Engineering Sciences and Fundamentals 2017

Core Programming Area at the 2017 AIChE Annual Meeting

Minneapolis, Minnesota, USA
29 October – 3 November 2017

Volume 1 of 2

ISBN: 978-1-5108-5790-2

Printed from e-media with permission by:

Curran Associates, Inc.
57 Morehouse Lane
Red Hook, NY 12571

Some format issues inherent in the e-media version may also appear in this print version.

Printed by Curran Associates, Inc. (2018)

For permission requests, please contact AIChE
at the address below.

AIChE
120 Wall Street, FL 23
New York, NY 10005-4020

Phone: (800) 242-4363
Fax: (203) 775-5177

www.aiche.org

Additional copies of this publication are available from:

Curran Associates, Inc.
57 Morehouse Lane
Red Hook, NY 12571 USA
Phone: 845-758-0400
Fax: 845-758-2633
Email: curran@proceedings.com
Web: www.proceedings.com

TABLE OF CONTENTS

VOLUME 1

(27a) Anisotropic Particle Interactions with Surfaces, Other Particles, and External Fields 1
Michael A. Bevan
(27b) Particle-Wavy Wall Interactions in a Nematic Liquid Crystal 2
Kathleen J. Stebe, Yimin Luo, Francesca Serra
(27c) Influence of Surface Roughness on Particle-Substrate Interactions: Double Layer Repulsion, Van Der Waals Attraction, and Depletion Interaction 3
Sven H. Behrens, Joanna W. Tsao
(27d) Functional Particles and Microcapsules 4
Simon Biggs, Alison Tasker
(27e) The Hindered Translational and Rotational Dynamics of Anisotropic Nanoparticles Diffusing Near a Solid-Liquid Interface Measured in Aqueous Solution 5
Christopher Bolton, Raymond R. Dagastine
(27f) Strong Deformation of the Electric Double Layer during Sedimentation or Electrophoresis of a Charged Particle 6
Aditya S. Khair
(27g) Understanding and Engineering Diffusiophoretic Suspensions 7
Todd M. Squires
(27h) Chemically-Driven Transport -- a Pervasive Mechanism That Is Seldom Recognized 8
Darrell Velegol
(66a) Understanding Catalyst Selectivity for CO2 Electro-Reduction 9
Thomas F. Jaramillo, Christopher Hahn, Drew Higgins, Jeremy T. Feaster, Stephanie Nitopi, Carlos Morales Guio, Lei Wang, Alan Landers, John Lin, Antaeres Antoniuk-Pablant
(66b) Ag-Sn Bimetallic Catalyst with a Core-Shell Structure for CO2 Reduction 10
Wesley Luc, Charles Collins, Feng Jiao
(66c) Electrochemical CO2 Reduction over Cuag Bimetallic Electrodes and Well-Defined Surface Alloys with Enhanced Oxygenate Selectivity 11
Ezra L. Clark, Christopher Hahn, Thomas F. Jaramillo, Alexis T. Bell
(66d) A Carbon Nanotube Supported Gold Catalyst for the Electroreduction of Carbon Dioxide (CO2) 12
Sumit Verma, Yuki Hamasaki, Chaerin Kim, Wenxin Huang, Shawn Lu, Molly Jhong, Tsuyohiko Fujigaya, Naotoshi Nakashima, Paul J. A. Kenis
(66e) Interplay of Mass Transfer and Local pH Effects in CO2 Reduction Electrocatalysis on Cu Nanowires 13
David Raciti, Chao Wang
(66f) Measurement of Intrinsic Activity of Electrocatalytic Reduction of CO2 over Cu 14
Aditya Prajapati, Meenesh R. Singh
(66g) Electrocatalytic Carbon Fixation on Molecular-Functionalized Surfaces 15
Ming Gong, Zhi Cao, Wei Liu, Christopher Chang
(66h) Electrochemical Reduction of CO2 on Metal Doped Graphene 16
Charlotte Kirk, Leanne Chen, Samira Siahrostami, Mohammadreza Karamad, Michal Bajdich, Johannes Voss, Jens Norskov, Karen Chan
(70a) A Generic Coarse-Grained Model of Influenza Budding: What Can We Learn? 17
Jesper J. Madsen, John M. A. Grime, Gregory A. Voth
(70b) Allosteric Effects of Gold Nanoparticles on Human Serum Albumin 18
Qing Shao, Carol K. Hall
(70c) Self-Assembly of Proteins: The Role of Shape and Specific Interaction 19
Jens Glaser, Sharon C. Glotzer
(70d) Predictive Design of Next-Generation Nanomaterials and Devices Via Bottom-up Approaches 20
Trung Nguyen
(70e) Solvation of Self-Assembled Complexes: Using Molecular Simulations to Probe Energetics, Structure, and Dynamics 21
Kevin R. Hinkle, Frederick R. Phelan Jr.
(70f) Level Set Strategy for Self Consistent Field Theory 22
Gaddiel Ouaknin
(70g) Simulations of Nonlinear Flows in Nonequilibrium Complex Liquids 23
Rui Zhang

(70h) Employing a Multipole Approximation in a Hybrid Fluid Via Relative Resolution 24
Aviel Chaimovich, Christine Peter, Kurt Kremer
(70j) Quantitatively Reliable Molecular Modeling and Simulation of Vapor-Liquid Equilibria 25
Martin T. Horsch
(80a) Production of High-Purity Pyrene Oligomers in a Supercritical Fluids Medium 26
Willam Lamie, Mark C. Thies
(80b) Development and Analysis of Green Pathways Isolation of Medically Active Components from Grape Pomace Biomass 27
Kyle Cogswell, Aydin K. Sunol
(80c) A Novel Supercritical CO2-Based Treatment for Decellularization That Maintains Mechanical and Structural Integrity 28
Dominic M. Casali, Rachel M. Handleton, Michael A Matthews
(80d) Supercritical Fluid Enhanced Swelling and Drug Impregnation of Biomedical Polymers 29
Kyle Cogswell, Aydin K. Sunol
(80e) Effect of Precursor on Crystal Structure of Titania Synthesis in Supercritical Sol - Gel Reaction 30
Motohiro Kinoshita, Yusuke Shimoyama
(80f) Accelerated Conversion of Magnesium Oxychloride to Chlorartinite Using Pressurized Carbon Dioxide and the Resulting Increased Water Resistance 31
Roque Gochez, Christopher L. Kitchens, Thomas Vreeland
(81a) High-Throughput Mechanical Stimulation and Functional Imaging in vivo 32
Hang Lu
(81b) The Steady Motion of a Train of Vesicles in a Cylindrical Channel of Arbitrary Cross Section 33
Joseph M. Barakat, Eric S. G. Shaqfeh
(81c) The Dissolution of a Sheared Water Drop in Bitumen Measured Using a Novel Microfluidic Platform 34
Sachin Goel, Samson Ng, Edgar Acosta, Arun Ramachandran
(81d) Adhesion and Break-up of Model Metastatic Cancer Cell Clusters during Passage through a Microfluidic Constriction 35
Yeng-Long Chen, Sam Au, Mehmet Toner
(81e) Chaotic Thermal Convection in Microfluidic Hydrothermal Pore Environments 36
Aashish Priye, Yuncheng Yu, Vijay Ravisankar, Yassin A. Hassan, Victor M. Ugaz
(81f) Using Droplet Microfluidics to Study Chemical Thermodynamics of Aqueous Atmospheric Aerosol Particles 37
Lucy Nandy, Cari S. Dutcher
(81g) Analysis of the Engulfment of a Foreign Particle By a Solidification Interface: A New Scaling for Flow and Interfacial Forces Involving SiC Particles during Silicon Crystal Growth 38
Jeffrey J. Derby, Yutao Tao, Christian Reimann, Jochen Friedrich, Thomas Jauss, Tina Sorgenfrei, Arne Croell
(81h) Hydrodynamic Mobility of Particles, Vesicles and Cancer Cells in a Square Microchannel 39
Shamim Ahmmed, Naureen Suteria, Valeria Garbin, Siva A. Vanapalli
(81i) Deformable Droplet Migration in a Narrow Microchannel at Finite Reynolds Numbers Via Dissipative Particle Dynamics (DPD) 40
Ryan L. Marson, Ronald G. Larson
(83a) Using Molecular Simulation to Understand the Interfacial Behavior of Ionic Liquids 41
Xiaoyu Liang, Jeffrey R. Errington
(83b) Capacitance and Ion Dynamics of Ionic Liquids Near Oxidized Graphene 42
Yu Zhang, Boris Dyatkin, Yury Gogotsi, Peter T. Cummings
(83c) Using Simulations and Experiments to Characterize Water Structure Near Mica Surfaces for Heterogeneous Ice Nucleation 43
Brittany Glatz, Jiarun Zhou, Sapna Sarupria
(83d) A Novel Density Gradient Theory for Surfactant Molecules - Applied to Oil/Water Interfaces 44
Xiaoqun Mu, Walter G. Chapman
(83e) Computational Chemistry Design of Liquid Crystal-Based Chemoresponsive Systems with Increased Water Tolerance 45
Tibor Szilvasi, Huaizhe Yu, Prabin Rai, Robert Twieg, Nicholas L. Abbott, Manos Mavrikakis
(83g) A Semi-Empirical Analytical Model for Determining the Thermodynamic Stability of Pickering Emulsion 46
Guolin Zhao, Junyin Xiao, Yanyang Wu, Shuangliang Zhao, Honglai Liu
(83h) Computational Method for Extracting Individual Adsorption Parameters from Experimental Dynamic Interfacial Tension of Mixed Surfactant Solutions 47
Fang Liu, Nelya Akhmetkhanova, Vincent Pauchard
(84a) Atomistic Simulations of Ordered and Disordered Carbons: Mimicking Versus Targeting 48
Roland J.-M. Pellenq

(84b) Graphene Oxide Membranes: A Molecular Simulation Approach 49
Christopher Williams, Paola Carbone, Flor R. Siperstein
(84c) Polyethylene Oxide (PEO) in a Polyethylene (PE) Framework: A Simple Model for Simulation Studies of Scaling and Solvent Effects on Polymers in an Open Framework 50
Kwong-Yu Chan, Liangxu Xie, Nicholas Quirke
(84d) Molecular Understanding and Design of Zwitterionic Materials 51
Shaoyi Jiang
(84e) Interfacial Transport of Protons on a 2-Dimensional Functionalized Graphane Surface 52
J. Karl Johnson, Abhishek Bagusetty, Bridget Derksen, Pabitra Choudhury
(84f) Gas Adsorption Behavior in Ionic Polyimide Composite Membranes 53
C. Heath Turner, Asghar Abedini, Ellis Crabtree, Jason E. Bara
(84g) Computational Simulation of Supported Nanocatalysts Under Realistic Conditions 54
Jian-Guo Wang
(90b) Continuous Synthesis of Ethyl Esters from Free Fatty Acids over Metal Oxides in Sub/Supercritical Ethanol 55
Jiuxu Liu, Yue Nan, Lawrence L. Tavlarides
(90c) Denitrogenation of Pyridine with Acetic Acid As Hydrogen Source Under Catalytic Hydrothermal Gasification Environment 56
Peng Zhang, Young Hwan Shin, Lance Schideman, Yuanhui Zhang, Wan-Ting Chen
(90d) The Non-Catalytic Decomposition of Lignin into Chemical and Fuel Intermediates 57
Sara Pourjafar, Wayne S. Seames
(90e) Hydrothermal Liquefaction of Food Waste and Remediation of Aqueous Byproducts 58
Alex D. Paulsen, Alex Maag, Ted J. Amundsen, Michael T. Timko, Paul E. Yelvington
(90f) Hydrothermal Liquefaction of Wastewater Treatment Microalgae in a Pilot-Scale Continuous Flow Reactor 59
Feng Cheng, Travis Le-Doux, Brian Treftz, Scott Woolf, Juanita Miller, Umakanta Jena, Catherine E. Brewer
(92a) The Importance of Classical Soft Matter Physics in the Development of New Nanomaterials 60
Matteo Pasquali
(92b) Rapid and Accurate Methods for Modeling Hydrodynamic Forces in Brownian Dynamics Simulations 61
Andrew Fiore, James Swan
(92c) Rotational Dynamics of Nanoparticles in Polymer Solutions and Melts 62
Lorena Maldonado-Camargo, Carlos Rinaldi
(92d) Morphology of Nanoparticle Aggregates in Flow through Beds Packed with Spheres: Self- and Flow-Induced Assembly Using Lattice Boltzmann Simulations 63
Ngoc Hong Pham, Dimitrios V. Papavassiliou
(92e) Shear Thickening of Colloidal Dispersions at High Particle Concentrations- Rheology and SANS Investigations into Particle-Scale Mechanisms 64
Kevin Whitcomb, Norman J. Wagner
(92f) Colloidal Hydrodynamics at a Fluid-Fluid Interface 65
Charles Maldarelli, Archit Dani
(92g) Effect of Surface Geometry on the Frictional Properties of Poly(dimethyl siloxane) 66
Yunhu Peng, Lilian Hsiao
(92h) Microrheology and Structural Reconfiguration of Artificial Biofluids Composed of Xanthan Gum in Salt Solutions 67
Mingyang Tan, Yating Mao, Britany M. Swann, Travis W. Walker
(92i) Probe Rheology Simulation Technique for Determination of Viscoelasticity of Complex Fluids 68
Pouria Nourian, Dinesh Sundaravadivelu Devarajan, Rajesh Khare
(93a) Approach to Contact, and Adhesion at Elastic and Structured Interfaces 69
Joelle Frechette
(93c) Surface Modification of Stainless Steels for Advanced Functionalities 70
Won Tae Choi, Victor Breedveld, Dennis W. Hess, Preet M. Singh
(93d) Understanding the Wetting Behavior of Water-Octane-Silica Systems Using Monte Carlo Simulation 71
Wenjing Guo, Jeffrey R. Errington
(93e) Time-Resolved in situ Studies of Zeolite Crystal Growth 72
Madhuresh K. Choudhary, Manjesh Kumar, Jeffrey D. Rimer
(93f) Understanding the Solid-Liquid Phase Transition during the Growth of Scintillator Single Crystals Via Computational Modeling and Neutron Imaging 73
Chang Zhang, Jeffrey H. Peterson, Jan Seebeck, Anton S. Tremsin, Didier Perrodin, Gregory A. Bizarri, Edith D. Bourret, Sven Vogel, Jeffrey J. Derby

(93g) Efficient Dispersion of Crude Oil By Blends of Food-Grade Surfactants: Toward Greener Oil Spill Treatments 89
David Riehm, Jasmin C. Athas, John Neilsen, Geoffrey D. Bothun, Vijay T. John, Srinivasa R. Raghavan, Alon McCormick

(93h) Comparison of Experimental and Predicted Adsorption Isotherms of Mixtures 90
Julian Butz, Sabine Enders

(93i) Towards Connecting the Microstructural and Structural Changes in Hierarchical Nanoporous Materials (e.g., clays) during Gas Adsorption (CO2, CH4, H2) Using in-Operando Multi-Scale X-Ray and Neutron Scattering Measurements 91
Greeshma Gadikota, Andrew J. Allen

(140a) Adsorption and Phase Behavior of Mixed Alkanes in Nano Slit Graphite Pores: An iSAFT Application 92
Jinlu Liu, Dilip Asthagiri, Walter G. Chapman

(140b) Oscillation and Enhancement of Gas Solubility in Nanopores 93
Xiaochen Yu, Liangliang Huang, Shuangliang Zhao, Honglai Liu, Keith E Gubbins

(140c) Effect of Interaction Position on Molecular Transport and Separation through a Multi-Site Nanopore 102
Shaghayegh Agah, Matteo Pasquali, Anatoly Kolomeisky

(140d) Strain-Induced Topological Defects and Configurational Transitions in Liquid Crystals 103
Monirosadat Sadati, Jose Martinez-Gonzalez, Khia Kurtenbach, Luis X. De Pablo, Harrison Shapiro, Samuel Morin, Ye Zhou, Juan J. De Pablo

(140e) Alkyl Chain Length and Nanoconfinement Effects on Dynamics of Imidazolium-Based Ionic Liquids 104
Yu Zhang, Naresh C. Otsi, Eugene Mamontov, Peter T. Cummings

(140f) A Molecular Simulation Study on CO2 and a Deep Eutectic Solvent in Slit Nanopores 105
Yan Shen, Francisco R. Hung

(140g) Molecular Simulation of Methane Adsorption Behavior in Nanopores for Shale Gas Development: With Comparison between Graphite and Kerogen Models 106
Jinrong Cao, Yunfeng Liang, Yoshihiro Masuda, Hiroaki Koga, Hiroyuki Tanaka, Toshifumi Matsuoka

(140h) Brownian Dynamics Simulations on Spontaneous Extension of DNA on Cationic Lipid Bilayers Along Grooved Structures 115
Chang Ming-Yi, Ching-Kuan Wang, Chih-Chen Hsieh

(141a) Design of Advanced Nanomaterials for Electrocatalysis 116
Yijin Kang

(141c) Layered Hybrid Structure of Cobalt/N-Doped Carbon Derived from Metal-Organic Frameworks for Electrocatalytic Hydrogen Evolution 117
Tan Huang, Jong-Min Lee

(141d) Comparison of Hydroxide-Mediated and Direct Mechanisms for Alkaline Hydrogen Electrocatalysis 118
Saad Intikhab, Jennifer Gallup, Joshua Snyder, Maureen H. Tang

(141e) Alkaline Hydrogen Evolution from Ni-Mo Intermetallics with High Mo Content 119
James R. McKone, Peter Csernica, Francis J. Disalvo, Hector D. Abruna

(141f) Increased Photocatalytic Activity of TiO2 Nanoparticles with Defects for Sustainable Hydrogen Production 120
Ashley M. Pennington, Fuat E. Celik

(141g) Transition Metal Nitride Core-Noble Metal Shell Nanoparticles As Highly CO Tolerant Catalysts 121
Aaron Garg, Maria Milina, Madelyn Ball, Sean T. Hunt, James A. Dumesic, Yuriy Roman-Leshkov

(141h) Understanding the Electrochemical Behavior of a Thin, Flexible Micro Fuel Cell 122
Seyed Reza Mahmoodi, Matthew Mayer, Ronald S. Besser

(147a) Metastable Phase Transitions 123
Pablo G. Debenedetti

(147b) Computational Investigation of Multipolar Colloidal Particle 124
David M. Rutkowski, Ryan C. Maloney, Orlin D. Velev, Sabine H. L. Klapp, Carol K. Hall

(147c) Interfacial Tensions from SAFT: Connecting Equations of State to Molecular Simulations 125
Erich A. Muller

(147d) Generalized Gibbs Free Energy of Confined Nanoparticles 126
Xiaohua Lu

(147e) Molecular Modeling of Polymeric Systems 127
Coray M. Colina

(147f) Effect and Regulation of Surface Wettability on Molecular Transport and Reaction 128
Shuangliang Zhao, Honglai Liu

(148a) Dynamics of Collective Endothelial Cell Migration in Response to Fluid Shear Stress 129
Alexander Dunn
(148b) Rupture of Cancer Cells Under Microcirculatory Conditions 130
Nabiollah Kamyabi, Siva A. Vanapalli
(148c) Transport of Nanoparticles in the Brain 131
Justin Rosch, C. B. Schaffer, William L. Olbricht
(148d) Towards Understanding Nanoparticle Diffusion in Synovial Fluid Analogues 132
Mythreyi Unni, Lorena Maldonado-Camargo, Kyle Allen, Carlos Rinaldi
(148e) Diffusion of Concentrated Macromolecules within Living Cells 133
Jiyuan Li, Xikai Jiang, Juan Hernandez-Ortiz, Olle G. Heinonen, Juan J. De Pablo
(148f) Platelet Margination, Adhesion, and Activation in Secondary Flows Are Necessary for Thrombus Propagation in an in vitro Model of Venous Thrombosis 134
Marcus Lehmann, Patrick Krohl, Keith B. Neeves
(148g) The Importance of Rheology in Blood Flow Modeling 135
Jeffrey S. Horner, Norman J. Wagner, Antony N. Beris, Donna S. Woulfe
(148h) In Vitro Measurement and Modelling of Platelet Adhesion on Von-Willebrand-Factor-Coated Surfaces in Channel Flow 144
Qin M. Qi, Irene K. Oglesby, Jonathan Cowman, Antonio J. Ricco, Dermot Kenny, Eric S. G. Shaqfeh
(148i) A Novel Viscoelastic Thrombogenesis Model from High Performance Lattice Boltzmann Method Yield-Stress Calculations Based on Intravital Images of Clot Formation in Live Mice 145
Vishnu Deep Chandran, Olufemi Kadri, Roman Voronov
(150a) Determination of the Zeta Potential of Planar Solids in Nonpolar Liquids 146
Paul J. Sides, Dennis Prieve, Benjamin A. Yezer
(150b) Toward Simulation-Based Design of Particle Handling Processes 147
Jennifer Sinclair Curtis
(150c) Sustainable Electrocatalytic Water Splitting with Earth-Abundant Materials 148
Hao Yuan, Richard Lunt, Robert Y. Ofoli
(150d) Human Tear Production from Capillary Wicking Dynamics 149
Clayton J. Radke
(150e) Structure, Elasticity, and Non-Equilibrium State Diagram of Depletion Gels 150
Eric M. Furst
(150f) Harnessing Solvation Forces for Dispersing Colloids in Ionic Liquids with Application in Human Exploration of Space 151
Jingsi Gao, Norman J. Wagner
(150g) Control of Interfacial Phase Separation By Electro-Autocatalysis 152
Martin Z. Bazant
(150h) Reduction of Prticle Jamming in Abrasive Slurries 153
Aditya A. Salunkhe, Rene M. Overney, John C. Berg
(152a) Digital Alchemy for Assembly Engineeringc 164
Sharon C. Glotzer
(152b) Improved Algebraic, Numerical, and Graphical Representations in Fluid Mechanics 165
Stuart W. Churchill, James C. Hill
(152c) The Scaling of Turbulence Near the Wall and the Churchill Turbulent Flux Correlation: Insights with Lagrangian Simulations 166
Dimitrios V. Papavassiliou, Quoc T. Nguyen, Chiranth Srinivasan
(152d) Flow Boiling Using a Piranha Pin Fin Heat Sink 167
Cory Woodcock, Xiangfei Yu, Yoav Peles, Joel L. Plawsky
(152e) Transport Problems in the Spirit of Stuart Churchill for Teaching and Research at the University of Michigan 168
Ronald G. Larson, Claudio Vilas Boas Favero
(160a) Layered Fluid-Fluid Interfaces Confined in Microfluidics 169
Bruno Pinho, Ryan L. Hartman
(160b) Focused DC and AC Electric Fields at Conic Pipettes: Nanoscale Thermal Hotspot and Nano-Droplet Generation with Universal Scalings 170
Hsueh-Chia Chang, Zehao Pan
(160c) Real-Time Monitoring of Complex Multiphase Behavior in a High Pressure High Temperature Microfluidic Chip 171
R. M. Ripken, J. G. E. Gardeniers, S. Le Gac
(160d) Motion of a Deformable Drop in Microchannels of Complex Shape 172
Robert H. Davis, Rocio Navarro, Alexander Zinchenko

(160k) Microfluidic Models of Porous Media to Examine Asphaltene Deposition in the Presence of Chemical Dispersants 173
Yu-Jiun Lin, Sibani Lisa Biswal
(160f) Microfluidic Device for the Continuous Measurement of Viscosity 174
Yunzi Li, Sarah E. Mena, Kevin R. Ward, Mark A. Burns
(160g) Simple Algebraic Formulas for the Practical Interpretation of Mercury Porosimetry Data 175
Zongyu Gu, Martin Z. Bazant
(160h) A Microfluidic Platform to Measure Dynamic Interfacial Tension of Complex Fluid Systems 176
Shweta Narayan, Davis B. Moravec, Brad G. Hauser, Andrew J. Dallas, Cari S. Dutcher
(160i) Controlled Liquid Entrapment through Photo-Patterned Obstacles and Patterned Surfaces 177
Ankur Gupta, Hyundo Lee, T. Alan Hatton, Patrick S. Doyle
(160j) Tears of Wine 178
Prerana Rathore, Vivek Sharma
(163a) Kinetics of (un)Binding between DNA-Functionalized Particles Using a Coarse-Grained Model with Explicit Nucleotide Representation 179
Tiara Ann Maula, Jeetain Mittal
(163c) Hydration Structure and Dynamics of Poly(2-methacryloyloxyethyl phosphorylcholine) 180
Christoph Klein, William L. Roussell, Christopher R. Iacovella, Clare McCabe, Peter T. Cummings
(163d) Polyelectrolyte Interactions: Simulation and Theory 181
Maria Sammalkorpi, Hanne Antila, Paul R. Van Tassel
(163e) Modeling Solute and Solvent Distributions in Functionalized Dendrimers from iSAFT Density Functional Theory 182
Yuchong Zhang, Walter G. Chapman
(163f) Dependence of Relaxations and Mechanical Properties on Molecule Shape in Dissipative Particle Dynamics 183
Michael L. Greenfield, Claire A. Lemarchand, Jesper S. Hansen
(163g) Thermodynamics of Self-Assembly of Perylene Derivatives 184
Jorg Baz, Niels Hansen
(163h) A Molecular Dynamics Study of Actinide Nanoclusters 185
Ken Newcomb, Edward J. Maginn
(176a) Solve this! Fundamental Approach to Problem Solving in Industrial Processes I (Invited Talks) 186
Zdravko Stefanov, Paul Chauvel, Jr., Eldad Herceg, Dana A. Livingston
(179a) Solubiltiy and Phase Equilibria for Optimizing the Processing of Cannabinoids and Terpenes in Cannabis and Hemp Extractions 187
Jerry W. King
(179b) Measurement and Modeling of Vinyl Fluoride Solubility in Aqueous Lithium Bis(trifluoromethylsulfonyl)Imide Solutions 188
David L. Minnick, William J. R. Gilbert, Alejandra M. Rocha, Mark B. Shiflett
(179c) Heat and Momentum Transport Properties of CO2-Expanded Liquids (CXLs): N-Hexane, n-Decane, and n-Tetradecane 189
Kourosh Kian, Aaron M. Scurto
(179d) Van Der Waals Phenomenological Transport Equation of State for Dense Viscosities of the Ultra-Deep Petroleum Reservoirs in the Gulf of Mexico 190
Akanni S. Lawal
(179e) Viscosity Models for Hydrocarbons at High Pressure: A State-of-the-Art Review 208
Isaac Gamwo, Hseen O. Baled, Mark A. McHugh, Robert M. Enick
(179f) Quantification of Uncertainties in Experimental Liquid Density and Viscosity at Hthp Conditions By Using the Corresponding States Law As Arbitrator 209
Akanni S. Lawal
(195a) Characterization of Orientation and Photochemical Function of Chlorophyll a Molecules in Self-Assembled Membranes 217
Shogo Taguchi, Keishi Suga, Keita Hayashi, Yukihiro Okamoto, Hidemi Nakamura, Hiroshi Umakoshi
(195b) Design of Lipid Membrane Surfaces As Organocatalyst for Michael Reactions in Aqueous Media 218
Masanori Hirose, Keishi Suga, Yukihiro Okamoto, Hiroshi Umakoshi
(195c) A Unified Structure-Property Relationship for Alkyl-Polyoxide Surfactants 219
Zachary R. Hinton, Nicolas J. Alvarez
(195d) Relating Rheology with Morphology: Cholesterol in a Model Lung Surfactant Monolayer 220
Steven Patton, Amit Kumar Sachan, Ian Williams, Todd M. Squires, Joesph A. Zasadzinski
(195e) A Matlab Based Tool Package for Interfacial Property Calculations 221
Xiaoqun Mu, Walter G. Chapman, Florian Frank, Faruk O. Alpak

(195i) Interaction Forces between Colliding Emulsion Drops (Oil or Water) Coated with Non-Ionic Surfactants (C12E5 and PGPR) 222
Srinivas Mettu, Joe Berry, Chu Wu, Raymond R. Dagastine
(204a) Prediction of Vapor Pressure and Critical Properties for Non-Electrolyte Organic Compounds from PR+Cosmosac Eos 223
Chien-Yi Li, Chieh-Ming Hsieh
(204b) Volumetric Properties of Saccharide in Aqueous Ionic Liquid Mixtures at Different Temperatures 224
Natalia D. F. Val, Heloisa E. Hoga, Ricardo B. Torres
(204c) Density and Derived Properties of Binary Mixtures Containing {Dimethyl Carbonate (DMC) + 1-Propanol} at T = (288.15 to 308.15) K and at P = (0.1 to 40) Mpa 225
Gustavo V. Olivieri, Ricardo B. Torres
(204e) A SAFT Equation of State Based on Triangular-Well Fluid Potential 226
Luis A. Galicia-Luna, Alfredo Pimentel-Rodas, Jaime A. Riera-Ortiz, Jose M. Rosete-Barreto
(204g) Modelling the Solubility of Naphthalene and Phenanthrene in Binary and Ternary Systems Containing Carbon Dioxide 227
Francisco Javier Veronico Sanchez, Miguel Gonzalo Arenas Quevedo, Octavio Elizalde-Solis
(204h) Molecular-Inspired Parameters Concealed in the Van Der Waals Attractive Force Revealed By First Principles, Statistical Mechanics and Perturbation Methods 228
Akanni S. Lawal
(204i) Solid-Liquid Equilibria for Selected Binary Mixtures Containing Diphenyl Carbonate 229
Hiroyuki Matsuda, Yuki Ohashi, Kiyofumi Kurihara, Katsumi Tochigi
(204j) Thermophysical Suitability of Terphenyl for Hydrogen Storage 230
Jonas Obermeier, Yannick Werner, Ferdinand Heusinger, Michael Geißelbrecht, Patrick Preuster, Peter Wasserscheid, Wolfgang Arlt, Karsten Muller
(204k) Atomic-Level Mechanistic Insights into Monolayer hBN Growth from Reactive Molecular Dynamics Simulations for Catalysis Applications 231
Bin Liu, Song Liu, Mingxia Zhou, James H. Edgar
(204l) Differential Retention and Release of CO2 and CH4 in Kerogen Nanopores: Implications for Gas Extraction and Carbon Sequestration 232
Tuan Ho, Louise J. Criscenti, Yifeng Wang
(204m) Investigation of Electronic Properties of Imidazolium-Based Ionic Liquids in the Presence of Iron Porphyrins for Understanding Their Biodegradability 233
Atiya Banerjee, Jindal K. Shah
(204n) Capturing the Membrane-Triggered Conformational Transition for Pore Forming Cytolysin (ClyA) Using Structure Based Models 234
Hemanth Giri Rao V. V., Rajat Desikan, Shachi Gosavi, K. G. Ayappa
(204o) Solid Form Transformation of Disodium Guanosine 5€²-Monophosphate: Thermodynamic Perspective 235
Qiao Chen, Fengxia Zou, Pengpeng Yang, Jinglan Wu, Wei Zhuang, Hanjie Ying
(204p) Estimation of Kinematic Viscosity for CO2 expanded Liquids By Asog-Visco Method 236
Toshitaka Funazukuri, Hiroyuki Matsuda, Kiyofumi Kurihara, Katsumi Tochigi, V. K. Rattan
(204q) On the Kac-Based Collision Models from Simplified Bernoulli till Its Intelligent Variants 237
Bijan Goshayeshi
(204r) Viscosity and Derived Properties of Binary Mixture Containing Acetonitrile + 1-Propanol at Different Temperatures and Atmospheric Pressure 238
Christian A. T. Campos, Heloisa E. Hoga, Ricardo B. Torres
(204s) Volumetric, Acoustic and Viscometric Properties of Binary Mixture of (n-butylammonium methanoate + 1-propanol) at Different Temperatures 239
Robert L. Fernandes, Heloisa E. Hoga, Ricardo B. Torres
(204t) Measurement of Diffusion Coefficients of Chromium(III) Acetylacetonate in Supercritical Carbon Dioxide at High Temperatures 240
Minoru Yamamoto, Junichi Sakabe, Minori Taguchi, Toshitaka Funazukuri
(204u) Simultaneous Determinations of Dynamic Viscosity and Density of Several Alcohols Using a Straight and Coil Capillary Viscometers at Temperatures between (313-353) K and Pressures up to 30 Mpa 241
Alfredo Pimentel-Rodas, Luis A. Galicia-Luna, Jose J. Castro-Arellano
(204v) Analysis of High-Speed Rotating Flow inside Gas Centrifuge Casing 242
Sahadev Pradhan
(204w) DSMC Simulations of High MACH Number Taylor-Couette FLOW 272
Sahadev Pradhan

(204x) Viscosity Prediction of the Carbon Dioxide Loaded Aqueous Solutions of Alkanolamines: Thermodynamic Approach 274
Naser S. Matin, Joseph E. Remias, Kunlei Liu
(204y) A Computational Study on the Transport of Actives through Skin Layers 275
Kishore Gajula, Rakesh Gupta, Dwadasi Balarama Sridhar, Beena Rai
(213a) Developing Chemical Engineering Acumen By Brewing Kicking Mule Beer 276
Matthew Armstrong, Joshua White, Geoffrey Bull, Jesse Hudgins
(213b) Beer Experiments - a Hopping Success 277
Claire F. Komives, Corey Lapeyri, Alexis Venegas, Diego Marquez, Joseph Pesek
(213c) Why Is the Mustard in the Fridge? A Fun Assignment on the Thermodynamics of Food Safety 278
Margot Vigeant
(213d) Improving Conceptual Knowledge and Retention in Introduction to Engineering Thermodynamics 279
Rachel Morrish
(213e) Tax Revenue Laffer Curve from Thermodynamics Perspective 280
Min Huang
(213f) The Art of the Game: Infusing Thermodynamics Learning with Classic Household Games 281
Reginald E. Rogers Jr.
(216a) Generalizing the Design Principles of Pt-Based Alloy Catalysts with Improved ORR Performance and Durability 282
Zhenhua Zeng, Jeffrey Greeley
(216b) Computational Insights into Thermal and Photoresponse of Pd/C and Pd/N Co-Doped TiO2 using DFT 283
Sai Phani Kumar Vangala, Parag Arvind Deshpande
(216c) Transition Metal-Doped C2N as Active Catalysts for the Oxygen Reduction Reaction 284
Anjli M. Patel, Ambarish R. Kulkarni, Jens Norskov
(216d) Electrochemical Synthesis of Hydrogen Peroxide Via Water Electrolysis: Challenges and Opportunities 285
Samira Siahrostami, Jens Norskov
(216e) Modeling Electrochemical Reactions: DFT-Based Models Including Explicit Solvation, Electrolyte, and Electrochemical Potential 286
Jason Goodpaster
(216f) Efficiently Determining Solvent Effects at Metal-Water Interfaces 287
Satish Iyemperumal, N. Aaron Deskins
(216h) Machine-Learning Model Development for Electrocatalyst Discovery 288
Zheng Li, Hongliang Xin
(218b) Climbing the Volcano: Active-Site Engineering at the Atomic Scale 289
Joseph H. Montoya
(218c) Developing Ab Initio Methodology for Advancing Catalytic Reactions 290
Eric Walker
(218d) First-Principles-Derived Structure-Energy Relationship for Surface Oxides 291
Alexander V. Mironenko, Dionisios G. Vlachos
(218e) First-Principles Based Design of Reaction Conditions for the Catalytic Conversion of Methane to Methanol over Cu-Exchanged SSZ-13 292
Florian Goltl, Manos Mavrikakis
(218f) Reaction Ensemble Monte Carlo Simulations of Xylene Isomerization Under Confinement 293
Ryan Gotchy Mullen, Edward J. Maginn
(218g) Understanding Separation and Catalysis in Nanoporous Materials 294
Peng Bai
(218h) Modeling Self-Assembly of Metal-Organic Frameworks with Enhanced Sampling Techniques 295
Yamil J. Colon, Ashley Guo, Lucas Antony, Kyle Hoffmann, Juan J. De Pablo
(218i) Computationally-Efficient High-Throughput Screening of Metal-Organic Frameworks for Hydrogen Storage 296
N. Scott Bobbitt, Arun Gopalan, Benjamin Bucior, Jiayi Chen, Randall Q. Snurr
(218j) Effects of Ion Self-Energy on the Double Layer Structure and Properties at the Dielectric Interface 297
Rui Wang
(225a) Discovery of the Universal Van Der Waals Generic Cubic Equation of State for the Continuity of Gaseous and Liquid States of Fluids and Mixtures 298
Akanni S. Lawal

(225b) Vapor-Liquid Equilibrium Data for the Systems H2S-MDEA-H2O and CH4-H2S-MDEA-H2O at High Solvent Concentrations and High Pressures 312
Eirini Skylogianni, Diego D. D. Pinto, Hanna K. Knuutila, Christophe Coquelet
(225c) Solubility of Timolol Maleate in Supercritical Carbon Dioxide with Water on Extraction from Silicone Hydrogel 315
Yusuke Shimoyama, Yuta Yokozaki
(225d) Experimental and Modelling Study of the Phase Behaviour of (CO2 + CH4 + Methylbenzene) at High-Pressure and High-Temperature Conditions 316
Saif Zs. Al Ghafri, J. P. Martin Trusler
(230a) Industrial-Scale Modeling of Herschel-Bulkley (Yield Stress) Fluids 317
John A. Thomas, Kevin Smith
(230b) Viscometer for on-Site Rheology Measurements: Development and Application 318
Richard K. Grenville, Jason J. Giacomelli, Wojciech Wyczalkowski, Addison Root
(230c) Using ?2rheology to Measure Rheological Properties during Repeated Phase Transitions of Hydrogenated Castor Oil 319
Matthew Wehrman, Seth Lindberg, Kelly M. Schultz
(230d) Model Systems for Mixed Particulate Fermentation Broth 320
Matthias Kraume, Lutz Bohm, Chrysoula Bliatsiou
(230e) Polycarboxylate Ether Cement Superplasticizers Containing Copolymer Backbones with Improved Rheology Performance 321
Xue Chen, Jennifer Lowe, Thomas Clark, Mike Radler
(234a) Microscale Acoustic Streaming Flows in Viscoelastic Fluids: Comparison of Experiment and Simulation 322
Behrouz Behdani, Ran Zhou, Cheng Wang, Joontaek Park
(234aa) Radioactive Particle Tracking (RPT) Techniquefor Pilot-Scale Bubble Column 323
Laith Sabri, Abbas Sultan, Muthanna Al-Dahhan
(234ab) Gamma-ray Computed Tomography (CT)Technique for Pilot-Scale Bubble Column Reactor 324
Abbas Sultan, Laith Sabri, Muthanna Al-Dahhan
(234c) Influence of Cell Lengths on Surface Drag in Motile Bacteria 325
Katie Ford, Pushkar Lele
(234d) Dynamics of Semiflexible Colloidal Particle Chains Under Rotating Magnetic Fields 326
Steve Kuei, Sibani Lisa Biswal
(234e) A Study of the Dynamics of Human Pedestrians Using Experiments and Simulations in the Indian Context 327
Indranil Saha Dalal, Anurag Tripathi, Amullya Kale, Ishan Prashant
(234f) Extensional Rheometry with a Handheld Mobile Device 328
Kristin A. Marshall, Aleesha M. Liedtke, Anika H. Todt, Travis W. Walker
(234g) Free Surface Flows and Extensional Rheology of Polymer Solutions 329
Jelena Dinic, Leidy N. Jimenez, Vivek Sharma
(234h) Rheological Behavior of Poly(vinyl alcohol) in Aqueous Solutions: Comparison and Assessment of Rheological Parameters Obtained By Empirical Correlations 330
Maria Veronica Carranza Oropeza, Luis Carrasco Venegas Sr., Sandy Candiotti Velasquez
(234i) The Phantom Generation of a Complex Nasal Geometry with Horizontal Cut for Lda Measurements 331
Manuel Berger, Martin Pillei, Andreas Mehrle, Wolfgang Recheis, Florian Kral, Wolfgang Freysinger, Michael Kraxner
(234j) Using Laos and Transient Data to €œFingerprint€ Human Blood Rheological Data 332
Tyler Helton, Matthew Armstrong
(234l) Comparison of Simple Rheological Models in Fitting and Predicting Steady State and Transient Blood Rheology 333
Michael Deegan, Evan Ousley, Matthew Armstrong
(234o) Development of Predictive Model for Sizes of Gas and Liquid Slugs Formed in Millimeter-Scale T-Shaped Channels 334
Gwangnoh Ahn, Osamu Tonomura, Satoshi Taniguchi, Aoyama Tomoya, Shinji Hasebe
(234p) Shape Evolution and Spreading of Liquid Droplets in Miscible Environments 335
Dan Walls, Simon Haward, Amy Shen, Gerald G. Fuller
(234q) Impact Forces of Inertia-Driven Liquid Drops 336
Ting-Pi Sun, Leonardo Gordillo, Xiang Cheng
(234r) Visualizing Nanoscopic Topography, Patterns, Flows, Thickness Transitions and Instabilities in Stratifying Freestanding Thin Films 337
Yiran Zhang, Subinuer Yilixiati, Vivek Sharma

(234s) Contrasting Drainage and Stratification in Horizontal Vs Vertical Micellar Foam Films 338
Subinuer Yilixiati, Ewelina Wojcik, Yiran Zhang, Vivek Sharma
(234t) Solvent Effects on the Crystallinity of Petroleum Asphaltenes 339
Yuan Yang, Thomas Headen, Michael P. Hoepfner
(234v) Role of Bifurcation Geometry on Stability of Thrombus 340
Hari Hara Sudhan Lakshmanan, Jeevan Maddala
(234w) Using ?2rheology to Measure Rheological Properties of Hydrogenated Castor Oil 341
Matthew Wehrman, Seth Lindberg, Kelly M. Schultz
(234x) CFD Simulation of Bubble Columns Operating in Heterogeneous Regime 342
Tatiana Matiazzo, Bruna L Mees, Jaci Carlo Schramm Camara Bastos, Henry F. Meier, Marcela Kotsuka Silva
(234z) Deposition and Oil-Brine Interfacial Rheology of Asphaltene-Stabilized Emulsions 343
Yu-Jiun Lin, Peng He, Zhuqing Zhang, Steve Kuei, Sibani Lisa Biswal
(257a) Vapor Recompression Mechanism Introducing in Internal Heat-Integrated Distillation Column: Impact of Internal Energy Driven Intermediate and Bottom Reboiler 344
Bandaru Kiran
(257b) Experimental Studies Controlling Trace Components in a Dividing Wall Distillation Column 345
Melissa Donahue, Michael Baldea, R. Bruce Eldridge
(257c) Novel Metal-Enzyme Catalyst for One-Pot Dynamic Resolution in a Spinning Cloth Disc Reactor 346
Parimala Shivaprasad, Matthew Jones, Emma Emanuelsson
(257d) Scale and Effects of Catalyst Deactivation in Enzymatic Catalyzed Reactive Distillation 355
Torben Egger, Georg Fieg
(257e) Use of Computational Fluid Dynamics (CFD) Simulation and Image Analysis Tool for Modelling Light in a Microalgal Photobioreactor 356
Arpit Mishra, Geetanjali Yadav, Parthasarathi Ghosh, Ramkrishna Sen
(261a) Interfacial Dynamics between Pathogenic Nanoparticles and Cell Membrane Surfaces 357
Susan Daniel
(261b) Bicontinuous Microemulsions in Homopolymer-Block Copolymer Blends 358
Frank S. Bates
(261c) Self-Assembly of Block Copolymers: From Nanostructure to Function to Applications 359
Paschalis Alexandridis
(282a) Rational Design and Synthesis of Surfactant-Free Pt-Based Ternary Nanoalloys As Superior Oxygen Reduction Reaction Electrocatalysts 360
Sheng Hu, Bamin Khomami, Dibyendu Mukherjee
(282c) Improved Activity and Durability of Ionic Liquid Composite Nanoporous Nanoparticle Electrocatalysts for Oxygen Reduction Reaction 361
Yawei Li, Joshua Snyder
(282d) Particle Atomic Layer Deposition for Stabilization of Oxygen Reduction Catalysts 362
William McNeary Iv, Audrey Linico, Megan Maguire, Alia M. Lubers, Alan W. Weimer
(282e) A Mechanistic Insight into the Electrocatalytic Behavior of Mixed Oxide Cathode Catalysts for Li-O2 Batteries 363
Samji Samira, Ayad Nacy, Eranda Nikolla
(282f) Characterization of Feni Alloy Nanoparticles for Oer As a Function of Iron-Nickel Composition 364
Prashant Acharya, Zachary Nelson, Lauren F. Greenlee
(282g) Active Sites in Nitrogen-Doped Carbon Nanostructures for Oxygen Reduction and Oxygen Evolution Reactions 365
Kuldeep Mamtani, Deeksha Jain, Anne Co, Umit S. Ozkan
(282h) Atomic Iron-Dispersed Electrocatalysts Derived from Metal-Organic Framework for Oxygen Reduction 366
Hanguang Zhang, Gang Wu
(286a) Nucleation Mechanism of Clathrate Hydrates of Soluble Guest Molecules 367
Ryan Defever, Sapna Sarupria
(286b) Investigation on Thermophysical Properties of Structure II Clathrate Hydrates Using Molecualr Dynamics Simulations 368
Bing Fang, Fulong Ning, Ping Cao, Jianyang Wu, Thijs J. H. Vlugt, Signe Kjelstrup
(286c) Prediction of Gas Hydrate Formation Conditions in Brine Systems with the Enrtl Model 379
Amadeu K. Sum, Yue Hu
(286d) Clathrate Hydrate Formation Using Fluorocarbons 380
M. Alejandra Rocha, Mark B. Shiflett
(286e) Methane Hydrate Formation Accelerated with Various Promoters in an Unstirred Reactor 381
Seungjun Baek, Yun-Ho Ahn, Junshe Zhang, Juwon Min, Huen Lee, Jae W. Lee

(286f) Effect of Thermodynamic Promoter (TBAB) in Gas Hydrates Dissociation Conditions of the Mixture H2O + CO2 + Undecane 382
Raul A. Santos-Serena, Pedro Esquivel-Mora, Jose J. Castro-Arellano, Luis A. Galicia-Luna
(286g) Study on Flow Behavior of CO2-in-Water (C/W) Emulsion in the Porous Medium Under Hydrate Stability Conditions 383
Yoshihiro Masuda, Keishi Usui, Yuta Tosuji, Shigemi Naganawa, Masahiro Yasue, Yunfeng Liang, Norio Tenma
(286h) Ionic Lqiuids As Dual Functional Inhibitors for Natural Gas Hydrates 384
Majeda Khraisheh
(289j) Complex Fluids in the Gut 396
Sujit S. Datta
(289b) Boundary Guidance in Self Propelled Colloidal Motors 397
Ali Mozaffari, Joel Koplik, Charles Maldarelli
(289c) Active Nematic Liquid Crystals with Variable Activity and Elasticity 398
Rui Zhang, Nitin Kumar, Jennifer Ross, Margaret L. Gardel, Juan J. De Pablo
(289d) Using a Stochastic Field Theory to Understand Active Colloidal Suspensions 399
Yuzhou Qian, Peter R. Kramer, Patrick T. Underhill
(289e) Imaging the Onset Kinetics of the Swarming Transition Using Light-Controlled Bacteria 400
Yi Peng, Yishu Tai, Kechun Zhang, Xiang Cheng
(289f) Surface Drag and Swarming in Motile Bacteria 401
Pushkar Lele, Katie Ford
(289g) Effects of Elasticity and Hydrodynamic Interactions on Swimmer Shape and Trajectories in a Coarse-Grained Model of Monotrichous Bacteria 402
Frank Nguyen, Michael D. Graham
(289h) Microscopic Dynamics of Bacterial €œSuperfluids€ Under Planar Oscillatory Shear 403
Shuo Guo, Devranjan Samanta, Yi Peng, Xinliang Xu, Xiang Cheng
(289i) Monte Carlo Simulaitons on the Aggregate Structures in a Supension Composed of Magnetic Cubic Particles on a Material Surface 404
Kazuya Okada, Akira Satoh
(296a) Pattern Formation from Interfacial Instabilities in Miscible Viscous Fluids 405
Irmgard Bischofberger, Qing Zhang
(296b) Electrokinetic Fingering: A Problem in Vector Laplacian Growth 406
Mohammad Mirzadeh, Martin Z. Bazant
(296c) Designing Functional Materials with Interfacial Instabilities in Thin Films 407
Pierre Brun, Joel Marthelot, Elizabeth Strong, Pedro Reis
(234n) Numerical Simulation on Flow Behaviour of Twin-Liquid Film over a Vertical Plate with an Open Window 408
Hanguang Xie, Jianguang Hu, Gance Dai
(296e) Droplet Wetting Transitions on Inclined Substrates in the Presence of External Shear and Substrate Permeability 409
Leonardo Espin, Satish Kumar
(296f) Measuring Contact-Line Mobility Using Drop Resonance 410
Yi Xia, Paul Steen
(296g) Electrostatic Suppression of the Leidenfrost State 411
Arjang Shahriari, Soumik Das, Vaibhav Bahadur, Roger T. Bonnecaze
(296h) Linear Stability of Layered Two-Phase Flow through Soft Gel Coated Walls 412
Dinesh N V S S R Bhagavatula, S. Pushpavanam
(305a) Inverse Design of Interactions for Assembly 413
Thomas M. Truskett
(305b) High-Throughput Computational Screening As a Tool for Understanding the Molecular Thermodynamics of Adsorption 414
Randall Q. Snurr
(305c) Awe-Somes: All Water Emulsion Bodies 415
Kathleen J. Stebe, Sarah Hann, Daeyeon Lee
(305d) Succession of Alkane Conformational Motifs Bound within Hydrophobic Nano-Capsule Assemblies 416
Hank Ashbaugh
(305e) Thermodynamics and Transport Properties of Fluids with Intermediate Range Order with Application to Protein Solutions and Biopharmaceuticals 417
P. Douglas Godfrin, Yun Liu, Nestor Valadez, Ramon Castaneda-Priego, Jonas Riest, Gerhard Naegele, Norman J. Wagner
(320d) 20 Years of Corrosion Sensing and Microvisualization of Corrosion Processes 418
William H. Smyrl

(320a) The Need for Engineering Methods for Electrochemical Systems 419
Richard Alkire
(320b) Characterizing Metal-Air Batteries Using Electrochemical Impedance Spectroscopy and Mass Spectrometry 420
Bryan D. McCloskey
(320c) Platinum-Free Fuel Cells for Affordable Zero-Emission Cars 421
Yushan Yan
(351a) Deciphering the Chemical and Physical Mechanisms Active in Elevated Temperature Photocatalytic Synthesis of Hydrocarbons and Ammonia 422
Siris Laursen, Samiksha Poudyal, Morghan Parker
(351b) A High-Throughput Computational Screening Approach for Solar Fuels Photoelectrocatalysis 423
Joseph H. Montoya, Kristin Persson
(351c) Computational Insights into Photo(electro)Chemical Nitrogen Fixation over Titania Catalysts 424
Andrew Medford, Benjamin Comer, Marta Hatzell
(351d) Engineering the Interface to Improve the Efficiencies of Insulator-Protected Semiconductors for Photoelectrochemical Cells 425
Joseph Quinn, Suljo Linic
(351e) The Role of Surface-Bound Dihydropyridine Analogs in Pyridine-Catalyzed CO2 Reduction over Semiconductor Photoelectrodes 426
Thomas P. Senftle, Emily A. Carter
(351f) Surface Plasmon-Assisted Photoelectrochemical CO2 Reduction on Well-Defined Nanostructured Silver Electrodes 427
Elizabeth Corson, Youngsang Kim, Erin Creel, Fen Qiu, Robert Kostecki, Jeffrey Urban, Bryan D. McCloskey
(351g) Artificial Leaf for Carbon Dioxide Photo-Reduction to Fuels 428
Mohammad Asadi, Pedram Abbasi, Kibum Kim, Amin Salehi-Khojin
(351h) Understanding Photoelectrochemistry on Epitaxial Oxides through Surface Electronic Structure 429
Kelsey A. Stoerzinger, Yingge Du, Scott A. Chambers
(352a) Rethinking Grid-Level Energy Storage with Minimal Architecture Zinc-Bromine Batteries 430
Kevin Knehr, Shaurjo Biswas, Hang Huynh, Daniel Steingart
(352b) Highly Energy Dense Cu-Intercalated Bi-Birnessite/Zn Battery 431
Gautam G. Yadav
(352c) Designing Electrolytes for Beyond Li-Ion Batteries Using Coupled High Throughput Ab Intio Calculations and MD Simulations 432
Nav Nidhi Rajput, Xiaohui Qu, Vijayakumar Murugesan, Karl Mueller, Kristin Persson
(352d) Rational Design of Solid-Liquid Interphases for Reactive Metal Batteries 433
Snehashis Choudhury, Lynden A. Archer
(352e) Molecular Design of Redox-Interfaces: Selective Electrochemical Separations and Beyond 434
Xiao Su
(352j) Atomistic modeling of metallic anodes in beyond Li-ion batteries 435
Jeffrey S. Lowe, Donald J. Siegel
(352g) H2 Production Via Photovoltaic-Electrolysis with over 30% Solar-to-Hydrogen Efficiency 436
Jesse D. Benck, Linsey C. Seitz, Jieyang Jia, Yijie Huo, Yusi Chen, Jia Wei Desmond Ng, Taner Bilir, James Harris, Thomas F. Jaramillo
(352h) Remote Control Electrodeposition: Design Criteria for Patterning on Substrates without Direct Electrical Connections 437
Trevor M Braun, Daniel T. Schwartz
(352i) Stalbe Electrochemical Growth in Viscoelastic Flow 438
Shuya Wei, Lynden A. Archer
(358a) Searching for Volume Diffusion 439
Narendra Singh, T. E. Schwartzentruber, Edward L. Cussler
(358b) Condensation on Highly Superheated Surfaces: Unstable Thin Films in a Wickless Heat Pipe 443
Thao T. T. Nguyen, Akshay Kundan, Jiaheng Yu, Peter C. Wayner Jr., Joel L. Plawsky
(358k) Demixing in Dilute Polydisperse Emulsions 444
Rodrigo Reboucas, Michael Loewenberg
(358c) Experimental Investigation of Air Ingress Scenario during Natural Circulation Cooling of a Very Hot Channel 453
Apoorva Rudra, Sanjoy Banerjee, Masahiro Kawaji
(358d) Hydrodynamic Study to Determine Volumetric Mass Transfer Coefficient for Cell Culture Applications 454
Sohail Rasool Lone, Vimal Kumar, Jeffrey Seay, Derek Englert, Hyun-Tae Hwang

(358e) Turbulent Flow of Diutan Biopolymer Solutions and Carbon Nanotube Suspensions in a 4.6 mm ID x 200 L/D Smooth Pipe 455
Preetinder S. Virk
(358g) Fluid Dynamics with Superimposed Mass Transfer of Single Bubbles in Reacting Liquids 456
David Merker, Lutz Bohm, Matthias Kraume
(358h) Experimental Temperature and Fluid Flow Measurements in Packed Beds with Magnetic Resonance Imaging 457
Matthew Skuntz, James E. Maneval, Dinal Perera, Joseph D. Seymour, Ryan Anderson
(358i) Experimental Study of Gas Liquid Mass Transfer Enhanced By Bottom Shear Turbulence 458
Tom Lacassagne, Serge Simoã«ns, Mahmoud El Hajem, Jean-Yves Champagne
(358j) Influence of Thermophysical Properties of Cryogenic Fluids on Growth and Collapse of Cavitating Bubbles 461
Arpit Mishra, Rajaram Lakkaraju, Arnab Roy, Parthasarathi Ghosh
(360a) Interfacial Activity of Silica and Ethyl Cellulose Particles in Fluid-Fluid Interfaces 462
Songcheng Wang, Yi Zhang, Jiarun Zhou, Ruiyang Zhao, Gregory Benz, Stephane Tcheimou, J. Carson Meredith, Sven H. Behrens
(360b) Adhesion of Explosive Particles to Functionalized Surfaces 463
Darby J. Hoss, Sanjoy Mukherjee, Bryan W. Boudouris, Stephen P. Beaudoin
(360j) Dynamic Collision of Water Drops Coated with Food Grade Surfactant Poly-Glycerol-Poly-Ricinoleate (PGPR) in Canola Oil 464
Srinivas Mettu, Chu Wu, Raymond R. Dagastine
(360d) Design of New Liquid Crystal-Based Systems with Improved Chemo-Responsiveness Towards the Detection of Nerve Gases 465
Tibor Szilvasi, Nanqi Bao, Nicholas L. Abbott, Manos Mavrikakis
(360e) Factors Affecting the Transient Stability of Solid Stabilized Emulsions 466
Ashwin Kumar Yegya Raman, Michael Miranda, Jarred Kelsey, Jeff White, Clint P. Aichele
(360f) Adsorption of Star Polymers to Fluid Interfaces 467
Yun-Ru Huang, Robert D. Tilton
(360g) Influence of Order within Non-Polar Monolayers on Hydrophobic Interactions 468
Nicholas L. Abbott
(360h) How Molecular Structure of Surfactants Determine the Dynamics and Viability of Wettability Alteration 469
Soumik Das, Quoc P. Nguyen, Roger T. Bonnecaze
(360i) Chiral Nonlinear Rheology of Phospholipid Monolayers 470
Joseph A. Zasadzinski, Todd M. Squires, Kyuhan Kim, Siyoung Choi
(360j) Visualizing Nanoscopic Topography and Patterns in Freely Standing Thin Films 471
Yiran Zhang, Subinuer Yilixiati, Vivek Sharma
(365a) Heavy Oil and Bitumen Complex Viscosity Simulation - Impacts of Phase Behavior on Rheology 472
John M. Shaw, Sepideh Mortazavi-Manesh, Mildred Becerra
(365b) Maxwell and Marco: Thermophysical Property Needs of Rate-Based Process Simulation Tools 473
Ross Taylor
(365c) The Expanded Fluid Concept for Transport Property Correlations 474
Harvey W. Yarranton, Francisco Ramos-Pallares, Shawn D. Taylor
(365d) Grid Evaluation of Pure-Compound Properties 475
Vladimir Diky, Andrei Kazakov, Kenneth Kroenlein
(365e) Water Content of Natural Gas Systems in Equilibrium with an Aqueous or a Hydrate Phase - Experimental Measurements and Molecular Modeling 476
Walter G. Chapman, Wael A. Fouad, Kyoo Song, Kenneth R. Cox
(365f) When Experimental and Predicted Data Are in Conflict, What Should We Trust? 477
Ala Bazyleva, Eugene Paulechka, Vladimir Diky, Joseph W. Magee, Kenneth Kroenlein
(369a) Global Strain-Field Mapping of a Carbon Nanotube-Laden Interface Using Digital Image Correlation 478
Sahil R. Vora, Brice Bognet, Huseini S. Patanwala, Charles Young, Virgile Daux, Anson Ma
(369b) Simulations of Transient Dynamics of Dense Suspensions 479
Rui Zhang, Endao Han, Heinrich Jaeger, Juan J. De Pablo
(369c) Interfacial Rheology of Coexisting Solid and Fluid Monolayers 480
Joesph A. Zasadzinski, Amit Kumar Sachan, Siyoung Choi, Kyuhan Kim, Ka Yee Lee, Luke Hwang, Todd M. Squires
(369d) Modulation of Dilatational Rheology with Interfacial Curvature and Phase Morphology 481
Amit Kumar Sachan, Joesph A. Zasadzinski

(369e) Surface Forces, Flows and Fluxes Underlying Nanoridge Formation and Instabilities in Stratifying, Micellar Freestanding Films 482
Yiran Zhang, Vivek Sharma
(369f) Wrinkling Instabilities in Thin Inhomogenously Stretched Viscous Sheets 483
Siddarth Srinivasan, Zhiyan Wei, L. Mahadevan
(369g) Fluid Instabilities, Density Stratification, and Lee Waves That Result in Growth Limitations and Morphological Instability of Liquid-Crystal Interfaces 484
Jeffrey H. Peterson, Jeffrey J. Derby
(369h) Liquid-Film Coating on Topographically Patterned Rotating Cylinders 485
Weihua Li, Marcio S. Carvalho, Satish Kumar
(369i) Drying of Multicomponent Thin Films on Substrates with Topography 486
Truong Pham, Xiang Cheng, Satish Kumar
(369j) Effect of Elasticity on Stability of Viscoelastic Liquid Curtain 487
Alireza Mohammad Karim, Wieslaw Suszynski, Lorraine F. Francis, Marcio S. Carvalho
(380a) Flow, Arrest and Yielding in Dense Colloidal Suspensions - Glasses Vs. Gels 488
Chinedum O. Osuji
(380b) Modeling a Hydrodynamic Instability in Freely Settling Colloidal Gels 489
Zsigmond Varga, Jennifer L. Hofmann, James Swan
(380c) Effects of Attractive Interparticle Interactions on Normal Stresses and Shear-Induced Migration of Colloids 490
Nayoung Park, Jacinta C. Conrad
(380d) Connecting Bidisperse and Polydisperse Suspension Rheology 491
Sidhant Pednekar, Jaehun Chun, Jeffrey F. Morris
(380e) The Steric Effect of a Solid Boundary on the Orientation Distribution of a Dilute Suspension of Rigid Rod-like Particles Under Shear Flow 492
Saman Monjezi, James Jones, Shawn Oettle, Joontaek Park
(380f) Particle Crystallinity in Automated Langmuir-Blodgett Deposition: Effect of Speed and Concentration 493
James F. Gilchrist, Kedar Joshi, Zhiqiao Zeng, Xue Li
(380g) A Hybrid Lattice Boltzmann-Random Walk Method for Multiphase Heat Transfer 494
Aaron Lattanzi, Xiaolong Yin, Christine Hrenya
(380h) Separation By Design - Towards Simulation Guided Engineering of Coiled Channels for Precise Particle Separation 495
Jakob D. Redlinger-Pohn, Federico Municchi, Stefan Radl
(380i) Simulations of Particle-Laden Flows in Microchannels 496
Wouter Van Aeken, Khurram Shahzad, Tom Van Gerven, Simon Kuhn
(392b) Prediction of Solid State Phase Diagrams Using Multistate Reweighting and Jacobian Mapping 497
Natalie Schieber, Eric Dybeck, Nathan Abraham, Michael Shirts
(392c) Evaluation of Defect Concentrations in Crystalline Systems Via Molecular Simulation 498
Apoorva Purohit, Andrew J. Schultz, Jeffrey Errington, David A. Kofke
(392d) Nature of the Instability of the Body-Centered-Cubic (bcc) Structure in Classical Hard Sphere Solids 499
Vadim B. Warshavsky, David M. Ford, Peter A. Monson
(392e) Entropy Driven Solid-Solid Transitions in Colloids 500
Chrisy Xiyu Du, Greg Van Anders, Richmond S. Newman, Sharon C. Glotzer
(392f) Calculation of Glass Transition Using Sanchez-Lacombe - Eos 501
Margarete Roericht, Kirstin Taufertshofer, Sabine Enders
(409a) Engineering Shape Entropy for Colloidal Crystal Design 502
Yina Geng, Greg Van Anders, Paul Dodd, Julia Dshemuchadse, Sharon C. Glotzer
(409b) Controlling Anisotropic Colloidal Assembly in External Fields 503
Isaac Torres-Diaz, Michael A. Bevan
(409c) Deposition Dynamics of Rod-Shaped Colloids during Transport in Porous Media 504
Huilian Ma, Ke Li
(409d) Engineering the Phase Behavior of Anisotropic Particles By Inverse Design 505
Chrisy Xiyu Du, Greg Van Anders, Paul Dodd, Julia Dshemuchadse, Sharon C. Glotzer
(409e) Self-Assembly of Magnetic Janus Colloids Via Brownian Dynamics Simulation 506
Gabriel Vega-Bellido, Ronal A. Delacruz-Araujo, Ilona Kretzschmar, Ubaldo M. Cordova-Figueroa
(409f) Brownian Dynamics of a Spherical Janus Particle Near a Boundary As a Tool to Investigate TIRM 507
Aidin Rashidi, Christopher L. Wirth
(409g) Examining the Stability of Amphiphilic Janus Particle-Laden Interfaces 508
Ellen M. Knapp, Ilona Kretzschmar, Raymond Tu

(409h) Self-Assembly of Amphiphilic Janus Nanoparticles 509
Alan Hanley, Allan E. David
(409i) Magnetic Janus Particle Aggregation Rate at Long Times 510
Thomas Long, Ilona Kretzschmar, Joel Koplik
(414a) Using ?2 rheology to Characterize Consecutive Phase Transitions in a Hydrogenated Castor Oil Colloidal Gel 511
Matthew Wehrman, Seth Lindberg, Kelly M. Schultz
(414b) Hydrodynamic Entrainment in Micro-Confined Suspensions and its Implications for Two-Point Microrheology 512
Christian Aponte-Rivera, Roseanna N. Zia
(414c) Shear-Induced Structural Ordering in Jammed Suspensions of Soft Particle Glasses 513
Fardin Khabaz, Tianfei Liu, Michel Cloitre, Roger T. Bonnecaze
(414d) Dynamic Simulation of Aging in a Hard-Sphere Colloidal Glass after Volume-Fraction Jumps 514
Jialun Wang, Xiaoguang Peng, Xi Li, Gregory B. McKenna, Roseanna N. Zia
(414e) A New Conformation Tensor Based Macroscopic Model for Emulsions with Particle Inertia 515
Paul M. Mwasame, Antony N. Beris, Norman J. Wagner
(414f) Polyelectrolyte - Particle Flocculation in Complex Aqueous Solutions and Mixed Hydrodynamic Fields 516
Athena E. Metaxas, Nikolas A. Wilkinson, Cari S. Dutcher
(414g) From Filaments to Coils: Controlling the Dynamics of Linked Colloidal Particle Chains 517
Steve Kuei, Sibani Lisa Biswal
(414h) A Combined Experimental and Numerical Analysis of DNA-Functionalized Colloidal Particle Deposition in a Channel Flow 518
Young Ki Lee, Christopher Porter, John C. Crocker, Scott L Diamond, Talid Sinno
(414i) Structural and Rheological Relaxation upon Flow Cessation in Colloidal Dispersions: Transient, Nonlinear Microrheology 519
Ritesh P. Mohanty, Roseanna N. Zia
(422a) Durable, Self-Hydrating, Tungsten Carbide-Based Composite Polymer Electrolyte Membrane Fuel Cells 520
Weiqing Zheng, Liang Wang, Stephen A. Giles, Yushan Yan, Ajay K. Prasad, Dionisios G. Vlachos
(422c) Teaching an Old Electrocatalyst to Do New Tricks 521
Daniel V. Esposito, Natalie Labrador, Eva Songcuan, Jacob Robinson, Bicheng Xu, Sophia Kurdziel
(422d) Using Surface Chemistry to Understand Aqueous-Phase Thermal Catalytic and Electrocatalytic Hydrogenation of Bio-Oil Model Substrates 522
Nirala Singh, Udishnu Sanyal, John L. Fulton, Oliver Gutierrez, Donald M. Camaioni, Charlie Campbell, Johannes A. Lercher
(422e) Design of Heterogeneous Catalysts for High Temperature Electrochemical Reduction of CO2 523
Juliana S. A. Carneiro, Xiang-Kui Gu, Zachary Kuczera, Eranda Nikolla
(422f) Enhancing Methanol Electrooxidation Activity Using Double Oxide Catalyst Support 524
Ahmed Jasim, Yangchuan Xing
(422g) Study of Impact of Electrode Catalyst Loadings on Thin Flexible Fuel Cell (TFFC) Performance 525
Seyed Reza Mahmoodi, Ronald S. Besser, Matthew Mayer
(422h) Transition Metal Modified Graphitic Mesoporous Carbon for PEMFC Application 526
Khondker Sultana, Dereje Worku, Md. Tashfin Zayed Hossain, Shamsuddin Ilias
(425a) Capillary Foams: How a Small Amount of Oil Can Make a Big Difference 527
Yi Zhang, Sven H. Behrens, J. Carson Meredith
(425b) Wetting on Strain Induced Microstructured Surfaces 528
Vartika Parihar, Soumen Das, Sunando Dasgupta
(425c) Water Wetting on Surfaces Under Fuel Oil Containing Surfactants and Its Implication for Coalescence Separation of w/o Emulsions 530
Qian Zhang, Yanxiang Li, Lixia Cao, Lei Li, Chuanfang Yang
(425e) Dynamic Surface Tension Measurements of Ionic Surfactants Using Maximum Bubble Pressure Tensiometry 531
Camilla U. Ortiz, Vivek Sharma
(425g) The Importance of Spreading Pressure on Adsorption Based Surface Energy Measurements; The Case of IGC 532
Eftychios Hadjittofis, Geoff G. Z. Zhang, Jerry Heng
(425h) Stabilization of Glycerol/Dodecanol Pickering Emulsions with Surface-Grafted Silica Nanoparticles 533
Guolin Zhao, Shuangliang Zhao, Junyin Xiao, Honglai Liu, Marc Pera-Titus

(425i) In situ Observations of Dynamics of Active Braze Joining 534
Anne M. Grillet, Robert M. Garcia, David A. Barringer, Adam D. Martinez
(425j) High Throughput Fabrication of Synthetic Asymmetric Bacterial Membranes 535
Sepehr Maktabi, Li Lu, Jeffrey W. Schertzer, Paul R. Chiarot
(431a) Design Insights for a Thermal Visbreaking Pilot Based on Microwave Heating 536
Shawn D. Taylor, Simon Andersen, Tracy Neitz, David O'Brien, Amin Saeedfar, Merlyn Pulikkathara, Wai-Ming Tam
(431b) On Cubic Eos Interaction Parameter Estimation for Long Chain n-Alkane + Aromatic Binary Mixtures 537
Sourabh Ahitan, Miao Luo, John B. McLaughlin, John M. Shaw
(431c) Uncertainty in Heat Exchanger Design for Diluted Bitumen 538
Harry Z Ha, Paul M. Mathias

VOLUME 2

(431d) Practical Aspects in Using Distillation VLE and Enthalpy Models 550
Daniel R. Summers, Senthil Krishnamoorthy
(431e) Bubble Pressure Measurement and Prediction for n-Alkane + Naphthenic Hydrocarbon Binary Mixtures 551
Sourabh Ahitan, John M. Shaw
(431f) When Is It Appropriate to Kiss? a Case Study on the Adverse Consequences of Overfitting GE Models 552
Paul M. Mathias
(433a) Photochemical Reduction of Enzymes for Light-Driven Chemical Reactions 553
Paul King
(433c) Termite-Inspired Electrochemical Processing of Lignocellulose to Chemicals and Fuels 554
Xiong Peng, Ying Liu, Andrea Kadilak, Sujan Shrestha, Leslie M. Shor, William E. Mustain
(433d) Nature-Inspired Flow Fields for PEM Fuel Cells 555
Jason Cho, Tobias P. Neville, Panagiotis Trogadas, Billy Wu, Dan Brett, Marc-Olivier Coppens
(435a) Numerical Analysis of the Flow, Transport, and Interfacial Phenomena Associated with Growth of Crystalline CZT Under Crucible Rotation 556
Mia S. Divecha, Jeffrey J. Derby
(435b) Modeling the Transient Shear Flow and Predicting Large Amplitude Oscillatory Shear (LAOS) Flow of a Thermoreversible Gel Using a Scalar Structure Parameter Thixotropic Model 557
Matthew Armstrong, Ryan P. Murphy, Norman J. Wagner, Antony N. Beris
(435c) Modeling and Uncertainty Quantification of Vapor Diffusion and Reactions in Polymer 568
Yunwei Sun, Hom Sharma, Elizabeth Glascoe
(435e) Modeling of Transport and Reaction in a Novel Hydride Vapor Phase Epitaxy System 569
Min Yao, James B. Rawlings, Thomas F. Kuech
(435f) Discretized Modeling of a Simple Motionless 3D Printer Based on Retarded Bending Motion and Electrically Controlled Patterning of Fiber Deposition 570
Mounica Jyothi Divvela, Yong Lak Joo
(435g) A Forward Modeling Approach for the Inverse Estimation of Transient Local Heat Fluxes 571
Jiu Luo, Ya-Qiao Wang, Jia-Li Luo, Dong-Chuan Mo, Yuan-Xiang Fu, Shu-Shen Lyu, Yi Heng, Hiroyuki Ozoe
(435h) Modeling of the Evaporation of the Polymer Slurry in the Porous Media 584
Shuji Hironaka, Gen Inoue, Jun Fukai, Yoshifumi Tsuge
(435i) Flow of a Power-Law Fluid across an Asymmetrically Confined Rotating Cylinder 592
Pooja Thakur, Naveen Tiwari, R. P. Chhabra
(435j) Binary Gas Mixture in a High Speed Channel 593
Sahadev Pradhan
(444a) The Shape Evolution of Pendant Droplets in Miscible Environments 640
Dan Walls, Simon Haward, Amy Shen, Gerald G. Fuller
(444b) Shape Evolution of Miscible Drops with Arbitrary Viscosity Ratio: Lagrangian-Eulerian Swarms of Stokeslets and Subgrid Resolution 641
Ludwig C Nitsche, Paola Leon Plata, Rafael G. Henriquez Rivera, Ying Liu
(444c) Coalescence of Drops Due to a Constant Force Interaction in a Viscous Quiescent Fluid 642
John M. Frostad, Alexandra Paul, L. Gary Leal
(444e) Nanoemulsion Formation: Controlling and Predicting Droplet Size 643
Ankur Gupta, T. Alan Hatton, Patrick S. Doyle

(444f) Using a Novel Approach to Model Drop Size Distribution: The Adaptive Multi-Size Group Method 644
Thomas Eppinger, Koichi Akano, Simon Lo, Alexander Vichansky, Ravindra Aglave
(444g) Describing Emulsions in Terms of Pseudo-Potential Lattice Boltzmann Modeling 645
Mohammad Pourtousi, Siddhartha Mukherjee, Pieter Berghout, Orest Shardt, Harry E. A. Van Den Akker
(444h) Comparison of the Rise Dynamics of Standard Bubbles and Oil-Coated Bubbles: Experiments and Simulations 646
Songcheng Wang, Manoj Kumar Tripathi, Kirti Chandra Sahu, J. Carson Meredith, Sven H. Behrens
(444j) A Quadrature-Based Model for Polydisperse Laminar Bubbly Flows 647
Jeffrey C Heylmun, Bo Kong, Alberto Passalacqua, Rodney O. Fox
(234y) Stability of Stagnation Point Flows of Newtonian and Complex Fluids 648
Noa Burshtein, Simon Haward, Amy Shen
(453a) Prediction of Mean Ionic Activity Coefficient of NaCl-H2O System at High Concentrations Using Molecular Dynamics Simulations 649
Nazir Hossain, Ashwin Ravichandran, Rajesh Khare, Chau-Chyun Chen
(453b) Thermodynamic Modeling of Lithium Salt Systems with the Electrolyte NRTL Model 650
Toni Kirkes, Chau-Chyun Chen
(453c) Why Tetrabutylphosphonium Hydroxide-Water Mixtures Can Dissolve Cellulose at Extremely High Water Mass Fractions 651
Brad Crawford, Ahmed E. Ismail
(453d) Phase Behavior of Compressed Gases in Ionic Liquids at the Liquid-Solid Transition Point 652
David L. Minnick, Brooks B. Danahy, Mark B. Shiflett
(453e) Is Difference in Hydrogen Bond Ability of Anions an Exclusive Descriptor of Nonideality in Binary Ionic Liquid Mixtures? 653
Utkarsh Kapoor, Jindal K. Shah
(453f) Polymer-Solvent Phase Behavior of Lignin with Hot Aqueous Solvent Systems 654
Junhuan Ding, Adam S. Klett, Jordan A. Gamble, Graham W. Tindall, Mark C. Thies, Mark E. Roberts
(453g) Application of Thermodynamic Models in the Context of Liquid Chromatography 655
Franziska Ortner, Chantal Ruppli, Marco Mazzotti
(453h) The Solubility and the Freezing Point Depression (FPD) Measurements of Na-1-Ethanethiolate, Na-1-Propanethiolate, Na-2-Propanethiolate, Na-1-Butanethiolate and Na-2-Methyl-2-Propanethiolate Salts in Pure Water 656
Javeed Awan
(464a) Computation of Amino Acid-Mineral Surface Thermodynamics and Interaction Modes via Importance Sampling 701
Kayvon Tabrizi, Michael S. Pacella, Jeffrey J. Gray
(464b) Research on the Mechanism of Spontaneous DNA Extension on Grooved Surface Covered with Cationic Lipid Bilayers 702
Tzu-Hsien Fan, Hou-Jun Guo, Chih-Chen Hsieh
(464c) Effect of Interchain Associations on Hybridization Activity of DNA Brushes 703
Hao-Chun Chiang, Rastislav Levicky
(464d) Understanding Marine Mussel Adhesion: Interfacial Energy from Microscopic to Macroscopic Length Scales 704
George Degen, Jacob Israelachvili
(464e) Tuning Underwater Adhesion with Cation-Pi Interactions 705
Matthew A. Gebbie, Jacob Israelachvili, J. Herbert Waite
(464f) Influence of the Surface Charge on the Brownian Motion of Colloidal Particles Near Surfaces with Biomimetic Characteristics 706
Juan Manuel Hernandez Meza, Angeles Ramirez-Saito, B. Jose Luis Arauz-Lara, Said E. Aranda Espinoza, Bernardo Yanez Soto
(464g) Charged Nanoparticles Interacting with Giant Vesicles Fabricated from Inverted-Headgroup Lipids 707
Lu Wang, Noah Malmstadt
(464h) Modulation of Phase Morphology of Surfactant Monolayer with Interfacial Curvature 708
Amit Kumar Sachan, Joseph A. Zasadzinski
(464i) Recognition of Target Cells By Vibrio cholerae outer Membrane Vesicles 709
Elnaz S. Rasti, Angela C. Brown
(468a) Applying Computational Tools of Polymer Field Theory to out-of-Equilibrium Polymer Solutions in Flow 710
Charles Young, Charles E. Sing

(468b) Capillary Thinning and Pinch-Off Dynamics and Extensional Rheology of Polyelectrolyte Solutions 711
Leidy N. Jimenez, Jelena Dinic, Nikhila Parsi, Vivek Sharma
(468c) Relaxation of Knotted Polymers 712
Vivek Narsimhan, Alexander Klotz, Patrick S. Doyle
(468d) On Synthetic Drilling Fluid Rheological Modelling 713
Silvio Baldino, Reza E. Osgouei, Evren Ozbayoglu, Stefan Z. Miska
(468e) Experimental Evidence for Shear Banding in Large Amplitude Oscillatory Shear of Entangled DNA Solution. 722
Seunghwan Shin, Kevin D. Dorfman, Xiang Cheng
(468f) PadÃ© Approximants for Shear Stress in Large-Amplitude Oscillatory Shear Flow 723
Chaimongkol Saengow, A. Jeffrey Giacomin, Nidal Khalaf, Martin Guay
(468g) The Effect of Branching on Shear Band Formation and Evolution in Wormlike Micelles (WLMs) 724
Michelle A. Calabrese, Simon Rogers, Lionel Porcar, Norman J. Wagner
(468h) Predictions for Non-Linear Flows of Polydisperse Blends Based on a Differential-Constitutive Analogue of the Double Reptation Model 725
Joseph Peterson, L. Gary Leal, Glenn H. Fredrickson
(468i) Marangoni Instability during the Non-Solvent Induced Phase Separation of a Ternary Polymer Solution 726
Douglas Tree, Glenn H. Fredrickson
(482a) Highly Stable Mn-Based Nanocarbon: A New Class of Bifunctional Electrocatalyst for Oxygen Reduction and Evolution Reactions 727
Shiva Gupta, Gang Wu
(482b) Density Functional Theory Study of the Oxygen Reduction Reaction in a Graphene Surface Modified with S or N 728
Elizabeth Montiel-Macias, Ysmael Verde-Gomez, Perla B. Balbuena
(482c) Highly Stable Nanocarbon Supports for Pt Cathode Catalysts in Polymer Electrolyte Fuel Cells 729
Mengjie Chen, Gang Wu
(482d) Approaching 2 W-cm-2 Aemfcs through Electrode Engineering and Controlling the Cell Water Content and Balance 730
Travis J. Omasta, Xiong Peng, William E. Mustain
(482e) Design and Engineering of Membraneless Electrolysis Devices 731
Jonathan Davis, Justin Bui, Daniel V. Esposito
(482g) Design of a Novel Electrochemical Membrane Reactor for Hydrogen Production Via the S-NH3 Cycle 732
Raul Marquez-Montes, Raul Orozco-Mena, Virginia Collins-Martinez, Eduardo Herrera-Peraza, David Chavez-Flores, Victor Ramos-Sanchez
(488a) Portable Smartphone-Enabled Capillary-Based Viscometer and Tensiometer 733
Jose C. Contreras-Naranjo, Xiaorui Dong, Victor M. Ugaz
(488b) An Experimental Study of a Liquid Drop Impinging on a Liquid Pool 734
Manas Ranjan Behera, Anirvan Dasgupta, Sudipto Chakraborty
(488c) Arresting Dissolution By Interfacial Rheology Design 736
Peter J. Beltramo, Manish Gupta, Alexandra Alicke, Jan Vermant
(488d) The Importance of the Amorphous Material Interface in the Measurement of Thermodynamic Parameters, Using Inverse Gas Chromatography 737
Eftychios Hadjittofis, Geoff G. Z. Zhang, Jerry Heng
(488e) Mechanisms of Dynamic Wetting Failure in the Presence of Soluble Surfactants 738
Chen-Yu Liu, Marcio S. Carvalho, Satish Kumar
(488f) Electrostatic Assist of Liquid Transfer in Printing Processes 739
Chung-Hsuan Huang, Satish Kumar
(488g) Analysis of Temperature Gradient Zone Melting and Annealing for Mitigation of Second-Phase Particles in Single Crystals 740
Kerry Wang, Nathan T. Morgan, Jeffrey J. Derby
(488h) In situ Partitioning Measurements of Multiple Fermentation End-Products into Lipid Membranes 741
Geoffrey D. Bothun, John Nunes, Carmen Scholz, Javier Sanchez
(488i) Synchronization and Traveling Waves within Arrays of Electrostatic Oscillators 742
Shashank Pandey, Yong Dou, Charles A. Cartier, Mikolaj Kowalik, Kyle J. M. Bishop
(494a) Probing Viscosity of Complex Emulsions with Micro-Rheometry 743
Archit Dani, Shweta Narayan, Cari S. Dutcher

(494b) The Study on Mechanism of Flow Distribution and Its Optimization in a Spray Column 744
Penghui Wang, Gance Dai
(480e) Transition between Turbulence Regimes in Particle-Laden Channel Flows 745
Jesse Capecelatro, Olivier Desjardins, Rodney O. Fox
(494d) Analysis of Interfacial Forces of a Slurry Bubble Column 746
Bruna L Mees, Tatiana Matiazzo, Jaci Carlo Schramm Camara Bastos, Henry F. Meier, Marcela Kotsuka Silva
(494e) Semi-Batch Liquid-Liquid Taylor Vortex Flow: Comparison of 3-D and 2-D Eulerian-Eulerian CFD Simulations and Experimental Measurements for Fluid Flow Patterns and Droplet Holdup 747
John Lavey, Arya Haghighat, Charlton Campbell, Xi Gao, Bo Kong, Meesha Legg, Michael Olsen, R. Dennis Vigil
(494g) Experiments on Bubble Behaviors in Three Phase Internal-Loop Airlift Reactor Using Optical Fiber Probe Measurements 748
Yongxiang Gao
(494h) Flow Properties of Gas Hydrate Slurry in Oil-Dominated Transportation Systems 749
Lin Ding, Bohui Shi, Yang Liu, Shangfei Song, Jing Gong
(494i) Relationship between Flow Pattern in Cyclone Classifier and Classification Process 750
Zhanpeng Sun, Guogang Sun, Yan Du
(511a) Computational Study of the Stability of the Miniprotein Trp-Cage in Extreme Environments 760
Betul Uralcan, Sang Beom Kim, Pablo G. Debenedetti
(511b) Engineering Robust Activity in Extremophilic Enzymes 761
Siva Dasetty, Weigao Wang, Mark Blenner, Sapna Sarupria
(511c) Effect of Manufacturing Conditions on the Fabrication of Protein/Polysaccharide Biocompatible Nanotubes (BNTs) 762
Luis F. Maldonado, Jozef Kokini
(511d) Towards a Thermodynamic Model for Predicting Coiled-Coil Protein Structures 763
Mojtaba Jokar, Korosh Torabi
(511e) Effect of Residual Secondary Structure Propensity on Liquid-Liquid Phase Separation of Tdp-43 764
Gregory L. Dignon, Gul H. Zerze, Nicolas Fawzi, Young C. Kim, Jeetain Mittal
(511f) Solvation Contribution to the Conformational Preferences of the Hydrated Peptides 765
Gul H. Zerze, Alexander Bourque, Jeetain Mittal
(511g) Influence of Backbone Hydrogen Bonding on the Stability of the WW Domain of the Protein Pin1 766
Daniel Markthaler, Niels Hansen
(511h) Molecular Simulations of the Force Spectroscopy of a Biomimetic Polymer 775
Aviel Chaimovich, Christian Leitold, Christoph Dellago
(511i) Insights into the Binding of Î²-Wrapins to Amyloidogenic Proteins Using Simulations and Experiments 776
Asuka A. Orr, Michael M. Wordehoff, Hamed Shaykhalishahi, Ewa A. Mirecka, Sai Vamshi R Jonnalagadda, Wolfgang Hoyer, Phanourios Tamamis
(512a) Monte Carlo Simulations Probing Interfacial Adsorption and Uptake in Surfactant Bilayers 777
J. Ilja Siepmann, Mona Minkara, Jingyi Chen, Cor J. Peters
(512c) Phase and Interfacial Behaviour of Aqueous-Two Phase Systems Based on Hyperbranched Polymers 778
Andres Kulaguin Chicaroux, Tim Zeiner
(512d) From Metallic Hydrogen to Polymeric Sulfur: Peculiar Thermodynamics of Polyamorphic Fluids 779
Lauren E. Amrhein, Mikhail A. Anisimov, Frederic Caupin, Michal Duska, Amanda Rosenbaum
(512e) Computational Studies of Mixtures of Refrigerants, Ionic Liquids and Deep Eutectic Solvents for Absorption Refrigeration Systems 780
Rubaiyet Abedin, Sharareh Heidarian, John C. Flake, Francisco R. Hung
(512f) Oil Desulfuriztion Using Deep Eutectic Solvents Via Liquid-Liquid Extraction 781
Samah E. E. Warrag, Nerea R. Rodriguez, Martin Van Sint Annaland, Maaike C. Kroon, Cor J. Peters
(512g) Relevance of Thermophysical Property Data in Rate-Based Gas Treating Simulations 782
Diego E. Cristancho, Jacob M. Crosthwaite, Allen Day, John Dowdle, Clare Worley
(512h) Identification of the Equilibrium Parameters for the Modeling of the Hydrogen Solubility in Heavy Crude Cuts Using the Matlab-Aspen Connection 793
Camilo Monroy-Pena, Johana Orjuela
(527a) Molecular Perspective on Protein-Protein Interactions at the Tight Junctions Interface 794
Flaviyan Jerome Irudayanathan, Shikha Nangia
(527b) Pore Formation By Aggregates of Antimicrobial Peptides in Dmpc Bilayers 795
Yuan Lyu, Maya Frityanti, Xiao Zhu, Ganesan Narsimhan

(527c) Single-Molecule Characterization of Protein Adsorption to Multivalent Glycan Polymers 796
Bhargava Nemmaru, Sonia K Brady, Matthew J. Lang, Shishir Chundawat

(527d) Controlling Local Hydrophobicity in Poly(ethylene glycol) Brushes with Poly(sulfobetaine) to Mediate the Conformation of Fibronectin on Biomaterial Surfaces 797
David Faulon Marruecos, Hye Hyun Kim, Michael Shirts, Daniel K. Schwartz, Joel L. Kaar

(527e) Adsorption of Recombinant Human Interleukin-1 Receptor Antagonist to Silicone Oil-Water Interfaces Leads to Gel Formation and Subsequent Surface-Induced Aggregation 798
Lea L. Sorret, Theodore W. Randolph

(527f) Fibrinogen Adsorption Onto Phospholipid Monolayers: Aging and Stiffening 799
Ian Williams, Todd M. Squires

(527g) Morphological and Mechanical Studies of Multicomponent Phospholipid Monolayers Containing 27-Hydroxycholesterol 800
Benjamin Stottrup, Amit Kumar Sachan, Joesph A. Zasadzinski

(527h) Understanding the Effect of Engineered Carbon Nanodiamonds on the Reversible Collapse of Lung Surfactant Monolayers 801
Aishik Chakraborty, Prajnaparamita Dhar

(527i) High Resolution Optical and Electrical Recording of Free Standing Lipid Bilayers 802
Peter J. Beltramo, Jan Vermant

(535a) Dynamics of Lipid Lamellae and Lamellar Vesicles in Shear and Extensional Flows 803
Subas Dhakal, Radhakrishna Sureshkumar

(535b) Probing the Structure of High Viscosity Complex Fluids at High Shear Rates 804
Javen Weston, Kathleen Weigandt, Steven D. Hudson

(535c) Mechanistic Constitutive Model for Wormlike Micelle Solutions with Flow-Induced Structure Formation 818
Sarit Dutta, Michael D. Graham

(535d) Rheology and Microstructure of Thermoreversible Micellar Crystals in Ionic Liquid 819
Ru Chen, Carlos R. Lopez-Barron, Norman J. Wagner

(535e) Use of Microfluidics and Advanced Microscopy to Study Shear-Induced Structuring in Wormlike Micellar Solutions 823
Carla Caiazza, Valentina Preziosi, Giovanna Tomaiuolo, Denis O'Sullivan, Vincenzo Guida

(535f) A Multiscale Tensor-Based Model for the Rheology of Aggregating Thixotropic Colloidal Suspensions 824
Paul M. Mwasame, Antony N. Beris, Norman J. Wagner

(535g) Effect of Shear Stress on Rheological Properties of Fibrous Colloidal Gels during Gel-Sol Transitions Induced By Osmotic Pressure Gradients 825
Matthew Wehrman, Seth Lindberg, Kelly M. Schultz

(535h) Microstructural Basis for Colloidal Gel and Glass Aging 826
Melissa B. Gordon, Christopher J. Kloxin, Norman J. Wagner

(535i) The Mechanism for Shear Thickening in Viscoelastic Suspensions 827
Mengfei Yang, Eric S. G. Shaqfeh

(535j) Homogenization of Self-Organizing Colloidal Particles in a Microscopy Chamber 828
Paul W. Todd, Michael (Andy) Kurk, Scott Moyers

(543a) Elevated Pressure Interfacial Dynamics of LS Surfactants at the Water-CO2 Interface 829
Zachary R. Hinton, Nicolas J. Alvarez

(543b) Atomistic Simulations of Micellization and Adsorption of Imidazoline-Based Surfactants Near Metal-Water Interfaces 830
Sumit Sharma, Yathish Kurapati

(543c) Surfactant Chemistry and Counter-Ion Clouds in Bile Salt Stabilized Swcnts 831
Ketan S. Khare, Frederick R. Phelan Jr.

(543d) Tailoring Inhibitors of Pathological Crystallization: New Platforms for Drug Design 832
Bryan G. Alamani, Doyoung Kim, Mangalaa Dinivahi, Diego Guala, Brandon Melendez, Jeffrey D. Rimer

(543e) Directed Crystal Nucleation and Growth of Blue Phases By Chemically Patterned Surfaces 833
Xiao Li, Jose Martinez-Gonzalez, Juan Hernandez-Ortiz, Ye Zhou, Monirosadat Sadati, Rui Zhang, Juan J. De Pablo, Paul F. Nealey

(543f) Transient Reorientation of Liquid Crystals Induced By Water Transport 834
Hadi Ramezani-Dakhel, Monirosadat Sadati, Rui Zhang, Mohammad Rahimi, Khia Kurtenbach, Nicholas L. Abbott, Benoit Roux, Juan J. De Pablo

(543h) DNA-Micropost Adsorption and Enhanced Transport in More Crowded Micropost Arrays 835
Yeng-Long Chen, Wei Chien, Fan-Tso Chien

(543i) A Miniaturized, Radial Langmuir Trough for Simultaneous Visualization and Dilatational Deformation of a Complex Fluid-Fluid Interface 836
Joseph R. Samaniuk

(554a) Electrocatalytic CO2 Reduction for Enhanced C-C Coupling 837
Youngkook Kwon
(554c) Reaction Path Discovery Under Potential Bias 838
Eric Walker, Paul M. Zimmerman
(554e) Electrocatalytic Carbon Fixation on Molecular-Material Hybrids 839
Zhi Cao, Christopher Chang
(554f) Investigation of the Morphological Factors That Drive Selectivity on Polycrystalline Cu during CO2 Electroreduction 840
Alexandros N. Karaiskakis, Elizabeth J. Biddinger
(554g) Modulation of Charge Transfer Kinetics By Two Orders of Magnitude at Back-Gated Monolayer MoS2 Electrodes 841
Yan Wang, C. Daniel Frisbie
(574a) Predicting Vapor-Liquid Equilibria with Augmented Ab Initio Potentials 842
Maryna Vlasiuk, Richard J. Sadus
(574b) New Vapor Pressure Prediction with Improved Thermodynamic Consistency Using the Riedel Equation 843
Joseph W. Hogge, Neil Giles, Richard L Rowley, Thomas A. Knotts Iv, W. Vincent Wilding
(574c) Solvent Screening and Pure Component Thermophysical Property Prediction for Pharmaceutical Process Design 844
Getachew S. Molla, Lukasz Ruszczynski, Jens Abildskov, Gurkan Sin
(574d) A Two-Structure Equation of State for Super Cooled and Stretched Light and Heavy Water 845
Michal Duska, Jan Hruby, Frederic Caupin, Mikhail A. Anisimov
(574e) Can Molecular Simulations Predict the Binary Interaction Parameters of the Activity Coefficient Models? 853
Ashwin Ravichandran, Rajesh Khare, Chau-Chyun Chen
(574f) A Wertheim Activity Coefficient Model for Associating Mixtures 854
Aseel M. Bala, William G. Killian, Jackson A. Storer, James E. Jackson, Paul M. Mathias, Navin Patel, Timothy C. Frank, Dung T. Vu, Eric L. Cheluget, Carl T. Lira
(574g) Understanding the Role of Intermolecular Forces in Molecular-Based Equations of State: 20 Years of the Soft-SAFT Equation 855
Lourdes F. Vega, Felix Llovell, Felipe J. Blas, Oriol Vilaseca, Joao A. P. Coutinho, Jordi Torne
(196b) Coarse-Grained Modeling of Polymer Materials in Molecular Simulations with SAFT-Î³ Mie Corresponding States Correlation Specified Potentials 856
Amulya Pervaje, Saad A. Khan, Erik E. Santiso
(574h) New SAFT Eos for Electrolyte Solution 857
Reza Shahryari, Mohammad Reza Dehghani
(575a) Controlling Vesicle Shape By Adsorption of a Semiflexible Polymer 858
Bing Li, Steven M. Abel
(575b) Polymer-Induced Restructuring in the Gut 859
Sujit S. Datta, Asher Preska Steinberg, Rustem Ismagilov
(575c) Multicomplex Formation in the Presence of Polymer Crowders Mimicking in-Cell Environment 860
Vahid Rahmanian, Young C. Kim, Jeetain Mittal
(575d) Effect of Palmitoylation in Membrane Proteins at the Blood-Brain Barrier Interface 873
Nandhini Rajagopal, Flaviyan Jerome Irudayanathan, Shikha Nangia
(575e) Determining the Limitations of Processive Polysaccharide Deconstruction By Glycoside Hydrolases 874
Suvamay Jana, Brandon C. Knott, Gregg T. Beckham, Christina M. Payne
(575f) Computational Design of Novel Self-Assembling Peptide Biomaterials Based on an Amyloid Forming Motif from the Adenovirus Fiber Shaft 875
Sai Vamshi R Jonnalagadda, Graziano Deidda, Eirini Ornithopoulou, Asuka Orr, Anna Mitraki, Phanourios Tamamis
(575g) Conformational Changes of 2'-Hydroxybiphenyl-2-Sulfinate Desulfinase 876
Yue Yu, Christina M. Payne
(575h) Chromatin Folding Via Coarse-Grained Multi-Scale Simulation 877
Joshua Lequieu, Andres Cordoba, Joshua Moller, Juan J. De Pablo
(577a) Quantitative Prediction of Industrial Turbulent Processes in Stirred Tank Reactors- Invited Presentation in Turbulent and Reactive Flows 878
Minye Liu
(577b) Passive Scalar Mixing in Anisotropic Turbulence from Line Sources 879
Quoc T. Nguyen, Dimitrios V. Papavassiliou

(577c) Dynamics and Structures of Viscoelastic Turbulence in Transitional Channel Flow 880
Ashwin Shekar, Sung-Ning Wang, Michael D. Graham
(577d) Polymer Effects on the Development and Bursting of Turbulent Vortices: Implication on High-Extent Drag Reduction 881
Lu Zhu, Xue Bai, Li Xi
(577f) Macroscale Insensitivity of Type B Drag Reduction By Two Biopolymers 882
Preetinder S. Virk
(577g) Euler-Euler and Euler-Lagrange Simulations of Heavy Particles Clustering in Homogeneously Sheared Gas 883
Mohamed H. Kasbaoui, Olivier Desjardins, Donald L. Koch
(577h) Breakage of Single Drops and Bubbles in a Turbulent 2-D Orifice Flow 884
Derrick I. Ko, Richard V. Calabrese
(577i) On the Influence of Spanwise Boundary Conditions on the Large Eddy Simulation of the Flow Past a Single Cylinder 885
Barbara L. Silva, Matheus R. Barbieri, Jonathan Utzig, Waldir Pedro Martignoni, Henry F. Meier
(588a) Out-of-Equilibrium Phase Behavior of Dielectric/Paramagnetic Nanoparticle Suspensions in Toggled Electric/Magnetic Fields 893
Zachary Sherman, James Swan
(588b) Investigation of the Motion of Patchy Particles at Liquid/Fluid Interfaces 894
Zohreh Jalilvand, Ilona Kretzschmar
(588c) Preparation and Characterization of Janus Dumbbells Nanomotors 895
Florian Guignard, Marco Lattuada
(588d) Magnetic Microlassos for Reversible Cargo Capture, Transport, and Release 896
Tao Yang, Tonguc Onur Tasci, Keith B. Neeves, Ning Wu, David W. M. Marr
(588e) Fly Larvae Mix to Increase Eating Rates 897
Olga Shishkov, David L. Hu
(588f) Programming the Dynamics of Active Colloids in 3D 898
Jingyun Lee, Stacey Wieseneck, Bhuvnesh Bharti
(588g) Tuning the Collective Behavior of Active Electrohydrodynamic Motors 899
Xingfu Yang, Ning Wu
(588h) Heterogeneous Active Matter Systems 900
Thomas Kolb
(588i) Directed Motion of Metallodielectric Particles By Contact Charge Electrophoresis 901
Yong Dou, Charles A. Cartier, Wenjie Fei, Shashank Pandey, Sepideh Razavi, Ilona Kretzschmar, Kyle J. M. Bishop
(603a) Molybdenum Dioxide Functionalized CoP Porous Nanocatalyst for Highly Efficient Hydrogen Evolution 902
Jun Wang, Liang-Xin Ding, Haihui Wang
(603b) Bijel-Derived Materials for Electrochemical Energy Storage and Conversion 903
Kyle McDevitt, Daniel R. Mumm, Ali Mohraz
(603c) Electrochmical Lithiation-Delithation of Sulfur in Sub-Nano Confinement 904
Chengyin Fu, Juchen Guo
(603d) Durable High Capacity Li-O2 Cathode Composed of Iron-Nitrogen-Doped Mesoporous Core-Shell Carbon Loaded with RuO2 Nanoparticles 905
Ming Zhou, Kwong-Yu Chan, Chi-Ying Vanessa Li
(603e) Magnetic Field-Induced Fabrication of Fe3O4@Graphene Nanocomposites for Enhanced Electrode Performance in Lithium-Ion Batteries 906
Huan Wang, Jingyi Xie, Marissa Follette, Tyson C. Back, Placidus B. Amama
(613a) Elucidating Mechanisms of Substrate Transport in Membrane Transporters 907
Diwakar Shukla, Balaji Selvam
(613b) Grafting Charged Species to Membrane-Embedded Scaffolds Dramatically Increases the Rate of Bilayer Translocation 908
Reid Van Lehn
(613c) Molecular Simulations of Nanoparticles Permeating Lipid Membranes for Drug Delivery Applications 909
Priyanka Oroskar, Cynthia J. Jameson, Sohail Murad
(613d) Transport Characteristics and Stability of Artificial Water Channels in Lipid and Biomimetic Membranes 910
Harish Vashisth
(613e) Solute Transport through Outer Membrane Porins of Gram-Negative Bacteria 911
Huilin Ma, Aliza Khan, Shikha Nangia

(613f) Could Class IIb Bacteriocins Induce Pore Formation? Investigation through Microsecond Long Atomistic Molecular Dynamic Simulation 912
Panagiota Kyriakou, Yiannis N. Kaznessis
(613g) Host-Pathogen Interactions at the Tight Junctions: Molecular Insights from Membrane Guided Self-Assembly 913
Flaviyan Jerome Irudayanathan, Xiaoyi Wang, Nan Wang, Shikha Nangia
(613h) Structure of Multi-Component, Gel-Phase Lipid Bilayers 914
Alexander Yang, Timothy C. Moore, Christopher R. Iacovella, Michael Thompson, Pallav Bulsara, David J. Moore, Clare McCabe
(613i) Multiplicity of States and Internal Structure of Supported Lipid Bilayers 915
Aleksey Vishnyakov, Ting Li, Alexander V. Neimark
(613j) Line Tension and Lipid Sorting Modulate Dynamics of Hemifusion Diaphragm Dissipation 916
Jasmine Gardner, Cameron F. Abrams
(614a) Understanding the Removal of Pharmaceuticals from Water Effluents By Adsorption in Activated Carbons - A Molecular Simulation Approach 917
Lourdes F. Vega, Daniel Bahamon
(614b) Insights into the Adsorption and Phase Behaviour of Fluids in Nanoporous Materials with Hierarchical Pore Structure: Towards an Advanced Textural Characterization 918
Matthias Thommes
(614c) Molecular Studies of Supercapacitors: Ionic Liquids Adsorbed into Porous Carbon Electrodes 919
Peter T. Cummings
(614d) Structure of Ice in Confinement; Water in Mesopores 920
Malgorzata Sliwinska-Bartkowiak
(614e) Towards Understanding the Role of Microstructure in Energetic Material Response: Coarse-Grain Modeling and Simulation 921
John K. Brennan
(614g) Adsorption and Transport in Multiscale Porous Media 922
Benoit Coasne
(629a) Redox-Triggered Mixing and Demixing of Surfactants within Assemblies Formed in Solution and at Surfaces 923
Nicholas L. Abbott
(629b) Static and Dynamic Signatures of Branching in Wormlike Micelles (WLMs) Via Advanced Techniques in Rheology and Neutron Scattering 924
Michelle A. Calabrese, Simon Rogers, Lionel Porcar, Norman J. Wagner
(629c) Self-Assembly of Achiral Surfactants Can Conduct Enantioselective C-C Bond Formation of Amino Acid Derivative 925
Hiroshi Umakoshi, Fumihiko Iwasaki, Keishi Suga, Yukihiro Okamoto
(629d) A Statistical Associating Fluid Theory (SAFT) Framework for Aqueous Nonionic Surfactant Systems 926
Aubrey Winiarski, Arthur S. Gow
(629e) First Proof of Self-Assembly of Block Copolymers in a Deep Eutectic Solvent 927
Dannie J. G. P. Van Osch, Marco M. R. M. Hendrix, Nicole M. W. Van Der Heijden, Jaap Van Spronsen, A. Catarina C. Esteves, Remco Tuinier
(629f) Fast and Slow Dynamical Processes in Simulations of Block Copolymer Micelles 928
Joshua Mysona, David Morse, Alon McCormick
(629g) Synthesis of Hydrophobically Modified Polybetaines (HMPB) and Study of Their Self-Assembly By Molecular Dynamics Simulations 929
Xiao Zhao, Ashwin Ravichandran, Sarkyt Kudaibergenov, Rajesh Khare, Nurxat Nuraje
(629h) A Hexagonal Columnar Liquid Crystal Phase Formation in Dilute Solutions of Carbon Nanotubes 930
Vida Jamali, Francesca Mirri, Paul Van Der Schoot, Fred Mackintosh, Matteo Pasquali
(629i) Computational Phase Space Screening of Isotropic Multi-Well Pair Potentials 931
Julia Dshemuchadse, Michael Engel, Pablo F. Damasceno, Carolyn L. Phillips, Sharon C. Glotzer
(629j) Methanol Self-Association and Preferential Solvation of Chelating Agents for the Extraction of Nuclear Fission Products in Supercritical CO2 932
Trent R. Graham, Daniel J. Pope, Aurora E. Clark, Steven R. Saunders
(654a) Effect of Salt Concentration on the Ability to Form Stable Close-Packed Vesicular Dispersions for Stabilizing Suspensions of Dense Particles Against Sedimentation 933
An-Hsuan Hsieh, Yung-Jih Yang, David . S. Corti, Elias I Franses
(654b) Toward a Better Understanding of Nanoassembly: Importance of Molecular Granularity on Colloidal Forces 934
Jaehun Chun, Christopher J. Mundy, Gregory K. Schenter, Dongsheng Li, James J. De Yoreo

(654c) Structural Insights into DNA-Stabilized Silver Clusters through Analytical Ultracentrifugation 935
Danielle Schultz, Nancy J. Lin, Jeffrey A. Fagan
(654d) Stabilization of Colloidal Suspensions with a Bimodal Distribution of Microparticles and Nanoparticles 936
Marzieh Moradi, Stuart J. Williams, Gerold A. Willing
(654e) Rheological and Electrical Percolation Behavior of Carbon Black Suspensions in Polar Aprotic Solvents 937
Norman J. Wagner, Jeffrey J. Richards, Paul Butler
(654f) Functional Hybrids Based on Assembly of 2D Materials and Polymeric Nanoparticles 938
Nader Taheri Qazvini, Kylie Zane, Brian Schwartz, Matthew V. Tirrell, Juan J. De Pablo
(654g) Convective Fluid Motions in Droplets Driven By Global and Local Chemical Gradients 939
Nan Shi, Todd M. Squires
(654h) Characterizing Gelatin Hydrogel Viscoelasticity with Diffusing Colloidal Probe Microscopy 940
Soheila Shabaniverki, Jaime Juarez
(654i) Self-Immobilized Cross-Linked Enzyme Aggregates for Biocatalysis 941
Yang Kun-Lin, Le Truc Nguyen
(669a) Impact of Molecular Structure on Surface and Thermal Properties of Amido-Amine Cationic Gemini Surfactants 942
Muhammad Shahzad Kamal, S. M. Shakil Hussain, Abdullah S. Sultan
(669b) Surfactant Effect on Hydrate Crystallization at Oil-Water Interface 943
Liat Rosenfeld, Kevin Dann
(669c) Surface Tension Behavior of Aqueous Solutions of a Propoxylated Surfactant and Interfacial Tension Behavior Against an Un-Preequilibrated Crude Oil 944
Jaeyub Chung, Bryan W. Boudouris, Elias I Franses
(669d) Eco-Friendly Sacrificial Amphiphiles As Chemical Herders for Oil Spill Remediation 945
Hao Zhou, George John, Charles Maldarelli
(669e) Interfacial Interactions in Oil/Brine Emulsions Stabilized By Combinations of Cellulose Nanocrystals and Emulsion Stabilizers 946
Sanjiv Parajuli, Chadwick E. Middleton, Andres E. Rodriguez Zambrano, Esteban E. Urena-Benavides
(669f) Measurement and Surface Complexation Modeling of Calcite Zeta Potential in Mixed Brines for Carbonate Wettability Characterization 947
Jin Song, Yongchao Zeng, Xindi Duan, Le Wang, Walter G. Chapman, Sibani Lisa Biswal, George J. Hirasaki
(669g) Dissolution and Restructuring of Calcite Surfaces in Contact with Electrolyte Solutions - Implications for Enhanced Oil Recovery 955
Kai Kristiansen, Szu-Ying Chen, Yair Kaufman, Nicholas Cadirov, Howard Dobbs, Alex Schrader, Dongjin Seo, J. Boles, Jacob Israelachvili
(669h) Marine Bacteria Adhesion to Oil/Water Interfaces 956
Michael Godfrin, Maswazi Sihlabela, Arijit Bose, Anubhav Tripathi
(669i) Oil Spill Dispersion with Food-Grade Amphiphiles: Formation of Oil-Bacteria Agglomerates during Oil Biodegradation 957
Geoffrey D. Bothun, Joseph Rocchio, Srinivasa R. Raghavan, Vijay T. John, Alon McCormick
(669j) The Role of Bacterial Biofilm in Particle-Stabilized Oil-in-Water Emulsions: Implications to the Biodegradation of Oil Spills 958
Marzhana Omarova, Lauren T Swientoniewski, Diane A. Blake, Geoffrey D. Bothun, Vijay T. John
(670a) Novel Cathode Materials for Rechargeable Aluminum Batteries 959
Robert J. Messinger, Jeffrey Xu, Ankur Jadhav
(670b) Reversibility of Zinc Metal Anode: Fundamentals and Engineering 960
Nian Liu
(670c) An Air Breathing Lithium-Oxygen Battery 961
Baharak Sayahpour, Mohammad Asadi, Pedram Abbasi, Larry A. Curtiss, Amin Salehi-Khojin
(670d) Appropriate Characterization Techniques for Lithium-Oxygen Batteries, and Implications for Understanding 2e- Vs. 4e- Oxygen Reduction Processes 962
Colin M. Burke, Bryan D. McCloskey
(670e) Mechanistic Insights into the Sodium-Oxygen Battery Cathode Electrochemistry 963
Jessica E. Nichols, Bryan D. McCloskey
(670f) Mechanistic Evaluation of Thermal Runaway in Potassium-Ion Batteries 964
Ryan A. Adams, Arvind Varma, Vilas G. Pol
(670g) Molten Salt Batteries: Mechanics and Electrolyte Transport 965
Christine Cardinal Roberts, Martin B. Nemer, Mark Stavig, Alexander Headley, Ryan Solich, Scott A. Roberts
(683a) Entropy-Driven, Two-Step Crystallization of Colloidal Clathrate Crystal 966
Sangmin Lee, Michael Engel, Sharon C. Glotzer

(683b) Unraveling the Coupling between Demixing and Crystallization in Mixtures 967
Jerome Delhommelle, Caroline Desgranges

(683c) La Mer Burst Nucleation and Growth: Assumptions, Models, and Data 968
Baron Peters

(683d) A Stochastic Model of Primary Nucleation of Polymorphs 969
Giovanni Maria Maggioni, Leonard Bezinge, Marco Mazzotti

(683e) Step Velocity of a Crystal Edge with Alternating Rows of Growth Units 970
Mark Joswiak, Baron Peters, Michael F. Doherty

(683f) Multiscale, Multiphysics, Mechanistic Model for Computation of Face-Specific Growth Rates 971
James Fell, Anish V. Dighe, Meenesh R. Singh

(683g) Anisotropic Growth Kinetics of Triblock Janus Colloids 972
Wesley F. Reinhart, Athanassios Z. Panagiotopoulos

(685a) Enhanced Sampling Methods for Modulating Density Fields 973
Zhitong Jiang, Suruchi Fialoke, Amish Patel

(685b) Mapped Averaging Methods for Accurate and Precise Evaluation of Free Energies and Other Properties By Molecular Simulation 974
Weisong Lin, Akshara Goyal, Sabry G. Moustafa, Andrew J. Schultz, David A. Kofke

(685c) Predicting the Free Energy Landscape of Multicomponent Fluids 975
Nathan A. Mahynski, Jeffrey R. Errington, Vincent K. Shen

(685d) A Novel Molecular Simulation Method for Liquid-Liquid Equilibria Predictions and in silico Screening of Desalination Solvents 976
Prashanth Chandran, Jindal K. Shah

(685e) Ssages: A Comprehensive Platform for Enhanced Sampling Simulations 977
Hythem Sidky, Yamil J. Colon, Benjamin J. Sikora, Cody Bezik, Federico Giberti, Ashley Guo, Julian Helfferich, Xikai Jiang, Joshua Lequieu, Jiyuan Li, Joshua Moller, Michael Quevillon, Mohammad Rahimi, Hadi Ramezani-Dakhel, Vikramjit Rathee, Daniel Reid, Emre Sevgen, Vikram Thapar, Michael Webb, Justin Wozniak, Xujun Zhao, Nicola J. Ferrier, Olle G. Heinonen, Giulia Galli, Francois Gygi, Juan J. De Pablo, Jonathan K. Whitmer

(685f) Multiscale Modeling of Multicompartment Micelle Nanoreactors 978
Connor Callaway, Parveen Sood, Seung Soon Jang

(685g) Reaction Ensemble Monte Carlo: Applications to Ionic Liquids 979
Ryan Gotchy Mullen, Edward J. Maginn

(685h) Temperature-Dependent Physicochemical Properties of Nitrotoluenes from Solvation Free Energies 980
Alauddin Ahmed, Stanley I. Sandler

(688a) Predicting the Solubility and Diffusivity of Gases (CO2, CH4, H2, Noble Gases) in Nano-Confined Interlayer Water and Bulk Water Using Molecular Dynamics Simulations 981
Greeshma Gadikota, Ian Bourg

(688b) Simulation of Tracer Particle Diffusion in Attractive and Repulsive Glassy Matrices 982
Ryan Roberts, Ryan Poling-Skutvik, Jacinta C. Conrad, Jeremy C. Palmer

(688f) [Invited Talk] Predictive Modeling of Adsorption and Reaction Equilibria in Nanoporous Materials 983
J. Ilja Siepmann, Evgenii Fetisov, Mansi S. Shah, Michael Tsapatsis

(688d) Characterizing Protein Hydration to Inform its Interactions 984
Erte Xi, Amish Patel

(688e) Confinement-Induced Supercriticality and Phase Equilibria of Hydrocarbons in Nanopores 985
Sheng Luo, Hadi Nasrabadi, Jodie L. Lutkenhaus

(688c) [Invited Talk] Simulations of Biomolecular Assembly Processes at Interfaces 986
Jeetain Mittal

(688g) Nucleation of Capillary Bridges and Bubbles 987
Jerome Delhommelle, Caroline Desgranges

(688h) Modeling Single Component Transport in Mesoporous Membranes Under Non-Equilibrium Conditions 988
Ashutosh Rathi, Eustathios Kikkinides, David M. Ford, Peter A. Monson

(704a) Optimizing the Formation of Solid Solutions with Colloids of Different Shapes 989
Fernando Escobedo

(704b) Programmed Assembly of Anisotropic Patchy Colloids By Nonlinear Learning and Landscape Engineering 990
Andrew W. Long, Andrew L. Ferguson

(704c) Evaporation-Induced Assembly of Colloidal Crystals 991
Michael P. Howard, Wesley F. Reinhart, Arash Nikoubashman, Athanassios Z. Panagiotopoulos

(704d) Amphiphile-Induced Reorganization of Liquid Crystals at Aqueous Interfaces 992
Hadi Ramezani-Dakhel, Mohammad Rahimi, Joel Pendery, Nicholas L. Abbott, Benoit Roux, Juan J. De Pablo

(704e) Investigating Thermodynamics and Kinetics of a Î²(16-22) Peptide Aggregation Using Coarse-Grained Simulations 993
Yiming Wang, Stefan Auer, Sergei V. Krivov, Carol K. Hall

(704f) Inverse Materials Design from Phase Transitions in Shape Space 994
Rose Cersonsky, Greg Van Anders, Paul Dodd, Sharon C. Glotzer

(704g) Concentration Effects in Simulations of Non-Ionic and Ionic Surfactant Micellization 995
Andrew P. Santos, Athanassios Z. Panagiotopoulos

(704h) Self-Assembly Simulations of Stratum Corneum Lipid Mixtures 996
Timothy C. Moore, Christopher R. Iacovella, Remco Hartkamp, Annette L. Bunge, Clare McCabe

(704i) Self-Assembly of Multi-Flavored DNA-Functionalized Particles into Binary Superlattices 997
Hasan Zerze, Nathan A. Mahynski, Evan Pretti, Vincent K. Shen, Jeetain Mittal

(708a) Accelerating Force Field Parameterization to Improve the Quantitative Predictability of Thermophysical Properties 998
Richard A. Messerly, Andrei Kazakov

(708b) Validation of Trimethylamine-N-Oxide (TMAO) Force Fields Based on Thermophysical Properties of Aqueous Tmao Solutions 999
Daniel Markthaler, Johannes Zeman, Jorg Baz, Jens Smiatek, Niels Hansen

(708d) Optimized Mie Potentials for Phase Equilibria: Application to Noble Gases, Alkanes, Alkynes and Their Mixtures 1008
Mohammad Barhaghi, Jason R. Mick, Jeffrey J. Potoff

(708e) Evaluation of Virial Coefficients and Their Temperature Derivatives for Multibody Potential Models 1009
Navneeth Gokul, Andrew J. Schultz, David A. Kofke, Hainam Do, Richard J. Wheatley

(708f) Prediction of Interaction Parameters for Coarse-Grained Models Using Ab Initio Calculations 1010
Jennifer A. Clark, Erik E. Santiso

(708g) Modifying Nonplanar Vibrational Modes of Aromatic Rings in Biomolecular Modeling 1011
Faramarz Joodaki, Lenore M. Martin, Michael L. Greenfield

(708h) A Polarizable Force Field of Inorganic Phosphates and Hydroxyapatite Based on the Classical Drude Oscillator 1012
Hadi Ramezani-Dakhel, Hui Li, Nader Taheri Qazvini, Juan J. De Pablo, Benoit Roux

(708i) Using Minimal Biasing Methods in Plumed and Lammps 1013
Andrew White

(713a) A General Route for Nanoemulsion Synthesis Using Low Energy Methods at Constant Temperature 1014
Ankur Gupta, Abu Zayed Md Badruddoza, Patrick S. Doyle

(713b) Advanced Emulsion-Templating of Microreactors for the Scalable Production of Semiconductor Nanowires 1015
Maritza Mujica, Michael A. Filler, Victor Breedveld, Sven H. Behrens

(713c) The Discovery of Novel Hydrophobic DES-Water Emulsions 1022
Dannie J. G. P. Van Osch, Nicole M. W. Van Der Heijden, Jaap Van Spronsen, A. Catarina C. Esteves, Remco Tuinier, Mark Vis

(713d) Controlling Foaming during Thermal Cracking - a Non-Silicon Based Antifoaming Agent 1023
Amaka Waturuocha, Michael Volk, Glixon Mavarez Nava, Dwijen Banerjee, Keith Wisecarver

(713e) Effect of Salt on Drainage Via Stratification in Micellar Foam Films 1024
Subinuer Yilixiati, Rabees Rafiq, Yiran Zhang, Vivek Sharma

(713f) Asphaltenes Adsorption at Water/Oil Interface, a Classical Surfactant Approach 1025
Shaghayegh Darjani, Fang Liu, Nelya Akhmetkhanova, Vincent Pauchard

(713g) Rapid Demulsification of Water-in-Oil Emulsions Using Silica Nanoparticles 1026
Ashwin Kumar Yegya Raman, Clint P. Aichele

(713h) Linking Adhesive Emulsions Behaviour in Microfluidic Devices to Direct Force Measurements between Drop Pairs 1027
Christopher Fewkes, Emily Jamieson, Joe Berry, Raymond R. Dagastine

(713i) Interfacial Routes to Gelation in Solid-Stabilized Emulsions 1028
Max Kaganyuk, Ali Mohraz

(713j) Predicting the Microstructure of an Interface Laden with Anisotropic Particles 1029
Sri Harsha Nuthalapati, Christopher L. Wirth

(718a) Controlling the Polymer/Gas Interface of the Ionic Polymer Phase of a PEM Fuel Cell Catalyst Layer during Membrane Electrode Assembly Fabrication 1030
Regis Dowd Jr., Trung Van Nguyen

(718b) Modeling the Effect of Charged Lithium-Metal Anode on Electrolyte Decomposition 1031
Luis E. Camacho-Forero, Perla B. Balbuena

(718c) Dynamic Measurement Method of the Emulsified Water Content Based on Interfacial Polarization 1032
Meiyi Qing, Huaqing Liang, Jinjun Zhang, Min Wei, Chenbo Ma
(718d) Stretchable and Soft Electroadhesion Enhanced By Liquid-Metal Subsurface Microstructures 1041
Sungjune Park, Jun Shintake, Ishan Joshipura, Michael D. Dickey, Kunal Mondal
(718e) Giant Electret Polarization in Electrochemically Deposited Hydroxyapatite Ceramic Coatings 1042
Matthew Yates, Xuefei Zhang, Wanaruk Chaimayo, Benjamin Miller
(718f) Electrochemical Charging of CdSe Quantum Dots: Effects of Adsorption Versus Intercalation 1043
Ajinkya Puntambekar, Qi Wang, Vidhya Chakrapani
(718g) Microstructured Conducting Polymer Swabs for Enhanced Trace Explosive Detection 1044
Jennifer S. Laster, Bryan W. Boudouris, Stephen P. Beaudoin
(718h) Atomistic Modeling of Solvent Decomposition in Magnesium Batteries 1045
Jeffrey S. Lowe, Donald J. Siegel
(719b) Understanding the Reduction Reaction Mechanisms of Sulfur-Based Cathodes: A Theoretical Approach from DFT and Reaxff Molecular Dynamics 1046
Saul Perez Beltran, Perla B. Balbuena
(719c) Reversible Aluminum Intercalation in Transition Metal Sulfides 1047
Linxiao Geng, Juchen Guo
(719d) Highly Structured Titanium Nitride As Novel Cathode Materials of Lithium Sulfur Batteries 1048
Wenduo Zeng, Mark Cheng, Simon Ng
(719e) Elucidating the Solvation Structure and Dynamics of Lithium Polysulfides Using Coupled High Throughput Simulations and Experiments 1058
Nav Nidhi Rajput, Vijayakumar Murugesan, Karl Mueller, Kristin Persson
(719f) Solvation Structure and Behavior of Lithium Polysulfide Species in Electrolytes of Lithium-Sulfur Batteries 1059
Ethan P. Kamphaus, Perla B. Balbuena
(719g) Design Stable Room-Temperature Metal-Sulfur Batteries 1060
Shuya Wei, Lynden A. Archer
(749a) Diverse Colloidal Crystals from DNA-Grafted Spheres Via Self-Assembly 1061
Yifan Wang, Ian Jenkins, James T. McGinley III, Talid Sinno, John C. Crocker
(749b) Material Design By DNA-Mediated Interactions between Colloids 1071
Runfang Mao, Jeetain Mittal
(749c) High Throughput Acoustically Driven Self-Assembly of Microfluidic Colloidal Crystals 1072
Meghana Akella, Jaime Juarez
(749d) The Role of Interaction Heterogeneity in Colloidal Crystallization 1073
Ian Jenkins, John C. Crocker, Talid Sinno
(749e) Self-Assembly of Open Structures Using Depletion 1074
Jens Glaser, Sharon C. Glotzer
(749f) Controlled Self-Assembly of Colloidal Discotic Liquid Crystals 1075
Zhengdong Cheng
(749g) Stratification Dynamics in Drying Colloidal Mixtures 1076
Michael P. Howard, Arash Nikoubashman, Athanassios Z. Panagiotopoulos
(749h) Theoretical and Experimental Investigation of Phase Separation in Noble Metal Nanoparticle Monolayers 1077
Steven Merz, Zachary Farrell, Sergei Egorov, David Green
(749i) Bending-Induced Buckling Instabilities in Self-Assembled Elastomeric Composite Films 1078
Peng Jiang, Sin-Yen Leo, Zhuxiao Gu
(749j) Molecular Modeling of Microstructure, Solubilization and Micro-Emulsion of Block Copolymer Micelles By iSAFT 1079
Shun Xi, Le Wang, Walter G. Chapman
(754a) Spreading of Nanoscale Droplets of Ionic Liquids on the Mica Surface 1080
Lei Li
(754b) Long Range Electrostatic Screening in Ionic Liquids 1081
Matthew A. Gebbie
(754c) Mesopore Confinement Effects on Ionic Liquid Hydration 1082
Yuxin He, Daudi Saang' Onyo, Folami Ladipo, Barbara L. Knutson, Stephen E. Rankin
(754d) Water Sorption in Ionic Liquids Characterized Using a Dynamic Vapor Sorption Analyzer (IGASorp) and High Pressure X-Ray Photoelectron Spectroscopy (APXPS) 1083
M. Alejandra Rocha, Alicia Broderick, John T. Newberg, Mark B. Shiflett
(754e) Property Modeling of Ionic Liquids for Gas Separation Processes 1084
Xinyan Liu, Xiaodong Liang, Xiangping Zhang, Suojiang Zhang, Rafiqul Gani

(754f) An Investigation of Cellulose Solubility in Ionic Liquids with Added Cosolvents 1085
Brooks D. Rabideau
(754g) Effects of Ionic Liquids and Liposomes on Enzymatic Cellulose Hydrolysis Process 1086
Kazuhiko Tanimura, Yoshiko Ooe, Keishi Suga, Yukihiro Okamoto, Makoto Yoshimoto, Hiroshi Umakoshi
(754i) Micellization of Poloxamer Block Copolymers in Ionic Liquids and Their Mixtures with Water 1087
Zhiqi He, Yingzhen Ma, Paschalis Alexandridis
Author Index

Some abstracts in this book refer to images that were not made available by the publisher

487399 Anisotropic Particle Interactions with Surfaces, Other Particles, and External Fields

Abstract:

Anisotropic colloidal particles are present in numerous natural and synthetic materials that are important to environmental, biological, and advanced material applications. Interactions of anisotropic particles with each other, surfaces, and external fields determines all of their behavior and properties. As such, it is essential to have physically realistic accurate models of anisotropic particle interactions to design, control, and optimize their behavior and properties in applications. In this work, we report closed-form analytical potentials for anisotropic particle interactions, which we also measure on the kT-scale via optical microscopy experiments. First, we present an interaction potential for arbitrarily oriented convex anisotropic particles with planar surfaces using DLVO half-space interactions in combination with the Derjaguin approximation. A relatively simple potential is obtained for all particles that can be described by the super-ellipsoid equation, which includes spheres, ellipsoids, cylinders, polygons, super-spheres, as well as many other shapes. We extend the analysis of particle-wall potentials to determine pair potentials between arbitrarily oriented anisotropic particles. Next, we report dipole-field and dipole-dipole potentials for super-ellipsoidal particles in AC electric fields. We demonstrate the validity of these potentials through measurements of single particles and quasi-2D phase behavior in combination with an inverse Monte Carlo analysis. Finally, we show that tuning electric field mediated potentials can be used to manipulate position, orientation, and assembly of super-ellipsoidal colloidal particles.

498465 Particle-Wavy Wall Interactions in a Nematic Liquid Crystal

Abstract:

We are developing strategies for directed assembly for particles in nematic liquid crystals (NLCs), in which elastic energy fields and topological constraints provide a rich landscape for structure formation. NLCs are comprised of elongated molecules that co-align to minimize their free energy; deviations from this co-alignment are costly. Thus, these systems are sensitive to boundary topography and colloidal inclusions. By confining a NLC near a wavy wall with homeotropic (perpendicular) anchoring and a given wavelength, we distort the NLC director field, and impose well defined zones of bend and splay that decay over distances comparable to the wavelength. A colloid with well-defined anchoring, placed within this director field, migrates and docks in preferred locations that depend on the nature of the colloidâ€™s anchoring, and its defect structure. We examine several aspects of the particle wall interaction for colloids, including metastable and stable states.Â Strategies to harness this landscape for preferred docking sites, structure formation and control are discussed.

497925 Influence of Surface Roughness on Particle-Substrate Interactions: Double Layer Repulsion, Van Der Waals Attraction, and Depletion Interaction

Abstract:

The interaction of colloidal particles with each other and with solid substrates has long been studied, because it determines dispersion stability and deposition kinetics; it also holds the key to directing any form of colloidal self-assembly. For particles in dilute dispersions interacting across simple electrolytes at low to moderate ionic strength, the relevant interaction forces are considered reasonably well understood; and still, discrepancies between theory and experiment are observed routinely, and theoretical predictions rarely prove quantitatively accurate. Neglected effects of surface roughness are often blamed for these problems, but efforts to account for roughness and test the proposed modeling strategies have been limited. One difficulty in assessing the influence of surface topography lies in varying surface roughness in a controlled way without concomitantly changing the surface chemistry. In our present study, the particle-substrate interaction of microparticles with tunable surface roughness at fixed surface chemical composition was probed using Total Internal Reflection Microscopy (TIRM). These measurements reveal a simultaneous enhancement of the electric double layer interaction and weakening of the van der Waals interaction caused by the surface roughness. We further used small particles with negligible scattering intensity in the evanescent wave to induce depletion interaction between the probe particle and the substrate. At a fixed size and concentration of the depletant particles we studied the effect of probe particle roughness up to roughness amplitudes comparable to the depletant size. As one might expect, the depletion interaction gets weaker with increased particle roughness, but to our knowledge, ours is the first quantitative study of this roughness effect, which is analogous to the effect of molecular-scale roughness on the solvent-mediated â€œstructural forcesâ€ reported from observations with the surface forces apparatus.

487957 Functional Particles and Microcapsules

Abstract:

Functional Particles & Microcapsules

Simon Biggs and Alison Tasker

School of Chemical Engineering, The University of Queensland, St. Lucia QLD 4072

E-mail: simon.biggs@uq.edu.au

Over the last few within our research group we have developed a number of bespoke microcapsule and particle designs tailored to the characteristics of the material to be encapsulated. In this presentation, examples of core-shell particles for use as semi-permeable and impermeable microcapsules for the encapsulation of large and small molecular weight materials will be discussed.

For semi-permeable capsules, I will show how we can use environment-responsive polymers to manage the permeability of polymer shells built around emulsion droplet precursors. I will also briefly introduce the potential of membrane emulsification techniques to manufacture these capsules on multi-litre scales.

Finally, I will discuss a recently developed method for the preparation of metal-coated emulsion droplets for the long-term retention of small, volatile encapsulated species. I will indicate the key aspects of the synthesis method for such metal shells and demonstrate our ability to control the thickness of the shell deposited on the surface of the droplets and exemplify the release properties of these microcapsules.

496109 The Hindered Translational and Rotational Dynamics of Anisotropic Nanoparticles Diffusing Near a Solid-Liquid Interface Measured in Aqueous Solution

Abstract:

We discuss a new method for measuring the three-dimensional translational and rotational dynamics of various nanoparticles (*e.g.* ZnO, gold and clay nanorods) diffusing in aqueous solution above a solid-liquid interface. The utility of conventional optical tools for probing these systems is limited by the proximity of an interface and the presence of particle anisotropy; yet it is the influence of these factors that makes such systems interesting, introducing asymmetric interfacial forces and separation-dependent hydrodynamic hindrance in each of the spatial modes of diffusion. We simultaneously record the spatially correlated scattering of multiple evanescent light sources by isolated anisotropic particles, and use this data to reconstruct instantaneous positions and orientations at millisecond time-intervals. By observing diffusion in each spatial mode over time we are able to quantify each translational and rotational diffusion coefficient as a function of interfacial separation, and show an asymptotic approach to the expected constant value in the bulk. These same data can also be used to measure equilibrium particle-interface interaction potentials with orientation and position effects optionally decoupled, completing a rich set of tools for probing anisotropic interactions. Aside from fundamental applications, this approach will be particularly useful for understanding and tuning the self-assembly of films and other structures incorporating anisotropic nanoparticles.

494224 Strong Deformation of the Electric Double Layer during Sedimentation or Electrophoresis of a Charged Particle

Abstract:

Electrokinetic effects arise from the deformation of electric double layers under electric fields or fluid flows. The majority of studies in this venerable field consider weak fields or flows, for which the double layer is only slightly perturbed from equilibrium. Here, we examine the strong deformation of a double layer during sedimentation or electrophoresis of a charged colloid with a fixed surface charge. The former is relevant to particle motion in ultracentrifuges and the latter to electrophoresis in doped nonpolar fluids. Specifically, we invoke the Debye-Huckel approximation to the electrokinetic equations for a particle with a small surface potential (relative to thermal voltage) and also consider the Debye length to be large compared to particle size. In this thick-double-layer regime, the particle appears as a point charge, and the compensating counter-ion charge in the cloud surrounding it obeys an advection-diffusion-screening equation, which encapsulates the effects of advection due to the particle motion; Brownian diffusion of ions; and electrostatic screening of the surface charge by the cloud. For weak advection the cloud is only slight distorted. In contrast, for strong advection the cloud can be completely stripped from the particle: consequently, electrokinetic effects vanish in this strong-advection regime. Thus, for sedimentation, the drag is limits to Stokes law for an uncharged particle. Likewise, the particle velocity for electrophoresis approaches Huckelâ€™s expression.

500072 Understanding and Engineering Diffusiophoretic Suspensions

Abstract:

The long history of diffusiophoresis owes much to Dennis Prieve and co-workers, who developed elegant theoretical expressions, membrane deposition experiments that supported these theories, and enabled quantitative prediction of diffusiophoretic phenomena for various purposes. Our lab has developed a microfluidic technique to design and control compositional gradients across suspensions, and to directly measure the diffusiophoretic migration that results from these gradients. This technique has revealed both quantitative agreement for existing theories, and qualitative surprises in new systems. We will describe diffusiophoretic migration under some unorthodox gradients - including solvents, ionic surfactants, zwitterions, and reacting solutes. We will present a generalized theory that reproduces Prieve & Anderson's original expressions for diffusiophoresis, but also reveals connections to other phoretic migrations, and which explain some of the surprises we have observed in our experiments. Finally, we will describe our efforts to engineer 'soluto-inertial' interactions in suspensions, in which we engineer "beacon" particles that emit long-lived, long-ranged chemical fluxes, driving neighboring particles into diffusiophoretic migration. Building on Prieve's fundamental insights about diffusiophoresis in this way, we envision versatile new capabilities for `smart suspensions'.

494240 Chemically-Driven Transport -- a Pervasive Mechanism That Is Seldom Recognized

Abstract:

Fluid and particle transport that is driven by gradients of pressure, gravity, or electro-magnetic potential is well-known and studied in many fields. A subtler type of transport, called diffusiophoresis, occurs in a gradient of chemical concentration, either electrolyte or non-electrolyte. Although this mechanism of particle transport is well-known, the diffusiophoresis mechanism is often considered to be an esoteric laboratory phenomenon. However, in this talk it is shown that concentration gradients can develop in a surprisingly wide variety of physical phenomena – imposed gradients, asymmetric reactions, dissolution, crystallization, evaporation, mixing, sedimentation, and others – so that diffusiophoresis is in fact a very common transport mechanism in complex fluids, in both natural and artificial systems. It is anticipated that in georeservoir extractions, physiological systems, drying operations, laboratory and industrial separations, crystallization operations, membrane processes, and many other situations, diffusiophoresis is already occurring – often without being recognized – and that opportunities exist for designing this transport to great advantage. In this talk methods are shown for identifying when diffusiophoresis might be important, and for analyzing transport in complex fluid systems, from both a modeling and experimental standpoint.

501759 UnderstandingÂ Catalyst Selectivity forÂ CO_2 Electro-Reduction

Abstract:

Developing new processes that can employ renewable electricity to produce fuels and chemicals using water and CO_2 as inputs could play an important role in a sustainable future. New catalysts are needed for these processes, particularly those that are selective to desired products, one of the principal challenges in the field. In this talk, we will discuss some of our latest efforts to understand catalyst selectivity for CO_2 electro-reduction, including in-depth studies of catalyst surface structure. We then leverage that knowledge to develop catalyst materials and interfaces with improved performance.

491315 Ag-Sn Bimetallic Catalyst with a Core-Shell Structure for CO_2 Reduction

Abstract:

Converting greenhouse gas carbon dioxide (CO_2) to value-added chemicals is an appealing approach to tackle CO_2 emission challenges. The chemical transformation of CO_2 requires suitable catalysts that can lower the activation energy barrier to minimize the energy penalty associated with the CO_2 reduction reaction. First-row transition metals are potential candidates as catalysts for electrochemical CO_2 reduction; however, their high oxygen affinity makes them easy to be oxidized, which could strongly affect the catalytic properties of metal-based catalysts.[1] Recently, we proposed a strategy to synthesize ultra-thin electrocatalysts using a core-shell nanostructure that contains a bimetallic core responsible for high electronic conductivity and an ultra-thin partially oxidized shell for catalytic CO_2 conversion.

In this presentation, we will show a model core-shell electrocatalyst that has a silver-tin bimetallic core and an ultra-thin partially oxidized tin oxide shell (denoted as $AgSn/SnO_x$) for CO_2 reduction to formate with high Faradaic efficiencies and high current densities.[2] A structure-activity correlation between the $AgSn/SnO_x$ core-shell structure and CO_2 electrocatalytic activity was established by studying a series of Ag-Sn catalysts with different Ag/Sn compositions. More importantly, a volcano-like correlation between the bulk composition and electrocatalytic performance was observed, in which the optimal thickness of the partially oxidized SnO_x shell was ~1.7 nm.

References:

1. Lu, Q., & Jiao, F.* Electrochemical CO_2 reduction: electrocatalyst, reaction mechanism, and process engineering. *Nano Energy* 29, 439-456 (2016).

2. Luc, W., Collins, C., Wang, S. W., Xin, H. L., He, K., Kang, Y. J. & Jiao, F.* Ag-Sn bimetallic catalyst with a core-shell structure for CO_2 reduction. *Journal of the American Chemical Society* 139, 1885-1893 (2017).

501199 Electrochemical CO_2 Reduction over Cuag Bimetallic Electrodes and Well-Defined Surface Alloys with Enhanced Oxygenate Selectivity

Abstract:

Atmospheric CO_2 is a potential source of renewable carbon for the production of fuels and chemicals. However, for this process to be sustainable the hydrogen required must be derived from water and the necessary energy must be supplied by the sun. One approach to this goal is the utilization of renewable electricity to drive the electrochemical CO_2 reduction reaction (CO_2RR). Previous research has shown that the overall efficiency of CO_2RR and the distribution of generated products depend primarily on the electrocatalyst used as the cathode. Cu is the only monometallic electrocatalyst capable of reducing CO_2 into potential fuels or multi-carbon chemicals with a total Faradaic efficiency (FE) in excess of 1%. The principal products formed over polycrystalline Cu are H_2, CH_4, C_2H_4, and C_2H_5OH. However, at potentials cathodic of -1 V vs RHE excessive H_2 evolution leads almost exclusively to the methanation of CO_2. Since multi-carbon products are more valuable precursors to fuels and chemicals than CH_4, recent studies have focused on identifying factors governing the multi-carbon product selectivity. These studies have revealed that the multi-carbon product selectivity is sensitive to the surface morphology, electrolyte composition, and the applied potential. However, these studies have not managed to substantially reduce the fraction of current lost to the parasitic evolution of H_2.

Herein we report our investigations of electrochemical CO_2 reduction over CuAg bimetallic electrodes and well-defined surface alloys, which we have found to exhibit an enhanced activity for the evolution of products derived from CO relative to H_2 compared to pure Cu. Furthermore, we have found that these electrocatalysts display an unusually high selectivity for the formation of carbonyl-containing compounds, such as acetaldehyde. While this bimetallic system was investigated on the hypothesis that CO spillover from phase-segregated domains of Ag to Cu would increase the rate of C-C coupling, further characterization by photoelectron spectroscopy and electrochemical cycling led to a very different conclusion. The revised hypothesis was confirmed by synthesizing well-defined Cu+Ag surface alloys, which reproduced the activity and selectivity trends observed over the bimetallic electrodes. The results are explained in terms of an observable shift in the electronic properties of Cu and the resulting impact that these modifications have on the adsorption energies of key reaction intermediates.

493375 A Carbon Nanotube Supported Gold Catalyst for the Electroreduction of Carbon Dioxide (CO_2)

Abstract:

The electrochemical reduction of carbon dioxide (CO_2) to value added chemicals (such as formic acid, carbon monoxide (CO), ethylene, and ethanol) has been proposed as a potential strategy for the utilization of excess anthropogenic CO_2 emissions.[1, 2] In particular, the electroreduction of CO_2 to CO seems interesting, as CO (either alone or in combination with H_2 *i.e.*, syngas) can be used as a platform chemical to produce a variety of hydrocarbons. However, even after a few decades of research, electrochemical systems that can produce CO *via* the electroreduction of CO_2 at industrially relevant rates (partial current density for CO >150 mA cm^{-2}) and low overall cell overpotentials (absolute value <1V),[3] remain difficult to develop.

In this presentation, we will report a supported gold (Au) catalyst that exhibits remarkably high activity (partial current density for CO ~158 mA cm^{-2}) at low cell overpotentials (~-0.94 V), for the electroreduction of CO_2 to CO. The catalyst was synthesized *via* an *in situ* chemical reduction of the Au precursor (chloroauric acid) on a polybenzimidazole wrapped carbon nanotube support. The electrochemical characterization was performed in a flow electrolyzer to avoid mass transfer limitations associated with the low solubility of CO_2 in aqueous solutions.[4] Furthermore, we will also discuss the effect of electrolyte composition (concentration and pH) with respect to increasing the activity and lowering the overpotential required for the electroreduction of CO_2 to CO on the supported Au catalyst. The mechanistic implications of such a behavior will also be discussed.

References:

[1] A.M. Appel, J.E. Bercaw, A.B. Bocarsly, H. Dobbek, D.L. DuBois, M. Dupuis, J.G. Ferry, E. Fujita, R. Hille, P.J.A. Kenis, C.A. Kerfeld, R.H. Morris, C.H.F. Peden, A.R. Portis, S.W. Ragsdale, T.B. Rauchfuss, J.N.H. Reek, L.C. Seefeldt, R.K. Thauer, G.L. Waldrop, *Chem. Rev.* 2013, 113, 6621-6658.

[2] H.R.M. Jhong, S. Ma, P.J.A. Kenis, *Curr. Opin. Chem. Eng.* 2013, 2, 191-199.

[3] S. Verma, B. Kim, H.R.M. Jhong, S. Ma, P.J.A. Kenis, *ChemSusChem* 2016, 9, 1972-1979.

[4] D.T. Whipple, E.C. Finke, P.J.A. Kenis, *Electrochem. Solid-State Lett.* 2010, 13, B109-B111.

487255 Interplay of Mass Transfer and Local pH Effects in CO_2 Reduction Electrocatalysis on Cu Nanowires

Abstract:

Besides the catalyst structures, the electrocatalysis for CO_2 reduction is also dependent on the electrolyte and CO_2 partial pressure, which can affect the catalytic activity and selectivity via mass transfer and/or local pH effects. While protons are readily available in aqueous solutions, the supply of reactants to the electrode surface is limited by the concentration and diffusivity of dissolved CO_2 molecules, which places a hurdle on the current density and efficiency of CO_2 reduction. The reduction of CO_2, as well as the evolution of H_2, produces hydroxide anions ($OH^{â€“}$) and causes rise of the local pH on the electrode surface.

Here we report on the study of mass transfer and local pH effects on CO_2 reduction by using Cu nanowires as the electrocatalysts. Highly dense Cu nanowires were synthesized by first thermal oxidation of Cu mesh in air, and then reduction by applying a cathodic electrochemical potential or thermal annealing in a reducing atmosphere. Our previous report has shown that the electrochemically reduced (ECR) Cu nanowires are highly active for CO_2 reduction, with favorable production of CO and formate at potentials more positive than -0.5 V (versus reversible hydrogen electrode) and C_2 species (ethanol, ethylene, and ethane) at more negative potentials. We have further performed more systematic studies in the high-overpotential region, i.e., potentials more negative than -0.5 V, where the mass transport of CO_2 and high local pH near the electrode surface are believed to play a significant role in determining the electrocatalytic performance. Simulations have been developed to understand the interplay between the mass transfer and local pH effects and elucidate their roles in the structure-performance relationships of the high-surface-area nanowire catalysts. Based on these results, discussions are provided for optimizing the electrochemical conditions for CO_2 reduction to produce valuable C_2 products.

497824 Measurement of Intrinsic Activity of Electrocatalytic Reduction of CO_2 over Cu

Abstract:

Electrocatalytic reduction of CO_2 offers a sustainable route to mitigate global warming and store intermittent solar and wind energy. Various electrocatalysts have been studied over the past three decades for the conversion of CO_2 to oxygenates such as CO, HCOOH, $C_2H_2O_4$, CH_3OH, and C_2H_5OH; and hydrocarbons such as CH_4, C_2H_4, and C_2H_6. Among these electrocatalysts, Cu is the only catalyst that can produce hydrocarbons. Several electrochemical cells have been proposed in the literature to measure activity and selectivity of the CO_2 reduction reaction. However, all of these measurements were affected by the poor mass transfer of CO_2 to the cathode at potentials more negative than -1 V vs RHE. The limiting factor in the electrochemical reduction of CO_2 is the depletion of dissolved CO_2 coupled with the increase in pH near the cathode. Here we propose a compact design of an electrochemical cell with the cathode being a Rotating Disk Electrode (RDE). The RDE sparged with gaseous CO_2 minimizes the cathode polarization and hence enable the measurement of intrinsic activity. The effect pH and applied potential on the intrinsic activity and selectivity of electrocatalytic CO_2 reduction reaction over Cu will be presented in this talk.

495014 Electrocatalytic Carbon Fixation on Molecular-Functionalized Surfaces

Abstract:

Global warming and related environmental issues have been exacerbated by the accelerated consumption of fossil fuels as energy resources. Utilizing renewable energy input to transform the abundant chemical feedstock carbon dioxide (CO_2) back into fuels and other value-added chemicals and materials can complete a potential carbon-neutral cycle to help the sustainability of our planet. To this end, we are developing hybrid homogeneous-heterogeneous catalysts for electrochemical carbon fixation based on the combination of inorganic molecules and surfaces. Our latest work on the development of systems capable of CO_2 and CO reduction to C-C coupled products will be discussed in this presentation.

494963 Electrochemical Reduction of CO_2 on Metal Doped Graphene

Abstract:

Single atom transition metals embedded at single vacancies of graphene provide a unique paradigm for catalytic reactions.[1,2] We present a density functional theory study of such systems for the electrochemical reduction of CO_2. Using an explicit description of the solvent at the electrochemical interface, the activation energies for key proton-electron transfer steps were determined. On transition metals, the first proton transfer ($*CO + H^+ + e^- \rightarrow *CHO$) has been shown to be the critical rate determining step in C_1 product formation. On these single atom sites, the corresponding barrier scales more favorably with the CO binding energy than for 211 and 111 transition metal surfaces, in the direction of improved activity.[3] Hydrogen evolution intermediates, on the other hand, are found to less stable than those on transition metals, which suggests a lower hydrogen evolution activity. Using a simple microkinetic model based on the calculated electrochemical barriers, we present a rate volcano for the production of methane from CO. All the considered transition metal dopants (Cr, Mo, W, Mn, Fe, Ru, Os, Co, Rh, Ir, Ni, Pd, Pt, and Cu) form stable doped-graphene structures, and a subset of these dopants (Ru, Fe, Ni, Pd, Pt, Rh, Co, and Ni) are not expected to be poisoned by oxide species. The presence of nitrogen near the active site was found to affect intermediate and transition state energies, but variations in these energetics followed the same scaling determined for these materials without nitrogen. We identify promising candidates with high activity, stability, and selectivity for the reduction of CO_2. This work highlights the potential of these systems as improved electrocatalysts over pure transition metals for CO_2 reduction.

1. Krasheninnikov, A.V; Lehtinen, P.O.; Foster, A.S.; Pyykkö P.; Nieminen, R.M. *Phys. Rev. Lett.* **102**, 126807 (2009).
2. Stolbov, S.; Ortigoza, M.A. *J. Chem. Phys.* **142**, 154703 (2015).
3. Liu, X.; Xiao, J.; Peng, H.; Hong, X.; Chan, K.; Nørskov, J.K. *Nat. Comm.* (2017).

501706 A Generic Coarse-Grained Model of Influenza Budding: What Can We Learn?

Abstract:

Viral release of influenza is believed to occur independently of the usual cellular machinery of endosomal sorting complexes required for transport (ESCRT); instead, scission of the extruding bud is facilitated by the M2 protein. [1] The shape of the M2 transmembrane domain is distinctly conical and the assumed mechanism by which scission proceeds is that M2 aggregates in the budding neck to cause constriction that eventually pinches off the enveloped virus.

Constructing a physiologically inspired, minimal physical model of viral budding that can elucidate molecular mechanisms is nontrivial. In the spirit of Einsteinâ€™s famous quote: â€œEverything should be made as simple as possible, but no simpler,â€ we will discuss model design principles, appropriate modelistic idealizations, technical challenges, and validation exemplified by the chosen application in virology.

REFERENCES

[1] Rossman et al. Cell 141, pp. 902-913, 2010

492019 Allosteric Effects of Gold Nanoparticles on Human Serum Albumin

Abstract:

The ability of nanoparticles to alter protein structure and dynamics plays an important role in their medical and biological applications. We investigate allosteric effects of gold nanoparticles on human serum albumin protein using molecular simulations. The extent to which bound nanoparticles influence the structure and dynamics of residues distant from the binding site is analyzed. The root mean square deviation, root mean square fluctuation and variation in the secondary structure of individual residues on a human serum albumin protein are calculated for four protein-gold nanoparticle binding complexes. The complexes are identified in a brute-force search process using an implicit-solvent coarse-grained model for proteins and nanoparticles. They are then converted to atomic resolution and their structural and dynamic properties are investigated using explicit-solvent atomistic molecular dynamics simulations. The results show that even though the albumin protein remains in a folded structure, the presence of a gold nanoparticle can cause more than 50% of the residues to decrease their flexibility significantly, and approximately 10% of the residues to change their secondary structure. These affected residues are distributed on the whole protein, even on regions that are distant from the nanoparticle binding site. Our simulations suggest that allosteric effects must be considered when designing and deploying nanoparticles in medical and biological applications that depend on protein-nanoparticle interactions.

500931 Self-Assembly of Proteins: The Role of Shape and Specific Interaction

Abstract:

Crystallization is routinely exploited for understanding protein structure, but the molecular details of how proteins crystallize remain poorly understood. Using computer simulations, we investigate the role of patchy interactions and protein shape in self-assembly. We demonstrate that for green fluorescent protein with its beta-barrel shape, a patchy sphere model is not sufficient to explain the experimentally observed unit cell, but a model that accounts for anisotropic shape can reproduce protein crystal growth. We explore the role of rigid molecular shape in simulations of hard proteins represented solely by their solvent excluded surfaces. Hard convex polyhedra are known to crystallize into a multitude of close-packed structures, but for concave shapes such as proteins much less is known. In our simulations we observe metastable order, which could shed light on the role of shape in protein crystal nucleation.

497188 Predictive Design of Next-Generation Nanomaterials and Devices Via Bottom-up Approaches

Abstract:

Nanomaterials and devices that resemble biological matter in their ability to reconfigure and adapt on demand have captured increasingly growing interest over the past decades. Towards this end, bottom-up approaches including self- and directed-assembly techniques have been shown as a promising means for engineering the underlying nanostructure of this exciting class of materials and devices. In these approaches, the underlying nanostructures are obtained from the self- or directed-organization of nano building blocks such as block copolymers, nanoparticles and colloids. The fundamental challenges to the bottom-up techniques are to design the optimal assembling units, to tailor their effective interactions and to find efficient assembly pathways. In this contribution, I will address the design rules for the building blocksâ€™ geometry and their interactions targeting several hierarchically assembled structures including terminal supraparticles, helical ribbons and various columnar morphologies. I will also demonstrate numerous unconventional pathways to assemble these bio-mimicking and reconfigurable nanostructures including interaction tuning to shape shifting. The ultimate goals of my research program are to design of adaptive, programmable nanomaterials that are of potential use in a host of nanotechnology applications including, but not limited to, optical devices, drug delivery, biosensing and energy storage and conversion. Also presented are the tools and methods I have been developing to improve the efficiency of the computational studies of interest, ranging from large-scale GPU-accelerated codes to enhanced sampling methods.

492812 Solvation of Self-Assembled Complexes: Using Molecular Simulations to ProbeÂ Energetics, Structure, and Dynamics

Abstract:

Single walled-carbon nanotubes (SWCNTs) possess unique properties that make them attractive for a number of applications. However, the difficulty in obtaining uniform samples in terms of size, chirality, and handedness inhibits their widespread use. Various separation techniques address this problem through use of surfactants in order to suspend the tubes in aqueous media and then apply other approaches such as aqueous two phase extraction (ATPE) or ion-exchange chromatography (IEX) to sort the dispersed SWCNTs by their physiochemical properties. Single-stranded DNA (ssDNA) is a very effective dispersant that shows sequence-specific behavior during separation. Previous experimental work has shown that this specificity can be harnessed to tune the separation in favor of particular nanotube chiralities. Currently, the nature of this specificity is not yet well understood and optimal ssDNA/SWCNT pairs must be searched by trial and error. In this study, we use replica exchange molecular dynamics (REMD) simulations to assemble the ssDNA-SWCNT complexes and then apply free energy perturbation methods to obtain an estimate for the solvation energy of the complex. Such systems require the construction of a novel free energy pathway to access the quantity of interest. Additionally, the structure and dynamics of the hydrating water has been analyzed to better understand the solvation behavior on a molecular level and to provide insight into the effects of ssDNA sequence. These investigations work towards our goal to provide better insight into the sequence/chirality specific separation mechanism, and to eventually develop a model that allows for the prediction of other pairs leading to the efficient sorting of a desired nanotube chirality.

495168 Level Set Strategy for Self Consistent Field Theory

Abstract:

We introduce level set methods to design new algorithms for self-consistent field theory (SCFT) in polymer physics. SCFT computes the structure and energy of inhomogeneous self-assembling polymers at thermodynamic equilibrium. Level set methods enable to represent the surface of an enclosing domain with an implicit function which in turn provides a powerful way to evolve a free surface. We introduce the concept of *shape derivative* into SCFT and rigorously derive expressions for the change of energy of a diblock copolymer melt with respect to its enclosing shape. The shape derivative is then used to embed SCFT into a variable shape simulation where the internal structure and the enclosing shape are coupled and evolve in tandem in order to reduce the energy. Finally an algorithm for the inverse geometric problem is presented. The algorithm finds a shape in order to obtain a desired internal structure of the confined polymeric material.

501391 Simulations of Nonlinear Flows in Nonequilibrium Complex Liquids

Abstract:

Complex liquids, such as active biofluids and colloidal suspensions, are important for biophysical and other industrial applications. Here we focus on dense suspensions of semi flexible bioplomers, such as microtubule and actin filaments. When the biopolymer concentration is high, it exhibits nematic phase. And if motor protein clusters and adenosine triphosphate (ATP) are added to the suspension, biopolymers are able to slide on each other due to the action of motors - such collective motion leads to a spontaneous, cohesive flow in the system. The nature of such flows is not fully elucidated. Here we rely on a continuum hydrodynamic model of active nematics to characterize such nonequilibrium liquids. The model is realized via a hybrid lattice Boltzmann solver. We first validate the model by comparing it to benchmark problems of passive nematic liquid crystal flows and active nematics experiments. We then extend it to include unequal elasticities and local concentration variations. The model is able to describe actin-based active nematics with variable activity and elasticity. We further apply it to explain the ordered flow emerged from confinements and anisotropic boundaries. Based on the above results, we propose that active nematics can be the candidate for active microfluidic devices. Finally, we investigate the transient flows in dense suspensions. Discrete element method is used to simulate a suspension of bi-disperse non-Brownian spherical particles. As the system is subject to a sudden simple shear, dynamic jamming happens as there is a propagating wave moving with a constant speed. We focus on how the inter particle frictional force and the short-range repulsive force determine the formation of the front and its velocity. Our simulations provide detailed structures of the suspension at transient state and therefore helpful for elucidating the transient dynamics of dense suspensions.

491508 Employing a Multipole Approximation in a Hybrid Fluid Via Relative Resolution

Abstract:

Recently, we presented a novel framework for multiscale simulations, termed "Relative Resolution" (RelRes) [1]. Inspired by another hybrid approach [2], our algorithm also contains both Fine-Grained (FG) and Coarse-Grained (CG) models in asingle fluid system. However, a unique feature of RelRes is that it switches between molecular resolution in terms of relative separation: While nearest neighbors are characterized by a FG (geometrically detailed) model, other neighbors are characterized by a CG (isotropically simplified) model. We in turn make important connections with classical theories in statistical mechanics [3]. Instead of performing an iterative procedure for relating the FG and CG models [4], we formulate for this task an analytical expression which is based on a multipole approximation at appropriate distances: This algebraic equation stems in the energy conservation between a molecular pair in an infinite limit, and it substantially enhances the computational efficiency of RelRes. With this hybrid formalism for fluid mixtures, we consequently examine several systems in molecular simulations (e.g., a liquid-liquid of oxygens-nitrogens, a liquid-gas of tetrachloromethanes-thiophene, etc.). We primarily focus on structural correlations and on thermal properties, correctly retrieving their corresponding behavior across state space. As a next step, we also proceed by implementing RelRes for polymers. In summary, we show here for fluids, that molecular resolution is inherently hybrid in terms of relative separation.

[1] A. Chaimovich, C. Peter, and K. Kremer. Relative resolution: A hybrid formalism for fluid mixtures. The Journal of Chemical Physics 143:243107, 2015.

[2] M. Praprotnik, L. Delle Site, and K. Kremer. Adaptive resolution molecular-dynamics simulation: Changing the degrees of freedom on the fly. The Journal of Chemical Physics 123:224106, 2005.
[3] B. Widom. Intermolecular forces and the nature of the liquid state. Science 157:375-382, 1967.
[4] M. G. Saunders and G. A. Voth. Coarse-graining methods for computational biology. Annual Review of Biophysics 42:73-93, 2013.

489615 Quantitatively Reliable Molecular Modeling and Simulation of Vapor-Liquid Equilibria

Abstract:

For a variety of low-molecular fluids, two-center Lennard-Jones plus point dipole (2CLJD) and two-center Lennard-Jones plus point quadrupole (2CLJQ) models, which describe bulk properties of pure-component vapor-liquid equilibria with a good accuracy, are available from the literature [1]. In the present work, these models are applied for simulating the vapor-liquid surface tension. Thereby, the literature models, which were adjusted to bulk properties, but not to interfacial properties, are validated against the surface tension of the real fluids.

By molecular dynamics simulation, employing a recently developed efficient long-range correction scheme [2, 3], it is found that on average, the literature models overestimate the vapor-liquid surface tension by 15 to 20% [3, 4]. To improve the overall accuracy of the models, optimization criteria based on the vapor pressure and the saturated liquid density, i.e., bulk properties which were previously taken into account during model parameterization, are combined with the vapor-liquid surface tension, which was not considered when the literature models were developed. This multicriteria optimization problem is addressed by constructing the Pareto front, i.e., the set of rational compromises between the competing objectives, from which models tailored to special needs can subsequently be chosen [5].

[1] J. Vrabec, J. Stoll, H. Hasse, J. Phys. Chem. B 105(48), 12126-12133, 2001.
[2] S. Werth, G. Rutkai, J. Vrabec, M. Horsch, H. Hasse, Mol. Phys. 112(17), 2227-2234, 2014.
[3] S. Werth, M. Horsch, H. Hasse, Mol. Phys. 113(23), 3750-3756, 2015.
[4] S. Werth, M. Horsch, H. Hasse, J. Chem. Phys. 144, 054702, 2016.
[5] K. Stöbener, P. Klein, M. Horsch, K. Küfer, H. Hasse, Fluid Phase Equilib. 411, 33-42, 2016.

498779 Production of High-Purity Pyrene Oligomers in a Supercritical Fluids Medium

Abstract:

Polycyclic aromatic hydrocarbons (PAHs) have been shown to be ideal precursors for liquid crystalline materials, due to their ability to form an ordered mesophase. This feature gives them excellent heat and charge-transfer properties, which are needed for many applications. In our group, we have focused our work on the PAH pyrene. In that work, pyrene was catalytically reacted to form pyrene oligomers, which were then fractionated for individual property testing. During that testing, we found that the pyrene trimer oligomer was able to form 100% mesophase, making it the lowest melting point mesophase ever discovered.

Unfortunately, the current method of reacting pyrene monomer to form trimer results in trimer yields that are fairly low; thus, the oligomeric product must be extracted and purified, which is time-consuming, expensive, and difficult to scale up. Thus, we are interested in methods in which the pyrene reaction will be cut off at the desired oligomer, resulting in a higher-purity oligomeric product.

Using our knowledge of the solubility of pyrene oligomers in supercritical (SC) toluene, we are reacting a pure pyrene with $AlCl_3$ catalyst in the presence of SC toluene. At appropriate temperatures and pressures, the SC toluene solubilizes the monomer and any dimer products from the reaction, while any trimer produced will precipitate out of solution because of its low solubility. Once precipitated from solution, the trimer reacts far less with the monomer and dimer reactants, meaning that the monomer and dimer will continue to react only with themselves. This process can be carried out until the final mixture is left with a much higher concentration of trimer. Preliminary results indicate that trimer concentrations can be at least doubled when pyrene oligomerization is carried out in the presence of supercritical toluene.

498796 Development and Analysis of Green Pathways Isolation of Medically Active Components from Grape Pomace Biomass

Abstract:

Growth in the medical sector, fueled by advancing technologies, has afforded people a greater than ever level of wellness. The use of supercritical fluids for processing continues to expand in the realm of mainstream manufacturing, benefitting greatly from co-solvents and biphasic pairings. Innovation has allowed for utilization of natural components in health fields. However, significant bottlenecks exist with regards to obtaining natural constituents via a green pathway when compared to organic solvent extraction. Progressive environmental regulation and a health conscious population necessitates a shift towards green technology.

Supercritical CO_2can be used exclusively or with chemical enhancement (co-solvents, biphasic systems, etc.) to obtain active components from a biological matrix. This project focuses principally on grape pomace, a waste product from juice and wine production consisting of stems, seeds, peels, and remaining fruit matter. The peels contain multiple different types of polyphenols useful for natural medicines. The seeds of the pomace also contain polyphenols and also oil useful for various applications. Utilization of this waste as a feedstock allows the juice and wine industries to function in a more sustainable and environmentally friendly manner while allowing for the extraction of valuable natural medicinal compounds.

Pressures up to 300 bar are used with temperatures up to 60° Celsius. Extraction may be further modified by polar co-solvents to increase the solubility of polar compounds in the CO_2 extraction media. Sequential extraction is also utilized to obtain 2 extract fractions, involving extraction via warm pressurized water and SC CO_2 phases. Multiple analysis techniques are required to quantify both the amount of extract and the composition. Liquid chromatography (LC) and gas chromatography (GC) paired with mass spec are industry standards for identifying the extract compositions. Physisorption is planned as a method that may demonstrate quantitatively how the biomass structure has changed due to extraction. Modeling is implemented to test conditions before apparatus modification and also to obtain basic results for those conditions not conducive to lab experimentation. Different CFD models are analyzed to describe the complex fluid behavior and wetting patterns in these extractions.

492907 A Â Novel Supercritical CO_2-Based Treatment for Decellularization That Maintains Mechanical and Structural Integrity

Abstract:

We have previously presented a novel, hybrid process for decellularizing a porcine aorta, while maintaining structural and mechanical integrity. The process comprises brief exposure to an aqueous solution (for cell disruption), following by washing with supercritical CO_2(to remove cellular debris and aqueous residue). In addition to achieving decellularization, we now demonstrate that the aorta so treated maintains favorable mechanical properties, and that the structural proteins collagen and elastin maintain structure that is much closer to the native tissue than aorta treated with a standard decellularization method.

Whether derived from synthetic or natural materials, TE scaffolds must be sterile, porous, mechanically strong, biocompatible, bioactive, and properly direct cell proliferation and differentiation. Additionally, the scaffold fabrication process and subsequent treatments may introduce several structural and biochemical deficiencies, including loss of mechanical strength, loss of surface activity, denaturation of extracellular matrix (ECM) proteins, scaffold dehydration, and residual cytotoxicity of some solvents, detergents, and/or crosslinking agents. All of these challenges require continual development of novel and innovative scaffold fabrication methods. In 2016, our lab presented a novel hybrid decellularization method that utilizes supercritical carbon dioxide ($scCO_2$) and sodium dodecyl sulfate (SDS) to decellularize a tissue. We demonstrated that the hybrid decellularization method was able to completely decellularize porcine aorta faster than a standard SDS method.

In this work, scaffolds of porcine aorta were treated with different methods: a standard aqueous method using surfactant solutions (control), a SC CO2 method using ethanol as an additive, and the hybrid of aqueous treatment followed by CO2 rinsing. The mechanical properties of decellularized scaffolds were evaluated by performing a uniaxial ring test. Arterial tissue is non-Hookean and does not have a linear modulus of elasticity, but contributions to modulus of elasticity from elastin ($E_{elastin}$, 0-8% strain) and collagen ($E_{collagen}$, 12-20% strain) can be determined from stress-strain curves. These values were calculated for the native tissue, hybrid treatment, and treatments with only SDS and only $scCO_2$. The hybrid treatment best maintained the mechanical properties of the scaffold, as substantial protein denaturing was observed during SDS treatment and excessive dehydration was observed during treatment with only $scCO_2$and ethanol.

The microscopic arrangement of ECM fibers (elastin and collagen) was visualized by staining with Massonâ€™s trichrome. For the hybrid treatment, complete decellularization is observed along with a distribution of collagen that closely resembles the native aorta. SDS treatment also removes cells, but disruption of collagen is observed.

Finally, a methylene blue assay was used to quantify residual SDS for the standard SDS and hybrid SDS/$scCO_2$ decellularization protocols. Just one hour of treatment with $scCO_2$ removed a similar amount of SDS compared with a whole day of washing with phosphate-buffered saline. Prolonging $scCO_2$treatment beyond one hour would likely reduce residual SDS concentration below the cytotoxic level of 0.002%.

Overall, scaffolds produced by the hybrid SDS/$scCO_2$ decellularization protocol have more desirable mechanical and biochemical properties compared to standard treatments.

498845 Supercritical Fluid Enhanced Swelling and Drug Impregnation of Biomedical Polymers

Abstract:

The world of biomedical implants is growing faster than ever due to technological advancement driven by societal need. The baby boomer generation is proceeding into their post retirement years when health and wellness begin to decline. The American Heart Association states that 70% of men and women in the 60-79 year-old age group have cardiovascular disease. While diet and exercise may aid in the improvement of health, medicines and biomedical devices are often required to ensure health. Often, medicines are not adequate for treatment and more invasive devices must be utilized. As an example, cardiac catheters and stents are common devices used in cardiovascular procedures to diagnose and treat cardiovascular issues respectively; and like all devices put into the body they are foreign objects which result in an immune response often resulting in inflammation and scar tissue formation. The ultimate goal should be to produce an implant which minimizes the likelihood of immune response to the device while minimizing the need for supplemental medications used to prevent irritation, scarring, rejection, or clotting about the device.

What if devices could be prepared with polymeric coatings embedded with localized immunosuppressing technology? The polymer may be impregnated with a corticosteroid such as dexamethasone in order to reduce the inflammatory response around the device. This stands to minimize inflammation during routine procedures and also may function to extend the longevity of an implanted device before it begins to induce scarring. This idea exists firmly in academia but faces particular hurdles in making it to the consumer. Impregnation of a polymer in a medically safe way is the largest concern. Swelling the polymer and impregnating it with the drug is an advantageous method. However, temperature and solvent choices are limited due to the limitations of the drug to be impregnated and solvent interactions with the body.

Supercritical fluids, namely supercritical carbon dioxide (SC-CO_2), are applied to drug impregnation of polymers. SC-CO_2 has a low critical temperature and reasonable pressure, it is nonflammable, and it readily leaves the system when pressure is relieved. Polymer solubility in SC-CO_2 is highly dependent on the type of polymer and chain length. Polymers that are hard enough to be used for biomedical purposes are not solubilized in SC-CO_2, but do swell to some extent in its presence. Corticosteroids and other inflammatory response drugs are relatively small molecules when compared to polymers and have the ability to be soluble in SC-CO_2 as a function of temperature, pressure, and presence of a co-solvent.

A Thar SC-CO_2 phase analyzer is used to swell batches of polymer rings like those used in various biomedical applications. It features video capture technologies, variable speed motorized stirring, and a varying volume cell. Various inserts have been fashioned to further modify the functionality of the cell. The polymeric rings are inspected via SEM before and after swelling to make note of structural changes. Adequate swelling must be observed such that the drug can be properly impregnated. This allows for the drug to have much better functionality due to not only having drug coated on the outer surface but also contained within polymeric pores allowing for diffusion from within.

501084 Effect of Precursor on Crystal Structure of Titania Synthesis in Supercritical Sol – Gel Reaction

Abstract:

Titania has been applied as a photocatalyst in water treatment technology. On the development of titania, there are the drawbacks, the calcination step for the crystallization could cause the reduction of the surface area. Recently, the carbon / titania composite fabrication have attracted attention on the enhancement of the formation of mixed crystal structure of anatase and rutile phase which suppresses the recombination of electron and hole (A. Chatterjee et al., J. Raman Spectrosc. 42, 1075 – 1080 (2011)). Typically, anatase crystal of the titania can be formed in the calculation at the lower temperature and then the rutile crystal phase at around600 . Graphite or carbon nanotube were used for incorporation of carbon source into titania. These composite are expected to increasing of the photocatalytic activity. However, they are expensive and synthesized in complex reaction process.

In our research group, titania particles with various morphologies were synthesized by a sol-gel reaction in supercritical carbon dioxide(M. Kinoshita et al., J. of Supercritical Fluids 116 (2016)). Supercritical carbon dioxide has been applied as both synthesis and drying solvents for the fabrication of nanostructured titania particles because of the unique properties, such as low surface tension, high solubility and diffusivity. We found that the titanium precursor affects the remained amount of carbon in the titania particles. It was confirmed that the titania from the lower reactive precursor contained the larger amount of carbon and expected to reduction of the calcination temperature for the formation of the mixed crystal structure of the anatase and rutile phases.

In this work, the effect of the titanium precursor on the particle fabrication is investigated. The sol-gel reaction for synthesis of titania particle was conducted using the three types of titanium precursors acetic acid (HAc) in in supercritical carbon dioxide. Titanium isopropoxide (TIP), titanium ethylhexoxide (TEO) and titanium diisopropoxide bis acetylacetonate (TDB) used as their precursors. A high-pressure vessel with 34 ml was used for the synthesis of titania particle in supercritical carbon dioxide. The known amounts of the precursor and HAc were loaded into the high-pressure vessel before supplying supercritical carbon dioxide into the system. The molar ratio of HAc to the precursor was 10. Carbon dioxide from the cylinder was passed through the cell including the silica gel in order to remove the trace amount of water. The purified carbon dioxide was liquefied at a chiller and pressurized by a feed pump. The pressurized carbon dioxide was supplied into the system in the air thermostat and achieved to the supercritical fluid. The supercritical carbon dioxide was installed into the high-pressure vessel in which the titanium precursor and HAc were loaded. The sol-gel reactions in supercritical carbon dioxide were conducted in the by a batch type operation at 313 to 333 K and 20.0 MPa. The mixture inside the high-pressure vessel was stirred by a magnetic stirrer. After the sol-gel reaction for 24 h, the contents in the vessel were dried by supercritical carbon dioxide for 6 h. The system was depressurized and then the dried titania particle was removed from the high-pressure vessel. The titania particles from the sol-gel reaction and drying using supercritical carbon dioxide was calcined at 973 K in air for 2 h for the crystallization. The synthesized titania particle was also analyzed by scanning electron microscope (SEM), x-ray diffraction (XRD), thermogravimetric analysis (TGA)

and second-ion mass spectroscopy (SIMS). The fibrous titania was fabricated using TIP precursor, whereas block and sphere ones were synthesized in case of TEO and TDB precursors. It is considered that the different intermediate during the sol-gel reaction was formed due to their difference of reaction rate. The amount of carbon in titania was measured by SIMS analysis. In case of TIP, the carbon was hardly observed inside the titania particles. On the other hand, the large amount of carbon was in the particles from TEO and TDB. It is thought that the sol-gel reaction did not sufficiently proceed and unreacted carbon group remained in titania due to their low reactivity of TEO or TDB. The calcinated titania from TDB or TEO has the low weight loss in the TGA analysis. The ratio of anatase and rutile was calculated from XRD analysis results. The ratio of rutile in the titania from TDB and TEO are larger than those from TIP precursor. From TGA abd XRD results, it was suggested that the large amount of the residual carbon in the titania could enhance the transition of the crystal structure from anatase to rutile phases.

498570 Accelerated Conversion of Magnesium Oxychloride to Chlorartinite Using Pressurized Carbon Dioxide and the Resulting Increased Water Resistance

Abstract:

Magnesium oxychloride cement (MOC) is a fire resistant structural material with broad applications in passive fire protection, especially for scenarios where high-strength or low carbon footprint materials are needed; however, despite many advantages for construction applications, MOC can show low water stability, where untreated MOC degrades over time to magnesium hydroxide when exposed to water, and thus has not found widespread adoption in US construction.

MOC water resistance can increase over time when exposed to air. Previous studies have found that chlorartinite formation, $Mg_2(CO_3)(OH)Cl$â–ª$2H_2O$, forms in the surface of this material as a result of exposure to CO_2. It has been suggested that chlorartinite increases the water stability due to the lower water solubility of this material, but this phenomenon has not been quantified before. This study has used quantitative X-ray diffraction data and scanning electron microscope images to examine and quantify the conversion of magnesium oxychloride to chlorartinite and the improved water stability of the cement. Samples were exposed to pressurized CO_2 to accelerate the formation of chlorartinite on the cement surface and chlorartinite formation was assessed. The extent of chlorartinite formation was shown to correlate with improved water stability. We hypothesize that a protective semi-insoluble chlorartinite layer forms on the surface of the magnesium oxychloride crystals and slows the degradation of the magnesium oxychloride when exposed to water. Untreated samples convert to magnesium hydroxide when subjected to an accelerated water resistance test (submersion in deionized water at 60 °C for 24 hours). Samples exposed to CO_2 show reduced MOC conversion. Chlorartinite was observed to convert to hydromagnesite, magnesium hydroxide and amorphous phase, but at a much slower rate.

501032 High-Throughput Mechanical Stimulation and Functional Imaging *in vivo*

Abstract:

Mechanosensation is required for multiple sensory modalities such as touch, hearing, and balance, and is linked to a multitude of disorders including deafness. Molecular mechanisms for mechanotransduction have been partially elucidated using a variety of model organisms, including *Caenorhabditis elegans*. Conventional mechanosensation experiments with *C. elegans* typically involve the manual delivery of a mechanical stimulus via an eyebrow hair or metal pick, and visual scoring of touch avoidance behavior, an assay subject to considerable variability between experimenters. Microfluidics has long been used for well-controlled and high-throughput experiments with small samples. In addition to enabling precise perturbations on the micron scale, microfluidic devices can easily be designed to work together with optical microscopy, allowing for imaging of fluorescent probes such as calcium indicators. For *C. elegans* experimentation particularly, microfluidics has been a widely adopted technology due to the match in length scale and compatibility with fluid handling. In this talk, I will present a microfluidic platform for delivering robust and precise mechanical stimuli to *C. elegans* by using pneumatically actuated structures. The device is fully automated, minimizing human variability and improving experimental throughput; it is fully compatible with fluorescent imaging of calcium dynamics of neurons. We show that this device can deliver mechanical stimuli with precise control in timing, location, and deformation. These characteristics allow us not only to recapitulate the well-known receptive field of the neurons in the gentle-touch circuit, but also to understand the dynamics of the harsh-touch neurons. We further developed this microfluidic system to interrogate behavior of mechanosensory neurons during development. Another direct application of the devices are in drug screens in a humanized worm model. Our microfluidic system directly addresses the current bottleneck of a lack of functional assay for such drug screens. The ability to automate and streamline the microfluidic system will greatly enhance our ability to identify potential therapeutics.

490791 The Steady Motion of a Train of Vesicles in a Cylindrical Channel of Arbitrary Cross Section

Abstract:

The three-dimensional, inertialess motion of a train of lipid-bilayer vesicles flowing through a cylindrical channel of variable cross section is simulated using the boundary integral equation method. Steady-state results are reported for vesicles positioned concentrically inside cylindrical channels of circular, square, and rectangular cross sections. The vesicle translational velocity U and excess channel pressure drop Δp^+ are found to depend strongly on the ratio of the vesicle radius to the hydraulic radius λ and the vesicle reduced volume υ. "Deflated vesicles" of lower reduced volume $\upsilon < 1$ have more streamlined bodies and, therefore, translate with greater velocity U relative to the mean velocity V of the flow. Increasing the vesicle size (λ) increases the wall friction force and extra pressure drop Δp^+, which in turn reduces the vesicle velocity U. Hydrodynamic interactions between vesicles in a periodic train are largely screened by the channel walls, in accordance with previous results for spheres and drops. Changing the channel cross section has little effect on the steady vesicle shape but changes the mass flux through a cross section, viz. a vesicle passing through a square channel has a higher translational velocity than in a circular channel of equal hydraulic radius. A simple correction factor proportional to the ratio of mean velocities in channels of different cross sections is proposed to unify the results. Nonlinear effects are observed when β – the ratio of membrane bending elasticity to viscous traction – is changed. Increasing β tends to streamline the vesicle shape – preventing the formation of "parachute-like shapes" with regions of high membrane curvature – which, in turn, tends to increase U and decrease Δp^+. This effect is dampened – and, under certain conditions, reversed – as confinement is increased. The impact of the viscosity contrast κ between the interior and exterior fluid is also explored. The simulation results show excellent agreement with experimental measurements of vesicles in microfluidic channels and circular capillaries. In addition, the simulation results are commensurate with a previously reported "small-gap theory" valid for large values of λ.

494181 The Dissolution of a Sheared Water Drop in Bitumen Measured Using a Novel Microfluidic Platform

Abstract:

It is well known that the dissolution rate and solubility of drops of one liquid emulsified in a second, immiscible liquid phase can be enhanced significantly by the presence of surfactants in the continuous phase. While methods for measuring solubility abound in the literature, measurement techniques for the dissolution rate are rare. In this study, we present a new, robust microfluidic platform for determining the dissolution dynamics of an individual drop in an immiscible suspending medium. We demonstrate this technique for emulsions of water in bitumen, which contains indigenous surfactants that enhance water solubilization. This is a difficult system for dissolution measurements due to the opacity of bitumen, but our shallow microfluidic channels allow us to clearly image the water-bitumen interface[1]. Moreover, water-in-bitumen emulsions are extremely stable[1,2], and any sub-micron emulsified water droplets, which are difficult to remove even by ultracentrifugation, are also counted as solubilized water in traditional solubility measurements such as Karl Fischer titration. Our single drop dissolution technique avoids this complication.

Depending on the bitumen composition, water chemistry and flow conditions, the time for water dissolution in bitumen is governed either by interfacial phenomena or by mass transfer between the bulk bitumen phase and the water-bitumen interface. In the latter case, we can also calculate the solubility of water in bitumen, provided an independent measure of the size of the water-carrying species is available. Our study has uncovered four key features of the dissolution of water in bitumen under flowing conditions. First, increasing bitumen concentration enhances the water solubility, but the dissolution process slows down. Second, both dissolution rate and water solubility increase with increasing diluent aromaticity. Third, for a fixed strain rate of the flow past the drop, as the pH of the water drop increases, the dissolution rate of water diminishes. Fourth, water solubility decreases when asphaltenes are removed from the bitumen system. Our microscale results can predict the time required to completely solubilize a spherical drop in a suspending medium for a bench scale experiment, if the mixing conditions are specified. These and other distinguishing characteristics of our study will be elucidated in the presentation.

References

1) Sonthalia, R., Ng, S. and Ramachandran, A., 2016. Formation of extremely fine water droplets in sheared, concentrated bitumen solutions via surfactant-mediated tip streaming.Â *Fuel*,Â *180*, pp.538-550.

2) Tchoukov, P., Yang, F., Xu, Z., Dabros, T., Czarnecki, J. and Sjoÿblom, J., 2014. Role of asphaltenes in stabilizing thin liquid emulsion films.Â *Langmuir*,Â *30*(11), pp.3024-3033.

495859 Adhesion and Break-up of Model Metastatic Cancer Cell Clusters during Passage through a Microfluidic Constriction

Abstract:

During the extravasation stage of cancer metastases, cancer cells and clusters exit the blood vessel though constrictions that may be an order of magnitude smaller than the cluster dimension. The physics of the extravasation process involves cell deformation and inter-cell adhesion during through confinement and constriction flow. Using advanced modeling coupling fluid flow and cell mechanics, we investigated how circulating tumor cell clusters are able to traverse a much smaller capillary. We found that depending on the inter-cellular adhesion, the cluster would either become stuck, pass through intact, or break up during passage [1].Â The model predictions capture the qualitative behavior of cancer cell cluster passage through microfluidic constrictions, suggesting possible pathway for future treatments.

[1]Â S.H. Au, B.D. Storey, J.C. Moore, Q. Tang,Y.-L. Chen, S. Javaid, A.F. Sarioglu, R. J. Sullivan, M.W. Madden, R. Oâ€™Keefe, D.M. Langenau, D.A. Haber, S. Maheswaran, S.L. Stott and M. Toner* (2016) Clusters of circulating tumor cells traverse capillary-sized vessels, Proc. Natl. Acad. Sci. U.S.A. **113**, 4947

497594 Chaotic Thermal Convection in Microfluidic Hydrothermal Pore Environments

Abstract:

A key unanswered question in the origin of life involves elucidating how complex macromolecules spontaneously emerged from elementary building blocks present at extremely dilute concentrations in the prebiotic ocean. Pore networks permeating mineral formations near undersea hydrothermal vents have emerged as potential hot spots for these kinds of biochemical processes, fueled by the recent discovery of Earthâ€™s oldest microfossils in hydrothermal systems and the existence of similar environments under the icy surface of Saturnâ€™s moon Enceladus. Here we explore how thermally-actuated flows that naturally occur within these micro-scale pore spaces display surprisingly complex 3D chaotic characteristics that can enable continuous transport of particulate and molecular species from the bulk fluid to discrete targeted locations along interfacial boundaries. Using coordinated simulations and experiments, we find that these flows are capable of accelerating transport to the surface by several orders of magnitude under hydrothermally relevant conditions. We also explore partitioning and assembly of macromolecular species into membrane-like films capable of maintaining localized pH gradients. These results suggest that microscale chaotic thermal convection may supply a previously unappreciated mechanism to explain prebiotic surface-catalyzed processes and establishment of early bioenergetic pathways central to the origin of life.

498317 Using Droplet Microfluidics to Study Chemical Thermodynamics of Aqueous Atmospheric Aerosol Particles

Abstract:

Accurate predictions of gas-liquid-solid equilibrium partitioning of atmospheric aerosols by thermodynamic modeling and measurements is critical for determination of atmospheric particle properties and processes at temperatures and relative humidities relevant to the atmosphere. Organic compounds like dicarboxylic acids that originate from biomass burning, and direct biogenic emission make up a significant fraction of the organic mass in atmospheric aerosol particles. In addition, inorganic compounds like ammonium sulfate and sea salt also exist in atmospheric aerosols, that results in a mixture of single, double or triple charged ions and non-dissociated organic acids. In this work, atmospheric aerosol droplet chemical mimics are generated in microfluidic channels and stored in passive traps until dehydration to study the influence of relative humidity and water loss on phase behavior. The solution volume of the droplet trapped changes with respect to time, and is calculated by image analysis and correlated with the concentration of the solution to determine water activities at each time interval. The research will help determine mixture efflorescence of droplets that include presence of other dissolved components and tend to decrease the water activity. The measurements will also be used to parameterize statistical thermodynamic modeling for predictions of thermodynamic properties of aqueous aerosols as a function of relative humidity.

488078 Analysis of the Engulfment of a Foreign Particle By a Solidification Interface: A New Scaling for Flow and Interfacial Forces Involving SiC Particles during Silicon Crystal Growth

Abstract:

The engulfment of foreign particles during solidification has been studied for over 50 years, yet classical theories been unable to adequately describe the behavior of silicon carbide (SiC) particles during silicon crystal growth. The engulfment of these particles during the growth of multicrystaline silicon (mc-Si) remains a vexing problem for the production of high-quality, low-cost photovoltaic devices. In particular, carbon contamination of the melt, from the original silicon feed and via gas phase reaction and transport from hot furnace internals, leads to the formation of silicon carbide particles of 1-100 microns in diameter that may be engulfed during crystal growth. These particles are problematic for subsequent processing of solar cells, leading to issues of lower cell efficiency via the formation of dislocations and shunts, wafer breakage, sawing defects, and even saw wire breakage.

The engulfment of a micron-sized particle during solidification is determined by a balance of repulsive van der Waals forces between the particle surface and the solidification interface and drag forces arising from the flow around the particle and into a thin liquid gap between particle and interface, typically on the order of 10 nanometers in thickness. When drag forces overcome repulsive forces, the particle is engulfed, otherwise it is steadily pushed ahead of the advancing interface. Since drag increases with particle size and velocity, there exists a critical velocity at which a particle of a certain size is engulfed. However, drag forces are strongly dependent upon the shape of the solid-liquid interface as the particle approaches. Thus, the critical velocity is affected by significant and nonlinear interactions involving heat transfer, premelting, and Gibbs-Thomson phenomena.

In this presentation, we discuss a significant and previously unascertained interaction between particle-induced interface deflection (originating from the thermal conductivity of the SiC particle being larger than that of the surrounding silicon liquid) and curvature-induced changes in melting temperature arising from the Gibbs-Thomson effect. For a particular range of particle sizes, the Gibbs-Thomson effect flattens the deflected solidification interface, thereby reducing drag on the particle and increasing its critical velocity for engulfment. We show via numerical calculations and analytical reasoning that these effects give rise to a new scaling of the critical velocity to particle size as $v_c \sim R^{-5/3}$, whereas all prior models have predicted either $v_c \sim R^{-1}$ or $v_c \sim R^{-4/3}$. This new scaling is needed to quantitatively describe the experimental observations for this system.

This research was supported in part by NASA NNX10AR70G; no official endorsement should be inferred.

501153 Hydrodynamic Mobility of Particles, Vesicles and Cancer Cells in a Square Microchannel

Abstract:

Understanding the transport of deformable particles such as vesicles and cells in moderately or tightly confined conduits is of interest in microfluidic technologies ranging from particle sorting to biomedical diagnostics. A key parameter of interest in these studies is the hydrodynamic mobility of the deformable particle, Î², which is defined as the ratio of the particle velocity to the surrounding fluid velocity. Here, we report experimental measurements of hydrodynamic mobility of particles, vesicles and tumor cells at low Reynolds number, under conditions of moderate geometric confinement in a square microchannel. For rigid and elastic particles, we find that the mobility as a function of particle confinement agrees well with the theoretical result of Hetsroni, Haber and Wacholder (J. of Fluid Mech., 41, 1970) for cylindrical conduits under Stokes flow, with the modification that the mean velocity is that of a square microchannel. With respect to vesicles, we find that they move faster than the particles probably due to their lower reduced volume and mobile interface. For cancer cells, we observe that their mobility lies between those of particles and vesicles. We will also discuss aspects such as the role of particle off-centeredness and nonspherical shape on hydrodynamic mobility.

501734 Deformable Droplet Migration in a Narrow Microchannel at Finite Reynolds Numbers Via Dissipative Particle Dynamics (DPD)

Abstract:

We report the results of a Dissipative Particle Dynamics (DPD) study on deformable droplet migration in a narrow microchannel at finite Reynolds numbers. We find that droplet migrates to a position between that of the well-known Segre and Silberberg position for that of a rigid particle due to inertial lift forces, and those reported for deformable droplets at zero Reynolds number by Chan and Leal. The ultimate, steady-state, position in our simulations is dependent upon the capillary number, the Reynolds number, the relative viscosities of the droplet and suspending fluid, and to some extent on the ratio of droplet size to channel height. We discuss the relevant dimensionless groups that affect the droplet position, the effects of inertia and inter-particle/inter-wall interactions, and establish connections to related previous studies in the literature. We also assess the conditions needed for convergence of the DPD results to the continuum limit in which the discreteness of the particles in the DPD simulations and size of the periodic simulation domain become negligible. The study not only improves our understanding of migration phenomena in microfluidic geometries, but also provides an assessment of the DPD method as a tool for predicting such phenomena in realistic experiments.

501717 Using Molecular Simulation to Understand the Interfacial Behavior of Ionic Liquids

Abstract:

Understanding the bulk and interfacial properties of ionic fluids is important for several industrial applications. In this presentation, we examine the properties of a relatively simple ionic system consisting of the restricted primitive model (RPM) at a structureless substrate that interacts with the fluid via an attractive van der Waals potential of variable strength. The model system exhibits several behaviors that are distinct from simple non-ionic fluids (e.g., Lennard Jonesium) at an equivalent structureless substrate. For example, the contact angle of the RPM system increases with increasing temperature over a relatively wide range of substrate strengths. In contrast, the contact angle of a non-ionic fluid generally decreases with increasing temperature. One of the consequences of this behavior is that the RPM system generally does not tend towards a wetting point with decreasing temperature. There are also interesting ramifications regarding the transition from the partial wetting to prewetting, to complete wetting regimes. We discuss now these trends are related to the microstructure of the fluid at the liquid-vapor and solid-liquid interfaces. The interface potential approach is used to quantitatively explore the interfacial properties of the RPM system, with interface potentials generated via Monte Carlo simulation.

499685 Capacitance and Ion Dynamics of Ionic Liquids Near Oxidized Graphene

Abstract:

Supercapacitors are being adopted for a wide range of electrical energy applications due to their relatively high energy and power densities. With the supercapacitor research community, graphene and room-temperature ionic liquids (RTILs) are commonly studied electrode and electrolyte materials respectively. However, surface chemistry, such as oxygen groups and defects, can be easily introduced to the graphene surface during synthesis. Nevertheless, there is limited knowledge concerning the influence of surface functional groups on energy density and power density of graphene-based supercapacitor.

In this work, we first investigate the capacitance of RTILs near oxidized graphene surface using molecular dynamics (MD) simulation. The structure of electrical double layers (EDL) is discussed and related to the capacitive performance. Cyclic voltammetry experiments further validate the influence of surface oxidization on capacitance. The diffusivities of ions in the EDL are examined by calculating their self-diffusion coefficients, which are presented as a function of the distance to the electrode surface. Ion diffusivities are compared between charged surface and neutral surface, as well as between pristine graphene and oxidized graphene. This integrated study provides a comprehensive understanding of both energy and power density of oxidized graphene.

498887 Using Simulations and Experiments to Characterize Water Structure Near Mica Surfaces for Heterogeneous Ice Nucleation

Abstract:

Understanding the role played by solid surfaces in ice nucleation is a significant step toward designing surfaces that can control ice nucleation. This is relevant to various fields ranging from atmospheric chemistry to transportation and food industries. Our overall goal is to uncover the effects of surface properties on interfacial water structure relevant to heterogeneous ice nucleation. Such an understanding can guide the development of predictive models correlating surface properties, and interfacial water structure and dynamics to the ice nucleation propensity of that surface. To this end, we use a combined experimental and simulation approach to investigate the effect of surface properties on interfacial water structure near mica surfaces. Mica is used because it can be made atomistically smooth in experiments thereby enabling a comparison between experiments and simulations. In our talk, we will discuss the change in interfacial water structure and dynamics with different surface ions. We will particularly investigate the effect of ion size, ion charge and spatial distribution. We will correlate these changes to the differences in ice nucleating behavior. This will provide better understanding of the mechanisms of heterogeneous ice nucleation.

501152 A Novel Density Gradient Theory for Surfactant Molecules â€“ Applied to Oil/Water Interfaces

Abstract:

In the upstream operations of the oil & gas industry, water and surfactant can be injected into the reservoir for enhanced oil recovery (EOR). The surfactant mobilizes trapped oil by reducing the interfacial tension (IFT) between oil and water, and thus increased the oil recovery efficiency. Predicting the IFT of water/oil/surfactant mixtures is essential for the design of EOR processes. To interpolate and extrapolate expensive experimental data, different theoretical and simulation methods have been developed to model the interfacial properties of reservoir mixtures. Among these methods, density gradient theory (DGT) has become a fast and accurate tool in the interfacial property prediction of many alkane mixtures. However, several physical and numerical issues of DGT restrict its application in more complex systems such as water containing and surfactant containing systems. In this paper, we presented a novel density gradient theory to predict the interfacial properties of water/oil/surfactant mixtures, and also extended our stabilized density gradient theory (SDGT) algorithm to fit with the new model. A characteristic of surfactant molecules is the ability to span the interface; however, conventional DGT uses a single position for the molecule, which limits its ability and accuracy to describe the surfactant behavior. In this work we redefined the DGT free energy to include head and tail components of a surfactant molecule by introducing a chain free energy functional. We also optimized the SDGT algorithm to speed up calculations with associating molecules (e.g., water and surfactant molecules). By taking advantage of the optimized SDGT algorithm, the new DGT was applied to calculate the IFT of several water/oil/surfactant mixtures. The numerical stability of the optimized SDGT algorithm was tested, and the IFT calculation results were quantitatively verified with laboratory data.

493484 Computational Chemistry Design of Liquid Crystal-Based Chemoresponsive Systems with Increased Water Tolerance

Abstract:

Nematic liquid crystals are prone to anchor to surfaces that can alter their overall orientation in the bulk liquid phase. This orientation change can be exploited to detect analytes which can replace liquid crystal molecules bound to the surface and thus it can force liquid crystals to restore their original orientation in the bulk.[1] Dimethyl-methylphosphonate (DMMP), a common precursor of nerve gases, is an important analyte which has been proved to replace liquid crystals.[2,3] To make one step further towards developing a sensor that can be used in practical applications it is important to exclude false positive signals arising from ubiquitous molecules such as water. Previously, we have developed theoretical models, based on quantum mechanics, to simulate the orientational transition of nematic liquid crystals[4,5] which provide the basis to describe the effect water in these complex environments. We have also developed an integrated theoretical and experimental approach where computational models are validated against well-defined experimental data and new predictions are tested experimentally.[4,5]

In this presentation, we present computational models to simulate the effect of humidity on potential liquid crystal-based chemical sensors. Using experimental data, we validate descriptors that can indicate the sensitivity of liquid crystal-based systems towards high relative humidity. Based on our models, we have identified new previously unknown liquid crystal molecules that can have higher water tolerance than traditional liquid crystal molecule 4-cyano-4â€™-pentylbiphenyl (5CB). Motivated by our calculations, we have successfully synthesized the predicted molecules and tested their chemo-responsiveness in humid environment. Our experimental results have confirmed our theoretical predictions about increased water tolerance while the detection of DMMP is still possible. These results indicate the importance of close collaboration between theory and experiments to facilitate the progress in complex liquid crystal material design for chemoresponsive applications.

1. Shah, R. R.; Abbott, N. L., *Science*, **2001**, 293, 1296.

2. Yang, K. L.; Cadwell, K.; Abbott, N. L., *Journal of Physical Chemistry B*, **2004**, 108, 20180.

3. Hunter, J.T.; Abbott, N.L., *Applied Materials and Interfaces*, **2013**, 6, 2362.

4. Roling L. T.; Scaranto, J.; Herron, J. A.; Yu, H.; Choi, S.; Abbott, N. L.; Mavrikakis, M., *Nature Communication*, **2016**, 7, 13338.

5. Szilvási, T.; Roling, L. T.; Yu, H.; Rai, P.; Choi, S.; Twieg, R. J.; Mavrikakis, M.; Abbott, N. L., *Chemistry of Materials*, **2017**, accepted.

494354 A Semi-Empirical Analytical Model for Determining the Thermodynamic Stability of Pickering Emulsion

Abstract:

The prediction of thermodynamic stability of Pickering emulsion is of fundamental importance for various practical applications. Whereas massive experimental and theoretical work has been contributed for examining the stability of functional solid nanoparticles at different liquid-liquid interfaces, most of these works are very specific, and no generic model has yet been reported. Here, we first propose a semi-empirical analytical model to determine the thermodynamic stability of Pickering emulsion. This model, simply relying on the Flory-Huggins parameters of different compositions as input, is developed by evaluating the contact angle of solid nanoparticle. The determination of contact angle is associated with the liquid-liquid interfacial tension and liquid-solid surface tensions, whose analytical relations in terms of Flory-Huggins parameters and solid particle size are devised with the help of Dissipative Particle Dynamics simulations. This semi-empirical analytical model, in good agreement with experiment data, provides a straightforward route for predicting the thermodynamic stability of Pickering emulsion, as the Flory-Huggins parameters of most chemicals are available in standard manuals.

500227 Computational Method for Extracting Individual Adsorption Parameters from Experimental Dynamic Interfacial Tension of Mixed Surfactant Solutions

Abstract:

Diffusion-controlled adsorption dynamics of surface-active agents at fluid/fluid interfaces is described by the Ward Tordai equation which must be solved numerically when a nonlinear adsorption isotherm is involved [1]. These calculations are generally performed when knowing the composition of the surfactant solution and the adsorption behavior of its individual components (in terms of adsorption isotherm and equation of state as well as the associated parameters such as surface excess coverage and adsorption coefficient). However, for naturally occurring mixtures of surface active molecules (such as asphaltenes in crude oils [2], Arabic gum in food products [3], proteins in human biological fluids [4]) such information is often not available. This can also be the case for synthetic surfactants coming as mixtures of oligomers such as Tritons [5]. To alleviate this limitation, it is here reported a computational method for extracting the individual properties of pseudo-components within a surfactant mixture from its dynamic interfacial tension. The number of pseudo-components (*n*) and their adsorption isotherm/equation of state are first chosen. The pseudo-components' properties (concentration, adsorption coefficient, surface excess coverage) are then initialized at random values to solve numerically the *n* corresponding Ward Tordai equations. The resulting evolution of surface coverage is subsequently converted into a dynamic interfacial tension curve, which can be compared to the experimental one by means of a weighted Sum of Squared Error (WSSE) accounting for the exponential decay form of most available data. During an iterative process using the quasi-Newton Broyden-Fletcher-Goldfarb-Shanno (BFGS) algorithm, the pseudo-components' properties can be optimized to minimize the WSSE. This method has been tested with the experimental dataset from Freer and Radke [6]. A ternary mixture model with 7 parameters (3 concentrations, 3 adsorption coefficients and 1 surface excess coverage) proved enough to capture the dynamic interfacial tension from a few seconds up to 24 hours adsorption. The sum of the pseudo-components' concentrations is found very close to the nominal concentration of the asphaltenes solution used in experiments. Most asphaltenes appear to have a low surface activity while a tiny fraction (<1%) appears to have an extremely high one, in line with previous results obtained by fractionation or washout experiments. The accuracy of the method can further be checked by using the same calculations to predict the dilatational rheology of the same interface after 24 hours. The result compares very well with corresponding experiments over 7 frequency decades.

[1] Miller, R. (1981). *Colloid and Polymer Science, 259*(11), 1124-1128.

[2] Liu, F., Darjani, S., Akhmetkhanova, N., Maldarelli, C., Banerjee, S., & Pauchard, V. (2017) *Langmuir, 33*(8), 1927–1942.

[3] Castellani, O., Guibert, D., Al-Assaf, S., Axelos, M., Phillips, G. O., & Anton, M. (2010). *Food hydrocolloids, 24*(2), 193-199.

[4] Trukhin, D. V., Sinyachenko, O. V., Kazakov, V. N., Lylyk, S. V., Belokon, A. M., & Pison, U. (2001). *Colloids and Surfaces B: Biointerfaces, 21*(1), 231-238.

[5] Fainerman, V. B., Lylyk, S. V., Aksenenko, E. V., Makievski, A. V., Petkov, J. T., Yorke, J., & Miller, R. (2009). *Colloids and Surfaces A: Physicochemical and Engineering Aspects, 334*(1), 1-7.

[6] Freer, E. M., & Radke, C. J. (2004). *Journal of Adhesion, 80*(6), 481-496.

505008 Atomistic Simulations of Ordered and Disordered Carbons: "Mimicking" Versus "Targeting"

Abstract:

Atomistic simulation techniques have become extremely valuable tools for the elucidation of the structure of disordered or (partially) ordered materials. These techniques not only allow for the production of large-scale three-dimensional representations of the materials from the atomic scale, but also to predict most of their properties, and in some cases, to assess their formation mechanisms. In this talk, I will focus on carbon materials that can be ordered as well as disordered. There two routes to obtain realistic molecular-scale reconstruction of porous carbons: mimicking their formation process and targeting a â€œfinalâ€ state by constraining/guiding the simulation using experimental texture information. Obviously the first is a priori more predictive than the second but requires dealing upfront with chemistry (through reactive force fields) than can be rather complex with kinetically hindered processes.

To illustrate the mimicking route, I will present (i) the Grand-Canonical numerical synthesis of ordered carbon phases by carbon vapor deposition on zeolites and MCM-type mesoporous oxides, (ii) the Replica Exchange MD simulation of the geological degradation of the main components of wood (lignin and cellulose that are both organic molecules) into disordered nanoporous carbons and kerogen phases with the production of a complex liquid phase. The so-called targeting route will be illustrated with the example of the kerogen phases using the Hybrid-Reverse Monte-Carlo approach with an extension towards their mechanical properties from elasticity to failure (fracture).

505007 Graphene Oxide Membranes: A Molecular Simulation Approach

Abstract:

Graphene oxide membranes have attracted a lot of attention from the scientific community due to their potential application in challenging separations. Nevertheless, to optimise a material for an engineering application, it is necessary to understand their structures and the features that can improve their performance. Following on Keith Gubbins philosophy that realistic models of materials at a molecular level can be obtained from mimicking some aspects of their synthesis using molecular simulations, we have developed a protocol to create model graphene oxide sheets and mimic the membrane formation procedure to obtain model materials, with properties comparable to those measured experimentally.

495702 Polyethylene Oxide (PEO) in a Polyethylene (PE) Framework: A Simple Model for Simulation Studies of Scaling and Solvent Effects on Polymers in an Open Framework

Abstract:

We report molecular dynamics simulations studies of a polyethylene oxide polymer in a cubic polyethylene framework. Radius of gyration, shape, and other structural parameters are computed. Self-diffusion and displacement relative to the framework are also computed to investigate mobility and transport behavior. To investigate scaling effects, polyethylene oxide polymers of three different lengths and framework of three different sizes are simulated. Explicit water molecules are introduced to investigate solvation effects in different dimensions of polymer and framework.

Confinement of the polymer in smaller framework lead to closer interactin of polymer and framework and a more elongated configuration. The elongated configuration is favorable for diffusion through small windows of the framework. A larger mean square displacement is observed for the smaller framework, particularly in the presence of water. Adjusting the framework interaction potential sees a critical threshold dividing normal diffusion behavivoir and a regime dominated by hopping between cavities of the framework.

Although the model framework contains only simple van der Waalsl groups, confinement and solvent effects on polymer observed agree qualitatively with those in experiments reported recently for polyeletrolyte threaded in metal organic frameworks [1,2]

[1] L. Gao, C.Y.V. Li, and K.Y. Chan, *Chem. Mater.* 27(10) (2015) 3601-3608.

[2] L. Gao, C.Y. V. Li, K.Y. Chan, and Z.N. Chen, *J. Am. Chem. Soc.* 136 (2014) 7209-7212.

505006 Molecular Understanding and Design of Zwitterionic Materials

Abstract:

An important challenge in many applications, ranging from medical devices to ship hulls, is the prevention of nonspecific protein adsorption on surfaces. To address this challenge, our goals are twofold. First, we strive to provide a fundamental understanding of nonfouling mechanisms at the molecular level using an integrated experimental and simulation approach. Both simulation and experimental results show that the strong hydration of zwitterionic materials is responsible for their excellent nonfouling properties (JACS, **127**, 14473, 2005). Further studies examine zwitterionic structures and reveal the role of these structures in nonfouling properties beyond hydration (Advanced Materials, **27**, 15, 2015). Second, we aim to develop biocompatible and environmentally benign ultra low fouling materials based on the molecular principles learned. We have demonstrated that zwitterionic and mixed charge materials and surfaces are highly resistant to nonspecific protein adsorption, cell adhesion and bacteria adhesion/biofilm formation from complex media. Recent results show that zwitterionic materials induce no capsule formation upon implantation (Nature Biotechnology, **31**, 553, 2013) and no immunological response in blood circulation (PNAS, **112**, 12046, 2015). Furthermore, these materials are shown to preserve protein (Nature Chemistry, 4, 59, 2012) and cell (Angewandte Chemie, **53**, 12729, 2014) bioactivity. Because of these unique properties, zwitterionic materials are excellent alternatives to widely used poly(ethylene glycol) (PEG)-based materials.

504992 Interfacial Transport of Protons on a 2-Dimensional Functionalized Graphane Surface

Abstract:

Proton transport is crucial for many industrial and biological processes. The movement of protons is most often facilitated by the presence of water because water forms an extended hydrogen bonded network allowing for fast hopping of protons from one molecule to another. We have developed a novel 2-dimensional material that is capable of facilitating fast proton transport in the complete absence of water. Anhydrous proton transport is important for applications such as high- and intermediate-temperature fuel cell membranes. Our material was designed and tested using first principles density functional theory calculations. The material is based on graphane, which is an sp3 analogue of graphene. We have functionalized the graphane surface with hydroxyl groups and have used ab initio molecular dynamics and transition state calculations to study the mechanism and barriers of proton transport. We examine the limit of 1-dimensional proton hopping along a line of OH groups and the limit of fully hydroxylated graphane, forming 2-dimensional hydrogen bonding networks. We report diffusivities and transition state theory calculations. We find that anhydrous functionalized graphane has higher transport rates than hydrated Nafion.

504993 Gas Adsorption Behavior in Ionic Polyimide Composite Membranes

Abstract:

Industrial gas separation and storage processes present many technical and economic challenges to the natural gas industry and to power plants. In particular, palliating CO_2 from emission sources is a critical need in industry, and it is necessary in order to meet current and future environmental regulations. Polyimides are at the forefront of advanced membrane materials for CO_2 capture and gas purification processes. Recently, â€œionic polyimidesâ€ (i-PIs) have been reported as a new class of condensation polymers which combine structural components of both ionic liquids (ILs) and polyimides through covalent linkages. In this study, we report CO_2 and CH_4 adsorption and structural analyses of an i-PI and an i-PI + IL composite containing [C_4mim][Tf_2N]. The combination of molecular dynamics (MD) and grand canonical Monte Carlo (GCMC) simulations are used to compute the gas solubility and the adsorption performance with respect to the density, fractional free volume (FFV), and surface area of the materials.

Our results highlight the polymer relaxation process, and its correlation to the gas solubility. In particular, the surface area can provide meaningful guidance with respect to the gas solubility, and it tends to be a more sensitive indicator of the adsorption behavior versus only considering the system density and FFV. For instance, as the polymer continues to relax, the density, FFV, and pore-size distribution remain relatively constant, while the surface area can continue to increase, enabling more adsorption. Structural analyses are also conducted to identify the nature of the gas adsorption once the ionic liquid is added to the polymer. The presence of the IL significantly displaces the CO_2 molecules from the ligand nitrogen sites in the neat i-PI to the imidazolium rings in the i-PI + IL composite. Whereas, the CH_4 molecules move from the imidazolium ring sites in the neat i-PI to the ligand nitrogen atoms in the i-PI + IL composite. These molecular details can provide critical information for the experimental design of highly selective i-PI materials, as well as provide additional guidance for the interpretation of the simulated adsorption systems.

493614 Computational Simulation of Supported Nanocatalysts Under Realistic Conditions

Abstract:

Supported noble metal catalysts have played an important role in the conversion of energy and resources. Computational simulations ranging from quantum chemistry density functional calculations, molecular dynamic have been widely used to investigate the reaction mechanisms and the relationship between structure and properties of nanocatalysts. However, there are still huge gaps between computational simulations and realistic experiments. During the recent years, we try to computational simulate the nanocatalysts by consideration of realistic conditions in order to reduce the pressure and materials gap.

489964 Continuous Synthesis of Ethyl Esters from Free Fatty Acids over Metal Oxides in Sub/Supercritical Ethanol

Abstract:

2017 AIChE Annual Meeting

Oct 29th to Nov 3rd, Minneapolis, MN

Continuous synthesis of ethyl esters from free fatty acids over metal oxides in sub/supercritical ethanol

Jiuxu Liu, Yue Nan and Lawrence L. Tavlarides

Department of Biomedical and Chemical Engineering, Syracuse University.

Email: lltavlar@syr.edu (Tavlarides); jliu23@syr.edu(Liu)

Abstract

In our previous study, we proposed a method for biodiesel synthesis by running the transesterification reactions under sub/supercritical conditions with trace amounts of homogeneous catalysts in order to achieve a high product yield and reduce the dosage of catalysts compared to the conventional technology. Another way to avoid using large amount of homogeneous catalysts (which bring burdens in product purification process) is switching to heterogeneous catalysts, acidic or basic. The common basic solid catalysts are hydrotalcites, dolomites, loaded and supported alkaline and alkaline metals and corresponding oxides, and mixed basic metal oxides. The basic solid catalysts reported in the literature face leaching problems. The active compounds can be detected in the product with high concentration, and the catalysts have to be regenerated after a few runs. The solid acid catalysts seem to be more promising compared to basic ones by considering the fact that they are insensitive to free fatty acid content, and esterification and transesterification occurs simultaneously. Commonly used acidic solids in biodiesel synthesis include acidic metal oxides, sulfonic ion-exchange resin, sulfated/ tungstated metal oxides, zeolites, heteropolyacids, amongst others. Similar with basic solids, the reported acidic catalysts deactivate along with reaction due to the leaching problem.

A literature review indicates that several compounds were found to be stable even at alcohol temperature higher than 180 °C including zirconia, titania, silica, and alumina. These simple metal oxides have been commonly used as catalyst supports due to their thermal and mechanical stability. However they did not show any catalytic ability under conventional conditions (T=70 °C, ambient pressure) due to their weak acid/base property.

In this study, we are investigating the possibility and mechanism of continuously synthesizing ethyl-ester biodiesel under sub and supercritical alcohol conditions from fatty acids, which largely exist in cheap feedstocks such as waste cooking oil, catalyzed by simple and stable metal oxides catalysts including ZrO_2, TiO_2 and Al_2O_3. The objectives include: a) kinetics of ethyl-ester biodiesel synthesis with oleic acid at SCTE conditions without the catalysts, b) effect of addition of the solid catalysts on reaction yields, and c) kinetics of the biodiesel synthesis over the selected metal oxides.

At this writing the first aim has been completed, which is the kinetic study of the non-catalytic esterification reaction of oleic acid with sub and supercritical ethanol. The reactions were conducted at temperatures from 175 to 325 °C, residence times from 1 to 8 min, and ethanol-to oleic acid molar ratio of 18:1. The pressure was kept as 200 bar. Under subcritical conditions, biodiesel yield slightly increased from 8% to 33% with temperature changing from 175 to 225 °C at 8 min residence time. Under supercritical conditions, a yield of 60% was reached at 275 °C and 8 min, and increased further to 80% at 325 °C and 8 min. A one-step reversible second-order reaction model was employed to best fit the data, and the activation energy for forward and reverse reaction step was calculated as 55 and 32 kJ/mol, respectively. A negative enthalpy change as -23 kJ/mol indicates that the reaction is exothermic. These kinetic parameters will be used when conducting the kinetic analysis of reaction with presence of the solid catalysts.

Experiments to complete the remaining objectives are in progress and will be reported.

501066 Denitrogenation of Pyridine with Acetic Acid As Hydrogen Source Under Catalytic Hydrothermal Gasification Environment

Abstract:

Biofuel production from microalgae via hydrothermal liquefaction (HTL) produces wastewater with high organic (COD: 40 ~ 100 g/L) and nitrogen content (TN: 15g/L). The majority of nitrogen from the biomass remains in the post-HTL wastewater (PHWW). The most commonly identified chemical compounds in the PHWW are volatile fatty acids (VFAs) and cyclic amines such as pyridines. Pyridines, due to the aromaticity, are chemically stable, and they are also identified as cytotoxicity-inducing chemicals that inhibit the growth of microalgae even in 1% PHWW dilution, when the PHWW is reused as a nutrient source. Hydrodenitrogenation is an effective way to convert the pyridines into inorganic form as ammonia, or organic form as amides, which are better forms of nitrogen for biomass growth, as well as to mitigate the toxicity. Thus it is a beneficial to investigate an effective process of hydrodenitrogenation for recycling the HTL wastewater in large quantity without much dilution needed.

The first step of hydrodenitrogenation of pyridine is hydrogenation of pyridines to its piperidine form, followed by ring-opening and then C-C bond cleavage for ammonia/amine and small hydrocarbon production. Conventional hydrodenitrogenation requires hydrogen gas at high pressure to realize this series of conversions. Due to the large quantity of VFAs in the PHWW, and that VFAs have been extensively studied as model compounds for high hydrogen production in catalytic gasification conditions, it is possible to use VFAs as hydrogen precursor in the hydrodenitrogenation process without apply hydrogen gas.

In this study, acetic acid and pyridine are chosen as model compounds to evaluate the feasibility of using VFA as hydrogen source for hydrodenitrogenation of pyridine. Denitrogenation/gasification will be carried out under 500C reaction temperature, 60 min reaction time, which is previously discovered optimal operation condition for PHWW gasification. Denitrogenation efficiency and gasification yield under different pyridine to acetic acid ratio will be evaluated. Possible reaction pathway, and reaction kinetics will be discussed.

497002 The Non-Catalytic Decomposition of Lignin into Chemical and Fuel Intermediates

Abstract:

The goal of this presentation . . .

In this presentation we will present the results of a study performed to explore the non-catalytic decomposition of lignin into chemical intermediates.

Motivation . . .

Lignin is one of the most abundant biopolymers. On average, 24 wt% of a plantâ€™s structure is composed of lignin. Lignin is produced as a low value by-product from wood pulping processes as well as from ligno-cellulosic biofuel facilities, such as those producing ethanol. Most produced lignin is burnt to recover its energy content. The quest to decompose lignin into fragments that could serve as the building blocks for useful fuel and chemical components has been an active area of research for many years. These efforts have primarily focused on catalytic pyrolytic or biochemical decomposition strategies. However, a commercially feasible approach has been elusive.

Presentation Contents â€“ a Novel Approach to Lignin Decomposition

This presentation will describe a novel, high temperature, high pressure continuous reaction scheme designed and built to study lignin decomposition. A dilute slurry of lignin entrained in a basic water solution was pressurized to near super-critical conditions and heated to its decomposition temperature in a continuous tubular reactor. The reactor outlet stream was cooled against room temperature water and the four resulting phases (gas, polar liquid, non-polar liquid, and solid) were collected and analyzed with a variety of methods. When these initial results are compared to recently published catalytic decomposition results, this novel approach appears to be attractive. A combination of low residence time with appropriate reaction temperature and pressure result in the cleavage of ether bonds and the production of phenolic hydroxyl groups and various organic acids (e.g. vanillic and homovanillic acid) while minimizing the formation of char.

490374 Hydrothermal Liquefaction of Food Waste and Remediation of Aqueous Byproducts

Abstract:

Thermochemical processing via hydrothermal liquefaction (HTL) is capable of processing a broad range of feedstocks, has a favorable fossil energy ratio, and produces an energy-dense liquid bio-oil product and hydrochar co-product with emerging markets. However, there is a well-known obstacle facing current HTL processes that has hampered the adoption of this technology, namely the HTL process produces a considerable amount of aqueous organic byproducts. As such, the carbon yield to bio-oil is reduced and the organic content of the aqueous waste stream exceeds the environmental standards for discharging the water.

The project team is developing HTL technologies to produce bio-oil from commercial and institutional food waste. Using a two-pronged approach, we have demonstrated an improved HTL oil yield while also reducing aqueous contaminants. First, we have successfully identified a stable and active heterogeneous HTL catalyst capable of improving bio-oil yields (dry mass basis) from 19% to 28% compared to traditional homogeneous base catalysts. IR spectra, GC/MS chromatograms, and model compound experiments revealed that the heterogeneous catalyst increased oil molecular weight, increased ketone and ether content, and decreased alcohol content, thereby driving products out of the aqueous phase and into the organic oil phase. Second, we have tested two active hydrothermal gasification (HTG) catalysts capable of converting 93%â€“98% of aqueous organics to permanent gases composed primarily of hydrogen, methane, and carbon dioxide. In addition to reducing aqueous contaminants and simplifying waste water disposal, the hydrogen-rich gas produced from HTG can be recycled to the HTL reactor to improve oil yield and reduce the required residence time.

488218 Hydrothermal Liquefaction of Wastewater Treatment Microalgae in a Pilot-Scale Continuous Flow Reactor

Abstract:

Hydrothermal liquefaction (HTL) is considered as one of most promising techniques for valorization of whole, wet algae biomass. HTL uses subcritical water (270-350 ºC and 8-18 MPa) to convert the organic biomass constituents into an energy-rich bio-crude oil that can be upgraded into liquid fuels. To date, many studies have used batch reactor; shifting to continuous reactors is necessary for progressing to commercial-scale processes due to the higher productivity and lower operating costs of steady-state, continuous systems.

This study reports the modification and preliminary testing of a pilot-scale continuous flow reactor (CFR) for hydrothermal liquefaction of wastewater microalgae at low solids loadings. The goal of this reactor is to perform continuous (>30 liters or >8 hours) HTL on algae slurries with solid algal contents up to 10 wt.%, and to produce char-free bio-crude oils using a dual high-pressure cylinder filter system with blow-down pots to remove solids. CFR systems must address several challenges: smooth flow of homogenous biomass slurry through unit operations, clogging, solid-liquid-gas separation/filtration, multiple unit operations, and process safety and control. Here we report on the progress of the reactor development relative to bio-crude oil production, reaction residence time, and operational stability. In addition to reactor development, HTL product characterization was also conducted using Fourier transform ion cyclotron resonance mass spectroscopy (FT-ICR MS) and fatty acid methyl ester (FAME) analysis by gas chromatography mass spectroscopy (GC/MS for bio-crude oil quality, and oxy-combustion calorimetry and elemental analysis (CHNS) for mass and energy balances.

501306 The Importance of Classical Soft Matter Physics in the Development of New Nanomaterials

Abstract:

Upon their introduction, certain classes of gas-synthesized nanomaterials such as carbon nanotubes (CNTs) appeared completely disjointed from the area of soft matterâ€"in part because forming stable fluid phases with such materials appeared nearly impossible. In this talk, I will discuss the importance of classical soft matter physics in the development of high-performance CNT materials. I will describe how classical concepts from colloids and polymer science, such as intrinsic viscosity, persistence length, liquid crystalline phase transitions, are key to understanding and controlling CNT fluid phases, and how this degree of control is now yielding a new class of soft materials that combines the most attractive traits of polymers (mechanical softness, easy flow processing) with the properties of conductors and semiconductors, that had remained so far elusive.

495047 Rapid and Accurate Methods for Modeling Hydrodynamic Forces in Brownian Dynamics Simulations

Abstract:

Complex fluids composed of colloids, surfactants or dissolved polymers, generically referred to as ``particles", are ubiquitous in nature and industry. The dynamics of these materials are controlled by solvent-mediated interactions, called hydrodynamic interactions (HIs), as well as the stochastic motion associated with Brownian diffusion. Constructing accurate computational models to simulate such systems is difficult because the structure and dynamics of colloidal materials are inherently multiscale. Practically, this means that very large numbers of discrete elements need to be simulated to capture the full range of dynamics in complex materials. Performing such large simulations is difficult because conventional simulation methods that incorporate HIs with Brownian motion exhibit a superlinear scaling of calculation time with the number of particles, which limits the size of simulations to at most a few thousand particles. For many self assembling materials -- colloidal, polymeric, micellar networks, for example -- millions of particles would be preferred. A further complication is that high accuracy hydrodynamic models, which are constructed mathematically via additional constraints on particle motion, require additional computational effort and further restrict accessible system sizes.

We will present a novel approach to performing Brownian dynamics simulations with hydrodynamic interactions and constraints that results in an algorithm with linear scaling in the number of particles modeled. This method, called the positively-split Ewald (PSE) algorithm, decomposes the operator describing the hydrodynamic interactions into a sparse local contribution, and a low-rank long wavelength contribution. Decomposing the interactions in this way allows the stochastic displacements to be efficiently sampled from the distribution defined by these two independent operators, leveraging their algebraic structures to accelerate the sampling to stochastic displacements. Hydrodynamic constraints, for instance those associated with lubrication forces, stresslets, and rigid particle assemblies, are readily incorporated in this algorithm and are used to build high accuracy simulations. The PSE approach enables a speedup of two orders of magnitudes over existing techniques. A high-performance implementation of this algorithm on graphics processing units (GPUs) enables simulations of millions of particles in reasonable wall clock times on consumer grade hardware. Applications of the method to dispersions of colloidal particles are discussed.

496371 Rotational Dynamics of Nanoparticles in Polymer Solutions and Melts

Abstract:

The dynamics of nanoparticles in complex fluids are of interest for a wide range of applications, such as high performance materials, where nanoparticles are used to tune mechanical properties or add functionality, and biomedical applications, where nanoparticles must navigate a complex, crowded, and confined environment. Polymer solutions and melts are good examples of complex fluids where improved understanding of nanoparticle dynamics could impact applications while improving our fundamental understanding of nanoparticle behavior. Prior studies have explored the translational diffusion of nanoparticles in polymer melts and polymer solutions, demonstrating so-called breakdown of the Stokes-Einstein equation for the diffusivity under certain conditions. However, the rotational diffusion of nanoparticles in such systems has remained largely unexplored.

Here we report on studies of the rotational diffusivity of inorganic nanoparticles coated with a polymer brush and dispersed in polymer solutions and melts. The nanoparticles consist of cobalt ferrite cores with narrow size distribution coated with poly(ethylene glycol) covalently grafted to the particles through a silane coupling agent. The counterpart polymer used in the solutions and melts was also poly(ethylene glycol). The rotational diffusivity of the nanoparticles was determined by measuring their response to alternating magnetic fields of low amplitude and variable frequency, in what we call Dynamic Magnetic Susceptibility (DMS) measurements. In the case of polymer solutions we explored the effects of ratio of particle radius to polymer radius of gyration and of polymer crowding by varying the molecular weight and concentration of the solution polymer. Our results suggest breakdown of the Stokes-Einstein relation for the rotational diffusivity of the nanoparticles when the radius of gyration of the polymer in solution is much greater than the particleâ€™s hydrodynamic radius. A simple model for rotational diffusion of a nanoparticle surrounded by a polymer depletion layer and embedded in a polymer solution was obtained and appears to explain the measurements well. In the case of polymer melts, the nanoparticles were dispersed in poly(ethylene glycol) melts of varying molecular weights and their rotational diffusivity was measured also by DMS measurements. The predictions of the Stokes-Einstein relation were found to be valid when the contour length of the polymer was shorter than the nanoparticleâ€™s hydrodynamic diameter. For larger polymer chains in the melt it appears as if the rotational hydrodynamic drag on the nanoparticles is no longer a function of melt polymer molecular weight. These observations suggest that rotational hydrodynamic drag on the nanoparticle is determined by the melt polymer segments in the immediate vicinity of the nanoparticles, and not by the motion of the melt chains as a whole.

496969 Morphology of Nanoparticle Aggregates in Flow through Beds Packed with Spheres: Self- and Flow-Induced Assembly Using Lattice Boltzmann Simulations

Abstract:

The aggregation of nanoparticles as they propagate through porous media is an important process for the environment, since the use of nanotechnology has led to increasing rates of nanoparticle manufacturing that can in turn lead to nanoparticles introduced into the subsurface and possibly into aquifers. In addition, controlled assembly of nanoparticles can lead to the manufacturing of nanoparticles of specific shape and functionality. Understanding the process of nanoparticle aggregation in porous media, and the effects of the flow in the process, can therefore be of practical interest. Electrostatic forces and van der Waals forces drive the nanoparticle aggregation process. The mean size of aggregates as a function of the monomer size and of different operating conditions, and the growth rate of the aggregates have mostly attracted research attention. However, the morphology of the aggregates is also important, as is the final shape of the particles. Aggregate morphology affects the way colloid aggregates pass through the pores, either enhancing or reducing the clogging of pores. In this study, the lattice Boltzmann method (LBM) is used to simulate flow through a porous medium, and a Lagrangian method is used to simulate the transport of individual nanoparticles in the flow field [1,2,3]. The packed sphere beds are represented by packing known-sized spheres in either ideal arrays (i.e., body centered cubic, face centered cubic, and simple cubic) or a random array, using the modified Lubachevsky-Stillinger algorithm [4]. After that, the velocity field of water in these beds is generated, using the LBM simulations. The Lagrangian trajectories of the nanoparticles are then calculated based on convection due to the flow field and due to forces exerted on the nanoparticles. In our simulation, the balance of drag force, random force, buoyancy force, van der Waals force, and electrostatic force is responsible for advancing the nanoparticles. At nano-scale distances (i.e., 0.75 nm), the nanoparticles are pulled strongly together by the van der Waals force, and the aggregates are formed. The addition of each nanoparticle on an aggregate changes the morphology of the nano-aggregate, and we can observe the formation of these aggregates at different hydrodynamic and electrolyte conditions while they propagate in the packed beds. The change in morphology is quantified with calculations of the transient fractal dimensions of the particles, changes in the size of the aggregates with time, and with changes in the asphericity of the particle aggregates. Aggregates that are elongated, like cylinders, or round, like spheres, can be formed under different flow conditions.

ACKNOWLEDGEMENTS

The computational support of XSEDE (CTS090017) is gratefully acknowledged.

REFERENCES

1. Pham, N., Swatske, D.E., Harwell, J.H., Shiau, B.-J., and D.V. Papavassiliou, "Transport of nanoparticles and kinetics in packed beds: A numerical approach with lattice Boltzmann simulations and particle tracking," *Int. J. Heat and Mass Transfer*, **72**, 319-328, 2014
2. Voronov, R.S., VanGordon, S., Sikavitsas, V.I., and D.V. Papavassiliou, "Efficient Lagrangian scalar tracking method for reactive local mass transport simulation through porous media," *Int. J. for Numerical Methods in Fluids*, **67**, 501-517, 2011
3. Pham, N.H., and D.V. Papavassiliou, "Nanoparticle transport in heterogeneous porous media with particle tracking numerical methods," *Computational Particle Mechanics*, **4**(1), 87-100, 2017
4. M. Skoge, A. Donev, F. H. Stillinger, and S. Torquato, "Packing hyperspheres in high-dimensional Euclidean spaces", *Physical Review E*, **74**, 041127, 2006

498285 Shear Thickening of Colloidal Dispersions at High Particle Concentrations- Rheology and SANS Investigations into Particle-Scale Mechanisms

Abstract:

There is substantial interest in the mechanism of shear thickening in concentrated dispersions, with experiments, models, and simulations including various interparticle forces. Shear thickening mechanisms include lubrication hydrodynamics coupled with DLVO potentials, short range nanoscale forces, particle compressibility (elastohydrodynamics), and the possibility of contact friction. The existence of the shear thickened state in colloidal dispersions has been mapped experimentally for a model system of stabilized colloidal silica particles in an organic suspending medium (Cwalina and Wagner, JOR, 2015) that suggests a maximum packing fraction in the shear thickened state of 0.54. Further, direct measurements of the microstructure under flow provide microstructural evidence connecting the measured negative first and second normal stress differences to the predominance of lubrication hydrodynamics in the shear thickened state (Gurnon and Wagner, JFM, 2015). In this work, we explore the properties of the shear thickened state for this same system at volume fractions exceeding this maximum packing fraction by steady and time-dependent rheological measurements of the shear and normal stress differences, as well as complementary neutron scattering measurements of the microstructure under flow and at rest. The transition from negative to positive first normal stress difference is examined with regards to the dilational properties of flows exceeding the maximum packing fraction in the shear thickened state, along with the stability of such flows. These results provide experimental evidence to directly test models and simulations for discontinuous shear thickening in colloidal dispersions.

500407 Colloidal Hydrodynamics at a Fluid-Fluid Interface

Abstract:

The self assembly of colloids at a fluid/fluid interface is a balance between different forces of interest. These typically include interactions because of particle shape and weight (capillary attraction), surface charge (electrostatic repulsions) and viscous dissipative forces which stem from relative motion between the particles and the adjoining fluids. As the particle size decreases from millimeter to colloidal scale, thermal forces and electrostatic interactions play a greater role in affecting the self assembly into ordered patterns and crystal structures. Such self-assembly processes, which initiate at the microscale are the starting point for different types of novel materials.
In this work, simulations involving two particles are implemented to describe the lateral organization of particles along the surface in regimes that includes capillary attraction of a monopolar and quadrupolar nature, electrostatic repulsion of dipolar nature, the hydrodynamic interaction between particles (incorporating the effect of the particle immersion depth) and thermal fluctuations. Clustering is observed at critically large values of the Péclet numbers, while smaller values exhibit higher probabilities of yielding states in which particles remain uncorrelated in space and more widely separated.

498785 Effect of Surface Geometry on the Frictional Properties of Poly(dimethyl siloxane)

Abstract:

Natural cartilage is durable and elastic, providing a low friction coefficient to moving joints under frequent applications of heavy loads. However, the physical mechanisms contributing to this low friction coefficient is not well understood. We hypothesize that the non-ideal surface geometry of cartilage gives rise to its low friction coefficient in certain directions of motion. We design soft poly(dimethyl siloxane) (PDMS) substrates with lithography to study the influence of surface geometry on their frictional and lubrication properties. The PDMS surfaces consist of stripes with controlled dimensions and spacings between each other. Tribological tests performed with a thin layer of aqueous glycerol solution at different concentrations between the PDMS substrates show that the friction coefficient is a function of the sliding speed. However, we find that the tribological behavior does not follow the type of Stribeck curve that is typically observed in soft materials with flat surfaces. The hysteresis from the surface geometry could be a reason for this deviation. To test this hypothesis, we develop a spring-dashpot model that represents viscoelastic contributions from the multiple surface length scales.

494126 Microrheology and Structural Reconfiguration of Artificial Biofluids Composed of Xanthan Gum in Salt Solutions

Abstract:

We use a particle tracking method to investigate the microrheology and the timescales of structural rearrangement of solutions of xanthan gum that are being formulated for use as synthetic biofluids. By suspending Brownian fluorescent particles inside the medium and tracking changes of their mean-squared displacement (MSD), we can identify timescales when the particles are sub-diffusive, purely diffusive, or super-diffusive in behavior.

Solutions of various concentrations of xanthan gum and salt are investigated. At high concentrations of xanthan gum, we observe that the particles are inhibited from free diffusion, indicative of a purely elastic response from the material, which is consistent with the bulk rheology. However, for lower concentrations, we observed that the particles showed a super-diffusive behavior on the order of hours to days. We hypothesize that this behavior is a result of the rearrangement of the structures of the solutions over this timescale, and we use the transient changes in the MSD to provide structural information of this reconfiguration process. After equilibrium is reached, consistent short-time sub-diffusive behavior and long-time diffusive behavior is observed. These results indicate that the particles are able to eventually escape the cages of xanthan gum, and the intersection of the two regions can be used to estimate the cage size and the relaxation time of the solutions.

499651 Probe Rheology Simulation Technique for Determination of Viscoelasticity of Complex Fluids

Abstract:

The viscoelastic properties of complex fluids are governed by the interplay of specific interactions and microscopic structure. The structural and dynamic heterogeneities in these systems lead to a local variation in their rheological properties. Over the last two decades, probe microrheology has emerged as an experimental technique that has the potential for determining local viscoelasticity of complex fluids. Recently, inspired by the experimental probe microrheology technique, we have developed an approach that combines molecular dynamics (MD) simulations with inertial generalized Stokes-Einstein relation (IGSER) to extract the viscoelastic properties of a complex medium. In our previous work, we showed that this method can be applied in both passive and active modes to polymer melt systems. The probe rheology simulation results for viscoelastic moduli so obtained were in good agreement with those obtained from nonequlibrium MD (NEMD) simulations and the Green-Kubo formalism. In order to test the generality of the approach, the probe rheology simulation technique is applied to other soft matter systems in this work. Specifically, results will be presented for the application of probe rheology technique to polymer solutions and colloidal suspensions. Probe rheology results for viscoelastic moduli will be compared with those obtained from the NEMD simulation.

488252 Approach to Contact, and Adhesion at Elastic and Structured Interfaces

Abstract:

Understanding and harnessing the coupling between lubrication pressure and elasticity provides materials design strategies for applications such as adhesives, coatings, microsensors, and biomaterials. This presentation will discuss our efforts to unnderstand the analogies between drops and soft materials in terms of lubrication forces, adhesion and deformation in viscous environments. We will show how elastic films deform due to viscous forces and how these deformations are similar to those observed with fluid droplets. We will discuss implications for adhesion and soft lubrication. In particular, our experiments highlight important length scales for hydrodynamic drainage of past soft or structured surfaces and the importance of elastic deformation during the approach of compliant surfaces.

499399 Surface Modification of Stainless Steels for Advanced Functionalities

Abstract:

Surface characteristics of stainless steels play a critical role in solid/liquid interfacial phenomena including wetting, fouling, and corrosion, which determine performance of stainless steels for target applications. In this presentation, we describe electrochemical method to control surface topography and chemistry of stainless steel 316L (SS316L), which is one of the most widely used stainless steels, to achieve advanced functionalities in aqueous environments. Potentiostatic polarization in a nitric acid solution was performed to modify SS316L surface properties. Application of a low anodic potential results in a selective grain boundary etching that yields significant microscale topography. With high anodic potential, porous nanoscale topography was achieved. Based upon the potential-topography relationship, hierarchical structure with both nanoscale and microscale features was designed to achieve super-hydrophobic SS316L surface. Nano-textured SS316L surface demonstrated anti-bacterial performance, which can be applied to biomedical devices such as body implants and surgical tools. Improved corrosion resistances of the modified SS316L surfaces were observed, which are attributed to the formation of superior chemical composition at the surfaces during the electrochemical surface modification. These results offer a simple strategy to control surface characteristics and attain/improve functionalities on the metals by using electrochemical method.

501665 Understanding the Wetting Behavior of Water-Octane-Silica Systems Using Monte Carlo Simulation

Abstract:

Understanding the interfacial properties of the hydrocarbon-water system is important for the design of many industrial applications, including enhanced oil recovery processes. Oil reservoirs consist of porous rock (minerals) in intimate contact with aqueous and oil phases. Reservoirs are often characterized by the phase that has a greater affinity for the rock surface. In the so-called "water-wet" case, water preferentially wets the mineral surface and oil is displaced from the interior of pores. For the "oil-wet" scenario, this relative preference is reversed. From an oil recovery perspective, one generally prefers the water-wet case, as oil is the more mobile phase and is driven to the surface as pressure is applied to the well. Under oil-wet conditions, the water needs to overcome a capillary pressure, which depends on two key interfacial properties: (1) the tension between the hydrocarbon-rich and water-rich phases and (2) the contact angle that a water-rich droplet forms on a mineral surface in a mother alkane-rich background. We discuss here the use of Monte Carlo (MC) molecular simulation to measure these properties for hydrocarbon/water mixtures in the presence of an atomistically detailed mineral surface. We focus on the evolution of the water contact angle on a silica surface in a mother n-octane phase. We find that the contact angle exhibits a minimum upon increasing the substrate hydrophilicity. At a molecular level, this minimum is connected to the manner in which water organizes at the silica-water interface.

495441 Time-Resolved *in situ* Studies of Zeolite Crystal Growth

Abstract:

Zeolites are widely used in commercial processes spanning from ion exchange in detergents to catalysis in the (petro)chemical industry. Understanding zeolite growth mechanisms at the molecular level aids the *a priori* selection of synthesis parameters to tailor the physicochemical properties of zeolites. Despite tremendous effort to elucidate the mechanisms of nucleation and crystal growth, these pathways in zeolite synthesis are not well understood. This is due in large part to the inherent complexity of zeolite crystallization and the synthesis conditions (i.e., high pH, high temperature, etc.) that render *in situ* characterization challenging. Recently, our group developed a system to carry out solvothermal *in situ* atomic force microscopy (AFM) wherein we can observe zeolite surfaces at near molecular resolution under realistic growth conditions.[1]

Here, we will present *in situ* AFM measurements of silicalite-1 (MFI) crystallization using zeolite growth modifiers (ZGMs) to modulate the shape of zeolite crystals. ZGMs are molecules or macromolecules that selectively bind to zeolite crystal surfaces and mediate the anisotropic rate(s) of growth to achieve desired crystal size and morphology. We have reported the efficacy of ZGMs in silicalite-1 bulk crystallization experiments.[2] Here, we will discuss *in situ*AFM studies of ZGM effects on silicalite-1 crystals wherein we observe differences in the relative rates of growth by two distinct pathways: classical processes involving molecule addition and nonclassical pathways involving the attachment of amorphous nanoparticle precursors. We show that these pathways can be influenced by the presence of ZGMs.

We will also present *in situ* AFM results of industrially relevant aluminum-containing zeolite LTA where we observe distinct growth regimes as a function of supersaturation. At high supersaturation, we observe the three-dimensional assembly and structural evolution of gel-like islands on LTA crystals leading to a mechanism that differs from kinetic roughening in classical theory. We also report a distinct switch in the growth mode at low supersaturation marked by two-dimensional nucleation of single layers with step heights corresponding to the composite building units of the crystal structure.

(1) Lupulescu, A. I.; Rimer, J. D.; *Science* 344 (**2014**)Â 729-732.

(2) Lupulescu, A. I.; Rimer, J. D.; *Angew. Chemie Int. Ed.* 51 (**2012**) 3345â€"3349.

Understanding the Solid-Liquid Phase Transition during the Growth of Scintillator Single Crystals Via Computational Modeling and Neutron Imaging

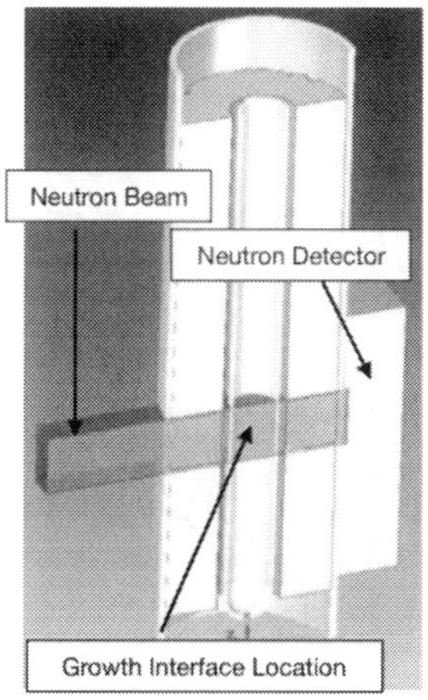

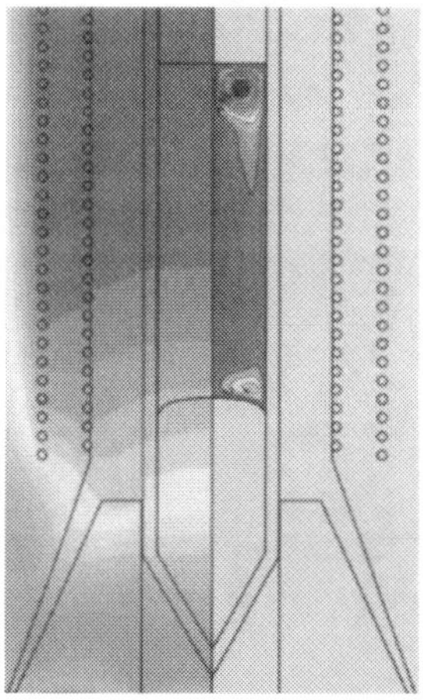

C. Zhang, J.H. Peterson, J. Seebeck, A.S. Tremsin,
D. Perrodin, G.A. Bizarri, E.D. Bourret,
A.S. Losko, S. Vogel, M. Bourke, and J.J. Derby

Department of Chemical Engineering &
Materials Science
University of Minnesota

2017 AICHE
Annual Meeting
Minneapolis, MN
Oct 30, 2017

Single Crystals and Their Applications

Si crystal:
Solar cells,
electronic devices...

Sapphire crystal:
Insulating substrates,
electronic wafers...

CZT crystal:
Radiation detectors,
optical devices...

Bridgman Crystal Growth Method

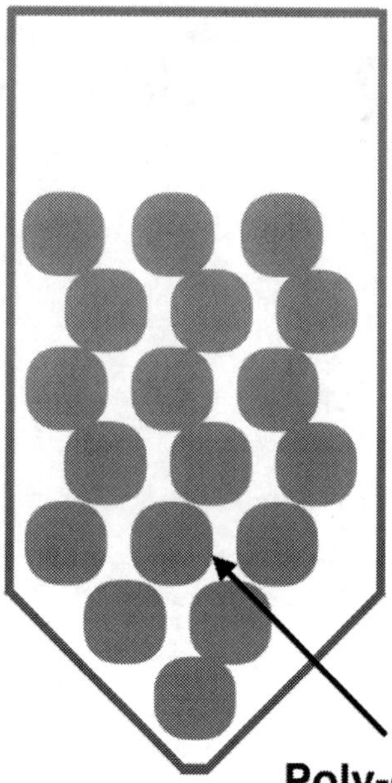

Bridgman Crystal Growth Method

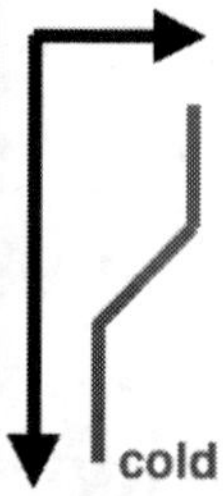

Bridgman Crystal Growth Method

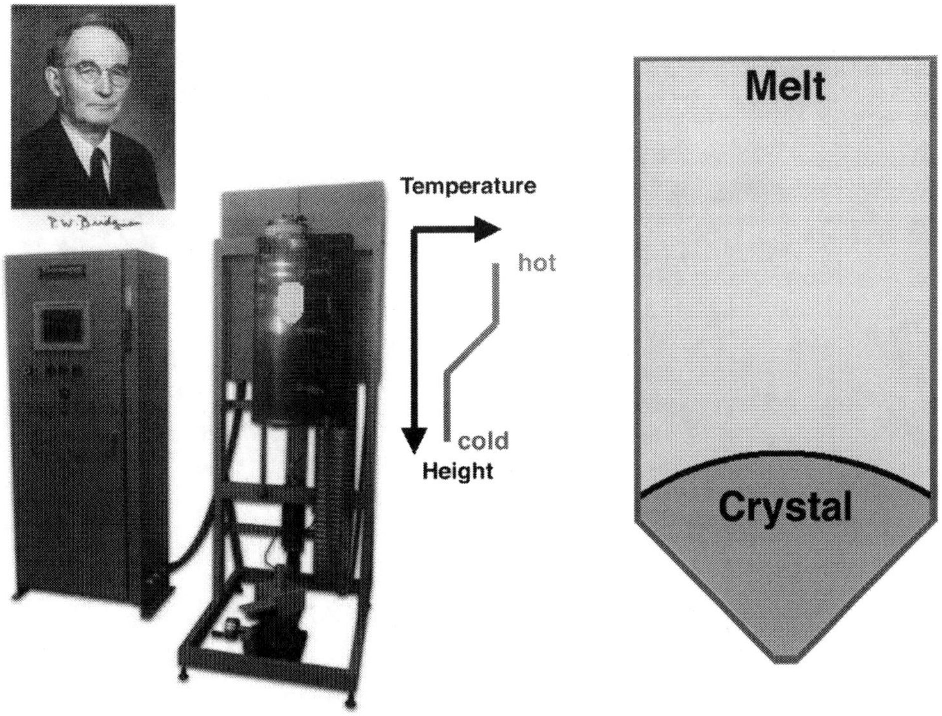

Bridgman Crystal Growth Method

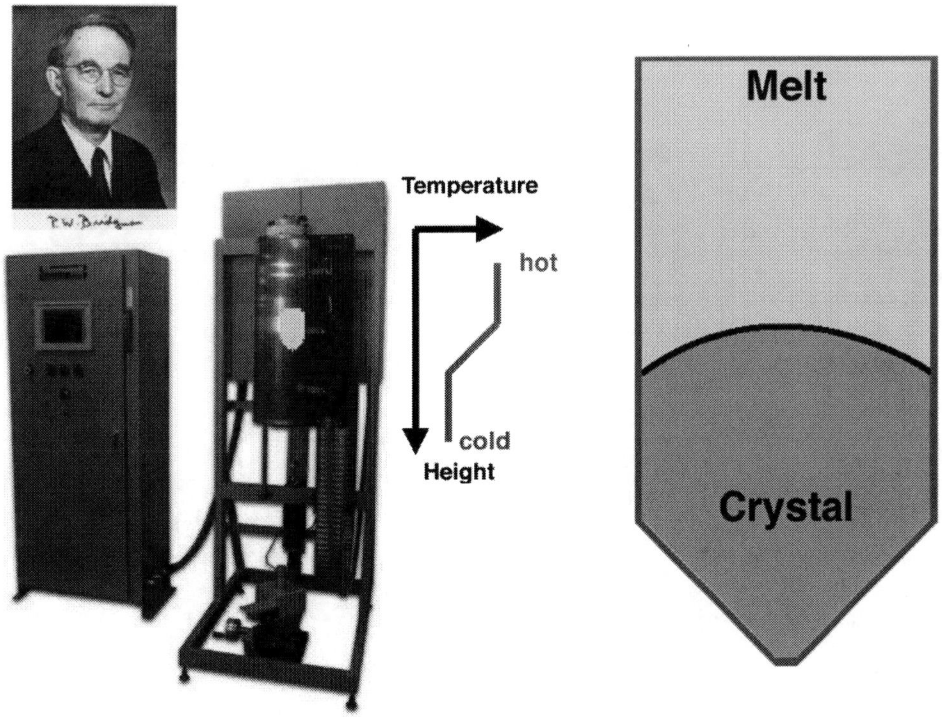

Bridgman Crystal Growth Method

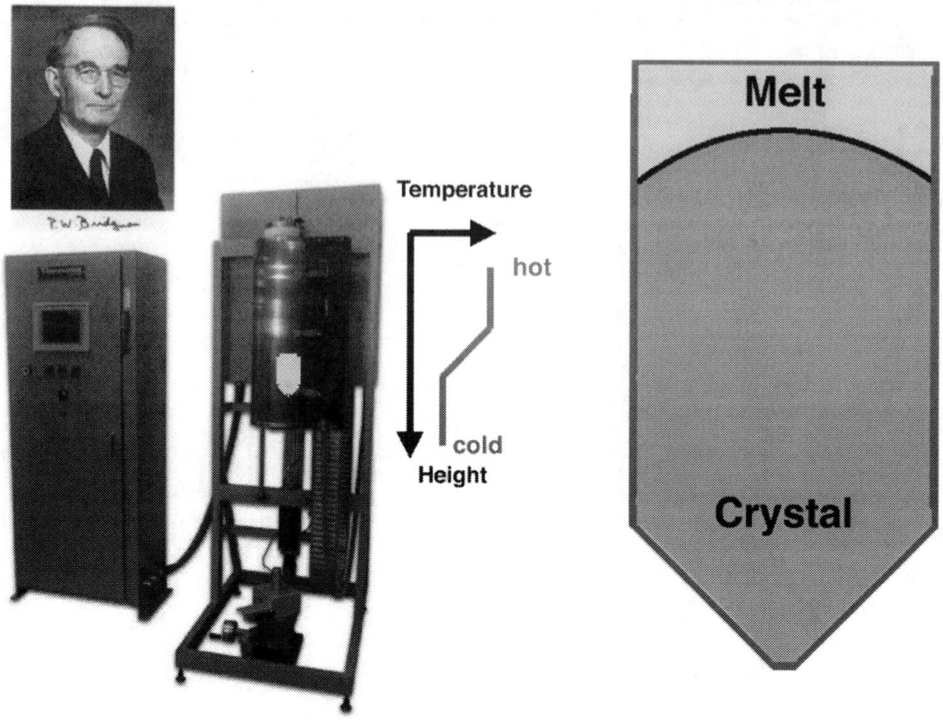

Bridgman Crystal Growth Method

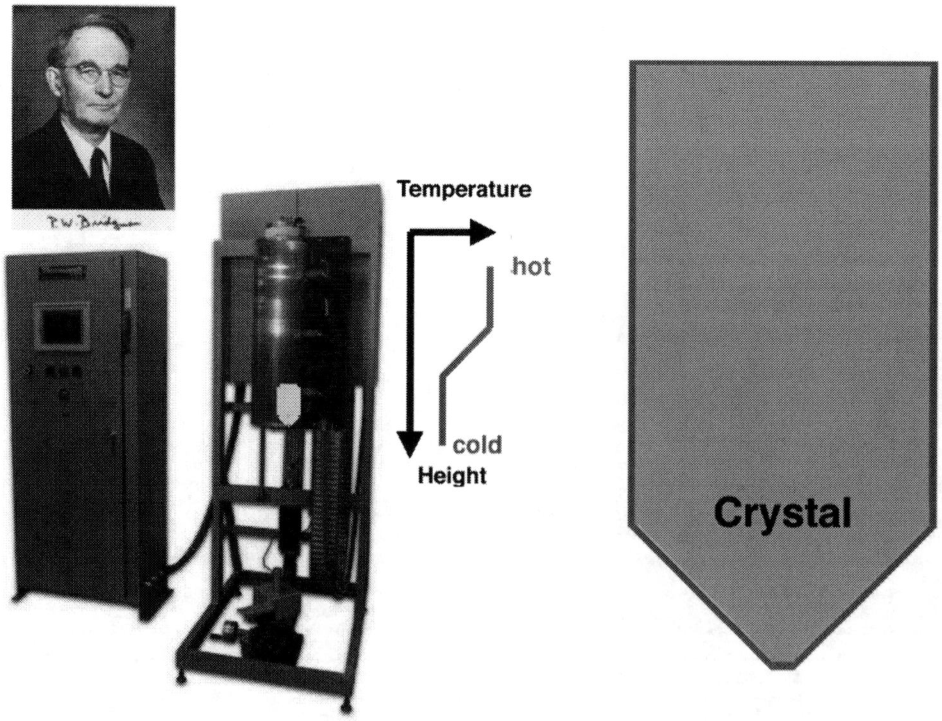

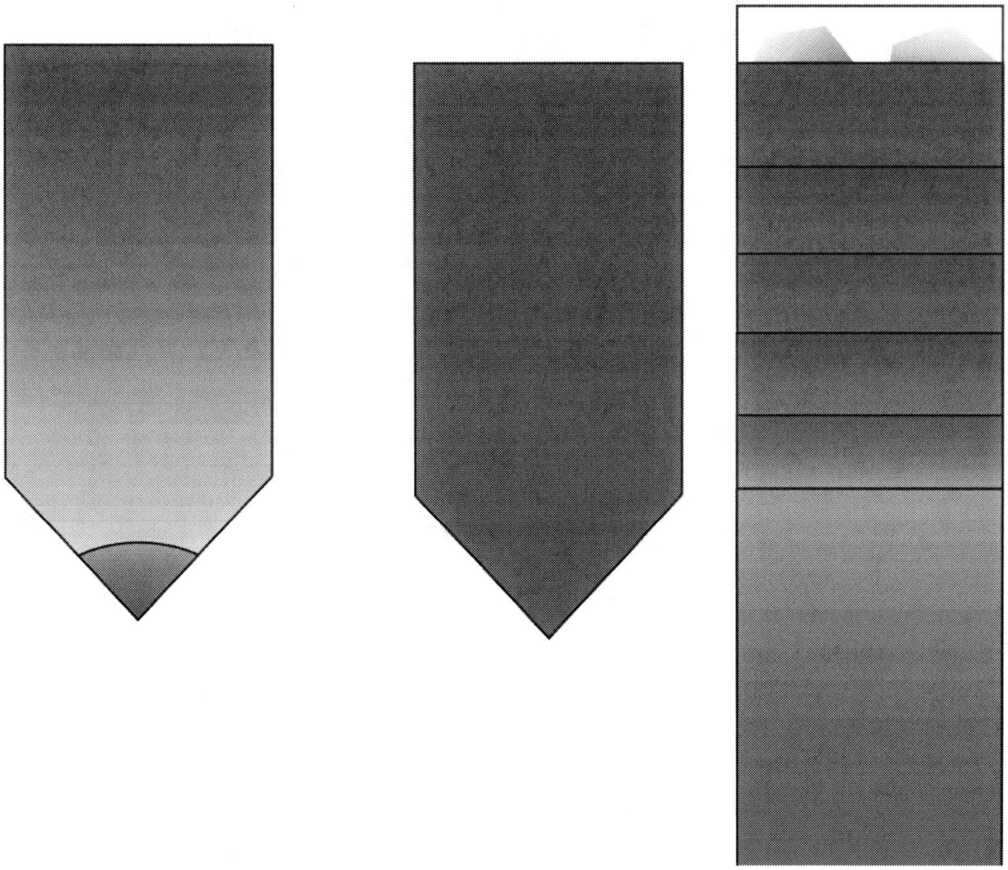

Physical Processes inside the Bridgman Method

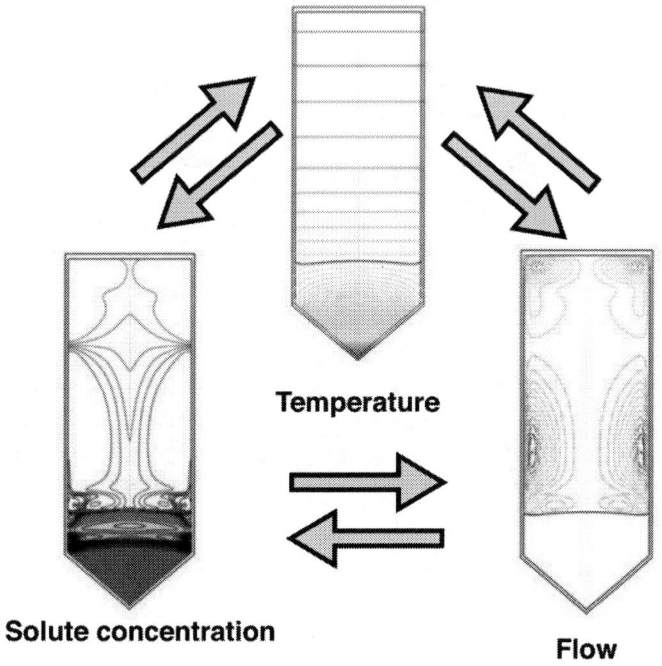

- The system is extremely nonlinear.
- The physical processes are highly coupled.

→ We are interested in controlling interface shape and understanding dopant segregation.

Historically, the Bridgman crystal growth process has been carried out blindly

In situ diagnostics are challenging due to:

•Extremely high temperatures for melt growth

•Refractory materials used for ampoule and furnace

•Very slow process dynamics

➡ We need a way to "see" what is happening to improve growth!

In our collaboration, neutron imaging is being used to see inside the Bridgman process

➡ We still need modeling to interpret what we see!

Together, neutron imaging and modeling provide new synergies to understand crystal growth

- *In situ,* real-time neutron imaging provides direct observation of large-scale crystal growth
- Models of the growth process provide basis for interpreting images
- Imaging information will enable model-based feedback control to improve growth outcomes

Vertical Bridgman growth of 12.3 mm-diameter, *Eu*-doped *BaBrCl*

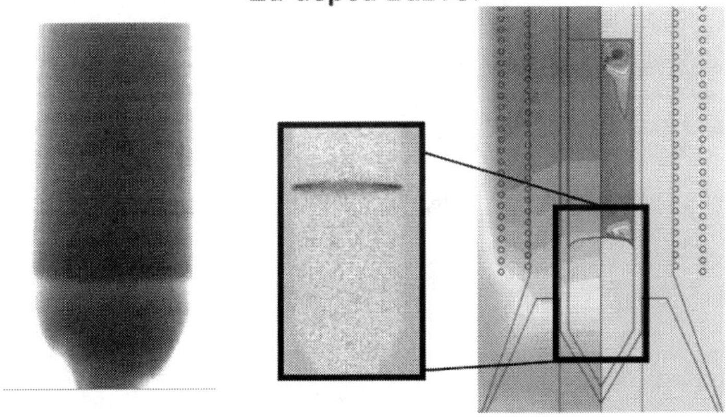

In situ, real-time neutron imaging — Frame difference contrast image — Crystal growth simulation

Energy-resolved neutron transmission was initially tested with "clamshell" furnace for the gradient freeze growth of BaBrCl:Eu

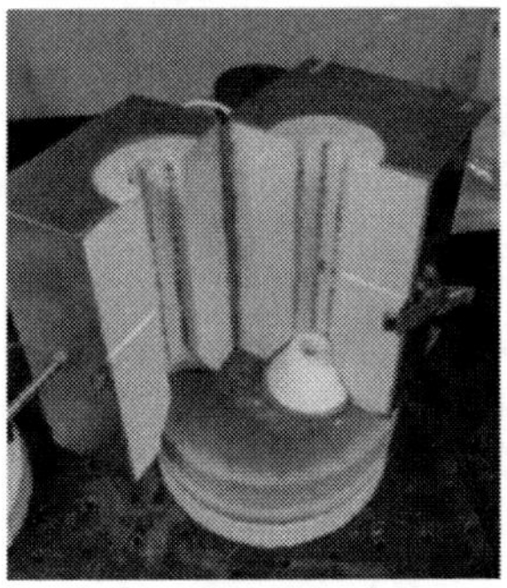

"Clamshell" Furnace

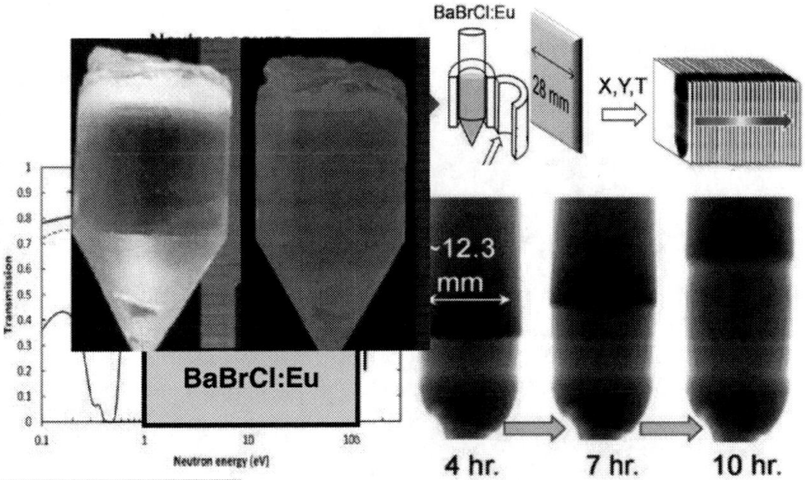

Element	σ_{abs} (barn)	σ_{scatt} (barn)
Ba	1.1	3.38
Br	6.9	5.9
Cl	33.5	16.8
Eu	4530	9.2

Neutron cross-section of selected elements

→ **We can visualize the crystal growth process by imaging the Eu concentration.**

A. S. Tremsin et al., J. Appl. Cryst. (2016). 49, 743–755

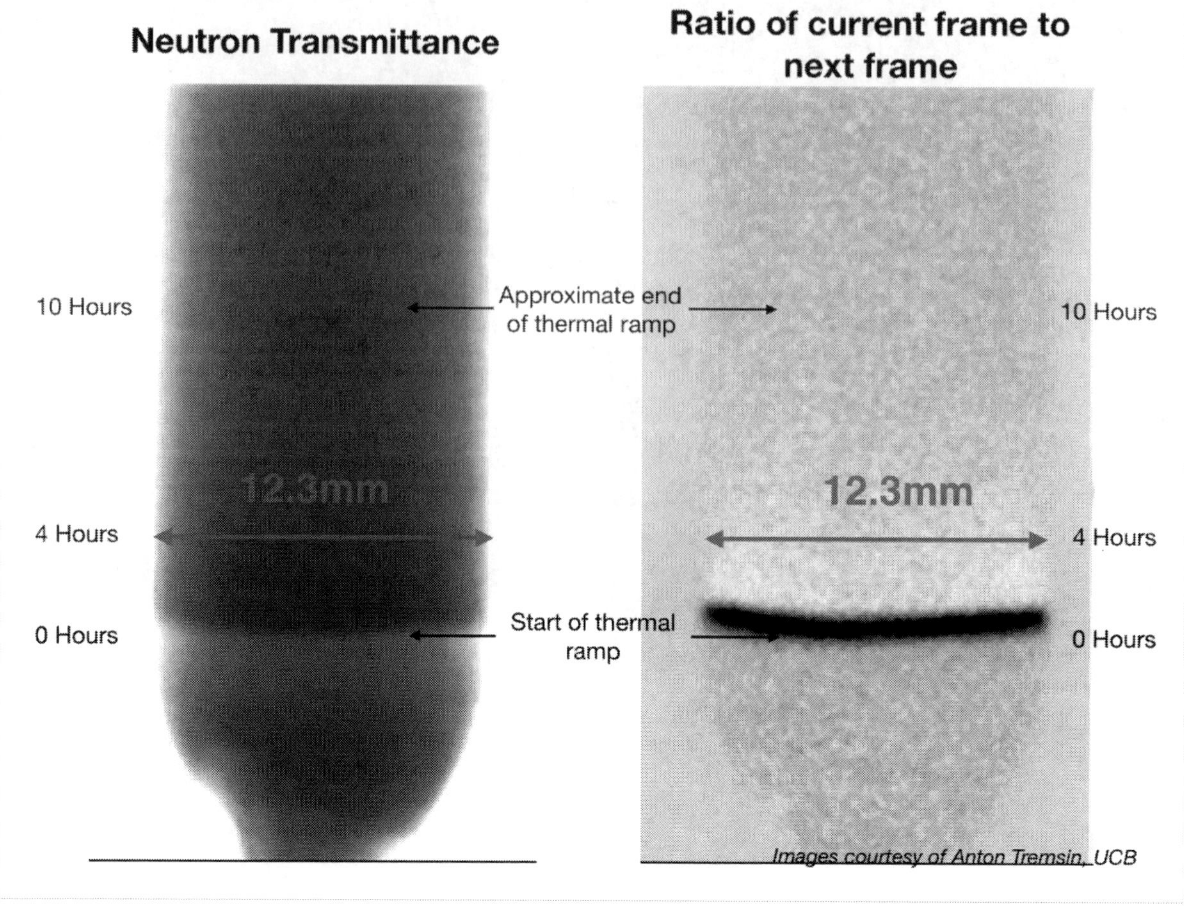
Neutron Transmittance
Ratio of current frame to next frame
10 Hours
Approximate end of thermal ramp
10 Hours
12.3mm
12.3mm
4 Hours
4 Hours
0 Hours
Start of thermal ramp
0 Hours
Images courtesy of Anton Tremsin, UCB

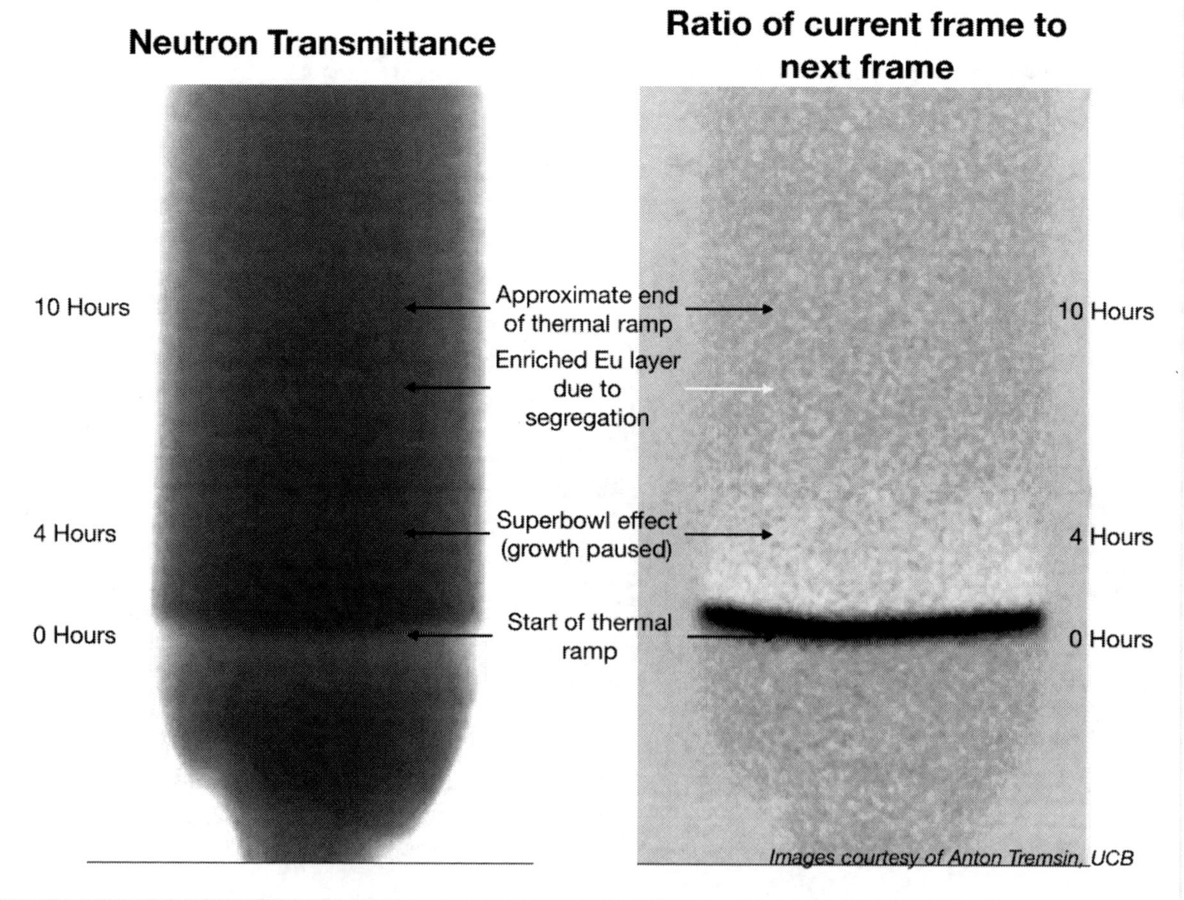
Neutron Transmittance
Ratio of current frame to next frame
10 Hours
Approximate end of thermal ramp
10 Hours
Enriched Eu layer due to segregation
4 Hours
Superbowl effect (growth paused)
4 Hours
0 Hours
Start of thermal ramp
0 Hours
Images courtesy of Anton Tremsin, UCB

Comparison between Experiment and Modeling Results:

Interface Shape

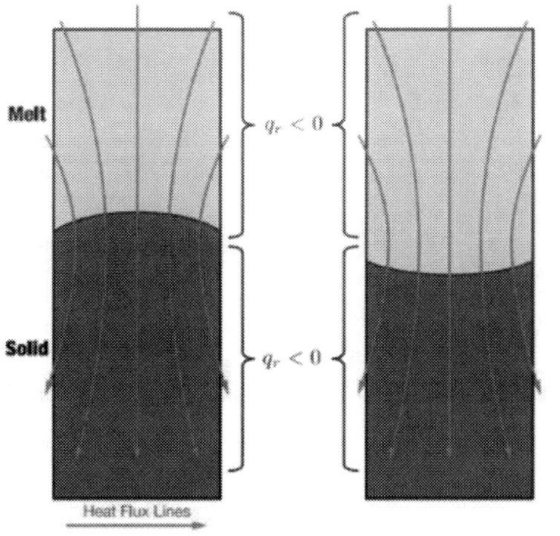

Melt-solid interface shape changes during growth, as observed in experiment and predicted by model

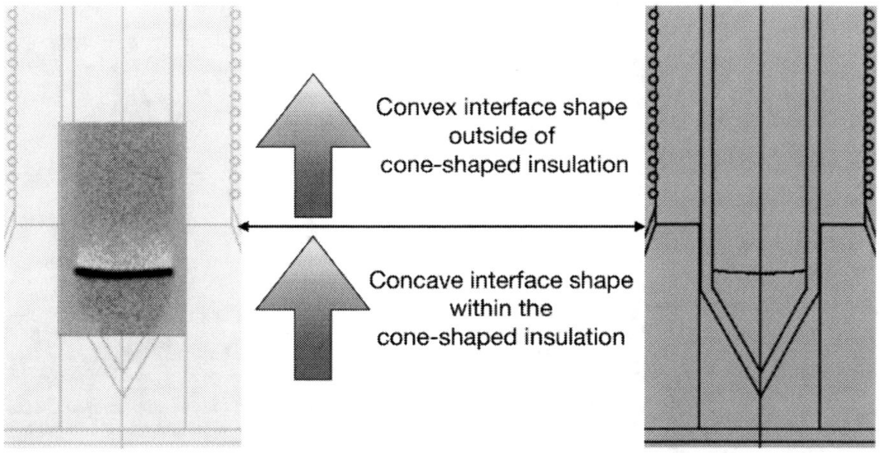

Can we understand why the interface changes its shape?

Theoretically, interface shape is affected by the direction of radial heat flux through melt above its position

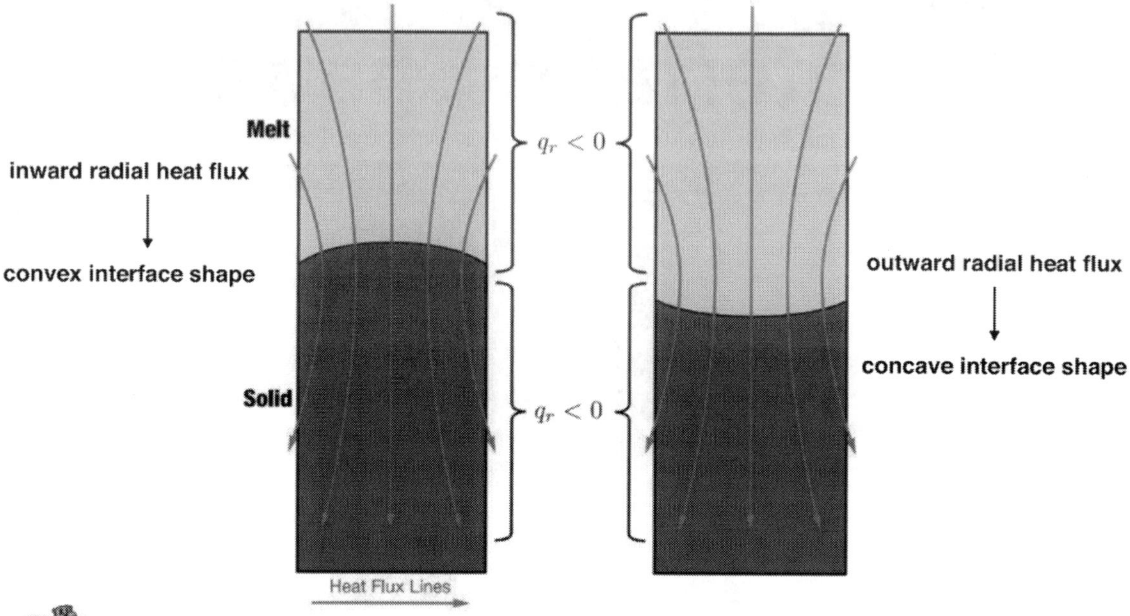

JH Peterson and JJ derby, Journal of Crystal Growth 468 (2017) 899

When growth is within cone, melt heat flux is directed outward near interface, consistent with concave shape

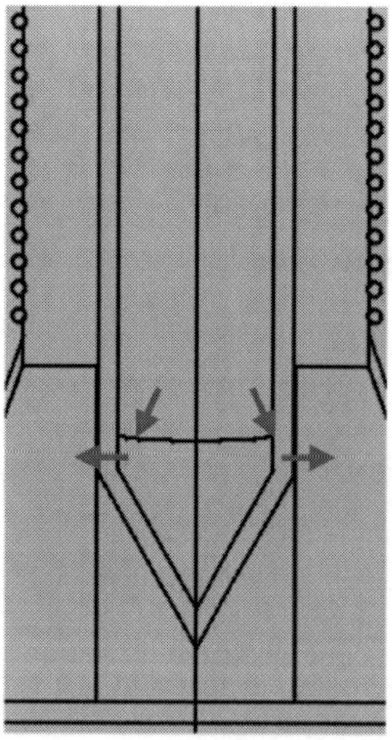

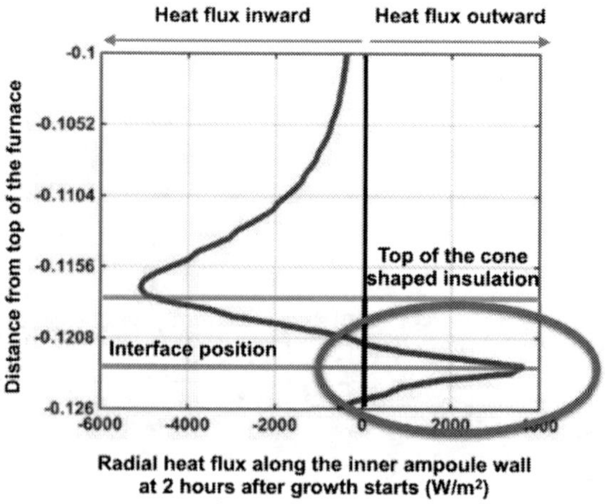

When growth is above cone, melt heat flux is directed inward near interface, consistent with convex shape

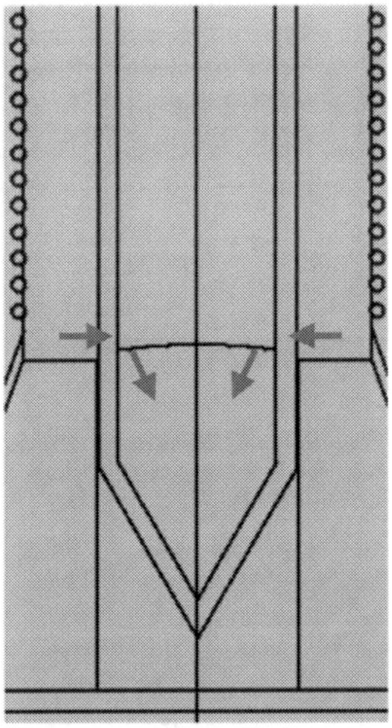

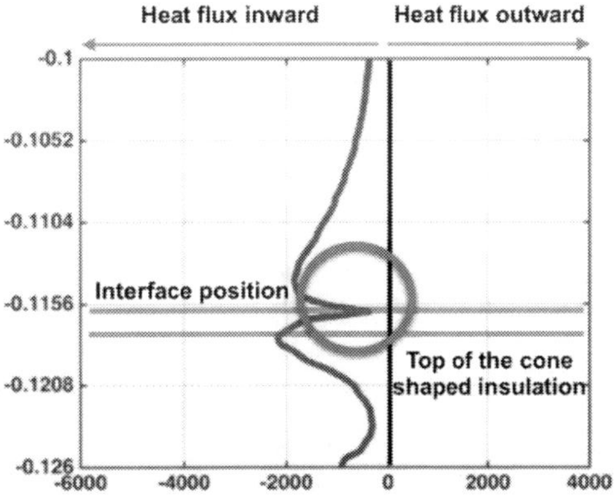

Radial heat flux along the inner ampoule wall at 10 hours after growth starts (W/m^2)

Modeling results agrees with theoretical prediction.

Comparison between Experiment and Modeling Results:

Eu composition

Classical segregation profiles

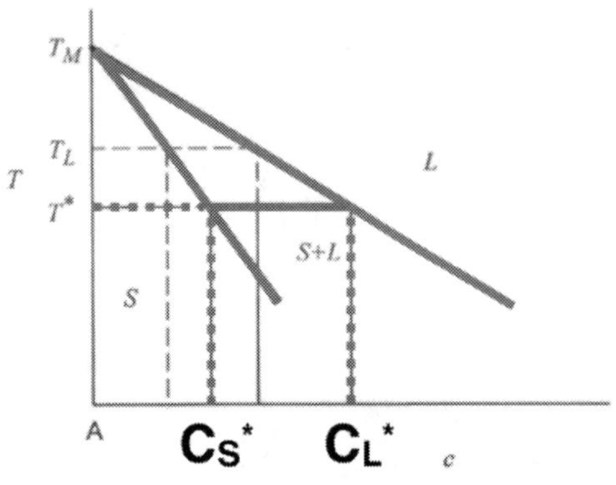

At melt-solid interface, the amount of solute which is partitioned to solid phase from melt phase is always in equilibrium:

$k = C_S^*/C_L^*$

http://coursenotes.mcmaster.ca/3E04/Plane_front_solidification.pdf

Crystal composition is determined by the segregation of species during growth: idealized cases of axial solute profiles

No mixing in the melt:
enriched solute layer built up at interface

Complete mixing in the melt:
uniform solute concentration in melt

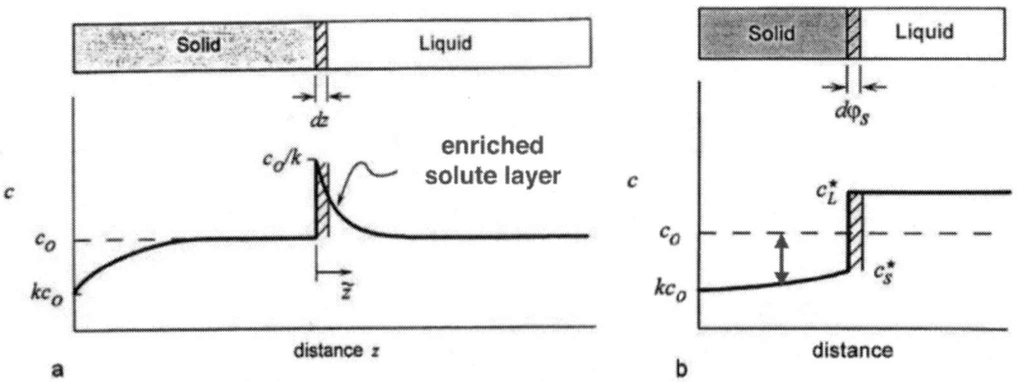

lower solute concentration in the solid

Does Eu dopant in our system follow these behaviors?

http://coursenotes.mcmaster.ca/3E04/Plane_front_solidification.pdf

We predict surprisingly complicated, multi-cellular flows that strongly impact Eu distribution during growth: Initial conditions

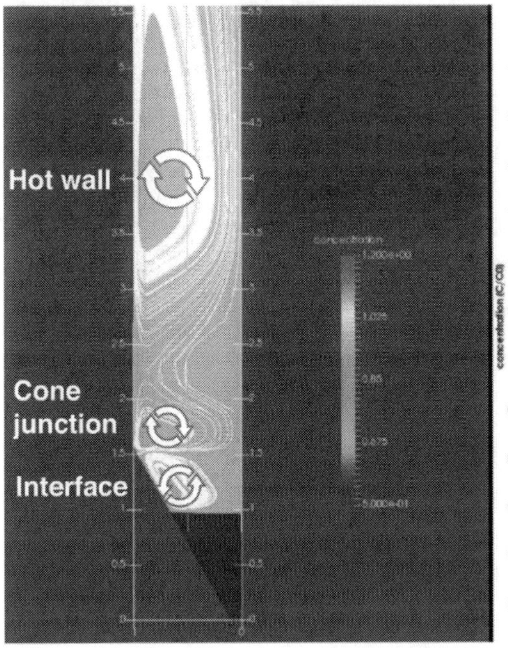

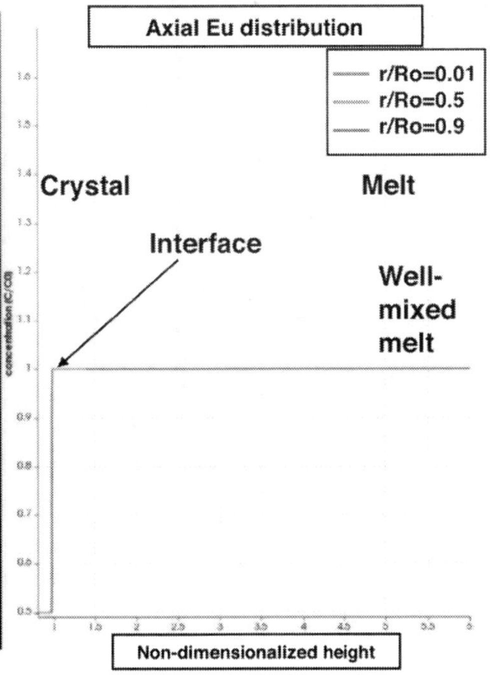

We predict surprisingly complicated, multi-cellular flows that strongly impact Eu distribution during growth: Transient evolution

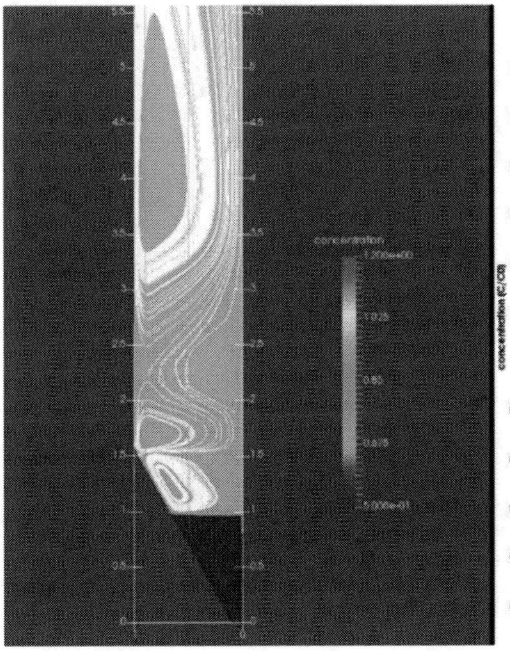

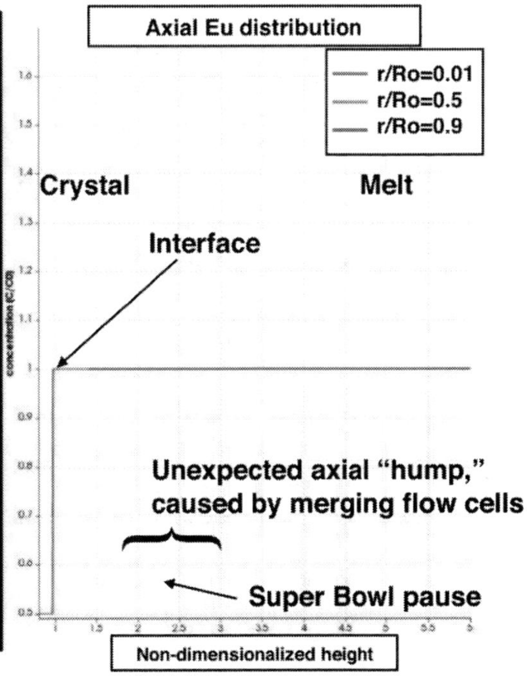

Eu distribution in grown crystal is well described by model and a mechanistic explanation of non-classical segregation outcome

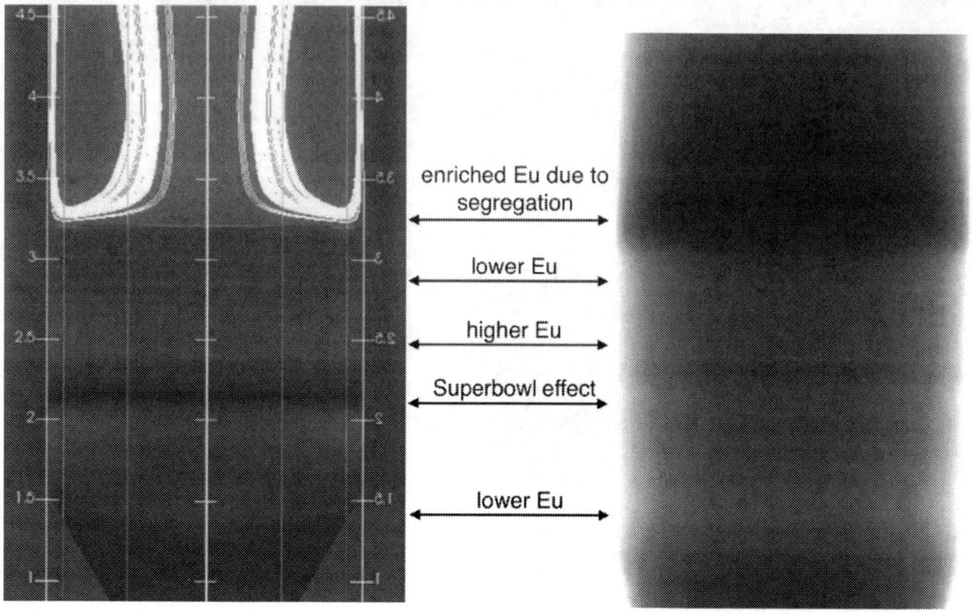

Final remarks

- These were the first experiments to demonstrate feasibility of neutron imaging of large-scale, melt crystal growth
- Computational modeling shows good agreement with experimental observations and provides a mechanistic explanation of outcomes
  - Shape of melt-crystal interface
  - Segregation of Eu in grown crystal
- Next phases of work include:
  - Investigation of residual stress and cracking issues (with dislocation dynamics simulation)

Neutron imaging and modeling, together, are promoting a deeper understanding of the solid-liquid phase transition during melt crystal growth!

Funded by DOE/NNSA/NA-22

Acknowledgements

University of Minnesota

J.J. Derby,
J.H. Peterson,
J. Seebeck,
K.P. Wang,
M.S. Divecha,
S. Dossa,
J. Roerig,
J. Brodzik

Lawrence Berkeley National Laboratory

D. Perrodin,
G.A. Bizarri,
E.D. Bourret

University of California

A.S. Tremsin

Los Alamos National Laboratory

A.S. Losko,
S. Vogel,
M. Bourke

Wuhan University

B. Gao

Thanks! Any questions?

Crystal Growth from Molten Material

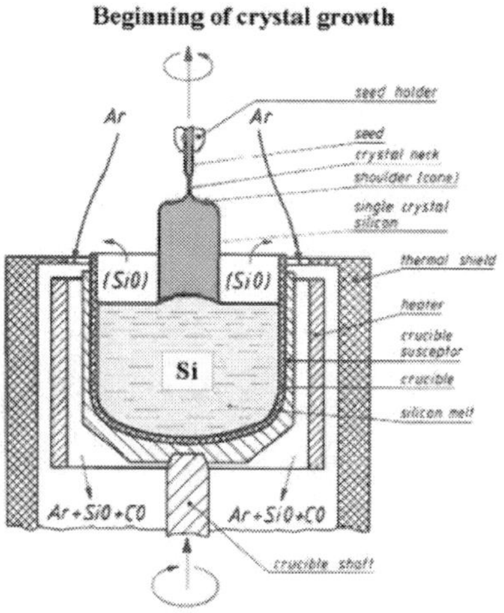

Czochralski method

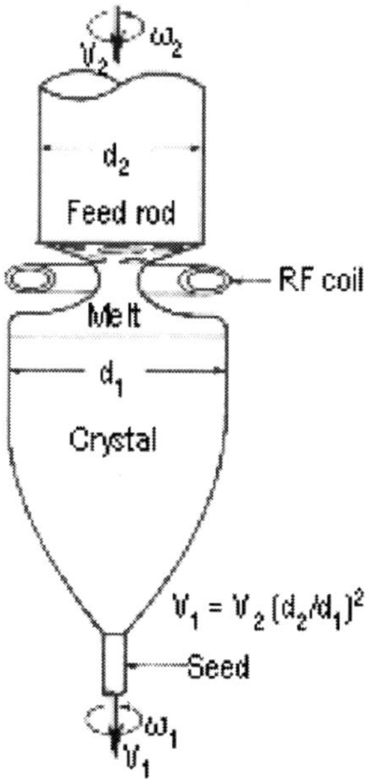

Float Zone method

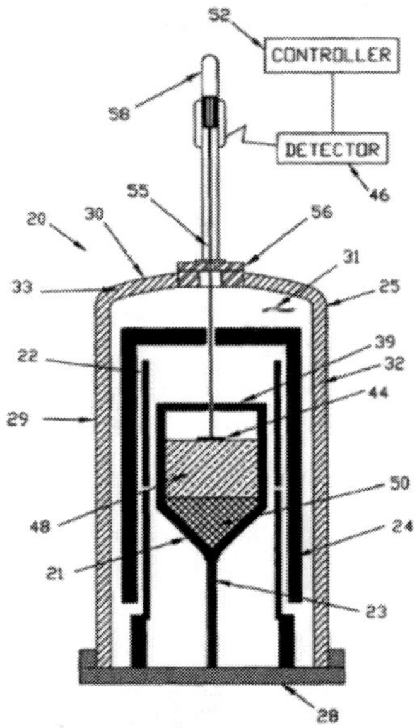

Bridgman method

496194 Efficient Dispersion of Crude Oil By Blends of Food-Grade Surfactants: Toward Greener Oil Spill Treatments

Abstract:

For decades, dispersants used for marine oil spills have been formulated with a blend of particular surfactants that are known to ensure effective dispersion at sea. After the Deepwater Horizon event, interest arose about what alternative surfactants might also be suitable to consider for future use. In this presentation we will very briefly summarize one promising avenue that has emerged.

Recent work has suggested that marine oil dispersants containing blends of the nontoxic surfactants lecithin (L) and Tween 80 (T) may be effective alternatives to traditional dispersant formulations containing blends of Tween 80, Span 80, and DOSS (dioctyl sodium sulfosuccinate). The dispersion effectiveness of lecithin-Tween 80 (L-T) blends in an ethanol solvent was measured for South Louisiana crude oil using baffled flask effectiveness testing, as a function of L:T ratio, surfactant:solvent ratio, solvent composition, and wt% dispersant in oil. The most effective L-T dispersants performed comparably to Tween 80-Span 80-DOSS dispersants, dispersing 75-90% of the test slick at 1.25-5 wt% dispersant in crude oil. Increasing surfactant:solvent ratio increased dispersant effectiveness even when dispersant dosage was proportionally reduced to keep total wt% surfactant in the oil constant. Replacing some of the ethanol solvent with octane or octanol also increased dispersant effectiveness, suggesting that ethanol's hydrophilicity lowered dispersant-oil miscibility, and that more hydrophobic solvents would increase dispersant effectiveness.

Two other important findings were that:

(1) lecithin-rich L-T dispersants were significantly more effective than Tween 80-rich L-T dispersants with lower or comparable interfacial tension, and

(2) all L-T dispersants produced much higher interfacial tension than Tween 80/Span 80/DOSS dispersants.

This suggests that interfacial phenomena other than interfacial tension, such as the formation of an interfacial gel or oil-dropletsâ€™ resistance to coalescence, may also influence L-T dispersantsâ€™ effectiveness.

489644 Comparison of Experimental and Predicted Adsorption Isotherms of Mixtures

Abstract:

Comparison of experimental and predicted adsorption isotherms of mixtures

Julian Butz, *Sabine Enders*

Karlsruhe Institute of Technology, Institute of Technical Thermodynamics and Refrigeration Engineering, Engler-Bunte-Ring 21, D-76131 Karlsruhe, Germany.

Email: julian.butz@kit.edu; sabine.enders@kit.edu

Key words: adsorption isotherms of mixtures, prediction

Abstract:

Multicomponent substances who build close-boiling system are a huge challenge for the separation through distillation. In such systems adsorption often is a lucrative separation alternative. The foundation for a good estimation of the outcome of the adsorption process is the adsorption equilibrium which can be described via adsorption isotherms. Unfortunately, measurements of adsorption isotherm for mixtures are still seldom while the measurement of pure component adsorption isotherms are a common but tedious method. Because of that it is suitable to predict the mixtures adsorption isotherms from only one adsorption isotherm for each pure component. The density functional approach [1] in combination with an convenient equation of state gives an excellent opportunity to do so. The solid-liquid interactions are modelled with a Steele 10-4-3 potential [2] in a slit pore while the liquid-liquid interactions are modelled with the PC-SAFT equation of state [3]. While the parameters of the PC-SAFT EOS for a lot pure components are listed in the literature [3] the parameters of the energy parameters of the Steele potential and the geometry parameters of the slit pore must be extracted from all pure component adsorption isotherms at one temperature. In terms of parameters this means one interaction solid-fluid energy parameter for every component $e_{sf,i}$ one interaction distance for the adsorbent s_{ss}, the mean pore width H as well as the mean pore volume V_p. For a pure component system that are 4 parameters while a ternary system needs 6 parameters. Having these parameters fitted to the pure components it is now possible to examine broad parameter studies. For the case of a varying feed composition in the adsorption process it is possible to pretend the adsorption isotherms for the possible feed configurations. Furthermore, a prediction of adsorption for temperatures yields a good opportunity to choose a good process window. It was found that a comparison to experimental multicomponent adsorption data [4] at different temperatures yields a very good compliance to the calculated predictions.

[1] C. Lastoskie, K.E. Gubbins, N. Quirke, J. Phys. Chem. 97 (1993) 4786.

[2] W.A. Steele, Surf. Sci. 36 (1973) 317.

[3] J. Gross, G. Sadowski, Ind. Eng. Chem. Res 40 (2001) 1244.

[4] R. Reich, W. T. Ziegler, K.A. Rogers, Ind. Eng. Chem. Process Des. Dev. 19 (1980) 336.

499546 Towards Connecting the Microstructural and Structural Changes in Hierarchical Nanoporous Materials (e.g., clays) during Gas Adsorption (CO_2, CH_4, H_2) Using *in-Operando* Multi-Scale X-Ray and Neutron Scattering Measurements

Abstract:

A fundamental understanding of the effect of gas adsorption and desorption behaviors at solid-water interfaces on the structures and microstructures of nano-porous materials has implications for engineered gas separations, and for engineering gas transport, recovery, and storage processes in subsurface environments. In this study, the effects of gas (e.g., CO_2, CH_4, H_2) adsorption and desorption behaviors on the structural and microstructural changes in nanoporous materials (e.g., clays such as Na-montmorillonite) with variable amounts of nano-confined interlayer water are investigated. The dynamic effects of gas adsorption and desorption on the molecular structure, the nano-confined interlayer water, the interlayer nanoporosity, and the meso-scale porosity are probed using multi-scale X-ray scattering methods which encompass Ultra-Small Angle, Small-Angle, and Wide-Angle X-ray Scattering (USAXS/SAXS/WAXS). Further, the application of contrast variation methods applied in Ultra-Small Angle and Small-Angle Neutron Scattering (USANS/SANS) methods allow us to investigate the displacement of nano-confined fluids during gas adsorption and desorption while quantifying the bulk gas uptake and release rates.

501740 Adsorption and Phase Behavior of Mixed Alkanes in Nano Slit Graphite Pores: An iSAFT Application

Abstract:

The promising development in shale gas/oil makes it critical to understand fluid properties in pores of shale kerogen. Confinement in nanopores shifts the fluid phase behavior making the exploration and production predictions challenging. As the pore shape, pore size and solid composition can all affect the fluid behavior, prediction of fluid properties in shale networks is still under study by many researchers. Molecular density functional theory (DFT) is and efficient and accurate approach to model the thermodynamic properties of confined fluid.

In this work, we have applied interfacial Statistical Associating Fluid Theory (iSAFT) to model alkane (C_1-C_8) adsorption/desorption and phase behavior in graphite slit pores for pure fluids and mixtures. iSAFT accurately predicts the effect of size and shape on component partitioning and phase behavior. The pure component parameters were determined by fitting to bulk saturated liquid density and vapor pressure data in certain temperature ranges. In quantifying the fluid-solid interactions, we have introduced the potential of mean force from molecular simulation to evaluate the external potential in the non-local DFT framework. The density distribution of molecules in adsorption of methane and heptane are in good agreement with molecular simulation. The predicted shift in critical properties of confined pure methane and ethane is of similar magnitude to that from simulation. For a binary system, a methane and ethane mixture at 250 K in a 5 nm pore was chosen as a showcase. The shifted bubble point, dew point and critical pressure show good agreement with molecular simulation, with the errors mainly arisen from the mean field approximation in treating dispersion energy. We predict the competitive adsorption of multicomponent systems and discuss the dominating factor that determines the pore selectivity to the heavy components in gas or liquid adsorption. With these studies, we confirm the validity of the simplest form of iSAFT in predicting confined molecule phase behavior and show the promising application in oil and gas industry with its advantage in being more computationally efficient than molecular simulation method.

Oscillation and Enhancement of Gas Solubility in Nanopores

Xiaochen Yu, Shuangliang Zhao*, Honglai Liu

State Key Laboratory of Chemical Engineering
East China University of Science and Technology (ECUST)

Oct. 30, 2017 Minneapolis

"This Rock Could Power the World"

Time Magazine, 04/2011

- **~50% U.S. natural gas supply, 2015**
- **~1115 TCF Recoverable in China**
- **Cleaner Fuel**

Energy consumption per capita

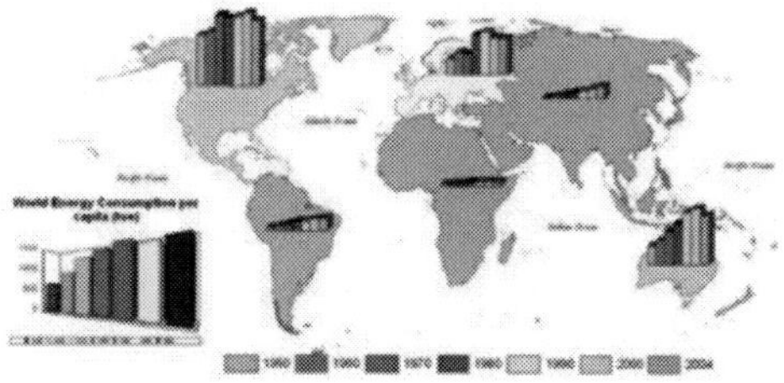

Thermodynamics of Fluids in Nanopores

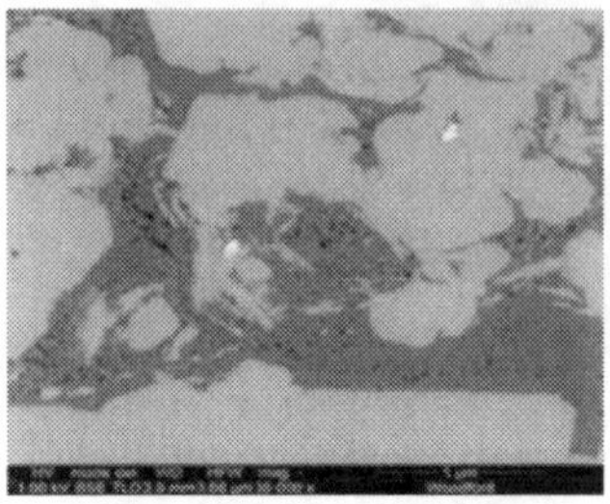

SEM image of Woodford Shale
(Rassenfoss 2011)

- Low Porosity (< 10%)
- Pores in *nm* range

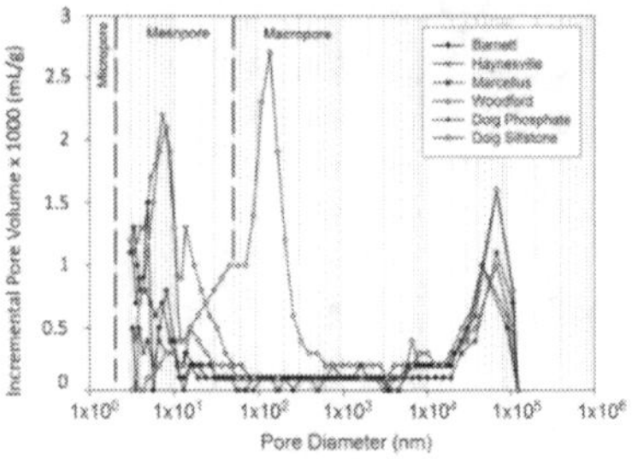

Pore size distributions of various shales
(Chalmers *et al.* 2012)

Key Point: What's gas solubility in shale rock?

- Conventional Rock

$$\Gamma = \rho_b V$$

- Shale Rock

- ➢ Tiny Pores (nm)
- ➢ Dissolution

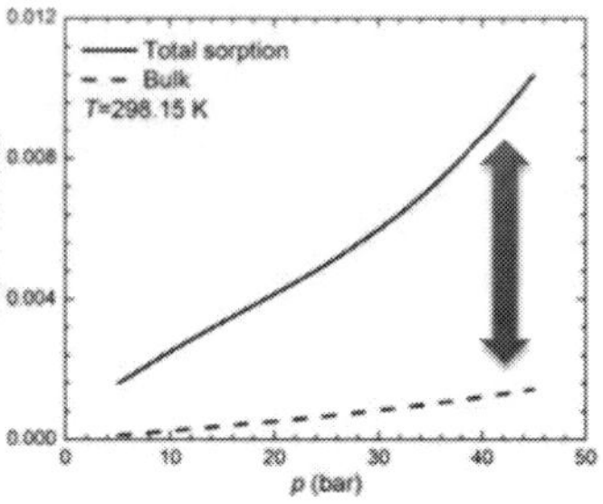

CO_2 sorption in Barnett Shale
(Vermylen 2011)

Gas Uptake in Solvents Confined in Mesopores: Adsorption versus Enhanced Solubility JPCL, 2013, 4(14):2274-2278.

Linh Ngoc Ho,[†] Stéphanie Clauzier,[†] Yves Schuurman,[†] David Farrusseng,[†] and Benoit Coasne*[,‡,§]

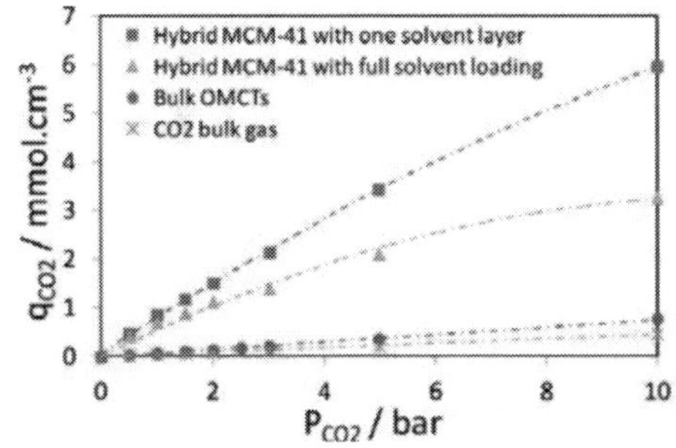

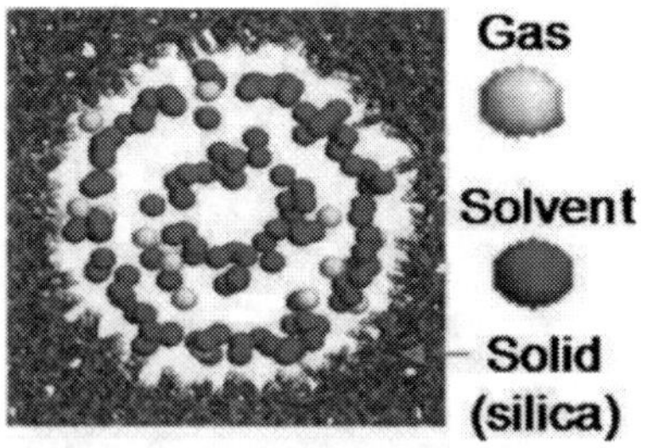

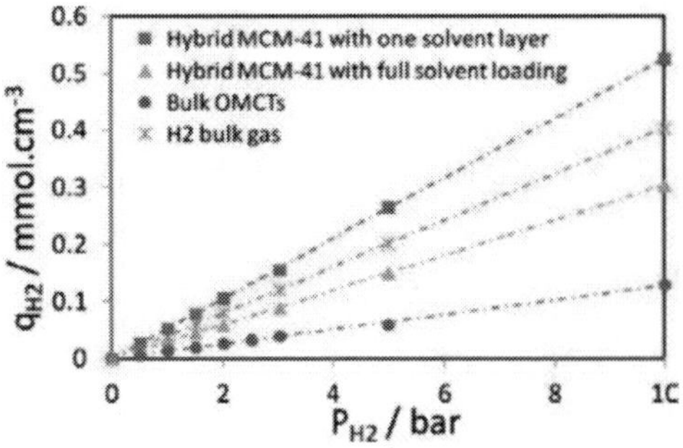

GCMC. Solubility of H_2 and CO_2 in OMCTS under confinements is higher than that in bulk phase. "Over-Solutility"

Effect of confinement on the hydration and solubility of NaCl in water

A. Malani [a,b], K.G. Ayappa [a], S. Murad [b,*]

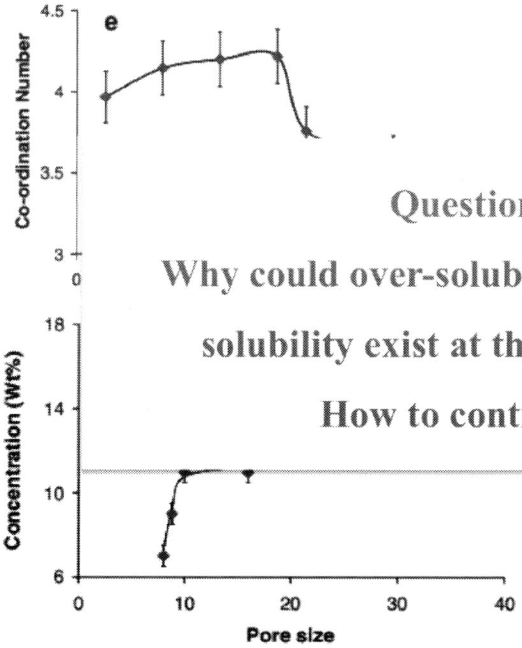

Question:

Why could over-solubility and under-solubility exist at the same time?

How to control it?

…l in water under confinements is lower than that in bulk system. "Under-Solubility".

Chemical Physics Letters, 431 (2006) 88-93

Method: Combination of EOS and DFT

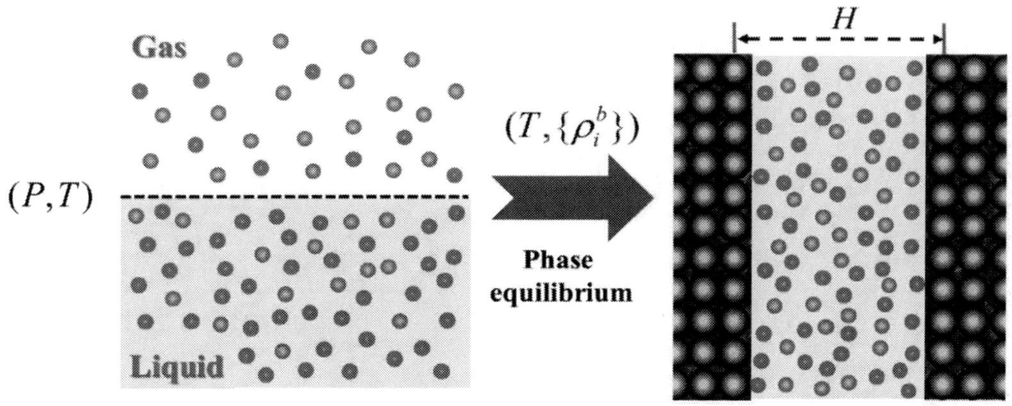

Equation of State (EOS) **Density Functional Theory (DFT)**

EOS for gas solubility under reservoir condition

DFT for gas solubility in confined solvents

Chemical Potential: EOS and Phase Equilibrium

The chemical potential of solute and solvent are determined by the gas-liquid (G/L) phase equilibrium.

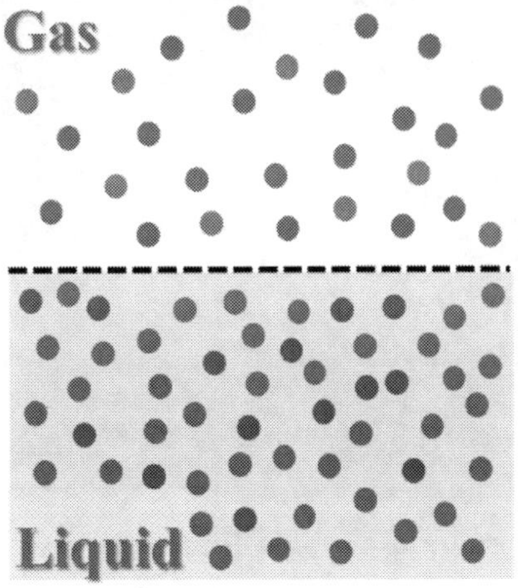

At given (P, T), the G/L phase equilibrium determined (i=1,2):

$$\mu_i^L(\rho_1^L,\rho_2^L)=\mu_i^G(\rho_1^G,\rho_2^G)$$
$$f(\rho_1^L,\rho_2^L)=\mathrm{P}$$
$$f(\rho_1^G,\rho_2^G)=\mathrm{P}$$

MBWR EOS

EOS and Phase Equilibrium: Solubility prediction

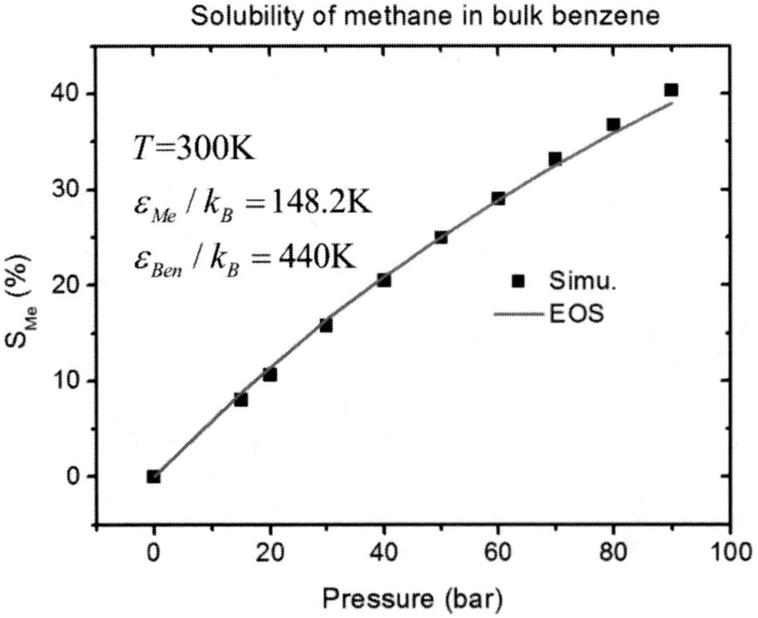

Solubility:

$$S_2=\frac{N_2}{N_1+N_2}\times 100\%$$

(Mol. Fraction)

Simu.: 5-10h

EOS: 1-5min

Classical DFT: Robust tool in chemical engineering

We have proposed a unified framework of density functional theory.

$$\frac{\delta\Omega[\rho(\mathbf{R})]}{\delta\rho(\mathbf{R})} = 0$$

angstrom → micron

Quantum DFT | Atomic DFT | Molecular DFT | Polymeric DFT

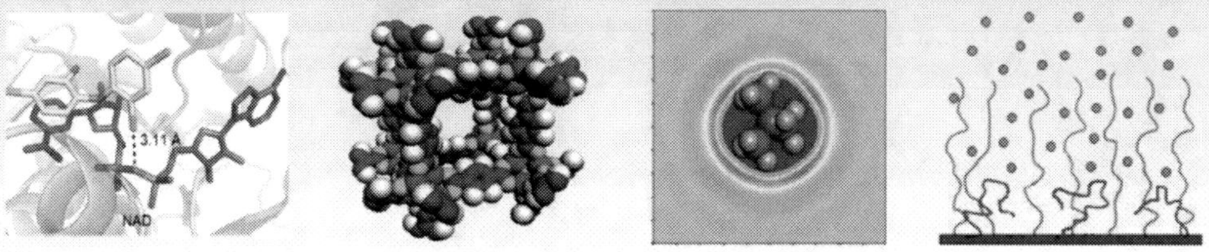

Affinity & Intensity | Adsorption & Separation | Structure & Energy | Mechanism & Regulation

Shuangliang Zhao, Yu Liu, Xueqian Chen, Yunxiang Lu, Honglai Liu*, and Ying Hu. Advance in Chemical Engineering, G.B. Marin, J.H. Li, Mesoscale Modeling in Chemical Engineering Part II, Chapter one, 1-83, 2016

Density Functional Theory

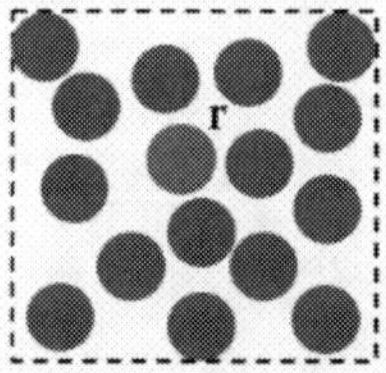

adsorption of an isolated site

Grand Potential

$$\Omega[\rho(\mathbf{r})] = F[\rho(\mathbf{r})] + \int \rho(\mathbf{r})(V_{ext}(\mathbf{r}) - \mu)d\mathbf{r}$$

Intrinsic Helmholtz free energy

$$F[\rho(\mathbf{r})] = F^{id}[\rho(\mathbf{r})] + F^{ex}[\rho(\mathbf{r})]$$

Ideal contribution (for F^{id})

Due to intermolecular interaction (for F^{ex})

$$\frac{\delta\Omega[\rho(\mathbf{r})]}{\delta\rho(\mathbf{r})} = 0$$

$$\rho(\mathbf{r}) = \rho_b \exp\left(\beta\mu^{ex} - \beta\frac{\delta F^{exc}[\rho(\mathbf{r})]}{\delta\rho(\mathbf{r})} - \beta V_{ext}(\mathbf{r})\right)$$

DFT verification

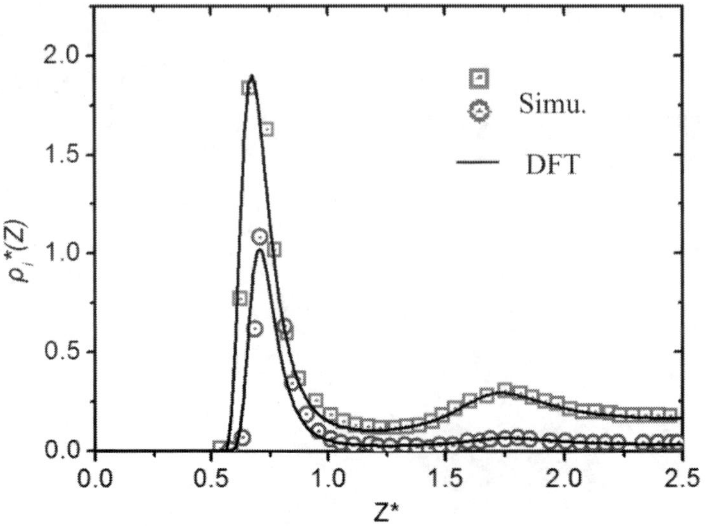

$\sigma_1 = 0.3405$ nm
$\sigma_2 = 0.363$ nm
$\varepsilon_1 / k = 109.8$K
$\varepsilon_2 / k = 163.1$K
$H^* = H / \sigma_1 = 5$
$T^* = k_B T / \varepsilon_1 = 2.0$
$\rho_0^* = 0.103,$
$x_1 = 0.109$

Simu: >24h

DFT : <20min

Accurate and Efficient!

PHYSICAL REVIEW E, VOLUME 64, 021206

Gas Solubility in Nanopores: Model

- **Slit pore**

Fluid-substrate interaction: Steel's 10-4-3 potential

$$u_{aw}(z) = 2\pi\rho_w\Delta_w\varepsilon_{aw}\sigma_{aw}^2\left[\frac{2}{5}\left(\frac{\sigma_{aw}}{z}\right)^{10} - \left(\frac{\sigma_{aw}}{z}\right)^4 - \left(\frac{\sigma_{aw}^4}{3\Delta_w(z+0.61\Delta_w)^3}\right)\right]$$

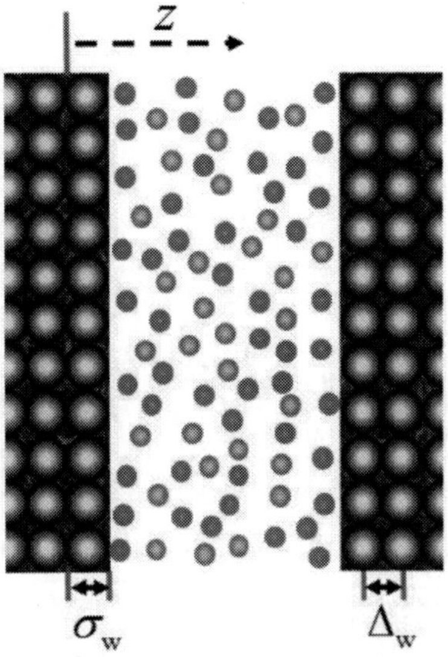

Wall	ρ_w/nm^{-3}	Δ_w/nm	σ_w/nm	ε/k_B/K
Graphite	114.0	0.335	0.34	28.0
Silica	44.2	0.220	0.27	230
Mica	25.4	0.287	0.35	940

Fluid	σ (nm)	ε/k (K)
Ar	0.3405	119.8
CCl_4	0.514	366
OMCTS	0.77	351.4
$C_6H_5NO_2$	0.514	425
C_6H_5Br	0.538	351.2

Gas Solubility in Nanopores: Pore size effect

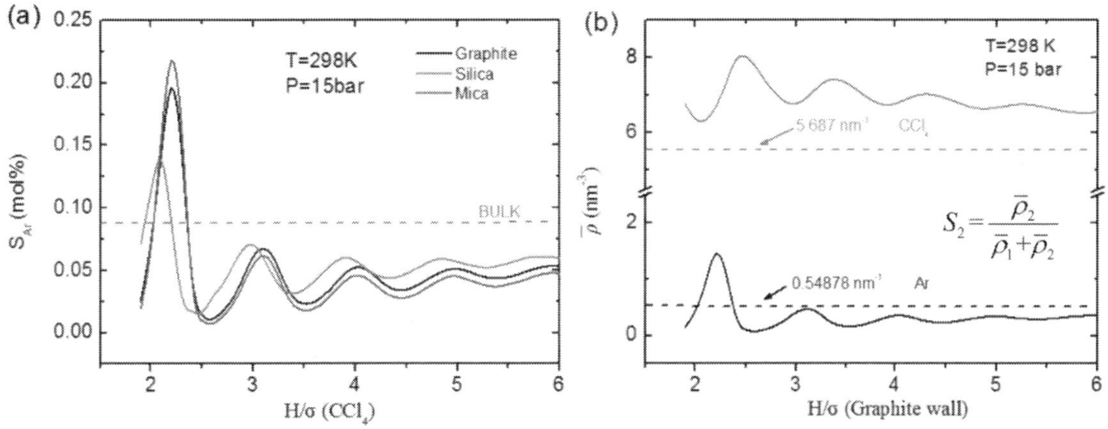

- **The solubility of Argon in confined CCl_4 is sensitive to the substrate**
- **The gas solubility in confined solvent oscillates with pore width, resulting from the size effect.**

Gas Solubility in Nanopores: Substrate Effect

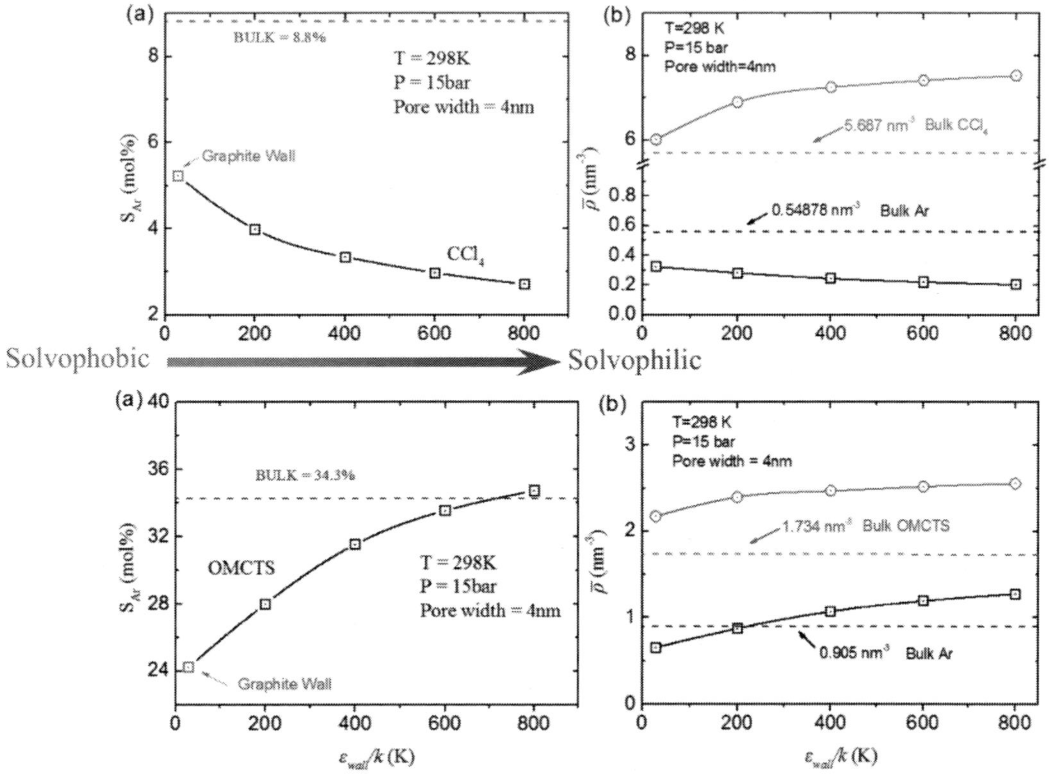

Gas Solubility in Nanopores: **Substrate Effect**

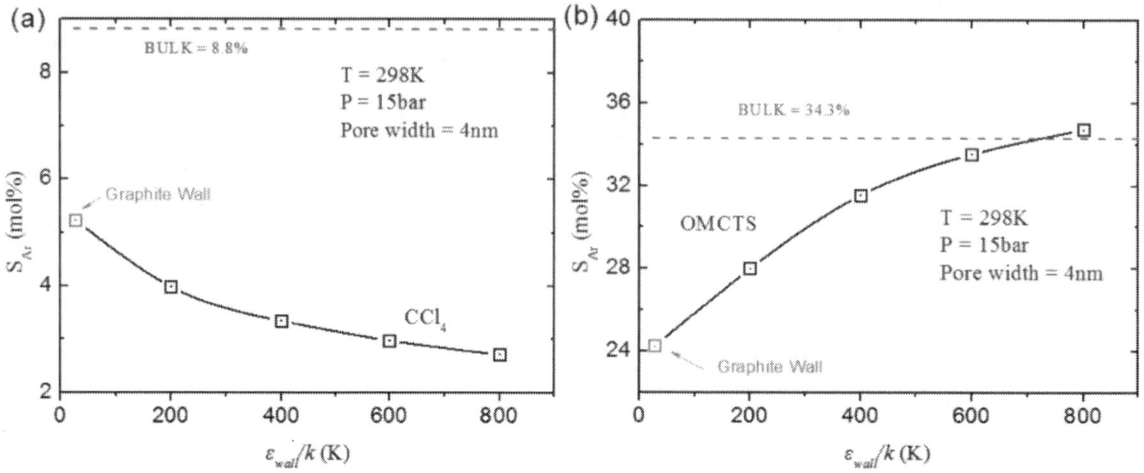

- **Solvophilic modification of the substrate surface may influence the solubility.**

Conclusion

- **The gas solubility in confined solvent oscillates with pore width, resulting from the size effect.**
- **Solvophilic modification of the substrate surface may change the solubility.**
- **Confinement Effect: Size Effect Vs. Adsorption Effect.**

Acknowledgement

- Honglai Liu

- Keith Gubbins

国家自然科学基金委员会 NSFC

973 CHINA

Skloche
化学工程联合
国家重点实验室

Thanks for your attentions !

501545 Effect of Interaction Position on Molecular Transport and Separation through a Multi-Site Nanopore

Abstract:

Mass transport in confinement spaces like nanopores has become a field of interest for many applications such as separation employing membranes, microfluidic systems, chemical analysis, and drug delivery. This is due to the possibility of having efficient, fast, and robust separation in transport through one dimensional membranes and pores. This type of transport differs in many ways from bulk transport due to specific interactions between the pore and molecules when the pore dimension is around nanometer. Several studies are focused on mimicking the physical and chemical properties of biological pores that exhibit real example of enhanced separation through nanopores. However, a theoretical model that describes the role of the molecule-pore interaction in the molecular transport through channels has not been developed yet. Here, we develop a theoretical approach to explain the origin of separation efficiency in molecular transport through channels. Our method utilizes discrete-state stochastic models that take into account all relevant chemical transitions. More specifically, we analyze channels with different binding sites employed for separating mixtures of two types of molecules. The effects of the symmetry and the strength of the molecular-pore interactions are examined. The most efficient separation is predicted when the interaction site is located near the entrance to the nanopore. However, based on the amount of two limiting rate steps that are introduced in the model, changing the position of interaction to the middle of the pore can enhance the separation. In general, the selectivity is higher for large entrance rates into the channel. The origin of the results is discussed as well.

498701 Strain-Induced Topological Defects and Configurational Transitions in Liquid Crystals

Abstract:

In this work we explore the effect of geometrical confinement on the orientational ordering and transition of topological defects in nematic and smectic liquid crystals. The geometrical confinement is induced by uniaxial and biaxial stretching of a polymer dispersed liquid crystal (PDLC) film, with homeotopic anchoring. Our results show that upon uniaxial stretching, the nematic liquid crystal (LC) in the prolate cavity of a stretched polymer film retains a radial configuration, and the defect position at the droplet center remains unchanged. In the radial smectic LC droplet, however, the topological defect undergoes a transition from a point to a line, whose length is a function of the aspect ratio. The formation of this topological defect can be attributed to the misorientation and dilation of the smectic layers. Upon biaxial stretching, liquid crystal droplets become increasingly oblate. In the oblate smectic LC droplet, a layering configuration undergoes a transition from radial to parallel ordering, starting from the central part of the droplet. On the other hand, in the oblate nematic LC droplet, a topological distortion competes with homeotropic alignment and radial ordering, causing a configurational director twist. At a particular strain level, the LC molecules in the twisted oblate align normal to the respective stretch directions. Landau-de Gennes simulations are used to gain additional insights into the morphological changes of the nematic LC under biaxial stretching. Consistent with our experimental results, the continuum simulations capture the biaxial stretch-induced twist configuration, and demonstrate a uniform configuration when the mechanical strain exceeds 20%.

499811 Alkyl Chain Length and Nanoconfinement Effects on Dynamics of Imidazolium-Based Ionic Liquids

Abstract:

Room-temperature ionic liquids (RTILs), due to their unique properties, such as wide electrochemical window and high thermal stability, are considered as excellent candidates as electrolytes for supercapacitors. While the high operating voltage window allows for a high energy density, the slow dynamics of RTILs may lead to a moderate power density, especially in the case of confinement. Understanding the constraints on the power density of supercapacitor based on nanoporous materials and RTILs is still limited.

In this work, molecular dynamics (MD) simulations were conducted to investigate the dynamics of a set of imidazolium-based RTILs with different side chain lengths. The effect of alkyl chain length is discussed by comparing the ion self-diffusion coefficients between different RTILs. The ion diffusivities are also compared with bulk values to evaluate the effects of nanoconfinment. Structural analysis of ions in bulk and confinement are made to further explain the effect of nanoconfinement. In addition, we rely on Quasi-elastic neutron scattering (QENS) experiments to validate the observed trends in MD simulations. This study helps to provide theoretical support for the design of energy storage devices that are better both in energy and power densities.

493119 A Molecular Simulation Study on CO_2 and a Deep Eutectic Solvent in Slit Nanopores

Abstract:

A molecular simulation study on CO_2 and a deep eutectic solvent in slit nanopores

Yan Shen and Francisco R. Hung

Department of Chemical Engineering, Northeastern University, Boston, MA 02115

Molecular dynamics (MD) simulations were performed to study systems of CO_2 and deep eutectic solvents (DESs) in the bulk and inside slit-like nanopores. The porous materials considered were graphite and rutile, while the DES studied was composed of choline chloride and ethylene glycol with a 1:2 molar ratio; different amounts of DES inside the nanopores were considered. Our work was motivated by previous studies exploring the use of bulk DESs in CO_2 capture[1, 2], as DESs share many of the physical and chemical properties of room temperature ionic liquids (ILs) while being considerably cheaper and (mostly) nontoxic. On the other hand, recent reports[3-5] have studied the â €œoversolubilityâ€ phenomenon, where gas molecules exhibit larger solubility in solvents that are confined within a nanoporous material, as compared to the gas solubility in the bulk solvents. Our results show that the properties of the DES-CO_2 systems in the bulk are significantly different from those of the same systems confined in the nanopores. For the confined systems, the chemical composition of the pore walls, and the amounts of DES inside the nanopores have a great impact on the behavior of CO_2 molecules. Detailed comparisons of the properties of all systems will be presented and discussed.

References:

1. García, G.; Atilhan, M.; Aparicio, S. *The Journal of Physical Chemistry C* **2015**, 119, (37), 21413-21425.

2. García, G.; Aparicio, S.; Ullah, R.; Atilhan, M. *Energy & Fuels* **2015**, 29, (4), 2616-2644.

3. Bratko, D.; Luzar, A. *Langmuir* **2008**, 24, (4), 1247-1253.

4. Clauzier, S.; Ho, L. N.; Pera-Titus, M.; Coasne, B.; Farrusseng, D. *Journal of the American Chemical Society* **2012**, 134, (42), 17369-17371.

5. Ho, N. L.; Perez-Pellitero, J.; Porcheron, F.; Pellenq, R. J. M. *The Journal of Physical Chemistry C* **2012**, 116, (5),

3600-3607.

Molecular Simulation of Methane Adsorption Behavior in Nanopores for Shale Gas Development: With Comparison between Graphite and Kerogen Models

Jinrong Cao
Department of Systems Innovation, the University of Tokyo, Tokyo, Japan
Yunfeng Liang and Yoshihiro Masuda
Center for Engineering, Research into Artifacts (RACE), the University of Tokyo, Kashiwa, Japan
Hiroaki Koga and Hiroyuki Tanaka
Japan Oil, Gas and Metals National Corporation (JOGMEC), Chiba, Japan
Toshifumi Matsuoka
Fukada Geological Institute, Tokyo, Japan

In shale reservoirs, amount of gas is stored in organic nanopores. The effects of confinement on gas molecular properties are significantly different from those in bulk system. Molecular simulation of methane adsorption behavior in nano-confinement space is investigated in this research. The simulation was performed at 10 MPa and 350 K in kerogen or graphite nanopores system. Molecular dynamics (MD) and grand canonical Monte Carlo (GCMC) methods were selected to conduct the simulation. The general aim of our researches is to select an appropriate method for investigation of gas adsorption behavior. Focus on the objective, three comparison studies were performed: (i) compare the adsorption behavior in kerogen nanopores using different cut-off values with dispersion correction and without dispersion correction, respectively; (ii) compare the adsorption behavior in graphite nanopores using MD and GCMC simulation methods; (iii) compare the adsorption behavior in kerogen and graphite nanopores using MD simulations. Our work improves understanding for gas adsorption in confinement space.

Keywords: Molecular Simulation, Nanopores, Methane Adsorption, Shale Gas

Introduction

Organic rich shale contains kerogen nanopores with characteristic lengths typically about tens of nm to about one hundred nm[1]. Most of the shale gas is adsorbed in nanoscale pores of kerogen. An adequate knowledge of gas adsorption behavior in the nanopores is necessary to accurately estimate gas-in-place for shale gas resource assessment and development. Molecular simulation is a powerful technique and gets increasingly important in this research direction[2-5]. Among all the molecular simulation techniques, molecular dynamics (MD) and grand canonical Monte Carlo (GCMC) simulations are two typical methods[6,7] for the shale gas-in-place calculations. However, the results are still controversial[2-5], a few issues need to be carefully discussed in order to better address the methane adsorption behavior for engineering applications. In this research, focusing on the dispersion interaction, we need to consider whether implement correction and how much the cut-off value. On the other hand, MD and GCMC simulations can, in principle, be used to study these problems consistently by using the same atomistic model with the same force field. Hence, it will be very valuable if we can examine two methods in detail by using the same atomistic models. In addition to that, attention also to be paid for the difference between graphite model and kerogen model.

Methodology

Molecular Dynamics Simulation

MD simulation is a method to calculate the force acting between all atoms of the system, it is a technique to develop the coordinates of the particles time by solving the Newton equation.[6] In this methodology, we describe the system at the atomic scale and simulate the motions of all the atoms in the system to understand and predict macroscopic properties of the system under desired conditions. In this study, MD simulations were performed by GROMACS 5.1.2 software package[8]. The simulations were performed in constant pressure-constant temperature (NPT) ensemble.

Grand Canonical Monte Carlo Simulation

For adsorption studies, a natural ensemble to use is the grand-canonical ensemble (or μ, V, T ensemble). GCMC simulation is a method that can be applied to study the adsorption of a bulk fluid of constant temperature and chemical potential on a solid adsorbent. The volume of the solid adsorbent is also constant throughout the simulation. The equilibrium conditions are that the temperature and chemical potential of the gas inside and outside the adsorbent must be equal. The pressure is a function of the chemical potential and is always possible to calculate the pressure of the gas that corresponds to a given chemical potential and vice versa. The temperature and chemical potential are imposed and the number of particles is allowed to fluctuate during the simulation. This makes the simulation different from the conventional ensembles, where the number of molecules is fixed.

In Monte Carlo simulation, a Markov chain of molecular configurations is generated. The Markov chain is produced using three trial moves:

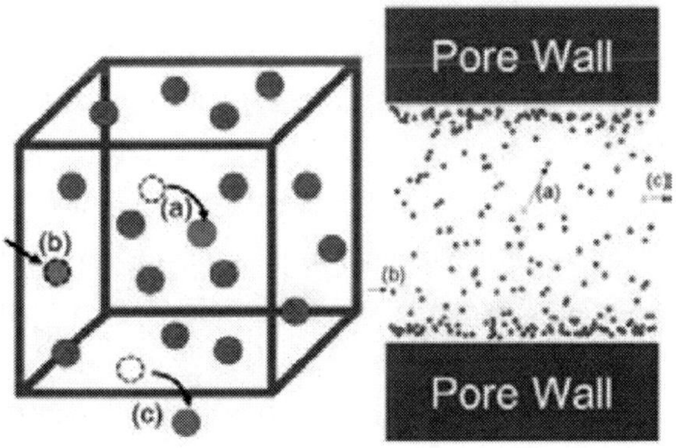

Fig. 1. Schematic of the GCMC trial moves: (a) particle displacement, (b) particle insertion and (c) particle removal.

(a) Particle displacement: A randomly selected particle is displaced by a predetermined distance. The move is accepted with a probability:

$$\mathcal{P} = \min[1, exp\{-\beta[\mathcal{U}(s'^N) - \mathcal{U}(s^N)]\}]$$

where $\beta = 1/[k_B T]$ (k_B is Boltzmann's constant and T is temperature), and $U(s'^N)$ and $U(s^N)$ are the potential energy functions of the new and old configurations, respectively.

(b) Particle insertion: A particle is inserted at a random position. The insertion is accepted with a probability:

$$\mathcal{P} = \min\left[1, \frac{v}{\Lambda^3 (N+1)} exp\{\beta[\mu - \mathcal{U}(N+1) + \mathcal{U}(N)]\}\right]$$

where V is the volume, Λ is the thermal de Broglie wavelength, μ is the chemical potential and N is the number of molecules inside the framework.

(c) Particle removal: A randomly selected particle is removed from the system. The removal of a particle is accepted with a probability:

$$\mathcal{P} = \min\left[1, \frac{\Lambda^3 N}{v} exp\{-\beta[\mu + \mathcal{U}(N-1) - \mathcal{U}(N)]\}\right]$$

In this study, GCMC simulations were performed by Towhee code 7.2.0[9].

Dispersion Correction

One common optimization in simulations with pairwise potentials is to divide the interactions into long and short range, which are treated differently. For long range Van der Waals interactions, we assume that the cut-off is so long that the repulsion term can safely be neglected, therefore only the dispersion term is taken into account.

Due to the nature of the dispersion interaction, energy and pressure corrections are both negative. For a homogeneous mixture we need to define an average dispersion constant:

$$\langle C_6 \rangle = \frac{2}{N(N-1)} \sum_{i}^{N} \sum_{j>i}^{N} C_6(i,j)$$

While the energy correction is usually small, it may be important for free energy calculations. The long-range contribution to the dispersion energy is:

$$V_{lr} = -\frac{2}{3}\pi N \rho C_6 r_c^{-3}$$

In contrast, the pressure correction is very large and cannot be neglected under any circumstances where a correct pressure is required, especially for any NPT simulations. The correction to the pressure is:

$$P_{lr} = -\frac{4}{3}\pi C_6 \rho^2 r_c^{-3}$$

In some cases, dispersion corrections appropriate for fluid systems (homogeneous system) are applied to the pressure and the potential energy. For the adsorption process, especially those which electrostatic interaction energy is not strong enough, dispersion correction played a significant role.

In addition to dispersion correction, so as to treat system using Lennard-Jones potentials, which is inhomogeneous outside of the cut-off distance, we can instead use the Particle-mesh Ewald (PME)[8] method as discussed for electrostatics above. In this case, the dispersion correction can be described explicitly for an inhomogeneous property.

Simulation Details

In our research, we have studied gas adsorption behavior inside the nanopores with graphite and full-atom kerogen slab models, respectively. **Fig. 2** shows the graphite and kerogen molecular model structure[4,10]. Snapshots for graphite and kerogen nanopores system also shown as in **Fig. 3**.

The optimized potential for liquid simulations (OPLS) – all-atom force field is used[11,12]. The pressure and temperature are fixed at 10 MPa and 350 K, while the pore size varies from 0.5 to 70 nm.

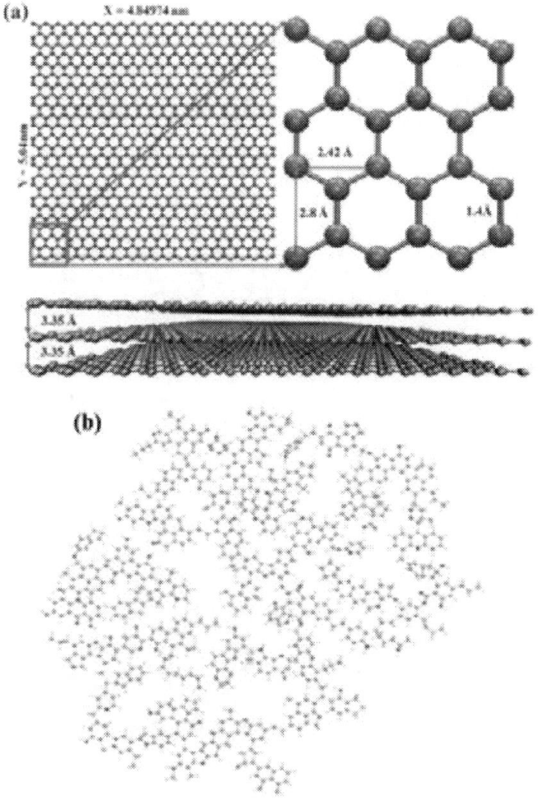

Fig. 2. Initial graphite (a) and kerogen (b) molecular model structure.

Fig. 3. Snapshots for graphite (a) and kerogen (b) nanopores system.

MD simulations were conducted under the condition where an anisotropic scaling without shear deformation was set to barostat for pressure control of the system. GCMC simulations require adsorbate chemical potential rather than gas pressure as input. So, we performed "bulk" GCMC simulations (i.e., a box without graphite or kerogen pore wall unit cell) to establish appropriate chemical potentials for the target pressure as **Fig 4**. We chose the corresponding chemical potential for the goal pressure 10 MPa in GCMC simulation.

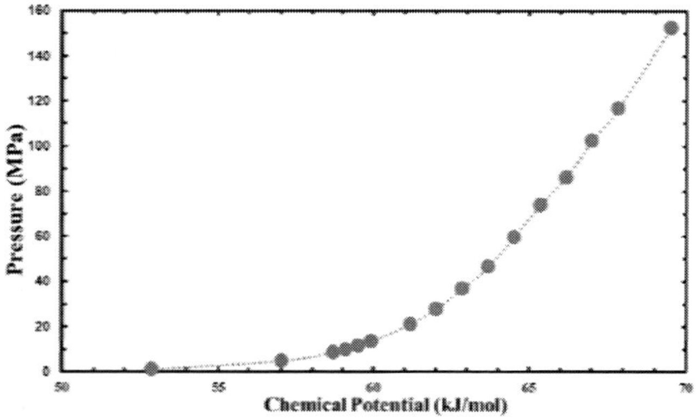

Fig. 4. Pressure (output) as a function of chemical potential (input) in GCMC simulations.

Results and Discussion

In this paper, we have focused on three points about the computational techniques. At first, we will introduce the simulations in bulk methane system with MD and GCMC methods.

Bulk Methane System

The simulated average bulk methane density at 10 MPa and 350 K from MD and GCMC simulations as in **Fig. 5**. It is found that the simulation results are essentially identical for two methods in the bulk system compared with NIST experimental data.

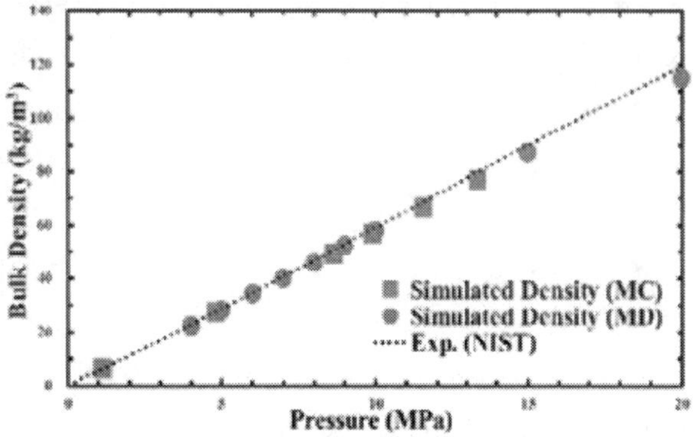

Fig. 5. Simulated bulk density as a function of pressure compared with experimental data.

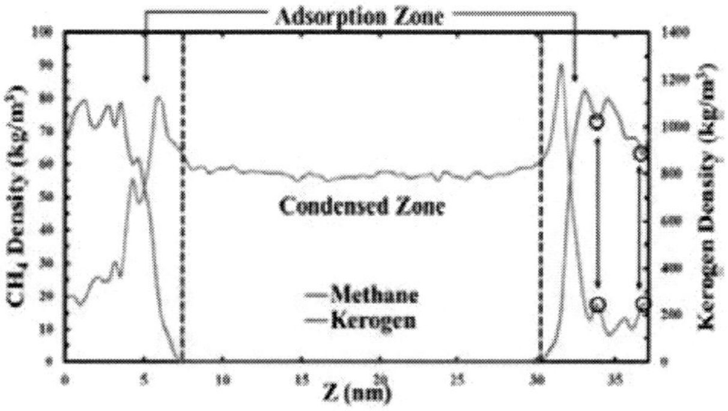

Fig. 6. Categorization of the simulation system into two zones: (1) the adsorption zone and (2) the "condensed" zone.

Long Range Van der Waals interactions

First of all, we discussed the limitation on the implementation of a long-range correction scheme due to the use of a cut-off for dispersion interactions.

This scheme can be derived analytically, and appropriately applied for a homogeneous system while saving the computational cost. However, there is no guarantee for this scheme to be applied in an inhomogeneous system, such as methane in kerogen nanopores system. For this point, we have selected different cut-off values to simulate methane adsorption in kerogen nanopores system (at 10 MPa and 350 K) with dispersion correction and without dispersion correction, respectively. In addition, Lennard-Jones PME scheme is also used to calculate the dispersion interactions as a benchmark for this system.

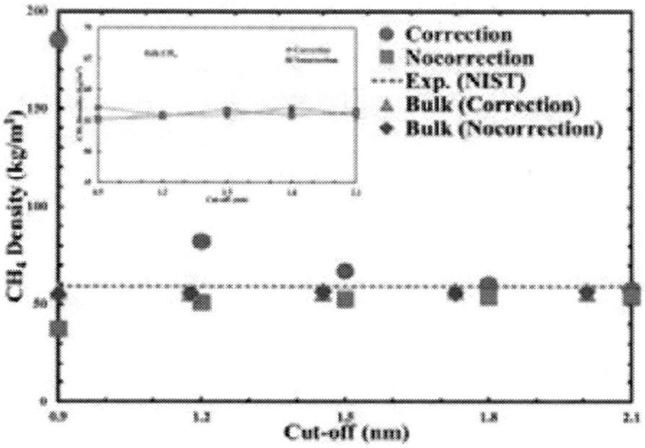

Fig. 7. "Bulk" CH_4 density with different cut-off values with dispersion correction and without dispersion correction in kerogen system and bulk system.

The above points were addressed by MD simulations. The simulation results with dispersion correction and without dispersion correction are compared. Here, we just calculated the average bulk density in the condensate zone as in **Fig. 6**. As **Fig. 7**, it is shown that the computational details, such as the cut-off values, can play a significant role in the calculation of methane density in the mesopores. In addition, the result of PME scheme is close to the simulation result without dispersion correction when we select larger cut-off (> 1.5 nm) as in **Fig. 8**.

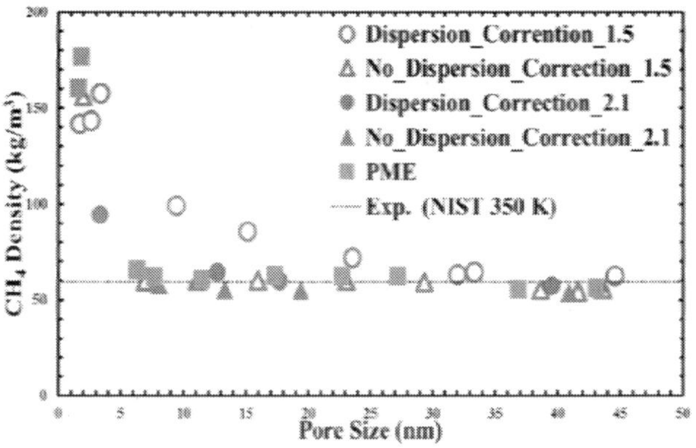

Fig 8. "Bulk" CH_4 density as a function of pore size in kerogen system.

Comparison between MD and GCMC Methods

Second, we compared the results from MD and GCMC simulations using graphite nanopores model at 10 MPa and 350 K, while the pore size varies from 0.5 to 80 nm as **Fig 9**.

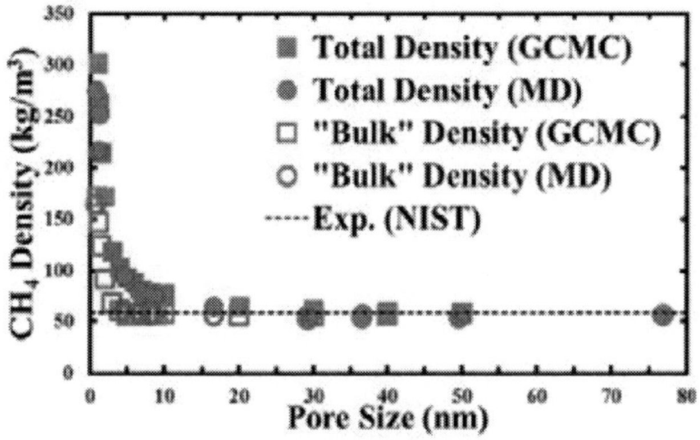

Fig. 9. CH_4 density as a function of pore size using MD and GCMC methods.

We have been employing GCMC simulations to compare the results obtained from MD simulations. The purpose is to examine the "pore pressure" of methane as used in MD simulations. We compared the simulated bulk CH_4 densities by MD and GCMC simulations as a function of pressure and compared with NIST experiment data, which is in good agreement with each other (**Fig. 5**).

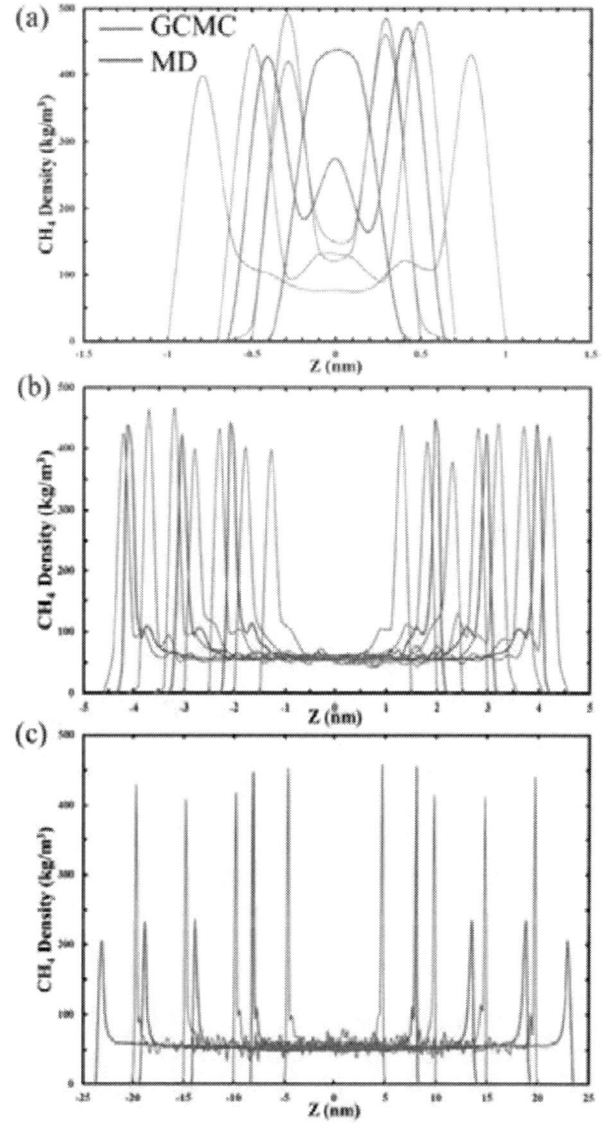

Fig. 10. Density profile for graphite systems using MD and GCMC methods at (a) ~2 nm pore size; (b) 2~10 nm pore size; (c) 10~50 nm pore size. Here, the grid size is 0.2 nm for density profile calculation.

Then, we simulated CH_4 adsorption behavior in graphite system with MD and GCMC simulations, respectively. As in **Fig. 9**, here, the "Bulk" density means the methane average density in the condensed zone. Total density is the methane density included the adsorbed methane on the pore wall. From **Fig. 10**, we can notice that when the pore size is smaller than ~2 nm for two methods, the average density ("Bulk" Density) in the mesopore is higher than average bulk density.

Graphite and Kerogen Models

At last, we have also discussed the limitation on the use of graphite as a kerogen model. For that, we have compared the results for gas density inside the nanopores with graphite and full-atom kerogen slab models, respectively. The pressure and temperature are fixed at 10 MPa and 350 K, while the pore size varies from 0.5 to 80 nm. The simulations have been performed with PME scheme for the dispersion interactions, and 1.5 cut-off value.

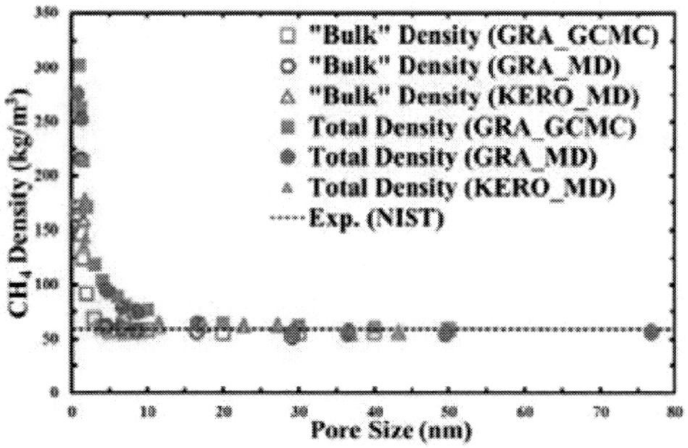

Fig. 11. CH_4 density as a function of pore size in kerogen and graphite models.

We conducted simulations with graphite and kerogen models using MD simulation. In kerogen model, the simulation results show that the methane density in the nanopores with pore sizes larger than ~ 2 nm is almost identical to that in bulk phase at the corresponding pressure and temperature. It also seems to present pore size effect at ~ 2 nm in graphite model, that is, when the mesopore ends up with a micropore size. The adsorption peak is higher in graphite system. This deviation between two models is presumably because there are micropores in kerogen slab (including the interfacial region), where the methane density can be influenced each other in the micropore and mesopore regions. However, care must be taken for the calculation of total adsorption excess, where both the adsorption peak and density deviation in the "bulk" region should be counted. In our research, it was found that the total adsorption excess deviates starting from about 10 nm.

One of the density profiles of two different models is shown in **Fig. 12**. Different from the graphite model, the CH_4 molecule can be adsorbed inside the kerogen bulk. Two small density peaks in kerogen region can be used to indicate the existence of micropores in kerogen slab. **Fig. 12** also shows the definition of mesopore region.

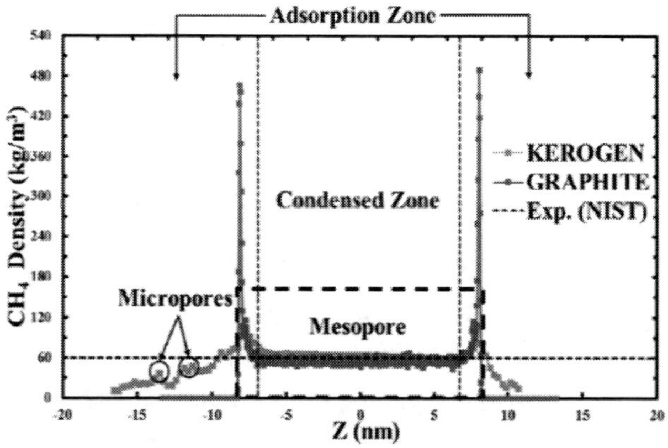

Fig. 12. CH_4 density distribution for two systems with ~16 nm mesopores.

Conclusions

We have focused on some points about the adsorption behaviors simulations.

First, we discussed the limitation on the implementation of a long-range correction scheme due to the use of a cut-off for dispersion interactions.

Second, we have compared the results from MD and GCMC simulations. The input is chemical potential in GCMC simulation, however, pressure along the normal direction as an input parameter in MD simulation. This difference should be verified. We found that the simulation results are essentially identical for two simulation methods.

Finally, we discussed the difference between graphite model and kerogen model. By using MD simulations, we have compared the results for gas density inside the nanopores with graphite and full-atom kerogen slab models, respectively. Under the pressure and temperature conditions we studied, we found that the methane density inside nanopores starts to deviate from the bulk density at around 2 nm for kerogen models, as well as for graphite models. However, if we focus on the total adsorption excess, it deviates starting from about 10 nm. This study solves the controversial raised in previous molecular scale simulations[2–5].

The knowledge that we obtained in this study will help us select an appropriate scheme for investigation of methane adsorption in kerogen nanopores. The results will contribute to the quantitative evaluation of methane adsorption under a certain temperature and pressure for certain kerogen constituents, and enable us to connect the nanoscale simulation and macroscopic experimental results. In the near future, we will calculate the adsorption isotherms using kerogen models and MD simulations.

Acknowledgment

This study has been financially supported by Japan Oil, Gas and Metals National Corporation (JOGMEC), and Arai Science & Technology Foundation. The computations were carried out on the Supercomputing Division, Information Technology Center at The University of Tokyo.

References

1. Milliken, K., Milliken, K.L., Rudnicki, M., Awwiller, D.N., and Awwiller, D. (2011), "Organic Matter-Hosted Pore System, Marcellus Formation (Devonian), Pennsylvania", *AAPG Bulletin* **97** (2): 177–200.
2. Ambrose, R. J., Hartman, R. C., Diaz-Campos, M., Akkutlu, I. Y. and Sondergeld, C. H. (2012), "Shale Gas-in-Place Calculations Part I: New Pore-Scale Considerations", *SPE Journal* **17** (1): 219–229. SPE-131772-PA.
3. Al Ismail, M.I., and Horne, R.N. (2014), "Modeling Adsorption of Gases in Nanoscale Pores Using Grand Canonical Monte Carlo Simulation Techniques", Paper SPE 170948 presented at *the SPE Annual Technical Conference and Exhibition*, Amsterdam, The Netherlands, 27-29 October.
4. Sawa, Y., Liang, Y., Murata, S., Matsuoka, T., Akai, T. and Takagi, S. (2015), "Pore-Filling Nature of CH_4 Adsorption Behavior in Kerogen Nano pores: A Molecular View Based on Full-Atom Kerogen Models", Paper SPE 176999 presented at *the SPE Asia Pacific Unconventional Resources Conference,*

Brisbane, Australia, 9-11 November.

5. Stukan, M., and Abdallah (2015), "Nano-Confined Adsorbed and Free Gas in Shale Reservoirs: A Molecular Dynamic Study", Paper SPE 172589 presented at *the SPE Middle East Oil & Gas Show and Conference*, Manama, Bahrain, 8-11 March.
6. Allen, M.P., and Tildesley, D.J. (2002), "*Computer Simulation of Liquids*", Oxford University Press, Oxford.
7. Frenkel, D. and Smit, B. (2002), "*Understanding Molecular Simulation: From Algorithms to Application, 2nd edition*", Academic Press, San Diego, CA.
8. Abraham, M. J., Murtola, T., Schulz, R., Páll, S., Smith, J. C., Hess, B., Lindahl, E. (2015) "GROMACS: High performance molecular simulations through multi-level parallelism from laptops to supercomputers", *SoftwareX* **1–2**:19–25, 2015.
9. Martin, M. G. (2013), "MCCCS Towhee: a tool for Monte Carlo molecular simulation", *Mol. Simulat.* **39**: 1212-1222.
10. Shinn, J. H. (1983) "From Coal to Single-stage and Two-stage Products: A Reactive Model of Coal Structure", *Fuel* **63** (9): 1187-1196.
11. Kaminski, G., Duffy, E. D., Matsui, T. and Jorgensen, W. L. (1994), "Free Energies of Hydration and Pure Liquid Properties of Hydrocarbons from the OPLS All-atom Model", *J. Phys. Chem.* **98**: 13077-13082.
12. Jorgensen, W.L., Maxwell, D.S., Tirado-Rives, J. (1996), "Development and Testing of the OPLS All-Atom Force Field on Conformational Energetics and Properties of Organic Liquids", *J. Am. Chem. Soc.* **110** (45): 11225–11236.

491172 Brownian Dynamics Simulations on Spontaneous Extension of DNA on Cationic Lipid Bilayers Along Grooved Structures

Abstract:

When DNA adsorbed on grooved glass covered with cationic lipid bilayers, they extend spontaneously along the roots of the grooves where the surface curvature is positive. The phenomenon is spontaneous and thus it suggests there exists a deep energy well for DNA. This energy well can be considered as a strip-like confinement for DNA, and the width of the strip can be inferred from the width of area with positive curvature. However, the experimentally determined relationship between the DNA extension and the strip width somewhat deviates from the theoretical predictions. We suspect the deviation is related to the shape of the energy well that is assumed to be a deep square in the theories. To verify our argument, we use Brownian dynamics with the bead-rod model to simulate DNA trapped by the energy well with different shapes, depths and widths. The shape of energy well was found to have a strong effect on DNA extension. Moreover, when the depth of energy well is fixed, DNA extension was found decreased with increasing the width of energy well. On the other hand, when the width of confinement is fixed, DNA extension gradually increases as the energy well gets deeper. We have also simulated the cases that the depth of energy well decreases with the increasing width of confinement to mimic the real situation in our experiments. The results are in qualitative agreement with the experimental observation. In conclusion, the simulation results support our argument that the energy well is not a deep square in our system. This finding also bridges the gap between the theoretical predictions and experimental observation for DNA behavior in strip confinement.

487709 Design of Advanced Nanomaterials for Electrocatalysis

Abstract:

Hydrogen is a promising energy carrier for the replacement of fossil fuels. The hydrogen cycle in the sustainable energy system involves four important reactions, namely, oxygen reduction reaction (ORR), hydrogen oxidation reaction (HOR), oxygen evolution reaction (OER) and hydrogen evolution reaction (HER). Our research emphasizes the catalysts for these key reactions. Here, we introduce a "model surface - nano verification - catalyst design" research system for discovery and production of advanced nanomaterials for sustainable energy and fuel.

491078 Layered Hybrid Structure of Cobalt/N-Doped Carbon Derived from Metal-Organic Frameworks for Electrocatalytic Hydrogen Evolution

Abstract:

This study reports in situ synthesis of Co/N-Carbon hybrid structures via the pyrolysis of layered cobalt based metal–organic frameworks. As excellent hydrogen evolution reaction (HER) catalysts, the 2D hybrid layered structure composed by ultrafine metallic cobalt nanoparticles and N-carbon, showed a small overpotential of 103 mV (vs RHE) at 10mA/cm2, high cycle durability and considerable long-term stability for HER. The superior electrochemical performances of Co/N-carbon for HER can be attributed to such 2D configuration which includes the low H atom binding energy of metallic cobalt, the high conductivity of carbon and the stability of carbon-encapsulated Co nanoparticles. Therefore, we have worked on the synthesis and characterization of the 2D hybrid structure of Co/N-Carbon derived from metal organic frameworks. The as-prepared 2-D sheet structures retained the well-defined morphology of the initial Co-based MOFs, and cobalt nanoparticles (ca. 7.2 nm) were well embedded on the N-carbon. Due to the integration of the superior features of ultrafine cobalt nanoparticles and N-carbon, the resultant Co/N-Carbon showed enhanced electrocatalytic activity and durability when compared to the commercial Pt/C electrocatalyst at the alkaline media. More importantly, Co/N-carbon's electrocatalytic activities in terms of overpotential at 10mA/cm2 for hydrogen evolution have been more advantageous over other nanocarbon-based electrocatalysts.

Acknowledgements: Financial support from Interdisciplinary Graduate School, Nanyang Technological University Singapore is gratefully acknowledged.

References

[1] Walter, M. G.; Warren, E. L.; McKone, J. R.; Boettcher, S. W.; Mi, Q.; Santori, E. A.; Lewis, N. S. Chem. Rev 2010, 110 (11), 6446-6473.

[2] Hou, Y.; Wen, Z.; Cui, S.; Ci, S.; Mao, S.; Chen, J. Adv. Funct. Mater 2015, 25 (6), 872-882.

492501 Comparison of Hydroxide-Mediated and Direct Mechanisms for Alkaline Hydrogen Electrocatalysis

Abstract:

It has long been recognized that the reaction rates of the hydrogen oxidation and hydrogen evolution reactions (HOR and HER) are slower in basic than acidic electrolytes, even though the surface intermediate of adsorbed hydrogen is independent of solution pH. Understanding the root of this observation is critical to designing catalysts for a multitude of electrochemical reactions with relevance to energy conversion and storage. In this work, we undertake a fundamental investigation to determine if adsorbed hydroxide plays an active or spectator role in HER/HOR kinetics.

One proposed explanation for the pH dependence of HER/HOR kinetics, suggested by the research groups of both Gasteiger and Yan, is that in base, hydroxide ions stabilize the Pt-H bond for stronger binding and slower catalysis [1,2]. Central to this hypothesis is an experimentally measured shift with pH in the peak potential for hydrogen underpotential deposition (H-UPD) on the (100) and (110) steps on the Pt surface. However, the Janik and Koper groups have found that the shift in H-UPD potential is caused not by changes to hydrogen adsorption, but instead by competitive adsorption of hydroxide in the presence of alkalai cations [3,4]. Markovic et al have proposed that such hydroxide adsorption is actually necessary to facilitate water recombination/dissociation in base, and that oxophilic sites such as Ni(OH)2 generate a bifunctional mechanism for efficient alkaline hydrogen electrocatalysis [5].

In this work, we investigate specifically the hypothesis that adsorbed hydroxide is an active participant in the alkaline HER and HOR by combining electroanalytical techniques with microkinetic modeling. We develop expressions for time-dependent current and voltage as a function of hydrogen and hydroxide binding strength for the indirect (hydroxide-mediated) and direct (hydroxide-as-spectator) Volmer steps in base. These expressions are compared to experimental cyclic voltammograms of H-UPD on polycrystalline platinum. The free energy of hydroxide adsorption is controlled experimentally by the electrolyte cation via changes in cation solvation strength [6]. Varying the sweep rate results in greater peak separation that can be used to extract rate constants for the adsorption reaction via traditional electroanalysis [7].

Our computations show that at both slow and fast scan rates, either the direct or indirect mechanism can describe the peak potential splitting and peak current. However, hydroxide binding strength affects the mechanisms differently. In the indirect mechanism, stronger hydroxide binding reduces overpotential and, therefore, peak splitting. In the direct mechanism, stronger hydroxide binding decreases available surface sites, leading to lower exchange current densities and greater overpotential. Comparison with experiment shows that on the (110) and (100) facets of polycrystalline platinum, moving from 0.1M KOH to 0.1M LiOH decreases the peak potential location, but increases peak potential splitting. This trend shows that stronger hydroxide binding has detrimental effects on H-UPD kinetics, and that the direct mechanism is therefore more likely to dominate observed behavior.

Our results strongly suggest that adsorbed hydroxide serves as a competitive spectator in the alkaline Volmer step, and that the bifunctional HOR/HER mechanism plays only a minor role at best. This study contributes to resolving a long-standing paradox in electrocatalysis and surface science by determining that oxophilicity is not an accurate descriptor for alkaline hydrogen electrocatalysts. Other parameters, such as water orientation and non-covalent interactions, must play a greater role in overall activity. Future work will discuss these and other factors in the relation between H-UPD kinetics and HER/HOR catalyst activity .

(1) Durst et al. Energy Environ. Sci. 2014, 2255.

(2) Sheng et al. Nat. Commun. 2015, 5848.

(3) McCrum et. al. J. Phys. Chem. C, 2015, 457.

(4) Van Der Niet et al. Catal. Today 2013, 105.

(5) Strmcnik et al, Nat. Chem. 2013, 1.

(6) Danilovic, N. et al. Electrocatalysis 3, 2012, 221.

(7) Angerstein-Kozlowska et al. J. Electroanal. Chem. Interfacial Electrochem. 1979, 1.

497033 Alkaline Hydrogen Evolution from Ni–Mo Intermetallics with High Mo Content

Abstract:

The transition from acidic to alkaline conditions promises considerable cost savings for water electrolysis systems based on polymer ion-exchange membrane electrolytes. Binary transition metal alloys containing Ni and Mo have been known for decades to exhibit excellent performance as alkaline water electrolyzer cathodes,[1] but the specific reason that these alloys are more active toward the hydrogen evolution reaction (HER) than either Ni or Mo alone remains the subject of some considerable debate.[2] Moreover, previously reported mass-specific activities for Ni–Mo alloys are still at least 100-fold lower than that of Pt, which makes them difficult to deploy in a practical electrolysis system that operates at suitably high energy efficiency and power density.

Prior work on Ni–Mo electrocatalysts has suggested that the alloy phase exhibits monotonically increasing HER activity as the bulk Mo mole fraction is increased.[3] However, the accessible composition range for binary Ni–Mo alloys is constrained by the solid solubility of Mo in the Ni fcc lattice, which is only ~25 mol% at 1000 °C. Surface-science experiments on bulk $Ni_{0.8}Mo_{0.2}$ alloys have indicated that the topmost atomic layer becomes considerably enriched in Mo, up to approximately equimolar Mo and Ni.[4] Thus, the question remains as to how much Mo is present in the most active Ni–Mo binary HER electrocatalyst surface, let alone what is the mechanism for enhanced activity.

In this presentation, I will discuss recent and ongoing work aimed at understanding the influence of bulk composition on the activity of nanostructured Ni–Mo alkaline HER catalysts.[5] We synthesized unsupported powders of Ni, $Ni_{0.92}Mo_{0.08}$ alloy, and the intermetallic compound Mo_7Ni_7. Mass-specific activity was found to be highest for the 8 mol% Mo alloy, but these data were clearly convoluted by catalyst morphologies, as the 50 mol% Mo intermetallic exhibited very low surface roughness. Attempts to account for electrochemically active surface area using a straightforward method based on electrochemical impedance spectroscopy (EIS) yielded unexpected results that conflicted with morphological analysis from electron microscopy. We have therefore postulated that Mo undergoes a pseudocapacitive redox transition near the onset potential for the HER, leading to erroneous results from EIS measurements. These results further implicate a plausible bifunctional catalytic mechanism that explicitly involves Mo redox transitions.

(1) Brown, D. E.; Mahmood, M. N.; Man, M. C. M.; Turner, A. K. *Electrochimica Acta* **1984**, *29* (11), 1551–1556.

(2) McKone, J. R.; Marinescu, S. C.; Brunschwig, B. S.; Winkler, J. R.; Gray, H. B. *Chem. Sci.* **2014**, *5* (3), 865–878.

(3) McKone, J. R.; Sadtler, B. F.; Werlang, C. A.; Lewis, N. S.; Gray, H. B. *ACS Catal.* **2013**, *3* (2), 166–169.

(4) Soriaga, M. P.; Baricuatro, J. H.; Cummins, K. D.; Kim, Y.-G.; Saadi, F. H.; Sun, G.; McCrory, C. C.; McKone, J. R.; Velazquez, J. M.; Ferrer, I. M.; I, C., Azhar. *Surf. Sci.* **2015**, *631*, 285–294.

(5) Csernica, P. M.; McKone, J. R.; Mulzer, C. R.; Dichtel, W. R.; Abruña, H. D.; DiSalvo, F. J. *ACS Catal.* **2017**, 3375–3383.

485062 Increased Photocatalytic Activity of TiO_2 Nanoparticles with Defects for Sustainable Hydrogen Production

Abstract:

Hydrogen gas is touted as a clean-burning energy carrier that can be used in fuel cells for stationary and transportation power at higher efficiency than conventional internal combustion engines. The primary source of H_2today is from nonrenewable natural gas-derived methane steam reforming (MSR). MSR suffers from being a highly endothermic and endergonic reaction (though less so than water splitting), requiring significant energy input as heat generated from burning half of the feed methane, limiting atom efficiency and emitting greenhouse gases. Photocatalytic MSR circumvents the high heat requirement and lowers the reaction temperature, thereby avoiding coke buildup problems as well.

Titanium dioxide (TiO_2) is of particular interest in photocatalysis due to its high photoactivity under ultraviolet illumination and its stability under catalytic reaction conditions. Together with supported metal nanoparticles, TiO_2has been used to catalyze a number of important chemical and photochemical reactions.

We have synthesized defect-rich anatase TiO_2 that shows unique selectivity for hydrogen formation from methane. We have also demonstrated that ultra-violet light illumination boosts the activity and selectivity of the catalyst for hydrogen formation. We have demonstrated hydrogen yield of 27% under steady state conditions using a nickel cocatalyst. Through a combination of experimental and computational techniques, we investigate the mechanisms of defect formation and how these defects activate methane and selectively catalyze hydrogen formation.

486781 Transition Metal Nitride Core-Noble Metal Shell Nanoparticles As Highly CO Tolerant Catalysts

Abstract:

Introduction

Core-shell nanoparticles (NPs) have become increasingly attractive as a catalytic design platform due to their unique ability to simultaneously enhance the activity, increase the durability, and reduce the loading of noble metal (NM) catalysts. However, these materials have been limited by the lack of independent control over the core and shell compositions, shell thickness, and particle size [1]. To address these issues, our group recently developed a novel method for the synthesis of transition metal carbide (TMC) core-NM shell (denoted NM/TMC) NPs with precise control over the core-shell architecture [2]. Group 4-6 TMCs are earth-abundant compounds with high sinter-resistance, electrical conductivity, and chemical stability, making them ideal candidates for core materials [3].

Transition metal nitrides (TMNs) share all of these favorable material properties of TMCs but have rarely been investigated in core-shell materials. Here, we present the ability to synthesize NM/TMN core-shell NPs with varying core and shell compositions and controllable shell thickness via nitridation of the corresponding NM/TMC. The impacts of titanium tungsten nitride (TiWN) and titanium tungsten carbide (TiWC) cores on the electronic properties of a Pt shell are compared, and the effect of these differences on catalytic properties was investigated by measuring the activity for the hydrogen oxidation reaction (HOR) under the presence of CO to assess the CO tolerance. The results illustrate how the use of nitride versus carbide cores offers an effective way to tune the reactivity of NM catalysts.

Materials and Methods

NM/TMC NPs were nitrided under NH_3 at 800°C for 3 h to form NM/TMN NPs. Synthetic details for NM/TMC are provided elsewhere [2]. All materials were supported on carbon. Electrochemical measurements were performed at 30°C on a rotating disk electrode with a Ag/AgCl reference and platinum counter electrode. All potentials are reported versus a reversible hydrogen electrode (RHE). For comparison, catalyst loadings were targeted to have roughly equal Pt surface areas. HOR scans were performed in H_2-sat 0.1 M $HClO_4$ at 2 mV/s and 1600 rpm with iR compensation. To test CO tolerance, the working electrode was first held at 0.025 V until the electrolyte equilibrated with 1000 ppm CO before starting the HOR scan.

Results and Discussion

Pt/TiWC NPs with varying Pt coverage were synthesized using the method previously reported by our group and transformed into Pt/TiWN NPs via nitridation at 800°C. The high sinter-resistance and strong interfacial bonds of Pt/TiWC prevent particle agglomeration and preserve the core-shell structure while the core is converted into a nitride, which is confirmed by XRD, STEM-EDX, TEM, XPS, and CHNS analysis (Figure 1A).

XPS spectra of the core-shell materials reveal shifts in the core electron binding energies, called core level shifts (CLSs), for both the W 4f and Pt 4f signals. These CLSs indicate a significant change in the Pt electronic structure, as shifts in the core level energies of metal overlayers are reflected in similar shifts in the valence d-band center, which is known to correlate strongly with adsorbate binding energies and thus the reactivity [4-5]. The XPS results suggest downward shifts in the Pt d-band center for the core-shell materials compared to a pure Pt surface and an even lower d-band center and weaker binding for Pt/TiWN compared to Pt/TiWC.

To probe the difference in adsorbate binding and reactivity among our core-shell and commercial Pt (denoted Pt_{comm}) catalysts, we used CO as a model adsorbate and measured the HOR activity in the presence of CO (Figure 1B). Strong CO binding is a common problem for Pt catalysts, poisoning the surface and impeding reactivity. In the presence of 1000 ppm CO, the HOR activity of Pt_{comm} is completely suppressed by CO adsorption and is not recovered until above 0.5 V when CO can be oxidized off the surface. In contrast, both core-shell materials retain HOR activity at potentials as low as 0.1 V, thus demonstrating dramatically improved CO tolerance and reduced CO bond strength. Furthermore, 2-3 monolayer (ML) Pt/TiWN displays even greater CO tolerance than 2-3 ML Pt/TiWC, indicative of weaker CO binding as predicted from the XPS CLSs.

Figure 1. A) STEM-EDX map of 2-3 ML Pt/TiWN. B) HOR polarization curves in H_2-sat 0.1 M $HClO_4$ at 1600 rpm with (solid line) and without (dotted line) 1000 ppm CO.

Significance

In addition to Pt/TiWC, the synthetic method can be applied to core-shell NPs of varying NM/TMC compositions, demonstrating a promising way of synthesizing new core-shell architectures with tunable electronic properties that can serve as next generation catalysts.

References

1. Yang, H. *Angew. Chem. Int. Ed.* 50, 2674-2676 (2011).

2. Hunt, S.T., Milina, M., Alba-Rubio, A.C., Hendon, C.H., Dumesic, J.A., and Román-Leshkov, Y. *Science* 352, 974-978 (2016).

3. Zhong, Y., Xia, X., Shi, F., Zhan, J., Tu, J., and Fan, H.J. *Adv. Sci.* 3, 1500286 (2016).

4. Weinert, M., and Watson, R.E. *Phys. Rev. B* 51, 17168-17180 (1995).

5. Kitchin, J.R., Nørskov, J.K., Barteau, M.A., and Chen, J.G. *J. Chem. Phys.* 120, 10240-10246 (2004).

502612 Understanding the Electrochemical Behavior of a Thin, Flexible Micro Fuel Cell

Abstract:

Portable fuel cell systems in the power range of few milliwatts to a few hundred watts have become the focus of vigorous research and development activities. Fuel cells promise to provide more reliable, longer portable power than batteries. Micro fuel cell (mFC) technology can take advantage of fuels with energy densities of an order of magnitude higher than the energy stored in batteries. The air breathing cathode reduces the size and weight of the cathodic plate, and simplifies system requirements since oxygen inlet is fully passive. On the other hand, working conditions for the air breathing cell are more demanding due to the absence of an air flow behind the cathode gas diffusion layer to assist the removal of water.

The importance of micro-structures on polymers for micro fuel cell fabrication is enormous, particularly when considered as a low-cost alternative to the silicon- or glass-based MEMS technologies, for disposable small electronic gadgets. Many polymer-based microfabrication techniques via microinjection molding, casting, and micro-hot embossing have been developed.

Hot embossing provides advantages such as a relatively low-cost for embossing tools, simple operation and higher accuracy in the replication of small features. The embossing master stamp can be a silicon wafer, glass, electroplated nickel mold or other stamp with microfeatures. The hot embossing process introduces less residual stress in the polymer because the polymer stretches for a very short distance from the substrate into microstructure during hot embossing. In addition, the temperature variation range for the polymer is smaller than that required in injection molding, thus can reduce shrinkage during cooling and the friction forces acting on the microfeatures during de-molding. However, still micro-hot embossing is facing challenge in terms of process feasibility, since it is difficult to make the polymer to fill completely into microfeatured geometry of high aspect ratio and it is also delicate to separate the embossed structures from the mold without breakage.

An air-breathing micro fuel cell with direct hydrogen flow through porous anode electrode is realized and reported in this study. The performance of the microdevice with embedded flow channels and electrodes is characterized in ambient conditions. To achieve this goal, we implement a SU-8 microchannel stamp to transfer a pattern into a Nafion 1110 membrane by hot embossing. Nafion 212 thermally seals the whole device as a blanket top layer. The fabrication process of microchannels and micro fuel cell construction is evaluated with optical and scanning electron microscopy (SEM) step by step. Variations of hydrogen feed rate on performance were investigated. The shared-anode characteristic design of the double-sided cell is further studied by separate performance and electrochemical impedance spectroscopy (EIS) measurements taken from each side and compared with data for the complete cell. The maximum power density per superficial (footprint) area is 68.4 mWcm-2 for the shared anode stack. In the present micro fuel cell architecture, the external package is constructed of Nafion polymer, which offers a thin, light and flexible energy source with low cost of manufacturing. The device performance offers a high volumetric and gravimetric energy density for portable applications.

498780 Metastable Phase Transitions

Abstract:

Several model systems exhibit a low-temperature first-order transition between two disordered phases that are metastable with respect to an ordered phase. The preponderance of experimental evidence is consistent with the possibility that water may behave in this way. The confluence of structural, kinetic and thermodynamic factors that make this interesting phase behavior possible is an active area of inquiry. I will present a theoretical and computational overview of metastable phase transitions, including discussions of finite size effects, the interplay of nucleation and relaxation times, the stability of interfaces, and metastable criticality.

498824 Computational Investigation of Multipolar Colloidal Particle

Abstract:

Colloids with anisotropic charge distributions hold promise for creating a number of useful new materials including optic materials with novel symmetries, materials for information storage, and dampers for controlling vibrations in structures. Experimentally manipulating properties of anisotropic particles in a controlled fashion can sometimes be difficult and is time consuming. The search for novel anisotropic colloidal materials can be enhanced and even guided by simulations of colloidal system assembly. We report the results of two projects. In the first we use discontinuous molecular dynamics (DMD) simulation to investigate how the internal charge separation of anisotropic, dipolar colloidal rods with 4:1 aspect ratio affects their phase behavior. The differences between the head-tail and side-side ordering displayed in these simulations could be exploited to design gels whose properties (especially those related to percolation: viscosity, conductivity, etc.) are responsive to external stimuli. In the second project we perform quasi-2D Monte Carlo simulations of systems of colloidal spheres with off-center, extended dipoles. These colloidal spheres are found to aggregate into cyclical structures whose interior angle is equal to the angle between the charges on a single dipolar sphere.

498803 Interfacial Tensions from SAFT: Connecting Equations of State to Molecular Simulations

Abstract:

A description of fluid systems with molecular-based algebraic equations of state (EoSs) and by direct molecular simulation is common practice in chemical engineering and the physical sciences, but the two approaches are rarely closely coupled. The key for an integrated representation is through a well-defined force field and Hamiltonian at the molecular level. We discuss the latest developments in a top-down representation of fluids, where we use an accurate EoS to link the macroscopic properties of the fluid and the force-field parameters, with a particular focus on a coarse-grained formulation of the statistical associating fluid theory (SAFT-Î³ Mie). The EoS is used to estimate the parameters of the Mie force field, which can then be used with confidence in direct molecular simulations to obtain structural and dynamical properties that are otherwise inaccessible from the EoS, including the interfacial tension, which is a property not included in the original parametrization.

A further theoretical framework is developed based on scaling arguments applied to the influence parameter of the van der Waals theory of inhomogeneous fluids. The molecular model stems from the application of the square gradient theory to the SAFT-Î³ Mie equation of state. The theory is validated against computer simulation results for homonuclear pearl-necklace linear chains made up to six Mie (Î»-6) beads with repulsive exponents spanning Î» = 8 to 44 by combining the theory with a corresponding states correlation to determine the intermolecular potential parameters. We show how this can be employed to provide a predictive tool to determine interfacial tensions for a wide range of molecules including hydrocarbons, fluorocarbons, polar molecules, among others. The proposed methodology is tested against comparable existing correlations in the literature, proving to be vastly superior, exhibiting an average absolute deviation of 2.2%

493586 Generalized Gibbs Free Energy of Confined Nanoparticles

Abstract:

The nanoparticles generally show abnormal properties compared to those in the bulk phase, and they exhibit significant potential in various applications such as catalysis and energy conversion. However, the theoretical work for describing the properties of nanoparticles is limited with poor prediction capacity. In this work, the Gibbs free energy was studied, from both marcoscope and microscope, predictive models were proposed to study the thermodynamic properties of nanoparticles with a generalized description of the Gibbs free energy considering the effects of surface-energy and the substrate contacted. The proposed model from the microscope was based on the corresponding states theory to describe the effect of the substrate on the Gibbs free energy of nanoparticles, in which the molecular parameter with a generalized parameter was obtained from the melting point of metals due to sufficient information. The comparison with the new measured experimental results proves the reliability of the model prediction.

498818 Molecular Modeling of Polymeric Systems

Abstract:

In this talk, I will present the results of three exciting opportunities to comprehensively investigate nano- materials through the use of molecular simulations. Tailoring and optimization of these materials in- clude: 1) shape-directed packing of rigid 2D-like polymer chains (PIM-TMN-Trip), that provides ul- trapermable polymers with highly interconnected micropores facilitating rapid and selective gas transport of polymers, 2) ultra-thin molecular layer-by-layer polyamide membranes which can play a crucial role in the design and understanding of the next generation of reverse osmosis membranes, and 3) bio-conjugates, protein-polymer systems, which will benefit from a predictive knowledge of how an attached polymer will interact with a protein and its environment.
In all cases, it is anticipated that experimentalists will avoid the peril of performing a large number of difficult synthetic chemistry that would result in no enhancement of the target property. Concurrent to the above goals, is the generation of large-data sets and the extraction of critical information from that data (e.g., structure factors to understand the intrinsic correlations between structure and properties/ behavior) to catalyze materials breakthroughs.

492444 Effect and Regulation of Surface Wettability on Molecular Transport and Reaction

Abstract:

Whereas massive experimental works have been contributed into the investigation of surface wettability effect on fluid transport and reaction, theoretical study on molecular scale is still few. The obstacle mainly originates from the lack of both microscopic description for the surface wettability and accurate theoretical method for quantifying the effect. Here, we demonstrate that the surface wettability effect on molecular transport and reaction can be quantitatively accessed from molecular perspective. The surface wettability is characterized with the ratio of fluid-fluid interaction to fluid-substrate interaction, whose relation to contact angle is given by an empirical yet analytical equation. This mesoscale relation, connecting macroscopic measurement and microscopic modeling, allows for seeking regulation method by identifying the thermodynamic and/or dynamic properties of fluid nearby substrate, and these properties can be well predicted with the help of modern statistical theories including classical density functional theories and molecular simulation. By discussing two specific examples associated with the flux through nanopores and conversion rate of a surface catalytic reaction, we show that this route is generic and has directive and practical importance to the design of efficient processes.

501332 Dynamics of Collective Endothelial Cell Migration in Response to Fluid Shear Stress

Abstract:

The endothelial cells that line blood and lymphatic vessels undergo complex, collective migration and rearrangement processes during embryonic development, and are known to be exquisitely responsive to fluid flow. At present, the molecular mechanisms by which endothelial cells sense fluid flow remain incompletely understood. Here, we report that both the G-protein-coupled receptor sphingosine 1-phosphate receptor 1 (S1PR1) and its ligand sphingosine 1-phosphate (S1P) are required for the collective upstream migration of human lymphatic endothelial cells. These findings are consistent with a model in which S1P and fluid flow act synergistically to guide development and remodeling in the lymphatic circulatory system.

493803 Rupture of Cancer Cells Under Microcirculatory Conditions

Abstract:

Survival of circulating tumor cells (CTCs) in blood circulation is essential for cancer metastasis. However, fluid forces in microcirculation have been shown to cause cell rupture, lending to the notion that such mechanical trauma of CTCs is responsible for metastatic inefficiency. Literature evidence for tumor cell death due to mechanical damage remains ambiguous and moreover the physical mechanisms of cell rupture are poorly understood. Here, we develop a microfluidic capillary model of microcirculation that contains an array of flow bifurcations and investigate rupture of tumor cells. We find that when a cancer cell arrives at a narrow bifurcation, it is strongly deformed and stretched into the bifurcating branches, eventually causing a cytosol-filled membrane-bound vesicle to be pinched-off. To identify the essential features of the mechanism underlying the rupture process, we used a library of drugs that (i) either enhance or inhibit the polymerization of actin and microtubules (ii) inhibit myosin II and decrease cortical tension, and (iii) condense and decondense nuclear chromatin. Results from these studies show that the mechanical properties of the membrane and cytoskeletal filaments play an important role in regulating the time-scale for rupture of tumor cells, but not the nucleus. In addition, the vesicle size was not sensitive to any of the drug treatments indicating that the cytosol content determines the vesicle volume. We develop a model that is a result of balance between external fluid pressure and stresses internal to the cell, which captures the key experimental observations. Taken together, our results support a picture, where under mechanical stress, the cytosol flows through the elastic cytoskeleton and fills the membrane that is detached from the cortex, similar to cellular blebbing. Finally, we find that a very small fraction of the ruptured tumor cells undergo apoptosis, suggesting that CTCs can survive the impact of fluid forces in microcirculation and that mechanical trauma is not a significant cause for metastatic inefficiency.

498114 Transport of Nanoparticles in the Brain

Abstract:

We report the results of experimental studies of the motion of fluid and nanoparticles in the brain. This is an important problem for drug delivery in neural tissue, and it is directly relevant to convective-enhanced delivery (CED), a treatment in which fluid containing therapeutic molecules is injected directly into the interstitium of the brain. In some strategies, the therapeutic molecules are packaged in nanoparticles and released over time after the nanoparticles have reached targeted tissue. Numerous studies show that the efficacy of CED therapy depends on a uniform spatial distribution of therapeutic throughout the afflicted tissue, which is difficult to achieve in clinical practice. Many investigators have found indirectly that fluid motion in CED is enhanced in perivascular spaces, which are thin regions of relatively high hydraulic conductivity immediately outside blood vessel walls.

Using two-photon excitation microscopy, we have measured the velocities of fluid and nanoparticles in real time during CED into the cortex of living, anesthetized rodents. Two-photon microscopy is especially useful in highly scattering materials such as tissue, and in this case the motion the motion of fluorescent nanoparticles and fluid can be detected deep into the cortex. During CED infusion at constant volumetric flow rate into the cortex, we have measured simultaneously the velocity of fluid and rigid fluorescent nanoparticles in the interstitial space and in the perivascular space. Nanoparticles ranged in size from 20 to 200 nm.

We have found that nanoparticle motion is significantly hindered in the interstitial space; the fluid velocity can exceed the nanoparticle velocity by more than an order of magnitude for larger nanoparticles. However, the difference in velocity is much less for nanoparticles in the perivascular space. Furthermore, the velocity of a nanoparticle in the perivascular space does not decrease with distance from the infusion source, suggesting that motion in the perivascular space resembles flow in a closed annular tube rather than motion from a point source in a porous medium. These results could help explain the distribution of nanoparticles in CED therapy and could be used to improve numerical simulations of the therapy.

495106 Towards Understanding Nanoparticle Diffusion in Synovial Fluid Analogues

Abstract:

The use of nanoparticles in complex biological environments has motivated study of their dynamics in concentrated mixtures of macromolecules. Synovial fluid, a lubricant found in joints, is one such complex fluid, with macromolecules like hyaluronic acid and lubricin as major constituents that define its structure and function. Fundamental knowledge of how nanoparticles diffuse in synovial fluid would help engineer better drug delivery agents and also improve basic understanding of the structure of synovial fluid. Here, we study the translation and rotation of magnetic nanoparticles in synovial fluid analogues – concentrated suspensions of hyaluronic acid.

Hyarulonic acid (HA) solutions at concentrations where the HA molecules are well separated (i.e., the dilute regime) and where they overlap (i.e., the semi-dilute regime) were prepared as a first approach to mimic the composition of synovial fluid. Cobalt ferrite particles of two hydrodynamic sizes (40 nm and 240 nm) with an outer coating of polyethylene glycol were suspended in these HA solutions to study the effect of particle size on hydrodynamic drag exerted by the complex fluid matrix. The dynamic response of the particles to an applied alternating magnetic field was analyzed to obtain their rotational diffusion coefficients and, through the Stokes-Einstein relation, the apparent viscosity of the fluid. We found that the average viscosities obtained from nanoscale measurements were similar to that of water and did not correspond to the viscosity of the bulk fluid as measured using a rheometer, for HA solutions in both the dilute and semi-dilute regimes. Control measurements of rotation of the particles in glycerol solutions indicated that the viscosity determined by rotation of the particles in simple fluids was similar to the bulk viscosity.

To complement these measurements, semi-quantitative estimates of the relative translational hydrodynamic drag on the nanoparticles was obtained through measurements of the rate of magnetic capture. Magnetic nanoparticles were subjected to an external magnetic field gradient and their translation was monitored by the amount of particles recovered as a function of time. The rate of translation of particles in solutions in the semi-dilute regime was found to be slower than for particles in solutions in the dilute regime, indicating greater viscous drag experienced by the particles in the more concentrated HA solutions. However, comparison to rates of capture in control glycerol solutions with similar viscosity suggests that the translational hydrodynamic drag experienced by the nanoparticles in the HA solutions is lower than expected based on their macroscopic viscosity.

These studies suggest that rotational motion of nanoparticles below 240 nm in diameter is unimpeded by the HA network in these synovial fluid analogues, whereas translation of the particles does appear to be impeded by the HA network, albeit the effect is much less pronounced than expected based on macroscopic rheological measurements. These observations provide insight into the dynamics of nanoparticles in complex biological fluids and can serve to develop novel nanoparticle-based drug delivery and diagnostic technologies applicable in joint diseases.

493889 Diffusion of Concentrated Macromolecules within Living Cells

Abstract:

The diffusion of concentrated macromolecules, protein, RNAs, etc., in living cells can have significant effects on the kinetics of biological processes, including the collapse of polypeptide chains, enzyme activity, and signal transduction near cell membranes. Such diffusion in a crowded environment has been measured by various experimental techniques, and appears to be greatly reduced compared to that in dilute conditions. In order to understand the origins of this reduction, Brownian dynamics (BD) simulations have been used to examine the motion of macromolecules. In past studies, however, the long-range and many-body hydrodynamic interactions (HI) under confinement have often been over-simplified, because of the complexity of the calculation and their considerable computational costs. Recently, we have developed a high-performance parallelized finite-element method based on the GGEM-IBM technique[1], which employs Fixmanâ€™s integration scheme without explicitly constructing a diffusion matrix or performing matrix decompositions. With this technique, which is of O(N), we have performed HI-BD simulations to study the effects of HI, domain sizes, particle concentrations, etc. on the diffusion of macromolecules in confined spherical cavities. We believe that these studies will help provide an improved description of macromolecular dynamics in crowded environments, including cells.

[1] X. Zhao, J. Li, X. Jiang, D. Karpeyev, O. Heinonen, B. Smith, J. P. Hernandez-Ortiz, J. J. de Pablo, Parallel O(N) Stokesâ€™ solver towards scalable Brownian dynamics in general geometries. *The Journal of Chemical Physics*. Submitted (2017)

501474 Platelet Margination, Adhesion, and Activation in Secondary Flows Are Necessary for Thrombus Propagation in an *in vitro* Model of Venous Thrombosis

Abstract:

Background: The histology of venous thrombi implies a role for blood flow in thrombus propagation. The alternating layered structure of red, fibrin-rich regions that begin at the vessel wall in the valve sinus followed by white, platelet-rich regions, called the lines of Zahn, suggests a mechanism of platelet accumulation to the initial fibrin-rich regions. The geometry of the valve sinus defined by a large cavity distal to the expansion created by the valve leaflets yields unique flow patterns. Flow through fixed venous valves of dogs shows flow separation that results in a large primary vortex adjacent to the valve cusps, and a secondary vortex in the deepest recess of the valve pocket. The fluid velocity in this secondary vortex is extremely slow and corresponds to the most hypoxic area of the valve sinus. Vortical flows in the valve sinus have also been observed by ultrasound in human venous valves. Platelets are essential for thrombus formation under flow and thrombus growth in murine models of venous thrombosis (VT). The murine models used in VT however do not capture the vortical flows and geometry that are characteristic of human valves where VT initiates. The objective of this study was to determine the importance of platelets on thrombus propagation an in vitro model of VT that mimics the hemodynamics of human venous valves. Methods: The flow chamber consists of a 150 μm x 150 μm channel that undergoes a 1:3 expansion, which is similar to the stenosis ratio of venous valves, with a varying undercut angle (90, 120, 135 or 150°) to mimic different positions of a valve leaflet. In order to match flow vortices in our system with human venous valves, we used the Reynolds number (Re) to scale our flow rates. The viscosity of blood is a as function of hematocrit and the channels size. Here, we use the semi-empirical equation from Pries et al.[1] to calculate the viscosity in order to appropriately match Re between flow conditions. Velocity fields and streamlines for the device were calculated using commercially available computational fluid dynamics software (COMSOL). The streaklines of platelet sized fluorescent beads (2 μm) mixed with red blood cells (RBC) in buffer were used to characterize the flow field for Re of 1-100. For experiments where blood clots are formed, tissue factor (TF) was adsorbed in the valve sinus in order to initiate the coagulation cascade. Human whole blood was separated into platelet rich plasma (PRP) and packed RBC and reconstituted to hematocrits of 0, 0.2, 0.4 or 0.6. Platelets were labeled with DiOC6 prior to and/or with annexin V (phosphatidylserine stain) after an experiment and exogenous Alexa 555-fibrinogen was added to visualize fibrin. Thrombus formation was measured by confocal microscopy as plasma, plasma with RBC, PRP, or reconstituted blood (PRP and RBC) was perfused over 30 minutes at flow rates that yield primary and secondary vortices similar to those measured in humans. Results: CFD simulations and streakline imaging showed flow separation for Re > 1 and the presence of a primary vortex for all expansion angles near the valve cusp. A secondary vortex was observed for expansion angles of 135° and 150° in the valve pocket. The size of the secondary vortex was more strongly dependent on expansion angle than Re. RBC were required for significant quantities of platelets and fluorescent beads to enter into the primary vortex, suggesting the need for platelet margination upstream of the expansion. Plasma or RBC suspensions without platelets resulted in a fibrin gel that was confined to the deepest part of the pocket that coincided with the secondary vortex. Both platelet and RBC were required for the thrombus to grow out of the valve sinus over and into the main flow channel. It appears that there is a critical platelet flux necessary to support additional coagulation, and that flux is dependent on RBC-platelet collisions. The thrombus that forms with reconstituted blood shows a layered structure whereby a dense layer of platelets in the primary vortex adhered to the initial fibrin gel formed in the secondary vortex. This dense platelet layer is followed by another fibrin rich layer. These data suggest that fibrin adhered platelets support coagulation beyond the near-wall region coated with TF. Indeed, platelet adhered at this interface are positive for the procoagulant lipid phosphatidylserine as measured by annexin V staining.

Conclusions: Immobilized TF can initiate coagulation in a model venous valve sinus within deep regions defined by a secondary vortex, but platelets and RBC are necessary to spatially propagate a thrombus. Platelets entrained in the primary vortex in the valve sinus, a process that is enhanced by RBC, adhere in a dense layer to fibrin. This platelet layer supports additional coagulation, and the formation of a fibrin rich thrombus that grows beyond the valve sinus.

[1] A. R. Pries, D. Neuhaus, and P. Gaehtgens, "Blood viscosity in tube flow: dependence on diameter and hematocrit," *American Journal of Physiology - Heart and Circulatory Physiology*, vol. 263, no. 6, pp. H1770–H1778, Dec. 1992.

The Importance of Rheology in Blood Flow Modeling

Jeffrey S. Horner, Antony N. Beris, Norman J. Wagner, and Donna S. Woulfe

jshorner@udel.edu
Chemical &Biomolecular Engineering

University of Delaware

Blood Makeup & Complex Rheology

University of Delaware

- Human Blood
  - Plasma (~55% by volume)
    - Fibrinogen (2-4 g/L)
    - Albumin (35-50 g/L)
  - Erythrocytes (~45% by volume)
  - Leukocytes (~0.25% by volume)
  - Platelets (~0.50% by volume)

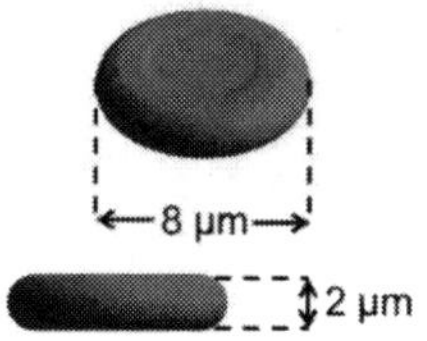

Shear Thinning

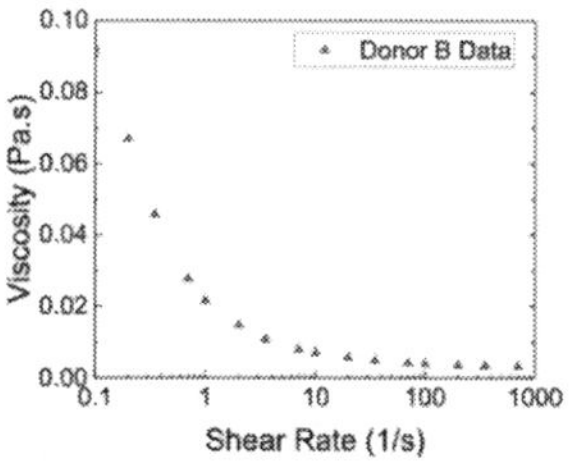

Viscoelasticity

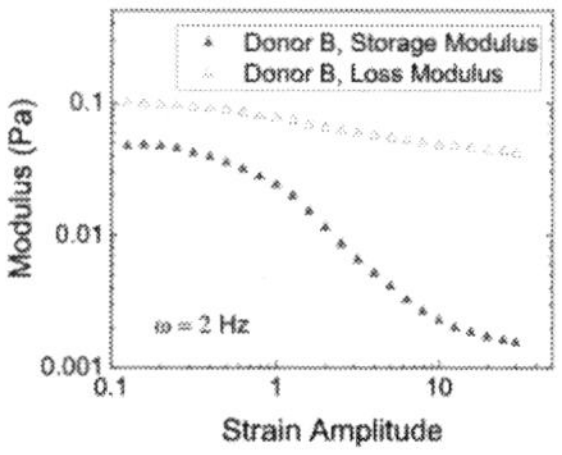

Thixotropy

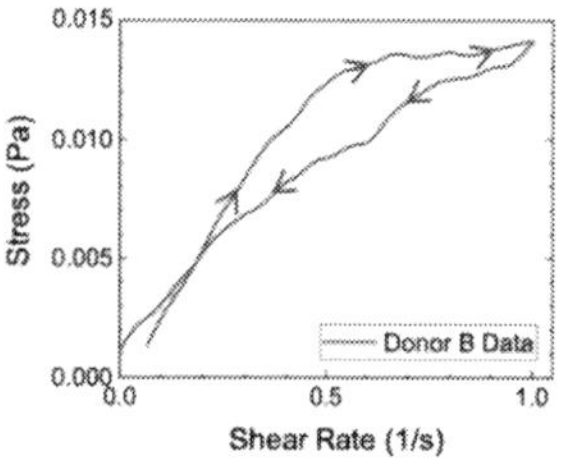

The Components of Blood and Their Importance, American Society of Hematology
https://www.youtube.com/watch?v=R-sKZWqsUpw

jshorner@udel.edu | 2 / 13

Motivation

UNIVERSITY OF DELAWARE

- Motivation
  - Blood rheology is critical for accurate blood flow simulations
  - Various diseases have been linked to changes in blood rheology
  - How do we relate blood physiology to the rheology?
  - Careful data on well characterized samples are required to understand and model the rheology
- Measurement Issues
  - Blood ex vivo aging
  - Thixotropy
  - Sample evaporation
  - Temperature/interfacial effects
  - Red blood cell sedimentation
  - Red blood cell syneresis/wall slip

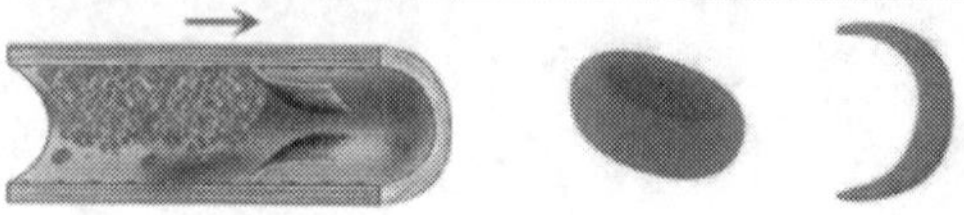

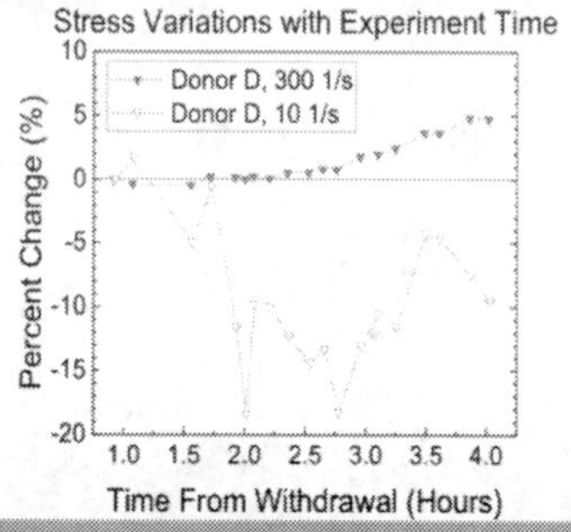

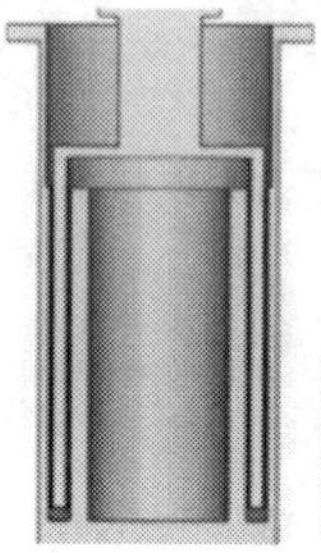

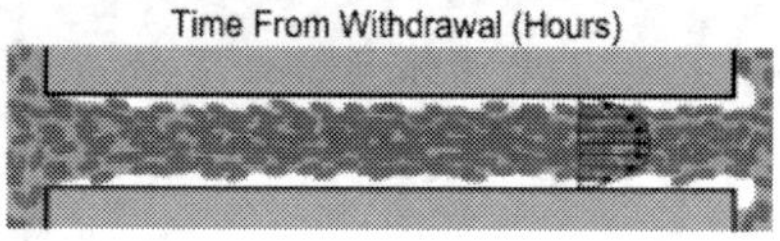

Maslak, *Blood*, (2005)
Sickle Cell Disease, KidsHealth, http://kidshealth.org/en/parents/sickle-cell-anemia.html
Horner, et al., *Clinical Hemorheology and Microcirculation*, (Submitted)
Fahraeus Effect, Wikipedia.org, https://en.wikipedia.org/wiki/F%C3%A5hr%C3%A6us_effect/

Steady Shear RBC Contributions

UNIVERSITY OF DELAWARE

- Rouleaux only form in the presence of proteins and break at high shear rates
- Shear thinning at high shear rates is due to red blood cell deformations
- For isolated red blood cells (C), steady shear rheology can be described by a Cross model

$$\sigma_{ss,C} = \eta_C(\dot{\gamma})\dot{\gamma} = \left(\frac{\mu_{0,C} - \mu_{\infty,C}}{1 + \tau_C|\dot{\gamma}|} + \mu_{\infty,C}\right)\dot{\gamma}$$

- Can use an extended White-Metzner model to represent the viscoelastic contribution of the red blood cells

$$\boldsymbol{\sigma_C} + \left(\frac{\eta_C(c_I)}{G_C}\right)\boldsymbol{\sigma_{C_{(1)}}} = \eta_C(c_I)\boldsymbol{\gamma_{(1)}}$$

$$\boldsymbol{\sigma_C} = G_C(\boldsymbol{c} - \boldsymbol{\delta})$$

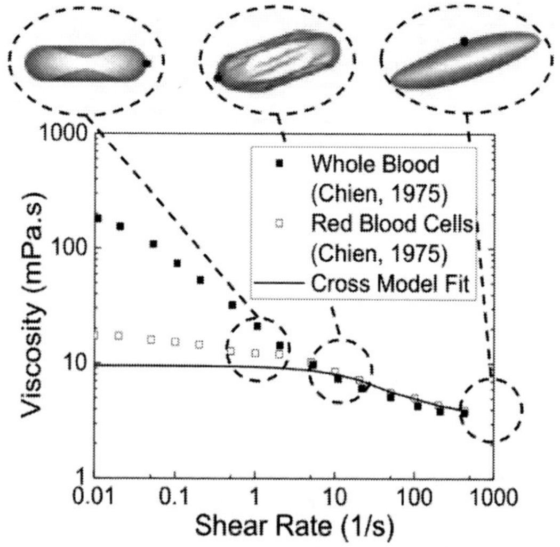

Chien, The Red Blood Cell, Vol. 2, Academic Press, (1975)
Omori et al., *Physical Review E*, (2012)
Souvaliotis and Beris, *Journal of Rheology*, (1992)

Steady Shear Rouleaux Contributions

UNIVERSITY OF DELAWARE

- At low shear rates, the red blood cells will reversibly form structures called rouleaux (R)
- The presence of rouleaux significantly increases the viscosity and results in a non-zero yield stress
- The rouleaux add an elastic stress component associated with the yield stress and viscous component

$$\sigma_{Tot} = \sigma_R + \sigma_C$$
$$\sigma_{ss,R} = \mu_R \lambda^3 \dot{\gamma} + \sigma_y \lambda$$

- The structural parameter (λ) evolves according to a kinetic rate equation

$$\frac{d}{dt}\lambda = \frac{1}{\tau_\lambda}\left((1-\lambda) + (1-\lambda)\tau_a|\dot{\gamma}_p| - \lambda(\tau_b \dot{\gamma}_p)^2\right)$$

No Structure $\leftarrow 0 \leq \lambda \leq 1 \rightarrow$ Fully Structured

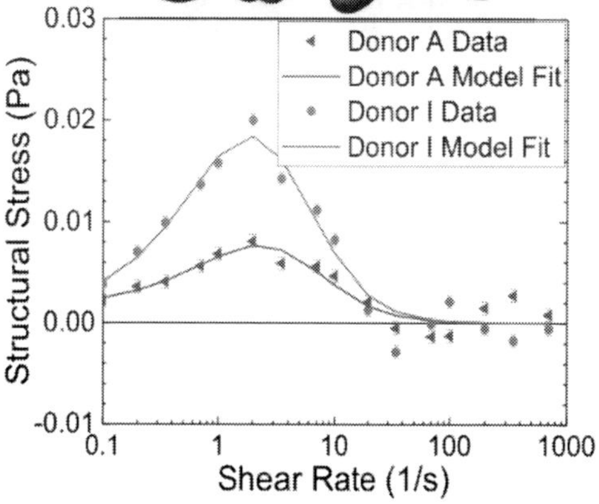

Apostolidis et al., *Journal of Rheology*, (2015)
Armstrong et al., *Journal of Rheology*, (2016)

Model Steady Shear Fits

UNIVERSITY OF DELAWARE

- The model may be solved to obtain the steady state solution:

$$\lambda_{SS} = \frac{1 + \tau_a|\dot{\gamma}|}{1 + \tau_a|\dot{\gamma}| + (\tau_b \dot{\gamma})^2}$$
$$\sigma_{ss} = \mu_R \lambda_{SS}^3 \dot{\gamma} + \sigma_y \lambda_{SS} + \left(\frac{\mu_{0,C} - \mu_{\infty,C}}{1 + \tau_C|\dot{\gamma}|} + \mu_\infty\right)\dot{\gamma}$$

- Newtonian at high shear rates
- Yield stress behavior at low shear rates

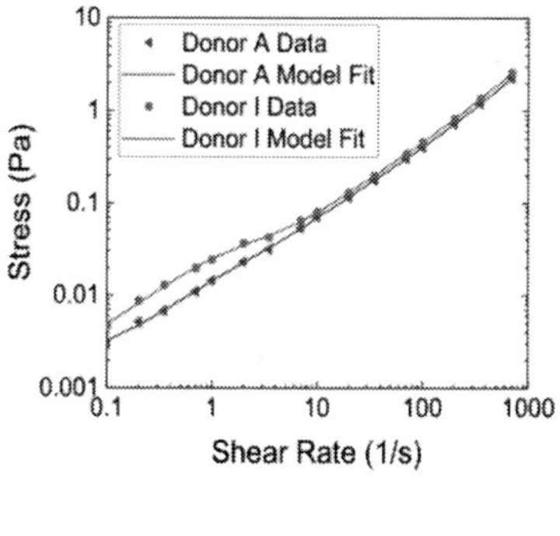

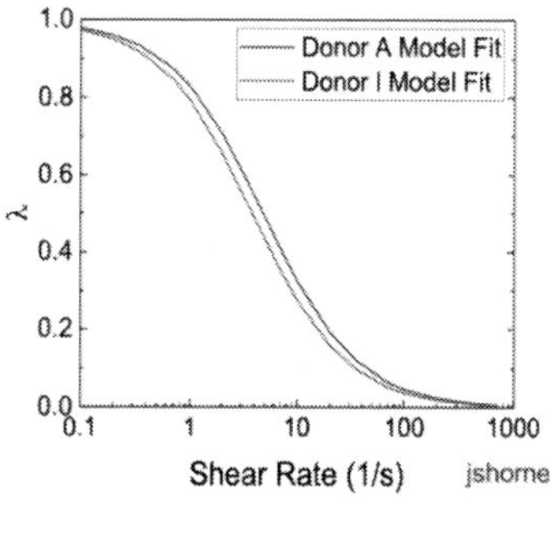

Correlation Analysis

- Several physiological properties of the blood samples were considered including hematocrit, fibrinogen, triglycerides, LDL cholesterol, HDL cholesterol (HDL), and total cholesterol (TC)
- Significant correlations ($p < 0.01$) were identified between:
  - The high shear model viscosity ($\mu_{\infty,C}$) and the ratio of cholesterol levels
  - The structural model viscosity (μ_R) and the fibrinogen content

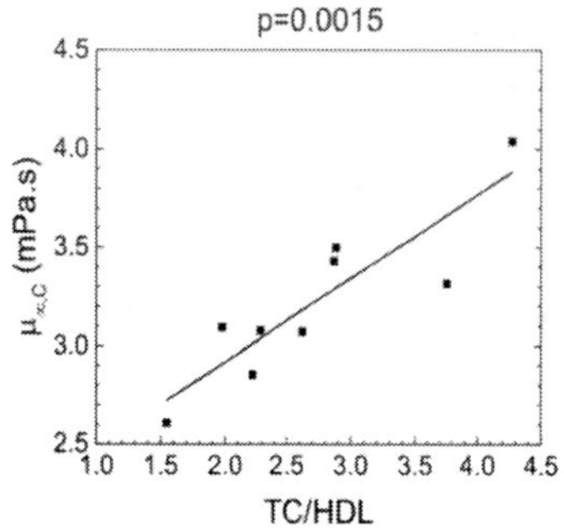

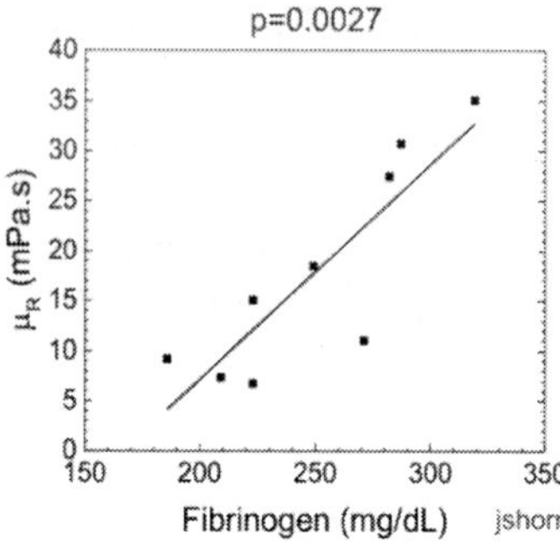

UD-LAOS

Unidirectional Large Amplitude Oscillatory Shear

$$\dot{\gamma}(t) = \gamma_0 \omega \cos(\omega t) + \gamma_0 \omega$$

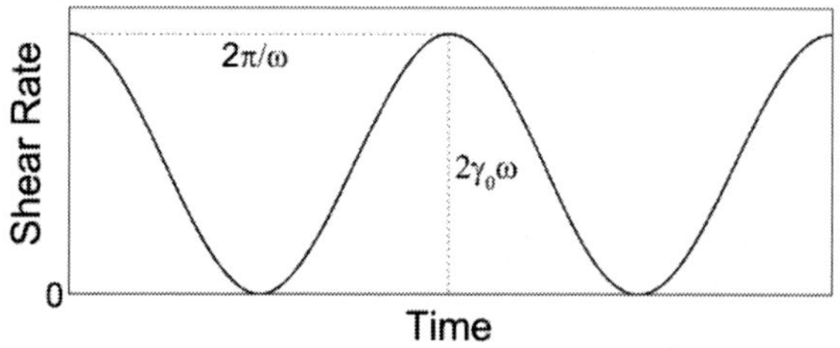

$$\gamma_r(t) = \gamma_0 \sin(\omega t)$$
$$\dot{\gamma}_r(t) = \gamma_0 \omega \cos(\omega t)$$

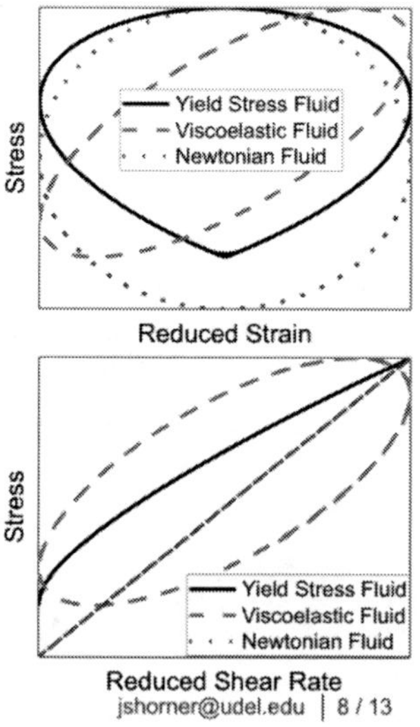

UD-LAOS Model Fits

UNIVERSITY OF DELAWARE

- The transient model parameters were fit to four different UD-LAOS experiments
- Two stress contributions
  - Viscoelastic at high shear rates
  - Yield stress at low shear rates
- Different qualitative and quantitative responses are seen at different conditions
- The model underpredicts stresses at low strain amplitude and high frequency

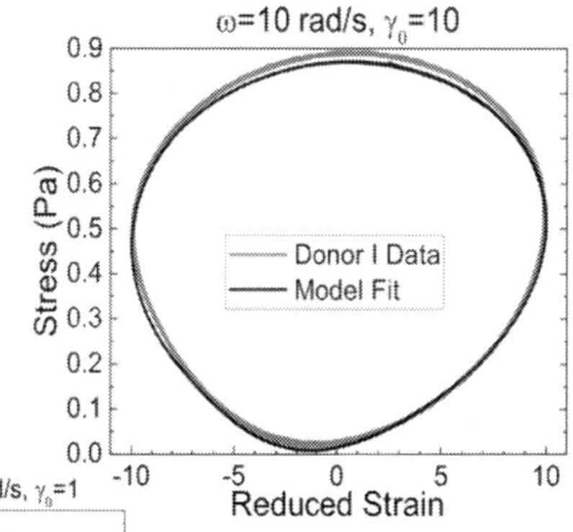

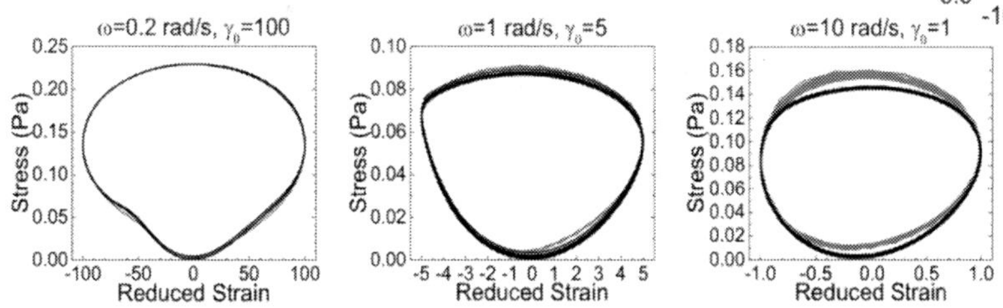

jshorner@udel.edu | 9 / 13

UD-LAOS Model Predictions

UNIVERSITY OF DELAWARE

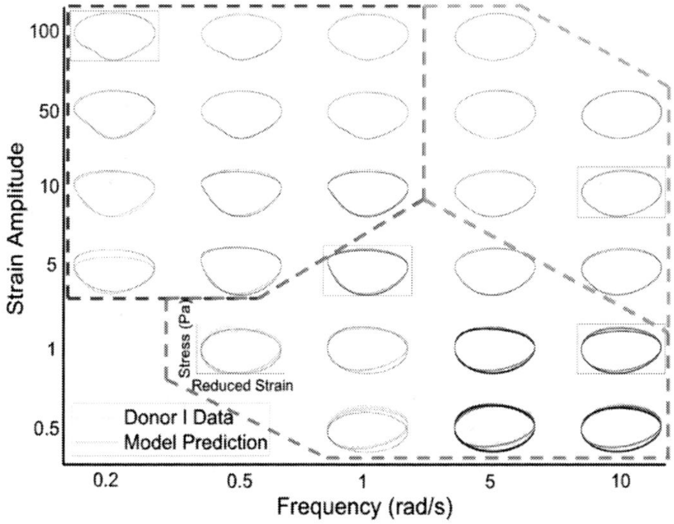

Viscous yield stress response

Transient structure reformation

Viscoelastic response

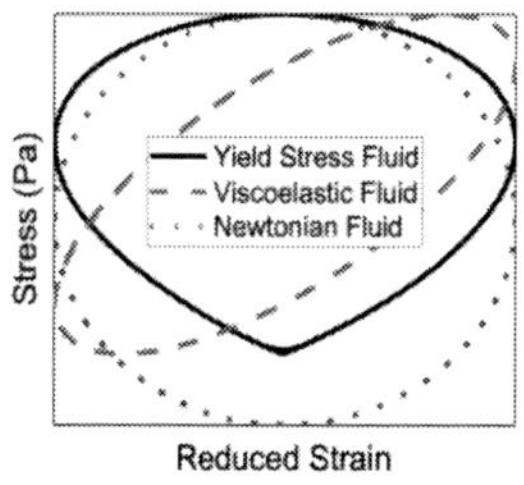

jshorner@udel.edu | 10 / 13

CFD Simulations

- Effects of rheology assessed on a mesh of the first bifurcation of the Left Coronary Artery (LCA) created using ANSYS Fluent

Newtonian

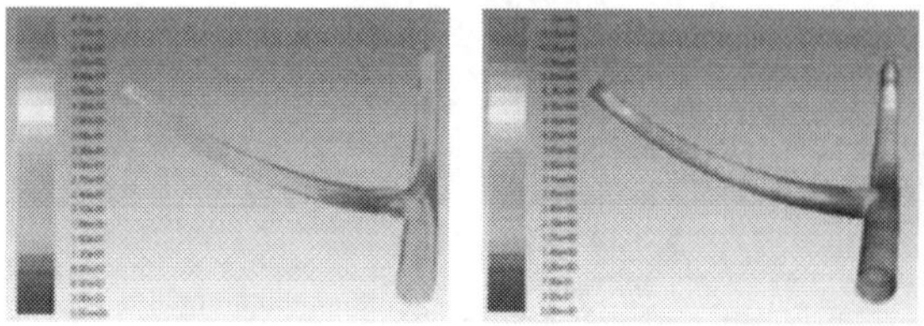

Velocity Pathlines (m/s) Wall Shear Stress (Pa)

Wall Shear Stress (Pa)	
Average	Minimum
2.49	0.0147

Steady State Model

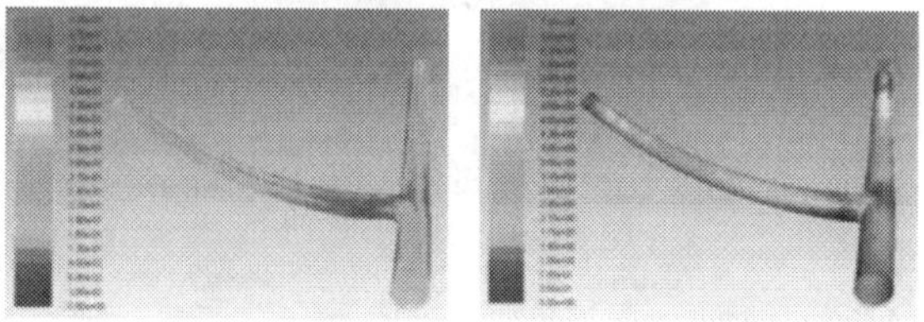

Velocity Pathlines (m/s) Wall Shear Stress (Pa)

Wall Shear Stress (Pa)	
Average	Minimum
2.89	0.0873

Apostolidis et al., *Journal of Non-Newtonian Fluid Mechanics*, (2016)

jshorner@udel.edu | 11 / 13

RBC Syneresis Future Work

- For small gaps, a red blood cell free layer will form near the walls
- Time scale for diffusion away from wall is long

$$\tau_D = \frac{L^2}{D} \sim \frac{O(-5)O(-5)}{O(-14)} = O(4)\ s$$

- Layer formation must be shear driven and possibly structure limited

$$N_c = -K_c a^2 (\phi^2 \nabla\dot{\gamma} + \phi\dot{\gamma}\nabla\phi)$$

$$N_\eta = -K_\eta \dot{\gamma}\phi^2 \left(\frac{a^2}{\eta}\right) \frac{d\eta}{d\phi} \nabla\phi$$

$$\frac{\partial\phi}{\partial t} = \nabla \cdot (N_c + N_\eta)$$

$$a = f(\lambda) \longrightarrow$$

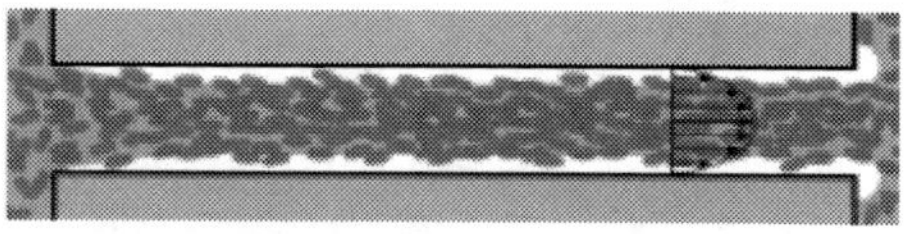

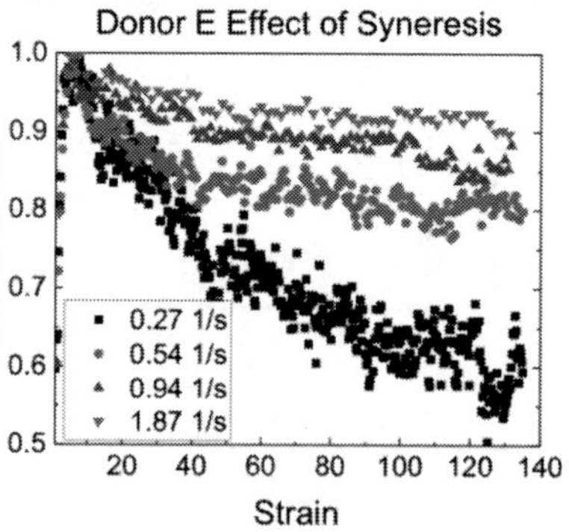

Fahraeus Effect, Wikipedia.org, https://en.wikipedia.org/wiki/F%C3%A5hr%C3%A6us_effect
Horner, et al., *Clinical Hemorheology and Microcirculation*, (Submitted)
Phillips et al., *Physics of Fluids A: Fluid Dynamics*, (1992)

jshorner@udel.edu | 12 / 13

Summary

- A new model for transient blood rheology was proposed which included a viscoelastic red blood cell stress component and a thixotropic rouleaux component
- Significant correlations were identified between the high shear viscosity and the cholesterol ratio and the low shear viscosity and the fibrinogen content
- The model provides a good fit to a range of physiologically relevant flow conditions
- Incorporation of the rheology into CFD blood flow simulations can result in significantly different wall shear stress values

Acknowledgments

- National Science Foundation
- Dr. Matthew J. Armstrong, United States Military Academy
- Dominic Gallo
- Willow Bowen
- Nurse Managed Primary Care Center

Human Blood Physiological Parameters

Plasma	
Specific Gravity	~1.03
Viscosity	~1.28 cP
pH	~7.46
Fibrinogen Content	~3 g/L
Albumin Content	~37 g/L
Total Protein Content	~73 g/L
Red Blood Cells	
Volume Fraction (Hematocrit)	42-47 %
Mean Corpuscular Volume	~80 fL
Thickness	1-2 μm
Diameter	~8 μm
Lifespan	~121 days
Specific Gravity	~1.09
Cytoplasm Viscosity	~13.2 cP
White Blood Cells	
Volume Fraction	~0.50 %
Diameter	7-10 μm
Lifespan	~7 days

Platelets	
Volume Fraction	~0.50 %
Diameter	2-3 μm
Lifespan	8-14 days
Blood Flow	
Capillary Velocity	~0.1 cm/s
Capillary Diameter	5-10 μm
Vena Cavae Velocity	~15 cm/s
Vena Cavae Diameter	2-3.5 cm
Aorta Velocity	~40 cm/s
Aorta Diameter	2-3 cm
Heart Rate	60-100 bpm
Volume per Heartbeat	~70 mL
Total Volume	~5.5 L
Circulation Time	~1 min

Steady Shear Rheology

- Significant shear thinning behavior
- Arises due to the formation of rouleaux
- Can be fit to the Casson model:

$$\sigma^{\frac{1}{2}} = \sigma_y^{\frac{1}{2}} + (\mu\dot{\gamma})^{\frac{1}{2}}$$

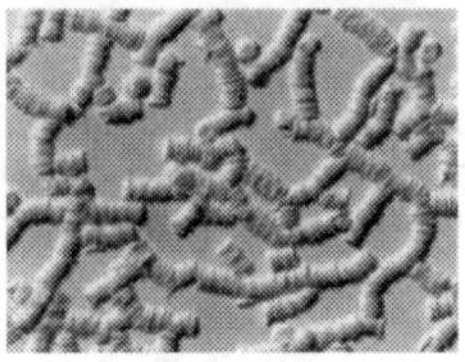

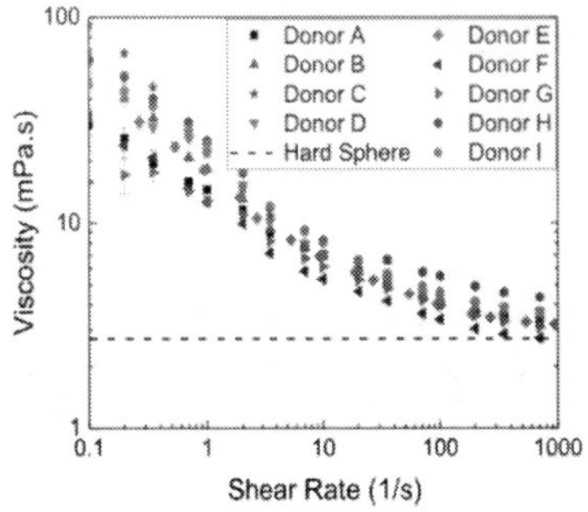

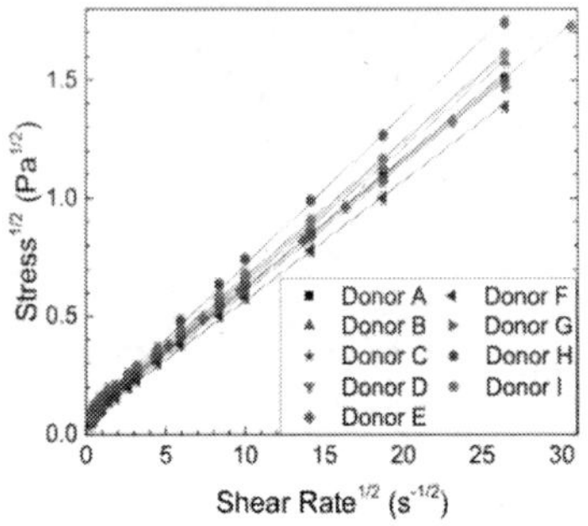

Baskurt et al., Red Blood Cell Aggregation, (2012)
Casson, Rheology of Disperse Systems, (1959)

jshorner@udel.edu | 15

Isolated Red Blood Cell Viscoelasticity

$$c_I = \mathrm{tr}(\boldsymbol{c}) = 3 + \frac{\sigma_{11}}{G_C} \overset{SS}{\Rightarrow} c_I = 3 + \frac{2\eta^2\dot{\gamma}^2}{G_C^2}$$

$$\boldsymbol{\sigma} + \left(\frac{\eta_C(c_I)}{G_C}\right)\boldsymbol{\sigma}_{(1)} = \eta_C(c_I)\boldsymbol{\gamma}_{(1)}$$

$$\boldsymbol{\sigma} = G_C(\boldsymbol{c} - \boldsymbol{\delta})$$

$$\boldsymbol{\sigma} = \begin{pmatrix} \sigma_{11} & \sigma_{12} & 0 \\ \sigma_{21} & 0 & 0 \\ 0 & 0 & 0 \end{pmatrix}$$

$$a = 1$$

$$b = G_C\tau_C\sqrt{\frac{c_I - 3}{2}} - \mu_{0,C}$$

$$c = -G_C\mu_{\infty,C}\tau_C\sqrt{\frac{c_I - 3}{2}}$$

$$\eta_C(c_I) = \frac{-b - \sqrt{b^2 - 4ac}}{2a}$$

Souvaliotis and Beris, *Journal of Rheology*, (1992)

jshorner@udel.edu | 16

Rouleaux Thixotropy

$$\sigma_R = G_R\gamma_e + \lambda^3\mu_R\dot{\gamma}_p$$

Shear Aggregation

$$\frac{d}{dt}\lambda = \frac{1}{\tau_\lambda}\left((1-\lambda) + (1-\lambda)\tau_a|\dot{\gamma}_p| - \lambda(\tau_b\dot{\gamma}_p)^2\right)$$

Brownian Aggregation　　Shear Breakup

$$G_R = \frac{1}{\tau_G}\left(\frac{\sigma_y}{\gamma_0\lambda} - G_R\right)$$

$$\dot{\gamma} = \dot{\gamma}_p + \dot{\gamma}_e \leftrightarrow \gamma = \gamma_p + \gamma_e$$

$$\dot{\gamma}_e = \begin{cases} \dot{\gamma}_p - \dfrac{\gamma_e}{\gamma_{max}}|\dot{\gamma}_P|, & \dfrac{d\gamma_{max}}{dt} \geq 0 \\ \dot{\gamma}_p - \dfrac{\gamma_e}{\gamma_{max}}|\dot{\gamma}_p| + \dfrac{\gamma_e}{\gamma_{max}}\dfrac{d\gamma_{max}}{dt}, & \dfrac{d\gamma_{max}}{dt} < 0 \end{cases}$$

$$\gamma_{max} = \gamma_0\lambda^2$$

λ = Nondimensional structural parameter

Kinematic hardening model proposed by Dimitriou et al.

Apostolidis et al., *Journal of Rheology*, (2015)
Armstrong et al., *Journal of Rheology*, (2016)
Fedosov and Gompper, *Proceedings of the 6th NIC Symposium*, (2012)
Dimitriou et al., *Journal of Rheology*, (2013)

jshorner@udel.edu | 17

Steady Shear RBC Normal Stress

- The normal stresses associated with the isolated red blood cells may be calculated for steady shear according to the model

$$\eta_{SS,C}(\dot{\gamma}) = \left(\frac{\mu_{0,C} - \mu_{\infty,C}}{1 + \tau_C|\dot{\gamma}|} + \mu_{\infty,C}\right)$$

$$\sigma_{SS,11,C} = 2\left(\frac{\eta^2_{SS,C}(\dot{\gamma})}{G_C}\right)\dot{\gamma}$$

$$\sigma_{22} = \sigma_{33} = 0$$

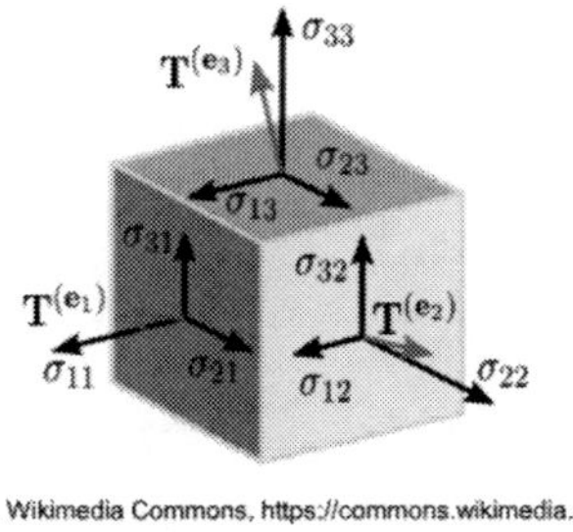

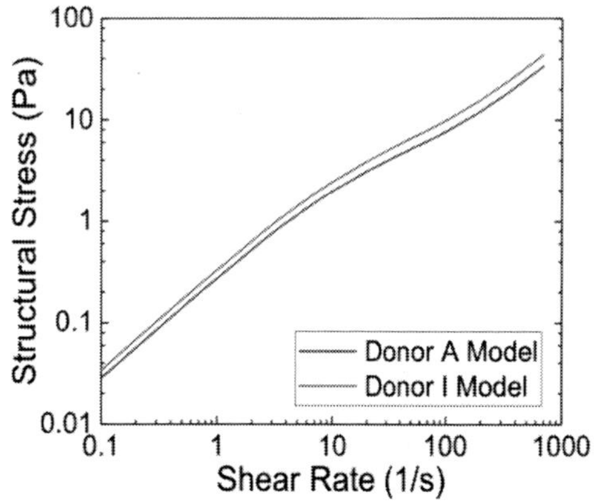

jshorner@udel.edu | 18 / 13

Components stress tensor, Wikimedia Commons, https://commons.wikimedia.org/wiki/File:Components_stress_tensor.svg

495678 *In Vitro* Measurement and Modelling of Platelet Adhesion on Von-Willebrand-Factor-Coated Surfaces in Channel Flow

Abstract:

Understanding the biophysics of platelet adhesion is a long-standing problem in the study of hemostasis and thrombosis. Blood coagulation is initiated by GPIb and GPIIbIIIa receptors on the platelet surface binding with von Willebrand factors (VWF) on the vascular walls. The facts that each platelet contains multiple copies of such receptors and that GPIb-VWF binding is fast and reversible complicate the reaction kinetics. Thus, a clear distinction must be drawn between the bond-level kinetics, i.e. receptor-ligand binding, and the platelet-level kinetics, i.e. the adhesion and detachment of a platelet from the vessel wall. There has not been a detailed study relating these two phenomena at their associated different length scales. Another challenge is that platelet adhesion is a shear-induced phenomenon. Thus, the accurate modeling must consider the fluid mechanics of blood flow.

In this talk, we introduce a multi-scale model for platelet adhesion in channel flow, which takes into account both the fluid mechanics and the biochemistry. Based on our existing model of the margination of platelets in a cellular suspension, we now consider near-wall platelets interacting with wall-tethered VWFs. Therefore, the concentration distribution of platelets in the cross-flow direction is influenced by red-blood-cell-platelet interactions, platelet-platelet interactions as well as the near-wall binding reactions. Our approach utilizes single molecule measurements of bond-level kinetics and couples them with the platelet effective reactive area determined from simulations. Thus, we are able to obtain platelet-level rate constants to compare with our experimental measurements. We utilize a microfluidic device with VWF-coated surfaces in 50-μm-high channels to mimic the actual flow conditions inside human arterioles. We carefully choose our control variables to be the shear rate, hematocrit, GPIb inhibitor dosage and GPIIbIIIa inhibitor dosage. We demonstrate good agreement between our model and experiments and investigate how the change of flow and platelet properties affect the adhesion process. These findings also provide implications for how platelet defects and abnormal flow conditions influence hemostasis and thrombosis.

501695 A Novel Viscoelastic Thrombogenesis Model from High Performance Lattice Boltzmann Method Yield-Stress Calculations Based on Intravital Images of Clot Formation in Live Mice

Abstract:

Thrombo-embolic infarction is the major cause of mortality and morbidity in the United States, causing over 1 million strokes, heart attacks and other life-threatening thrombotic events each year in the United States. Conversely, deficiencies in these processes result in severe bleeding risks. Unfortunately, despite tremendous efforts in understanding thrombosis, the viscoelastic properties of thrombi that are responsible for embolism are not well understood. The ability to access hydrodynamics stresses at which thrombus structure yield to deformation in a blood vessel can provide meaningful information on when the thrombus is likely to embolize. Lack of an accurate method/computational model for estimating the stresses acting on a thrombus in vivo prompted us in developing a robust approach that combines intravital imaging and simulation. Interestingly, the viscoelastic behavior exhibited by the thrombus resembles that of a Bingham fluid - a material that behaves as a rigid body at low stresses but flows as a viscous fluid when it experiences stresses in excess for its "critical yield stress". Thus, we ventured to measure the critical yield stress at which the thrombi yield behavior occurs.

All experiments were performed at Dr. Skip Brass's Lab, University of Pennsylvania. To that end, thrombi were induced by means of a laser injury in cremaster muscle microcirculation of live mice and tracked via fluorescence markers using confocal microscopic imaging. The images are stabilized using Fiji plugins StackReg and Image Stabilization to fix translation. An in-house Matlab® code is used to produce 3D reconstruction by interpolating the thrombus height to width ratio and fitting a parabola across the edge in the direction perpendicular to blood flow. An in-house D3Q15 Lattice Boltzmann Method code is used to simualte convective blood flow on supercomputers, due to the large simulation sized. The simulations are performed for a constant pressure drop across the channel and the pressure drop was estimated based on the average blood velocity obtained in vivo by the Doppler velocimetry. A blood viscosity of $\mu \approx 0.03$ is assumed based on literature, and no-slip boundary condition is applied at the wall faces.

Our results indicate that although each clot evolves uniquely, all of the data can be collapsed on a single dimensionless scale. The non-dimensionalization of the data is done by taking the critical time at which the clot's structure yields to the surround blood flow as a reference. At this special point, the clot stops undergoing uniform growth, yields to the surrounding blood flow and rearranges its shape to minimize the drag on its surface. Consequently, we present results for both wild type mice and various mutants on this master scale.

To our knowledge, this is the first image-based in-vivo assessment of blood clots viscoelastic nature. Collectively, the outcome of this work is expected to provide insight towards an understanding of clot generation, its structure, stability, thromboembolism and coagulopathy. Moreover, they can assist in

creating simpler thrombogenesis models that can help improve the understanding of risk factors associated with blood clotting, and ideally help researchers to reduce risks of occlusion and embolism in patients.

502096 Determination of the Zeta Potential of Planar Solids in Nonpolar Liquids

Abstract:

An apparatus and a method for determining the zeta potential of planar solid surfaces in contact with nonpolar liquids are described. Rotation of disk-shaped mica, silica, and ITO samples in surfactant-doped dodecane generated streaming potential that was measured between a glassy carbon electrode near the sample at its axis and a second glassy carbon electrode far from the sample. After rotation began, the measured potential changed rapidly at first but asymptotically approached a steady value over a timescale from fractions of a second to seconds. The potential difference between the rest potential and the asymptote was proportional to the 3/2 power of the rotation rate, which agrees with the theory of streaming potential for the rotating disk. Also, the potential measured at the axis had the opposite polarity of potential measured by the same sensor at the diskâ€™s periphery, which is expected from the same theory used to determine the zeta potential. This evidence indicated that the difference between the potential before rotation and its asymptotic value during rotation was streaming potential. The zeta potential of freshly cleaved mica in dodecane doped with 0.5% wt. AOT was +100 mV. The zeta potential of fused silica in dodecane doped with OLOA 11000 varied from -10 mV to -25 mV agreeing qualitatively with zeta determined from electrophoretic mobility measurements of silica particles. The magnitude of the zeta potential of ITO coated glass was smaller than 25 mV in solutions of OLOA 11000 in dodecane, which means that the diffuse layer capacitance near the ITO equals the permittivity divided by the Debye length, to a good approximation.

501709 Toward Simulation-Based Design of Particle Handling Processes

Abstract:

Particulate processes pervade the petroleum, chemical, mining, pharmaceutical, and agricultural industries. Many of these processes have significant opportunities for optimization and productivity enhancements. Reliable particle modeling and simulation tools can improve critical understanding and design of particle transport processes; this talk will provide several examples in this regard. A recent approach to modeling particle flow employs the discrete element method (DEM). In DEM, the dynamics of individual powder particles are described rather than treating the particle phase as a continuum via CFD. DEM models are especially useful for investigating phenomena occurring at the particle-length scale. Although DEM models are computationally intensive, they are capable of developing constitutive relations for the particle phase that are needed for continuum-based CFD simulations.

Specifically, this presentation will focus on DEM modeling of complex particulate flows for non-spherical particles which are described via linked and overlapping spheres. In addition, both rigid and flexible non-spherical particles, including particle breakage, will be discussed. Finally, we consider the effect of cohesion of non-spherical particles via liquid bridging. The particle models are validated via experiments - shear cells testing, angle of repose testing, hopper discharge, particle packing, and particle breakage due to compaction and flow in a FT4 rheometer. These aspects of particle flow (asphericity, flexibility, breakage and cohesion) are of great importance in many practical applications and, hence, are the subjects of research at the forefront of current particle modeling efforts.

503436 Sustainable Electrocatalytic Water Splitting with Earth-Abundant Materials

Abstract:

Water splitting has potential for sustainable solar energy storage in the form of fuels and organic materials. Its practical use requires highly effective electrocatalysts synthesized from earth-abundant materials such as manganese and nickel. We have developed a synthesis method for manganese-based (MnO_x) catalytic films *in situ*, with steady and excellent water oxidation performance[1]. The method involves two series of cyclic voltammetry (CV) over different potential ranges, followed by calcination to increase crystallinity. We have incorporated the catalysts into TiO_2 to build a composite photoanode with excellent activity under ultraviolet (UV) illumination, with current work focused on improving performance by enabling solar energy collection over a wider range of wavelengths. We have also developed an electrodeposition method to synthesize $Ni/Ni(OH)_2$ electrocatalysts *in situ* on conductive surfaces for the hydrogen evolution reaction[2]. The new method uses two cycles of CV over a single potential range, producing a film with a large number of $Ni(OH)_2/Ni$ interfaces that significantly shifts performance towards that of platinum.

1. H Yuan, RR Lunt, GJ. Blanchard, and RY Ofoli. Synthesis of MnO_x water oxidation catalyst on fluorine-doped tin oxide with a dual-series cyclic voltammetry method, *ChemElectroChem*, 3(5): 709-712 (2016).
2. H Yuan, R.R Lunt, JI Thompson, and RY Ofoli Electrodeposition of $Ni/Ni(OH)_2$ Catalytic films for the hydrogen evolution reaction produced by using cyclic voltammetry. *ChemElectroChem*, DOI: 10.1002/celc.201600572 (2016)

484240 Human Tear Production from Capillary Wicking Dynamics

Abstract:

A Schirmer tear test (STT) commonly gauges human tear production, especially when dry-eye symptoms are present. In a STT, the rounded tip of a standardized paper strip (Whatman #41) is inserted into the lower fornix of the eye and the wetted length extending out from the lower lid is recorded after 5 min of eye closure. Longer wetted lengths suggest higher tear production rates. To date, however, there is no methodology to transform STT transient wetting lengths into basal tear production rates, e.g., in Î¼L/min. We develop a physicochemical wicking model to elucidate wetting kinetics in a Schirmer strip. Careful consideration of the initial depletion of tear in the closed-eye tear prism reveals an initial fast increase in wetted length followed by slower growth. Evaporation from the exposed portion of the strip is carefully considered. Excellent agreement of the proposed model is achieved with available experimental observation.

Based on our proposed numerical model, we suggest a new clinical methodology in which the Schirmer strip is sheathed against evaporation by transparent tape and the wetted length of a distance-marked strip is measured at three time points near the current 5 min of insertion. Based the proposed wicking model, the slope of the resulting best-fit straight line gives basal tear production rate. Preliminary clinical evaluation of ten college-age subjects shows subject variation in tear production rates between 1 and 4 Î¼L/min, all that lie within the range of currently accepted values. Our proposed modeling effort provides a simple new tool for diagnosing dry eye.

496567 Structure, Elasticity, and Non-Equilibrium State Diagram of Depletion Gels

Abstract:

Depletion gels, soft solids that form in a mixture of colloidal particles and non-adsorbing polymer, are ubiquitous in industries where fine (colloidal) solids are dispersed in polymers, and have often served as model systems to understand the fundamental origins and rheology of particulate gels, including their elasticity and yielding. Using a recent model system that enables the rheology, structure, and particle interactions to be measured in concert [1], we show that the elastic modulus of a depletion gel is governed by the spatial organization and interactions of particle clusters. Each cluster acts as a mechanical unit that propagates the elastic deformation. As the attractive strength between particles increases, the gel elastic modulus increases, but this change cannot be accounted for by the immediate increase in bond stiffness between particles and clusters alone. The modulus is, however, consistent with an increasing number of cluster-cluster contacts. Two principal length scales thus emerge: at the particle level, the internal cluster structure becomes less dense with increasing attractive strength, and extends from the attractive glass line of colloids with short range attraction into the lower density gel region. However, the overall size of the clusters, the length scale over which stress is transmitted during elastic deformation, does not depend on the magnitude of the attractive strength.

1. Hsiao, L. C., Solomon, M. J., Whitaker, K. A. & Furst, E. M. A model colloidal gel for coordinated measurements of force, structure, and rheology. J. Rheol. 58, 1485â€“1504 (2014).

497886 Harnessing Solvation Forces for Dispersing Colloids in Ionic Liquids with Application in Human Exploration of Space

Abstract:

Ionic liquids are proposed for many applications due to their remarkable properties. Of particular interest here are space-based applications such as for improving astronaut survivability through enhanced ballistic, puncture and abrasion resistance of space suits and micrometeorite and orbital debris (MMOD) shielding for spacecraft. Specifically, ionic liquids are proposed as the dispersing phase of shear thickening colloidal dispersions because of their stability over the broad range of temperatures and low volatility. However, dispersing particles in ionic liquids can be challenging because the high ionic strength of ionic liquids screens the electrostatic stabilizing forces that are typically important for dispersing particles in polar solvents. Recently, (Gao *et al.* ACS Nano 2015), we created stable nanoparticle dispersions in the ionic liquid $[C_4mim][BF_4]$ by inducing solvation layering, whereby a 5 nm layer or organized IL forms around the particle due to strong hydrogen bonding between anion $[BF4]^-$ and the fluorinated alcohol functionalized particle surface. However, in agreement with theory, the presence of this steric solvation layer prevented shear thickening by suppressing hydrocluster formation. To achieve a stable dispersion that also exhibits shear thickening, commercial silica particles with an alcohol- functionalized particle surface were dispersed in the ionic liquid $[C_4mim][BF_4]$. It was expected that the alcohol coating will exhibit weaker hydrogen bonding with the anion $[BF_4]^-$, leading to a thinner solvation layer that is still sufficient for dispersion, but enables hydrocluster formation at high shear rates. Dynamic light scattering (DLS), small angle neutron scattering (SANS) and rheology were employed to determine the solvation layer thickness and microstructure of dispersions. Analysis of SANS spectra across a broad range of particle concentrations was used to develop a quantitative model for the inter-particle interactions including the thickness of the solvation layer. The effects of temperature on microstructure and shear thinning and shear thickening rheology are investigated.Â Dispersions lose their stability and transit from stable dispersion to â€œsoftâ€ gel due to the loss of solvation layers thickness with increasing temperature. This work provides guidance on formulating colloidal dispersions in ionic liquids and may have implications for environmental and energy engineering as ionic liquids are candidates for remediation, separation, and recycling of nuclear waste.

490360 Control of Interfacial Phase Separation By Electro-Autocatalysis

Abstract:

Motivated by electrochemical control of interfacial phase separation, a general theory of nonequilibrium thermodynamic stability is developed for driven reactive mixtures. Pattern formation is determined by the competition of cooperative diffusion and driven autocatalysis. Novel features arise for electrochemical interfaces, related to controlled total current (galvanostatic operation), concentration-dependent exchange current (Butler-Volmer kinetics), and negative differential reaction resistance (Marcus kinetics). The theory shows how spinodal decomposition can be controlled by solo-autocatalytic charge transfer, with only a single Faradaic reaction. Experimental evidence is presented for intercalation and electrodeposition in rechargeable batteries, and further applications arise in solid state ionics, electrophoretic optics, electrochemical precipitation, and biological pattern formation.

All the best !

1

UNIVERSITY *of* WASHINGTON

The Use of Boundary Lubricants for the Reduction of Particle Friction in Abrasive Particle Slurries

Aditya Salunkhe
Rene Overney
John Berg

10/30/2017

2

UNIVERSITY of WASHINGTON

Application of abrasive slurries

- Metal machining, polishing

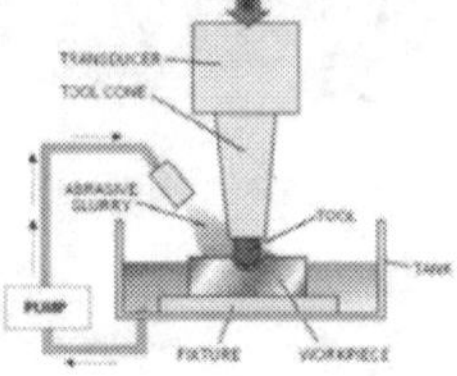

- Chemical mechanical planarization (CMP)

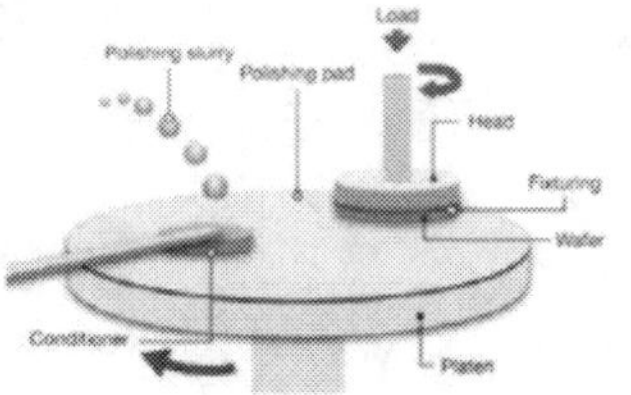

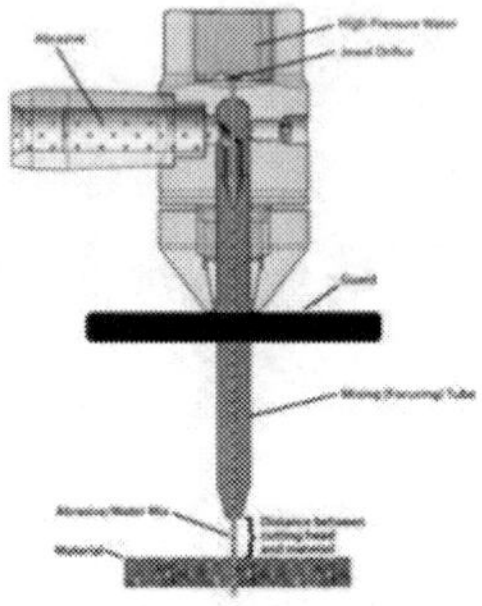

- Water-jet cutting

3

UNIVERSITY of WASHINGTON

Abrasives used in jet cutting

Garnet (silicate material)

220 mesh (30-75μm diameter)

Jagged shape

Moh's hardness 7.0 - 7.5

Specific gravity: 4.0

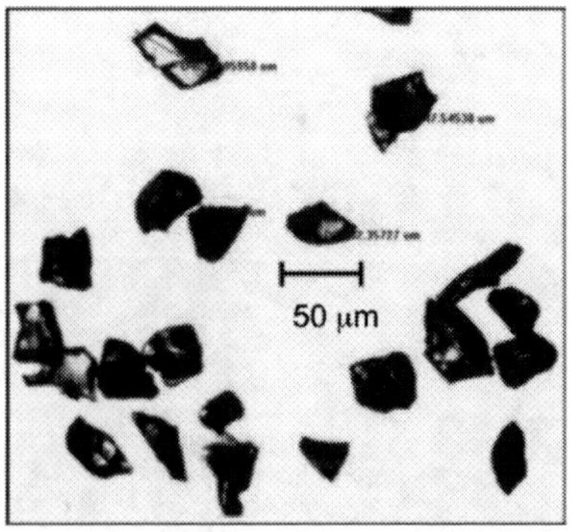

4

Abrasive Water Jets *vs*. Slurry Jets

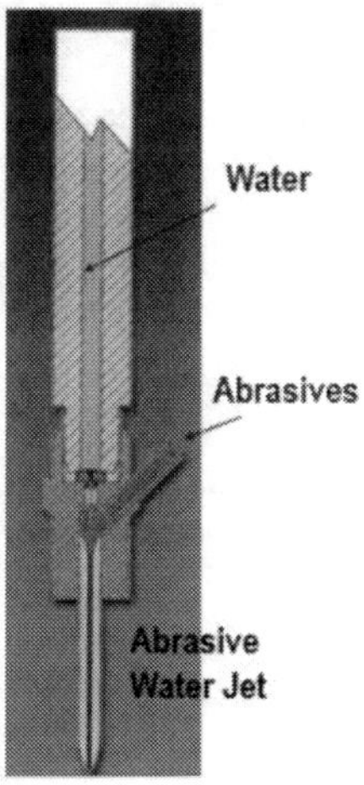

$P \rightarrow$ 90,000 psi

$d \rightarrow$ 800 μm

$v \rightarrow$ 1,140 m/s

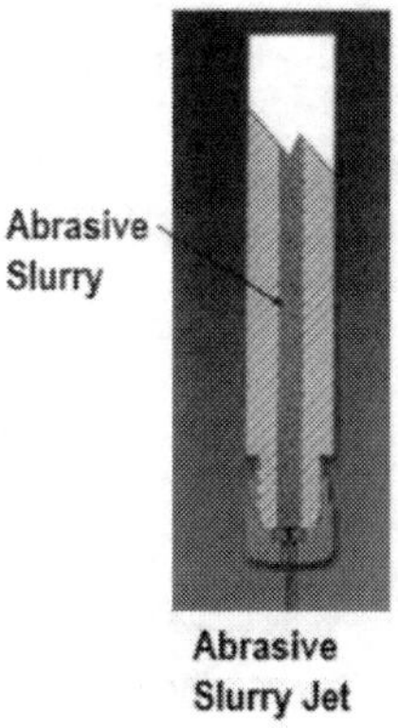

Dr. Mohamed Hashish
Flow International Corp.
Kent, WA
"The father of abrasive slurry jets"

Abrasive Slurry Jets (ASJ)

- Advantages
  - Efficient power transfer to abrasive particles
  - High concentration of abrasive particles in jet
- Disadvantages
  - Particle settling
  - Severe shear thickening, jamming

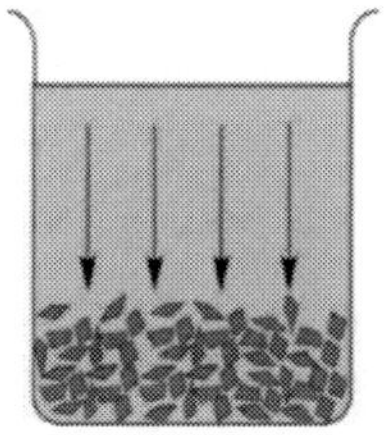

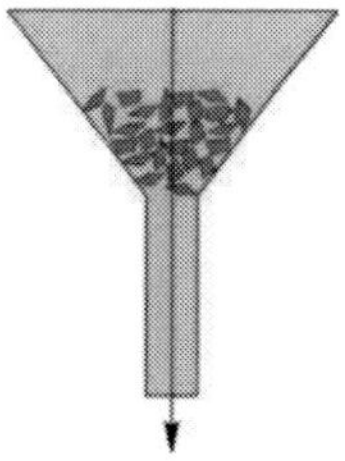

Possible solution: boundary lubrication

Dense slurry
$\phi_{rlp} = 0.575$

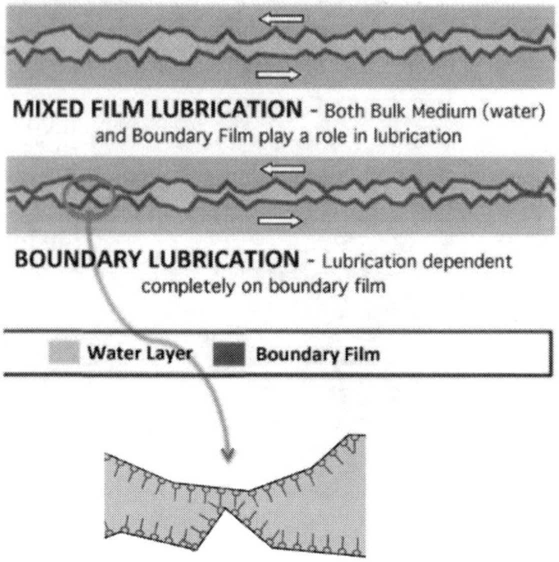

Objectives

- Evaluate candidate boundary lubricants for garnet surfaces in water using Atomic Force Microscopy (AFM) in lateral force mode (LFM)
- Examine rheometric behavior of dense garnet particle dispersions of bare and surface-treated particles

W UNIVERSITY *of* WASHINGTON

Candidates

The challenge: Garnet (≈ silica) in neutral-to-higher pH water has a low interfacial energy, unfavorable to physical adsorption.

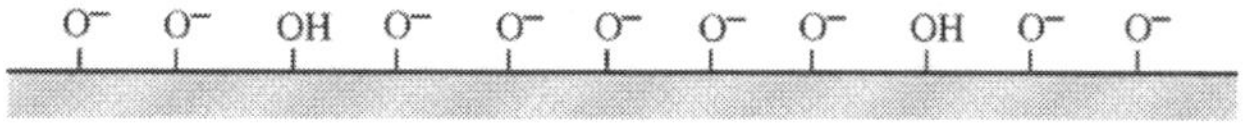

1. Cationic surfactant, e.g. HTAB

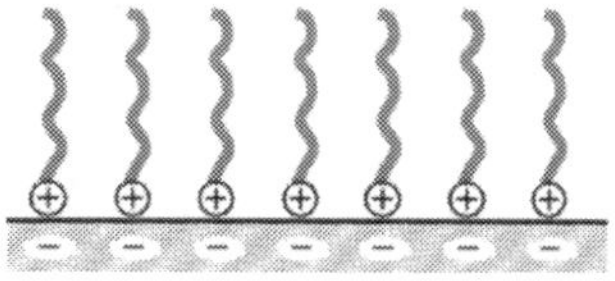

2. H-bonding nonionic, e.g. Pluronic® L62

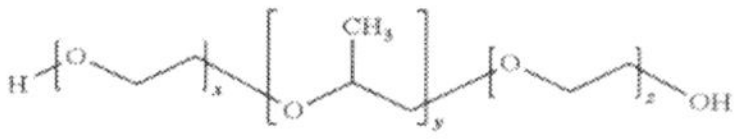

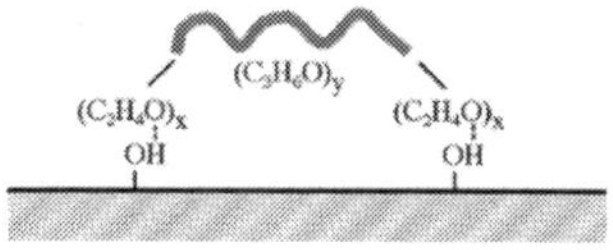

9

W UNIVERSITY *of* WASHINGTON

3. Silane surface functionalization, e.g. OTS

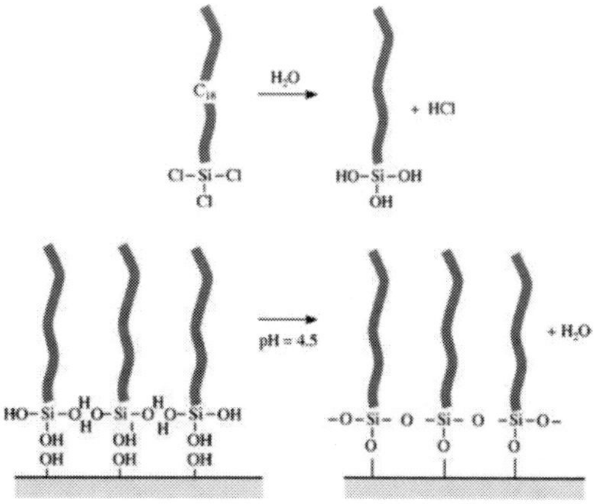

10

Lateral Force Microscopy (LFM)

- Contact Mode Imaging
- Measures the deflection of the cantilever in the horizontal direction
- The cantilever bends laterally due to a friction force between the tip and the sample surface

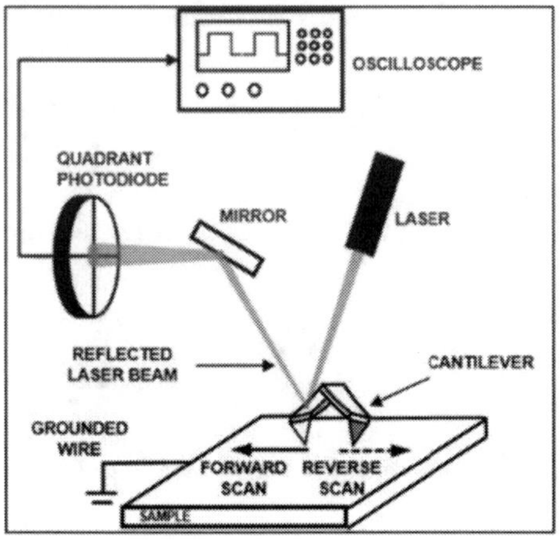

LFM Results for HTAB

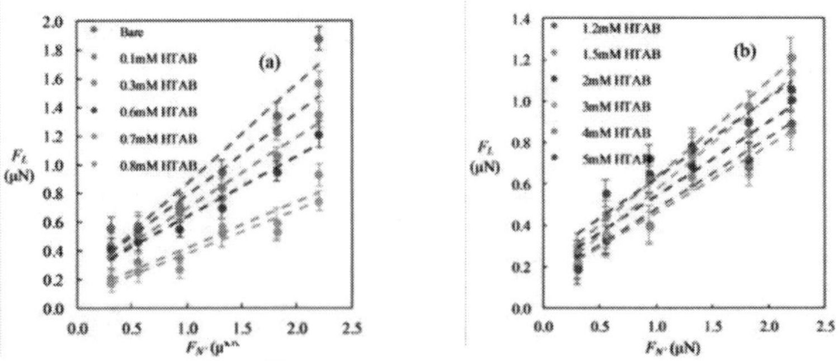

Coefficient of friction

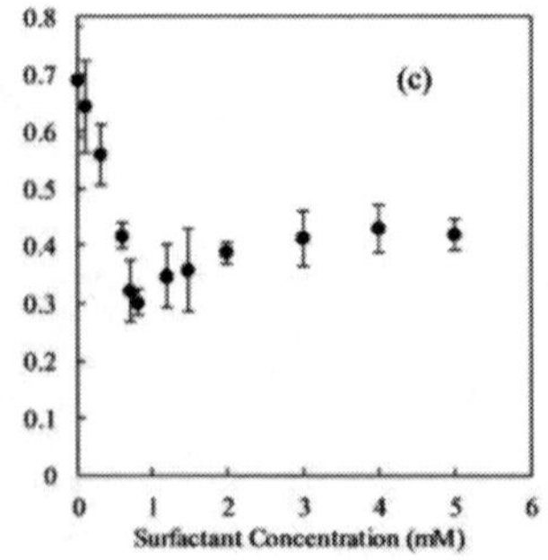

μ_{bare} = 0.69

μ_{min} = 0.29

UNIVERSITY of WASHINGTON

Interpretation

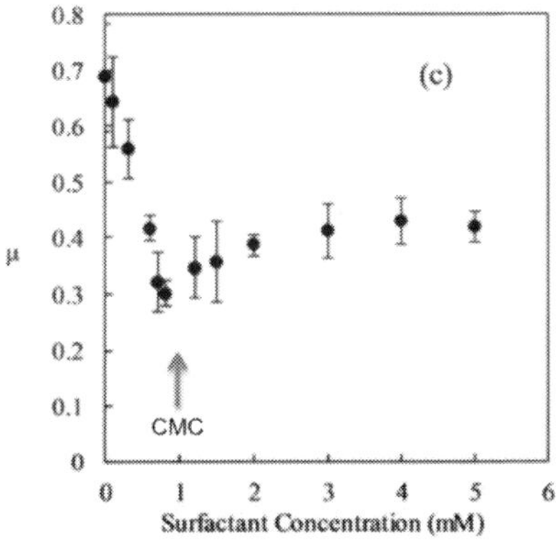

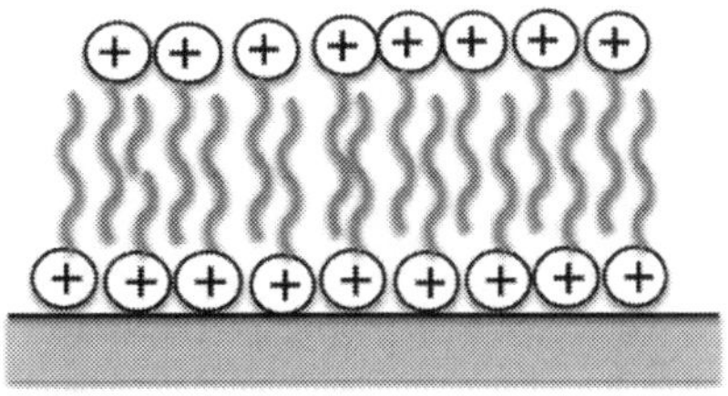

Conc. < CMC

Conc. ≈ CMC (0.95 mM)

Conc. > CMC

13

UNIVERSITY of WASHINGTON

LFM Results for Pluronic® L62

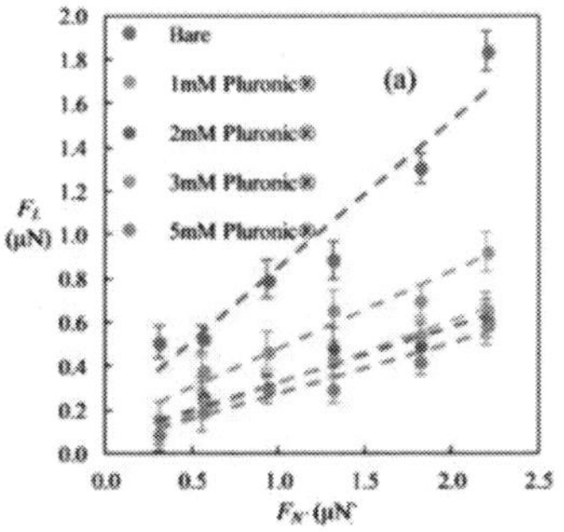

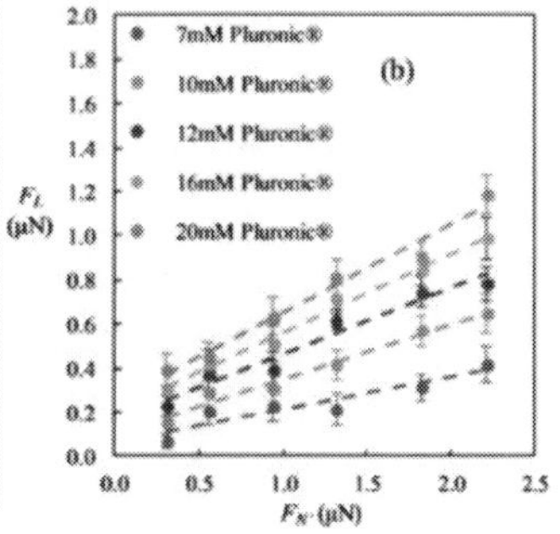

Coefficient of friction

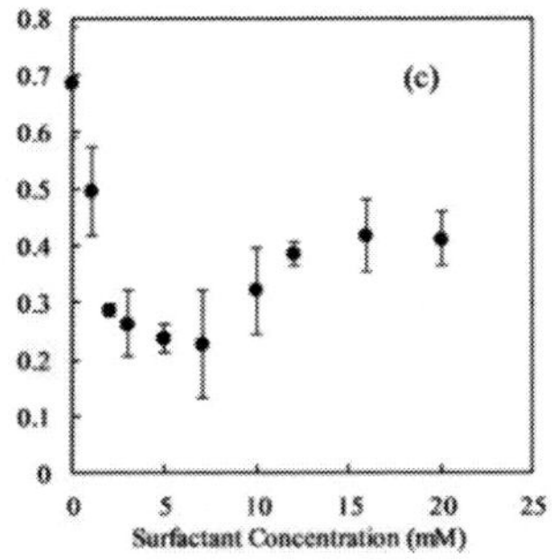

μ_{bare} = 0.69

μ_{min} = 0.21

14

W UNIVERSITY *of* WASHINGTON

Interpretation

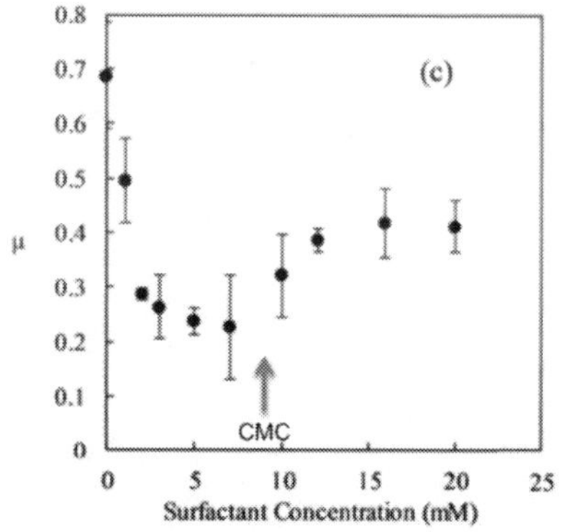

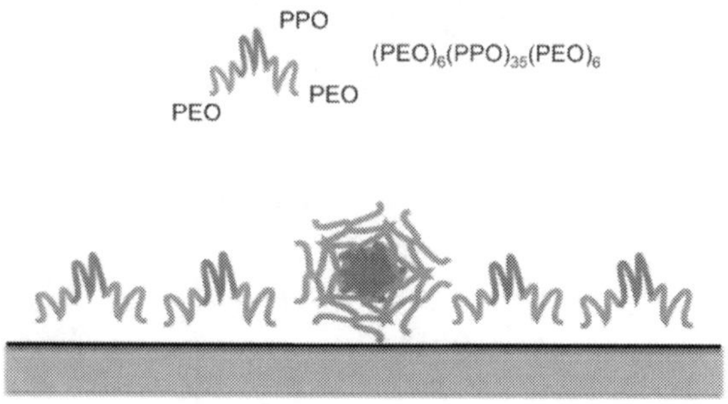

Conc. < CMC

Conc. ≈ CMC (9.6 mM)

Conc. > CMC

15

W UNIVERSITY *of* WASHINGTON

LFM Results for OTS Treatment

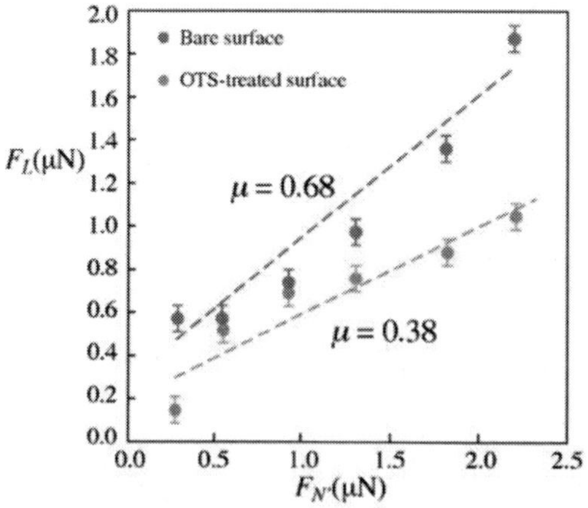

16

Vane tool Couette rheometry

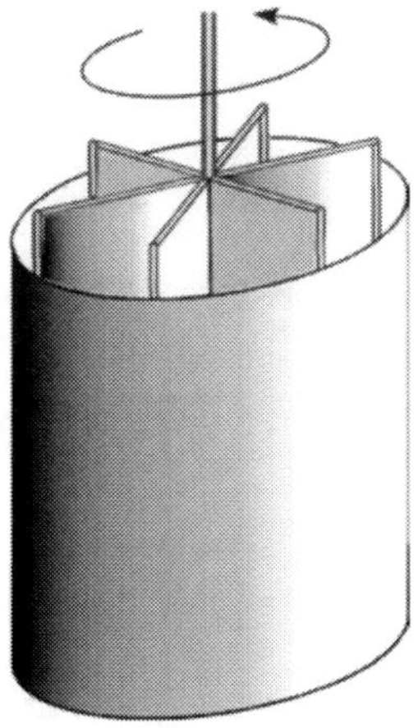

Viscosity of dense dispersion (ϕ = 0.575) *vs* shear rate

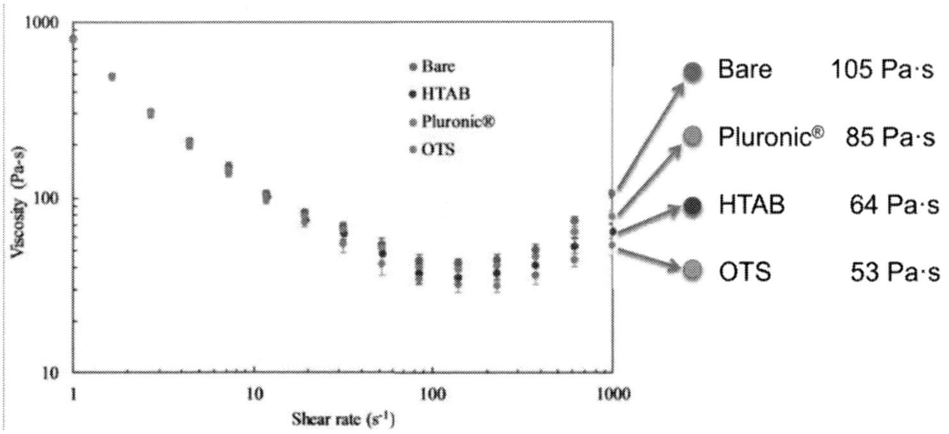

Jamming tests

Slurry in the needle

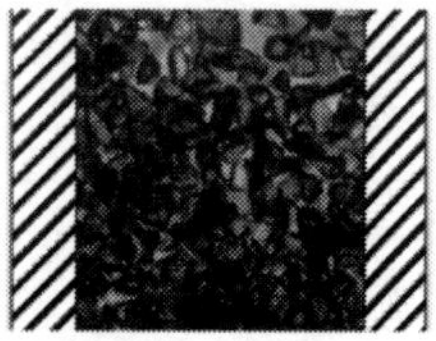

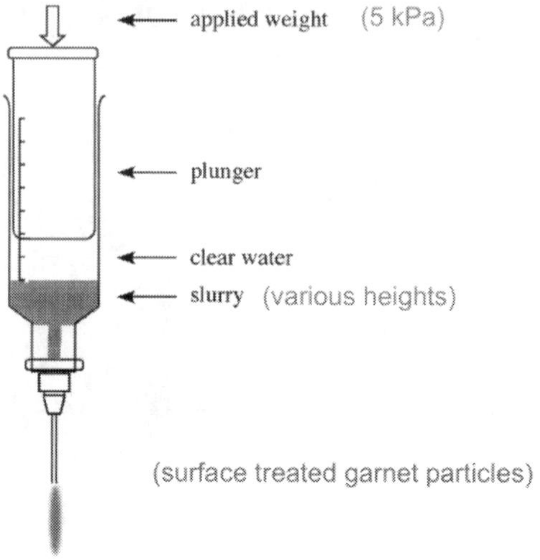

W UNIVERSITY of WASHINGTON

Jamming Results

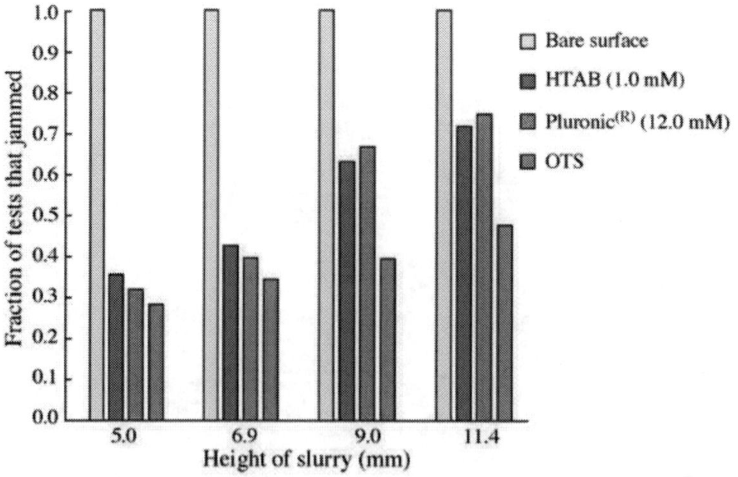

CONCLUSIONS

- LFM tests indicate that pre-treatment of garnet surfaces with HTAB or Pluronic® L62 surfactants, or functionalization with OTS produced decreases in friction by factors of ≈ 2 to 3.
- Surfactants produced maximum friction reduction at concentrations near their respective *CMC*'s.
- Vane tool rheometry showed increasing reduction in shear thickening with surface treatment as shear rate was increased. OTS treatment decreased viscosity at 1000 s^{-1} by a factor of ≈ 2.
- Jamming tests showed significant reduction in jamming with surface treatments.

W UNIVERSITY of WASHINGTON

Thanks to…

- Dr. Mohammed Hashish, Flow International Corp.
- Nick Sage
- Jacob Hatzinger
- UW Center for Surfaces, Polymers & Colloids

503670 Digital Alchemy for Assembly Engineeringc

Abstract:

Back in the day, Stuart Churchill was a pioneer in the use of computers for modeling chemical and thermal processes. Since that time, the role of computing in chemical engineering has become indispensable. In this talk, we present new computational methods we call "digital alchemy" that allow for predictive modeling of thermodynamic assembly, and permit the design of building blocks optimized for assembly into target structures with desired properties. In this approach, standard thermodynamic ensembles are generalized to include alchemical variables and their thermodynamic conjugates, allowing for rapid, real-time computational investigation of large, complex design spaces.

503671 Improved Algebraic, Numerical, and Graphical Representations in Fluid Mechanics

Abstract:

The concept of equivalents is applied to fully developed flows through straight channels, and two applications are introduced: the equivalents that produce equal rates of flow through channels of different cross-section, and the diminishment that occurs between the regimes of laminar and turbulent flow. The validity of the eddy diffusivity and the invalidity of the mixing-length as measures of the shear stress due to turbulence have perhaps never before been demonstrated so clearly. The possible choices of primary variables and the consequences are also examined. The objective of these explorations is an improved understanding of the basic fluid mechanics by teachers, students, and industrial practitioners.

503672 The Scaling of Turbulence Near the Wall and the Churchill Turbulent Flux Correlation: Insights with Lagrangian Simulations

Abstract:

The evaluation of mean velocity and mean temperature in wall-bounded turbulent flows has been based on the "classical" theory of turbulence and the law of the wall. The viscous wall parameters, namely the friction velocity, the friction temperature, and the kinematic viscosity of the fluid have been used to scale not only the mean values of velocity, but also other mean turbulent quantities close to the wall, including the Reynolds stresses and the turbulent velocity and temperature fluctuations. Stuart Churchill, as he questioned the validity of the Reynolds analogy between momentum and heat or mass transfer in his works [1], he realized that extending the scaling of momentum to the scaling of scalar quantities in turbulence was not a sound approach. He proposed a scaling of turbulence that depends on the turbulent momentum flux and the turbulent heat flux. In his AIChE Institute Lecture (in Miami, 1998), Churchill presented a summary of his work on this topic, and closed form solutions for the mean velocity and temperature [2]. The turbulent Prandtl number, the most significant parameter for the modeling of turbulent transport not only in theory but also in modern CFD commercial software packages, was also interpreted physically based on these ideas [3]. We have used direct numerical simulation results and results from Lagrangian simulations [4] for the transport of passive scalars in turbulent channel flows to explore Churchill's ideas on turbulent flow scaling and to develop an updated theory for the scaling of heat transport in turbulent flows [5-7]. His passion to remove empiricisms and approximations from the development of predictive correlations in turbulent transport has been an inspiration. In this talk, we will revisit the Churchill scaling in view of our simulation results, and will discuss the counterintuitive finding that wall turbulence can be used to separate rather than mix substances under special circumstances [8].

1. Churchill, S. W., 1996, "A Critique of Predictive and Correlative Models for Turbulent Flow and Convection," Ind. Eng. Chem. Res., 35, pp. 3122–3140.
2. Churchill, S.W., 2000, "Progress in the Thermal Sciences: AICHE Institute Lecture," AIChE Journal, 46(9), pp. 1704-1722.
3. Churchill, S. W., 2002, "A reinterpretation of the Turbulent Prandtl Number," Ind. Eng. Chem. Res., 41, pp. 6393–6401.
4. Papavassiliou, D.V, 2002, "Turbulent transport from continuous sources at the wall of a channel," Int. J. Heat Mass Transfer, 45(17), pp. 3571-3583.
5. Le, P.M., and Papavassiliou, D.V., 2008, "On the scaling of heat transfer using thermal flux gradients for fully developed turbulent channel and Couette flows," Int. Commun. Heat Mass Transfer, 35(4), pp. 404-412.
6. Srinivasan, C., and Papavassiliou, D. V., 2011, "Prediction of Turbulent Prandtl Number in Wall Flows With Lagrangian Simulations," Ind. Eng. Chem. Res., 50(15), pp. 8881–8891.
7. Srinivasan, C., and Papavassiliou, D.V., 2013, "Heat transfer scaling for wall bounded turbulent flows," Applied Mechanics Review 65(3), Art. 031002.
8. Nguyen, Q., Srinivasan, C., and Papavassiliou, D.V., 2015, "Flow induced separation in wall turbulence", Phys Rev E, 91, Art. 033019.

503674 Flow Boiling Using a Piranha Pin Fin Heat Sink

Abstract:

Flow Boiling using a Piranha Pin Fin Heat Sink

Corey Woodcock, Xiangfei Yu, Yoav Peles[1] and Joel Plawsky

Rensselaer Polytechnic Institute; [1]University of Central Florida

The *Piranha Pin Fin* (*PPF*) microdevice is a heat exchanger engineered to dissipate extreme thermal loads using either dielectric fluids such as HFE7000 or water. Fabrication of the device is CMOS compatible and the first generations of the device have been fashioned from silicon substrates. The PPF operates on the principle that heat transfer would be more efficient if heated fluids could be removed nearly as fast as they are created. This process would keep the driving force as high as possible. Thus the heat exchanger is built as two layers; a primary substrate that dissipates the bulk of the heat and shunts the heated fluids to a booster section that continues to heat the fluid and extends the total surface area of the device. Three generations of devices were fabricated. The first two generations were single-layer systems designed to help understand the operation of the primary layer and optimize the shape and location of the pin fins. The third generation was the first, two-layer system to be fabricated and tested.

Figure 1 Representative PPF array

Correlations for the heat transfer coefficient, h_{tp}, based on the Boiling Number and the Weber Number in the PPF design was developed and validated with Gen II. The discrepancy of heat transfer coefficient between predicted and measured values was within 20%. The ratio of two-phase to single-phase pressure drop was correlated with the Boiling Number (Bo) and the Jakob Number (Ja). The discrepancy of the two-phase multiplier between prediction and experimental measurements were within 12%.

The third generation *PPF* heat sinks were the first series to provide both a primary substrate populated with *PPFs*, as well as a booster substrate populated with solid micro pin fins. The substrates are micro-aligned and fusion-bonded to provide a full three dimensional (3D) architecture with bulk silicon properties. Experimental results from the Gen III *PPF* microdevices provide a proof of concept of the true *PPF* heat sink concept with fluid venting into multiple substrates. At low flowrates (G = 670 [kg/s/m^2]) the Gen III *PPF* heat sink has been shown to dissipate heat loads approaching 1.1 [kW/cm^2] at an exit quality of 95%, while at higher flowrates (G = 3,300 [kg/s/m^2]) the thermal limit of the heat sink remains unknown due to exceeding the available power supply at >1.4 [kW/cm^2].

503673 Transport Problems in the Spirit of Stuart Churchill for Teaching and Research at the University of Michigan

Abstract:

Stuart Churchillâ€™s solutions to important heat transport problems have greatly influence both pedagogy and research. The correlation of Churchill and Chu for the natural convection heat transfer to or from a vertical plate, over an astonishing range of Rayleigh numbers is based partly on a similarity solution of a type common in heat, mass, and momentum transfer for problems in chemical engineering. One such problem is asphaltene or wax transport and deposition onto the wall of an oil pipeline, which can be solved using similar methods, as will be discussed. In addition, such problems, and the wealth of their applications make for excellent motivations and examples in the teaching of graduate or undergraduate transport, as will be discussed in the talk.

494194 Layered Fluid-Fluid Interfaces Confined in Microfluidics

Abstract:

Fluid-fluid interactions are present in a vast number of multiphase processes in biological[1,2] and chemical fields[3,4]. Understanding fluid-fluid mass transport at a interface can help: (i) optimize the manufacture of chemicals, commodities, and pharmaceuticals[5–8], (ii) monitor methane dissolved in groundwater caused by fracturing activities[9,10], (iii) study methane hydrate formation[11], and (iv) investigate methane release from melting permafrost [12,13]. Aside from the mentioned applications, preliminary studies of the methane/water, and non-polar/aqueous film thickness by direct probe has not yet been investigated on a flow microscale (scale of porous or grooved surfaces). In fluid-fluid mass transfer limited systems, the overall performance is closely related to how molecules are transported[14], and how they are affected by attractive and repulsive forces and reactive states present near the interface[15,16]. On the surroundings of an interface, intermolecular and viscous forces play an important role in defining the concentration profiles' shape[17,18]. At the molecular level, the shape is often a manifestation of the excess of free energy, leading to an increase or decrease of the local density[19,20]. The molecular interactions are normally manifested macroscopically as the balance of forces tangent to the surface[19].

The nature of these can be attractive or repulsive towards the interface. Understanding how these forces intervene in mass transfer is fundamental to better understand liquid-vapor processes. One of today's challenges is related to scale-up molecular simulations (MD) to real world scenarios[21,22].

A microfluidic apparatus (semi-flow) associated with in situ Raman spectroscopy is reported for the investigation of toluene-, diethyl ether-, and xylenes- and methane-water interactions. The fluid/fluid phases interfaces exist in static contact by force balance; one is the moving phase across the microfluidic channel, other is trapped inside a μReservoir. The behavior of confined interfaces was characterized using high-resolution *in situ*1D, 2D, and 3D Raman spectroscopy.

It was found that rarefaction, mixture, thin film, and shockwave layers are present. For non-polar/aqueous interfaces, the Schlieren pattern was observed for all solvent pairs. The bulk-to-bulk region was characterized by water rarefaction (17.5 to 23.2 μm), mixing (3.4 to 6.4 μm), and hydrophobic shockwaves (34.9 to 38.8 μm). For gas/liquid interfaces, the apparent CH_4 and H_2O concentrations are reported for different Reynolds numbers (*Re*). The mixture interface is comprised different thicknesses, varying from 19 to 57 μm. The results indicate that the mixture layer thickness (δ) increases with *Re*. Remarkably, traditional transport film theory for unconfined interfaces does not explain this phenomena at confined interfaces. At the interface, it is expected molecular diffusivity enhancement, and/or an evaporation film.

Reference:

1. M. J. Syu, *Appl. Microbiol. Biotechnol.*, 2001, **55**, 10–18.

2. C. W. Li, J. Ciston, and M. W. Kanan, *Nature*, 2014, **508**, 504–507.

3. L. X. Dang and T. M. Chang, *J. Phys. Chem. B*, 2002, **106**, 235–238.

4. L. X. Dang and T. M. Chang, *J. Chem. Phys.*, 1997, **106**, 8149–8159.

5. K. F. Jensen, *Chem. Eng. Sci.*, 2001, **56**, 293–303.

6. P. Claus, D. Hönicke, and T. Zech, *Catal. Today*, 2001, **67**, 319–339.

7. I. Pitault, P. Fongarland, D. Koepke, M. Mitrovic, D. Ronze, and M. Forissier, *Chem. Eng. Sci.*, 2005, **60**, 6240–6253.

8. T. Gervais and K. F. Jensen, *Chem. Eng. Sci.*, 2006, **61**, 1102–1121.

9. R. W. Howarth, A. Ingraffea, and T. Engelder, *Nature*, 2011, **477**, 271–275.

10. R. D. Vidic, S. L. Brantley, J. M. Vandenbossche, D. Yoxtheimer, and J. D. Abad, *Science*, 2013, **340**, 1235009–1:9.

11. A. Vysniauskas and P. R. Bishnoi, *Chem. Eng. Sci.*, 1983, **38**, 1061–1072.

12. E. G. Nisbet, E. J. Dlugokencky, and P. Bousquet, *Science*, 2014, **343**, 493–495.

13. T. R. Christensen, T. Johansson, H. Jonas Å Kerman, M. Mastepanov, N. Malmer, T. Friborg, P. Crill, and B. H. Svensson, *Geophys. Res. Lett.*, 2004, **31**, L04501.

14. J. P. Lopes, S. S. S. Cardoso, and A. E. Rodrigues, *Chem. Eng. J.*, 2011, **176–177**, 3–13.

15. H. Brenner and L. G. Leal, *AIChE J.*, 1978, **24**, 246–254.

16. I.-F. W. Kuo and C. J. Mundy, *Science*, 2004, **303**, 658–660.

17. A. L. Dessimoz, L. Cavin, A. Renken, and L. Kiwi-Minsker, *Chem. Eng. Sci.*, 2008, **63**, 4035–4044.

18. B. Pinho and R. L. Hartman, *React. Chem. Eng.*, 2016.

19. A. Marchand, J. H. Weijs, J. H. Snoeijer, and B. Andreotti, *Am. J. Phys.*, 2011, **79**, 999.

20. S. Iatsevitch and F. Forstmann, *J. Chem. Phys.*, 1997, **107**, 6925–6935.

21. Y. Li, J. Xu, and D. Li, *Microfluid. Nanofluidic*, 2010, **8**, 1011–1031.

22. D. M. Holland, D. A. Lockerby, M. K. Borg, W. D. Nicholls, and J. M. Reese, *Microfluid. Nanofluidics*, 2015, **18**, 461–474.

494233 Focused DC and AC Electric Fields at Conic Pipettes: Nanoscale Thermal Hotspot and Nano-Droplet Generation with Universal Scalings

Abstract:

We report two new phenomena due to focused electric fields at a conic nanopipette and suggest how they can be harnessed for molecular assays:

(1) Ohmic heating due to a DC ionic current through a conic nanopipette or a conic nanopore is shown to produce a nanoscale thermal hotspot whose dimension is independent of the nanopipette length[1]. As in DC Taylor cones, the conic geometry allows for a local analysis of the thermal temperature that is independent of far field conditions. The singular Ohmic heating rate in the bulk fluid at the tip is exactly balanced by the singular conductive heat loss, as the area per unit volume also blows up with the same power in the same limit. Using matched asymptotics, we obtain an estimate of the thermal hot spot temperature and size as a function of a single dimensionless parameter Z that scales quadratically with respect to voltage and cone angle. We use this self-similar local solution to collapse measured quadratic increase in the current at high voltages due to thermal effects on the fluid viscosity and ion conductivity. Temperature approaching 100 degrees C can be reached with modest voltages like 20 V. Nanothermophoresis of molecules is a possible application of this nanofluidic phenomenon due to focused DC field.

(2) We extend our early work on AC spray of liquid droplets into air[2] to generation of water droplets into an immiscible oil phase. The higher viscosity of the oil phase produces significant convection within the water phase. As a result, dynamic AC charge accumulation seen in our previous work was eliminated. Without charge polarization, charged AC cone, nanojets and nanosprays are not observed in oil. A conic pipette is also used to enhance capillary forces, such that the AC Maxwell pressure cannot pull out a jet but pinches off individual droplets at the tip. The frequency of the AC field is further tuned to the inverse RC time of a desirable pinching droplet size and the voltage is adjusted so that the electric Maxwell pressure matches the capillary pressure of the same droplet. Validated size calibrations and phase boundaries are offered in terms of the electric Bond number and a dimensionless frequency. Unlike conventional flow focusing design, which relies on oil/water flow rate ratio for size tuning, we can generate monodispersed water droplets from 1 to 100 Î¼m in diameter by simply tuning the electric field. A digital droplet PCR technology based on this phenomenon is shown to allow nucleic acid quantification over a large dynamic range.

References:

[1] Pan, Zehao, Ceming Wang and Chang, H.-C. "Universal Scaling of Robust Thermal Hot Spot and Ionic Current Enhancement by Focused Ohmic Heating in a Conic Nanopore." , Phys Rev Lett, 117, 134301(2016).

[2] Chetwani, N., Maheshwari, S. and Chang, H.-C."Universal Cone Angle of AC Electrosprays due to Net Charge Entrainment", Phys Rev Lett, 101, 204501(2008).

494582 Real-Time Monitoring of Complex Multiphase Behavior in a High Pressure High Temperature Microfluidic Chip

Abstract:

Real-Time Monitoring of Complex Multiphase Behavior in a High Pressure High Temperature Microfluidic Chip

Renée M. Ripken, Johannes G.E. Gardeniers and Séverine Le Gac

University of Twente, The Netherlands

Here, we determined the phase diagrams of aqueous solutions of model biomass compounds in a novel high-pressure-high-temperature microfluidic set-up. We find that the experimental boiling temperature is significantly higher than the theoretical value due to local pressure effects, the energy barrier for bubble nucleation and superheating effects. Furthermore, this set-up provides exquisite control on the point of nucleation by accurate control of temperature.

Good control on the multiphase behavior of aqueous biomass mixtures is of extremely high importance for the biomass conversion industry. Previously, we developed a theoretical model of the saturated vapor pressure of water/biomass model substrate solutions as a function of the temperature and the substrate mole fraction using the AspenPlus software with a Redlich-Kwong-Soave Boston-Mathias (RKS-BM) model. [1] Here, we experimentally validate the calculated vapor/liquid equilibria (VLE) for temperatures between 25 and 200 °C and for pressures between 1 and 10 bar. For this purpose, a microfluidic platform is considered ideal, as it provides excellent heat transfer properties so that thermodynamic equilibrium is reached fast.

A glass-silicon microfluidic device with a meandering channel (length: 0.4 or 0.2 m; depth: 250 µm and width 250 or 500 µm) was fabricated using standard microfabrication techniques. A dedicated chip-holder was designed to include the device and heater on the silicon side. Three K-type thermocouples allowed controlling the temperature: two between the heater and the silicon, and one in a 300 µm-wide and 450-µm deep groove on the glass side to determine the heat loss over the device. The pressure was controlled with an adjustable back pressure regulator and pressure meter. Aqueous biomass solutions (10 wt% of ethylene glycol, glycerol, xylose or xylitol) were perfused at a flow rate of 10 µL/min using a syringe pump. The temperature was increased until gas bubbles were formed as a result of the phase transition.

A significant difference was observed between the theoretical and experimental saturation temperatures; typically experimental values were degrees higher; sometimes saturation could not be achieved. A local pressure increase due to both the thermal expansion of the liquid and the Laplace pressure, superheating effects and possible fouling of the channel wall by hydrocarbon deposits are considered possible explanations for these findings.

The observed multiphase flow pattern indicated that good control of the point of nucleation could be achieved by taking advantage of the pressure drop across the channel and accurate control of the temperature; a higher temperature resulted in nucleation further upstream in the channel. As the void fraction increases downstream, the flow regime changes from a bubbly flow to Taylor flow to annular flow. All three flow regimes were observed simultaneously in a single microfluidic device.

This high-pressure-high-temperature set-up will be further developed to study the mass and heat transfer and kinetics of biomass conversion for sustainable hydrogen production.

References and acknowledgements

1. "Influence of the water phase state on the thermodynamics of aqueous phase reforming for hydrogen production" R.M. Ripken, J. Meuldijk, J.G.E. Gardeniers and S. Le Gac, ChemSusChem, under revision

The authors want to thank Remco Sanders, Stefan Schlautmann and Sip-Jan Boorsma (University of Twente, The Netherlands) for the help with the development of the experimental set-up and microreactor fabrication.

This work was supported by the Netherlands Center for Multiscale Catalytic Energy Conversion (MCEC), an NWO Gravitation programme funded by the Ministry of Education, Culture and Science of the government of the Netherlands.

494807 Motion of a Deformable Drop in Microchannels of Complex Shape

Abstract:

A theoretical model and a robust algorithm have been developed to simulate tight low-Reynolds number motion of a three-dimensional, freely suspended drop in plane microchannels of complex shape (assuming the channel depth but not necessarily the width to be large compared to the drop size). The channel profile may consist of an arbitrary number of straight line segments with sharp corners in an arbitrary configuration. This geometry provides a suitable model for drop transport in many microfluidic devices with multiple branch bifurcations. One example is the pinched-flow fractionation (PFF) device [1] consisting of two inlet channels at an angle with different flow rates, followed by a plane-parallel and widening sections, and the outlet. The drops fed into one of the input branches (at dilute concentration) are pressed by a buffer fluid (entering the second branch) to the wall of the pinched area, and eventually separate into the outlet according to droplet size. Another, simpler example is drop motion in a T-bifurcation with sharp corners. Here, the fluid flux $Q=Q_1 + Q_2$ through the upstream branch is split into the fluxes through the side (Q_1) and the downstream (Q_2) branches, with a specified partition ratio Q_1/Q_2; of primary interest is to determine the corresponding partition of the drop phase fluxes. Solid sphere transport through a PFF channel and a T-junction has been previously simulated [2]. The main goal of the present study is to determine how different these phenomena are for strongly deformable drops, and explore the effects of the capillary number and drop-to-medium viscosity ratio on the drop transport. In particular, for a T-junction and a large enough solid sphere (still well fitting the junction branches), it was demonstrated [2] that the particle-corner interactions (when the particle rolls around the corner before separating) can make the side branch totally inaccessible to particles, regardless of their initial location in the inlet, even for relatively strong fluid suction through the side branch. The case of a sufficiently deformable drop is much different in that the drop does not hit a corner but wraps around it with some clearance. The occurrence of particle-corner contacts did not allow us [2] to simulate the true temporal dynamics of particle motion (only the particle trajectories could be simulated). In contrast, for a sufficiently deformable drop capable of staying away from sharp corners, both the drop trajectories and temporal dynamics are captured by the present simulations.

The drop size is comparable (or close) to the width of the narrowest branch sections, but is generally much smaller than the overall channel domain size. This difference in the scales makes it difficult to address the problem by a standard boundary-integral (BI) algorithm (or other approaches) for the entire domain due to necessary resolution on the drop and the local vicinity as the drop proceeds through the channel, resulting in a very large total number of BI elements. Instead, we adapt the idea of the "moving frame" BI method [2]. Namely, the drop is embedded in a dynamically constructed computational cell (moving frame, MF) and the 3D BI problem is solved in the cell only, with the outer boundary conditions for the fluid velocity provided by the 2D flow that would exist in the channel without the drop. The MF cell is obtained by intersecting a cubic box around the drop with the entire channel domain, and it can acquire quite complex shapes as the drop proceeds. Parts of the MF boundary may be inside the flow domain, while other parts

may belong to the channel walls, so our method captures strong drop-wall hydrodynamical interactions. The cutting box has an intermediate size, several times larger than the non-deformed drop but typically much smaller than the overall channel domain, thus resulting in many-fold computational savings compared to the standard approach. Robustness of the proposed algorithm is further demonstrated by an example of tight drop motion through a multiple bifurcation system (akin to those in experiments [3]).

[1] Nakashima M., Yamada M. and Seki M. 2004 "Pinched flow fractionation (PFF) for continuous particle separation in a microfluidic device", in proceedings of 17thIEEE International Conference on Micro Electro Mechanical Systems, Maastricht, The Netherlands (Piscataway, NJ), pp. 33-36.

[2] Zinchenko A.Z., Ashley J.F. and Davis R.H. 2012 "A moving-frame boundary-integral method for particle transport in microchannels of complex shape". Phys. Fluids, vol. 24, 043302.

[3] Roberts B.W. and Olbricht W.L. 2006 "The distribution of freely suspended particles at microfluidic bifurcations". AIChE J., vol. 52, pp.199-206.

511033 Microfluidic Models of Porous Media to Examine Asphaltene Deposition in the Presence of Chemical Dispersants

Abstract:

Microfluidic devices offer direct visualization of oil-flow processes. Here we utilize a porous media micro model to characterize how chemical dispersants alter the deposition of asphaltenes in porous media. The chemical dispersants alter the effective size and intermolecular interactions which allow asphaltenes aggregates to become â€œsofterâ€ , which in turn, affects how readily they can be sheared in porous media flows.

496441 Microfluidic Device for the Continuous Measurement of Viscosity

Abstract:

Viscosity is an important variable in chemical processes that is commonly used to monitor the extent of a reaction, the quality of an end product, and the design and optimization of flow systems, among other applications. The accurate measurement of viscosity usually involves sample collection for batch testing that can take hours before analysis is completed. If viscosity is being used as a quality control parameter, the delay in response can prevent actions to correct the process and may not even indicate the current state of the system. We have developed a microfluidic device that can measure viscosity of aqueous solutions in a fast and continuous way.

The device operates by pumping the sample and an immiscible oil phase into a cross junction where droplets are created. At low capillary numbers, the regime our device operates, the mechanism that controls the droplet formation is flow rate breakup, which relates the length of the droplets to their viscosity. Previously we have presented that our device has a linear relationship between droplet length and viscosity, with the results being in agreement with cone-and-plate rheometer measurements. In this talk we will present an optimized design that works with a variety of Newtonian and non-Newtonian fluids including Boger fluids and Xanthan gum solutions. We have also used our device for continuous measurement of viscosity by monitoring the coagulation process in blood at different shear rates. Each device costs only a few dollars to fabricate and can respond to changes in viscosity in the order of seconds. Additionally it can operate using small sample volumes of only nanoliters per hour. These advantages make our technology appealing in chemical industry and biomedical devices that require viscosity measurements in a continuous, inexpensive, fast and reliable way.

498786 Simple Algebraic Formulas for the Practical Interpretation of Mercury Porosimetry Data

Abstract:

Mercury porosimetry is a standard laboratory characterization technique for porous materials. It is used to probe microstructure information of such materials, notably their pore-size distribution. The porous sample, first evacuated in a vacuum, is exposed to mercury under quasistatically increasing pressure, causing mercury to progressively intrude into the pore space of the sample due to capillary effects. Subsequently, the pressure can be quasistatically reduced, leading to mercury extrusion. We record the volume of mercury in the sample as a function of the imposed pressure, which often follows distinct paths for intrusion and extrusion. In practice, mercury porosimetry experiments are commonly interpreted using the "Washburn equation" [1], which uses the intrusion data alone to derive a pore-size distribution, hence disregarding the hysteretic nature of the measurements. Pore-network simulations are based on more realistic pore-scale models and predict hysteresis, but are arguably too sophisticated for the routine analysis of porosimetry data. In this work, we propose a set of simple algebraic formulas for interpreting hysteretic intrusion-extrusion porosimetry data. The model is intended as an incremental improvement over Washburn's approach. It contains three physically meaningful fitting parameters, which captures connectivity effects, contact-angle hysteresis, and mercury entrapment in a minimal fashion. We show the model is easy to apply in practical applications, leads to better estimates of the pore-size distribution, and readily reveals qualitative features of the porous material.

[1] E. W. Washburn, Note on a method of determining the distribution of pore sizes in a porous material, P. Natl. Acad. Sci. USA (1921) 115-116.

499022 A Microfluidic Platform to Measure Dynamic Interfacial Tension of Complex Fluid Systems

Abstract:

Liquid-liquid emulsions are central to many applications, including food processing, oil recovery, liquid filtration as well as environmental science. The accurate and rapid determination of interfacial tension is vital for characterizing their formation and stability behavior. Microfluidics has emerged as a powerful tool for measuring such fundamental properties in a cost-effective and reproducible manner, with low sample volumes and high throughput[1]. Such measurements employ optical microscopy and high-speed imaging to record droplet shape deformation in a contraction-expansion geometry, and further use the parameters extracted from image analysis to calculate interfacial tension using existing theory[2-4]. While applicable to many systems, the underlying assumptions of the theory can limit its application to particularly complex emulsions.

In this work, microfluidic tensiometry experiments are implemented to characterize interfacial properties of water droplets in complex, low viscosity diesel-fuel systems. Diesel fuel systems are multicomponent mixtures containing additives which significantly lower the interfacial tension and hinder the coalescence of water droplets dispersed in the fuel. The dynamic interfacial behavior exhibited by these systems has remained largely unexplored, despite the significant impact that they have on coalescence. This work uses microfluidic drop-shape analysis in contraction-expansion geometries to study how liquid properties, flow conditions, and device geometries can impact the measurement of dynamic interfacial tension. Empirical measurements are also compared to those based on theory. This analysis will enable accurate prediction of interfacial tension in complex, low viscosity liquid-liquid systems, using high-throughput microfluidics. The results of this work reveal new insights into the dynamics of additive transport to fuel-water interfaces.

References:

(1) Cabral, João T., and Steven D. Hudson. "Microfluidic approach for rapid multicomponent interfacial tensiometry." *Lab on a Chip* 6.3 (2006): 427-436.

(2) Taylor, G. I. "The viscosity of a fluid containing small drops of another fluid." *Proceedings of the Royal Society of London. Series A, Containing Papers of a Mathematical and Physical Character* 138.834 (1932): 41-48.

(3) Taylor, G. I. "The formation of emulsions in definable fields of flow." *Proceedings of the Royal Society of London. Series A, Containing Papers of a Mathematical and Physical Character* 146.858 (1934): 501-523.

(4) Rallison, J. M. "The deformation of small viscous drops and bubbles in shear flows." *Annual Review of Fluid Mechanics* 16.1 (1984): 45-66.

499038 Controlled Liquid Entrapment through Photo-Patterned Obstacles and Patterned Surfaces

Abstract:

We present a novel method to trap liquid (either oil or water) in a microfluidic channel by sequentially injecting two immiscible liquids over photo-patterned obstacles or patterned surfaces. By using geometrical arguments, we develop an analytical solution for liquid-liquid interface evolution, and demonstrate that the wetting and shape of the patterned objects govern the trapping process. The solution is consistent with experiments for circular, square, half-moon and raindrop shaped obstacles, and allows us to understand the relative importance of shape and wetting. We also predict the condition for formation of a liquid bridge for two circular obstacles separated by a distance. Based on our understanding of trapping process over isolated obstacles, we trap liquid over sawtooth and sinusoidal patterned surfaces. By varying the amplitude and frequency of the patterned surface, we are able to control and predict the amount of liquid trapped. Our platform can have diverse applications in diagnostics, drug discovery, reaction engineering, material synthesis and printing processes.

500815 Tears of Wine

Abstract:

`Tears of wine' refer to the rows of wine-drops that spontaneously emerge within a glass of strong wine. Evaporation-driven Marangoni flows near the meniscus of water-alcohol mixtures drive liquid upward forming a thin liquid film, and a rim or ridge forms near the moving contact line. Eventually the rim undergoes an instability forming drops, that roll back into bulk reservoir forming so called tears or legs of wine. Most studies in literature argue the evaporation of more volatile, lower surface tension component (alcohol) results in a concentration-dependent surface tension gradient that drives the climbing flow within the thin film. Though it is well-known that evaporative cooling can create temperature gradients that could provide additional contribution to the climbing flows, the role of thermocapillary flows is less well-understood. Furthermore, the patterns, flows and instabilities that occur near the rim, and determine the size and periodicity of tears, are not well-studied. Using experiments and theory, we visualize and analyze the formation and growth of tears of wine. The sliding drops, released from the rim towards the bulk reservoir, show oscillations and a cascade of fascinating flows that are analyzed for the first time.

502752 Kinetics of (un)Binding between DNA-Functionalized Particles Using a Coarse-Grained Model with Explicit Nucleotide Representation

Abstract:

Recent technological advances have required the manufacture of materials at increasingly precise scales. Developing such materials, requiring finely ordered structures up to the nanometer scale, is infeasible using traditional methods. This has prompted interest in self assembling materials like DNA-functionalized particles (DFPs). In this research, we use a coarse-grained DFP model which preserves key chemical and structural properties of DNA by explicitly representing nucleotide bases to study the kinetics of two complementary particles. We explore the effects of temperature, DNA surface density, number of sticky beads, and location of the sticky beads on the binding lifetimes of particles, with a particular focus on how these factors can be used as tuning parameters for the design of new materials. Our results should help tune assembly conditions and obtain desired crystal lattices unhindered by kinetically arrested amorphous or undesired structures.

501522 Hydration Structure and Dynamics of Poly(2-methacryloyloxyethyl phosphorylcholine)

Abstract:

We explore the hydration structure and dynamics of poly(2-methacryloyloxyethyl phosphorylcholine)50, or pMPC50, in solution and grafted to silica using molecular dynamics simulations with an emphasis on isolating the contributions of the key chemical moieties. Surface bound films of this biocompatible polymer have been shown experimentally to produce tribological properties that surpass those of the human synovial joint [1] and are being developed for use in artificial joints [2]. The mechanism by which such materials are thought to provide ultra-low friction coefficients has been termed "hydration lubrication" [3]. In a simplified monolayer analogue to the brush structures, water hydrates surface bound MPC monomers and enables such materials to utilize the hydration lubrication mechanism as monolayers experience shear; a clear cross-over, as a function of compression, is found between regimes dominated by hydration lubrication and those where MPC-MPC interactions dominate [4]. Previous results [5] showed that the charged choline groups of individual monomers tend to fold onto neighboring monomers yielding solvent facing, highly hydrated phosphoryl groups.

Here, we further analyze the effect of the two outer chemical moieties, the choline and phosphoryl groups. For bulk solutions, we compare the behavior of pMPC to poly(2-methacryloyloxyethyl phosphate) and show that the removal of the choline group reduces hydration around the polymer. When bound to a substrate, the monolayer analogue to the experimental system did not permit significant inter-chain associations via the choline groups. Simulations of much larger pMPC50 brush layers demonstrate that the choline moiety serves to stabilize the material through non-covalent cross-links between individual strands in the layer. Preliminary results strengthen the previous hypothesis [5] that the choline moiety is responsible for stabilizing the polymer while the phosphoryl moiety contributes to the hydration lubrication mechanism.

[1] M. Chen et al., "Lubrication at Physiological Pressures by Polyzwitterionic Brushes." Science 2009, 323, 1698–1701.

[2] M. Kyomoto et al., "Biomimetic hydration lubrication with various polyelectrolyte layers on cross-linked polyethylene orthopedic bearing materials." Biomaterials 2012, 33, 4451-9.

[3] J. Klein, "Hydration lubrication." Friction 2013, 1, 1-23.

[4] C.Klein, C.R. Iacovella, P.T. Cummings, C. McCabe, "Tunable transition from hydration to monomer-supported lubrication in zwitterionic monolayers revealed by molecular dynamics simulation" Soft Matter 2015, 11, 3340-3346.

[5] W.L. Roussell, C. Klein, C.R. Iacovella, P.T. Cummings, C. McCabe, "Molecular Origins of the Ultra-Low Friction Exhibited by Biocompatible Zwitterionic Polymer Brushes" Young Scientist, May 2015

497706 Polyelectrolyte Interactions: Simulation and Theory

Abstract:

Polyelectrolyte interactions govern many natural and technological processes. Recent experiments and simulations suggest like-charged polymers to attract one another under certain conditions, owing to preferential counter-ion positioning between the two chains. Such attraction may have significant implications to nucleic acid delivery and polyelectrolyte film formation. We address here the problem of like-charged polymer attraction using a coarse grained model, consisting of two similarly charged aligned cylinders, oppositely charged spherical counter-ions, and spherical salt species. Employing Monte Carlo simulation, we show polymer-polymer attraction to correlate strongly with a Coulomb coupling parameter, defined as the ratio of i) the electrostatic interaction energy of two counter-ions at the cylinder surface and ii) the thermal energy. A simple statistical mechanical model is proposed, where ions â €œcondensedâ€ to the cylinder surface are treated as a highly correlated 2-D one-component plasma, and ions away from the surface are treated in a mean field description. We discuss these results in light of recent experimental findings of significantly enhanced polyelectrolyte adsorption to electrified interfaces.

498527 Modeling Solute and Solvent Distributions in Functionalized Dendrimers from iSAFT Density Functional Theory

Abstract:

A new density functional theory (DFT) formalism is designed to study the phase behavior of dendrimers in inhomogeneous systems. Dendrimers are highly branched and globular macromolecules. They are of great research interest because of their potential applications such as in drug delivery and catalysis. However, it is a challenge to model the structure or phase behavior of dendrimer systems by theory or simulation because of their complex architectures and usually big sizes. The new DFT formalism proposed by us is based on inhomogeneous statistical associating fluid theory (iSAFT) and it is capable of predicting the radius of gyration and radial density profile for all types of dendrimers. The effect of implicit solvent on dendrimers predicted by the theory agrees very well with simulation results. Further the theory allows us to study the behavior of dendrimers in explicit solvents. To test applicability of the theory for drug delivery applications, we have modelled the solute and solvent distributions in functionalized dendrimers and arrived at similar conclusions as

experimentalists. We believe this theory to be a powerful tool to model dendrimer systems with advantages over simulation (less computational expense) and other theories (more detailed information).

499316 Dependence of Relaxations and Mechanical Properties on Molecule Shape in Dissipative Particle Dynamics

Abstract:

Dissipative Particle Dynamics (DPD) has been established as an effective way to reach long time scales in molecular simulations. Strong forces at short separations are neglected in the method, and consequently each DPD particle represents a local collection of neighboring atoms and molecules. Our long-term interest is simulating amorphous mixtures of molecules that relax over a range of time scales; DPD is of interest for averaging over sufficient numbers of relaxations. To quantify relaxation dynamics in DPD, we constructed systems that share common interparticle interactions yet differ in bonding geometry. We chose systems of spheres in which some were bonded into simple geometries -- dimers, trimers, triangles, squares -- to focus on how physical shape affects relaxations and mechanics. At fixed density, the pressure varied by about ± 3% as a consequence of bonding between the extremes of 0 and 100% unbonded spheres. Larger bonded entities diffused more slowly; at low temperatures they showed an anomalous region ($r^{\wedge 2} \sim t^n$, $n<1$) over moderate time scales prior to reaching a diffusive regime. Diffusion coefficient and rotational relaxation times, in dimensionless simulation units, changed less than predicted by an Arrhenius dependence over temperatures of $k_BT/a = 0.15$ to 1.2, where a is the parameter that quantifies conservative particle-particle repulsive forces. The difference from Arrhenius (i.e. "strong") or fragile liquid molecular-scale behavior indicates that no single set of mass, length, and time units can interrelate DPD and molecular results for these liquids. Comparing stress contributions to modulus and viscosity indicates that kinetic energy, random force, and dissipative force contributions are independent of particle-particle bonding. Only the conservative-conservative and bond forces lead to viscosities that increase with decreasing temperature and with the presence of bonds. Near k_BT/a â‰ˆ 1, viscosity rises with temperature, as in a gas. Relaxations of correlation functions that are derived form generalized hydrodynamics also support gas-like mechanical behavior near this temperature.

492610 Thermodynamics of Self-Assembly of Perylene Derivatives

Abstract:

Functional organic materials composed of dye-based molecules that stack into various structures by non-covalent intermolecular forces have fascinating optical and mechanical properties. These novel materials are of increasing importance for many high-technology applications. Here is a need for a rational design of such materials based on the engineering of dye-dye interactions and the prediction of how these interactions impact the properties of nano- or bulk state materials. The required insight into the underlying physical processes can be derived from investigations on dye assemblies that constitute the intermediate state of matter between monomeric dyes and supramolecular materials. Perylene di-imide derivatives (PDIs) emerged as a prototype class of molecules for the elucidation of the transition from monomeric to bulk materials via the supramolecular state [1,2]. Their self-assembly into complex structures is determined by size and shape of the monomeric unit, system composition and thermodynamic boundary conditions. Given this diversity of factors, a molecular-level understanding of the various driving forces is essential for the rational material design.

In this project the free energy for the formation of dimers and higher aggregates are calculated for prototypical molecular building blocks carrying different functional groups in different solvent environments using molecular dynamics free-energy simulations based on force fields inherited from the area of biomolecular simulations. The temperature dependence of the free energy gives access to the entropic contribution while the change in potential energy during assembly provides the enthalpic contribution.

We discuss the influence of the molecular architecture and the solvent composition on the thermodynamic fingerprint of the aggregation and investigate the ability of classical molecular dynamics simulations to be used in rational materials design.

[1] F. Würthner, C. R. Saha-Möller, B. Fimmel, S. Ogi, P. Leowanawat, D. Schmidt, *Chem. Rev.* **116**, 962 (2016).

[2] E. Krieg, M. M. C. Bastings, P. Besenius, B. Rybtchinski, *Chem. Rev.* **116**, 2414 (2016).

498840 A Molecular Dynamics Study of Actinide Nanoclusters

Abstract:

A Molecular Dynamics Study of Actinide Nanoclusters

Ken Newcomb, Edward J. Maginn

The actinides are an important class of materials to modern society, due to their role in nuclear power generation, nuclear waste treatment, and national security. Recent experimental work has led to the characterization of a new class of materials: actinide nanoclusters[1]. UO_2 groups self-assemble in aqueous solution to form the closed clusters, ranging in size from 1.5 to 3.0 nm in diameter. The nanoclusters could play an important role in nuclear fuel reprocessing, since they can be readily separated from aqueous solution using ultrafiltration. We have developed a classical force field for the simulation of U_{20} nanoclusters in water, parameterized from highly accurate quantum-mechanical calculations. The clusters carry a large negative charge, which causes cations in solution to occupy locations within and around the cluster. These cations are known to play a key role in cluster aggregation. but the exact positions and quantity of cations within the cluster are hard to determine experimentally. By performing molecular dynamics simulations, the free energy profile associated with bringing various cations (Li^+, Na^+, K^+, Rb^+, Cs^+) from the bulk into the clusters has been calculated, revealing a process hindered by two significant barriers. Using this method, the equilibrium distribution of cations around the cluster was determined.

References:

1) Burns, P. C., Kubatko, K., Sigmon, G., Fryer, B. J., Gagnon, J. E., Antonio, M. R., & Soderholm, L. (2005). Actinyl Peroxide Nanospheres. *Angewandte Chemie Angew. Chem., 117*(14), 2173-2177. doi:10.1002/ange.200462445

509054 Solve this! Fundamental Approach to Problem Solving in Industrial Processes I (Invited Talks)

Abstract:

Solving complex plant problems is an everyday exercise that affects all projects and budgets, and it is very rarely introduced to students at any level. In fact, by the time young engineers are few years into working in industry, most have lost the engineering skills that they developed in school. We hope to encourage future engineers to retain those skills and learn the importance of plant problem-solving.

501857 Solubiltiy and Phase Equilibria for Optimizing the Processing of Cannabinoids and Terpenes in Cannabis and Hemp Extractions

Abstract:

The extraction and processing of the components associated with medicinal cannabis and hemp extracts is rather daunting since cannabis plant materials can have in excess of 400 molecular compounds. Two molecular groups consisting of 12 major cannabinoids and 10-15 terpenoids components are of particular interest and the focus of using pressurized fluid extractions for their processing. Pressurized fluids such as carbon dioxide and butane coupled with conventional liquid solvents such as ethanol and water are routinely used to produce cannabis concentrates in which the pressurized fluid is used in both the sub- as well as the supercritical state (CO_2), or as a near critical fluid in the case of butane. Optimization of extraction and fractionation conditions is difficult due to the lack of fundamental thermodynamic solubility data and phase equilibria. Sources of useful data have been employed toward this end using extraction data available for essential oil components in liquefied- and supercritical- carbon dioxide as well as the meager literature data available at lower pressures (<200 bar) for important cannabinoids such as cannabinol (CBD) and tetrahydrocannabinol (THC). Solubility data from the Delft study show that CBD and THC solubility increases with pressure, differentially for the major cannabinoid components allowing for their fractionation at these low pressures as well as exhibiting the â€œcrossover regionâ€ between 110-140 bar and 40-60°C· Â Solubility parameter theory has also been employed to rationalize experimental extraction data by using group contribution calculations based on the molecular structure of the major cannabinoid and terpene components for both critical fluids and liquid solvent media - particularly with respect to their sequential use in extractions in both the near critical and supercritical state. For the major cannabinoid components found in cannabis and hemp extracts, solubility parameter theory predicts solubility maximum occurring to occur at quite high extraction pressures equivalent to solubility parameters of 22-25 $MPa^{1/2}$, or equivalent to the solubility parameter of ethanol. Many of the difficult to isolate terpenoid components show solubility minimum in CO_2 at 80-90 bar and 60°C (limonene and beta-carophyllene) and their solubility increases in both sub- and supercritical CO_2 with both an increase or decrease in temperature. At higher extraction pressures and temperatures in CO_2, terpenoids should exhibit solubility maxima at lower pressures than those for the acidic and neutral cannabinoid compounds, thereby permitting the fractionation at either subcritical or supercritical CO_2 conditions. Major terpenoid components such as beta-carophyllene found in cannabis can be fractionated from plant wax material optimally below 40°C and a CO_2 pressure of 90 bar since the wax components exhibit a solubility maxima under these conditions. Separator conditions can be adjusted between 30-70 bar and 0-40°C to separate the essential oil components, i.e., the terpenes from the plant wax material The presence of water in the plant material can also influence the terpenoid composition in the resultant extract at conditions close to P_c of CO_2 and the temperature range of 40-50°C. Aside from predicting optimal extraction conditions for cannabinoid and terpene constituents, the above approaches will also be used along with equation-of-state calculations for selecting conditions for precipitation of these solutes in separator vessels as commonly employed in processing of cannabis and hemp extracts. In the case

of the cannabinoid components, CBD, delta-8-THC, and delta-9-THC, both sub- and super-critical CO_2 conditions are used for the adjustment of precipitation conditions in separator vessels between 29-50°C and 55-70 bar. An example of the sequential extraction of terpenes and cannabinoids from cannabis using subcritical (69 bar/26°C) and supercritical (240 bar/40-45°C) CO_2 will be shown, and conditions suggested for stripping ultra-low levels of pesticides from cannabis and hemp matrices.

484103 Measurement and Modeling of Vinyl Fluoride Solubility in Aqueous Lithium Bis(trifluoromethylsulfonyl)Imide Solutions

Abstract:

This presentation will highlight experimental (*P-x*) phase equilibrium results for the solubility of gaseous vinyl fluoride (VF) in lithium bis(trifluoromethylsulfonyl)imide ($Li^+Tf_2N^-$) + water (H_2O) solutions at pressures up to 40 bar. Vinyl fluoride solubility will be explored as a function of pressure and temperature at constant $Li^+Tf_2N^-$ loadings (20.15, 40.0, and 80.0 wt.%) and as a function of pressure and salt composition at constant temperatures (10°C, 30°C, and 100°C). The experimental results will demonstrate that VF is highly soluble in aqueous $Li^+Tf_2N^-$ salt solutions and that solubility is directly proportional to salt loading. A temperature dependent Peng-Robinson equation of state with a van der Waals 1-parameter mixing rule will be utilized to model the experimental results. In addition this presentation will highlight a few examples where $Li^+Tf_2N^-$ + H_2O solutions could be utilized as a new solvent for fluoromonomer polymerizations and as a new working fluid for absorption-cooling applications.

504027 Heat and Momentum Transport Properties of CO_2-Expanded Liquids (CXLs): N-Hexane, n-Decane, and n-Tetradecane

Abstract:

Many important processes exploit liquids saturated with significant quantities of gases at elevated pressures for separations, reactions, and materials processing; e.g. CO_2 capture and sequestration; enhanced oil recovery ; Gas- or CO_2 -expanded Liquids; particle formation from Gas or Supercritical Anti-Solvent methods; etc. To engineer these systems, the thermodynamic and transport properties are required. This contribution is one of the first reports for the full heat transport properties (thermal conductivity, thermal diffusivity, heat capacity, and dynamic and kinematic viscosity) of gas-saturated liquids at elevated pressures for the binary systems of CO_2 and the n-alkanes, n-hexane, n-decane, or n-tetradecane, at 25°, 40°, and 55°C and pressures up to 106 bars. Equation of state modeling of literature vapor-liquid equilibrium data was used to determine the compositions at the conditions of the experimental heat transport measurements. All measured properties decrease with increasing pressure (i.e. increasing composition of CO_2) and tend to be relatively linear until a CO_2 composition of approximately 70%mole. The Prandtl number, Pr, is calculated and the effect of CO_2 composition on heat transfer engineering is discussed.

Van der Waals Phenomenological Transport EOS for Dense Viscosities of the Ultra-Deep Petroleum Reservoirs in the Gulf of Mexico

Akanni S. Lawal
Petroleum Engineering Department
Texas Tech University, P. O. Box 43111
Lubbock, Texas 79423

Take Away From This Presentation

- **There is Phenomenological Similarity between the Graphs of P-V-T and T-η-P**
- **No Agreement on Whether Viscosity η is Due to Attractive or Repulsive Forces (Brush: *Chem. Rev.* 1962)**
- **Viscosity η Correlating Methods Rely on Judicious Choice of Density ρ Correlations**
- **NONE of the η Correlating Methods Reduces to $f(T_R, \eta_R, P_R)$**
- **Transport EOS Require NO Density ρ or Cp Correlation**
- **NO Molecular Descriptors For High-Molecular-Weight Fluids**
- **NO Tb, Tc, Pc, Vc, Zc, ω, PNA for Reference Standard High-Temperature High-Pressure (HTHP) Liquid Viscosity**

2017 AIChE Annual Meeting, Minneapolis

2

iC4 Viscosity (Gonzalez-Lee, JCED 1966)

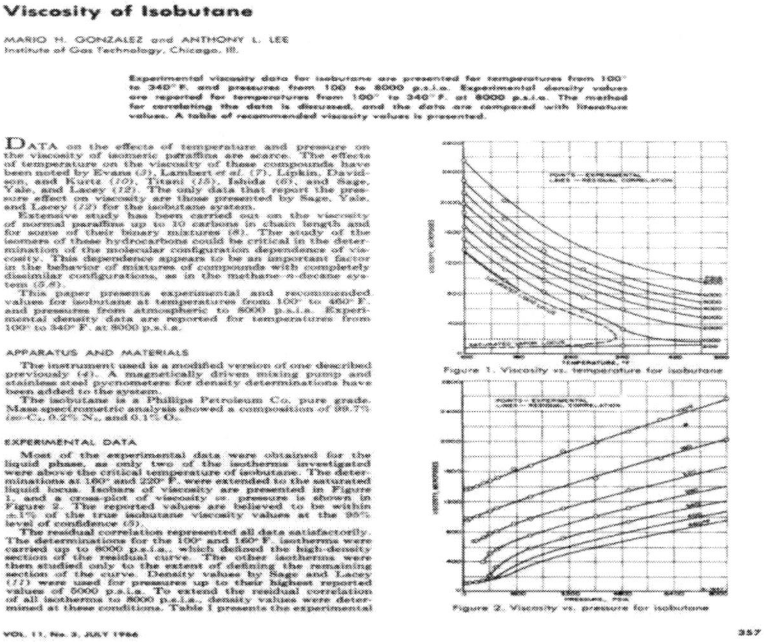

Viscosity of Isobutane

MARIO H. GONZALEZ and ANTHONY L. LEE
Institute of Gas Technology, Chicago, Ill.

Experimental viscosity data for isobutane are presented for temperatures from 100° to 340° F. and pressures from 100 to 8000 p.s.i.a. Experimental density values are reported for temperatures from 100° to 340° F. at 8000 p.s.i.a. The method for correlating the data is discussed, and the data are compared with literature values. A table of recommended viscosity values is presented.

DATA on the effects of temperature and pressure on the viscosity of isomeric paraffins are scarce. The effects of temperature on the viscosity of these compounds have been noted by Evans (3), Lambert et al. (7), Lipkin, Davidson, and Kurtz (10), Titani (15), Ishida (6), and Sage, Yale, and Lacey (12). The only data that report the pressure effect on viscosity are those presented by Sage, Yale, and Lacey (12) for the isobutane system.

Extensive study has been carried out on the viscosity of normal paraffins up to 10 carbons in chain length and for some of their binary mixtures (8). The study of the isomers of these hydrocarbons could be critical in the determination of the molecular configuration dependence of viscosity. This dependence appears to be an important factor in the behavior of mixtures of compounds with completely dissimilar configurations, as in the methane-n-decane system (5,8).

This paper presents experimental and recommended values for isobutane at temperatures from 100° to 460° F. and pressures from atmospheric to 8000 p.s.i.a. Experimental density data are reported for temperatures from 100° to 340° F. at 8000 p.s.i.a.

APPARATUS AND MATERIALS

The instrument used is a modified version of one described previously (4). A magnetically driven mixing pump and stainless steel pycnometers for density determinations have been added to the system.

The isobutane is a Phillips Petroleum Co. pure grade. Mass spectrometric analysis showed a composition of 99.7% iso-C_4, 0.2% N_2, and 0.1% O_2.

EXPERIMENTAL DATA

Most of the experimental data were obtained for the liquid phase, as only two of the isotherms investigated were above the critical temperature of isobutane. The determinations at 160° and 220° F. were extended to the saturated liquid locus. Isobars of viscosity are presented in Figure 1, and a cross-plot of viscosity vs. pressure is shown in Figure 2. The reported values are believed to be within ±1% of the true isobutane viscosity values at the 95% level of confidence (5).

The residual correlation represented all data satisfactorily. The determinations for the 100° and 160° F. isotherms were carried up to 8000 p.s.i.a., which defined the high-density section of the residual curve. The other isotherms were then studied only to the extent of defining the remaining section of the curve. Density values by Sage and Lacey (11) were used for pressures up to their highest reported values of 5000 p.s.i.a. To extend the residual correlation of all isotherms to 8000 p.s.i.a., density values were determined at these conditions. Table I presents the experimental

Figure 1. Viscosity vs. temperature for isobutane

Figure 2. Viscosity vs. pressure for isobutane

VOL. 11, No. 3, JULY 1966 357

iC4 Viscosity Data Conform to the Application of the Cubic Equations of State

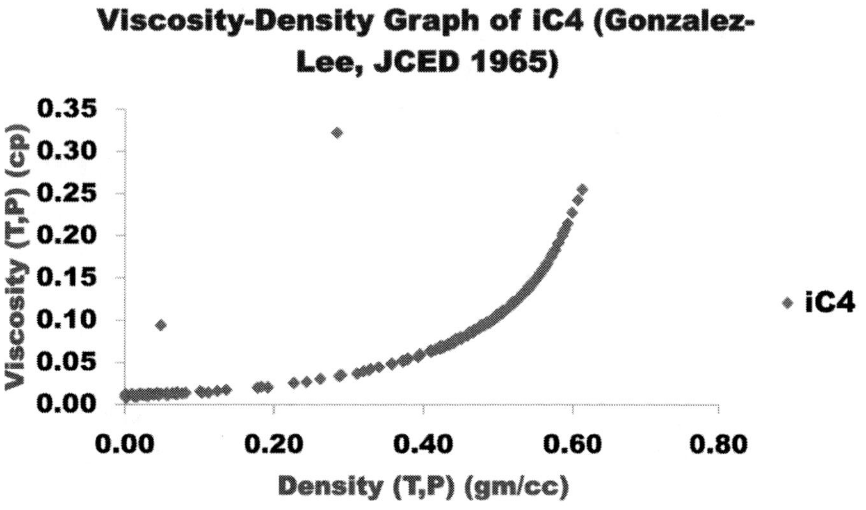

iC4 Viscosity Data DO NOT Conform to the Corresponding State Principles

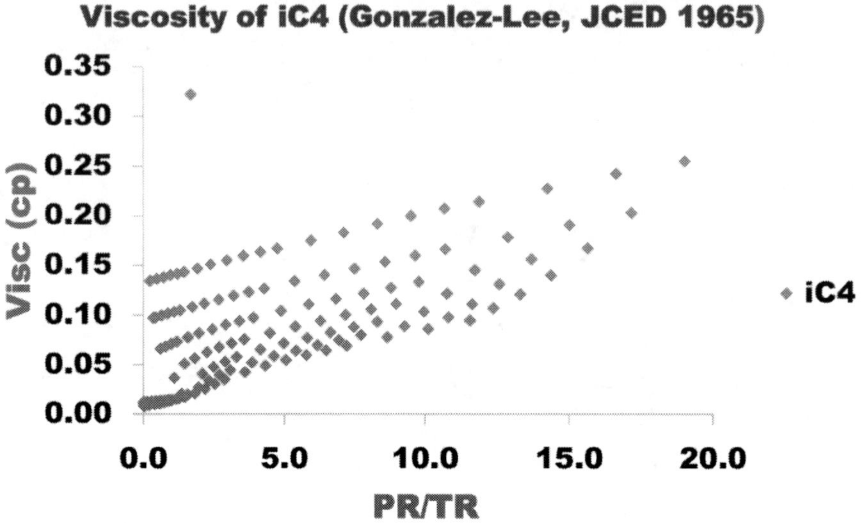

n-C5 Viscosity (Lee-Ellington, JCED 1965)

Viscosity of *n*-Pentane

A. L. LEE and R. T. ELLINGTON[1]
Institute of Gas Technology, Chicago, Ill.

Experimental *n*-pentane viscosity data are presented for temperatures from 100° to 340° F. and pressures from 200 to 3000 p.s.i.a. The methods for correlation are discussed, and the data are compared with literature values. A table of recommended viscosity values is presented.

RESULTS of the ethane, propane, and *n*-butane viscosity investigations have been presented (4, 6, 16). A detailed study of the viscosity behavior of ethane, propane, and *n*-butane in the vicinity of their critical points (15), and that of binary mixtures of methane–*n*-butane has also been reported (3, 10).

Experimental results of this investigation were used to test two correlations. Recommended values for *n*-pentane viscosity are presented for temperatures from 100° to 460° F. and pressures from atmospheric to 8000 p.s.i.a.

APPARATUS AND MATERIALS

The viscometer used for this investigation has been described in detail (3, 5, 7). The *n*-pentane used was Phillips Petroleum Co. pure grade, certified 99 mole % minimum purity.

EXPERIMENTAL DATA

All the data are for the liquid phase, and they were obtained only to the extent necessary to obtain a residual viscosity correlation and test the generalized 9-parameter equation of Lee and coworkers (10). In previous work (4, 6, 16), the residual concept was satisfactory for representing the viscosity behavior of light hydrocarbons. Therefore, extensive data were not required for all regions, particularly for high density regions.

Isobars of *n*-pentane viscosity are presented in Figure 1, and a crossplot of viscosity vs. pressure is also presented in Figure 2. These plots are based on the authors' experimental data and correlations to be discussed. Experimental

[1] Present address: Sinclair Research, Inc., Tulsa, Okla.

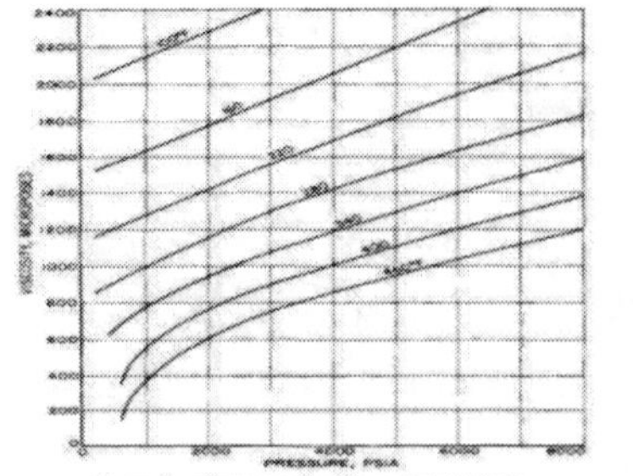

Figure 1. *n*-Pentane viscosity vs. temperature

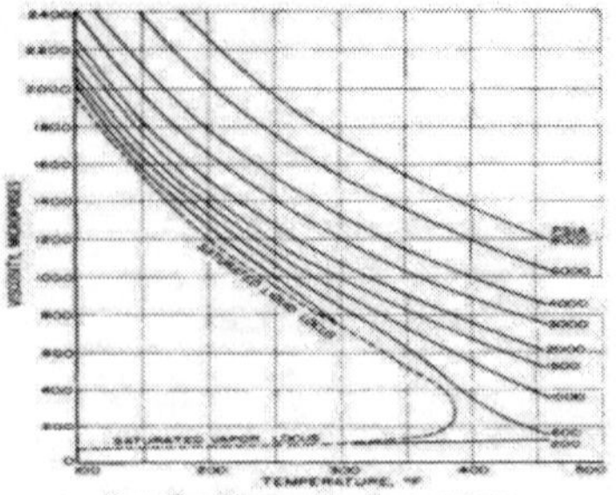

Figure 2. *n*-Pentane viscosity vs. pressure

VOL. 10, No. 2, APRIL 1965 101

nC5 Viscosity Data Conform to the Application of Cubic Equations of State

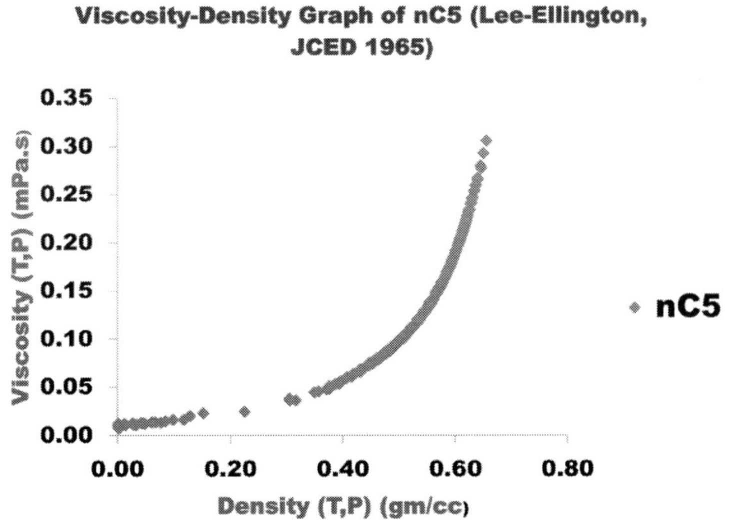

2017 AIChE Annual Meeting, Minneapolis

7

nC5 Viscosity Data DO NOT Conform to the Corresponding State Principles

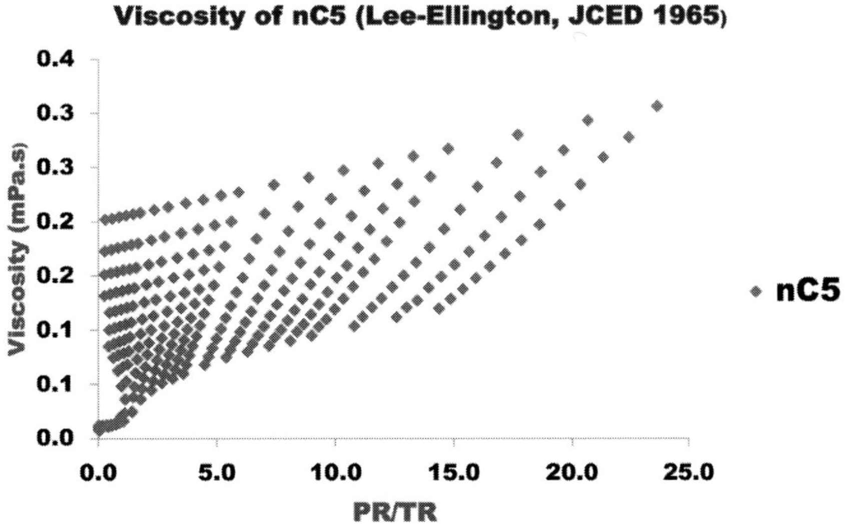

2017 AIChE Annual Meeting, Minneapolis

8

Viscosity Data of Isopropyl Ether Conform to the Application of Cubic Equations of State

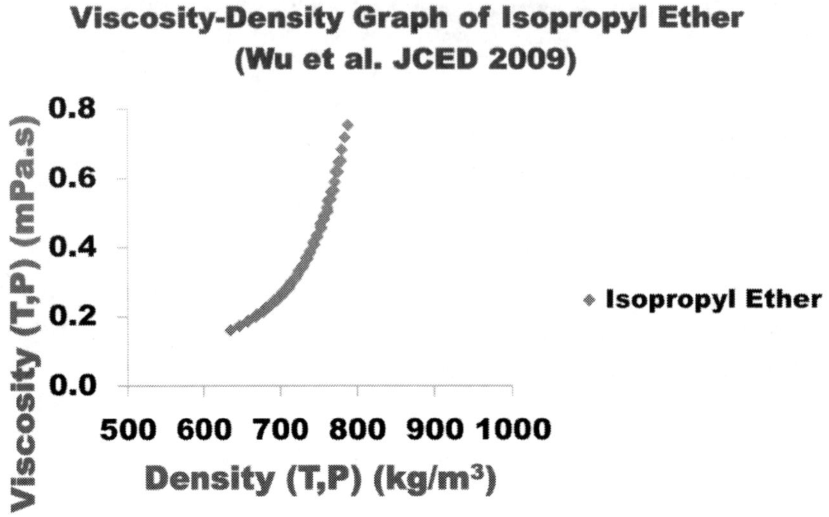

2017 AIChE Annual Meeting, Minneapolis

9

Viscosity Data of Isopropyl Ether DO NOT Conform to the Corresponding State Principles

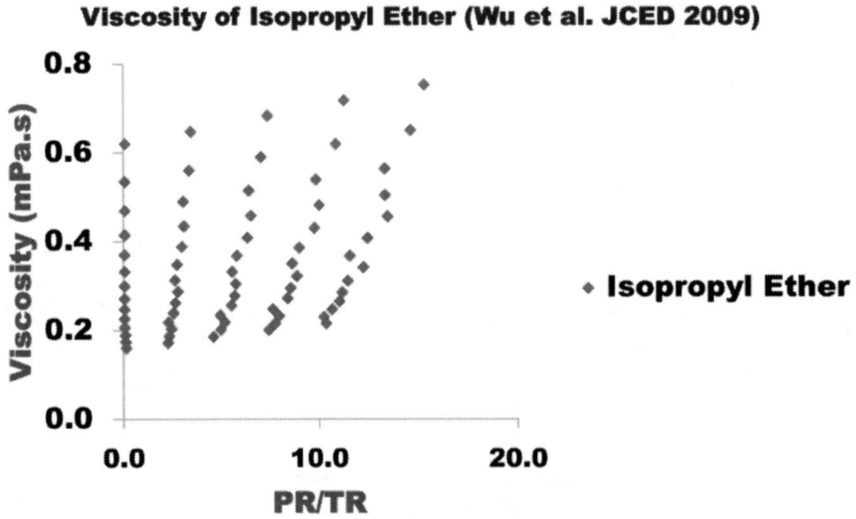

2017 AIChE Annual Meeting, Minneapolis

10

Viscosity Data of n-Butyl Ether Conform to the Application of Cubic Equations of State

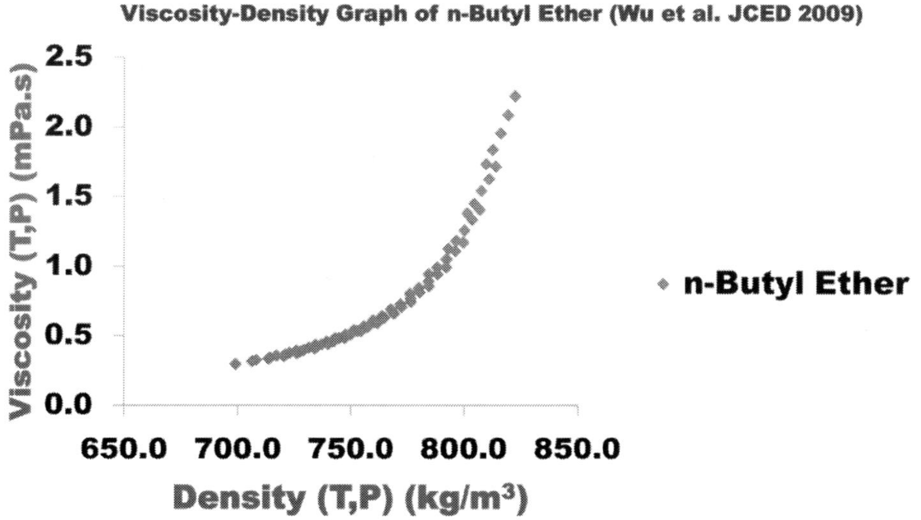

2017 AIChE Annual Meeting, Minneapolis

11

Viscosity Data of n-Butyl Ether DO NOT Conform to the Corresponding State Principles

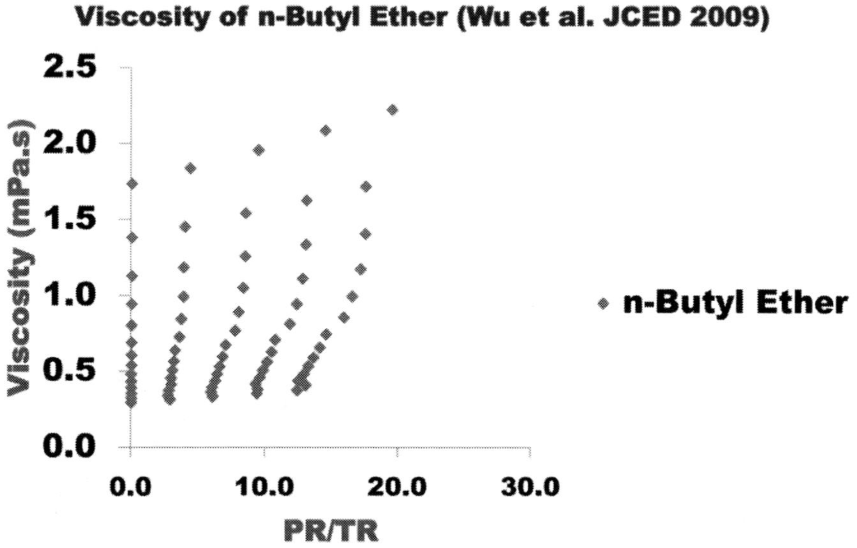

2017 AIChE Annual Meeting, Minneapolis

12

100 Years Survey of the Notable Transport Cubic Equations of State

- **Phillips:** VdW-based Viscosity of CO_2 (Proc. R. Soc. 87A 1912)
- **Phillips:** VdW-based VEOS (MIT J. 1922)
- **Brush:** Theories of Liquid Viscosity (Chem. Rev. 1962)
- **Lohrenz-Bray-Clark: Residual VEOS** (JPT 1964)
- **Little & Kennedy:** VdW-based VEOS (SPE J. 1968)
- **Lawal:** LLS-based VEOS (SPE/DOE Paper 14926, 1986)
- **Guo et al.:** PR-based VEOS, (FPE 1997); (J. Pet. Sci. Eng. 2001)
- **Tian & Wang:** PR-based VEOS (FPE 2006)
- **Wu et al.** PR-based VEOS (FPE 2014)
- **Bonyadi & Rostami:** SRK-based VEOS (FPE 2017)

Selection Criteria for Van der Waals Cubic Equations of State Procedure

- Cubic EOS Applicable to the Entire PVT States
- **Cubic EOS Applicable to Gas-Liquid Critical Point**
- Uses Measurable Macroscopic Properties: Pc, Tc, Zc, ω
- **Conforms to Corresponding States Law:** $f(P_R, \Phi_R, T_R) = 0$

Testable Hypothesis for Gas-Liquid Viscosity and Dilute Gas and Dense Gas Viscosities

- Brush (**Chem. Rev., 1962; 584 Refs.**):

  "…No General Agreement on Whether Viscosity is Essentially Due to Attractive or Repulsive Forces"
- No Statistical Theory for Dynamic Viscosity and Thermal Conductivity
- **Density and Viscosity are due to Molecular Motion & Interaction (Microscale)**
- Density and Viscosity are Function of Thermodynamic State of Fluids

Thermodynamic Equation of State and the Transport Cubic Equation of State

- **Thermodynamic State Equation (LLS EOS)**

$$P = \frac{RT}{v - b} - \frac{a(T)}{v^2 + \alpha b v - \beta b^2}$$

- **Viscosity Equation of State (LLS VEOS)**

$$T = \frac{R^{vdw}P}{\eta - b} - \frac{a(P)}{\eta^2 + \alpha b \eta - \beta b^2}$$

Universal Van der Waals Transport Cubic Equation of State for Transport Properties

- Based on the Framework of the Lawal-Lake-Silberberg (LLS) Cubic Equation of State

Universal Transport Property Cubic Equation:

$$T = \frac{\left(\frac{T_c\phi_c}{P_cZ_c}\right)P}{\phi - b(T)} - \frac{a(T_c)\,\gamma(P_R,\omega)}{\phi^2 + \alpha b(T)\phi - \beta[b(T)]^2}$$

$$\boldsymbol{\phi \equiv D_e, \eta, \left(\frac{\eta}{\rho}\right), \lambda, \left(\frac{\lambda}{\rho C_p}\right), \left(\frac{\eta C_p}{\lambda}\right), \left(\frac{\lambda}{\eta C_v}\right)}$$

- **Thermal Diffusivity, αT, Prandtl Number, NPr, Eucken Factor, Eu**

ϕ_c = Gas and Liquid Critical Property

Gas Constant and Parameters of Transport Cubic Equation of State

$$\mathbf{R_{vdw}} = \frac{\mathbf{T_c\phi_c}}{\mathbf{Z_cP_c}};\ R_{vdw} = \left(\sum_{i}^{n} x_i \left(\frac{T_c\phi_c}{Z_cP_c}\right)^{1/3}\right)^3$$

$$a(P) = \Omega_a \frac{R_{vdw}^2 P_c^2}{T_c}\xi(P)$$

$$\xi(P) = \begin{cases} \ln(P_R), & \text{if } P_R \neq 1 \\ 1 & \text{if } P_R = 1 \end{cases}$$

Dilute-Gas, Dense-Gas and Super-Critical Temperature Functions

$$b(T) = \begin{cases} b(T_c) = \Omega_b \frac{R_{vdw} P_c}{T_c} \\ b(T_c) T_R^{\theta_b} \ \langle \text{dilute gas} \rangle \\ b(T_c) T_R^{-\varphi_b} \ \langle \text{dense gas} \rangle \end{cases}$$

$$\mathbf{b_m} = \begin{cases} \left(\sum_{i=1}^{n} x_i [b(T_c)]_i^{1/3}\right)^3 & \langle \text{VLE} \rangle \\ \left(\sum_{i=1}^{n} x_i [b(T)]_i^{1/3}\right)^3 & \langle \text{Dilute or Dense gas} \rangle \end{cases}$$

2017 AIChE Annual Meeting, Minneapolis

19

Mixture Transport Equation of State

$$T = \frac{R_{vdw} P}{\eta - b_m} - \frac{a_m}{\eta^2 + \alpha_m b_m \eta - \beta_m b_m^2}$$

$$\mathbf{a_m} = \sum_{i=1}^{n} \sum_{j=1}^{n} x_i x_j \left(\frac{[\mathbf{a(P)}]_i + [\mathbf{a(P)}]_j}{2} \right) \mathbf{a}_{ij}$$

$$\alpha_m = \sum_{i=1}^{n} \sum_{j=1}^{n} x_i x_j \alpha_i^{1/2} \alpha_j^{1/2} \alpha_{ij}$$

$$\beta_m = \sum_{i=1}^{n} \sum_{j=1}^{n} x_i x_j \beta_i^{1/2} \beta_j^{1/2} \beta_{ij}$$

2017 AIChE Annual Meeting, Minneapolis

20

Methods of Pure Substance Critical Viscosity Estimations & Correlations

- Rackett Equation (JCED 1973): $\mathbf{V_L = V_c Z_c^{(1-T_R)^{2/7}}}$
- Rackett Viscosity Analogy (2003 Spring AIChE by ASL): $\boldsymbol{\eta_L = \eta_c Z_c^{3.5 \ln T_R}}$
- Back-Out Critical Viscosity: $\boldsymbol{f(T_R, P_R, \eta_R, Z_c, \Omega_w, \Omega_b, \Omega_a, \alpha, \beta) = 0}$
- Critical Viscosity Correlation: $\boldsymbol{\frac{\eta_c}{Z_c} = a_0 \left(\frac{M^{1/2}}{T_c^{1/6} P_c^{2/3}}\right)^{a_1}}$
- Critical Viscosity Tabulation: $\boldsymbol{\eta_c = b + \frac{R_{vdw} T_c}{P_c} - \left(\frac{\eta_c - b}{P_c}\right)\left[\frac{a(T_c)}{\eta_c^2 + \alpha b \eta_c - \beta b^2}\right]}$

2017 AIChE Annual Meeting, Minneapolis

21

Availability of HTHP Experimental Liquid Density and Viscosity Data

- **Research Groups (World Wide): Australia, Spain, Portugal, Japan, Greece, Germany, United Kingdom, Schlumberger, Edmonton (Canada) and Wu (China)**
- **USA Research Team: NETL, University of Pittsburgh, Virginia Commonwealth University (VCU) labs, Laesecke, NIST Boulder, Bair, Georgia Tech**
- **Texas Tech University (TTU) CAPRS Research Lab**

- **Estimation of Oil Remaining in West Texas Permian Basin (68% of Texas Oil or 22% of US Oil @ 160 Mile Radius of Lubbock)**
- **Estimation of BP Oil Spill into Gulf of Mexico (Environmental impact)**
- **CO2 Sequestration in Depleted Oil Reservoir: On- and Off-shores**

2017 AIChE Annual Meeting, Minneapolis

22

Need Characterization of Standard Reference Liquid Viscosity/ Density

Substance	Formula	Mw	Tb	Tc	Pc	Vc	Zc	ACF
DEHS								
DEHP	$C_{24}H_{38}O_4$	390.60						
DIDP	$C_{28}H_{46}O_4$	446.7						
PFPE	$C_{32}F_{66}O_{10}$	3100						
PFPE		8600						
Squalane	$nC_{30}H_{62}$	422.0						
Krytox GPL102	[(CF(CF3) CF2O]	1720.0						
diPE/C9	$C_{64}H_{118}O_{13}$	1095.6						
TOTM	$C_{33}H_{54}O_6$	546.8						

Viscosity Data of Squalane Conform to the Application of Cubic Equations of State

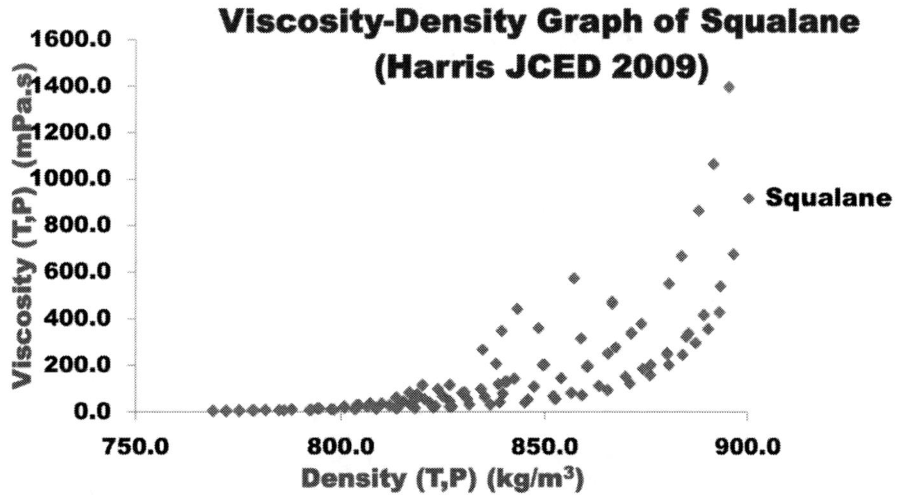

Viscosity Data of Squalane DO NOT Conform to the Corresponding State Principles

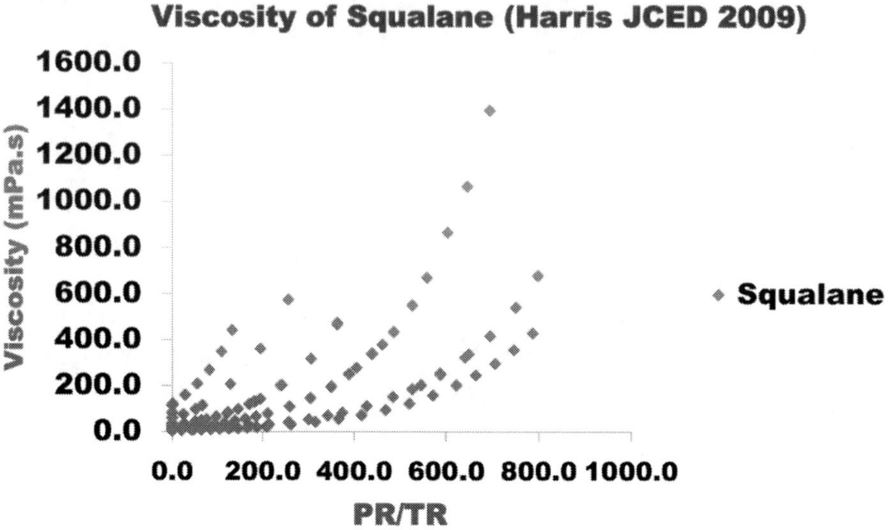

Viscosity Data of DIDP Conform to the Application of Cubic Equations of State

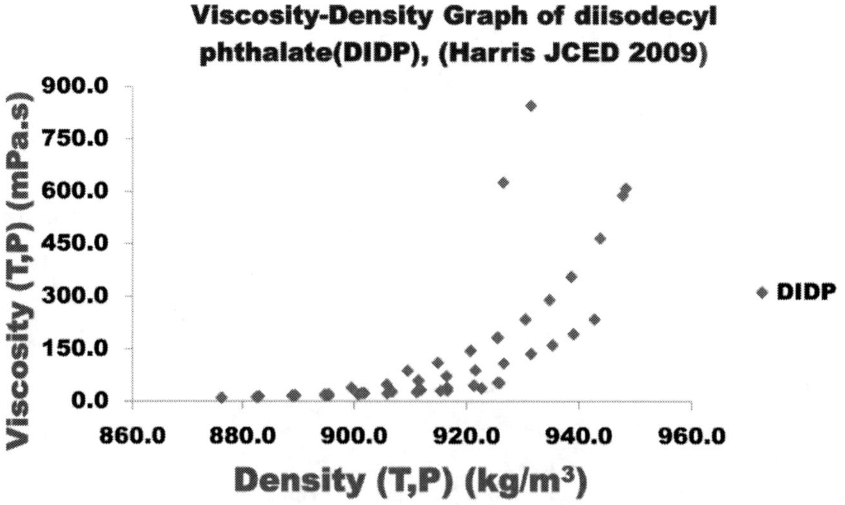

Viscosity Data of DIDP DO NOT Conform to the Corresponding State Principles

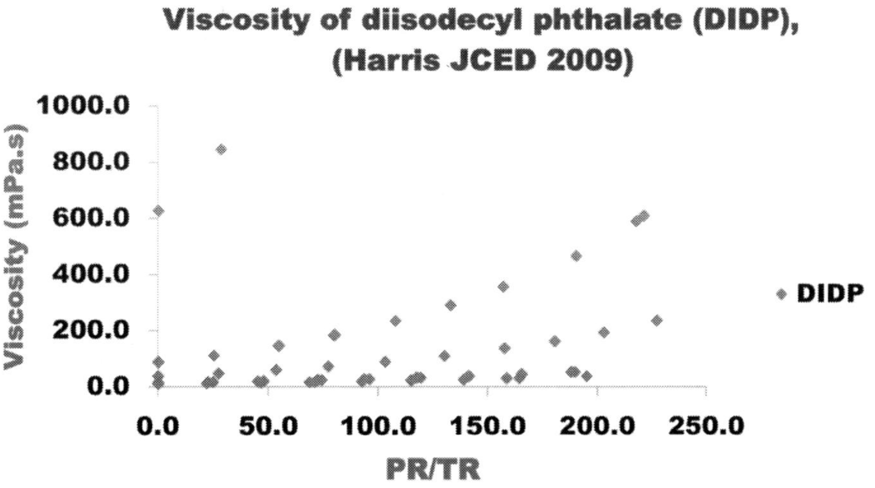

2017 AIChE Annual Meeting, Minneapolis

27

Viscosity Data of DIDP DO NOT Conform to the Corresponding State Principles

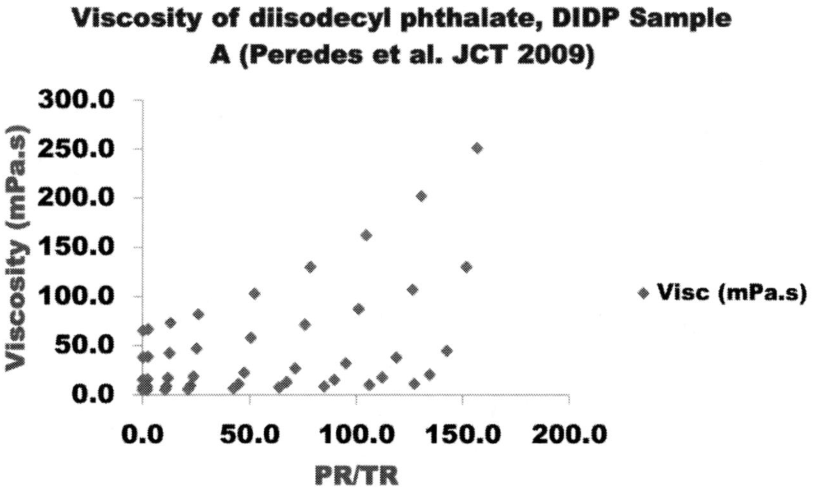

2017 AIChE Annual Meeting, Minneapolis

28

Composite Squalane Viscosity Data Conform to the Application of Cubic Equations of State

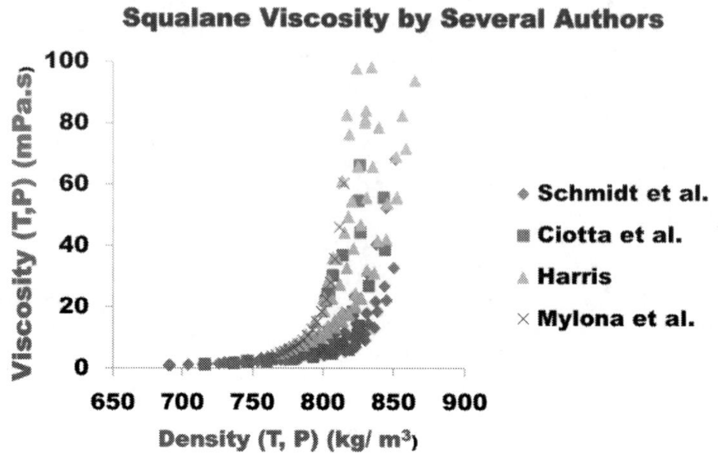

2017 AIChE Annual Meeting, Minneapolis

29

Composite Squalane Viscosity Data DO NOT Conform to the Corresponding State Principles

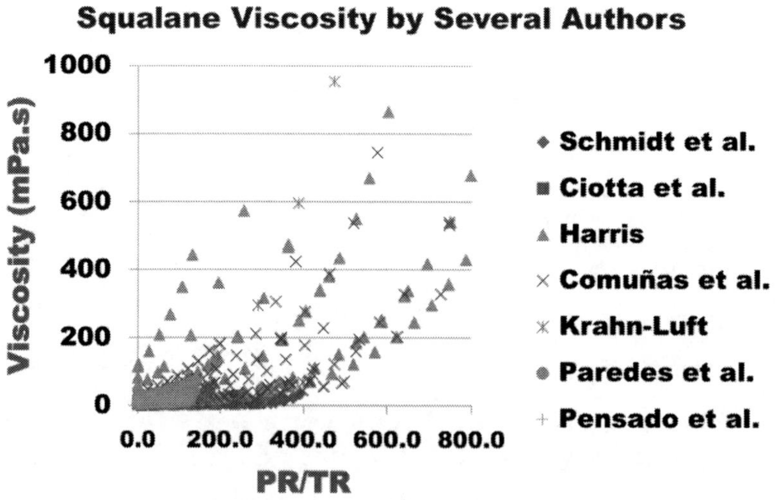

2017 AIChE Annual Meeting, Minneapolis

30

Composite Squalane Density Data Conform to the Application of Cubic Equations of State

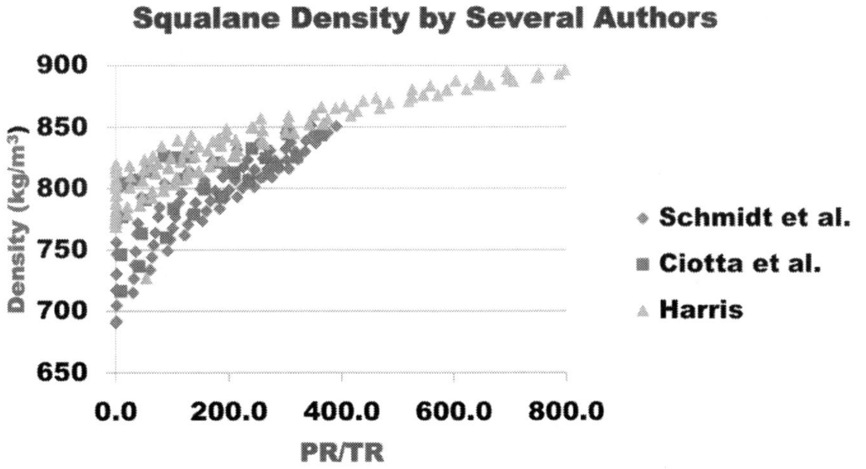

Results: Critical Viscosity of Squalane

Authors	# of Data	Av. η_c (cp) of Data
Krahn- Luft (1994)	18	4.67
Bair (2002)	32	954.24*
Pensado et al. (2006)	84	7.66
Paredes et al. (2009)	84	7.99
Harris (2009)	151	8.15
Ciotta et al. (2009)	32	6.42
Mylona et al. (2013)	19	atm
Comunas et al. (2013)	54	atm
Comunas et al. (2014)	176	3.85
Schmidt et al. (2015)	86	1.91

Squalane Viscosity (Krahn-Luft, JCED 1994)

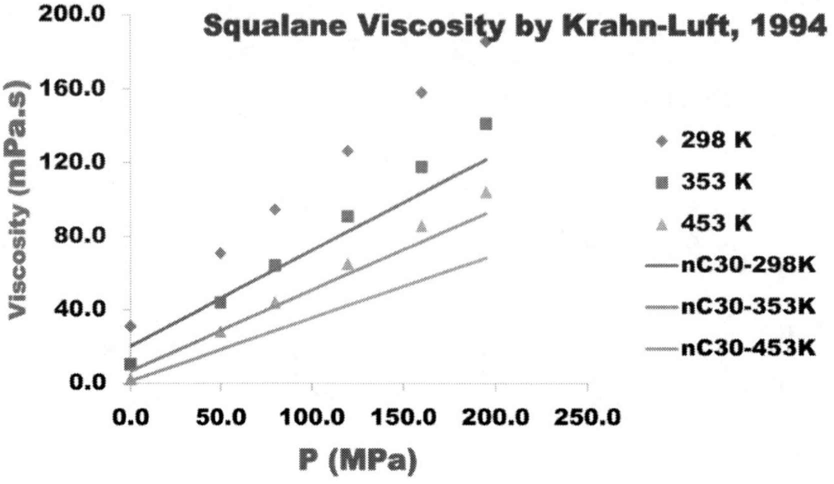

Conclusion and Recommendation

- VdW 1873 Theory is Satisfactory for Transport Equation of State
- **Transport Equation of State Reduces to the Corresponding States Chart**
- Need Comprehensive Tabulation of Gas-Liquid Critical Property for Transport Properties: De, η, λ, kinematic Viscosity, Thermal Diffusivity, Prandtl Number, Eucken Factor, etc.
- **Need Comprehensive Characterization Scheme for High-Molecular-Weight Fluids, Natural Products, Esters**
- Squalane Viscosity Data do not Conform to the Corresponding State Law???

Nitrogen Dioxide Viscosity Data DO NOT Conform to the Corresponding State Principles

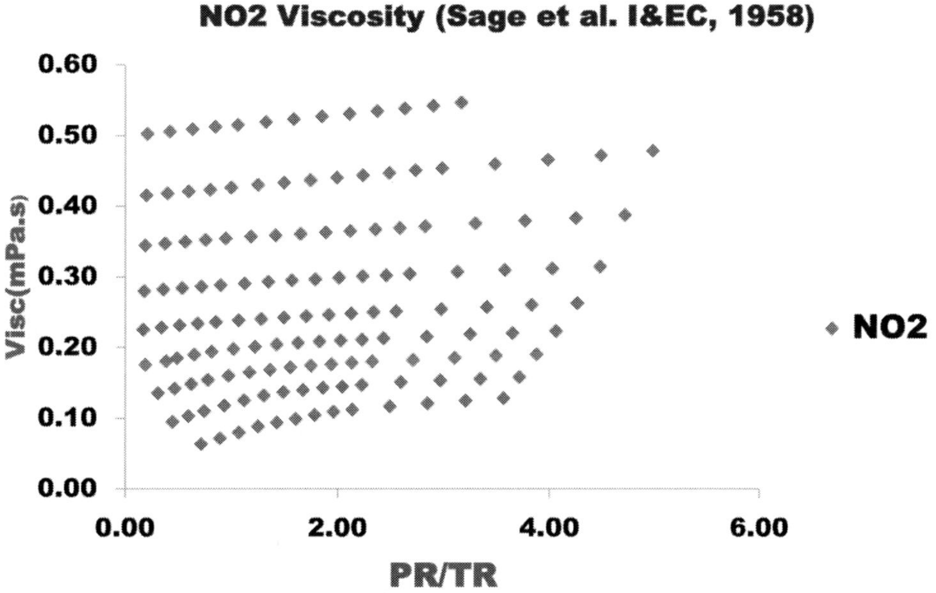

2017 AIChE Annual Meeting, Minneapolis

35

489830 Viscosity Models for Hydrocarbons at High Pressure: A State-of-the-Art Review

Abstract:

Viscosity is a fundamental property required in many petroleum reservoir applications, including the estimation of reserves, the calculation of flow rates in porous media or wellbores, and the determination of the mobility ratio of a displacement process. Direct measurements of this property are usually expensive and time-consuming. Therefore, reliable predictive methods are often employed to obtain the viscosity.

In this work, viscosity models for pure hydrocarbons are comprehensively reviewed. The models assessed in this study are suitable for use in reservoir simulators and are capable of estimating the viscosity of multi-component mixtures with changing composition. The performance of several selected viscosity models representative of various predictive and correlative approaches is discussed and evaluated by comparison to experimental viscosity data at ambient and extremely high-temperature, high-pressure (HTHP) conditions up to 300 °C and 300 MPa, conditions that are typically encountered in ultra-deep petroleum reservoirs, such as those found beneath the Gulf of Mexico. The applicability of each method is discussed and recommended models and approaches are identified.

Quantification of Uncertainties in Experimental Liquid Density and Viscosity at HTHP Conditions By Using the Corresponding States Law As Arbitrator

Akanni S. Lawal
Petroleum Engineering Department
Texas Tech University, P. O. Box 43111
Lubbock, Texas 79423

Take Away From This Presentation

- **Corresponding State Law Provides Approaches to Quantify Uncertainties in the Measured HTHP Viscosity and Density Data**
- Reformed Van der Waals Cubic Equation Conforming to Molecular Theory of the Corresponding States Law is NOW Available for ρ and η
- Possibility for Calibrating HTHP Viscosity Chart as Function of the Reduced Temperature and Pressure, $f(T_R, \eta_R, P_R)$
- **HTHP Viscosity Data DO NOT Conform to the Corresponding State Law**
- **HTHP Density Data CONFORM to the Corresponding State Law**
- **As Yet, No Unique Critical Viscosity Possibility for HTHP Viscosity Data**
- **Molecular Descriptors Needed for Tb, Tc, Pc, Vc, Zc, ω of High-Mw Fluids**

The Guiding Principles for Analyzing Experimentally Measured Data

- "It is better to be *roughly right* than *precisely wrong*." — **Alan Greenspan**

- **The Silberberg KISS Principle: Keep It Short and Simple**
- **Simplest Explanation for Deviations is Always the Correct Answer**
- **EYE-Test: Display Data via Charts to Light-up the Non-Obvious**
- **Avoid Rows/ Columns of Numbers and Statistical Jiggle Bell**

2017 AIChE Annual Meeting, Minneapolis

3

Research Questions Regarding the Measured HTHP Viscosity/ Density Data

- **What is the Quantification of Uncertainty in the World-Wide HTHP Viscosity Measurements Due to the Various Viscometer Designs used in the World-Wide Laboratories?**
- **What is the Effect of the Viscosity Calibration Constant, K on the Estimated HTHP Viscosity Data (or Pressure Dependent Viscosity Measurements)?**
- **What is the Pressure Limit of the Constancy of the Viscometer Calibration Constant, K (due to the Possibility that Calibration Constant K Changes with Time Due to Wear)?**
- **What is the Effect of the Estimated Density (ρ_{fluid}, ρ_{sinker}) Variation with T & P on the Calibrants and Substance under Study which may be significantly differ in Density?**

2017 AIChE Annual Meeting, Minneapolis

4

Universal Law of Corresponding States

- CSP Method:

  Tool to Assess Accuracy and Consistency of Measured Data

- Molecular Theory: $\phi(T_R, P_R, Z, Z_c, \beta_s) = 0$; β_s = Shape Factor
- VdW (Theory) Molecular Properties: $Z_c, \Omega_w, \Omega_b, \Omega_a$
- VdW Equation Corresponding States: $f(P_R, V_R, T_R) = 0$
- VdW Reformed Equation (LLS EOS):

$$f(P_R, V_R, T_R, Z_c, \Omega_w, \Omega_b, \Omega_a, \alpha, \beta) = 0$$

The VdW Reduced Corresponding States Cubic Equation of State

- Thermodynamic Reduced Equation

$$P_R = \left(\frac{T_R}{Z_c V_R - [\Omega_b \phi(T_R)]}\right) - \left(\frac{\Omega_a \gamma(T_R)}{Z_c^2 V_R^2 + \alpha[\Omega_b \phi(T_R)] Z_c V_R - \beta[\Omega_b \phi(T_R)]^2}\right)$$

- Transport Reduced Equation

$$T_R = \left(\frac{P_R}{Z_c \eta_R - [\Omega_b \varphi(T_R)]}\right) - \left(\frac{\Omega_a \phi(P_R)}{Z_c^2 \eta_R^2 + \alpha[\Omega_b \varphi(T_R)] Z_c \eta_R - \beta[\Omega_b \varphi(T_R)]^2}\right)$$

- Dimensional Similitude

$$f(P_R, T_R, \phi_R, Z_c, \Omega_w, \Omega_b, \Omega_a, \alpha, \beta) = 0$$

nC5 Viscosity Data Conform to the Application of Cubic Equations of State

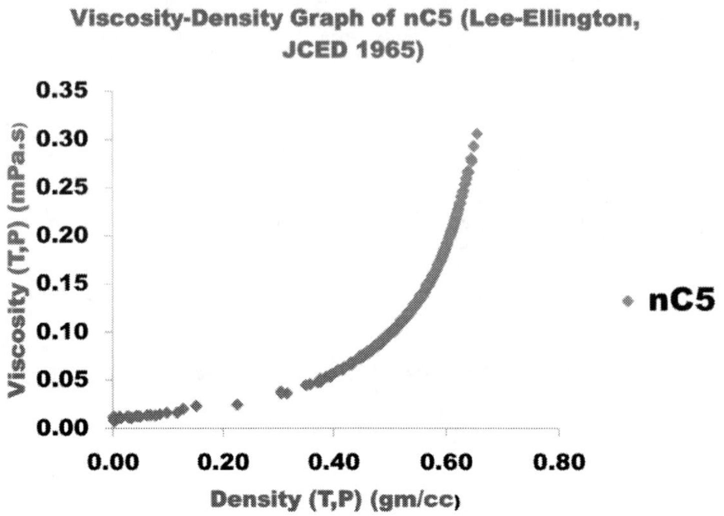

2017 AIChE Annual Meeting, Minneapolis

7

Statistical Measure for the HTHP Viscosity and Density Data

$$\text{BIAS} = \frac{1}{N}\sum_{i=1}^{N}\left(\frac{\rho^{\text{expt}} - \rho^{\text{cal}}}{\rho^{\text{cal}}}\right)_i$$

$$\text{BIAS} = \left(\frac{\rho(\text{T,P})}{\rho(\text{T},0.1)}\right) \text{ Versus Pressure}$$

$$\text{BIAS} = \left(\frac{\eta(\text{T,P})}{\eta(\text{T},0.1)}\right) \text{ Versus Pressure}$$

- **Regard ρ(T,0.1) and η(T, 0.1) as Standard States for Normalizing Measured Data to Conform with the Corresponding State Law**

2017 AIChE Annual Meeting, Minneapolis

8

Notable Viscometer Calibrating Equation and Viscometer Constant

- Baled et al. (Fuel 2017):

$$\eta = \frac{k(\rho_b - \rho_{fl})\text{Sin }\theta}{v}$$

- Harris (*JCED* 2015 v60 1510)

$$\eta(p,T) = \frac{t(1-\rho/\rho_s)}{K[1+2\alpha_p(T-T_{ref})][1-2\beta_T(P-P_{ref})/3]}$$

- **Required Calibration: $K(T, 0.1)$; $\eta(T, 0.1)$**

2017 AIChE Annual Meeting, Minneapolis

9

Visualization Bias in the Viscometer/ Densimeter Calibration Constant

- **Independent Variables: T and P**
- Viscometer K BIAS:

  $\frac{k(\mathrm{T,P})}{k(\mathrm{T,0.1})} \cong \frac{\eta(\mathrm{T,P})}{\eta(\mathrm{T,0.1})}$ Vs P $\langle$*horizontal line parallel to abscissa*$\rangle$

- **Densimeter BIAS:**

  $\frac{\rho(\mathrm{T,P})}{\rho(\mathrm{T,0.1})}$ **Vs P** $\langle$***horizontal line parallel to abscissa***$\rangle$

2017 AIChE Annual Meeting, Minneapolis

10

Bias Plot of iC8 & nC8 (Baled et al., Fuel 2016)

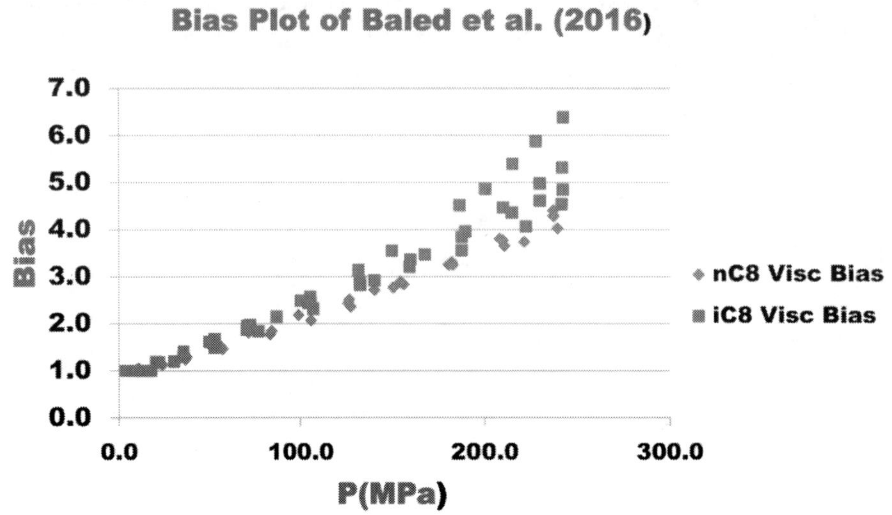

2017 AIChE Annual Meeting, Minneapolis

11

Bias Graphs of Density and Viscosity

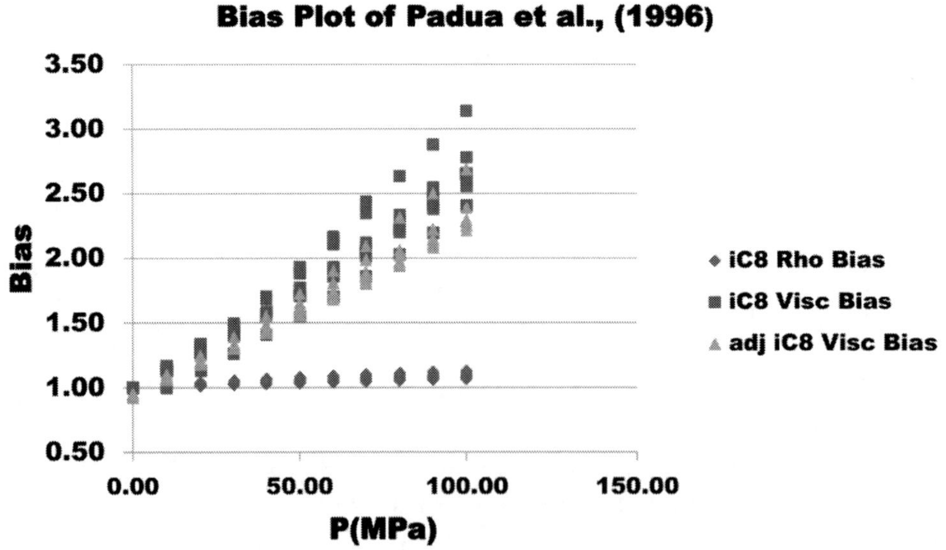

2017 AIChE Annual Meeting, Minneapolis

12

6th Iterations of Bias Method (Padua et al.)

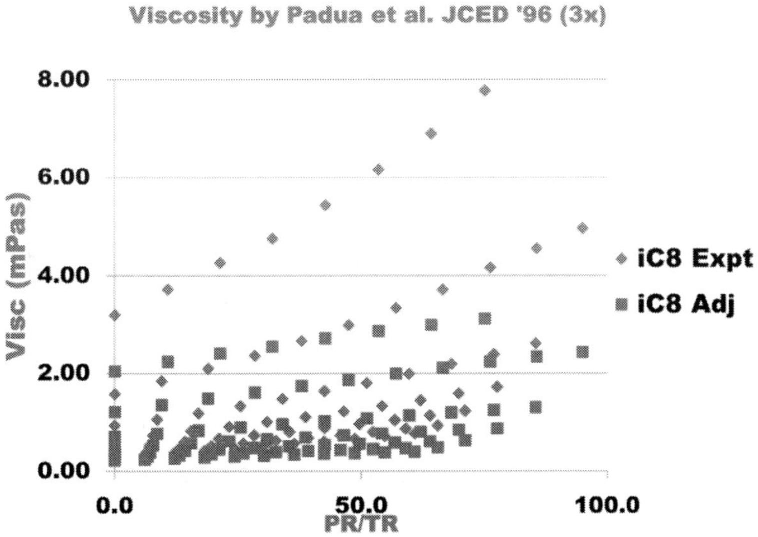

2017 AIChE Annual Meeting, Minneapolis

13

Eliminating the Effect of Viscometer Calibration Constant

$$\langle \eta(\mathrm{T}_f, \mathrm{P}) \rangle_{\mathrm{corrected}} = \eta(\mathrm{T}_f, \mathrm{P})_{\mathrm{expt}} - \left(\eta(\mathrm{T}_f, \mathrm{P})_{\mathrm{expt}} \left[\frac{\left(\frac{\eta(\mathrm{T}_f, \mathrm{P})_{\mathrm{expt}}}{\eta(\mathrm{T}_f, 0.1)} \right)_i}{\sum_{i=1}^{m} \left(\frac{\eta(\mathrm{T}_f, \mathrm{P})_{\mathrm{expt}}}{\eta(\mathrm{T}_f, 0.1)} \right)_i} \right] \right)$$

2017 AIChE Annual Meeting, Minneapolis

14

Conclusion and Recommendation

- **Corresponding State Law Provides Approaches to Quantify Uncertainties and Corrected HTHP Viscosity η and Density ρ**
- HTHP Viscosity Data **do not Conform** to the Corresponding State Law
- HTHP Density Data **Conform** to the Corresponding State Law
- **No Unique Critical Viscosity Possibility for HTHP Viscosity Data**
- **Need Comprehensive Characterization Scheme for High-Molecular-Weight Fluids; Tb, Tc, Pc, Vc, Zc**

2017 AIChE Annual Meeting, Minneapolis

15

495827 Characterization of Orientation and Photochemical Function of Chlorophyll *a* Molecules in Self-Assembled Membranes

Abstract:

The long chained DMPC (1,2-dimyristoyl-*sn*-glycero-3-phosphocholine) and the short chained DHPC (1,2-dihexanoyl-*sn*-glycero-3-phosphocholine) mixtures provided bicelle (disk like membrane) structures at high phospholipid concentration [1]. The morphology of DMPC/DHPC self-assembly changed to vesicle in diluted conditions. Self-assemblies of phospholipids, the bicelle and the vesicle, were employed to organize photosynthesis pigments, chlorophyll *a* (Chl*a*) molecules, in aqueous solution. In nature, Chl*a* act as reaction center (RC) in light harvesting (LH), wherein they are lined up around apoprotein rings. These organized pigments transfer the absorbed light energy efficiently to the RC. Inspired from the orientation of Chl*a* molecules in natural LH systems; we here investigated the behavior of inserted Chl*a* molecules in phospholipid membranes. The behaviors of a Chl*a*-incorporated phospholipid membranes were evaluated by optical techniques. In this study, we revealed the phase diagram of DMPC/DHPC mixture in dilution process based on membrane properties such as turbidity, membrane fluidity and membrane polarity of that. The designed assembly grounded the phase diagram (DMPC/DHPC = 3:2 including 1.2mol% Chl*a*) resulted J-aggregate, and the Chl*a*-Chl*a* interaction was also maintained at diluted condition. In addition, the photo-function of the vesicle with a Chl*a* was examined the influence of the membrane properties on the photoreduction with methyl viologen (MV) [2]. The J-aggregate of Chl*a* on DMPC/DHPC = 3:2 efficiently catalyzed photoreduction of methyl viologen, suggesting the importance of controlled orientation of pigments by used self-assembly membrane. Preparing highly efficient LH membrane with Chl*a* can contribute to develop the photochemical energy conversion system.

[1] L. van Dam, G. Karlsson, K. Edwards, Biochim. Biophys. Acta 2004, 1664, 241. [2] M. Kobayashi, S. Ohashi, K. Iwamoto, Y. Shiraiwa, Y. Kato, and T. Watanabe, Biochim. Biophys. Acta 2007, 1767, 596.

496007 Design of Lipid Membrane Surfaces As Organocatalyst for Michael Reactions in Aqueous Media

Abstract:

L-proline is known to catalyze Michael addition reaction with enantioselectivity, whereas its catalytic activity is decreased by surrounding water. Herein, we developed the liposomes which can conduct Michael addition reaction of trans-β-nitrostyrene and acetone in aqueous media. In the case of L-Pro-catalyzed Michael addition, the membrane fluidity and polarity are major controlling factors for this reaction [1]. The highest conversion and rate constant of the reaction within the liposomes was achieved with the 1,2-dioleoyl-sn-glycero-3-phosphocholine (DOPC)/1,2-dipalmitoyl-3-trimethylammoniumpropane (DPTAP) system. The catalytic activity of L-Pro in the liposome suspension was found to be comparable to that in a DMSO system. The reaction rate constant was found to be controlled by both the phase state of the liposome membrane and the surface charge on the membrane. Greater enantioselectivity was achieved in the presence of the liposomes than in DMSO solution, with corresponding enantiomeric excess values of 97.6% for the DOPC/DPTAP liposome system. Furthermore, the amphiphilic amine molecules (stearylamine)-modified liposomes also showed catalytic activities with enantioselectivity. The hydrophobic region of the liposome membrane, which is a relatively stable self-organizing system, can serve as an effective "platform" for molecular recognition and selective conversion in aqueous media.

[1] M. Hirose *et al.,* Liposome Membrane as a Platform for the L-Pro-Catalyzed Michael Addition of trans-β-Nitrostyrene and Acetone. *Langmuir,* **2015**, *31* (47), 12968–12974.

499071 A Unified Structure-PropertyÂ Relationship for Alkyl-PolyoxideÂ Surfactants

Abstract:

Polyoxide (e.g. polyethylene glycol) surfactants are the most well studied and understood nonionic surfactants and are used in a variety of applications as detergents, additives, etc. While there is a constant debate over the roles of surfactant structure on interfacial performance, no complete model has been presented. In this work, we will combine previously documented data with new measurements to develop a predictive model of surface properties of alkyl-polyoxide surfactants. From this model, one can predict the critical aggregation concentration, equilibrium surface tensions, surface packing density, and kinetic ratio simply from surfactant structure parameters.

501234 Relating Rheology with Morphology: Cholesterol in a Model Lung Surfactant Monolayer

Abstract:

The addition of cholesterol to lung surfactant (LS) model monolayers decreases viscosity and affects domain morphology. Recent viscosity measurements of the DPPC:HD:cholesterol system, often used as a model for LS studies, showed this trend with cholesterol. For cholesterol concentrations from zero to four percent, increased packing density corresponds to an increase in viscosity. However, samples with eight mole percent cholesterol were shown to decrease in viscosity as packing density increases, a surprising result. We present a morphological study to complement recent viscosity measurements of the DPPC:HD:cholesterol system. We performed confocal imaging on DPPC:HD:cholesterol monolayers in a Langmuir trough to identify morphological changes that correspond with the observed changes in viscosity. Characteristic features at various cholesterol concentrations are identified, and we observe a possible three-phase system in six percent and eight percent cholesterol samples. We also consider the role of experimental parameters such as camera gain, saturation, and dye quenching in the interpretation of these phase-separated systems. The resulting images shed new light on a potential mechanism for viscosity regulation in LS.

501577 A Matlab Based Tool Package for Interfacial Property Calculations

Abstract:

In the upstream operations of the oil & gas industry, the interfacial tension (IFT) of reservoir fluids greatly affects the efficiency of oil recovery process. Predicting the IFT of reservoir fluids is essential for the design of enhanced oil recovery (EOR) processes. To interpolate and extrapolate expensive experimental data, different theoretical and simulation methods have been developed to model the interfacial properties of reservoir mixtures. Among these methods, density gradient theory (DGT) has become a fast and accurate tool in the interfacial property prediction of many oil mixtures and water/oil mixtures.

In our work, the stabilized density gradient theory (SDGT) algorithm was developed to optimize the solving process of the DGT equations. We have also modified the DGT equations so that it can handle molecules with chain structures, such as surfactant molecules. By taking advantage of the SDGT algorithm and the modified DGT functions, we were able to conduct interfacial property calculations of oil mixtures, water/oil mixtures and water/oil/surfactant mixtures.

The Matlab based tool package was developed by integrating the prediction models we have developed into a user-friendly interface, which can be readily used by engineers to conduct fast and accurate reservoir property simulations. Some features of the package include phase equilibrium calculation, free energy analysis, IFT calculation and results plotting & exporting. The advanced numerical methods were implemented to speed up the calculation. The user-friendly interface and the userâ€™s manual make it easy to use in practical calculations.

507742 Interaction Forces between Colliding Emulsion Drops (Oil or Water) Coated with Non-Ionic Surfactants ($C_{12}E_5$ and PGPR)

Abstract:

We measured the interaction forces between colliding emulsion drops (Oil or Water) coated with non-ionic surfactants $C_{12}E_5$ (penta-ethylene-glycol-mono-dodecyl-ether) and PGPR (polyglycerol polyricinoleate). Non-ionic surfactants are often used in combination with ionic surfactants in detergency, oil recovery and mineral flotation applications because they increase surface activity of surfactant mixtures. The emulsion systems tht we studied are oil-in-water (O/W) and water-in-oil (W/O) in the cases of $C_{12}E_5$ and PGPR respectively. The oils used are Tetradecane and Canola oil for O/W and W/O emulsions respectively. The interaction force between Tetradecane drops were measured in the range of pre-micellar compositions of $C_{12}E_5$ in various salt solutions. The Tetradecane drops were charged even though the surfactant was non-ionic in nature. The surface potential of the oil drops decreased with increase in $C_{12}E_5$ concentration in agreement with the experimentally measured electrophoretic mobilities and zeta potentials. The surface potential decreased with increasing electrolyte concentration that confirmed the presence of a charged oil-water interface. We also measured the non-equilibrium film drainage between tetradecane drops coated with $C_{12}E_5$ as a function of magnitude of the force of interaction and velocity. We found that the tetradecane drops were stable against coalescence even when the drops were deformed on the order of their radii. This observation is in contrast to some existing experiments in literature.

Unlike the case of O/W emulsion with $C_{12}E_5$, in the case of W/O emulsion where water drops were emulsified into Canola Oil (with PGPR), we found that the interaction forces were dominated by steric forces. PGPR is a fat soluble non-ionic surfactant widely used in chocolate industry. Hydrophile lipophile balance (HLB) number for PGPR is between 2 and 4; hence it is known to be the best food grade emulsifier for stabilizing water-in-oil emulsions. Interfacial tension measurement and adsorption isotherm fitting showed that the water-canola oil interface gets saturated with PGPR for concentrations above 0.2 wt%. In order to probe meta-stability of emulsified water drops coated with PGPR for concentrations below 0.2 wt%, we measured the interaction force between the drops as a function of PGPR concentration and velocity of collisions. We also measured forces between hydrophilic rigid solid surfaces coated with PGPR in canola oil in order to get qualitative and quantitative understanding of repulsive forces resulting from adsorption of PGPR. There was a long range attraction between the hydrophilic rigid surfaces for PGPR concentrations below 0.02 wt% due to insufficient coverage of PGPR. Steric repulsion was observed for PGPR concentration of 0.05 wt% and above. Estimated thickness of adsorbed brush layer of PGPR on hydrophilic surfaces was found to increase from 10 to 64 nm when PGPR concentration increased from 0 to 0.2 wt%. In the case of force between PGPR coated water drops in canola oil, hysteresis between the approach and retraction curves increased with increase in the maximum force to which the water drops are pushed to. We observed that during the retract phase of drop collision there was a kink in the force curves and sudden jump out. We attributed such a force jump outs to bridging and extension of adsorbed PGPR molecules on water drops. In order to rule out the role of water transport in the form of PGPR reverse micelles to observed force jump outs, we carried out water drop collision experiments with 1 wt% salt in water. Salt was introduced to arrest the water transport across oil layer. However, we found similar force jump outs even in the presence of salt. This observation indicated that bridging and extension of PGPR molecules was the mechanism behind the force jumps observed during retraction phase of drop collision.

The above finding relevant to O/W and W/O emulsions have significant implications in the stability of emulsions in food, personal care and detergent industries.

489315 Prediction of Vapor Pressure and Critical Properties for Non-Electrolyte Organic Compounds from PR+Cosmosac Eos

Abstract:

Thermodynamic properties, such as vapor pressures, are important information for both process design and optimization in chemical industry and estimation of gas/particle partitioning of organic compounds in environmental applications. The PR+COSMOSAC EOS has been developed to estimate the energy and volume parameters of the PR EOS from quantum mechanical and COSMO solvation calculations with the only input of the molecular structure. This method has been shown to provide reasonable prediction for vapor pressures of pure substances and fluid phase equilibrium of mixtures without the issue of missing parameters. In this study, the temperature-dependence of the dispersion contribution was revised to improve the accuracy in predicting vapor pressure of pure substances, especially at temperatures lower than the normal boiling temperature of the substance. Furthermore, a hydrogen-bonding correction term was newly introduced for amines, accompanying with two new dispersion parameters. The revised model has a total of 16 global parameters and three parameters for each element (one atomic radius and two dispersion parameters). All the dispersion parameters involved in the dispersion contribution were re-optimized by regressing vapor pressures of a few selected pure substances. A total of 1125 pure substances are included to investigate the accuracy of the revised model. The overall average logarithmic deviation in pressure (ALD-P) is 0.195 from the revised model which is better than that from the original model (ALD-P = 0.241). The accuracy of the revised PR+COSMOSAC EOS in predicting critical properties is underway. In general, the PR+COSMOSAC EOS could be useful when no experimental data is available.

494995 Volumetric Properties of Saccharide in Aqueous Ionic Liquid Mixtures at Different Temperatures

Abstract:

Saccharides and their derivatives as the most abundant class of biomolecules are the receptors of biologically active compounds (enzymes, drugs, etc.) and they are able to stabilize the native state of proteins/enzymes. In the present study, experimental density data of aqueous solutions of 1-butyl-3-methylimidazolium hydrogen sulfate ionic liquid containing saccharide have been determined at different temperatures and composition. The saccharide studied was D(+)-glucose (molalities saccharides range from 0.015 - 1.7 molâˆ™kg^{-1}) in water and in (0.025; 0.050; 0.075 and 0.1 molâˆ™kg^{-1}) aqueous ionic liquid solution at temperatures of *T* = (283.15, 288.15, 293.15, 298.15 and 303.15) K. Density of solutions have been determined by using a vibrating-tube densimeter manufactured by Anton Paar (Model 4500). Experimental results have been used to determine the apparent molar volumes, apparent molar volumes at infinite dilution and partial molar volumes of transfer of the saccharide from water to aqueous ionic liquid solutions. The results lead to believe that possible hydrophilic-ionic interactions among the hydrophilic groups (-OH, -C=O, -O-) of the saccharide and the ions of ionic liquid can be predominant in the systems studied.

495066 Density and Derived Properties of Binary Mixtures Containing {Dimethyl Carbonate (DMC) + 1-Propanol} at T = (288.15 to 308.15) K and at P = (0.1 to 40) Mpa

Abstract:

Excess properties have been a qualitative and quantitative way to predict deviation from ideality of binary liquid solution. Moreover, excess properties have been used to develop models and propose new solution theories. As a continuation of an experimental study and modeling of excess molar properties of binary systems containing alcohols [1-2], in the present work, experimental data of density of binary liquid mixtures containing {dimethyl carbonate (DMC) + 1-propanol have been determined over the entire composition range, at temperatures from (288.15 to 308.15) K and at pressure range between (0.1 and 40) MPa. Density measurements of pure components and their respective solutions were determined by using a vibration-tube densimeter (Model DMA 4500) coupled with a high-pressure cell (DMA HP), both manufactured by Anton Paar. The results of densities were used to calculate excess molar volume and other volumetric properties, such as partial molar volume, excess partial molar volume and respective partial molar volume at infinite dilution, and apparent molar volume of each compound. Density data were also fit to Tait equation and used to determine the isothermal compressibility and the thermal expansion coefficient. The excess molar volume values were positive for the system studied. It was possible to observe that the absolute values of the excess molar volume decrease with increasing in pressure and increase with increasing in temperature. In order to complement the thermodynamic analysis and understand the possible interactions between the components present in the mixtures, Fourier Transformed Infrared (FT-IR) spectroscopy study was realized. The results were interpreted in terms of intermolecular interactions and structural effects between the components present in the systems. The thermodynamic and spectroscopic results suggest that the structural effects and dispersion forces must outweigh other possible effects present in the mixtures studied. For the best of our knowledge, no data of volumetric properties at high-pressure for the systems studied in this work were found in the literature.

References

[1] S.S. Bittencourt, H.E. Hoga R.B. Torres, J.V.H. d´Angelo. Thermodynamic and spectroscopic properties of binary mixtures of *n*-butylammonium butanoate ionic liquid with alcohols at T = (293.15 to 313.15) K. Journal of Chemical Thermodynamics, 2017. 105: p. 238-252.

[2] H.E. Hoga R.B. Torres. Volumetric and viscometric properties of binary mixtures of methyl tert-butyl ether (MTBE) + alcohol at several temperatures and p = 0.1 MPa: Experimental Results and application of the ERAS Model. Journal of Chemical Thermodynamics, 2011. 43: p. 1104-1134.

501311 A SAFT Equation of State Based on Triangular-Well Fluid Potential

Abstract:

In this works, the main objective is to develop the frame work for a Statistical Associated Fluid Theory based upon the Triangular-Well (TW) fluid potential proposed by Betancourt-CÃ¡rdenas et al. [Mol. Phys. 105 (2007) 2987 and Mol. Phys. 106 (2008) 113] based on the perturbation theory by Barker-Henderson. The TW potential is more realistic than the well-know square-well (SW) potential commonly used in the popular SAFT and PC-SAFT, besides the TW keep a significative analytical simplicity compared to others potentials. TW showsÂ a great potential to be used for the molecular based equations of state as it can be observe in the followings works Zhou [J. Chem. Phys. 130 (2009), 014502], Koyuncu [Mol. Phys. 109 (2011), 565] y Rivera et al. [Mol. Phys. 110 (2012), 1317]. The formulation by Betancourt-Cardenas used a second order perturbation theory with a macroscopic compressibility approximation. We present vapourâ€“liquid phase diagrams and thermodynamic properties such as the density as a function of the pressure at different temperatures and for several values of the potential range. The preliminary results are promising of good agreement with data reported in the literature.

482607 Modelling the Solubility of Naphthalene and Phenanthrene in Binary and Ternary Systems Containing Carbon Dioxide

Abstract:

Experimental solid solubility data are necessary to develop new supercritical fluid processes. Solubility data for different solids in supercritical fluids are important in the petroleum, food and pharmaceutical industries. Because of the limited amount of experimental data for solid + supercritical fluid systems [1], it is important to have an experimental method to determine reliable properties, as well as models for correlation and prediction.

In this work, we present the results of correlation of solubility data of the binary system carbon dioxide + naphthalene and the ternary system carbon dioxide + naphthalene + phenanthrene. These solubility data were measured using an apparatus based on dynamic methods in order to test the newly developed apparatus in our laboratory; the solubilities of a solid in a binary system and in a ternary system were measured based on the dynamic analytic method. The developed equipment and the experimental methodology used were suitable for determining solubilities of solid in carbon dioxide over a wide range of pressure.

We used two approaches for modelling our obtained solubility data: density-based empirical models and a cubic equation of state. The first approach does not need properties of the solutes, while the latter approach does. Modelling by an equation of state requires critical and physical properties of the solutes and the solvent, as well as the selection of a proper mixing rule. For the binary system, the models tested in this work were the empirical models presented by Chrastil [2], Kumar and Johnston (K-J) [3], Bartle [4], Mendez-Santiago and Teja (M-S-T) [5], and the modification of Chrastil model by Garlapati and Madras [6]. These models relate the solubility of a solute to the temperature and density of the supercritical solvent. For the ternary system, we used the same models for the binary systems and also the modifications of Chrastil [7] and Mendez-Santiago and Teja [8] models.

The equation of state and the mixing rules used for both the binary and the ternary systems were the Peng-Robinson [9] equation of state and the van der Waals mixing rules, with 1 or 2 parameters. The average absolute relative deviation (AARD) was used to evaluate the results of the correlation.

For naphthalene in the binary and ternary systems, we had ranges of values for the AARD % of (3.14-18.48) with the use of empirical models. Using the Peng-Robinson equation of state, we had minimal AARD % values of 2.55. The addition of the co-volume parameter (lij) to the van der Waals mixing rule improved the correlation in all cases. This means that there might be complex interactions in the solid-supercritical fluid equilibrium. The energy parameters (kij) show little dependence with temperature for all cases.

The AARD % values had a range of (5.34-15.03), using empirical models for phenanthrene in the ternary system. The modelling of the ternary system with an equation of state gave values of AARD % up to 27 for naphthalene and 48 % for phenanthrene. This modelling was performed using optimised kij parameters for the two solutes with the solvent (naphthalene + carbon dioxide and phenanthrene + carbon dioxide), while the parameter for the interactions between naphthalene and phenanthrene was considered an adjustable parameter [10]. It is necessary to perform the modelling with another approach in order to obtain lower values of AARD.

We concluded that the use of empirical models is good for the modelling of solubility of solids in supercritical solvents, because they need the density of the solvent and give good results in correlation. However, the use of an equation of state improves the correlation when using a proper mixing rule for binary systems. In the case of ternary systems, it is necessary to use a different approach than the conventional one. Nevertheless, an equation of state cannot be a model that is required to give accurate results, because it is not designed to model solid compounds.

[1] Skerget, M.; Knez, Z.; Knez-Hrn, M. Solubility of Solids in Sub- and Supercritical Fluids: a Review. J. Chem. Eng. Data 2011, 56, 694–719.

[2] Chrastil, J. Solubility of Solids and Liquids in Supercritical Gases. J. Phys. Chem. 1982, 86, 3016–3021.

[3] Kumar, S. K.; Johnston, K. P. Modelling Solubility of Solids in Supercritical Fluids with Density as the Independent Variable. J. Supercrit. Fluids 1988, 1, 15–22.

[4] Bartle, K. D.; Clifford, A. A.; Jafar, S. A.; Shilstone, G. F. Solubilities of Solids and Liquids of Low Volatility in Supercritical Carbon Dioxide. J. Phys. Chem. Ref. Data 1991, 20, 713–756.

[5] Mendez-Santiago, J.; Teja, A. S. The Solubility of Solids in Supercritical Fluids. Fluid Phase Equilib. 1999, 158, 501–510.

[6] Garlapati, C.; Madras, G. Solubilities of Palmitic and Stearic Fatty Acids in Supercritical Carbon Dioxide. J. Chem. Thermodyn. 2010, 42, 193–197.

[7] J.C. González, M.R. Vieytes, A.M. Botana, J.M. Vieites, L.M. Botana, Modified mass action law-based model to correlate the solubility of solids and liquids in entrained supercritical carbon dioxide, J. Chromatogr. A 910 (2001) 119-125.

[8] M. Sauceau, J.J. Letourneau, D. Richon, J. Fages, Enhanced density-based models for solid compound solubilities in supercritical carbon dioxide with cosolvents, Fluid Phase Equilib. 208 (2003) 99-113.

[9] Peng, D. Y.; Robinson, D. B. A new two-constant equation of state. Ind. Eng. Chem. Fundam. 1976, 15, 59–64.

[10] Wu, H.; Zhu, J.; Wang, Y.; Chang, Ch.; Jin, J. Measurement and modeling for solubility of 3-hydroxybenzaldehyde and its mixture with 4-hydroxybenzaldehyde in supercritical carbon dioxide. Fluid Phase Equilib. 2016, 409, 271-279.

Molecular-Inspired Parameters Concealed in the Van der Waals Attractive Force Revealed By First Principles, Statistical Mechanics and Perturbation Methods

Akanni S. Lawal (akannislawal@aol.com) , Petroleum Engineering Department
Reservoir and Petrophysical Research Laboratory, Texas Tech University, P. O. Box 43111, Lubbock, Texas 79423

ABSTRACT

- Methodology of Interpreting the Molecular Significance of the Additional Empirical Parameters the VDW 1873 Cubic Equation of State
- The Reformed VDW 1873 Cubic Equation which has been named the Lawal-Lake-Silberberg (LLS) Equation of State is used as Prototype
- Sensitivity Analysis and Perturbation Techniques Used for Interpreting the Molecular Significance of Empirical Parameters in the VDW 1873 Cubic Equation

WEAKNESS OF VAN DER WAALS EQUATIONS

- Inaccurate Saturated Liquid Densities
- Inaccurate Substance Critical Volumes (Densities)
- Inaccurate HTHP Volumetric (Density) Properties
- Inaccurate Gas-Liquid Critical Properties
- Inaccurate for Polar (Hydrogen-Bonded) Substances
- Inaccurate Third + Higher Virial Coefficients
- Inaccurate Second-order Derivative of Thermo-Properties
- Inaccurate Van der Waals Repulsive Term

TESABLE HYPOTHESIS FOR VDW PARAMETERS

- Could the Absence of Empirical Parameters Replicating Molecular Structure and Molecular Shape in the Van der Waals Theory Be Responsible?

THE VAN DER WAALS OBJECTIVE

- The VdW Ultimate Objective**: "*to determine the relation between p, v and T for a substance,*" and thus, "*...determine the equation of state of the molecule itself.*"

**Van der Waals 1910 Nobel Prize Lecture at the Nobel organization website http://www.nobelprize.org/nobel_prizes/physics/laureates/1910/waals-lecture.html

PARAMETERS OFTHE LAWAL-LAKE-SILBERBERG EOS

- Molecular Critical Parameters

$$Z_c = \varphi(\alpha,\beta),\quad v_c = \phi_v(\alpha,\beta)b,\ P_c = \phi_p(\alpha,\beta)\frac{a}{b^2},\ T_c = \phi_T(\alpha,\beta)\frac{a}{Rb}$$

$$\Omega_w = \phi_o(\alpha,\beta),\quad \Omega_a = \phi_1(\alpha,\beta),\quad \Omega_b = \phi_2(\alpha,\beta)$$

MOLECULAR STRUCTURAL PARAMETERS

- Dimensionless Molecular Parameters

$$Z_c = \frac{P_c v_c}{RT_c},\ \Omega_w = \frac{b}{v_c},\ \Omega_a = \frac{aP_c}{R^2T_c^2},\ \Omega_b = \frac{bP_c}{RT_c}$$

$$Z_c = \frac{P_c v_c}{RT_c},\ \Omega_w = \frac{b_m}{v_c},\ A = \frac{a_m P_c}{R^2T_c^2},\ B = \frac{b_m P_c}{RT_c}$$

VDW PARAMETERS AS FUNCTIONS OF (α, β)

$$\Omega_w = f_w(\alpha,\beta):\ (\alpha^2-\alpha\beta+\beta)\Omega_w^3 + 3(\alpha-\beta)\Omega_w^2 + 3\Omega_w - 1 = 0$$

$$\Omega_a = f_a(\alpha,\beta):\ \Omega_a = 3Z_c^2 + \alpha\Omega_b + (\alpha+\beta)\Omega_b^2 \ \text{ or }\ \Omega_a = [1+(\Omega_w-1)Z_c]^3$$

VIRIAL COEFFICIENTS AND BOYLE TEMPERATURE

- Second Virial Coefficient

$$B = \Omega_b(\alpha,\beta)T_R^{\theta_b} - \Omega_a(\alpha,\beta)T_R^{-\theta_a}$$

- Third-Virial Coefficient

$$C = \left[\Omega_b(\alpha,\beta)T_R^{\theta_b}\right]^2 - \Omega_a(\alpha,\beta)T_R^{-\theta_a}\left[\alpha\Omega_b(\alpha,\beta)T_R^{\theta_b}\right]$$

- Boyle Temperature

$$T_B = T_c\left[\frac{\Omega_a(\alpha,\beta)}{\Omega_b(\alpha,\beta)}\right]^{f(\theta)},\ \text{or}\ \ T_c\Phi_B(\alpha,\beta)$$

PERTURBATION-SENSITIVITY METHODS

- Sensitivity of Parameter α to Molecular Structure

$$\left(\frac{\partial Z_c}{\partial\alpha}\right)_\beta,\ \left(\frac{\partial\Omega_w}{\partial\alpha}\right)_\beta,\ \left(\frac{\partial\Omega_a}{\partial\alpha}\right)_\beta,\ \left(\frac{\partial\Omega_b}{\partial\alpha}\right)_\beta,\ \left(\frac{\partial P_c}{\partial\alpha}\right)_\beta,\ \left(\frac{\partial T_c}{\partial\alpha}\right)_\beta,\ \left(\frac{\partial B}{\partial\alpha}\right)_\beta,\ \left(\frac{\partial C}{\partial\alpha}\right)_\beta$$

- Sensitivity of Parameter β to Molecular Shape $\left(\frac{\partial v_c}{\partial\beta}\right)_\alpha,\ \left(\frac{\partial T_B}{\partial\beta}\right)_\alpha$

VDW PARAMETERS AS FUNCTIONS OF (α, β)

$$Z_c = f_z(\alpha,\beta):\ \vartheta_4 Z_c^3 + \vartheta_3 Z_c^2 + \vartheta_2 Z_c + \vartheta_1 = 0$$
$$\vartheta_4 = (\alpha^3 + 6\alpha^2 + 12\alpha + 8)$$
$$\vartheta_3 = -(12\alpha^2 + 12\alpha - 9\alpha\beta + 9\beta + 3)$$
$$\vartheta_2 = (6\alpha^2 + 3\alpha - 6\alpha\beta + 6\beta)$$
$$\vartheta_1 = -(\alpha^2 - \alpha\beta + \beta)$$

INTERPRETATION OF MACROSCOPIC PARAMETERS

- Interpretation of Macroscopic Properties

(Leland, AIChE J. *12*(6), 1227, 1966; *14* (4), 568, 1968)

- **Zc, Pc, Tc, B, C ~ Reflect Molecular Structure of Substances**

- Interpretation of Multiplier of Covolume Parameter

(Leland-Mueller, I&E C., *51*, 597, 1959; I&E C. Fund., *9*, 537, 1970)

- **$V_c = \phi b$, ϕ denotes shape Factor of Potential Function**

- Interpretation of 3rd Virial Coefficient & Boyle Temp. (T_B)

(Chueh-Prausnitz, AIChE J., 13(5), 896, 1967)

B, C, T_B ~ Reflects Molecular Structure of Substances

VDW PARAMETERS AS FUNCTIONS OF (α, β)

$$\Omega_b = f_b(\alpha,\beta):\ \theta_4\Omega_b^3 + \theta_3\Omega_b^2 + \theta_2\Omega_b + \theta_1 = 0$$
$$\theta_4 = (\alpha^3 + 6\alpha^2 + 12\alpha + 8)$$
$$\theta_3 = -3(\alpha^2 - 5\alpha + 9\beta - 5)$$
$$\theta_2 = (3\alpha + 6)$$
$$\theta_1 = -1$$

INTERPRETATION OF PARAMETERS IN VDW EQUATION

$$P = \frac{RT}{v-b} - \frac{a}{v^2 + \alpha bv - \beta b^2}$$

Intermolecular Potential Parameter (a)

Molecular Size Parameter (b)

Molecular Structure Parameter (α)

Molecular Shape Parameter (β)

CONCLUSION

- Resolved the Overlooked of Structure-Shape (S-S) Parameters of Substances in the VDW 1873 Theory
- Statistical Mechanics Needed for Verification of S-S Parameters at the Microscopic Level
- Structure and Shape Parameters Resolved ALL the Weaknesses of the VDW 1873 EOS

2017 AIChE Annual Meeting, Minneapolis, MN

504934 Solid-Liquid Equilibria for Selected Binary Mixtures Containing Diphenyl Carbonate

Abstract:

Polycarbonate (PC) is a widely used engineering plastic which has excellent features, such as transparency, impact resistance, thermal stability, dimensional stability, and flame resistance. Most PC production has been via the phosgene process, which uses carbon monoxide and chloride as raw materials. However, this phosgene process has a number of environmental drawbacks. In particular, highly toxic phosgene is used as a monomer and a large quantity of the low-boiling-point solvent methylene chloride is required for polymerization. Also, a very large quantity of water must be used to remove phosgene and chlorine (from methylene chloride) from the product PC. To overcome these environmental problems, a new process for PC production starting from carbon dioxide and bisphenol-A has been developed [1], and has attracted much attention as it offers much greener and more sustainable chemistry. Diphenyl carbonate (DPC) is an important intermediate material in this new process. The purity of the DPC directly influences the downstream PC quality, with high-purity DPC being critical for PC optical media applications. One of the options for purification of DPC is a crystallization process [2-4]. In the design and development of this crystallization process, knowledge of solid-liquid equilibrium (SLE) data is important.

The objective of this work to determine the SLE data for binary systems containing DPC. In this study, we have focused on the transesterification reaction for production of DPC. This reaction is based on alcohol, dialkyl carbonate, phenol, and DPC. Thus, we have measured the SLE for tselected binary mixtures: ethanol or 1-propanol + DPC at atmospheric pressure by visually observing the melting points using the synthetic method. The experimental SLE data of the three binary mixtures were represented by a polynomial equation of the composition suggested by Ott and Goates [5] and the NRTL model [6].

References

[1] N. Miyake, Expected Material for the Future 5 (2010) 46-51 .

[2] S. Kuze, R. Okumura, Y. Suwabe, Preparation of Polycarbonate by Melt Method without Discoloration. Eur. Patent EP0561363, 1993.

[3] D. W. Wei, Y. H. Pei, Ind. Eng. Chem. Res. 47 (2008) 8953-8956.

[4] D. W. Wei, Y. H. Pei, J. Chem. Eng. Data 53 (2008) 2710-2711.

[5] J. B. Ott, J. R. Goates, J. Chem. Thermodyn. 15 (1983) 267-278.

[6] H. Renon, J. M. Prausnitz, AIChE J. 14 (1968) 135-144.

489626 Thermophysical Suitability of Terphenyl for Hydrogen Storage

Abstract:

Liquid Organic Hydrogen Carrier (LOHC) are an interesting option for the storage of hydrogen via the reversible hydrogenation of aromatic compounds. Terphenyls have might by a new class of materials in this respect.

Terphenyl usually consist of a mixture of three isomers, which can be hydrogenated reversibly to their respective saturated derivatives.

However, despite its promising chemical and thermodynamic properties its physical properties, especially its high melting point, is disadvantageous compared to conventional LOHC materials like dibenzyltoluene. Nevertheless, mixtures of isomers or with small amounts of other LOHC materials can achieve reasonable physical properties.

In this contribution the feasibility of hydrogen storage using terphenyl as a LOHC is evaluated with respect to its physical properties. The focus is set on the effect on the ease of handling. Hence, important properties are melting point and viscosity. The properties are analyzed and compared to other LOHCs like Nâ€‘ethylcarbazole and dibenzyltoluene.

494801 Atomic-Level Mechanistic Insights into Monolayer hBN Growth from Reactive Molecular Dynamics Simulations for Catalysis Applications

Abstract:

Atomic-level mechanistic insights into monolayer hBN growth from reactive molecular dynamics simulations for catalysis applications

Bin Liu, Mingxia Zhou, Song Liu, and James H. Edgar

Department of Chemical Engineering, Kansas State University, Manhattan, Kansas 66506

Hexagonal boron nitride (hBN) is an important 2D material. Structurally analogous to graphene, hBN is mechanically and chemically stable, and exhibits exceptional optical and electronic properties for applications such as, ultraviolet light emitters, neutron detectors, catalyst supports, and so on. Understanding of the material growth mechanism is one of the key steps towards optimizing and tailoring material synthesis for desired applications.

In this talk, reactive molecular dynamics (MD) simulations â€“ based on the newly developed ReaxFF force field â€“ have been employed to understand the mechanism of hBN growth on nickel substrates. In order to elucidate the growth pathways, NVT-MD simulations, from stoichiometric elemental B and N species deposited onto a Ni substrate, were performed in the temperature range of 1100-1500 K. The elementary nucleation and growth process of a monolayer hBNâ€¨structure on the nickel substrate from such elemental B and N sources has been revealed at the atomistic scale: the nucleation initiates from the growth of linear BN chains, which evolve into branched and then hexagonal lattices as the centers for growth. Subsequent DFT calculations are able confirm the energetics of the simulated structure evolution, and also validate the self-consistency within our multiscale modeling framework. Based on the established modeling framework, this talk will also present the recent work further exploring the hBN growth mechanism by varying the stoichiometric ratios of the deposited B and N.

Furthermore, I will also demonstrate - using density functional theory (DFT) calculation - that transition metal (i.e., Pt) catalysts supported on defective hBN monolayer can enhance the activity of CO oxidation reaction. Bader charge analyses were conducted to show that charge transfers associated with the Pt nanoparticles (NPs) bound at the N or B-type vacancies can impact on the binding energies of CO and O_2, on Pt NPs, and thus the reactivity.

496686 Differential Retention and Release of CO_2 and CH_4 in Kerogen Nanopores: Implications for Gas Extraction and Carbon Sequestration

Abstract:

Kerogen is the largest pool of organic matter on earth. During kerogen maturation, a considerable amount of CO2 is generated concurrently with petroleum products. Surprisingly, no one has reported any significant amount of CO2 for shale gas plays across the US. One explanation is that once CO2 is expelled from kerogen, it can be removed from the gas phase by mineral precipitation and dissolution. Therefore, fundamental investigation of the retention and expulsion of CO2 in kerogen is important to understand the fate of CO2 in sedimentary basins. In this talk we show that kerogen retains more CO2 than CH4 and that a large amount of the CO2 generated during kerogen maturation will not be released until the reservoir pressure decreases below CO2 supercritical pressure. Our results, derived from careful selection and validation of a simulation model, provide fundamental insight into the fate of CO2 generated in kerogen and potentially in organic matter in general. This research also investigates the introduction of CO2 into kerogen nanopores to mimic the potential use of CO2 to enhance gas recovery and CO2 sequestration scenarios. CO2 can migrate into kerogen nanopores even when the nanopore openings are blocked with water, and displace a large quantity of CH4.
Sandia National Laboratories is a multi-mission laboratory managed and operated by Sandia Corporation, a wholly owned subsidiary of Lockheed Martin Corporation, for the U.S. Department of Energyâ€™s National Nuclear Security Administration under contract DE-AC04-94AL85000.

498509 Investigation of Electronic Properties of Imidazolium-Based Ionic Liquids in the Presence of Iron Porphyrins for Understanding Their Biodegradability

Abstract:

Imidazolium-based ionic liquids have been shown to be highly effective in a wide variety of chemical engineering applications. These include extractive distillation, catalysis and electrochemistry among many others. This is promoted by multiple factors including their design flexibility, low vapor pressure and favorable solvability. Although being extremely useful in nature, they have been shown to be partially soluble in water and often leave a footprint on the ecosystem. This might percolate through the elements and be detrimental to the quality of essential natural resources. Hence, it becomes imperative to think about a rational design of these liquids that may prove to be environmentally friendly. Though there have been significant investigations into the biodegradability of this class of ionic liquids in the experimental direction, computational attempts have rather lesser in number. A lack of understanding of the degradation phenomena of the above said ionic liquids at the molecular level is the motivation behind this research.

With respect to degradation, oxidation has been proposed to be one of the prime mechanisms through which degradation may occur in nature. Former studies illustrate that cytochrome P-450 and its variants are instrumental in catalyzing oxidation in a variety of different aerobic and anaerobic environments. Fe- porphyrin (FeP) represents the core of this enzyme and an important part of its active site. In the present work, the electronic properties of imidazolium cations were studied in the presence of FeP in complex. *Ab initio* calculations were performed on these systems varying the alkyl chain length of imidazolium cations.

A quantum mechanical treatment of the overall system was adopted considering these cations as potential substrates to the FeP receptor. To reflect the suggestions of previous quantitative structure biodegradability relationships (QSBR), variations in frontier orbital energies i.e. Highest Occupied and Lowest Unoccupied energy levels (HOMO and LUMO) were investigated in detail. To account for the effect of geometry on the energetics, three different conformations, namely, tail up, tail down and interplanar conformations were considered. Different aspects concerning the HOMO-LUMO energies across different alkyl chain lengths and conformations will be discussed to elucidate the effect of receptor on the dynamics of the local electronic environment of the cations. Variations in key thermodynamic entities of the system, such as Gibbs free energy and entropy give us important information regarding imidazolium cation-FeP interactions, which would also be included in the results.

497647 Capturing the Membrane-Triggered Conformational Transition for Pore Forming Cytolysin (ClyA) Using Structure Based Models

Abstract:

Pore forming proteins lyse target cells by rapid and unregulated pore formation on the plasma membrane and are widely implicated in a variety of virulent bacterial diseases. Cytolysin a (ClyA) is anÂ Î±-pore forming proteins which undergoes a large conformation change upon binding to the target membrane. The rate determining step of this process is the conversion of the water soluble Clya monomer into a membrane inserted protomer. We elucidate the mechanism of this conformational transition using molecular dynamics simulations of coarse-grained structure based (GO) models of Clya and a membrane. We show that a membrane is necessary for the conformational conversion because membrane protein interactions counteract the loss of the many intra-protein hydrophobic interactions that stabilize theÂ membrane-inserting segments in the Clya monomer. Of the two membrane-inserting segments, the flexible and highly hydrophobic Î²-tongue inserts into the membrane, while the insertion of Helix A1 is membrane assisted. We conclude that the Î²-tongue is designed to behave as a rapidÂ membrane sensor, while Helix A1 improves target selectivity for cholesterol containing cell membranes. We connect these results with recent single molecule experiments carried out in our laboratory and comment on the implications of the study on the pore forming pathway for ClyA.

505275 Solid Form Transformation of Disodium Guanosine 5′-Monophosphate: Thermodynamic Perspective

Abstract:

Disodium guanosine 5′-monophosphate (5′-$GMPNa_2$) is one of the four common ribonucleotides that constitute ribonucleic acids (RNAs), containing pentose phosphate and amidogen groups, which is widely used in food, pharmaceutical and fine chemistry and other industries. The mainly solid forms of 5′-$GMPNa_2$ are amorphous form and crystal polymorphs (I, II, III and IV-form), among all the solid forms, the I-form and amorphous form are the common forms. However amorphous form is often prepared during crystallization process with immature processing conditions, while the crystal products is with poor quality. Some researches have reported that the amorphous form of 5′-$GMPNa_2$ can transform to crystal form with the driving force of solubility difference of the two different forms. Hence the solubility of polymorphous of 5′-$GMPNa_2$ is extremely significant for control course of crystallization and study mechanism of polymorphous transformation. In this work, we are trying to elaborate the transformation mechanism of 5′-$GMPNa_2$ between amorphous form and crystal form from thermodynamic perspective.

A series of means (TG-DSC, PXRD, SEM) were employed to characterize 5′-$GMPNa_2$, and to prove that the different solid form of experimental materials are different in thermodynamic properties, structure arrangement and crystal habit. The solubility of crystalline and amorphous form of 5′-$GMPNa_2$ in binary isopropanol-water and ethanol-water solvent mixture with the temperature ranging from 298.15 K to 318.15 K and pH ranging from 8.5 to 10.0 were studied, meanwhile the residual solids were measured by PXRD and SEM. The solubility data shows that 5′-$GMPNa_2$ is more soluble in ethanol than isopropanol at same mole fraction and temperature, mainly because that 5′-$GMPNa_2$ is polyhydroxy compound,which can easily dissolve in solvent with larger polarity. The PXRD together with solubility data indicates that the transformation of solid forms 5′-$GMPNa_2$ is existed during the solubility experiment but have different direction in different solution environment. In isopropanol - water system, crystalline form can transform to amorphous under the high temperature and low proportion isopropanol, while amorphous form has the tendency to transform to crystalline form under the low temperature and high proportion isopropanol with thermodynamic viewpoint; However in ethanol - water system, amorphous form can transform spontaneously to crystalline form when ethanol proportion great than or equal to 20%, in addition amorphous form has the tendency to transform to crystalline when ethanol proportion less than or equal to 10% with thermodynamic viewpoint theoretically.

Reasons for transformation of 5′-$GMPNa_2$ in different solvent environment can summarize as the different adsorption capacity of solvent between 5′-$GMPNa_2$ contributes different direction of structure rearrangement. Since adsorption capacity between 5′-$GMPNa_2$ and ethanol is more stronger than isopropanol, so isopropanol is more easily dropped out from 5′-$GMPNa_2$ than ethanol during the formative process of 5′-$GMPNa_2$ cluster, which results that clusters have large disordering in isopropanol and need more time to rearrange in isopropanol. That's why that crystalline form transform to amorphous form in isopropanol-water system, while crystalline form transform to amorphous from in ethanol-water system during the solubility experiment process.

503529 Estimation of Kinematic Viscosity for CO_2 expanded Liquids By Asog-Visco Method

Abstract:

A mixture of CO_2 and a co-solvent is widely employed as a solvent or mobile phase for supercritical fluid extraction, chemical reaction, supercritical fluid chromatography etc. In particular, CO_2 expanded liquid (CXL), a liquid phase in a mixture of pressurized CO_2 gas and a liquid organic solvent at gasâ€"liquid phase equilibrium, CXL is expected to provide advantages in these processes because the pressure is lower than the critical pressure and their unique properties can be modified significantly with the small changes in temperature and/or pressure. Kinematic viscosity is one of important transport properties, and is required for designing chemical process as well as estimating dimensionless numbers such as Reynolds number, Schmidt number, Prandtl number [1,2]. A more accurate estimation method for kinematic viscosity is increasingly demanded. The authors have proposed the estimation method, that is, the ASOG-VISCO method by which the excess Gibbs free energy at transition state in the Eyring kinetic equation is obtained by the group contribution method. In fact, the parameters for functional groups such as CH_2, ArCH, CyCH, OH, H_2O, CO, COO, CCl_3, and CCl_4 were determined using the literature data on kinematic viscosities experimentally measured, and those for binary and ternary mixtures consisting of hydrocarbons, alcohols, water, ketones, esters, trichlorides and carbon tetrachloride were estimated with accuracies of 4.2 and 5.0 % for binary and ternary mixtures, respectively [3,4]. In this presentation to estimate kinematic viscosities for binary mixtures using the ASOG-VISCO model, the group parameters were determined from kinematic viscosities of binary mixtures for CO_2 and compounds having functional groups such as CH_2, OH and CO under gas-liquid equilibria conditions. The model was applied to estimate the kinematic viscosities for binary mixtures of saturated carbon dioxide and organic compounds such as paraffins, alcohols, and ketones. The predicted results have been compared with those using the McAllister model [5].

Reference:

1) B. E. Poling, J. M. Prausnitz, J. P. Oâ€™Connell: The Properties of GASES AND LIQUIDS, fifth edition, McGraw Hill (2001)

2) T. Funazukuri, Y. Ishiwata, N. Wakao: AIChE J., 38, 1761 (1992)

3) A. Murata, K. Tochigi, H. Yamamoto: Molecular Simulation, 15, 451 (2004)

4) K. Tochigi, K. Yoshino, V. K. Rattan: Int. J. Thermophys., 26, 413 (2005)

5) R. A. McAllister: AIChE J., 6, 427 (1960)

483976 On the Kac-Based Collision Models from Simplified Bernoulli till Its Intelligent Variants

Abstract:

On the Kac-based collision models from Simplified Bernoulli till its Intelligent variants

Bijan Goshayeshi

High Performance Computing (HPC) Laboratory, Department of Mechanical Engineering, Ferdowsi University of Mashhad, P.O. Box: 91775-1111, Mashhad, Iran

Key words

Direct Simulation Monte Carlo method (DSMC), kinetic theory, rareﬁed gas ﬂows, Simplified Bernoulli-trials (SBT) scheme, Intelligent Simplified Bernoulli-trials (ISBT) scheme, Transient Adaptive Subcell (TAS), Nearest Neighbor (NN).

Abstract

Graeme Bird [1] developed the Direct Simulation of Monte Carlo (DSMC) in the 1960s, and nowadays this method has gained popularity in simulation of rarefied gas dynamics and micro gas flow problems. In addition to the requirement of employing multidimensional computational mesh, the DSMC method uses a finite set of particles or simulators, denoted by their positions and velocities, to model the advection and collision terms of the Boltzmann equation. The requirement of these large computational resources have been a prohibitive barrier in the DSMC analysis of massive computational two- and three-dimensional problems in rarefied gas dynamics. This work is dedicated to review some recent advancements in reducing the DSMC computational requirements by using Kac model -based collision schemes [2-4] in the Direct Simulation of Monte Carlo. In general, two major concepts exist for obtaining collision schemes, and here we focus our attention to the one based on the Kac stochastic model. The common advantage of these schemes is that their algorithms avoid the repeat collisions, and can be used to reduce the number of particles as a portion of computational resources. The paper reviews this conception since Yanitiskiy first introduced the Bernoulli Trials collision scheme (BT), to the introduction of Simplified Bernoulli Trials (SBT), proposed by Stefanov. We also present a new intelligent variant of the SBT collision scheme, entitled ISBT, which enters some semi-deterministic elements in the SBT scheme, that ensure selection of closer placed collision pairs.

References

[1] Bird, G. A. (1994) Molecular Gas Dynamics and the Direct Simulation of Gas Flows, Oxford :Clarendon Press.

[2] Stefanov, S. K. (2011). Particle Monte Carlo algorithms with small number of particles in grid cells. In Numerical Methods and Applications, 110-117P. Tata, C. Toto, *A theory of many things*, Fluid Dyn. Res. **44**, 031202 (2012)

[3] Goshayeshi, B., Roohi, E., & Stefanov, S. (2015). DSMC Simulation of hypersonic flows using an improved SBT-TAS technique. Journal of Computational Physics.

[4] Goshayeshi, B., Roohi, E., Stefanov, S. (2015). A novel Simplified Bernoulli Trials collision scheme in the DSMC with intelligence over particle distances, Physics of Fluids, To appear in Vol. **27**(10).

494968 Viscosity and Derived Properties of Binary Mixture Containing Acetonitrile + 1-Propanol at Different Temperatures and Atmospheric Pressure

Abstract:

Experimental data of viscosity and therefore its derivatives properties of liquid systems as a function of composition and temperature are important due to the fact that the viscosity is a property frequently used in designs of industrial processes. These properties are required in many calculations in Chemical Engineering involving fluid flow, heat and mass transfer. Acetonitrile is widely used as a reaction solvent for the production of antibiotics. It is also one of the most used solvents in the butadiene extraction. Once highly purified, it can be used as a solvent in high performance liquid chromatography. As a continuation of our study involving excess properties containing acetonitrile [1-4], in the present work, viscosities and density of binary mixture of acetonitrile + 1-propanol have been determined over the entire composition range at different temperatures and atmospheric pressure. The temperatures studied were 288.15, 293.15, 298.15, 303.15 K and 308.15 K. Viscosity and density measurements of pure liquid and mixtures were performed using a Stabinger viscometer (Anton Paar, Model SVM 3000/G2). The experimental data have been used to calculate deviation in viscosity and excess Gibbs energy of activation for viscous flow, and correlated to the Redlich-Kister equation. For the studied systems, deviation in viscosity and excess Gibbs energy of activation for viscous flow were negative over the entire composition range and become less negative with increase in temperature. The experimental viscosity results were also used to test the applicability of the empirical equations proposed by Grunberg-Nissan, Hind-McLaughlin, Katti-Chaudhri, Heric-Brewer and Van der Wyk. The results obtained are discussed in terms of alcohol self-association and cross-association between acetonitrile and 1-propanol molecules.

References

[1] S.S. Bittencourt, R.B. Torres. Volumetric properties of binary mixtures of (acetonitrile + amines) at several temperatures with application of the ERAS model. Journal of Chemical Thermodynamics, 2016. 93: p. 222-241.

[2] R.B. Torres, A.Z. Francesconi, P.L.O. Volpe. Volumetric properties of binary mixtures of acetonitrile and alcohols at different temperatures and atmospheric pressure. Journal of Molecular Liquids, 2007. 131: p. 139-144.

[3] R.B. Torres, A.Z. Francesconi, P.L.O. Volpe. Thermodynamics of binary liquid mixtures: application of the Prigogine-Flory-Patterson theory to excess molar volumes of acetonitrile + 1-alkanols systems. Journal of Molecular Liquids, 2004. 110: p. 81-85.

[4] R.B. Torres, A.Z. Francesconi, P.L.O. Volpe. Experimental study and modeling using the Eras-model of the excess molar volume of acetonitrile + 1-alkanols mixtures at different temperatures and atmospheric pressure. Fluid Phase

Equilibria, 2003. v. 210: p. 287-306.

495030 Volumetric, Acoustic and Viscometric Properties of Binary Mixture of (*n*-butylammonium methanoate + 1-propanol) at Different Temperatures

Abstract:

In this work, experimental data of density, speed of sound and viscosity for binary mixture of {*n*-butylammonium methanoate + 1-propanol} have been determined as a function of composition at T = (293.15 - 313.15) K. The results were used to calculate the excess molar volume, V^E, deviation in isentropic compressibility, Î”k_S , deviation in viscosity, Î”Î·, and excess Gibbs free energy of activation of viscous flow, Î”G^{*E}, and correlated to the Redlich-Kister polynomial equation. Ionic liquids were synthesized by the method of acid-base neutralization reaction using *n*-butylamine and methanoic acid. *N*-Butylamine (Merck, purity > 99.0%) and methanoic acid (Merck, purity > 98.0) were used without further purification. Ionic Liquids were characterized by using ^{1}H-NMR, ^{13}C-NMR and FTIR spectroscopy techniques. The binary mixtures were prepared by weighing of the pure components. The densities and speed of sound of pure components and their solutions were measured using a commercial density and speed of sound analyzer manufactured by Anton Paar (Model DSA5000). The viscosities were measuring using a Stabinger viscometer manufactured by Anton Paar (Model SVM3000/G2). The analysis of ^{1}H-NMR and ^{13}C-NMR were obtained in an Avance III 600 HD Bruker spectrometer operating at 600 MHz for protons and 150 MHz for carbon. The IR spectra were obtained in Fourier transform infrared spectroscopy Nicolet with attenuated total reflectance (FTIR-ATR) (Model 6700). For all systems studied, the values of excess molar volumes, deviation in isentropic compressibility and deviation in viscosity were negative over the entire composition range. The values of excess Gibbs energy of activation for viscous flow were positive over the entire composition range. The results obtained were discussed in terms of intermolecular interactions, particularly hydrogen-bonding interactions between like and unlike molecules.

References

[1], S.S. Bittencourt, H.E. Hoga, R.B. Torres, J.V.H Dâ€™Angelo. Thermodynamic and spectroscopic properties of binary mixtures of *n*-butylammonium butanoate ionic liquid with alcohols at T = (293.15 - 313.15) K. The Journal of Chemical Thermodynamics, v. 105, p. 238-252, 2017.

496275 Measurement of Diffusion Coefficients of Chromium(III) Acetylacetonate in Supercritical Carbon Dioxide at High Temperatures

Abstract:

Recently, supercritical fluid deposition (SFD) has become focus of attention as well as various chemical processes with supercritical CO_2 such as extraction, dyeing, chemical reaction etc. SFD is useful for plating with metals on high-aspect-ratio substrates due to negligible surface tension of $scCO_2$. These methods are expected for materials of nano-devices instead of using aqueous solutions which results in destructing trenches on the substrates when evaporating solvents having high surface tension such as an aqueous solution. Thus, SFD could be a promising process, and the physical properties, especially diffusion coefficient, are required for design reactors at high temperatures [1, 2]. While the Taylor dispersion method is common to measure diffusion coefficients in $scCO_2$, it is not suitable with high accuracies because of peak tailing at high temperatures. Consequently, measurements of diffusion coefficients in $scCO_2$ at high temperatures were scarcely reported.

In this work, diffusion coefficients of chromium(III) acetylacetonate, $Cr(acac)_3$, together with retention factors were measured above 70 Â°C using the chromatographic impulse response method, which is a kind of transient response methods and two parameters: diffusion coefficient and retention factor are simultaneously determined [3]. Validating a hydrodynamic equation [4] as a predictive correlation for determined diffusion coefficients of $Cr(acac)_3$, average absolute relative deviation (AARD) increased with increasing temperature. The diffusion coefficients at higher temperatures deviate from the reference values in low temperature $scCO_2$. It is implied that the hydrodynamic equation based on the Stokes-Einstein equation, derived from those in liquid phases, is no longer valid in supercritical fluids at high temperatures and low pressures. Moreover, solubilities of the solutes in $scCO_2$ can be estimated using an Ekartâ €™s equation [5] from retention factors determined. The estimated solubilities provided high accuracies with AARD < 10% at all temperatures.

Bibliography

[1] Y. Zhao et al., J. of Supercritical Fluids, 120 (2017) 209.
[2] S. Yoda et al., J. of Supercritical Fluids, 123 (2017) 82.
[3] T. Funazukuri et al., Ind. Eng. Chem. Res., 39 (2000) 4462.
[4] C.Y. Kong et al., J. Chromatogr. A, 1035 (2004) 177.
[5] M. Ekart et al., AIChE J., 39 (1993) 235.

501060 Simultaneous Determinations of Dynamic Viscosity and Density of Several Alcohols Using a Straight and Coil Capillary Viscometers at Temperatures between (313-353) K and Pressures up to 30 Mpa

Abstract:

In this work, simultaneous determinations of viscosity and density of propanol, butanol and pentanol using two different capillaries viscometers (straight and coil) were performed. The goal of this analysis is to determine the influence that have: a) the configuration of the capillary tube (straight and coil) considering the existing centripetal force in the coil tube in the determination of liquid viscosities and densities of pure fluids; b) the length of the capillary (0.755 m and 5.5 m for straight and coil tubes, respectively). For this reason, three n-alcohols were used as the test fluids, determining the deviation between the data obtained with both tubes and the international literature. Moreover, for both cases the experimental uncertainty was estimated. The results show that, for the operating conditions used, the coil capillary viscometer has the smallest deviation from literature data (for viscosity: 0.35 % coil capillary and 0.5 % straight capillary; for density: 0.04 % coil capillary and 0.05 % straight capillary). This can be explained because there is a long length difference between the two tubes (5.5 m and 0.755 m for coil and straight tube, respectively). In the estimation of relative combined expanded uncertainty, there is no noticeable difference between the two capillaries (for viscosity: 0.7 % and 0.72 % for coil and straight capillary, respectively; for density: 0.07 % and 0.071 % for coil and straight capillary). It is noteworthy that the effect of centripetal force in the calculation of the uncertainty was evaluated to be 0.03%. All measurements were carried out at temperature up to 353 K and pressures up to 30 MPa

Analysis of high-speed rotating flow inside gas centrifuge casing

Sahadev Pradhan

Department of Chemical Engineering, Indian Institute of Science, Bangalore 560 012, India

The generalized analytical model for the radial boundary layer inside the gas centrifuge casing in which the inner cylinder is rotating at a constant angular velocity Ω_i while the outer is stationary, is formulated for studying the secondary flow due to wall thermal forcing, inflow/outflow of light gas along the boundaries, as well as the combination of the above two external forcing. The analytical model includes the sixth order differential equation for the radial boundary layer at the cylindrical surface in terms of master potential (χ), which is derived from the equations of motion in an axisymmetric ($r - z$) plane. The linearization approximation is used, where the equations of motion are truncated at linear order in the velocity and pressure disturbances to the base flow. Additional assumptions in the analytical model include constant temperature in the base state (isothermal compressible Couette flow), but there is no limitation on the Mach number. The discrete eigenvalues and eigenfunctions of the linear operators (sixth-order in the radial direction for the generalized analytical equation) are obtained. The solutions for the secondary flow is determined in terms of these eigenvalues and eigenfunctions. Next we perform the direct simulation Monte carlo (DSMC) simulations inside the gas centrifuge casing for a range of Mach number, Knudsen number, and aspect ratio to understand the detailed flow phenomenan and heat transfer mechanism which could help to improve the efficiency of the centrifugal separation process.

Key words: High speed rotating flow, Gas centrifuge casing, DSMC Simulation, Rarefied gas flow.

---

1. Introduction

Centrifugal gas separation processes effect separation by utilizing the difference in the mole fraction in a high speed rotating cylinder caused by the difference in molecular mass, and consequently the centrifual force density. These have been widely used in isotope separation because chemical separation methods cannot be used to separate isotopes of the same chemical species. An important application is the separation of trace radioactive isotopes of metals (molybdenum, tellurium, iridium, cadmium, technitium, indium, etc.) for medical imaging, as well as use in non-destructive testing using gamma cameras. These are also used in the nuclear industry in the separation of uranium isotopes for power generation, as well as zinc isotopes for corrosion resistance.

In a gas centrifuge, the inner cylinder rotates at very high-speed, while the outer is stationary. The region between these two cylinders communicates with the center of the machine and so contains essentially light gas. By introducing the inert Helium gas inside the casing the thermal convection can be increased and thereby enhance the efficiency of the wall thermal drive. For this reason, it is important to analyse the flow field and temperature field inside the casing in terms of residual pressure in the casing. The vacuum inside the casing is created due to the rotation of the inner cylinder itself, by use of a molecular pump.

Here we have attempt to study the cylindrical Couette flow in the casing between the rotating inner cylinder and the fixed outer cylinder, using a combination of analytical model and molecular simulations based on direct simulation Monte Carlo (DSMC) method. At moderate residual casing pressure (by introducing the inert Helium gas inside the casing to enhance the thermal convection), we formulate the generalized analytical model from the linearized Navier-Stokes equations. The analytical results have been compared with the results of DSMC simulations in order to quantitatively estimate the accuracy of the analytical model.

Stefanov & Cercignani (1993) discussed about the computations concerning Taylor's instability of Couette flow, a well-known phenomenon in hydrodynamics. They mentioned the classical case of this flow which occurs when we consider two coaxial cylinders of infinite length and the inner cylinder rotates while the outer one is at rest. At a certain critical value of the angular velocity of the rotating cylinder, the Couette flow becomes unstable and transforms into the so-called Taylor-Couette flow. This flow is characterized by a system of Taylor cells in the form of toroidal vortices, two neighbouring vortices rotating in opposite directions. Their results refer to Knudsen numbers of order 10^{-2} and different Mach numbers, based on the velocity of the inner cylinder. The ranges of these numbers are chosen in such a way as to indicate the onset of instability in the cylindrical Couette flow of a rarefied gas. When the Taylor-Couette flow appears to be fully developed they extended the study to higher Mach numbers and exhibit results that seem to indicate transition to chaotic behaviour. Again the real physics of the transition to turbulence escapes these computations because of their two dimensional nature.

They have mentioned that the Knudsen number to be considered in the analysis is either based on the radius of the inner cylinder or based on the annular gap width; but what is the Knudsen number to be taken into account in order to describe the rarefaction effects in the instability that leads to Taylor cells. From their results one sees that the Knudsen number based on the distance over which there is an important gradient in a cell is about one fifth of the gap's width and a factor 5 in the Knudsen number completely changes the order of magnitude of the perturbations, which can no longer be regarded as 'small'. The issues that appear in this connection are so many that only a study of comparable complexity for the Navier-Stokes equations (including compressibility at very large Mach numbers and slip effects) could give us an answer. A study that takes into account compressibility, but not slip, has also been carried out. The results of the paper indicate that the rarefaction effects play a very important role.

Another issue that has established in their work, concerning the ability of the DSMC technique to faithfully resolve vortex motion, owing to the lack of accurate conservation of angular momentum in collisions. The first paper to cast doubts on this ability is applied both molecular dynamics and the direct simulation Monte Carlo (DSMC) method to the case of a rarefied flow past an impulsively started inclined flat plate. A clear vortex structure was obtained in the wake region of the molecular dynamics calculation, but the wake in the DSMC method was relatively devoid of structure. As pointed out by Bird (1993), the approximations associated with either method cannot be tested for the unsteady flow past a plate because this problem places excessive demands on computer time. He therefore introduced the forced vortex flow produced by a moving wall in a two-dimensional cavity as an alternative test case and came to the conclusion that as long as the cell size requirement is met, the nonconservation of angular momentum in collisions does not appear to have a significant effect on the results of direct simulation Monte Carlo calculations. Bird (1993) also examined the values of the parameters in Meiburg

(1986) calculations and showed that the density of the gas was too high to employ the DSMC method, because the mean free path was of the order of the molecular diameter and the size of the cell is too large to analyse the vortical wake structure. The issue of nonconservation of angular momentum was addressed by Nanbu (1988), who showed that if the cell is sufficiently small, the total angular momentum of the molecules in a cell is almost conserved in the Monte Carlo calculations.

Manela & Frankel (2007) consider the linear temporal stability of a Couette flow of a Maxwell gas within the gap between a rotating inner cylinder and a concentric stationary outer cylinder both maintained at the same temperature. The neutral curve was obtained for arbitrary Mach (Ma) and small Knudsen (Kn) numbers by use of a 'slip-flow' continuum model and was verified via comparison to direct simulation Monte Carlo results. At subsonic rotation speeds they find that, for the radial ratios considered, that the neutral curve nearly coincides with the constant-Reynolds-number curve pertaining to the critical value for the onset of instability in the corresponding incompressible flow problem. With increasing Mach number, transition was deferred to larger Reynolds numbers. It was remarkable that for a fixed Reynolds number, instability was always eventually suppressed beyond some supersonic rotation speed. To clarify this they examine the variation with increasing Ma of the reference Couette flow and analyse the narrow-gap limit of the compressible Taylor-Couette flow problem. The results of these suggest that, as in the incompressible problem, the onset of instability at supersonic speeds was still essentially determined through the balance of inertial and viscous-dissipative effects. Suppression of instability is brought about by increased rates of dissipation associated with the elevated bulk-fluid temperatures occurring at supersonic speeds. A useful approximation was obtained for the neutral curve throughout the entire range of Mach numbers by an adaptation of the familiar incompressible stability criteria with the critical Reynolds (or Taylor) numbers now based on average fluid properties. The narrow-gap analysis further indicates that the resulting approximate neutral curve obtained in the (Ma, Kn) plane consists of two branches: (i) the subsonic part corresponding to a constant ratio (Ma/Kn) (i.e. a constant critical Reynolds number) and (ii) a supersonic branch which at large Ma values corresponds to a constant product (Ma $\times$ Kn). Finally, their analysis helps to resolve some conflicting views in the literature regarding apparently destabilizing compressibility effects

The Taylor-Couette (TC) instability in a fluid between rotating concentric cylinders giving rise to a secondary vortical flow ('Taylor vortices') is a classical problem in hydrodynamic stability theory (Chandrasekhar (1961); Drazin (1981); Koschmieder (1993)). The problem has been investigated extensively for incompressible fluids. Rayleigh (1916) formulated a stability criterion for the mean viscous flow which is based on the inviscid perturbation equations. Subsequent analyses have considered the viscous perturbation equations where transition to instability is governed by Re_c, a critical value of the Reynolds number (Re), depending on the ratios of cylinders' radii and rotation speeds. Taylor (1923) studied the narrow-gap approximation. He found that, while in this limit Re_c, the transition to instability is governed by a finite critical value of the Taylor (Ta) number. Others have analysed the problem for arbitrary gap widths and have demonstrated that viscosity stabilizes the flow in comparison with the inviscid Rayleigh criterion. Only a relatively few studies have so far addressed the compressible TaylorCouette problem. This flow problem is governed by additional parameters, including the temperature ratio of the cylinders and the Mach (Ma) number. Kuhlthau (1960) conducted experiments with dry air to study the effects of transonic Mach numbers (Ma) on the Taylor-Couette instability. During each set of experiments the rotation speed (Ma) was maintained constant while the average density was increased (i.e. Kn decreased). The onset of instability was identified through a sharp increase in the torque measured on the outer (stationary) cylinder. The experiments were carried out for Ma in the range (0.7-1.5) showing an increase in the critical Knudsen number (Kn_c) with increasing Ma. The value of the critical Reynolds number (Ma/Kn_c) was found to be nearly the same as in the corresponding incompressible problem.

Kao (1992) assumed axisymmetric perturbations and studied the linear stability problem for the ratio of the radii of the outer and inner cylinders ($R_R = R_o/R_i = 2$). Their results apparently suggested that increasing Ma had a destabilizing effect in the sense that it decreased Re_c relative to its incompressible value. Hatay (1993) considered various parameter combinations and non-axisymmetric perturbations. Their results essentially agreed with those of Kao (1992) in that the effect of increasing Ma, which was stabilizing for narrow gaps, apparently became destabilizing for wider gaps. In both of these studies the Reynolds number was defined on the basis of the local gas density at the inner cylinder. Unlike the average density (which is determined by the total mass of fluid within the gap), the local density is not a priori prescribed in experiments or simulations. Furthermore, owing to the appearance of both elevated bulk-fluid temperatures and large radial pressure gradients at supersonic speeds, the local and average densities significantly differ from each other. Consequently, comparison of the critical values of the Reynolds numbers based on the local density with their incompressible counterparts may not provide an unequivocal indication of stabilizing or destabilizing effect of increasing Ma. Rather, it will be established that the effects of increasing Ma are adequately correlated and interpreted in terms of modified Reynolds number (Re_m), and Taylor number Ta_m which are based on average fluid properties. For sufficiently wide gaps, instability phenomena in the TaylorCouette problem occur at Re = 100 ((Chandrasekhar (1961)). Both Kao and Chow (1992) and Hatay et al. (1993) have studied the problem for Ma = 4. The Knudsen number (corresponding to the upper end is Kn & O(10). Thus, the very use of the continuum approach may not be consistent (Cercignani 2000). Furthermore, with the advent of microfluidic systems (e.g. micro-gas turbines; see Epstein 2004), the TC problem in rarefied gases have attracted considerable interest in recent years. The axisymmetric problem has principally been studied by means of the direct simulation Monte Carlo (DSMC) method (Bird 1994). The numerical simulations follow the evolution of the system through its terminal state which, in turn, serves to classify the system response as stable or unstable. Riechelmann Nanbu (1993) studied the problem for a Maxwell gas. They justified their application of DSMC to this problem by demonstrating a close agreement with the experimental results of Kuhlthau (1960). Stefanov Cercignani (1993) considered the problem for a hard-sphere gas and found that, contrary to the above-mentioned claims of Kao Chow (1992) and Hatay et al. (1993), increasing Ma had a stabilizing effect for RR = 2. Stefanov Cercignani (1993) related this disagreement to the effects of rarefaction which were not considered by the former authors. They also pointed out the potential usefulness of a stability analysis based on the slip-flow continuum model. Golshtein Elperin (1995) and Usami (1995) carried out DSMC calculations for parameter combinations within the instability domain and studied the form of vortices obtained. Bird (1998) investigated the time evolution of the vortices and presented results of three-dimensional DSMC calculations. Aoki, Sone Yoshimoto (1999) and Yoshida Aoki (2005) applied the DSMC method to study the influence on the neutral curve of varying both temperature and velocity ratios. All of the above-mentioned DSMC studies demonstrate that TC instability is a small O(100) Knudsen phenomenon. However, the artificial noise inherent in these simulations makes it difficult to identify and characterize the final states clearly, particularly for parameter combinations in the vicinity of the transition to instability. Furthermore, these simulations become extremely time-consuming in the continuum limit, obstructing accurate delineation of the domain of instability. Consequently, explicit results in the literature have been presented only for a limited number of parameter combinations.

Recently, Yoshida Aoki (2006) have studied the corresponding linear stability problem based on the BhatnagarGrossKrook (BGK) model of the Boltzmann equation for RR = 2. A close agreement was found with continuum NavierStokes calculations. The present contribution is intended to complement the above studies by considering the corresponding linear hydrodynamic temporal stability problem based on the NavierStokes slip-flow continuum model for (arbitrarily) small Knudsen numbers. Our main objective is to gain some insight into the physical mechanism causing suppression of the TC instability at moderately large supersonic speeds of rotation.

In §2, the linearized mass, momentum and energy conservation equations for an axisymmetric flow in a cylindrical coordinate system are presented and different dimensionless numbers are defined. The equations for perturbations to the base state (isothermal compressible cylindrical Couette flow) are then used to get the equation for the master potential in §3. In §4, the radial boundary layer equations are solved to obtain the generalized analytical equation. In §5, a brief summary on molecular simulation based on direct simulation Monte Carlo (DSMC) method is presented. In §6, secondary flow due to wall thermal forcing, inflow/outflow of light-gas along the boundaries, as well as combination of those is studied and analytical results have been compared with those of DSMC simulations. The important conclusions of the analysis are summarized in §7.

2. Formulation

he configuration consists of right circular concentric cylinders of radius R_i and R_o for the inner and outer cylinders respectively, and height Z. The inner cylinder is rotating at a constant angular velocity Ω_i while the outer is kept stationary. The fluid (rarefied gas) is confined between the concentric cylinders which are maintained at the uniform temperature T_0. A cylindrical coordinate system with origin at the bottom of the cylinder along the axis of rotation, with radial, meridional and axial coordinates r, θ, and z respectively, is used. Throughout this analysis, we assume that the flow is axisymmetric, so there is no dependence on the meridional coordinate θ. Gravitational acceleration is neglected, since it has a negligible effect for high-speed gas flows. The compressible mass, momentum and energy equations in the cylindrical coordinate system are provided in the appendix for reference. There are six variables in these equations, the density ρ, the three components of the velocity v_r, v_θ, v_z, the pressure p and the temperature T. In equations A 1 to A 5, $\mathcal{S}_m$ is the mass source density (mass input per unit volume per unit time), f_r, f_θ and f_z are the force densities in r, θ and z directions respectively, and $\mathcal{S}_e$ is the energy source density (energy input per unit volume per unit time).

The pressure is related to the density by the ideal gas equation of state,

$$\rho = \frac{pm}{k_B T}, \tag{2.1}$$

where m is the mass of a molecule and k_B is the Boltzmann constant.

If the temperature on the walls of the cylinders is uniform, and there is no inflow or outflow of gas, a steady solution for the equations of motion is one where the gas undergoes isothermal compressible Couette flow with angular velocity of the inner cylinder Ω_i at constant temperature. The temperature $\bar{T}$ and the components of the velocity $\bar{\mathbf{v}}$ in the base state in this steady state are

$$\bar{v}_z = 0, \quad \bar{v}_r = 0, \quad \bar{T} = T_0, \tag{2.2}$$

where T_0 is the constant temperature. The angular velocity is obtained by solving the θ momentum equation,

$$\frac{\partial}{\partial r}\left(\frac{1}{r}\frac{\partial}{\partial r}\left(r\bar{v}_\theta\right)\right) = 0. \tag{2.3}$$

The above equation is easily solved to obtain,

$$\bar{v}_\theta = br + \frac{a}{r}, \tag{2.4}$$

where

$$b = -\frac{\Omega_i R_i^2}{R_o^2 - Ri^2},$$
$$a = \frac{\Omega_i R_i^2 R_o^2}{R_o^2 - Ri^2}. \tag{2.5}$$

The pressure is obtained by solving the radial momentum equation,

$$\frac{d\bar{p}}{dr} = \bar{\rho}\frac{\bar{v}_\theta^2}{r} = \frac{\bar{p}m}{k_B T_0}\frac{\bar{v}_\theta^2}{r}, \tag{2.6}$$

where $\bar{\rho}$ is the density at steady state, together with the ideal gas equation of state,

$$\bar{\rho} = \frac{\bar{p}m}{k_B T_0}, \tag{2.7}$$

where m is the mass of a molecule and k_B is the Boltzmann constant. The above equation is easily solved to obtain

$$\frac{\bar{p}_o}{\bar{p}_i} = \frac{\bar{\rho}_o}{\bar{\rho}_i} = \exp\left(\frac{m}{k_B T_0}\left(\frac{b^2}{2}\left(R_o^2 - R_i^2\right) + 2ab\log_e\frac{R_o}{R_i} - \frac{a^2}{2}\left(\frac{1}{R_o^2} - \frac{1}{R_i^2}\right)\right)\right),$$

where $\bar{p}_w$ and $\bar{\rho}_w$ are the pressure and density at the wall, and Ma is the Mach number defined below.

The volume-averaged gas density (total mass of the gas within the annular gap divided by volume) for the base state is denoted by $\bar{\rho}$, and defined by

$$\bar{\rho} = \frac{1}{\pi\left(R_R^2 - 1\right)}\int_1^{R_R} \rho\, r dr, \tag{2.9}$$

where R_R is the ratio of the radii of the outer and inner cylinders $R_R = R_o/R_i$.

It is useful to discuss the different dimensionless numbers used in the analysis. The Mach number obtained from the equation for the density is

$$\mathrm{Ma} = \sqrt{\frac{m\Omega_i^2 R_i^2}{k_B T_0}} \propto \frac{R_i\Omega_i}{c_s}, \tag{2.10}$$

where Ω_i and R_i are the rotational speed and radius of the inner cylinder, $(R_i\Omega_i)$ is the characteristic flow velocity, and $c_s = \sqrt{\gamma\, kT_0/m}$ is the speed of sound. The Reynolds number, based on annular gap width $(R_o - R_i)$, and rotational speed of the inner cylinder $(\Omega_i R_i)$, is defined by

$$\mathrm{Re} = \frac{\bar{\rho}\Omega_i R_i\left(R_o - R_i\right)}{\bar{\mu}}, \tag{2.11}$$

where $\bar{\mu}$ is the viscosity of the fluid in the unperturbed state. The viscosity of a dilute gas scales as

$$\bar{\mu} \sim \frac{\sqrt{mk_B T_0}}{d^2}, \tag{2.12}$$

where m is the molecular mass and d is the molecular diameter. Using the above expression for the viscosity, it can easily be verified that the Reynolds number can be rewritten as

$$\mathrm{Re} \sim \frac{\mathrm{Ma}}{\mathrm{Kn}}, \tag{2.13}$$

where the Knudsen number, based on annular gap width $(R_o - R_i)$ is defined by

$$\begin{aligned}\mathrm{Kn} &= \left(m/\sqrt{2}\pi\bar{\rho}d^2\left(R_o - R_i\right)\right), \\ &= \lambda/\left(R_o - R_i\right),\end{aligned} \tag{2.14}$$

where $\lambda = (m/\sqrt{2}\pi\bar{\rho}d^2)$ is the mean free path based on the average density within the annular gap. The other dimensionless groups used are the Taylor number, Prandtl number and the ratio of specific heats,

$$\mathrm{Ta} = \frac{4\bar{\rho}^2\Omega_i^2 R_i^4}{\bar{\mu}^2\left(1 - \left(\frac{R_i}{R_o}\right)^2\right)^2}, \tag{2.15}$$

$$\mathrm{Pr} = \frac{C_p\bar{\mu}}{K}, \tag{2.16}$$

$$\gamma = \frac{C_p}{C_v}. \tag{2.17}$$

3. Equations of motion in axisymmetric coordinates

The disturbance to the base state, which is a isothermal compressible cylindrical Couette flow, due to a temperature gradient at the walls, or due to a small mass flux through the cylinder, is incorporated by perturbing the base velocity and pressure profile as follows:

$$\rho = \bar{\rho} + \hat{\rho}, \tag{3.1}$$

$$\begin{aligned}p &= \bar{p} + \hat{p}, \\ T &= \bar{T} + \hat{T}, \\ v_r &= \hat{v}_r, \\ v_\theta &= \left(br + \frac{a}{r}\right) + \hat{v}_\theta, \\ v_z &= \hat{v}_z.\end{aligned} \tag{3.2}$$

The above forms for the dynamical variables are inserted into the mass, momentum and energy conservation equations, and linearized in the perturbation (hatted) quantities. Steady state is assumed, so that all the time derivatives are zero. After subtracting the corresponding base state equations, we obtain the following equations for the perturbations:

$$\frac{1}{r}\frac{\partial}{\partial r}(r\bar{\rho}\hat{v}_r) + \frac{\partial}{\partial z}(\bar{\rho}\hat{v}_z) = \mathcal{S}_m, \tag{3.3}$$

$$-\hat{\rho}b^2 r - \frac{2\hat{\rho}ab}{r} - \frac{\hat{\rho}a^2}{r^3} - \frac{2}{r}\left(br + \frac{a}{r}\right)\bar{\rho}\hat{v}_\theta = -\frac{\partial\hat{p}}{\partial r} + \bar{\mu}\left(\frac{\partial}{\partial r}\left(\frac{1}{r}\frac{\partial}{\partial r}(r\hat{v}_r)\right) + \frac{\partial^2\hat{v}_r}{\partial z^2}\right) + \left(f_r + \frac{\bar{\mu}}{3}\frac{\partial\mathcal{S}_m}{\partial r}\right), \tag{3.4}$$

$$\frac{2}{r}\left(br + \frac{a}{r}\right)\bar{\rho}\hat{v}_r = \bar{\mu}\left(\frac{\partial}{\partial r}\left(\frac{1}{r}\frac{\partial}{\partial r}(r\hat{v}_\theta)\right) + \frac{\partial^2\hat{v}_\theta}{\partial z^2}\right) + f_\theta, \tag{3.5}$$

$$0 = -\frac{\partial \hat{p}}{\partial z} + \bar{\mu}\left(\frac{1}{r}\frac{\partial}{\partial r}\left(r\frac{\partial \hat{v}_z}{\partial r}\right) + \frac{\partial^2 \hat{v}_z}{\partial z^2}\right) + \left(f_z + \frac{\bar{\mu}}{3}\frac{\partial \mathcal{S}_m}{\partial z}\right), \tag{3.6}$$

$$-\frac{\bar{\rho}}{r}\left(br + \frac{a}{r}\right)^2 \hat{v}_r = -\bar{p}\mathcal{S}_m + \bar{K}\left(\frac{1}{r}\frac{\partial}{\partial r}\left(r\frac{\partial \hat{T}}{\partial r}\right) + \frac{\partial^2 \hat{T}}{\partial z^2}\right) + \mathcal{S}_e. \tag{3.7}$$

Here, $\bar{\mu}$ and $\bar{K}$ are the viscosity and thermal conductivity of the gas in the base state, where the temperature is T_0. Note that in the ideal gas limit, the viscosity and thermal conductivity are functions of the temperature alone, and they do not depend on the density. In equation 3.4, the term on the left is the contribution to the r momentum equation due to the centrifugal force, while in equation 3.5, the left side represents the contribution to the θ momentum equation due to the Coriolis force. In the energy equation 3.7, the term on the left is the pressure work done due to the compression or expansion of the gas, and we have used the equation of state 2.7 and the mass conservation equation 3.3 to simplify this term. The perturbation to the equation of state is

$$\hat{\rho} = \left(\frac{m}{k_B T_0}\right)\hat{p} - \bar{\rho}\left(\frac{\hat{T}}{T_0}\right). \tag{3.8}$$

First, we eliminate the density in the r momentum equation, 3.4, using the equation of state, equation 3.8, to obtain

$$-b^2 r\frac{m\hat{p}}{k_B T_0} + b^2 r\frac{\bar{\rho}\hat{T}}{T_0} - \frac{2ab}{r}\frac{m\hat{p}}{k_B T_0} + \frac{2ab}{r}\frac{\bar{\rho}\hat{T}}{T_0} - \frac{a^2}{r^3}\frac{m\hat{p}}{k_B T_0} + \frac{a^2}{r^3}\frac{\bar{\rho}\hat{T}}{T_0} - \frac{2}{r}\left(br + \frac{a}{r}\right)\bar{\rho}\hat{v}_\theta + \frac{\partial \hat{p}}{\partial r}$$
$$= \bar{\mu}\left(\frac{\partial}{\partial r}\left(\frac{1}{r}\frac{\partial}{\partial r}(r\hat{v}_r)\right) + \frac{\partial^2 \hat{v}_r}{\partial z^2}\right) + \left(f_r + \frac{\bar{\mu}}{3}\frac{\partial \mathcal{S}_m}{\partial r}\right). \tag{3.9}$$

The identity,

$$\bar{\rho}\frac{\partial}{\partial r}\left(\frac{\hat{p}}{\bar{\rho}}\right) = \frac{\partial \hat{p}}{\partial r} - \frac{\hat{p}}{\bar{\rho}}\frac{\partial \bar{\rho}}{\partial r}$$
$$= \frac{\partial \hat{p}}{\partial r} - \frac{\hat{p}m}{k_B T_0}\frac{1}{r}\left(br + \frac{a}{r}\right)^2, \tag{3.10}$$

can be inserted into the momentum conservation equation, 3.9, to obtain

The velocity components $\hat{v}_r$ and $\hat{v}_z$ can be expressed in terms of a stream function using the mass conservation equation 3.3,

$$r\bar{\rho}\hat{v}_r = -\frac{\partial \psi}{\partial z} + \int_{r_0}^{r} \mathcal{S}_m\, r dr, \tag{3.12}$$

$$r\bar{\rho}\hat{v}_z = \frac{\partial \psi}{\partial r}, \tag{3.13}$$

where the difference in the stream function between two radial locations gives the axial mass flux in the region between these two locations,

$$\psi(a, z) - \psi(r, z) = \int_r^a \bar{\rho}\hat{v}_z\, r\, dr. \tag{3.14}$$

Expressing in terms of the stream function, the equation ?? is,

$$\frac{1}{r}\frac{\partial}{\partial r}\left(\frac{1}{\bar{\rho}r}\frac{\partial}{\partial r}\left(r\frac{\partial}{\partial r}\right) + \frac{1}{\bar{\rho}}\frac{\partial^2}{\partial z^2}\right)\left(\frac{1}{\bar{\rho}r}\frac{\partial \psi}{\partial r}\right)$$

$$+\left(\frac{1}{\bar{\rho}r}\frac{\partial}{\partial r}\left(\frac{1}{r}\frac{\partial}{\partial r}\left(r\frac{\partial}{\partial z}\right)\right)+\frac{1}{\bar{\rho}r}\frac{\partial^3}{\partial z^3}\right)\left(\frac{1}{\bar{\rho}r}\left(\frac{\partial\psi}{\partial z}-\int_{r_0}^{r}\mathcal{S}_m\,rdr\right)\right)$$

$$-\frac{1}{\bar{\mu}}\frac{\partial}{\partial z}\left(\frac{2}{r^2}\left(br+\frac{a}{r}\right)\hat{v}_\theta-\frac{1}{r^2}\left(br+\frac{a}{r}\right)^2\frac{\hat{T}}{T_0}\right)=\frac{1}{\bar{\mu}\bar{\rho}r}\left(\frac{\partial f_r}{\partial z}\right)-\frac{1}{\bar{\mu}r}\frac{\partial}{\partial r}\left(\frac{f_z}{\bar{\rho}}\right).\quad(3.15)$$

The stream function ψ is expressed in terms of the 'master potential' χ:

$$\psi=\frac{1}{r}\frac{\partial\chi}{\partial r}.\quad(3.16)$$

This is inserted into the momentum conservation equation ??, to obtain

$$\frac{1}{r}\frac{\partial}{\partial r}\left(\frac{1}{\bar{\rho}r}\frac{\partial}{\partial r}\left(r\frac{\partial}{\partial r}\right)+\frac{1}{\bar{\rho}}\frac{\partial^2}{\partial z^2}\right)\left(\frac{1}{\bar{\rho}r}\frac{\partial}{\partial r}\left(\frac{1}{r}\frac{\partial\chi}{\partial r}\right)\right)$$
$$+\left(\frac{1}{\bar{\rho}r}\frac{\partial}{\partial r}\left(\frac{1}{r}\frac{\partial}{\partial r}\left(r\frac{\partial}{\partial z}\right)\right)+\frac{1}{\bar{\rho}r}\frac{\partial^3}{\partial z^3}\right)\left(\frac{1}{\bar{\rho}r^2}\frac{\partial}{\partial r}\left(\frac{\partial\chi}{\partial z}\right)-\frac{1}{\bar{\rho}r}\int_{r_0}^{r}\mathcal{S}_m\,rdr\right)$$
$$-\frac{1}{\bar{\mu}}\frac{\partial}{\partial z}\left(\frac{2}{r^2}\left(br+\frac{a}{r}\right)\hat{v}_\theta-\frac{1}{r^2}\left(br+\frac{a}{r}\right)^2\frac{\hat{T}}{T_0}\right)=\frac{1}{\bar{\mu}\bar{\rho}r}\left(\frac{\partial f_r}{\partial z}\right)-\frac{1}{\bar{\mu}r}\frac{\partial}{\partial r}\left(\frac{f_z}{\bar{\rho}}\right).\quad(3.17)$$

4. Axial flow in radial boundary layer

Further development of the equation for the master potential depends on whether the boundary layer considered is in the radial or the axial direction. When the boundary layer is in the radial coordinate, the radial derivatives of $\hat{v}_\theta$ and $\hat{T}$ on the right sides of equations 3.5 and 3.7 are balanced by the left sides. The angular velocity $\hat{v}_\theta$ in equation 3.17 can be expressed in terms of the master potential as follows. The angular momentum conservation equation, (3.5), can be rewritten as,

$$\frac{2\bar{\rho}}{r}\left(br+\frac{a}{r}\right)\hat{v}_r=\bar{\mu}\left(\frac{1}{r^2}\frac{\partial}{\partial r}\left(r^3\frac{\partial}{\partial r}\left(\frac{\hat{v}_\theta}{r}\right)\right)+\frac{\partial^2\hat{v}_\theta}{\partial z^2}\right)+f_\theta.\quad(4.1)$$

In this equation, the velocity $\hat{v}_r$ is expressed in terms of the master potential using equation (3.12) and equation (3.16):

$$\frac{\partial}{\partial z}\left(-\frac{2}{r^3}\left(br+\frac{a}{r}\right)\frac{\partial\chi}{\partial r}\right)+\frac{2}{r^2}\left(br+\frac{a}{r}\right)\int_{r_0}^{r}\mathcal{S}_m\,rdr=\bar{\mu}\left(\frac{1}{r^2}\frac{\partial}{\partial r}\left(r^3\frac{\partial}{\partial r}\left(\frac{\hat{v}_\theta}{r}\right)\right)+\frac{\partial^2\hat{v}_\theta}{\partial z^2}\right)+f_\theta.\quad(4.2)$$

The above equation is multiplied throughout by r^2, and integrated once with respect to the radial coordinate, to obtain

$$-\frac{2}{r}\left(br+\frac{a}{r}\right)\frac{\partial\chi}{\partial z}+\frac{2}{r}\left(br+\frac{a}{r}\right)\frac{\partial\chi}{\partial z}\bigg|_{r=r_0}+\frac{2}{r}\left(br+\frac{a}{r}\right)\int_{r_0}^{r}r\left[\int_{r_0}^{r}\mathcal{S}_m\,rdr\right]dr$$
$$=\bar{\mu}\left(r^3\frac{\partial}{\partial r}\left(\frac{\hat{v}_\theta}{r}\right)\bigg|_r-r^3\frac{\partial}{\partial r}\left(\frac{\hat{v}_\theta}{r}\right)\bigg|_{r=r_0}\right)$$
$$+\bar{\mu}\int_{r_0}^{r}r^2\,dr\left(\frac{\partial^2\hat{v}_\theta}{\partial z^2}\right)+\int_{r_0}^{r}f_\theta r^2\,dr.\quad(4.3)$$

Here r_0 is a reference location which will be discussed a little later. The above equation is differentiated with respect to the z coordinate, and multiplied by $\left(\frac{2}{r}\left(br+\frac{a}{r}\right)/\bar{\mu}^2r^4\right)$, to obtain

$$\frac{4\left(br+\frac{a}{r}\right)^2}{\bar{\mu}^2 r^6}\frac{\partial^2 \chi}{\partial z^2}-\frac{4\left(br+\frac{a}{r}\right)^2}{\bar{\mu}^2 r^6}\int_{r_0}^{r} r\left[\int_{r_0}^{r}\frac{\partial \mathcal{S}_m}{\partial z}\, r dr\right] dr + R_\theta = -\frac{2}{\bar{\mu} r}\left(br+\frac{a}{r}\right)\frac{\partial}{\partial z}\left(\frac{1}{r}\frac{\partial}{\partial r}\left(\frac{\hat{v}_\theta}{r}\right)\right)$$
$$+C_\theta+\frac{2}{\bar{\mu}^2 r^5}\left(br+\frac{a}{r}\right)\frac{\partial}{\partial z}\left(\int_{r_0}^{r} f_\theta\, r^2\, dr\right), \quad (4.4)$$

where

$$R_\theta = \frac{2}{\bar{\mu} r^5}\left(br+\frac{a}{r}\right)\int_{r_0}^{r} r^2\, dr\left(\frac{\partial^2 \hat{v}_\theta}{\partial z^2}\right), \quad (4.5)$$

and the function C_θ, which appears as a boundary term in equation 4.3, is,

$$C_\theta = \frac{2}{\bar{\mu} r^5}\left(br+\frac{a}{r}\right)\frac{\partial}{\partial z}\left(r^3\frac{\partial}{\partial r}\left(\frac{\hat{v}_\theta}{r}\right)\right)\Bigg|_{r=r_0}-\frac{4}{\bar{\mu}^2 r^6}\left(br+\frac{a}{r}\right)^2\left(\frac{\partial^2 \chi}{\partial z^2}\right)_{r=r_0}. \quad (4.6)$$

The temperature perturbation $\hat{T}$ in equation 3.17 can be expressed in terms of the master potential from the energy equation, 3.7. Expressing the radial velocity $\hat{v}_r$ in terms of the master potential using equation 3.12 and equation 3.16, we obtain

$$\frac{1}{r^3}\left(br+\frac{a}{r}\right)^2\frac{\partial}{\partial r}\left(\frac{\partial \chi}{\partial z}\right)-\frac{1}{r^2}\left(br+\frac{a}{r}\right)^2\int_{r_0}^{r}\mathcal{S}_m\, r dr = \bar{K}\left(\frac{1}{r}\frac{\partial}{\partial r}\left(r\frac{\partial \hat{T}}{\partial r}\right)+\frac{\partial^2 \hat{T}}{\partial z^2}\right)+\mathcal{S}_e. \quad (4.7)$$

The above equation can be multiplied by r, integrated once with respect to the radial coordinate with lower limit r_0 and upper limit r, to obtain

$$\frac{1}{r^2}\left(br+\frac{a}{r}\right)^2\left(\frac{\partial \chi}{\partial z}\right)-\frac{1}{r^2}\left(br+\frac{a}{r}\right)^2\int_{r_0}^{r} r\left[\int_{r_0}^{r}\mathcal{S}_m\, r dr\right] dr = \bar{K}\left(r\frac{\partial \hat{T}}{\partial r}\right)\Bigg|_r-\bar{K}\, r\frac{\partial \hat{T}}{\partial r}\Bigg|_{r=r_0}$$
$$+\int_{r_0}^{r} r\, dr\left(\frac{\partial^2 \hat{T}}{\partial z^2}\right)+\int_{r_0}^{r}\mathcal{S}_e\, r\, dr. \quad (4.8)$$

This is then multiplied by $\left(\frac{1}{r^2}\left(br+\frac{a}{r}\right)^2/\bar{K}T_0 r^2\right)$, and differentiated once with respect to z, to obtain

$$\frac{\left(br+\frac{a}{r}\right)^4}{\bar{K}\bar{\mu}T_0 r^6}\frac{\partial^2 \chi}{\partial z^2}-\frac{\left(br+\frac{a}{r}\right)^4}{\bar{K}\bar{\mu}T_0 r^6}\int_{r_0}^{r} r\left[\int_{r_0}^{r}\frac{\partial \mathcal{S}_m}{\partial z}\, r dr\right] dr + R_T + C_T = \frac{\frac{1}{r^2}\left(br+\frac{a}{r}\right)^2}{\bar{\mu}}\frac{1}{r}\frac{\partial}{\partial r}\frac{\partial}{\partial z}\left(\frac{\hat{T}}{T_0}\right)$$
$$+\frac{\left(br+\frac{a}{r}\right)^2}{\bar{K}\bar{\mu}T_0 r^4}\left(\frac{\partial}{\partial z}\int_{r_0}^{r}\mathcal{S}_e r\, dr\right), \quad (4.9)$$

where

$$R_T = \frac{\left(br+\frac{a}{r}\right)^2}{\bar{K}\bar{\mu}T_0 r^4}\int_{r_0}^{r} r\, dr\left(\frac{\partial^3 \hat{T}}{\partial z^3}\right). \quad (4.10)$$

and the term appearing due to the constant of integration in equation 4.8 is

$$C_T = \frac{\left(br+\frac{a}{r}\right)^2}{T_0 r^4}\frac{\partial}{\partial z}\left(r\frac{\partial \hat{T}}{\partial r}\Bigg|_{r=r_0}\right)-\frac{\left(br+\frac{a}{r}\right)^4}{\bar{K}\bar{\mu}T_0 r^6}\left(\frac{\partial^2 \chi}{\partial z^2}\right)_{r=r_0}, \quad (4.11)$$

where r_0 is a reference location to be discussed a little later.

We take the derivative $(1/r)(\partial/\partial r)$ of the equation for the master potential, 3.17, and then insert equations 4.4 and 4.9, to obtain

$$
\begin{aligned}
&\frac{1}{r}\frac{\partial}{\partial r}\left(\frac{1}{r}\frac{\partial}{\partial r}\left(\frac{1}{\bar{\rho}r}\frac{\partial}{\partial r}\left(r\frac{\partial}{\partial r}\right)+\frac{1}{\bar{\rho}}\frac{\partial^2}{\partial z^2}\right)\left(\frac{1}{\bar{\rho}r}\frac{\partial}{\partial r}\left(\frac{1}{r}\frac{\partial\chi}{\partial r}\right)\right)\right)\\
&+\frac{1}{r}\frac{\partial}{\partial r}\left(\frac{1}{\bar{\rho}r}\frac{\partial}{\partial r}\left(\frac{1}{r}\frac{\partial}{\partial r}\left(r\frac{\partial}{\partial z}\right)\right)+\frac{1}{\bar{\rho}r}\frac{\partial^3}{\partial z^3}\right)\left(\frac{1}{\bar{\rho}r^2}\frac{\partial}{\partial r}\left(\frac{\partial\chi}{\partial z}\right)-\frac{1}{\bar{\rho}r}\int_{r_0}^{r}\mathcal{S}_m r\,dr\right)\\
&+\left(\frac{\left(br+\frac{a}{r}\right)^4}{\bar{K}\bar{\mu}T_0r^6}+\frac{4\left(br+\frac{a}{r}\right)^2}{\bar{\mu}^2r^6}\right)\frac{\partial^2\chi}{\partial z^2}+\left(R_\theta+R_T\right)+\left(C_\theta+C_T\right)=\\
&\frac{1}{r}\frac{\partial}{\partial r}\left(\frac{1}{\bar{\mu}\bar{\rho}r}\left(\frac{\partial f_r}{\partial z}\right)-\frac{1}{\bar{\mu}r}\frac{\partial}{\partial r}\left(\frac{f_z}{\bar{\rho}}\right)\right)+\frac{2\left(br+\frac{a}{r}\right)}{\bar{\mu}^2r^5}\frac{\partial}{\partial z}\left(\int_{r_0}^{r}f_\theta r^2\,dr\right)\\
&-\frac{\left(br+\frac{a}{r}\right)^2}{\bar{K}\bar{\mu}T_0r^4}\frac{\partial}{\partial z}\int_{r_0}^{r}\mathcal{S}_e r\,dr+\left(\frac{\left(br+\frac{a}{r}\right)^4}{\bar{K}\bar{\mu}T_0r^6}+\frac{4\left(br+\frac{a}{r}\right)^2}{\bar{\mu}^2r^6}\right)\int_{r_0}^{r}r\left[\int_{r_0}^{r}\frac{\partial\mathcal{S}_m}{\partial z}\,rdr\right]dr. \qquad (4.12)
\end{aligned}
$$

The radial coordinate is expressed in terms of the scaled radial coordinate ξ, defined by

$$
\xi=\frac{\mathrm{Ma}^2}{2}\left(1-\left(\frac{(r-R_i)}{R_o-R_i}\right)^2\right). \qquad (4.13)
$$

With this transformation, it can easily be verified that

$$
\frac{1}{r-R_i}\frac{\partial}{\partial r}=-\frac{\mathrm{Ma}^2}{\left(R_o-R_i\right)^2}\frac{\partial}{\partial\xi}, \qquad (4.14)
$$

where $\delta=\left(R_o-R_i\right)/\mathrm{Ma}$ is the boundary layer thickness in the radial coordinate. When the equation for the master potential is expressed in terms of this scaled coordinate, we obtain

$$
\begin{aligned}
&\frac{\mathrm{Ma}^{12}}{\bar{\rho}^2\left(R_o-R_i\right)^{10}}\frac{\partial}{\partial\xi^2}\left(\frac{\partial}{\partial\xi}\left(\left(1-\frac{2\xi}{\mathrm{Ma}^2}\right)\frac{\partial}{\partial\xi}\left(\frac{\partial^2\chi}{\partial\xi^2}\right)\right)\right)+\frac{\mathrm{Ma}^8}{\bar{\rho}^2\left(R_o-R_i\right)^8}\frac{\partial^2}{\partial\xi^2}\left(\frac{\partial^2}{\partial\xi^2}\,\frac{\partial^2\chi}{\partial z^2}\right)+\\
&\frac{\mathrm{Ma}^8}{\bar{\rho}^2\left(R_o-R_i\right)^8}\frac{\partial}{\partial\xi}\left(\frac{\partial^2}{\partial\xi^2}\left(\frac{\partial}{\partial\xi}\frac{\partial^2\chi}{\partial z^2}\right)\right)+\frac{\mathrm{Ma}^4}{\bar{\rho}^2\left(R_o-R_i\right)^6}\frac{\partial}{\partial\xi}\left(\frac{1}{\left(1-\left(2\xi/\mathrm{Ma}^2\right)\right)}\frac{\partial}{\partial\xi}\frac{\partial^4\chi}{\partial z^4}\right)\\
&+\left(\frac{\left(b\left(R_i+\sqrt{\left(1-\frac{2\xi}{\mathrm{Ma}^2}\right)}\left(R_o-R_i\right)\right)+\frac{a}{\left(R_i+\sqrt{\left(1-\frac{2\xi}{\mathrm{Ma}^2}\right)}\left(R_o-R_i\right)\right)}\right)^4}{\bar{K}\bar{\mu}T_0\left(R_i+\sqrt{\left(1-\frac{2\xi}{\mathrm{Ma}^2}\right)}\left(R_o-R_i\right)\right)^6}\right.\\
&\left.+\frac{4\left(b\left(R_i+\sqrt{\left(1-\frac{2\xi}{\mathrm{Ma}^2}\right)}\left(R_o-R_i\right)\right)+\frac{a}{\left(R_i+\sqrt{\left(1-\frac{2\xi}{\mathrm{Ma}^2}\right)}\left(R_o-R_i\right)\right)}\right)^2}{\bar{\mu}^2\left(R_i+\sqrt{\left(1-\frac{2\xi}{\mathrm{Ma}^2}\right)}\left(R_o-R_i\right)\right)^6}\right)\frac{\partial^2\chi}{\partial z^2}
\end{aligned}
$$

$$+(R_\theta + R_T) + (C_\theta + C_T) = (S.15)$$

where S represents the source term in the equation for the master potential obtained from the mass, momentum and energy sources.

The above equation can be reduced to dimensionless form by dividing by the prefactor of the first term on the left,and defining a scaled master potential. The scaling for master potential (χ) is best inferred by balancing the first term on the left and the term on the right in equation 4.2, which provides $\chi^* = \left(\chi / \bar{\mu} \, (R_o - R_i)^2 \, Z\right)$. Using this scaled master potential, we obtain

$$\frac{\partial^2}{\partial\xi^2}\left(\frac{\partial}{\partial\xi}\left(\left(1-\frac{2\xi}{\mathrm{Ma}^2}\right)\frac{\partial}{\partial\xi}\left(\frac{\partial^2\chi^*}{\partial\xi^2}\right)\right)\right)+\frac{1}{\mathrm{Ma}^4}\frac{(R_o-R_i)^2}{Z^2}\frac{\partial^2}{\partial\xi^2}\left(\frac{\partial^2}{\partial\xi^2}\frac{\partial^2\chi^*}{\partial\eta^2}\right)+$$
$$\frac{1}{\mathrm{Ma}^4}\frac{(R_o-R_i)^2}{Z^2}\frac{\partial}{\partial\xi}\left(\frac{\partial^2}{\partial\xi^2}\left(\frac{\partial}{\partial\xi}\frac{\partial^2\chi^*}{\partial\eta^2}\right)\right)+\frac{1}{\mathrm{Ma}^8}\frac{(R_o-R_i)^4}{Z^4}\frac{\partial}{\partial\xi}\left(\frac{1}{(1-(2\xi/\mathrm{Ma}^2))}\frac{\partial}{\partial\xi}\frac{\partial^4\chi^*}{\partial\eta^4}\right)$$
$$+\left(\frac{\mathrm{Re}^2\mathrm{Ma}^2\mathrm{Pr}(1-\gamma^{-1})}{(1-2\xi/\mathrm{Ma}^2)}+\frac{4\mathrm{Re}^2}{(1-2\xi/\mathrm{Ma}^2)^2}\right)\frac{1}{\mathrm{Ma}^{12}}\frac{(R_o-R_i)^2}{Z^2}\frac{\partial^2\chi^*}{\partial\eta^2}$$
$$+(R^*_\theta + R^*_T) + (C^*_\theta + C^*_T) = S^*, \quad (4.16)$$

where the Reynolds number, Mach number and the Prandtl number are given in equations 2.11, 2.10 and 2.16 respectively, and

$$R^*_\theta = \frac{2}{\mathrm{Ma}^{14}}\frac{(R_o-R_i)^4}{Z^4}\left(\left(1-2\xi/\mathrm{Ma}^2\right)^{-2}\int_{\xi_0}^{\xi} d\xi \left(1-2\xi/\mathrm{Ma}^2\right)^{1/2}\left(\frac{\partial^3\hat{v}^*_\theta}{\partial\eta^3}\right)\right), \quad (4.17)$$

$$C^*_\theta = \frac{2}{\mathrm{Ma}^{10}}\frac{(R_o-R_i)^2}{Z^2}\left(\left(1-2\xi/\mathrm{Ma}^2\right)^{-2}\left(\left.\frac{\partial\hat{v}^*_\theta}{\partial\xi}\right|_{\xi=\xi_0}\right)\right), \quad (4.18)$$

$$R^*_{\hat{T}} = -\frac{1}{\mathrm{Ma}^{14}}\frac{(R_o-R_i)^4}{Z^4}\left(\left(1-2\xi/\mathrm{Ma}^2\right)^{-1}\int_{\xi_0}^{\xi} d\xi\left(\frac{\partial^3\hat{T}^*}{\partial\eta^3}\right)\right), \quad (4.19)$$

$$C^*_{\hat{T}} = -\frac{1}{\mathrm{Ma}^{10}}\frac{(R_o-R_i)^2}{Z^2}\left(\left(1-2\xi/\mathrm{Ma}^2\right)^{-1}\left.\frac{\partial}{\partial\eta}\frac{\partial\hat{T}^*}{\partial\xi}\right|_{\xi=\xi_0}\right), \quad (4.20)$$

where $\hat{v}^*_\theta = \hat{v}_\theta / \left(\left(b\,(R_o-R_i) + \frac{a}{(R_o-R_i)}\right)\right)$ and $\hat{T}^* = \hat{T}/T_0$ and S^* is the scaled source term.

The above equation for the master potential, 4.16, can be used to assess the approximations made in the analytical equation for the master potential which, in the present notation, is

$$\frac{\partial^2}{\partial\xi^2}\left(\frac{\partial}{\partial\xi}\left(\frac{\partial}{\partial\xi}\left(\frac{\partial^2\chi^*}{\partial\xi^2}\right)\right)\right)+\left(\frac{\mathrm{Re}^2\mathrm{Ma}^2\mathrm{Pr}(1-\gamma^{-1})}{(1-2\xi/\mathrm{Ma}^2)}+\frac{4\mathrm{Re}^2}{(1-2\xi/\mathrm{Ma}^2)^2}\right)\frac{1}{\mathrm{Ma}^{12}}\frac{(R_o-R_i)^2}{Z^2}\frac{\partial^2\chi^*}{\partial\eta^2} = \mathcal{S}^*, \quad (4.21)$$

where

$$\mathcal{S}^*(\xi,\eta) = \frac{\mathrm{Re}}{\mathrm{Ma}^{10}}\frac{(R_o-R_i)}{Z}\frac{(R_o-R_i)^2}{\mathrm{Ma}^4\bar{K}T_0}\left(1-2\xi/\mathrm{Ma}^2\right)^{-1}\int_{\xi_0}^{\xi}\frac{\partial(\mathcal{S}_e-\bar{p}\mathcal{S}_m)}{\partial\eta}\,d\xi^*$$

$$
\begin{aligned}
&- \frac{2\mathrm{Re}}{\mathrm{Ma}^{10}} \frac{(R_o - R_i)}{Z} \frac{(R_o - R_i)}{\mathrm{Ma}^4 \Omega \bar{\mu}} (1 - 2\xi/\mathrm{Ma}^2)^{-2} \int_{\xi_0}^{\xi} \frac{\partial f_\theta}{\partial \eta} \left(1 - \frac{2\xi^*}{\mathrm{Ma}^2}\right)^{\frac{1}{2}} d\xi^* \\
&- \left(\frac{\mathrm{Re}^2 \mathrm{Ma}^2 \mathrm{Pr}(1-\gamma^{-1})}{(1 - 2\xi/\mathrm{Ma}^2)} + \frac{4\mathrm{Re}^2}{(1 - 2\xi/\mathrm{Ma}^2)^2} \right) \frac{1}{\mathrm{Ma}^{16}} \frac{R}{\rho_w \Omega Z} \int_{\xi_0}^{\xi} \left[\int_{\xi_0}^{\xi^*} \frac{\partial \mathcal{S}_m(\xi^{**}, \eta)}{\partial \eta} d\xi^{**} \right] d\xi^* \\
&+ \frac{(R_o - R_i)^4}{\mathrm{Ma}^8 Z^2 \bar{\mu}} \left(\frac{\partial}{\partial \xi} \left(e^{\xi} \frac{\partial}{\partial \xi} \left(\frac{\partial}{\partial \xi} \left(\left(1 - \frac{2\xi}{\mathrm{Ma}^2}\right)^{\frac{1}{2}} \frac{\partial}{\partial \eta} \right) \right) \right) + \frac{R^2}{Z^2} \frac{\partial}{\partial \xi} \left(\frac{e^{\xi}}{(1 - 2\xi/\mathrm{Ma}^2)^{\frac{1}{2}}} \frac{\partial^3}{\partial \eta^3} \right) \right) \\
&\times \left[\frac{e^{\xi}}{(1 - 2\xi/\mathrm{Ma}^2)^{\frac{1}{2}}} \int_{\xi_0}^{\xi} \mathcal{S}_m \, d\xi^* \right] \\
&- \frac{\mathrm{Re}}{\mathrm{Ma}^{10}} \left[\frac{1}{\rho_w \Omega^2 Z} \frac{\partial}{\partial \xi} \left(\frac{1}{(1 - 2\xi/\mathrm{Ma}^2)^{\frac{1}{2}}} \frac{\partial (e^{\xi} f_r)}{\partial \eta} \right) + \frac{1}{\rho_w \Omega^2 R} \frac{\partial^2 (e^{\xi} f_z)}{\partial \xi^2} \right]. \qquad (4.22)
\end{aligned}
$$

In the above equation, the contribution to S^* due to the fourth term on the right side of the above equation (fourth and fifth lines of equation 4.22) is $O(\mathrm{Re}^{-1}(R/Z))$ smaller than that due to the third term on the right side (third line in equation 4.22). Therefore, this can be neglected in the leading approximation. The magnitude of the other terms do depend on the source strengths, and so these cannot be neglected in general. However, equation 4.22 provides the conditions under which the source strengths are significant, that is, for $\mathcal{S}_m \sim (\rho_w \Omega Z/R)$, $f_r \sim (\rho_w \Omega^2 R/\mathrm{Re})$, $f_\theta \sim (\Omega \bar{\mu} Z/R^2 \mathrm{Re})$, $f_z \sim (\rho_w \Omega^2 R/\mathrm{Re})$ and $S_e \sim (\bar{K} T_0 Z/R^3 \mathrm{Re})$.

The following terms in equation (4.16) are neglected in the "analytical equation" (4.21).

(*a*) The second, third and fourth terms on the left side of equation 4.16 are neglected, because the parameter $((R_o - R_i)^2/Z^2 \mathrm{Ma}^4) \ll 1$. Clearly, this is valid for high aspect ratio cylinders, $Z \gg (R_o - R_i)$, and also when the Mach number is large.

(*b*) The sixth term on the left is also neglected in comparison to the fifth term on the left side. It can easily be verified, from equations 4.1 and 4.7 that these terms are also $O((R_o - R_i)^2/\mathrm{Ma}^4 Z^2)$ smaller than the fifth term on the left side of equation 4.16.

(*c*) The seventh term on the left side of equation 4.16 is also neglected. The value of this term depends on the location of the boundary surface $\xi = \xi_0$.

(*d*) Within the first term on the left of equation 4.16, a term proportional to (ξ/Ma^2) is neglected. Similarly, within the fifth term on the left, it is assumed that $(1 + \xi/\mathrm{Ma}^2) \approx 1$. These imply that the Mach number $\mathrm{Ma} \gg 1$.

(*e*) The first and fifth terms on the left side of equation 4.16 are retained in the "analytical equation" (4.21), implying that either $(\mathrm{Re}^2/\mathrm{Ma}^{12})((R_o - R_i)^2/Z^2) \sim 1$, or $(\mathrm{Re}^2 \mathrm{Pr}(1-\gamma^{-1})/\mathrm{Ma}^{10})(R^2/Z^2) \sim 1$.

The limitation $\mathrm{Ma} \gg 1$ and $\mathrm{Re} \gg 1$ imposed by condition (d) above can be relaxed if we formulate a "generalized analytical equation",

$$
L_6^{\mathrm{Ma}}(\chi^*) + \frac{\mathrm{Re}^2}{\mathrm{Ma}^{12}} \frac{(R_o - R_i)^2}{Z^2} \frac{\partial^2 \chi^*}{\partial \eta^2} = S^*, \qquad (4.23)
$$

where the scaled source term (S^*) was defined in equation 4.22, the linear differential operator L_6^{Ma} in the ξ coordinate is defined by

$$
L_6^{\mathrm{Ma}} = \left(\frac{\mathrm{Ma}^2 \mathrm{Pr}(1-\gamma^{-1})}{(1 - 2\xi/\mathrm{Ma}^2)} + \frac{4}{(1 - 2\xi/\mathrm{Ma}^2)^2} \right)^{-1} \left(\frac{\partial^2}{\partial \xi^2} \left(\frac{\partial}{\partial \xi} \left(\left(1 - \frac{2\xi}{\mathrm{Ma}^2}\right) \frac{\partial}{\partial \xi} \left(\frac{\partial^2}{\partial \xi^2} \right) \right) \right) \right). \qquad (4.24)
$$

The operator has been written with the superscript ^{Ma} to indicate that there is an explicit dependence on the Mach number. This operator also depends on the Prandtl number Pr and the ratio of specific heats, but these are gas properties and not flow properties, and are both $O(1)$ for an ideal gas. Therefore, the the only flow dependence in the operator L_6^{Ma} enters through the Mach number, while the Prandtl number and ideal gas constant are determined by the nature of the gas. Here, we will assume an ideal gas of hard spheres.

It is easily verified that this operator is also self-adjoint, when the inner product is defined as,

$$(A, B) = \int_{\xi_i}^{\xi_o} d\xi w(\xi) A(\xi) B(\xi), \tag{4.25}$$

where the weighting function $w(\xi)$ is,

$$w(\xi) = \left(\frac{\mathrm{Ma}^2 \mathrm{Pr}(1 - \gamma^{-1})}{(1 - 2\xi/\mathrm{Ma}^2)} + \frac{4}{(1 - 2\xi/\mathrm{Ma}^2)^2} \right), \tag{4.26}$$

when the boundary conditions are appropriately specified as discussed next. Therefore, the eigenvalues of this operator are real, and the eigenfunctions are orthogonal. We refer to equation (4.23) as the 'generalized' analytical equation, to distinguish this from the analytical equation (4.21).

Since the equation (4.16) is a sixth order differential equation, it is necessary to specify three boundary conditions at each boundary. Here we will consider the homogeneous form of the analytical and generalized analytical equation and we will investigate the secondary flow due to wall thermal gradients and external inflow/outflow of light-gas into the rotating cylinder.

The following expressions for the velocity components and the axial flow rate are useful for specifying the boundary conditions,

$$\hat{v}_r = \frac{\left(b\,(R_o - R_i) + \frac{a}{(R_o - R_i)} \right) \mathrm{Ma}^2}{\mathrm{Re}} \left(1 - \frac{2\xi}{\mathrm{Ma}^2} \right)^{-1/2} \frac{\partial^2 \chi^*}{\partial \xi \partial \eta}, \tag{4.27}$$

$$\hat{v}_z = \frac{Z \left(1 - \frac{2\xi}{\mathrm{Ma}^2} \right) \left(b\,(R_o - R_i) + \frac{a}{(R_o - R_i)} \right) \mathrm{Ma}^4}{(R_o - R_i)\,\mathrm{Re}} \frac{\partial^2 \chi^*}{\partial \xi^2}, \tag{4.28}$$

$$\begin{aligned} F_z &= 2\pi \int_{R_i}^{R_o} r\, dr\, \bar{\rho} \hat{v}_z \\ &= \frac{2\pi \bar{\rho}\,(R_o - R_i)^2\, Z \mathrm{Ma}^2}{\mathrm{Re}} \left. \frac{\partial \chi^*}{\partial \xi} \right|_{\xi_i}^{\xi_o}, \end{aligned} \tag{4.29}$$

where F_z is the total flow rate in the axial direction.

For the flow in the annulus between two cylinders where R_o and R_i are the radii of the outer and inner cylinders respectively, the boundary conditions are as follows:

At the inner wall $r = R_i$ ($\xi = \xi_i$), the zero radial velocity boundary condition requires that $(\partial \psi / \partial z) = 0$, or

$$\frac{\partial \chi^*}{\partial \xi} = 0. \tag{4.30}$$

The axial velocity condition at the inner wall can be written as

$$\frac{\partial^2 \chi^*}{\partial \xi^2} = V_{z,i}^*, \tag{4.31}$$

where $V^*_{z,i} = (\hat{v}_{z,i}\mathrm{Re}\,(R_o - R_i)) / \left(Z\left(1 - \frac{2\xi}{\mathrm{Ma}^2}\right)\left(bR_i + \frac{a}{R_i}\right)\mathrm{Ma}^4\right)$ is the scaled axial velocity at the inner wall. It might be expected that the axial velocity should be zero at the wall from the no-slip condition, but in the DSMC simulations, we find that the axial velocity is not zero at the wall, and there is a slip velocity. This slip velocity $V^*_{z,i}$ is included in the boundary condition 4.33.

At the outer wall $r = R_o\,(\xi = \xi_o)$, the zero radial velocity boundary condition requires that $(\partial\psi/\partial z) = 0$, or

$$\frac{\partial \chi^*}{\partial \xi} = 0. \tag{4.32}$$

The axial velocity condition at the outer wall can be similarly written as

$$\frac{\partial^2 \chi^*}{\partial \xi^2} = V^*_{z,o}, \tag{4.33}$$

where

$$V^*_{z,o} = (\hat{v}_{z,o}\mathrm{Re}\,(R_o - R_i)) / \left(Z\left(1 - \frac{2\xi}{\mathrm{Ma}^2}\right)\left(bR_o + \frac{a}{R_o}\right)\mathrm{Ma}^4\right), \tag{4.34}$$

is the scaled axial velocity at the inner wall.

The third boundary condition at the outer wall emerges from the temperature at the outer surface of the cylinder. This boundary condition is formulated equation 3.17 for the master potential, in which we neglect the axial gradients of the master potential in comparison to the radial derivatives. Since the velocity perturbation $\hat{v}_\theta$ is zero at the cylindrical surface, the boundary condition reduces to,

$$\frac{1}{r}\frac{\partial}{\partial r}\left(\frac{1}{\bar{\rho} r}\frac{\partial}{\partial r}\left(r\frac{\partial}{\partial r}\right)\right)\left(\frac{1}{r}\frac{\partial}{\partial r}\left(\frac{1}{\bar{\rho} r}\frac{\partial \chi}{\partial r}\right)\right) = -\frac{\Omega^2}{\bar{\mu}T_0}\left(\frac{\partial \hat{T}}{\partial z}\right). \tag{4.35}$$

The temperature boundary condition at the outer wall (from equation 4.35) is expressed in terms of scaled coordinates as,

$$L_5^{\mathrm{Ma}}(\chi^*) = -\frac{\mathrm{Re}^2}{\mathrm{Ma}^{10}}\frac{(R_o - R_i)^2}{Z^2}\left(\frac{1}{T_0}\frac{\partial \hat{T}}{\partial \eta}\right), \tag{4.36}$$

where the operator L_5^{Ma} is,

$$L_5^{\mathrm{Ma}} = \frac{\partial}{\partial \xi}\left(\frac{\partial}{\partial \xi}\left(\left(1 - \frac{2\xi}{\mathrm{Ma}^2}\right)\frac{\partial}{\partial \xi}\left(e^{\xi}\frac{\partial^2}{\partial \xi^2}\right)\right)\right). \tag{4.37}$$

Note that in general, equation 4.36 provides the boundary condition for a linear combination of the perturbation in $\hat{v}_\theta$ and $\hat{T}$, and not their individual boundary conditions. Since $\hat{v}_\theta$ is zero at the outer surface, equation 4.36 reduced to a boundary condition for $\hat{T}$.

5. DSMC simulation

The direct simulation Monte Carlo method (DSMC) is a numerical method for modeling rarefied gas flows. Here, the Boltzmann equation for the probability distribution function for gas molecules is solved using a probabilistic (Monte Carlo) simulation technique. The Boltzmann equation is more fundamental than the Navier-Stokes equations, since the it describes the evolution of the velocity distribution function of the gas molecules. The Navier-Stokes equations are obtained by taking the first, second and third moments of the Boltzmann equation with respect to the molecular velocities. While taking moments, the assumption of 'local equilibrium' (velocity distribution function close to the Maxwell-Boltzmann distribution) is made, thereby restricting the Navier-Stokes equations to near-equilibrium flows. Since there is no assumption of local equilibrium in the Boltzmann equation, it is valid even for far-from equilibrium flows, provided the molecular chaos approximation is satisfied.

The DSMC method uses 'simulated molecules' which represent a large number of real gas molecules. A basic assumption is that the inter-molecular collisions and molecular motion between collisions can be decoupled over time scales comparable to the time between collisions. The translation of molecules between collisions is deterministically modeled, while the inter-molecular collisions and molecular-surfaces are incorporated using probabilistic models. The most common model is the hard-sphere model, but other models such as the variable hard sphere and variable soft sphere models are also used in order to accurately reproduce the density and temperature dependence of the viscosity. The hydrodynamic fields (density, velocity, temperature) are obtained by averaging over the molecular distribution functions in the DSMC simulations.

At the boundaries, it is necessary to prescribe interaction laws between the molecules and the solid surfaces. Accurate specification of the interaction laws requires a detailed knowledge of the molecule-surface interaction. However, there are standard interaction laws based on the surface velocity and temperature which are sufficient for predicting the hydrodynamic fields accurately. The most commonly used is the 'diffuse reflection' conditions, where a molecule incident on the surface gets reflected with a Gaussian velocity distribution with the same mean velocity and temperature as that of the surface. Another commonly used condition is the specular reflection condition, that the velocity normal to the surface is reversed, while that tangential to the surface is unchanged. It is important to note that these boundary conditions are different from the no-slip conditions in the continuum models. In high Knudsen number flows, the diffuse and specular reflection conditions do predict a slip velocity at the surface.

From kinetic theory of dilute gases, the viscosity of a hard-sphere gas in the dilute limit is given by

$$\mu = \frac{5\sqrt{mk_BT}}{16\sqrt{\pi}d^2}, \tag{5.1}$$

where k_B is the Boltzmann constant, while the speed of sound c_s is given by the relation

$$mnc_s^2 = \gamma p = \gamma nk_BT, \tag{5.2}$$

where m is the molecular mass, n is the number density, and p is the pressure. From this the Mach number(Ma), Reynolds number(Re), cylinder radius(R), molecular diameter(d) and volume fraction of the molecules(ϕ) are related by the following equation:

$$\text{Re} = \frac{96}{5} \times \frac{1}{\sqrt{\pi}} \text{Ma}\, \phi\, \frac{R}{d}, \tag{5.3}$$

where the volume fraction of the molecules is given by

$$\phi = \frac{\pi}{6} d^3 n, \tag{5.4}$$

where n is the number density of the molecules. The important point to note, from 5.3, is that the Reynolds number is proportional to the Mach number times $\phi(R/d)$, where R is the macroscopic scale. Therefore if we are keeping the volume fraction (density) a constant, increasing (decreasing) the Reynolds number at constant Mach number corresponds to increasing (decreasing) the system size R in relation to the molecular diameter.

The details of the DSMC simulation procedure are discussed in Bird (1963) and Bird (1994), and so we do not repeat the discussion here, but restrict attention to the parameters in the simulations. The number of cells along the radial direction is 50, while there are 200 cells along the axial direction. The cell width is no greater than one-third of the local mean free path. The number of subcells per cell is taken as 4 for both radial and axial directions. The number of simulated particles per cell is taken as 100. In our simulations, the number of real molecules per simulated particles varied between 1×10^{16} and 1×10^{18} so that the statistical fluctuations not becomes large.

The hard sphere (HS) molecular model is used in the present simulations, with the molecular mass 6.63×10^{-26} kg, and molecular diameter $= 4.17 \times 10^{-10}$ m. The cylindrical surfaces are specified as being diffusely reflecting with the thermal accommodation coefficient equal to 1, and the end plates are considered as specularly reflecting. The time step in the simulation is chosen to be small compared to the time between collisions. The mean collision time is calculated as, $t = (\lambda / c_{av})$, where λ is the mean free path and c_{av} is the average molecular velocity. The minimum value of mean free path, calculated from the gas density at the wall, is $\lambda = 2.2609 \times 10^{-4}$ m. The average molecular velocity is calculated as $\sqrt{k_B T/m} = 249.92$ m/sec. This gives the mean collision time (t) $= 9.04649 \times 10^{-7}$ sec. The time step used in the simulations is 5.0×10^{-8} sec, which is less than mean collision time.

In the analysis, Mach number (Ma) varied in the range $1 - 5$, Reynolds number (Re) in the range $1 - 100$ and aspect ratio $(Z/(R_o - R_i)) = 4$.

6. Results

The direct simulation Monte Carlo method (DSMC) is used here to simulate the gas flow in two-dimensional axisymmetric $r - z$ plane. The flow geometry and coordinate system are shown in figure 3 (a), and characteristic stream line plot in figure 3 (b), for wall thermal forcing. In case of wall thermal forcing, a linear temperature profile is applied along the outer-cylinder stationary sidewall. This results in an upward secondary flow along the wall. Simulations were carried out for Mach numbers in the range $1 - 5$, and Reynolds numbers in the range $1 - 100$. Three different aspect ratios $Z/(R_o - R_i)$ $= 2$, 4 and 10 are considered in the analysis (where Z is the height and $(R_o - R_i)$ is the annular-gap (gap between the rotating inner cylinder and the concentric stationary outer cylinder). Only the comparisons for $Z/(R_o - R_i) = 2$, 4, 8 are shown here. This because the smallest aspect ratio provides the most stringent test.

6.1. *Linear wall temperature profile*

First, we study the steady state hydrodynamic fields, the temperature, density, velocity distribution (v_r, v_θ, v_z), axial mass flux (ρv_z), shear-stress (τ), and heat-flux (q) inside the gas-centrifuge casing for Mach number (Ma) = 5, Reynolds number (Re) in the range 1-100, and aspect ratio $(Z/(R_o - R_i))$ in the range 2-8 using DSMC simulations.

The streamlines in the flow domain for Mach number (Ma) = 5, Knudsen Number Kn = 0.01, Reynolds number (Re) in the range 1-100, and aspect ratio $(Z/(R_o - R_i))$ in the range 2-8, is shown in figure 4 (a, b, c), and figure ?? (a, b, c). The flow is induced by the wall thermal forcing with a linear temperature profile along the sidewall of the stationary outer cylinder. The formation of Taylor vortices are clearly observed in the simulations.

The velocity vector in the flow domain for Mach number (Ma) = 5, Knudsen Number Kn = 0.01, Reynolds number (Re) in the range 1-100, and aspect ratio $(Z/(R_o - R_i))$ in the range 2-8, is shown in figure 5 (a, b, c), and figure ?? (a, b, c). The flow is induced

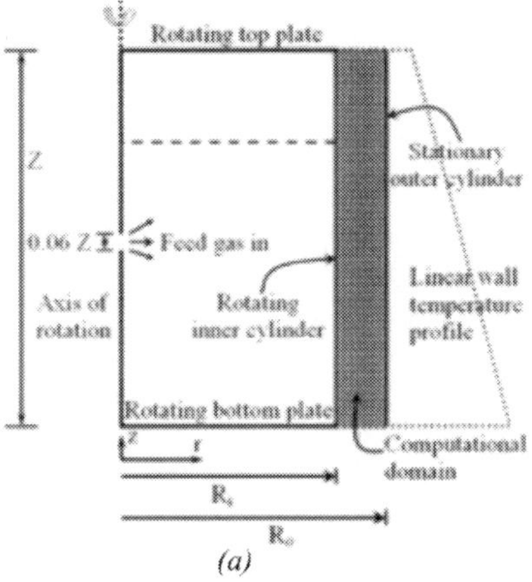

FIGURE 1. Computational domain and coordinate system (a), and characteristic stream line plot for Ma = 5, Re = 100, F_T = 0.2, aspect ratio $Z/(R_o - R_i) = 4$ (b), between the rotating inner cylinder and the concentric stationary outer cylinder, with a linear temperature profile along the sidewall of the stationary outer cylinder. The lines show the analytical solution for the generalized analytical model (−−), and the results of DSMC simulation (—). Here, $F_T = (Z/T_0)(\partial T/dz)$.

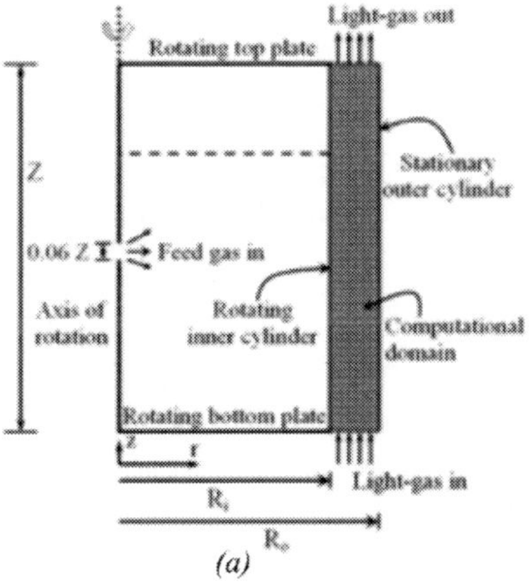

FIGURE 2. Computational domain and coordinate system (a), and characteristic stream line plot for Ma = 5, Re = 100, $F^* = 0.2$, aspect ratio $Z/(R_o - R_i) = 4$ (b), between the rotating inner cylinder and the concentric stationary outer cylinder, for inflow/outflow of light-gas along the boundaries. The lines show the analytical solution for the generalized analytical model (−−), and the results of DSMC simulation (—). Here, $F^* = F/(H\Omega)$.

by the wall thermal forcing with a linear temperature profile along the sidewall of the stationary outer cylinder.

The steady state density contours in the flow domain for Mach number (Ma) = 5, Knudsen Number Kn = 0.01, Reynolds number (Re) in the range 1-100, and aspect ratio $(Z/(R_o - R_i))$ in the range 2-8, is shown in figure ?? (a, b, c), and figure ?? (a, b, c). Due to high surface speed of the inner cylinder, a significant heating of the gas is taking place. The gas temperature increases until the heat transfer to the surface equals the work done in moving the surface.

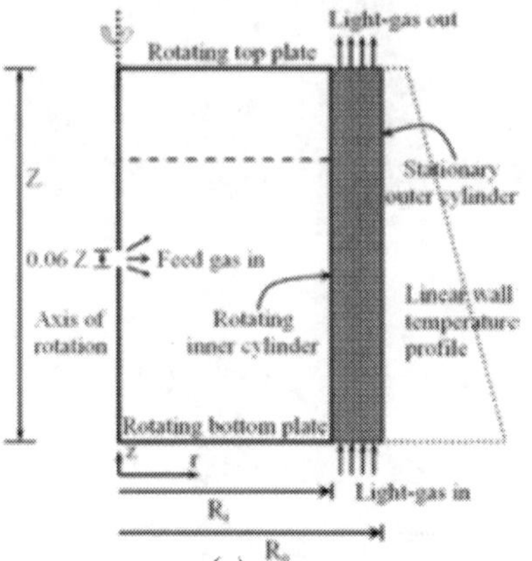

FIGURE 3. Computational domain and coordinate system (a), and characteristic stream line plot for Ma = 5, Re = 100, $F_T/F^* = 0.2$, aspect ratio $Z/(R_o - R_i) = 4$ (b), between the rotating inner cylinder and the concentric stationary outer cylinder, with a combination of linear temperature profile along the sidewall of the stationary outer cylinder, and inflow/outflow of light-gas along the boundaries. The lines show the analytical solution for the generalized analytical model (−−), and the results of DSMC simulation (—).

The steady state number density contours in the flow domain for Mach number (Ma) = 5, Knudsen Number Kn = 0.01, Reynolds number (Re) in the range 1-100, and aspect ratio $(Z/(R_o - R_i))$ in the range 2-8, is shown in figure ?? (a, b, c), and figure ?? (a, b, c).

The steady state axial-velocity contours in the flow domain for Mach number (Ma) = 5, Knudsen Number Kn = 0.01, Reynolds number (Re) in the range 1-100, and aspect ratio $(Z/(R_o - R_i))$ in the range 2-8, is shown in figure 8 (a, b, c), and figure ?? (a, b, c).

The steady state axial-mass flux contours in the flow domain for Mach number (Ma) = 5, Knudsen Number Kn = 0.01, Reynolds number (Re) in the range 1-100, and aspect ratio $(Z/(R_o - R_i))$ in the range 2-8, is shown in figure 9 (a, b, c), and figure ?? (a, b, c).

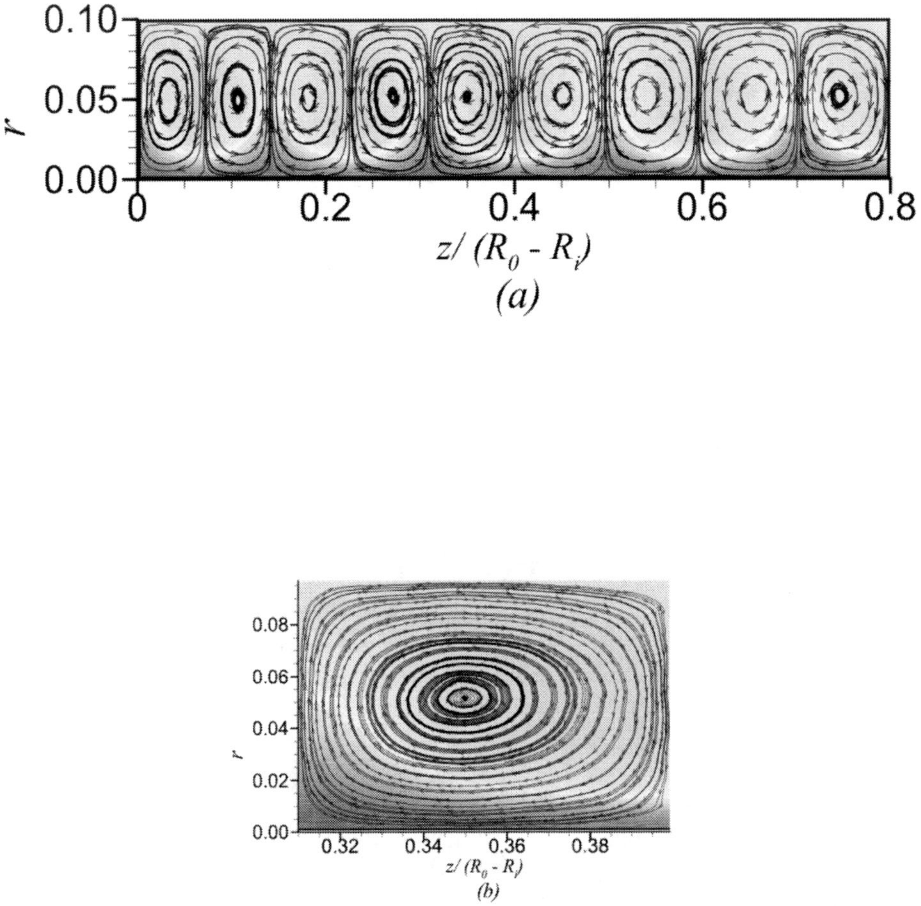

FIGURE 4. The streamlines in the flow domain for Mach number Ma = 5, Knudsen Number Kn = 0.01, and for different aspect ratio : $Z/(R_o - R_i) = 8$ (a), $Z/(R_o - R_i) = 4$ (b), $Z/(R_o - R_i) = 2$ (c). These results for the scaled wall temperature gradient along the sidewall of the stationary outer cylinder $(F_T) = 0.2$.

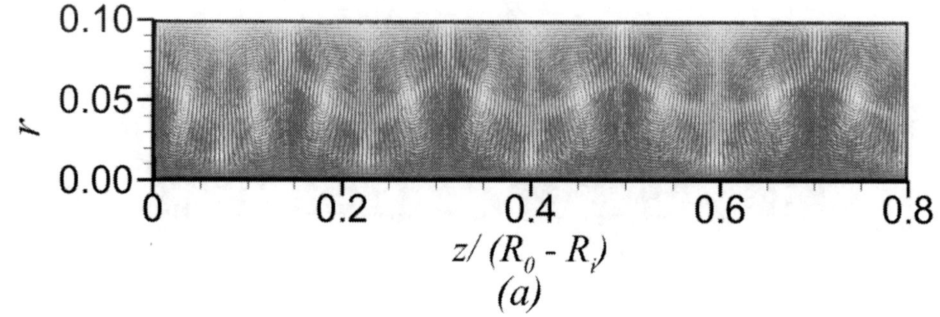

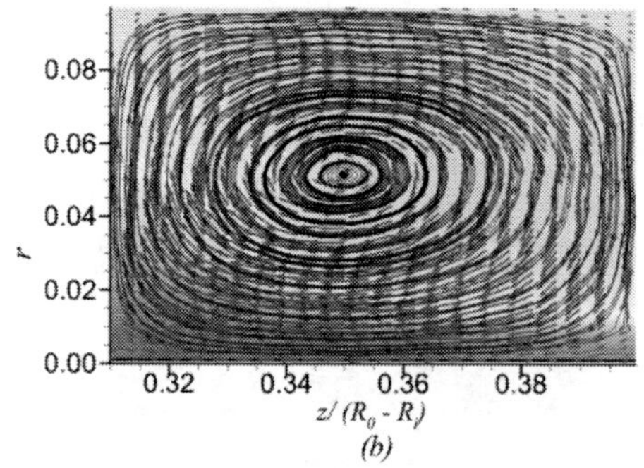

FIGURE 5. The velocity vector field in the flow domain for Mach number Ma = 5, Knudsen Number Kn = 0.01, and for different aspect ratio : $Z/(R_o - R_i) = 8$ (a), $Z/(R_o - R_i) = 4$ (b), $Z/(R_o - R_i) = 2$ (c). These results for the scaled wall temperature gradient along the sidewall of the stationary outer cylinder $(F_T) = 0.2$.

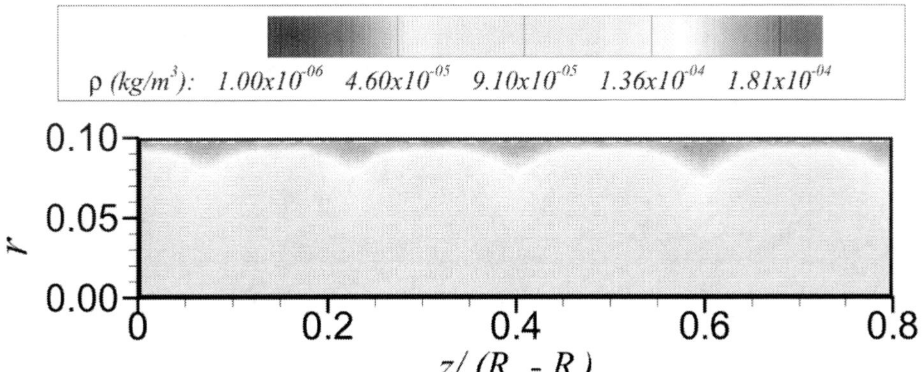

(a)

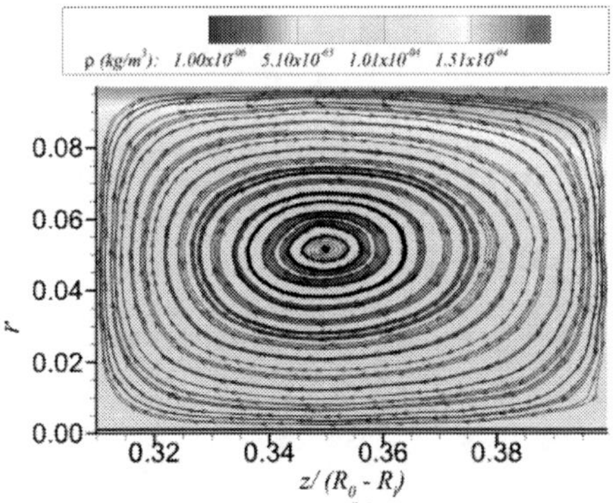

(b)

FIGURE 6. The density contours in the flow domain for Mach number Ma = 5, Knudsen Number Kn = 0.01, and for different aspect ratio : $Z/(R_o - R_i) = 8$ (a), $Z/(R_o - R_i) = 4$ (b), $Z/(R_o - R_i) = 2$ (c). These results for the scaled wall temperature gradient along the sidewall of the stationary outer cylinder $(F_T) = 0.2$.

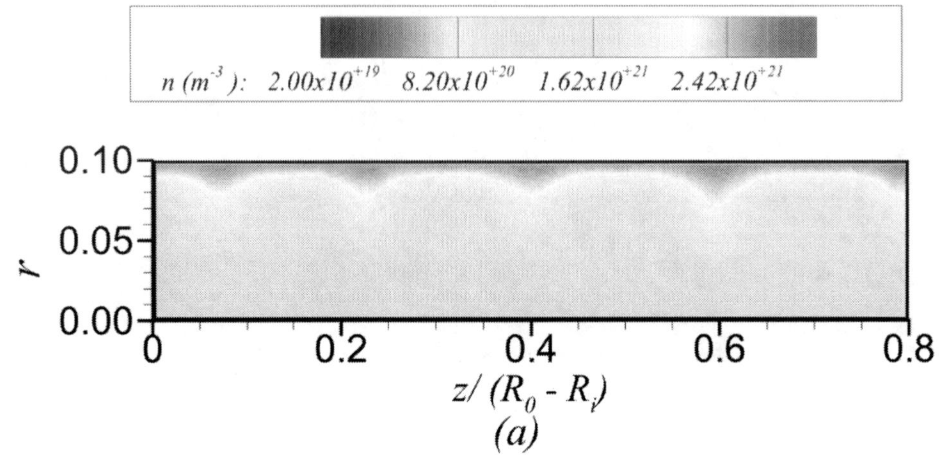

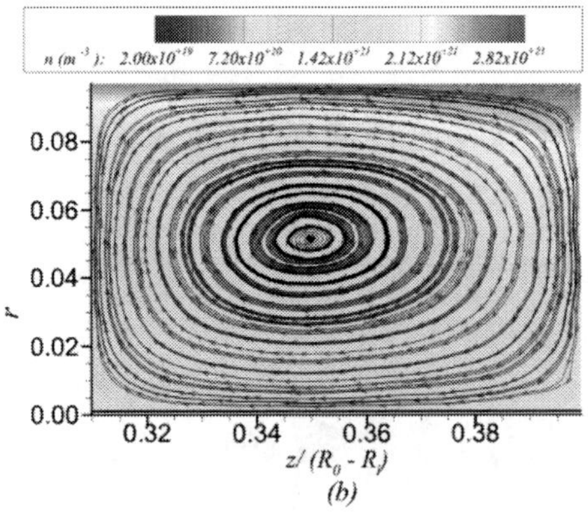

FIGURE 7. The number-density contours in the flow domain for Mach number Ma = 5, Knudsen Number Kn = 0.01, and for different aspect ratio : $Z/(R_o - R_i) = 8$ (a), $Z/(R_o - R_i) = 4$ (b), $Z/(R_o - R_i) = 2$ (c). These results for the scaled wall temperature gradient along the sidewall of the stationary outer cylinder $(F_T) = 0.2$.

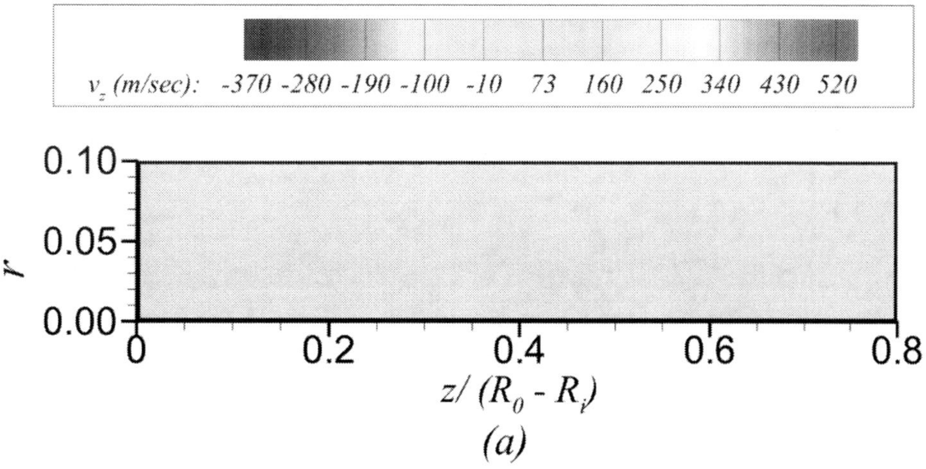

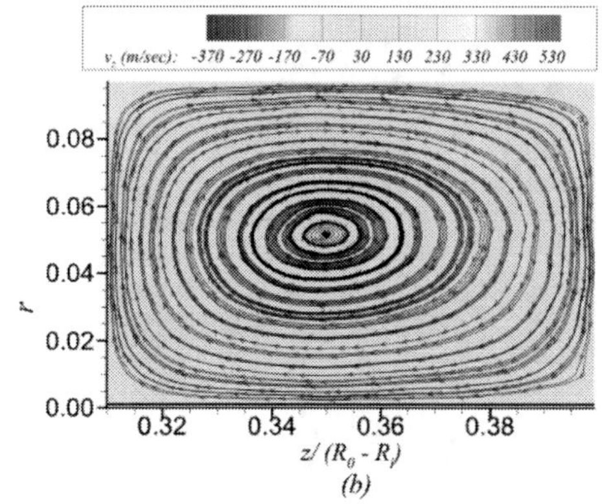

FIGURE 8. The axial velocity contours in the flow domain for Mach number Ma = 5, Knudsen Number Kn = 0.01, and for different aspect ratio : $Z/(R_o - R_i) = 8$ (a), $Z/(R_o - R_i) = 4$ (b), $Z/(R_o - R_i) = 2$ (c). These results for the scaled wall temperature gradient along the sidewall of the stationary outer cylinder $(F_T) = 0.2$.

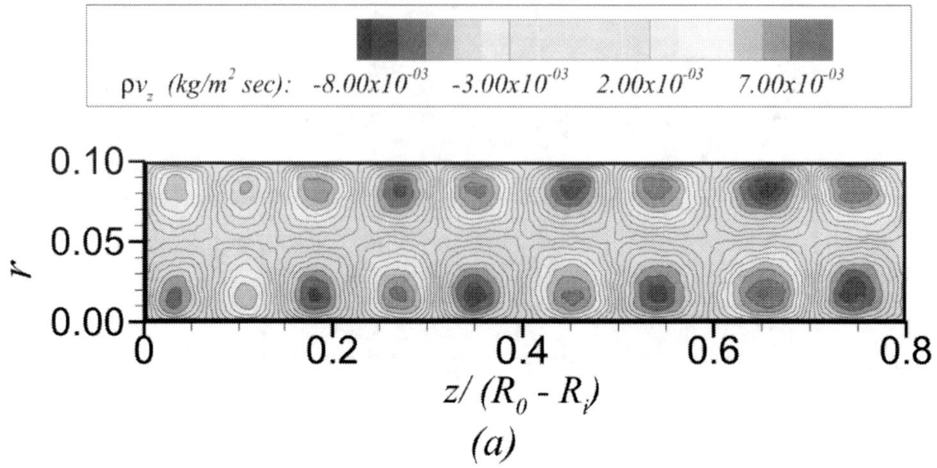

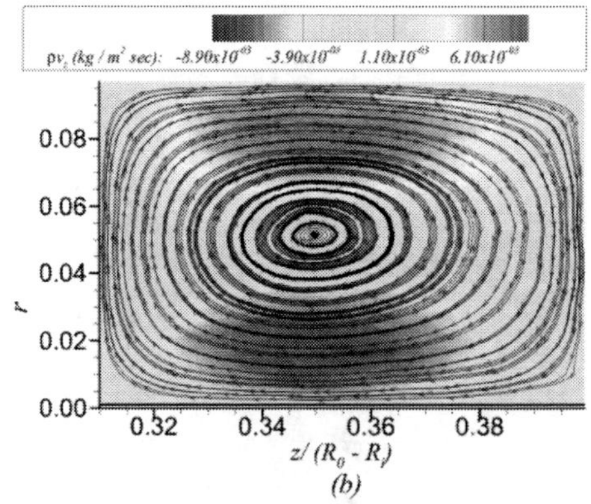

FIGURE 9. The axial mass flux contours in the flow domain for Mach number Ma = 5, Knudsen Number Kn = 0.01, and for different aspect ratio : $Z/(R_o - R_i) = 8$ (a), $Z/(R_o - R_i) = 4$ (b), $Z/(R_o - R_i) = 2$ (c). These results for the scaled wall temperature gradient along the sidewall of the stationary outer cylinder $(F_T) = 0.2$.

7. Conclusions

The objective of the present work is to carry out a comprehensive study of the comparison of the results obtained from analytical equation, and generalized analytical equation, derived in §2, for rarefied gas flow between two cylinders, where the inner cylinder is rotating at a constant angular velocity Ω_i while the outer is kept stationary, with molecular simulations based on the direct simulation Monte Carlo (DSMC) method, for the flow field and temperature field. Several situations have been considered, including wall thermal forcing with a linear temperature profile along the sidewall of the stationary outer cylinder, inflow/outflow of light-gas along the boundaries, and the combination of the two. Simulations have been carried out for Mach number (Ma) in the range $1-5$, Reynolds number (Re) in the range $1-100$ and aspect ratio $(Z/(R_o - R_i))$ in the range $2-10$.

The major advantage of the present formulation is that it is not restricted to the asymptotic limit of high Mach number. Due to this, the model is not restricted to just boundary layer flows at the curved surface of the cylinder. Therefore, the model can be used for much slower flows, and this provides a way to bridge the existing gap between the very high speed (high Mach number) flows, and incompressible flows where the incompressible Navier-Stokes equations can be used.

Next, we compare the analytical results with those of DSMC simulations and found excellent agreement (with a difference of less than 15%) between the predictions of the analytical model and the DSMC simulations, provided the boundary conditions in the analytical model are accurately specified.

Appendix A. Conservation equations in cylindrical coordinates:

The mass momentum and energy conservation equations for an axisymmetric flow in a cylindrical coordinate system are,

$$\frac{\partial \rho}{\partial t} + \nabla.(\rho \mathbf{v}) = \mathcal{S}_m, \tag{A 1}$$

$$\rho\left(\frac{\partial v_r}{\partial t} + v_r\frac{\partial v_r}{\partial r} - \frac{v_\theta^2}{r} + v_z\frac{\partial v_r}{\partial z}\right) = -\frac{\partial p}{\partial r} + \left[\frac{1}{r}\frac{\partial \tau_{rr}}{\partial r} - \frac{\tau_{\theta\theta}}{r} + \frac{\partial \tau_{rz}}{\partial z}\right] + f_r, \tag{A 2}$$

$$\rho\left(\frac{\partial v_\theta}{\partial t} + v_r\frac{\partial v_\theta}{\partial r} + \frac{v_r v_\theta}{r} + v_z\frac{\partial v_\theta}{\partial z}\right) = \left[\frac{1}{r^2}\frac{\partial (r^2\tau_{r\theta})}{\partial r} + \frac{\partial \tau_{\theta z}}{\partial z}\right] + f_\theta, \tag{A 3}$$

$$\rho\left(\frac{\partial v_z}{\partial t} + v_r\frac{\partial v_z}{\partial r} + v_z\frac{\partial v_z}{\partial z}\right) = -\frac{\partial p}{\partial z} + \left[\frac{1}{r}\frac{\partial (r\tau_{rz})}{\partial r} + \frac{\partial \tau_{zz}}{\partial z}\right] + f_z, \tag{A 4}$$

$$\rho C_v\left(\frac{\partial T}{\partial t} + v_r\frac{\partial T}{\partial r} + v_z\frac{\partial T}{\partial z}\right)$$

$$= -p(\nabla.\mathbf{v}) + \tau_{rr}\frac{\partial v_r}{\partial r} + \tau_{\theta\theta}\frac{v_r}{r} + \tau_{zz}\frac{\partial v_z}{\partial z}$$
$$+\tau_{r\theta}\left(r\frac{\partial}{\partial r}\left(\frac{v_\theta}{r}\right)\right) + \tau_{\theta z}\left(\frac{\partial v_\theta}{\partial z}\right) + \tau_{rz}\left(\frac{\partial v_z}{\partial r} + \frac{\partial v_r}{\partial z}\right)$$
$$+\frac{1}{r}\frac{\partial}{\partial r}\left(rK\frac{\partial T}{\partial r}\right) + \frac{\partial}{\partial z}\left(K\frac{\partial T}{\partial z}\right) + S_e, \tag{A 5}$$

where

$$\nabla.\mathbf{v} = \frac{1}{r}\frac{\partial(rv_r)}{\partial r} + \frac{\partial v_z}{\partial z}, \tag{A 6}$$

$$\tau_{rr} = \mu\left(2\frac{\partial v_r}{\partial r} - \frac{2}{3}\nabla.\mathbf{v}\right), \tag{A 7}$$

$$\tau_{\theta\theta} = \mu\left(-\frac{2}{3}\nabla.\mathbf{v}\right), \tag{A 8}$$

$$\tau_{zz} = \mu\left(2\frac{\partial v_z}{\partial z} - \frac{2}{3}\nabla.\mathbf{v}\right), \tag{A 9}$$

$$\tau_{r\theta} = \mu\left(r\frac{\partial}{\partial r}\left(\frac{v_\theta}{r}\right)\right), \tag{A 10}$$

$$\tau_{rz} = \mu\left(\frac{\partial v_r}{\partial z} + \frac{\partial v_z}{\partial r}\right), \tag{A 11}$$

$$\tau_{\theta z} = \mu\left(\frac{\partial v_\theta}{\partial z}\right). \tag{A 12}$$

Equation A 1 is the mass conservation equation, equations A 2, A 3 and A 4 are the momentum conservation equations in the r, θ and z directions, and equation A 5 is the energy conservation equation.

REFERENCES

AOKI, E. & SUZUKI, M. 1985 Numerical studies on gas flow around scoop of a gas centrifuge. *In Proc. Sixth Workshop on Gases in Strong Rotation, Tokyo (ed. Y. Takashima)*, 11.

ANDRADE, D. A. & BASTOS, J. L. F. 1998 A Gas centrifuge thermal hydrodynamical model. *In Proc. of the Sixth International Workshop on Separation Phenomena in Liquids and Gases.*, 92.

AOKI, K. & YOSHIDA, H. & NAKANISHI, T. & GARCIA, A. L. 2003 Inverted velocity profile in the cylindrical Couette flow of a rarefied gas. *Physical Review E.* **68 (016302).**

AVERY, D. G. & DAVIES, E. 1973 Uranium enrichment by gas centrifuge *Mills and Boon Ltd. London.*

BABARSKY, R. J. & WOOD, H. G. 1990 Approximate eigensolutions for non-axisymmetric rotating compressible flows. *Comp. Meth. Appl. Mech. and Engg.* **81**, 317.

BABARSKY, R. J. & HERBST, W. I. & WOOD, H. G. 2002 A new variational approach to gas flow in a rotating system. *Phys. Fluids.* **14(10)**, 3624.

BARK, F. H. & BARK, T. H. 1976 On vertical boundary layers in a rapidly rotating gas. *J. Fluid Mech.* **78**, 749.

BERGER, M. H. 1987 Finite element analysis of flow in a gas-filled rotating annulus. *International Journal for Numerical Methods in Fluids.* **7**, 215.

BIRD, G. A. 1963 Approach to translational equilibrium in a rigid sphere gas. *Phys. Fluids.* **6**, 1518.

BIRD, G. A. 1981 Monte-Carlo simulation in an engineering context. *In S. S. Fischer, editor, Rarefied Gas Dynamics, proceedings of the 12th international symposium - part-1.*, 239.

BIRD, G. A. 1987 Direct simulation of high-vorticity gas flows. *Phys. Fluids.* **30**, 364.

Bird, G. A. 1994 Molecular gas dynamics and the direct simulation of gas flows. *Clarendon Press. Oxford.*

Bourn, R. & Peterson, T. D. & Wood, H. G. 1999 Solution of the pancake model for flow in a gas centrifuge by means of a temperature potential. *Comp. Meth. Appl. Mech. and Engg.* **178**, 183.

Brauckmann, H. J. & Salewski, M. & Eckhardt, B. 2016 Momentum transport in Taylor-Couette flow with vanishing curvature. *J. Fluid Mech.* **790**, 419.

Brouwers, J. J. H. 1978 On compressible flow in a gas centrifuge and its effect on maximum separative power. *Nucl. Technology.* **39**, 311.

Cercignani, C. & Lampis, M. 1971 Kinetic models for gassurface interactions. *Transport Theory Stat. Physics.* **1**, 101.

Cercignani, C. 2000 Rarefied gas dynamics. From basic concepts to actual calculations. *Cambridge University Press.*

Chapman, S. & Cowling, T. G. 1970 The Mathematical Theory of Non-Uniform Gases, 2nd edition. *Cambridge, Cambridge University Press.*

Chen, G. & Boyd, I. D. 1996 Statistical error analysis for the Direct Simulation Monte Carlo technique. *J. Comput. Phys.* **126**, 434.

Copper, G. R. & Morton, J. B. 1988 Driven Wave Motions in a Rotating Gas. *J. Fluid Struct* **2**, 453.

Fan, J. & Shen, C. 2001 Statistical simulation of low-speed rarefied gas flows. *J. Comp. Phys.* **167**, 393.

Garcia, A. L. & Wagner, W. 2000 Time step truncation error in direct simulation Monte Carlo. *Phys. Fluids* **12**, 2621.

Greenspan, H. P. 1968 The theory of Rotating Fluids. *Cambridge, Cambridge University Press.*

Gunzburger, M. D. & Wood, H. G. 1982 A finite element method for the Onsager pancake equation. *Comp. Meth. Appl. Mech. Engg.* **31**, 43.

Gunzburger, M. D. & Wood, H. G.& Jordan, J. A. 1984 A finite element method for gas centrifuge problem. *SIAM J. Sci. Stat. Comp.* **5**.

Hadjiconstantinou, N. 2000 Analysis of discretization in the direct simulation Monte Carlo. *Phys. Fluids.* **12**, 2634.

Harada, I. 1980 A Numerical Study of Wikely Compressible Rotating Flows in a Gas Centrifuge. *Nuclear Sc. and Engg.* **73**, 225.

Harada, I. 1980 Computation of Strongly Compressible Rotating Flows. *J. Comput. Phys.* **38**, 335.

Hittinger, M. & Holt, M. & Soubbaramayer, B. 1981 Numerical solution of the flow field near a gas centrifuge scoop. *In Proc. Fourth Workshop on Gases in Strong Rotation, Oxford (ed. E.Ratz)*, 22.

Hittinger, M. & Holt, M. & Soubbaramayer, B. & Cortet, C. 1983 Simulation of gas flow in front of the scoop of a gas centrifuge. *In Proc. Fifth Workshop on Gases in Strong Rotation, Charlottesrille (ed. H.G.Wood)*, 515.

Jung, E. 1983 An analytic solution of the linearized flow equations using the method of eigenvalues. *In Proc. Fifth Workshop on Gases in Strong Rotation, Charlottesrille (ed. H.G.Wood)*, 247.

Kai, T. 1983 Analysis of Separation Performance of Gas Centrifuge. *In Proc. Fifth Workshop on Gases in Strong Rotation, Charlottesrille (ed. H.G.Wood)*, 1.

Kumaran, V. & Pradhan, S. 2014 The generalized Onsager model for a binary gas mixture. *J. Fluid Mech.* **753**, 307.

Larignon, B. & Marr, K. & Goldstein, D. B. 2016 Monte Carlo and Navier-Stokes Simulations of Compressible Taylor-Couette Flow. *AIAA*.

Macrossan, M. N. 2001 DSMC : A fast simulation method for rarefied flow. *J. Comp. Phys.* **173**, 600.

Matsuda, T. 1976 A New proposal of gas centrifuge with desirable counter-current. *J. Nucl.Sci. Tech.* **13**, 98.

Matsuda, T. & Hashimoto, K. 1978 The Structure of the Stewartson layers in a gas centrifuge. Part 1: Insulated end plates. *J. Fluid Mech.* **85**, 433.

Matsuda, T. & Nakagawa, K. 1983 A new type of boundary layer in a rapidly rotating gas. *J. Fluid Mech.* **126**, 431.

MANELA, A. & FRANKEL, I. 2007 On the compressible TaylorCouette problem. *J. Fluid Mech.* **588**, 59.

MASLEN, S. H. & MOORE, F. K. 1956 On strong transverse waves without shocks in a circular cylinder. *J. Aeronautical Sciences* **23**, 583.

MATSUDA, T. & NAKAGAWA, K. 1983 A new type of boundary layer in a rapidly rotating gas. *J. Fluid Mech.* **126**, 431.

MATSUDA, T. & TAMURA, N. & SAWADA , K. 1989 Three-dimensional numerical simulation of flows past scoops in a gas centrifuge. *J. Fluid Mech.* **201**, 203.

MIKAMI, H. 1973 Thermally induced flow in a gas centrifuge. *J. Nucl. Sci. Tech.* **10**, 580.

MIKAMI, H. 1981 Rotating supersonic flow about scoop inlet using an unsteady implicit technique. *In Proc. Fourth Workshop on Gases in Strong Rotation,Oxford (ed. E.Ratz)*, 94.

MIKAMI, H. 1985 Two dimensional numerical calculation of flow fields about scoop inlet by the piecewise linear method. *In Proc. Sixth Workshop on Gases in Strong Rotation,Tokyo (ed. Y.Takashima)*, 67.

NAKAYAMA, W. & TORII, T. 1974 Numerical analysis of separative power of isotope centrifuges. *J. Nucl. Sci. Tech.* **11**, 495.

NANBU. 1980 Direct simulation scheme derived from the Boltzmann equation I. Multicomponent gases. *J. Phys. Soc. of Japan.* **45**, 2042.

OLANDER, D. R. 1981 The theory of uranium enrichment by the gas centrifuge. *Progress in Nuclear Energy.* **8**, 1.

OMNES, P. 2007 Numerical and physical comparisons of two models of a gas centrifuge. *Comp. Fluids.* **36** ,1028.

PRADHAN, S. & KUMARAN, V. 2011 The generalized Onsager model for the secondary flow in a high-speed rotating cylinder. *J. Fluid Mech.* **686**, 109.

RIBANDO, R. J. 1984 A Finite difference solution of Onsager's model for flow in a gas centrifuge. *Comp. Fluids.* **12 (3)**, 235.

RIEFFEL, M. A. 1999 A method for estimating the computational requirements of DSMC simulations. *J. Comp. Phys.* **149**, 95.

ROBERTS, W. W. 1985 Three dimensional stratified gas flows past impact probes and scoops : N-body Monte Carlo calculations. *In Proc. Sixth Workshop on Gases in Strong Rotation, Tokyo (ed. Y.Takashima.)*, 115.

ROBERTS, W. W. & HAUSMAN, M. A. 1988 Hypersonic, stratified gas flows past an obstacle : Direct Simulation Monte Carlo calculations. *J. Comp. Phys.* **77**, 283.

ROBLIN, P. & DONEDDU, F. 2000 Direct Monte-Carlo Simulations In a Gas Centrifuge. *In Proc. 22nd International Symposium on Rarefied Gas Dynamics. (ed. T.J.Bartel & M.C.Gallis.)*, 169.

SAKURAI, T. 1981 Linearized thin-wing theory of gas-centrifuge scoops. *J. Fluid Mech.* **103**, 257.

STEFANOV, S. & CERCIGNANI, C. 1993 Monte Carlo simulation of the Taylor-Couette flow of a rarefied gas. *J. Fluid Mech.* **256**, 199.

SUN, Q. & BOYD, I. D. 2002 A Direct Simulation Method for Subsonic Microscale Gas Flows . *J. Comp. Phys.* **179**, 400.

THOMPSON, J. J. & CHEN, Y. P. 1980 Eigen vector solution of the linearized equations of thermal convection in a closed gas centrifuge.Part-1 : Problem formulation and solution by collocation. *Nuclear Engineering and Design.* **58**, 97.

VIECELLI, J. A. 1983 Exponential difference operator approximation for the sixth order Onsager equation. *J. Comp. Phys.* **50**, 162.

VOLOSCIUK, K. 1981 Application of the vortex transport equations to the calculation of 2 and 3 dimensional extraction chamber flows. *In Proc. Fourth Workshop on Gases in Strong Rotation, Oxford (ed. E.Ratz)*, 604.

WAGNER, W. 1992 A convergence proof for Bird's Direct Simulation Monte Carlo method for the Boltzmann equation. *J. Statistical Physics.* **66**, 1011.

WALZ, A. & VOLOSCIUK, K. & SCHUTZ, H. 1983 Numerical investigations of the flow field near a model of a scoop using the vortex transport equations. *In Proc. Fifth Workshop on Gases in Strong Rotation,Charlottesrille (ed. H.G.Wood)*, 425.

WOOD, H. G. & MORTON, J. B. 1980 Onsager's pancake approximation for the fluid dynamics of a gas centrifuge. *J. Fluid Mech.* **101**, 1.

WOOD, H. G. & SANDERS, G. 1983 Rotating compressible flows with internal sources and sinks. *J. Fluid Mech.* **127**, 299.

WOOD, H. G. & JORDAN, J. A. & GUNZBURGER, M. D. 1984 The effect of curvature on the flow field in rapidly rotating gas centrifuges. *J. Fluid Mech.* **140**, 373.

WOOD, H. G. & BABARSKY, R. J. 1992 Analysis of a rapidly rotating gas in a pie-shaped cylinder. *J. Fluid Mech.* **239**, 249.

DSMC SIMULATION OF HIGH MACH NUMBER TAYLOR-COUETTE FLOW

S. Pradhan

Department of Chemical Engineering, Indian Institute of Science, Bangalore- 560 012, India

ABSTRACT: The main focus of this work is to characterize the Taylor-Couette flow of ideal gas between two coaxial cylinders at Mach number $Ma = (U_w / \sqrt{kb\ T_w / m})$ in the range $0.01 < Ma < 30$, and Knudsen number $Kn = (1 / (\sqrt{2} \pi d^2\ n_d\ (r_2 - r_1)))$ in the range $0.001 < Kn < 10$, using two-dimensional (2D) direct simulation Monte Carlo (DSMC) simulations [1, 2, 3, 4, 5, 6, 7, 8, 9, 10]. Here, r_1 and r_2 are the radius of inner and outer cylinder respectively, U_w is the circumferential wall velocity of the inner cylinder, T_w is the wall temperature, n_d is the number density of the gas molecules, m and d are the molecular mass and diameter, and kb is the Boltzmann constant. The cylindrical surfaces are specified as being diffusely reflecting with the thermal accommodation coefficient equal to one.

In the present analysis of high Mach number compressible Taylor-Couette flow using DSMC method, wall slip in the temperature and the velocities are found to be significant ((Pradhan & Kumaran, *J. Fluid Mech.*, vol. 686, 2011, pp. 109-159); (Kumaran & Pradhan, *J. Fluid Mech.*, vol. 753, 2014, pp. 307- 359)). Slip occurs because the temperature/velocity of the molecules incident on the wall could be very different from that of the wall, even though the temperature/velocity of the reflected molecules is equal to that of the wall. Due to the high surface speed of the inner cylinder, significant heating of the gas is taking place. The gas temperature increases until the heat transfer to the surface equals the work done in moving the surface. The highest temperature is obtained near the moving surface of the inner cylinder at a radius of about ($1.26\ r_1$).

In a compressible Taylor-Couette flow we examine the result that the splitting of the Taylor vortices takes place proportional as $(L/(r_2 - r_1))$. The resolution suggested by our simulation is that even though the Mach number based on the wall velocity and temperature is large, the local Mach number based on the local dissipation velocity in regions of high shear decreases due to an increase in temperature. Due to this, the ratio of the mean free path and characteristics flow scale $(\lambda/(r_2 - r_1))$ appears to taper off in the high Mach number limit.

A modification of the velocity profile in the viscous rotating boundary layer near the wall, which takes into account temperature and density variations, is derived. The variation of the velocity and temperature is predicted under the assumption that the increase in temperature across the viscous rotating boundary layer is larger than the wall temperature. It is found that the scaling laws do depend on the molecular model, through the dependence of viscosity and thermal conductivity on the temperature. The predicted law, are found to be in good agreement with simulations, for two different molecular models, the hard-sphere and the variable hard-sphere.

KEY WORDS: Compressible Taylor-Couette flow, Rarefied gas flow, Direct Simulation Monte Carlo (DSMC) simulations.

REFERENCES:

[1] Pradhan, S. & Kumaran, V. 2011 The generalized Onsager model for the secondary flow in a high-speed rotating cylinder. J. Fluid Mech. 686, 109 - 159.

[2] Kumaran, V. & Pradhan, S. 2014 The generalized Onsager model for a binary gas mixture. J. Fluid Mech. 753, 307 - 359.

[3] Bird, G.A. 1994 Molecular gas dynamics and the direct simulation of gas flows. *Clarendon Press. Oxford.*

[4] Bird, G. A. 1998 Recent advances and current challenges for DSMC. Comput. Math. Appl. 35,1-14.

[5] Cercignani, C. 2000 Rarefied Gas Dynamics. Cambridge University Press.

[6] Chapman, S. & Cowling, T. 1970 The Mathematical Theory of Non-Uniform Gases. 3rd edn. Cambridge University Press.

[7] Rayleigh, Lord 1916 On the dynamics of revolving fluids. Sci. Pap. 6, 447–453.

[8] Usami, M. 1995 Direct simulation Monte Carlo on Taylor vortex flow. In Rarefied Gas Dynamics (ed. J. Harvey & G. Lord), vol. 1, pp. 389–395. Oxford University Press.

[9] Taylor, G. I. 1923 Stability of a viscous liquid contained between two rotating cylinders. Phil. Trans. R. Soc. A 223, 289–343.

[10] Stefanov, S. & Cercignani, C. 1993 Monte Carlo simulation of the Taylor–Couette flow of a rarefied gas. J. Fluid Mech. 256, 199–213.

DSMC SIMULATION RESULTS:

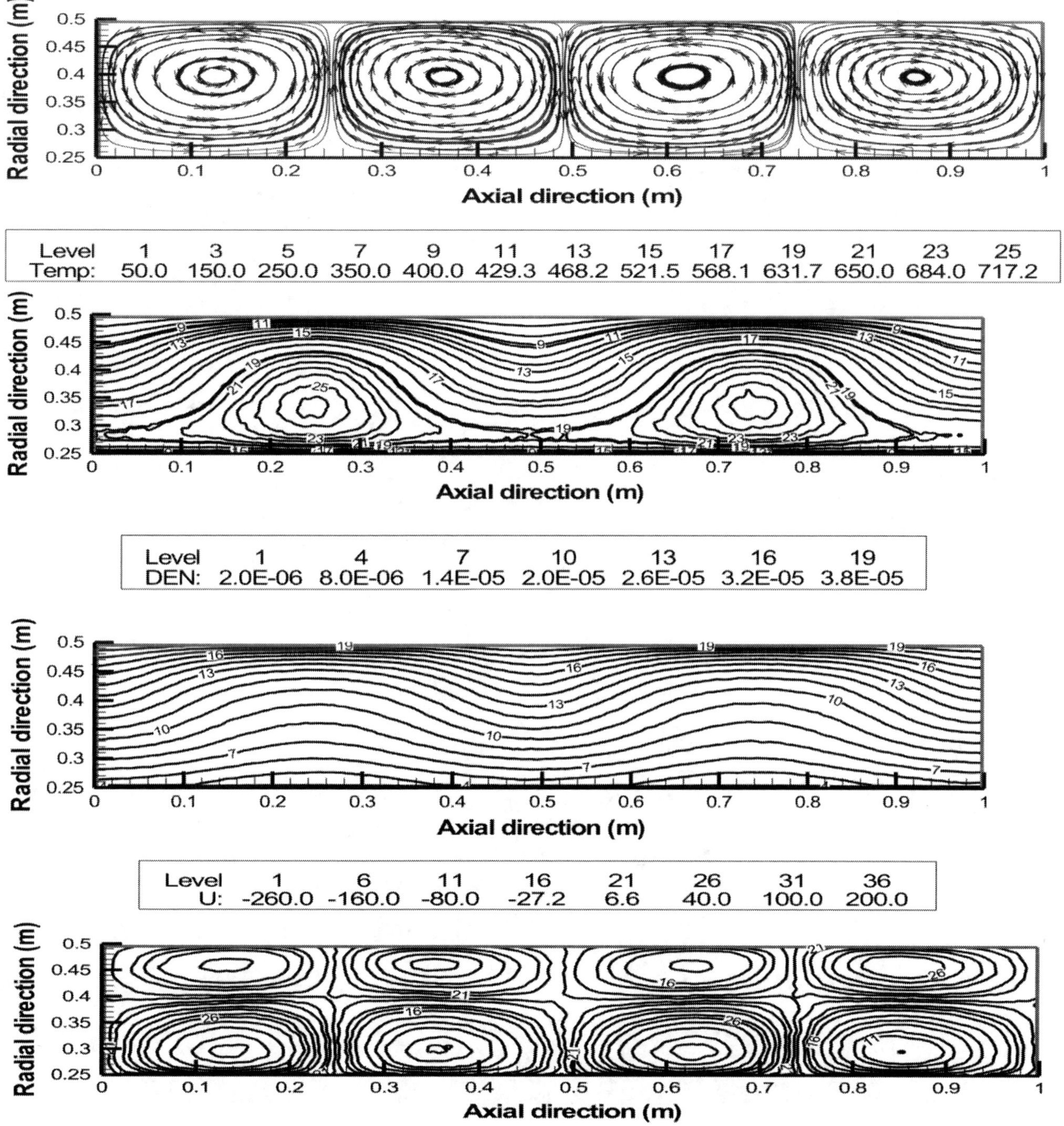

Figure- 1 : DSMC simulation results for the Taylor-Couette flow : (a) Streamlines in the flow domain, (b) Temperature contours, (c) Density contours, and (d) Axial-velocity contours in the flow domain.

Viscosity Prediction of the Carbon Dioxide Loaded Aqueous Solutions of Alkanolamines: Thermodynamic Approach

Naser S. Matin, Joseph E. Remias and Kunlei Liu
Center for Applied Energy Research, University of Kentucky, Lexington, KY 40511, United States

Abstract: The Eyring absolute rate theory (EART) based methodology has been used for estimation of the dynamic viscosity of CO_2 loaded aqueous solutions of a representative amine-monoethanolamine (MEA). The electrolyte-NRTL model is applied to estimate the activation free energy of the solution and consequently enable prediction of the dynamic viscosity of the electrolyte CO_2:MEA:H_2O system. According to EART for the dynamic viscosity of asolution,[1]

$$\eta/\eta_0 = (V_0/V)\exp(\Delta F^{\neq}/RT) \quad (1)$$

η, η_0, V, V_0, R and $\Delta F^{\neq}$, are the dynamic viscosity of solution and pure solvent, the solution and pure solvent molar volume, gas constant and activation free energy for the flow process, respectively.

The $\Delta F^{\neq}$ was generally related to the Gibbs free energy of mixing, $f(\Delta G^{\neq}_{mix})$.

$$\Delta G^{\neq}_{mix} = \Delta G^{\neq}_{id,mix} + \Delta G^{\neq}_{ex} \quad (2)$$

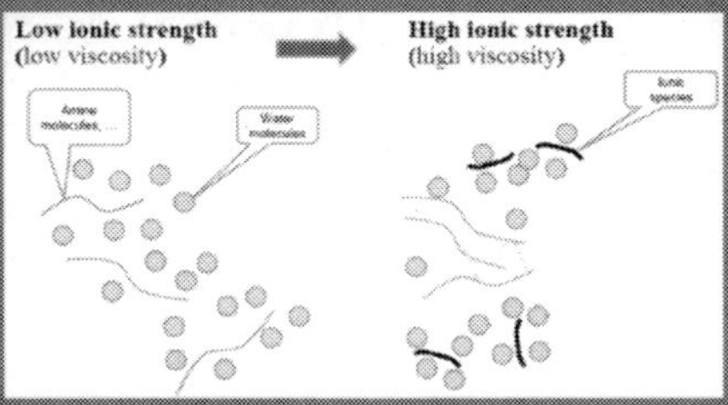

Solution pH and Ionic Strength, MEA 30%

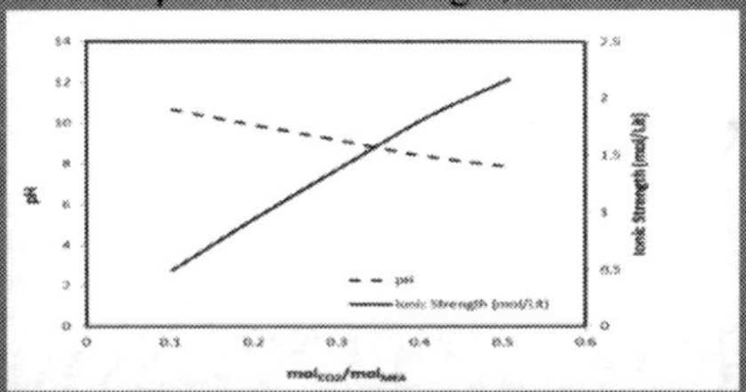

(1) Martins, et al., Ind. Eng. Chem. Res. 40 (2001), 1271.

Effects of viscosity on the liquid side mass transfer coefficient

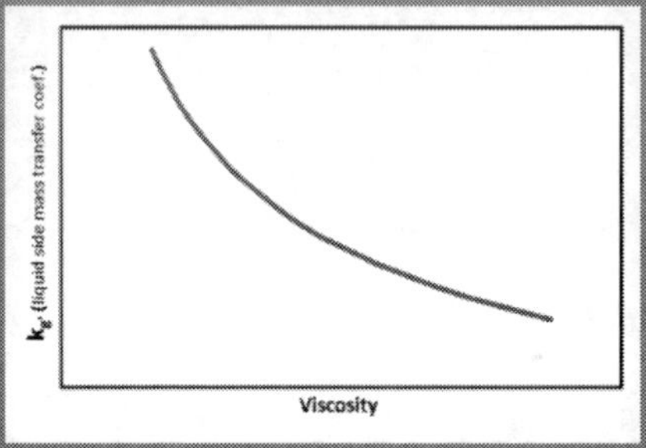

In equation (2), $\Delta G^{\neq}_{id,mix}$, can be calculated through the system speciation which is available from thermodynamic modeling. (e.g. Aspen Plus®). $\Delta G^{\neq}_{ex}$, is related to the solution components (species) mole fractions and their corresponding activity coefficients.

For a multicomponent mixture, Eq (1) can be written as,

$$\ln(\eta V) = \sum_i X_i \ln(\eta_i) + \ln(\sum_i X_i V_{i,0}) + \frac{\Delta F^{\neq}}{RT}$$

X_i, $V_{i,0}$ and η_i are the mole fraction of component i, molar volume and viscosity of the pure chemical i, respectively.

Results: Using the Electrolyte-NRTL model for excess Gibbs energy, the MEA solution viscosity calculated through model equation based on EART approach. The results revealed that the positive sign for the $\Delta F^{\neq}/RT$ term in Eq. (1), yields 4 to 5 times lower viscosity values compared to the experimental data. While, the negative sign, i.e. $-\Delta F^{\neq}/RT$, provides consistent estimation with the experimental viscosity data.[2]

Conclusion:
The EART approach based model:

- Employs thermodynamic framework for viscosity prediction.
- Displays the importance of the Gibbs free energy of mixing in solution viscosity.
- Needs a comprehensive thermodynamic modeling for the system.
- Can estimate the viscosity of the solution mixtures at different concentrations, temperatures, CO_2 loadings, and operating pressures.

Acknowledgments: The Carbon Management Research Group (CMRG) members, including Duke Energy, Electric Power Research Institute (EPRI), Kentucky Department of Energy Development and Independence (KY-DEDI), and Louisville Gas & Electric (LG&E) and Kentucky Utilities (KU) for their financial support.

MEA 20 wt%

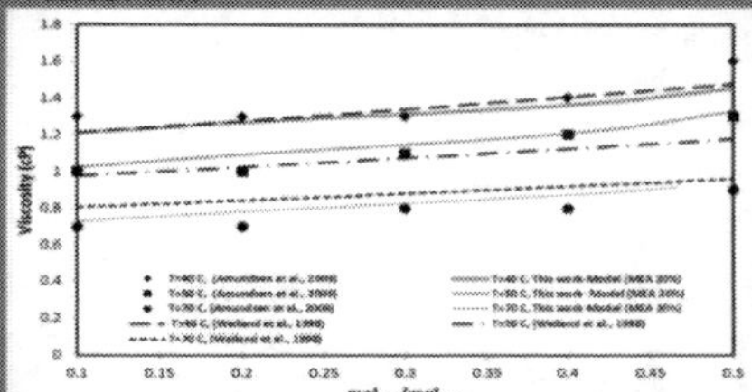

MEA 30 wt%

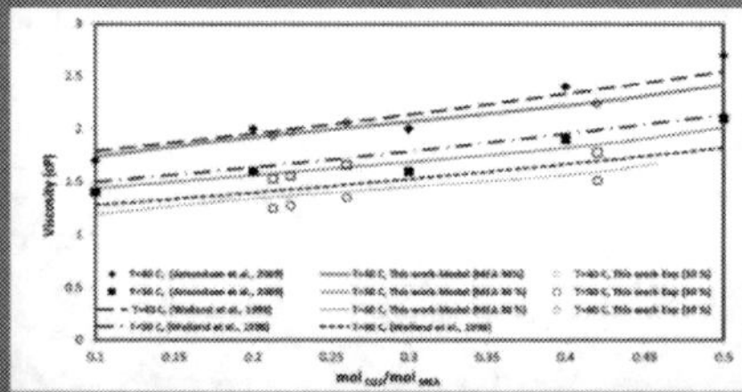

(2) Matin, et al., Ind. Eng. Chem. Res., 52 (2013), 16979.

Center for Applied Energy Research

caer.uky.edu

1

492293 A Computational Study on the Transport of Actives through Skin Layers

Abstract:

The accurate prediction of dermal uptake of actives is relevant to both transdermal drug delivery as well as topical application of cosmetics. The current industry standard, however, both in pharma and cosmetics, is to conduct detailed *in-vitro* and *in-vivo* trials. These obviously incur huge expenses thereby leading to a very few successful candidates that are finally approved by regulatory authority (FDA). The plasma uptake of drug through skin is limited by the systematic absorption of drugs through distinct layers of skin. The drug permeation through skin mostly happens through passive diffusion. The drug diffusion coefficient and partition coefficient in these layers is mainly responsible for the fate of the drug inside the skin layer.

Skin is composed of many layer in series, mainly SC (stratum corneum), VE (viable epidermis) and dermis. Out of these three layers, the first layer, SC, is responsible for barrier function of the skin. The SC is made up of corneocyte and lipid matrix, which are arranged in brick and mortar fashion respectively. Dermis consists of microscopic blood vessels through which drug enters into systemic circulation.

The detailed modelling of all the skin layers would give an insight into the transport mechanism of drugs in skin layers and would help in obtaining the desired total drug uptake. In this study, we have integrated the models of all the skin layers, and performed finite element simulations to obtain the dermal uptake. The heterogeneity of SC was captured using bricks and mortar model, the diffusion coefficient which governs the transport process in SC was obtained accurately from molecular dynamics simulations[1] instead of depending on experiments or from fitted data. The VE layer and subsequent dermis layer were approximated with compartmental models. Diffusion coefficients of the drug in the VE and dermis were obtained from available empirical correlations.[2] Finite element method framework was used to predict the time and space dependent concentration profiles of drug in different layers of skin. Permeation of the drug in all these layers were linked in the integrated model. The model was validated with experimentally available release profile.[3,4] The concentration depth profiles in VE and dermis layers were also compared with the available experimental data. Our integrated model provides flexibility in analyzing diffusion process in individual skin layers as well as whole skin. This model shall be used as guiding tool for performing targeted experiments in designing a suitable delivery system.

References:

1. Gupta, Rakesh, D. B. Sridhar, and Beena Rai. "Molecular Dynamics Simulation Study of Permeation of Molecules through Skin Lipid Bilayer." The Journal of Physical Chemistry B 120.34 (2016): 8987-8996.

2. Bunge, Annette L., and Robert L. Cleek. "A new method for estimating dermal absorption from chemical exposure: 2. Effect of molecular weight and octanol-water partitioning." Pharmaceutical research 12.1 (1995):

   88-95.

3. Kushner, Joseph, et al. "Firstâ€principles, structureâ€based transdermal transport model to evaluate lipid partition and diffusion coefficients of hydrophobic permeants solely from stratum corneum permeation experiments." Journal of pharmaceutical sciences 96.12 (2007): 3236-3251.

4. Dias, M., et al. "Topical delivery of caffeine from some commercial formulations." International journal of pharmaceutics 182.1 (1999): 41-47.

West Point Kicking Mule Brewery

Cadets Jesse Hudgins, Evan Ousley, Denis Glinski

CPT Joshua White, LTC Matthew Armstrong, LTC Corey James, LTC Geoffrey Bull

Department of Chemistry and Life Science

West Point, New York

ABSTRACT

The Kicking Mule Brewing Co. enables members of our chemical engineering club students and beer enthusiasts alike to combine chemical engineering acumen, knowledge and skills by developing a tangible product. Currently, Kicking Mule Brewing Co. aspires to hone in on the scientific aspects of brewing to enter into competitive beer brewing competitions including reaction engineering, heat transfer, and separations. The West Point Chemical Engineering program is process oriented and the Kicking Mule Brewery allows Cadets hands on experience with a batch reaction process. Through a rigorous on-site brewing process, with subject matter experts, our end goal is enhancement of student chemical engineering knowledge with respect to batch reactors and mixtures, heat exchange processes, chemical engineering process control, and bio-chemical reaction rates while simultaneously providing quality products to The United States Corps of Cadets.

PROJECT GOALS

- Produce a West Point beer using raw ingredients, a self-made recipe, and experimental procedures.
- Apply chemical engineering principles such has heat exchange, thermodynamics, batch reactions, mass flow, kinetics and process control.
- Determine reaction rates through ethanol production.
- Combine chemistry, biological processes, process control and chemical engineering principles in a tangible and fun way.
- Process parameters for varied flavors and palette while monitoring cost and profit.

$$(C_6H_{10}O_5)_x \xrightarrow{\text{enzymes}} C_6H_{12}O_6 \xrightarrow{\text{enzymes}} 2CH_3CH_2OH + 2CO_2$$

Starch — Glucose — Ethanol

Figure 1. Fermentation equation

PRECURSORS

- Fermentation requires sugars in order for yeast to produce alcohol (Figure 1).
- In order to make glucose (our reactant), malted barley at different roasts are built into a recipe for desired flavor profile and alcohol content.
- Sugars for fermentation are cooked out of the malted barley after the grain has been milled.

WORT

- Wort is the sugary precursor to beer. This includes all sugar for the reaction as well as hops for bitterness and aroma.
- We create the wort from milled grain. Cooking the grain in an aqueous batch reaction (the mash) activates enzymes that break down long chain sugars into short chain fermentables.
- During the boil, hops are added into the kettle to add bitterness and temper the sweetness.
- Our 3-vessel electric system is a professional pilot system to brew beer from grain. It has a PID controlled heat exchanger to keep the mash at the exact desired temperature.

PICTURE REFERENCES

- Image 1 - http://2012books.lardbucket.org/books/introduction-to-chemistry-general-organic-and-biological/s17-04-reactions-that-form-alcohols.html
- Image 3 - http://www.blichmannengineering.com/products/therminator

EQUIPMENT

- ½ bbl (15.5 gal) three tank, electric heat exchange recirculating mash system (Figure 5).
- Temp controlled Fermenter (25 gal).
- Heat Exchanger and Pump.
- Grain mill.
- Assorted hoses, and instrumentation.

Figure 2: Temp controlled fermenter

PROCESS CONTROL

- PID control is utilized to maintain strict adherence to the set point, which is set by the brewer and recipe dependent.
- Modeling of the control structure and system response allows a comparison of modeled data with real data.

Figure 3: Results of system temperature with PID control.

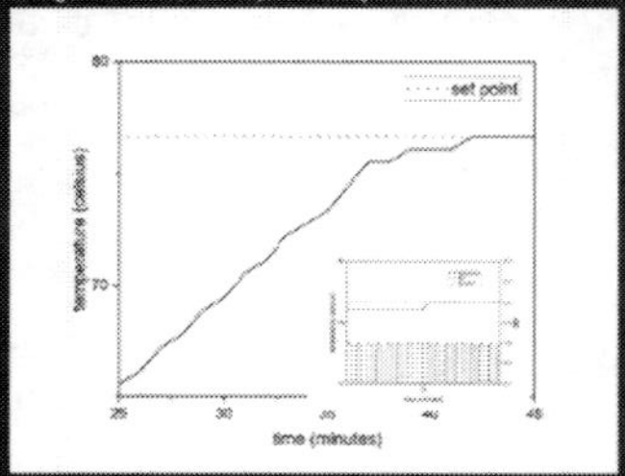

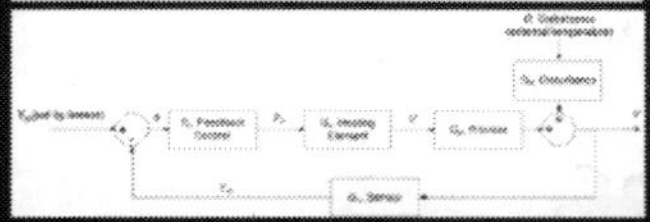

Figure 4: Block diagram of system feedback control.

Figure 5: ½ bbl E-HERMS system

FUTURE RESEARCH

- Optimizing our system and techniques to obtain the highest efficiency of sugar extraction and bring down cost.
- Continue to collect and analyze kinetic data to obtain best-fit parameters.
- Yeast counting to get biomass information to solve the remaining kinetic differential equation.

HEAT EXCHANGE

- Utilizing a countercurrent plate heat exchanger, the wort is chilled from boiling to ≈70°F (30°C) utilizing tap water.
- Adjusting flow rates of the wort with a pump, and the countercurrent chilling water with the faucet, we can achieve optimal temperature and flow. This information can be used to calculate heat transfer information of the heat exchanger.
- The heat exchanger was modeled in ChemCAD for energy balance. Correct modeling (Figure 6) of wort properties and sizing of the exchanger will be a future project.

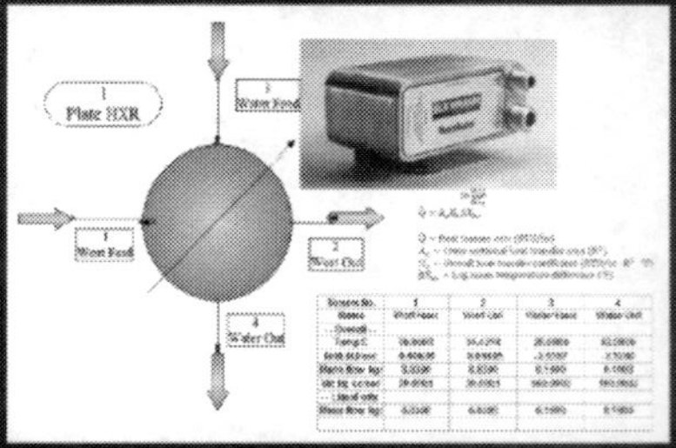

Figure 6: ChemCAD model of plate-type heat exchanger.

REACTION RATES AND FERMENTATION

- Different fermentation temperatures will alter reaction rates.
- The higher the temperature of the fermenter, the higher the reaction rate and the more vigorous the fermentation.
- We collected data on several fermentations and the results to start modeling kinetic data.
- Data was collected using a refractometer, hydrometer and a GC mass spec.

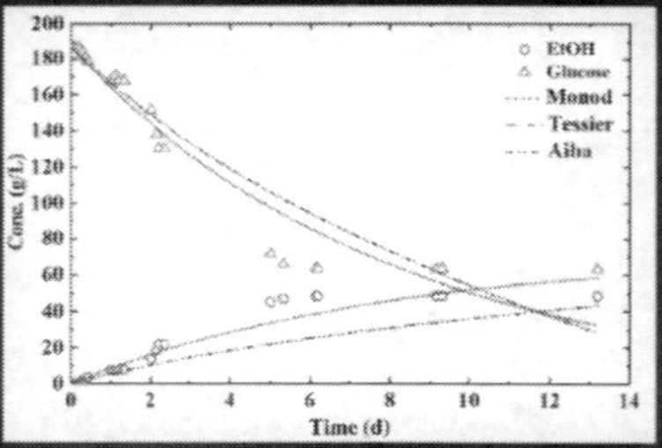

Figure 7: Kinetic model of fermentation

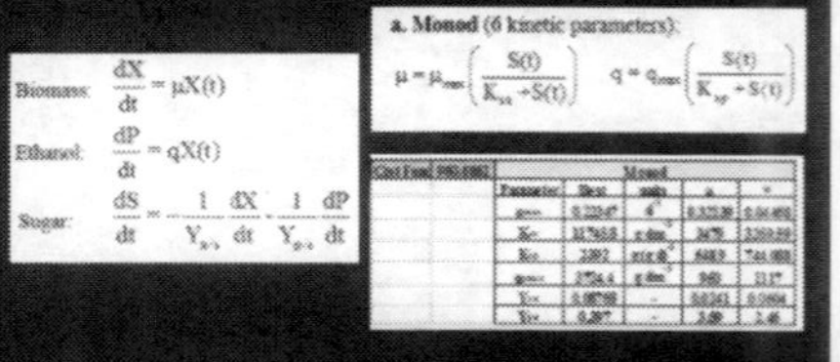

$$\mu = \mu_{max}\left(\frac{S(t)}{K_{s\mu}+S(t)}\right) \qquad q = q_{max}\left(\frac{S(t)}{K_{sp}+S(t)}\right)$$

$$\text{Biomass: } \frac{dX}{dt} = \mu X(t)$$

$$\text{Ethanol: } \frac{dP}{dt} = qX(t)$$

$$\text{Sugar: } \frac{dS}{dt} = -\frac{1}{Y_{x/s}}\frac{dX}{dt} - \frac{1}{Y_{p/s}}\frac{dP}{dt}$$

Figure 8: Governing equations/parameter values for kinetics

PACKAGING

- After fermentation is complete, the uncarbonated beer is filtered and transported to a keg.
- Once carbonated and refrigerated, the beer is now ready to serve.
- The beer is sold by DCA* at the on-post First Class Club which gives the cadets a real-world economics model to determine profitability and viability of the process.

ACKNOWLEDGMENTS

•Directorate of Cadet Activities for funding and a venue to sell Kicking Mule Beer.

499372 Beer Experiments - a Hopping Success

Abstract:

In the Biochemical Engineering Laboratory (CHE 194) course, students learn bioprocess engineering experimental techniques for eight weeks including microbial fermentation, protein purification, ultrafiltration, chromatography and enzyme kinetics.

Following the modular labs, a true inquiry lab experience is had by the students. Student teams submit a proposal for an experiment of their choice, using a previously published rubric for design of experiments. Five weeks is allotted for the inquiry experiment so they have sufficient time to repeat parts of the experiment that didnâ€™t work the first time. Students rate the experience highly on the end-of-semester evaluations.

Beer is a popular subject for many chemical engineering students. An experiment has been added into the chromatography module giving the students the opportunity to calibrate the high pressure liquid chromatograph (HPLC) instrument using a standard for alpha acids and then they can bring their favorite beers and test them for alpha acid content. Beer hops confer alpha acids into the beverage that have some health benefits, such as antioxidant activity, as well as anti-microbial activity for preserving the beer early during the fermentation when the alcohol level is low. Hops are added during the boil because the solubility of the alpha acids is higher in hot water than cold. Popular now is also the dry-hopping process whereby hops are added after the boil and the ethanol produced during the fermentation enables the solubilization of the alpha acids. Students have done several experiments related to beer over the years. The alpha acid HPLC module will be presented here as well as the inquiry experiment design and two of the most interesting beer experiments designed and executed by students over the years.

496488 Why Is the Mustard in the Fridge?Â a Fun Assignment on the Thermodynamics of Food Safety

Abstract:

Activity is often introduced in chemical engineering thermodynamics courses in the context of modeling non-ideal solutions in vapor-liquid equilibrium, but this is only the first of many useful applications. One of these applications is estimation of osmotic pressure, often in the context of drinking water purification or creation of isotonic medical saline solution. In both of these examples, the activity of water in solution (a_w) is the key value in the analysis. For both my thermodynamics and applied food science courses, I created an assignment that focuses on yet another important application of a_w: food safety. Prior to refrigeration, extremes of osmotic pressure were used to prevent or reduce microbial spoilage of foods through addition of salt, sugar, acid, or a combination thereof. This assignment, which works both in-class and as homework, starts with published food labels and a few assumed molecular masses and results in fairly accurate determinations of whether or not a given food should be â€œshelf stableâ€ without refrigeration. The assignment can also be extended to ask students to invent a recipe for preserving fruits and vegetables that result in answers remarkably similar to traditional versions of jam, raisins, and pickles.

499784 Improving Conceptual Knowledge and Retention in Introduction to Engineering Thermodynamics

Abstract:

By the end of our sophomore level Introduction to Engineering Thermodynamics course (CBEN 210), students should be able to demonstrate both procedural and conceptual knowledge. Procedural knowledge is the ability to solve domain problems and usually yields a final numerical answer.[1] Conceptual knowledge is understanding of the principles governing a domain.[2] Data collected over the course of two years in CBEN 210 showed that students consistently performed better on procedural problems than conceptual problems. Conceptual knowledge can be more challenging because it requires organizing key principles, like content experts, and it is frequently coupled with misconceptions, which are hard to undo.[3] In this study, we identified three conceptual topics that students performed poorly on and initiated three in-class interventions to address the deficiency. The three topic areas were isenthalpic flow through a valve, isentropic work, and entropy generation. The interventions were all interactive activities that tied the thermodynamic concept to a real-word application. For example, isenthalpic flow through a valve was examined by the students as a technique to measure steam quality in a process line. The impact of these interventions will be measured with a concept inventory exam administered 8 -12 months following enrollment in CBEN 210. Since our chemical engineering students take a second course in thermodynamics, their ability to retain knowledge beyond the end of the sophomore level class is important. The test group contains approximately 70 students and will be compared to a control group from the previous year that did not receive the interventions. Correlations between thermodynamic knowledge retention and CBEN 210 performance, class timing (8 or 12 months), and other factors such as GPA will be examined.

[1] R. Taraban *et al*, *J. Eng. Ed.*, **96**, 57, 2007.

[2] R. A. Streveler *et al*, *J. Eng. Ed.*, **97**, 279, 2008

[3] P. N. Van Meter *et al*, *J Eng. Ed.*, **105**, 245, 2016.

494839 Tax Revenue Laffer Curve from Thermodynamics Perspective

Abstract:

It has been a long tradition of economists and scientists making parallel analogy between economics and thermodynamics, establishing phenomenological theory of economic systems. In this contribution, the proposed economic temperature and economic entropy identification and subsequently developed economic production and growth models were used to obtain the quantitative relationship between tax revenue, payroll tax as well as corporate/personal income tax, and the wealth creation efficiency of an economic entity. The derived expression for corporate/personal income tax exhibits a quadratic dependence of wealth creation efficiency and tax rate, which agrees well with the famous tax revenue Laffer curveâ€"There are always two tax rates that yield the same revenues. The expression also indicates that the wealth creation efficiency improvement plays a pivotal role in the pay off of the revenue lost by prior tax rate cut. The derived mathematical revenue expression together with the proposed monetary rule may be used for future fiscal policy design and optimization.

492365 The Art of the Game: Infusing Thermodynamics Learning with Classic Household Games

Abstract:

We all understand the drab that comes with learning Chemical Engineering Thermodynamics in general. Now try teaching it at 8am and one can immediately see the conundrum that presents itself. Altering the classroom learning environment in unusual ways can spark new life into students who do not necessarily appreciate thermodynamics for what it is worth in their later courses. In this presentation, the use of classic household games to change the dynamics of the classroom will be discussed. From MONOPOLY to Blackjack, creating an incentive to learn Thermodynamics (and other concepts) leads to a transformation of the student that promotes growth and turns the classroom into a fun learning environment, even at 8am in the morning.

496817 Generalizing the Design Principles of Pt-Based Alloy Catalysts with Improved ORR Performance and Durability

Abstract:

Since the discovery of Pt skin type Pt_3Ni catalysts with superior performance towards oxygen reduction reaction,[1] there has been a blossoming of Pt-based alloy catalysts for ORR and various other applications.[2,3] In last couple years, the focus has slowly shifted from catalysts with monolayer skins to those with thicker Pt overlayers for the sake of improving the stability.[4-6] But the decrease in the activity has been observed with increasing overlayer thickness. Thus, how to increase the stability of catalysts, with no harm on the activity, or even simultaneously increase activity, have been the focus of the current catalyst development, but few systematic design principles have been established to guide these efforts. It is worsened by a lacking of atomic-scale understanding of water/electrode interface and consequently the bonding nature of ORR intermediates. For example, it is still in the debate regarding whether hydrated OH or non-hydrated OH is the plausible ORR intermediate.

In the present talk, we will start by determining the atomic-scale structure of oxygenated species at water/electrode interfaces under electrochemical ORR conditions, by combining a careful calibration of Density Functional Theory (DFT)-determined energetics, the liquid water/Pt interfaces generated from *ab-initio* molecular dynamics (AIMD) simulations and a detailed simulation of X-ray Photoelectron Spectroscopy (XPS) signatures.[7] In the second part, we will investigate the structure and the stability of Pt overlayers with various thickness. We will further analyze their activity based on a scheme established through above calibration. Then, general design principles will be proposed for searching for ORR catalysts with improved stability and activity. We will close by comparing the predicted stability and activity of various catalysts with those available in the literature and from our collaborators.

[1] V.R. Stamenkovic, B. Fowler, B.S. Mun, G.F. Wang, P.N. Ross, C.A. Lucas, N.M. Markovic, Science 315 (2007) 493.

[2] J. Greeley, I.E.L. Stephens, A.S. Bondarenko, T.P. Johansson, H.A. Hansen, T.F. Jaramillo, J. Rossmeisl, I. Chorkendorff, J.K. Nørskov, Nat. Chem. 1 (2009) 552.

[3] P. Strasser, S. Koh, T. Anniyev, J. Greeley, K. More, C. Yu, Z. Liu, S. Kaya, D. Nordlund, H. Ogasawara, M.F. Toney, A. Nilsson, Nat Chem 2 (2010) 454.

[4] M. Escudero-Escribano, P. Malacrida, M.H. Hansen, U.G. Vej-Hansen, A. Velázquez-Palenzuela, V. Tripkovic, J. Schiøtz, J. Rossmeisl, I.E.L. Stephens, I. Chorkendorff, Science 352 (2016) 73.

[5] N. Todoroki, H. Watanabe, T. Kondo, S. Kaneko, T. Wadayama, Electrochim. Acta 222 (2016) 1616.

[6] M. Asano, R. Kawamura, R. Sasakawa, N. Todoroki, T. Wadayama, ACS Catal. 6 (2016) 5285.

[7] Z. Zeng, J. Greeley, Nano Energy 29 (2016) 369.

490188 Computational Insights into Thermal and Photoresponse of Pd/C and Pd/N Co-Doped TiO_{2}Â using DFT

Abstract:

TiO_2 has attracted increasing attention during the past few decades as a promising catalytic and photocatalytic material due to its unique physicochemical properties. Specifically, in heterogeneous catalysis, it has been used as a catalytic material for treatment of environmental pollutants (greenhouse gases), photocatalytic reduction of CO_2 with water vapour and photodegradation of dyes. In recent years, doping of transition metal and nonmetal ions in TiO_2 has been reported to enhance the catalytic and photocatalytic performance of TiO_2. The main objective of this work is to study the effect of co-doping on the structure-property relationships of anatase TiO_2 using density functional theory (DFT+U) approach. Palladium ion and carbon (C)/Nitrogen (N) were co-doped in TiO_2 bulk and surface (001) and (100) planes in order to investigate the thermal and photoresponse of TiO_2. Differential bond length formations and associated net oxygen activations (NOA) were determined for the bulk and surface exposed Pd/C and Pd/N doped TiO_2. Effects of anion vacancies on the bond length distributions and net oxygen activations were found for the co-doped structures. NOA was found to be higher for Pd/N co-doped catalysts with 18.8 % and 28% in case of bulk and surface exposed codoped TiO_2 when compared to 17 % and 25 % in case of bulk and surface exposed Pd/C co-doped TiO_2. Band structure analyses were performed in order to examine the photoresponse of co-doped TiO_2. Band gaps obtained from the band structure analyses were in a range of 1.9-3 eV for Pd/C-doped TiO_2 whereas 2.7-3.1 eV for Pd/N-doped TiO_2, respectively, thereby establishing the importance of codoping for thermal and photocatalytic behaviour of TiO_2-based materials.

491746 Transition Metal-Doped C_2N as Active Catalysts for the Oxygen Reduction Reaction

Abstract:

Meeting ever-increasing worldwide energy demands in a sustainable way is one of the most pressing challenges of our times. The oxygen reduction reaction (ORR) plays a critical role in pollution-free energy conversion technologies, such as hydrogen fuel cells and metal-air batteries. Platinum (Pt) is one of the best known catalysts for ORR, but the high cost of this precious metal and the sluggish ORR kinetics inhibit the widespread growth and implementation of fuel cell technologies. [1,2]

Two-dimensional (2D) transition metal-doped carbon nitride catalysts have emerged as abundant and low-cost alternatives to Pt for ORR. [3] A more recent member of this family, C_2N, [4] exhibits several promising features for ORR, including a conductive framework and the potential to create and tune active sites through controlled transition metal doping. [5]

In this work, we use density functional theory (DFT) to investigate the ORR mechanism for C_2N catalysts doped with various transition metals. We find that the choice of metal dopant significantly affects the coverages of the ORR intermediates (*OOH, *O, *OH) on this material. These coverages directly influence the predicted ORR mechanism and overpotentials, which determine catalytic activity. By including explicit solvent layers in our predictions, we also investigate solvation effects on the binding energies. By studying the ORR mechanism and activity of various metal-doped C_2N catalysts, we identify critical material properties that influence the performance of single site 2-D catalysts for electrochemical reactions.

References:

[1] Xia, W., Mahmood, A., Liang, Z., Zou, R., & Guo, S. (2016). Earthâ€Abundant Nanomaterials for Oxygen Reduction.Â *Angewandte Chemie International Edition*,Â *55*(8), 2650-2676.

[2] United States Department of Energy. (2016). Comparison of Fuel Cell Technologies. https://energy.gov/eere/fuelcells/comparison-fuel-cell-technologies.

[3] Zheng, Y. et al. (2017). Molecule-Level g-C_3N_4 Coordinated Transition Metals as a New Class of Electrocatalysts for Oxygen Electrode Reactions. *Journal of the American Chemical Society*, 139(9), 3336-3339.

[4] Mahmood, J., et. al. (2015). Nitrogenated holey two-dimensional structures. *Nature communications*, *6*, 6486.

[5] Li, X., Zhong, W., Cui, P., Li, J., & Jiang, J. (2016). Design of Efficient Catalysts with Double Transition Metal Atoms on C_2N Layer. *The Journal of Physical Chemistry Letters*, *7*(9), 1750-1755.

496687 Electrochemical Synthesis of Hydrogen Peroxide Via Water Electrolysis: Challenges and Opportunities

Abstract:

Electrochemical production of hydrogen peroxide (H_2O_2) from water oxidation could provide a way to produce a chemically valuable product from an earth abundant material. Today there are only a few experimental studies that produce measurable amount of hydrogen peroxide through electrochemical oxidation of water.[1,2] A major challenge is that the electrocatalyst must suppress the thermodynamically favored four-electron pathway (O_2 evolution) and direct selectivity towards two-electron path (H_2O_2 evolution).[3] We have recently developed a complete understanding of water oxidation incorporating the formation of hydroxyl radical, a one-electron water oxidation reaction.[4] Based on thermodynamic analysis, we demonstrated leads to OH radical, H_2O_2 and O_2 evolution and rationalized product selectivity in several different experimental reports.

In this work, by combining the above knowledge and density functional theory (DFT) calculations we search for other materials to identify suitable electrocatalyst with high selectivity towards two-electron path (H_2O_2 evolution).

References

(1) Fuku, K.; Sayama, K.; Fujishima, A.; Honda, K.; Maeda, K. *Chem. Commun.* **2016**, *52* (31), 5406–5409.

(2) Izgorodin, A.; Izgorodina, E.; MacFarlane, D. R. *Energy Environ. Sci.* **2012**, *5* (11), 9496.

(3) Viswanathan, V.; Hansen, H. A.; Nørskov, J. K. *J. Phys. Chem. Lett.* **2015**, *6* (21), 4224–4228.

(4) Siahrostami, S.; Li, G.-L.; Viswanathan, V.; Nørskov, J. K. *J. Phys. Chem. Lett.* **2017**, 1157–1160.

497976 Modeling Electrochemical Reactions:Â DFT-Based Models Including Explicit Solvation, Electrolyte, and Electrochemical Potential

Abstract:

Modeling Electrochemical Reactions: DFT-Based Models Including

Explicit Solvation, Electrolyte, and Electrochemical Potential

Jason D. Goodpaster*

Department of Chemistry, University of Minnesota Twin Cities, Minneapolis, Minnesota 55455, USA

**Corresponding author: jgoodpas@umn.edu*

I report an electrochemical model based on periodic Kohn-Sham Density Functional Theory (DFT) that includes the effects of the electrochemical potential, solvent, and electrolyte. In this model, the electrode surface and all adsorbates are treated using DFT. In the region close to the surface, the solvent is treated explicitly, whereas in the region far from the surface the solvent is treated using a continuum dielectric. The electrolyte ions are implicitly represented by a linearized Poisson-Boltzmann model.[1,2] The electrode potential was calculated from the Fermi energy. The Fermi energy can be varied by changing the number of electrons in the simulation cell; therefore, the electrochemical potential was set by calculating the number of electrons the simulation cell required for a specific Fermi energy.[3] Using this model, free energy profiles for a reaction can be calculated as a function of the applied potential. The free energy of activation for each elementary step can then used to determine the rate coefficient for that step and combined with microkinetic models to determine the rate of product formation.

I applied this model to investigate the mechanistic pathways by which CO_2 on copper and silver surfaces are electrochemically reduced.[3,4] I compare this model to vacuum DFT, implicit solvation DFT, explicit solvation DFT, and electrochemical potential DFT models. Our results demonstrated that including electrochemical potential is critical to accurate calculations of free energy barriers and kinetics for electrochemical reactions. Additionally, we show that for some reactions explicit solvation plays an important role in solvating adsorbed intermediates. Finally, I discuss progress in using these free energies for microkinetic modeling.

References:

1. K. Letchworth-Weaver and T. A. Arias Phys. Rev. B 86, 075140 (2012).
2. K. Mathew, R. Sundararaman, K. Letchworth-Weaver, T. A. Arias, and R. G. Hennig. J. Chem. Phys. 140, 084106 (2014).
3. J. D. Goodpaster, Alexis T. Bell, and M. Head-Gordon J. Phys. Chem. Lett. 7, 1471 (2016).

4. J. D. Goodpaster, Alexis T. Bell, and M. Head-Gordon *in prep.*

491234 Efficiently Determining Solvent Effects at Metal-Water Interfaces

Abstract:

Accurately modeling interfaces between liquid and solid phase is one important frontier in computational catalysis. Explicitly modeling these interfaces can be challenging due to the large number of molecules needed, as well as the large number of possible configurations. Implicit or continuum solvent models can be much quicker, but have only recently been included in plane wave density functional theory (DFT) codes. In contrast, implicit solvation methods are much more mature in molecular DFT codes. How implicit solvation methods perform in modeling periodic catalytic systems is still unclear. In this work we systematically evaluated how interfaces between water can metals can be treated using implicit solvation methods.

In total we considered 41 different adsorbates and used three different solvation methods implemented in the NWChem, VASP, and JDFTx codes. Agreement between the three programs was strong. We classified the adsorbates into five different groups, and found that trends within each group were consistent. For instance, adsorbed atomic or weakly polar species showed almost negligible solvation effects, while adsorbed aromatic compounds were strongly stabilized by the presence of water solvent (up to 0.44 eV). We also modeled four prototypical reactions in vacuum and water: oxygen reduction reaction, formic acid oxidation, C-C cleavage, and water-gas shift. Reaction energies changed up to 0.23 eV when solvent was present. We analyzed the contributing energy terms to the solvation energy of adsorbed species, and found that relevant desciptors (e.g. Bader charge of adsorbates, dipole moments) are correlated to these solvation energy terms. We furthermore created several artificial neural networks, where we evaluated potential descriptors for predicting solvation energies of adsorbates. Our best neural network indicated that Bader charge of adsorbed species, solvation energies of gas species, gas dipole moments, and molecular surface area of adsorbates are all simple descriptors that can be used to predict solvation energies of adsorbates. Our work provides guidelines on when solvation effects may be important for surface chemistry, and also provides valuable insights on modeling such effects in an efficient manner using implicit solvation.

498550 Machine-Learning Model Development for Electrocatalyst Discovery

Abstract:

Machine-Learning Model Development for Electrocatalyst Discovery

Zheng Li, and Hongliang Xin*

*hxin@vt.edu

Session: Catalysis and Reaction Engineering Division- Electrocatalysis and Photoelectrocatalysis

Transition metal alloys, metal oxides and molecular complexes have shown great promise for catalyzing many chemical reactions. For a given type of catalyst, its surface reactivity can be tailored by varying geometric characteristics, metal ligands, organic functional groups and extrinsic factors (e.g., solvation, support) in the vicinity of active sites [1]. With hierarchical complexities in catalyst design, a priori estimation of chemical reactivity of surface atoms is attractive. However, it is very time-consuming and costly, if at all possible, to search for highly optimized catalyst structures by the empirical testing and/or quantum-chemical calculations. Therefore, we tackle this problem by developing the predictive, data-driven machine-learning-augmented models (e.g., artificial neural network, kernel ridge regression) that can unravel the complex, non-linear interactions between the electronic structure properties with reactivity descriptors, thus enabling fast and accurate predictions of surface reactivity of a broad range of material space without resorting to the expensive quantum-chemical calculations.

Since the efficiency and accuracy of machine-learning models are strongly dependent on the input features, we develop various types of data representation of catalyst surfaces. For metal alloys and metal oxides, we engineer numerical representation of surface metal atoms by using geometry and electronic structure based primary features [2-4] including the local electronegativity and the d-band features that are dependent on the surroundings of an adsorption site, together with the intrinsic properties of active metal atoms including the electronegativity, ionic potential, electron affinity, etc. For the description of molecular complexes, we apply the Coulomb matrix [5], which is a simple matrix form crystal structure representation based on the molecular complex structure coordination and atoms physical properties. Another type of representation method for molecular complexes is based on the similarity indexing of molecular complexes with the associating functional groups and structure components. The machine-learning-augmented models exhibit outstanding prediction power in screening for the bimetallic, metal oxide and molecular complexes surfaces with desired adsorption properties. Compared with the traditional high-throughput computational and experimental trial-and-error approach, the machine-learning models have great potential in accelerating the discovery of new catalytic materials with desired properties. Sensitivity analysis of the machine-learning models can shed light on the nature of chemical bonding and its governing factors, providing chemical insights into catalytic processes and fundamental knowledge base of molecule-catalyst interactions.

[1] Vojvodic, Aleksandra, and Jens K. Nørskov. 2015. "New Design Paradigm for Heterogeneous Catalysts." *National Science Review* 2 (2): 140–49.

[2] Z. Li, X. Ma, and H. Xin, Feature engineering of machine-learning chemisorption models for catalyst design, Catalysis Today. 280, Part 2 (2017) 232–238.

[3] X. Ma, Z. Li, L.E.K. Achenie, and H. Xin, Machine-Learning-Augmented Chemisorption Model for CO2 Electroreduction Catalyst Screening, J. Phys. Chem. Lett. 6 (2015) 3528–3533.

[4] Z. Li, S. Wang, W. Chin, and H. Xin, Rapid Screening of Bifunctional Alloy Catalysts Enabled by Machine Learning, Submitted to Journal of Materials Chemistry A. (2017).

[5] M. Rupp, A. Tkatchenko, K.-R. Müller, and O.A. von Lilienfeld, Fast and Accurate Modeling of Molecular Atomization Energies with Machine Learning, Phys. Rev. Lett. 108 (2012) 58301.

499574 Climbing the Volcano: Active-Site Engineering at the Atomic Scale

Abstract:

In recent years, molecular simulations of catalyst surfaces have yielded considerable insights into how the electronic structure of catalyst materials influences activity. One of the most notable examples is that of the scaling relations, or correlations between adsorption energies of molecules onto surfaces. Using these correlations, models for turnover rates, coverages, and theoretical overpotentials (in the case of electrocatalysis) may be constructed, often yielding "volcano" relationships which may define fundamental limits on the ideal performance of catalysts. Some of the clearest examples of these limits are present in electrocatalysis of earth-ambient molecules into fuels and chemicals, which include reactions like water oxidation and CO2 electroreduction, in which even the most efficient catalysts require significant overpotentials to achieve moderate turnover rates.

In this work, I identify materials for which assumptions of scaling-relation-based models might be strategically broken in order to circumvent fundamental limits on activity. These include bifunctional materials, in which adsorbates interact with multiple sites on a surface and may be selectivity stabilized by distinct moities in alloys and other bulk binary compounds. Electrolytes, interfaces, ligand-functionalized surfaces and confined nanostructures may also achieve a similar selective stabilization or destabilization, being perturbed off of the traditional scaling relation in order to achieve higher activity. Lastly, I describe a high-throughput methodology developed using software infrastructure from the Materials Project which may aid in the more rapid identification of materials which meet these criteria.

496619 Developing Ab Initio Methodology for Advancing Catalytic Reactions

Abstract:

Ab initio investigations have advanced the molecular-scale understanding of catalytic reactions. Current methods allow for predictions of catalytic reaction behavior, although reliability of calculations is a barrier to widespread adoption. The growing string method automates molecular-scale reaction path discovery and is a rapid, unbiased approach. It is enhanced for electrochemical systems which experience electrical potential bias. Although calculations without potential bias treatment are capable to make meaningful predictions for electrochemical systems, a more reliable approach considers the charge of the electrocatalyst surface and solvent effects on adsorbates. This improved growing string method is demonstrated on copper and silver for elementary reactions of electrochemical carbon dioxide reduction. In a second example of ab initio method strengthening, a study of platinum nanoparticles supported by titanium-dioxide catalyzing the water-gas shift reaction is presented. This study incorporates uncertainty quantification and the Bayesian inference from experiments to complement calculations. A microkinetic model connects the atomic-scale calculations with the reactor-scale kinetic experiments. Through the Bayesian inference process, the uncertainties of the ab initio calculations are refined, and a conclusion about the catalytically active site is achieved.

493253 First-Principles-Derived Structure-Energy Relationship for Surface Oxides

Abstract:

Metal-supported, 2-dimensional surface oxide monolayers represent an emerging class of catalysts for a variety of applications, ranging from biomass conversion to electrocatalysis. A tremendous challenge with modeling such materials is their atomic structure, which is typically unknown. Commonly employed structure prediction methods (simulated annealing, evolutionary algorithms) require accurate force fields, which are either expensive (density functional theory), or too complex and have transferability issues (ReaxFF).

Herein, we show that the systematic reduction of Kohn-Sham equations via the Harris energy functional leads to a simple structure-energy relationship involving only interatomic distances and coordination numbers for systems, in which angular dependence of energy is not important. The model reproduces the functional dependence of the metal cohesive energy on geometric characteristics, and its connection to molecular orbital theory allows generalizations to surface oxides. We assess the model performance for Co_xO_y/Pt(111) monolayer structures and investigate the transferability of its parameters.

491413 First-Principles Based Design of Reaction Conditions for the Catalytic Conversion of Methane to Methanol over Cu-Exchanged SSZ-13

Abstract:

Due to its perfect symmetry, methane is the most stable hydrocarbon and its selective conversion to methanol is one of the long-standing challenges in catalysis. Cu containing zeolites are able to catalyze this reaction under relatively mild conditions but a variety of $Cu_xO_yH_z$ clusters have been suggested as active sites. Results in the literature indicate that the type of active site does not only vary with the zeolite framework, but different types of active sites have been observed for the same zeolite structure prepared by different groups or under varying conditions and clear rules for the formation of these sites have not been established so far.

In this contribution, we construct a thermodynamic model for Cu in zeolite SSZ-13 based on RPA Density Functional Theory, and show that under almost all conditions the formation of a distribution of $Cu_xO_yH_z$ clusters is favorable. We identify the chemical potential of O and H (Î¼^O and Î¼^H), the local Al distribution, the reference state and the temperature as important parameters in predicting the exact stoichiometry of these sites. We furthermore model the $Cu_xO_yH_z$ distribution of several systems studied in the literature; comparing these results to experimental studies for the catalytic conversion of methane to methanol significantly narrows down candidates for the most active sites. In a subsequent step we then use our model to predict conditions, which lead to equally as good or better catalytic performance.

The systematic work presented here shows that the Cu speciation in zeolites is condition-dependent and that there is not one single Cu species present. Furthermore, we establish basic rules for choosing the proper conditions to optimize the performance of this catalytic system.

498697 Reaction Ensemble Monte Carlo Simulations of Xylene Isomerization Under Confinement

Abstract:

Xylene exists as three isomers: meta- (m-), ortho- (o-), and para- (p-). m-Xylene is the most abundant isomer in bulk fluids at equilibrium, but p-xylene is a valuable feedstock in the production of terephthalic acid. p-Xylene can be selectively produced using ZSM-5, a zeolite with narrow channels that prevent m- and o-xylene from escaping interior pores. Using reaction ensemble Monte Carlo (RxMC), we demonstrate that confining a xylene fluid to the interior of a carbon nanotube shifts the equilibrium composition as a function of nanotube diameter.

We adapt the original RxMC reaction move to align both the position and orientation of inserted product molecules and deleted reactant molecules. The accuracy and efficiency of this move is demonstrated for xylene isomerization in vapor, liquid, and supercritical phases. Classical RxMC requires the ideal gas free energy of reaction as an input. We compare three methods for computing this free energy: using tabulated enthalpies and entropies of formation, using the harmonic oscillator and rigid rotor approximation, and using QM/MM alchemical transformation combined with the multistate Bennett acceptance ratio (MBAR). We find that the tabulated free energies of reaction give the best agreement with experimental equilibrium compositions in bulk fluids. RxMC simulations in a carbon nanotube with an

consistent with the production of p-xylene in ZSM-5. We also show that o-xylene becomes the dominant isomer in

nanotubes of this diameter, effectively allowing o-xylene to fit into cavities inaccessible to the other isomers.

501357 Understanding Separation and Catalysis in Nanoporous Materials

Abstract:

The sustainable production of energy and chemicals requires efficient separation and catalytic systems, and an improved understanding of the principles driving the various physical and chemical processes is clearly critical to our success in this endeavor. This talk will highlight applications using crystalline nanoporous materials, primarily zeolites, and show how molecular modeling can be used both to gain a microscopic understanding into phenomena difficult to probe experimentally and to predict accurate data at a scale beyond the reach of the traditional trial-and-error approach. Examples discussed include the aqueous-phase adsorption of oxygenated compounds, the unique adsorption/diffusion behaviors in hierarchical materials and ultrathin membranes, and the molecular structure and framework confinement effects on hydrocarbon cracking. New insights and methodological developments from these studies provide the basis for an effort in which a large number of materials were screened with the purpose of finding better candidates for two energy-related applications, ethanol/water separation and hydrocarbon iso-dewaxing, and some of our predictions were subsequently verified by experiments. Together they demonstrate the possibility of using theory and computation in a predictive mode to expedite the pace of materials discovery for complicated, realistic, and industrially relevant applications.

493009 Modeling Self-Assembly of Metal-Organic Frameworks with Enhanced Sampling Techniques

Abstract:

Self-assembly is a process by which building blocks form functional materials. In principle, instructions can be imparted onto the building blocks to engineer and design the final structure. One class of materials which takes advantage of this building block approach is metal-organic frameworks (MOFs), nanoporous crystalline materials formed in solution through coordination-driven self-assembly of inorganic nodes and organic linkers. Experimental observations of MOF self-assembly reveal complex and interesting phenomena where synthetic conditions (solvent, temperature, concentration of building blocks) alter the final MOF structure as well as intermediate metastable states. However, computational efforts to elucidate MOF self-assembly are scarce. For this study, we focus on the self-assembly of MOF-5.

We employ a combination of enhanced sampling techniques to study MOF self-assembly for different scenarios and system sizes. We model MOF-5 with an atomistic model as either a discrete unit cell or four unit cells in explicit solvent. To validate the model, we perform replica exchange with solute scaling (REST2) simulations and calculate the free energy profile using the average distance and average angle between nodes as collective variables. The free energy minima observed in these systems are close to the values in the experimental unit cell, therefore validating that the chosen model represents at least a metastable state.

To model self-assembly, we perform finite temperature string (FTS) method calculations using the average distance and angle between nodes and the total coordination between nodes and linkers. The process was modeled for both single and four unit cells systems from two starting points: fully disassembled and amorphous. The self-assembly starting from the disassembled state was found to be downhill in free energy for both the single unit cell and four unit cells systems. In contrast, the self-assembly starting from the amorphous state reveals energetic barriers on the way to the final MOF structure. Finally, we discuss the mechanism of self-assembly of MOF-5 for the scenarios and system sizes considered.

492643 Computationally-Efficient High-Throughput Screening of Metal-Organic Frameworks for Hydrogen Storage

Abstract:

Metal-organic frameworks (MOFs) are porous materials consisting of metal nodes connected with organic linkers. These materials can have high surface area and porosity, which make them interesting candidates for applications in gas separation and storage. Since MOFs can be made from many different nodes, linkers, and functional groups, there are a huge number of possible MOF structures. High-throughput computational screening can efficiently characterize large numbers of materials to identify the best candidates for applications such as gas separation and storage. [1,2]

We have used high-throughput screening techniques paired with grand canonical Monte Carlo simulations to calculate the hydrogen storage capacity for tens of thousands of MOFs. We will discuss commonalities in the structural and textual properties among the top-performing candidates from this screening and further explore how the information gleaned from high-throughput screening can inform the design of new materials that can be synthesized for hydrogen storage applications.

Due to the significant computational cost of performing GCMC simulations for a large number of MOFs, we will introduce an efficient screening method that can estimate the storage capacity for thousands of structures in a small fraction of the time required for full GCMC simulations. This allows us to quickly identify the most promising candidates for more detailed study, and this technique can be applied to other gases, such as methane, as well as a wide range of temperatures and pressures.

1. Y.J. Colon, R.Q. Snurr, “High-throughput computational screening of metal-organic frameworks.” *Chem. Soc. Rev.* 43, 5735-5749 (2014)
2. N. S. Bobbitt, J. Chen, R. Q. Snurr, "High-Throughput Screening of Metal–Organic Frameworks for Hydrogen Storage at Cryogenic Temperature." *The Journal of Physical Chemistry C* 120(48), 27328-27341 (2016)

502903 Effects of Ion Self-Energy on the Double Layer Structure and Properties at the Dielectric Interface

Abstract:

A systematic, unified and predictive theoretical framework that is able to capture all the essential physics of the interfacial behaviors of ions, such as the Hofmeister series effect, Jones-Ray effect and the salt effect on the bubble coalescence remains an outstanding challenge. The most common approach is the Poisson-Boltzmann (PB) theory; however, there are many systems for which the PB theory fails to offer even a qualitative explanation, especially for ions distributed in the vicinity of an interface with dielectric contrast between the two medium. A key factor missing in the PB theory is the self-energy of the ion. We develop a self-consistent theory that treats the electrostatic self-energy (including both the short-range Born solvation energy and the long-range image charge interactions), the nonelectrostatic contribution of the self-energy, the ion-ion correlation and the screening effect systematically in a single framework. The theory gives a continuous self-energy across the interface, which allows ions on the water side and the vapor/oil side of the interface to be treated in a unified framework. Using the theory, we demonstrate three essential effects, the image charge effect, the inhomogeneous screening effect and the specific ion effect, for the ions near the dielectric interfaces. First, we show that the image charge repulsion creates a depletion boundary layer that cannot be captured by a regular perturbation approach. The image force qualitatively alters the double layer structure and properties, and gives rise to many non-PB effects, such as like-charge attraction and charge inversion. Then, we show that the double layer structure and interfacial properties is drastically affected by the inhomogeneous screening due to the nonuniform ion distribution particularly at high salt concentrations when the bulk Debye screening length is comparable to the Bjerrum length. The characteristic length of the depletion layer scales with the Bjerrum length instead of the Debye screening length, resulting in a linear increase of the negative adsorption of ions with concentration, in agreement with experiments. Finally, we study the self-energy of a single ion across the dielectric interface. Using intrinsic parameters of the ions, such as the valency, radius, and polarizability, we predict the specific ion effect on the interfacial affinity of halogen anions at the water/air interface, and the strong adsorption of hydrophobic ions at the water/oil interface, in agreement with experiments and atomistic simulations.

Discovery of the Universal Van der Waals Generic Cubic Equation of State for the Continuity of Gaseous and Liquid States of Fluids and Mixtures

Akanni S. Lawal
Petroleum Engineering Department
Texas Tech University, P. O. Box 43111
Lubbock, Texas 79423

Take Away From This Presentation

- **VdW Theory is Correct BUT Chosen Parameters of EOS are the Culprits**
- Critical Indexes & Cross-Over Theories do NOT lead to Critical Properties
- **Human Error is due to:**
  - **Inaccurate Saturated Liquid Density [Discriminant $\neq f(Z_c)$]**
  - **Inaccurate HTHP Phase Equilibria [Discriminant $\neq f(x)$]**
  - **Inaccurate HTHP Densities & Virial Coeff. [$b \neq b(T)$]**
  - **Inaccurate Reduced VdW Corresponding States [$f(P_R, V_R, T_R) \neq \varphi(Z_c, \Omega_w, \Omega_b, \Omega_a)$]**
  - **Inaccurate HTHP Thermodynamic Properties [$b \neq b(T)$]**
  - **Inaccurate Critical Properties [$\lim_{\substack{T\to T_c \\ P\to P_c}} v \to v_c$ & $\lim_{\substack{T\to T_c \\ P\to P_c}} Z \to Z_c$]**

2017 AIChE Annual Meeting, Minneapolis

2

Discovery of the Universal VdW Generic Cubic Equation of State – Terminology

- Discovery: **Means Looking at the Same Problem Like Everybody Else BUT Thinking Differently**
- Universal VdW Equation Means: **Using the** Same Set **of the VdW Molecular Properties (Zc, Ω_w, Ω_a, Ω_b) to Predict**
  - *Volumetric Properties, Virial Coefficients, Compressibility Factor, Vapor Pressure, Phase Densities, Phase Boundaries, Phase Equilibria, Critical Properties, Gas-Liquid Critical Points for Thermodynamic & Transport Properties*
  - ***Thermodynamic Properties (Isothermal Compressibility, Volume Expansivity, Enthalpy, Entropy, Internal Energy, Isobaric and Isochoric Heat Capacities, Speed of Sound, Joule-Thomson Coefficient, Gibbs Free Energy and Helmholtz Energy)***
  - *Transport Properties (Dynamic Viscosity (η), Kinematic Viscosity (η/ρ), Thermal Conductivity (λ), Thermal Diffusivity ($\lambda/\rho C_p$), Prandtl Number ($\eta C_p/\lambda$), Eucken Factor ($\lambda/\eta C_v$) and Diffusivity or Diffusion Coefficient)*

Discovery of the Universality of the Van der Waals Generic Cubic State Equation

VdW PVT Cubic Equation:

$$\mathbf{P} = \frac{\mathbf{RT}}{v - \mathbf{b}} - \frac{\mathbf{a(T_c)\,\gamma(T_R, \omega)}}{v^2 + \alpha \mathbf{b} v - \beta \mathbf{b}^2}$$

Thermodynamic Property Cubic Equation:

$$\mathrm{P} = \frac{\left(\frac{P_c \phi_c}{T_c Z_c}\right) T}{\phi - b} - \frac{a(T_c)\,\gamma(T_R, \omega)}{\phi^2 + \alpha b \phi - \beta b^2}; \quad \phi \equiv \alpha_T, k_p, IFT, H, S, U, G, C_p, C_v, \mu_{JT}, u_s$$

Transport Property Cubic Equation:

$$\mathrm{T} = \frac{\left(\frac{T_c \phi_c}{P_c Z_c}\right) P}{\phi - b(T)} - \frac{a(T_c)\,\gamma(P_R, \omega)}{\phi^2 + \alpha b(T) \phi - \beta [b(T)]^2}$$

$$\boldsymbol{\phi} \equiv \mathbf{D}_e, \boldsymbol{\eta}, \left(\frac{\boldsymbol{\eta}}{\boldsymbol{\rho}}\right), \boldsymbol{\lambda}, \left(\frac{\boldsymbol{\lambda}}{\boldsymbol{\rho} \mathbf{C_p}}\right), \left(\frac{\boldsymbol{\eta} \mathbf{C_p}}{\boldsymbol{\lambda}}\right), \left(\frac{\boldsymbol{\lambda}}{\boldsymbol{\eta} \mathbf{C_v}}\right)$$

The Determination of Gas and Liquid Critical Properties for Thermodynamic and Thermophysical Properties

- Apply the Standard States of Thermodynamic Properties to Establish their Critical Properties
- In the Absence of the Tabulated Values of Standard States for Thermodynamic Properties, Use Methods of Statistical Mechanics (see Lemmon et al. 2015)

$$\left(\frac{P\phi^o}{ZT}\right)_{\text{std state}} = \left(\frac{P_c\phi_c}{Z_cT_c}\right)_{\text{crt.pt.}}$$

$$P = 1\text{bar};\ T = 298\text{ K};\ Z = 1$$

$$\phi^o = H^o, S^o, U^o, G^o, C_p^o, C_v^o, u_s^o, \text{etc.}$$

Van der Waals 1910 Nobel Lecture

$$\left(P + \frac{a}{v^2}\right)(v - b) = RT$$

- Van der Waals (Theory) Property: $Z_c = \frac{3}{8};\ \Omega_w = \frac{1}{3};\ \Omega_b = \frac{1}{8};\ \Omega_a = \frac{27}{64}$
- Van der Waals Stipulation: $\Omega_b = \Omega_w Z_c;\ \ \Omega_a = [1 + (\Omega_w - 1)Z_c]^3$
- "...the weak point of my theory is" $v_c = 3b$ or $\Omega_w \equiv \frac{b}{v_c} = \frac{1}{3};\ \langle \sim \pi! \rangle$
- "...to determine the relation between *p, v* and *T* for a substance"
- "As regards... the mutual attraction of the molecules... the last word has not yet been spoken;" "Temperature Functions Restricted to Parameters (*a, b*)"
- "...I was quite convinced of the real existence of molecules,...I never regarded them as a figment of my imagination..."

Van der Waals Rejected non-Molecular Inspired Parameters in the 1873 Theory

Van der Waals (*1873 PhD Thesis*)

$$\left(P + \frac{a}{v^2}\right)(v - b) = \mathrm{RT}$$

Clausius (*Ann. Phys.* v9, 1880)

$$\left(P + \frac{a}{T(v + c)^2}\right)(v - b) = \mathrm{RT}$$

Berthelot (*Comptes Rendus*, v130, 1900)

$$\left(P + \frac{a}{v^2 + nbv + mb^2}\right)(v - b) = \mathrm{RT}$$

Lawal (*I &E Chem. Res.*, v26, 1987)

$$\left(P + \frac{a}{v^2 + \alpha bv + \beta b^2}\right)(v - b) = \mathrm{RT}$$

Johannes van der Waals (1837-1923) - LAST WORDS ON CUBIC EQUATIONS

Van der Waals (1910 Nobel Lecture)

"... **And here I have come to the weak point in the study of the equation of state.** I still wonder whether there is a better way. **In fact** this question continually obsesses me, **I can never free myself from it,** it is with me even in my dreams..."

Objective of this Presentation

-to Show a Better Way of Resolving the Weaknesses of the Van der Waals Cubic Equations of State by Reflecting Structure and Shape Parameters **in the VdW 1873 CEOS:** Using the Lawal-Lake-Silberberg (LLS) Cubic Equation of State

Van der Waals Cubic Equation by First Principles - Reconstruction

$$P = \frac{RT}{v-b} - \frac{a}{v^2}$$

$$P = \frac{RT}{v-b} - \frac{a}{(v+s_0 b)(v+s_1 b)} \quad \langle shift\ the\ Root\ from\ the\ pole \rangle$$

$$P = \frac{RT}{v-b} - \frac{a}{v^2 + (s_0+s_1)bv + (s_0 s_1)b^2} \quad \langle Quadratic\ Expansion \rangle$$

$$(\partial P/\partial v)_T = 0\ and\ \left(\partial^2 P/\partial v^2\right)_T = 0 \quad \langle Applying\ Critical\ Constraints \rangle$$

$$\phi(v_c, b, s_0, s_1) = 0\ or\ \phi\left(\frac{b}{v_c}, s_0, s_1\right) = 0 \quad \langle Asymptotic\ Critical\ Limit \rangle$$

$$P = \frac{RT}{v-b} - \frac{a}{v^2 + \alpha bv - \beta b^2} \quad \langle Applying\ Dimensional\ Similitude \rangle$$

$$\sqrt{\alpha^2 + 4\beta} \quad \langle Discriminant\ of\ the\ LLS\ EOS \rangle$$

Molecular-Inspired EOS Parameters Interpretation by Macroscopic Modeling

$$Z_c = f_z(\alpha, \beta);\quad \Omega_w = f_w(\alpha, \beta);\quad \Omega_b = f_b(\alpha, \beta);\quad \Omega_a = f_a(\alpha, \beta)$$

$$P_c = \frac{\Omega_b^2}{\Omega_a}\frac{a}{b^2} \equiv \phi(\alpha, \beta)\frac{a}{b^2};\ T_c = \frac{\Omega_b}{\Omega_a}\frac{a}{Rb} \equiv \varphi(\alpha, \beta)\frac{a}{Rb};\ V_c = \frac{Z_c}{\Omega_b} b \equiv \emptyset(\alpha, \beta) b$$

$$B_R = \Omega_b - \Omega_a \frac{1}{T_R};\quad C_R = \Omega_b^2;\quad D_R = \Omega_b^3 \quad \langle f(\alpha, \beta) \rangle$$

$$T_{Boyle} = T_c \frac{\Omega_a}{\Omega_b} \equiv T_c \varphi(\alpha, \beta)$$

Molecular Significance of Virial Coefficients & Macroscopic Properties

- **Molecular Explanation of Macroscopic Critical Properties** **(Leach-Leland, *AIChE J., 14*(4), 568,** 1968**; Leland, *AIChE J.*, *12*(6), 1227,** 1966); **(Leland-Mueller, *IEC*, *51*(4), 597,** 1959**; Leach et al., *AIChE J.*, *14*, 568,** 1968**; Fisher-Leland, *IECFund.*, *9*(4), 537,** 1970) **(Tahery, *JPC B*, *111*,** 2007)
  - Z_c, T_c, P_c ~ reflects *molecular structure* of Pure Substances
  - $v_c = \phi b$, *ϕ denotes shape factor* of Potential Function
  - *B, C, D, E ~ Two, Three, Four, Five Body Molecular Interaction*

- **Molecular Interpretation of 3rd Virial Coefficient** (C) **and Boyle Temperature (Chueh-Prausnitz, *AIChE J.*, *13*(5), 896,** 1967)
  - C ~ f_d(Polarizability, Molecular Size, Molecular Shape)
  - T_B ~ reflects *molecular structure of Individual* Pure Substances

EOS Molecular-Inspired Parameters Interpretation – Schematic Illustration

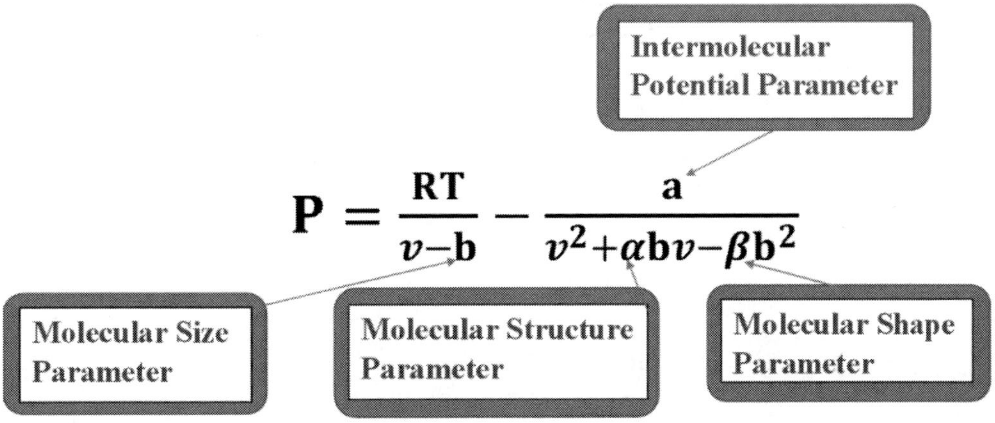

Discriminant of the Lawal-Lake-Silberberg Cubic Equation Changes with Substance Zc-Factor and Mixture Composition for the VLE

$$P = \frac{RT}{v - b} - \frac{a(T)}{v^2 + \alpha b v - \beta b^2}, \langle \text{discriminant} = \sqrt{\alpha^2 + 4\beta} \rangle$$

$$\ln\left(\frac{f}{P}\right) = (Z - 1) - \ln(Z - B) - \left\{\frac{A}{B(\alpha^2+4\beta)^{0.5}}\right\} \ln\left[\frac{Z+\left\{\alpha+(\alpha^2+4\beta)^{0.5}\right\}\frac{B}{2}}{Z+\left\{\alpha-(\alpha^2+4\beta)^{0.5}\right\}\frac{B}{2}}\right]$$

$$\ln\left(\frac{f_i}{x_i P}\right) = (Z - 1)\left\{3\left[\frac{b_i}{b_m}\right]^{1/3} - 2\right\} - \ln(Z - B) - F_a + F_{\alpha\beta} + F_\alpha + F_\beta + \ldots$$

$$F_a = \left\{\frac{A}{B(\alpha_m^2+4\beta_m)^{0.5}}\right\} \ln\left[\frac{Z+\left\{\alpha_m+(\alpha_m^2+4\beta_m)^{0.5}\right\}\frac{B}{2}}{Z+\left\{\alpha_m-(\alpha_m^2+4\beta_m)^{0.5}\right\}\frac{B}{2}}\right]\left\{\frac{2\sum_{k=1}^{n} x_k a_i^{0.5} a_k^{0.5} a_{ik}}{a_m} - 3\left[\frac{b_i}{b_m}\right]^{1/3} + 2\right\}$$

2017 AIChE Annual Meeting, Minneapolis 13

Peng-Robinson Discriminant Responsible for Inaccurate Liquid Density & Phase Equilibria

$$P = \frac{RT}{v - b} - \frac{a(T)}{v^2 + 2bv - b^2}, \langle \text{discriminant} = 2\sqrt{2} \rangle$$

$$\ln\frac{f}{P} = (Z - 1) - \ln(Z - B) - \frac{A}{2\sqrt{2}B}\left(\ln\left[\frac{Z + (2 + 2\sqrt{2})\frac{B}{2}}{Z - (2 - 2\sqrt{2})\frac{B}{2}}\right]\right)$$

$$\ln\frac{f_i}{x_i P} = \frac{b_i}{b_m}(Z - 1) - \ln(Z - B) - \frac{A}{2\sqrt{2}B}\left(\frac{2\sum_i^n x_i a_{ik}}{a_m} - \frac{b_i}{b_m}\right)\left(\ln\left[\frac{Z + 2.414B}{Z - 0.414B}\right]\right)$$

2017 AIChE Annual Meeting, Minneapolis 14

Effect of Pure Substance Zc-Factor on the Discriminant of CEOS

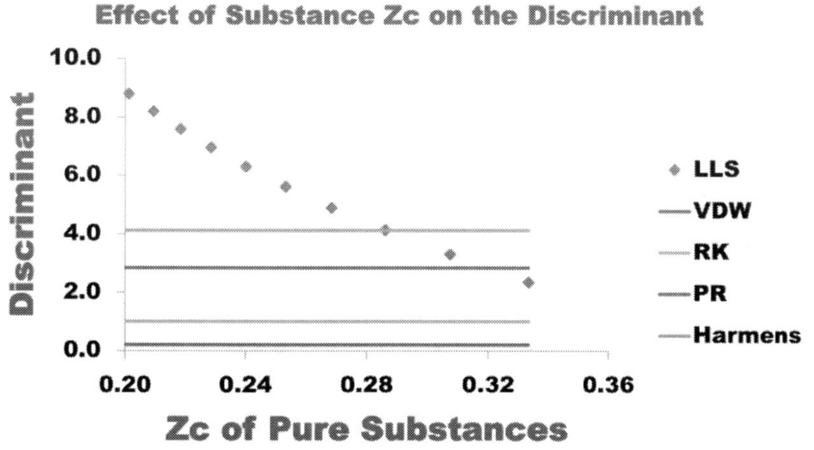

Effect of the Discriminant of Cubic EOS on Binary Mixture Composition

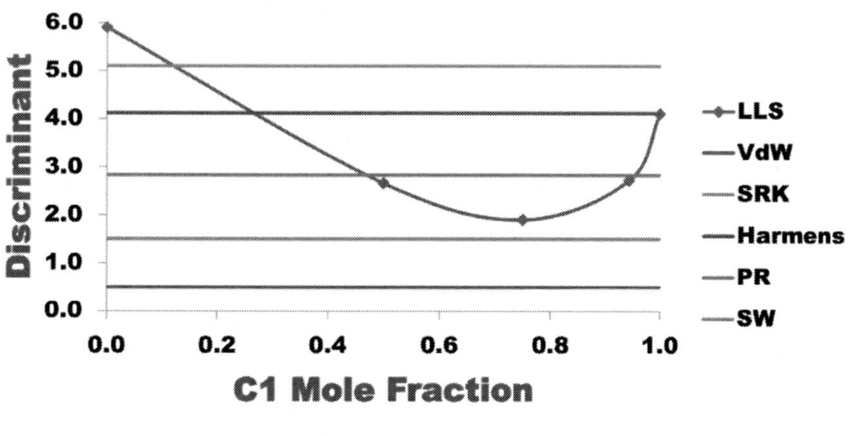

Prediction of Pure Substances Zc-Factor

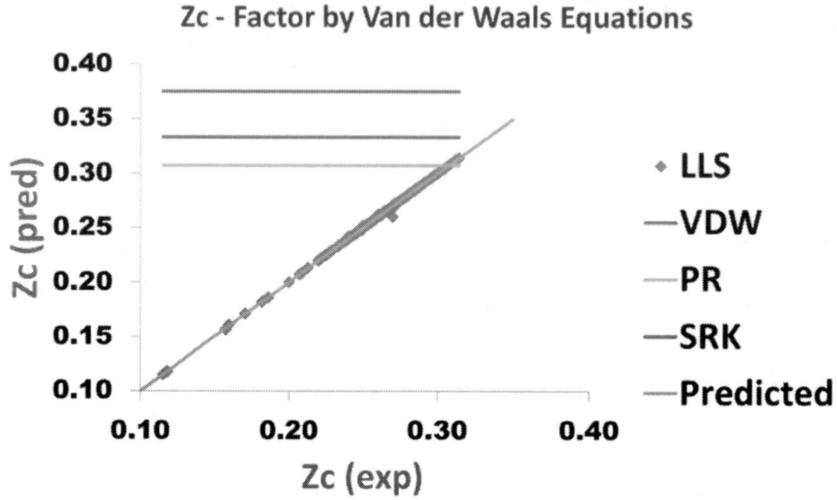

2017 AIChE Annual Meeting, Minneapolis

17

Prediction of Pure Substances Vc

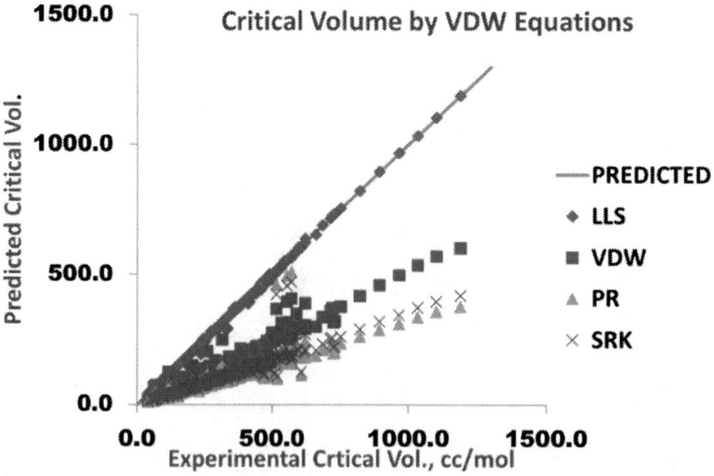

2017 AIChE Annual Meeting, Minneapolis

18

3rd Virial Coefficient LACK Structure and Shape Parameters in the VdW 1873 Cubic EOS

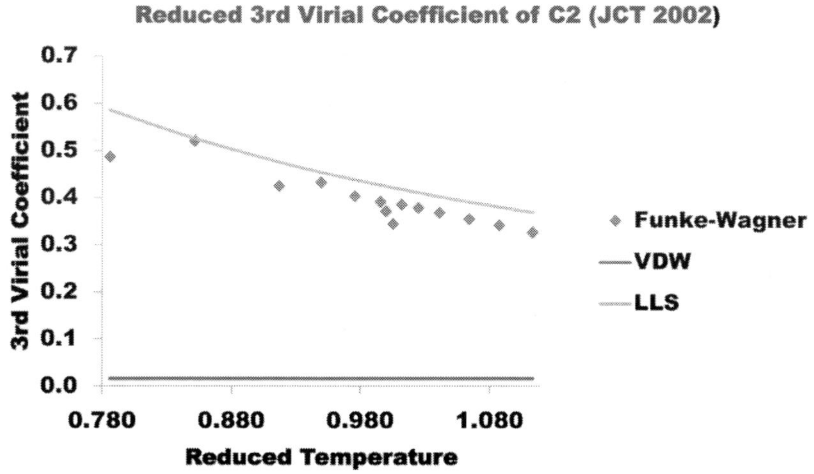

2017 AIChE Annual Meeting, Minneapolis

19

3rd Virial Coefficient LACK Structure and Shape Parameters in the VdW 1873 Cubic EOS

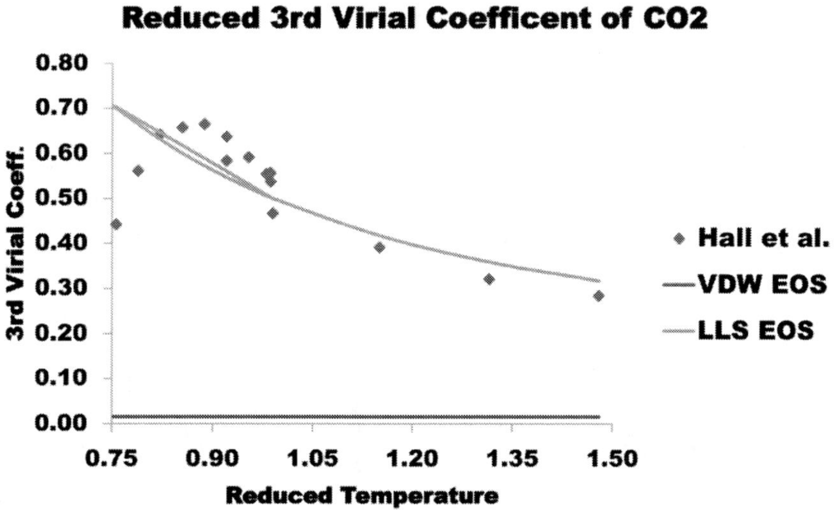

2017 AIChE Annual Meeting, Minneapolis

20

3rd Virial Coefficient LACK Structure and Shape Parameters in the VdW 1873 Cubic EOS

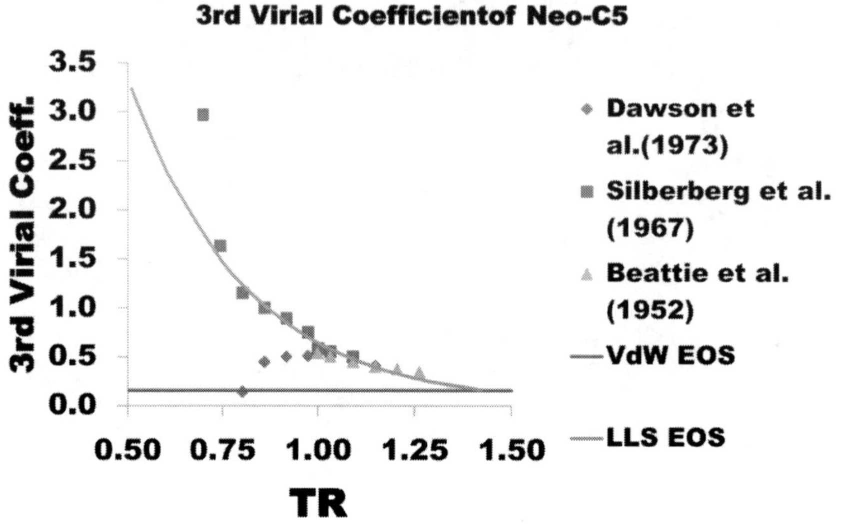

2017 AIChE Annual Meeting, Minneapolis

21

HTHP Liquid Volume of Pure Substances Suggests Covolume Temperature in the VdW Cubic Equations

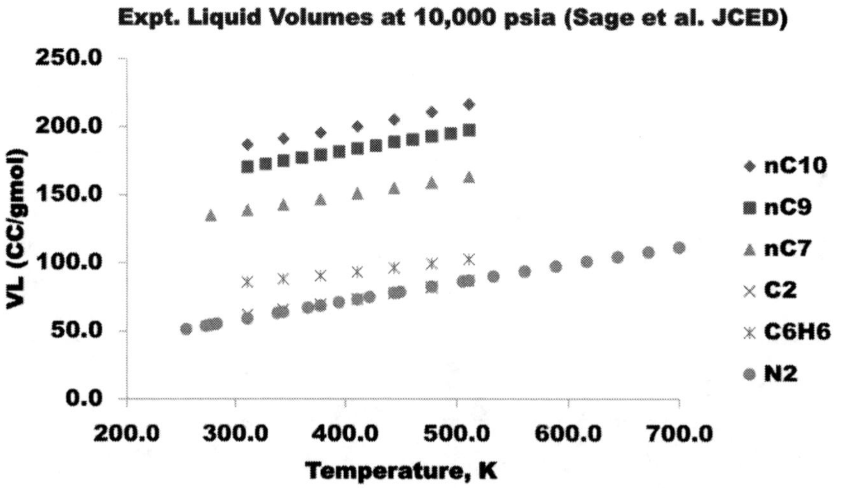

2017 AIChE Annual Meeting, Minneapolis

22

HTHP Liquid Volume of Pure Substances Suggests Covolume Temperature in the VdW Cubic Equations

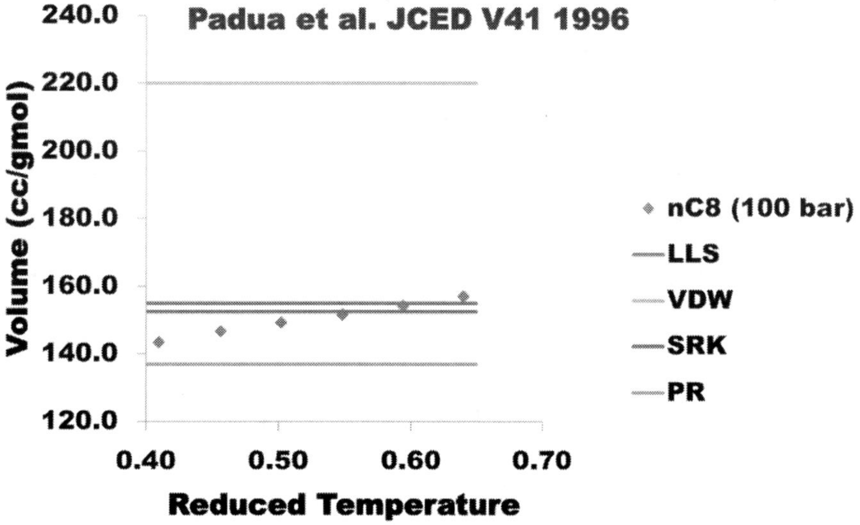

HTHP Liquid Volume of Pure Substances Suggests Covolume Temperature in the VdW Cubic Equations

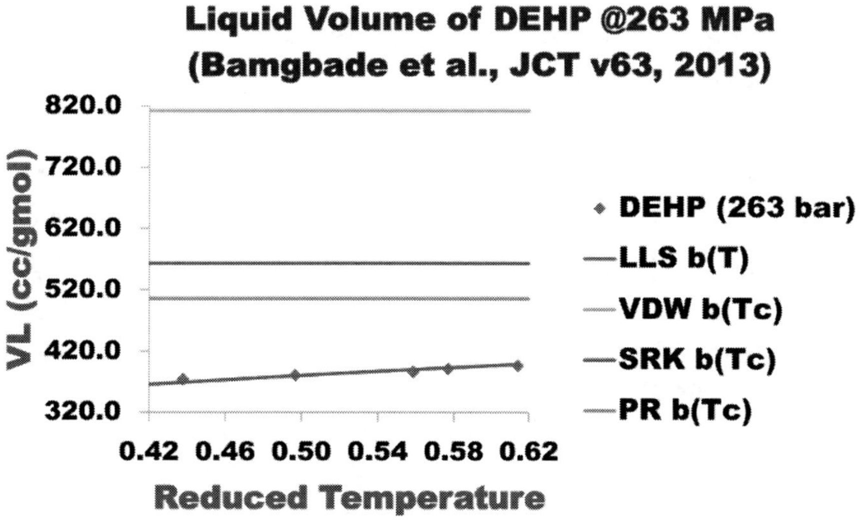

Results – Saturated Liquid Density

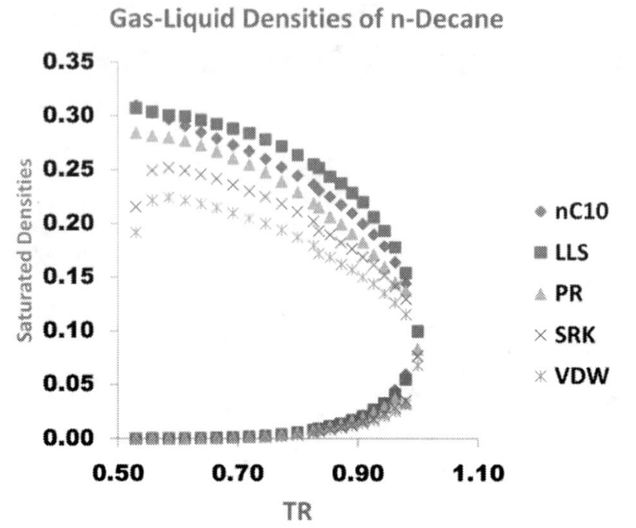

Results – Binary Critical Loci

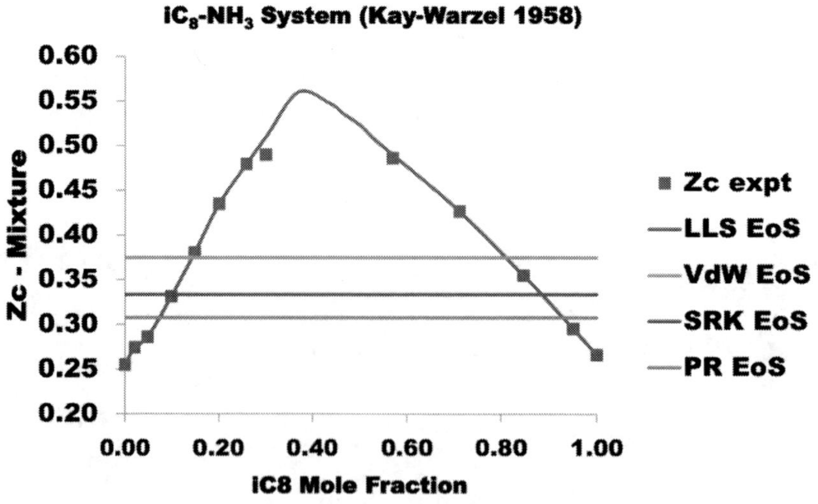

Conclusion and Recommendation

- VdW Theory is Satisfactory But the VdW 1873 Equation is Problematic
- Four-Parameter Cubic EOS is Required to Resolve the Entire PVT States, including Gas-Liquid Critical Point
- Thermodynamic and Transport Property Equations Established with the Defined VdW Gas Constant R^{vdw}
- Need to Generalize BIJ: a_{ij}, α_{ij}, β_{ij}, B_{ij}, C_{ij}, P_{cij}, T_{cij}, V_{cij}
- Need Comprehensive Tabulation of Gas-Liquid Critical Property for Thermodynamic Properties: α_T, k_P, H, S, U, G, Cp, Cv, μ_{JT} etc.
- Need Comprehensive Characterization Scheme for High-Molecular-Weight Fluids

2017 AIChE Annual Meeting, Minneapolis 27

492263 Vapor-Liquid Equilibrium Data for the Systems H_2S-MDEA-H_2O and CH_4-H_2S-MDEA-H_2O at High Solvent Concentrations and High Pressures

Abstract:

Vapor-Liquid Equilibrium Data for the Systems H_2S-MDEA-H_2O and CH_4-H_2S-MDEA-H_2O at High Solvent Concentrations and High Pressures

Eirini Skylogianni[1], Diego D. D. Pinto[1], Hanna K. Knuutila[1], Christophe Coquelet[2]

[1] Norwegian University of Science and Technology (NTNU), Trondheim, Norway

[2] Mines ParisTech, Fontainebleau, France

INTRODUCTION

Trouble-free operations and stringent environmental requirements in the oil and gas industry demand the removal of, among other, acidic constituents from natural gas. The presence of carbon dioxide (CO_2) and hydrogen sulfide (H_2S) increases corrosion, creates safety hazards for operations and results in an export gas of lower value. The most commonly used technology for acid gas removal is chemical absorption into aqueous amine-based solvents. When the selective removal of H_2S over CO_2 is desired, tertiary amines are preferred since hydrogen sulfide reacts instantaneously through a proton transfer while carbon dioxide cannot react directly with the amine (Versteeg et al., 1996). In our work, aqueous solutions of N-methyldiethanolamine (MDEA) are studied. This amine demonstrates high selectivity for H_2S over CO_2, degradation resistance, lower losses by vaporization because of its low vapor pressure and lower energy requirements during regeneration thanks to its low enthalpy of reaction (Kohl and Nielsen, 1997).

OBJECTIVE

Vapor-liquid equilibrium (VLE) data over a wide range of pressures, temperatures and concentrations of the various solvents used for acid gas removal are essential for the development of the thermodynamic models employed during the design and operation of the absorption and regeneration units. Plenty of equilibrium data are already available in the literature for the system CO_2-MDEA-H_2O, at various MDEA concentrations, pressure and temperature conditions since the use of amine-based solvents is also the dominant technology for CO_2 capture from the flue gases emitted by fossil fuel power plants. Moreover, the effect of the co-existence of carbon dioxide and hydrogen sulfide to the acid gas solubility into an MDEA solution has also been studied. However, few experimental measurements of the solubility of H_2S into aqueous amine solutions have been performed. In addition to this, to the authors' best knowledge, there are not any available data for high MDEA concentrations (MDEA > 50 wt. %). The objective of this work is, therefore, to provide new experimental data for the system H_2S-MDEA-H_2O at pressures, temperatures and concentrations that have not been studied so far in the literature with focus on high pressures, as high-pressure processes are relevant to the offshore/subsea operations and long-distance transportation of natural gas. When needed, methane will be used for pressurization, as it is the main component present in natural gas. Therefore experimental data for the system H_2S-CH_4-MDEA-H_2O will be also provided. Such data can be used to validate and/or extend existing models in terms of concentration and/or pressure and predict the thermodynamic behavior of the system accurately.

APPARATUS

A high-pressure equilibrium cell described by Dicko et al. (2010) is used for the experimental measurements of vapor-liquid equilibrium (Figure 1). A static-analytic method is employed utilizing two ROLSI® capillary samplers for fluid sampling (ARMINES, 2003), one for the vapor phase and one for the liquid phase. The VLE apparatus is designed for measurements with acid gases and organic sulfur compounds with different solvents and it allows for measurements at a pressure range of 0 – 10000 kPa and temperature range of 243 – 473 K. The analysis of the vapor and liquid phase samples is performed by means of gas chromatography. The apparatus is connected to a gas chromatograph which allows the analysis of the two phases samples and so the estimation of the respective molar compositions.

Figure 1: Flow diagram of the equipment. d.a.u.: data acquisition unit; DDD: displacement digital display; DM: degassed mixture; DT: displacement transducer; EC: equilibrium cell; GC: gas chromatograph; LB: liquid bath; LS: liquid sampler; LV_i: loading valve; MR: magnetic rod; P: propeller; PP: platinum probe; PN: pressurized nitrogen; PT: pressure transducer; PTh: pressure transducer for high pressure values; PTl: pressure transducer for low pressure values; SD: stirring device; SM: sample monitoring; ST: sapphire tube; TR: thermal regulator; Th: thermocouple; V_i: valve; VP: vacuum pump; VS: vapor sampler; VVCM: variable volume cell for mixture.

RESULTS

In this paper, we are going to present experimental vapor-liquid equilibrium data for the system H_2S-MDEA-H_2O and CH_4-H_2S-MDEA-H_2O at different loadings of the solution and at conditions relevant to the operating conditions of the absorber and the stripper (up to 100 bar and up to 120°C). A 50 wt.% MDEA solvent will be used for validation by comparing with available data from the literature, as well as two highly concentrated solvents, i.e. 70 wt. % and 90 wt. % MDEA, will be studied. A comparison of the new data and the predicted values from a thermodynamic model will be also presented.

REFERENCES

ARMINES, 2003. Procédé et Dispositif Pour Prélever Des Microéchantillons D'un Fluide Sous Pression Contenu Dans Un Container. 2 853 414.

Dicko, M., Coquelet, C., Jarne, C., Northrop, S., Richon, D., 2010. Acid gases partial pressures above a 50 wt% aqueous methyldiethanolamine solution: Experimental work and modeling. Fluid Phase Equilibria 289, 99–109. doi:10.1016/j.fluid.2009.11.01

Kohl, A.L., Nielsen, R.B., 1997. Chapter 2 - Alkanolamines for Hydrogen Sulfide and Carbon Dioxide Removal, in: Gas Purification. Gulf Professional Publishing, Houston, pp. 40–186.

Versteeg, G.F., Van Dijck, L.A.J., Van Swaaij, W.P.M., 1996. On the kinetics between CO2 and alkanolamines both in aqueous and non-aqueous solutions. An overview. Chem. Eng. Commun. 144, 133–158.

500838 Solubility of Timolol Maleate in Supercritical Carbon Dioxide with Water on Extraction from Silicone Hydrogel

Abstract:

Development of ocular drug release system using hydrogel has been demanded toward to the achievement of the high bioavailability and sustainable release on the ocular treatment. The fabrication techniques and process for the ocular drug release system using hydrogel have been reported the fabrication method or process by many research groups, such as the polymerization of the hydrogel with drugs and impregnation of the hydrogel with drugs. Impregnation type method has a merit as simple the fabrication method and process compared with those by the polymerization. For the hydrophilic drug, it is important to reduce the diffusion rate into the ocular surface. The reduction of the ocular drug diffusivity inside the silicone hydrogel release system by addition of vitamin E has been reported. Vitamin E is a fat-soluble compound and not toxic to the human body. Supercritical carbon dioxide has been applied for fabrication of drug / polymer composite as impregnation media because of high solubility and diffusivity, low surface tension and no toxicity of supercritical carbon dioxide. The supercritical carbon dioxide has also a possibility to be an impregnation solvent for loading of vitamin E into the hydrogel. In our previous work, the vitamin E loading process into the silicone hydrogel with timolol maleate using supercritical carbon dioxide has been developed for the fabrication of the timolol maleate release system. However, the loading of vitamin E in supercritical carbon dioxide caused that timolol maleate from the silicone hydrogel was extracted into the supercritical carbon dioxide and then reducing the drug loading amount. Knowledge of the solubility of timolol maleate in supercritical carbon dioxide for the process design of the vitamin E loading. The solubility of timolol maleate in supercritical CO_2 was measured in different three conditions, in pure CO_2, CO_2 saturated with water and CO_2 with water phase in this work. This knowledge would be useful to know how to reduce timolol maleate extraction. In addition, the extraction of timolol maleate from the silicone hydrogel into supercritical carbon dioxide with water was investigated. The silicone hydrogel containing timolol maleate was impregnated with vitamin E in supercritical carbon dioxide with water for the fabrication of the timolol maleate release system. The release of timolol maleate from the hydrogel with vitamin E are studied and the diffusivity of timolol maleate in the hydrogel with vitamin E are also discussed. The solubility of timolol maleate in supercritical carbon dioxide and supercritical carbon dioxide saturated with water, and supercritical carbon dioxide with water phase were experimentally determined at 313 K and 8 to 18 MPa. The solubility in supercritical carbon dioxide saturated with water is slightly higher than those in pure carbon dioxide without because of the molecular interaction between timolol maleate and water in supercritical carbon dioxide. Two phase system of supercritical carbon dioxide water phase gives the reduction of the solubilities of timolol maleate in supercritical carbon dioxide largely due to the ionization of timolol maleate in the water phase. This result of the quite low solubility of timolol maleate would succeed in the restrained extraction during the supercritical impregnation. Consequently, the large amount of timolol maleate is loaded into the silicone hydrogel with vitamin E treated in supercritical CO_2 saturated with water. The silicone hydrogel impregnated with vitamin E in supercritical CO_2 saturated with water results in the diffusivity of timolol maleate lower than that without vitamin E.

485294 Experimental and Modelling Study of the Phase Behaviour of (CO_2 + CH_4 + Methylbenzene) at High-Pressure and High-Temperature Conditions

Abstract:

1. Introduction

Knowledge and understanding of the phase behavior of CO_2-hydrocarbon mixtures at reservoir conditions is essential for the design, construction and operation of carbon capture and storage (CCS) and enhanced oil recovery (EOR) processes. For an optimal design and operation of these processes, numerical simulations are developed which play a key role in predicting the flow of CO_2 from the injection wells into the storage formation and determining the long-term evolution of the CO_2 plume after the termination of injection as well as predicting the required operating conditions for the maximum recovery of heavy oil. These models, however, require detailed and accurate thermophysical-property data for CO_2 and its mixtures with the reservoir fluids it encounters, in particular their phase behaviour. Providing such experimental data is expensive and time consuming. Therefore there is a great deal of interest in developing improved models and computational packages to predict the phase behavior and thermophysical properties of such mixtures with the least inputs of experimental data. However, to validate and improve those models, experimental data, in particular data for ternary and multi-component CO_2-hydrocarbon mixtures, are still required to tune the interaction parameters of such models, to help in developing predictive approaches, and to assess the predictive capabilities of all models.

2. Experimental Work

Various techniques have been employed to determine the phase behaviour of multicomponent mixtures under high-pressure and high-temperature conditions. In this work, a new experimental apparatus [1] was used to measure the phase equilibria of systems containing $CO_2$and hydrocarbons at reservoir temperatures and pressures. The apparatus involved a high-pressure high-temperature variable-volume view cell, with wetted parts fabricated from Hastelloy C-276, driven by a computer-controlled servo motor system, and equipped with a sapphire window for visual observation. Two calibrated syringe pumps with a maximum service pressure of 70 MPa were used for quantitative fluid injection. A low-dead-volume pressure transducer was used for the pressure measurements while the cell temperature was controlled by a heating jacket, and the temperature was measured by means of a Pt100 sensor. The phase behavior of the mixture in the cell was observed with the aid of a CCD camera operating with both front and rear illumination. The maximum operating pressure and temperature of the entire setup were 40 MPa and 473.15 K, respectively.

The apparatus was previously validated by comparison with published isothermal vapor-liquid equilibrium data for the binary system (CO_2 + heptane). Measurements were made on the ternary mixture (CO_2 + CH_4 + methylbenzene) over the temperature range (323 to 423) K at pressures up to 35 MPa. In this case, the molar ratio of $CO_2$to methane in the ternary system was fixed at three different values (0.25, 0.5 and 0.75) and the measurements were made on three isotherms (323, 373 and 423) K for each fixed value.

3. Modeling Work

In this work, we explore the predictive capability of SAFT-g-Mie [2] to model the phase equilibria of this mixture. The Statistical Associating Fluid Theory, stemming from the first order perturbation theory of Wertheim [3], was implemented in this work with a group contribution approach and the generalized Mie potential to represent segment-segment interactions. In the resulting SAFT-g-Mie, complex molecules are represented by fused segments representing the functional groups from which the molecule may be assembled. All interactions, both like and unlike, as implemented in this work were determined from experimental data of systems comprising the constituent groups. We also explore the predictive capability of, PPR78 [4], which is based on the Peng-Robinson 78 EoS [5]. In the PPR78 approach, the binary parameters are obtained from a group-contribution scheme.

The current work provides new experimental data, detailed assessment of predictive models as well as tuning the model parameters through regress analysis and fitting procedures.

References

1. Al Ghafri SZ, Maitland GC, Trusler JPM, Experimental and modeling study of the phase behavior of synthetic crude oil + CO_2, Fluid Phase Equilib., 365 (2014)
2. V. Papaioannou, T. Lafitte, C. Avendano, C.S. Adjiman, G. Jackson, E.A. Muller, A. Galindo, Group contribution methodology based on the statistical associating fluid theory for heteronuclear molecules formed from Mie segments, J. Chem.Phys, 140 (2014)
3. M.S. Wertheim, Fluids with highly directional attractive forces. II. Thermodynamic perturbation theory and integral equations, J. Stat. Phys, 35 (1984)
4. J.-N. Jaubert, F. Mutelet, VLE predictions with the Pengâ€"Robinson equation of state and temperature dependent kij calculated through a group contribution method, Fluid Phase Equilib., 224 (2004)
5. D.-Y. Peng, D.B. Robinson, A new two-constant equation of state, Ind. Eng. Chem. Fun, 15 (1976)

Â

Acknowledgment

We gratefully acknowledge the funding of QCCSRC provided jointly by Qatar Petroleum, Shell, and the Qatar Science and Technology Park, and their permission to publish this research.

503576 Industrial-Scale Modeling of Herschel-Bulkley (Yield Stress) Fluids

Abstract:

The suitability of lattice-Boltzmann-based CFD for modeling Herschel-Bulkley fluid is examined. Under the Herschelâ€“Bulkley model, the strain experienced by a fluid is related to the stress via a yield shear stress, flow consistency, and flow index. Although common in industrial practices, the yield stress-induced discontinuity in the constitutive relationship is difficult to model using time-averaged, RANS-based CFD. The time-accurate nature of the lattice-Boltzmann approach, combined with the high spatial resolution it affords, enables a DNS approach to modeling momentum transport through such systems. Here we assess the industrial suitability of this lattice-Boltzmann DNS approach for modeling Herschelâ€“Bulkley fluids, as applied to draining storage tanks, flow through pipes, and flow inside agitated systems.

499620 Viscometer for on-Site Rheology Measurements:Â Development and Application

Abstract:

It is vitally important to have accurate rheological measurements for fluids when designing equipment for processes requiring mixing of viscous, non-Newtonian fluids. In many cases, it is convenient to send samples to a laboratory and have the measurements made there but there are certain processes in which the properties of the fluid change so rapidly in time that representative measurements can only be made at the site where the process is carried out.

This presentation will discuss the development of a portable viscometer which can be easily transported to a site and used to measure rheology. A case study will be presented demonstrating how the viscometer has been used to investigate problems in a high solids concentration slurry mixing process.

502086 Using Î¼²rheology to Measure Rheological Properties during Repeated Phase Transitions of Hydrogenated Castor Oil

Abstract:

Rheological modifiers are often used as ingredients in home care goods. These materials undergo phase changes during manufacturing and regular use. Rheological and material properties can be altered by repeated phase transitions, and it is important to know the extent to which microstructural changes affect the performance of the end use product. In this work, we investigate a hydrogenated castor oil (HCO) gel, a rod-like colloid (aspect ratio 50-2500) that forms a stable aqueous suspension with the addition of a surfactant and undergoes a phase transition due to an osmotic pressure gradient. Î¼²rheology, the combination of microfluidics and multiple particle tracking microrheology (MPT), is used to measure material properties during transition. A new microfluidic device design enables the exchange of fluids while minimizing shear stress on the sample. In MPT, fluorescent probe particles are embedded in the sample and their Brownian motion is recorded using video microscopy. The particle trajectories are then related to rheological properties using the Generalized Stokes-Einstein Relation. MPT data are compared to phases previously determined by Wilkins et. al., 2009. The phases of rod-like systems are network, where the particles have a sample spanning structure; a bundled phase where rods do not create a network; and a transitional phase where the colloidal particles partially lose scaffold structure and begin forming bundles. MPT results show that a 4wt% HCO gel undergoes a repeatable phase transition between the network and transitional phases due to the osmotic gradient, but never reaches the bundled phase. A 0.125wt% solution of colloidal rods that was prepared using a large amount of shear undergoes repeatable phase transitions but does not form the network structure measured in the 4wt% gel. This is due to the shear added during sample preparation permanently breaking the network structure. This work shows that sample preparation can have irreversible effects on the microstructure of the gels, which can cause changes in end use products made with rheological modifiers.

499275 Model Systems for Mixed Particulate Fermentation Broth

Abstract:

Model systems for mixed particulate fermentation broth

C Bliatsiou, L Böhm, M Kraume

Chair of Chemical and Process Engineering, Technische Universität Berlin, Germany

FH6-1, Fraunhoferstrasse 33-36, 10587 Berlin, Germany. Tel: +493031472791, email: lutz.boehm@tu-berlin.de

The project which is part of the Priority Programme "SPP1934 DiSPBiotech" is concerned with the experimental and numerical characterization of local mechanical stresses occurring in gassed stirred fermenters and their effect on biological agglomerates. The findings thus obtained are practically tested on the fermentation of Aspergillus niger. Furthermore, the results are used for the identification and development of low-shear stirrers.

To create controlled and repeatable conditions in experiments, suitable model systems which represent the biological broth are of high interest. The material properties such as density and rheology but also characteristics of the agglomerates depend on numerous parameters such as biological behavior, the mechanical and operational process settings as well as time. Even for such well known processes as the fermentation of A.niger, material property data can rarely be found. It is known that, depending on the conditions, the broth can have Newtonian (e.g., for pellet growth; Fig.1a) or non-Newtonian characteristics (incl. shear thinning, shear thickening, viscoelastic, e.g., for mycelia growth; Fig.1b) or even changing rheology over time due to increasing biomass concentration in the batch process. The same is valid for the characteristics of the agglomerates with rather large round particles in case of pellets down to smaller, in parts branched, filamentous shapes. This broad variety of conditions in real fermentation broth cannot all be covered by one model system. The requirements that the model solution should fulfill are challenging. It should be able to mimic all the material properties of the real system while covering the requirements of the applied measurement techniques as well. To name one example, e.g., for optical measurements, the continuous phase must be transparent.

Here, material and particle information is collected from literature and generated by own work and together with project partners within the frame of the Priority Programme. This information will be used to find proper model solutions. For a Newtonian rheology of the continuous phase, aqueous solutions with different additives can be used. Furthermore, for high viscous conditions, silicone oil or glycerol are valid options. To generate a non-Newtonian behavior carboxyl cellulose, xanthan gum or polyacrylamid solutions show a shear-thinning and, in parts, additionally viscoelastic behavior while still being transparent.

Covering the task of finding a model system for the dispersed phase as found in real systems with a comparable morphology and morphological behavior during the application of mechanical stress is an even the more complicated task. Suitable options are, e.g., a second immiscible liquid phase (Fig.1a) or blue-clay flocs (Fig.1b). The latter one, especially, is highly promising. Still it must be kept in mind that the dispersed phase can influence the rheology of the whole mixture, as well. To find a guideline for choosing the best model solution representing a particulate fermentation broth, these options must be investigated with according techniques for the determination of material properties, such as rheometers. Here, the accuracy of the measurements is of high importance especially for liquid-solid-mixtures as the solids can settle which changes the properties of the sample over time. As mentioned above, in the frame of this project, the morphology and morphological behavior during the application of mechanical stress, as found in stirred tanks, is of highest interest. This will be investigated in the stirred tank by using an endoscope measurement techniques together with an automated image analysis delivering the particle properties.

Figure 1. Images of a) pellet-like growth of A.niger (upper left) and an according potential model system (liquid-liquid dispersion) and b) of branched filamentous agglomerates of the fungi (upper left) with an according image of blue-clay flocs in a stirred tank

Acknowledgments

Financial support by the Deutsche Forschungsgesellschaft (DFG) in the frame of the Priority Programme SPP1934 DiSPBiotech is gratefully acknowledged. Special acknowledgements go to the working group of Prof. Krull from Technische Universität Braunschweig for the support in the fermentation and imaging of A.niger.

489218 Polycarboxylate Ether Cement Superplasticizers Containing Copolymer Backbones with Improved Rheology Performance

Abstract:

Polycarboxylate ether (PCE) superplasticizers are used in cement to reduce the water level necessary to achieve flow and the desired open time for workability, thus reducing the impact of waterâ€™s deleterious effect on concrete strength. Standard PCEs used in cements are graft copolymers consisting of a polyacrylic acid (PAA) backbone with acrylic ester units containing polyethylene glycol (PEG) side chains. Unfortunately, while PAA-based PCEs show equal or better performance initially, the alkalinity of the cement leads to rapid hydrolysis and loss of slump flow. We have found that using an acrylic acid copolymer backbones to produce PCEs provides excellent flow and open time characteristics. The acrylic acid copolymers are statistical copolymers of acrylic acid and hydrophobic acryloyl morpholine (ACMO). This acrlylic acid and acryloyl morpholine copolymer based PCE has shown extended open time for workability, indicating better stability in cement formulation than PAA based PCE.

491129 Microscale Acoustic Streaming Flows in Viscoelastic Fluids: Comparison of Experiment and Simulation

Abstract:

Micro-bubble streaming induces vortices of secondary flows around a vibrating bubble in a microfluidic device. It has been applied to micro-scale fluid manipulations such as mixing enhancement, fluid pumping and micro-particle sorting. Although microbubble streaming has been widely investigated both experimentally and numerically in Newtonian fluids, there is little information in the literature on such flows in viscoelastic fluids specifically on the numerical aspect. The Oldroyd-B family of viscoelastic fluids provides a linear constitutive model to describe the behavior of the dilute polymeric solutions based on the dumbbell model. This study focuses on the investigation of the flow around a sessible vibrating micro-bubble attached to the wall of a micro-channel in two cases: a Newtonian fluid and a viscoelastic fluid. In the experimental investigation, aqueous glycerol and dilute solution of polyethylene oxide (PEO) in aqueous glycerol are respectively utilized as the Newtonian and viscoelastic fluids in the micro-channel. The patterns of the flow field at different frequencies are visualized by high-speed imaging and particle-lines. In the numerical solution, finite element method is employed to simulate flow patterns around the vibrating micro-bubble and to probe the effect of the elasticity on how the bubble vibrates. Both numerical and experimental results suggest a significant change in the number and the position of vortices in the steady streaming flow pattern around the micro-bubble.

509815 Radioactive Particle Tracking (RPT) Techniquefor Pilot-Scale Bubble Column

Abstract:

Radioactive Particle Tracking (RPT) Technique for Pilot-Scale Bubble Column

Laith S. Sabri*1, Abbas J. Sultan*2, Muthanna H. Al-Dahhan*3

*Multiphase Reactors and Applications Laboratory (mReal). *Department of Chemical and Biochemical*

Engineering, Missouri University of Science and Technology. Rolla, MO 65409-1230. USA

1lssf25@mst.edu, 2ajshw9@mst.edu, 3aldahhanm@mst.edu,

Abstract

The multiphase flow reactors are extensively used in chemical and biochemical industries. Where perfect mixing and ideal mass and heat transfer are importance. In this work, pilot plant scale of bubble column reactor was used for air-water system. The local hydrodynamics can considerably affect the reactor conduct. Due to the local hydrodynamics parameters are the main factor for multiphase reactor design. Consequently, effective reactor design and successful scale-up require a suitable understanding of the local hydrodynamics for multiphase flow through the reactor running stages. Unfortunately, these parameters need to be further developed and evaluated to account of 3D flow field in each position inside the reactor in industry scale (pilot plant) which is unclear yet and need more information for deeply understanding. Therefore, In our laboratory (Multiphase Reactors and Applications Laboratory MReal), an advanced diagnostic techniques for industrial scale have been well established and used extensively for monitoring the flow field in a non-invasive manner in opaque multiphase pilot plant reactor systems in column diameter 18-inch (0.46 m) Plexiglas reactor with height 144 inches (3.66m). A non-invasive advanced measurement technique called Radioactive particle tracking (RPT) (see Figure 1) for large scale columns has been implemented to provide detailed and in 3D hydrodynamic parameters (velocity components, turbulent parameters, shear stresses, normal stresses, kinetic energy, and eddy diffusivity). For present study, quantifies of the flow characterization in the time series of lagrangian tracer particle trajectory are available for fluctuating liquid velocity fields at various superficial gas velocities (10 and 45 cm/sec). As a conclusion, the experimental results provide benchmark data for simulation, design, scale-up, and performance of the pilot plant scale bubble column reactors. The experimental results and conclusions will present at the conference.

Figure 1. Radioactive Particle Tracking (RPT) Technique for Pilot-Scale of Multiphase flow system

509816 Gamma-ray Computed Tomography (CT)Technique for Pilot-Scale Bubble Column Reactor

Abstract:

Gamma-ray Computed Tomography (CT) Technique for Pilot-Scale Bubble Column Reactor

Abbas J. Sultan*1, Laith S. Sabri*2, Muthanna H. Al-Dahhan*3

*Multiphase Reactors and Applications Laboratory (mReal). *Department of Chemical and Biochemical*

Engineering, Missouri University of Science and Technology, Rolla, MO 65409-1230. USA

1ajshw9@mst.edu, 2lssf25@mst.edu , 3aldahhanm@mst.edu,

Introduction

Cross-sectional gas holdup distribution in pilot-scale bubble column without and with a bundle of cooling tubes is crucial for design, scale-up and performance of these multiphase flow system. There is still a lack of knowledge and understanding of the cross-sectional gas holdup distribution in large scale reactors. Therefore, in this work an advanced gamma-ray computed tomography (CT) technique (see figure 1) has been utilized for the first time to visualize and quantify the impact of size, and configurations of the intense heat exchanging tubes (internals) on the cross-sectional distribution and diameter profiles of the gas holdup at different superficial gas velocities covers the bubbly to the turbulent flow regime (0.05-0.45 m/s). Two different sizes of internals (0.5 and 1-inch in diameter) with different configurations, including hexagonal and circular shape arrangements for internals, have been investigated in the 6-inch (0.14 m) and 18-inch (0.46 m) bubble columns using the air-water system. The vertical internals arrangements were blocked 25% of the total cross-sectional area of the column to mimic the occluded cross-sectional area of heat exchanger tubes that are used in industrial such as Fischer-Tropsch (FT) synthesis. As a conclusion, the experimental results provide benchmark data for simulation, design, scale-up, and performance of the pilot plant scale bubble column reactors. The experimental results and conclusions will present at the conference.

Figure 1 Advanced Non-Invasive Gamma-ray Computed Tomography (CT) technique.

(a) Pilot-Scale Bubble column without Internals (b) Pilot-Scale bubble column with internals

501055 Influence of Cell Lengths on Surface Drag in Motile Bacteria

Abstract:

Bacterial surface motility is an important factor in host-colonization and infections. Bacteria swim at low Reynold's numbers, where viscous forces dominate. Consequently, the cell length significantly impacts the surface drag experienced by motile cells near substrates, with longer cells experiencing higher drag. We are interested in determining how cells with widely differing cell lengths overcome drag. To study these strategies, we employed mutants that exhibited variable cell morphologies and tested motility on agar media. For each mutant, the surface drag was varied. Our measurements indicated a positive correlation between motility and cell lengths. We will discuss our findings and present additional evidence for this surprising observation. We anticipate that these studies will provide key insight into pathogen-surface interactions.

501668 Dynamics of Semiflexible Colloidal Particle Chains Under Rotating Magnetic Fields

Abstract:

Several naturally occurring and practically useful phenomena, such as flagellar propulsion and fluid manipulation via artificial cilia, are products of the unique dynamics of semi-flexible filaments as they respond to external forces in a fluid environment. In order to harness these phenomena and design optimal synthetic systems, we must be able to predict the eventual dynamics of fiber as a function of the applied driving forces and material properties. However, in the intermediate semi-flexible regime, where many fibers of interest, such as microtubules, carbon nanotubes, and artificial cilia reside, there appear to be deviations from the scalings expected of rigid and highly flexible fibers. We investigate these dynamics via a tunable experimental model, consisting of chains of linked, paramagnetic colloidal particles. When these chains are driven by an external, rotating magnetic field, rich dynamical regimes are observed, including non-reciprocal, wagging orbits and potentially bifurcating regimes. Using a combination of experimental filaments, numerical simulations, and theoretical calculations, we classify and predict these dynamics as a function of the dimensionless Mason and magnetoelastic numbers of the system.

493666 A Study of the Dynamics of Human Pedestrians Using Experiments and Simulations in the Indian Context

Abstract:

The world population, currently at about 7.5 billion, has experienced dramatic growth rates since the industrial revolution. We are poised to add another 2 billions by 2050, with a significant portion residing in the developing countries. Even with the existing numbers, India and other developing countries are witnessing unprecedented levels of population density. The management of these large volumes of crowds, especially in mass gatherings and public spaces, is an extremely challenging problem. One should also remember the fact that, in many cases, several well-populated public spaces are interconnected with a steady flow of crowds between them, which creates a network and further complicates the problem. Quite unsurprisingly, we witness a significant number of disasters like stampede, accompanied by losses of hundreds of lives every year. To prevent such unfortunate scenarios, we need to optimally design public spaces such that the chances of these disasters are minimized. These should ideally incorporate the fundamental nature of interactions, in a mathematically usable form, between human beings in such scenarios. Through significant improvements over the past decade, researchers have made significant progress towards obtaining physical models for the forces between human beings. There are several such 'social force' based models for human pedestrian interactions that have been proposed in literature. However, none of these were parameterized with existing scenarios of a developing country like India, where the population densities are an order of magnitude larger relative to the western nations. Higher densities, coupled with cultural and social differences, could give rise to significantly different forces between pedestrians in India. To the best of our knowledge, no such studies exist in the Indian context.

In this study, we try to understand and model the pedestrian interactions in Indian public spaces. We perform experiments in the "ultra-dilute" regime, which consist of a pair of pedestrians moving towards each other, to understand the validity of the existing 'social force' models in the Indian context. Also, we perform some experiments on single isolated pedestrians, which provided us with the characteristics of the motion in the absence of other interactions. In these experiments, the pedestrians mostly consisted of student volunteers at IIT Kanpur. We employ standard image processing techniques to obtain the trajectories of the pedestrians for further analysis. These trajectories provide the position, velocity and acceleration of the pedestrians at all times during the experiment. Fourier transforms of the data from experiments reveal the dynamics at various length and time scales, for both the single and a pair of pedestrians. For comparison of these with the predictions of the 'social force' models, we perform simulations of a pedestrian pair moving towards each other. These simulations modelled each pedestrian as a particle, which exert forces upon each other that are estimated from the relevant social force model. In the process, we also compare the predictions of two promising social force models existing in literature to highlight the similarities and differences. Further, by comparing the dynamics obtained from simulations to our experiments with volunteers, we identify some severe limitations of these social force models. Some of the qualitative trends obtained from the experiments, especially the velocity and accelerations of the pedestrians as they approach each other, are in sharp contrast with those from simulations. Our observations from experiments are quite different from the fundamental characteristics that the social force models are expected to capture, thereby raising doubts about whether they can be used in the current form for Indian scenarios.

493307 Extensional Rheometry with a Handheld Mobile Device

Abstract:

The on-site characterization of complex fluids is important for a number of academic and industrial applications. Consequently, a need exists to develop portable rheometers that can provide in-the-field diagnostics and serve as tools for rapid quality assurance. With the advancement of smartphone technology and the widespread global ownership of smart devices, mobile applications are attractive as platforms for rheological characterization. The present work investigates the use of a smartphone device for the extensional characterization of a series of Boger fluids composed of glycerol/water and poly(ethylene oxide), taking advantage of the increasing high-speed video capabilities (currently up to 240 Hz capture rate at 720p) of smartphone cameras. We report a noticeable difference in the characterization of samples with slight variations in polymer concentration and discuss current device limitations. Potential benefits of a handheld extensional rheometer include its use as a point-of-care diagnostic tool, especially in developing communities, as well as a simple and inexpensive tool for assessing product quality in industry.

500870 Free Surface Flows and Extensional Rheology of Polymer Solutions

Abstract:

Free-surface flows -- jetting, spraying, atomization during fuel injection, roller-coating, gravure printing, several microfluidic drop/particle formation techniques, and screen-printing -- all involve the formation of axisymmetric fluid elements that spontaneously break into droplets by a surface-tension-driven instability. The growth of the capillary-driven instability and pinch-off dynamics are dictated by a complex interplay of inertial, viscous and capillary stresses for simple fluids. Additional contributions by elasticity, extensibility and extensional viscosity play a role for complex fluids.~We show that visualization and analysis of capillary-driven thinning and pinch-off dynamics of the columnar neck in an asymmetric liquid bridge created by dripping-onto-substrate (DoS) can be used for characterizing the extensional rheology of complex fluids. Using a wide variety of complex fluids, we show the measurement of the extensional relaxation time, extensional viscosity, power-law index and shear viscosity. Lastly, we elucidate how polymer composition, flexibility, and molecular weight determine the thinning and pinch-off dynamics of polymeric complex fluids.

501966 Rheological Behavior of Poly(vinyl alcohol) in Aqueous Solutions: Comparison and Assessment of Rheological Parameters Obtained By Empirical Correlations

Abstract:

Aqueous solutions of poly(vinyl alcohol) at different concentrations (4%, 6%, 8% and 10%) and temperatures (20°C and 35°C) have been used as model of fluid to assess the behavior of various rheological parameters using five models of empirical correlations, such as, Ferrys, Roberston-Stiff, Williamson, Ellis de Haven and Sisko. In this way, regression analysis on experimental data were developed to associate the rheological behavior with the model parameters. The rheological models chosen were compared in their coefficient of multiple correlation to fit the experimental data with the selected models. We observed through statistical analysis that each model presents different behaviors in relation to differential and dynamic viscosity. Further, it is shown that the choice of rheology model has influence on the prediction of the behavior of the aqueous solution of poly(vinyl alcohol) at different concentrations. Also, the results demonstrated that the precision of the calculated rheological parameters does not depend on the initial parameter values. This paper highlight the importance of understand the rheological behavior of a fluids through various models of empirical correlations and include directions for future research.

492262 The Phantom Generation of a Complex Nasal Geometry with Horizontal Cut for Lda Measurements

Abstract:

The validation of Computational Fluid Dynamic (CFD) simulations using the non-invasive flow measurement technique of Laser Doppler Anemometry (LDA) needs a model (phantom) that guarantees optical accessibility in order to result in the desired velocity vector field. The Region Of Interest (ROI) is the human nose based on a Computer Tomography (CT) dataset. Segmentation selects the relevant voxels for the breathing process (voxels with the smallest X-ray density = air) to create a 3D surface geometry [1] which is used for LDA as well as CFD.

Optical accessibility for LDA measurements is guaranteed by a transversal cut through the inferior turbinates as a part of the complex nasal structure, which is further on sealed by a glass pane to ensure appropriate airflow. Three reference points attached to the model allow aligning the LDA measurement system in a way that assures that both coordinate systems are identical.

For the investigation two models with different 3D printing systems are used: The first of which is Polyjet Connex 350 developed by Stratasys which features a spatial resolution within the range of micrometers â€“ all cavities are filled with a support material. Second the PLA filament printer 3D20 by Dremel with a resolution of about 0.05 mm is used. The latter brings about the advantage that no support structures for the phantom are required. The defined research goal is to unveil the better suited printing system.

In order to remove the support material when using the Connex 350 created model coarse and easy accessible structures are cleared away by a water jet. However, nearly all important structures are hidden which makes a subsequent bath in an ultrasonic tank with 5% sodium hydroxide in water necessary. It should be mentioned that there is a critical point in time when fine and filigree structures begin to lose strength, requiring careful stoppage of the process in order to not harm the model.

The complete vanishing of all support material is controlled by LDA measurements of the cross-section on the inferior turbinate for both models. There is a higher velocity of about 7 m/s observed in the Polyjet model. Since flow rate is constant and velocity only depends on the crosssection, higher velocity results from still existing non-removable support material. The reason is the fulfillment of the continuity equation. Even though the 3D print with the higher resolution seems to be the preferred one, in this application the filament model is more suitable for the measurement.

The next steps are to evaluate all possible measurable regions to have good basis for the validation of CFD simulations.

[1] Newman, T. S., Yi, H., A survey of the marching cubes algorithm In: Computer Graphics, Vol. 30, Nr. 5, Oktober, p. 854-879, 2006.

497481 Using Laos and Transient Data to "Fingerprint" Human Blood Rheological Data

Abstract:

American Institute of Chemical Engineering

Tyler T. Helton and Matthew J. Armstrong

Department of Chemistry and Life Science, United States Military Academy, West Point, NY 10996

Using LAOS and Transient Data to "Fingerprint" Human Blood Rheological Data

Complex, viscoelastic materials exhibit a change in viscosity depending on the extent of shearing forces placed upon them. Among these types of materials is blood, a well documented viscoelastic material. An underdeveloped area of study in this field analyzing these mechanical measurements of blood for use in clinical diagnoses. Rheologists have long studied a unique condition of shear flow called large amplitude oscillatory shear (LAOS). The challenge of this condition is its nonlinearity, stemming from the difficulty of modeling the unique internal structure of each material. There is a possibility that blood could be classified using its own unique LAOS signature, and with the addition of large quantities of data for comparison, could be used as a method of diagnoses.

Blood is a great example of a non-Newtonian fluid. It is a suspension red blood cells (RBCs), white blood cells, proteins, hormones, platelets, and other ionic solutes in the blood plasma. This internal microstructure affects the dynamic viscosity of blood when forced into flow conditions. The interactions of the microstructure play a major role in determining the viscosity of blood. RBC aggregation, deformability, and membrane rigidity all contribute to the blood's viscosity [1]. Plasma proteins, most notably fibrinogen, contribute to the interactions among RBCs as well as platelets [2].

Recent work with blood using LAOS has shown that its unique rheological signature can be used as a psuedo- "fingerprint" [3-7], able to clearly show and delineate elastic and viscous regions of the LAOS cycle. Applying large amplitude oscillatory shear (LAOS) to complex fluids induces nonlinear rheological responses, that, with proper LAOS analysis technique, can be used to sensitively probe the underlying microstructure and its dynamics. Using the previously published blood LAOS data of Sousa *et al.* [10] we compare LAOS using three published LAOS analysis frameworks, including the aforementioned by Bharadwaj and Ewoldt as well as a novel Series of Physical Processes (SPP) framework [8,9] by Rogers. The techniques will be compared and contrasted and respective strengths and weakness discussed.

For many years, the standard means of analyzing LAOS has been the use of discrete Fourier transforms in order to break down the sinusoidal stress signal, $\sigma(t)$, into elastic and viscous components respectively

[14,15]. These components can be used to analyze the extent of the elastic and viscous contributions to the stress signal. A new method of analyzing LAOS data, (SPP), uses all the components of LAOS data used by the rheometer and treats the now three dimensional data as a space curve. SPP also allows for a comparison between the G'and G" analog at every given point. With this advantage, the materials movement between the elastic and viscous regimes can be analyzed. We demonstrate the unique fingerprinting technique using published LAOS [10] and transient data [17].

References

[1] O.K. Baskurt, H.J. Meiselman, Seminars in Thrombosis and Hemostasis. 29(5) (2003) 435-450.

[2] G.D.O. Lowe. Clinical Science. 71 (1986) 137-146.

[3] C.J. Dimitriou, R.H. Ewoldt and G.H. McKinley. J. Rheol. 571(1), 27-70.

[4] B.C. Blackwell and R.H. Ewoldt. JNNFM 227, (2016) 80-89.

[5] B.C. Blackwell and R.H. Ewoldt. JNNFM 208-209, (2014) 27-41.

[6] R.H. Ewoldt and N.A. Bharadwaj. Rheol. Acta. (2013) 52:201-219.

[7] N.A. Bharadwaj and R.H. Ewoldt. J.Rheol. 59(2), (2015) 557-592.

[8] S.A. Rogers. J. Rheol. 56(5), (2012) 1129-1151.

[9] S.A. Rogers and M.P. Lettinga. J. Rheol. 56(1), (2012) 1-25.

[10] P. C. Sousa. Biorheology 50 (2013), 269 – 282.

[11] Moreno et al., Korea-Australia Rheology Journal Vol. 27, No.1 (2015) 1 – 10.

[12] J. Mewis and N. J. Wagner, Colloidal Suspension Rheology, Cambridge Univ. Press, 2012.

[13] O.K. Baskurt, H.J. Meiselman, Seminars in Thrombosis and Hemostasis. 29(5) (2003) 435-450.

[13] G.D.O. Lowe. Clinical Science. 71 (1986) 137-146.

[14] R.H. Ewoldt, A.E. Hasoi, G.H. McKinley. J. Rheol. 52(6). (2008) 1427-1458.

[15] M.J. Armstrong. PhD Thesis, University of Delaware (2015).

[16] S. Rogers. Rheol. Acta. (2017).

[17] Bureau et al. Biorheology (1980).

497503 Comparison of Simple Rheological Models in Fitting and Predicting Steady State and Transient Blood Rheology

Abstract:

American Institute of Chemical Engineering

Michael Deegan, Evan Ousley and Matthew J. Armstrong

Department of Chemistry and Life Science, United States Military Academy, West Point, NY 10996

Comparison of Simple Rheological Models in Fitting and Predicting Steady State and Transient Blood Rheology

Many complex materials have thixotropic behaviors, displaying time dependent viscous and elastic properties that are a function of microstructure. Some examples of thixotropic materials are aqueous nuclear waste, crude oil, and paints. Another example is blood. Modeling the behavior of these materials allows us to better understand how they can be used effectively in their respective industry [1,2]. We have characterized human blood using published steady state data and different models from literature. First, we looked at data from previous rheological experiments (Moreno and Sousa) [3,4] done on blood and modeled them using nine different simple models. Using the best model, we predicted small amplitude oscillatory shear (SAOS) with our own parameter determination technique and used two published transient models to predict large amplitude oscillatory shear (LAOS) and SAOS.

We characterize blood to develop better models, and evolve strategies to give way to better understanding of blood to facilitate benchmarking blood's "normal" rheological fingerprint, with a view toward a methodology to diagnose pathologies based on rheological deviations from blood's baseline mechanical properties. Based on the biochemistry of blood, its fluid mechanics, and rheological properties, pathological blood will have different rheological properties than healthy blood [7,8]. By modeling healthy blood, we can establish a baseline that will then allow us to characterize and diagnose pathological blood based on its flow behavior, as was seen in the Moreno paper with high cholesterol [3].

Applying large amplitude oscillatory shear (LAOS) to complex fluids (including blood) induces nonlinear rheological responses, that with proper modeling, can be used to sensitively probe the underlying microstructure and its dynamics. We demonstrate this with published blood steady state and transient data, using a the best simple models [3,4,5,6]. We will first demonstrate the state of blood modeling using simple models comparing and contrasting relevant models and model features. With the best models we will fit to steady state data, and compare predictions of transient, and LAOS data with the published data of Sousa *et al.* [4] and Bureau *et al.* [9], to show weaknesses in current best models, and propose potential model improvements. In addition we propose modifications to current transient models that have shown potential

to fit these transient experiments with human blood.

References

[1] Merrill, Edward. Rheology of Blood. Physiological Reviews. (1969). USA.

[2] Baskurt OK, Meiselman HJ, Semin Thromb Hemost· Blood rheology and hemodynamics. Department of Physiology, Akdeniz University Faculty of Medicine, Antalya, Turkey. 2003 Oct;29 (5):435-50.

[3] Moreno, Leonardo and Fausto Caldera, *et al.* Effect of Cholesterol and Triglycerides Levels on the Rheological Behavior of Human Blood. Korea-Australia Rheology Journal. (2015).

[4] Sousa, P.C. and J Carneiro, et al. Shear Viscosity and Nonlinear Behavior of Whole Blood Under Large Amplitude Oscillatory Shear. Biorheology. (2013).

[5] Apostolidis, A., Armstrong, M., and Beris, A. Modeling of Human Blood Rheology in Transient Shear Flows. University of Delaware. (2014).

[6] Akaike, Hirotugu. A New Look at the Statistical Model of Identification. IEE Transactions of Automatic Control (1974).

[7] Baskurt, Oguz Kerim. Pathophysiological Significance of Blood Rheology. Turk J Med Sci. 33 (2003) 347-355.

[8] Fedosov, Dmitry A., Wenxiao Pan, et al. Predicting human blood viscosity in silico. New York

[9] Bureau et al. Biorheology (1980).

492604 Development of Predictive Model for Sizes of Gas and Liquid Slugs Formed in Millimeter-Scale T-Shaped Channels

Abstract:

Gas-liquid slug flow in micro/millimeter-scale channels is attractive in a broad range of applications because of its high mass and heat transfer rates [1][2]. In almost all applications, the sizes of slugs need to be known precisely, since the mass transfer characteristic depends on the slug length and velocity [1]. To estimate the slug length from design and operating conditions or efficiently design devices that can generate slugs with a desirable size, a number of models have been developed so far for micro/millimeter-scale T-shaped channels, which are often used as generators of gas-liquid slugs, but parameters of such models need to be empirically determined through many experimental trials [3]. To reduce the experimental efforts, the physically sound model has been developed [4] but are limited to micrometer-scale T-shaped channels. In this study, the applicability of the existing physically sound model is experimentally investigated in the case of millimeter-scale T-shaped channels under various channel sizes, flow rates, and channel materials. The results show that the estimation accuracy of the existing model deteriorates when the feed flow rate ratio of gas to liquid becomes small and when the channel width ratio of the gas to the liquid flow becomes small. Taking these results, the combination of physical and empirical models is newly developed. Our developed model will ensure the high accuracy in the design of the millimeter-scale T-shaped channels with gas-liquid slug flows.

[1] G. Berčič and A. Pintar, "The role of gas bubbles and liquid slug lengths on mass transport in the Taylor flow through capillaries," Chem. Eng. Sci., vol. 52, no. 21–22, pp. 3709–3719, 1997.

[2] G. Dummann, U. Quittmann, L. Groschel, D. W. Agar, O. Worz, and K. Morgenschweis, "The capillary-microreactor: A new reactor concept for the intensification of heat and mass transfer in liquid-liquid reactions," Catal. Today, vol. 79–80, pp. 433–439, 2003.

[3] P. Garstecki, M. J. Fuerstman, H. A. Stone, and G. M. Whitesides, "Formation of droplets and bubbles in a microfluidic T-junction - scaling and mechanism of break-up, " Lab Chip, 6 pp. 437-446, 2006

[4] V. van Steijn, C. R. Kleijn, and M. T. Kreutzer, "Predictive model for the size of bubbles and droplets created in microfluidic T-junctions.," Lab Chip, vol. 10, no. 19, pp. 2513–2518, 2010.

500883 Shape Evolution and Spreading of Liquid Droplets in Miscible Environments

Abstract:

The spreading of liquids is a classical problem in interfacial fluid mechanics and, historically, the examination has been limited to immiscible systems. We have reported previously on our experimental studies and observations of the spreading of sessile drops in miscible environments, which have distinctly different shape evolution and power law dynamics from sessile drops that spread in immiscible environments. We have extended this experimental work to include the shape evolution of pendant drops existing in a miscible environment. By examining pendant drops, the need to account for surface energies arising from a solid-fluid interface, as in the case of a sessile drop, is eliminated. We have complemented these experimental studies with a theoretical scaling analysis as well as numerical simulation.

As time evolves, diffusion across the miscible liquid-liquid boundary proceeds due to the chemical potential difference between the two initially distinct, homogeneous phases. Diffusion, in turn, imparts a time-dependence to the properties of the liquids in the diffusive region – notably the density, viscosity, and interfacial tension – that influence the shape evolution. It was found for these droplets in miscible environments that gravitational forces dominate the shape evolution process. The presence of diffusion sets up a fluid flow of free convection in the case of a pendant drop in a miscible environment; in the case of a sessile drop in a miscible environment, free convection occurs in tandem with spreading along the solid substrate.

A series of liquid pairs (corn syrup-water, glycerol-water, glycerol-ethanol, tricresyl phosphate-ethanol, silicone oil-silicone oil) and volumes of droplets have been studied. Solving the convection-diffusion and Stokes equations numerically and in tandem has been used to simulate these systems, which quantitatively match observations of the experiments. This poster will present the sessile and pendant drop work together, spanning the experimental, theoretical, and numerical work.

Figure 1: Side view of a miscible sessile drop of corn syrup immersed in water.

Figure 2: Image sequence taken in time of a corn syrup pendant drop immersed in water. A strand emanates from the apex of the drop and continues to flow as the entire drop descends and elongates.

Figure 3: Numerical simulation of a pendant drop in a miscible environment evolving in time.

489083 Impact Forces of Inertia-Driven Liquid Drops

Abstract:

Liquid drop impacts on solid surface are ubiquitous relevant to many important natural and industrial processes. Although the dynamics of liquid-drop impacts have been extensively studied by high-speed photography, comparatively fewer experiments have been conducted to explore the impact forces induced by liquid drops. Particularly, the temporal variation of impact forces during impact has not been systematically investigated. Here, by synchronizing high-speed imaging with fast force measurements, we investigate the detailed temporal variation of impact forces induced by inertiaâ€“driven liquid-drop impacts. Our experiments on the rise of impact forces at early times unambiguously demonstrate the existence of self-similar inertial fronts during the initial impact of liquid drops. Moreover, we systematically study how the intrinsic properties of drops affect impact forces. We find that, for low-viscosity liquid drops, the temporal variation of impact forces is highly asymmetric. As viscosity increases, impact forces become more symmetric approaching those induced by the impact of solid spheres. Our study reveals an important yet unexplored aspect of drop impacts and illustrates the transition between liquid-drop and solid-sphere impacts from the perspective of impact forces.

500903 Visualizing Nanoscopic Topography, Patterns, Flows, Thickness Transitions and Instabilities in Stratifying Freestanding Thin Films

Abstract:

Freely standing thin liquid films containing supramolecular structures including micelles, nanoparticles, polyelectrolyteâ€"surfactant complexes, and smectic liquid crystals undergo drainage via stratification. The layer-by-layer removal of these supramolecular structures manifests as stepwise thinning over time and a coexistence of domains and nanostructures of discretely different thickness. The layering of supramolecular structures in confined thin films contributes additional non-DLVO, supramolecular oscillatory surface forces to disjoining pressure, thus influencing both drainage kinetics and stability of thin films. Understanding and characterizing the spontaneous creation and evolution of nanoscopic topography of stratifying, freely standing thin liquid films have been long-standing challenges due to the absence of experimental techniques with the requisite spatial (thickness <10 nm) and temporal resolution (<1 ms). Using Interferometry Digital Imaging Optical Microscopy (IDIOM) protocols developed herein, we visualize and characterize size, shape, and evolution kinetics of nanoscopic mesas, terraces, and ridges. The exquisite thickness maps created using IDIOM protocols provide much needed and unprecedented insights into the role of supramolecular oscillatory surface forces in driving growth of such nanostructures as well as in controlling properties and stability of freely standing thin films and, more generally, of colloidal dispersions like foams.

500967 Contrasting Drainage and Stratification in Horizontal Vs Vertical Micellar Foam Films

Abstract:

Understanding and controlling the drainage kinetics of thin films is an important problem that underlies the stability, lifetime and rheology of foams and emulsions. In foam films formed with micellar solutions, the surfactant is present as interfacially-adsorbed layer at both liquid-air interfaces, as well as in bulk as self-assembled supramolecular structures called micelles. Ultrathin micellar films exhibit stratification due to confinement-induced structuring and layering of micelles. Stratification in micellar foam films is manifested as stepwise thinning over time, and it leads to the coexistence of flat domains with discretely different thicknesses. In this contribution we use Interferometry Digital Imaging Optical Microscopy (IDIOM) protocols to visualize and analyze thickness transitions and variations associated with stratification in micellar foam films made with sodium dodecyl sulfate (SDS) . We contrast the drainage and stratification dynamics in horizontal and vertical foam films. We experimentally study the emergence of thickness fluctuations near the borders and within thinning films, and study how buoyancy, capillarity and gravity driven instabilities and flows, are affected by variation in bulk and interfacial physicochemical properties dependent on the choice of constituents.

501544 Solvent Effects on the Crystallinity of Petroleum Asphaltenes

Abstract:

Petroleum asphaltenes are the most aromatic and heaviest fraction of crude oil and they can destabilize and precipitate on surfaces with changes in composition, temperature, and pressure [1]. The molecular structure of asphaltenes is that of a large aromatic core along with alkyl stabilizing side chains. Asphaltenes have been shown in primarily exist as stable semi-crystalline aggregates, called nanoaggregates (~2nm diameter), made of ~3-8 molecules in crude oils and good solvents. Previous research indicated that the aromatic areas of asphaltenes' molecules stack at a similar separation distance to graphite in the solid phase. However, it is reported that the correlations between asphaltene molecules can be influenced by the liquid surroundings. In this study, neutron diffraction has been used to determine the influence of the surrounding liquid on the crystallinity of asphaltene nano-aggregates (stacked graphite-like structure) in mixtures organic solvents (e.g., toluene).

This talk will present first of their kind neutron diffraction results that were generated for samples containing 5 vol. % asphaltenes dissolved in deuterated solvents and anti-solvents on the NOMAD instrument at the Spallation Neutron Source at Oak Ridge National Lab. Moreover, the crystallinity of different subfractions of asphaltenes (extracted from crude oil using different fraction of the precipitant) were studied using neutron diffraction technique on the NIMROD instrument at the Rutherford Appleton Laboratory. Results were instrument, solvent background and inelasticity effect corrected to isolate the scattering signal of the structures formed by asphaltene molecules.

In the neutron diffraction results, the peak corresponding to aromatic core stacking ([002] graphite peak) was observed in agreement with previous X-ray results. The [002] graphite peak is dominated by the correlation between asphaltene structures and represents the crystallinity of asphaltene nano-aggregates. The results showed the crystallinity is approximately constant with changes in the concentrations of asphaltenes or precipitant, indicating a highly stable aggregated state. The characteristics of other diffraction peaks varied with different kinds of solvents, which indicate different solvation mechanisms. The results at different temperatures will also be presented (25 – 240). The novel results generated with neutron diffraction provide necessary experimental data as a benchmark for molecular simulations and reinforce the need to consider the local solvent matrix on the structure and interactions of asphaltenes and other self-assembling hydrocarbon systems. This work also demonstrates the capability of neutron diffraction technique to probe assembly structures at Angstrom length scale of dispersed semi-crystalline materials.

[1] Hoepfner, M. P. and Fogler, H. S., "Multiscale Scattering Investigations of Asphaltene Cluster Breakup, Nanoaggregate Dissociation, and Molecular Ordering," Langmuir, 29 (49), 2013.

498108 Role of Bifurcation Geometry on Stability of Thrombus

Abstract:

An injury or a high shear gradient triggers the formation of a thrombus in a blood vessel. Though thrombus formation under flow involves understanding reaction kinetics and hydrodynamics in a dynamically changing geometry, the stability of thrombus is described through hydrodynamic forces around it. Platelet aggregation rate on a thrombus has a non-linear dependence on shear rate exhibiting a zero-aggregation rate at a critical shear rate. Complete occlusion or partial occlusion of a thrombus depends on the shear rate around the thrombus, controlled through fluid dynamics and geometry of the vasculature. Bifurcations are repeated structures in human vasculature geometry regulating blood flow and hydrodynamic forces in the blood vessel. The flow through one branch of a bifurcation is regulated by the other branch and vice-versa. The critical shear rate behaviour results in a critical bifurcation radius ratio determining the dynamics of occlusion of an evolving thrombus. In this study, we quantify the role of bifurcation geometry on thrombus evolution and stability through finite element simulations. Our theory is further validated using microfluidic blood flow experiments.

504134 Using Î¼²rheology to Measure Rheological Properties of Hydrogenated Castor Oil

Abstract:

Rheological modifiers are often used as ingredients in home care goods. These materials undergo phase changes during manufacturing and regular use. Rheological and material properties can be altered by repeated phase transitions, and it is important to know the extent to which microstructural changes affect the performance of the end use product. In this work, we investigate a hydrogenated castor oil (HCO) gel, a rod-like colloid (aspect ratio 50-2500) that forms a stable aqueous suspension with the addition of a surfactant and undergoes a phase transition due to an osmotic pressure gradient. Î¼²rheology, the combination of microfluidics and multiple particle tracking microrheology (MPT), is used to measure material properties during transition. A new microfluidic device design enables the exchange of fluids while minimizing shear stress on the sample. In MPT, fluorescent probe particles are embedded in the sample and their Brownian motion is recorded using video microscopy. The particle trajectories are then related to rheological properties using the Generalized Stokes-Einstein Relation. MPT data are compared to phases previously determined by Wilkins et. al., 2009. The phases of rod-like systems are network, where the particles have a sample spanning structure; a bundled phase where rods do not create a network; and a transitional phase where the colloidal particles partially lose scaffold structure and begin forming bundles. MPT results show that a 4wt% HCO gel undergoes a repeatable phase transition between the network and transitional phases due to the osmotic gradient, but never reaches the bundled phase. A 0.125wt% solution of colloidal rods that was prepared using a large amount of shear undergoes repeatable phase transitions but does not form the network structure measured in the 4wt% gel. This is due to the shear added during sample preparation permanently breaking the network structure. This work shows that sample preparation can have irreversible effects on the microstructure of the gels, which can cause changes in end use products made with rheological modifiers.

498269 CFD Simulation of Bubble Columns Operating in Heterogeneous Regime

Abstract:

Bubble columns are widely used for gas-liquid contact, applied in several chemical process, they are often used for promoting high heat and mass transfers rates combined with low operating and implantation costs. For most industrial applications the multiphase flow occurs with high superficial gas velocities consequently, the column operates at a heterogeneous regime. The main difference between the homogeneous and heterogeneous regimes lies within the superficial velocity magnitude, while low superficial velocities characterize homogeneous regime with low presence of breakage and coalescence the heterogeneous flow presents a higher superficial velocity and grater turbulence, and so the breakage and coalescence phenomena have a significant importance. The heterogeneous regime flow characteristically allows the achievement of higher heat and mass transfer rates, which makes it industrially relevant. Bubble columns have a large application as well as a great potential use as multiphase contactors and multiphase reaction systems at several industrial areas. Therefore, the correct scale-up of this equipment is crucial to the plants operation. The successful scale-up is highly dependent of the correct prediction of the mixing and of fluid dynamics. A brief literature assessment allows the conclusion that besides the fact that heterogeneous regime is largely verified in industrial operations few are the studies focused on such regime. Therefore, several issues remain unanswered so far. In this way, the investigation of bubble columns fluid dynamics is proposed for the heterogeneous regime. Numerical simulations were conduct with a CFD commercial code using an Euler-Euler methodology. The biggest problem of fluid dynamics simulation for this kind of systems by CFD technics is the dynamic bubble behavior. For that, a population balance is solved applying a class method. The numerical results of holdup and bubble velocities were compared with literature experimental data of a bubble column operating with high superficial velocity.

506178 Deposition and Oil-Brine Interfacial Rheology of Asphaltene-Stabilized Emulsions

Abstract:

Asphaltene-related flow assurance problems are prevalent in the oil production process, and are at the heart of issues such as the plugging of pipelines, the damage of rock formations, and the formation of viscous emulsions. Understanding of the interfacial behavior of asphaltenes is required to design the effective solutions to these problems. In this study, we elucidate the deposition behavior of asphaltene-stabilized water/brine-in-oil emulsions in microfluidics and further use the double-wall ring (DWR) rheometer to characterize the rheological responses of the asphaltene interface. Firstly, various oil-to-water ratios (water concentration: 0.03 wt% to 20 wt%) are studied; we find that lower ratios give rise to higher permeability reduction, and that the deposition mechanism changes from adhesion to bridging. Moreover, less deposition is found with more salts in the adhesion-dominated stage, while more severe pressure drops are observed in the bridging-dominated stage, which is caused by larger emulsions in the presence of salts. Also, interfacial rheological elasticity of the interface reduces, and the disjoining pressure decreases with higher salinity. This combining effect results in larger emulsions. This alteration causes stronger erosion in the adhesion stage and facilitates blockage in the bridging stage. The zeta potential can be neutralized by salts, suggesting the possible screening effect of charged asphaltenes at the interface on the electrostatic attraction, and the inter-droplet disjoining pressure.

502794 Vapor Recompression Mechanism Introducing in Internal Heat-Integrated Distillation Column: Impact of Internal Energy Driven Intermediate and Bottom Reboiler

Abstract:

A novel combination of internally heat-integrated distillation column (HIDiC) and vapor recompression column (VRC) with intermediate reboiler (IR) is proposed. Supplying heat at the highest temperature point (i.e., column bottom) of the VRC scheme is not thermodynamically favorable and, therefore, we aim to install the IR for better distribution of heat along the column length, thereby reducing the compressor work. Introducing IR in the combined HIDiC-VRC system formulates an open-loop variable manipulation policy to evaluate the comparative impact of internal and external heat sources on bottom liquid reboiling. With internal energy driven bottom reboiler, we further investigate the hybrid HIDiCVRCIR column with proposing the two modes of compressor arrangement, namely parallel and series. Finally, a multicomponent distillation system is exampled to show the promising potential of the proposed HIDiC-VRCIR configurations in improving the energetic and economic performance over the HIDiC-alone and HIDiC-VRC schemes with reference to a conventional standalone column.

496355 Experimental Studies Controlling Trace Components in a Dividing Wall Distillation Column

Abstract:

Experimental Studies Controlling Trace Components in a Dividing Wall Distillation Column

Melissa Donahue, Michael Baldea, R. Bruce Eldridge

McKetta Department of Chemical Engineering and

James R. Fair Process Science and Technology Center

The University of Texas at Austin

An example of one of the prominent technologies in the field of process intensification is dividing wall distillation columns or DWCs. As part of an ongoing experimental study of dividing wall columns at The University of Texas at Austin[1], the design and selection of control strategies for a pilot plant DWC will be presented. The pilot plant column is six inches in diameter and approximately 30 feet tall. The column has 24 theoretical stages of structured packing and a welded wall in the horizontal and vertical middle of the column[1]. To investigate the operational flexibility of this fixed design DWC, a fourth trace component is added to the feed. A sensitivity analysis studying the minimum energy usage and optimal distribution of impurities is conducted using a dynamic process simulator. These operating targets and conventional distillation temperature control techniques are used to design control configurations that are then implemented on the pilot scale dividing wall column. The control configurations are tested against changes in set point and feed flow and composition disturbances.

(1) Blevins, T.; Downs, J. J.; Donahue, M. M.; Roach, B. J. Use Model Predictive Control to Achieve Real-Time Management of a DWC. *Hydrocarbon Processing* **2015**.

UNIVERSITY OF BATH

CASCADE REACTION IN A SPINNING MESH DISC REACTOR USING A NOVEL METAL- ENZYME CATALYST

PARIMALA SHIVAPRASAD
Department of Chemical Engineering
ps784@bath.ac.uk

Emma Emanuelsson*, Darrell Patterson*, Matthew Jones**
* Department of Chemical Engineering, University of Bath
** Department of Chemistry, University of Bath

Bath - A University in a UNESCO World Heritage City

UNIVERSITY OF BATH

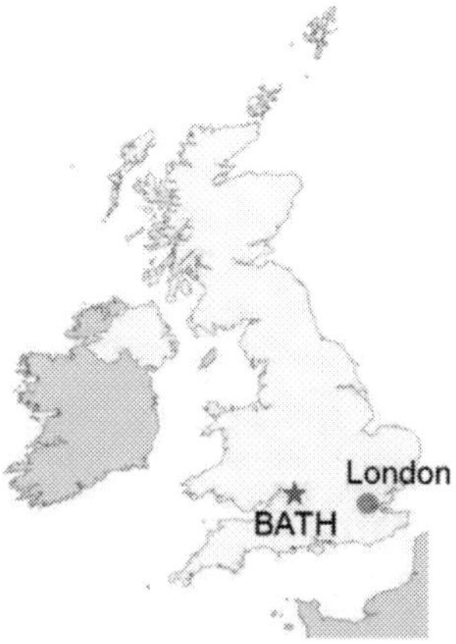

Outline

Background

The Idea

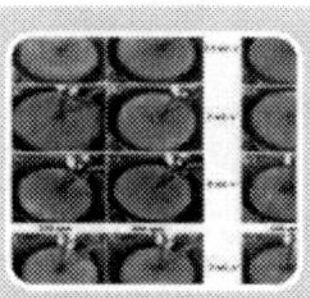

Results

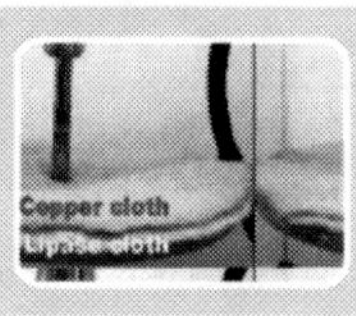

The Path Forward

Problem

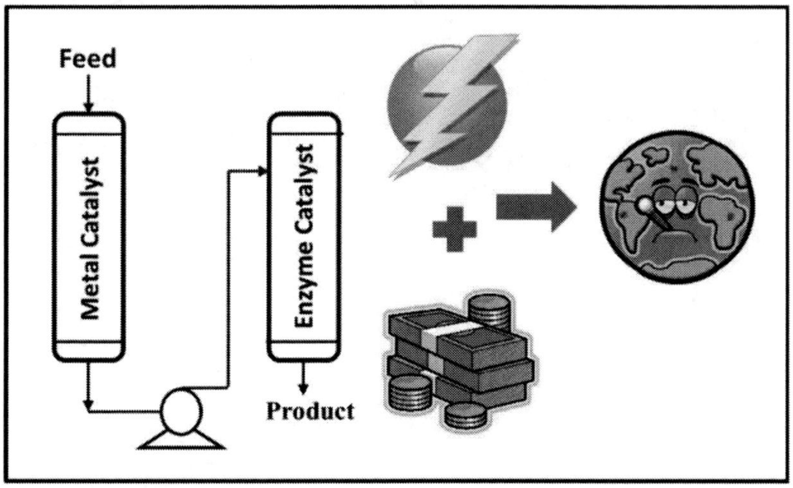

Multi-step processes lead to higher recovery and energy costs.

Solution

Imagine a reactor that could do a cascade of reactions without the need for product purification in between!

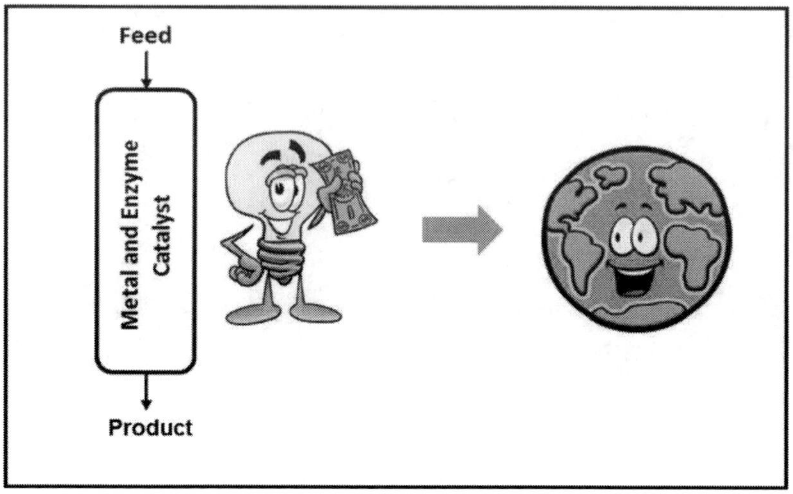

Catalyst immobilisation on wool

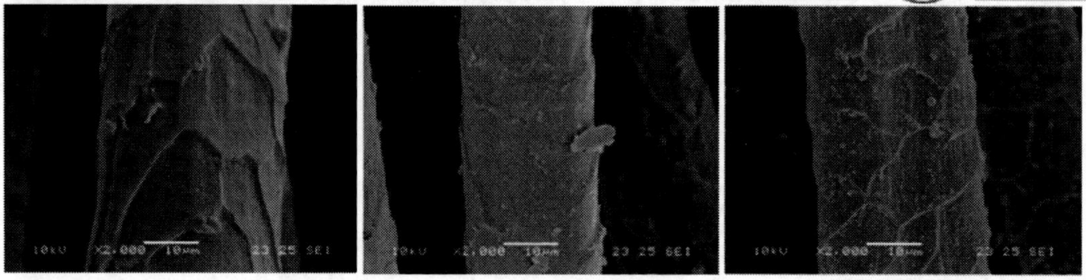

Shivaprasad, P., *et al.(2017) Chemical Engineering and Processing: Process Intensification.*

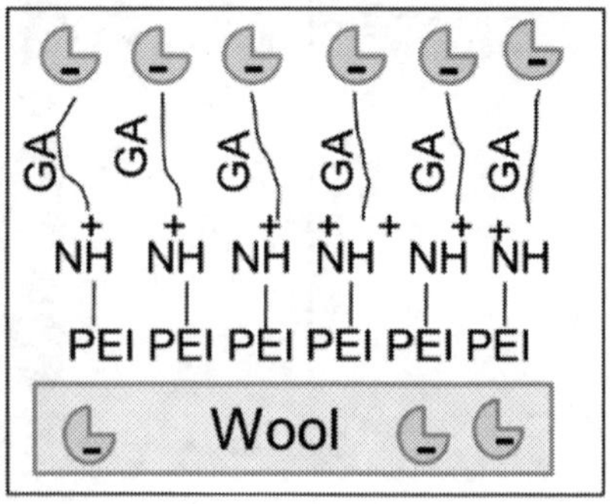

Feng, X., *et al.(2012) Colloids and Surfaces B: Biointerfaces*, 102, pp. 526-533

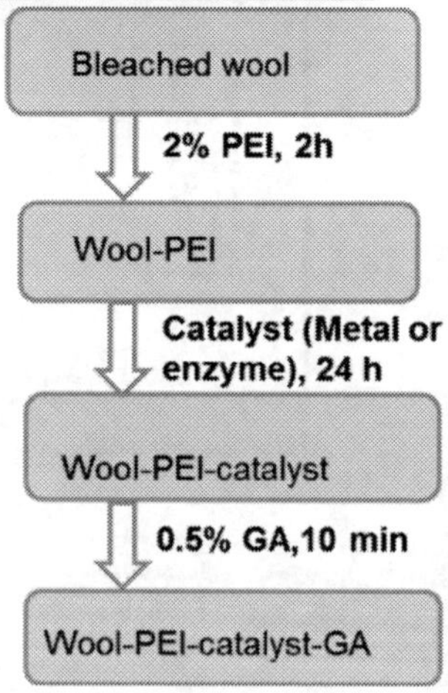

The Idea: Process Intensification

UNIVERSITY OF BATH

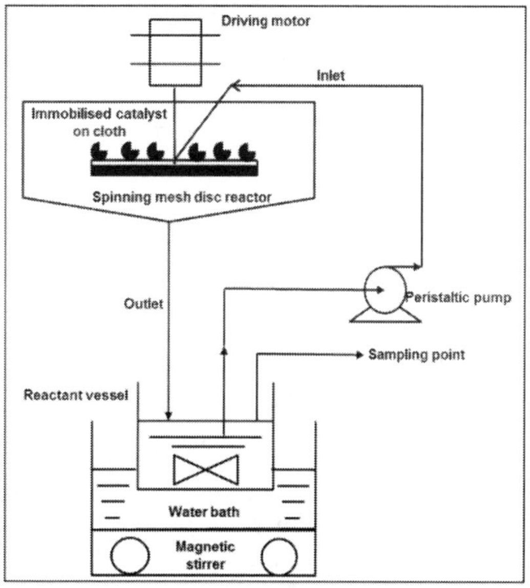

Spinning Mesh Disc Reactor (SMDR)

The SMDR: Characteristics

UNIVERSITY OF BATH

Reactor characterisation for enzymatic hydrolysis of tributyrin

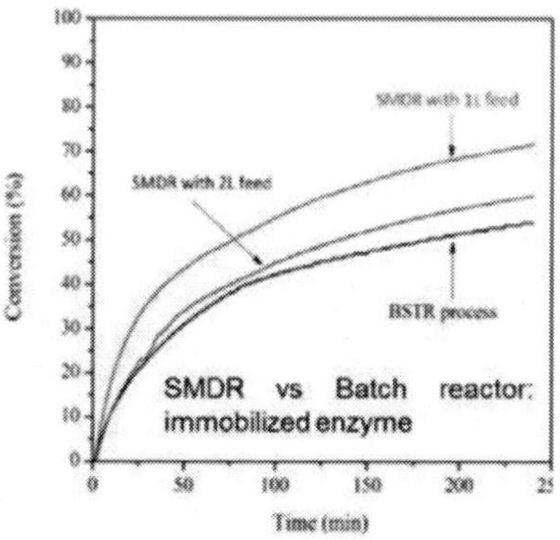

(a)

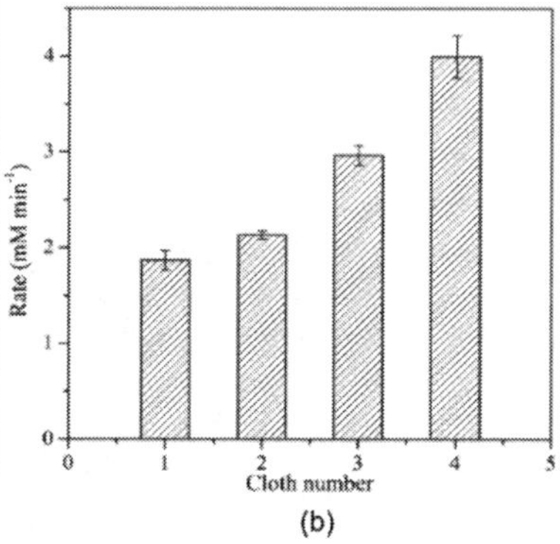

(b)

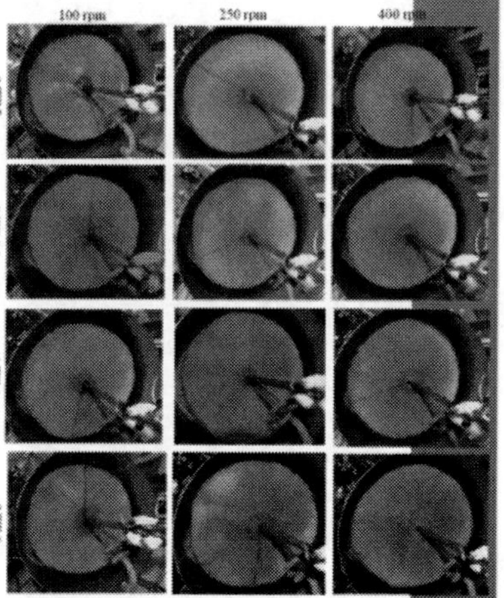

(c)

Spinning Mesh Disc Reactor

- ✓ Process Intensification
- ✓ High Reusability of enzyme
- ✓ Scale-up

Feng *et al.* (2013) *Chemical Engineering Journal*, 221, 407-417

Towards a one pot cascade reaction

UNIVERSITY OF BATH

Dynamic Kinetic Resolution (DKR) of Henry Reaction

Perfecting individual reactions in the SMDR

STEP 1: Copper catalysed Henry reaction

STEP 2: Enzyme catalysed kinetic resolution of racemic alcohol

One-pot cascade reaction

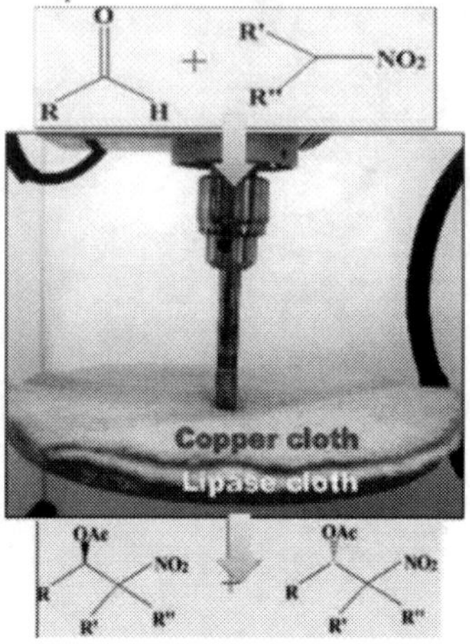

Copper-lipase catalyst in tandem

The Henry reaction

UNIVERSITY OF BATH

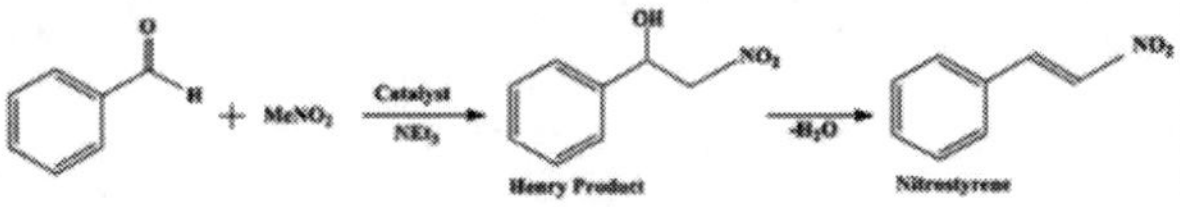

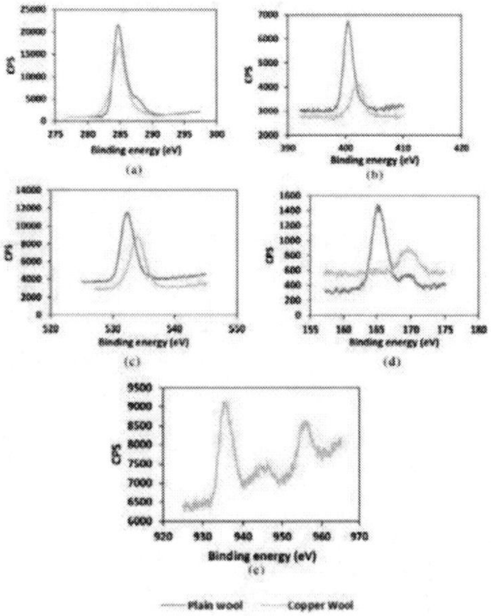

Process intensification of catalysed henry reaction using copper-wool catalyst in a spinning mesh disc reactor

Parimala Shivaprasad[a], Matthew David Jones[b], Darrell Alec Patterson[a], Emma Anna Carolina Emanuelsson[a,*]

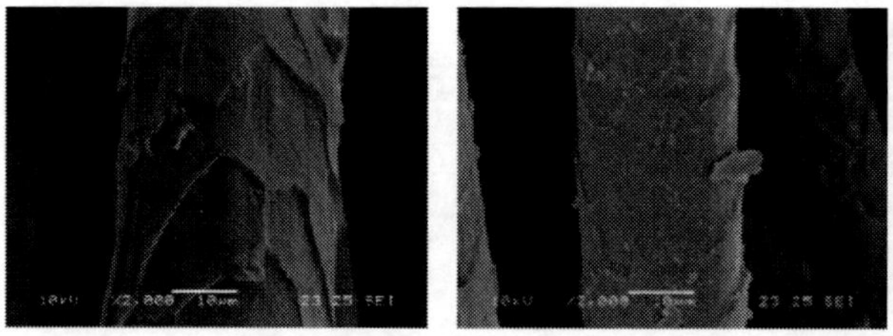

Copper triflate was successfully immobilised on wool

The Henry reaction

Henry reaction with free and immobilised catalyst in batch

Catalyst	Time (h)	% Conversion
Free copper triflate	4	38
(homogeneous batch)	6	42
	24	57
Copper cloth	4	41
(heterogeneous batch)	6	63
	24	85

Henry reaction using the copper cloth in the SMDR

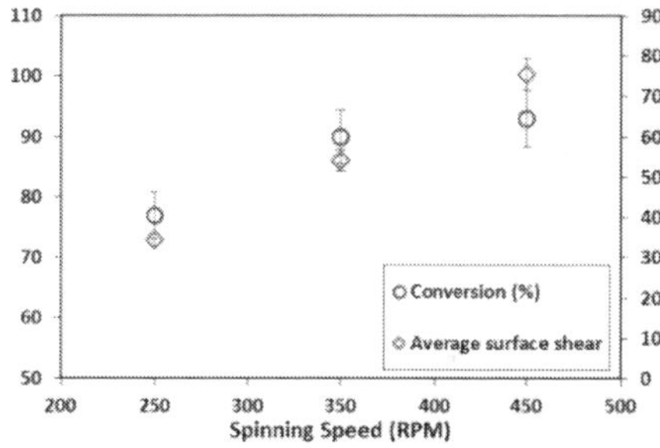

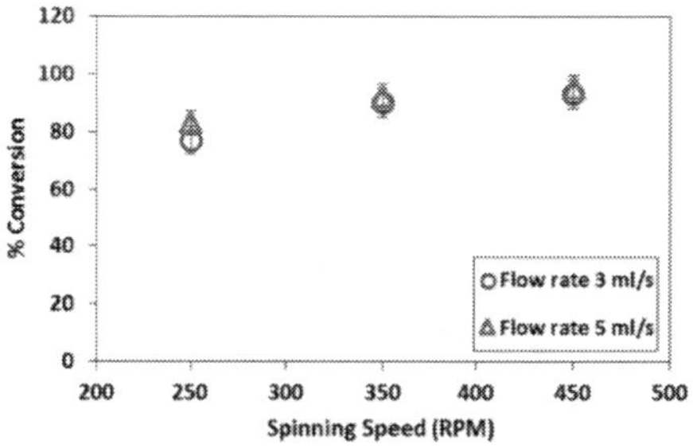

- ✓ Copper immobilised on wool showed a higher conversion in batch.
- ✓ Higher reaction rate was observed in the SMDR using the copper cloth
- ✓ The SMDR was successfully applied for a solvent based reaction.

Shivaprasad, P., *et al.*(2017) *Chemical Engineering and Processing: Process Intensification.*

Wool as a catalyst

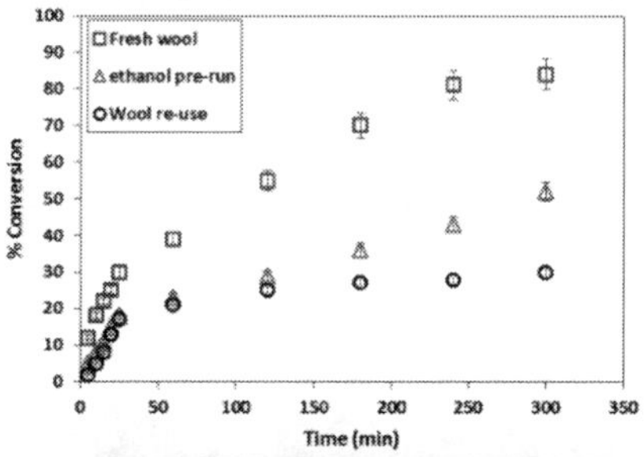

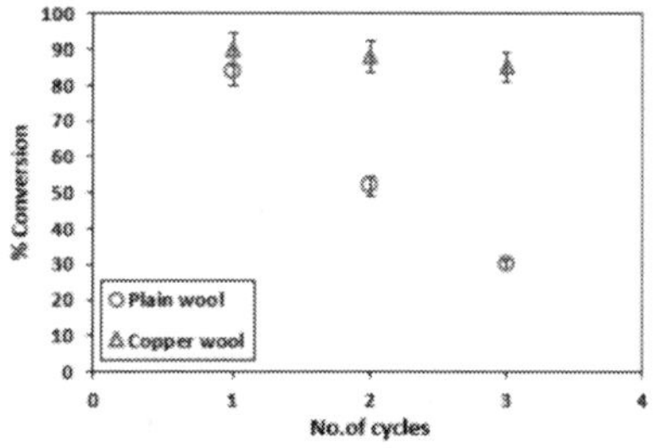

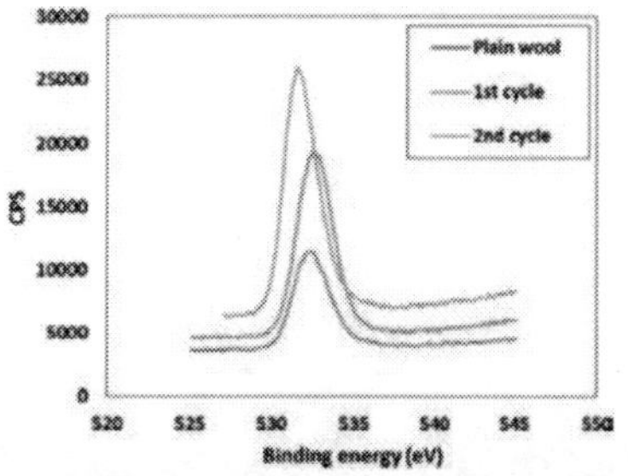

- ✓ Wool was found to catalyse the Henry reaction but was not re-usable for multiple cycles.
- ✓ Copper triflate immobilised on wool was more robust than plain wool.

Shivaprasad, P., *et al.*(2017) *Chemical Engineering and Processing: Process Intensification.*

Enzymatic kinetic resolution in the SMDR

UNIVERSITY OF BATH

Kinetic resolution of 1-phenyl ethanol catalysed by lipase

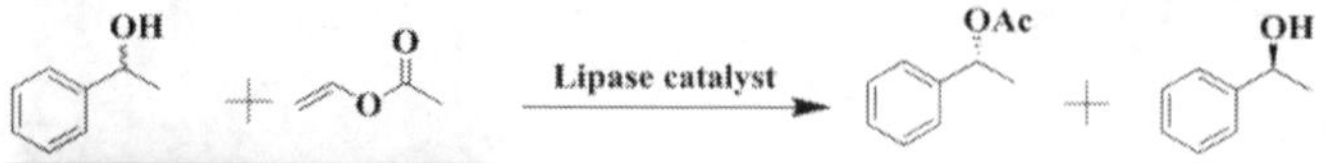

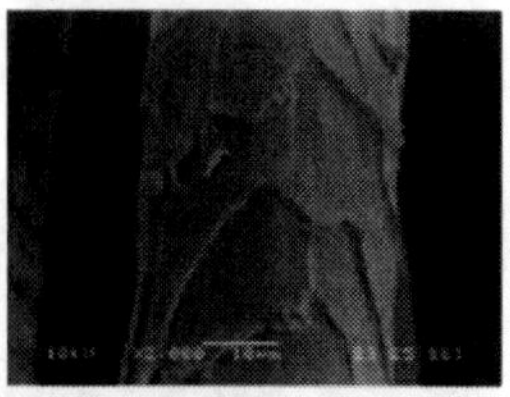

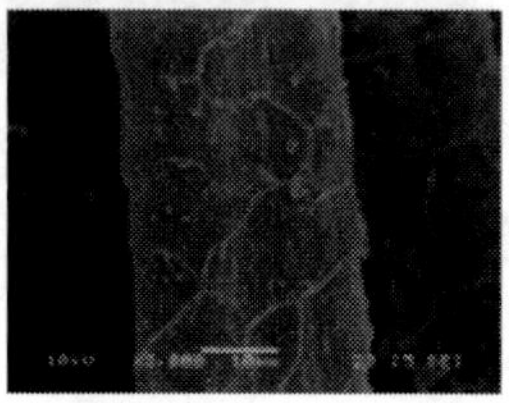

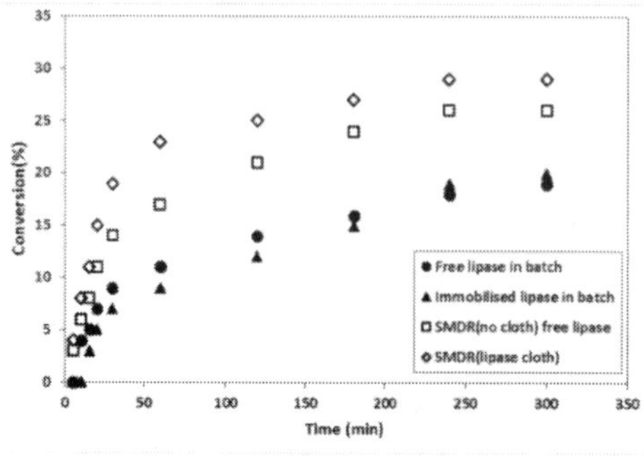

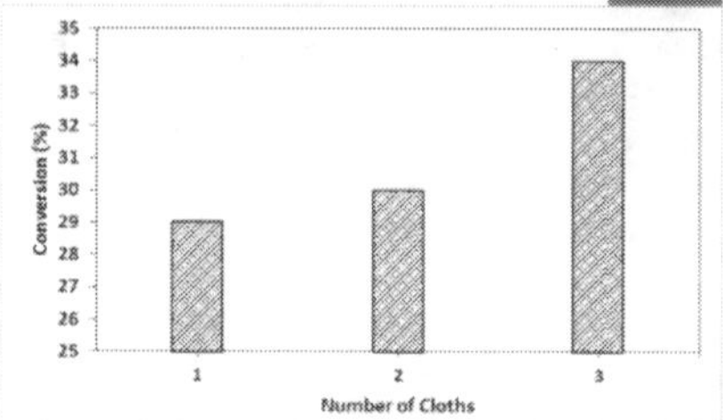

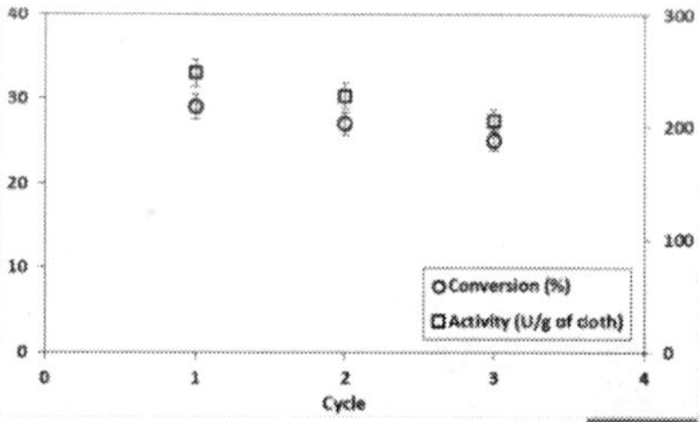

- ✓ Lipase immobilise on wool was used for the KR reaction in both batch and SMDR
- ✓ The reaction rate in the SMDR was higher for both free and immobilised lipase compared to the batch reaction.

TOWARDS CASCADE REACTIONS

UNIVERSITY OF BATH

Can copper and lipase immobilised on wool catalyse the reaction in tandem?

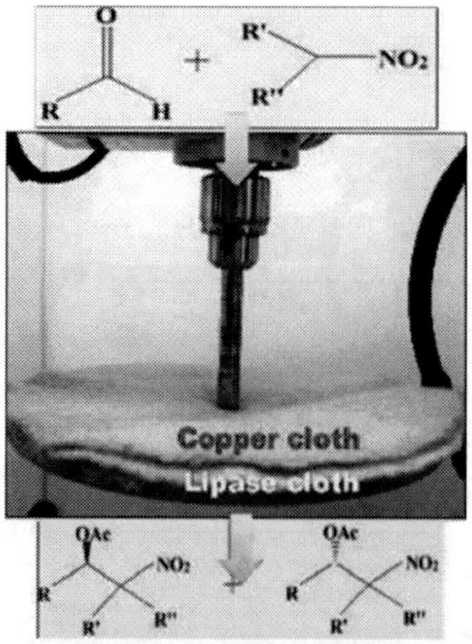

Challenges

- ✓ Solvent compatibility
- ✓ Robustness of the metal and the enzyme catalyst in one-pot
- ✓ Reaction kinetics

SUMMARY

Process intensification compared to more conventional reactors under equivalent operating conditions:

- enhanced reaction rate
- enhanced enzyme stability

The SMDR was used to demonstrate traditional organic synthesis reactions:

- The Henry reaction
- Enzyme catalysed kinetic resolution

Future work

One-Pot Cascade reaction

- Different enzyme and catalyst

Spinning Mesh Disc Reactor – further characterisation

- Increasing cloth size vs stacking cloths?
- A prototype for 'Perfect Reactor' – The future of process intensification

Acknowledgement

UNIVERSITY OF BATH

- Emanuelsson and Patterson Research Group
- Dr. Xudong Feng and Kemi Lawal
- The Santander mobility fund
- University of Bath and European Union's Seventh Framework Programme for research funding.

QUESTIONS?

484756 Scale and Effects of Catalyst Deactivation in Enzymatic Catalyzed Reactive Distillation

Abstract:

Increasing global competition and environmental legislation require the chemical sector to produce more cost efficient and reduce emissions. Process integration reduces the energy consumption of a process as well as its investment costs. An example for an integrated process is the combination of reaction and separation in form of the reactive distillation (RD) or reactive dividing wall column (RDWC). However, many of the industrially employed catalysts also enhance the production of unwanted side products that require additional, expensive purification steps. The employment of enzymatic catalysts is a promising approach to greatly increase the selectivity and reduce the required temperatures for chemical reactions. As enzymes are often product inhibited, the in situ product removal of a reactive distillation allows fully taking advantage of the enzymeâ€™s potential. Studies on laboratory and pilot plant scale columns have already shown the feasibility of this approach[1,2]. For the next step, the industrial application in a continuous reactive distillation, the stability of the enzymes is a fundamental condition, but yet untested under operating conditions. Therefore, knowledge about the scale of deactivation, the influencing factors and how to operate a column to minimize deactivation and maximize the production time before a standstill is required, are key questions still to be answered.

Our research aims to systematically investigate the behavior of eRDWCs. The first step has been to show the general feasibility of a successful operation with a continuous eRDWC 12 m high pilot plant.[1] Focus of this work was to develop a start-up strategy and to show that the column can be operated successfully for different operating conditions. Using the acquired experimental data a comprehensive simulation model was validated. The next step is a comprehensive approach to answer the questions about enzyme stability that hinder the industrial application. To analyze the scale and driving forces of enzyme deactivation two different experimental set-ups have been built. First, the eRDWC pilot plant and additionally a fixed-bed reactor have been employed in long term studies to test enzyme deactivation. After examining the experimental results on the quantity and driving forces of deactivation the second part of the project uses the developed mathematical model to analyze the influence on deactivation. The combined results from experimental and simulation studies allow a much deeper understanding about how to minimize enzyme deactivation in reactive distillation and additionally how to achieve the required conversion and purities for reduced enzyme activities.

This presentation will give a comprehensive overview about the scale and underlying causes of enzyme deactivation. A short introduction will explain how an eRDWC works and when to employ it. The simulation model and pilot plant will be presented. Selected experimental results about enzyme activity from the eRDWC and fixed-bed reactor experiments will be shown. From this data conclusions will be drawn about scale and influencing factors. Finally, it will be explained how to limit the enzyme deactivation by employing a smart process control and how to maintain the product specifications if deactivation occurs.

References

1.Egger T, Fieg G. Enzymatic catalyzed reactive dividing wall column: Experiments and model validation. AIChE J. 2016. doi:10.1002/aic.15598 .

2.Heils R, Jensen J, Wichert S, et al. Enzymatic Reactive Distillation: Kinetic Resolution of rac -2-Pentanol with Biocatalytic Coatings on Structured Packings. Ind. Eng. Chem. Res. 2015;54(38):9458-9467. doi:10.1021/acs.iecr.5b02802 .

498215 Use of Computational Fluid Dynamics (CFD) Simulation and Image Analysis Tool for Modelling Light in a Microalgal Photobioreactor

Abstract:

Light, one of the most important parameter for the growth of microalgal cells, is converted into chemical energy by the process of oxygenic photosynthesis. Microalga is a valuable feedstock for the production of industrially important metabolites. The energy for algal growth is obtained from the visible light within the photosynthetically active radiation (PAR) ranging from 400 to 700 nm wavelength. The high density of microalgal cultures is usually attained by cultivation in the closed photo-bioreactors (PBR) as compared to open systems due to better process control, mixing, light distribution etc. However, there is a need for optimizing the cultivation processes that should support both autotrophic and mixotrophic cultivation. Light in particular is known to directly influence the algal biomass production and product yield (Kumar, Sirasale, & Das, 2013). The conventional algal PBRs suffers from self-shading effect which is caused when microalga attains high cell density, thereby reducing the light penetration inside PBR and leading to development of dark zones. This is mainly because gradient of incident light intensity varies along the radius of PBR. Thus, local light intensity distribution at any given point inside the PBR is dependent on the path length, light penetration and absorption coefficient that becomes much acute especially during scaling-up of the process. Existing PBR designs do not have the desired efficiencies, and lead to high material and energy costs. Moreover, predicting the hydrodynamics (fluid flow pattern) of a PBR in laboratory experiments is difficult and cumbersome. Therefore, novel PBR designs need to be developed and benchmarked against existing designs.
Light utilization in the microalgal PBRâ€™s is usually measured by incident light intensity on the surface of the PBR or the irradiating surface area per unit culture volume. Average light intensity received by the microalgal cells is used for measuring the light utilization or photosynthetic efficiency of photosynthetic microorganisms. Efficient mixing allows cells to continuously pass through various zones inside a PBR such as complete dark, photo-limiting (cell growth increases with the light intensity), saturating and photo-inhibition (increase in light intensity leads to growth inhibition) and gives them flashing light effect. Despite several simulation studies on PBRs, a comprehensive analysis and evaluation of light penetration in PBRs has been limited.
The present study discusses use of CFD simulation for designing a novel airlift PBR with improved hydrodynamics of fluid flow. The novel PBR will be used for the cultivation of microalgae. The designed reactor performance in terms of light distribution will be compared with the conventional bubble column PBR. Since the application of multiphase CFD is complicated due to difficulty in describing the variety of interactions in these systems. The hydrodynamics of the reactor will be simulated using CFD and later will be used for microalgal cultivation. The present work describes the use of image analysis technique for the measurement and tracking of cell growth at different growth phases of a live bio-culture in the indigenously designed novel PBR. The light penetrability of incident light intensity was assessed in a Chlorella sp. culture throughout a batch experiment. The light penetration profile could be generated following modified Beer-Lambertâ€™s law by spectrophotometric method. Digital images of the top view of the PBRâ€™s were taken and processed using image processing tool in the MATLAB software. This was used to estimate the light intensity distribution in the externally radiating column PBR across the radial path length. Light gradient pattern inside the algal suspension was assessed by combining fluid dynamics simulations of three-dimensional turbulent single-phase fluid flow with statistical particle tracking and signal analysis. (Perner-Nochta & Posten, 2007). The experimental results and CFD simulations exhibited a good fit. Also, the light penetration was mapped with the image analysis tool for a better understanding of light availability and detection of dark zones inside a working PBR.

503392 Interfacial Dynamics between Pathogenic Nanoparticles and Cell Membrane Surfaces

Abstract:

Small vesicles derived from live, biological cell membranes are ubiquitous in nature and found among both mammalian and bacterial cells. These vesicles range from the nano- to micro- scale in diameter and can serve a variety of purposes. Examples of such particles include exosomes, for membrane trafficking of proteins; microvesicles (MVs), derived from cancer cells; outer membrane vesicles (OMVs), produced by bacteria like *E. Coli*; and membrane-enveloped viruses, borne from infected mammalian cells. What is common to all these particles is that they are essentially cell membranes encapsulating an aqueous compartment filled with biological materials like proteins and RNA. Another common theme is that these particles are thought to be critical in cellular reprogramming. While this cellular reprogramming can serve physiological roles it also enables pathological changes. In the case of cancer, for example, vesicle-mediated cellular reprogramming leads to the formation of tumor-promoting microenvironments, biofilms in the case of bacteria, and the spread of infection in the case of viruses. Hence, gaining an understanding of how these particles interact with â€œhostâ€ cell surfaces and which specific material properties of host cell surfaces facilitate these interactions is important for the development of strategies to interrupt these outcomes. Improved understanding of how interfacial properties of cellular membranes modulate the dynamic interactions between particles and host cell surfaces, in turn, makes it possible to use that knowledge for beneficial purposes, like controlling the regrowth of damaged tissue, or designing next-generation particles for targeted drug delivery. But, our understanding of how microvesicles and cell surfaces interact is very limited. That is, how the particles interact with the cell surface and deliver their message is unknown in most cases, with one exception: viruses. Virus infection has been studied in much detail over decades. However, only recently has it become possible to study molecular-scale interactions of viruses with host cell membranes and the resulting consequences on virus entry and delivery of viral genomes across membranes. Detailed studies of these interactions are enabled by modern tools like microfluidics, biomimetic membrane materials, and sophisticated microscopy. Several groups, including the Daniel group, have pioneered single particle tracking (SPT) to directly observe virus binding to, and fusion with, host membranes for the successful delivery of viral RNA across cell membranes. Because individual virions can be monitored with SPT, details about on and off binding rates, binding residence times, and intermediate fusion steps like membrane merging and pore formation can be timed and analyzed for individuals and populations. This quantitative approach to characterizing the dynamic interactions of virus with host cell surface has provided critical new insights about virus infection, its dependence on the material properties of the membranes (*e.g.*, composition, fluidity, charge, *etc.*), and pinpoints new targets for therapeutic intervention. Our recent work leverages the tools and biomimetic membrane materials we have pioneered for the study of dynamic interactions of viruses with host cell surfaces to understand the dynamic interactions of other kinds of microvesicular particles with their target cell surfaces. In this plenary talk, I will give an overview of our work on understanding pathogenic nanoparticle interactions with cell surfaces and the exciting, new directions possible for study, enabled by high resolution microscopy techniques and quantitative analysis.

499151 Bicontinuous Microemulsions in Homopolymer-Block Copolymer Blends

Abstract:

A bicontinuous microemulsion (BmE) is a familiar state of structured soft material enabled by the interfacial action of properly configured surfactants in combination with immiscible oil-water mixtures. Similar interpenetrating domains with dimensions ranging from 10 nanometers to 1 micron can be created by blending appropriate amounts of two thermodynamically incompatible homopolymers with the corresponding diblock copolymer. The resulting phase prism, containing three composition variables and temperature, is characterized by various nanoscale ordered phases and macroscopic phase separation at high and low copolymer concentrations, respectively. Increasing temperature leads to disorder (DIS) and a state of phase mixing. We have explored the phase behavior of a model ternary system comprised of poly(cyclohexylethylene) (PCHE), poly(ethylene) (PE) and the associated PCHE-PE diblock copolymer. These materials were synthesized by anionic polymerization of styrene and butadiene followed by heterogeneous catalytic hydrogenation. The BmE state exists within a relatively narrow portion of the phase prism bounded by ordered lamellar (LAM) and the 2-phase region at concentrations of PCHE and PE that result in zero mean interfacial curvature. Small-angle neutron and x-ray scattering (SANS and SAXS) and transmission electron microscopy (TEM) have been employed to probe the structure of the BmE and the line of congruent LAM-DIS transitions, which define the peak temperature separating the LAM and DIS states, terminating at the BmE. Dynamic mechanical spectroscopy (DMS) experiments have established the viscoelastic behavior of the mixtures as a function of temperature and composition. These results will be discussed in the context of self-consistent mean-field theory (SCFT), highlighting the role of fluctuation effects.

* Research conducted in collaboration with Robert Hickey, Timothy Gillard, Matthew Irwin, David Morse, and Timothy Lodge.

500855 Self-Assembly of Block Copolymers: From Nanostructure to Function to Applications

Abstract:

Self-assembly leads to diverse and interesting properties and function which emanate from the nano- and meso-scale organization of constituents such as surfactants, lipids, and polymers. Block copolymers, in particular, are prime examples of tunable materials. Selective solvents provide valuable degrees of freedom for controlling the self-assembled morphology and, hence, structure/property relationships and can dramatically affect the local mobility. Incorporation of "hard" nanoparticles into a "soft" matrix can modify dramatically the structure and dynamics, and also confer novel properties (optical, electrical, catalytic).

The presentation will highlight the interplay between (A) fundamental aspects (thermodynamics, interactions, structure, dynamics) of complex fluids and soft materials based on the self-assembly of block copolymers in selective solvents, and (B) applications of such self-assembled systems in the (a) structuring of waterborne fluids and surfaces with properties tailored for pharmaceutics and biomedical applications, (b) environmentally benign synthesis of metal and semiconductor nanoparticles in a size- and shape-controlled manner, and (c) formulation of polymer gels with ionic liquids as potential electrolytes for energy applications.

The utilization of block copolymers of the poly(ethylene oxide)-block-poly(propylene oxide) (PEO-PPO) family, commercially available as Pluronics or Poloxamers and approved for pharmaceutical use, enables the direct translation of fundamental knowledge to practice.

500365 Rational Design and Synthesis of Surfactant-Free Pt-Based Ternary Nanoalloys As Superior Oxygen Reduction Reaction Electrocatalysts

Abstract:

Efficient and low cost electrocatalysts are indispensable for electrochemical oxygen reduction reactions (ORR) during proton exchange membrane fuel cell (PEMFC) operations. Yet, the challenge for solution phase synthesis of nanocatalysts with low precious metal loading, high durability and catalytic activities remains in tailoring their architectures and compositions without using unwanted chemicals and surfactants/ligands that are known to impede their active catalytic sites. We present to synthesize ternary nanoalloys (NAs) of Pt with transition metals (Co, Cu, Ni, Ti, Ru, Mn) as ORR electrocatalysts using our recently developed laser ablation synthesis in solution-galvanic replacement reaction (LASiS-GRR) technique as a facile and surfactant-free nanomanufacturing route. The specific choice of the elemental compositions is driven by the respective target metal/metal salt redox potential gaps as well as the target metal and/or, metal oxide solubility in desired acids. The high-energy thermodynamics of the LASiS-GRR process enables control of the sizes, elemental compositions and distributions of the ternary NAs through systematic tuning of initial metal salt concentration, pH, laser fluence and ablation time. Specifically, the PtCuCo NAs synthesized with an elemental composition of 72:12:16 (Pt:Co:Cu) indicates the best ORR catalytic activity. The NAs largely possess shell-core structure with the shell comprising mostly of Pt and minor amount Cu, along with a uniformly alloyed PtCuCo core. Mass and specific activities for ORR performance of the NAs indicate a 4 and 6.5-fold improvements respectively over the corresponding activities of commercial Pt/C. We attribute the enhanced activity to 1) our surfactant/ligand-free synthesis technique that prevents catalytic site degradations; and 2) minor alloying of the second transition metal Cu that shifts back the Pt d-band center to an optimal position between that of Pt and the PtCo binary NAs, thereby tuning their binding affinities for both oxygen and oxygenated species. Finally, this work establishes the versatility of the LASiS-GRR technique through the synthesis of other ternary NAs (PtRuNi, PtCoMn, PtNiTi) that also exhibit reasonably good ORR activities.

496706 Improved Activity and Durability of Ionic Liquid Composite Nanoporous Nanoparticle Electrocatalysts for Oxygen Reduction Reaction

Abstract:

The development of efficient and stable catalysts for the cathodic oxygen reduction reaction (ORR) is one of the greatest challenges for the exploitation of polymer electrolyte membrane fuel cells (PEMFCs). Dealloying has shown increasing utility in the field of electrocatalysis as a tool for the synthesis and development of nanoporous materials possessing high surface-to-volume ratios with controlled morphology and compositional gradient (core-shell structure) [1]. After electrochemical dealloying, the open, bicontinuous, three-dimensional nanoporous nanoparticle electrocatalysts exhibit dramatically enhanced electrocatalytic properties [2,3]. We have taken advantage of the free volume within these nanoporous nanoparticles and created a composite catalyst architecture through the incorporation of a hydrophobic, protic ionic liquid (IL), specifically [MTBD][beti]. Modification of the interface through the formation of IL thin film on the surface of an electrocatalyst yields enhanced catalytic rates by chemically biasing reactants to the surface and changing the degree of solvation/stabilization of adsorbed intermediates.

In the development of efficient electrocatalysts for ORR, durability is too often ignored in the pursuit of higher activities. Here we studied the manner by which ILs impact the stability of 3-dimensional, nanoporous nanoparticle electrocatalysts. In addition to the standard mechanisms of catalyst degradation including Pt dissolution/Ostwald ripening and coalescence/aggregation, nanoporous materials can lose electrochemically active surface area (ECSA) through surface smoothening driven coarsening. We use a combination of in-situ and ex-situ experimental techniques, under accelerated degradation conditions, to develop insight into the structural and compositional evolution of nanoporous PtNi nanoparticles (np-NiPt) formed through the dealloying of $Pt_{20}Ni_{80}$precursor nanoparticles, and demonstrate that facilitated diffusion of stable surface atoms under relevant electrochemical conditions will drive evolution of the particle morphology from a porous to a nonporous structure. With a better understanding of the interplay between nanoporous structure coarsening and transition metal loss, we have developed strategies to mitigate coarsening through incorporating ILs at the metal/electrolyte interface to directly mitigate electrochemically enhanced surface diffusion, which leads to ECSA loss. The hydrophobicity of these ILs is known to change the interaction of water with the surface, positively shifting the onset potential of Pt oxidation [3]. The interfacial IL acts to limit the charge dependent formation of a surface metal/electrolyte anion complex which is responsible for the potential dependent enhancement in surface diffusion. The oxygen reduction reaction activity, ECSA and residual transition metal content of these complex composite nanoporous electrocatalysts was tracked as a function of cycle number during accelerated degradation testing providing insight into the mechanism of catalyst degradation and the effect of ILs on catalyst stability.

References

[1] Snyder, J.; McCue, I.; Livi, K. & Erlebacher, J., *J. Am. Chem. Soc.* **134**,8633-8645 (2012).

[2] Snyder, J.; Fujita, T.; Chen, M. W. & Erlebacher, J., *Nature Mater.* **9**,904-907 (2010).

[3] Snyder, J.; Livi, K. & Erlebacher, J., *Adv. Funct. Mater.* **23**, 5494-5501 (2013).

489132 Particle Atomic Layer Deposition for Stabilization of Oxygen Reduction Catalysts

Abstract:

As a highly efficient, portable, non-greenhouse gas emitting source of electrical power, the proton exchange membrane (PEM) fuel cell holds great potential as a replacement for the internal combustion engine. One of the most significant development challenges in the commercialization of the PEM fuel cell is the long-term durability of the oxygen reduction reaction (ORR) catalyst. Cathode potential cyclingâ€"resulting from the variable voltage loads imposed during vehicular operationâ€"is known to promote agglomeration and growth of the Pt nanoparticle catalyst, which in turn degrades the power output of the fuel cell. In this work, atomic layer deposition (ALD) was used to deposit protective TiO_2 films onto both commercial and ALD-synthesized Pt/C catalyst in order to increase its electrochemical durability. Deposition half-reactions during fluidized bed particle ALD were observed via in-situ mass spectrometry, and chemisorption analysis was used to quantify changes in Pt surface area following TiO_2 deposition. Rotating disk electrode analysis indicated that electrochemical surface area (ECSA) and activity were slightly depressed by the addition of TiO_2 ALD layers. Activity retention during electrochemical durability testing was modestly improved, with TiO_2-coated catalysts retaining upwards of 60% of their initial activity. Examination of catalysts before and after testing also showed that TiO_2 ALD barriers were effective at preventing Pt agglomeration under fuel cell operating conditions. Further testing of the catalysts in membrane electrode assemblies was used to quantify effects of TiO_2 ALD on the hydrophobicity of the catalysts, which has been previously demonstrated as an important factor in fuel cell device operation with ALD-fabricated catalysts.

500977 A Mechanistic Insight into the Electrocatalytic Behavior of Mixed Oxide Cathode Catalysts for Li-O_2 Batteries

Abstract:

The recent advancements in renewable energy generation will require investments in the development of energy storage technologies. Generally, the generated renewable energy can be stored in multiple forms, such as chemically (i.e., water splitting to form hydrogen fuel), mechanically (i.e., flywheel that converts electrical to rotational energy), or electrochemically (i.e., batteries that can store electrical energy). Among the many different electrochemical energy storage technologies, lithium-oxygen (Li-O_2) batteries exhibit the highest theoretical energy density (11,680 Wh/kg) [1]. Unlike the current state-of-the-art Li-ion batteries, where the technology has been explored for many years and reached a ceiling performance, Li-O_2 batteries are still far from reaching the theoretical energy density, thus are still under wide investigation [2]. In order to improve the performance of these systems, a fundamental understanding of the reaction mechanism involved in both the oxygen reduction reaction (ORR) and oxygen evolution reaction (OER) during discharge and charge respectively, at the cathode of these systems is necessary. It is in part due to the reported discharge product, Li_2O_2, being a high bandgap insulator, thereby making the electron transfer during OER extremely challenging [1]. We have shown previously that the performance of the Li-O_2 battery cathodes can be improved upon addition of non-precious metal oxide catalysts, known as Ruddlesden-Popper (R-P) oxides [3]. To optimize the performance of these catalytic materials, a fundamental understanding of the mechanism that leads to this improvement is required. In this contribution, we combine electrochemical and spectroscopic techniques to unravel the complex chemistry involved in Li-O_2 cathodes containing R-P oxide electrocatalysts. The effect of the nature of the discharged product species during ORR and its effect on lowering the overpotential losses during OER will be discussed. We finalize by devising ways of enhancing the performance of R-P oxides for oxygen reduction and evolution in Li-O_2 batteries.

References

1. Girishkumar, G., et al., *Lithium - Air Battery: Promise and Challenges.* Journal of Physical Chemistry Letters, 2010. **1**(14): p. 2193-2203.

2. Armand, M. and J.M. Tarascon, *Building better batteries.* Nature, 2008. **451**(7179): p. 652-657.

3. Nacy, A., X.F. Ma, and E. Nikolla, *Nanostructured Nickelate Oxides as Efficient and Stable Cathode Electrocatalysts for Li-O-2 Batteries.* Topics in Catalysis, 2015. **58**(7-9): p. 513-521.

494499 Characterization of Feni Alloy Nanoparticles for Oer As a Function of Iron-Nickel Composition

Abstract:

Iron-doped nickel hydroxide catalysts are now recognized as the leading catalyst composition for alkaline electrochemical water splitting, and in particular, the anodic reaction of oxygen evolution [1]. The iron-nickel bimetallic composition of these hydroxide catalysts has been demonstrated to be critical for optimal catalyst performance [2]. However, few studies thus far have investigated if the same trend in catalyst activity is observed when catalyst morphology is scaled from bulk films to the nanoscale [3], where the overall field of iron-nickel hydroxide/oxide catalysts is in its infancy in terms of developing, synthesizing, and fully characterizing these nanoparticulate bimetallic hydroxides. Further, preliminary stability studies performed on iron-nickel [4] and iron-cobalt [5] thin films suggests that an evaluation of the stability of the bimetallic catalysts as a function of composition will be critical to identifying the best-performing catalyst compositions to pursue. In particular, Burke et al. report that while iron atomic compositions of less than 54 % resulted in high catalytic activities, the activities of these iron-cobalt bimetallic catalysts decreased by up to 62 % over two hours [5]. In the iron-nickel catalyst field, little work has evaluated how catalyst stability is affected by bimetallic composition in either thin film or nanocatalyst morphologies. Thus far, most studies have primarily focused on how the iron-nickel bimetallic composition affects catalyst performance from the perspective of lowering the overpotential, thereby lowering the kinetic limitation to the oxygen evolution reaction [2, 6].

In this talk, our ongoing work to develop and characterize a suite of bimetallic iron-nickel alloy hydroxide nanoparticle catalysts for the oxygen evolution reaction will be discussed. In particular, results will be presented for three iron-nickel compositions (1:5, 1:1, and 5:1 mol:mol Fe:Ni) and trends in both overpotential and stability will be discussed. Nanoparticle characterization via high resolution transmission electron microscopy, elemental analysis, and synchrotron-based x-ray absorption spectroscopy will be presented and discussed as a function of bimetallic composition. The role of nanoparticle structure in controlling activity and stability will be discussed, and electrochemical performance data will be used to demonstrate that both activity (i.e., overpotential) and stability (i.e., degradation rate as mV/hr) are critical to evaluate as we develop these catalysts for water splitting.

References

[1] D. Friebel, M.W. Louie, M. Bajdich, K.E. Sanwald, Y. Cai, A.M. Wise, M.-J. Cheng, D. Sokaras, T.-C. Weng, R. Alonso, R.C. Davis, J.R. Bargar, J.K. Norskov, A. Nilsson, A.T. Bell, Identification of highly active Fe sites in (Ni,Fe)OOH for electrocatalytic water splitting, J. Am. Chem. Soc., 137 (2015) 1305â€"1313.

[2] L. Trotochaud, S.L. Young, J.K. Ranney, S.W. Boettcher, Nickel-iron oxyhydroxide oxygen-evolution electrocatalysts: The role of intentional and incidental iron incorporation, J. Am. Chem. Soc., 136 (2014) 6744-6753.

[3] J.A. Bau, E.J. Luber, J.M. Buriak, Oxygen Evolution Catalyzed by Nickelâ€"Iron Oxide Nanocrystals with a

Nonequilibrium Phase, ACS Applied Materials & Interfaces, 7 (2015) 19755-19763.

[4] M. Gong, Y. Li, H. Wang, Y. Liang, J.Z. Wu, J. Zhou, J. Wang, T. Regier, F. Wei, H. Dai, An Advanced Niâ€"Fe Layered Double Hydroxide Electrocatalyst for Water Oxidation, Journal of the American Chemical Society, 135 (2013) 8452-8455.

[5] M.S. Burke, M.G. Kast, L. Trotochaud, A.M. Smith, S.W. Boettcher, Cobalt-iron (oxy)hydroxide oxygen evolution electrocatalysts: The role of structure and composition on activity, stability, and mechanism, J. Am. Chem. Soc., 137 (2015) 3638-3648.

[6] M.W. Louie, A.T. Bell, An investigation of thin-film Ni-Fe oxide catalysts for the electrochemical evolution of oxygen, J. Am. Chem. Soc., 135 (2013) 12329-12337.

495038 Active Sites in Nitrogen-Doped Carbon Nanostructures for Oxygen Reduction and Oxygen Evolution Reactions

Abstract:

Regenerative polymer electrolyte membrane (PEM) fuel cell technology has a promising future for sustainable power generation. In regenerative PEM fuel cells, oxygen reduction (ORR) and oxygen evolution (OER) are two important reactions that take place in the fuel cell mode and electrolyzer mode, respectively. State-of-the-art catalysts used for these reactions – Pt for ORR and Ir (or Ru) for OER – are expensive and found in limited reserves. Also, both these catalysts are active only for one of the two reactions, thus necessitating the development of cheaper materials that can serve as efficient bifunctional catalysts for both ORR and OER in regenerative PEM fuel cells. We have reported nitrogen doped carbon nanostructures, denoted as 'CN_x' [1], to be active, selective and stable catalysts for ORR in acidic medium. Our current focus is to further evaluate the applicability of CN_x as bifunctional ORR and OER catalysts, and also understand the nature of active sites in these materials.

Electrocatalytic activity measurements using rotating ring disk electrode (RRDE) show significant ORR and OER activity of CN_x compared to state-of-the-art Pt and Ir-based catalysts. Analysis of bifunctional electrocatalytic activity demonstrates much better bifunctional characteristics of CN_x compared to Ir/C and Pt/C [2].

Use of poisoning probes is a useful methodology to systematically examine the nature of active sites in catalytic materials. However, due to the absence of a metal centered active site, CN_x materials are not susceptible to poisoning by CO, H_2S or CN^- [3, 4]. We have recently identified phosphate anions ($H_2PO_4^-$) as probe molecules for poisoning CN_x catalysts in an acidic ORR environment [5]. In-situ RRDE measurements show a linear decrease in ORR activity with increasing $H_2PO_4^-$ anion concentration. Transmission IR spectroscopy and Raman spectroscopy reveal the presence of $H_2PO_4^-$ species on H_3PO_4–soaked CN_x. A linear decrease in ORR kinetic current with a decrease in pyridinic nitrogen content obtained using X-ray photoelectron spectroscopy (XPS) is observed. This poisoning phenomenon is consistent with two possible active site models: (i) pyridinic N as the active site which is rendered inactive by protonation [6, 7], and (ii) carbon sites adjacent to pyridinic N as the active site (i.e. pyridinic N as a marker for the active site) wherein poisoning is caused by a site-blocking effect due to adsorption of $H_2PO_4^-$ species on carbon. OER activity has also been found to increase with increasing pyridinic nitrogen site density in CN_x samples.

References

[1] P.H. Matter, U.S. Ozkan, Non-metal catalysts for dioxygen reduction in an acidic electrolyte, Catal. Lett., 109 (2006) 115-123.

[2] K. Mamtani, D. Jain, D. Dogu, A. C. Co, U. S. Ozkan, Insights into Oxygen Reduction Reaction (ORR) and Oxygen Evolution Reaction (OER) Active Sites for Nitrogen-doped Carbon Nanostructures (CNx) in Acidic Media, Applied Catalysis B: Environmental, Under Review, (2017).

[3] D. von Deak, D. Singh, J.C. King, U.S. Ozkan, Use of carbon monoxide and cyanide to probe the active sites on nitrogen-doped carbon catalysts for oxygen reduction, Appl. Catal. B-Environ., 113-114 (2012) 126-133.

[4] D. von Deak, D. Singh, E.J. Biddinger, J.C. King, B. Bayram, J.T. Miller, U.S. Ozkan, Investigation of sulfur poisoning of CNx oxygen reduction catalysts for PEM fuel cells, J. Catal., 285 (2012) 145-151.

[5] K. Mamtani, D. Jain, D. Zemlyanov, G. Celik, J. Luthman, G. Renkes, A.C. Co, U.S. Ozkan, Probing the Oxygen Reduction Reaction Active Sites over Nitrogen-Doped Carbon Nanostructures (CN x) in Acidic Media Using Phosphate Anion, ACS Catalysis, 6 (2016) 7249-7259.

[6] G. Liu, X. Li, P. Ganesan, B.N. Popov, Studies of oxygen reduction reaction active sites and stability of nitrogen-modified carbon composite catalysts for PEM fuel cells, Electrochim. Acta, 55 (2010) 2853-2858.

[7] X. Li, G. Liu, B.N. Popov, Activity and stability of non-precious metal catalysts for oxygen reduction in acid and alkaline electrolytes, J. Power Sources, 195 (2010) 6373-6378.

497178 Atomic Iron-Dispersed Electrocatalysts Derived from Metal-Organic Framework for Oxygen Reduction

Abstract:

To significantly reduce the cost of proton exchange membrane (PEM) fuel cells, current Pt must be replaced by platinum-metal-group (PGM)-free catalysts for the oxygen reduction reaction (ORR) in acid. We report here a new class of high-performance atomic iron-rich catalyst via chemical doping of iron ions into zinc-zeolitic imidazolate framework (ZIF), a type of metal-organic framework (MOF). The novel synthetic chemistry enables accurate size control of Fe-doped ZIF catalyst particles with a wide range from 20 to 1000 nm without changing chemical property. This provides a great opportunity to increase the density of active sites. An optimal size at 50 nm exhibited the highest activity; further reducing size to 20 nm results in particle agglomeration with decreased active sites. In addition to the size effect, we carefully explored the thermal activation mechanism to provide insight into the origin of active site formation. A temperature of 800oC was identified as the critical point to generate active sties with new bond formation associated with nitrogen doping, carbon, and iron. 1100 oC was found for the best activity with maximum density of active sites along with optimal porosity, graphitic nitrogen doping, and carbon structures. We also found the control of optimal iron doping is crucial for performance enhancement without the formation of inactive iron cluster. Overall, the best performing catalyst is enriched in atomic iron sites embedded into carbon structures, achieving a new performance milestone for the ORR including a half-wave potential of 0.84 V vs RHE in 0.5 M H2SO4 and only 20 mV loss after 10,000 cycles in O2 saturated electrolyte, which has been approaching to Pt in the challenging acidic media. Thus, the chemically Fe doped MOF catalysts are able to further narrow down the performance gap with Pt catalysts in more challenging acidic media, holding great promise to advance PGM-free catalyst as a replacement of Pt for PEM fuel cells.

499431 Nucleation Mechanism of Clathrate Hydrates of Soluble Guest Molecules

Abstract:

Clathrate hydrates are a solid phase composed of guest molecules entrapped in a lattice of polyhedral water cages. The guest molecules are generally light gases, such as methane or carbon dioxide. Currently active research on applications of clathrate hydrates ranges from energy and flow assurance to gas storage. Many such applications of clathrate hydrates require a firm understanding of the mechanism of hydrate formation.

The birth of the hydrate phase occurs from a supersaturated liquid solution of water and guest molecules through an activated process called nucleation. Hydrate nucleation is inherently a molecular-level phenomenon, occurring on length- and time-scales that are difficult to probe with current experimental techniques. Though the length and time-scales of hydrate nucleation appear well-suited to study through molecular simulation, in most cases the associated activation barrier is large, and nucleation is a rare event in simulations. Correspondingly, in straightforward simulations most of the computational power is expended waiting for the nucleation event to happen rather than simulating the nucleation events of interest. To overcome these challenges, we use a technique called forward flux sampling (FFS) to generate over one thousand nucleation trajectories.

We focus on the nucleation of clathrate hydrates of soluble guest molecules. The water and guest molecules are described by the coarse-grained monatomic water and associated guest models. Accurate committor probabilities were calculated for over 150 configurations generated with FFS. One and two parameter reaction coordinate models were then created from several common order parameters and variants thereof. In this talk, we present results indicating that order parameters that rely on water structure are generally more effective at describing the liquid-to-solid transition than order parameters relying on guest structure. An analysis of over 200 configurations near the transition state reveals that the transition state consists of a small number of empty, face-sharing cages surrounded by guest molecules. These surrounding guest molecules begin to form a network of connected precursors to occupied cages. Motivated by these findings, we discuss results from further simulations which reveal that two or more face-sharing cages significantly affect the structure of the surrounding guest molecules. We conclude by investigating how our proposed nucleation mechanism applies to mixed hydrate systems with two types of guest molecules.

AIChE
American Institute of Chemical Engineers

Modeling Thermodynamic Properties of Propane or Tetrahydrofuran Mixed with Carbon Dioxide or Methane in Structure-II Clathrate Hydrates

Bin Fang, **Fulong Ning***, Pinqiang Cao, Li Peng, Jianyang Wu, Zhun Zhang, Thijs J.H. Vlugt , Signe Kjelstrup

China University of Geosciences in Wuhan

29 Oct 2017

CONTENTS

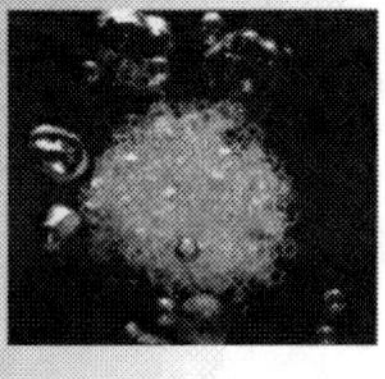

- BACKGROUND
- SIMULATION MODEL
  - Initial structure construction
  - Parameter definition
  - Simulation procedure
  - Calculation methods
- RESULTS AND ANALYSIS
  - Lattice parameter and isobaric thermal expansion coefficient
  - Lattice parameter and isothermal compressibility
  - Specific heat capacity at constant pressure
- CONCLUSIONS

■BACKGROUND

Thermal expansion coefficient--basic information for risk assessment studies of the mechanical stability of the earth strata hosting hydrates. (Ning et al., 2012)

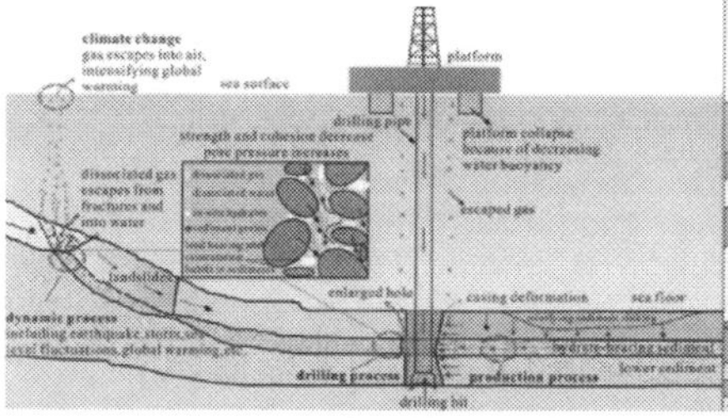

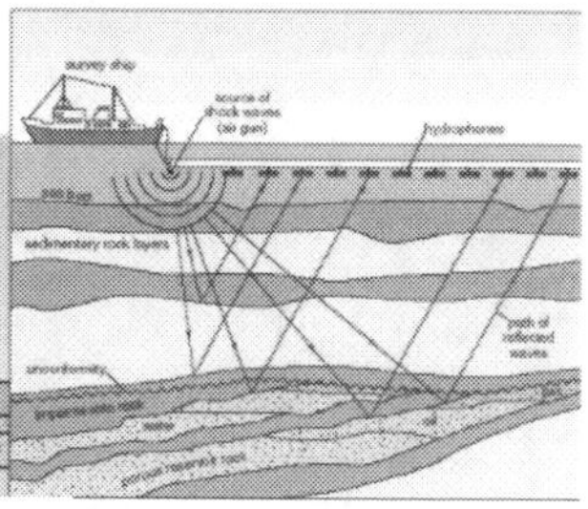

Seismic waves are usually employed in order to detect natural gas hydrates in sedimentary layers, and the wave speed is connected with the elastic constants and ***compressibility*** of the media through which they propagate. (Yuan et al.,1999)

Simultaneous determination of thermal conductivity, thermal diffusivity and specific heat in sI methane hydrate

W. F. Waite,[1] L. A. Stern,[2] S. H. Kirby,[2] W. J. Winters[1] and D. H. Mason[1]

On localized length scales, thermal diffusivity and specific heat of CH_4 hydrate are important not only for modelling controlled production of CH_4 from hydrate (Ji *et al.* 2003; Pooladi-Darvish 2004; Hong & Pooladi-Darvish 2005; Kurihara *et al.* 2005; Moridis *et al.* 2005), but for hazard mitigation in conventional hydrocarbon extraction, which can destabilize hydrate underlying production equipment (Briaud & Chaouch 1997; Moridis & Kowalsky 2006).

The heat capacity is not only used for the modelling of natural gas production from hydrate-bearing sediments, but also for hazard mitigation in conventional hydrocarbon extraction. (Waite et al., 2007)

the understanding of thermodynamic properties such as the thermal expansion coefficient, the isothermal compressibility and the heat capacity at constant pressure of the gas hydrates are essential.

■BACKGROUND

- System selection

◆ sII structure is one of the three typical hydrate structures occurring in the permafrost region and in deep oceans as well as in pipelines.

◆ In our project, structure-II hydrates with propane or tetrahydrofuran (THF) mixed with carbon dioxide or methane guest molecules were selected.

- C_3H_8 hydrate is the major component of structure II hydrates in sediments
- THF hydrate is a promising analog material for CO_2 recovery from flue gas as well as for hydrogen storage.
- the small cages of sII hydrates are often occupied by relative small molecules (also called secondary "help-gases") like CH_4 and CO_2 molecules.

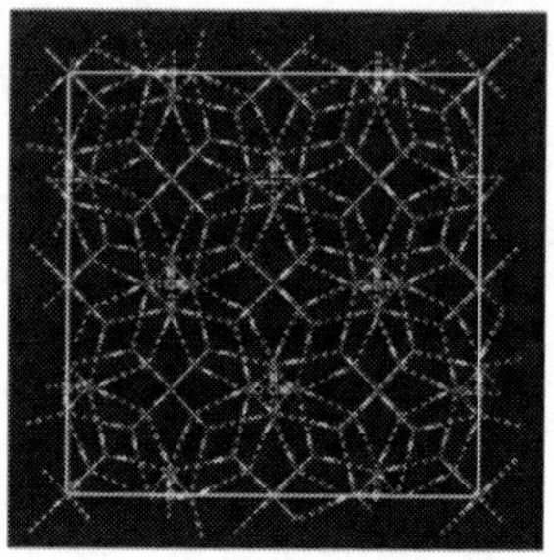

Cubic structure II (sII) clathrate hydrate
denoted as $16M_S8M_L136H_2O$ has 16 small pentagonal dodecahedron cages and eight hexakaidecahedron large cages.

■SIMILATION MODEL

➢ Initial structure construction

- 1. The positions of the water oxygen atoms of the initial hydrate configuration were taken from x-ray crystallography, and the water hydrogen atoms were inserted to adjust the orientation for yield to Bernal-Fowler rule and minimize the net unit cell dipole moment and potential energy
- 2. All guest molecules were placed at the center of the water cages, assuming that the propane and THF molecules both full occupied the large cages, and that the methane, carbon dioxide molecules fully lived in the small cages
- 3. Annealing simulations were additionally performed for hydrates with THF

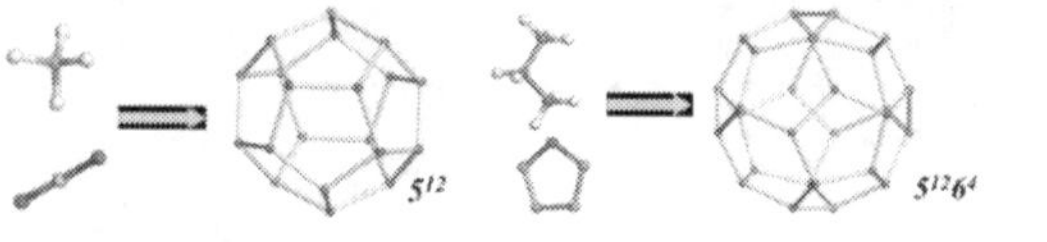

➢ Parameter definition

1. Force field
- C_3H_8/CH_4 opls-aa
- THF/CO_2 TraPPE
- H_2O TIP4P/2005 (all simulations)
  TIP4PQ/2005
  (only for isobaric thermal expansion)

2. Temperature and pressure range
- Temperature 5-287.15K
- Pressure 10-100MPa

3. Simulation tool
- Gromacs package version 4.5.6

4. NPT ensemble

■SIMILATION MODEL

➢ Simulation procedure

1. Costandy correction (for lattice parameter)
- Step one

  $C(T)\ (=\rho^{(CS)}{}_{empty}-\rho^{(PI)}{}_{empty})$ $\quad C(T)=a_0+a_1T+a_2T^2$

  with $a_0=0.04672$, $a_1=-1.68913\times10^{-4}$, $a_2=2.17391\times10^{-7}$

  C(T) is the density difference of empty sII structure between quantum path-integral simulation and classical simulation at temperature of 77, 100 and 125K, therefore, the density differences were fitted to second-order polynomials.

- Step two

  classical MD simulations for all sII hydrates were performed using **TIP4PQ/2005** water model as well as the other guest molecular force fields described above.

- Step three

  $\rho_{(corr)}=\rho-C(T)$

  $a_{corr}=(V_{corr})^{1/3}$

  $$\rho=\frac{n_wM_w+\sum_i^2\theta_i v_i M_g}{N_{A}V} \qquad (1)$$

  the hydrate density is corrected, and then obtained value for ρ and re-solve it in order to calculate the corrected value for unit cell volume (V) and the lattice constant.

➢ Thermodynamic properties calculation methods

α_P is the thermal expansion coefficient, K^{-1};
k_T is the isothermal compressibility coefficient, MPa^{-1};
C_P is the heat capacity at constant pressure, J/K

- numerical differentiation method

$$\alpha_P=\frac{1}{V}\left(\frac{\partial V}{\partial T}\right)_P \qquad (2) \qquad\qquad \kappa_T=-\frac{1}{V}\left(\frac{\partial V}{\partial P}\right)_T \qquad (3)$$

- fluctuation theory method

$$\alpha_P=\frac{\left[\langle V\cdot U\rangle-\langle V\rangle\cdot\langle U\rangle+P\left(\langle V^2\rangle-\langle V\rangle^2\right)\right]_{NPT}}{k_BT^2\langle V\rangle} \qquad (4)$$

$$\kappa_T=\frac{1}{k_BT\langle V\rangle}\left(\langle V^2\rangle-\langle V\rangle^2\right)_{NPT} \qquad (5)$$

$$C_P=\frac{ik_BN}{2}+\frac{1}{k_BT^2}\left[\begin{pmatrix}\langle U^2\rangle-\langle U\rangle^2+\\ 2P\left(\langle V\cdot U\rangle-\langle V\rangle\cdot\langle U\rangle\right)\\ +P^2\left(\langle V^2\rangle-\langle V\rangle^2\right)\end{pmatrix}\right]_{NPT} \qquad (6)$$

■RESULTS AND ANALYSIS

➢ Lattice parameter and isobaric thermal expansion coefficient

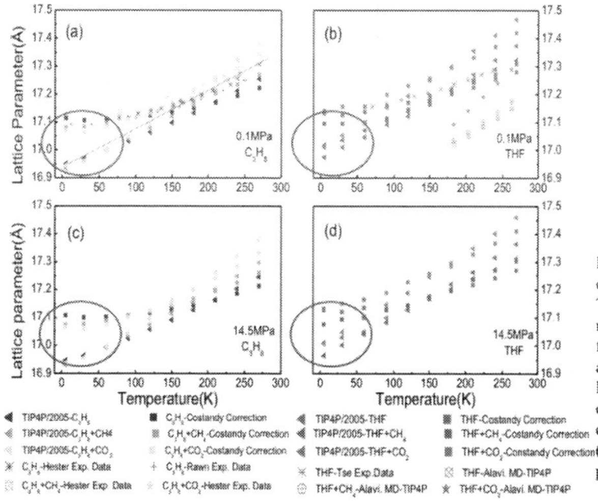

Figure 1 Lattice parameters for sII clathrate hydrates of the C_3H_8(a/c), THF(b/d) with different second-help molecules in the small cages as a function of temperature at atmospheric pressure (a/b) or 14.5 MPa (c/d) compared with the experimental values. Results are obtained by MD simulations, with (square points) and without (triangular points) the Costandy correction.

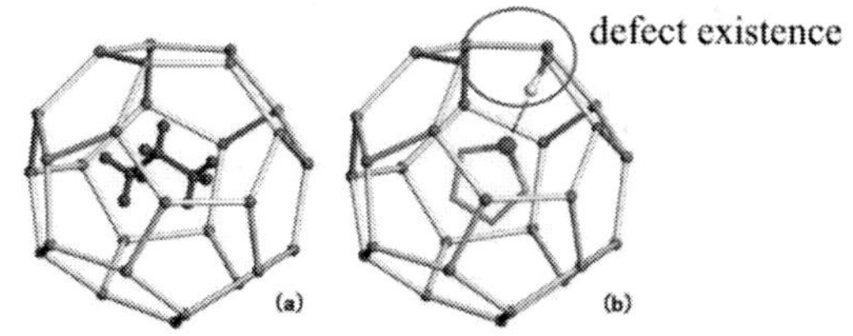

Figure 2. Schematic representation of H_2O molecule cavities occupied by non-hydrogen bond forming guest C_3H_8(a) and hydrogen bond forming guest THF(b).

■RESULTS AND ANALYSIS

➢ isobaric thermal expansion coefficient (0.1MPa or 14.5MPa)

$$V = B_2T^2 + B_1T + B_0$$

Table ½ Coefficients of polynomial function for the unit volume of sII hydrates at atmospheric pressure or 14.5MPa

	TIP4P/2005				TIP4PQ/2005 (and corrected)			
	B_2 (×10⁻⁴)	B_1	B_0	Determination Coeff.	B_2 (×10⁻⁴)	B_1	B_0	Determination Coeff.
C_3H_8	9.9244	0.7331	4865.9839	0.9998	0.0019	0.1551	5011.6093	0.9985
$C_3H_8+CH_4$	9.5216	0.9731	4850.7726	0.9999	0.0017	0.1615	4981.8777	0.9997
$C_3H_8+CO_2$	18.6	0.9979	4851.5848	0.9989	0.0025	0.2638	4964.4122	0.9971
THF	7.2386	0.8599	4917.7378	0.9965	0.0013	0.1343	5025.8891	0.9811
THF+CH_4	7.9961	1.1019	4886.4697	0.9994	0.0022	0.1365	4889.0561	0.9891
THF+CO_2	9.3672	1.2577	4921.6374	0.9997	0.0016	0.5090	5026.1473	0.9951

	TIP4P/2005				TIP4PQ/2005 (and corrected)			
	B_2 (×10⁻⁴)	B_1	B_0	Determination Coeff.	B_2 (×10⁻⁴)	B_1	B_0	Determination Coeff.
C_3H_8	9.6815	0.7267	4861.4437	0.9997	0.0018	0.1642	5005.8863	0.9984
$C_3H_8+CH_4$	9.2894	0.9656	4846.9991	0.9999	0.0016	0.1661	4975.5149	0.9997
$C_3H_8+CO_2$	17.1	1.0437	4846.3404	0.9992	0.0025	0.2266	4960.7774	0.9965
THF	5.3119	0.8860	4912.2907	0.9962	0.0018	0.0013	5020.3309	0.9807
THF+CH_4	6.4517	1.1533	4877.3176	0.9997	0.0016	0.2971	4980.3069	0.9965
THF+CO_2	6.8359	1.3178	4915.3161	0.9995	0.0016	0.4942	5025.4776	0.9984

units B_2: Å³/K²; B_1: Å³/K; B_0: Å³

■RESULTS AND ANALYSIS

➢ isobaric thermal expansion coefficient (0.1MPa or 14.5MPa)

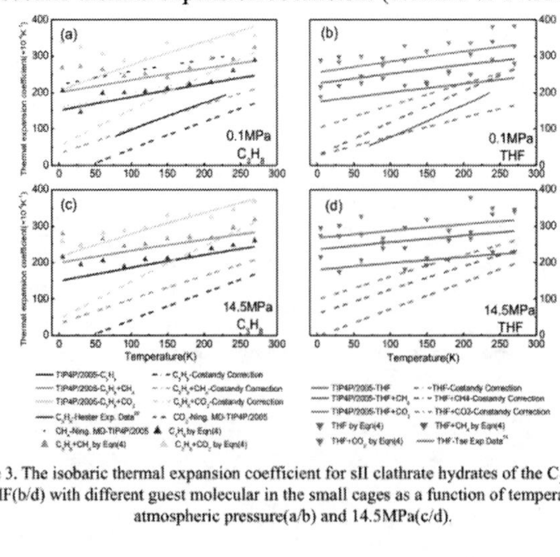

Figure 3. The isobaric thermal expansion coefficient for sII clathrate hydrates of the C_3H_8(a/c) or THF(b/d) with different guest molecular in the small cages as a function of temperature at atmospheric pressure(a/b) and 14.5MPa(c/d).

■RESULTS AND ANALYSIS

➢ Lattice parameter and Isothermal compressibility

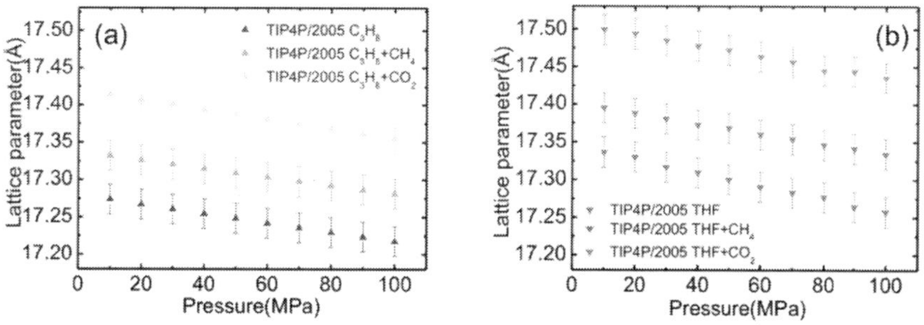

Figure 4. Lattice parameters for sII clathrate hydrates of C_3H_8(a) and THF(b) with different guest molecules in the small cages, as a function of pressure at 287.15K.

■RESULTS AND ANALYSIS

➢ Isothermal compressibility
(287.15K)

$$V = B_2 P^2 + B_1 P + B_0$$

$$V_{C_3H_8} = 1.8935\times10^{-4}P^2 - 0.58P + 5159.9018$$

$$V_{C_3H_8+CH_4} = 2.1732\times10^{-4}P^2 - 0.5305P + 5211.8002$$

$$V_{C_3H_8+CO_2} = 2.1092\times10^{-4}P^2 - 0.6130P + 5286.8328$$

$$V_{THF} = 6.9879\times10^{-4}P^2 - 0.869P + 5219.6656$$

$$V_{THF+CH_4} = 5.2064\times10^{-4}P^2 - 0.6673P + 5270.0361$$

$$V_{THF+CH_4} = 3.8109\times10^{-4}P^2 - 0.70134P + 5366.2880$$

$$10 \le P \le 100MPa$$

■RESULTS AND ANALYSIS

➢ Isothermal compressibility
(287.15K)

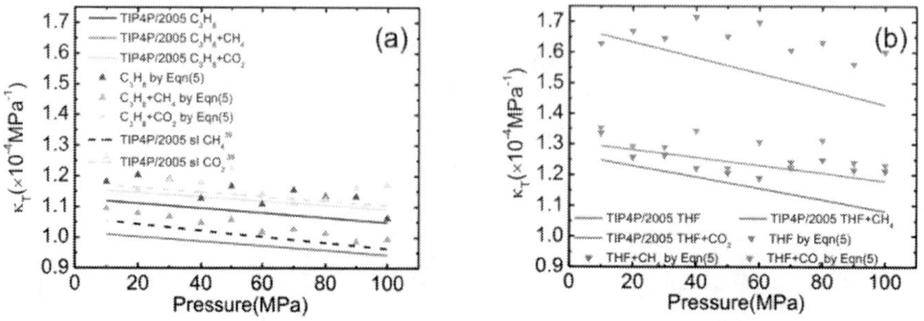

Figure 5. The isothermal compressibility of sII hydrates of C_3H_8(a) and THF(b) with different guest molecules in the small cages as a function of pressure at 287.15 K

Question one: Why THF hydrates is easier to compress?

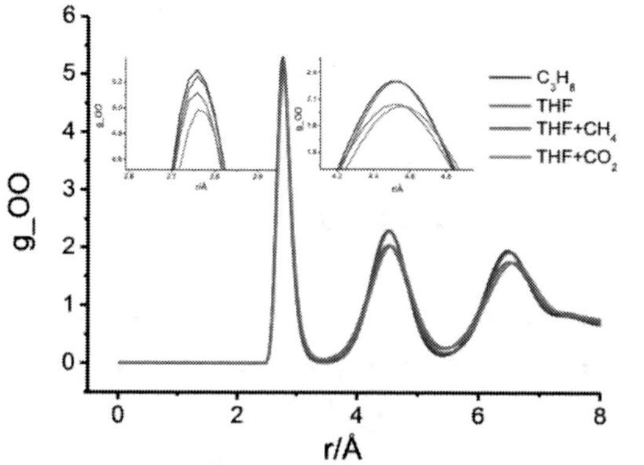

Figure 7. Radial distribution function (RDF) for the oxygen-oxygen atoms of water at 287.15K and 14.5MPa for different pure and mixed hydrates.

Question two: Why the binary hydrates with CO_2 would be more compressible than those with CH_4 as the help gas?

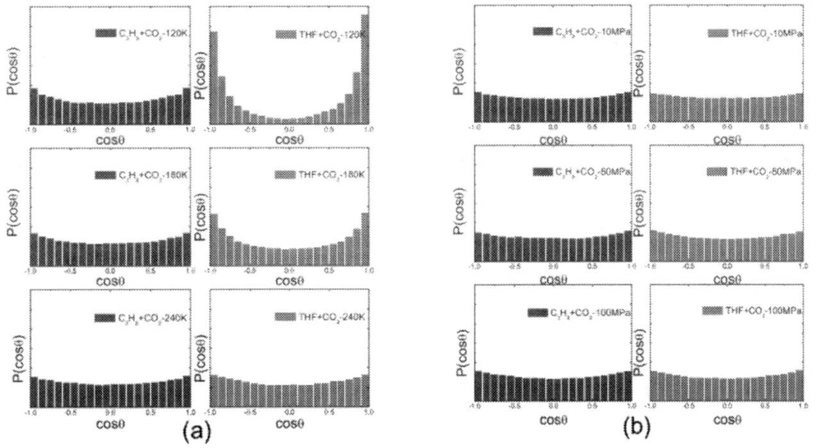

Figure 6. The distribution of cosine values of rotation angle of linear CO_2 molecule relative to the initial direction both in C_3H_8 with CO_2 (blue) and THF with CO_2 (red) binary hydrates at different temperatures (a) and pressures (b) (counted 2000 snapshots from each trajectories). Here, θ represent the rotation angle of CO_2 linear molecule relative to the initial direction.

■RESULTS AND ANALYSIS

➤ Specific heat capacity at constant pressure

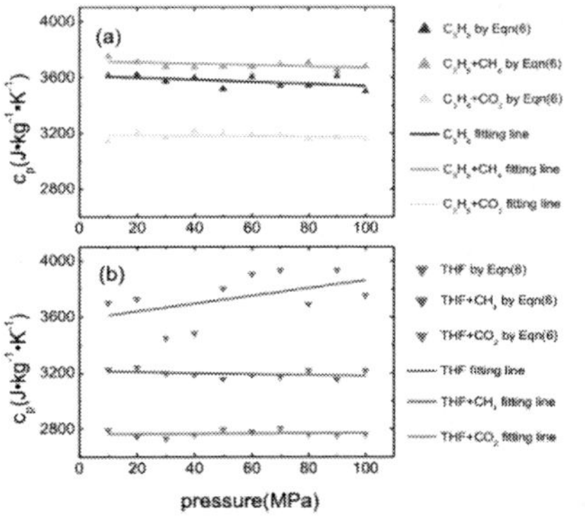

$$c_{P\,C_3H_8} = -0.7563P + 3613.2690$$
$$c_{P\,\mathrm{THF}} = 2.7669P + 3585.7779$$
$$c_{P\,C_3H_8+CH_4} = -0.4813P + 3717.0008$$
$$c_{P\,\mathrm{THF}+CH_4} = -0.3824P + 3215.7374$$
$$c_{P\,C_3H_8+CO_2} = -0.1763P + 3189.2362$$
$$c_{P\,\mathrm{THF}+CO_2} = -0.0559P + 2764.3700$$
$$(10 \leq P \leq 100\mathrm{MPa})$$

Figure 8. Pressure/temperature dependence of the specific heat capacity at constant pressure of C_3H_8(a), THF(b) pure and with CH_4 or CO_2 binary hydrates at 287.15K.

■RESULTS AND ANALYSIS

➤ Specific heat capacity at constant pressure

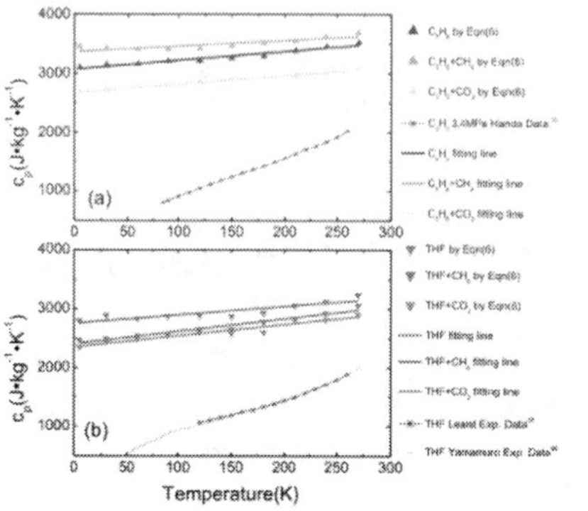

$$c_{P\,C_3H_8} = 1.5018T + 3071.3113$$
$$c_{P\,\mathrm{THF}} = 2.1112T + 2412.7323$$
$$c_{P\,C_3H_8+CH_4} = 0.9386T + 3368.7513$$
$$c_{P\,\mathrm{THF}+CH_4} = 1.4161T + 2760.1949$$
$$c_{P\,C_3H_8+CO_2} = 1.4115T + 2675.4371$$
$$c_{P\,\mathrm{THF}+CO_2} = 1.9013T + 2358.7803$$
$$(5 \le T \le 270K)$$

Figure 9. Pressure/temperature dependence of the specific heat capacity at constant pressure of C_3H_8(a), THF(b) pure and with CH_4 or CO_2 binary hydrates at 14.5MPa

Question one: Why the specific heat capacities of the C_3H_8 hydrates are slight larger than those of the THF hydrates?

Table. 3 Mean hydrogen-bond length (Å) in water in the large cage at 14.5MPa

	180K	240K
THF	1.8466	1.8895
C_3H_8	1.8297	1.8768

the water-water contribution to $\triangle cp$ arises from the change in the hydrogen bond network of water shell around solutes, and manifests that increasing H-bond length can decrease the water heat capacity contribution. THF can form hydrogen bonds with host water molecules and change the hydrogen bond network of their hydrates. (Madan et al.,1997) So we concluded that the slightly large is from guest- host interaction.

Discussion: the relationship between the number of freedom degrees of water and specific heat capacity

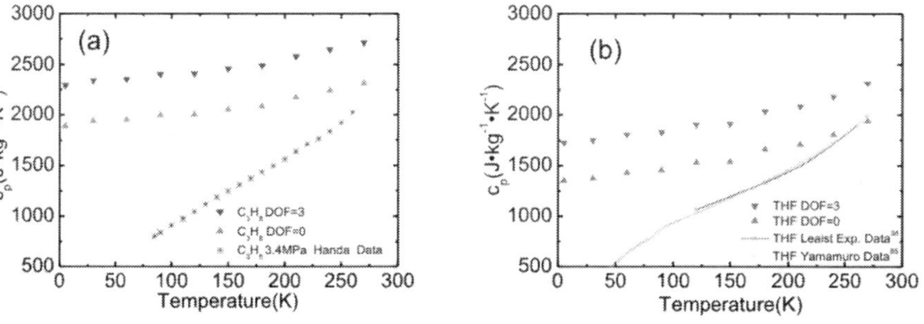

Figure 9. Temperature dependence of specific heat capacity of C_3H_8(a) and THF(b) pure hydrates at 14.5MPa using DOF=0 and 3.

■CONCLUSION

- The correction for lattice parameters of hydrates in low temperature region is significant improvement of the results compared to experimental data, and provide a basis for further study of the effect of temperature and guest molecule types on the expansion at low temperature by MD simulations.
- the lattice parameter at a constant pressure or a constant temperature varies as a function of the guest type and guest coupling interaction.
- The thermodynamic properties of THF hydrates are similar to that of C_3H_8 hydrates and CH_4 hydrates except for compressibility, which also provides the possibility for the application of THF hydrates instead of other hydrates in the experiment.
- Small molecules, such as CH_4 and CO_2, play an obvious role in the stabilization of THF hydrates structure with defect. Moreover, when the CO_2 molecules replace CH_4 molecules in the small cages of sII structure, the lattice parameters, isothermal compressibility and thermal expansion of the hydrates increase, and the bulk modulus and specific heat capacity decrease accordingly. And this effect has nothing to do with crystal structure and other guest type in the hydrates.

Thank you for your attention

NSFC
National Natural Science Foundation of China

中国地质大学
CHINA UNIVERSITY OF GEOSCIENCES

青岛海洋科学与技术国家实验室
Qingdao National Laboratory for Marine Science and Technology

491422 Prediction of Gas Hydrate Formation Conditions in Brine Systems with the Enrtl Model

Abstract:

As oil and gas fields and developments delve into deeper water, the harsh conditions in terms of high pressure and salinity requires reliable and accurate prediction of hydrate phase equilibrium conditions in salt aqueous solutions with concentration up to saturation over a wide range of pressures. This study develops a predictive method to estimate hydrate suppression temperature in high salinity systems. The simple but robust prediction is based on the fundamental principle of freezing point depression, which for hydrate is equivalent to the suppression temperature from the uninhibited (salt-free) systems. The water activity, highly dependent on the salt species and concentration, is the key factor for determining the hydrate inhibition strength. Therefore, in this developed predictive method, the hydrate suppression temperature with any saline systems can be accurately and directly predicted with the calculation of water activity using the electrolyte-NRTL (eNRTL) model. The presentation will show the reliability of the predictions as well as how it can be generally used for the prediction of the hydrate suppression temperature for any brine system.

492780 Clathrate Hydrate Formation Using Fluorocarbons

Abstract:

Global demand for freshwater continues to increase and as a response to this pressing need, hundreds of desalination plants have been constructed over in the past decade. Most plants utilize reverse osmosis (RO) or multi-stage flash distillation (MSF) as the separation technology. However, pumping large volumes of water to high pressures for RO and the significant thermal heating requirement for MSF makes these technologies expensive. Clathrate hydrates serve as an economic alternative desalination technology if it can be operated near ambient conditions. Fluorocarbon refrigerants have shown promising results as clathrate formers for producing clathrates at such conditions. However, concerns regarding global warming potential and ozone depletion have limited the extent to which some fluorocarbons can be used.

Understanding the conditions in terms of temperature and pressure at which clathrates form between fluorocarbons and water is an important aspect in evaluating optimum clathrate formers. Our studies use accurate gravimetric and volumetric techniques to measure sorption of different fluorocarbon gases into water to determine phase transition conditions. Combining these two methods allows us to investigate the solubility of the refrigerant in water and the rate of formation of the clathrates.

492061 Methane Hydrate Formation Accelerated with Various Promoters in an Unstirred Reactor

Abstract:

Recently, many fundamental researches and technologies have focused on the natural gas storage in artificially synthesized clathrate hydrate because of its several advantages of cost-effectiveness, non-explosiveness, and eco-friendly method. For practical applications of hydrate-based natural gas storage, rapid hydrate formation without mechanical agitation is one of key factors to be solved. Zhang et al. reported that sodium dodecyl sulfate (SDS) plays an important role in increasing both nucleation and growth rate of hydrate in the unstirred batch reactor.[1] Kang et al. also used a water-insoluble heavy guest molecule of iodomethane which shows rapid hydrate formation with a high conversion rate by renewing liquid-liquid interfaces without mechanical agitation.[2] Veluswamy et al. introduced some additives and evaluated kinetic performance of methane hydrate formation in several systems.[3-4] Here, we suggest a new concept of cyclopentane (CP) hydrate seeds as a kinetic hydrate promoter; the induction time of pure sI methane hydrate can be eliminated in the presence of small amounts of CP hydrate seeds and SDS in an unstirred reaction system.

1. Zhang, J. S.; Lee, S.; Lee, J. W., Kinetics of Methane Hydrate Formation from Sds Solution. *Ind Eng Chem Res* **2007**, *46*, 6353-6359.

2. Kang, H.; Ahn, Y. H.; Koh, D. Y.; Baek, S.; Lee, J. W.; Lee, H., Rapid Clathrate Hydrate Formation Using a Heavy Guest Molecule with Sodium Dodecyl Sulfate. *Ind Eng Chem Res* **2016**, *55*, 6079-6084.

3. Veluswamy, H. P.; Kumar, A.; Kumar, R.; Linga, P., An Innovative Approach to Enhance Methane Hydrate Formation Kinetics with Leucine for Energy Storage Application. *Appl Energ* **2017**, *188*, 190-199.

4. Veluswamy, H. P.; Hong, Q. W.; Linga, P., Morphology Study of Methane Hydrate Formation and Dissociation in the Presence of Amino Acid. *Cryst Growth Des* **2016**, *16*, 5932-5945.

501347 Effect of Thermodynamic Promoter (TBAB) in Gas Hydrates Dissociation Conditions of the Mixture $H_2O + CO_2$ + Undecane

Abstract:

In México, there are reservoirs with high content of water (20 % or more) like Cantarell where hydrate formation can be simple. For this reason, the oil industry record an increase of produced water during the extraction of these resources in oil production wells, so it is necessary to develop a low cost effective process for separating water from oil. A novel alternative that can be considered for the separation of these phases is the gas hydrates formation. Gas hydrates are ice-like crystalline structures, mostly formed by water and small gas molecules with low molecular weight.

In this work, the experimental dissociation conditions for the gas hydrate formed in the mixture: $H_2O + C_{11}H_{24} + CO_2$ + TBAB in a wide composition range of TBAB (0.05, 0.10, 0.20, 0.30, 0.40) were determined in the temperature range from 274.15 to 291.15 K and a pressure up to 4.0 MPa. The composition is expressed as mass fraction, the combined expanded uncertainties of the dissociation conditions were estimated to be less than 0.012 MPa, 0.020 K and 0.0015 for pressure, temperature and mass fraction, respectively. The goal of this study is to know the effect of the thermodynamic promoter in the hydrate formation process. The experimental equipment used in this work is based on the static-synthetic method using the isochoric technique.

500038 Study on Flow Behavior of CO_2-in-Water (C/W) Emulsion in the Porous Medium Under Hydrate Stability Conditions

Abstract:

CO_2-in-Water (C/W) emulsion injection method (a method of injecting CO_2 droplets mixed in water as an emulsion state) is expected to be a useful technique for recovering methane gas from methane hydrate (MH). In this technique, CO_2 hydrate particles are formed from C/W emulsion injected into aquifer layers in the MH zone. The heat of CO_2 hydrate formation can be used to increase the temperature of sediment and supply energy to dissociate MH. In this work, we have developed a numerical program for predicting flow behavior of C/W emulsion with hydrate formation in the porous medium. After comparing the calculated pressure and temperature behavior with experimental data of C/W emulsion injection into the core, injection schemes of C/W emulsion to make the heat of CO_2 hydrate formation maximum were investigated by numerical case studies.

In the simulation program, five governing equations are solved: Three continuity equations of H_2O component, liquid CO_2 component, and total CO_2 component (liquid and dissolved CO_2) in C/W emulsion; one equation of describing transport of CO_2 hydrate particles; and one energy balance equation considering heat convection/conduction and the heat of hydrate formation/CO_2 dissolution into water. These equations are formulated based on the following assumptions: (1) C/W emulsion flows as a homogeneous phase before hydrate formation. (2) CO_2 droplets are converted into hydrate particles at a given rate after induction time (thin films are formed around CO_2 droplets). (3) Hydrate particles are transported in the pore by emulsion flow and dispersion, where some particles are trapped on the pore surface and the pore throat of the porous medium. (4) The permeability of the porous medium decreases with increasing saturation of trapped hydrate particles. In the numerical procedure, five independent variables of pressure, temperature, concentration of H_2O and liquid CO_2 in C/W emulsion, and saturation of flowing hydrate particles are solved by five governing equations.

To examine the validity of numerical calculation, a number of experiments were conducted to inject C/W emulsion (CO_2 vol. concentration of 10-30%) into 0.1 mm glass-bead sintered water-saturated cores under controlled pressure and temperature conditions of 7 MPa, 7°C and 5 MPa, 2°C. The length of the core was 10 cm, and injection pressure and core surface temperatures at the locations of 2.5 cm, 5 cm, and 7.5 cm from inlet were sequentially recorded during the experiment. In most experiments, we observed that inlet pressure rose steeply to lead the core plugged a certain time after the increase in temperatures by hydrate formation.

After confirming that simulation could reproduce the tendency of pressure and temperature behavior during the experiments, numerical case studies were carried out to find appropriate injection schemes of C/W emulsion maximizing the heat of CO_2 hydrate formation until the plug of the core. The following conclusions were obtained. (1) Heat of 10-16 MJ/m of the porous medium is generated during C/W emulsion until the plug of the core. This corresponds to the heat value that can dissociate about 10 % of MH. (2) The lower the CO_2 concentration of C/W emulsion, the smaller the heat generation rate of hydrate formation, but the time to core plugging is much delayed compared with the case of high CO_2 concentration. As a result, the use of lower CO_2 concentration generated the maximum heat. This study showed that C/W emulsion injection method would be effective to increase gas recovery efficiency from MH if we select an appropriate injection scheme.

Ionic Liquids as Dual Functional Inhibitors for Natural Gas Hydrates (NGH).

Tausif AlTamash[a], Fahed Qureshi[a], Majeda Khraisheh[a1], Mohammad Ali Saad [a], Mert Atilhan [b]

[a] Department of Chemical Engineering, Qatar University, P.O.Box 2713, Doha, Qatar.

[b] Department of Chemical Engineering, Texas A&M University, College Station, TX, USA.

Abstract.

The thermodynamic inhibition (TI) and kinetic inhibition (KI) effect of pyrrolidinium based Ionic liquids (ILs) 1-Methyl-1-Propyl-pyrrolidinium Chloride [PM-Py][Cl] and 1-Methyl-1-Propyl-pyrrolidinium Triflate [PM-Py][Triflate] have been tested with and without the addition of synergents /additives such as polyethylene oxide (PEO) and Caprolactum (Cap) on the methane rich mixture (Methane C_1: 85.24 %, Ethane C_2: 10.03 %, Carbon dioxide CO_2: 2.49%, Nitrogen N_2 : 2.22 and n-hexane C_{6+} : 0.02 %) using the rocking cell assembly at different concentrations (1-5wt%) and pressure range of 40-120 bars. The experimental results, show that the selected ILs act as thermodynamic and kinetic inhibitors both simultaneously. The addition of synergent / additives in equal ratio with ILs helps to improve their kinetic and thermodynamic inhibition performance significantly.

Keywords: Ionic liquids, Synergents, Hydrates, Natural Gas Mixture

Introduction

Natural Gas hydrates (NGH) are ice-like crystalline structures that are formed of gas and water molecules.[1] NGH mainly have of three distinctive structures: structure I (sI), structure II and structure H (sH).[2] The NGH surface is known to be hydro-philic and a thin layer of liquid water film resides on the NGH surface that helps hydrate particles to bridge together via capillary force leading to agglomeration [3] and blockages in offshore deep water sea-lines.[4] Therefore, NGH are considered to be a major flow assurance threat to oil and gas industry.

[1] Corresponding author Tel.:+974 4403 4993; Fax: +974 4403 4131
Mobile: +974 30999747

Hydrates are threat to flow assurance, but they also have many positive applications. These positive applications include CO_2 sequestration[5], gas storage and transportation[6], gas separation in mines [7], desalination[8] and separation of unwanted components from bio-gases.[9] The incidents of NGH formation has been reported in Caspian Sea and North Sea where petroleum crude is pumped at the high pressures and low temperature conditions causing blockages in the subsea lines.[10] The formation of NGH has also been reported in Gulf of Mexico where temperature goes down to 2-4 ^{0}C at pressures up to 30 MPa. [11] In the middle-eastern countries the formation of NGH is likely to occur during winter seasons. [12] Therefore, over a year an expenditure of worth billion dollars is made on NGH prevention measures. [13]

The use of chemical inhibitors is considered to be one of the most viable method to prevent NGH formation in oil and gas sector. These inhibitors are classified into three types such as thermodynamic hydrate inhibitors (THI), kinetic hydrate inhibitors (KHI) and ant-agglomerants (AG). THI function by shifting hydrate dissociation temperature to lower values and KHI function by delaying the hydrate formation time.[14] The AG does not inhibit hydrate formation, but prevent the growth of hydrate crystals. [15]

The commercial THI like Methanol and Mono-ethylene glycol (MEG) perform well, but these inhibitors are required in bulk quantities (> 30 wt%) and cannot be easily disposed of into the environment. Therefore, the industrial and academic research interest is shifting toward the low dosage and environmentally friendly inhibitors that can prevent hydrate formation.

Ionic liquids (ILs) also known as ionic fluids are classified as organic salt that have low melting point and tend to exist in a liquid state at low or ambient temperature. [16] The ILs are widely used in different green chemical processes due to their negligible vapour pressure and low viscosity. Recently, ILs have been identified as the dual functional inhibitors as they have tendency to act as both KHI and THI at the same time.[17] The thermodynamic inhibition (TI) and kinetic inhibition (KI) effectiveness of the IL varies with their concentration and length of alky chain.

Xiao et al. [18] showed that imidazolium-based ILs with halides tend to acts as a dual functional inhibitors for methane sI crystals. The ILs possess strong electrolytic charges and strength to form hydrogen bonding with water molecules that helps to reduce hydrate suppression temperature and also delay the hydrate formation time. Kim et al. [19], formulated a pyrrolidinium cation-based ILs with tetra fluoroborate $[BF_4]^-$ anion and found that these ILs reduced hydrate suppression temperature and also delayed hydrate formation time. Shin et al. [20], studied the thermodynamic inhibition effect of pyrrolidinium cation based ILs on CO_2 hydrates and found that the selected ILs acted as a THI. Rasoolzadeh, et al. [21] studied the KI effect of three imidazolium based ILs at different concentration (1-20 wt%) on methane hydrate formation and found that the ILs with shorter alkyl chain and stronger electrostatic forces tend to be more effective as KHI. It was also observed that the effect of cation is more significant than the effect of anions in the selected ILs. Zare, et al. [22] studied the KI and TI effect of two imidazolium based ILs on methane hydrate formation in the presence of ethylene glycol methyl ether (EGME). It was found that one of the ILs acted as KHI and THI both as the same time and the other IL acted as the hydrate promoter.

In order to stimulate the inhibiton effectiveness of the hydrate inhibitors, the effect of mixing additives with the hydrate inhibitors like poly-vinylpyrrolidine (PVP) and polyvinyl caprolactum (PVCap) has been investigated in different literatures. [23] It has been found that the addition of these synergistic compounds helps to enhance the inhibition effectiveness of the hydrate inhibitors.[24] So the objective of this work is investigate the KI and TI effectiveness of ILs 1-Methyl-1-Propyl-pyrrolidinium Chloride [PMPy][Cl] and 1-Methyl-1-Propyl-pyrrolidinium Triflate [PMPy][Triflate] in a methane rich (MR) mixture at different pressures (40-120 bars) and concentrations (1-5wt%) in the presence of additive such as polyethylene oxide (PEO) and Caprolactum (VCap). Most part of this work has already been published in the ACS journal 'Energy & Fuel' with slight editions.[25]

Experimental Procedure

Materials

The ILs [PM-Py][Cl] and [PM-Py][Triflate] were acquired from Iolitec ionic liquid technologies GmbH (≥ 98 %) and the additives Polyethylene oxide (PEO) and Caprolactum (Cap) were acquired from Sigma Aldrich Korea (≥ 98 %) [Table 1].

Table 1. Ionic Liquids (ILs) and Synergents (Syn) used in this work

Ionic Liquids (ILs)	Structure	Synergents	Structure
1-Methyl-1-Propylpyrrolidinium Chloride [PMPy][Cl]	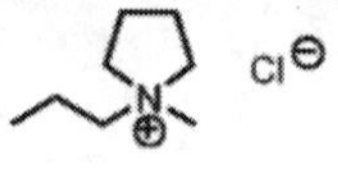	VinylCaprolactum (VCap)	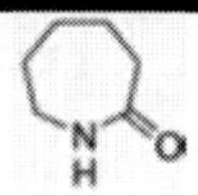

1-Methyl-1-Propylpyrrolidinium Triflate [PMPy][Triflate]	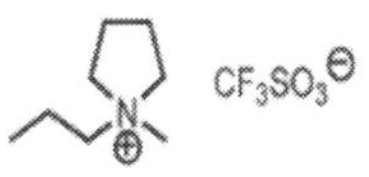	Polyethylene Oxide (PEO)	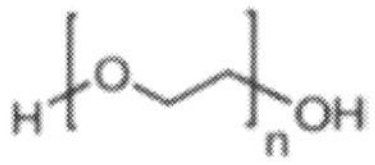

These ILs and additives sample were diluted with de-ionized water to prepare sample solutions for the experiments. The MR mixture used for these experiments was purchased from Buzwair Industrial Gas Factories (Doha-Qatar) [Table 2].

Table 2. Composition of the MR mixture as per GC analysis in Mol %

Components	Composition (Mol %)
C_1	85.24
C_2	10.03
C_{6+}	0.02
CO_2	2.49
N_2	2.22

Procedure

All the experimental work was conducted using the Rocking cell assembly (RC-5) purchased from PSL system-technik GmbH (Germany). The assembly consist of 5 stainless steel cells that are attached to a rig submerged in the water bath. These cells have the volume of 40.13 cm^3 and consist of a stainless steel ball that moves back and forth to provide the necessary agitation. A 15 ml of the sample solution is added to each of these cells and then the cells are pressurized chronologically from 40-120 bars.

Before starting of the experiments, the RC-5 is maintained steady at the temperature of 20 °C for one hour. When the RC-5 gets steady at 20 °C the experiment process is initiated using isochoric pressure search method.[26] The submerged cells in the bath are then gradually cooled from 20 °C to 2 °C within the period of 9 hours, followed by an isothermal step of 24 hours at 2 °C and then heating the cells at the rate of 0.1 °C per hour to the temperature 20 °C. The extraction of hydrate dissociation points has been conducted using the method proposed by Kim, et al. [27] The hydrate formation initiates during sub-cooling and the hydrate growth is indicated through the sharp pressure drop that takes place during isothermal step.[28]

Results & Discussion

Thermodynamic Hydrate Inhibition using IL/ IL + Additives

The initial experiments were conducted using 1 wt% of selected ILs on MR mixture. But no significant TI effect was observed at this concentration. So the further experiments were conducted using 5wt% of ILs. The 5 wt% IL [PMPy][Cl] and 5 wt% IL [PMPy][Triflate] provided a noticeable TI effect with hydrate suppression temperature of 1.8 °C ± 0.2 °C observed at 40 bars for [PMPy][Cl]. It is reported in previous works[23,32] that when the NGH formation occurs the water molecules form clusters around guest molecule and then these clusters grow in size and become more structured at low temperatures. The ILs tend to disrupts the hydrogen bonding between the water molecules and their TI effectiveness increases with their respective concentration. The same effect was also observed by Keshavarz et al. [29] and Partoon et. al [30] who found that by increasing the mass fraction of IL [Bmim][BF_4^-] and [Bmim][Cl^-] the overall TI effect enhances. In terms of effectiveness as THI the selected ILs can be listed as:

[PM-Py][Cl] > [PM-Py][Triflate]

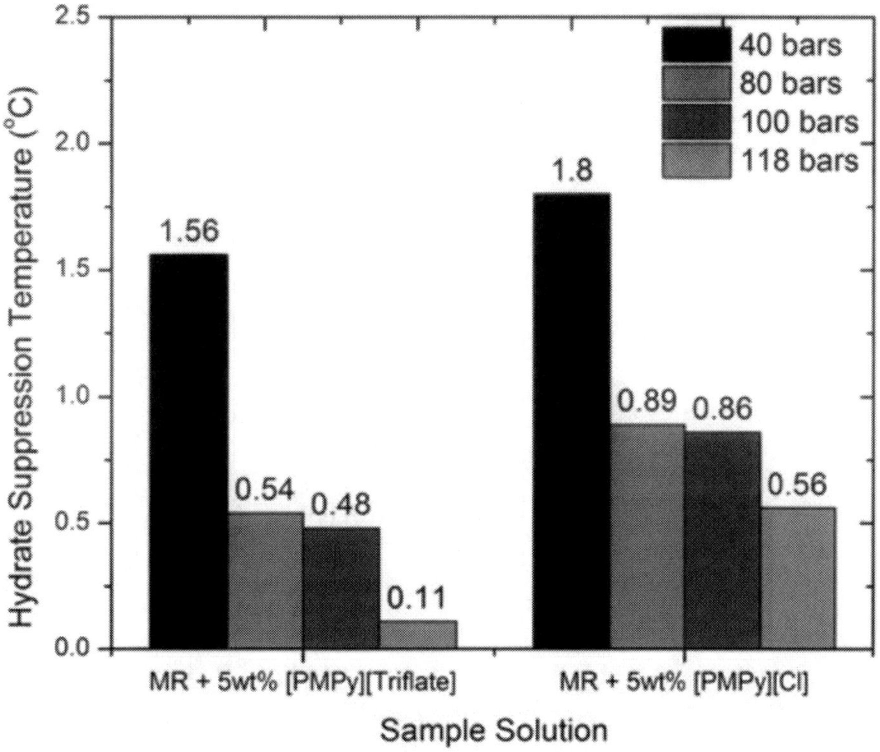

Figure 1: Showing the thermodynamic inhibition effect of ILs 5wt% [PMPy][Cl] and 5wt% [PMPy][Triflate] on the methane rich (MR) mixture hydrate formation at different pressures. The figure is adapted from the published work.[25]

In order to improve the TI effectiveness of the selected ILs the additives such as Caprolactum (Cap) and polyethylene oxide (PEO) were added in ratio 1:1 with the ILs. No previous work has reported the effect of adding the additives with the selected ILs previously. It was found that by the addition of 5wt% Cap with the selected ILs the hydrate suppression temperature increases to 2.2 °C ± 0.2 °C (40 bars). The additive PEO did not provided any enhancement in the TI effect of the selected ILs. So in terms of effectiveness as a THI the ILs + additives mixtures can be listed as:

[PMPy][Cl] + VCAP > [PMPy][Cl] + PEO

Kinetics of Hydrate Inhibition using ILs / ILs + Additives

In order to confirm the dual functional effect of the selected ILs the kinetic inhibition effectiveness of the selected ILs was tested at different concentration with and without the presence of additives such as PEO and VCAP. At the concentration of 1wt% both the ILs showed a slight KI effect confirming that they have a tendency to act as a KHI. As the concentration of IL was increased to further this KI effect got more significant and the delay in hydrate formation time of up to 1.3 hrs ± 0.5 hrs was observed at moderate pressures (< 60 bars) for IL **[PMPy][Cl]** and **[PMPy][Triflate].** This confirms that the selected ILs have a tendency to act as a dual functional inhibitors.

The delay in hydrate induction/formation time provided by the ILs was not significant. So the additives PEO and VCAP were added with the selected ILs to improve their overall KI effect. It was found that the KI effectiveness of the selected ILs increases significantly in the presence of additive PEO. The PEO was found to be an effective Kinetic synergent for the both ILs. The hydrate induction time at lower pressures (~37 bars) for [PMPy][Cl] was found to 7.6± 0.5 hrs, followed by [PMPy][Cl]+VCAP with induction time 8.3 ± 0.5 hrs and [PMPy][Cl]+PEO with induction time 22.7 ± 0.5 hrs. So in terms of effectiveness as KHI the ILs + additive mixtures at lower pressures (~37 bars) can be listed as:

[PMPy][Cl]+PEO > [PMPy][Cl]+VCAP > [PMPy][Cl]

Similarly the hydrate induction time at lower pressures (~38 bars) for [PMPy][Triflate] was found to 7.3± 0.5 hrs, followed by [PMPy][Triflate]+VCAP with induction time 8.5 ± 0.5 hrs and [PMPy][Triflate]+PEO with induction time 20 ± 0.5 hrs. So in terms of effectiveness as KHI the ILs + additive mixtures at lower pressures (~37 bars) can be listed as [Figure 2]:

[PMPy][Triflate]+PEO > [PMPy][Triflate]+VCAP > [PMPy][Triflate]

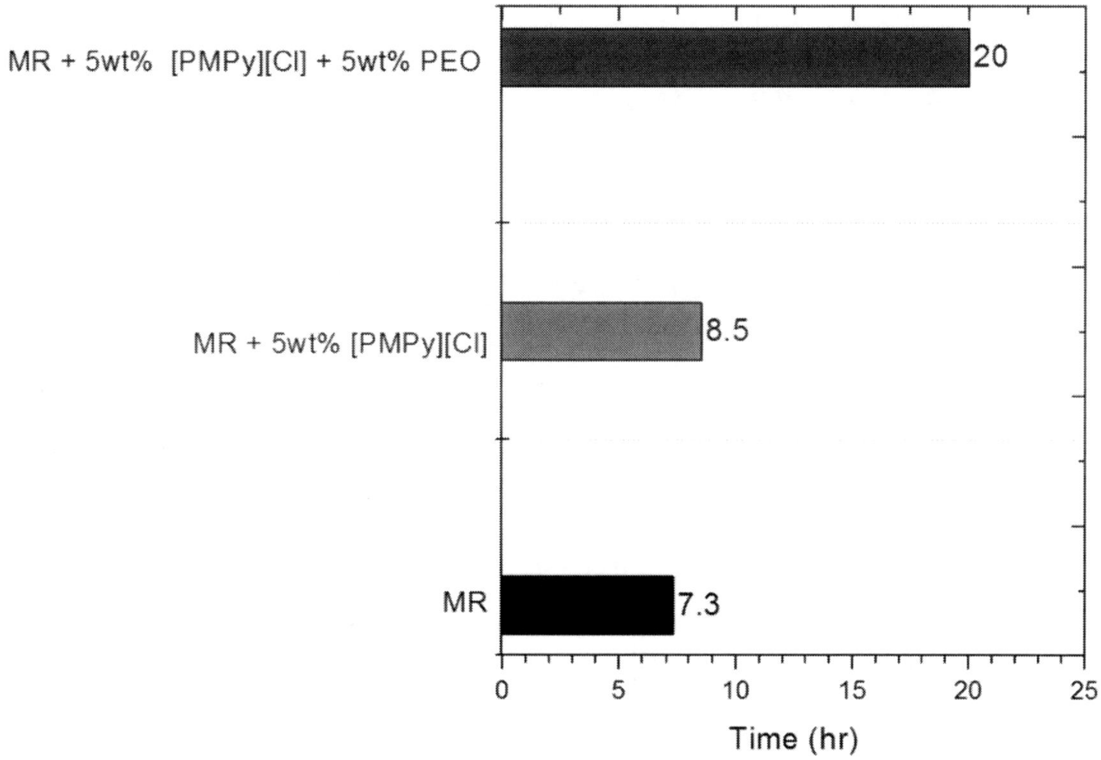

Figure 2: The hydrate induction time in the presence of IL 5wt% [PMPy][Cl] and 5wt% [PMPy][Cl] + 5wt% PEO for the methane rich (MR) mixture at 38 bars.

This shows that the addition of PEO with the ILs in similar ratio can help to increase their inhibition effectiveness significantly. The synergistic effect of addition of additive 0.5 wt% PVCap with ILs [EMIM][BF4], [BMP][BF4] and [HEMP][BF4] was investigated by Kang et al[8]. It was found that the addition of 0.5 wt% PVCap with these ILs help to delay hydrate formation by 1-3 hrs at pressures of 70 bars. This indicate that the addition of such additives effect the hydrate nucleation rather than the growth of hydrate crystals. Daraboina, et al. [13] also found that the addition of PEO with the commercial inhibitor Luvicap helps to enhance the KI effectiveness of the Luvicap. Hammouda, et al. [31], stated that by when PEO is added with the water, the PEO forms hydrophilic and hydrophobic interactions with water molecules. The presence of hydroxyl group in PEO allows it to form hydrophilic interactions with water molecule via hydrogen bonding. This allows it to disrupt the water molecules structure before the process of hydrate nucleation occurs. In addition to that, the hydrophobic group helps to repel away the water molecules and it may associate with inhibitor molecules and vary their conformation.

Conclusion

The experimental results show that the selected ILs have a tendency to act as a dual function inhibitors by themselves. The IL [PMPy][Cl] provided hydrate suppression temperature within the range of 0.8-1.8 K (± 0.2) and delayed hydrate formation by 0.8-1.3 hrs (± 0.5). The IL [PMPy][Cl] was found to be more effective than the IL [PMPy][Triflate].

In order to improve the TI and KI effectiveness of the ILs the additives such as PEO and VCAP were added with them in equal ratio and the experiments were repeated. The new set of experiments how that the additive VCAP helped to improve the hydrate suppression temperature from 1.8 K to 2.2 K at 40 bars. This shows that the additive VCAP acts likes a thermodynamic synergent. Similarly by the addition of PEO, the hydrate induction time got delayed from 7.6 ± 0.5 hrs to 22.7 ± 0.5 hrs for the IL [PMPy][Cl] at 40 bars.

To sum up, in this work it has been found that the use of additives/ synergents can help to improve the kinetic and thermodynamic inhibition effectiveness of the pyrrolidinium based ILs. This motivates us to look for new combination of ILs + synergents to prepare mixtures that help to mitigate the hydrate formation.

Acknowledgement:

Author Fahed Qureshi would like to acknowledge GSRA. This work was made possible by GSRA # 2-1-0603-14012 from the Qatar National Research Fund (a member of Qatar Foundation). The statements made herein are solely the responsibility of the authors.

References:

1. Sohn, Y. h.; Kim, J.; Shin, K.; Chang, D.; Seo, Y.; Aman, Z. M.; May, E. F., Hydrate plug formation risk with varying watercut and inhibitor concentrations. *Chemical Engineering Science* **2015,** *126*, 711-718.
2. Kim, E.; Lee, S.; Lee, J. D.; Seo, Y., Influences of large molecular alcohols on gas hydrates and their potential role in gas storage and CO2 sequestration. *Chemical Engineering Journal* **2015,** *267*, 117-123.
3. Delavar, H.; Haghtalab, A., Thermodynamic modeling of gas hydrate formation conditions in the presence of organic inhibitors, salts and their mixtures using UNIQUAC model. *Fluid Phase Equilibria* **2015,** *394*, 101-117.
4. Sun, M.; Firoozabadi, A., Gas hydrate powder formation – Ultimate solution in natural gas flow assurance. *Fuel* **2015,** *146*, 1-5.
5. Linga, P.; Daraboina, N.; Ripmeester, J. A.; Englezos, P., Enhanced rate of gas hydrate formation in a fixed bed column filled with sand compared to a stirred vessel. *Chemical Engineering Science* **2012,** *68* (1), 617-623.
6. Prasad, P. S. R.; Chari, V. D., Preservation of methane gas in the form of hydrates: Use of mixed hydrates. *Journal of Natural Gas Science and Engineering* **2015,** *25*, 10-14.
7. Zhong, D.-L.; Daraboina, N.; Englezos, P., Recovery of CH4 from coal mine model gas mixture (CH4/N2) by hydrate crystallization in the presence of cyclopentane. *Fuel* **2013,** *106*, 425-430.
8. Kang, K. C.; Linga, P.; Park, K.-n.; Choi, S.-J.; Lee, J. D., Seawater desalination by gas hydrate process and removal characteristics of dissolved ions (Na+, K+, Mg2+, Ca2+, B3+, Cl−, SO42−). *Desalination* **2014,** *353*, 84-90.
9. Kondo, W.; Ohtsuka, K.; Ohmura, R.; Takeya, S.; Mori, Y. H., Clathrate-hydrate formation from a hydrocarbon gas mixture: Compositional evolution of formed hydrate during an isobaric semi-batch hydrate-forming operation. *Applied Energy* **2014,** *113*, 864-871.
10. Sami, N. A.; Das, K.; Sangwai, J. S.; Balasubramanian, N., Phase equilibria of methane and carbon dioxide clathrate hydrates in the presence of (methanol+MgCl2) and (ethylene glycol+MgCl2) aqueous solutions. *The Journal of Chemical Thermodynamics* **2013,** *65*, 198-203.
11. Zhao, X.; Qiu, Z.; Huang, W., Characterization of kinetics of hydrate formation in the presence of kinetic hydrate inhibitors during deepwater drilling. *Journal of Natural Gas Science and Engineering* **2015,** *22*, 270-278.
12. Jordan, M. M.; Weathers, T. M.; Jones, R. A.; Sutherland, L.; Giddings, D. M.; Abla, M., The Impact of Kinetics Hydrate Inhibitors Within Produced Water on Water Injection/Disposal Wells. In *SPE International Symposium and Exhibition on Formation Damage Control, 26-28 February.*, Society of Petroleum Engineers: Lafayette, Louisiana, USA 2014.
13. Daraboina, N.; Malmos, C.; von Solms, N., Synergistic kinetic inhibition of natural gas hydrate formation. *Fuel* **2013,** *108*, 749-757.
14. Riza Turkmen, I.; Upadhyay, N.; Adham, S.; Gharfeh, S., Evaluation of analytical methods for determination of kinetic hydrate inhibitor (KHI) in produced waters. *Journal of Petroleum Science and Engineering* **2015,** *126*, 63-68.
15. Xiao, C.; Adidharma, H., Dual function inhibitors for methane hydrate. *Chemical Engineering Science* **2009,** *64* (7), 1522-1527.
16. Richard, A. R.; Adidharma, H., The performance of ionic liquids and their mixtures in inhibiting methane hydrate formation. *Chemical engineering science* **2013,** *87*, 270-276.
17. (a) Del Villano, L.; Kelland, M. A., An investigation into the kinetic hydrate inhibitor properties of two imidazolium-based ionic liquids on Structure II gas hydrate. *Chemical Engineering Science* **2010,** *65*

(19), 5366-5372; (b) Li, X.-S.; Liu, Y.-J.; Zeng, Z.-Y.; Chen, Z.-Y.; Li, G.; Wu, H.-J., Equilibrium hydrate formation conditions for the mixtures of methane+ ionic liquids+ water. *Journal of Chemical & Engineering Data* **2010,** *56* (1), 119-123.

18. Xiao, C.; Wibisono, N.; Adidharma, H., Dialkylimidazolium halide ionic liquids as dual function inhibitors for methane hydrate. *Chemical Engineering Science* **2010,** *65* (10), 3080-3087.

19. Kim, K. S.; Kang, J. W.; Kang, S. P., Tuning ionic liquids for hydrate inhibition. *Chemical communications* **2011,** *47* (22), 6341-3.

20. Shin, B. S.; Kim, E. S.; Kwak, S. K.; Lim, J. S.; Kim, K.-S.; Kang, J. W., Thermodynamic inhibition effects of ionic liquids on the formation of condensed carbon dioxide hydrate. *Fluid Phase Equilibria* **2014,** *382*, 270-278.

21. Rasoolzadeh, A.; Javanmardi, J.; Eslamimanesh, A.; Mohammadi, A. H., Experimental study and modeling of methane hydrate formation induction time in the presence of ionic liquids. *Journal of Molecular Liquids* **2016,** *221*, 149-155.

22. Zare, M.; Haghtalab, A.; Ahmadi, A. N.; Nazari, K.; Mehdizadeh, A., Effect of imidazolium based ionic liquids and ethylene glycol monoethyl ether solutions on the kinetic of methane hydrate formation. *Journal of Molecular Liquids* **2015,** *204*, 236-242.

23. (a) Kelland, M. A., History of the Development of Low Dosage Hydrate Inhibitors. *Energy & Fuels* **2006,** *20* (3), 825-847; (b) Daraboina, N.; Linga, P.; Ripmeester, J.; Walker, V. K.; Englezos, P., Natural gas hydrate formation and decomposition in the presence of kinetic inhibitors. 2. Stirred reactor experiments. *Energy & Fuels* **2011,** *25* (10), 4384-4391; (c) Townson, I.; Walker, V. K.; Ripmeester, J. A.; Englezos, P., Bacterial inhibition of methane clathrate hydrates formed in a stirred autoclave. *Energy & Fuels* **2012,** *26* (12), 7170-7175; (d) Del Villano, L.; Kommedal, R.; Kelland, M. A., Class of kinetic hydrate inhibitors with good biodegradability. *Energy & Fuels* **2008,** *22* (5), 3143-3149.

24. (a) Kelland, M. A.; Moi, N.; Howarth, M., Breakthrough in Synergists for Kinetic Hydrate Inhibitor Polymers, Hexaalkylguanidinium Salts: Tetrahydrofuran Hydrate Crystal Growth Inhibition and Synergism with Polyvinylcaprolactam. *Energy & Fuels* **2013,** *27* (2), 711-716; (b) Magnusson, C. D.; Kelland, M. A., Study on the Synergistic Properties of Quaternary Phosphonium Bromide Salts with N-Vinylcaprolactam Based Kinetic Hydrate Inhibitor Polymers. *Energy & Fuels* **2014,** *28* (11), 6803-6810; (c) Chua, P. C.; Kelland, M. A., Tetra (iso-hexyl) ammonium Bromide□ The Most Powerful Quaternary Ammonium-Based Tetrahydrofuran Crystal Growth Inhibitor and Synergist with Polyvinylcaprolactam Kinetic Gas Hydrate Inhibitor. *Energy & Fuels* **2012,** *26* (2), 1160-1168.

25. Qureshi, M. F.; Atilhan, M.; Altamash, T.; Tariq, M.; Khraisheh, M.; Aparicio, S.; Tohidi, B., Gas Hydrate Prevention and Flow Assurance by Using Mixtures of Ionic Liquids and Synergent Compounds: Combined Kinetics and Thermodynamic Approach. *Energy & Fuels* **2016,** *30* (4), 3541-3548.

26. Tariq, M.; Rooney, D.; Othman, E.; Aparicio, S.; Atilhan, M.; Khraisheh, M., Gas Hydrate Inhibition: A Review of the Role of Ionic Liquids. *Industrial & Engineering Chemistry Research* **2014,** *53* (46), 17855-17868.

27. Kim, K.-S.; Kang, J. W.; Kang, S.-P., Tuning ionic liquids for hydrate inhibition. *Chemical Communications* **2011,** *47* (22), 6341-6343.

28. Mohammad-Taheri, M.; Zarringhalam Moghaddam, A.; Nazari, K.; Gholipour Zanjani, N., The role of thermal path on the accuracy of gas hydrate phase equilibrium data using isochoric method. *Fluid Phase Equilibria* **2013,** *338*, 257-264.

29. Keshavarz, L.; Javanmardi, J.; Eslamimanesh, A.; Mohammadi, A. H., Experimental measurement and thermodynamic modeling of methane hydrate dissociation conditions in the presence of aqueous solution of ionic liquid. *Fluid Phase Equilibria* **2013,** *354*, 312-318.

30. Partoon, B.; Wong, N. M. S.; Sabil, K. M.; Nasrifar, K.; Ahmad, M. R., A study on thermodynamics effect of [EMIM]-Cl and [OH-C2MIM]-Cl on methane hydrate equilibrium line. *Fluid Phase Equilibria* **2013,** *337*, 26-31.

31. Hammouda, B.; Ho, D.; Kline, S., SANS from poly (ethylene oxide)/water systems. *Macromolecules* **2002,** *35* (22), 8578-8585.

511620 Complex Fluids in the Gut

Abstract:

The gut governs digestion and nutrient absorption, is a promising target for drug delivery, and teems with active micro-organisms that can have remarkably strong effects on host health. Many of the materials that transit through it regularly are complex fluids composed of ingested/secreted polymers and active microbes. Despite their importance, however, little is known about the interactions between the gut and the complex fluids within it. By combining animal experiments, biophysical processing and characterization, and microscopy, we show how polymers abundant in the gut can alter the structure of the gut and its contents. These alterations could impact how microbes transit through the gut, potentially take up residence within it, and potentially infect it. Our results provide physical principles by which we can better understand the hydrodynamics of the gut. More broadly, this research stimulates new findings and questions at the interface of engineering, physics, and biology.

499713 Boundary Guidance in Self Propelled Colloidal Motors

Abstract:

Self-propelled, chemically-powered colloidal locomotors are engines which convert fuel from the environment into mechanical propulsion. They are designed to transverse small scale landscapes in a range of applications involving micropumping, sensing and cargo transport. Precise navigation across prescribed pathways is usually required, and for this purpose onboard steering mechanisms can be incorporated. Here we demonstrate, through numerical simulation, how boundaries can simply and passively guide locomotors to move along desired pathways.

As a model system we choose an engine design in which a spherical Janus colloid coated with a symmetrical catalyst cap converts fuel into a product solute on one side of the colloid. The solute is repelled from the colloid through a short range interaction in a thin layer around the particle which creates a slip velocity that propels the swimmer in the direction opposite to the cap (self-diffusiophoresis).

As a first example of boundary guidance, we have previously shown that diffusiophoretically-driven Janus colloid locomotors reaching a planar wall can, for a range of active catalyst areas, rotate to a configuration in which the active side is partially inclined from the wall and the locomotor skims at a constant separation distance. Here we demonstrate more advanced and intricate examples of boundary guidance: We examine the motion of a spherical Janus locomotor in a two dimensional channel, and the trajectory of spheroidal locomotors around a spherical obstacle. Locomotors with small catalyst areas in a channel repeatedly reflect off of the channel wall until they become rotated to a symmetrical cap orientation which focuses them to move at the channel centerline. For larger coverages, the swimmers approach one wall of the channel and skim directly without reflection. For ellipsoidal and spherical locomotors approaching a spherical obstacle, locomotors becomes locked in an orbit around the obstacle for large obstacles, but, for smaller obstacles, only partly navigate around the obstacle before leaving the orbit and moving away from the obstacle.

499779 Active Nematic Liquid Crystals with Variable Activity and Elasticity

Abstract:

Actomyosin has long been believed to be an active contractile gel. Here we show for the first time that when actin filaments are sufficiently short, a two-dimensional dense sheet of actin-myosin solution will exhibit a nematic phase and spontaneously develop extensile flow. As short actin filaments slide on each other due to the action of motor protein clusters, such collective motion leads to a sustainable, cohesive flow that drives the proliferation, motion, and annihilation of topological defects, reminiscent of active nematic liquid crystals. Our hybrid lattice Boltzmann simulation based on the hydrodynamic model of active nematics can well capture the experimental observations that the orientation and velocity correlation lengths scale inversely with the activity parameter(s?). We further show that by visualizing the +1/2 defect's morphology, one can deduce the ratio of the splay and bend elastic constants of the material. As the system is driven out of equilibrium by the action of myosin clusters, the defect morphology transitions from a U-shape to a V-shape, implying an increase of the apparent bend-to-splay ratio - such change in a material's apparent elasticity is an indicator of how far the system is away from equilibrium. We also investigate how defect velocity and defect-defect interaction depend on the local director field, its material constants, and its activity parameters. Both experiments and simulations demonstrate that one can switch the interaction of $\pm1/2$ defect pairs between repulsion and attraction by simply tuning the activity parameter. We therefore propose that by controlling the spontaneous flow of active nematics, one can assemble and transport defect-carried cargos.

495385 Using a Stochastic Field Theory to Understand Active Colloidal Suspensions

Abstract:

Even without external forcing active systems are out of equilibrium, which gives rise to interesting properties in both small and large concentrations of the particles. These properties have been observed in experiments as well as simulation/modeling approaches. It is important to understand how hydrodynamic interactions between active colloids cause and/or alter the suspension properties including enhanced transport and mixing. One of the most successful approaches has been a mean field theory. However, in some situations the mean field theory makes predictions that differ significantly from experiments and direct (agent or particle based) simulations. There are also some quantities that cannot be calculated by the mean field theory. In this talk, we will describe our new approach which uses a stochastic field to overcome the limitations of the mean field assumption. It allows us to calculate how interactions between organisms alter the correlations and mixing even in conditions where there is no large-scale group behavior.

486722 Imaging the Onset Kinetics of the Swarming Transition Using Light-Controlled Bacteria

Abstract:

Active fluids are a novel class of nonequilibrium soft materials, which are composed of a large number of self-propelled particles. These particles collectively form coherent clusters at high densities and low noise levels, as illustrated vividly by the striking patterns of flocking birds, schooling fishes and swarming bacteria. Although the disorder-coherent transition of active fluids has been extensively studied, its very nature is still on heated debate and the transition kinetics has not been explored. Here, using a mutated E. coli strain, whose locomotion can be reversibly controlled by light, we experimentally study the swarming transition in active fluids and explore its kinetic pathway. Particularly, we trigger bacterial swarming by tuning the light intensity and image the dynamics of coherent clusters in concentrated bacterial suspensions. We systematically map the phase diagram of bacterial swarming with respect to the concentration n and the faction of tumblers f. We find that the transition points follow n*f=12. The continuous nature of the phase transition is revealed by the absence of hysteresis loop and zero incubation times in the swarming transition. We further investigate the formation of swarming cluster, dynamics of correlation length and energy spectrum. Our study sheds light on the phase transition of active matter and the emergent properties of many-body nonequilibrium systems.

496230 Surface Drag and Swarming in Motile Bacteria

Abstract:

Escherichia coli swim in liquid medium with the aid of flagellar filaments that switch erratically between clockwise (CW) and counter-clockwise (CCW) directions of rotation. When grown on semi-soft agar substrates, cells undergo subtle morphological changes and transition into a swarming state (a type of 2D group-motility). Presumably, the transition is initiated by mechanical forces arising from contact with the surface. In *E. coli*, the cell lengths double during swarming. Longer cells counter the increase in surface drag by increasing the number of filaments to generate additional thrust. We are interested in determining the interplay between cell lengths, filament forms, the direction of flagellar rotation and hydrodynamic interactions with the substrates during swarming.

We will present our latest findings that suggest that CCW-biased mutants fail to swarm but motility can be rescued by simply wetting the agar surface, in agreement with earlier reports. However, swarming in CW-biased mutants cannot be rescued by increasing surface wetness. To determine the underlying reason for this difference, we employed biophysical characterization assays and measured the torque generated by individual flagellar motors in the CCW and CW mutants. Results suggested anisotropic torque generation in the two directions, consistent with our previous observations in a different species [1]. However the anisotropy by itself was inadequate in explaining the inhibition of swarming in the CW-mutants. Fluorescent-labeling experiments were carried out to determine the predominant polymorphic flagellar forms in the two types of mutants. We will discuss the insights revealed by the fluorescent experiments and the resulting effects on the thrust developed by each cell. The implications of these phenotypes in swarm-inititation will also be discussed.

[1] P. P. Lele, T. Roland, A. Shrivastava, Y. Chen, H. C. Berg, â€œDirectional dependence of stall torques in *C. crescentus* motorsâ€, *Nature Physics* (2016), 12, 175-178.

497190 Effects of Elasticity and Hydrodynamic Interactions on Swimmer Shape and Trajectories in a Coarse-Grained Model of Monotrichous Bacteria

Abstract:

The locomotion of singly-flagellated (monotrichous) bacteria in viscous fluid is a quintessential example of movement at low Reynolds number. Observations of such organisms have shown that both the hook protein (connecting cell body and flagellum) and the flagellum itself may buckle when a critical flexibility is exceeded, relative to the swimmer's generated propulsion. The former result is demonstrated with a simple linked rigid swimmer. The latter phenomena is considered at varying levels of description. First, we consider the elastic stability of a pinned flagellum in three separate cases with different dynamics and equilibrium shape, finding in each case that the flagellum indeed buckles above a critical flexibility. Next, we consider the case of an elastic flagellum pushing a cell body, examining the resultant trajectories and flow fields. Relatively rigid flagella produce similar results to the linked rigid swimmer, but very flexible flagella render propulsion ineffective post-buckling. These considerations may provide insight on how swimmers move through complex environments and how to expand potential application.

487175 Microscopic Dynamics of Bacterial “Superfluids” Under Planar Oscillatory Shear

Abstract:

Bacterial suspensions show interesting rheological behaviors including a remarkable "superfluid" flows with vanishing apparent viscosity. To reveal the microscopic dynamics of bacterial "superfluids", we systematically studied the shear profiles of *E-coli* suspensions using ultra-fast confocal microscopy. The flow fields of bacterial suspensions under planar oscillatory shears of different shear rates are measured along both the flow-vorticity and the flow-gradient planes. We found that concentrated bacterial suspensions exhibit strong non-homogeneous flow profiles at low shear rates, where shear rates vanish away from the moving shear plate. We characterized the shape of the nonlinear shear profiles at different applied shear rates and bacterial concentrations and activities. The shear profiles follow a simple scaling relation with the applied shear rates and the enstrophy of suspensions. We demonstrated that this scaling relation can be quantitatively understood by considering the ensemble average of the unusual shear banding predicted by the theory of active liquid crystals. Our experiments reveal a profound influence of shear flows on the locomotion of bacteria and provide new insights into the dynamics of active fluids.

484051 Monte Carlo Simulaitons on the Aggregate Structures in a Supension Composed of Magnetic Cubic Particles on a Material Surface

Abstract:

A magnetic particle suspension has been actively investigated in a various fields such as the fluid engineering, the magnetic material engineering, and the colloid and interface science. In addition, the fields of the biomedical engineering and the environmental resource engineering may be new challenging areas as an application of suspensions composed of magnetic particles. In contrast to the ordinary ferromagnetic rod-like particles, hematite particles have a characteristic feature in that the particles have a significantly weak magnetization, which may be a merit for obtaining a stable suspension in a relatively straightforward manner. In addition to spindle-like hematite particles, modern synthesis technologies enable the generation of disk-like and cube-like hematite particles. In the present study, we concentrate on a suspension composed of magnetic cubic particles on a material surface from the viewpoint of applying a magnetic cubic particle suspension to the development of the technology of surface modification. We here treat a variety of situations of non-equal number of particles with the magnetic moments aligning with the two diagonal directions, i.e., the directions toward and away from the material surface. It is assumed that the cubic particles are magnetized in the diagonal direction of the particle, which may be a typical characteristic of hematite cubic particles. Also, the cubic particles are assumed to move on the surface plane in a quasi-2D system in thermodynamic equilibrium with the face of each cubic particle in contact with the bottom material surface under the situation of a strong gravitational field. Monte Carlo simulations have been performed to obtain the results regarding the characteristics of the aggregates and the order parameters of the system for a variety of situations of the composition ratio of upward and downward particles, the magnetic particle-particle interaction and the magnetic field strength. We summarize the main results that were obtained from Monte Carlo simulations in the following. The regime in the particle aggregates is strongly dependent on the composition ratio of these cubic particles. Specifically, the size of closely-packed aggregates becomes smaller, and thin linear clusters are formed in the situation of a strong magnetic field with decreasing composition ratio. An external magnetic field does not function as more significantly inducing the formation of thick and closely-packed clusters for smaller values of the composition ratio. A decrease in the composition ratio comes to dull the regime change in the particle aggregates for a change in the magnetic particle-particle interaction strength. From these results, we understand that a decrease in the composition ratio leads to a suppression of the growth of large closely-packed clusters, which clearly suggests a possibility that the composition ratio may be used as a technique for controlling the cluster formation and the size of closely-packed clusters of cubic hematite particles in the situation of a strong external magnetic field.

493135 Pattern Formation from Interfacial Instabilities in Miscible Viscous Fluids

Abstract:

The invasion of one fluid into another of higher viscosity is unstable in a quasi-two dimensional geometry. This viscous-fingering instability typically produces complex patterns that are characterized by repeated branching of the evolving structure, which leads to the common morphologies of fractal or dense-branching growth. However, for two miscible fluids with a high viscosity ratio between the inner and the outer fluid, an entirely different type of pattern formation occurs. Here, structures form that, most remarkably, all expand at nearly the same rate in all directions. This leads to patterns that do not change shape as they develop. This type of growth is called *proportionate growth* and it has not previously been identified in a physical system despite its common occurrence in the biological world. Such growth relies on a high degree of regulation of the growth rate of widely separated features. We discuss the characteristics of this growth and its relation to the common patterns of fractal, dense-branching and dendritic growth.

491328 Electrokinetic Fingering: A Problem in Vector Laplacian Growth

Abstract:

Pattern formation is ubiquitous in various physical and chemical processes and has fascinated many scientists over the past couple of decades. In many instances, interfacial instabilities play a crucial role in creating these patterns and controlling their spatiotemporal evolution. Perhaps one of the most well-known examples is the striking figures generated when a high-viscosity fluid is displaced by a low-viscosity fluid. This "viscous fingering" phenomenon was originally described by Saffman and Taylor and is one of the limiting factors in determining the extraction efficiency in Enhanced Oil Recovery (EOR). More generally, viscous fingering belongs to the class of "Laplacian Growth" phenomena which includes many other notable examples such as crystal growth, solidification, and dendritic growth in metal electrodepostion. In this framework, the interfacial velocity is related to the gradient of a harmonic function (e.g. pressure, temperature, or concentration) and the instability is attributed to the curvature-dependence of the gradient.

More generally, however, the interface could be driven by multiple competing forces. In this talk, we will consider a class of problems which we term "Vector Laplacian Growth". In these problems, fluxes are linearly proportional to the gradient of a vector-valued potential function through a generalized mobility tensor. Specifically, we will study the linear electrokinetic response of two immiscible electrolytes in a Hele-Shaw cell where the interface motion is set by both the pressure-driven and electro-osmotic flows. By using linear stability analysis, as well as nonlinear numerical simulations, we will show that the coupling between pressure and potential fields strongly affects the interfacial stability and, under certain conditions, can suppress viscous fingering. Remarkably, our theory suggests that for a given pair of electrolytes, the ratio of electric current to the flowrate controls the onset of instability, suggesting new directions for further experimental investigations. We believe that our findings could have implications for electrically enhanced oil recovery application and might help with understanding other similar multi-field driven interfacial instabilities.

490575 Designing Functional Materials with Interfacial Instabilities in Thin Films

Abstract:

We harness interfacial instabilities in thin liquid films to fabricate lens-shaped solid structures. A liquid elastomeric polymer is spin-coated on a flat substrate to create a uniform film, which is then turned upside-down, leading to the formation of a pattern of drops that emerges through the Rayleigh-Taylor instability. As the polymer cures, this array of liquid drops solidifies, thereby permanently structuring the geometry of the originally fluid system. The drops arrange in a lattice pattern with a wavelength that is well captured by stability analysis so that tuning the properties of the polymer enables us to tailor the morphology of the drops. Upon curing, the drops can then be peeled off from the substrate. Carefully designing a layout of surface defects, etched on the substrate onto which the thin film is initially deposited, is used as seed perturbations that trigger the instability in a controlled manner. This perturbation field significantly augments the monodispersity of the final solid structures, thereby making the fabrication process both robust and scalable. We demonstrate that by using an optically clear polymer (e.g. PDMS) during fabrication, the solid drops that are produced act as working lenses that can readily turn any smartphone into a microscope.

494222 Numerical Simulation on Flow Behaviour of Twin-Liquid Film over a Vertical Plate with an Open Window

Abstract:

Liquid film, as a basic contact way between gas and liquid, is commonly encountered in chemical engineering process. Based on the features of film formation, it can typically be divided into two types: wall-bounded film[1], which is supported by a solid wall, and free film, which flows without contact with any solid frame and has two apparent free interfaces. Different from traditional liquid film investigation, coexistence and interaction of wall-bounded film and confined free film[2] were found on some chemical engineering equipment, such as falling-film heat exchangers with bundles of horizontal tubes, rotating disc reactors with large windows on the wall and random or structured packings with holes.

VOF method was used to simulate flow behaviour of such a novel film (twin-liquid film) over a vertical plate with an open window. According to its development of flow structure in stream wise, three regions are divided: Region 1, at inlet of window, velocity profile rearrangement within falling liquid film in its transition from wall-bounded film to free film flow are shown; Region 2, at middle of window, liquid film falls freely under gravity and stationary varicose waves occur; Region 3, at outlet of window, liquid film impinges on the lower wall and flows back on the plate. The existence of confined free film apparently changes film thickness and velocity distribution of wall-bounded film around while the latter increases stability and waviness of confined free film in the window. Compared with traditional wall-bounded film, twin-liquid film owns a larger mean velocity, thinner film thickness, stronger fluctuations and vorticity on the free surfaces, which are all considered to improve mass transfer efficiency. Considering geometry effects of window: with increasing window depth, amplitude of waves in the window increases first and then fluctuation of free surface becomes dynamic and irregular; with increasing window height, typical flow characteristics of wall-bounded film, wavy twin-liquid film and smooth free film are seen one after the other; with increasing window width, waves on confined free film become three dimensional and wave intensity decreases. Considering effects of operation condition: for varicose waves in the window, the wave length decreases with increasing *Re* and *Ca*, while the wave amplitude increases with increasing *Re* and decreasing *We*. Complement of experimental observations show a good agreement with corresponding three-dimensional simulations and offer more three-dimensional flow characteristics of twin-liquid film. The conception of twin-liquid film will expand traditional liquid film investigation field and offers an effective way to improve heat and mass transfer in chemical engineering process.

Figure 1 *Re*=150, *Ca*=0.0175, *D*=0.28 *H*, Case A: (a) Liquid fraction (0-1);

(b) Pressure (-300-200 Pa); (c) X-velocity (0-0.9 m/s); (d) and (e) streamline (0-0.9 m/s)

Figure 2 Comparison of twin-liquid film (left) and wall-bounded film (right) at *Re*=150 and *Ca*=0.0175. (a) Liquid fraction (0-1); (b) Y-velocity (-0.3-0.3 m/s); (c) X-velocity (0-0.9 m/s) (d) Pressure (-300-200 Pa); (e) Vorticity (0-7000/s); (f) Swirling strength (0-3000/s)

Figure 3 Comparison between 3D simulations and experimental results (water, *Re*=150)

References

1. Batchelor GK, Moffatt HK, Worster M. *Perspectives in fluid dynamics: a collective introduction to current research*: Cambridge University Press; 2002.

2. Hu J, Yang X, Dai G. Numerical investigation on hydrodynamics of vertically confined free film. *The Canadian Journal of Chemical Engineering*. 2016;94(2):340-348.

484939 Droplet Wetting Transitions on Inclined Substrates in the Presence of External Shear and Substrate Permeability

Abstract:

Understanding the gravity-driven motion of droplets on inclined substrates in the presence of external shear and substrate permeability is important for applications such as spray coating and filtration. In this work, we use a lubrication-theory-based model to study how external shear and substrate permeability affect droplet wetting transitions. A nonlinear evolution equation for the droplet height as a function of time and two spatial variables is derived and numerically solved. The contact-line region is described using a precursor film and disjoining pressure. Depending on its direction, external shear can either suppress or drive wetting transitions, but does not appear to significantly change the critical droplet speeds associated with these transitions. Substrate permeability generally suppresses wetting transitions due to liquid absorption and does appear to significantly affect these critical droplet speeds. The strong influence of substrate permeability and external shear on droplet wetting transitions indicates that it will be important to account for these effects when developing accurate models for industrial applications (L. Espin and S. Kumar, Phys. Rev. Fluids 2 (2017) 014004).

497213 Measuring Contact-Line Mobility Using Drop Resonance

Abstract:

Contact-line *mobility* characterizes how fast a liquid can coat a solid support. During rapid movement of liquid across a solid, the contact angle changes dynamically in response to changes in contact line speed. Mobility captures this far-from-equilibrium behavior. For sufficiently rapid movement of thin liquids in contact with low-hysteresis supports, motions are largely inviscid. In this regime, surface tension and liquid inertia compete to influence the interface shape near the three-phase contact line. We use a mechanical oscillator to induce resonance in a supported liquid drop and resolve contact line motions using high-speed imaging. The mobility parameter is then extracted from measurements of contact line displacement, contact line speed and contact angle. The introduced procedure enables the resonant-drop dynamical goniometer (RDG) and recovers the traditional contact-angle versus contact-line speed diagram as a limiting case. Advantages of the RDG are summarized.

498648 Electrostatic Suppression of the Leidenfrost State

Abstract:

It is a common sight in the kitchen to observe droplets of water dancing on a hot frying pan. The evaporating liquid droplet floats above the hot surface, its weight supported by the slightly higher pressure in the thin layer of its vapor between the droplet and the surface. This homely observation known as the Leidenfrost effect has serious ramifications for process phenomena like steam generation, thermal desalination and quenching of metals[1]. Here, we investigate the fundamental mechanism and identify the competing forces underlying the limits of electrostatic suppression of the Leidenfrost state. We consider liquids with finite electrical conductivity, wherein the fluid remains equipotential and the entire potential difference falls across the vapor gap. We postulate and analyze several competing forces on the perturbed interface of the droplet as the voltage approaches its threshold value. The attraction (interface destabilizing) force is due to the electrostatic attraction between the charged fingers of a disturbance on the interface and the electrically grounded surface. There are two repulsive or stabilizing effects acting on these fingers. Firstly, there are capillary forces that smooth out disturbances to the interface, especially the high wave number disturbances. Secondly, as a finger gets closer to the interface, the rate of vaporization of the liquid increases locally and creates an additional stabilizing pressure to suppress the finger. If the applied voltage is large enough, the electrostatic forces dominate and the fingers will grow and bridge the vapor gap completely, enabling liquid-solid contact. We observe and analyze the electrohydrodynamic instabilities occurring above the threshold voltage that result in a wavy liquid-vapor pattern with a characteristic wavelength. The critical voltage required for Leidenfrost suppression is estimated and is found to increase with both the droplet volume and the surface temperature[2]. There is very good agreement between theoretical predictions and experimental measurements.

[1] D. Quere, Leidenfrost dynamics, Annu. Rev. Fluid Mech. **45**, 197 (2013).

[2] Shahriari, A., Das, S., Bahadur, V., & Bonnecaze, R. T. (2017). Analysis of the instability underlying electrostatic suppression of the Leidenfrost state. Physical Review Fluids, 2(3), 034001.

499073 Linear Stability of Layered Two-Phase Flow through Soft Gel Coated Walls

Abstract:

Linear stability of layered two-phase flow through soft gel coated walls

Two phase stratified flows between soft-gel coated walls arise in flows in micro-channels. It is important to understand the elastohydrodynamic coupling between the soft-gel layers and the fluids to design lab on a chip devices. In this work, a theoretical model is developed in a generalized framework and different types of instabilities that arise due to the fluid flow are understood. The solid membrane is modelled as an incompressible linear viscoelastic material. To simplify the analysis, inertia in the solid is neglected as a first step. Linear stability analysis is carried around the base state velocity of the fluid and displacement field of the solid. The flow is perturbed by a small disturbance and a normal mode analysis is carried out to study the growth rate of the disturbance. An eigenvalue problem in formulated using Chebyshev spectral method and is solved to obtain the growth rate of the disturbance. The effect of different parameters such as thickness of the flexible membrane, Reynolds number, viscosity ratio, density ratio, Capillary number and Weissenberg number on the stability characteristics of the flow is studied in detail. Dispersion curves are obtained which explain the stability of the flow. A detail energy analysis is carried out to determine different ways through which energy transfers from the base flow to the disturbed flow. Three instability modes are identified a liquid-liquid long wave mode, a liquid-liquid short wave mode and a gel-liquid mode. Parameters for which the liquid-liquid long wave mode is unstable for flow between rigid plates gets stabilized by introducing a soft-gel layer near to the more viscous fluid. For parameters for which the liquid-liquid long wave mode is stable for flow between rigid plates gets destabilized by introducing a soft-gel layer near to the less viscous fluid.

498346 Inverse Design of Interactions for Assembly

Abstract:

Nanometer-scale, colloidally-stable particles dispersed in a fluid can be driven to assemble into a wide variety of different structures depending on the control parameters of the system and the nature of the effective interparticle interactions. In many cases, the relevant interactions are tunable via external fields, physical or chemical modification of the particle surfaces or changes in the composition of the suspending solvent. In this talk, we discuss some of the theoretical challenges associated with using inverse methods to guide the design of such interactions for assembly of colloidal mesophases with targeted structure and the opportunities that new machine-learning based simulation approaches may provide for addressing them.

498131 High-Throughput Computational Screening As a Tool for Understanding the Molecular Thermodynamics of Adsorption

Abstract:

High-throughput computational screening, in which molecular simulations are performed for large numbers of systems, could transform how research is performed in a variety of areas. In this talk, I will present examples of this paradigm emerging in the area of nanoporous materials, especially metal-organic frameworks (MOFs). MOFs are well suited to computational screening because they are synthesized in a â€œbuilding-blockâ€ approach from inorganic nodes and organic linkers and there exist a large number of known and potential materials.Â By selecting appropriate building blocks, the structural and chemical properties of the resulting materials can be finely tuned, and this makes MOFs promising materials for applications such as gas storage, chemical separations, sensing, drug delivery, and catalysis. This talk will focus on efforts to design or screen MOFs for gas storage and separations to 1) find top candidates for a given application and 2) to develop insights and structure/property relationships. The latter can, in turn, help improve our understanding of the molecular thermodynamics of these systems.

498261 Awe-Somes: All Water Emulsion Bodies

Abstract:

We are developing encapsulation methods via interfacial complexation of polyelectrolytes and other charged species in all aqueous two phase systems to make multi-functional **A**ll **W**ater **E**mulsion bodies- AWE-somes. Such capsules can be particularly interesting for sequestration of delicate components, including proteins and microbes, which should not be placed in contact with oils or hydrophobic media. These platforms have several exciting applications, and open interesting questions for fundamental study.

Here we discuss the example of the PEG-Dextran-water system, which separates into PEG-rich and dextran-rich phases. The interfacial tension between the phases is quite low. Furthermore, many molecules, including polyelectrolytes, partition freely between the two phases. These factors make interfacial structure formation especially challenging. We develop strategies to build membranes from complementary polyelectrolytes in each phase by balancing their rates of transport to the interface. To impart additional functionality, we develop methods to include charged nanoparticles (NPs) in such membranes. Here, NPs can be selected that preferentially partition into one of the phases, facilitating interfacial transport, and creating an osmotic stress imbalance between the two phases that allows us to form encapsulated multiple emulsions. These AWE-somes, with internal structures reminiscent of membraneless organelles, provide a rich platform for separation, partitioning, reaction, and transport, suggesting AWE-somes might be developed into protocell systems.

499527 Succession of Alkane Conformational Motifs Bound within Hydrophobic Nano-Capsule Assemblies

Abstract:

Alkane encapsulation experiments within dimeric octa-acid cavitand capsules in water reveal a succession of packing motifs from extended, to helical, to hairpin, to spinning top structures with increasing chain length. Here, we report a molecular simulation study of alkane conformational preferences within these host-guest assemblies to uncover the factors stabilizing distinct conformers. The simulated alkane conformers follow the trends inferred from ^{1}H-NMR experiments, while guest proton chemical shifts evaluated from Gauge Invariant Atomic Orbital calculations provide further evidence our simulations capture guest packing within these assemblies. Analysis of chain length and dihedral distributions indicates that packing under confinement to minimize non-polar guest and host interior contact with water largely drives the transitions. Mean intramolecular distance maps and transfer free energy differences suggest the transition between the extended and helical motifs is a continuous process accompanied by increases in the guest gauche population to allow accommodation within the complex, while breaks observed between the helical/hairpin and hairpin/spinning top motifs are indicative of true transitions. Our results represent the first bridging of empirical and simulation data for flexible guests encapsulated within confined nano-spaces, and constitute an effective strategy by which guest packing motifs within artificial or natural compartments can be rationalized and/or predicted a priori.

498136 Thermodynamics and Transport Properties of Fluids with Intermediate Range Order with Application to Protein Solutions and Biopharmaceuticals

Abstract:

Colloidal liquids interacting with short range attraction and long range repulsion, such as proposed for some protein solutions, have been found to exhibit novel states consisting of equilibrium particle clusters. Monte Carlo simulations are performed for two physically meaningful inter-particle potentials across a broad range of interaction parameters, temperatures and volume fractions to locate the conditions where clustered states are found. A corresponding states phase behavior is identified when normalized by the critical point of an appropriately selected reference attractive fluid. Clustered fluid states and cluster percolated states are found exclusively within the two phase region of the state diagram for a reference attractive fluid, confirming the underlying intrinsic relation between clustered states and bulk phase separation. Clustered and cluster percolated states consistently exhibit an intermediate range order peak in their structure factors with a magnitude above 2.7, leading to a semi-empirical rule for identifying clustered fluids in scattering experiments. (Soft Matter, 2014)

The glass transition of colloidal dispersions interacting with both a short-ranged attraction and long-ranged repulsion is studied using highly purified lysozyme solutions. Newtonian liquid behavior is observed at all conditions while measurements of the dynamics in the short-time limit show features typical of glassy colloidal systems at high protein concentrations. This interesting behavior is due to the competition of the attraction and repulsion that produces a heterogeneous microstructure only at intermediate range length scales. The results demonstrate that theories for the macroscopic properties of systems with competing interactions need to include intermediate range order. (Phys. Rev. Lett. 2015)

Recently, reversible cluster formation has been identified as an underlying cause of anomalously large solution viscosities observed in some concentrated monoclonal antibody (mAb) formulations, which poses a major challenge to the use of subcutaneous injection for some mAbs. A fundamental understanding of the structural and dynamic origins of high viscosities in concentrated mAb solutions is thus of significant relevance to mAb applications in human health care as well as of scientific interests. Here, we present a detailed investigation of an IgG1 based mAb to relate the short time dynamics and microstructure to significant viscosity changes over a range of pharmaceutically relevant physiochemical conditions. The combination of light scattering, small angle neutron scattering and neutron spin echo measurement techniques conclusively demonstrates that, upon adding Na_2SO_4, these antibodies form strongly bounded reversible dimers at dilute concentrations, which interact with each other to form loosely bounded, large, transient clusters when concentrated. This hierarchical structure formation in solution causes a significant increase in the solution viscosity. (J. Chem. Phys. B, 2015)

Finally, simulations and theory have been performed to explore the dynamics of colloids with intermediate range order. The results are compared with recent experiments on globular proteins. The dynamical signatures of intermediate range order and cluster formation are presented and discussed.

506667 20 Years of Corrosion Sensing and Microvisualization of Corrosion Processes

Abstract:

Localized corrosion processes continue to be a major cause for degradation failure in damaging fluids. Initiation stages of response remain difficult to characterize, arising in part because of the difficulty in being able to predict the location and onset of the attack. Propagation stages are seldom constant but change with time due to competing processes of dissolution and product deposition at the corrosion sites. The progress of the attack changes the local environmental concentrations which are especially important at buried interfaces that are isolated from bulk conditions. The present paper will describe some of the recent advances in optical, electrochemical, and microscopy that are directed at providing local information on â€œprecursor sitesâ€ and vulnerable areas on metal and covering oxide surfaces. Previous studies of coupled reactions on reactive surfaces provide a general basis for the work to be described here.

The objectives of the experimental techniques are: (1) to measure the local corrosion rate; (2) to locate breakdown sites, and; (3) to study the fundamental properties of passive films and how they are related to the stability of the films and damaging environments. The latter is of particular interest since it will allow one to predict why a particular precursor site is more vulnerable to breakdown as compared to neighboring areas.

In locating precursor sites, entire regions of the corroding surface can be imaged concurrently or point-by-point using a scanned probe or beam. Both classes will be discussed with their capabilities. The techniques used in our laboratory are: (1) the Quartz Crystal Microbalance (QCM); (2) Optical Fiber Micromirrors (FOP); (3) Confocal Laser Scanning Microscopy (CLSM); (4) Photoelectrochemical Microscopy (PEM); (5) Scanning Electrochemical Microscopy (SECM); (6) Scanning Photoelectrochemical and Electrochemical Microscopy (SPECM); (7) Near Field Scanning Optical and Spectroscopic Microscopy (NSOSM). The experimental setup for each and their capability help to reveal the advantages that are offered by them.

Optical sensing by the QCM and optical fiber micromirrors have been used to measure corrosion rates, with sensitivity of fractions of an atomic layer. Similar to them but with the capability to image a larger reactive surface is the CLSM to determine local attack rates at various surface sites. The NSOSM is valuable to obtain both vertical and lateral super resolution images of reacting surfaces of about 60 nanometers. The techniques support the development of miniature corrosion sensors devoted to monitoring reactive surfaces in remote and occluded regions, coupled with wireless transmission. The wireless corrosion sensors must be (1) small, (2) inexpensive, (3) tailored to evaluate corrosion rates, and (4) robust.

493414 The Need for Engineering Methods for Electrochemical Systems

Abstract:

Electrochemical phenomena control the existence and movement of charged species in the bulk as well as across interfaces between ionic, electronic, semiconductor, photonic, and dielectric materials. The pervasive occurrence of electrochemical phenomena may be seen in many nanoscale and biological systems, energy systems, microelectronic and photovoltaic devices, and in green processes as well as in natural systems.

Future trends in electrochemical engineering will be influenced by the need to control processes and insure product quality at the molecular scale. Transfer of molecular-scale understanding and discoveries into new and improved products and processes requires integration of system behavior across a range of length- and time-scales. Engineering approaches are needed that couple traditional current- and potential-distribution approaches to molecular-scale events in order to accurately describe and design systems to meet the needs of the next century.

The existing technology base of the electrochemical field is massive and of long-standing. For economic reasons, most traditional large-scale electrolytic processes are driven to transport-limited rates. For many decades, the traditional electrochemical engineering literature therefore focused strongly on understanding how ohmic and mass transport processes influence the potential field between electrodes and the current distribution along electrode surfaces.

Mathematical modeling of transport and reaction processes in electrochemical systems has advanced steadily for many decades. Continuum models based on differential-algebraic equations dominate the extensive literature. Modern sophisticated models typically include simultaneous consideration of thermodynamic, kinetic, and transport phenomena in adjoining phases as well as reactions and other processes at the interface between them. Such models can provide a rational basis for complex *engineering applications* involving design, optimization, scale-up, optimization, and process control. For many applications, commercial software is available.

Because of the large variety of phenomena that can be taken into account, engineering models can also be used to advance *fundamental understanding* of such complicated processes as shape evolution of irregular surfaces including dendrites, complex reaction sequences involving additives, phase transformations and surface films, and multiphase phenomena associated with bubbles and wetting, among many other examples.

In addition, there are by now many examples of close integration of numerical and experimental approaches from the start of a project. By pursuing both approaches simultaneously from the outset, these can serve to guide experimental design, to interpret sensitivity data for assessing which among competing effects are the most important, to construct chemical mechanisms and estimate the parameters, to assess uncertainty, and predict behavior of well-characterized model systems.

During the past several decades, the electrochemical field has benefited from a suite of remarkable new tools that provide the ability to create precisely characterized systems for fundamental study; to monitor behavior at

unprecedented levels of sensitivity, atomic resolution, and chemical specificity; and to predict behavior with new theories and improved computational abilities. These capabilities have revolutionized fundamental scientific understanding of interfacial and catalytic processes, as well as contributed to the present rapid pace of discovery of novel technological materials and devices for which product quality is determined at the molecular scale.

To drive these new electrochemical discoveries into technology innovation, new engineering methods are needed which take advantage of advances in molecular simulation methods, sensor technology and data-storage, computer speed and memory, and numerical algorithms. New engineering methods are needed for analysis of extremely large experimental data sets, for efficient algorithms which enable optimization and quantification of uncertainty, and for easy-to-use multi-scale methods that link small-scale events to macroscopic processes. The development and reduction to routine use, of such methods will provide engineering tools needed for the next-generation approach to design and control of electrochemical systems, as well as opening the way to exploiting and controlling self-assembly during processing.

492730 Characterizing Metal-Air Batteries Using Electrochemical Impedance Spectroscopy and Mass Spectrometry

Abstract:

Metal-air batteries have extraordinarily high theoretical specific energy densities and therefore have been explored as potential alternatives to current state-of-the-art Li-ion batteries. However, electronic passivation of the cathode and electrolyte stability have, among other challenges, precluded their use as practical energy storage devices. My tutorial will discuss characterization techniques, namely electrochemical impedance spectroscopy and mass spectrometry for gas consumption and evolution analysis, that have been employed to quantitatively understand these limitations and trace their origins. Although this presentation will primarily focus on application of these techniques to metal-air batteries, the discussion will be of interest to anyone studying electronic or ionic transport in porous electrodes or electrochemical systems where a gas is a product, reactant, or both.

501146 Platinum-Free Fuel Cells for Affordable Zero-Emission Cars

Abstract:

One of the grand challenges facing humanity today is the development of an alternative energy system that is safe, clean, and sustainable and where combustion of fossil fuels no longer dominates. A distributed renewable electrochemical energy and mobility system (DREEMS) could meet this challenge. At the foundation of this new energy system, we have chosen to study a number of electrochemical devices including fuel cells, electrolyzers, and flow batteries. For all these devices electrocatalysis and polymer electrolytes play a critical role in controlling their performance, cost, and durability, and thus their economic viability. In this presentation, I will focus on our recent work on hydroxide exchange membrane fuel cells (HEMFCs) which can work with nonprecious metal catalysts and inexpensive hydrocarbon polymer membranes. I will briefly show the roadmap we have developed for this technology, the progress we have made in developing the most stable membranes and the most active nonprecious metal catalysts. I will discuss in detail why hydrogen oxidation reactions are slower in base than in acid for precious metal catalysts.

499699 Deciphering the Chemical and Physical Mechanisms Active in Elevated Temperature Photocatalytic Synthesis of Hydrocarbons and Ammonia

Abstract:

The photocatalytic synthesis of hydrocarbons, oxygenates, and NH_3 from CO_2, N_2, NO/NO_2, and H_2O is a truly sustainable route to fuels and building block chemicals production. Unfortunately, the community still lacks the ability to rationally improve overall rates of reaction and control product distributions due to a lack of reaction mechanism understanding. Endemic issues with organic contamination in experimental studies has further contributed to this lack of progress. Our combined computational/experimental studies of photocatalytic CO_2 reduction by H_2O in organic-free reaction conditions at an appreciable temperature range (gas phase reaction at ~100-400°C+) have shed significant light upon the surface chemical and electronic properties that affect these reaction mechanisms. Moreover, studies have also isolated new H-transfer mechanisms. Both of these developments have the power to significantly aid in the design of these materials. The new H-transfer mechanisms and an understanding of the role of surface chemistry have further enabled control over the selectivity of surface-bound atomic H between hydrogenation and H_2 evolution -- a critical efficiency aspect of the system. Additionally, key carbon-containing reaction intermediates that allow for carbon-carbon coupling have been isolated as well. Catalysts shown to be active at elevated temperatures indicate semiconductors with a high Debye temperature, robust bulk bonding, and elevated surface chemical reactivity may lend additional tunable features for photocatalytic synthesis mechanisms. With the ultimate aim of producing higher hydrocarbons, oxygenates, and ammonia photocatalytically from simple thermodynamic-minimum reactants, these studies have mapped the course forward for materials selection and mechanistic understanding.

502400 A High-Throughput Computational Screening Approach for Solar Fuels Photoelectrocatalysis

Abstract:

In recent years, theoretical overpotential has become increasingly popular as a metric for characterizing the suitability of oxygen evolution electrocatalysts using density functional theory (DFT). In addition, DFT has become an invaluable tool for characterizing photoabsorbing materials in a number of high-throughput screening studies aimed at designing materials for solar fuels production from stability and bandstructure criteria. In this work, we present an approach which integrates calculations of theoretical overpotential with those of band-gap and band-edge positions on mixed metal oxides. Given that the prospect of discovering lone materials fulfilling the criteria of stability, catalytic activity, and photoabsorption efficiency, we also report on how materials with suitable surface chemistry may be more effectively paired with suitable photoabsorbing substrates using interfacial matching. Lastly, we benchmark our approach comparing potential combined metrics with experimentally measured photoelectrocatalytic activity.

499265 Computational Insights into Photo(electro)Chemical Nitrogen Fixation over Titania Catalysts

Abstract:

Photocatalytic nitrogen fixation is a form of artificial photosynthesis capable of enabling distributed production of fixed nitrogen at near-ambient conditions from the nitrogen, water, and photons that are readily available from the environment. The process has been demonstrated over titania catalysts through both reductive and oxidative pathways, although the currently observed nitrogen fixation rates are too low to be practical for fertilizer synthesis and there is little fundamental understanding of the reaction mechanism or chemical phenomena that enable the process. Interestingly, the process occurs under environmental conditions and has been speculated to play a significant role in the global nitrogen cycle owing to the prevalence of TiO_2 in naturally occurring sands and man-made pigments. Despite the wide-ranging potential impacts of this reaction, first demonstrated nearly 75 years ago, the process has received relatively little attention and has not been studied with modern techniques. This talk will briefly review the historical context and current progress toward photocatalytic nitrogen fixation, followed by a theoretical perspective into the thermodynamic driving forces and likely reaction mechanisms, with an emphasis on the potential for developing photoelectrochemical processes. The computational results will be complemented by recent experiments demonstrating the feasibility of the reaction.

501308 Engineering the Interface to Improve the Efficiencies of Insulator-Protected Semiconductors for Photoelectrochemical Cells

Abstract:

Recent improvements in solar fuels production efficiencies have been enabled by the development of protection schemes for low bandgap semiconductors that would otherwise be unstable in the harshly oxidative environment of an electrochemical cell. One common protection strategy incorporates an insulator layer placed between a metal electrocatalyst and semiconductor. Recent research has shown that the characteristics of the insulator layer have a large effect on the overall efficiency of a photoelectrochemical cell. In this contribution, we will provide experimental evidence that demonstrates how by tuning the properties of the insulator layer we can better control charge transfer across the interface in order to improve efficiencies. We will also present a model to describe the insulator layerâ€™s effect on charge transfer mechanisms. Provided with these modeling insights, we will discuss which experimental systems can benefit from the most from insulator tuning in order toÂ expand the flexibility of material selection for a photoelectrochemical device.

492457 The Role of Surface-Bound Dihydropyridine Analogs in Pyridine-Catalyzed CO_2 Reduction over Semiconductor Photoelectrodes

Abstract:

The (photo)electro-catalytic reduction of CO_2 to useful fuels via energy from sunlight has received significant attention as a promising route for generating carbon-neutral fuels and value-added chemicals. Experimental evidence suggests that pyridine (Py) is an effective co-catalyst during CO_2 reduction over GaP, CdTe, and $CuInS_2$ semiconductor electrodes. Identifying the role of Py during catalytic reduction is essential for optimizing the design of such photocatalytic processes. We propose a general reaction mechanism for the Py-catalyzed reduction of CO_2 over GaP(111), CdTe(111), and $CuInS_2$(112) photoelectrode surfaces. This mechanism proceeds via formation of a surface-bound dihydropyridine (DHP) analog, which is a newly postulated intermediate in the Py-catalyzed mechanism. Using density functional theory, we calculate the standard reduction potential related to the formation of the DHP analog, which demonstrates that it is thermodynamically feasible to form this intermediate on all three investigated electrode surfaces under photo-electrochemical conditions. Hydride transfer barriers from the intermediate to CO_2 demonstrate that the surface-bound DHP analog is as or even more effective at reducing CO_2 to HCOO- as the $DHP_{(aq)}$ molecule in solution. This intermediate is predicted to be both stable and active on many varying electrodes, therefore pointing to a mechanism that can be generalized across a variety of semiconductor surfaces and explains the observed electrode dependence of the photocatalysis.

490917 Surface Plasmon-Assisted Photoelectrochemical CO_2 Reduction on Well-Defined Nanostructured Silver Electrodes

Abstract:

Plasmonic hot-carriers and the strong local electric fields produced by plasmon excitation may open new pathways for CO_2 reduction, resulting in increased product selectivity and lower overpotential. Plasmon decay generates â€œhotâ€ electrons which can be transferred selectively to a surface molecule for direct reduction. Simultaneously, the enhanced electric fields can alter the electronic coupling with surface adsorbed molecules, thereby changing the binding energy of these species and the catalytic properties of the plasmonic metals.[1-3] Plasmon-assisted photoelectrochemical CO_2 reduction was demonstrated when optically excited surface plasmons at the roughened surface of a Ag electrode in CO_2-containing solutions produced strong enhancement of the cathodic current.[4-5]

Here we reveal the mechanism of plasmonic hot-carriers interacting with CO_2 and associated adsorbates on plasmonically-active Ag nanopyramid catalysts. Photocurrent measurements and photoelectrochemical spectroscopy were performed on the plasmonic Ag nanopyramid photocatalysts in electrolyte saturated with CO_2 and Ar. We observed that the photocatalytic reaction is much stronger in the presence of CO_2 and is resonantly enhanced at a specific potential. In Ar-saturated electrolyte, no characteristic feature of photocurrent was observed. These results indicate that the hot-carriers created by the localized surface plasmon resonance effect are reacting specifically with CO_2 molecules and CO_2 reduction intermediates. Further, the photocurrent onset potential was significantly reduced with respect to the dark current. Our findings suggest that the plasmonic hot-carriers behave as an effective co-catalyst with Ag and may play a critical role in discovering a CO_2reduction process that forms readily usable hydrocarbon fuels.

Scanning electron microscopy (SEM) was used to confirm the shape, size, and distribution of the silver nanopyramids. Photocurrent measurements were performed in a custom glass photoelectrochemical cell with a quartz window that allows for front- or back-illumination of the electrode. The two electrode chambers are separated by a membrane and can achieve a gas-tight seal that allows for precise pressure measurements. The cell was coupled to a gas chromatograph (GC) that allows for detection and quantification of gaseous products. *Ex-situ* proton nuclear magnetic resonance spectroscopy (^1H NMR) and high performance liquid chromatography (HPLC) were used to quantify liquid products.

References

[1] A. O. Govorov, et. al., *Nano Today* 9, 85 (2014)
[2] R. Sundararaman, et. al., *Nature Comm.* 5, 5788 (2014)
[3] S. Mukherjee, et. al., *J. Am. Chem. Soc.* 136, 64 (2014)
[4] R. Kostecki, et. al., *J. Chem. Phys. Lett.* 194, 386 (1992)

[5] R. Kostecki, et. al., *J. Appl. Phys.* 77, 4701 (1995)

This material is based upon work performed by the Joint Center for Artificial Photosynthesis, a DOE Energy Innovation Hub, supported through the Office of Science of the U.S. Department of Energy under Award Number DE-SC0004993.
The work is supported by the National Science Foundation Graduate Research Fellowship under Grant No. 1106400.
Work at the Molecular Foundry was supported by the Office of Science, Office of Basic Energy Sciences, of the U.S. Department of Energy under Contract No. DE-AC02-05CH11231.

490812 Artificial Leaf for Carbon Dioxide Photo-Reduction to Fuels

Abstract:

An artificial leaf that mimics photosynthesis process is a promising approach to directly convert solar energy into energy-rich chemicals[1–4]. Thus far, this concept has been successfully developed for the hydrogen generation process[2–4]. Here, we have developed an artificial leaf for the conversion of carbon dioxide (CO_2) into synthesis gas in an acidic electrolyte using the energy from a sun simulator. A configuration of the artificial leaf was achieved by using cobalt, as an oxygen-evolving catalyst and molybdenum disulfide (MoS_2) catalyst for CO_2 reduction reaction deposited on the illuminating and back-substrate surfaces of a photovoltaic cell, respectively. We have performed detail experimental studies such as linear sweep voltammetry (LSV) and chronoamperometry (CA), in-situ differential electrochemical mass spectroscopy (DEMS), and gas chromatography (GC) followed by some characterization techniques e.g. scanning electron microscopy (SEM), X-ray photoelectron spectroscopy (XPS) and energy dispersive spectroscopy (EDS) to elucidate the performance and stability of our developed artificial leaf. The experimental analyses indicate that our custom made artificial leaf cell can directly convert the solar energy to synthesis gas with a catalytic conversion efficiency of 75%. Our results reveal that the operation of the artificial leaf cell is not restricted by the CO_2 reduction reaction according to the remarkable activity of MoS_2 NFs catalysts[5]. However, it is rather limited by the oxygen evolution reaction (water splitting) at the anode side. This finding could potentially open a new route in the solar-driven CO_2 conversion field where the CO_2 reduction reaction is known to be the bottle neck of this process.

References:

1. Asadi, M. *et al.* Nanostructured transition metal dichalcogenide electrocatalysts for CO_2 reduction in ionic liquid. *Science.* **353,**467–470 (2016).

2. Kanan, M. W. & Nocera, D. G. In Situ Formation of an Water Containing Phosphate and Co^{2+}. *Science.* **321,**1072 (2008).

3. Reece, S. Y. *et al.* Wireless Solar Water Splitting Using Silicon-Based Semiconductors and Earth-Abundant Catalysts. *Science.* **334,**645–648 (2011).

4. Qiu, F. *et al.* Photocatalytic Hydrogen Generation by CdSe/CdS Nanoparticles. *Nano Lett.* **16,**5347–5352 (2016).

5. Abbasi, P. *et al.* Tailoring the Edge Structure of Molybdenum Disulfide toward Electrocatalytic Reduction of Carbon Dioxide. *ACS Nano.* **11**, 453–460 (2017).

500536 Understanding Photoelectrochemistry on Epitaxial Oxides through Surface Electronic Structure

Abstract:

The intermittent nature of renewable energy sources requires a clean, scalable means of converting and storing energy. One earth abundant storage option is water electrolysis, storing energy in the bonds of O_2 and H_2. Photoelectrochemical (PEC) cells based on semiconductor/liquid interfaces can convert sunlight to chemical fuels without external circuitry, such as "splitting" water into O_2 and H_2 upon illumination of a photoabsorber and catalyst. Tuning the location of the band edges of a semiconductor tailors the wavelength of photoabsorption and the charge transfer to adsorbed species. The efficacy of conversion depends in part on the rectifying properties of semiconductor–electrolyte junctions, as the band bending present in the semiconductor space-charge region from Fermi level equilibration drives the separation of electron–hole pairs. However, the relationship between band structure, band bending/flattening, photocatalytic activity, and surface functionality (such as hydroxylation properties and hydrophilicity) is not well understood. Improving our understanding of the semiconductor/electrolyte interface is crucial in the design of efficient photoelectrode materials for harvesting solar energy in a PEC system.

In the particular case of the semiconductor Ge, favorable conduction band alignment with the hydrogen evolution reaction (HER) suggests the potential for efficient solar-to-hydrogen conversion with visible light. However, material instability in aqueous environments and the potential for photocorrosion limits its utility in PEC applications. In order to protect the photoabsorber surface while maintaining favorable alignment of the conduction band, we epitaxially grow $SrTiO_3$ on top of p-Ge. We will present studies of such model oxide photoelectrodes grown by molecular beam epitaxy (MBE) on single crystal substrates that display a known crystallographic orientation, surface area, path for charge transport, and strain. Photoelectrochemical measurements on these heterostructures can establish the intrinsic activity of oxide catalysts in a way that cannot be realized with polydisperse nanoparticle systems. Insight into the band bending between the substrate and oxide overlayer, as well as at the semiconductor surface, can be obtained from X-ray photoelectron spectroscopy (XPS).[1] Measurement of XPS at ambient pressures (AP-XPS) can further elucidate the relationship between adsorbates and surface band bending.[2] This fundamental insight will build understanding necessary for the design of active, earth-abundant photocatalysts that can be integrated into PEC devices for efficient conversion of solar energy into chemical fuels.

References

1. S.A. Chambers, Y. Du, R.B. Comes, S.R. Spurgeon, P.V. Sushko, *Applied Physics Letters,* **110**, 082104 (2017).

2. K.A. Stoerzinger, R. Comes, S.R. Spurgeon, S. Thevuthasan, K. Ihm, E.J. Crumlin, S.A. Chambers. *J. Phys. Chem. Lett.* **8**, 1038 (2017).

497185 Rethinking Grid-Level Energy Storage with Minimal Architecture Zinc-Bromine Batteries

Abstract:

One of the major factors limiting a deeper market penetration of intermittent, renewable energy sources (i.e. wind and solar) is a lack of reliable, cost-effective energy storage technologies. Of the existing technologies, electrochemical energy storage in the form of batteries is a promising candidate due to high efficiencies (up to 95%) and modular designs that can be easily installed next to generation sites. Despite this important application for batteries, significant research efforts still focus on maximizing energy and power density for use in portable applications like consumer electronics and automobiles. One major exception are redox flow batteries, which are designed for grid-level applications, where low cost and long lifetime are the key design criteria. Flow batteries are slowly entering the market; however, in most cases, their cost is still prohibitive. The prohibitive cost is largely due to the use of expensive passive components like membranes and pumps needed to manage the flowing, reacting fluid. To address this issue, our lab has recently developed a minimal architecture zinc-bromine battery (MA-ZBB), which utilizes the same chemistry from commercial zinc-bromine flow batteries, but eliminates the need for any passive components, significantly reducing the cost (projection of < $90/kWh cell cost) [1].

The MA-ZBB is composed of two carbon based electrodes, which are spatially separated in the vertical direction, in an aqueous electrolyte containing zinc-bromide salt. At the negative electrode on top of the cell, zinc ions are reduced (Zn) and oxidized (Zn2+) on a carbon cloth during charge and discharge, respectively. At the positive electrode, which is located on the bottom of the cell, bromide ions are oxidized (Br2) and reduced (Br-) within a carbon foam electrode. The electrodes are vertically separated to utilize the variations in density between bromine and the aqueous electrolyte as an advantage to prevent bromine, which may escape the carbon foam electrode, from reaching with the zinc electrode. This design is key for limiting self-discharge associated with the unwanted chemical reaction of bromine with zinc.

To date, we have demonstrated an energy efficiency of over 60% for over 1000 cycles at lab scale. In this work, we present recent efforts at characterizing the efficiency losses in the system to identify methods for improving the electrochemical performance. Galvanostatic intermittent titration technique (GITT) and electrochemical impedance spectroscopy (EIS) are utilized to determine the relative importance of ohmic resistance, charge transfer (reaction) resistance, and mass transport limitations on the performance of the lab-scale cell. In addition, variations in cell and electrode design are investigated to maximize the capacity and energy efficiency of the system. Finally, insights from these studies are used to develop a prototype MA-ZBB cell with improved performance (>75% energy efficiency) that is used to study the technologyâ€™s viability for grid-level energy storage applications.

[1] Shaurjo Biswas, Aoi Senju, Robert Mohr, Thomas Hodson, Nivetha Karthikeyan, Kevin. W. Knehr, Andrew G. Hsieh, Xiaofang Yang, Bruce E. Koel, Daniel A. Steingart, Energy Environ. Sci., 10 (2017) 114.

501791 Highly Energy Dense Cu-Intercalated Bi-Birnessite/Zn Battery

Abstract:

Manganese dioxide (MnO_2) and zinc (Zn) are one of the most abundant, safest and cheapest materials available. Together, they are found in common household batteries like Duracell, Energizer, etc. as small cylindrical alkaline cells. These cells or batteries are used as primary batteries, i.e., as single use batteries, where the entire capacity of the battery is delivered once and then discarded. The disadvantage of primary batteries is that it takes a lot of energy to produce the battery than the energy that can be actually obtained from it, and also, it creates environmental waste. However, the manufacturing of primary cells has still been rampant due to the cost of manufacturing MnO_2-Zn cells being very cheap. In terms of improving the overall energy efficiency, reducing waste and maintaining its cost advantage, it makes good sense, economically and environmentally, to make MnO_2-Zn cells rechargeable. However, the main deterrent to this direction has been the fundamental material and chemical problems of the main raw components, i.e., MnO_2 and Zn.

Manganese dioxide can theoretically deliver a capacity of approximately 617mAh/g. It delivers this capacity through a 2 electron electrochemical reaction (each electron providing around 308mAh/g). MnO_2 has been found to be rechargeable when the capacity has been limited to around 5-10% of the 617mAh/g. It suffers a crystal structure breakdown as more of the capacity is accessed, and it inherently forms electrochemical irreversible phases. If the entire 2 electron capacity can be accessed then theoretically it can reach energy density numbers near lithium-ion batteries. Similar problems are associated with the zinc electrode, where higher utilization of its capacity causes dendrite formation, shape change and formation of inactive zinc oxides that ultimately lead to electrode failure. These are the main deterrent to a cheap and safe battery that could be a disruptive technology in the energy storage field.

In this presentation, we report the breakthrough of reversibly accessing the 2nd electron capacity of MnO_2 by using its layered polymorph called birnessite mixed with bismuth oxide (Bi-birnessite) and intercalating the layers with Cu ions (1). Bi-birnessite's undergo conversion reactions in alkaline electrolyte and ultimately form electro-inactive hausmannite (Mn_3O_4) because of its poor charge transfer characteristics. Intercalating the layers of Bi-birnessite with Cu ions is shown to improve its charge transfer characteristics dramatically and regenerate its layered structure reversibly for thousands of cycles as shown in Figure 1. We also present a case of Cu-intercalated Bi-birnessite's applicability in practical batteries by cycling the material at high areal capacities (10-29mAh/cm^2) for thousands of cycles at C-rates that are of interest in the battery community.

For true applicability in practical energy dense batteries its pairing with a Zn anode is essential. The use of Zn anodes has also presented problems as it is the source of zincate ions in electrolyte that react with the cathode, MnO_2, to form electro-inactive phase called haeterolite ($ZnMn_2O_4$). The best reported cycle life data for high depth-of-discharge (DOD) birnessite cathodes with Zn anodes had been 50 cycles till our recent publication, which showed over 90 cycles achieving 140Wh/L.

In this presentation, we also report the effect of zincate ions on the Cu-intercalated Bi-birnessite cathodes beyond 100 cycles (2). The Cu-intercalated Bi-birnessite cathodes when paired with Zn anodes are shown to deliver 160Wh/L and cycle reversibly for over 100 cycles. The Cu ions play an important role in mitigating the detrimental effect of zincate ions in the 100 cycles; however, the zincate ions eventually poison the cathode to form $ZnMn_2O_4$. The mechanism through which $ZnMn_2O_4$ is formed is presented in detail with the aid of electroanalytical and spectroscopic methods. A solution of trapping the zincate ions is also presented, where the membrane that is used successfully traps the zincate ions from interacting with the cathode and thus, extend cycle life to over 900 cycles as shown in Figure 2. This is the best reported cycle life data with a manganese dioxide cathode accessing the near 2nd electron capacity paired with Zn anodes.

Figure 1. (a) Volumetric energy density of a Cu-intercalated Bi-birnessite/Zn cell. Inset shows the first 5 cycles of the cell. (b) Energy density comparison of different energy storage systems. (c) Specific capacity (mAh/g) vs cycle number for the Cu-intercalated Bi-birnessite against a sintered Ni counter electrode. Insets show specific cycle discharge curves for different wt.% loadings of MnO_2.

Figure 2. A Cu-intercalated Bi-birnessite cathode paired with Zn anode achieving cycle life greater than 900 cycles with specialized membrane.

References:

1] Yadav, G. G.; Gallaway, J. W.; Turney, D. E.; Nyce, M.; Huang, J.; Wei, X.; Banerjee, S. *"Regenerable Cu-intercalated MnO_2 layered cathode for highly cyclable energy dense batteries"* Nat. Commun. **8**, 14424 (2017).

2] Yadav, G. G.; Wei, X; Huang, J.; Gallaway, J. W.; Turney, D. E.; Nyce, M.; Secor, J.; Banerjee, S., paper submitted (2017).

498514 Designing Electrolytes for Beyond Li-Ion Batteries Using Coupled High Throughput *Ab Intio* Calculations and MD Simulations

Abstract:

Transformative outcomes for transportation and electrical grid require high performance, low cost energy storage systems beyond Li-ion technology and innovations in electrodes and electrolytes, alike. Multivalent batteries, chemical transformation and redox flow batteries are promising candidates for next generation electricity storage technologies. An automatic High-throughput computational infrastructure has been constructed for the electrolyte genome project supported by the US Joint Center for Energy Storage Research (JCESR) coupled with experimental analysis with the aim to design electrolytes for beyond Li-ion battery technology.[1] A non-aqueous multivalent (e.g., Mg^{2+}, Ca^{2+} and Zn^{2+}) metal cell is one of the potential candidates for a post-lithium-ion battery. The theoretical volumetric capacity of a metal anode coupled with the lack of dendrite formation at a multivalent metal anode provide an attractive opportunity in energy storage.[2] However, development and commercialization of multivalent batteries are fraught with complications. Thus a fundamental understanding of molecular level properties of these electrolytes is required to improve the electrochemical stability and the charge transfer properties. In this work, we present a multi-scale modelling approach for multivalent salts in various solvents. We uncover a novel effect between concentration dependent ion pair formation and anion stability at reducing potential, e.g., at the metal anode.[3] We find that both Mg and Zn electrolytes are highly prone to ion pair formation, even at modest concentrations, for a wide range of solvents with different dielectric constants, which have implications for dynamics as well as charge transfer.[3,4] Specifically, we observe that, at Mg metal potentials, the ion pair undergoes partial reduction at the Mg cation center (Mg^{2+}â†'Mg^{+}), which competes with the charge transfer mechanism and can activate the anion to render it susceptible to decomposition.[3] It was observed that Zn electrolytes are also prone to similar $TFSI^{-}$ anion decomposition but with much higher transition state barrier, which results in better reversible Zn deposition on Zn metal anode in non-aqueous Zn electrolytes as compared to Mg electrolytes. Another category of battery that has gained much attention recently includes chemical transformation systems such as Li-S batteries, which stores energy through chemical bonds instead of intercalation. Here, we study the effect of salt anion and the solvent on the solvation structure and dynamics of Li-polysulfide in the solution. It is observed that the polysulfide chain length has a significant effect on the ion-ion and ion-solvent interaction as well as on the diffusion coefficient of the ionic species in solution. In particular, optimal solvent/salt ratio is essential to control the solubility and conductivity as the addition of Li salt increases the solubility but decreases the mobility of the ionic species.[5] The third energy storage concept is of redox flow batteries, which have shown outstanding promise for grid-scale energy storage with improved grid stability. Here we study ionic liquid tethered ferrocene catholyte used as redox center utilizing the Fc/Fc^{+} reaction for improving the performance of non-aqueous redox flow battery materials. We observed that solubility of Fc/Fc^{+} redox center can be increased up to eight-fold in a mixture of carbonate electrolytes through the modification Fc with a polar tetra alkylammonium functional group containing $TFSI^{-}$ anion. It was also observed that at solubility limit, the precipitation of solute is initiated[6-7]

through agglomeration of contact-ion pairs due to overlapping solvation shells. This works shows that the combination of multi-scale modeling with experimental techniques provides unprecedented insight into the origin of the electrochemical, structural, and transport properties of electrolytes, which is crucial in designing electrolytes for beyond Li-ion batteries.

References

1. Qu, X.; Jain, A.; Rajput, N. N.; Cheng, L.; Zhang, Y.; Ong, S. P.; Brafman, M.; Maginn, E.; Curtiss, L. A.; Persson, K. A., The Electrolyte Genome project: A big data approach in battery materials discovery. *Computational Materials Science* **2015,** *103*, 56-67.

2. Muldoon, J.; Bucur, C. B.; Gregory, T., Quest for nonaqueous multivalent secondary batteries: magnesium and beyond. *Chem. Rev.* **2014,** *114* (23), 11683-11720.

3. Rajput, N. N.; Qu, X.; Sa, N.; Burrell, A. K.; Persson, K. A., The Coupling between Stability and Ion Pair Formation in Magnesium Electrolytes from First-Principles Quantum Mechanics and Classical Molecular Dynamics. *J. Am. Chem. Soc.* **2015,** *137* (9), 3411-3420.

4. Han, S.-D.; Rajput, N. N.; Qu, X.; Pan, B.; He, M.; Ferrandon, M. S.; Liao, C.; Persson, K. A.; Burrell, A. K., Origin of Electrochemical, Structural, and Transport Properties in Nonaqueous Zinc Electrolytes. *ACS Applied Materials & Interfaces* **2016.**

5. Rajput, N. N.; Murugesan, V.; Shin, Y.; Han, K. S.; Lau, K. C.; Chen, J.; Liu, J.; Curtiss, L. A.; Mueller, K. T.; Persson, K. A., Elucidating the Solvation Structure and Dynamics of Lithium Polysulfides Resulting from Competitive Salt and Solvent Interactions. *Chem. Mater.* **2017**.

6. Han, K. S.; Rajput, N. N.; Wei, X.; Wang, W.; Hu, J. Z.; Persson, K. A.; Mueller, K. T., Diffusional motion of redox centers in carbonate electrolytes. *J. Chem. Phys.* **2014,** *141* (10), 104509.

7. Han, K. S.; Rajput, N. N.; Vijayakumar, M.; Wei, X.; Wang, W.; Hu, J.; Persson, K. A.; Mueller, K. T., Preferential Solvation of an Asymmetric Redox Molecule. *J. Phys. Chem. C* **2016,** *120* (49), 27834-27839.

498451 Rational Design of Solid-Liquid Interphases for Reactive Metal Batteries

Abstract:

Advancement in energy storage technologies is necessary owing to the growing demand for portable consumer electronics and also for the global consensus of energy policy shift to renewable resources. In this regard, metal based batteries comprising of a reactive metallic anode (like Li, Na, Al) have gained significant attention because of their promise of improving the anode-specific capacity by 10-fold compared to the current state-of-art Li-ion battery that uses a graphitic anode. However, in these metal batteries, the issue of sudden short-circuits by dendrite growth as well as rapid fade in battery capacity due to internal side reactions limit their practical usage. Such failures modes have been characterized to occur due to the uneven deposition of metal ions, however, a striking fact in the electrodeposition literature is that, certain metals like, Magnesium do not form dendrites. In this study, we use theoretical and experimental tools to compare the surface properties of Mg with that of alkali metals to rationally design the solid-liquid interphase of lithium and sodium electrodes. We, specifically, use Density Functional Theory (DFT) calculations to quantify the diffusion energy barrier of ions on Mg, Li, Na surfaces and interestingly it is seen that the diffusion barrier of Mg (0.02eV/atom) is several fold lower than Li (0.14eV/atom) or Na (0.16eV/atom) metals. In fact, the diffusion barrier of Li_2CO_3, Li_2O (the commonly found compounds in lithium interface) is even higher, thus indicating that electrodeposition in such batteries can lead to the formation of dendrites. However, in quest for finding stable interfaces, we observed that most metal halides (LiF, LiBr, NaF etc.) have much lower diffusion barrier. In other words, lithium or sodium anode with halide-rich interface can lead to stable electrodeposition similar to Mg deposition. We further utilize the energy values from DFT calculation in Classical Nucleation Theory (CNT) to determine the microscopic deposit size for different metals and interfaces at various operating conditions (like current density, overpotential and temperature). The predictions from the coupled DFT-CNT model were validated using ex-situ scanning electron microscopy as well as in-situ X-ray imaging. The nucleation pattern, indeed, showed a strike difference between carbonate-rich and halide-rich lithium interfaces, which were utilized to calculate the experimental diffusion barriers. In order to understand the macroscopic implications of the interfacial ion transport, we performed in-situ optical microscopy of electrodeposition to directly visualize the morphology of dendrite growth in lithium and sodium batteries, without and with halide generating additives. Electrodeposition in usual electrolytes led to accelerated dendrite growth capable of causing short-circuits, while with halide-rich interfaces a more compact and smooth deposition was observed. Finally, the concept of halide-protected anodes was utilized in lithium and sodium metal batteries with high energy density cathodes that demonstrated extended capacity retention and longer cycle life.

497461 Molecular Design of Redox-Interfaces: Selective Electrochemical Separations and Beyond

Abstract:

Molecular level design is crucial for the development of new, more efficient electrochemical interfaces for a wide range of industrial and environmental applications. In particular, selective separations remain some of the most important processes in the chemical and biochemical industries, and are crucial for water purification and environmental remediation. Although there has been great interest in electrochemical systems for energy storage and electrocatalysis, they are still of limited use in the separations field due to a lack of molecular-selectivity, and high energetic costs. Here, I present my research in developing redox-active interfaces for selective electrochemical separations.

Redox-active species offer an attractive materials platform, especially organometallic compounds (metallopolymers and associated metal-organic complexes), due to fine control of electronic properties through chemical design. First, I discuss the engineering of specific Faradaic-driven interactions between our redox-centers and target compounds. We investigate the selective sorption and release of anions, cations, and even proteins, based solely on electrochemical control. In parallel, we unravel the intermolecular mechanisms through a combination of electronic structure calculations and spectroscopy, and show how molecular tuning can further enhance interactions, to address tough challenges such as fine chemical separations.

Second, we propose asymmetric Faradaic systems as a next generation configuration for electrochemical processes, since asymmetric systems show much higher current and energy efficiencies. We focus on counter-electrode design, in which the Faradaic process at the cathode operates in tandem with the equivalent process at the anode; ultimately enhancing ion-selective performance, lowering potential windows and preventing parasitic side-reactions. Finally, we explore the nano-scale film properties and physico-chemical behavior of our redox-films, and point to emerging directions beyond separations, especially novel materials design.

Fundamentally, the concepts explored have broad implications in electrochemical sensing, electrocatalysis and interfacial electrosorption. For chemical engineering, these findings demonstrate the capability of redox-based technologies for both environmental and chemical process separations, and eventually lead towards process intensification.

509240 Atomistic modeling of metallic anodes in beyond Li-ion batteries

Abstract:

The lithium-ion (Li-ion) battery has revolutionized the consumer electronics industry in the past two decades, achieving moderate battery lifetimes and higher energy storage densities. However, we have begun to reach the limits to improvements in Li-ion batteries. Further, in order to reach the energy demands posed by the transportation industry, batteries that possess energy densities much larger than current state-of-the-art Li-ion technologies are highly desirable. This research investigates two â€˜beyond Li-ionâ€™ battery technologies that boast theoretical energy densities more than five times larger than Li-ion batteries: batteries employing (i) metallic Mg anodes and (ii) metallic Li anodes. We used density functional theory to probe atomistic interactions on each of these metallic surfaces. For the metallic Mg anode, we identified likely reactions between the anode surface and the battery solvent. The composition of the model Mg anode surface was modified to mimic realistic electrode surfaces. We found that solvent decomposition products and reaction energies were highly dependent on the model anode surface composition, with the pristine Mg anode displaying a greater tendency for solvent decomposition. For the metallic Li anode, we have modeled the interface between Li metal and its native oxide using a crystalline Li/Li_2O slab model and an oxidized slab model in which the native oxide layer was dynamically grown on the Li metal slab. We show the effect of this native oxide layer on the transport properties in each model. Taken together, these studies provide a more complete picture of the atomic-scale factors governing the successful implementation of metallic anodes for beyond Li-ion batteries.

497130 H_2 Production Via Photovoltaic-Electrolysis with over 30% Solar-to-Hydrogen Efficiency

Abstract:

Solar water splitting is a promising technique for producing clean, renewable H_2. Currently, no solar water splitting devices are economically competitive with alternative technologies such as steam methane reforming. A great deal of effort has been invested into developing devices with high solar to hydrogen (STH) efficiency, because modeling studies have predicted that device efficiency is a primary factor that affects the overall H_2 cost.

We report a photovoltaic-electrolysis system with the highest STH efficiency for any water splitting technology reported to date, to the best of our knowledge. This system consists of two polymer electrolyte membrane electrolyzers in series with one triple-junction InGaP/GaAs/GaInAsSb photovoltaic cell, which produces a high enough voltage to drive both electrolyzers with no additional energy input. The solar concentration is is adjusted such that the maximum power point of the photovoltaic is well matched to the operating capacity of the electrolysers to optimize the system efficiency.

Our system achieves a 48 hour average STH efficiency of 30%. This high efficiency demonstrates that photovoltaic-electrolysis systems are a promising solution for cost-effective solar energy storage. We will discuss the implications of these results for the comparitive advantages and disadvantages of competing electrochemical H_2 production technologies and suggest directions for further research in this area.

496306 Remote Control Electrodeposition: Design Criteria for Patterning on Substrates without Direct Electrical Connections

Abstract:

Recent advances in additive manufacturing technologies have been primarily driven by the availability of simple, integrated software/hardware platforms capable of rendering computer aided design files into fabricated objects. These easy-to-use systems facilitate a wider range of users, allowing highly technological hardware (such as 3D-printing) to be used both by hobbyists and trained technicians [1]. Despite recent advances in software integration and commercialization of 3D-printing, a similar transition to fully software-reconfigurable electrodeposition-based prototyping has yet to emerge. The industrial electrodeposition patterning standard, through-mask plating, requires a mask and a series of deposition and etch steps to fully develop metal patterns. Even more sophisticated electrochemical fabrication technology, such as EFAB or MICA, require successive deposition, planarization, and chemical etch steps to build 3D objects layer-by-layer. The reconfigurability of these technologies is limited by their reliance on masks and chemical etching. However, direct-write electrodeposition can bridge the advantages of electrodeposition-based micro/nanofabrication with user friendly software inherent with current reconfigurable additive manufacturing technologies. Previously, localized electrodeposition using microelectrodes and microjet nozzles has demonstrated direct-write metal patterning dictated by the geometry of the electrode or nozzle [2-3]. Impinging jet electroplating was further expanded by Dr. Daniel Schwartz and coworkers at the University of Washington by implementing full software control of electrodeposition and mass transport conditions with a system called Electrochemical Printing (EcP) [4].

More recently, our lab modified the EcP system to operate in a bipolar manner, enabling direct-write electrodeposition and patterning without needing direct electrical connections to the substrate [5]. This configuration, called a scanning bipolar cell (SBC), is advantageous for direct electrodeposition on surfaces that are difficult to electrically connect to, and allows for further automation of the software/hardware. Bipolar electrochemistry involves spatially segregated, equal and opposite oxidation and reduction reactions on a single conductive substrate. In the SBC, reduction of metal ions to metal occurs beneath a rastering microjet nozzle and oxidation of an electron-donating species takes place across the far-field area of the conductive substrate. The constraint of connecting to the substrate in traditional electroplating techniques is thus eliminated, albeit through more intricate electrolyte design.

In this presentation, we outline the design guidelines for bipolar electrodeposition of metals across a wide range of nobilities, and discuss how the coupling of kinetic reversibility, thermodynamics of the bipolar pair, and ionic transport through the electrolyte affects the spatiotemporal stability of electrodeposited patterns [6]. The kinetic reversibility and thermodynamic equilibrium potential of the reduction reaction dictates the selection of a suitable electron-donating oxidation chemistry. Metal deposition chemistries that are irreversible, such as nickel deposition, are paired with electron-donating species that have fast kinetics and a more positive equilibrium potential. This type of bipolar couple is thermodynamically uphill, and the reactions are driven by inputted current supplied by the SBC. For reversible metal deposition chemistries, such as silver deposition, previously patterned material can electrochemically erased during subsequent patterning. Being reversible, any silver metal deposited on the substrate can act as the equal and opposite oxidation chemistry for the bipolar reaction. This challenge is overcome by designing pseudo-stable baths, similar to those used in electroless deposition, which protect the deposited metal from oxidation. For silver deposition, Fe^{2+} oxidation to Fe^{3+} is used as the oxidation chemistry. This electrolyte is designed so the equilibrium potential for the Fe^{2+}/Fe^{3+} couple is marginally less (about 20 mV) than that of the $Ag^{+}/Ag_{(s)}$ couple, creating an environment where Fe^{2+} oxidation is preferential to $Ag_{(s)}$ etching.

Electrodeposition manufacturing processes are unique for their ability to control vertical material resolution at the nanometer scale. Incorporating micropipettes and nanopipettes through direct-write electroplating with EcP or the SBC adds control of lateral resolution ideal for additive manufacturing and patterning. The highest value commercial application for metal patterning in this domain is in integrated circuit and printed circuit board manufacturing. Bipolar electrodeposition with the SBC offers a solution for automated direct-write patterning by eliminating the need to electrically connect to the multifaceted substrates in these industries. Further research in this area will explore scale-down of the SBC nozzles to sub-micron dimensions as well as the impact of substrate surface area on the spatiotemporal behavior of bipolar electrodeposition.

References:

[1] M. Vaezi, et al., *Int. J. Adv. Manuf. Technol.*, **67**, 1721 (2013).

[2] J. D. Madden et al., *J. Microelectromech. Syst.*, **5**, 24 (1996).

[3] R. Von Gutfeld et al., *Appl. Phys. Lett.*, **43**, 876 (1983).

[4] J. D. Whitaker et al., *J. Micromech. Microeng.*, **15**, 1498 (2005).

[5] T. M. Braun et al., *J. Electrochem. Soc.*, **162**, D180 (2015).

[6] T. M. Braun et al., *J. Electrochem. Soc.*, **163**, D3014 (2016).

501965 Stalbe Electrochemical Growth in Viscoelastic Flow

Abstract:

Electrodeposition is used in various manufacturing processes for creating metal, colloid, and polymer coatings on conductive substrates. The process also plays an important role in electrochemical storage technologies based on rechargeable batteries, where it must be carefully managed to facilitate stable and safe operations at low operating temperatures, high rates, and over many cycles of charge and discharge. A successful electrodeposition processes requires fast transport of charged species (e.g. ions, particles, & polymers) and stable redox reactions at liquid-solid interfaces. In almost all used liquid electrolytes, deposition is subject to a variety of hydrodynamic and morphological instabilities that lead to complex transport phenomena in the electrolyte and unstable deposition, including formation of ramified structures known as dendrites on the substrate/electrode. This talk considers the stability of electrodeposition of reactive metals on planar electrodes with an emphasis on its role in enabling next-generation batteries in which metallic lithium or sodium serves as the electrode. Such batteries have been argued to offer step-change improvements in electrochemical storage technology over todayâ€™s state-of-the art lithium ion batteries and are under active investigation worldwide for high-energy, portable energy storage solutions in multiple fields. Beginning with analyses of ion transport processes and stability of electrodeposition in a structured electrolyte, the talk will discuss the fundamental fluid properties such as viscosity, elasticity, electro-convection etc. of the electrolytes that affect the stability of deposition and electrode-electrolyte interphases for stabilizing deposition of reactive metals. The talk will also discuss contemporary efforts to evaluate these approaches using electrochemical and visualization studies.

358a Searching For Volume Diffusion

Narendra Singh,
T. Schwartzentruber,
and E L Cussler

U MN

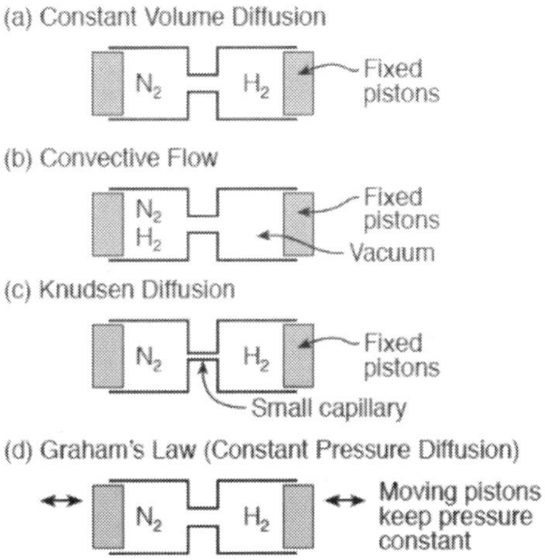

Searching for Volume Diffusion

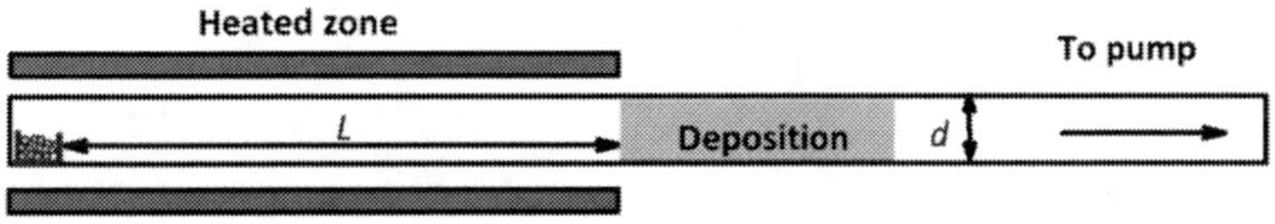

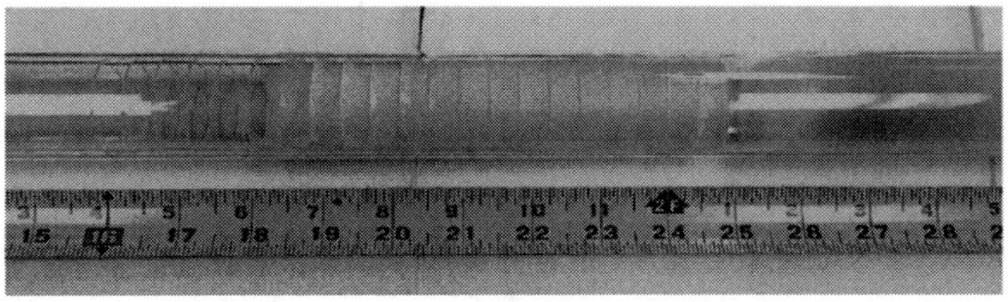

Sublimation Varies with (Diameter)^2

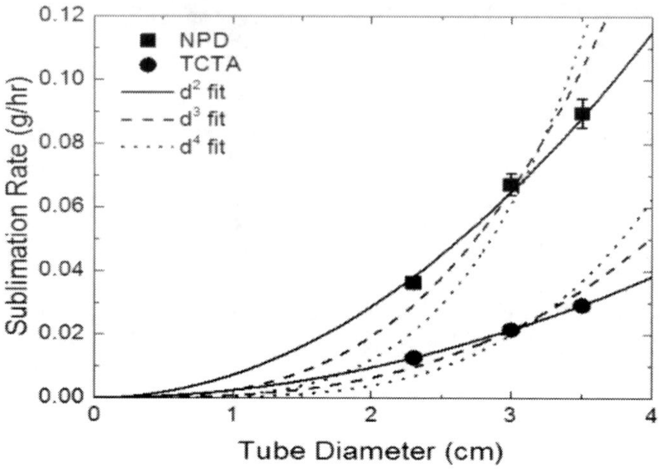

Sublimation/Area Varies with (Diameter)$^?$

Resistances Parallel and Series

Flow

$$j = j_v + \frac{1}{\frac{1}{j_D} + \frac{1}{j_{Kn}}} = \left[\frac{pd^2}{3z\mu} + \frac{1}{\frac{1}{D} + \frac{1}{D_{Kn}}} \right] \frac{\Delta c}{L} = D' \frac{\Delta c}{L}$$

Volume Knudsen Length

Experiment (Knudsen, 1909) vs. Theory (2015)

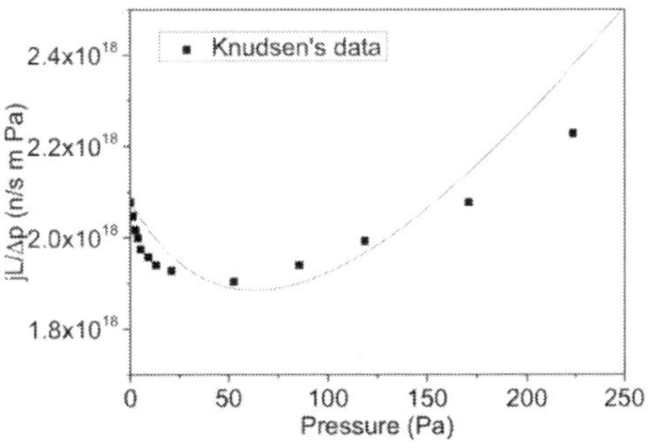

Experiment:	Theory:
$a' = 7.0\times10^{15}$	$a = 2.4\times10^{15}$
$b' = 2.1\times10^{18}$	$b = 2.1\times10^{18}$
$c' = 0.7\times10^{-2}$	$c = 1.1\times10^{-2}$

Velocity with Deposition

Velocity = Velocity(radius)

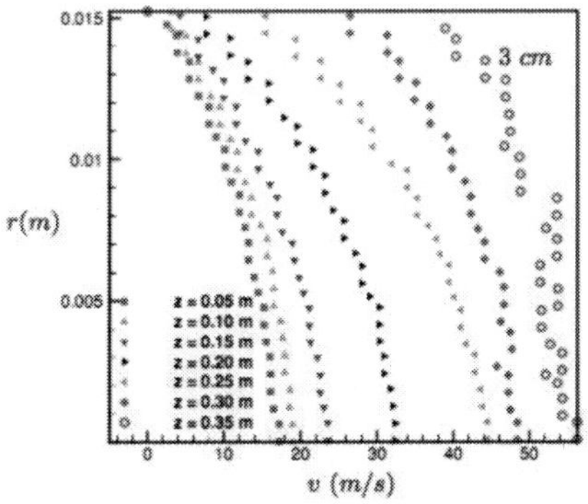

Hagen – Poiseuille test

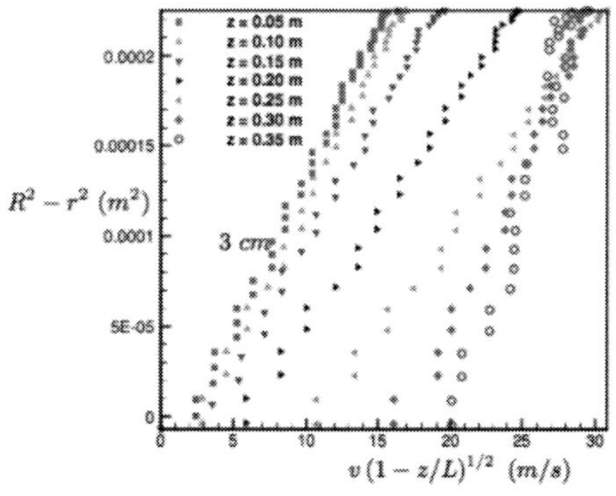

Velocity without Deposition

Velocity = Velocity(radius) Hagen – Poiseuille test

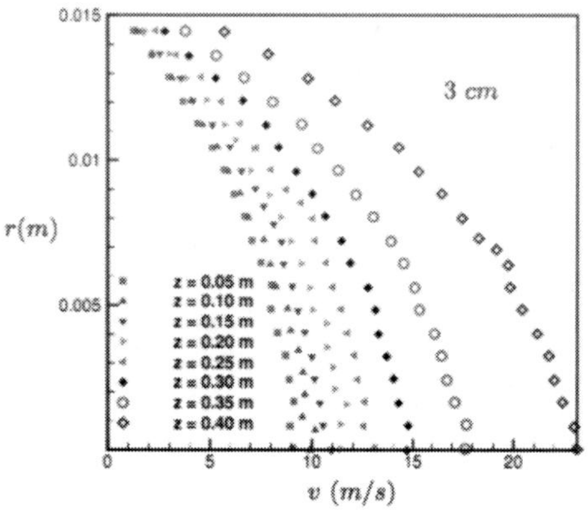

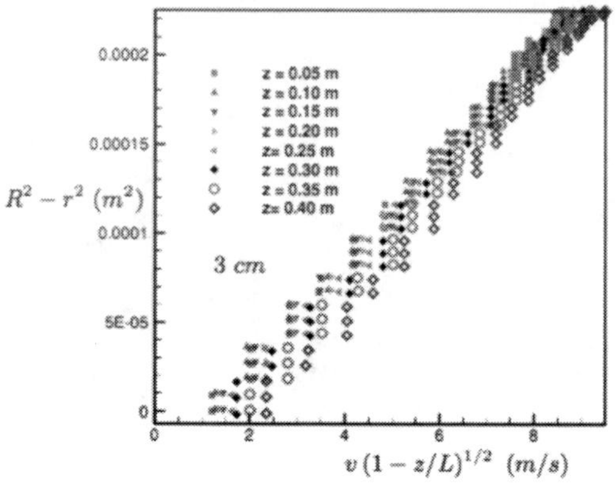

Diffusion of What?

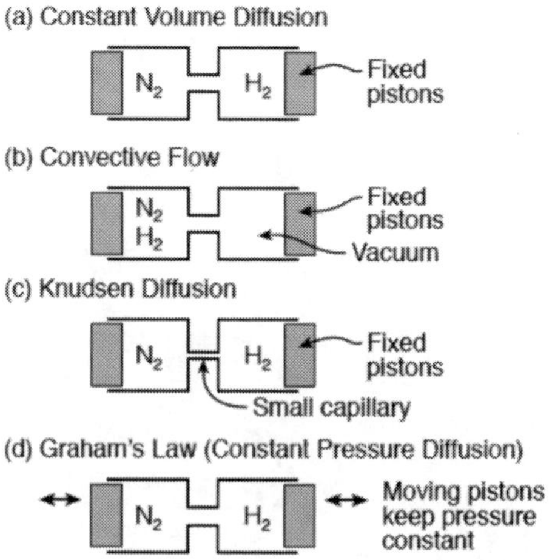

501204 Condensation on Highly Superheated Surfaces: Unstable Thin Films in a Wickless Heat Pipe

Abstract:

The Constrained Vapor Bubble (CVB) heat pipe is a wickless heat pipe that is designed to produce a simple, light, and reliable heat transfer system that can be used for cooling critical components of spacecraft. The CVB system consists of a relatively simple setup - a quartz cuvette with sharp corners partially filled with pure pentane as the working fluid. A full-scale fluid experiment was conducted on the International Space Station to provide a better understanding of how the microgravity environment might alter the physical and interfacial forces driving evaporation and condensation. Along with temperature and pressure measurements, the Light Microscopy Module (LMM) was used to determine the two-dimensional thickness profile of the menisci formed on the wall surfaces as well as at the corners of the cuvette. Interfacial forces dominate in these extremely small Bond number systems. The transport processes were found to be complex despite being conceptually simple.

Traditional heat pipes are divided into three zones: evaporation at the heated end, condensation at the cooled end, and intermediate/adiabatic in between. The microgravity experiments reported herein show the situation may be dramatically more complicated. Beyond a threshold heat input, there was a transition from evaporation at the heated end to large-scale condensation, even as surface temperatures exceeded the boiling point by 160 K. The hotter the surface, the more vapor was condensed onto it. The condensation process at the heated end is initiated by thickness and temperature disturbances in the thin liquid film that wet the solid surface. Those disturbances effectively leave the vapor â€œsuperheatedâ€ in that region. Condensation is amplified and sustained by the high Marangoni stresses that exist near the heater end that drive liquid to cooler regions of the device. This led to flooding of the heater end of the device and the forming of a large liquid drop on the wall surface. This condensation phenomenon at the heater end also occurs in a 1-g environment. However, since the return flow of liquid is meager in 1-g, there is little vapor to condense at the heater end and the phenomenon appears as if pentane were a partially wetting fluid. Thus, it can be easily mistaken for the effects of surface contamination.

This material is based on the work supported by the National Aeronautics and Space Administration (NASA) under Grant No. NNX13AQ78G and the National Science Foundation under Grant No. CBET-1603318.

Demixing in Dilute Polydisperse Emulsions and Suspensions

Rodrigo Reboucas and Michael Loewenberg
Yale University
Abstract ID: 509578

17AIChE
Annual Meeting, Minneapolis, MN

10/31/2017

Motivation

Fig.1 - Blood flow
Ref.: A. Kumar & M. D. Graham (2012).

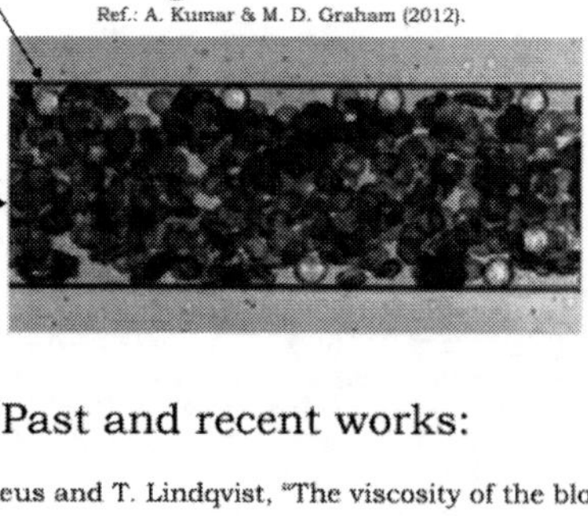

Fig.2 - Drug delivery
Ref.: Muller et al (2016).

- Past and recent works:

R. Fahraeus and T. Lindqvist, "The viscosity of the blood in narrow capillary tubes," Am. J. Physiol. 96, 562–568 (1931)

V. Narsinham, H. Zhao and E. S. G. Shaqfeh, "Coarse-grained theory to predict the concentration distribution of red blood cells in wall-bounded Couette flow at zero Reynolds number", Physics of Fluids (2013)

R. H. Riveira, X. Zhang and M. D. Graham, "Mechanistic theory of margination and flow-induced segregation in confined multicomponent suspensions: Simple shear and Poiseuille flows", Physical Review Fluids 1 (2016)

Outline

1. **Hydrodynamic pair interactions**

2. Analytic trajectory integration

3. Droplet transport near a solid wall

Hydrodynamic Pair Interactions

- Smooth, spherical and rigid particles

➡ Reversible

- Roughness
- Deformation
- Porosity

➡ Irreversible

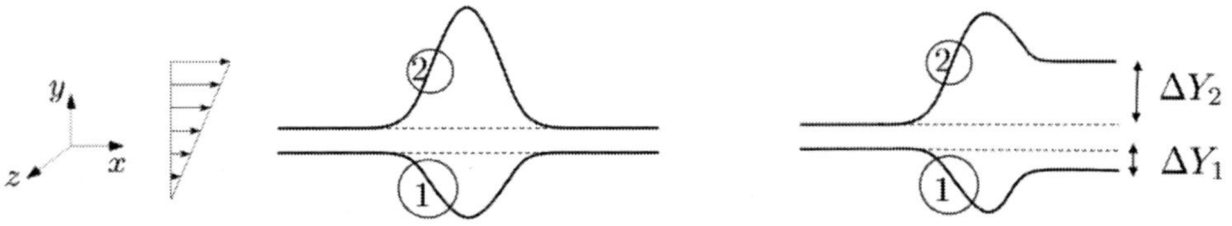

Shear flow

$\Delta Y_i = 0 \quad i = 1, 2$

$\Delta Y_i \neq 0 \quad i = 1, 2$

↳ **Particle Flux**

Rough Particles

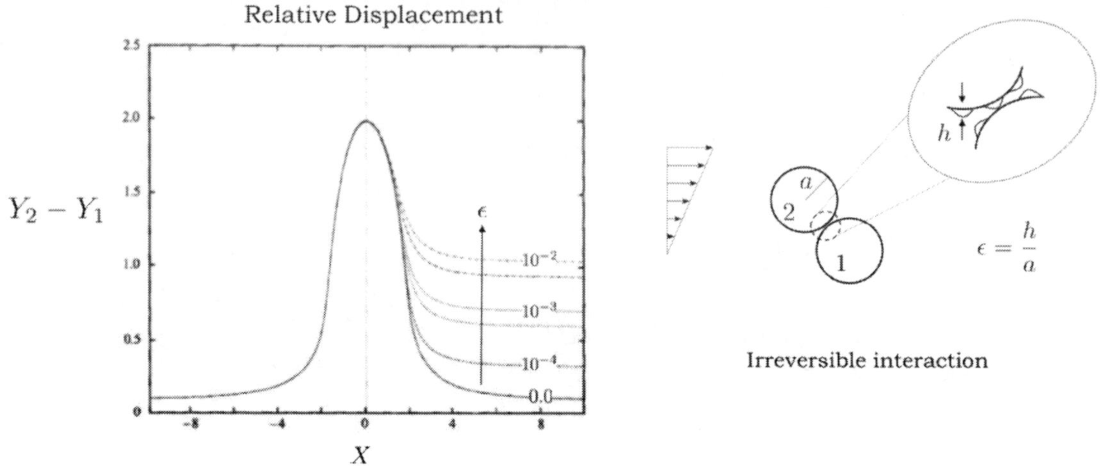

Ref.: Figure – Adapted from Da Cunha & Hinch (1996).

Deformable Drops

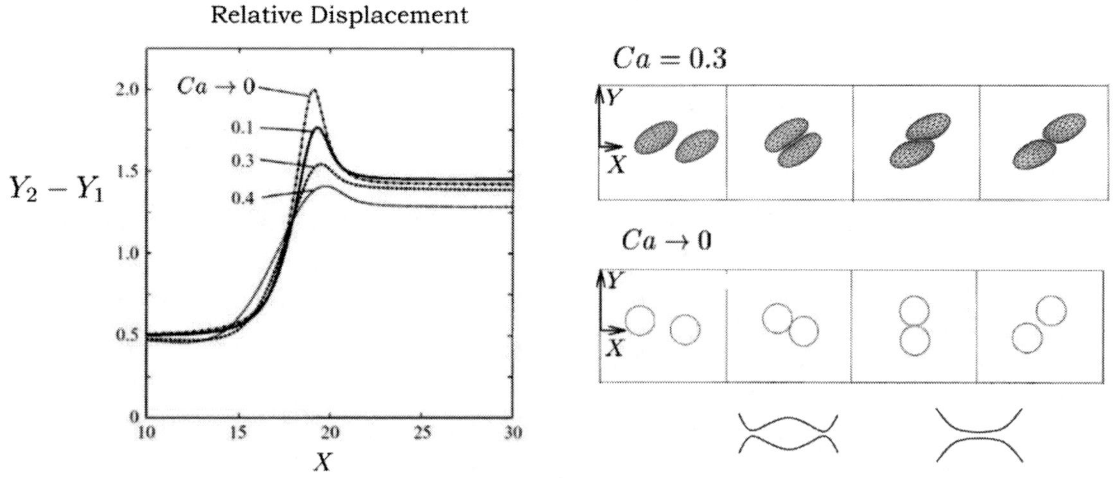

Ref.: Figure – Adapted from Loewenberg & Hinch (1997).

Outline

1. Hydrodynamic pair interactions

2. Analytic trajectory integration

3. Droplet transport near a solid wall

7

Trajectory Integration

$$\mathbf{V}_i = \mathbf{U}_\infty + \left[\mathcal{A}_i(r)\tfrac{\mathbf{rr}}{r^2} + \mathcal{B}_i(r)\left(\mathbf{I} - \tfrac{\mathbf{rr}}{r^2}\right)\right] \cdot \mathbf{E} \cdot \mathbf{r} \quad (i = 1, 2)$$

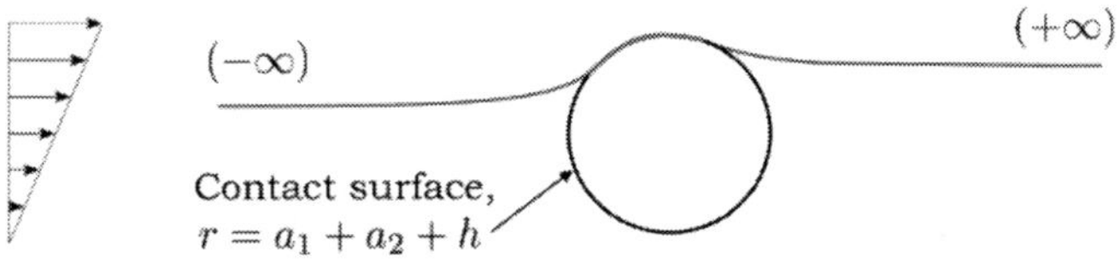

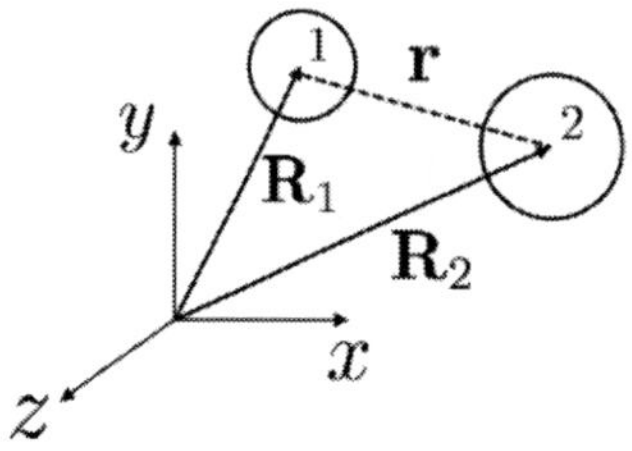

- Analytic trajectory integration:

$$\Delta\mathbf{R}_1 = \mathbf{I}_1(\mathcal{A}_1, \mathcal{A}_2, \mathcal{B}_1, \mathcal{B}_2)$$

$$\Delta\mathbf{R}_2 = \mathbf{I}_2(\mathcal{A}_1, \mathcal{A}_2, \mathcal{B}_1, \mathcal{B}_2)$$

8

Displacements

λ : viscosity ratio

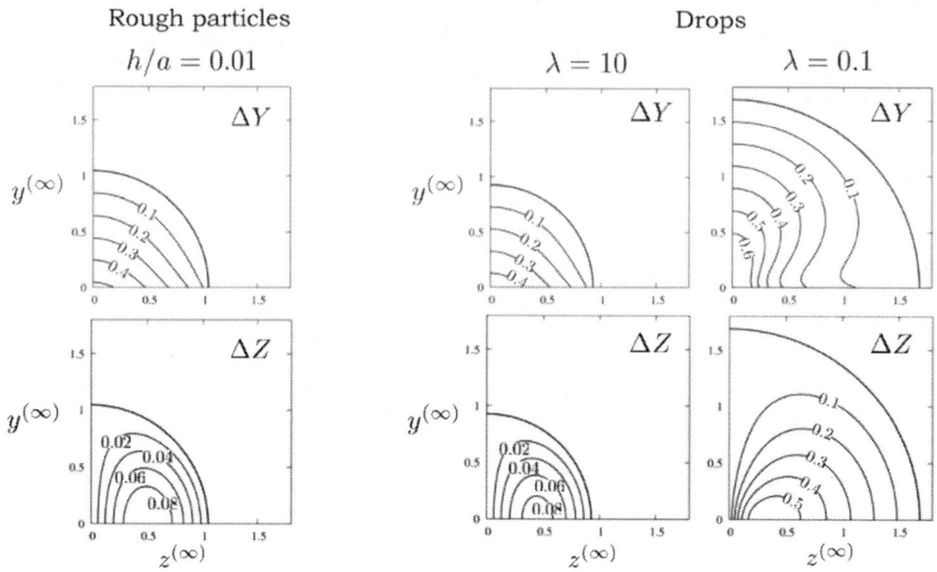

9

Outline

1. Hydrodynamic pair interactions

2. Analytic trajectory integration

3. Droplet transport near a solid wall

10

Droplet transport near a solid boundary

- Wall Migration

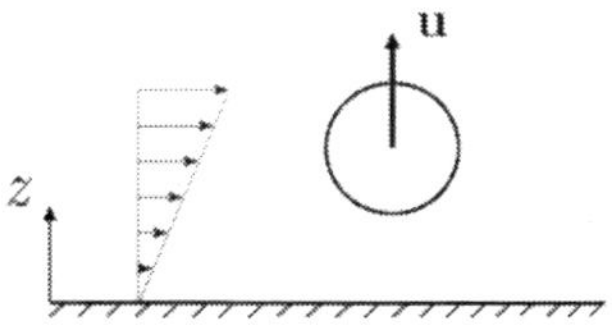

- Hydrodynamic Diffusion

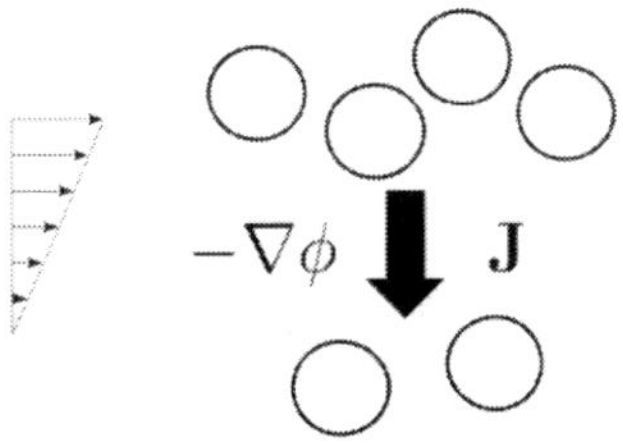

11

Wall Migration

- Deformation-induced velocity

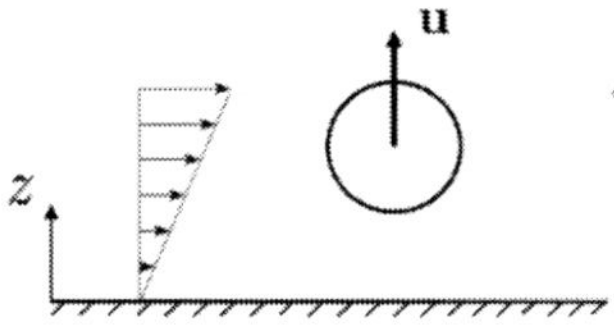

$$u = \frac{\mu \dot{\gamma}^2}{\sigma} g(\lambda) \frac{a^4}{z^2}$$

Ref.: C. E. Chaffey, H. Brenner & G. S. Mason (1965)

12

Hydrodynamic Diffusion:

$$J_i = \underbrace{\left(\sum_j D_{ij}^{(1)}\right) \frac{\partial n_i}{\partial z}}_{\text{Down gradient}} + \underbrace{\sum_j D_{ij}^{(2)} \frac{\partial n_j}{\partial z}}_{\text{Induced Migration}}$$

Down Gradient

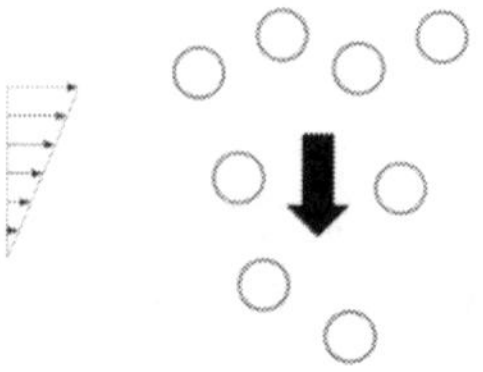

Induced Diffusion

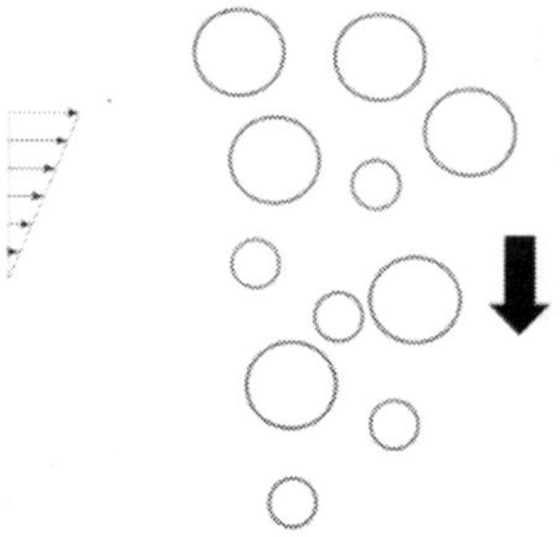

$$D_{ij}^{(1)} = -n_j \dot{\gamma} \iint_{-\infty}^{+\infty} \left| y^{(\infty)} \right| \frac{1}{2} (\Delta_{ij})^2 \, dy^{(\infty)} dz^{(\infty)}$$

$$D_{ij}^{(2)} = -n_i \dot{\gamma} \iint_{-\infty}^{+\infty} \left| y^{(\infty)} \right| \left[\frac{1}{2} (\Delta_{ij})^2 + y^{(\infty)} \Delta_{ij} \right] dy^{(\infty)} dz^{(\infty)}$$

Δ_{ij} : Pair displacement

13

Droplet transport near a solid boundary

- Wall Migration

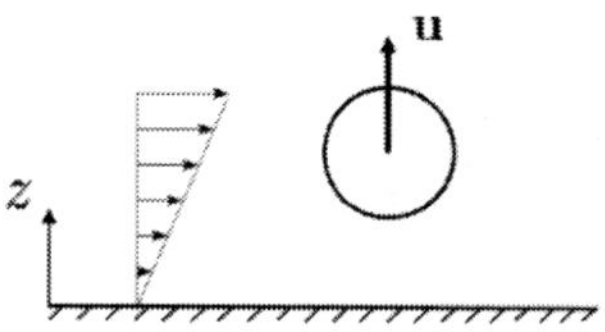

- Hydrodynamic Diffusion

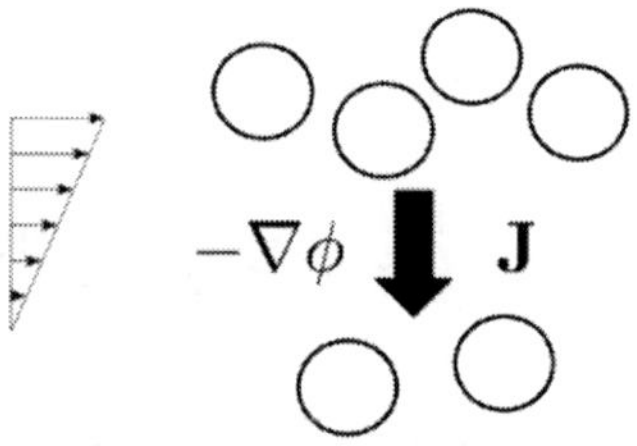

$$\frac{D_\infty}{\mathcal{D}} \gg 1 \qquad \boxed{J_i + n_i u_i = 0} \quad \Rightarrow \quad L = \frac{Ca}{\phi_\infty} a$$

14

Results

Monodisperse

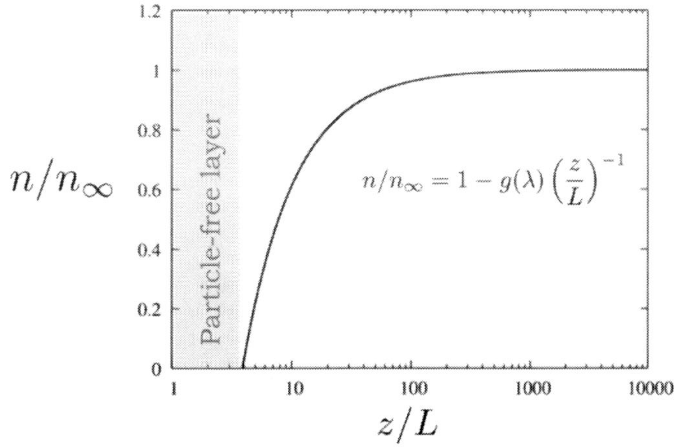

15

Segregation:

$$x_i = \frac{\phi_i}{\phi}$$

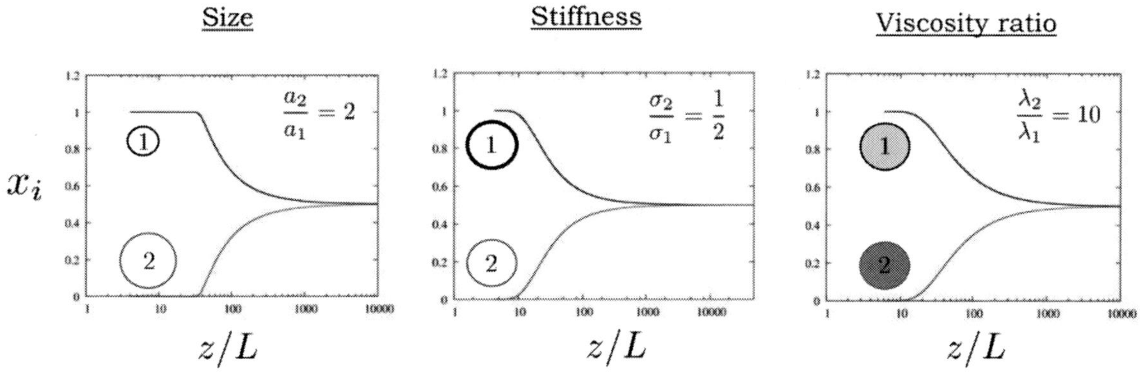

16

Final Remarks

- Irreversible pair interactions ➡ $\Delta_{ij} \neq 0$
- Analytic trajectory integration
- Drop transport near boundary:
  - ✓ Segregation: size, viscosity ratio and surface tension

Acknowledgements: CAPES NSF

17

498488 Experimental Investigation of Air Ingress Scenario during Natural Circulation Cooling of a Very Hot Channel

Abstract:

Experimental Investigation of Air ingress scenario during Natural Circulation Cooling of a Very Hot Channel

Apoorva V. Rudra *[a], Narbeh Artoun [b], Dinesh Kalaga [b], Sanjoy Banerjee [a, c], Masahiro Kawaji [b, c]

[a] Department of Chemical Engineering, City College of the City University of New York, New York

[b] Department of Mechanical Engineering, City College of the City University of New York, New York

[c] The Energy Institute, City College of the City University of New York, New York

After the severe accidents at Fukushima Daiichi Nuclear Power Plants, the need for understanding the response of nuclear facilities to externally or environmentally initiated events, such as strong earthquakes, tsunamis, and flooding has become significantly important [1, 2]. VHTRs are one of the important Gen-IV reactor designs which have passive safety features. Although these designs ensure passive cooling in case of loss of forced circulation or pressure [3], it is still unknown how they will behave under all conceivable accident scenarios. One such accidental scenario is air ingress scenario in which a crack/breakage might occur in the reactor due to an externally or environmentally initiated event leading to a pressure front. Due to pressure difference, air from the outside can enter the reactor loop and cause graphite oxidation. It could also induce secondary effects such as increased temperatures due to an exothermic reaction which lowers the graphite's thermal conductivity, thus causing a degradation in conduction heat transfer.

In this study, we present the experimental measurements of the coolant flow and heat transfer behavior in a flow channel in a graphite block simulating a prismatic core of a Very High Temperature Gas-Cooled Reactor (VHTR). Natural circulation experiments are conducted using a helium-nitrogen mixture representing a helium-air mixture initially present in the lower plenum. A helium analyzer is used to investigate the nitrogen transport from the lower to upper plena as a function of time.

The natural circulation flow rate is measured and the flow rate data is related to the graphite temperatures, operating pressure, and overall concentration of nitrogen in helium. The amount of nitrogen gas injected into the lower plenum and thus the overall nitrogen concentration in the gas mixture is systematically changed to determine its effect at different graphite temperature settings in the riser and downcomer. To what extent the local helium concentrations will differ between the lower and upper plena during both steady and transient natural circulation is determined in order to elucidate the air transport process following a postulated accident involving air ingress into the lower plenum. In order to measure the natural circulation flow rate, a mass flow measurement system was designed. The natural circulation flow and heat transfer data along with the viscosity effects are analyzed to identify possible changes in heat transfer rates from the graphite in the riser.

[1] ScienceDaily. "Fukushima at Increased Earthquake Risk, Scientists Report." *ScienceDaily*. 13 Feb. 2012. Web. 6 Nov. 2012. <http://www.sciencedaily.com/releases/2012/02/120214100819.htm>

[2] Alvarez, Robert. "Spent Nuclear Fuel Pools in the U.S.: Reducing the Deadly Risks of Storage." *Institute for Policy Studies*. May 2011. <http://www.ips-dc.org/files/3200/spent_nuclear_fuel_pools_in_the_US.pdf>

[3] Vilim, R.B., Pointer, W.D., and Wei, T.Y.C., 2006, "Prioritization of VHTR System Modeling Needs Based on Phenomena Identification, Ranking and Sensitivity Studies," ANL-GenIV-071.

491554 Hydrodynamic Study to Determine Volumetric Mass Transfer Coefficient for Cell Culture Applications

Abstract:

The stirred tank reactors are most commonly used both in the laboratory and process industry to identify key factors that influence reactor performance. Particularly for bioreactors, the volumetric mass transfer coefficient ($k_L a$) of oxygen is used as one of the important parameters for determining efficiencies of reactors and for successful scale-up. A number of methods have been developed to determine the volumetric mass transfer coefficient such as dynamic method, steady-state method, gaseous oxygen balance and carbon dioxide balance. In the present work, using the dynamic â€œgassing out-gassing-inâ€ method, the effect of various parameters such as impeller agitation rate, air flow rate, and impeller type (e.g. Rushton Turbine, Pitched Blade Turbine, and Dislocated Rushton Turbine) on the volumetric mass transfer coefficient has been investigated in a stirred tank bioreactor. It was found that the volumetric mass transfer coefficient increases with both the agitation rate and the air flow rate. In addition, the power input per unit volume was shown to decrease with the air flow rate or aeration. Furthermore, it was found that the dislocated Rushton turbine can achieve higher volumetric mass transfer coefficient due to enhanced gas dispersion, as compared to the standard Rushton Turbine. Finally, several empirical correlations have been developed for the evaluation of $k_L a$. This study can be used for cell cultivation systems/applications/reactors.

486656 Turbulent Flow of Diutan Biopolymer Solutions and Carbon Nanotube Suspensions in a 4.6 mm ID x 200 L/D Smooth Pipe

Abstract:

Turbulent flow of dilute aqueous solutions of a polysaccharide biopolymer, Diutan, and dilute aqueous suspensions of multi-walled carbon nanotubes was explored experimentally in a 4.6 mm ID x 200 L/D smooth pipe. The test pipe was seamless stainless steel with electropolished bore, of internal diameter and bore roughness (D, k) = (4.60 mm, 0.18 microns), and comprized 7 segments, each of L/D = 30 with a pressure tap near its downstream end. It was installed in a single-pass, progressing cavity pump-driven flow system, fed from a 200 liter tank. Pipe Reynolds numbers varied from 8000 to 80000 and pipe wall shear stresses from 8 to 600 Pa. Friction factors for deionized water in all pipe segments adhered to the Prantdl-Karman law within ±0.2 units of $1/\surd f$, indicative of fully-developed, hydrodynamically smooth turbulent flow.

The Diutan biopolymer, MW ~ 5.0E6, had end-to-end length Le ~ 6.0 um, formal contour length Lc ~ 11.7 um and chain diameter Dch ~ 1.5 nm. Aqueous Diutan solutions of concentrations C from 1 to 100 wppm exhibited Type B drag reduction, characteristic of extended macromolecules, yielding turbulent flow segments roughly parallel to, but displaced upwards from, the Prandtl-Karman law, the more so with increasing concentration. At fixed $Re\surd f$ = 5000, flow enhancements relative to solvent $S' = [(1/\surd f)_p - (1/\surd f)_n]_{Re\surd f}$ increased almost linearly with increasing concentration, with intrinsic flow enhancements $[S'] = Lim_{c\to 0}[S'/c] = 0.10\pm0.02$.

Industrial grade multi-walled carbon nanotubes, abbr MWCNT, of average diameter and length (Dch, Le) = (12 nm, 30 um), were used. Suspensions of concentration C from 1 to 1000 wppm were made in deionized water containing 900 wppm of polyvinylpyrolidone surfactant. On account of their scarcity, nanotube suspensions were tested in brief "scanning" runs of 30 s duration at each of four Re = (8300, 16700, 33000, 67000). First, two scans at C = 0, that is, of the 900 wppm PVP surfactant solution alone, served to establish a basis, yielding P-K results of $1/\surd f$ (11.2±0.1, 12.3±0.1, 13.3±0.1, 14.5±0.1) at $Re\surd f$ (790, 1440, 2670, 4890) respectively. Thereafter, scans of MWCNT suspensions at each of c = (1.0, 10, 100, 1000) wppm provided sets of $1/\surd f$ (11.2±0.2, 12.2±0.1, 13.2±0.1, 14.4±0.1) at $Re\surd f$ (770, 1450, 2650, 4850) respectively that were virtually indistinguishable from the basis scans at c = 0 wppm. Suspensions of C = 1 to 1000 wppm MWCNTs in 900 wppm PVP surfactant thus exhibited Newtonian turbulent flow behavior.

Under similar flow conditions, solutions of Diutan biopolymer, with Le ~ 6 um and aspect ratio Le/Dch ~ 4000, exhibit Type B turbulent drag reduction with intrinsic slip [S'] = 0.10, whereas suspensions of multi-walled carbon nanotubes, with Le ~ 30 um and aspect ratio Le/Dch ~ 2500, both the same order of magnitude as Diutan, exhibit Newtonian turbulent flow, without any hint of drag reduction. It would be most interesting to divine what skeletal attribute(s) allow(s) drag reduction by Diutan biopolymer but

forbid(s) drag reduction by carbon nanotubes.

497890 Fluid Dynamics with Superimposed Mass Transfer of Single Bubbles in Reacting Liquids

Abstract:

The design of reactors with a gaseous phase dispersed in a liquid phase (e.g. bubble columns) is often based on rather roughly estimated parameters. The DFG priority program SPP 1740 "Influence of local transport processes on chemical reactions in bubbly flows" was set up to get a deeper insight in such systems. The interdisciplinary program allows a close cooperation between chemists, process engineers and mathematicians for both experimental work and simulation e.g. using CFD. Aim of the SPP 1740 is to develop a fundamental understanding of the interactions in gas/liquid-reactions, especially concerning fluid dynamics, mass transfer and reaction rate and their influence on yield and selectivity.

In this sub-project, the local and integral mass transfer is determined qualitatively and quantitatively. A considerable amount of literature has been published on the analysis of volume specific mass transfer coefficient or integral mass transfer in gas/liquid systems. The main influencing factors were found to be the pressure, temperature, gas and liquid velocity, liquid properties and column geometry.

In this work, single bubble investigations were performed in two different setups: a rising cell with a length of 2 meters and a counter flow cell. These setups allow different contact times to be applied, varying between a couple of seconds depending on the rise velocity of the bubble and up to several minutes in case of the counter flow setup. Additionally the overall pressure and temperature can be changed. The bubble size and the applied gases and liquids have been varied during the investigations.

To get a profound understanding of the local effects influencing the fluid dynamics and mass transfer during the rise of a bubble, the rising test cell uses two high-speed cameras on a carriage, to track and observe the bubble during its ascent. The velocity of the carriage is regulated in real-time by the position of the bubble in the captured picture. The view of the second camera has an angle of 90°, thus allowing the measurement of the trajectory, velocity, size and shape of a free rising bubble. The contact time is limited by the height of the column (H = 2 m). Two reactive systems were investigated. Carbon dioxide / water and nitric oxide bubbles in a solution of iron(ii) sulfate heptahydrate ($FeSO_4$) and EDTA (ethylenediaminetetraacetic acid). This chemical reaction has been chosen due to the color change of the reaction product in order to visualize local effects and the wake of the bubble. The reaction rate can be adjusted by the choice of ligand (e.g. EDTA).Depending on the concentration of the educts, the mass transfer can be enhanced by the chemical reaction in the liquid phase. This enhancement can be determined qualitatively.

In the counter-current measuring cell the preparation of the liquid phase can also play a significant role. Due to the dissolved gases in the liquid, the mass transfer from the continuous phase into the gas bubble must also be taken into account.

Acknowledgement

The authors gratefully acknowledge the financial support provided by the German Research Foundation within the Priority Program "Reactive Bubbly Flows", SPP 1740.

499791 Experimental Temperature and Fluid Flow Measurements in Packed Beds with Magnetic Resonance Imaging

Abstract:

Packed bed units are of industrial importance in the development of thermal energy storage systems such as those associated with concentrated solar-power operations. Such units experience a temperature front that propagates through the bed and control of this front is key to improved efficiency of operations. A challenge in this field is measuring high-accuracy temperature profiles inside these systems. Also, assessing local heat-transfer processes are hampered by limited information on local flow conditions. To determine the spatial and temporal temperature and velocity profiles, magnetic resonance imaging (MRI) is used with encapsulated phase change materials as particles and a fluorinated fluid as heat transfer fluid. The encapsulated phase change materials generate signal as they change phase so that the melt front at a fixed melting temperature is detected in the images. Thus, the propagation of the melt front can serve as a proxy to the local temperature conditions. Additionally, MRI measurements using ^{19}F NMR provide local fluid-phase velocity profiles that can identify flow maldistribution and help assess local heat-transfer environments by providing one or more components of the local velocity vector. Results are presented for a range of flow rates (10-600 mL/hr) using encapsulated eicosane particles for the solid phase ($T_{melt} = 37^oC$, $d_p = 5$ mm) and Flourinert FC-43 for the fluid phase. Preliminary results show qualitative agreement with discrete element method based computational fluid dynamics studies.

492355 Experimental Study of Gas Liquid Mass Transfer Enhanced By Bottom Shear Turbulence

Abstract:

Introduction

The study of mass transfer at the interface between a gas and a liquid is of great importance in many environmental phenomena and industrial processes. In the industry, the physics of mass transfer defines the efficiency of diphasic mixing and degassing in stirred tanks, bubble columns, or even more complex mixing tank geometries. From an environmental point of view, the same mechanisms also control the dissolution of atmospheric gases into the oceans, rivers and lakes, and have therefore a huge impact on climate and biodiversity.

Gas dissolution at flat interface is controlled by gas and liquid turbulence characteristics [1]. Three possible turbulence sources may be considered [2]: gaseous phase turbulence, similar to that of wind above oceans or lakes; liquid turbulence caused by bottom shear found in rivers and stirred cells, and convective turbulence created by temperature gradients between the free surface and the bottom of the fluid.

To implement reliable models of mass transfer into CFD codes, a better understanding of the mechanisms ruling the transfer, like the influence of interface's shape or flow regime, is still needed.

The aim of the present work is to study experimentally the mass transfer of a low solubility gas at a flat gas-liquid interface, and how it is enhanced by bottom shear turbulence. The dissolution of carbon dioxide (CO_2) into water is first studied. The same experimental procedure will then be applied to the observation of CO_2 dissolution into dilute polymer solutions with viscoelastic, shear-thinning behavior.

Experimental setup

The experimental bench is a prismatic tank with transparent Plexiglas® walls and lid allowing flow visualization. Liquid phase turbulence is generated by an oscillating grid appartus similar to the one used by Hopfinger and Toly [3].

Carbon dioxide (99,99% purity) is introduced above the liquid at atmospheric pressure. The overall carbon dioxide mass transfer is monitored by a pH sensor placed at the bottom of the tank.

Two non-intrusive optical techniques are used simultaneously to measure both velocity and dissolved gas concentration fields at a good spatial resolution (typically 0.4 mm for the velocity field and up to 0.008 mm for the concentration field out of the Laser sheet thickness). The Region of Interest (ROI) is a vertical plane underneath the interface at the center of the tank.

Stereoscopic Particle Image Velocimetry (SPIV) allows for the measurement of the three components of the velocity field by stereoscopic reconstruction of particle images of two identical cameras looking at the same region of the flow. As the measurements are made inside the liquid phase, a Scheimpflung arrangement for the lenses and the use of a liquid prism are necessary [4].

An original Quenched Planar Laser Induced Fluorescence (Q-PLIF) technique is used for the determination of the dissolved gas concentration field [5]. A fluorescent dye (here Fluorescein disodium salt) is previously mixed into the liquid at low concentration (typically of 10^{-7}M). The fluorescence of this dye strongly depends on the pH, which is itself modified by the acid-base reactions of dissolved carbon dioxide in aqueous media. A rigorous calibration of the relations between carbon dioxide concentration and pH, and between pH and fluoresced intensity enables the transformation of fluorescence intensity fields into dissolved gas concentration fields, and consequently the measurement of local gas concentration.

External timing of the two techniques based on the grid's oscillation, and a common spatial calibration make it possible to correlate spatially and temporally the two measured fields and to compute the mass fluxes in the ROI.

Results and discussions

From PIV measurements, it has been shown that for a given grid oscillation frequency, velocity magnitude and fluctuations decrease toward the free surface. Vertical distribution of turbulence can be related to the grid oscillation frequency f, the stroke S, the mesh parameter M (distance between two consecutive grid bars), and the distance to the interface, as shown by Hopfinger and Toly [3]. Turbulence was also found to become almost homogeneous with no periodic oscillation of the velocity field in an horizontal plane at a given distance from the grid's average position [6].

When approaching the interface, the root mean square of velocity fluctuations profile is modified. The vertical component, which is the strongest fluctuation in the bulk fluid, becomes lower than the two horizontal components. All fluctuations decrease rapidly from a given depth to the interface, verifying at least from a statistical point of view, the concept of a viscous sublayer expressed in Hunt [7].

The dissolved-gas fields obtained by Q-PLIF show the existence of a thin subsurface layer in which the concentration rapidly decreases from the surface saturation concentration. Different types of scalar structures modifying the shape of this sublayer are observable.

- **Renewal events**correspond to "fresh" low concentration fluid being brought to the interface by an upwelling eddy to replace saturated fluid consequently enhancing the diffusion and the mass transfer

- **Injection events** or "peeling" [8] which consist in high concentration filaments being brought down by swirling velocity structures down to the bulk fluid where it is mixed.

Using coupled measurements, the correspondence between scalar and velocity structures can easily be done. But since the velocity field under the interface tends to be highly three dimensional, 3D velocity structures appear and are also found to be correlated with some complex injection and renewal events [6].

Current work and perspectives

From the previously cited measurements, it is possible to compute the correlation between two velocity components fluctuations and between velocity and concentration fluctuations, i.e. the Reynolds shear stresses and the turbulent contribution to the overall mass flux. The conditioning of this fluxes using quadrant analysis [9] based on the vertical velocity fluctuation and the concentration, and a further extension to an octant analysis [10] including the horizontal velocity fluctuation should give a better insight about the events that dominate turbulent mass transfer.

The next step of the study consists in the addition of a small concentration of polymer to the liquid phase. It has been known for a long time that such additives tend to lower the flow friction at rigid boundaries and modify turbulence dissipation [11], [12]. It has also been observed that mixing of passive scalar [13] and gas liquid mass transfer [14] in such media differ from the Newtonian case, but to the author's knowledge, no fundamental study has yet focused on the turbulent mass transfer events at a gas liquid interface for these complex fluids.

[1] D. E. Turney and S. Banerjee, "Air–water gas transfer and near-surface motions," *J. Fluid Mech.*, vol. 733, pp. 588–624, Oct. 2013.

[2] H. Herlina, "Gas transfer at the air-water interface in a turbulent flow environment," Universitätsverlag Karlsruhe, Karlsruhe, 2005, PhD Thesis.

[3] E. J. Hopfinger and J.-A. Toly, "Spatially decaying turbulence and its relation to mixing across density interfaces," *J. Fluid Mech.*, vol. 78, no. 1, pp. 155–175, Nov. 1976.

[4] A. K. Prasad and K. Jensen, "Scheimpflug stereocamera for particle image velocimetry in liquid flows," *Appl. Opt.*, vol. 34, no. 30, p. 7092, Oct. 1995.

[5] P. Valiorgue, N. Souzy, M. EL Hajem, H. B. Hadid, and S. Simoëns, "Concentration measurement in the wake of a free rising bubble using planar Laser-induced fluorescence (PLIF) with a calibration taking into account fluorescence extinction variations," *Exp. Fluids*, vol. 54, no. 4, pp. 1–10, Apr. 2013.

[6] T. Lacassagne, M. EL Hajem, F. Morge, S. Simoens, and J.-Y. Champagne, "Study of Gas Liquid Mass Transfer in a Grid Stirred Tank," *Oil Gas Sci. Technol. – Rev. D'IFP Energ. Nouv.*, vol. 72, no. 1, p. 7, Jan. 2017.

[7] J. C. R. Hunt, "Turbulence Structure and Turbulent Diffusion Near Gas-Liquid Interfaces," in *Gas Transfer at Water Surfaces*, W. Brutsaert and G. H. Jirka, Eds. Springer Netherlands, 1984, pp. 67–82.

[8] null Herlina and G. H. Jirka, "Experiments on gas transfer at the air–water interface induced by oscillating grid turbulence," *J. Fluid Mech.*, vol. 594, pp. 183–208, Jan. 2008.

[9] E. A. Variano and E. A. Cowen, "Turbulent transport of a high-Schmidt-number scalar near an air–water interface," *J. Fluid Mech.*, vol. 731, pp. 259–287, Sep. 2013.

[10] J.-Y. Vinçont, S. Simoëns, M. Ayrault, and J. M. Wallace, "Passive scalar dispersion in a turbulent boundary layer from a line source at the wall and downstream of an obstacle," *J. Fluid Mech.*, vol. 424, pp. 127–167, Dec. 2000.

[11] K. R. Sreenivasan and C. M. White, "The onset of drag reduction by dilute polymer additives, and the maximum drag reduction asymptote," *J. Fluid Mech.*, vol. 409, pp. 149–164, Apr. 2000.

[12] M. Q. Nguyen, A. Delache, S. Simoëns, W. J. T. Bos, and M. EL Hajem, "Small scale dynamics of isotropic viscoelastic turbulence," *Phys. Rev. Fluids*, vol. 1, no. 8, p. 83301, Dec. 2016.

[13] T. D. Nguyen, "Influences des propriétés non-Newtoniennes sur un mélange de scalaire passif," Lyon, INSA, 2013, PhD Thesis.

[14] V. R. Ranade and J. J. Ulbrecht, "Influence of polymer additives on the gas-liquid mass transfer in stirred tanks," *AIChE J.*, vol. 24, no. 5, pp. 796–803, Sep. 1978.

498024 Influence of Thermophysical Properties of Cryogenic Fluids on Growth and Collapse of Cavitating Bubbles

Abstract:

Influence of Thermophysical Properties of Cryogenic Fluids on Growth and Collapse of Cavitating Bubbles

A distinctive feature of the dynamics of liquids is the possibility of the coexistence of a vapor phase with the liquid phase. Such formation and disappearance of a vapor phase in liquid are known as cavitating flow [Rayleigh, 1917]. Cavitating flow has been significantly governed by the physical-chemical properties of the liquid. The investigation needs to be performed to study the influence of thermo-physical properties of cryogenic fluid on the cavitating flows.

In Cavitating flow, plenty of nuclei available for the inception of cavitation. The fluid is in a precarious imbalance due to repetitive growth and collapse of cavitating bubbles until an equilibrium condition is achieved. Thus, the primary focus is on an exhaustive investigation of bubble growth and collapse. This event rate is dominated by two mechanisms: a momentum for the impulsion/expulsion of ambient liquid, and a heat transfer to initiate phase change at the interface. These mechanisms are known as the inertia (momentum) control and the heat transfer control, respectively. As the cavitating bubble grows, latent heat is extracted from the surrounding liquid to the interface, creating a thermal boundary layer. The consequence is a small local decrease of the liquid temperature, which results in a slight drop in the vapor pressure. This pressure depression suppresses the further development of the bubble because a greater pressure drop is needed to maintain the process. This phenomenon is known as 'thermal suppression of cavitation,' as it plays a moderating role in the development of cavitation. The influence of thermodynamic effects can usually be neglected in fluids for which the critical-point temperature is much higher than the working temperature. But these effects become significant when the critical-point temperature is close to the temperature of the fluid, as in the case of cryogenic fluids [Ruggeri & Moore, 1968].

Cryogenic fluids are characterized by steep density variation and a small latent heat of vaporization as compared to room temperature fluids. The vapor pressure of cryogenic liquid is highly sensitive to temperature change as shown in figure 1(b). As a consequence, the liquid surrounding the vapor cavity undergoes substantial evaporative cooling. [Balje, 1981].

Figure 1 Variation in thermophysical properties for Liquid Nitrogen (NIST: REFPROP)

(a) Liquid to vapor density ration vs Temperature and (b) Vapor pressure vs. Temperature

Previous studies have concentrated on the consequences of the thermodynamic effect rather than on the investigation of the mechanism itself. However, on the local scale, each bubble undergoes a considerable rise of temperature during the collapse while significant local cooling of the liquid is also expected within the cavitation areas. The behavior of bubble breakdown process is governed by different thermophysical properties of fluids viz. vapor pressure, viscosity, and surface tension. The thermodynamic effects associated with the dynamically changing interfacial structures occurs due to coalescence and break-up of bubbles, significant discontinuities of the fluid properties and complicated flow field near the interface results in due to the prevailing phase change process. The repeating transient collapses of cavitating bubbles in high-pressure regions influences the shape and stability of the bubbles, which can respond with the formation of a micro-jet. These re-entrant jets penetrate the bubble induce the process of cavitation erosion.

Therefore, for modeling inertial and heat transfer controlled bubble growth in cryogenic fluids due to multiphase malevolent cavitation process and to investigate the effect of thermo-physical properties of cryogenic fluids, demands a computational approach which can simultaneously capture different elementary physical phenomenon (i.e., nucleation, evaporation/condensation, coalescence/fission, bubble-turbulence interaction, and so on). In this paper, a multiphase formulation which solves energy equation in conjunction with the mass and momentum conservation that accounts for variable thermodynamic properties of the cryogenic fluid and nucleation transport equation has been used to investigate the fundamental dynamics of growing or collapsing spherical and non-spherical bubbles in an infinite domain of cryogenic liquid subjected to large pressure variations.

494012 Interfacial Activity of Silica and Ethyl Cellulose Particles in Fluid-Fluid Interfaces

Abstract:

Like surfactants, particles with appropriate wettability can adsorb at fluid-fluid interfaces. However, there has been a long-standing debate on whether particles can lower the interfacial tension in the same fashion as surfactants do. In our work, we used the pendant drop method to investigate the interfacial activity of isotropic silica nanoparticles at air-water and hexadecane-water interfaces and of ethyl cellulose particles at the trimethylolpropane trimethacrylate (TMPTMA)-water interface. Our results suggest that simple, non-amphiphilic particles can change the interfacial tension as long as they do adsorb (make physical contact with the interface) and have strong affinity to the interface (high desorption energy). We also proposed a simple model that does not explicitly account for any particle-particle interaction but is seen to predict surprisingly well the effective interfacial tension due to particle adsorption at fluid-fluid interfaces. Our work also provides a convenient way to estimate the packing density of particles in fluid-fluid interfaces by combining dynamic interfacial tension measurements with information about the particlesâ€™ wetting properties.

498571 Adhesion of Explosive Particles to Functionalized Surfaces

Abstract:

Adhesion of Explosive Particles to Functionalized Surfaces

Darby J. Hoss, Sanjoy Mukherjee, Bryan W. Boudouris, and Stephen P. Beaudoin

Charles D. Davidson School of Chemical Engineering, Purdue University, 480 Stadium Mall Drive West Lafayette, IN 47907-2100, USA

The improved elucidation of the adhesion behavior of energetic materials will improve munitions formulation and the performance of collection materials (*e.g.*, swabs or traps) used to detect trace explosive residue. Resolving the role that surface chemistry plays in the adhesion of energetic materials is a critical step towards the rational design of these materials. In particular, the use of specific chemical functional groups at the interface allows for the modulation of the adhesion force, and this key parameter will directly impact the performance and reliability of munitions and collection materials. In order to evaluate the effects of specific functional groups on adhesion, atomic force microscopy (AFM) pull-off force measurements were performed using nitrate-based energetic (and non-energetic) particles against self-assembled monolayers (SAMs) of representative chemical functionalities, and these SAMs on gold substrates were selected to evaluate surface chemistries due to their reproducibility and facile production. Gold films prepared by the template-stripped method produced sufficiently smooth surfaces for repeatable force measurements. Microparticles composed of either: (1) 2,4,6-Trinitrotoluene (TNT), (2) pentaerythritol tetranitrate (PETN), (3) 1,3,5-Trinitroperhydro-1,3,5-triazine (RDX), and (4) silica were evaluated against methyl, carboxyl, amine, and hydroquinone surface end-groups. In addition to the experimental results, stabilization energies for the optimized most-stable configurations for a coupled receptor-analyte system were determined using density functional theory (DFT).

For PETN and RDX, the adhesion force of the functional groups determined by AFM can be ordered as follows in terms of the strength of the adhesion force: amine > carboxyl > hydroquinone > methyl. This trend is in agreement with the trend calculated in DFT simulation. For TNT, the adhesion force of the functional groups determined by AFM can be ordered: amine > carboxyl > hydroquinone > methyl. However, DFT calculations showed the highest interaction with hydroquinone. This is due to a possible p-p stacking geometry between the hydroquinone and TNT that is readily accessible in the computational calculations; however, experimentally, and in practical applications, this well-aligned p-p stacking interaction is unlikely to occur. The electron donating amine functional group shows enhanced adhesion specific to the electron withdrawing nitro sites of the energetics relative to the other functional groups. Additionally, the methyl groups yielded the lowest adhesion force across all particles. Overall, this study suggests that electron donating functional groups are the most promising class of chemical moieties to increase adhesion to common energetics. Ultimately, these results will facilitate the rational design of energetic particle collection materials and munitions through chemical-tailoring in order to enhance performance.

511763 Dynamic Collision of Water Drops Coated with Food Grade Surfactant Poly-Glycerol-Poly-Ricinoleate (PGPR) in Canola Oil

Abstract:

The dynamic collision of emulsified water drops in the presence of non-ionic surfactants plays a crucial role in many practical applications. We measured the interaction forces between colliding water drops coated with food grade non-ionic surfactant PGPR (polyglycerol polyricinoleate) in Canola oil at concentrations well below typically used to form stable emulsions. PGPR is a fat soluble non-ionic surfactant widely used in chocolate industry as it is known to be the best emulsifier for stabilizing water-in-oil emulsions. Even though the use of PGPR is widespread in food industry, there is a lack of understanding of fundamental mechanisms (and forces) that contribute to the stability of emulsions. The experiments indicated that the interaction forces were dominated by steric forces. Interfacial tension measurement and adsorption isotherm fitting showed that the water-canola oil interface is almost saturated with PGPR for concentrations above 0.2 wt%. To probe meta-stability of emulsified water drops for concentrations below 0.2 wt%, we measured the interaction force as a function of PGPR concentration and velocity of collisions. The hysteresis between the approach and retraction curves increased with increase in the maximum force to which the water drops are pushed to. We observed that during the retract phase of drop collision there was a kink in the force curves and sudden jump out. We attributed such a force jump outs to bridging and extension of adsorbed PGPR molecules on water drops. In order to rule out the role of water transport in the form of PGPR reverse micelles to observed force jump outs, we carried out water drop collision experiments with 1 wt% salt in water. Salt was introduced to arrest the water transport across oil layer. However, we found similar force jump outs even in the presence of salt. This observation indicated that bridging and extension of PGPR molecules was the mechanism behind the force jumps observed during retraction phase of drop collision. The Stokes-Reynolds-Young-Laplace (SRYL) model was used to calculate theoretical force curves. For higher collisions velocities, the agreement between the calculated and experiment data was acceptable, but the SRYL model failed to describe the extension or kink feature observed at slower velocities below. Both the AFM data and the comparison to the model calculation indicated the presence of a short-range attractive force, not of a hydrodynamic origin, attributed to the bridging and extension of PGPR molecules on the surface of water drops below saturation of the interface.

493476 Design of New Liquid Crystal-Based Systems with Improved Chemo-Responsiveness Towards the Detection of Nerve Gases

Abstract:

Surface anchored nematic liquid crystals can respond to external chemical stimuli by changing their bulk orientation.[1] This orientation change is the foundation for detecting chemical analytes by measuring the orientational response of the liquid crystal. Several analytes have already been identified that can be detected by liquid crystals but organophosphates, especially dimethyl-methylphosphonate (DMMP), a simulant of sarin nerve gas, has become the most important test subject.[2,3] One of the most important efforts in this area is to increase the chemo-responsiveness of liquid crystal-based systems to detect analytes such as DMMP as fast as possible. We have recently developed computational models, based on quantum mechanics, to describe and predict the orientational transition of liquid crystal-based systems.[4] We have also adopted a combined approach using theory and experiments together to test theoretical models against experimental results and verify predictions by synthesis and characterization.[5]

In this presentation, we will show improved models and simple descriptors, which can predict anchoring behavior and orientational transition of nematic liquid crystals in good agreement with experimental results. Exploiting the simple descriptors we have executed a high throughput screening to find liquid crystal-based systems with improved chemo-responsiveness. We show that the modeled parameters of the liquid crystal-based systems are strongly correlated with each other. Therefore it is difficult to predict new and better systems based on limited number of experiments and wide-range, extensive screening is needed for that which can only be realized efficiently using theory. We have evaluated several surprising, non-intuitive theoretical predictions, which were subsequently verified by experiments. We present liquid crystal-based systems with even an order of magnitude better chemo-responsiveness than current state-of-the-art systems. These results prove that the close integration between theory and experiments is a very promising path for expediting the design of new liquid crystal-based chemo-responsive systems.

1. Shah, R. R.; Abbott, N. L., *Science*, **2001**, 293, 1296.
2. Hunter, J.T.; Abbott, N.L., *Applied Materials and Interfaces*, **2013**, 6, 2362
3. Yang, K. L.; Cadwell, K.; Abbott, N. L., *Journal of Physical Chemistry B*, **2004**, 108, 20180.

4. Roling L. T.; Scaranto, J.; Herron, J. A.; Yu, H.; Choi, S.; Abbott, N. L.; Mavrikakis, M., *Nature Communication*, **2016**, 7, 13338.

5. Szilvási, T.; Roling, L. T.; Yu, H.; Rai, P.; Choi, S.; Twieg, R. J.; Mavrikakis, M.; Abbott, N. L., *Chemistry of Materials*, **2017**, accepted.

499093 Factors Affecting the Transient Stability of Solid Stabilized Emulsions

Abstract:

Emulsions are commonly found in energy and pharmaceutical industries. The presence of solid particle is known to stabilize emulsions. Fundamental insight on the transient stability of concentrated emulsions stabilized using solid particles is critical to manage these processes. This work focuses on developing a fundamental relationship between the stability of emulsions and the interfacial properties. In this work, nuclear magnetic resonance (NMR) has been utilized to characterize concentrated emulsions over time. A series of experiments were conducted on emulsions prepared using model oils with well-characterized surfactants and solid nanoparticles to elucidate the impact of stabilizer type on dynamic emulsion behavior. It was found that the presence of salts enhanced the stability of water-in-oil emulsions stabilized using silica nanoparticles while an increase in pH of the water phase led to destabilization of solid stabilized water-in-oil emulsions. In addition, the effect of salinity and pH on oil-in-water emulsions was also investigated. Dynamic oilâ^'water interfacial tension and rheological measurements were performed to characterize the interfacial and rheological properties of concentrated water-in-oil emulsions stabilized using solid particles. In addition, the difference in interfacial and rheological properties of water-in-oil emulsions stabilized using asymmetric and completely spherical silica nanoparticles were investigated.

489936 Adsorption of Star Polymers to Fluid Interfaces

Abstract:

Star polymers and similar nanoparticulate brushes can be effective and efficient emulsifiers and foaming agents. When environmentally responsive polymer arms are incorporated in the star polymers, the resulting emulsions can be rendered environmentally responsive. These attributes suggest possibly advantageous application potential for star polymers in commercial formulations, yet little is understood about their fundamental interfacial characteristics and quantitative structure/activity relationships are only beginning to emerge for this novel class of surface active materials. This presentation considers adsorption of a variety of star polymers at fluid interfaces, including air/water interfaces and two types of oil/water interfaces, one representing a poor solvent / water interface and the other a good solvent / water interface. Interfacial tension reduction and dynamic dilatational elastic modulus are measured by pendant drop tensiometry methods, and the extent of adsorption is independently determined by ellipsometry. Both large, multi-arm star polymers and smaller star polymers containing 14 polymer arms per star are examined. For the latter, variants are considered where the arms are block copolymers that present nonionic blocks, weakly cationic pH-responsive blocks, and/or zwitterionic blocks. Star polymer morphology and composition have strong effects on interfacial packing and the resulting interfacial tension reduction and dilatational elasticity characteristics. Differences in the latter may correlate with emulsifying and/or foaming effectiveness for different star polymers.

494055 Influence of Order within Non-Polar Monolayers on Hydrophobic Interactions

Abstract:

This presentation will describe the results of an experimental investigation of the influence of molecular-level order (crystallinity) within non-polar monolayers on hydrophobic interactions. The measurements were performed using gold film-supported monolayers formed from alkanethiols ($CH_3(CH_2)_nSH$, with n ranging from 3 to 17), which we confirmed by using polarization-modulation infrared reflection-adsorption spectroscopy to exhibit chain-length-dependent order (methylene peak moves from 2926 to 2919 cm^{-1}, corresponding to a transition from liquid- to quasi-crystalline-like order) in the absence of substantial changes in chain-density (constant methyl peak intensity). By using monolayer-covered surfaces immersed in either aqueous triethanolamine (TEA, 10 mM, pH 7.0) or pure methanol, we quantified hydrophobic and van der Waals contributions to adhesive interactions between identical pairs of surfaces (measured using an atomic force microscope) as a function of the length and order of the aliphatic chains within the monolayers. In particular, we measured pull-off forces arising from hydrophobic adhesion to increase in a non-linear manner with chain-length (abrupt increase between $n = 5$ and 6 from 2.1 ± 0.3 nN to 14.1 ± 0.7 nN) and to correlate closely with a transition from a liquid-like to crystalline-like monolayer phase. In contrast, adhesion in methanol increased gradually with chain-length from 0.3 ± 0.1 nN to 3.2 ± 0.3 nN for $n=3$ to 7, and then did not change further with an increase in chain-length. These results lead to the hypothesis that order within non-polar monolayers influences hydrophobic interactions. Additional support for this hypothesis was obtained from measurements reported in this paper using long-chain alkanethiols (ordered) and alkenethiols (disordered). The results are placed into the context of recent spectroscopic studies of hydrogen bonding of water at non-polar monolayers. Overall, our study provides new insight into factors that influence hydrophobic interactions at non-polar monolayers.

497345 How Molecular Structure of Surfactants Determine the Dynamics and Viability of Wettability Alteration

Abstract:

How Molecular Structure of Surfactants Determine the Dynamics and Viability of Wettability Alteration

Only about 35% of oil in place is recovered from carbonate reservoirs through primary and secondary flooding because of oil wet surfaces and resulting unfavorable capillary pressures. Surfactants can improve oil recovery significantly by altering wettability of surfaces. However, the process of selecting an effective surfactant for wettability alteration is dependent on a number of factors, including mineral type, temperature, salinity, and nature of adsorbed oil and ultimately how the molecular structure of the surfactant interacts with all of these. Here we present an experimental study of the effectiveness of nonionic surfactants with different hydrophobic groups and different lengths of hydrophilic ethylene oxide oligomers. The surfactants selected alter the wettability of the rock primarily by acting on the water-rock and oil-rock interfaces and have minimal effect on the interfacial tension of the oil-water interface. The dynamics of wettability alteration is measured by the evolution of contact angles of oil-drops on initially oil-wet surfaces and film coverage for surfactant-induced dewetting of oil-films. The hydrophobic groups are found to determine which surfactants primarily act on rock interfaces. The rate and degree of wettability alteration are strongly controlled by the length of the ethylene oxide oligomers. Shorter oligomers lead to faster and greater wettability alteration and are least sensitive to the effects of temperature and salt concentration. Experimental observations are efficiently summarized by a few thermodynamic and kinetic parameters. A theory is presented to describe the dynamics and provide a framework to extract rate constants for the adsorption and desorption of the surfactants to the rock interfaces. These rate constants and the surfactants effect on contact angle are used in a pore-scale model for imbibition in the presence of surfactant to determine which surfactants structures are practically viable for wettability alteration in oil-wet carbonate reservoirs.

500734 Chiral Nonlinear Rheology of Phospholipid Monolayers

Abstract:

Chiral objects, which break mirror symmetry, range in scale from spiral galaxies to hurricanes to DNA. Macroscopic examples include snail shells , flowers and bacterial flagella. The molecular building blocks of biology are chiral, including 19 of 20 amino acids, nucleotides, and many sugars and lipids. Natural phospholipids have a chiral carbon in the sn-2 position and are exclusively R-enantiomers. This chirality is manifested in the counter-clockwise spiral patterns of condensed-phase domains in monolayers and bilayers, in which the chiral orientational order extends tens of microns. Here, we demonstrate that the mechanical response of two-dimensional, condensed-phase monolayers of 1,2-dipalmitoyl-sn-glycero-3-sn-phosphocholine (R-DPPC) is chiral as well, with nonlinear rheological properties that differ between monolayers sheared clockwise (C) or counter-clockwise (CC). The nonlinear elastic modulus and yield stress of R-DPPC monolayers are greater when sheared clockwise, i.e. against the natural CC winding of individual DPPC domains, than when sheared counter-clockwise. These differences disappear for racemic mixtures. Direct visualization of the deforming domains reveals the dramatic consequences of this chiral rheology. Domains fracture when sheared in the clockwise direction, but deform in a ductile, plastic fashion under strong counter-clockwise shearing. The macroscopic consequences of the underlying molecular chirality are remarkable given the single-component, non-cross-linked nature of DPPC monolayers.

Monolayer viscous and elastic properties are key to the stability, function and dynamics of lung surfactant monolayers. Normal breathing necessitates that the lung surfactant monolayer maintain a low surface tension as alveoli are inflated every few seconds during normal respiration, requiring that the surfactant rapidly spread to cover the expanding interface. At the same time, the lung surfactant monolayer must remain within the deep lung, despite the large surface tension gradients acting to pull it toward the trachea where the surface tension is higher. In technology, applications of foams, emulsions and dispersions can depend critically on the rheology of the surfactant monolayers used to stabilize them. On a more fundamental level, surface viscosity and elasticity can be exquisitely sensitive to molecular packing, correlation lengths, and interactions in monolayers that are not easily resolved in pressure-area isotherms or fluorescence images as we show here.

500577 Visualizing Nanoscopic Topography and Patterns in Freely Standing Thin Films

Abstract:

Thin liquid films containing micelles, nanoparticles, polyelectrolyte-surfactant complexes and smectic liquid crystals undergo thinning in a discontinuous, step-wise fashion. The discontinuous jumps in thickness are often characterized by quantifying changes in the intensity of reflected monochromatic light, modulated by thin film interference from a region of interest. Stratifying thin films exhibit a mosaic pattern in reflected white light microscopy, attributed to the coexistence of domains with various thicknesses, separated by steps. Using Interferometry Digital Imaging Optical Microscopy (IDIOM) protocols developed in the course of this study, we spatially resolve for the first time, the landscape of stratifying freestanding thin films. In particular, for thin films containing micelles of sodium dodecyl sulfate (SDS), discontinuous, thickness transitions with concentration-dependent steps of 5-25 nm are visualized and analyzed using IDIOM protocols. We distinguish nanoscopic rims, mesas and craters and show that the~non-flat features are sculpted by oscillatory, periodic, supramolecular structural forces that arise in confined fluids

489427 Heavy Oil and Bitumen Complex Viscosity Simulation â€“ Impacts of Phase Behavior on Rheology

Abstract:

Complex viscosity, including both elastic and viscous contributions to flow properties, is a key property for heavy oil and bitumen production, transport and refining process design. Values vary by more than eight orders of magnitude under industrially relevant conditions and have proven difficult to estimate. There is also a growing recognition that heavy hydrocarbon resources are multiphase and that they exhibit complex rheological behaviors, with shear rate, shear history, and shear method dependent rheological responses particularly at low temperatures. The complexity of their responses to shear along with speciation and rheological measurement reproducibility uncertainties present significant simulation challenges. In this work, an approach that builds on the understanding of the responses of polymers to steady and oscillatory shear, as well as expected responses of dispersed phases, is presented and applied to heavy oils and bitumen. The challenge in this work is to identify, to the greatest extent possible, common parameters applicable to hydrocarbon resources as diverse as Maya crude oil (Mexico), Athabasca bitumen (Canada), and Safaniya vacuum residue (Saudi Arabia) and their nanofiltered and chemically partitioned fractions with pentane asphaltene contents ranging from 0 to 100 wt. %, that are rooted in their known phase behaviors. A workflow resulting in semi-quantitative agreement between measured and computed complex viscosity outcomes over eight orders of magnitude is presented and discussed. Four hydrocarbon resource specific parameters, linked to solid-liquid transition boundaries for maltene-rich and asphaltene-rich fractions, are needed to describe rheological responses. Impacts of non-Newtonian effects are isolated.

488313 Maxwell and Marco: Thermophysical Property Needs of Rate-Based Process Simulation Tools

Abstract:

Process engineers have been using the well-known equilibrium stage column model for more than a century. More recently, some engineers, including the late Marco Satyro have adopted a more fundamental approach based on the use of the Maxwell-Stefan equations to describe mass transfer. However, any illusion that these so-called rate-based models do away with the need for thermodynamics should be dispelled immediately. In fact, the reverse is true: the need for accurate thermodynamic models is greater than ever with these models supplying not only first order properties such as equilibrium ratios and enthalpies, but second order properties as well since thermodynamics plays a key role in the prediction of diffusion coefficients, an essential component to rate-based models and an area where thermodynamics is somewhat less definitive. In addition, rate based models place increased emphasis on transport properties, something that is entirely irrelevant to a true equilibrium stage model (but is necessary for actual process equipment design). This talk presents one long time rate-based modelers perspective on what is needed for superior process simulation models to be developed.

488376 The Expanded Fluid Concept for Transport Property Correlations

Abstract:

The Expanded Fluid concept is straightforward: as a fluid expands, the distance between the molecules in the fluid increases and the properties which are related mainly to intermolecular forces change in a systematic manner that can be correlated. In other words, properties such as viscosity can be correlated to density. This simple principle is at the heart of many viscosity correlations such as the corresponding states viscosity model. The corresponding states approach, which uses the critical point as a reference, begins to fail for viscous fluids that are near their glass transition at ambient conditions such as heavy oils. The Expanded Fluid model, first proposed by Yarranton and Satyro (2009), uses a fitted compressed state density (the density at which the viscosity tends to infinity) as a reference point. The model is structured as a departure from an independently determined gas viscosity so that the complete fluid phase region is mapped; the critical point viscosity is predicted from the change in density.

In this contribution, the Expanded Fluid model is briefly reviewed including its pure component form and its application to mixtures and characterized crude oils. The Expanded Fluid principle is then applied to thermal conductivity to develop a new model based on the same compressed state densities used in the Expanded Fluid viscosity model. Similar mixing rules are developed and a similar extension to characterized oils. Finally, the relationship between diffusivity and viscosity is demonstrated to illustrate how the three transport properties, viscosity, thermal conductivity, and diffusivity, all relate to fluid expansion.

Reference: Yarranton, H.W., Satyro, M.A., *Ind. Eng. Chem. Res.*, 48(7), 2009, 3640-3648.

500012 Grid Evaluation of Pure-Compound Properties

Abstract:

Holistic and internally consistent evaluation of property data is the essential idea behind the NIST ThermoData Engine (TDE, Standard Reference Database 103b). Developed at the Thermodynamics Research Center (TRC), TDE enforces the thermodynamic consistency of all recommendations for a given compound, either implicitly with an equation of state or explicitly so models agree within uncertainties. However, evaluating the properties of a compound without considering the behaviors of related compounds creates a situation where recommendation quality become susceptible to reporting errors or natural experimental variability, as well as inconsistencies of prediction methods should method selection vary across different compounds. Discrepancies that exceed engineering need requirements are often observed between different prediction schemes or between experiments and predictions; thus heterogenous data scenarios can interfere with good chemical design.

An automated method for identifying inconsistencies in a family of compounds has been developed. It is based on building of all possible series involving the compound of interest, where the same change in a molecular structure is repeated. As the analysis of a molecular structure is complex, descriptors expressing molecular structure are used as surrogate variables. For example, such a descriptor might be a decomposition for a group-contribution prediction method or a component used in a molecular structure similarity assessment. Then a one-dimension data analysis is applied to each pair of vectors representing the molecular descriptor and property value. Extending this analysis to multiple dimensions is also possible.

Once this web of relationships has been defined, analysis can be performed for any property that is expected to show regular variation with structure (with exceptions for smallest members and odd-even effects). For that purpose, a grid involving compounds with the most reliable values, as defined by expert analysis, can be built and gradually extended with the use of predicted values if necessary. The whole grid can then be iteratively refined as confidence can be gained in newly analyzed data. The properties of which consistency is most important are critical parameters and normal boiling points or acentric factors. The proposed method correlates with previously published work, where reparameterization of prediction methods is performed for each compound on the basis of the data for the set of most similar compounds with known properties. Its application should increase the reliability and stability of property data evaluation for pure compounds.

495496 Water Content of Natural Gas Systems in Equilibrium with an Aqueous or a Hydrate Phase â€“ Experimental Measurements and Molecular Modeling

Abstract:

Accurate determination of hydrate formation conditions and water content of natural gas systems is critical for cryogenic processes used in the production of natural gas liquids (NGL), liquefied natural gas (LNG) processing and carbon dioxide (CO_2) capture. Water content of three sweet and sour natural gas mixtures was measured using a Panametrics moisture analyzer at low temperatures and pressures up to 40 MPa. A molecular model based on the statistical associating fluid theory (SAFT) was developed to predict water content of pure alkanes, carbon dioxide (CO_2), hydrogen sulfide (H_2S), nitrous oxide (N_2O), nitrogen (N_2) and argon (Ar). The model application was also extended to model sour natural gas mixtures in equilibrium with an aqueous phase. Finally, the SAFT-based model was combined with a modified van der Waals and Platteeuw (vdWP) theory to predict water content of systems in equilibrium with hydrate. The theory was found to be in good agreement with experimental data found over a wide range of conditions, thus exhibiting the predictive ability of the proposed model.

499230 When Experimental and Predicted Data Are in Conflict, What Should We Trust?

Abstract:

Process simulation has become an essential part of development, optimization, and risk assessment in chemical engineering. To build adequate computer models of chemical, separation, and transport processes, engineers need reliable property data. Process simulations are usually developed on the basis of experimental results, and predicted values are involved when direct experimental data are missing. It is common that discrepancies between experimental and predicted values exceed the claimed uncertainties; it has long been an axiom that experimental measurements are fundamentally more trustworthy and experimental errors consistently lie within the boundaries of typical uncertainties for particular methods and samples. However, statistical analysis of modern publications shows a high probability of large deviations from expected behavior (*e.g.*, comparison with literature data or trends for similar compounds, self-consistency within reported data) or questionable experimental methods that make many published data suspect, adoption of which can lead engineers to draw misleading or even dangerous conclusions from process simulations. In general, chemical engineers should realize that a published experimental value cannot be blindly trusted. Additional comparison frequently reveals that even smooth data in self-consistent reports may be erroneous. On the other hand, rigorously developed prediction schemes are usually developed to follow patterns in chemical series. While each prediction method has its bias arising from the underlying physical model and parameterization used and should only be applied within the chemical domain it was developed for, it represents the aggregate of a range of experimental results and theoretical expectations. Therefore, predictions are often a more conservative choice and robust against crude errors. It is also much easier to verify if a prediction procedure was correctly followed and to repeat the analysis. However, most prediction methods are ultimately based on experimental data and depend on their quality. Consequently, neither experimental nor predicted data should be considered in a vacuum, and only through mutual validation of experimental and predicted data can insight regarding true values be gained. The activities of the Thermodynamic Research Center (TRC) at NIST include collecting and evaluating published experimental property data as well as development and validation of property prediction methods. Quality assessment of newly measured data, which is implemented at TRC, is a complex process involving analysis of experimental details (sample, method, instrumentation, *etc.*), validation against independent additional measurements (preferably measured with a different method), consistency between different properties, and agreement with theoretical calculations or correlation with similar compounds. This detailed analysis requires a comprehensive property database and a variety of computational and correlation methods individual for each property. Comprehensive coverage of the entire scope of properties required by chemical engineers is a challenging goal, but with the TRC constantly increasing its data archive and continuing its development of new validation procedures, that goal grows more achievable each day. Typical data scenarios, the current coverage by TRC procedures, future plans, and some suggestions for practical engineers will be discussed.

495463 Global Strain-Field Mapping of a Carbon Nanotube-Laden Interface Using Digital Image Correlation

Abstract:

Identifying the correct stress-strain relationship experimentally is important to understanding the mechanical response of an interface and provides the basis for the theoretical development and experimental validation of any constitutive models. Langmuir-Pockels (LP) trough is one of the most commonly used tools for studying an interface. In a typical LP trough experiment, as the interface is compressed by a pair of barriers, a Wilhelmy microbalance is used to measure the corresponding "surface pressure". However, the as-measured surface pressure is based on a vertical force balance and thus contains both surface energy and rheological contributions. Decoupling these contributions is non-trivial. Further, despite the relatively simple experimental setup, a mixed deformation field is created, further complicating the interpretation of the experimental results. Most, if not all, existing studies assume a 1D uniaxial compression during a LP-trough compression experiment. To examine this assumption, we custom-built a glass-bottomed LP trough equipped with a camera to capture a series of optical images as an interface is compressed. Carbon nanotubes (CNTs) were chosen as the model system as they formed a "speckle pattern" when spread onto an air-water interface. Based on the change in this speckle pattern, the displacement and strain fields were calculated using digital image correlation (DIC) analysis. Our experimental findings clearly show, for the first time, the development of a non-uniform and complex 2D strain field during compression. Although the compressive strain averaged over the whole trough area closely resembles the 1D uniaxial compression strain, the 1D compression assumption underestimates the local strain by about 36% at a compression area of 25 cm^2 at the center of the trough, where the surface stresses are measured. The DIC-corrected strain data were subsequently analyzed with the surface stress data to quantify the surface shear and dilatational moduli of the CNT-laden interface. This is the first study in applying the DIC technique to map out the global strain field as a particle-laden interface is compressed. The method may also be applicable to other systems with similar optical texture, allowing the correct identification of stress-strain relationship of an interface. This work is supported by NSF Career Award (# 1253613), GE fellowship, and Anton Paar fellowship.

500880 Simulations of Transient Dynamics of Dense Suspensions

Abstract:

Shear thickening of dense suspensions has been intensively studied due to its importance in many industrial applications. However, less is known about the transient phenomena such as impact activated solidification or dynamic shear jamming. Here we rely on LAMMPS-based discrete particle simulations to understand the dynamics of particulate suspensions, which are comprised of bi-disperse non-Brownian spherical particles at high volume fraction. Near-field hydrodynamic forces, short-range repulsions and frictional forces are considered in our model. As the volume fraction increases, we observe a transition from continuous shear thickening (CST) to discontinuous shear thickening (DST), which is consistent with experimental observations. We then apply our simulation tool to study dynamic shear jamming of dense suspensions in a quasi-one dimensional system. The particles are confined by two solid boundaries (top and bottom) that are sufficiently separated, and periodic boundary conditions are applied in the other directions. At the beginning of the simulation both boundaries are stationary, and then suddenly the bottom boundary starts to move with a constant speed. As the boundary moves, we observe a propagating front that moves with a constant speed, and the measured front speed is consistent with that reported in a recent experiment. We further investigate how the frictional force and the short-range inter-particle force determine the formation of the jamming front and its velocity. Our simulations provide detailed structures of the suspension at transient state, and are therefore helpful for elucidating the transient dynamics of dense suspensions.

491492 Interfacial Rheology of Coexisting Solid and Fluid Monolayers

Abstract:

Biologically relevant monolayer and bilayer films often consist of micron-scale high viscosity domains in a continuous low viscosity matrix. Here we show that this morphology can cause the overall monolayer fluidity to vary by orders of magnitude over a limited range of monolayer composition. Modeling the system as a two-dimensional suspension in analogy to classic three-dimensional suspensions of hard spheres in a liquid solvent explains the rheological data with no adjustable parameters. In monolayers with ordered, highly viscous domains dispersed in a continuous low viscosity matrix, the surface viscosity increases as a power law with the area fraction of viscous domains. Changing the phase of the continuous matrix from a disordered fluid phase to a more ordered, condensed phase dramatically changes the overall monolayer viscosity. Small changes in the domain density and/or continuous matrix composition can alter the monolayer viscosity by orders of magnitude.

500955 Modulation of Dilatational Rheology with Interfacial Curvature and Phase Morphology

Abstract:

Adsorption of complex molecular or particulate species on curved interfaces of bubbles, foams or drops can render the interface into a complex two-dimensional (2D) viscoelastic fluid. Here, we show that the dilatational viscoelastic response of an insoluble 2-D fluid film of Survanta surfactant is influenced by the curvature of interface, phase-morphology of the film and its ageing time. Survanta is a clinically approved multicomponent natural lung surfactant, which is used for treating fatal neonatal respiratory distress syndrome in preterm babies. We investigated the interfacial rheological behavior of Survanta surfactant adsorbed and coated on curved air-aqueous interfaces created in our in-house built micro-bubble dilatational rheometer, where we apply pure dilatational, small amplitude, oscillatory surface area deformations on the interface at frequencies 0.25 to 0.5 Hz near equilibrium regime and calculate complex dilatational modulus 'ε' of the interfacial film. We find that the dilatational modulus of 2-D fluid film of Survanta connected with the bulk possessing surfactant vesicles goes up surprisingly as we increase the curvature of (bubble) interface from ~0.003 μm^{-1} to 0.01 μm^{-1}. Our unique approach of combining 3-D imaging capabilities of confocal fluorescence microscopy with the dilatational rheometer setup reveals that the Survanta film which shows coexisting 'solid-like' ordered, circular domains flowing easily in the disordered, 'liquid-like' phase on a low curvature interface (~0.003 μm^{-1}) shifts its phase morphology to ordered phase organizing into a rigid, immobile framework of stripes with liquid phase organizing into isolated discontinuous patches at relatively higher curvature bubble interface (~0.01 μm^{-1}). Modulation of dynamics with the background curvature and phase morphology is found reversible up to certain extent. Further, ageing of the film at equilibrium yields increasing values of complex dilatational modulus and elastic modulus, and is found associated with the area fraction of solid phase being altered at equilibrium.

These findings on this complex 2D fluid film can be extrapolated to understand rheology of complex 3D fluids as well as how lung surfactants might be involved in avoiding Laplace instability among alveoli of different radii. Moreover, it offers insights for creating smart interfacial films of technological and commercial importance.

500405 Surface Forces, Flows and Fluxes Underlying Nanoridge Formation and Instabilities in Stratifying, Micellar Freestanding Films

Abstract:

Controlling and predicting the stability and lifetime of freestanding films, including foam and emulsion films, is crucial for many industrial and biological applications. Freestanding films (thickness < 100 nm), stabilized by surfactants above the critical micelle concentration, exhibit stratification or stepwise thinning. Stratification proceeds by formation of thinner domains that grow at the expense of surrounding films. In this article, we address several longstanding challenges related to the experimental characterization and theoretical description of thickness variations, forces, fluxes and flows underlying stratification. We show that nanoridges form and grow at the moving front around expanding domains, and we visualize their shape evolution using interferometry, digital imaging and optical microscopy (IDIOM) protocols with an unprecedented spatio-temporal resolution (thickness < 10 nm, time < 1 ms). We develop a theoretical model for drainage via stratification under the influence of supramolecular oscillatory surface forces arising from the confinement-induced layering of micelles, and we show that the nanoridge growth and domains expansion dynamics can be modeled quantitatively. Using IDIOM protocols, we also visualize the break-up of these nanoridges into mesas and describe how instability growth and evolution is determined by the interplay of capillarity and disjoining pressure.

488568 Wrinkling Instabilities in Thin Inhomogenously Stretched Viscous Sheets

Abstract:

Ultra-thin glass sheets with thicknesses that are < 100 μm, a necessity for many modern haptic technologies, are commonly manufactured by the glass redraw process. In this technique, molten glass is simultaneously heated in a furnace and drawn via a tensile force to obtain thin viscous fluid sheets of a required thickness. However, an injudicious choice of stretching rate or applied heating profile can give rise to undesired out-of-plane buckling and wrinkling instabilities that adversely affect the flatness of the sheet.

Here, we present a mathematical framework to characterize the shape and stability of a thin viscous fluid sheet that is inhomogeneously stretched in an imposed non-uniform temperature field. We determine that the sheet can become unstable in two regions that are upstream and downstream of the heating zone, and where the minimum in-plane stress is negative. We characterize the shape and growth rates of the most unstable modes in both regions for various operating conditions, and show that the behavior of the instability changes from a stationary mode to a pair of unstable traveling modes upon varying the heating zone width, corresponding to a transition from a stationary bifurcation to a Hopf bifurcation.

Finally, we investigate the effect of surface tension and show that the wrinkling instability can be entirely suppressed when the surface tension is large enough relative to the magnitude of the in-plane stress. We present an operating diagram that indicates regions of the parameter space that result in a required outlet sheet thickness upon stretching, while simultaneously minimizing or suppressing the out-of-plane buckling; a result that is relevant for the glass redraw method used to create ultrathin glass sheets.

489184 Fluid Instabilities, Density Stratification, and Lee Waves That Result in Growth Limitations and Morphological Instability of Liquid-Crystal Interfaces

Abstract:

It has long been known that bulk crystal growth processes are strongly affected by fluid flows arising from radial temperature and density gradients. In this presentation, we present new insight to flows that arise from the interactions of a destabilizing radial temperature profile and a stabilizing vertical gradient. This interaction explains a decades-long issue that has plagued the growth of the II-VI semiconductor, cadmium telluride, from a liquid solvent in a system known as the traveling heater method (THM).

Historically, the THM has suffered from growth rates that are an order of magnitude slower than other directional solidification techniques. We demonstrate that the growth rate limit in the THM system is intrinsically tied to a phenomenon known as lee waves, most typically observed in atmospheric flows. On the downwind side of a mountain, the denser air at low altitude acts as a spring to resist the downward flow of air coming over the mountain. The spring-like effect causes the flow to enter into an oscillation that can extend hundreds of miles downwind of the mountain itself. These standing waves are then called lee waves due to their downwind position and can often be observed in the formation of wave clouds that appear to hover motionless at regular intervals.

In our analysis of flows in the THM, we examine the role of buoyancy, not only as the primary driving force for flow in the system but also through its stabilizing influence that results in the creation of lee waves. First, we present results from a finite-element model of the THM that demonstrate the strong coupling between mass, momentum, and heat transport. Model predictions provide quantitative evidence for the formation of lee waves in the liquid zone of the THM.

Furthermore, we show how these lee waves lead to anomalous mass transfer in the melt zone, promoting excess solvent underneath the peak of the standing wave at the phase change interface. These regions of high solvent concentration depress the local melting temperature via conitutional effects, leading to undercooling of the liquid. We then discuss how the linear stability analysis of Mullins and Sekerka [1] predicts that this undercooling will cause the normally stable, planar phase-change interface to become unstable and develop a cellular structure. Thus, the outcome of the lee-wave structure is that stable growth is limited to rates far less than expected. Additionally, we will discuss possible process improvements that could either eliminate or help manage the undercooling in front of the solidification interface.

[1] W. W. Mullins and R. F. Sekerka, "Stability of a planar interface during solidification of a dilute binary alloy", *Journal of Applied Physics* **35**(2), pp. 444–451 (1964).

This work has been supported in part by the National Science Foundation, under DMR-1007885, and no

official endorsement should be inferred.

484941 Liquid-Film Coating on Topographically Patterned Rotating Cylinders

Abstract:

The coating of discrete objects having surface topography is an important step in the manufacturing of a broad variety of products. To develop fundamental understanding of this problem, we study liquid-film flow on rotating cylinders patterned with sinusoidal topographical features. The Stokes equations, augmented with a term accounting for centrifugal forces, are solved in a rotating reference frame using the Galerkin finite-element method (GFEM). A nonlinear evolution equation for the film thickness based on lubrication theory is also solved numerically and its predictions are compared to those from the GFEM calculations. When gravitational effects are negligible and the rotation rate is sufficiently low, liquid accumulates over the pattern troughs before merging to form multiple larger drops (located over troughs) whose number at steady state depends on the topography wavelength and rotation rate. When the rotation rate is sufficiently high, similar merging events occur but liquid accumulates over the pattern crests at steady state. When gravitational forces become significant, it is possible to obtain a coating that closely conforms to the surface topography. The GFEM calculations are in agreement with predictions from the lubrication model provided that the free-surface curvatures are sufficiently small. For sufficiently large pattern amplitude and film thickness, the GFEM calculations show that recirculation regions inside the troughs can appear and vanish as the cylinder rotates due to the variation of gravitational forces around the cylinder surface. This phenomenon, along with flow reversal over the crests, may strongly influence mixing, mass transport, and heat transport (W. Li, M. S. Carvalho, and S. Kumar, Phys. Rev. Fluids 2 (2017) 024001).

484946 Drying of Multicomponent Thin Films on Substrates with Topography

Abstract:

Drying of multicomponent thin liquid films is an important step in coating and printing processes. In many cases, the substrate on which the film rests may possess topography, either intended or unintended. We present a lubrication-theory-based model describing the fundamentals of drying on such substrates. The film consists of volatile solvent and additional non-volatile components such as colloidal particles, surfactants, and non-volatile solvents. A system of one-dimensional partial differential equations accounting for the film height, depth-averaged concentration of bulk non-volatile components, and interfacial concentration of insoluble surfactant is derived. Evaporation is included using the well-known one-sided description, and the governing equations are solved with finite-difference methods to study various limiting cases. The results highlight the influence of evaporation rate, and thermal, surfactant, and solutal Marangoni flows on the film thickness and colloidal particle distribution. We find that in a realistic region of parameter space, the addition of a non-volatile solvent yields a film that conforms to the substrate topography.

499975 Effect of Elasticity on Stability of Viscoelastic Liquid Curtain

Abstract:

Curtain coating is one the preferred methods for high-speed precision application of single-layer and multi-layer coatings in industry. Despite the extensive variety of applications of curtain coating, its operation is challenging and uniform coating is only obtained in a certain range of operating parameters, called the coating window. The two main physical mechanisms that limit curtain coating are the breakup of the liquid curtain, below a critical flow rate, and the catastrophic event of air entrainment, which occurs above a certain web speed. The rheological characteristics of the coating liquid play an important role on these mechanisms, but the fundamental understanding of the role of rheology is still not complete. In this work, we analyze the relative importance of shear and extensional viscosity on both curtain breakup and dynamic contact line instability (i.e. air entrainment). Aqueous solutions of polyethylene oxide (PEO) and polyethylene glycol (PEG) of different molecular weights were used as model liquids to obtain fluids with different levels of extensional thickening behavior.

500135 Flow, Arrest and Yielding in Dense Colloidal Suspensions â€“ Glasses Vs. Gels

Abstract:

Rheology modification in complex fluids is important for a wide range of applications including formulation of engineering fluids and consumer products. The ways in which such fluids evolve after cessation of flow and during yielding under flow have important implications for the properties and performance of the fluid in its intended application. Here we consider flow properties, dynamical arrest, and yielding in systems with repulsive and attractive interactions, i.e. systems which form soft colloidal glasses and colloidal gels, respectively. Rheological studies chart the evolution from metastable fluid-like states into arrested solid-like states as a function of time following cessation of flow, or a thermal quench. We observe distinct signatures that are associated with dynamical arrest in both cases. For colloidal glasses, arrest occurs due to a sudden onset of elasticity in which the viscosity of the system goes through a maximum as the elastic contribution becomes dominant at large times. Notably, the temporal width of the peak in the viscous modulus of the system is a strong function of stress, which signals that the stress modifies the distribution of arrest timescales in the system. In colloidal gels, arrest proceeds via the formation of a critical gel. The scaling exponents describing the approach to the critical state can be captured through a systematic set of time-resolved measurement, revealing strong similarities to the critical behavior of many commonly studied polymer gels. We examine the yielding response in a model suspension as the system is deliberately tuned from a soft colloidal glass to an attractive colloidal gel by adjusting ionic strength. The colloidal glass is marked by rapid recovery of rheological properties following yielding, whereas the gel is subject to much longer transients and exhibits path-dependence in its rheological properties due to the sensitivity of the aggregation state to shear flow. The implications of dynamical arrest and yielding are discussed in the context of formulating complex fluids and rheology modification for various applications. A novel approach to rheology modification is discussed, based on the deliberate design of systems featuring hierarchical dynamics.

494733 Modeling a Hydrodynamic Instability in Freely Settling Colloidal Gels

Abstract:

Attractive colloidal dispersions, suspensions of fine particles which aggregate and frequently form a space spanning elastic gel are ubiquitous materials used as thermal insulators, catalytic electrodes in fuel cells, and in many consumer care products and food stuffs. The colloidal networks in these materials can exist in a mode of free settling when the network weight exceeds its compressive yield stress. An equivalent state occurs when the network is held fixed in place and used as a filter through which the suspending fluid is pumped. In either scenario, hydrodynamic instabilities leading to loss of network integrity occur. Recent observations using ghost particle velocimetry [Secchi et al., Soft Matter, 2014] have shown that the loss of integrity is associated with the formation of eroded channels (so-called streamers) through which the fluid flows rapidly. In this talk, we use Brownian dynamics simulations of sedimenting and hydrodynamically interacting colloids in dilute colloidal gels to examine the initiation and propagation of this instability. In simulations, we measure the evolution of the network settling rate and identify a critical point in time beyond which the velocity grows rapidly. The rapid increase of the streamer volume in the network is shown to coincide with increasing settling rate. A simple phenomenological model is developed that describes dynamically the radial growth of a streamer due to erosion of the network by rapid fluid back flow. The model exhibits a finite-time blow-up - the onset of catastrophic failure in the gel - due to activated breaking of the inter-colloid bonds. The model relates blow-up time to the relevant dimensionless groups describing the network including: the ratio of buoyant forces to network strength, the particle volume fraction, and the strength of inter-particle bonds relative to the thermal forces acting on the particles. Simulations are used to test the model and show good agreement with the predicted dependence of the blow-up time on the physical properties of the network. The model dynamics are also shown to accurately replicate ghost particle velocimetry measurements of streamer growth. Finally, we summarize our findings in a stability-state diagram that provides insight into engineering strategies for avoiding settling instabilities in networks meant to have long shelf-lives.

501092 Effects of Attractive Interparticle Interactions on Normal Stresses and Shear-Induced Migration of Colloids

Abstract:

Particles in flowing suspensions can move across streamlines to form regions with concentrations higher than the bulk through shear-induced migration.[1, 2] This phenomenon has been observed in pipe or channel flow of suspensions, and during rheological measurements in Couette and cone-and-plate geometries. Controlling the microstructure of particles during flow hence requires understanding the mechanisms that drive particle migration. For suspensions of hard sphere particles, the migration is thought to arise from normal stress differences in the sample.[2] How attractive interactions between particles affect the normal stresses and/or the shear-induced migration is less studied. Here, we developed a model system in which to systematically measure normal stresses in colloid-polymer depletion mixtures, which are commonly used as simple models of attractive suspensions. We synthesized colloidal particles of a copolymer of trifluoroethyl methacrylate and tert-butyl methacrylate,[3] which were nearly refractive-index and density-matched to 80 wt% glycerol in water with Tris buffer added to control the pH. To controllably vary the strength of attraction between the particles, we added poly(acrylic acid) at three concentrations to induce depletion interactions. The first normal stress differences, measured for suspensions formulated at constant particle volume fraction Î¦ = 0.40 and various concentrations of polymer, approximately represent the contributions of the polymer and colloidal suspensions. We also examined the resulting migration of the particles in a cone-and-plate geometry after rheology measurements at high shear rates to determine the effect of normal stress differences on the migration of attractive particles. These measurements provide insight into the physical processes underlying migration in complex mixtures, which is of practical significance for controlling microstructure during flow-based processing routes in particulate suspensions.

1. D. Leighton and A. Acrivos, *Journal of Fluid Mechanics*, 1987, **181**, 415-439.

2. M. Frank, D. Anderson, E. R. Weeks and J. F. Morris, *Journal of Fluid Mechanics*, 2003, **493**, 363-378.

3. T. E. Kodger, R. E. Guerra and J. Sprakel, *Scientific Rep*orts 2015, **5**, 14635.

501287 Connecting Bidisperse and Polydisperse Suspension Rheology

Abstract:

In this work, we consider spherical particles which are distributed in size, as often seen in practical applications involving suspensions. Polydisperse suspensions display lower viscosities as compared to monodisperse suspensions at the same solid volume fraction. The reduced viscosity associated with polydispersity can be explained by an increase with dispersion in particle size of the jamming fraction, or maximum flowable fraction, of solids, Î¦$_m$. In fact, we show by numerical simulation that the classical Krieger viscosity relationship in which the effective viscosity diverges as (1-Î¦/Î¦$_m$)$^{-2}$ provides a collapse of data for suspensions having contrasting particle size distributions. Prior work [1] has shown that statistically equivalent bidisperse and normal and log-normal suspensions (to third moment of the size distribution) have similar Î¦$_m$. Using these arguments, this study draws parallels between flowing bidisperse and higher order polydisperse suspensions in the high shear limit demonstrating close agreement of both viscosity and normal stress response. Polydisperse suspensions and the complexities associated with understanding them can hence in theory be replaced by relatively well-understood bidisperse suspensions of equivalent rheology.

[1] Desmond, Kenneth W., and Eric R. Weeks. "Influence of particle size distribution on random close packing of spheres." *Physical Review E* 90.2 (2014): 022204.

491136 The Steric Effect of a Solid Boundary on the Orientation Distribution of a Dilute Suspension of Rigid Rod-like Particles Under Shear Flow

Abstract:

We performed numerical analysis to study the orientation distribution of thin rigid rod-like particles under shearing flow near a solid boundary. The focus of this study is the effect of excluded volume effect under various flow conditions. More specifically, the channel height is assumed to be larger than a rod length where the center-of-mass position of a rod from a wall is changed after each collision. Therefore, we classified the rod collision events according to the shear and the Brownian contributions to obtain the rod distribution correctly. Comapring with similar previous studies, which only considered 2D rotation and additional explicit parameter for the rod collision events, our approach is improved by considering the rod rotation in 3D without any adjustable parameters. The results, which will be presented, are the orientation distributions in terms of various angles, average orientation moments, and Jeffery orbit constants. We will also discuss the relation between our results and the average diffusivity, the center-of-mass distributions of rods in a bounded shear flow, and application to shape-based particle separations.

501478 Particle Crystallinity in Automated Langmuir-Blodgett Deposition: Effect of Speed and Concentration

Abstract:

The promise of scalable nanomanufacturing is to translate the ability to create nanoscale functionality, often in the form of 3D nanostructures, to processes that provide industrial-scale qualtities of material. Few efforts have succeeded in this effort. In the laboratory, colloidal crystals have long been used as templates for well-ordered structures enabling control of mass, heat, photon, and electron transport at the nanoscale. Automated Langmuir-Blodgett (ALB) roll-to-roll coating of particles into well-ordered monolayers that can be used to fabricate colloidal crystals. Previous work has characterized the process parameters with regard to coating speed and general coating morphology. This work directly measures the crystallinity of micron-scale microsphere coatings and correlates their structure back to the process parameters, primarily coating speed, mass delivery rate, and underflow recirculation rate. In addition, the forumlation of the suspension in the addition of nanoparticles much smaller than the coated particles is altered based on observation of coating enhancement in laboratory scale convective deposition. Overall, slower speed coatings and the addition of nanoparticles roughly 10:1 in radius to the microspheres tends to increase the crystallinity of the colloidal microstructure. The relationship of process parameters to structure are somewhat clear, however the role of nanoparticles in enhancing crystallization is confusing because only a small fraction of nanoparticles from suspension are translated to the final coating.

502558 A Hybrid Lattice Boltzmann-Random Walk Method for Multiphase Heat Transfer

Abstract:

The transport phenomena within multiphase flows plays a key role in the operation of catalytic reactors, calciners, fluidized beds, and porous media. However, a fundamental understanding of the physics within these complicated systems still remains elusive. Direct numerical simulation has been shown to be an effective method for obtaining high fidelity predictions of the heat transfer within multiphase flows. Most commonly, particle-resolved immersed boundary [4, 5] or thermal lattice Boltzmann [3] methods are employed. Here we utilize a hybrid framework where the velocity field is resolved by the lattice Boltzmann method and temperature field by random walk particle tracking (Brownian tracers). Due to the simplicity of the algorithm and boundary conditions, random walk particle tracking offers an efficient means of simulating scalar transport within complex geometries. However, interfaces pose a significant challenge to the random walk framework [1, 2], due to the discontinuous change in material properties – e.g., a particle and fluid with different diffusivities. We derive a technique for handling discontinuities in the diffusivity, arising at a particle-fluid interface, and implement said method within the tracer algorithm. Verification of the code is completed against a host of systems: dispersion within a stagnant particle-fluid assembly, a hot sphere in contact with an unbounded fluid, a cooling sphere in contact with an unbounded fluid, and unbounded flow past a hot sphere. Predictions made by the new code are found to be in agreement with analytical solutions and previous works.

[1] Bekri, S., Adler, P.M., 2002. Dispersion in multiphase flow through porous media. International Journal of Multiphase Flow 28, 665–697. doi:10.1016/S0301-9322(01)00089-1

[2] Hoteit, H., Mose, R., Younes, A., Lehmann, F., Ackerer, P., 2002. Three-Dimensional Modeling of Mass Transfer in Porous Media Using the Mixed Hybrid Finite Elements and the Random-Walk Methods. Mathematical Geology 34, 435–456. doi:10.1023/A:1015083111971

[3] Kruggel-Emden, H., Kravets, B., Suryanarayana, M.K., Jasevicius, R., 2016. Direct numerical simulation of coupled fluid flow and heat transfer for single particles and particle packings by a LBM-approach. Powder Technology 294, 236–251. doi:10.1016/j.powtec.2016.02.038

[4] Sun, B., Tenneti, S., Subramaniam, S., 2015. Modeling average gas–solid heat transfer using particle-resolved direct numerical simulation. International Journal of Heat and Mass Transfer 86, 898–913. doi:10.1016/j.ijheatmasstransfer.2015.03.046

[5] Tavassoli, H., Kriebitzsch, S.H.L., van der Hoef, M.A., Peters, E.A.J.F., Kuipers, J.A.M., 2013. Direct numerical simulation of particulate flow with heat transfer. International Journal of Multiphase Flow 57, 29–37. doi:10.1016/j.ijmultiphaseflow.2013.06.009

494427 Separation By Design - Towards Simulation Guided Engineering of Coiled Channels for Precise Particle Separation

Abstract:

Coiling a straight channel around an axis, i.e. forming a torus, gives rise to additional fluid motion, changing the trajectories of suspended particles. Most significant differences between straight and coiled channel flow are (i) the deflection of the velocity maximum to the outer bent, and (ii) the occurrence of secondary motion within the cross-sectional plane caused by centrifugal forces. Consequently, suspended particles are subject to a non-trivial interplay of various hydrodynamic forces, most notably drag and lift, resulting in an interesting orbiting behavior of suspended particles. Most important, such behavior can be used for size-based particle separation or fractionation (Yoon 2009, Di Carlo 2009). Previous studies found the process to be sensitive to the Reynolds number, and the channel's curvature, as well as the channel cross sectional shape. However, most studies followed a purely experimental pathway, and captured the particle motion by imaging methods (Guan 2013, Martel 2013). Unfortunately, experimental methods do not allow the precise measurement of certain flow details (e.g., hydrodynamic forces acting on the particles) and hence often fail to identify the true mechanism that leads to, e.g., particle separation. To highlight on such mechanisms, we follow a bottom-up approach focusing on the hydrodynamic forces acting on particles of different size. We use Euler-Lagrange simulations to quantify the relevance of these forces, and how they affect particle orbiting behavior.

The flow field was directly simulated using OpenFOAM® (Greenshields, 2016). Particles were then placed into the developed fluid flow field. Fluid-particle forces were calculated using CFDEM®coupling (www.cfdem.com), and the particle motion was calculated using LIGGGHTS® (Kloss 2012). In a sub-set of numerical experiments, we performed computationally extremely demanding particle-resolved suspension flow simulations to guide our choice of proper hydrodynamic interaction models with a special focus on lift forces. An array of particle-unresolved simulations was performed to study cases of different channel aspect ratio, curvature, and Reynolds number. Forces acting on the particles were recorded and carefully analyzed, i.e., they were correlated with certain features of the fluid flow field, as well as the particles' position. Finally, we condense our findings in general guidelines for the design of coiled channels to separate particles of given size.

Acknowledgements

The authors gratefully acknowledge support from NAWI Graz by providing access to dcluster.tugraz.at.

References

D. Di Carlo, "Inertial microfluidics", Lab Chip 9, 2009, 3038 – 3046.

D.H. Yoon, et al., "Size-Selective Separation of Micro Beads by Utilizing Secondary Flow in a Curved Ractangular Microchannel", Lab Chip 9, 2009, 87-90.

G. Guan, et al., "Spiral Microchannel with Rectangular and Trapezoidal Cross-Sections for Size-Based Particle Separation", Scientific Reports 3, 2013, 1475.

J.M. Martel, Mehmet Toner, „Particle Focusing in Curved Microfluidic Channels", Scientific Reports 3, 2013, 3340.

C. J. Greenshields, 2016, OpenFOAM User Guide, CFD Direct Ltd.

C. Kloss, et al., "Models, algorithms and validation for OpenSource DEM and CFD-DEM", Prog. Comput. Fluid Dyn. 12, 2012, 140–152.

485500 Simulations of Particle-Laden Flows in Microchannels

Abstract:

Microreaction engineering offers many advantages over conventional batch reaction systems commonly used in the pharmaceutical and specialty chemicals industries. The increased surface-to-volume ratio enhances the mass and heat transfer coefficients, thus improving efficiency and control. Despite the advantages of microreactors, their small dimensions also pose challenges. For example, the presence of particulate matter can quickly lead to clogging and permanently damage microchannels.

We developed a numerical tool to predict particle formation and agglomeration to study clogging in microsystems. This tool can be used to gain deeper insight into the underlying processes governing clog formation and help identify critical parameters. In addition, this tool enables process parameter-testing for adaption to microstructured devices.

For this, we used a coupled computational fluid dynamics (CFD) / discrete element method (DEM) approach. The DEM solver (LIGGGHTS) handles the equation of motion for each particle, while the CFD solver (OpenFOAM) solves the Navier-Stokes equations describing the fluid flow. The coupling software (CFDEM coupling) computes momentum exchange between particles and fluid. In the DEM solver, we implemented a contact model based on the JKR model [1] to accurately resolve particle-particle and particle-wall collisions.

As a test case, we investigated the precipitation and agglomeration of hydroxyapatite in a microchannel. Using the material properties of hydroxyapatite particles and the process parameters set in previous experimental studies [2], we investigated their effect on agglomeration size and distribution in time and position. The Youngâ€™s modulus was seen to have a great effect on whether stable agglomerates were formed.

Our previous experimental study [2] also showed an effect of dispersed inert gas bubbles on the polydispersity. Towards this end, we also address the three phase modelling (gas, particle and fluid) on the microscopic scale. For this, we developed a void fraction model which enables to simultaneously simulate large size bubbles and small size particles, respectively. Using the developed algorithm, we will predict particle deposition on bubbles and their behaviour in microfluidic devices. In addition, results in terms of bubble-particle aggregates percentage, and spatial and temporal evaluation of bubble-particle aggregates in 3D will be presented.

References

1. Johnson, K. L., K. Kendall, and D. Roberts. Surface Energy and the Contact of Elastic Solids. Proceedings of the Royal Society A: Mathematical, Physical and Engineering Sciences, 1971. 324(1558): p.301- 313.

2. Castro, F., S. Kuhn, K. Jensen, A. Ferreira, F. Rocha, A. Vicente, and J. A. Teixeira. Process intensification and optimization for hydroxyapatite nanoparticles production. Chemical Engineering Science, 2013. 100: p. 352-359.

490765 Prediction of Solid State Phase Diagrams Using Multistate Reweighting and Jacobian Mapping

Abstract:

In materials with multiple metastable crystalline packings (polymorphs) the specific polymorphic phase affects many properties of the material. These include optical properties of dyes, charge transport in semiconductors, detonation properties of energetic materials, and bioavailability in small molecule organic pharmaceuticals. For this reason, effective prediction of the solid state phase diagrams of small molecules is important in the efficient development of materials.

In this study, we present a new approach for the predicting solid phase diagrams using the calculation of the relative Gibbs free energy of polymorphs at a range of temperatures and pressures. Sampling is required in each polymorph at a range of temperatures and pressures. Multistate reweighting is used to calculate free energy differences between temperature and pressure points within a polymorph. These results are then combined with a reference Gibbs free energy obtained with a pseudo-supercritical path to obtain a surface representing the Gibbs free energy differences between polymorphs at each point. Determining the line of intersection of these surfaces, which can be done by several different adaptive methods. gives the coexistence lines in the phase diagram and the most stable in each region is the lowest free energy structure. With this method, the uncertainty in each coexistence point returned by the method can be efficiently and reliably estimated.

The computational scaling of multistate reweighting with system size is a common concern. In addition to developing this method, the size dependence of the uncertainty in the phase diagram has been determined, and thus the scaling of the method with increasing system sizes. This method scales slightly better than N, where N is the number of molecules in the system. To increase the efficiency by decreasing the number of sampled states required, we have also used configuration mapping, and demonstrated the large increase in efficiency this provides for the Lennard Jones system. Configuration mapping allows direct energy reevaluation of a trajectory in at other temperatures and pressures in order to increase phase space overlap. Increasing this overlap allows increased spacing of temperatures and pressures, significantly increasing the computational efficiency. We have shown the feasibility of these approaches by producing the solid phase diagrams of two different systems, benzene and Lennard-Jones spheres.

497007 Evaluation of Defect Concentrations in Crystalline Systems Via Molecular Simulation

Abstract:

Point defects â€“ particularly vacancies and substitutions â€“ are thermodynamically stable features of crystals, and consequently they should be expected in some concentration when using any self-assembly synthesis method. Such defects have a disproportionate impact on the electronic and photonic properties and mechanical performance of the material. Thus, it is important to assess defect concentration and distribution in the equilibrium crystal, and this is done through reference to the free energy. Calculation of defect free energies is significantly complicated by the need to match the lattice-site concentration, which is determined by thermodynamic equilibrium, to the size of the simulation cell. If this is not done carefully, then one results with behavior that appears to be thermodynamically inconsistent.[1,2]

We have developed several approaches to address this problem efficiently in molecular simulations. Each method may be most appropriate for different molecular systems and modeling goals. We evaluate defect behavior the demonstrate various cases, and show how these features affect the system properties. We also consider the behavior of point defects in relation to other types of defects, such as stacking faults.

(1) Swope, W. C. and Andersen, H. C. Thermodynamics, statistical thermodynamics, and computer simulation of crystals with vacancies and interstitials, *Phys. Rev. A* 1992, *46*, 4539-4548.

(2) Pronk, S. and Frenkel, D. Point defects in hard-sphere crystals, *J. Phys. Chem. B* 2001, *105*, 6722-6727.

496345 Nature of the Instability of the Body-Centered-Cubic (bcc) Structure in Classical Hard Sphere Solids

Abstract:

The body-centered cubic (bcc) solid phase of hard spheres is known to be mechanically unstable and also less thermodynamically stable than the face-centered cubic (fcc) and hexagonal close packed (hcp) solids. Yet interest in the bcc hard sphere solid remains, especially in relation to the development of perturbation theories for bcc solids of other model potentials. We employed Monte Carlo simulation in the canonical ensemble, with and without single occupancy cell (SOC) constraints, to explore the nature of the instability in the bcc structure of hard spheres. Starting from a perfect bcc lattice, we observed that the system rapidly evolved into one of four distorted structures, which we named bcc-d*i* with *i* ranging from 0 to 3, which then persisted for the duration of the simulation. These different structures appeared over a range of densities and were characterized by patterns in the radial distribution function and bond order parameter distribution. The bcc-d*i* were formed by small, well-defined average displacements of the particle centers away from the bcc lattice sites, but none of them comprises a repeating crystalline lattice. None of the traditionally employed SOC constraints, *e.g*. Wigner-Seitz, octahedral, or spherical, was sufficient to prevent the bcc structure from distorting into one of the bcc-d*i*. Finally, we also ran constant-pressure simulations with box shape fluctuations and observed that the bcc solids evolved into defective fcc structures, passing through one of the bcc-d*i*structures as an intermediate. We conclude that the perfect bcc hard sphere solid is a highly unstable structure surrounded by a set of metastable distorted, noncrystalline structures, and thus is unsuitable for use as a reference system for bcc solids of other interaction potentials.

498034 Entropy Driven Solid—Solid Transitions in Colloids

Abstract:

In classical, equilibrium statistical mechanics, entropy-driven order remains one of the most enigmatic phenomena. Although there is considerable work on entropy-driven fluid-solid transitions, the multiplicity of crystals that form in systems of hard, anisotropically shaped colloids suggests the possibility of studying entropy-driven solid-solid phase transitions. Here, we introduce a family of minimal model systems that exhibit solid—solid phase transitions that are driven by changes in the shape of colloidal particles. We carry out a detailed investigation of the thermodynamics of a series of isochoric, diffusionless solid—solid phase transitions within a single shape family, and find transitions that require thermal activation, or are "discontinuous", and transitions that occur without thermal activation, or are "continuous". Our results have direct implications for designing reconfiguration in soft materials, and our approach opens new avenues for the detailed study of the basic physics of solid-solid transitions, with potential applications in other areas of physics.

489628 Calculation of Glass Transition Using Sanchez-Lacombe - Eos

Abstract:

Calculation of glass transition using Sanchez-Lacombe - EOS

Margarete Roericht, Kirstin Taufertshöfer, Sabine Enders

Karlsruhe Institute of Technology, Institute of Technical Thermodynamics and Refrigeration Engineering, Engler-Bunte-Ring 21, 76131 Karlsruhe, Germany.

Email: margarete.roericht@kit.edu, kirstin.taufertshoefer@kit.edu; sabine.enders@kit.edu

Key words: Glass transition temperature of polymers and polymer blends, Sanchez-Lacombe Model, generalized entropy theory

Abstract:

The development of reliable approaches for the prediction of the thermodynamic properties of polymer systems is crucial for the rational design of polymer materials and polymer processing in a wide range of polymer applications (for example energy storage, electronics, food, controlled drug delivery systems, medical devises). Glasses are disordered materials that lack the periodicity of crystals but behave mechanically like solids.

The assumption that the glass transition is basically thermodynamic in nature was used to derive a theoretical framework for the calculation of the glass transition temperature (T_g) of polymers, copolymers and polymer blends [1,2,3]. The theory is a synthesis of a thermodynamic equation of state suitable for polymers, the generalized entropy theory for glass-formation in polymer materials, and the rigorous Kirkwood-Buff theory for concentration fluctuations in binary mixtures [2,3].

In the present contribution this theoretical framework is applied in order to compare the theoretical results with experimental data taken from the literature. We apply the Sanchez-Lacombe equation of state (SL-EOS) [4], because the pure-component parameters for several different polymers (polystyrene, (PS); poly(vinyl methyl ether), (PVME); poly(methyl methacrylate), (PMMA); poly(*p*-phenylene oxide), (PPO)) as well as for the statistical copolymer consisting of styrene and acrylonitrile (PSAN) are available. Unfortunately, the parameters for polyacrylonitrile (PAN) are not available, because no PVT data can be measured with high accuracy. The reason for this finding is the thermal instability of this polymer. Therefore, the pure-component parameter for PAN were estimated using PS and PSAN data simultaneously.

Using the above mentioned theoretical framework [1] the T_g of pure polymers was calculated as function of the molecular weight, and compared to experimental data. The analysis shows, that one adjustable parameter per pure polymer is required to match experimental T_g data. Regarding the molecular weight dependency, the theory predicts correctly that the T_g tends towards a limiting value for high molecular weight polymers. This concept can also be applied to statistical copolymers (i.e. PSAN), where the T_g is predicted as function of the chemical composition in excellent agreement with experimental data taken from the literature [5].

This theoretical method can also be applied for polymer blends, where the Kirkwood-Buff formalism serves for the determination of the concentration fluctuation. These calculations were performed for blends made of PS and PVME, where the T_g was investigated as function of blend composition [6]. Blends made of PS and PPO serve as an additional example, where experimental data are taken from [7]. In both cases, the T_g runs monotonically from the T_g of one pure polymer to the T_g of the other pure polymer. In contrast, the T_g as function of blend composition made of PSAN + PMMA [8] runs through a maximum.

The obtained values agree very nicely with experimental data [6,7,8].

References

[1] J. Dudowicz, K.F. Freed, J.F. Douglas, Adv. Chem. Physics 137 (2008) 125.

[2] J. Dudowicz, J.F. Douglas, K.F. Freed, J. Chem. Phys. 140 (2014) 244905.

[3] J. Dudowicz, J.F. Douglas, K.F. Freed, J. Chem. Phys. 141 (2014) 234903.

[4] I.C. Sanchez, R.H. Lacombe, Macromolecules 11 (1978) 1145.

[5] L. Fan, D. Zhao, C. Bian, Y. Wang, G. Liu, Polym. Bull. 67 (2011) 1311.

[6] E. Leroy, A. Alegria, J. Colmenero, Macromolecules 35 (2002) 5587.

[7] W.M. Prest, R.S. Porter, J. Polym. Sci. A-2 10 (1972) 1639.

[8] G. Wen, L. An, J. Appl. Polym. Sci. 90 (2003) 959.

498480 Engineering Shape Entropy for Colloidal Crystal Design

Abstract:

Engineering Shape Entropy for Colloidal Crystal Design

The inverse design of target material structures is a fundamental challenge. Here, we demonstrate the direct inverse design of anisotropic particles for target crystal structures using entropy alone. Our approach does not require any geometric ansatz. Instead, it efficiently samples 92- or 188-dimensional building-block parameter spaces to determine thermodynamically optimal shapes. We present detailed data for optimal particle characteristics and parameter tolerances for six target structures. Our results demonstrate a general, rational, and precise method for engineering new colloidal materials, and will guide nanoparticle synthesis to realize these materials.

492452 Controlling Anisotropic Colloidal Assembly in External Fields

Abstract:

The ability to assemble anisotropic nano- and micro- colloidal particles into hierarchically ordered structures that are also potentially reconfigurable provides the basis for exotic materials (e.g. photonic, negative refraction) and controllable devices in emerging technologies (e.g. optical computing, sub-diffraction limit imaging, invisibility cloaking). However, current capabilities to produce such ordered materials with a sufficiently low defect density have limited the development of the science and applications of such materials. These limitations are due to fundamental problems with designing, controlling, and optimizing the thermodynamics and kinetics of assembly processes for complex colloidal components. As a result, there is an interest in understanding how thermal motion, particle interactions, patterned surfaces, and external fields can be optimally coupled to robustly control the assembly of colloidal components into hierarchically structured functional meta-materials.

Our approach to this problem is to use inhomogeneous AC electric fields to manipulate the position, orientation, and assembly of tri-axial super-ellipsoidal colloids. We begin by first developing rigorous potentials to capture the position and orientation dependent DLVO particle-substrate and particle-particle, dipole-field and dipole-dipole interactions. Microscopy results are used to validate these theoretical expressions by showing measured and modeled potential energy landscapes for up to twelve unique states for single tri-axial particles (including all combinations of position and orientation). We then show how assemblies of many particles can be used to obtain a variety of equilibrium microstructures of super-ellipsoidal colloids. Finally, with the ability to measure, model, and tune anisotropic colloidal interactions and dynamics in inhomogeneous electric fields, we demonstrate how time-varying fields can be used to manipulate non-equilibrium pathways for the assembly, disassembly, reconfiguration, and repair of hierarchically ordered anisotropic colloidal microstructures.

500660 Deposition Dynamics of Rod-Shaped Colloids during Transport in Porous Media

Abstract:

The transport and deposition of colloids in saturated porous media is essential to various environmental processes, especially to produce safe drinking water. Almost all colloids in the environment possess certain degrees of inhomogeneity on their surfaces and are non-spherical in shape. However, existing theories often simplify natural colloids as "homogeneous spheres", which has been increasingly demonstrated inadedquate in describing their transport behavior. In this work, we develop a particle transport model that incorporates the anisotropic features of colloids, i.e. non-spherical shape and heterogeneous patches over particle surface. We will present results obtained during our systematic investigation of rod-shaped colloids with surface charge heterogeneity in Happel sphere-in-cell model (i.e., a well-known model to represent porous media in classic colloid filtration theory).

To improve the classic colloid filtration theory, we performed a series of simulations for ellipsoidal particles with different sizes (e.g., 50 nm – 10 μm), different surface charge heterogeneity coverage (e.g., 0-100%) and aspect ratios (e.g., 1 - 8, where 1 defines spherical particles) under a wide range of flow velocities that represent groundwater flow conditions. The deposition dynamics of rods were compared with that of spheres of equivalent volume under identical conditions. Under favorable conditions (the energy barriers to deposition are absent), for large particle sizes, rods exhibited enhanced attachment than spheres; whereas for small particle sizes, rods showed less retention than spheres. This crossover size strongly depends on flow velocity and porosity. The non-spherical shape of rod particles causes them to respond to hydrodynamic fluid drag very differently in comparison with their counterpart spheres. We found that the flow effect due to shape dominates other factors, e.g. Brownian motion effect for small particle sizes, or gravitational effect for large particle sizes; and can explain the above-observed deposition behavior for rods. Under unfavorable conditions (the energy barriers to deposition are present), surface charge heterogeneity coverage not only affects the extent of attachment, but the orientations of the attached particles. Trajectories of adhered colloids illustrated that they tend to orient to maximize the attraction between colloid surface heterogeneities and the collector surface for firm attachment. In addition, rod particles require less heterogeneous patches for attachment to happen, and for the attachment to behave like favorable conditions, compared to spheres. These results demonstrate that particle shape and surface heterogeneity are very important factors in governing particle transport and deposition.

498191 Engineering the Phase Behavior of Anisotropic Particles By Inverse Design

Abstract:

Anisotropic particles demonstrate a rich set of phase transitions through variation in temperature, pressure, solvent quality, and particle properties. Recent developments in anisotropic particle synthesis have shown promise for using these particles as building blocks for reconfigurable materials. However, due to the large design space, it can be challenging to find appropriate building blocks for targeted behaviors. Here, instead of mapping out the phase behavior for different types of particles, we use the inverse method of Digital Alchemy to find optimal particle shapes for targeted phase behaviors. We present a theoretical design framework along with computer simulation results that give candidate particles for reconfigurable colloidal materials.

490468 Self-Assembly of Magnetic Janus Colloids Via Brownian Dynamics Simulation

Abstract:

Magnetic colloids are of great interest for materials design due to the ease of controlling their structure after tuning their magnetic interactions. Previous studies of anisotropic magnetic particles have focused on discovering their ground state configurations for a few magnetic interaction strengths, however, they have not delved into the dynamics behind their aggregation. Understanding the evolution of these systems is useful for predicting their behavior when employed in realistic conditions. In this contribution, the self-assembly of colloidal magnetic Janus particles with a laterally displaced (or shifted), permanent dipole in a quasi-two-dimensional system is studied using Brownian dynamics simulations. Herein, the rate of formation of clusters and their structures are quantified for all values of dipolar shift from the particle center, which is made non-dimensionalized using the particleâ€™s radius so that it takes values ranging from 0 to 1, and examined under different magnetic interaction strengths relative to Brownian motion. For dipolar shifts of less than 0.1 (shifts close to particle center), chain-like structures are formed which grow at long times following a power law, while particles of shift higher than 0.2 generally aggregate in ring-like clusters and as such clusters experience limited growth. In the case of shifts between 0.4 and 0.5, the particles tend to aggregate in clusters of 3 to 6, while for all shifts higher than 0.6 clusters scarcely contain more than 3 particles due to the antiparallel dipole orientations which are most stable at those shifts. The strength of the magnetic interactions hastens the rate at which clusters are formed, however the effect it has on cluster size is lessened by increases in the shift of the dipoles. These results contribute to better understand the dynamics of magnetic Janus particles and can help the synthesis of functionalized materials for specific applications like drug delivery.

492838 Brownian Dynamics of a Spherical Janus Particle Near a Boundary As a Tool to Investigate TIRM

Abstract:

Colloidal particles dispersed in a liquid interact via surface forces to form microstructure that may have a profound impact on the mechanics and rheology of that liquid. Total Internal Reflection Microscopy (TIRM) has been used for decades to measure very weak ~kT scale colloidal interactions, including electrostatic double layer repulsion, van der Waals attraction, steric repulsion, depletion attraction, hydrodynamic friction, and even the Casimir force, between a sphere and flat plate. Unfortunately, one major restriction currently imposed by TIRM is that the particle needs to be chemically uniform and geometrically spherical. In response to this need, we are currently developing the theoretical and experimental tools to conduct TIRM of anisotropic particles. This talk will summarize our efforts to use Brownian Dynamics simulations to conduct *in silico* TIRM of a Janus sphere very near a boundary. We systematically altered a variety of system and experimental parameters, including the zeta potential of each half of the Janus sphere, sampling rate, measurement duration, and salt concentration, to test the impact and optimize conditions. We simulated the position and rotational trajectories near a boundary and subsequently calculated potential energy landscapes for the Janus particles. Results show that the Janus sphere will sample a potential energy landscape with a transition region at the location of the boundary between the two Janus halves, which depends on the relative zeta potential magnitude. Additionally, simulations show that an experiment may require more than 10^6 observations to obtain a suitable potential energy landscape. These results demonstrated the utility of Brownian dynamics simulations as a tool to probe TIRM for anisotropic particles.

499117 Examining the Stability of Amphiphilic Janus Particle-Laden Interfaces

Abstract:

Recent advances in fabrication techniques have allowed for synthesis of anisotropic Janus particles. Studies of these particles have revealed multitude of exciting properties such as their ability to self-assemble and self-propel. Owing to their colloidal nature and amphiphilicity, Janus particles are believed to stabilize interfaces. Here, we explore the behavior of gold-capped silica Janus particles at the air-water and liquid-liquid interface. More specifically, the work investigates how the degree of amphiphilicity of gold-capped silica Janus particles affects their stability at the air-water interface. Amphiphilicity is varied by modification of the gold surface with alkane thiols of varying chain lengths. Stability of the particle-laden interface is studied under successive compressions and expansions of the air/water interface on a Langmuir trough. Samples of the interface are deposited on silicon wafers and examined using scanning electron microscopy to connect particle amphiphilicity, orientation at the interface, and stability of the monolayer under compression/expansion cycles. Additionally, internally fluorescent Janus particles have been synthesized for the first time to enable studies via confocal microscopy. In particular, this synthesis route is of interest for applications of Janus particles at liquid-liquid interfaces, such as in bijel synthesis.

501380 Self-Assembly of Amphiphilic Janus Nanoparticles

Abstract:

Janus nanoparticles, due to their ability to self-assemble, have been of interest to researchers for over 30 years. Janus nanoparticles are defined by regions of differing properties on a nanoparticle surface. By changing the materials on the Janus nanoparticle surfaces, we can alter the characteristics of the self-assembled structures. In our research, we have modified nanoparticle surfaces with hydrophobic and hydrophilic compounds of varying molecular size to create amphiphilic Janus nanoparticles. Size ratio between the hydrophobic and hydrophilic molecules is altered between Janus nanoparticle samples to examine the difference in self-assembly behavior in selected solvents. To date, we have examined Janus nanoparticles composed of either iron oxide or silica cores. Hydrophilic compounds such as PEG (multiple Dalton weights) and fluorescent molecules have been utilized as well as various long chain hydrocarbons for the hydrophilic region. Using characterization methods including Dynamic Light Scattering, Fluorescent Spectroscopy, and Transmission Electron Microscope imaging, we are able to examine the surface of our Janus nanoparticles as well as self-assembled structures. By altering the size of the surface compounds as well as the size ratio of hydrophobic to hydrophilic compounds, we believe we can control the size of the self-assembled structures; this work provides a foundation for the development of self-assembled coatings composed of Janus nanoparticles, which have potential uses in improve water-resistant coatings and antimicrobial protective layers.

495687 Magnetic Janus Particle Aggregation Rate at Long Times

Abstract:

Magnetic Janus particles are spherical colloids coated asymmetrically on a single hemisphere with a thin iron oxide cap (cap volume approximately 1% of particle volume). The iron oxide gives the particles a magnetic dipole moment shifted away from the center of the particle. When placed in a uniform magnetic field, these particles aggregate into aligned chains, macroscopically creating a magnetorheological fluid. The rate of this aggregation is an essential component in understanding the fluid's properties.

At long times, the system switches from long chains mainly aggregating with small chains to long chains aggregating with each other and small chains at equal probabilities. The average chain length at these long times follows a power law relationship with time:

$< l > \sim t ^\wedge z$

Here, we quantify when the system reaches these "long times" by drawing analogies with polymer chain growth statistics. "long time" data from experiments and simulations are used to find the z-coefficient. Contrary to previous results, the coefficient is found not to be limited to the range $0.6<z<0.7$, but instead ranges from $0.4<z<0.8$ and is found to depend on concentration and dipole strength.

497803 Using μ^2rheology to Characterize Consecutive Phase Transitions in a Hydrogenated Castor Oil Colloidal Gel

Abstract:

The gelation and degradation of colloidal particles is of importance in the design of these materials for use in commercial products, such as consumer and home care products. Of concern in the design is whether the gradient that induces phase change can overcome any processing history, particularly due to shear stress. In this work, we characterize a hydrogenated castor oil (HCO) colloidal scaffold using μ^2rheology, multiple particle tracking microrheology (MPT) in a microfluidic device. HCO is a colloidal fiber that gels and degrades in response to osmotic pressure gradients. When an HCO gel is contacted with water it will degrade and when the surrounding environment is changed to a gelling solution, composed of glycerine and surfactant, the material will gel. Using μ^2rheology, we determine the scaffold properties and structure during consecutive phase changes on a single sample. The microfluidic device design is a two-level device with two chambers: sample and suction chambers. Channels that deliver solvent from the second level into the sample chamber are spaced 60° around the chamber. This creates equal pressure around the HCO sample locking it in place when solvent is exchanged. MPT measures gelation and degradation of the HCO scaffold in the microfluidic device. In MPT, the Brownian motion of probe particles embedded in the material is measured and related to rheological properties using the Generalized Stokes-Einstein Relation. μ^2rheology measurements of consecutive phase changes of HCO are taken starting with both a solution of HCO (0.125 wt%) and a gel (4 wt%). The HCO gel has minimal shear added prior to measurement. The HCO solution is created by diluting HCO in water and adding large amounts of shear. Single samples are successfully measured through 4 – 9 phase transitions. MPT measurements suggest that the structure of the HCO gel is dependent on the starting materials shear history. When gelled the sheared HCO solution cannot form a tightly associated, entangled network structure. During degradation, the samples that begin as an HCO gel cannot completely degrade into single colloids in solution. From these results, we conclude that the osmotic gradient is not a strong enough force to break or reform entanglements. Therefore, equilibrium structures depend on the shear history of the starting material, which can have important implications in end use products made with colloidal gel scaffolds.

500846 Hydrodynamic Entrainment in Micro-Confined Suspensions and its Implications for Two-Point Microrheology

Abstract:

We study hydrodynamic entrainment in spherically confined colloidal suspensions of hydrodynamically interacting particles as a model system for intracellular and other micro-confined biophysical transport. Modeling of transport and rheology in such inhomogeneous soft materials requires an accurate description of the microscopic forces driving particle motion, such as entropic and hydrodynamic forces, and of particle interactions with nearby boundaries. Experimentally, information about the rheological properties of micro-confined biophysical systems is often measured via two-point microrheology, where it is assumed that particle entrainment exhibits the same qualitative features as that of particles in unbound domains; the effects of confining boundaries are thus neglected. In the present work, we carry out dynamic simulations of concentrated, spherically confined colloids as a model system to study the effect of 3D confinement on entrainment and rheology. We show that entrainment between two tracer particles exhibits qualitatively different functional dependence on inter-particle separation compared to an unbound suspension. We develop a scaling theory that collapses the concentrated mobility of spherically confined suspensions for all volume fractions and particle to cavity size ratios onto a master curve. For widely separated particles, the master curve can be predicted via a Greenâ€™s function, which suggests a framework with which to conduct two-point microrheology measurements near confining boundaries.Â The implications of these results for two-point microrheology experiments in micro-confined biophysical systems, such as the interior of eukaryotic cells, are discussed.

497137 Shear-Induced Structural Ordering in Jammed Suspensions of Soft Particle Glasses

Abstract:

We computationally show that jammed suspensions of soft particles can form a variety of ordered microstructures and varying macroscopic properties as a function of the degree of polydispersity and volume fraction of particles. A three-dimensional particle-dynamics simulation of the micromechanics of the flow (Seth *et al.* Nat. Mater. **10**, 838 (2011) is employed to elucidate the effect of shear flow on the microstructure of suspensions that are composed of polydispersed particles. Simulations show that polydispersed jammed glasses in shear flow can undergo a structural transitions regardless of the volume fraction of particles. During the transition, the microstructure changes from a glass to a layer-like phase that results in a reduction of the shear stress and elastic energy. The onset shear rate of this transition depends on the volume fraction and degree of polydispersity. Furthermore, there is an induction strain prior to this transition which forms layer-like structures parallel to the flow-vorticity plane. The length of the induction strain decreases with shear rate, which suggests that the structural transition is a shear-activated process. In the case of monodispersed and suspensions with a low degree of polydispersity, shear flow rearranges particles in a hexagonal close-pack (HCP) phase at higher shear rates, while the structure becomes face-centered cubic (FCC) phase at lower shear rates. Simulation tools are used to describe the origin of these phase transitions, and a state diagram of these suspensions that can be utilized to predict the state of a given jammed suspension in shear flow is provided.

500874 Dynamic Simulation of Aging in a Hard-Sphere Colloidal Glass after Volume-Fraction Jumps

Abstract:

We study the colloidal glass transition from a micromechanical perspective via large-scale dynamic simulation. Many molecular liquids transition from a liquid to a crystalline state upon cooling. If the cooling is sufficiently rapid, however, the material can undergo a phase transition whereby it solidifies without crystallization; the amorphous structure is retained but the dynamics become extremely slow (Angell, 1995). The new "state" is path dependent, i.e. is not an equilibrium phase: material properties such as enthalpy continue to evolve in time. The material relaxes over time toward an "intransient" state, as measured by the Kovacs signature experiments, i.e. the intrinsic isotherms, the asymmetry of approach and the memory effect (Kovacs, 1963). The reasons for this path- and time-dependence are not fully understood in molecular glass-formers, owing partly to molecular relaxation that is so fast that the transition is conflated with the subsequent aging process, making it difficult to capture structural changes during the transition. Colloidal glass-formers provide a natural way to model such behavior, owing to the disparity in colloidal versus solvent time scales that allows tracking of particle dynamics and structure, and how these evolve with liquid-to-glassy macroscopic behavior. To shed light on the ambiguity of the glass transition and the glassy state, we study via large-scale dynamic simulation the structural relaxation, dynamics and rheology of a hard-sphere colloidal glass after various volume-fraction jumps. We execute such volume-fraction jumps with a range of quench rates, where particle size increases at constant system volume. Here we focus on post-jump particle dynamics, where we implement the protocols of the McKenna-Kovacs signature experiments (intrinsic iso-volume fraction) to study the approach to the stationary state. During and following each jump, the positions, velocities, and particle-phase stress are tracked and utilized to characterize relaxation time scales. The impact of both quench depth and quench rate on arrested dynamics and "state" variables is explored. A comparison to the experiments (Peng and McKenna, 2016; Li, Peng and McKenna, 2017) shows qualitative agreement. In addition, we expand our view to various structural signatures, and rearrangement mechanism is proposed. The results provide insight into not only the existence of an "ideal" glass transition, but also the role of structure in such a dense amorphous system.

498526 A New Conformation Tensor Based Macroscopic Model for Emulsions with Particle Inertia

Abstract:

Dispersion are common in numerous industrial processes justifying the large amount of research dedicated to their study. At the macroscopic level of description, a conformation tensor, **C,** can be advantageously used to represent their microstructure. In dilute emulsions with droplet morphology, the dynamics of droplets can be simply represented through a contravariant conformation tensor of constant determinant, det(**C**)=1 [Maffetone Minale (1998)]. Such models, developed when particle inertia is neglected, have been previously postulated by various authors e.g. Maffetone Minale (1998) and Wetzel and Tucker (2001).

This work makes a new advance in emulsion modeling through the development of a new, thermodynamically consistent, conformation tensor-based model that is able to describe unique signature rheological features of emulsion droplets seen only in the presence of particle (micro) inertia. This is motivated by recent microscopic studies of the rheological behavior of dilute emulsions at finite Reynolds number. Li and Sarkar (2005) have carried out simulations of single droplets in simple shear flows at finite particle Reynolds number. Unlike zero particle Reynolds flows where the droplets are observed to orient in the flow direction, in the presence of micro-inertia, characterized by particle Reynolds numbers of O(1), the droplets orient increasingly in the velocity gradient direction. Furthermore, negative first normal stress differences and positive second normal stress differences are seen. The emergence of negative first normal stress differences is also characterized by orientation of the major axis of the droplets at angles, θ, greater than 45^0 above the flow direction. The findings of Li and Sarkar (2005) have been further validated by recent independent asymptotic microscopic theory by Raja et al. (2010). Raja et al. (2010) associate the emergence of negative first normal stresses to a critical Ohnesorge number. This dimensionless group describes the competing effect of viscous forces on one hand and inertia and interfacial forces on the other.

Our work builds upon these recent microscopic developments to develop a new macroscopic conformation tensor-based model that incorporates micro-inertia effects. It achieves that goal through the use of the non-equilibrium thermodynamics bracket framework [Beris & Edwards (1994)]. They key outcome of the developments here is the revelation of a new dissipative term arising from the presence of micro-inertia leading to a new co-deformational time derivative. The new time derivative that results includes an additional parameter, ζ, that is related to the Ohnesorge number. This parameter is beyond the standard non-affine parameter, ξ , that arises in the Gordon-Schowalter derivative used in zero Reynolds number emulsion flow models. For specific values of the new parameter, ζ, the macroscopic model now allows us to predict negative first normal stresses. Furthermore, the droplet morphology associated with the negative first normal stresses is oriented increasingly in the velocity gradient direction with $\theta>45^o$, corresponding to behavior seen in simulation studies [Li and Sarkar (2005)]. The key development of this work is the use of microscopic theory [Raja et al. (2010)] to guide the development of a macroscopic emulsion model that allows for the effects of micro-inertia to be incorporated into macroscopic, continuum level, models.

References

1. Maffettone, P.L. and Minale, M., 1998. Equation of change for ellipsoidal drops in viscous flow. *Journal of Non-Newtonian Fluid Mechanics, 78*(2), pp.227-241.

2. Li, X. and Sarkar, K., 2005. Effects of inertia on the rheology of a dilute emulsion of drops in shear. *Journal of Rheology, 49*(6), pp.1377-1394.

3. Raja, R.V., Subramanian, G. and Koch, D.L., 2010. Inertial effects on the rheology of a dilute emulsion. *Journal of Fluid Mechanics, 646*, pp.255-296.

4. Beris, A.N. and Edwards, B.J., 1994. Thermodynamics of flowing systems: with internal microstructure.

5. Wetzel , E. D. & Tucker R III, C. L. 2001 Droplet deformation in dispersions with unequal viscosities and zero interfacial tension. *J. Fluid Mech.* **426**, 199-228.

499368 Polyelectrolyte – Particle Flocculation in Complex Aqueous Solutions and Mixed Hydrodynamic Fields

Abstract:

Water-soluble polymers with ionizable groups, or polyelectrolytes, are used in a number of applications, such as drinking water treatment, to form mesoscale structures ('flocs') with solid particulates in solution. There are several known potential mechanisms for floc assembly, including polymer bridging and patch flocculation, but the impact of transient chemical and hydrodynamic conditions on assembly kinetics and final floc microstructure remains largely uncertain. Studying assembly kinetics and floc microstructure is particularly complex for systems with anisotropic particulates, such as clays, due to the variety of solution-dependent clay aggregate sizes and morphologies. Here, we explore assembly dynamics of anisotropic Na-bentonite clay particles with cationic polyacrylamide, a polymer commonly used in water treatment, in a variety of chemical solution conditions. We will highlight recent advancements in polymer-particle floc structure visualization that allows for improved understanding of the process variable effects on flocculation (Wilkinson et al., 2016). A Taylor-Couette cell with the novel ability to radially inject fluid into the annulus will be used to study how the hydrodynamics affect assembly and structure of these materials during the flocculation process. This work sheds more light on the complexities of polymer-anisotropic particle flocculation, towards improving dosing and treatment optimization for more efficient water treatment.

501277 From Filaments to Coils: Controlling the Dynamics of Linked Colloidal Particle Chains

Abstract:

The deformations and dynamics of semi-flexible filaments in response to external forces and their surrounding fluid leads to non-reciprocal orbits, which have significant potential for application in microfluidic scale propulsion and fluid manipulation. In order to harness these phenomena, we must understand the competition of various elastic, hydrodynamic, and external forces on the filament, particularly in the semi-flexible regime. We thus investigate these dynamics experimentally through the use of a tunable, colloidal particle chain system; when the paramagnetic chains are driven by an external, rotating magnetic field, multiple dynamical regimes are observed. By pairing these experiments with Brownian dynamics simulations, we elucidate, predict, and begin to optimize semi-flexible fiber behavior as a function of the dimensionless Mason and magnetoelastic numbers of the system.

501115 A Combined Experimental and Numerical Analysis of DNA-Functionalized Colloidal Particle Deposition in a Channel Flow

Abstract:

Aggregating particulate flows are at the heart of an extremely broad range of phenomena in science, medicine and engineering, including biofilm formation in pipes and cellular aggregation in blood flow. These phenomena also are extremely popular targets for computer simulations, and impressive algorithmic developments over the years have enabled sophisticated, multiscale simulations of aggregating particulate flows in a variety of settings [e.g., 1,2]. Although various numerical models for particulate flow have been shown to capture many qualitative phenomena correctly, in most cases their quantitative predictive ability is not well established. Often, this is due to the sheer complexity of the various coupled phenomena at play. One example in which this complexity is highly evident is the aggregation of platelets flowing over an injury in a blood vessel [1]. Here, each platelet represents a complex particle with time-dependent shape, activity (or â€˜stickinessâ€™) that is also able to communicate with its local environment via an intricate sequence of biochemical pathways. While microfluidic models of platelet aggregation greatly simplified the setting in which analysis may be performed [3], the complexity of platelet biology is still present and confounds model testing and validation.

Here, we describe a simplified microfluidic model system that removes much of this complexity, while retaining some of the essential physics related to particle flow hydrodynamics. In particular, DNA-functionalized spherical particles [4] are used to simulate activated platelets, which interact with, and stick to, a patch on the microfluidic channel that presents a brush of complementary DNA single-strands to flowing particles. The evolution of the particle areal density on the adhesive patch is measured experimentally under varying conditions of shear rate and particle density. The results are first interpreted using a simple random-sequential adsorption model, which is shown to capture the data well if all the details of the hydrodynamic interactions are lumped into a single effective parameter. High-resolution images of the adhered particles are then analyzed using various structural measures, such as the radial distribution function, to assess the impact of the fluid flow on the adhered particle distributions.

More detailed simulations that explicitly account for the fluid mechanics are then employed to develop predictive and transferable simulations of the particle flow and adhesion process. We employ a variant of the immersed boundary technique based on a coupled Lattice Boltzmann Method/Brownian dynamics framework, and find that the fluid flow around adhered particles must be carefully resolved to provide accurate descriptions of even relatively simple measures such as the areal density as a function of time. Given the prohibitive computational expense associated with such simulations over the long times characteristic of the experiments, we also investigate an alternative approach in which high resolution particle trajectories are used to populate a database and corresponding regression model that are used to inform Brownian dynamics simulations.

[1] M. H. Flamm et al., Blood **120**, 190 (2012).

[2]. J. K. W. Chesnutt and J. S. Marshall, Comput Fluids **38**, 1782 (2009).

[3] T. V. Colace, G. W. Tormoen, O. J. T. McCarty, and S. L. Diamond, Annu Rev Biomed Eng. **15**, 283 (2013).

497468 Structural and Rheological Relaxation upon Flow Cessation in Colloidal Dispersions: Transient, Nonlinear Microrheology

Abstract:

We study the impact of particle roughness, Brownian motion, and hydrodynamic interactions on the relaxation in colloidal dispersions by theoretical study of structure and viscosity evolution following the cessation of an imposed flow. In particular, we focus on the disparity in timescales over which microscopic forces â€“ hydrodynamic, Brownian and interparticle â€“ act and influence the structural and rheological behavior utilizing the theoretical framework of active microrheology. In our recent work, we studied the formation of structure during flow startup and its impact on transient evolution of viscosity, finding that the transient rheology is a manifestation of a time evolving balance between microscopic forces, which leads to many interesting rheological behaviors, including transient viscosity overshoots, even for dilute colloidal dispersions. However, open questions remain about how the microscopic forces affect the structure and rheology after the external flow is removed, and how the structural relaxation and the probe motion varies with the external flow strength and strength of hydrodynamic interactions upon flow cessation. We formulate a time dependent Smoluchowski equation that governs the spatiotemporal evolution of the bath particle configuration around a probe particle, and solve it numerically for arbitrary strengths of pre-cessation external flow, and arbitrary strengths of hydrodynamic interactions. Upon flow cessation the stresses arising directly due to the external/hydrodynamic forces dissipate instantaneously, as expected, while the entropic Brownian and interparticle stresses take a finite time to relax. These entropic stresses relax via two processes: (i) the probe springs back immediately after flow cessation, and (ii) Brownian motion drives the bath particles towards an isotropic configuration relieving the concentration gradients in the structure. The rheological behavior shows at least two relaxation modes: an early-time fast relaxation, where the majority of the stresses relax, and a later time slow relaxation. We investigate the relaxation rates for arbitrary strengths of pre-cessation external flow and hydrodynamic interactions, and find that these factors have opposing effects on the relaxation process. While increasing pre-cessation flow strength enhances the relaxation rate, hydrodynamic interactions slow down relaxation as the relative motion between particles decreases. The connection between structural and rheological relaxation is studied, and the dissipation of stored entropic energy upon flow cessation is discussed.

498622 Durable, Self-Hydrating, Tungsten Carbide-Based Composite Polymer Electrolyte Membrane Fuel Cells

Abstract:

Proton conductivity of the polymer electrolyte membranes in fuel cell (PEMFC) dictates the performance and requires sufficient water management. Here, we introduce a simple scalable method to produce well-dispersed transition metal carbide (TMC) nanoparticles with narrow size distribution (3-5 nm). This protocol can be extended to the synthesis of other early transition metal carbides. Extensive characterizations including HRTEM, STEM tomography, SEM, XPS, XRD and TG are conducted.

WC nanoparticles incorporated into a recast Nafion membrane demonstrate significant improvement of the PEMFC power density compared to the baseline Nafion membrane. Based on both theoretical and experimental results, the WC composite membrane catalyzes the reaction of crossover H_2/O_2 and provides sufficient water management within the membrane, improving the proton conductivity. Tomographic studies using FIB-SEM indicate large pinholes forming around Pt nanoparticles and provide support that the membrane/Pt catalyst facilitates the production of radicals that leach from Nafion. Superior to Pt, incorporated WC nanoparticles in the membrane result in considerable durability of the membrane. DFT calculations provide insights into the mechanisms by which WC improves performance.

492100 Teaching an Old Electrocatalyst to Do New Tricks

Abstract:

Electrocatalysis underlies several key clean energy technologies such as fuel cells and electrolyzers that are expected to be key components of a sustainable energy future. However, continued improvements in electrocatalyst activity, durability, and selectivity must be achieved in order to make this future a reality. In this study, we explore the use of electrocatalyst architectures based on metallic electrocatalysts that have been encapsulated by ultra-thin silicon oxide overlayers. These oxide layers are synthesized with a room temperature process and deposited as continuous layers onto metallic thin films and nanoparticles. The catalytic metal used in these studies is Platinum (Pt), an "old catalyst" material that is one of the most commonly used electrocatalysts for electrolyzers and fuel cells. Inspired by past studies on glass microelectrodes[1] and a recent study from our lab on oxide-coated nanoparticles,[2] we show that well-defined oxide overlayers deposited on smooth metal substrates can be highly permeable to certain electroactive species and thereby enable efficient and selective electrocatalysis at the metal oxide / metal interface. By systematically varying the thickness of the oxide overlayer, we explore its transport properties and influence on the performance of thin film hydrogen evolution electrocatalysts. Our studies show that the properties of an "old catalyst" like Pt can be greatly altered, and perhaps tuned, through the use of the ultra-thin metal oxide overlayers, which can be thought of as nanomembranes that can selectively control transport of reactants and products. These results highlight the potential of this membrane-coated electrocatalyst architecture to serve as a tunable, efficient, and stable platform for a variety of electrochemical reactions.

References

1. J. Velmurugan, D. Zhan, and M. V. Mirkin, *Nature Chemistry*, 2, 498-502, 2010.

2. N. Y. Labrador, X. Li, Y. Liu, J. T. Koberstein, R. Wang, H. Tan, T. P. Moffat, and D. V. Esposito, *Nano Letters*, 16, 6452-6459, 2016.

493787 Using Surface Chemistry to Understand Aqueous-Phase Thermal Catalytic and Electrocatalytic Hydrogenation of Bio-Oil Model Substrates

Abstract:

One of the major challenges of the future is matching the increasing demand for energy using renewable resources. The energy density of hydrocarbons makes them ideal for transportation fuels, especially for aviation, where electric vehicles are not available. As the amount of energy transitions to renewable energy such as solar and wind, which are intermittent and potentially more decentralized, developing ways to store this energy also becomes important. Catalysis can play an integral role in the chemical processes required to produce renewable transportation fuels and store renewable energy. In this talk, I will discuss developments in catalysts used in a method to convert waste biomass to renewable fuels. The feedstock of this process is bio-oil, a product of pyrolysis or hydrothermal liquefaction of biomass. Bio-oil (modeled here using phenol and benzaldehyde) can be hydrogenated and then deoxygenated to yield fuel grade hydrocarbons.[1] Although hydrogen generally comes from steam methane reforming, which emits carbon dioxide, renewable electricity can be used to drive carbon-neutral (i) electrochemical hydrogen production (from water) which is then used for thermal catalytic hydrogenation (TCH), or (ii) direct electrochemical hydrogenation (ECH) of the bio-oil. Electrochemical routes are more economically competitive in decentralized applications and can make use of intermittent renewable electricity. The challenge of both electrochemical hydrogen production coupled with TCH and ECH is developing hydrogenation catalysts with high rates for commercial applications. The target is to achieve comparable rates (at approximately 80 °C) to existing commercial electrochemical processes, on the order of hundreds of mA per square centimeter (geometric current density) or turnover frequencies near 10 s^{-1}. In this work we discuss some advancements in the understanding of the aqueous-phase hydrogenation of phenol and benzaldehyde on different platinum group metals, and how this leads to improved rates for ECH and TCH.

For phenol hydrogenation, ECH and TCH on rhodium or platinum have similar activation barriers due to a shared rate determining step of adsorbed H adatoms reacting with adsorbed phenol. Using a flow cell designed for X-Ray Absorption Spectroscopy (XAFS), we probe Pt nanoparticles under reaction conditions for both near-edge (XANES) and extended X-ray absorption fine structure (EXAFS). This allows us to look at the catalyst state (from XANES) and surface coverages (from Pt-C scattering that we show is due to the adsorbed reactant species) under operating conditions. This spectroscopy, coupled with kinetic measurements, affords us a better understanding of the surface chemistry and its dependence on hydrogen pressure and applied potential. Using this understanding, we demonstrate methods to prevent catalyst deactivation at temperatures at 80 °C.[2] We also show how modifying the adsorption energy of the adsorbed species controls the reaction rate, and discuss how other reaction conditions (ex. pH) affect the surface coverages and rates. These results lead us to propose material properties that may result in more active electrocatalysts for phenol TCH or ECH, and how to reach reaction rates that may be necessary for a commercial ECH electrolyzer system.

For benzaldehyde, on metals such as platinum and rhodium, TCH and ECH have different benzaldehyde rate order dependence, unlike what is observed for phenol. This may be due to a difference in the reaction mechanism for the ECH and TCH of benzaldehyde on Pt and Rh, or surface coverages under operating conditions. Using XAFS (as well as traditional cyclic voltammetry), we try to understand how the surface coverage of the species may be responsible for the observed reaction rates and orders. We connect this with kinetic measurements and adsorption energies in an attempt to understand how the surface chemistry controls the reaction rates that are observed.

References

(1) Zhao, C.; He, J.; Lemonidou, A. A.; Li, X.; Lercher, J. A. *J. Catal.* **2011**, *280*(1), 8.

(2) Singh, N.; Song, Y.; Gutiérrez, O. Y.; Camaioni, D. M.; Campbell, C. T.; Lercher, J. A. *ACS Catal.* **2016**, *6*, 7466.

498568 Design of Heterogeneous Catalysts for High Temperature Electrochemical Reduction of CO_2

Abstract:

The high levels of carbon dioxide (CO_2) emissions, consequential from the extensive use of fossil fuels, are major contemporary challenges. An approach to alleviate this process is to activate the reverse chemical pathways in which CO_2 is reduced into high-energy molecules. The electrochemical reduction of CO_2 to CO has attracted increasing attention, since CO represents a valuable intermediate to the production of synthetic fuels using established processes, such as Fischer-Tropsch. The use of solid oxide electrolysis cells (SOECs) to electrochemically reduce CO_2 to CO is an attractive technology, given that this solid-state electrochemical systems can, in principle, facilitate the CO_2 reduction with potentially very high rates due to the favorable kinetics at high operating temperatures. While electrochemical reduction of CO_2 using SOECs offers a great deal of promise, these systems are still limited by the high activation overpotential losses induced by the sluggish kinetics of CO_2 reduction at the cathode. In the present work, we combine experimental and theoretical techniques to understand the chemical/electrochemical process that governs CO_2 electrolysis, and develop a structure/performance relation in order to identify the optimum electrocatalyst for this process. Controlled electrochemical experiments show that upon modification of the electrode metal electrocatalyst, significant decrease in the overpotential is achieved.

REFERENCES

1. Ebbesen, S. D., Jensen, S. H., Hauch, A., & Mogensen, M. B. (2014). High temperature electrolysis in alkaline cells, solid proton conducting cells, and solid oxide cells. *Chemical reviews*, 114(21), 10697-10734.

2. Gu, X. K., Carneiro, J., Nikolla, E. (2017). Electrolysis on Carbon Dioxide. In Catalysis (pp. 94-116). The Royal Society of Chemistry.

3. Gu, X. K., & Nikolla, E. (2017). First-Principles Study of High Temperature CO_2 Electrolysis on Transition Metal Electrocatalysts. *Industrial & Engineering Chemistry Research*, under review 2017.

494155 Enhancing Methanol Electrooxidation Activity Using Double Oxide Catalyst Support

Abstract:

Electrooxidation of methanol was investigated on platinum catalyst supported on a double oxide nanocomposite of tin oxide clusters and carbon-doped titanium dioxide nanocoatings on carbon nanotubes. The electrocatalyst demonstrated a higher forward peak current density in methanol eectrooxidation than the electrocatalyst with a single tin oxide or just carbon as support. Formation kinetics of the oxygenated species (-OH) groups was dramatically enhanced in the catalyst with double oxide support. This leads to a much lower onset potential for adsorbed CO oxidation with an enhanced bifunctional effect in tin oxide. Doping with carbon in the titanium oxide enables backward electron donation, making supported platinum more electron negative and stable. Binding energy shifts in oxidation states was observed for the double oxide support catalyst and is attributed to strong metal support interactions. This study demonstrated a synergistic effect of combining bifunctional and electronic mechanisms with double oxide support in the enhancement of methanol electrooxidation.

501794 Study of Impact of Electrode Catalyst Loadings on Thin Flexible Fuel Cell (TFFC) Performance

Abstract:

Thin Flexible Fuel Cell (TFFC) is a flat and flexible device without any rigid support capable of being integrated into portable electronics. TFFC uses hydrogen and oxygen as fuel to provide high volumetric and gravimetric energy densities capable of competing with battery technology as a power source. The group is currently investigating a TFFC with an air-breathing design that maximizes the cathode interface with ambient air. The current challenges of making TFFC commercialized are the cost and performance. Enhanced cell performance is critical for a competitive cost of the device. The group investigated the impact of catalyst loading on the cell performance to increase the overall performance of the device. The catalyst loading of the cathode electrode was varied while keeping the anode catalyst loading constant and vice versa. The effects of the various catalyst loadings were characterized using various analytical techniques. Polarization curves and Electrochemical Impedance Spectroscopy (EIS) were used to separate and quantify the sources of losses for the TFFC. Power curves provided the maximum power density of the device. Cyclic voltammetry (CV) was implemented to characterize the electrochemically active surface area (ECSA) and utilization of the various catalyst layers. Rotating Disc Electrodes (RDE) tests were performed for a direct assessment of the catalyst. We will report the results of these assessments and summarize the resulting guidance for optimizing catalyst loading on similar fuel cell structures.

501059 Transition Metal Modified Graphitic Mesoporous Carbon for PEMFC Application

Abstract:

Proton exchange membrane fuel cell (PEMFC) has drawn considerable attention due to its high power density, zero emission and fast startup features. In PEMFC, platinum metal is the catalyst of choice in anode and cathode electrodes. The cost of Pt is a significant issue in commercialization of PEMFC. Further due to its sensitivity to CO, PEMFC requires high purity hydrogen as fuel. Conventional catalyst support (activated carbon) lacks in adequate porous structure, which results in underutilization of Pt metal particles. To better utilize the Pt-catalyst, we synthesized high surface area metal modified mesoporous carbon using hard template-based (SBA-15) method. Highly porous structure of this carbon allows well-dispersed Pt metal loading ensuring its proper utilization. Transition metals (Ni, Fe and Co) and Ni-Fe alloy were used for catalytic graphitization of the carbon. Resulting carbon showed graphitic structure in X-ray diffraction (XRD) and Raman spectroscopy. Electrocatalysts synthesized using these metal modified carbons were tested on the fuel cell. The structural characteristics of these graphitic mesoporous carbon and their performance in PEMFC will be presented in this paper.

492994 Capillary Foams: How a Small Amount of Oil Can Make a Big Difference

Abstract:

Today, humanity faces unprecedented challenges in managing the word's critical energy, water, food, and healthcare needs. While there are many aspects to these challenges, one issue that is certain to play a pivotal role in any solution is new materials. Directed assembly provides a "bottom-up" approach to create novel, functional materials with well-defined properties and functionality from nanoscale building blocks. Here, I will present a new class of colloidal materials, capillary foams, through directed assembly in a colloidal four-phase systems. Capillary foams are formed by frothing a suspension of colloidal particles in the presence of a small amount of oil (less than 1 wt%). I will discuss our fundamental understanding of the formation mechanism and explain the huge difference that a small amount of oil can make in these foams. The presence of so many phases (and their interfaces) offers a wealth of processing opportunities for the development of new materials with interesting chemical or mechanical properties and transport characteristics. Looking beyond the fundamental studies, I will discuss some of the application opportunities for capillary foams, for instance in the fabrication of loadbearing, lightweight, or intensely colored porous materials, enhanced oil recovery, advanced froth flotation, wastewater treatment, and oil spill remediation. Capillary foams can be used for engineering a wide range of functional materials targeting energy and environment problems that can better serve our society.

491004 Wetting on Strain Induced Microstructured Surfaces

Abstract:

Wetting on Strain Induced Microstructured Surfaces

Vartika Parihar[a], Soumen Das[b] and Sunando Dasgupta[a]

[a]Department of Chemical Engineering, IIT Kharagpur 721302, India

[b]School of Medical Science and Technology, IIT Kharagpur 721302, India

E-mail: pari92vartika@gmail.com, sou@smst.iitkgp.ernet.in, sunando@che.iitkgp.ernet.in

Abstract

Wetting characteristics on hydrophobic micro-structured surfaces have remained a subject of eminent significance due to its characteristic properties and relevance in a plethora of industrial applications and scientific studies, such as anti-biofouling, self-cleaning and microfluidics etc[1]. Numerous processes for the fabrication and characterization of such specialty surfaces have been proposed and practiced[2]. However, most of them have limitations, i.e. they are inherently complex, intrinsically infrastructure dependent i.e. they require elaborate micro-fabrication facilities, and Most importantly, it is limited by the feature size of the available mask or the stamp[3]. Fabrication of microstructured surfaces with the controlled tuning of feature size has rarely been reported[4]. In this work, a new method for the fabrication of controlled dual scale roughened surfaces is reported involving relatively simpler processes compared to the existing processes[5]. The study enumerates a straightforward, masks-less process for the creation of microstructured surfaces, comprising of longitudinal wrinkles with controlled feature sizes. The microstructures result in the different wetting behavior of a sessile droplet with possible advantages in various applications. These wrinkles have been characterized by two morphological parameters, namely the amplitude and wavelength and they are found to be strong functions of the applied pre-strain. Wetting states over such wrinkled surfaces are observed to be governed by the underlying pattern of wrinkles. Modulation of the pre-strain leads to contrasting wetting regimes of Cassie-Baxter for high values of strain and Wenzel state for lower values of strain. The resulting equilibrium contact angles have been measured as a function of the buckle wavelength and amplitude. Thus, we have demonstrated a new and simple design paradigm, consisting of parallel grooves of appropriate aspect ratio with controlled anisotropic wetting.

The wrinkle process comprises stretching of a soft Polydimethylsiloxane(PDMS) (10:1) substrate and subsequent physical vapor deposition (sputtering)[6], resulting in a hard metal (Nichrome) layer on the surface. Upon relaxation, sinusoidal wrinkles with a uniform wrinkle wavelength (λ) and amplitude (A) are formed due to the buckling instability (mismatch of strain in between the two materials) that relieves stresses in the form of sinusoidal undulation. Subsequently, after preparation of the surfaces, the topography is transferred on a PDMS (10:1) spin-coated, glass slide for enhancing the hydrophobicity on surfaces with anisotropic wetting behavior. The final structure of the fabricated micro wrinkles on a PDMS substrate is obtained as shown in the 3D Optical Surface Profiler (3D-OSP) image (Figure 1).

The variations in the buckle wavelength and amplitude are studied after complete relaxation of initial applied pre-strains on different samples. Effect of this varying topography on the wetting behavior is also studied by measuring the contact angles of 2μl DI water droplet for different substrates by using a Rame-Hart Goniometer. The contact angles measured on the substrate are shown along with the images of the sessile droplets in Figure 2. The images of the droplets are captured in two prime directions, i.e. along the wrinkles and across the wrinkles, demonstrating the strong anisotropic wetting behavior of the surfaces with nearly superhydrophobic behavior in a direction perpendicular to the direction of wrinkles on surfaces at high values of pre-strains.

Keywords: Microstructures, Superhydrophobic, Microfluidics, Contact Angle, Anisotropic Wetting.

Figure 1. 3D optical surface profilometry of the wrinkled PDMS surface for 50 % applied pre-strain; (a) 3D image, and (b) corresponding surface profile.

Figure 2. Contact angles subtended by a DI water droplet of 2 μl volume on different surfaces. **(a-c)** shows contact angles measured along the direction of wrinkles on surfaces with pre-strains of 20%, 50%, and 80% respectively; whereas, **(d-f)** shows contact angles measured perpendicular to the direction of wrinkles on surfaces with pre-strains of 20%, 50%, and 80% respectively.

References:

1. Y. Y. Yan, N. Gao, and W. Barthlott (2011), "Mimicking natural superhydrophobic surfaces and grasping the wetting process: A review on recent progress in preparing superhydrophobic surfaces," *Adv. Colloid Interface Sci.*, vol. 169, no. 2, pp. 80-105.

2. B. N. Sahoo and B. Kandasubramanian (2014), "Recent progress in fabrication and characterisation of hierarchical biomimetic superhydrophobic structures," *RSC Adv.*, vol. 4, no. 42, pp. 22053-22093.

3. S. Chang and J. Yoon (2004), "arrays fabricated by a 3D diffuser lithography," *Opt. Express*, vol. 12, no. 25, pp. 6366-6371.

4. S. L. Zhang *et al.* (2002), "Variation of Raman feature on excitation wavelength in silicon nanowires," *Appl. Phys. Lett.*, vol. 81, no. 23, pp. 4446-4448.

5. N. Bowden, S. Brittain, A. G. Evans, J. W. Hutchinson, and G. M. Whitesides (1998), "Spontaneous formation of ordered structures in thin films of metals supported on an elastomeric polymer," *Nature*, vol. 393, pp. 146-149.

6. G. Bräuer, B. Szyszka, M. Vergöhl, and R. Bandorf (2010), "Magnetron sputtering - Milestones of 30 years," *Vacuum*, vol. 84, no. 12, pp. 1354-1359.

486104 Water Wetting on Surfaces Under Fuel Oil Containing Surfactants and Its Implication for Coalescence Separation of w/o Emulsions

Abstract:

Several chemicals were deposited on glass slide as thin layers by spin-coating and their response to water wetting in both air and under fuel oils with various surfactants were studied. One chemical was found to behave dramatically differently than others and its unknown chemistry was thus revealed. This chemistry, when applied to a porous filter media for emulsified water separation from surfactant-rich fuel oils, gave rise to enhanced separation efficiency. Molecular structure analysis indicates this chemistry is unique as it provides in one molecule a hard hydrophobic segment and a soft hydrophilic segment, both of which are extremely important and favorable for coalescence separation. Such chemistry also leads to surprisingly intensive but reversible adsorption of surfactants from the fuel oil. As a result, when a new water droplet approaches the coating surface sitting with an early droplet or film, the surfactant molecules freshly assembling around the new water/oil interface are pulled away immediately, yielding ways for droplet confluence.

500755 Dynamic Surface Tension Measurements of Ionic Surfactants Using Maximum Bubble Pressure Tensiometry

Abstract:

Dynamic surface tension refers to the time dependent variation in surface tension, and is intimately linked with the rate of mass transfer of a surfactant from liquid sub-phase to the interface. The diffusion- or adsorption-limited kinetics of mass transfer to interfaces is said to impact the so-called foamability and the Gibbs-Marangoni elasticity of surfaces. Dynamic surface tension measurements carried out with conventional methods like pendant drop analysis, Wilhelmy plate, etc. are limited in their temporal resolution (> 50 ms). In this study, we describe design and application of maximum bubble pressure tensiometry for the measurement of dynamic surface tension effects at extremely short (1-50 ms) timescales. Using experiments and theory, we discuss the overall adsorption kinetics of charged surfactants, paying special attention to the influence of added salt on dynamic surface tension.

487747 The Importance of Spreading Pressure on Adsorption Based Surface Energy Measurements; The Case of IGC

Abstract:

In his work on "The Spreading of Fluids on Glass",[1] published in 1919, Sir William Bate Hardy argues that "Whether primary or secondary spreading does or does not occur on a fluid face depends mainly upon the relative value of the surface tensions, but on a clean solid face it must depend wholly upon the value of vapour tension". This constitutes the first scientific statement highlighting the importance of spreading pressure in the formation of solid-fluid interfaces. Nonetheless, the concept of spreading pressure has been discussed, by Hardy and other prominent members of the then scientific community, well before 1919, in the context of its importance in the spreading of oils on water.[2-4]

In 1937 Bangham and Razouk apply Gibb's isotherm in gas-solid adsorption, in a work where they address the importance of spreading pressure.[5-7] However, in adsorption based techniques developed for surface energy measurements, in the years followed the publication of this work, the importance of spreading pressure was omitted, on the basis that its value can be considered negligible. One such technique is Inverse Gas Chromatography (IGC), a widespread technique used since 70's.[8,9]

This work shows that for low surface energy materials, such as organic solids and polymers, the influence of spreading pressure is sufficiently high, that it can lead to peculiar results if it is not taken into consideration. The example of two of Carbamazepine's polymorphs, the P-Monoclinic and the Triclinic, is employed to show that in the absence of the spreading pressure correction, the results would suggest an increase in the surface energy of the material, with an increasing temperature; a peculiar result for the specific material.

Using tailored experiments a roadmap is developed for the spreading pressure correction of the data. Upon correction the trend of the surface energy becomes the one expected from relevant theory, showing a decrease of about 7 mJ/m^2 for an increase in temperature from 25 ^{o}C to 60 ^{o}C. These findings reveal that appreciating the importance of spreading pressure is paramount for accurate surface energy measurements. Furthermore, the experiments suggest that a careful selection of the probes used in the measurements is important when the measurements are performed at temperatures well beyond ambient.

Computational models are employed to showcase that the surface energy distributions obtained from the spreading pressure adjusted results are much more realistic and in better agreement with the surface energy measurements obtained via wettability experiments on the individual facets of crystals of the P-Monoclinic polymorph.[10-12] This does not only validate the importance of computational models for the elucidation of adsorption measurements, but it also stretches the importance of cross verification of the results with complimentary experimental techniques.

This work exposes the importance of spreading pressure on adsorption based surface energy measurements

of organic materials and polymers. It highlights the need for complimentary experimental techniques, as also of computational modelling, for the in-depth understanding of interfacial phenomena at the low surface energy limit. It, finally, provides a framework, for researchers, to revisit some of their old data and examine them experimentally and/or computationally in the light of the new findings.

References

1. Hardy WB. The spreading of fluids on glass. *Philosophical Magazine*. **1919**, 38, 49-55.

2. Hardy WB. The tension of composite fluid surfaces and the mechanical stability of films of fluid. *Proceedings of the Royal Society of London Series A, Mathematical and Physical Sciences*. **1912**, 86, 610-635.

3. Hardy WB. The influence of chemical constitution upon interfacial tension. *Proceedings of the Royal Society of London Series A, Mathematical and Physical Sciences*. **1913**, 88, 303-313.

4. Hardy WB. The tension of composite fluid surfaces. No. II. *Proceedings of the Royal Society of London Series A, Mathematical and Physical Sciences*. **1913**, 88, 313-333.

5. Bangham DH, Razouk RI. Adsorption and the wettability of solid surfaces. *Transactions of the Faraday Society*. **1937**, 33, 1459-1463.

6. Bangham DH. The gibbs adsorption equation and adsorption on solids. *Transactions of the Faraday Society*. **1937**, 33, 805-811.

7. Bangham DH, Razouk RI. The wetting of charcoal and the nature of the adsorbed phase formed from saturated vapours. *Transactions of the Faraday Society*. **1937**, 33, 1463-1472.

8. Braun J, Guillet JE. Determination of crystallinity of olefin copolymers and polyolefin powders by inverse gas chromatography. *Polymer Chemistry*. **1975,** 13, 1119-1131.

9. Ho R, Heng JYY. A review of inverse gas chromatography and its Development as a tool to Characterize Anisotropic surface properties of pharmaceutical solids. *KONA Powder and Particle*. **2013**, 30, 164-180.

10. Jefferson EA, Williams DR, Heng JYY. Computing the surface energy distributions of heterogeneous crystalline powders. *Journal of Adhesion Science and Technology*. **2011,** 25, 339-355.

11. Smith RR, Williams DR, Burnett DJ, Heng JYY. A new method to determine dispersive surface energy site distributions by inverse gas chromatography. *Lagmuir*. **2014,** 30, 8029-8035.

12. Yao Z, Heng JYY, Lanceros-Méndez S, et al. Study on the surface properties of colored talc filler (CTF) and mechanical performance of CTF/acrylonitrile-butadiene-styrene composite. *Journal of Alloys and Compounds*. **2016,** 676, 513-520.

494331 Stabilization of Glycerol/Dodecanol Pickering Emulsions with Surface-Grafted Silica Nanoparticles

Abstract:

The conduct of Pickering interfacial catalysts (PIC) into glycerol/dodecanol nonaqueous biphasic system can significantly promote the reaction efficiency, while the design of optimal catalytic nanoparticles, which will stay farthest at the interface within the reaction process, remains a challenging task. Here by using dissipative particle dynamics simulations, the stabilization of glycerol-dodecanol systems with the addition of different functional silica nanoparticles was investigated. For mimicking the experimental operation, the silica nanoparticle in our study was surface modified by grafting alkyl chains and sulfonic group (catalytic site) for adjusting the emulsifying capability and catalytic performance. By accessing both the thermodynamic and structural properties of the interface systems, we identified the optimal surface modification including the chain length of alkyl chains and the molar ratio between sulfonic and alkyl moieties. In particular, by examining the surface tensions and interfacial contact angle between the nanoparticles and the glycerol/dodecanol interface, we found that the SiO_2-C_3 nanoparticle displayed excellent capacity to stabilize the emulsion with a wide range of alkyl grafting density ranging from 1.0 â€“ 2.2 group/nm^2, which was confirmed by the experimental result. In addition, the local molecular concentrations in the directions normal to and parallel with the solid particle surface were analyzed. The analysis indicated that the reaction would mostly take place in the glycerol phase close to the interface, and that the chain length of C_3 was appropriate for retaining sufficiently high number densities of both glycerol and dodecanol molecules in the vicinity of the catalytic sites, and thus facilitated the reactions. Finally, the effect of the product yield was conduct, and it was observed that the size of emulsion droplet decreased as increasing the product yield rate. This work provides useful insight from microscopic perspective toward the design of highly efficient PIC process for glycerol/dodecanol reaction.

499164 *In situ* Observations of Dynamics of Active Braze Joining

Abstract:

Brazing is an important joining technology for manufacturing hermetic seals between ceramic and metal surfaces. Active braze alloys make use of an active element such as titanium or zirconium that reacts with the ceramic to form an interfacial layer which improves the wettability of the metal alloy on the ceramic. The process of brazing is a complex competition between thermal transport, phase changes and chemical reactions between the molten braze and the bonding surfaces. Most active brazing operations take place at high temperatures up to 1000°C which makes studying defect formation and troubleshooting the mechanisms quite challenging. Literature studies of brazing have traditionally relied primarily on post mortem analysis of final joints to investigate the braze process. Recently, we have developed a method to visualize the dynamics of an active brazing process *in situ*. By substituting optically transparent sapphire for traditional alumina ceramic substrate, the braze alloy can be observed during joining within a high temperature vacuum chamber. We have successfully demonstrated the experiment with a silver-zirconium active braze alloy. Several interesting dynamics have been observed including phase transitions and wetting phenomena.Â We will compare in situ images with traditional post mortem analysis of the final joints.

Sandia National Laboratories is a multi-mission laboratory managed and operated by Sandia Corporation, a wholly owned subsidiary of Lockheed Martin Corporation, for the U.S. Department of Energyâ€™s National Nuclear Security Administration under contract DE-AC04-94AL85000.

495823 High Throughput Fabrication of Synthetic Asymmetric Bacterial Membranes

Abstract:

Giant unilamellar vesicles (GUVs) are of great interest as versatile model systems to emulate the function and fundamental properties of the plasma membrane in living cells. The double-membrane cellular envelope in Gram-negative bacteria enables them to endure harsh environments and represents a barrier to many clinically available antibiotics. The outer membrane (OM), which is exposed to the environment, is believed to be the first point of contact involved in fundamental bacterial processes such as signaling, pathogenesis, and motility. In this work, we report on a microfluidic technique for synthesizing monodispersed asymmetric GUVs possessing the Gram-negative bacterial OM lipid composition. As in the cytoplasmic membrane, the OM in Gram-negative bacteria has a phospholipid-rich inner leaflet; however, the outer leaflet is predominantly composed of lipopolysaccharide (LPS).

Our continuous microfluidic fabrication technique generates 50 to 100 μm diameter water-in-oil-in-water double emulsions at high-throughput. The water-oil and oil-water interfaces facilitate the self-assembly of phospholipid and LPS molecules to create the inner- and outer-leaflet of lipid bilayer, respectively. Bilayer membrane asymmetry is evaluated using a fluorescence quenching assay. Our approach addresses many of the deficiencies found in existing biotechnological techniques for building vesicles possessing strong membrane asymmetry. The effect of cholesterol and divalent metal ions (*i.e.* Mg^{2+}) are investigated on the membrane stability. The GUVs built using this strategy are collected off-chip and transferred to an aqueous solution to be studied as bacterial model membranes in biological processes.

496813 Design Insights for a Thermal Visbreaking Pilot Based on Microwave Heating

Abstract:

The transport of heavy oil by pipeline typically requires a significant reduction in the oil viscosity, which is often accomplished with a diluent. It is tempting to reduce or eliminate the need for diluent by performing some form of mild thermal cracking or viscosity breaking (visbreaking) of the heavy oil at the well-site prior to transport. The objective of this study was to evaluate the application of electromagnetic (EM) energy to a typical visbreaking (heater-soaker) reactor design where flowing heavy oil was heated in a microwave resonant cavity. The design concept was tested on laboratory scale reactor and it was demonstrated that the microwave resonant cavity could heat the heavy oil while maintaining efficient transfer of microwave energy to heating (over 80%). This heating significantly changed the oil composition and properties, including an order of magnitude reduction in viscosity [1,2].

The study also addressed two unique challenges in scaling results from the laboratory scale microwave reactor to a 1000 bbl/day pilot design: i) defining the dimensions of the resonant cavity to handle the 1000 bbl/day flow rate while maintaining the desired heating efficiency, and ii) identifying the tolerance of the microwave resonant cavity performance to produced water in the heavy oil feed (measured by resonance frequency and energy adsorption distribution). The first challenge was met by reducing the resonant frequency of the reactor by increasing the reactor tube diameter and using a modular heater-soaker unit design [3]. The second challenge was tackled using CST^{TM} simulation to model the EM behavior in the presence of water saturated heavy oil, emulsion and steam at reactor conditions. It was demonstrated that the reactor could tolerate a maximum water content in the range of 1 wt% before the reactor performance degraded.

[1] Taylor S.D., Andersen, S.I., Tam, W.M., Neitz, T., Oâ€™Brien, D., Pulikkathara, M., and Saeedfar, A., â€œA Thermal Visbreaking Prototype Based on Microwave Heatingâ€, 65th Canadian Chemical Engineering Conference, Calgary, Canada, 4-7 October 2015.

[2] T. Neitz, W-M. Tam, B. Clark, S.D. Taylor, D. O'Brien, Y. Cheng, A. Saeedfar, "Apparatus and Method Employing Microwave Resonant Cavity Heating of Hydrocarbon Fluid", (2013) WO 2014/189533.

[3] D. O'Brien, A. Saeedfar, S.D. Taylor, B. Clark, W.-M. Tam, M. Pulikkathara, T. Neitz, S. Andersen "Apparatus and Method Employing Microwave Resonant Cavity Heating for Visbreaking of Hydrocarbon Fluid", (2013) WO 2014/171960.

486906 On Cubic Eos Interaction Parameter Estimation for Long Chain n-Alkane + Aromatic Binary Mixtures

Abstract:

Cubic equations of state are widely used for phase equilibrium calculations. They are simple and reliable, especially for hydrocarbon mixtures. Unfortunately, common choices such as the Peng-Robinson and Soave-Redlich-Kwong equations of state, systematically predict both qualitatively and quantitatively incorrect phase behaviors for long chain n-alkane + aromatic and naphthenic mixtures unless negative binary interaction parameter (k_{ij}) values are used [1, 2]. Binary interaction parameters are typically obtained by fitting vapour-liquid equilibrium (VLE) data for binary mixtures. In the absence of data, interaction parameter values are obtained from generalized correlations typically leading to positive values [3,4] or are set to zero. Experimental VLE data for these mixtures are scarce and therefore reliable methods for estimating k_{ij} values in the absence of experimental data are needed to support industrial process design and optimization calculations. In this study, methods for estimating k_{ij} values for cubic EOS for binary mixtures of aromatics with long chain n-alkanes are evaluated. Options explored include obtaining k_{ij} values by regressing liquid phase activity coefficients from an *a priori* predictive model, the COnductor like Screening MOdel for Real Solvents (COSMOS-RS) [5] and the PC-SAFT EOS [6] (based on statistical associating fluid theory), and a predictive group-contribution k_{ij} estimation approach – PPR78 (predictive 1978, Peng–Robinson EOS) [7]. Calculated temperature-dependent interaction parameter values for 15 representative long chain n-alkane + aromatic binary mixtures are benchmarked against Peng-Robinson EOS k_{ij} values fitted to high precision VLE data [8]. The resulting skew and dispersion of predicted vs experimental k_{ij} values is discussed and best practices for estimating cubic EOS k_{ij} values for such mixtures are described.

References:

[1] Ahitan, S., Satyro, M. A., Shaw, J. M. Systematic misprediction of n-alkane + aromatic and naphthenic hydrocarbon phase behavior using common equations of state. J. Chem. Eng. Data 2015, 60, 3300-3318.

[2] Ahitan, S., Shaw, J. M. Quantitative comparison between predicted and experimental binary n-alkane+ benzene phase behaviors using cubic and PC-SAFT EOS. Fluid Phase Equilib. 2016, 428, 4-17.

[3] Gao, G.; Daridon, J.; Saint-Guirons, H.; Xans, P.; Montel, F. A simple correlation to evaluate binary interaction parameters of the Peng-Robinson equation of state: binary light hydrocarbon systems. Fluid Phase Equilibria 1992, 85-93.

[4] Chueh, P. L.; Prausnitz, J. M. Vapor-liquid equilibria at high pressures: Calculation of partial molar volumes in nonpolar liquid mixtures. AIChE J. 1967, 6, 1099-1107.

[5] Klamt, A.; Eckert, F. COSMO-RS: a novel and efficient method for the a priori prediction of

thermophysical data of liquids. Fluid Phase Equilibria 2000, 1, 43-72.

[6] Gross, J.; Sadowski, G. Perturbed-Chain SAFT: An Equation of State Based on a Perturbation Theory for Chain Molecules. Ind. Eng. Chem. Res. 2001, 4, 1244-1260.

[7] Jaubert, J.; Vitu, S.; Mutelet, F.; Corriou, J. Extension of the PPR78 model (predictive 1978, Peng–Robinson EOS with temperature dependent kij calculated through a group contribution method) to systems containing aromatic compounds. Fluid Phase Equilib. 2005, 1–2, 193-211.

[8] Liu, Q. MSc. Thesis, University of Alberta (in progress 2017).

Uncertainty in Heat Exchanger Design for Processing Bitumen

Harry Z. Ha

Paul M. Mathias

Fluor Enterprise

AIChE 2017 Fall Conference

Minneapolis, MN, USA

FLUOR.

Uncertainty in Heat Exchanger Design for Diluted Bitumen

FLUOR.

- The design of chemical processes today invariably uses computer simulations, and many technologists have recognized the need to understand and evaluate the impact of uncertainties in the underlying data on process design and plant operability.
- Transport properties have a significant impact on fluid heat transfer, consequently the design of the heat exchangers.
- The impact is especially significant when the heat exchanger is designed for the service of bitumen processing.
- Minimizing the uncertainty in transport property estimates will reduce the design margin in equipment design, consequently the project cost can be reduced.

1

Typical Viscosity Behavior of Oil Sand Bitumen

FLUOR

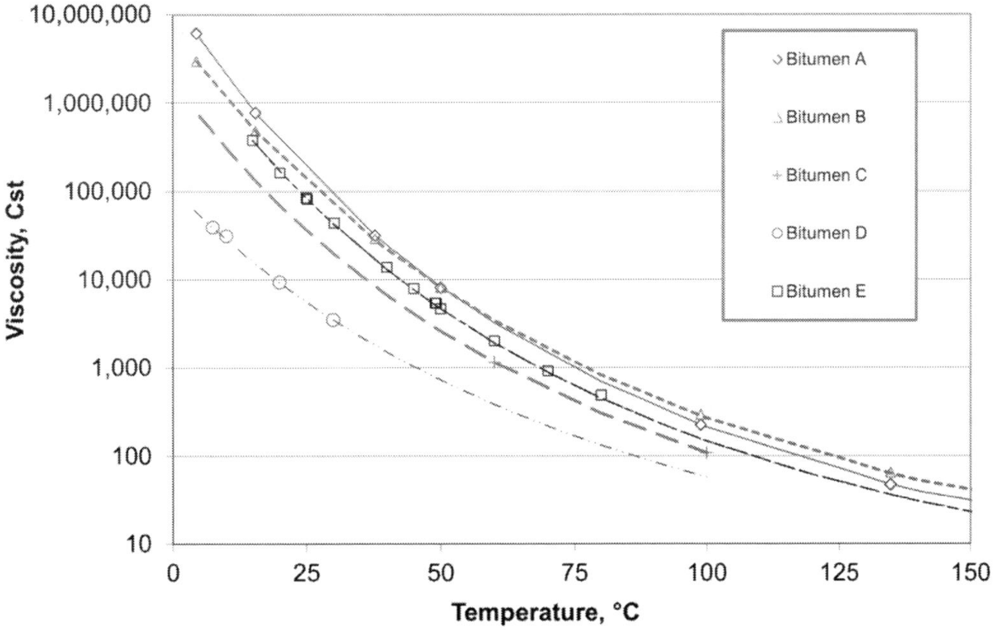

2

Predictions Errors in Dilbit Mixing Rules

FLUOR

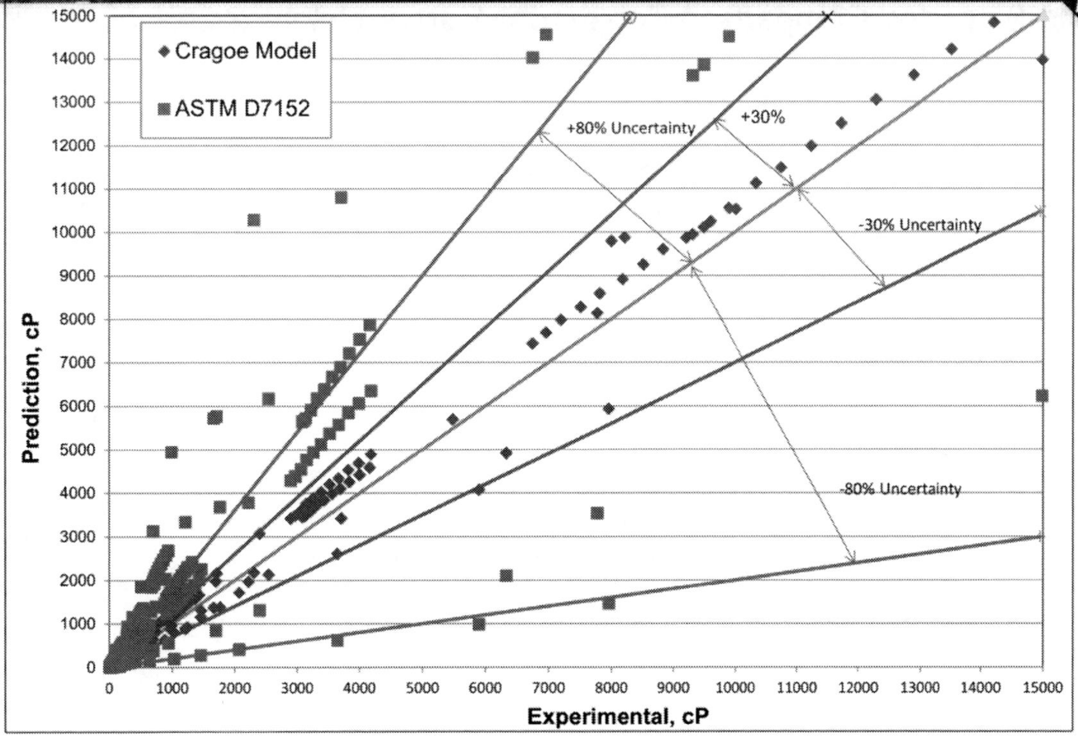

3

Prediction Errors of Viscosity of Diluted Bitumen

FLUOR

- Extracted Comparison Table from Marco A. Satyro's paper:

mixture	ideal mixing rules ($\alpha_{ij} = 0$) AARD (%)	ideal mixing rules ($\alpha_{ij} = 0$) MARD (%)	mixing rules with correlated α_{ij} AARD (%)	mixing rules with correlated α_{ij} MARD (%)
Athabasca 1/toluene	43	60	6	16
Athabasca 1/xylene	33	50	10	22
Athabasca 2/pentane	17	37	12	26
Athabasca 3/decane	8	35	9	38
Athabasca 4/tetradecane	24	48	9	32
Cold Lake/toluene	17	53	10	25
McKay River/hexane	19	37	16	25
McKay River/decane	22	48	24	53
McKay River/toluene	68	200	37	73
McKay River/mixture 1	17	34	16	28
McKay River/mixture 2	34	74	19	38

F. Ramos-Pallares, F. F. Schoeggl, S. D. Tsylor, M. A. Satyro, and H. W. Yarronto, Energy & Fuel, 30, 3575-3595, 2016.

4

Prediction Errors of Viscosity of Diluted Bitumen

FLUOR

- Extracted Comparison Table from Marco A. Satyro's paper:

Summary or errors for Cold Lake bitumen/toluene mixtures.

Fluid	% AARD$\beta_{ij} = 0$	% AARD$\beta_{ij} = -0.0538$
Bitumen	2.61	2.61
Mix 1 (1.61% toluene)	55.3	26.4
Mix 2 (4.71% toluene)	1147	18.2
Mix 3 (9.55% toluene)	143,311	30.7
Total average	251.5	15.5 ★

M. A. Satyro, and H. W. Yarronto, Fluid Phase Equilibria, 298, 1-11, 2010.

5

Uncertainty in Viscosity Calculation of Bitumen / Dilbit

FLUOR

- Viscosity correlation of bitumen is an ad hoc practice - accuracy depending on selection of model or model fitting
- Prediction performance differs significantly – no universal model that works accurately for all the samples
- Average Relative Deviations can be as high as ±200% if the prediction model is not validated.
- Generally the viscosity uncertainty can be managed within ±30% if the correct model or methods are used.
- Correspondingly, the exchanger design can be substantially impacted by errors in estimated fluid viscosities

6

Viscosity Impact on Exchanger Design – HTRI Study Case

FLUOR

- A case study is performed using hot bitumen cooler as an example – glycol water as coolant
- Coolant is on tube side and process properties are kept the same for all the cases studied
- The case design has 12 shells (3 parallel and 4 series)
- Hot bitumen is cooled from 130°C to 72°C on shell side
- Baseline case: average bitumen viscosity = 35 cSt
  - μ_{in} = 14 cSt and μ_{out} = 56 cSt
- A 30% step change in bitumen viscosity to explore the sensitivity in exchanger design at different viscosity ranges
- All other properties are kept the same for all cases to isolate the impact of the viscosity on heat transfer

7

Hot Bitumen Coolers in Oil Sand Project

Hot Bitumen Coolers in Oil Sand Project

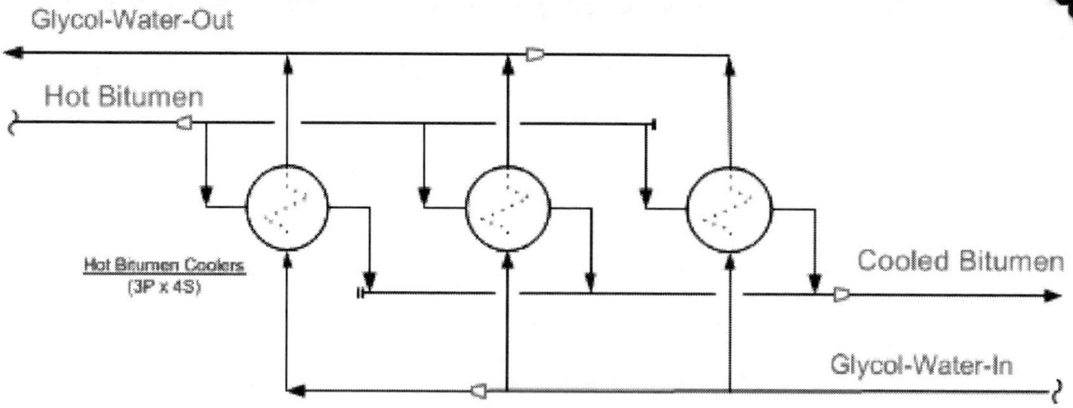

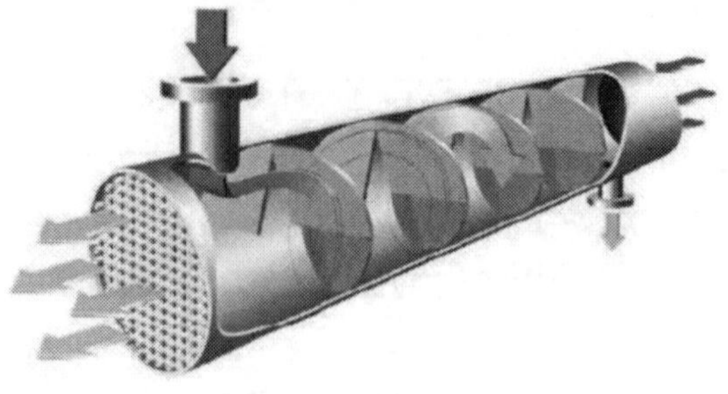

9

Viscosity Impact on Exchanger Design – HTRI Study Case

FLUOR

- A case study is performed using hot bitumen cooler as an example – glycol water as coolant
- Coolant is on tube side and process properties are kept the same for all the cases studied
- The case design has 12 shells (3 parallel and 4 series)
- Hot bitumen is cooled from 130°C to 72°C
- Baseline case: average bitumen viscosity = 35 cSt
  - μ_in = 14 cSt and μ_out = 56 cSt
- A 30% step change in bitumen viscosity to explore the sensitivity in exchanger design at different viscosity ranges
- All other properties are kept the same for all cases to isolate the impact of the viscosity on heat transfer

10

Viscosity Impact on Exchanger Design – HTRI Study Case

FLUOR

Perturbation Ratio - 30% Step Change	Average Bitumen Viscosity, cP	Req. A, Ft2	Over-Design% per 30% Step Change	Exchangers Equip. Cost $	Eng. Capital Cost $	$ change / 30% Step Change
3.71	129.9	65,700	-9.5%	$ 2,467,500	$ 9,870,000	$ 852,184
2.86	99.9	60,027	-4.7%	$ 2,254,454	$ 9,017,816	$ 406,002
2.20	76.9	57,325	-3.4%	$ 2,152,953	$ 8,611,814	$ 279,184
1.69	59.1	55,466	-3.8%	$ 2,083,157	$ 8,332,630	$ 304,976
1.30	45.5	53,436	-3.5%	$ 2,006,913	$ 8,027,654	$ 274,315
1.00	35.0	51,610	Baseline	$ 1,938,335	$ 7,753,339	Baseline
0.70	24.5	48,815	5.4%	$ 1,833,346	$ 7,333,383	$ (419,956)
0.49	17.1	46,190	5.4%	$ 1,734,744	$ 6,938,976	$ (394,406)
0.34	12.0	43,888	5.0%	$ 1,648,297	$ 6,593,186	$ (345,790)
0.24	8.4	42,458	3.3%	$ 1,594,610	$ 6,378,441	$ (214,745)
0.17	5.9	41,564	2.1%	$ 1,561,017	$ 6,244,069	$ (134,372)

11

Viscosity Impact on Exchanger Design – HTRI Study Case

FLUOR

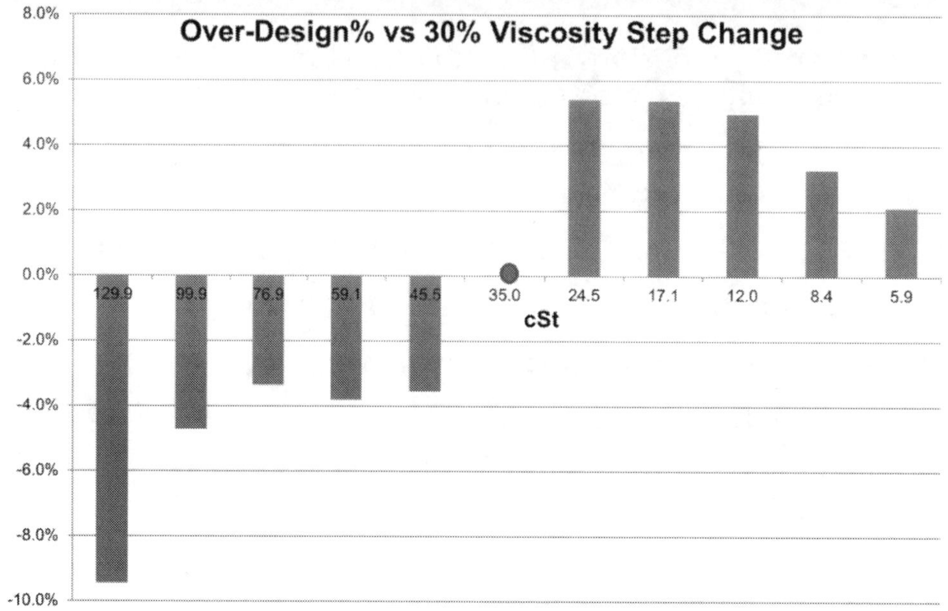

12

Viscosity Impact on Exchanger Design – HTRI Study Case

FLUOR

%Change over Baseline	Average DilBit Viscosity, cP	Req. A, Ft2	Over-Design% to Baseline	Exchangers Equip. Cost $	Eng. Capital Cost $	$ change to Baseline
271%	129.9	65,700	-27.3%	$ 2,467,500	$ 9,870,000	$ 2,116,661
186%	99.9	60,027	-16.3%	$ 2,254,454	$ 9,017,816	$ 1,264,478
120%	76.9	57,325	-11.1%	$ 2,152,953	$ 8,611,814	$ 858,475
69%	59.1	55,466	-7.5%	$ 2,083,157	$ 8,332,630	$ 579,291
30%	45.5	53,436	-3.5%	$ 2,006,913	$ 8,027,654	$ 274,315
0.00	35.0	51,610	0.0%	$ 1,938,335	$ 7,753,339	0
-30%	24.5	48,815	5.4%	$ 1,833,346	$ 7,333,383	$ (419,956)
-51%	17.1	46,190	10.5%	$ 1,734,744	$ 6,938,976	$ (814,362)
-66%	12.0	43,888	15.0%	$ 1,648,297	$ 6,593,186	$ (1,160,152)
-76%	8.4	42,458	17.7%	$ 1,594,610	$ 6,378,441	$ (1,374,897)
-83%	5.9	41,564	19.5%	$ 1,561,017	$ 6,244,069	$ (1,509,269)

13

Thermal Conductivity Impact on Exchanger Design – Uncertainty in Heavy Oil Prediction

FLUOR

- Thermal conductivity predictions for hydrocarbon liquid are typically empirical correlations based on experimental data
- J. R. Fair - "Advanced Process Engineering"
  - Overall ±20% error resulting in ~ ±10% equipment sizing
- For heavy oil like bitumen, the prediction error can be as high as 87% based on a comparison study over 52 Exp. data:

Prediction Model	Sato & Riedel (1977)	Riazi & Faghrl (1985)	Lakshimi & Prasad (1992)	Aboul & Moharam (1999)	API Proce. 12A3.2 (1997)	Latini et al. (1983)
AARD	11.0%	87.5%	16.8%	19.6%	27.0%	46.5%

- Typically thermal conductivity is over-estimated by simulators for heavy oil and resulting in under-sizing of the exchangers, based on our studies for the oil sand projects.

- J. R. Fair, AIChE Monograph Series, 76(13), 1980.

14

Thermal Conductivity Impact on Exchanger Design – HTRI Study Case

FLUOR

- A case study is performed using hot bitumen cooler as an example – glycol water as coolant
- The same system as viscosity study, but change TC only

Perturbation Ratio - 10% Step Change	Average TC, Btu/h-ft-F	Over-Design% per 10% Step Change	Exchangers Equipment Cost $	Engineering Capital Cost $	$ change / 10% Step Change
1.50	0.104	-3.44%	$ 2,043,090	$ 8,172,360	$ (291,165)
1.40	0.097	-3.52%	$ 2,115,881	$ 8,463,525	$ (308,931)
1.30	0.090	-3.71%	$ 2,193,114	$ 8,772,456	$ (337,554)
1.20	0.083	-3.79%	$ 2,277,503	$ 9,110,010	$ (359,268)
1.10	0.076	-4.00%	$ 2,367,320	$ 9,469,278	$ (394,722)
1.00	0.069	0.00%	$ 2,466,000	$ 9,864,000	Baseline
0.90	0.062	4.38%	$ 2,574,096	$ 10,296,384	$ 432,384
0.80	0.056	4.55%	$ 2,691,302	$ 10,765,209	$ 468,825
0.70	0.049	4.91%	$ 2,823,560	$ 11,294,241	$ 529,032
0.60	0.042	5.19%	$ 2,970,130	$ 11,880,519	$ 586,278
0.50	0.035	5.67%	$ 3,138,413	$ 12,553,653	$ 673,134

15

Thermal Conductivity Impact on Exchanger Design – HTRI Study Case

FLUOR

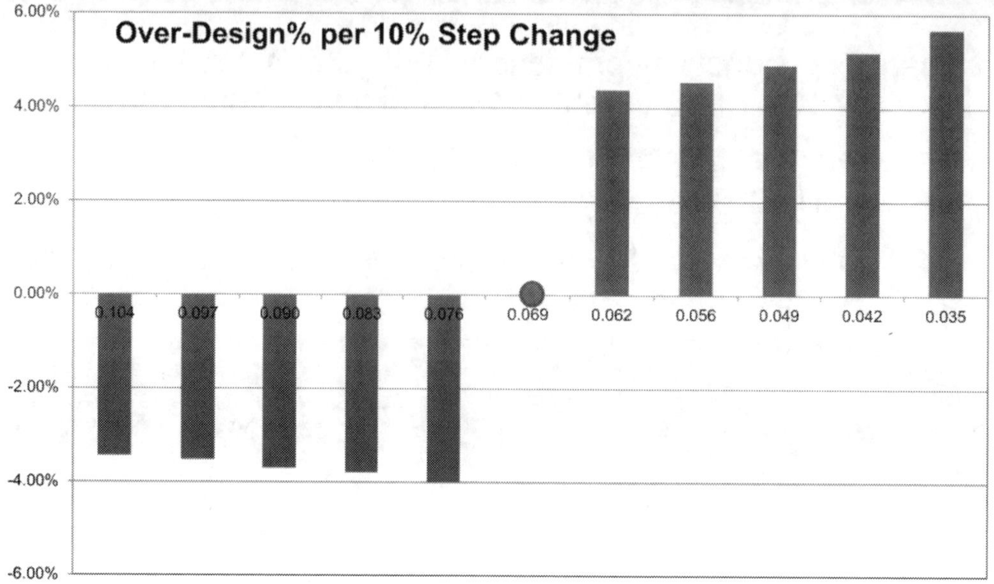

16

Thermal Conductivity Impact on Exchanger Design – HTRI Study Case

FLUOR

- A case study is performed using hot bitumen cooler as an example – glycol water as coolant
- The relative $ Changes over the baseline

%Change over Baseline	Average TC, Btu/h-ft-F	Over-Design% to Baseline	Exchangers Equipment Cost $	Engineering Capital Cost $	$ change over Baseline
50%	0.104	-18.5%	$ 2,043,090	$ 8,172,360	$ (1,691,640)
40%	0.097	-15.0%	$ 2,115,881	$ 8,463,525	$ (1,400,475)
30%	0.090	-11.5%	$ 2,193,114	$ 8,772,456	$ (1,091,544)
20%	0.083	-7.8%	$ 2,277,503	$ 9,110,010	$ (753,990)
10%	0.076	-4.0%	$ 2,367,320	$ 9,469,278	$ (394,722)
0.00	0.069	0	$ 2,466,000	$ 9,864,000	0.00
-10%	0.062	4.4%	$ 2,574,096	$ 10,296,384	$ 432,384
-20%	0.056	8.9%	$ 2,691,302	$ 10,765,209	$ 901,209
-30%	0.049	13.9%	$ 2,823,560	$ 11,294,241	$ 1,430,241
-40%	0.042	19.0%	$ 2,970,130	$ 11,880,519	$ 2,016,519
-50%	0.035	24.7%	$ 3,138,413	$ 12,553,653	$ 2,689,653

17

What can be done to reduce the uncertainty in Exchanger Design for Bitumen service?

FLUOR

- Both viscosity and TC prediction models need to be carefully selected for heavy oil service
- Preferable use models developed based on heavy oil data
- Validate the model performance with available experimental data
- Design the heat exchangers with the low viscous fluid (preferable in the < 100 cSt zone)
- Low viscosity impact and low uncertainty for less viscous fluid
- Showcase: shift bitumen cooler service into Dilbit cooler service by adding diluent upstream of the hot bitumen coolers

18

Hot Bitumen Coolers in Oil Sand Project

FLUOR

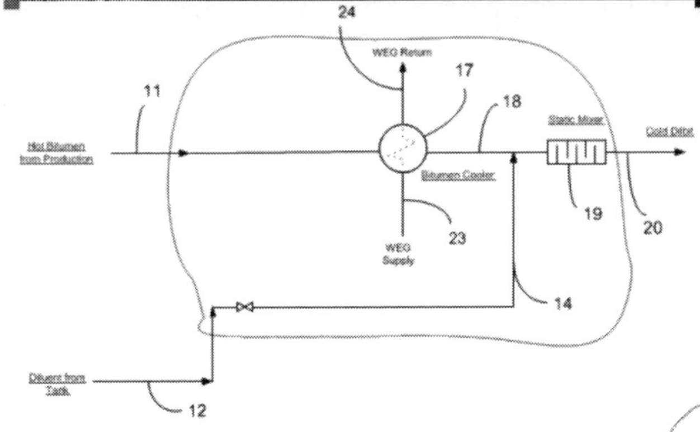

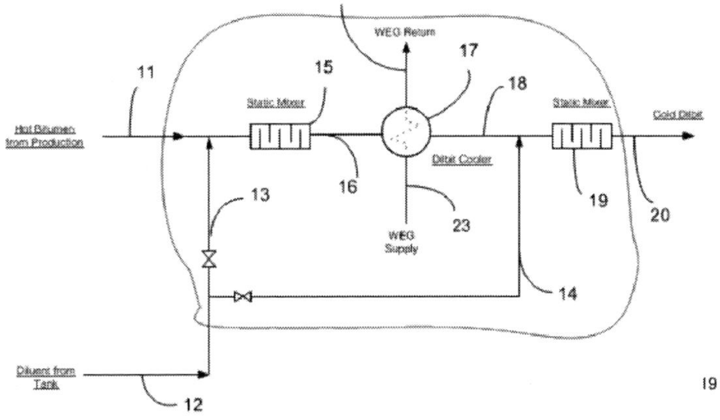

19

Hot Bitumen Cooling Process in Oil Sand Bitumen Production

FLUOR

- Current bitumen cooling process without pre-dilution:
  - Bitumen coolers processing ~200,000 BPSD bitumen
  - Helix exchangers are used to handle the viscous fluid (less ΔP)
  - WEG inlet T >= 50°C to avoid bitumen gel formation
  - High MTD, high fouling, and worse heat transfer due to high viscous
  - High uncertainty in viscosity prediction
- Proposed new bitumen cooling configuration with pre-dilution
  - Pre-dilute 53% of total amount of diluent required
  - Standard heat exchangers can be used (less viscous)
  - Colder WEG supply T can be used (no concern for gel formation)
  - Lower MTD, low fouling, and better heat transfer due to less viscous
  - Low uncertainty in viscosity prediction

20

Hot Bitumen Coolers in Oil Sand Project

FLUOR

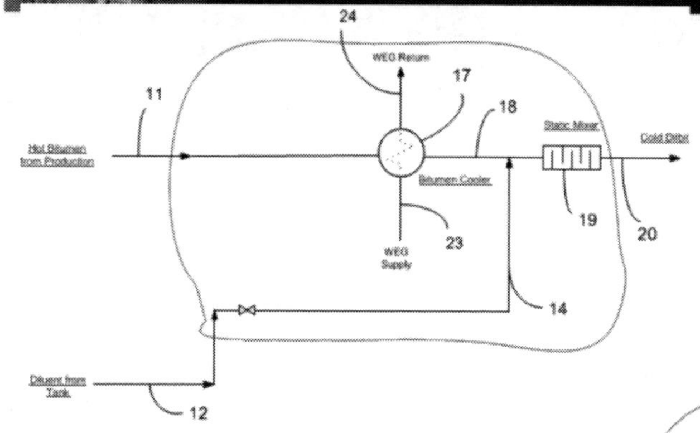

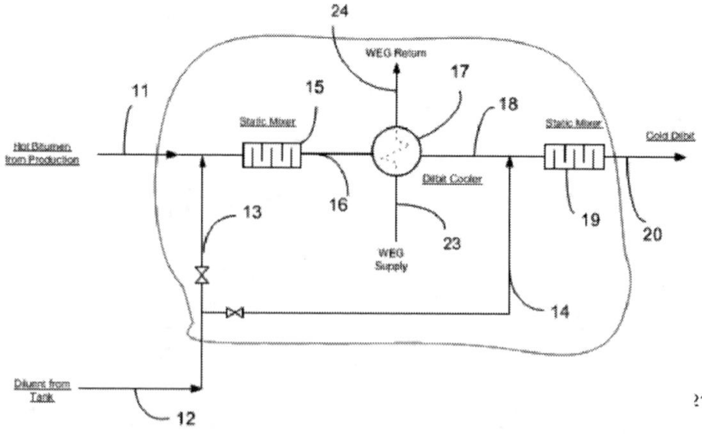

21

Comparison of Economic Study of Bitumen with / without Pre-dilution

FLUOR

Parameters \ Case	Current Plant Design (No dilulent Premixing)	Premixed Bitumen Cooler (Worst Case)	Savings in $CAN by Premixing of Diluent
Hot fluid viscosity at inlet / outlet	47.1 cP / 781.2 cP	20.0 cP / 100 cP	
Train#1 Bitumen Coolers duty	28.1 MW	28.1 MW	
No of Shell for Train#1 Bitumen Cooler	16 (4P x 4S)	12 (3P x 4S)	
Cost of Train#1 Bitument Cooler	$3,290,000	$2,467,500	$822,500
Train#2 Bitumen Coolers duty	18.4 MW	18.4 MW	
No of Shell for Train#2 Bitumen Cooler	12 (3P x 4S)	8 (2P x 4S)	
Cost of Train#2 Bitument Cooler	$2,280,000	$1,520,000	$760,000
Comparison Basis for Bitumen Cooler	Helix Exhanger, WEG Inlet/Out T = 50°C/93°C	Std. Exchanger, WEG Inlet/Out T = 38°C/81°C	
Primary WEG Air Cooler, Duty (# Bays)	53.8 MW (10 bays)	47.0 MW (8 bays)	
Secondary WEG Air Cooler, Duty (# Bays)	10.9 MW (5 bays)	17.5 MW (8 bays)	
Total of WEG Air Cooler	$3,000,000	$3,200,000	-$200,000
Total Equipment Cost Saving	/	/	$1,382,500
Total Engineering Cost Saving	/	/	$5,530,000

22

Conclusions

FLUOR

- Transport properties (viscosity and thermal conductivity) have strong impacts on heat exchanger design.
- The impacts are especially significant for heavy oil like bitumen, where the uncertainty in property estimate is also high.
- For bitumen processing, the uncertainty in viscosity prediction can be within ±30% if choosing proper models, resulting in ±10% margin required in exchanger design.
- For bitumen processing, the uncertainty in thermal conductivity prediction can be within ±10% if correct correlations are used, resulting in ±5% margin required in exchanger design.
- The exchanger design uncertainty can be improved by designing the heat exchanger in a less viscous zone if possible.
- Significant cost saving in engineering by reducing uncertainty in transport property predictions / resulting in less design margin req.

23

AUTHOR INDEX

Abbasi, Pedram 428, 961
Abbott, Nicholas L. 45, 465, 468, 834, 923, 992
Abedin, Rubaiyet 780
Abedini, Asghar 53
Abel, Steven M. 858
Abildskov, Jens....844
Abraham, Nathan 497
Abrams, Cameron F. 916
Abruna, Hector D....119
Acharya, Prashant 364
Acosta, Edgar 34
Adams, Ryan A. 964
Agah, Shaghayegh....102
Aglave, Ravindra 644
Ahitan, Sourabh....537, 551
Ahmed, Alauddin 980
Ahmmed, Shamim 39
Ahn, Gwangnoh 334
Ahn, Yun-Ho 381
Aichele, Clint P. 466, 1026
Akano, Koichi....644
Akella, Meghana 1072
Akhmetkhanova, Nelya 47, 1025
Al Ghafri, Saif Zs. 316
Alamani, Bryan G. 832
Al-Dahhan, Muthanna....323, 324
Alexandridis, Paschalis....359, 1087
Alicke, Alexandra....736
Alkire, Richard....419
Allen, Andrew J....91
Allen, Kyle 132
Alpak, Faruk O. 221
Alvarez, Nicolas J. 219, 829
Amama, Placidus B. 906
Amrhein, Lauren E. 779
Amundsen, Ted J. 58
Andersen, Simon 536
Anderson, Ryan....457
Anisimov, Mikhail A. 779, 845
Annaland, Martin Van Sint....781
Antila, Hanne 181
Antoniuk-Pablant, Antaeres 9
Antony, Lucas 295
Aponte-Rivera, Christian 512
Arauz-Lara, B. Jose Luis....706
Archer, Lynden A. 433, 438, 1060
Arlt, Wolfgang....230
Armstrong, Matthew 276, 332, 333, 557
Asadi, Mohammad....428, 961
Ashbaugh, Hank....416
Asthagiri, Dilip 92
Athas, Jasmin C. 89
Au, Sam....35
Auer, Stefan....993
Awan, Javeed....656
Ayappa, K. G. 234
Back, Tyson C. 906
Badruddoza, Abu Zayed Md 1014
Baek, Seungjun 381
Bagusetty, Abhishek 52
Bahadur, Vaibhav 411
Bahamon, Daniel 917
Bai, Peng 294
Bai, Xue 881
Bajdich, Michal 16
Bala, Aseel M. 854
Balarama Sridhar, Dwadasi 275
Balbuena, Perla B. 728, 1031, 1046, 1059
Baldea, Michael 345
Baldino, Silvio 713
Baled, Hseen O. 208
Ball, Madelyn 121
Banerjee, Atiya 233
Banerjee, Dwijen 1023
Banerjee, Sanjoy 453
Bao, Nanqi 465
Bara, Jason E. 53
Barakat, Joseph M. 33
Barbieri, Matheus R. 885
Barhaghi, Mohammad 1008
Barringer, David A. 534
Bastos, Jaci Carlo Schramm Camara 342, 746
Bates, Frank S. 358
Baz, Jorg 184, 999
Bazant, Martin Z. 152, 175, 406
Bazyleva, Ala 477
Beaudoin, Stephen P. 463, 1044
Becerra, Mildred 472
Beckham, Gregg T. 874
Behdani, Behrouz 322
Behera, Manas Ranjan 734
Behrens, Sven H. 3, 462, 527, 646, 1015
Bell, Alexis T. 11
Beltramo, Peter J. 736, 802
Beltran, Saul Perez 1046
Benck, Jesse D. 436
Benz, Gregory 462
Berg, John C. 153
Berger, Manuel 331
Berghout, Pieter 645
Beris, Antony N. 135, 515, 557, 824
Berry, Joe 222, 1027
Besser, Ronald S. 122, 525
Bevan, Michael A. 1, 503
Bezik, Cody 977
Bezinge, Leonard 969
Bhagavatula, Dinesh N V S S R 412
Bharti, Bhuvnesh 898
Biddinger, Elizabeth J. 840
Biggs, Simon 4

AUTHOR INDEX

Bilir, Taner 436
Bischofberger, Irmgard 405
Bishop, Kyle J. M. 742, 901
Biswal, Sibani Lisa 173, 326, 343, 517, 947
Biswas, Shaurjo 430
Bizarri, Gregory A. 73
Blake, Diane A. 958
Blas, Felipe J 855
Blenner, Mark 761
Bliatsiou, Chrysoula 320
Bobbitt, N. Scott 296
Bognet, Brice 478
Bohm, Lutz 320, 456
Boles, J. 955
Bolton, Christopher 5
Bonnecaze, Roger T. 411, 469, 513
Bose, Arijit 956
Bothun, Geoffrey D. 89, 741, 957, 958
Boudouris, Bryan W. 463, 944, 1044
Bourg, Ian 981
Bourque, Alexander 765
Bourret, Edith D. 73
Brady, Sonia K 796
Braun, Trevor M 437
Breedveld, Victor 70, 1015
Brennan, John K. 921
Brett, Dan 555
Brewer, Catherine E. 59
Broderick, Alicia 1083
Brown, Angela C. 709
Brun, Pierre 407
Bucior, Benjamin 296
Bui, Justin 731
Bull, Geoffrey 276
Bulsara, Pallav 914
Bunge, Annette L. 996
Burke, Colin M. 962
Burns, Mark A. 174
Burshtein, Noa 648
Butler, Paul 937
Butz, Julian 90
Cadirov, Nicholas 955
Caiazza, Carla 823
Calabrese, Michelle A. 724, 924
Calabrese, Richard V 884
Callaway, Connor 978
Camacho-Forero, Luis E. 1031
Camaioni, Donald M. 522
Campbell, Charlie 522
Campbell, Charlton 747
Campos, Christian A. T. 238
Cao, Jinrong 106
Cao, Lixia 530
Cao, Ping 368
Cao, Zhi 15, 839
Capecelatro, Jesse 745
Carbone, Paola 49
Carneiro, Juliana S. A. 523
Carter, Emily A. 426
Cartier, Charles A. 742, 901
Carvalho, Marcio S. 485, 487, 738
Casali, Dominic M. 28
Castaneda-Priego, Ramon 417
Castro-Arellano, Jose J. 241, 382
Caupin, Frederic 779, 845
Celik, Fuat E. 120
Cersonsky, Rose 994
Chaimayo, Wanaruk 1042
Chaimovich, Aviel 24, 775
Chakraborty, Aishik 801
Chakraborty, Sudipto 734
Chakrapani, Vidhya 1043
Chambers, Scott A. 429
Champagne, Jean-Yves 458
Chan, Karen 16
Chan, Kwong-Yu 50, 905
Chandran, Prashanth 976
Chandran, Vishnu Deep 145
Chang, Christopher 15, 839
Chang, Hsueh-Chia 170
Chapman, Walter G. 44, 92, 182, 221, 476, 947, 1079
Chauvel, Paul 186
Chavez-Flores, David 732
Cheluget, Eric L. 854
Chen, Chau-Chyun 649, 650, 853
Chen, Jiayi 296
Chen, Jingyi 777
Chen, Leanne 16
Chen, Mengjie 729
Chen, Qiao 235
Chen, Ru 819
Chen, Szu-Ying 955
Chen, Wan-Ting 56
Chen, Xue 321
Chen, Yeng-Long 35, 835
Chen, Yusi 436
Cheng, Feng 59
Cheng, Mark 1048
Cheng, Xiang 336, 400, 403, 486, 722
Cheng, Zhengdong 1075
Chhabra, R. P. 592
Chiang, Hao-Chun 703
Chiarot, Paul R. 535
Chicaroux, Andres Kulaguin 778
Chien, Fan-Tso 835
Chien, Wei 835
Cho, Jason 555
Choi, Siyoung 470, 480
Choi, Won Tae 70
Choudhary, Madhuresh K. 72

AUTHOR INDEX

Choudhury, Pabitra 52
Choudhury, Snehashis 433
Chun, Jaehun 491, 934
Chundawat, Shishir 796
Chung, Jaeyub 944
Churchill, Stuart W. 165
Clark, Aurora E. 932
Clark, Ezra L. 11
Clark, Jennifer A. 1010
Clark, Thomas 321
Cloitre, Michel 513
Co, Anne 365
Coasne, Benoit 922
Cogswell, Kyle 27, 29
Colina, Coray M. 127
Collins, Charles 10
Collins-Martinez, Virginia 732
Colon, Yamil J. 295, 977
Comer, Benjamin 424
Conrad, Jacinta C. 490, 982
Contreras-Naranjo, Jose C. 733
Coppens, Marc-Olivier 555
Coquelet, Christophe 312
Cordoba, Andres 877
Cordova-Figueroa, Ubaldo M. 506
Corson, Elizabeth 427
Corti, David . S. 933
Coutinho, Joao A. P. 855
Cowman, Jonathan 144
Cox, Kenneth R. 476
Crabtree, Ellis 53
Crawford, Brad 651
Creel, Erin 427
Criscenti, Louise J. 232
Cristancho, Diego E. 782
Crocker, John C. 518, 1061, 1073
Croell, Arne 38
Crosthwaite, Jacob M. 782
Csernica, Peter 119
Cummings, Peter T. 42, 104, 180, 919
Curtis, Jennifer Sinclair 147
Curtiss, Larry A. 961
Cussler, Edward L. 439
Dagastine, Raymond R. 5, 222, 464, 1027
Dai, Gance 408, 744
Dalal, Indranil Saha 327
Dallas, Andrew J. 176
Damasceno, Pablo F. 931
Danahy, Brooks B. 652
Dani, Archit 65, 743
Daniel, Susan 357
Dann, Kevin 943
Darjani, Shaghayegh 1025
Das, Soumen 528
Das, Soumik 411, 469
Dasetty, Siva 761
Dasgupta, Anirvan 734
Dasgupta, Sunando 528
Datta, Sujit S. 396, 859
Daux, Virgile 478
David, Allan E. 509
Davis, Jonathan 731
Davis, Robert H. 172
Day, Allen 782
De Pablo, Juan J. 103, 133, 295, 398, 479, 833, 834, 877, 938, 977, 992, 1012
De Pablo, Luis X. 103
De Yoreo, James J. 934
Debenedetti, Pablo G. 123, 760
Deegan, Michael 333
Defever, Ryan 367
Degen, George 704
Dehghani, Mohammad Reza 857
Deidda, Graziano 875
Delacruz-Araujo, Ronal A. 506
Delhommelle, Jerome 967, 987
Dellago, Christoph 775
Derby, Jeffrey J. 38, 73, 484, 556, 740
Derksen, Bridget 52
Desgranges, Caroline 967, 987
Deshpande, Parag Arvind 283
Desikan, Rajat 234
Desjardins, Olivier 745, 883
Deskins, N. Aaron 287
Devarajan, Dinesh Sundaravadivelu 68
Dhakal, Subas 803
Dhar, Prajnaparamita 801
Diamond, Scott L. 518
Dickey, Michael D. 1041
Dighe, Anish V. 971
Dignon, Gregory L. 764
Diky, Vladimir 475, 477
Ding, Junhuan 654
Ding, Liang-Xin 902
Ding, Lin 749
Dinic, Jelena 329, 711
Dinivahi, Mangalaa 832
Disalvo, Francis J. 119
Divecha, Mia S. 556
Divvela, Mounica Jyothi 570
Do, Hainam 1009
Dobbs, Howard 955
Dodd, Paul 502, 505, 994
Doherty, Michael F. 970
Donahue, Melissa 345
Dong, Xiaorui 733
Dorfman, Kevin D. 722
Dou, Yong 742, 901
Dowd Jr., Regis 1030
Dowdle, John 782

AUTHOR INDEX

Doyle, Patrick S. 177, 643, 712, 1014
Dshemuchadse, Julia 502, 505, 931
Du, Chrisy Xiyu 500, 505
Du, Yan 750
Du, Yingge 429
Duan, Xindi 947
Dumesic, James A. 121
Dunn, Alexander 129
Duska, Michal 779, 845
Dutcher, Cari S. 37, 176, 516, 743
Dutta, Sarit 818
Dyatkin, Boris 42
Dybeck, Eric 497
Edgar, James H. 231
Egger, Torben 355
Egorov, Sergei 1077
El Hajem, Mahmoud 458
Eldridge, R. Bruce 345
Elizalde-Solis, Octavio 227
Emanuelsson, Emma 346
Enders, Sabine 90, 501
Engel, Michael 931, 966
Englert, Derek 454
Enick, Robert M. 208
Eppinger, Thomas 644
Errington, Jeffrey 498
Errington, Jeffrey R. 41, 71, 975
Escobedo, Fernando 989
Espin, Leonardo 409
Espinoza, Said E. Aranda 706
Esposito, Daniel V. 521, 731
Esquivel-Mora, Pedro 382
Esteves, A. Catarina C. 927, 1022
Fagan, Jeffrey A. 935
Fan, Tzu-Hsien 702
Fang, Bing 368
Farrell, Zachary 1077
Favero, Claudio Vilas Boas 168
Fawzi, Nicolas 764
Feaster, Jeremy T. 9
Fei, Wenjie 901
Fell, James 971
Ferguson, Andrew L. 990
Fernandes, Robert L. 239
Ferrier, Nicola J. 977
Fetisov, Evgenii 983
Fewkes, Christopher 1027
Fialoke, Suruchi 973
Fieg, Georg 355
Filler, Michael A. 1015
Fiore, Andrew 61
Flake, John C. 780
Follette, Marissa 906
Ford, David M. 499, 988
Ford, Katie 325, 401
Fouad, Wael A. 476
Fox, Rodney O. 647, 745
Francis, Lorraine F. 487
Frank, Florian 221
Frank, Timothy C. 854
Franses, Elias I 933, 944
Frechette, Joelle 69
Fredrickson, Glenn H. 725, 726
Freysinger, Wolfgang 331
Friedrich, Jochen 38
Frisbie, C. Daniel 841
Frityanti, Maya 795
Frostad, John M. 642
Fu, Chengyin 904
Fu, Yuan-Xiang 571
Fujigaya, Tsuyohiko 12
Fukai, Jun 584
Fuller, Gerald G. 335, 640
Fulton, John L. 522
Funazukuri, Toshitaka 236, 240
Furst, Eric M. 150
Gadikota, Greeshma 91, 981
Gajula, Kishore 275
Galicia-Luna, Luis A. 226, 241, 382
Galli, Giulia 977
Gallup, Jennifer 118
Gamble, Jordan A. 654
Gamwo, Isaac 208
Gani, Rafiqul 1084
Gao, Jingsi 151
Gao, Xi 747
Gao, Yongxiang 748
Garbin, Valeria 39
Garcia, Robert M. 534
Gardel, Margaret L. 398
Gardeniers, J. G. E. 171
Gardner, Jasmine 916
Garg, Aaron 121
Gebbie, Matthew A. 705, 1081
Geißelbrecht, Michael 230
Geng, Linxiao 1047
Geng, Yina 502
Ghosh, Parthasarathi 356, 461
Giacomelli, Jason J. 318
Giacomin, A. Jeffrey 723
Giberti, Federico 977
Gilbert, William J. R. 188
Gilchrist, James F. 493
Giles, Neil 843
Giles, Stephen A. 520
Glascoe, Elizabeth 568
Glaser, Jens 19, 1074
Glatz, Brittany 43
Glotzer, Sharon C. 19, 164, 500, 502, 505, 931, 966, 994, 1074

AUTHOR INDEX

Gochez, Roque....31
Godfrin, Michael....956
Godfrin, P. Douglas....417
Goel, Sachin....34
Gogotsi, Yury....42
Gokul, Navneeth....1009
Goltl, Florian....292
Gong, Jing....749
Gong, Ming....15
Goodpaster, Jason....286
Gopalan, Arun....296
Gordillo, Leonardo....336
Gordon, Melissa B....826
Gosavi, Shachi....234
Goshayeshi, Bijan....237
Gow, Arthur S....926
Goyal, Akshara....974
Graham, Michael D....402, 818, 880
Graham, Trent R....932
Gray, Jeffrey J....701
Greeley, Jeffrey....282
Green, David....1077
Greenfield, Michael L....183, 1011
Greenlee, Lauren F....364
Grenville, Richard K....318
Grillet, Anne M....534
Grime, John M. A....17
Gu, Xiang-Kui....523
Gu, Zhuxiao....1078
Gu, Zongyu....175
Guala, Diego....832
Guay, Martin....723
Gubbins, Keith E....93
Guida, Vincenzo....823
Guignard, Florian....895
Guio, Carlos Morales....9
Guo, Ashley....295, 977
Guo, Hou-Jun....702
Guo, Juchen....904, 1047
Guo, Shuo....403
Guo, Wenjing....71
Gupta, Ankur....177, 643, 1014
Gupta, Manish....736
Gupta, Rakesh....275
Gupta, Shiva....727
Gutierrez, Oliver....522
Gygi, Francois....977
Ha, Harry Z....538
Hadjittofis, Eftychios....532, 737
Haghighat, Arya....747
Hahn, Christopher....9, 11
Hall, Carol K....18, 124, 993
Hamasaki, Yuki....12
Han, Endao....479
Handleton, Rachel M....28
Hanley, Alan....509
Hann, Sarah....415
Hansen, Jesper S....183
Hansen, Niels....184, 766, 999
Harris, James....436
Hartkamp, Remco....996
Hartman, Ryan L....169
Hasebe, Shinji....334
Hassan, Yassin A....36
Hatton, T. Alan....177, 643
Hatzell, Marta....424
Hauser, Brad G....176
Haward, Simon....335, 640, 648
Hayashi, Keita....217
He, Peng....343
He, Yuxin....1082
He, Zhiqi....1087
Headen, Thomas....339
Headley, Alexander....965
Heidarian, Sharareh....780
Heinonen, Olle G....133, 977
Helfferich, Julian....977
Helton, Tyler....332
Hendrix, Marco M. R. M....927
Heng, Jerry....532, 737
Heng, Yi....571
Herceg, Eldad....186
Hernandez-Ortiz, Juan....133, 833
Herrera-Peraza, Eduardo....732
Hess, Dennis W....70
Heusinger, Ferdinand....230
Heylmun, Jeffrey C....647
Higgins, Drew....9
Hill, James C....165
Hinkle, Kevin R....21
Hinton, Zachary R....219, 829
Hirasaki, George J....947
Hironaka, Shuji....584
Hirose, Masanori....218
Ho, Tuan....232
Hoepfner, Michael P....339
Hoffmann, Kyle....295
Hofmann, Jennifer L....489
Hoga, Heloisa E....224, 238, 239
Hogge, Joseph W....843
Horner, Jeffrey S....135
Horsch, Martin T....25
Hoss, Darby J....463
Hossain, Md. Tashfin Zayed....526
Hossain, Nazir....649
Howard, Michael P....991, 1076
Hoyer, Wolfgang....776
Hrenya, Christine....494
Hruby, Jan....845
Hsiao, Lilian....66

AUTHOR INDEX

Hsieh, An-Hsuan 933
Hsieh, Chieh-Ming 223
Hsieh, Chih-Chen 115, 702
Hu, David L. 897
Hu, Jianguang 408
Hu, Sheng 360
Hu, Yue 379
Huang, Chung-Hsuan 739
Huang, Liangliang 93
Huang, Min 280
Huang, Tan 117
Huang, Wenxin 12
Huang, Yun-Ru 467
Hudgins, Jesse 276
Hudson, Steven D. 804
Hung, Francisco R. 105, 780
Hunt, Sean T. 121
Huo, Yijie 436
Hussain, S. M. Shakil 942
Huynh, Hang 430
Hwang, Hyun-Tae 454
Hwang, Luke 480
Iacovella, Christopher R. 180, 914, 996
Ilias, Shamsuddin 526
Inoue, Gen 584
Intikhab, Saad 118
Irudayanathan, Flaviyan Jerome 794, 873, 913
Ismagilov, Rustem 859
Ismail, Ahmed E. 651
Israelachvili, Jacob 704, 705, 955
Iwasaki, Fumihiko 925
Iyemperumal, Satish 287
Jackson, James E. 854
Jadhav, Ankur 959
Jaeger, Heinrich 479
Jain, Deeksha 365
Jalilvand, Zohreh 894
Jamali, Vida 930
Jameson, Cynthia J. 909
Jamieson, Emily 1027
Jana, Suvamay 874
Jang, Seung Soon 978
Jaramillo, Thomas F. 9, 11, 436
Jasim, Ahmed 524
Jauss, Thomas 38
Jena, Umakanta 59
Jenkins, Ian 1061, 1073
Jhong, Molly 12
Jia, Jieyang 436
Jiang, Peng 1078
Jiang, Shaoyi 51
Jiang, Xikai 133, 977
Jiang, Zhitong 973
Jiao, Feng 10
Jimenez, Leidy N. 329, 711
John, George 945
John, Vijay T. 89, 957, 958
Johnson, J. Karl 52
Jokar, Mojtaba 763
Jones, James 492
Jones, Matthew 346
Jonnalagadda, Sai Vamshi R. 776, 875
Joo, Yong Lak 570
Joodaki, Faramarz 1011
Joshi, Kedar 493
Joshipura, Ishan 1041
Joswiak, Mark 970
Juarez, Jaime 940, 1072
Kaar, Joel L. 797
Kadilak, Andrea 554
Kadri, Olufemi 145
Kaganyuk, Max 1028
Kale, Amullya 327
Kamal, Muhammad Shahzad 942
Kamphaus, Ethan P. 1059
Kamyabi, Nabiollah 130
Kang, Yijin 116
Kapoor, Utkarsh 653
Karaiskakis, Alexandros N. 840
Karamad, Mohammadreza 16
Karim, Alireza Mohammad 487
Kasbaoui, Mohamed H. 883
Kaufman, Yair 955
Kawaji, Masahiro 453
Kazakov, Andrei 475, 998
Kaznessis, Yiannis N. 912
Kelsey, Jarred 466
Kenis, Paul J. A. 12
Kenny, Dermot 144
Khabaz, Fardin 513
Khair, Aditya S. 6
Khalaf, Nidal 723
Khan, Aliza 911
Khan, Saad A. 856
Khare, Ketan S. 831
Khare, Rajesh 68, 649, 853, 929
Khomami, Bamin 360
Khraisheh, Majeda 384
Kian, Kourosh 189
Kikkinides, Eustathios 988
Killian, William G. 854
Kim, Chaerin 12
Kim, Doyoung 832
Kim, Hye Hyun 797
Kim, Kibum 428
Kim, Kyuhan 470, 480
Kim, Sang Beom 760
Kim, Young C. 764, 860
Kim, Youngsang 427
King, Jerry W. 187

AUTHOR INDEX

King, Paul 553
Kinoshita, Motohiro 30
Kiran, Bandaru 344
Kirk, Charlotte 16
Kirkes, Toni 650
Kitchens, Christopher L. 31
Kjelstrup, Signe 368
Klapp, Sabine H. L. 124
Klein, Christoph 180
Klett, Adam S. 654
Klotz, Alexander 712
Kloxin, Christopher J. 826
Knapp, Ellen M. 508
Knehr, Kevin 430
Knott, Brandon C. 874
Knotts, Thomas A. 843
Knutson, Barbara L. 1082
Knuutila, Hanna K. 312
Ko, Derrick I. 884
Koch, Donald L. 883
Kofke, David A. 498, 974, 1009
Koga, Hiroaki 106
Kokini, Jozef 762
Kolb, Thomas 900
Kolomeisky, Anatoly 102
Komives, Claire F. 277
Kong, Bo 647, 747
Koplik, Joel 397, 510
Kostecki, Robert 427
Kowalik, Mikolaj 742
Kral, Florian 331
Kramer, Peter R. 399
Kraume, Matthias 320, 456
Kraxner, Michael 331
Kremer, Kurt 24
Kretzschmar, Ilona 506, 508, 510, 894, 901
Krishnamoorthy, Senthil 550
Kristiansen, Kai 955
Krivov, Sergei V. 993
Kroenlein, Kenneth 475, 477
Krohl, Patrick 134
Kroon, Maaike C. 781
Kuczera, Zachary 523
Kudaibergenov, Sarkyt 929
Kuech, Thomas F. 569
Kuei, Steve 326, 343, 517
Kuhn, Simon 496
Kulkarni, Ambarish R. 284
Kumar, Manjesh 72
Kumar, Nitin 398
Kumar, Satish 409, 485, 486, 738, 739
Kumar, Vimal 454
Kundan, Akshay 443
Kun-Lin, Yang 941
Kurapati, Yathish 830
Kurdzieł, Sophia 521
Kurihara, Kiyofumi 229, 236
Kurk, Michael (Andy) 828
Kurtenbach, Khia 103, 834
Kwon, Youngkook 837
Kyriakou, Panagiota 912
Labrador, Natalie 521
Lacassagne, Tom 458
Ladipo, Folami 1082
Lakkaraju, Rajaram 461
Lakshmanan, Hari Hara Sudhan 340
Lamie, Willam 26
Landers, Alan 9
Lang, Matthew J. 796
Lapeyri, Corey 277
Larson, Ronald G. 40, 168
Laster, Jennifer S. 1044
Lattanzi, Aaron 494
Lattuada, Marco 895
Laursen, Siris 422
Lavey, John 747
Lawal, Akanni S. 190, 209, 228, 298
Le Gac, S. 171
Leal, L. Gary 642, 725
Le-Doux, Travis 59
Lee, Daeyeon 415
Lee, Huen 381
Lee, Hyundo 177
Lee, Jae W. 381
Lee, Jingyun 898
Lee, Jong-Min 117
Lee, Ka Yee 480
Lee, Sangmin 966
Lee, Young Ki 518
Legg, Meesha 747
Lehmann, Marcus 134
Leitold, Christian 775
Lele, Pushkar 325, 401
Lemarchand, Claire A. 183
Leo, Sin-Yen 1078
Lequieu, Joshua 877, 977
Lercher, Johannes A. 522
Levicky, Rastislav 703
Li, Bing 858
Li, Chien-Yi 223
Li, Chi-Ying Vanessa 905
Li, Dongsheng 934
Li, Hui 1012
Li, Jiyuan 133, 977
Li, Ke 504
Li, Lei 530, 1080
Li, Ting 915
Li, Weihua 485
Li, Xi 514
Li, Xiao 833

AUTHOR INDEX

Li, Xue....493
Li, Yanxiang....530
Li, Yawei....361
Li, Yunzi....174
Li, Zheng....288
Liang, Huaqing....1032
Liang, Xiaodong....1084
Liang, Xiaoyu....41
Liang, Yunfeng....106, 383
Liedtke, Aleesha M. 328
Lin, John....9
Lin, Nancy J. 935
Lin, Weisong....974
Lin, Yu-Jiun....173, 343
Lindberg, Seth....319, 341, 511, 825
Linic, Suljo....425
Linico, Audrey....362
Lira, Carl T. 854
Liu, Bin....231
Liu, Chen-Yu....738
Liu, Fang....47, 1025
Liu, Honglai....46, 93, 128, 533
Liu, Jinlu....92
Liu, Jiuxu....55
Liu, Kunlei....274
Liu, Minye....878
Liu, Nian....960
Liu, Song....231
Liu, Tianfei....513
Liu, Wei....15
Liu, Xinyan....1084
Liu, Yang....749
Liu, Ying....554, 641
Liu, Yun....417
Livingston, Dana A. 186
Llovell, Felix....855
Lo, Simon....644
Loewenberg, Michael....444
Lone, Sohail Rasool....454
Long, Andrew W. 990
Long, Thomas....510
Lopez-Barron, Carlos R. 819
Lowe, Jeffrey S. 435, 1045
Lowe, Jennifer....321
Lu, Hang....32
Lu, Li....535
Lu, Shawn....12
Lu, Xiaohua....126
Lubers, Alia M. 362
Luc, Wesley....10
Lunt, Richard....148
Luo, Jia-Li....571
Luo, Jiu....571
Luo, Miao....537
Luo, Sheng....985
Luo, Yimin....2
Lutkenhaus, Jodie L. 985
Lyu, Shu-Shen....571
Lyu, Yuan....795
Ma, Anson....478
Ma, Chenbo....1032
Ma, Huilian....504
Ma, Huilin....911
Ma, Yingzhen....1087
Maag, Alex....58
Mackintosh, Fred....930
Maddala, Jeevan....340
Madsen, Jesper J. 17
Magee, Joseph W. 477
Maggioni, Giovanni Maria....969
Maginn, Edward J. 185, 293, 979
Maguire, Megan....362
Mahadevan, L. 483
Mahmoodi, Seyed Reza....122, 525
Mahynski, Nathan A. 975, 997
Maktabi, Sepehr....535
Maldarelli, Charles....65, 397, 945
Maldonado, Luis F. 762
Maldonado-Camargo, Lorena....62, 132
Malmstadt, Noah....707
Maloney, Ryan C. 124
Mamontov, Eugene....104
Mamtani, Kuldeep....365
Maneval, James E. 457
Mao, Runfang....1071
Mao, Yating....67
Markthaler, Daniel....766, 999
Marquez, Diego....277
Marquez-Montes, Raul....732
Marr, David W. M. 896
Marruecos, David Faulon....797
Marshall, Kristin A. 328
Marson, Ryan L. 40
Marthelot, Joel....407
Martignoni, Waldir Pedro....885
Martin, Lenore M. 1011
Martinez, Adam D. 534
Martinez-Gonzalez, Jose....103, 833
Masuda, Yoshihiro....106, 383
Mathias, Paul M. 538, 552, 854
Matiazzo, Tatiana....342, 746
Matin, Naser S. 274
Matsuda, Hiroyuki....229, 236
Matsuoka, Toshifumi....106
Matthews, Michael A....28
Maula, Tiara Ann....179
Mavrikakis, Manos....45, 292, 465
Mayer, Matthew....122, 525
Mazzotti, Marco....655, 969
McCabe, Clare....180, 914, 996

AUTHOR INDEX

McCloskey, Bryan D. 420, 427, 962, 963
McCormick, Alon 89, 928, 957
McDevitt, Kyle 903
McGinley III, James T. 1061
McHugh, Mark A. 208
McKenna, Gregory B. 514
McKone, James R. 119
McLaughlin, John B. 537
McNeary, William 362
Medford, Andrew 424
Mees, Bruna L 342, 746
Mehrle, Andreas 331
Meier, Henry F. 342, 746, 885
Melendez, Brandon 832
Mena, Sarah E. 174
Meredith, J. Carson 462, 527, 646
Merker, David 456
Merz, Steven 1077
Messerly, Richard A. 998
Messinger, Robert J. 959
Metaxas, Athena E. 516
Mettu, Srinivas 222, 464
Meza, Juan Manuel Hernandez 706
Mick, Jason R. 1008
Middleton, Chadwick E. 946
Milina, Maria 121
Miller, Benjamin 1042
Miller, Juanita 59
Min, Juwon 381
Ming-Yi, Chang 115
Minkara, Mona 777
Minnick, David L. 188, 652
Miranda, Michael 466
Mirecka, Ewa A. 776
Mironenko, Alexander V. 291
Mirri, Francesca 930
Mirzadeh, Mohammad 406
Mishra, Arpit 356, 461
Miska, Stefan Z. 713
Mitraki, Anna 875
Mittal, Jeetain 179, 764, 765, 860, 986, 997, 1071
Mo, Dong-Chuan 571
Mohanty, Ritesh P. 519
Mohraz, Ali 903, 1028
Molla, Getachew S. 844
Moller, Joshua 877, 977
Mondal, Kunal 1041
Monjezi, Saman 492
Monroy-Pena, Camilo 793
Monson, Peter A. 499, 988
Montiel-Macias, Elizabeth 728
Montoya, Joseph H. 289, 423
Moore, David J. 914
Moore, Timothy C. 914, 996
Moradi, Marzieh 936
Moravec, Davis B. 176
Morgan, Nathan T. 740
Morin, Samuel 103
Morris, Jeffrey F. 491
Morrish, Rachel 279
Morse, David 928
Mortazavi-Manesh, Sepideh 472
Moustafa, Sabry G. 974
Moyers, Scott 828
Mozaffari, Ali 397
Mu, Xiaoqun 44, 221
Mueller, Karl 432, 1058
Mujica, Maritza 1015
Mukherjee, Dibyendu 360
Mukherjee, Sanjoy 463
Mukherjee, Siddhartha 645
Mullen, Ryan Gotchy 293, 979
Muller, Erich A. 125
Muller, Karsten 230
Mumm, Daniel R. 903
Mundy, Christopher J. 934
Municchi, Federico 495
Murad, Sohail 909
Murphy, Ryan P. 557
Murugesan, Vijayakumar 432, 1058
Mustain, William E. 554, 730
Mwasame, Paul M. 515, 824
Mysona, Joshua 928
Nacy, Ayad 363
Naegele, Gerhard 417
Naganawa, Shigemi 383
Nakamura, Hidemi 217
Nakashima, Naotoshi 12
Nan, Yue 55
Nandy, Lucy 37
Nangia, Shikha 794, 873, 911, 913
Narayan, Shweta 176, 743
Narsimhan, Ganesan 795
Narsimhan, Vivek 712
Nasrabadi, Hadi 985
Nava, Glixon Mavarez 1023
Navarro, Rocio 172
Nealey, Paul F. 833
Neeves, Keith B. 134, 896
Neilsen, John 89
Neimark, Alexander V. 915
Neitz, Tracy 536
Nelson, Zachary 364
Nemer, Martin B. 965
Nemmaru, Bhargava 796
Neville, Tobias P. 555
Newberg, John T. 1083
Newcomb, Ken 185
Newman, Richmond S. 500
Ng, Jia Wei Desmond 436

AUTHOR INDEX

Ng, Samson 34
Ng, Simon........1048
Nguyen, Frank........402
Nguyen, Le Truc 941
Nguyen, Quoc P.........469
Nguyen, Quoc T........166, 879
Nguyen, Thao T. T. 443
Nguyen, Trung 20
Nichols, Jessica E. 963
Nikolla, Eranda 363, 523
Nikoubashman, Arash 991, 1076
Ning, Fulong 368
Nitopi, Stephanie........9
Nitsche, Ludwig C........641
Norskov, Jens 16, 284, 285
Nourian, Pouria........68
Nunes, John 741
Nuraje, Nurxat 929
Nuthalapati, Sri Harsha 1029
Obermeier, Jonas........230
O'Brien, David 536
Oettle, Shawn 492
Ofoli, Robert Y.........148
Oglesby, Irene K. 144
Ohashi, Yuki........229
Okada, Kazuya........404
Okamoto, Yukihiro........217, 218, 925, 1086
Olbricht, William L. 131
Olivieri, Gustavo V. 225
Olsen, Michael........747
Omarova, Marzhana........958
Omasta, Travis J.........730
Onyo, Daudi Saang' 1082
Ooe, Yoshiko 1086
Orjuela, Johana........793
Ornithopoulou, Eirini........875
Oropeza, Maria Veronica Carranza........330
Oroskar, Priyanka 909
Orozco-Mena, Raul........732
Orr, Asuka........776, 875
Ortiz, Camilla U.........531
Ortner, Franziska 655
Osgouei, Reza E........713
Osuji, Chinedum O. 488
O'Sullivan, Denis........823
Otsi, Naresh C. 104
Ouaknin, Gaddiel........22
Ousley, Evan........333
Overney, Rene M. 153
Ozbayoglu, Evren........713
Ozkan, Umit S. 365
Ozoe, Hiroyuki 571
Pacella, Michael S. 701
Palmer, Jeremy C........982
Pan, Zehao 170
Panagiotopoulos, Athanassios Z.....972, 991, 995, 1076
Pandey, Shashank 742, 901
Papavassiliou, Dimitrios V........63, 166, 879
Parajuli, Sanjiv........946
Parihar, Vartika 528
Park, Joontaek........322, 492
Park, Nayoung........490
Park, Sungjune 1041
Parker, Morghan........422
Parsi, Nikhila 711
Pasquali, Matteo 60, 102, 930
Passalacqua, Alberto........647
Patanwala, Huseini S........478
Patel, Amish........973, 984
Patel, Anjli M........284
Patel, Navin........854
Patton, Steven 220
Pauchard, Vincent........47, 1025
Paul, Alexandra........642
Paulechka, Eugene 477
Paulsen, Alex D.........58
Payne, Christina M. 874, 876
Pednekar, Sidhant 491
Peles, Yoav 167
Pellenq, Roland J.-M.........48
Pendery, Joel........992
Peng, Xiaoguang........514
Peng, Xiong........554, 730
Peng, Yi 400, 403
Peng, Yunhu 66
Pennington, Ashley M. 120
Pera-Titus, Marc 533
Perera, Dinal........457
Perrodin, Didier........73
Persson, Kristin 423, 432, 1058
Pervaje, Amulya........856
Pesek, Joseph 277
Peter, Christine........24
Peters, Baron 968, 970
Peters, Cor J. 777, 781
Peterson, Jeffrey H.........73, 484
Peterson, Joseph 725
Pham, Ngoc Hong........63
Pham, Truong........486
Phelan Jr., Frederick R........21, 831
Phillips, Carolyn L. 931
Pillei, Martin........331
Pimentel-Rodas, Alfredo........226, 241
Pinho, Bruno........169
Pinto, Diego D. D. 312
Plata, Paola Leon........641
Plawsky, Joel L.........167, 443
Pol, Vilas G. 964
Poling-Skutvik, Ryan 982
Pope, Daniel J. 932

AUTHOR INDEX

Porcar, Lionel....724, 924
Porter, Christopher....518
Potoff, Jeffrey J....1008
Poudyal, Samiksha....422
Pourjafar, Sara....57
Pourtousi, Mohammad....645
Pradhan, Sahadev....242, 272, 593
Prajapati, Aditya....14
Prasad, Ajay K....520
Prashant, Ishan....327
Pretti, Evan....997
Preuster, Patrick....230
Preziosi, Valentina....823
Prieve, Dennis....146
Priye, Aashish....36
Pulikkathara, Merlyn....536
Puntambekar, Ajinkya....1043
Purohit, Apoorva....498
Pushpavanam, S....412
Qazvini, Nader Taheri....938, 1012
Qi, Qin M....144
Qian, Yuzhou....399
Qing, Meiyi....1032
Qiu, Fen....427
Qu, Xiaohui....432
Quevedo, Miguel Gonzalo Arenas....227
Quevillon, Michael....977
Quinn, Joseph....425
Quirke, Nicholas....50
Rabideau, Brooks D....1085
Raciti, David....13
Radke, Clayton J....149
Radl, Stefan....495
Radler, Mike....321
Rafiq, Rabees....1024
Raghavan, Srinivasa R....89, 957
Rahimi, Mohammad....834, 977, 992
Rahmanian, Vahid....860
Rai, Beena....275
Rai, Prabin....45
Rajagopal, Nandhini....873
Rajput, Nav Nidhi....432, 1058
Ramachandran, Arun....34
Raman, Ashwin Kumar Yegya....466, 1026
Ramezani-Dakhel, Hadi....834, 977, 992, 1012
Ramirez-Saito, Angeles....706
Ramos-Pallares, Francisco....474
Ramos-Sanchez, Victor....732
Randolph, Theodore W....798
Rankin, Stephen E....1082
Rao, Hemanth Giri V....234
Rashidi, Aidin....507
Rasti, Elnaz S....709
Rathee, Vikramjit....977
Rathi, Ashutosh....988
Rathore, Prerana....178
Rattan, V. K....236
Ravichandran, Ashwin....649, 853, 929
Ravisankar, Vijay....36
Rawlings, James B....569
Razavi, Sepideh....901
Reboucas, Rodrigo....444
Recheis, Wolfgang....331
Redlinger-Pohn, Jakob D....495
Reid, Daniel....977
Reimann, Christian....38
Reinhart, Wesley F....972, 991
Reis, Pedro....407
Remias, Joseph E....274
Ricco, Antonio J....144
Richards, Jeffrey J....937
Riehm, David....89
Riera-Ortiz, Jaime A....226
Riest, Jonas....417
Rimer, Jeffrey D....72, 832
Rinaldi, Carlos....62, 132
Ripken, R. M....171
Rivera, Rafael G. Henriquez....641
Roberts, Christine Cardinal....965
Roberts, Mark E....654
Roberts, Ryan....982
Roberts, Scott A....965
Robinson, Jacob....521
Rocchio, Joseph....957
Rocha, Alejandra M....188
Rocha, M. Alejandra....380, 1083
Rodriguez, Nerea R....781
Roericht, Margarete....501
Rogers Jr., Reginald E....281
Rogers, Simon....724, 924
Roman-Leshkov, Yuriy....121
Root, Addison....318
Rosch, Justin....131
Rosenbaum, Amanda....779
Rosenfeld, Liat....943
Rosete-Barreto, Jose M....226
Ross, Jennifer....398
Roussell, William L....180
Roux, Benoit....834, 992, 1012
Rowley, Richard L....843
Roy, Arnab....461
Rudra, Apoorva....453
Ruppli, Chantal....655
Ruszczynski, Lukasz....844
Rutkowski, David M....124
Sabri, Laith....323, 324
Sachan, Amit Kumar....220, 480, 481, 708, 800
Sadati, Monirosadat....103, 833, 834
Sadus, Richard J....842
Saeedfar, Amin....536

AUTHOR INDEX

Saengow, Chaimongkol 723
Sahu, Kirti Chandra 646
Sakabe, Junichi 240
Salehi-Khojin, Amin 428, 961
Salunkhe, Aditya A. 153
Samaniuk, Joseph R. 836
Samanta, Devranjan 403
Samira, Samji 363
Sammalkorpi, Maria 181
Sanchez, Francisco Javier Veronico 227
Sanchez, Javier 741
Sandler, Stanley I. 980
Santiso, Erik E. 856, 1010
Santos, Andrew P. 995
Santos-Serena, Raul A. 382
Sanyal, Udishnu 522
Sarupria, Sapna 43, 367, 761
Satoh, Akira 404
Saunders, Steven R. 932
Sayahpour, Baharak 961
Schaffer, C. B. 131
Schenter, Gregory K. 934
Schertzer, Jeffrey W. 535
Schideman, Lance 56
Schieber, Natalie 497
Scholz, Carmen 741
Schrader, Alex 955
Schultz, Andrew J. 498, 974, 1009
Schultz, Danielle 935
Schultz, Kelly M. 319, 341, 511, 825
Schwartz, Brian 938
Schwartz, Daniel K. 797
Schwartz, Daniel T. 437
Schwartzentruber, T. E. 439
Scurto, Aaron M. 189
Seames, Wayne S. 57
Seay, Jeffrey 454
Seebeck, Jan 73
Seitz, Linsey C. 436
Selvam, Balaji 907
Sen, Ramkrishna 356
Senftle, Thomas P. 426
Seo, Dongjin 955
Serra, Francesca 2
Sevgen, Emre 977
Seymour, Joseph D. 457
Shabaniverki, Soheila 940
Shah, Jindal K. 233, 653, 976
Shah, Mansi S. 983
Shahriari, Arjang 411
Shahryari, Reza 857
Shahzad, Khurram 496
Shao, Qing 18
Shapiro, Harrison 103
Shaqfeh, Eric S. G. 33, 144, 827
Shardt, Orest 645
Sharma, Hom 568
Sharma, Sumit 830
Sharma, Vivek 178, 329, 337, 338, 471, 482, 531, 711, 1024
Shaw, John M. 472, 537, 551
Shaykhalishahi, Hamed 776
Shekar, Ashwin 880
Shen, Amy 335, 640, 648
Shen, Vincent K. 975, 997
Shen, Yan 105
Sherman, Zachary 893
Shi, Bohui 749
Shi, Nan 939
Shiflett, Mark B. 188, 380, 652, 1083
Shimoyama, Yusuke 30, 315
Shin, Seunghwan 722
Shin, Young Hwan 56
Shintake, Jun 1041
Shirts, Michael 497, 797
Shishkov, Olga 897
Shivaprasad, Parimala 346
Shor, Leslie M. 554
Shrestha, Sujan 554
Shukla, Diwakar 907
Siahrostami, Samira 16, 285
Sides, Paul J. 146
Sidky, Hythem 977
Siegel, Donald J. 435, 1045
Siepmann, J. Ilja 777, 983
Sihlabela, Maswazi 956
Sikora, Benjamin J. 977
Silva, Barbara L. 885
Silva, Marcela Kotsuka 342, 746
Simoã«ns, Serge 458
Sin, Gurkan 844
Sing, Charles E. 710
Singh, Meenesh R. 14, 971
Singh, Narendra 439
Singh, Nirala 522
Singh, Preet M. 70
Sinno, Talid 518, 1061, 1073
Siperstein, Flor R. 49
Skuntz, Matthew 457
Skylogianni, Eirini 312
Sliwinska-Bartkowiak, Malgorzata 920
Smiatek, Jens 999
Smith, Kevin 317
Smyrl, William H. 418
Snurr, Randall Q. 296, 414
Snyder, Joshua 118, 361
Solich, Ryan 965
Song, Jin 947
Song, Kyoo 476
Song, Shangfei 749

AUTHOR INDEX

Songcuan, Eva 521
Sood, Parveen 978
Sorgenfrei, Tina 38
Sorret, Lea L. 798
Soto, Bernardo Yanez 706
Squires, Todd M. 7, 220, 470, 480, 799, 939
Srinivasan, Chiranth 166
Srinivasan, Siddarth 483
Stavig, Mark 965
Stebe, Kathleen J. 2, 415
Steen, Paul 410
Stefanov, Zdravko 186
Steinberg, Asher Preska 859
Steingart, Daniel 430
Stoerzinger, Kelsey A. 429
Storer, Jackson A. 854
Stottrup, Benjamin 800
Strong, Elizabeth 407
Su, Xiao 434
Suga, Keishi 217, 218, 925, 1086
Sultan, Abbas 323, 324
Sultan, Abdullah S. 942
Sultana, Khondker 526
Sum, Amadeu K. 379
Summers, Daniel R. 550
Sun, Guogang 750
Sun, Ting-Pi 336
Sun, Yunwei 568
Sun, Zhanpeng 750
Sunol, Aydin K. 27, 29
Sureshkumar, Radhakrishna 803
Suszynski, Wieslaw 487
Suteria, Naureen 39
Swan, James 61, 489, 893
Swann, Britany M. 67
Swientoniewski, Lauren T 958
Szilvasi, Tibor 45, 465
Tabrizi, Kayvon 701
Taguchi, Minori 240
Taguchi, Shogo 217
Tai, Yishu 400
Tam, Wai-Ming 536
Tamamis, Phanourios 776, 875
Tan, Mingyang 67
Tanaka, Hiroyuki 106
Tang, Maureen H. 118
Taniguchi, Satoshi 334
Tanimura, Kazuhiko 1086
Tao, Yutao 38
Tasci, Tonguc Onur 896
Tasker, Alison 4
Taufertshofer, Kirstin 501
Tavlarides, Lawrence L. 55
Taylor, Ross 473
Taylor, Shawn D. 474, 536
Tcheimou, Stephane 462
Tenma, Norio 383
Thakur, Pooja 592
Thapar, Vikram 977
Thies, Mark C. 26, 654
Thomas, John A. 317
Thommes, Matthias 918
Thompson, Michael 914
Tilton, Robert D. 467
Timko, Michael T. 58
Tindall, Graham W. 654
Tirrell, Matthew V. 938
Tiwari, Naveen 592
Tochigi, Katsumi 229, 236
Todd, Paul W. 828
Todt, Anika H. 328
Tomaiuolo, Giovanna 823
Tomoya, Aoyama 334
Toner, Mehmet 35
Tonomura, Osamu 334
Torabi, Korosh 763
Torne, Jordi 855
Torres, Ricardo B. 224, 225, 238, 239
Torres-Diaz, Isaac 503
Tosuji, Yuta 383
Tree, Douglas 726
Treftz, Brian 59
Tremsin, Anton S. 73
Tripathi, Anubhav 956
Tripathi, Anurag 327
Tripathi, Manoj Kumar 646
Trogadas, Panagiotis 555
Truskett, Thomas M. 413
Trusler, J. P. Martin 316
Tsao, Joanna W. 3
Tsapatsis, Michael 983
Tsuge, Yoshifumi 584
Tu, Raymond 508
Tuinier, Remco 927, 1022
Turner, C. Heath 53
Twieg, Robert 45
Ugaz, Victor M. 36, 733
Umakoshi, Hiroshi 217, 218, 925, 1086
Underhill, Patrick T. 399
Unni, Mythreyi 132
Uralcan, Betul 760
Urban, Jeffrey 427
Urena-Benavides, Esteban E. 946
Usui, Keishi 383
Utzig, Jonathan 885
Val, Natalia D. F. 224
Valadez, Nestor 417
Van Aeken, Wouter 496
Van Anders, Greg 500, 502, 505, 994
Van Den Akker, Harry E. A. 645

AUTHOR INDEX

Van Der Heijden, Nicole M. W. 927, 1022
Van Der Schoot, Paul....................930
Van Gerven, Tom....................496
Van Lehn, Reid....................908
Van Nguyen, Trung....................1030
Van Osch, Dannie J. G. P. 927, 1022
Van Spronsen, Jaap....................927, 1022
Van Tassel, Paul R. 181
Vanapalli, Siva A. 39, 130
Vangala, Sai Phani Kumar....................283
Varga, Zsigmond....................489
Varma, Arvind....................964
Vashisth, Harish....................910
Vega, Lourdes F. 855, 917
Vega-Bellido, Gabriel....................506
Velasquez, Sandy Candiotti....................330
Velegol, Darrell....................8
Velev, Orlin D. 124
Venegas Sr., Luis Carrasco....................330
Venegas, Alexis....................277
Verde-Gomez, Ysmael....................728
Verma, Sumit....................12
Vermant, Jan....................736, 802
Vichansky, Alexander....................644
Vigeant, Margot....................278
Vigil, R. Dennis....................747
Vilaseca, Oriol....................855
Virk, Preetinder S. 455, 882
Vis, Mark....................1022
Vishnyakov, Aleksey....................915
Vlachos, Dionisios G. 291, 520
Vlasiuk, Maryna....................842
Vlugt, Thijs J. H. 368
Vogel, Sven....................73
Volk, Michael....................1023
Vora, Sahil R. 478
Voronov, Roman....................145
Voss, Johannes....................16
Voth, Gregory A. 17
Vreeland, Thomas....................31
Vu, Dung T. 854
Wagner, Norman J. 64, 135, 151, 417, 515, 557, 724, 819, 824, 826, 924, 937
Waite, J. Herbert....................705
Walker, Eric....................290, 838
Walker, Travis W. 67, 328
Walls, Dan....................335, 640
Wang, Chao....................13
Wang, Cheng....................322
Wang, Ching-Kuan....................115
Wang, Haihui....................902
Wang, Huan....................906
Wang, Jialun....................514
Wang, Jian-Guo....................54
Wang, Jun....................902
Wang, Kerry....................740
Wang, Le....................947, 1079
Wang, Lei....................9
Wang, Liang....................520
Wang, Lu....................707
Wang, Nan....................913
Wang, Penghui....................744
Wang, Qi....................1043
Wang, Rui....................297
Wang, Songcheng....................462, 646
Wang, Sung-Ning....................880
Wang, Weigao....................761
Wang, Xiaoyi....................913
Wang, Yan....................841
Wang, Ya-Qiao....................571
Wang, Yifan....................1061
Wang, Yifeng....................232
Wang, Yiming....................993
Ward, Kevin R. 174
Warrag, Samah E. E. 781
Warshavsky, Vadim B. 499
Wasserscheid, Peter....................230
Waturuocha, Amaka....................1023
Wayner Jr., Peter C. 443
Webb, Michael....................977
Wehrman, Matthew....................319, 341, 511, 825
Wei, Min....................1032
Wei, Shuya....................438, 1060
Wei, Zhiyan....................483
Weigandt, Kathleen....................804
Weimer, Alan W. 362
Werner, Yannick....................230
Weston, Javen....................804
Wheatley, Richard J. 1009
Whitcomb, Kevin....................64
White, Andrew....................1013
White, Jeff....................466
White, Joshua....................276
Whitmer, Jonathan K. 977
Wieseneck, Stacey....................898
Wilding, W. Vincent....................843
Wilkinson, Nikolas A. 516
Williams, Christopher....................49
Williams, Ian....................220, 799
Williams, Stuart J. 936
Willing, Gerold A. 936
Winiarski, Aubrey....................926
Wirth, Christopher L. 507, 1029
Wisecarver, Keith....................1023
Wojcik, Ewelina....................338
Woodcock, Cory....................167
Woolf, Scott....................59
Wordehoff, Michael M. 776
Worku, Dereje....................526
Worley, Clare....................782

AUTHOR INDEX

Woulfe, Donna S. 135
Wozniak, Justin 977
Wu, Billy 555
Wu, Chu 222, 464
Wu, Gang 366, 727, 729
Wu, Jianyang 368
Wu, Jinglan 235
Wu, Ning 896, 899
Wu, Yanyang 46
Wyczalkowski, Wojciech 318
Xi, Erte 984
Xi, Li 881
Xi, Shun 1079
Xia, Yi 410
Xiao, Junyin 46, 533
Xie, Hanguang 408
Xie, Jingyi 906
Xie, Liangxu 50
Xin, Hongliang 288
Xing, Yangchuan 524
Xu, Bicheng 521
Xu, Jeffrey 959
Xu, Xinliang 403
Yadav, Gautam G. 431
Yadav, Geetanjali 356
Yamamoto, Minoru 240
Yan, Yushan 421, 520
Yang, Alexander 914
Yang, Chuanfang 530
Yang, Mengfei 827
Yang, Pengpeng 235
Yang, Tao 896
Yang, Xingfu 899
Yang, Yuan 339
Yang, Yung-Jih 933
Yao, Min 569
Yarranton, Harvey W. 474
Yasue, Masahiro 383
Yates, Matthew 1042
Yelvington, Paul E. 58
Yezer, Benjamin A. 146
Yilixiati, Subinuer 337, 338, 471, 1024
Yin, Xiaolong 494
Ying, Hanjie 235
Yokozaki, Yuta 315
Yoshimoto, Makoto 1086
Young, Charles 478, 710
Yu, Huaizhe 45
Yu, Jiaheng 443
Yu, Xiangfei 167
Yu, Xiaochen 93
Yu, Yue 876
Yu, Yuncheng 36
Yuan, Hao 148
Zambrano, Andres E. Rodriguez 946
Zane, Kylie 938
Zasadzinski, Joesph A. 220, 470, 480, 481, 708, 800
Zeiner, Tim 778
Zeman, Johannes 999
Zeng, Wenduo 1048
Zeng, Yongchao 947
Zeng, Zhenhua 282
Zeng, Zhiqiao 493
Zerze, Gul H. 764, 765
Zerze, Hasan 997
Zhang, Chang 73
Zhang, Geoff G. Z. 532, 737
Zhang, Hanguang 366
Zhang, Jinjun 1032
Zhang, Junshe 381
Zhang, Kechun 400
Zhang, Peng 56
Zhang, Qian 530
Zhang, Qing 405
Zhang, Rui 23, 398, 479, 833, 834
Zhang, Suojiang 1084
Zhang, Xiangping 1084
Zhang, Xuefei 1042
Zhang, Yi 462, 527
Zhang, Yiran 337, 338, 471, 482, 1024
Zhang, Yu 42, 104
Zhang, Yuanhui 56
Zhang, Yuchong 182
Zhang, Zhuqing 343
Zhao, Guolin 46, 533
Zhao, Ruiyang 462
Zhao, Shuangliang 46, 93, 128, 533
Zhao, Xiao 929
Zhao, Xujun 977
Zheng, Weiqing 520
Zhou, Hao 945
Zhou, Jiarun 43, 462
Zhou, Ming 905
Zhou, Mingxia 231
Zhou, Ran 322
Zhou, Ye 103, 833
Zhu, Lu 881
Zhu, Xiao 795
Zhuang, Wei 235
Zia, Roseanna N. 512, 514, 519
Zimmerman, Paul M. 838
Zinchenko, Alexander 172
Zou, Fengxia 235

Engineering Sciences and Fundamentals 2017

Core Programming Area at the 2017 AIChE Annual Meeting

Minneapolis, Minnesota, USA
29 October – 3 November 2017

Volume 2 of 2

ISBN: 978-1-5108-5790-2

Printed from e-media with permission by:

Curran Associates, Inc.
57 Morehouse Lane
Red Hook, NY 12571

Printed by Curran Associates, Inc. (2018)

For permission requests, please contact AIChE
at the address below.

AIChE
120 Wall Street, FL 23
New York, NY 10005-4020

Phone: (800) 242-4363
Fax: (203) 775-5177

www.aiche.org

Additional copies of this publication are available from:

Curran Associates, Inc.
57 Morehouse Lane
Red Hook, NY 12571 USA
Phone: 845-758-0400
Fax: 845-758-2633
Email: curran@proceedings.com
Web: www.proceedings.com

TABLE OF CONTENTS

VOLUME 1

(27a) Anisotropic Particle Interactions with Surfaces, Other Particles, and External Fields 1
Michael A. Bevan

(27b) Particle-Wavy Wall Interactions in a Nematic Liquid Crystal 2
Kathleen J. Stebe, Yimin Luo, Francesca Serra

(27c) Influence of Surface Roughness on Particle-Substrate Interactions: Double Layer Repulsion, Van Der Waals Attraction, and Depletion Interaction 3
Sven H. Behrens, Joanna W. Tsao

(27d) Functional Particles and Microcapsules 4
Simon Biggs, Alison Tasker

(27e) The Hindered Translational and Rotational Dynamics of Anisotropic Nanoparticles Diffusing Near a Solid-Liquid Interface Measured in Aqueous Solution 5
Christopher Bolton, Raymond R. Dagastine

(27f) Strong Deformation of the Electric Double Layer during Sedimentation or Electrophoresis of a Charged Particle 6
Aditya S. Khair

(27g) Understanding and Engineering Diffusiophoretic Suspensions 7
Todd M. Squires

(27h) Chemically-Driven Transport -- a Pervasive Mechanism That Is Seldom Recognized 8
Darrell Velegol

(66a) Understanding Catalyst Selectivity for CO2 Electro-Reduction 9
Thomas F. Jaramillo, Christopher Hahn, Drew Higgins, Jeremy T. Feaster, Stephanie Nitopi, Carlos Morales Guio, Lei Wang, Alan Landers, John Lin, Antaeres Antoniuk-Pablant

(66b) Ag-Sn Bimetallic Catalyst with a Core-Shell Structure for CO2 Reduction 10
Wesley Luc, Charles Collins, Feng Jiao

(66c) Electrochemical CO2 Reduction over Cuag Bimetallic Electrodes and Well-Defined Surface Alloys with Enhanced Oxygenate Selectivity 11
Ezra L. Clark, Christopher Hahn, Thomas F. Jaramillo, Alexis T. Bell

(66d) A Carbon Nanotube Supported Gold Catalyst for the Electroreduction of Carbon Dioxide (CO2) 12
Sumit Verma, Yuki Hamasaki, Chaerin Kim, Wenxin Huang, Shawn Lu, Molly Jhong, Tsuyohiko Fujigaya, Naotoshi Nakashima, Paul J. A. Kenis

(66e) Interplay of Mass Transfer and Local pH Effects in CO2 Reduction Electrocatalysis on Cu Nanowires 13
David Raciti, Chao Wang

(66f) Measurement of Intrinsic Activity of Electrocatalytic Reduction of CO2 over Cu 14
Aditya Prajapati, Meenesh R. Singh

(66g) Electrocatalytic Carbon Fixation on Molecular-Functionalized Surfaces 15
Ming Gong, Zhi Cao, Wei Liu, Christopher Chang

(66h) Electrochemical Reduction of CO2 on Metal Doped Graphene 16
Charlotte Kirk, Leanne Chen, Samira Siahrostami, Mohammadreza Karamad, Michal Bajdich, Johannes Voss, Jens Norskov, Karen Chan

(70a) A Generic Coarse-Grained Model of Influenza Budding: What Can We Learn? 17
Jesper J. Madsen, John M. A. Grime, Gregory A. Voth

(70b) Allosteric Effects of Gold Nanoparticles on Human Serum Albumin 18
Qing Shao, Carol K. Hall

(70c) Self-Assembly of Proteins: The Role of Shape and Specific Interaction 19
Jens Glaser, Sharon C. Glotzer

(70d) Predictive Design of Next-Generation Nanomaterials and Devices Via Bottom-up Approaches 20
Trung Nguyen

(70e) Solvation of Self-Assembled Complexes: Using Molecular Simulations to Probe Energetics, Structure, and Dynamics 21
Kevin R. Hinkle, Frederick R. Phelan Jr.

(70f) Level Set Strategy for Self Consistent Field Theory 22
Gaddiel Ouaknin

(70g) Simulations of Nonlinear Flows in Nonequilibrium Complex Liquids 23
Rui Zhang

(70h) Employing a Multipole Approximation in a Hybrid Fluid Via Relative Resolution 24
Aviel Chaimovich, Christine Peter, Kurt Kremer
(70j) Quantitatively Reliable Molecular Modeling and Simulation of Vapor-Liquid Equilibria 25
Martin T. Horsch
(80a) Production of High-Purity Pyrene Oligomers in a Supercritical Fluids Medium 26
Willam Lamie, Mark C. Thies
(80b) Development and Analysis of Green Pathways Isolation of Medically Active Components from Grape Pomace Biomass 27
Kyle Cogswell, Aydin K. Sunol
(80c) A Novel Supercritical CO2-Based Treatment for Decellularization That Maintains Mechanical and Structural Integrity 28
Dominic M. Casali, Rachel M. Handleton, Michael A Matthews
(80d) Supercritical Fluid Enhanced Swelling and Drug Impregnation of Biomedical Polymers 29
Kyle Cogswell, Aydin K. Sunol
(80e) Effect of Precursor on Crystal Structure of Titania Synthesis in Supercritical Sol - Gel Reaction 30
Motohiro Kinoshita, Yusuke Shimoyama
(80f) Accelerated Conversion of Magnesium Oxychloride to Chlorartinite Using Pressurized Carbon Dioxide and the Resulting Increased Water Resistance 31
Roque Gochez, Christopher L. Kitchens, Thomas Vreeland
(81a) High-Throughput Mechanical Stimulation and Functional Imaging in vivo 32
Hang Lu
(81b) The Steady Motion of a Train of Vesicles in a Cylindrical Channel of Arbitrary Cross Section 33
Joseph M. Barakat, Eric S. G. Shaqfeh
(81c) The Dissolution of a Sheared Water Drop in Bitumen Measured Using a Novel Microfluidic Platform 34
Sachin Goel, Samson Ng, Edgar Acosta, Arun Ramachandran
(81d) Adhesion and Break-up of Model Metastatic Cancer Cell Clusters during Passage through a Microfluidic Constriction 35
Yeng-Long Chen, Sam Au, Mehmet Toner
(81e) Chaotic Thermal Convection in Microfluidic Hydrothermal Pore Environments 36
Aashish Priye, Yuncheng Yu, Vijay Ravisankar, Yassin A. Hassan, Victor M. Ugaz
(81f) Using Droplet Microfluidics to Study Chemical Thermodynamics of Aqueous Atmospheric Aerosol Particles 37
Lucy Nandy, Cari S. Dutcher
(81g) Analysis of the Engulfment of a Foreign Particle By a Solidification Interface: A New Scaling for Flow and Interfacial Forces Involving SiC Particles during Silicon Crystal Growth 38
Jeffrey J. Derby, Yutao Tao, Christian Reimann, Jochen Friedrich, Thomas Jauss, Tina Sorgenfrei, Arne Croell
(81h) Hydrodynamic Mobility of Particles, Vesicles and Cancer Cells in a Square Microchannel 39
Shamim Ahmmed, Naureen Suteria, Valeria Garbin, Siva A. Vanapalli
(81i) Deformable Droplet Migration in a Narrow Microchannel at Finite Reynolds Numbers Via Dissipative Particle Dynamics (DPD) 40
Ryan L. Marson, Ronald G. Larson
(83a) Using Molecular Simulation to Understand the Interfacial Behavior of Ionic Liquids 41
Xiaoyu Liang, Jeffrey R. Errington
(83b) Capacitance and Ion Dynamics of Ionic Liquids Near Oxidized Graphene 42
Yu Zhang, Boris Dyatkin, Yury Gogotsi, Peter T. Cummings
(83c) Using Simulations and Experiments to Characterize Water Structure Near Mica Surfaces for Heterogeneous Ice Nucleation 43
Brittany Glatz, Jiarun Zhou, Sapna Sarupria
(83d) A Novel Density Gradient Theory for Surfactant Molecules - Applied to Oil/Water Interfaces 44
Xiaoqun Mu, Walter G. Chapman
(83e) Computational Chemistry Design of Liquid Crystal-Based Chemoresponsive Systems with Increased Water Tolerance 45
Tibor Szilvasi, Huaizhe Yu, Prabin Rai, Robert Twieg, Nicholas L. Abbott, Manos Mavrikakis
(83g) A Semi-Empirical Analytical Model for Determining the Thermodynamic Stability of Pickering Emulsion 46
Guolin Zhao, Junyin Xiao, Yanyang Wu, Shuangliang Zhao, Honglai Liu
(83h) Computational Method for Extracting Individual Adsorption Parameters from Experimental Dynamic Interfacial Tension of Mixed Surfactant Solutions 47
Fang Liu, Nelya Akhmetkhanova, Vincent Pauchard
(84a) Atomistic Simulations of Ordered and Disordered Carbons: Mimicking Versus Targeting 48
Roland J.-M. Pellenq

(84b) Graphene Oxide Membranes: A Molecular Simulation Approach 49
Christopher Williams, Paola Carbone, Flor R. Siperstein
(84c) Polyethylene Oxide (PEO) in a Polyethylene (PE) Framework: A Simple Model for Simulation Studies of Scaling and Solvent Effects on Polymers in an Open Framework 50
Kwong-Yu Chan, Liangxu Xie, Nicholas Quirke
(84d) Molecular Understanding and Design of Zwitterionic Materials 51
Shaoyi Jiang
(84e) Interfacial Transport of Protons on a 2-Dimensional Functionalized Graphane Surface 52
J. Karl Johnson, Abhishek Bagusetty, Bridget Derksen, Pabitra Choudhury
(84f) Gas Adsorption Behavior in Ionic Polyimide Composite Membranes 53
C. Heath Turner, Asghar Abedini, Ellis Crabtree, Jason E. Bara
(84g) Computational Simulation of Supported Nanocatalysts Under Realistic Conditions 54
Jian-Guo Wang
(90b) Continuous Synthesis of Ethyl Esters from Free Fatty Acids over Metal Oxides in Sub/Supercritical Ethanol 55
Jiuxu Liu, Yue Nan, Lawrence L. Tavlarides
(90c) Denitrogenation of Pyridine with Acetic Acid As Hydrogen Source Under Catalytic Hydrothermal Gasification Environment 56
Peng Zhang, Young Hwan Shin, Lance Schideman, Yuanhui Zhang, Wan-Ting Chen
(90d) The Non-Catalytic Decomposition of Lignin into Chemical and Fuel Intermediates 57
Sara Pourjafar, Wayne S. Seames
(90e) Hydrothermal Liquefaction of Food Waste and Remediation of Aqueous Byproducts 58
Alex D. Paulsen, Alex Maag, Ted J. Amundsen, Michael T. Timko, Paul E. Yelvington
(90f) Hydrothermal Liquefaction of Wastewater Treatment Microalgae in a Pilot-Scale Continuous Flow Reactor 59
Feng Cheng, Travis Le-Doux, Brian Treftz, Scott Woolf, Juanita Miller, Umakanta Jena, Catherine E. Brewer
(92a) The Importance of Classical Soft Matter Physics in the Development of New Nanomaterials 60
Matteo Pasquali
(92b) Rapid and Accurate Methods for Modeling Hydrodynamic Forces in Brownian Dynamics Simulations 61
Andrew Fiore, James Swan
(92c) Rotational Dynamics of Nanoparticles in Polymer Solutions and Melts 62
Lorena Maldonado-Camargo, Carlos Rinaldi
(92d) Morphology of Nanoparticle Aggregates in Flow through Beds Packed with Spheres: Self- and Flow-Induced Assembly Using Lattice Boltzmann Simulations 63
Ngoc Hong Pham, Dimitrios V. Papavassiliou
(92e) Shear Thickening of Colloidal Dispersions at High Particle Concentrations- Rheology and SANS Investigations into Particle-Scale Mechanisms 64
Kevin Whitcomb, Norman J. Wagner
(92f) Colloidal Hydrodynamics at a Fluid-Fluid Interface 65
Charles Maldarelli, Archit Dani
(92g) Effect of Surface Geometry on the Frictional Properties of Poly(dimethyl siloxane) 66
Yunhu Peng, Lilian Hsiao
(92h) Microrheology and Structural Reconfiguration of Artificial Biofluids Composed of Xanthan Gum in Salt Solutions 67
Mingyang Tan, Yating Mao, Britany M. Swann, Travis W. Walker
(92i) Probe Rheology Simulation Technique for Determination of Viscoelasticity of Complex Fluids 68
Pouria Nourian, Dinesh Sundaravadivelu Devarajan, Rajesh Khare
(93a) Approach to Contact, and Adhesion at Elastic and Structured Interfaces 69
Joelle Frechette
(93c) Surface Modification of Stainless Steels for Advanced Functionalities 70
Won Tae Choi, Victor Breedveld, Dennis W. Hess, Preet M. Singh
(93d) Understanding the Wetting Behavior of Water-Octane-Silica Systems Using Monte Carlo Simulation 71
Wenjing Guo, Jeffrey R. Errington
(93e) Time-Resolved in situ Studies of Zeolite Crystal Growth 72
Madhuresh K. Choudhary, Manjesh Kumar, Jeffrey D. Rimer
(93f) Understanding the Solid-Liquid Phase Transition during the Growth of Scintillator Single Crystals Via Computational Modeling and Neutron Imaging 73
Chang Zhang, Jeffrey H. Peterson, Jan Seebeck, Anton S. Tremsin, Didier Perrodin, Gregory A. Bizarri, Edith D. Bourret, Sven Vogel, Jeffrey J. Derby

(93g) Efficient Dispersion of Crude Oil By Blends of Food-Grade Surfactants: Toward Greener Oil Spill Treatments 89
David Riehm, Jasmin C. Athas, John Neilsen, Geoffrey D. Bothun, Vijay T. John, Srinivasa R. Raghavan, Alon McCormick

(93h) Comparison of Experimental and Predicted Adsorption Isotherms of Mixtures 90
Julian Butz, Sabine Enders

(93i) Towards Connecting the Microstructural and Structural Changes in Hierarchical Nanoporous Materials (e.g., clays) during Gas Adsorption (CO2, CH4, H2) Using in-Operando Multi-Scale X-Ray and Neutron Scattering Measurements 91
Greeshma Gadikota, Andrew J. Allen

(140a) Adsorption and Phase Behavior of Mixed Alkanes in Nano Slit Graphite Pores: An iSAFT Application 92
Jinlu Liu, Dilip Asthagiri, Walter G. Chapman

(140b) Oscillation and Enhancement of Gas Solubility in Nanopores 93
Xiaochen Yu, Liangliang Huang, Shuangliang Zhao, Honglai Liu, Keith E Gubbins

(140c) Effect of Interaction Position on Molecular Transport and Separation through a Multi-Site Nanopore 102
Shaghayegh Agah, Matteo Pasquali, Anatoly Kolomeisky

(140d) Strain-Induced Topological Defects and Configurational Transitions in Liquid Crystals 103
Monirosadat Sadati, Jose Martinez-Gonzalez, Khia Kurtenbach, Luis X. De Pablo, Harrison Shapiro, Samuel Morin, Ye Zhou, Juan J. De Pablo

(140e) Alkyl Chain Length and Nanoconfinement Effects on Dynamics of Imidazolium-Based Ionic Liquids 104
Yu Zhang, Naresh C. Otsi, Eugene Mamontov, Peter T. Cummings

(140f) A Molecular Simulation Study on CO2 and a Deep Eutectic Solvent in Slit Nanopores 105
Yan Shen, Francisco R. Hung

(140g) Molecular Simulation of Methane Adsorption Behavior in Nanopores for Shale Gas Development: With Comparison between Graphite and Kerogen Models 106
Jinrong Cao, Yunfeng Liang, Yoshihiro Masuda, Hiroaki Koga, Hiroyuki Tanaka, Toshifumi Matsuoka

(140h) Brownian Dynamics Simulations on Spontaneous Extension of DNA on Cationic Lipid Bilayers Along Grooved Structures 115
Chang Ming-Yi, Ching-Kuan Wang, Chih-Chen Hsieh

(141a) Design of Advanced Nanomaterials for Electrocatalysis 116
Yijin Kang

(141c) Layered Hybrid Structure of Cobalt/N-Doped Carbon Derived from Metal-Organic Frameworks for Electrocatalytic Hydrogen Evolution 117
Tan Huang, Jong-Min Lee

(141d) Comparison of Hydroxide-Mediated and Direct Mechanisms for Alkaline Hydrogen Electrocatalysis 118
Saad Intikhab, Jennifer Gallup, Joshua Snyder, Maureen H. Tang

(141e) Alkaline Hydrogen Evolution from Ni-Mo Intermetallics with High Mo Content 119
James R. McKone, Peter Csernica, Francis J. Disalvo, Hector D. Abruna

(141f) Increased Photocatalytic Activity of TiO2 Nanoparticles with Defects for Sustainable Hydrogen Production 120
Ashley M. Pennington, Fuat E. Celik

(141g) Transition Metal Nitride Core-Noble Metal Shell Nanoparticles As Highly CO Tolerant Catalysts 121
Aaron Garg, Maria Milina, Madelyn Ball, Sean T. Hunt, James A. Dumesic, Yuriy Roman-Leshkov

(141h) Understanding the Electrochemical Behavior of a Thin, Flexible Micro Fuel Cell 122
Seyed Reza Mahmoodi, Matthew Mayer, Ronald S. Besser

(147a) Metastable Phase Transitions 123
Pablo G. Debenedetti

(147b) Computational Investigation of Multipolar Colloidal Particle 124
David M. Rutkowski, Ryan C. Maloney, Orlin D. Velev, Sabine H. L. Klapp, Carol K. Hall

(147c) Interfacial Tensions from SAFT: Connecting Equations of State to Molecular Simulations 125
Erich A. Muller

(147d) Generalized Gibbs Free Energy of Confined Nanoparticles 126
Xiaohua Lu

(147e) Molecular Modeling of Polymeric Systems 127
Coray M. Colina

(147f) Effect and Regulation of Surface Wettability on Molecular Transport and Reaction 128
Shuangliang Zhao, Honglai Liu

(148a) Dynamics of Collective Endothelial Cell Migration in Response to Fluid Shear Stress 129
Alexander Dunn
(148b) Rupture of Cancer Cells Under Microcirculatory Conditions 130
Nabiollah Kamyabi, Siva A. Vanapalli
(148c) Transport of Nanoparticles in the Brain 131
Justin Rosch, C. B. Schaffer, William L. Olbricht
(148d) Towards Understanding Nanoparticle Diffusion in Synovial Fluid Analogues 132
Mythreyi Unni, Lorena Maldonado-Camargo, Kyle Allen, Carlos Rinaldi
(148e) Diffusion of Concentrated Macromolecules within Living Cells 133
Jiyuan Li, Xikai Jiang, Juan Hernandez-Ortiz, Olle G. Heinonen, Juan J. De Pablo
(148f) Platelet Margination, Adhesion, and Activation in Secondary Flows Are Necessary for Thrombus Propagation in an in vitro Model of Venous Thrombosis 134
Marcus Lehmann, Patrick Krohl, Keith B. Neeves
(148g) The Importance of Rheology in Blood Flow Modeling 135
Jeffrey S. Horner, Norman J. Wagner, Antony N. Beris, Donna S. Woulfe
(148h) In Vitro Measurement and Modelling of Platelet Adhesion on Von-Willebrand-Factor-Coated Surfaces in Channel Flow 144
Qin M. Qi, Irene K. Oglesby, Jonathan Cowman, Antonio J. Ricco, Dermot Kenny, Eric S. G. Shaqfeh
(148i) A Novel Viscoelastic Thrombogenesis Model from High Performance Lattice Boltzmann Method Yield-Stress Calculations Based on Intravital Images of Clot Formation in Live Mice 145
Vishnu Deep Chandran, Olufemi Kadri, Roman Voronov
(150a) Determination of the Zeta Potential of Planar Solids in Nonpolar Liquids 146
Paul J. Sides, Dennis Prieve, Benjamin A. Yezer
(150b) Toward Simulation-Based Design of Particle Handling Processes 147
Jennifer Sinclair Curtis
(150c) Sustainable Electrocatalytic Water Splitting with Earth-Abundant Materials 148
Hao Yuan, Richard Lunt, Robert Y. Ofoli
(150d) Human Tear Production from Capillary Wicking Dynamics 149
Clayton J. Radke
(150e) Structure, Elasticity, and Non-Equilibrium State Diagram of Depletion Gels 150
Eric M. Furst
(150f) Harnessing Solvation Forces for Dispersing Colloids in Ionic Liquids with Application in Human Exploration of Space 151
Jingsi Gao, Norman J. Wagner
(150g) Control of Interfacial Phase Separation By Electro-Autocatalysis 152
Martin Z. Bazant
(150h) Reduction of Prticle Jamming in Abrasive Slurries 153
Aditya A. Salunkhe, Rene M. Overney, John C. Berg
(152a) Digital Alchemy for Assembly Engineeringc 164
Sharon C. Glotzer
(152b) Improved Algebraic, Numerical, and Graphical Representations in Fluid Mechanics 165
Stuart W. Churchill, James C. Hill
(152c) The Scaling of Turbulence Near the Wall and the Churchill Turbulent Flux Correlation: Insights with Lagrangian Simulations 166
Dimitrios V. Papavassiliou, Quoc T. Nguyen, Chiranth Srinivasan
(152d) Flow Boiling Using a Piranha Pin Fin Heat Sink 167
Cory Woodcock, Xiangfei Yu, Yoav Peles, Joel L. Plawsky
(152e) Transport Problems in the Spirit of Stuart Churchill for Teaching and Research at the University of Michigan 168
Ronald G. Larson, Claudio Vilas Boas Favero
(160a) Layered Fluid-Fluid Interfaces Confined in Microfluidics 169
Bruno Pinho, Ryan L. Hartman
(160b) Focused DC and AC Electric Fields at Conic Pipettes: Nanoscale Thermal Hotspot and Nano-Droplet Generation with Universal Scalings 170
Hsueh-Chia Chang, Zehao Pan
(160c) Real-Time Monitoring of Complex Multiphase Behavior in a High Pressure High Temperature Microfluidic Chip 171
R. M. Ripken, J. G. E. Gardeniers, S. Le Gac
(160d) Motion of a Deformable Drop in Microchannels of Complex Shape 172
Robert H. Davis, Rocio Navarro, Alexander Zinchenko

(160k) Microfluidic Models of Porous Media to Examine Asphaltene Deposition in the Presence of Chemical Dispersants 173
Yu-Jiun Lin, Sibani Lisa Biswal
(160f) Microfluidic Device for the Continuous Measurement of Viscosity 174
Yunzi Li, Sarah E. Mena, Kevin R. Ward, Mark A. Burns
(160g) Simple Algebraic Formulas for the Practical Interpretation of Mercury Porosimetry Data 175
Zongyu Gu, Martin Z. Bazant
(160h) A Microfluidic Platform to Measure Dynamic Interfacial Tension of Complex Fluid Systems 176
Shweta Narayan, Davis B. Moravec, Brad G. Hauser, Andrew J. Dallas, Cari S. Dutcher
(160i) Controlled Liquid Entrapment through Photo-Patterned Obstacles and Patterned Surfaces 177
Ankur Gupta, Hyundo Lee, T. Alan Hatton, Patrick S. Doyle
(160j) Tears of Wine 178
Prerana Rathore, Vivek Sharma
(163a) Kinetics of (un)Binding between DNA-Functionalized Particles Using a Coarse-Grained Model with Explicit Nucleotide Representation 179
Tiara Ann Maula, Jeetain Mittal
(163c) Hydration Structure and Dynamics of Poly(2-methacryloyloxyethyl phosphorylcholine) 180
Christoph Klein, William L. Roussell, Christopher R. Iacovella, Clare McCabe, Peter T. Cummings
(163d) Polyelectrolyte Interactions: Simulation and Theory 181
Maria Sammalkorpi, Hanne Antila, Paul R. Van Tassel
(163e) Modeling Solute and Solvent Distributions in Functionalized Dendrimers from iSAFT Density Functional Theory 182
Yuchong Zhang, Walter G. Chapman
(163f) Dependence of Relaxations and Mechanical Properties on Molecule Shape in Dissipative Particle Dynamics 183
Michael L. Greenfield, Claire A. Lemarchand, Jesper S. Hansen
(163g) Thermodynamics of Self-Assembly of Perylene Derivatives 184
Jorg Baz, Niels Hansen
(163h) A Molecular Dynamics Study of Actinide Nanoclusters 185
Ken Newcomb, Edward J. Maginn
(176a) Solve this! Fundamental Approach to Problem Solving in Industrial Processes I (Invited Talks) 186
Zdravko Stefanov, Paul Chauvel, Jr., Eldad Herceg, Dana A. Livingston
(179a) Solubiltiy and Phase Equilibria for Optimizing the Processing of Cannabinoids and Terpenes in Cannabis and Hemp Extractions 187
Jerry W. King
(179b) Measurement and Modeling of Vinyl Fluoride Solubility in Aqueous Lithium Bis(trifluoromethylsulfonyl)Imide Solutions 188
David L. Minnick, William J. R. Gilbert, Alejandra M. Rocha, Mark B. Shiflett
(179c) Heat and Momentum Transport Properties of CO2-Expanded Liquids (CXLs): N-Hexane, n-Decane, and n-Tetradecane 189
Kourosh Kian, Aaron M. Scurto
(179d) Van Der Waals Phenomenological Transport Equation of State for Dense Viscosities of the Ultra-Deep Petroleum Reservoirs in the Gulf of Mexico 190
Akanni S. Lawal
(179e) Viscosity Models for Hydrocarbons at High Pressure: A State-of-the-Art Review 208
Isaac Gamwo, Hseen O. Baled, Mark A. McHugh, Robert M. Enick
(179f) Quantification of Uncertainties in Experimental Liquid Density and Viscosity at Hthp Conditions By Using the Corresponding States Law As Arbitrator 209
Akanni S. Lawal
(195a) Characterization of Orientation and Photochemical Function of Chlorophyll a Molecules in Self-Assembled Membranes 217
Shogo Taguchi, Keishi Suga, Keita Hayashi, Yukihiro Okamoto, Hidemi Nakamura, Hiroshi Umakoshi
(195b) Design of Lipid Membrane Surfaces As Organocatalyst for Michael Reactions in Aqueous Media 218
Masanori Hirose, Keishi Suga, Yukihiro Okamoto, Hiroshi Umakoshi
(195c) A Unified Structure-Property Relationship for Alkyl-Polyoxide Surfactants 219
Zachary R. Hinton, Nicolas J. Alvarez
(195d) Relating Rheology with Morphology: Cholesterol in a Model Lung Surfactant Monolayer 220
Steven Patton, Amit Kumar Sachan, Ian Williams, Todd M. Squires, Joesph A. Zasadzinski
(195e) A Matlab Based Tool Package for Interfacial Property Calculations 221
Xiaoqun Mu, Walter G. Chapman, Florian Frank, Faruk O. Alpak

(195i) Interaction Forces between Colliding Emulsion Drops (Oil or Water) Coated with Non-Ionic Surfactants (C12E5 and PGPR) 222
Srinivas Mettu, Joe Berry, Chu Wu, Raymond R. Dagastine
(204a) Prediction of Vapor Pressure and Critical Properties for Non-Electrolyte Organic Compounds from PR+Cosmosac Eos 223
Chien-Yi Li, Chieh-Ming Hsieh
(204b) Volumetric Properties of Saccharide in Aqueous Ionic Liquid Mixtures at Different Temperatures 224
Natalia D. F. Val, Heloisa E. Hoga, Ricardo B. Torres
(204c) Density and Derived Properties of Binary Mixtures Containing {Dimethyl Carbonate (DMC) + 1-Propanol} at T = (288.15 to 308.15) K and at P = (0.1 to 40) Mpa 225
Gustavo V. Olivieri, Ricardo B. Torres
(204e) A SAFT Equation of State Based on Triangular-Well Fluid Potential 226
Luis A. Galicia-Luna, Alfredo Pimentel-Rodas, Jaime A. Riera-Ortiz, Jose M. Rosete-Barreto
(204g) Modelling the Solubility of Naphthalene and Phenanthrene in Binary and Ternary Systems Containing Carbon Dioxide 227
Francisco Javier Veronico Sanchez, Miguel Gonzalo Arenas Quevedo, Octavio Elizalde-Solis
(204h) Molecular-Inspired Parameters Concealed in the Van Der Waals Attractive Force Revealed By First Principles, Statistical Mechanics and Perturbation Methods 228
Akanni S. Lawal
(204i) Solid-Liquid Equilibria for Selected Binary Mixtures Containing Diphenyl Carbonate 229
Hiroyuki Matsuda, Yuki Ohashi, Kiyofumi Kurihara, Katsumi Tochigi
(204j) Thermophysical Suitability of Terphenyl for Hydrogen Storage 230
Jonas Obermeier, Yannick Werner, Ferdinand Heusinger, Michael Geißelbrecht, Patrick Preuster, Peter Wasserscheid, Wolfgang Arlt, Karsten Muller
(204k) Atomic-Level Mechanistic Insights into Monolayer hBN Growth from Reactive Molecular Dynamics Simulations for Catalysis Applications 231
Bin Liu, Song Liu, Mingxia Zhou, James H. Edgar
(204l) Differential Retention and Release of CO2 and CH4 in Kerogen Nanopores: Implications for Gas Extraction and Carbon Sequestration 232
Tuan Ho, Louise J. Criscenti, Yifeng Wang
(204m) Investigation of Electronic Properties of Imidazolium-Based Ionic Liquids in the Presence of Iron Porphyrins for Understanding Their Biodegradability 233
Atiya Banerjee, Jindal K. Shah
(204n) Capturing the Membrane-Triggered Conformational Transition for Pore Forming Cytolysin (ClyA) Using Structure Based Models 234
Hemanth Giri Rao V. V., Rajat Desikan, Shachi Gosavi, K. G. Ayappa
(204o) Solid Form Transformation of Disodium Guanosine 5€²-Monophosphate: Thermodynamic Perspective 235
Qiao Chen, Fengxia Zou, Pengpeng Yang, Jinglan Wu, Wei Zhuang, Hanjie Ying
(204p) Estimation of Kinematic Viscosity for CO2 expanded Liquids By Asog-Visco Method 236
Toshitaka Funazukuri, Hiroyuki Matsuda, Kiyofumi Kurihara, Katsumi Tochigi, V. K. Rattan
(204q) On the Kac-Based Collision Models from Simplified Bernoulli till Its Intelligent Variants 237
Bijan Goshayeshi
(204r) Viscosity and Derived Properties of Binary Mixture Containing Acetonitrile + 1-Propanol at Different Temperatures and Atmospheric Pressure 238
Christian A. T. Campos, Heloisa E. Hoga, Ricardo B. Torres
(204s) Volumetric, Acoustic and Viscometric Properties of Binary Mixture of (n-butylammonium methanoate + 1-propanol) at Different Temperatures 239
Robert L. Fernandes, Heloisa E. Hoga, Ricardo B. Torres
(204t) Measurement of Diffusion Coefficients of Chromium(III) Acetylacetonate in Supercritical Carbon Dioxide at High Temperatures 240
Minoru Yamamoto, Junichi Sakabe, Minori Taguchi, Toshitaka Funazukuri
(204u) Simultaneous Determinations of Dynamic Viscosity and Density of Several Alcohols Using a Straight and Coil Capillary Viscometers at Temperatures between (313-353) K and Pressures up to 30 Mpa 241
Alfredo Pimentel-Rodas, Luis A. Galicia-Luna, Jose J. Castro-Arellano
(204v) Analysis of High-Speed Rotating Flow inside Gas Centrifuge Casing 242
Sahadev Pradhan
(204w) DSMC Simulations of High MACH Number Taylor-Couette FLOW 272
Sahadev Pradhan

(204x) Viscosity Prediction of the Carbon Dioxide Loaded Aqueous Solutions of Alkanolamines: Thermodynamic Approach 274
Naser S. Matin, Joseph E. Remias, Kunlei Liu
(204y) A Computational Study on the Transport of Actives through Skin Layers 275
Kishore Gajula, Rakesh Gupta, Dwadasi Balarama Sridhar, Beena Rai
(213a) Developing Chemical Engineering Acumen By Brewing Kicking Mule Beer 276
Matthew Armstrong, Joshua White, Geoffrey Bull, Jesse Hudgins
(213b) Beer Experiments - a Hopping Success 277
Claire F. Komives, Corey Lapeyri, Alexis Venegas, Diego Marquez, Joseph Pesek
(213c) Why Is the Mustard in the Fridge? A Fun Assignment on the Thermodynamics of Food Safety 278
Margot Vigeant
(213d) Improving Conceptual Knowledge and Retention in Introduction to Engineering Thermodynamics 279
Rachel Morrish
(213e) Tax Revenue Laffer Curve from Thermodynamics Perspective 280
Min Huang
(213f) The Art of the Game: Infusing Thermodynamics Learning with Classic Household Games 281
Reginald E. Rogers Jr.
(216a) Generalizing the Design Principles of Pt-Based Alloy Catalysts with Improved ORR Performance and Durability 282
Zhenhua Zeng, Jeffrey Greeley
(216b) Computational Insights into Thermal and Photoresponse of Pd/C and Pd/N Co-Doped TiO2 using DFT 283
Sai Phani Kumar Vangala, Parag Arvind Deshpande
(216c) Transition Metal-Doped C2N as Active Catalysts for the Oxygen Reduction Reaction 284
Anjli M. Patel, Ambarish R. Kulkarni, Jens Norskov
(216d) Electrochemical Synthesis of Hydrogen Peroxide Via Water Electrolysis: Challenges and Opportunities 285
Samira Siahrostami, Jens Norskov
(216e) Modeling Electrochemical Reactions: DFT-Based Models Including Explicit Solvation, Electrolyte, and Electrochemical Potential 286
Jason Goodpaster
(216f) Efficiently Determining Solvent Effects at Metal-Water Interfaces 287
Satish Iyemperumal, N. Aaron Deskins
(216h) Machine-Learning Model Development for Electrocatalyst Discovery 288
Zheng Li, Hongliang Xin
(218b) Climbing the Volcano: Active-Site Engineering at the Atomic Scale 289
Joseph H. Montoya
(218c) Developing Ab Initio Methodology for Advancing Catalytic Reactions 290
Eric Walker
(218d) First-Principles-Derived Structure-Energy Relationship for Surface Oxides 291
Alexander V. Mironenko, Dionisios G. Vlachos
(218e) First-Principles Based Design of Reaction Conditions for the Catalytic Conversion of Methane to Methanol over Cu-Exchanged SSZ-13 292
Florian Goltl, Manos Mavrikakis
(218f) Reaction Ensemble Monte Carlo Simulations of Xylene Isomerization Under Confinement 293
Ryan Gotchy Mullen, Edward J. Maginn
(218g) Understanding Separation and Catalysis in Nanoporous Materials 294
Peng Bai
(218h) Modeling Self-Assembly of Metal-Organic Frameworks with Enhanced Sampling Techniques 295
Yamil J. Colon, Ashley Guo, Lucas Antony, Kyle Hoffmann, Juan J. De Pablo
(218i) Computationally-Efficient High-Throughput Screening of Metal-Organic Frameworks for Hydrogen Storage 296
N. Scott Bobbitt, Arun Gopalan, Benjamin Bucior, Jiayi Chen, Randall Q. Snurr
(218j) Effects of Ion Self-Energy on the Double Layer Structure and Properties at the Dielectric Interface 297
Rui Wang
(225a) Discovery of the Universal Van Der Waals Generic Cubic Equation of State for the Continuity of Gaseous and Liquid States of Fluids and Mixtures 298
Akanni S. Lawal

(225b) Vapor-Liquid Equilibrium Data for the Systems H2S-MDEA-H2O and CH4-H2S-MDEA-H2O at High Solvent Concentrations and High Pressures 312
Eirini Skylogianni, Diego D. D. Pinto, Hanna K. Knuutila, Christophe Coquelet
(225c) Solubility of Timolol Maleate in Supercritical Carbon Dioxide with Water on Extraction from Silicone Hydrogel 315
Yusuke Shimoyama, Yuta Yokozaki
(225d) Experimental and Modelling Study of the Phase Behaviour of (CO2 + CH4 + Methylbenzene) at High-Pressure and High-Temperature Conditions 316
Saif Zs. Al Ghafri, J. P. Martin Trusler
(230a) Industrial-Scale Modeling of Herschel-Bulkley (Yield Stress) Fluids 317
John A. Thomas, Kevin Smith
(230b) Viscometer for on-Site Rheology Measurements: Development and Application 318
Richard K. Grenville, Jason J. Giacomelli, Wojciech Wyczalkowski, Addison Root
(230c) Using ?2rheology to Measure Rheological Properties during Repeated Phase Transitions of Hydrogenated Castor Oil 319
Matthew Wehrman, Seth Lindberg, Kelly M. Schultz
(230d) Model Systems for Mixed Particulate Fermentation Broth 320
Matthias Kraume, Lutz Bohm, Chrysoula Bliatsiou
(230e) Polycarboxylate Ether Cement Superplasticizers Containing Copolymer Backbones with Improved Rheology Performance 321
Xue Chen, Jennifer Lowe, Thomas Clark, Mike Radler
(234a) Microscale Acoustic Streaming Flows in Viscoelastic Fluids: Comparison of Experiment and Simulation 322
Behrouz Behdani, Ran Zhou, Cheng Wang, Joontaek Park
(234aa) Radioactive Particle Tracking (RPT) Techniquefor Pilot-Scale Bubble Column 323
Laith Sabri, Abbas Sultan, Muthanna Al-Dahhan
(234ab) Gamma-ray Computed Tomography (CT)Technique for Pilot-Scale Bubble Column Reactor 324
Abbas Sultan, Laith Sabri, Muthanna Al-Dahhan
(234c) Influence of Cell Lengths on Surface Drag in Motile Bacteria 325
Katie Ford, Pushkar Lele
(234d) Dynamics of Semiflexible Colloidal Particle Chains Under Rotating Magnetic Fields 326
Steve Kuei, Sibani Lisa Biswal
(234e) A Study of the Dynamics of Human Pedestrians Using Experiments and Simulations in the Indian Context 327
Indranil Saha Dalal, Anurag Tripathi, Amullya Kale, Ishan Prashant
(234f) Extensional Rheometry with a Handheld Mobile Device 328
Kristin A. Marshall, Aleesha M. Liedtke, Anika H. Todt, Travis W. Walker
(234g) Free Surface Flows and Extensional Rheology of Polymer Solutions 329
Jelena Dinic, Leidy N. Jimenez, Vivek Sharma
(234h) Rheological Behavior of Poly(vinyl alcohol) in Aqueous Solutions: Comparison and Assessment of Rheological Parameters Obtained By Empirical Correlations 330
Maria Veronica Carranza Oropeza, Luis Carrasco Venegas Sr., Sandy Candiotti Velasquez
(234i) The Phantom Generation of a Complex Nasal Geometry with Horizontal Cut for Lda Measurements 331
Manuel Berger, Martin Pillei, Andreas Mehrle, Wolfgang Recheis, Florian Kral, Wolfgang Freysinger, Michael Kraxner
(234j) Using Laos and Transient Data to €œFingerprint€ Human Blood Rheological Data 332
Tyler Helton, Matthew Armstrong
(234l) Comparison of Simple Rheological Models in Fitting and Predicting Steady State and Transient Blood Rheology 333
Michael Deegan, Evan Ousley, Matthew Armstrong
(234o) Development of Predictive Model for Sizes of Gas and Liquid Slugs Formed in Millimeter-Scale T-Shaped Channels 334
Gwangnoh Ahn, Osamu Tonomura, Satoshi Taniguchi, Aoyama Tomoya, Shinji Hasebe
(234p) Shape Evolution and Spreading of Liquid Droplets in Miscible Environments 335
Dan Walls, Simon Haward, Amy Shen, Gerald G. Fuller
(234q) Impact Forces of Inertia-Driven Liquid Drops 336
Ting-Pi Sun, Leonardo Gordillo, Xiang Cheng
(234r) Visualizing Nanoscopic Topography, Patterns, Flows, Thickness Transitions and Instabilities in Stratifying Freestanding Thin Films 337
Yiran Zhang, Subinuer Yilixiati, Vivek Sharma

(234s) Contrasting Drainage and Stratification in Horizontal Vs Vertical Micellar Foam Films 338
Subinuer Yilixiati, Ewelina Wojcik, Yiran Zhang, Vivek Sharma
(234t) Solvent Effects on the Crystallinity of Petroleum Asphaltenes 339
Yuan Yang, Thomas Headen, Michael P. Hoepfner
(234v) Role of Bifurcation Geometry on Stability of Thrombus 340
Hari Hara Sudhan Lakshmanan, Jeevan Maddala
(234w) Using ?2rheology to Measure Rheological Properties of Hydrogenated Castor Oil 341
Matthew Wehrman, Seth Lindberg, Kelly M. Schultz
(234x) CFD Simulation of Bubble Columns Operating in Heterogeneous Regime 342
Tatiana Matiazzo, Bruna L Mees, Jaci Carlo Schramm Camara Bastos, Henry F. Meier, Marcela Kotsuka Silva
(234z) Deposition and Oil-Brine Interfacial Rheology of Asphaltene-Stabilized Emulsions 343
Yu-Jiun Lin, Peng He, Zhuqing Zhang, Steve Kuei, Sibani Lisa Biswal
(257a) Vapor Recompression Mechanism Introducing in Internal Heat-Integrated Distillation Column: Impact of Internal Energy Driven Intermediate and Bottom Reboiler 344
Bandaru Kiran
(257b) Experimental Studies Controlling Trace Components in a Dividing Wall Distillation Column 345
Melissa Donahue, Michael Baldea, R. Bruce Eldridge
(257c) Novel Metal-Enzyme Catalyst for One-Pot Dynamic Resolution in a Spinning Cloth Disc Reactor 346
Parimala Shivaprasad, Matthew Jones, Emma Emanuelsson
(257d) Scale and Effects of Catalyst Deactivation in Enzymatic Catalyzed Reactive Distillation 355
Torben Egger, Georg Fieg
(257e) Use of Computational Fluid Dynamics (CFD) Simulation and Image Analysis Tool for Modelling Light in a Microalgal Photobioreactor 356
Arpit Mishra, Geetanjali Yadav, Parthasarathi Ghosh, Ramkrishna Sen
(261a) Interfacial Dynamics between Pathogenic Nanoparticles and Cell Membrane Surfaces 357
Susan Daniel
(261b) Bicontinuous Microemulsions in Homopolymer-Block Copolymer Blends 358
Frank S. Bates
(261c) Self-Assembly of Block Copolymers: From Nanostructure to Function to Applications 359
Paschalis Alexandridis
(282a) Rational Design and Synthesis of Surfactant-Free Pt-Based Ternary Nanoalloys As Superior Oxygen Reduction Reaction Electrocatalysts 360
Sheng Hu, Bamin Khomami, Dibyendu Mukherjee
(282c) Improved Activity and Durability of Ionic Liquid Composite Nanoporous Nanoparticle Electrocatalysts for Oxygen Reduction Reaction 361
Yawei Li, Joshua Snyder
(282d) Particle Atomic Layer Deposition for Stabilization of Oxygen Reduction Catalysts 362
William McNeary Iv, Audrey Linico, Megan Maguire, Alia M. Lubers, Alan W. Weimer
(282e) A Mechanistic Insight into the Electrocatalytic Behavior of Mixed Oxide Cathode Catalysts for Li-O2 Batteries 363
Samji Samira, Ayad Nacy, Eranda Nikolla
(282f) Characterization of Feni Alloy Nanoparticles for Oer As a Function of Iron-Nickel Composition 364
Prashant Acharya, Zachary Nelson, Lauren F. Greenlee
(282g) Active Sites in Nitrogen-Doped Carbon Nanostructures for Oxygen Reduction and Oxygen Evolution Reactions 365
Kuldeep Mamtani, Deeksha Jain, Anne Co, Umit S. Ozkan
(282h) Atomic Iron-Dispersed Electrocatalysts Derived from Metal-Organic Framework for Oxygen Reduction 366
Hanguang Zhang, Gang Wu
(286a) Nucleation Mechanism of Clathrate Hydrates of Soluble Guest Molecules 367
Ryan Defever, Sapna Sarupria
(286b) Investigation on Thermophysical Properties of Structure II Clathrate Hydrates Using Molecualr Dynamics Simulations 368
Bing Fang, Fulong Ning, Ping Cao, Jianyang Wu, Thijs J. H. Vlugt, Signe Kjelstrup
(286c) Prediction of Gas Hydrate Formation Conditions in Brine Systems with the Enrtl Model 379
Amadeu K. Sum, Yue Hu
(286d) Clathrate Hydrate Formation Using Fluorocarbons 380
M. Alejandra Rocha, Mark B. Shiflett
(286e) Methane Hydrate Formation Accelerated with Various Promoters in an Unstirred Reactor 381
Seungjun Baek, Yun-Ho Ahn, Junshe Zhang, Juwon Min, Huen Lee, Jae W. Lee

(286f) Effect of Thermodynamic Promoter (TBAB) in Gas Hydrates Dissociation Conditions of the Mixture H2O + CO2 + Undecane 382
Raul A. Santos-Serena, Pedro Esquivel-Mora, Jose J. Castro-Arellano, Luis A. Galicia-Luna

(286g) Study on Flow Behavior of CO2-in-Water (C/W) Emulsion in the Porous Medium Under Hydrate Stability Conditions 383
Yoshihiro Masuda, Keishi Usui, Yuta Tosuji, Shigemi Naganawa, Masahiro Yasue, Yunfeng Liang, Norio Tenma

(286h) Ionic Lqiuids As Dual Functional Inhibitors for Natural Gas Hydrates 384
Majeda Khraisheh

(289j) Complex Fluids in the Gut 396
Sujit S. Datta

(289b) Boundary Guidance in Self Propelled Colloidal Motors 397
Ali Mozaffari, Joel Koplik, Charles Maldarelli

(289c) Active Nematic Liquid Crystals with Variable Activity and Elasticity 398
Rui Zhang, Nitin Kumar, Jennifer Ross, Margaret L. Gardel, Juan J. De Pablo

(289d) Using a Stochastic Field Theory to Understand Active Colloidal Suspensions 399
Yuzhou Qian, Peter R. Kramer, Patrick T. Underhill

(289e) Imaging the Onset Kinetics of the Swarming Transition Using Light-Controlled Bacteria 400
Yi Peng, Yishu Tai, Kechun Zhang, Xiang Cheng

(289f) Surface Drag and Swarming in Motile Bacteria 401
Pushkar Lele, Katie Ford

(289g) Effects of Elasticity and Hydrodynamic Interactions on Swimmer Shape and Trajectories in a Coarse-Grained Model of Monotrichous Bacteria 402
Frank Nguyen, Michael D. Graham

(289h) Microscopic Dynamics of Bacterial €œSuperfluids€ Under Planar Oscillatory Shear 403
Shuo Guo, Devranjan Samanta, Yi Peng, Xinliang Xu, Xiang Cheng

(289i) Monte Carlo Simulaitons on the Aggregate Structures in a Supension Composed of Magnetic Cubic Particles on a Material Surface 404
Kazuya Okada, Akira Satoh

(296a) Pattern Formation from Interfacial Instabilities in Miscible Viscous Fluids 405
Irmgard Bischofberger, Qing Zhang

(296b) Electrokinetic Fingering: A Problem in Vector Laplacian Growth 406
Mohammad Mirzadeh, Martin Z. Bazant

(296c) Designing Functional Materials with Interfacial Instabilities in Thin Films 407
Pierre Brun, Joel Marthelot, Elizabeth Strong, Pedro Reis

(234n) Numerical Simulation on Flow Behaviour of Twin-Liquid Film over a Vertical Plate with an Open Window 408
Hanguang Xie, Jianguang Hu, Gance Dai

(296e) Droplet Wetting Transitions on Inclined Substrates in the Presence of External Shear and Substrate Permeability 409
Leonardo Espin, Satish Kumar

(296f) Measuring Contact-Line Mobility Using Drop Resonance 410
Yi Xia, Paul Steen

(296g) Electrostatic Suppression of the Leidenfrost State 411
Arjang Shahriari, Soumik Das, Vaibhav Bahadur, Roger T. Bonnecaze

(296h) Linear Stability of Layered Two-Phase Flow through Soft Gel Coated Walls 412
Dinesh N V S S R Bhagavatula, S. Pushpavanam

(305a) Inverse Design of Interactions for Assembly 413
Thomas M. Truskett

(305b) High-Throughput Computational Screening As a Tool for Understanding the Molecular Thermodynamics of Adsorption 414
Randall Q. Snurr

(305c) Awe-Somes: All Water Emulsion Bodies 415
Kathleen J. Stebe, Sarah Hann, Daeyeon Lee

(305d) Succession of Alkane Conformational Motifs Bound within Hydrophobic Nano-Capsule Assemblies 416
Hank Ashbaugh

(305e) Thermodynamics and Transport Properties of Fluids with Intermediate Range Order with Application to Protein Solutions and Biopharmaceuticals 417
P. Douglas Godfrin, Yun Liu, Nestor Valadez, Ramon Castaneda-Priego, Jonas Riest, Gerhard Naegele, Norman J. Wagner

(320d) 20 Years of Corrosion Sensing and Microvisualization of Corrosion Processes 418
William H. Smyrl

(320a) The Need for Engineering Methods for Electrochemical Systems 419
Richard Alkire
(320b) Characterizing Metal-Air Batteries Using Electrochemical Impedance Spectroscopy and Mass Spectrometry 420
Bryan D. McCloskey
(320c) Platinum-Free Fuel Cells for Affordable Zero-Emission Cars 421
Yushan Yan
(351a) Deciphering the Chemical and Physical Mechanisms Active in Elevated Temperature Photocatalytic Synthesis of Hydrocarbons and Ammonia 422
Siris Laursen, Samiksha Poudyal, Morghan Parker
(351b) A High-Throughput Computational Screening Approach for Solar Fuels Photoelectrocatalysis 423
Joseph H. Montoya, Kristin Persson
(351c) Computational Insights into Photo(electro)Chemical Nitrogen Fixation over Titania Catalysts 424
Andrew Medford, Benjamin Comer, Marta Hatzell
(351d) Engineering the Interface to Improve the Efficiencies of Insulator-Protected Semiconductors for Photoelectrochemical Cells 425
Joseph Quinn, Suljo Linic
(351e) The Role of Surface-Bound Dihydropyridine Analogs in Pyridine-Catalyzed CO2 Reduction over Semiconductor Photoelectrodes 426
Thomas P. Senftle, Emily A. Carter
(351f) Surface Plasmon-Assisted Photoelectrochemical CO2 Reduction on Well-Defined Nanostructured Silver Electrodes 427
Elizabeth Corson, Youngsang Kim, Erin Creel, Fen Qiu, Robert Kostecki, Jeffrey Urban, Bryan D. McCloskey
(351g) Artificial Leaf for Carbon Dioxide Photo-Reduction to Fuels 428
Mohammad Asadi, Pedram Abbasi, Kibum Kim, Amin Salehi-Khojin
(351h) Understanding Photoelectrochemistry on Epitaxial Oxides through Surface Electronic Structure 429
Kelsey A. Stoerzinger, Yingge Du, Scott A. Chambers
(352a) Rethinking Grid-Level Energy Storage with Minimal Architecture Zinc-Bromine Batteries 430
Kevin Knehr, Shaurjo Biswas, Hang Huynh, Daniel Steingart
(352b) Highly Energy Dense Cu-Intercalated Bi-Birnessite/Zn Battery 431
Gautam G. Yadav
(352c) Designing Electrolytes for Beyond Li-Ion Batteries Using Coupled High Throughput Ab Intio Calculations and MD Simulations 432
Nav Nidhi Rajput, Xiaohui Qu, Vijayakumar Murugesan, Karl Mueller, Kristin Persson
(352d) Rational Design of Solid-Liquid Interphases for Reactive Metal Batteries 433
Snehashis Choudhury, Lynden A. Archer
(352e) Molecular Design of Redox-Interfaces: Selective Electrochemical Separations and Beyond 434
Xiao Su
(352j) Atomistic modeling of metallic anodes in beyond Li-ion batteries 435
Jeffrey S. Lowe, Donald J. Siegel
(352g) H2 Production Via Photovoltaic-Electrolysis with over 30% Solar-to-Hydrogen Efficiency 436
Jesse D. Benck, Linsey C. Seitz, Jieyang Jia, Yijie Huo, Yusi Chen, Jia Wei Desmond Ng, Taner Bilir, James Harris, Thomas F. Jaramillo
(352h) Remote Control Electrodeposition: Design Criteria for Patterning on Substrates without Direct Electrical Connections 437
Trevor M Braun, Daniel T. Schwartz
(352i) Stalbe Electrochemical Growth in Viscoelastic Flow 438
Shuya Wei, Lynden A. Archer
(358a) Searching for Volume Diffusion 439
Narendra Singh, T. E. Schwartzentruber, Edward L. Cussler
(358b) Condensation on Highly Superheated Surfaces: Unstable Thin Films in a Wickless Heat Pipe 443
Thao T. T. Nguyen, Akshay Kundan, Jiaheng Yu, Peter C. Wayner Jr., Joel L. Plawsky
(358k) Demixing in Dilute Polydisperse Emulsions 444
Rodrigo Reboucas, Michael Loewenberg
(358c) Experimental Investigation of Air Ingress Scenario during Natural Circulation Cooling of a Very Hot Channel 453
Apoorva Rudra, Sanjoy Banerjee, Masahiro Kawaji
(358d) Hydrodynamic Study to Determine Volumetric Mass Transfer Coefficient for Cell Culture Applications 454
Sohail Rasool Lone, Vimal Kumar, Jeffrey Seay, Derek Englert, Hyun-Tae Hwang

(358e) Turbulent Flow of Diutan Biopolymer Solutions and Carbon Nanotube Suspensions in a 4.6 mm ID x 200 L/D Smooth Pipe 455
Preetinder S. Virk
(358g) Fluid Dynamics with Superimposed Mass Transfer of Single Bubbles in Reacting Liquids 456
David Merker, Lutz Bohm, Matthias Kraume
(358h) Experimental Temperature and Fluid Flow Measurements in Packed Beds with Magnetic Resonance Imaging 457
Matthew Skuntz, James E. Maneval, Dinal Perera, Joseph D. Seymour, Ryan Anderson
(358i) Experimental Study of Gas Liquid Mass Transfer Enhanced By Bottom Shear Turbulence 458
Tom Lacassagne, Serge Simoã«ns, Mahmoud El Hajem, Jean-Yves Champagne
(358j) Influence of Thermophysical Properties of Cryogenic Fluids on Growth and Collapse of Cavitating Bubbles 461
Arpit Mishra, Rajaram Lakkaraju, Arnab Roy, Parthasarathi Ghosh
(360a) Interfacial Activity of Silica and Ethyl Cellulose Particles in Fluid-Fluid Interfaces 462
Songcheng Wang, Yi Zhang, Jiarun Zhou, Ruiyang Zhao, Gregory Benz, Stephane Tcheimou, J. Carson Meredith, Sven H. Behrens
(360b) Adhesion of Explosive Particles to Functionalized Surfaces 463
Darby J. Hoss, Sanjoy Mukherjee, Bryan W. Boudouris, Stephen P. Beaudoin
(360j) Dynamic Collision of Water Drops Coated with Food Grade Surfactant Poly-Glycerol-Poly-Ricinoleate (PGPR) in Canola Oil 464
Srinivas Mettu, Chu Wu, Raymond R. Dagastine
(360d) Design of New Liquid Crystal-Based Systems with Improved Chemo-Responsiveness Towards the Detection of Nerve Gases 465
Tibor Szilvasi, Nanqi Bao, Nicholas L. Abbott, Manos Mavrikakis
(360e) Factors Affecting the Transient Stability of Solid Stabilized Emulsions 466
Ashwin Kumar Yegya Raman, Michael Miranda, Jarred Kelsey, Jeff White, Clint P. Aichele
(360f) Adsorption of Star Polymers to Fluid Interfaces 467
Yun-Ru Huang, Robert D. Tilton
(360g) Influence of Order within Non-Polar Monolayers on Hydrophobic Interactions 468
Nicholas L. Abbott
(360h) How Molecular Structure of Surfactants Determine the Dynamics and Viability of Wettability Alteration 469
Soumik Das, Quoc P. Nguyen, Roger T. Bonnecaze
(360i) Chiral Nonlinear Rheology of Phospholipid Monolayers 470
Joseph A. Zasadzinski, Todd M. Squires, Kyuhan Kim, Siyoung Choi
(360j) Visualizing Nanoscopic Topography and Patterns in Freely Standing Thin Films 471
Yiran Zhang, Subinuer Yilixiati, Vivek Sharma
(365a) Heavy Oil and Bitumen Complex Viscosity Simulation - Impacts of Phase Behavior on Rheology 472
John M. Shaw, Sepideh Mortazavi-Manesh, Mildred Becerra
(365b) Maxwell and Marco: Thermophysical Property Needs of Rate-Based Process Simulation Tools 473
Ross Taylor
(365c) The Expanded Fluid Concept for Transport Property Correlations 474
Harvey W. Yarranton, Francisco Ramos-Pallares, Shawn D. Taylor
(365d) Grid Evaluation of Pure-Compound Properties 475
Vladimir Diky, Andrei Kazakov, Kenneth Kroenlein
(365e) Water Content of Natural Gas Systems in Equilibrium with an Aqueous or a Hydrate Phase - Experimental Measurements and Molecular Modeling 476
Walter G. Chapman, Wael A. Fouad, Kyoo Song, Kenneth R. Cox
(365f) When Experimental and Predicted Data Are in Conflict, What Should We Trust? 477
Ala Bazyleva, Eugene Paulechka, Vladimir Diky, Joseph W. Magee, Kenneth Kroenlein
(369a) Global Strain-Field Mapping of a Carbon Nanotube-Laden Interface Using Digital Image Correlation 478
Sahil R. Vora, Brice Bognet, Huseini S. Patanwala, Charles Young, Virgile Daux, Anson Ma
(369b) Simulations of Transient Dynamics of Dense Suspensions 479
Rui Zhang, Endao Han, Heinrich Jaeger, Juan J. De Pablo
(369c) Interfacial Rheology of Coexisting Solid and Fluid Monolayers 480
Joesph A. Zasadzinski, Amit Kumar Sachan, Siyoung Choi, Kyuhan Kim, Ka Yee Lee, Luke Hwang, Todd M. Squires
(369d) Modulation of Dilatational Rheology with Interfacial Curvature and Phase Morphology 481
Amit Kumar Sachan, Joesph A. Zasadzinski

(369e) Surface Forces, Flows and Fluxes Underlying Nanoridge Formation and Instabilities in Stratifying, Micellar Freestanding Films ... 482
Yiran Zhang, Vivek Sharma
(369f) Wrinkling Instabilities in Thin Inhomogenously Stretched Viscous Sheets ... 483
Siddarth Srinivasan, Zhiyan Wei, L. Mahadevan
(369g) Fluid Instabilities, Density Stratification, and Lee Waves That Result in Growth Limitations and Morphological Instability of Liquid-Crystal Interfaces ... 484
Jeffrey H. Peterson, Jeffrey J. Derby
(369h) Liquid-Film Coating on Topographically Patterned Rotating Cylinders ... 485
Weihua Li, Marcio S. Carvalho, Satish Kumar
(369i) Drying of Multicomponent Thin Films on Substrates with Topography ... 486
Truong Pham, Xiang Cheng, Satish Kumar
(369j) Effect of Elasticity on Stability of Viscoelastic Liquid Curtain ... 487
Alireza Mohammad Karim, Wieslaw Suszynski, Lorraine F. Francis, Marcio S. Carvalho
(380a) Flow, Arrest and Yielding in Dense Colloidal Suspensions - Glasses Vs. Gels ... 488
Chinedum O. Osuji
(380b) Modeling a Hydrodynamic Instability in Freely Settling Colloidal Gels ... 489
Zsigmond Varga, Jennifer L. Hofmann, James Swan
(380c) Effects of Attractive Interparticle Interactions on Normal Stresses and Shear-Induced Migration of Colloids ... 490
Nayoung Park, Jacinta C. Conrad
(380d) Connecting Bidisperse and Polydisperse Suspension Rheology ... 491
Sidhant Pednekar, Jaehun Chun, Jeffrey F. Morris
(380e) The Steric Effect of a Solid Boundary on the Orientation Distribution of a Dilute Suspension of Rigid Rod-like Particles Under Shear Flow ... 492
Saman Monjezi, James Jones, Shawn Oettle, Joontaek Park
(380f) Particle Crystallinity in Automated Langmuir-Blodgett Deposition: Effect of Speed and Concentration ... 493
James F. Gilchrist, Kedar Joshi, Zhiqiao Zeng, Xue Li
(380g) A Hybrid Lattice Boltzmann-Random Walk Method for Multiphase Heat Transfer ... 494
Aaron Lattanzi, Xiaolong Yin, Christine Hrenya
(380h) Separation By Design - Towards Simulation Guided Engineering of Coiled Channels for Precise Particle Separation ... 495
Jakob D. Redlinger-Pohn, Federico Municchi, Stefan Radl
(380i) Simulations of Particle-Laden Flows in Microchannels ... 496
Wouter Van Aeken, Khurram Shahzad, Tom Van Gerven, Simon Kuhn
(392b) Prediction of Solid State Phase Diagrams Using Multistate Reweighting and Jacobian Mapping ... 497
Natalie Schieber, Eric Dybeck, Nathan Abraham, Michael Shirts
(392c) Evaluation of Defect Concentrations in Crystalline Systems Via Molecular Simulation ... 498
Apoorva Purohit, Andrew J. Schultz, Jeffrey Errington, David A. Kofke
(392d) Nature of the Instability of the Body-Centered-Cubic (bcc) Structure in Classical Hard Sphere Solids ... 499
Vadim B. Warshavsky, David M. Ford, Peter A. Monson
(392e) Entropy Driven Solid-Solid Transitions in Colloids ... 500
Chrisy Xiyu Du, Greg Van Anders, Richmond S. Newman, Sharon C. Glotzer
(392f) Calculation of Glass Transition Using Sanchez-Lacombe - Eos ... 501
Margarete Roericht, Kirstin Taufertshofer, Sabine Enders
(409a) Engineering Shape Entropy for Colloidal Crystal Design ... 502
Yina Geng, Greg Van Anders, Paul Dodd, Julia Dshemuchadse, Sharon C. Glotzer
(409b) Controlling Anisotropic Colloidal Assembly in External Fields ... 503
Isaac Torres-Diaz, Michael A. Bevan
(409c) Deposition Dynamics of Rod-Shaped Colloids during Transport in Porous Media ... 504
Huilian Ma, Ke Li
(409d) Engineering the Phase Behavior of Anisotropic Particles By Inverse Design ... 505
Chrisy Xiyu Du, Greg Van Anders, Paul Dodd, Julia Dshemuchadse, Sharon C. Glotzer
(409e) Self-Assembly of Magnetic Janus Colloids Via Brownian Dynamics Simulation ... 506
Gabriel Vega-Bellido, Ronal A. Delacruz-Araujo, Ilona Kretzschmar, Ubaldo M. Cordova-Figueroa
(409f) Brownian Dynamics of a Spherical Janus Particle Near a Boundary As a Tool to Investigate TIRM ... 507
Aidin Rashidi, Christopher L. Wirth
(409g) Examining the Stability of Amphiphilic Janus Particle-Laden Interfaces ... 508
Ellen M. Knapp, Ilona Kretzschmar, Raymond Tu

(409h) Self-Assembly of Amphiphilic Janus Nanoparticles 509
Alan Hanley, Allan E. David
(409i) Magnetic Janus Particle Aggregation Rate at Long Times 510
Thomas Long, Ilona Kretzschmar, Joel Koplik
(414a) Using ?2 rheology to Characterize Consecutive Phase Transitions in a Hydrogenated Castor Oil Colloidal Gel 511
Matthew Wehrman, Seth Lindberg, Kelly M. Schultz
(414b) Hydrodynamic Entrainment in Micro-Confined Suspensions and its Implications for Two-Point Microrheology 512
Christian Aponte-Rivera, Roseanna N. Zia
(414c) Shear-Induced Structural Ordering in Jammed Suspensions of Soft Particle Glasses 513
Fardin Khabaz, Tianfei Liu, Michel Cloitre, Roger T. Bonnecaze
(414d) Dynamic Simulation of Aging in a Hard-Sphere Colloidal Glass after Volume-Fraction Jumps 514
Jialun Wang, Xiaoguang Peng, Xi Li, Gregory B. McKenna, Roseanna N. Zia
(414e) A New Conformation Tensor Based Macroscopic Model for Emulsions with Particle Inertia 515
Paul M. Mwasame, Antony N. Beris, Norman J. Wagner
(414f) Polyelectrolyte - Particle Flocculation in Complex Aqueous Solutions and Mixed Hydrodynamic Fields 516
Athena E. Metaxas, Nikolas A. Wilkinson, Cari S. Dutcher
(414g) From Filaments to Coils: Controlling the Dynamics of Linked Colloidal Particle Chains 517
Steve Kuei, Sibani Lisa Biswal
(414h) A Combined Experimental and Numerical Analysis of DNA-Functionalized Colloidal Particle Deposition in a Channel Flow 518
Young Ki Lee, Christopher Porter, John C. Crocker, Scott L Diamond, Talid Sinno
(414i) Structural and Rheological Relaxation upon Flow Cessation in Colloidal Dispersions: Transient, Nonlinear Microrheology 519
Ritesh P. Mohanty, Roseanna N. Zia
(422a) Durable, Self-Hydrating, Tungsten Carbide-Based Composite Polymer Electrolyte Membrane Fuel Cells 520
Weiqing Zheng, Liang Wang, Stephen A. Giles, Yushan Yan, Ajay K. Prasad, Dionisios G. Vlachos
(422c) Teaching an Old Electrocatalyst to Do New Tricks 521
Daniel V. Esposito, Natalie Labrador, Eva Songcuan, Jacob Robinson, Bicheng Xu, Sophia Kurdziel
(422d) Using Surface Chemistry to Understand Aqueous-Phase Thermal Catalytic and Electrocatalytic Hydrogenation of Bio-Oil Model Substrates 522
Nirala Singh, Udishnu Sanyal, John L. Fulton, Oliver Gutierrez, Donald M. Camaioni, Charlie Campbell, Johannes A. Lercher
(422e) Design of Heterogeneous Catalysts for High Temperature Electrochemical Reduction of CO2 523
Juliana S. A. Carneiro, Xiang-Kui Gu, Zachary Kuczera, Eranda Nikolla
(422f) Enhancing Methanol Electrooxidation Activity Using Double Oxide Catalyst Support 524
Ahmed Jasim, Yangchuan Xing
(422g) Study of Impact of Electrode Catalyst Loadings on Thin Flexible Fuel Cell (TFFC) Performance 525
Seyed Reza Mahmoodi, Ronald S. Besser, Matthew Mayer
(422h) Transition Metal Modified Graphitic Mesoporous Carbon for PEMFC Application 526
Khondker Sultana, Dereje Worku, Md. Tashfin Zayed Hossain, Shamsuddin Ilias
(425a) Capillary Foams: How a Small Amount of Oil Can Make a Big Difference 527
Yi Zhang, Sven H. Behrens, J. Carson Meredith
(425b) Wetting on Strain Induced Microstructured Surfaces 528
Vartika Parihar, Soumen Das, Sunando Dasgupta
(425c) Water Wetting on Surfaces Under Fuel Oil Containing Surfactants and Its Implication for Coalescence Separation of w/o Emulsions 530
Qian Zhang, Yanxiang Li, Lixia Cao, Lei Li, Chuanfang Yang
(425e) Dynamic Surface Tension Measurements of Ionic Surfactants Using Maximum Bubble Pressure Tensiometry 531
Camilla U. Ortiz, Vivek Sharma
(425g) The Importance of Spreading Pressure on Adsorption Based Surface Energy Measurements; The Case of IGC 532
Eftychios Hadjittofis, Geoff G. Z. Zhang, Jerry Heng
(425h) Stabilization of Glycerol/Dodecanol Pickering Emulsions with Surface-Grafted Silica Nanoparticles 533
Guolin Zhao, Shuangliang Zhao, Junyin Xiao, Honglai Liu, Marc Pera-Titus

(425i) In situ Observations of Dynamics of Active Braze Joining 534
Anne M. Grillet, Robert M. Garcia, David A. Barringer, Adam D. Martinez
(425j) High Throughput Fabrication of Synthetic Asymmetric Bacterial Membranes 535
Sepehr Maktabi, Li Lu, Jeffrey W. Schertzer, Paul R. Chiarot
(431a) Design Insights for a Thermal Visbreaking Pilot Based on Microwave Heating 536
Shawn D. Taylor, Simon Andersen, Tracy Neitz, David O'Brien, Amin Saeedfar, Merlyn Pulikkathara, Wai-Ming Tam
(431b) On Cubic Eos Interaction Parameter Estimation for Long Chain n-Alkane + Aromatic Binary Mixtures 537
Sourabh Ahitan, Miao Luo, John B. McLaughlin, John M. Shaw
(431c) Uncertainty in Heat Exchanger Design for Diluted Bitumen 538
Harry Z Ha, Paul M. Mathias

VOLUME 2

(431d) Practical Aspects in Using Distillation VLE and Enthalpy Models 550
Daniel R. Summers, Senthil Krishnamoorthy
(431e) Bubble Pressure Measurement and Prediction for n-Alkane + Naphthenic Hydrocarbon Binary Mixtures 551
Sourabh Ahitan, John M. Shaw
(431f) When Is It Appropriate to Kiss? a Case Study on the Adverse Consequences of Overfitting GE Models 552
Paul M. Mathias
(433a) Photochemical Reduction of Enzymes for Light-Driven Chemical Reactions 553
Paul King
(433c) Termite-Inspired Electrochemical Processing of Lignocellulose to Chemicals and Fuels 554
Xiong Peng, Ying Liu, Andrea Kadilak, Sujan Shrestha, Leslie M. Shor, William E. Mustain
(433d) Nature-Inspired Flow Fields for PEM Fuel Cells 555
Jason Cho, Tobias P. Neville, Panagiotis Trogadas, Billy Wu, Dan Brett, Marc-Olivier Coppens
(435a) Numerical Analysis of the Flow, Transport, and Interfacial Phenomena Associated with Growth of Crystalline CZT Under Crucible Rotation 556
Mia S. Divecha, Jeffrey J. Derby
(435b) Modeling the Transient Shear Flow and Predicting Large Amplitude Oscillatory Shear (LAOS) Flow of a Thermoreversible Gel Using a Scalar Structure Parameter Thixotropic Model 557
Matthew Armstrong, Ryan P. Murphy, Norman J. Wagner, Antony N. Beris
(435c) Modeling and Uncertainty Quantification of Vapor Diffusion and Reactions in Polymer 568
Yunwei Sun, Hom Sharma, Elizabeth Glascoe
(435e) Modeling of Transport and Reaction in a Novel Hydride Vapor Phase Epitaxy System 569
Min Yao, James B. Rawlings, Thomas F. Kuech
(435f) Discretized Modeling of a Simple Motionless 3D Printer Based on Retarded Bending Motion and Electrically Controlled Patterning of Fiber Deposition 570
Mounica Jyothi Divvela, Yong Lak Joo
(435g) A Forward Modeling Approach for the Inverse Estimation of Transient Local Heat Fluxes 571
Jiu Luo, Ya-Qiao Wang, Jia-Li Luo, Dong-Chuan Mo, Yuan-Xiang Fu, Shu-Shen Lyu, Yi Heng, Hiroyuki Ozoe
(435h) Modeling of the Evaporation of the Polymer Slurry in the Porous Media 584
Shuji Hironaka, Gen Inoue, Jun Fukai, Yoshifumi Tsuge
(435i) Flow of a Power-Law Fluid across an Asymmetrically Confined Rotating Cylinder 592
Pooja Thakur, Naveen Tiwari, R. P. Chhabra
(435j) Binary Gas Mixture in a High Speed Channel 593
Sahadev Pradhan
(444a) The Shape Evolution of Pendant Droplets in Miscible Environments 640
Dan Walls, Simon Haward, Amy Shen, Gerald G. Fuller
(444b) Shape Evolution of Miscible Drops with Arbitrary Viscosity Ratio: Lagrangian-Eulerian Swarms of Stokeslets and Subgrid Resolution 641
Ludwig C Nitsche, Paola Leon Plata, Rafael G. Henriquez Rivera, Ying Liu
(444c) Coalescence of Drops Due to a Constant Force Interaction in a Viscous Quiescent Fluid 642
John M. Frostad, Alexandra Paul, L. Gary Leal
(444e) Nanoemulsion Formation: Controlling and Predicting Droplet Size 643
Ankur Gupta, T. Alan Hatton, Patrick S. Doyle

(444f) Using a Novel Approach to Model Drop Size Distribution: The Adaptive Multi-Size Group Method 644
Thomas Eppinger, Koichi Akano, Simon Lo, Alexander Vichansky, Ravindra Aglave
(444g) Describing Emulsions in Terms of Pseudo-Potential Lattice Boltzmann Modeling 645
Mohammad Pourtousi, Siddhartha Mukherjee, Pieter Berghout, Orest Shardt, Harry E. A. Van Den Akker
(444h) Comparison of the Rise Dynamics of Standard Bubbles and Oil-Coated Bubbles: Experiments and Simulations 646
Songcheng Wang, Manoj Kumar Tripathi, Kirti Chandra Sahu, J. Carson Meredith, Sven H. Behrens
(444j) A Quadrature-Based Model for Polydisperse Laminar Bubbly Flows 647
Jeffrey C Heylmun, Bo Kong, Alberto Passalacqua, Rodney O. Fox
(234y) Stability of Stagnation Point Flows of Newtonian and Complex Fluids 648
Noa Burshtein, Simon Haward, Amy Shen
(453a) Prediction of Mean Ionic Activity Coefficient of NaCl-H2O System at High Concentrations Using Molecular Dynamics Simulations 649
Nazir Hossain, Ashwin Ravichandran, Rajesh Khare, Chau-Chyun Chen
(453b) Thermodynamic Modeling of Lithium Salt Systems with the Electrolyte NRTL Model 650
Toni Kirkes, Chau-Chyun Chen
(453c) Why Tetrabutylphosphonium Hydroxide-Water Mixtures Can Dissolve Cellulose at Extremely High Water Mass Fractions 651
Brad Crawford, Ahmed E. Ismail
(453d) Phase Behavior of Compressed Gases in Ionic Liquids at the Liquid-Solid Transition Point 652
David L. Minnick, Brooks B. Danahy, Mark B. Shiflett
(453e) Is Difference in Hydrogen Bond Ability of Anions an Exclusive Descriptor of Nonideality in Binary Ionic Liquid Mixtures? 653
Utkarsh Kapoor, Jindal K. Shah
(453f) Polymer-Solvent Phase Behavior of Lignin with Hot Aqueous Solvent Systems 654
Junhuan Ding, Adam S. Klett, Jordan A. Gamble, Graham W. Tindall, Mark C. Thies, Mark E. Roberts
(453g) Application of Thermodynamic Models in the Context of Liquid Chromatography 655
Franziska Ortner, Chantal Ruppli, Marco Mazzotti
(453h) The Solubility and the Freezing Point Depression (FPD) Measurements of Na-1-Ethanethiolate, Na-1-Propanethiolate, Na-2-Propanethiolate, Na-1-Butanethiolate and Na-2-Methyl-2-Propanethiolate Salts in Pure Water 656
Javeed Awan
(464a) Computation of Amino Acid-Mineral Surface Thermodynamics and Interaction Modes via Importance Sampling 701
Kayvon Tabrizi, Michael S. Pacella, Jeffrey J. Gray
(464b) Research on the Mechanism of Spontaneous DNA Extension on Grooved Surface Covered with Cationic Lipid Bilayers 702
Tzu-Hsien Fan, Hou-Jun Guo, Chih-Chen Hsieh
(464c) Effect of Interchain Associations on Hybridization Activity of DNA Brushes 703
Hao-Chun Chiang, Rastislav Levicky
(464d) Understanding Marine Mussel Adhesion: Interfacial Energy from Microscopic to Macroscopic Length Scales 704
George Degen, Jacob Israelachvili
(464e) Tuning Underwater Adhesion with Cation-Pi Interactions 705
Matthew A. Gebbie, Jacob Israelachvili, J. Herbert Waite
(464f) Influence of the Surface Charge on the Brownian Motion of Colloidal Particles Near Surfaces with Biomimetic Characteristics 706
Juan Manuel Hernandez Meza, Angeles Ramirez-Saito, B. Jose Luis Arauz-Lara, Said E. Aranda Espinoza, Bernardo Yanez Soto
(464g) Charged Nanoparticles Interacting with Giant Vesicles Fabricated from Inverted-Headgroup Lipids 707
Lu Wang, Noah Malmstadt
(464h) Modulation of Phase Morphology of Surfactant Monolayer with Interfacial Curvature 708
Amit Kumar Sachan, Joseph A. Zasadzinski
(464i) Recognition of Target Cells By Vibrio cholerae outer Membrane Vesicles 709
Elnaz S. Rasti, Angela C. Brown
(468a) Applying Computational Tools of Polymer Field Theory to out-of-Equilibrium Polymer Solutions in Flow 710
Charles Young, Charles E. Sing

(468b) Capillary Thinning and Pinch-Off Dynamics and Extensional Rheology of Polyelectrolyte Solutions 711
Leidy N. Jimenez, Jelena Dinic, Nikhila Parsi, Vivek Sharma
(468c) Relaxation of Knotted Polymers 712
Vivek Narsimhan, Alexander Klotz, Patrick S. Doyle
(468d) On Synthetic Drilling Fluid Rheological Modelling 713
Silvio Baldino, Reza E. Osgouei, Evren Ozbayoglu, Stefan Z. Miska
(468e) Experimental Evidence for Shear Banding in Large Amplitude Oscillatory Shear of Entangled DNA Solution. 722
Seunghwan Shin, Kevin D. Dorfman, Xiang Cheng
(468f) PadÃ© Approximants for Shear Stress in Large-Amplitude Oscillatory Shear Flow 723
Chaimongkol Saengow, A. Jeffrey Giacomin, Nidal Khalaf, Martin Guay
(468g) The Effect of Branching on Shear Band Formation and Evolution in Wormlike Micelles (WLMs) 724
Michelle A. Calabrese, Simon Rogers, Lionel Porcar, Norman J. Wagner
(468h) Predictions for Non-Linear Flows of Polydisperse Blends Based on a Differential-Constitutive Analogue of the Double Reptation Model 725
Joseph Peterson, L. Gary Leal, Glenn H. Fredrickson
(468i) Marangoni Instability during the Non-Solvent Induced Phase Separation of a Ternary Polymer Solution 726
Douglas Tree, Glenn H. Fredrickson
(482a) Highly Stable Mn-Based Nanocarbon: A New Class of Bifunctional Electrocatalyst for Oxygen Reduction and Evolution Reactions 727
Shiva Gupta, Gang Wu
(482b) Density Functional Theory Study of the Oxygen Reduction Reaction in a Graphene Surface Modified with S or N 728
Elizabeth Montiel-Macias, Ysmael Verde-Gomez, Perla B. Balbuena
(482c) Highly Stable Nanocarbon Supports for Pt Cathode Catalysts in Polymer Electrolyte Fuel Cells 729
Mengjie Chen, Gang Wu
(482d) Approaching 2 W-cm-2 Aemfcs through Electrode Engineering and Controlling the Cell Water Content and Balance 730
Travis J. Omasta, Xiong Peng, William E. Mustain
(482e) Design and Engineering of Membraneless Electrolysis Devices 731
Jonathan Davis, Justin Bui, Daniel V. Esposito
(482g) Design of a Novel Electrochemical Membrane Reactor for Hydrogen Production Via the S-NH3 Cycle 732
Raul Marquez-Montes, Raul Orozco-Mena, Virginia Collins-Martinez, Eduardo Herrera-Peraza, David Chavez-Flores, Victor Ramos-Sanchez
(488a) Portable Smartphone-Enabled Capillary-Based Viscometer and Tensiometer 733
Jose C. Contreras-Naranjo, Xiaorui Dong, Victor M. Ugaz
(488b) An Experimental Study of a Liquid Drop Impinging on a Liquid Pool 734
Manas Ranjan Behera, Anirvan Dasgupta, Sudipto Chakraborty
(488c) Arresting Dissolution By Interfacial Rheology Design 736
Peter J. Beltramo, Manish Gupta, Alexandra Alicke, Jan Vermant
(488d) The Importance of the Amorphous Material Interface in the Measurement of Thermodynamic Parameters, Using Inverse Gas Chromatography 737
Eftychios Hadjittofis, Geoff G. Z. Zhang, Jerry Heng
(488e) Mechanisms of Dynamic Wetting Failure in the Presence of Soluble Surfactants 738
Chen-Yu Liu, Marcio S. Carvalho, Satish Kumar
(488f) Electrostatic Assist of Liquid Transfer in Printing Processes 739
Chung-Hsuan Huang, Satish Kumar
(488g) Analysis of Temperature Gradient Zone Melting and Annealing for Mitigation of Second-Phase Particles in Single Crystals 740
Kerry Wang, Nathan T. Morgan, Jeffrey J. Derby
(488h) In situ Partitioning Measurements of Multiple Fermentation End-Products into Lipid Membranes 741
Geoffrey D. Bothun, John Nunes, Carmen Scholz, Javier Sanchez
(488i) Synchronization and Traveling Waves within Arrays of Electrostatic Oscillators 742
Shashank Pandey, Yong Dou, Charles A. Cartier, Mikolaj Kowalik, Kyle J. M. Bishop
(494a) Probing Viscosity of Complex Emulsions with Micro-Rheometry 743
Archit Dani, Shweta Narayan, Cari S. Dutcher

(494b) The Study on Mechanism of Flow Distribution and Its Optimization in a Spray Column 744
Penghui Wang, Gance Dai
(480e) Transition between Turbulence Regimes in Particle-Laden Channel Flows 745
Jesse Capecelatro, Olivier Desjardins, Rodney O. Fox
(494d) Analysis of Interfacial Forces of a Slurry Bubble Column 746
Bruna L Mees, Tatiana Matiazzo, Jaci Carlo Schramm Camara Bastos, Henry F. Meier, Marcela Kotsuka Silva
(494e) Semi-Batch Liquid-Liquid Taylor Vortex Flow: Comparison of 3-D and 2-D Eulerian-Eulerian CFD Simulations and Experimental Measurements for Fluid Flow Patterns and Droplet Holdup 747
John Lavey, Arya Haghighat, Charlton Campbell, Xi Gao, Bo Kong, Meesha Legg, Michael Olsen, R. Dennis Vigil
(494g) Experiments on Bubble Behaviors in Three Phase Internal-Loop Airlift Reactor Using Optical Fiber Probe Measurements 748
Yongxiang Gao
(494h) Flow Properties of Gas Hydrate Slurry in Oil-Dominated Transportation Systems 749
Lin Ding, Bohui Shi, Yang Liu, Shangfei Song, Jing Gong
(494i) Relationship between Flow Pattern in Cyclone Classifier and Classification Process 750
Zhanpeng Sun, Guogang Sun, Yan Du
(511a) Computational Study of the Stability of the Miniprotein Trp-Cage in Extreme Environments 760
Betul Uralcan, Sang Beom Kim, Pablo G. Debenedetti
(511b) Engineering Robust Activity in Extremophilic Enzymes 761
Siva Dasetty, Weigao Wang, Mark Blenner, Sapna Sarupria
(511c) Effect of Manufacturing Conditions on the Fabrication of Protein/Polysaccharide Biocompatible Nanotubes (BNTs) 762
Luis F. Maldonado, Jozef Kokini
(511d) Towards a Thermodynamic Model for Predicting Coiled-Coil Protein Structures 763
Mojtaba Jokar, Korosh Torabi
(511e) Effect of Residual Secondary Structure Propensity on Liquid-Liquid Phase Separation of Tdp-43 764
Gregory L. Dignon, Gul H. Zerze, Nicolas Fawzi, Young C. Kim, Jeetain Mittal
(511f) Solvation Contribution to the Conformational Preferences of the Hydrated Peptides 765
Gul H. Zerze, Alexander Bourque, Jeetain Mittal
(511g) Influence of Backbone Hydrogen Bonding on the Stability of the WW Domain of the Protein Pin1 766
Daniel Markthaler, Niels Hansen
(511h) Molecular Simulations of the Force Spectroscopy of a Biomimetic Polymer 775
Aviel Chaimovich, Christian Leitold, Christoph Dellago
(511i) Insights into the Binding of Î²-Wrapins to Amyloidogenic Proteins Using Simulations and Experiments 776
Asuka A. Orr, Michael M. Wordehoff, Hamed Shaykhalishahi, Ewa A. Mirecka, Sai Vamshi R Jonnalagadda, Wolfgang Hoyer, Phanourios Tamamis
(512a) Monte Carlo Simulations Probing Interfacial Adsorption and Uptake in Surfactant Bilayers 777
J. Ilja Siepmann, Mona Minkara, Jingyi Chen, Cor J. Peters
(512c) Phase and Interfacial Behaviour of Aqueous-Two Phase Systems Based on Hyperbranched Polymers 778
Andres Kulaguin Chicaroux, Tim Zeiner
(512d) From Metallic Hydrogen to Polymeric Sulfur: Peculiar Thermodynamics of Polyamorphic Fluids 779
Lauren E. Amrhein, Mikhail A. Anisimov, Frederic Caupin, Michal Duska, Amanda Rosenbaum
(512e) Computational Studies of Mixtures of Refrigerants, Ionic Liquids and Deep Eutectic Solvents for Absorption Refrigeration Systems 780
Rubaiyet Abedin, Sharareh Heidarian, John C. Flake, Francisco R. Hung
(512f) Oil Desulfuriztion Using Deep Eutectic Solvents Via Liquid-Liquid Extraction 781
Samah E. E. Warrag, Nerea R. Rodriguez, Martin Van Sint Annaland, Maaike C. Kroon, Cor J. Peters
(512g) Relevance of Thermophysical Property Data in Rate-Based Gas Treating Simulations 782
Diego E. Cristancho, Jacob M. Crosthwaite, Allen Day, John Dowdle, Clare Worley
(512h) Identification of the Equilibrium Parameters for the Modeling of the Hydrogen Solubility in Heavy Crude Cuts Using the Matlab-Aspen Connection 793
Camilo Monroy-Pena, Johana Orjuela
(527a) Molecular Perspective on Protein-Protein Interactions at the Tight Junctions Interface 794
Flaviyan Jerome Irudayanathan, Shikha Nangia
(527b) Pore Formation By Aggregates of Antimicrobial Peptides in Dmpc Bilayers 795
Yuan Lyu, Maya Frityanti, Xiao Zhu, Ganesan Narsimhan

(527c) Single-Molecule Characterization of Protein Adsorption to Multivalent Glycan Polymers 796
Bhargava Nemmaru, Sonia K Brady, Matthew J. Lang, Shishir Chundawat

(527d) Controlling Local Hydrophobicity in Poly(ethylene glycol) Brushes with Poly(sulfobetaine) to Mediate the Conformation of Fibronectin on Biomaterial Surfaces 797
David Faulon Marruecos, Hye Hyun Kim, Michael Shirts, Daniel K. Schwartz, Joel L. Kaar

(527e) Adsorption of Recombinant Human Interleukin-1 Receptor Antagonist to Silicone Oil-Water Interfaces Leads to Gel Formation and Subsequent Surface-Induced Aggregation 798
Lea L. Sorret, Theodore W. Randolph

(527f) Fibrinogen Adsorption Onto Phospholipid Monolayers: Aging and Stiffening 799
Ian Williams, Todd M. Squires

(527g) Morphological and Mechanical Studies of Multicomponent Phospholipid Monolayers Containing 27-Hydroxycholesterol 800
Benjamin Stottrup, Amit Kumar Sachan, Joesph A. Zasadzinski

(527h) Understanding the Effect of Engineered Carbon Nanodiamonds on the Reversible Collapse of Lung Surfactant Monolayers 801
Aishik Chakraborty, Prajnaparamita Dhar

(527i) High Resolution Optical and Electrical Recording of Free Standing Lipid Bilayers 802
Peter J. Beltramo, Jan Vermant

(535a) Dynamics of Lipid Lamellae and Lamellar Vesicles in Shear and Extensional Flows 803
Subas Dhakal, Radhakrishna Sureshkumar

(535b) Probing the Structure of High Viscosity Complex Fluids at High Shear Rates 804
Javen Weston, Kathleen Weigandt, Steven D. Hudson

(535c) Mechanistic Constitutive Model for Wormlike Micelle Solutions with Flow-Induced Structure Formation 818
Sarit Dutta, Michael D. Graham

(535d) Rheology and Microstructure of Thermoreversible Micellar Crystals in Ionic Liquid 819
Ru Chen, Carlos R. Lopez-Barron, Norman J. Wagner

(535e) Use of Microfluidics and Advanced Microscopy to Study Shear-Induced Structuring in Wormlike Micellar Solutions 823
Carla Caiazza, Valentina Preziosi, Giovanna Tomaiuolo, Denis O'Sullivan, Vincenzo Guida

(535f) A Multiscale Tensor-Based Model for the Rheology of Aggregating Thixotropic Colloidal Suspensions 824
Paul M. Mwasame, Antony N. Beris, Norman J. Wagner

(535g) Effect of Shear Stress on Rheological Properties of Fibrous Colloidal Gels during Gel-Sol Transitions Induced By Osmotic Pressure Gradients 825
Matthew Wehrman, Seth Lindberg, Kelly M. Schultz

(535h) Microstructural Basis for Colloidal Gel and Glass Aging 826
Melissa B. Gordon, Christopher J. Kloxin, Norman J. Wagner

(535i) The Mechanism for Shear Thickening in Viscoelastic Suspensions 827
Mengfei Yang, Eric S. G. Shaqfeh

(535j) Homogenization of Self-Organizing Colloidal Particles in a Microscopy Chamber 828
Paul W. Todd, Michael (Andy) Kurk, Scott Moyers

(543a) Elevated Pressure Interfacial Dynamics of LS Surfactants at the Water-CO2 Interface 829
Zachary R. Hinton, Nicolas J. Alvarez

(543b) Atomistic Simulations of Micellization and Adsorption of Imidazoline-Based Surfactants Near Metal-Water Interfaces 830
Sumit Sharma, Yathish Kurapati

(543c) Surfactant Chemistry and Counter-Ion Clouds in Bile Salt Stabilized Swcnts 831
Ketan S. Khare, Frederick R. Phelan Jr.

(543d) Tailoring Inhibitors of Pathological Crystallization: New Platforms for Drug Design 832
Bryan G. Alamani, Doyoung Kim, Mangalaa Dinivahi, Diego Guala, Brandon Melendez, Jeffrey D. Rimer

(543e) Directed Crystal Nucleation and Growth of Blue Phases By Chemically Patterned Surfaces 833
Xiao Li, Jose Martinez-Gonzalez, Juan Hernandez-Ortiz, Ye Zhou, Monirosadat Sadati, Rui Zhang, Juan J. De Pablo, Paul F. Nealey

(543f) Transient Reorientation of Liquid Crystals Induced By Water Transport 834
Hadi Ramezani-Dakhel, Monirosadat Sadati, Rui Zhang, Mohammad Rahimi, Khia Kurtenbach, Nicholas L. Abbott, Benoit Roux, Juan J. De Pablo

(543h) DNA-Micropost Adsorption and Enhanced Transport in More Crowded Micropost Arrays 835
Yeng-Long Chen, Wei Chien, Fan-Tso Chien

(543i) A Miniaturized, Radial Langmuir Trough for Simultaneous Visualization and Dilatational Deformation of a Complex Fluid-Fluid Interface 836
Joseph R. Samaniuk

(554a) Electrocatalytic CO2 Reduction for Enhanced C-C Coupling 837
Youngkook Kwon
(554c) Reaction Path Discovery Under Potential Bias 838
Eric Walker, Paul M. Zimmerman
(554e) Electrocatalytic Carbon Fixation on Molecular-Material Hybrids 839
Zhi Cao, Christopher Chang
(554f) Investigation of the Morphological Factors That Drive Selectivity on Polycrystalline Cu during CO2 Electroreduction 840
Alexandros N. Karaiskakis, Elizabeth J. Biddinger
(554g) Modulation of Charge Transfer Kinetics By Two Orders of Magnitude at Back-Gated Monolayer MoS2 Electrodes 841
Yan Wang, C. Daniel Frisbie
(574a) Predicting Vapor-Liquid Equilibria with Augmented Ab Initio Potentials 842
Maryna Vlasiuk, Richard J. Sadus
(574b) New Vapor Pressure Prediction with Improved Thermodynamic Consistency Using the Riedel Equation 843
Joseph W. Hogge, Neil Giles, Richard L Rowley, Thomas A. Knotts Iv, W. Vincent Wilding
(574c) Solvent Screening and Pure Component Thermophysical Property Prediction for Pharmaceutical Process Design 844
Getachew S. Molla, Lukasz Ruszczynski, Jens Abildskov, Gurkan Sin
(574d) A Two-Structure Equation of State for Super Cooled and Stretched Light and Heavy Water 845
Michal Duska, Jan Hruby, Frederic Caupin, Mikhail A. Anisimov
(574e) Can Molecular Simulations Predict the Binary Interaction Parameters of the Activity Coefficient Models? 853
Ashwin Ravichandran, Rajesh Khare, Chau-Chyun Chen
(574f) A Wertheim Activity Coefficient Model for Associating Mixtures 854
Aseel M. Bala, William G. Killian, Jackson A. Storer, James E. Jackson, Paul M. Mathias, Navin Patel, Timothy C. Frank, Dung T. Vu, Eric L. Cheluget, Carl T. Lira
(574g) Understanding the Role of Intermolecular Forces in Molecular-Based Equations of State: 20 Years of the Soft-SAFT Equation 855
Lourdes F. Vega, Felix Llovell, Felipe J. Blas, Oriol Vilaseca, Joao A. P. Coutinho, Jordi Torne
(196b) Coarse-Grained Modeling of Polymer Materials in Molecular Simulations with SAFT-Î³ Mie Corresponding States Correlation Specified Potentials 856
Amulya Pervaje, Saad A. Khan, Erik E. Santiso
(574h) New SAFT Eos for Electrolyte Solution 857
Reza Shahryari, Mohammad Reza Dehghani
(575a) Controlling Vesicle Shape By Adsorption of a Semiflexible Polymer 858
Bing Li, Steven M. Abel
(575b) Polymer-Induced Restructuring in the Gut 859
Sujit S. Datta, Asher Preska Steinberg, Rustem Ismagilov
(575c) Multicomplex Formation in the Presence of Polymer Crowders Mimicking in-Cell Environment 860
Vahid Rahmanian, Young C. Kim, Jeetain Mittal
(575d) Effect of Palmitoylation in Membrane Proteins at the Blood-Brain Barrier Interface 873
Nandhini Rajagopal, Flaviyan Jerome Irudayanathan, Shikha Nangia
(575e) Determining the Limitations of Processive Polysaccharide Deconstruction By Glycoside Hydrolases 874
Suvamay Jana, Brandon C. Knott, Gregg T. Beckham, Christina M. Payne
(575f) Computational Design of Novel Self-Assembling Peptide Biomaterials Based on an Amyloid Forming Motif from the Adenovirus Fiber Shaft 875
Sai Vamshi R Jonnalagadda, Graziano Deidda, Eirini Ornithopoulou, Asuka Orr, Anna Mitraki, Phanourios Tamamis
(575g) Conformational Changes of 2'-Hydroxybiphenyl-2-Sulfinate Desulfinase 876
Yue Yu, Christina M. Payne
(575h) Chromatin Folding Via Coarse-Grained Multi-Scale Simulation 877
Joshua Lequieu, Andres Cordoba, Joshua Moller, Juan J. De Pablo
(577a) Quantitative Prediction of Industrial Turbulent Processes in Stirred Tank Reactors- Invited Presentation in Turbulent and Reactive Flows 878
Minye Liu
(577b) Passive Scalar Mixing in Anisotropic Turbulence from Line Sources 879
Quoc T. Nguyen, Dimitrios V. Papavassiliou

(577c) Dynamics and Structures of Viscoelastic Turbulence in Transitional Channel Flow 880
Ashwin Shekar, Sung-Ning Wang, Michael D. Graham
(577d) Polymer Effects on the Development and Bursting of Turbulent Vortices: Implication on High-Extent Drag Reduction 881
Lu Zhu, Xue Bai, Li Xi
(577f) Macroscale Insensitivity of Type B Drag Reduction By Two Biopolymers 882
Preetinder S. Virk
(577g) Euler-Euler and Euler-Lagrange Simulations of Heavy Particles Clustering in Homogeneously Sheared Gas 883
Mohamed H. Kasbaoui, Olivier Desjardins, Donald L. Koch
(577h) Breakage of Single Drops and Bubbles in a Turbulent 2-D Orifice Flow 884
Derrick I. Ko, Richard V. Calabrese
(577i) On the Influence of Spanwise Boundary Conditions on the Large Eddy Simulation of the Flow Past a Single Cylinder 885
Barbara L. Silva, Matheus R. Barbieri, Jonathan Utzig, Waldir Pedro Martignoni, Henry F. Meier
(588a) Out-of-Equilibrium Phase Behavior of Dielectric/Paramagnetic Nanoparticle Suspensions in Toggled Electric/Magnetic Fields 893
Zachary Sherman, James Swan
(588b) Investigation of the Motion of Patchy Particles at Liquid/Fluid Interfaces 894
Zohreh Jalilvand, Ilona Kretzschmar
(588c) Preparation and Characterization of Janus Dumbbells Nanomotors 895
Florian Guignard, Marco Lattuada
(588d) Magnetic Microlassos for Reversible Cargo Capture, Transport, and Release 896
Tao Yang, Tonguc Onur Tasci, Keith B. Neeves, Ning Wu, David W. M. Marr
(588e) Fly Larvae Mix to Increase Eating Rates 897
Olga Shishkov, David L. Hu
(588f) Programming the Dynamics of Active Colloids in 3D 898
Jingyun Lee, Stacey Wieseneck, Bhuvnesh Bharti
(588g) Tuning the Collective Behavior of Active Electrohydrodynamic Motors 899
Xingfu Yang, Ning Wu
(588h) Heterogeneous Active Matter Systems 900
Thomas Kolb
(588i) Directed Motion of Metallodielectric Particles By Contact Charge Electrophoresis 901
Yong Dou, Charles A. Cartier, Wenjie Fei, Shashank Pandey, Sepideh Razavi, Ilona Kretzschmar, Kyle J. M. Bishop
(603a) Molybdenum Dioxide Functionalized CoP Porous Nanocatalyst for Highly Efficient Hydrogen Evolution 902
Jun Wang, Liang-Xin Ding, Haihui Wang
(603b) Bijel-Derived Materials for Electrochemical Energy Storage and Conversion 903
Kyle McDevitt, Daniel R. Mumm, Ali Mohraz
(603c) Electrochmical Lithiation-Delithation of Sulfur in Sub-Nano Confinement 904
Chengyin Fu, Juchen Guo
(603d) Durable High Capacity Li-O2 Cathode Composed of Iron-Nitrogen-Doped Mesoporous Core-Shell Carbon Loaded with RuO2 Nanoparticles 905
Ming Zhou, Kwong-Yu Chan, Chi-Ying Vanessa Li
(603e) Magnetic Field-Induced Fabrication of Fe3O4@Graphene Nanocomposites for Enhanced Electrode Performance in Lithium-Ion Batteries 906
Huan Wang, Jingyi Xie, Marissa Follette, Tyson C. Back, Placidus B. Amama
(613a) Elucidating Mechanisms of Substrate Transport in Membrane Transporters 907
Diwakar Shukla, Balaji Selvam
(613b) Grafting Charged Species to Membrane-Embedded Scaffolds Dramatically Increases the Rate of Bilayer Translocation 908
Reid Van Lehn
(613c) Molecular Simulations of Nanoparticles Permeating Lipid Membranes for Drug Delivery Applications 909
Priyanka Oroskar, Cynthia J. Jameson, Sohail Murad
(613d) Transport Characteristics and Stability of Artificial Water Channels in Lipid and Biomimetic Membranes 910
Harish Vashisth
(613e) Solute Transport through Outer Membrane Porins of Gram-Negative Bacteria 911
Huilin Ma, Aliza Khan, Shikha Nangia

(613f) Could Class IIb Bacteriocins Induce Pore Formation? Investigation through Microsecond Long Atomistic Molecular Dynamic Simulation 912
Panagiota Kyriakou, Yiannis N. Kaznessis
(613g) Host-Pathogen Interactions at the Tight Junctions: Molecular Insights from Membrane Guided Self-Assembly 913
Flaviyan Jerome Irudayanathan, Xiaoyi Wang, Nan Wang, Shikha Nangia
(613h) Structure of Multi-Component, Gel-Phase Lipid Bilayers 914
Alexander Yang, Timothy C. Moore, Christopher R. Iacovella, Michael Thompson, Pallav Bulsara, David J. Moore, Clare McCabe
(613i) Multiplicity of States and Internal Structure of Supported Lipid Bilayers 915
Aleksey Vishnyakov, Ting Li, Alexander V. Neimark
(613j) Line Tension and Lipid Sorting Modulate Dynamics of Hemifusion Diaphragm Dissipation 916
Jasmine Gardner, Cameron F. Abrams
(614a) Understanding the Removal of Pharmaceuticals from Water Effluents By Adsorption in Activated Carbons - A Molecular Simulation Approach 917
Lourdes F. Vega, Daniel Bahamon
(614b) Insights into the Adsorption and Phase Behaviour of Fluids in Nanoporous Materials with Hierarchical Pore Structure: Towards an Advanced Textural Characterization 918
Matthias Thommes
(614c) Molecular Studies of Supercapacitors: Ionic Liquids Adsorbed into Porous Carbon Electrodes 919
Peter T. Cummings
(614d) Structure of Ice in Confinement; Water in Mesopores 920
Malgorzata Sliwinska-Bartkowiak
(614e) Towards Understanding the Role of Microstructure in Energetic Material Response: Coarse-Grain Modeling and Simulation 921
John K. Brennan
(614g) Adsorption and Transport in Multiscale Porous Media 922
Benoit Coasne
(629a) Redox-Triggered Mixing and Demixing of Surfactants within Assemblies Formed in Solution and at Surfaces 923
Nicholas L. Abbott
(629b) Static and Dynamic Signatures of Branching in Wormlike Micelles (WLMs) Via Advanced Techniques in Rheology and Neutron Scattering 924
Michelle A. Calabrese, Simon Rogers, Lionel Porcar, Norman J. Wagner
(629c) Self-Assembly of Achiral Surfactants Can Conduct Enantioselective C-C Bond Formation of Amino Acid Derivative 925
Hiroshi Umakoshi, Fumihiko Iwasaki, Keishi Suga, Yukihiro Okamoto
(629d) A Statistical Associating Fluid Theory (SAFT) Framework for Aqueous Nonionic Surfactant Systems 926
Aubrey Winiarski, Arthur S. Gow
(629e) First Proof of Self-Assembly of Block Copolymers in a Deep Eutectic Solvent 927
Dannie J. G. P. Van Osch, Marco M. R. M. Hendrix, Nicole M. W. Van Der Heijden, Jaap Van Spronsen, A. Catarina C. Esteves, Remco Tuinier
(629f) Fast and Slow Dynamical Processes in Simulations of Block Copolymer Micelles 928
Joshua Mysona, David Morse, Alon McCormick
(629g) Synthesis of Hydrophobically Modified Polybetaines (HMPB) and Study of Their Self-Assembly By Molecular Dynamics Simulations 929
Xiao Zhao, Ashwin Ravichandran, Sarkyt Kudaibergenov, Rajesh Khare, Nurxat Nuraje
(629h) A Hexagonal Columnar Liquid Crystal Phase Formation in Dilute Solutions of Carbon Nanotubes 930
Vida Jamali, Francesca Mirri, Paul Van Der Schoot, Fred Mackintosh, Matteo Pasquali
(629i) Computational Phase Space Screening of Isotropic Multi-Well Pair Potentials 931
Julia Dshemuchadse, Michael Engel, Pablo F. Damasceno, Carolyn L. Phillips, Sharon C. Glotzer
(629j) Methanol Self-Association and Preferential Solvation of Chelating Agents for the Extraction of Nuclear Fission Products in Supercritical CO2 932
Trent R. Graham, Daniel J. Pope, Aurora E. Clark, Steven R. Saunders
(654a) Effect of Salt Concentration on the Ability to Form Stable Close-Packed Vesicular Dispersions for Stabilizing Suspensions of Dense Particles Against Sedimentation 933
An-Hsuan Hsieh, Yung-Jih Yang, David . S. Corti, Elias I Franses
(654b) Toward a Better Understanding of Nanoassembly: Importance of Molecular Granularity on Colloidal Forces 934
Jaehun Chun, Christopher J. Mundy, Gregory K. Schenter, Dongsheng Li, James J. De Yoreo

(654c) Structural Insights into DNA-Stabilized Silver Clusters through Analytical Ultracentrifugation 935
Danielle Schultz, Nancy J. Lin, Jeffrey A. Fagan
(654d) Stabilization of Colloidal Suspensions with a Bimodal Distribution of Microparticles and Nanoparticles 936
Marzieh Moradi, Stuart J. Williams, Gerold A. Willing
(654e) Rheological and Electrical Percolation Behavior of Carbon Black Suspensions in Polar Aprotic Solvents 937
Norman J. Wagner, Jeffrey J. Richards, Paul Butler
(654f) Functional Hybrids Based on Assembly of 2D Materials and Polymeric Nanoparticles 938
Nader Taheri Qazvini, Kylie Zane, Brian Schwartz, Matthew V. Tirrell, Juan J. De Pablo
(654g) Convective Fluid Motions in Droplets Driven By Global and Local Chemical Gradients 939
Nan Shi, Todd M. Squires
(654h) Characterizing Gelatin Hydrogel Viscoelasticity with Diffusing Colloidal Probe Microscopy 940
Soheila Shabaniverki, Jaime Juarez
(654i) Self-Immobilized Cross-Linked Enzyme Aggregates for Biocatalysis 941
Yang Kun-Lin, Le Truc Nguyen
(669a) Impact of Molecular Structure on Surface and Thermal Properties of Amido-Amine Cationic Gemini Surfactants 942
Muhammad Shahzad Kamal, S. M. Shakil Hussain, Abdullah S. Sultan
(669b) Surfactant Effect on Hydrate Crystallization at Oil-Water Interface 943
Liat Rosenfeld, Kevin Dann
(669c) Surface Tension Behavior of Aqueous Solutions of a Propoxylated Surfactant and Interfacial Tension Behavior Against an Un-Preequilibrated Crude Oil 944
Jaeyub Chung, Bryan W. Boudouris, Elias I Franses
(669d) Eco-Friendly Sacrificial Amphiphiles As Chemical Herders for Oil Spill Remediation 945
Hao Zhou, George John, Charles Maldarelli
(669e) Interfacial Interactions in Oil/Brine Emulsions Stabilized By Combinations of Cellulose Nanocrystals and Emulsion Stabilizers 946
Sanjiv Parajuli, Chadwick E. Middleton, Andres E. Rodriguez Zambrano, Esteban E. Urena-Benavides
(669f) Measurement and Surface Complexation Modeling of Calcite Zeta Potential in Mixed Brines for Carbonate Wettability Characterization 947
Jin Song, Yongchao Zeng, Xindi Duan, Le Wang, Walter G. Chapman, Sibani Lisa Biswal, George J. Hirasaki
(669g) Dissolution and Restructuring of Calcite Surfaces in Contact with Electrolyte Solutions - Implications for Enhanced Oil Recovery 955
Kai Kristiansen, Szu-Ying Chen, Yair Kaufman, Nicholas Cadirov, Howard Dobbs, Alex Schrader, Dongjin Seo, J. Boles, Jacob Israelachvili
(669h) Marine Bacteria Adhesion to Oil/Water Interfaces 956
Michael Godfrin, Maswazi Sihlabela, Arijit Bose, Anubhav Tripathi
(669i) Oil Spill Dispersion with Food-Grade Amphiphiles: Formation of Oil-Bacteria Agglomerates during Oil Biodegradation 957
Geoffrey D. Bothun, Joseph Rocchio, Srinivasa R. Raghavan, Vijay T. John, Alon McCormick
(669j) The Role of Bacterial Biofilm in Particle-Stabilized Oil-in-Water Emulsions: Implications to the Biodegradation of Oil Spills 958
Marzhana Omarova, Lauren T Swientoniewski, Diane A. Blake, Geoffrey D. Bothun, Vijay T. John
(670a) Novel Cathode Materials for Rechargeable Aluminum Batteries 959
Robert J. Messinger, Jeffrey Xu, Ankur Jadhav
(670b) Reversibility of Zinc Metal Anode: Fundamentals and Engineering 960
Nian Liu
(670c) An Air Breathing Lithium-Oxygen Battery 961
Baharak Sayahpour, Mohammad Asadi, Pedram Abbasi, Larry A. Curtiss, Amin Salehi-Khojin
(670d) Appropriate Characterization Techniques for Lithium-Oxygen Batteries, and Implications for Understanding 2e- Vs. 4e- Oxygen Reduction Processes 962
Colin M. Burke, Bryan D. McCloskey
(670e) Mechanistic Insights into the Sodium-Oxygen Battery Cathode Electrochemistry 963
Jessica E. Nichols, Bryan D. McCloskey
(670f) Mechanistic Evaluation of Thermal Runaway in Potassium-Ion Batteries 964
Ryan A. Adams, Arvind Varma, Vilas G. Pol
(670g) Molten Salt Batteries: Mechanics and Electrolyte Transport 965
Christine Cardinal Roberts, Martin B. Nemer, Mark Stavig, Alexander Headley, Ryan Solich, Scott A. Roberts
(683a) Entropy-Driven, Two-Step Crystallization of Colloidal Clathrate Crystal 966
Sangmin Lee, Michael Engel, Sharon C. Glotzer

(683b) Unraveling the Coupling between Demixing and Crystallization in Mixtures 967
Jerome Delhommelle, Caroline Desgranges
(683c) La Mer Burst Nucleation and Growth: Assumptions, Models, and Data 968
Baron Peters
(683d) A Stochastic Model of Primary Nucleation of Polymorphs 969
Giovanni Maria Maggioni, Leonard Bezinge, Marco Mazzotti
(683e) Step Velocity of a Crystal Edge with Alternating Rows of Growth Units 970
Mark Joswiak, Baron Peters, Michael F. Doherty
(683f) Multiscale, Multiphysics, Mechanistic Model for Computation of Face-Specific Growth Rates 971
James Fell, Anish V. Dighe, Meenesh R. Singh
(683g) Anisotropic Growth Kinetics of Triblock Janus Colloids 972
Wesley F. Reinhart, Athanassios Z. Panagiotopoulos
(685a) Enhanced Sampling Methods for Modulating Density Fields 973
Zhitong Jiang, Suruchi Fialoke, Amish Patel
(685b) Mapped Averaging Methods for Accurate and Precise Evaluation of Free Energies and Other Properties By Molecular Simulation 974
Weisong Lin, Akshara Goyal, Sabry G. Moustafa, Andrew J. Schultz, David A. Kofke
(685c) Predicting the Free Energy Landscape of Multicomponent Fluids 975
Nathan A. Mahynski, Jeffrey R. Errington, Vincent K. Shen
(685d) A Novel Molecular Simulation Method for Liquid-Liquid Equilibria Predictions and in silico Screening of Desalination Solvents 976
Prashanth Chandran, Jindal K. Shah
(685e) Ssages: A Comprehensive Platform for Enhanced Sampling Simulations 977
Hythem Sidky, Yamil J. Colon, Benjamin J. Sikora, Cody Bezik, Federico Giberti, Ashley Guo, Julian Helfferich, Xikai Jiang, Joshua Lequieu, Jiyuan Li, Joshua Moller, Michael Quevillon, Mohammad Rahimi, Hadi Ramezani-Dakhel, Vikramjit Rathee, Daniel Reid, Emre Sevgen, Vikram Thapar, Michael Webb, Justin Wozniak, Xujun Zhao, Nicola J. Ferrier, Olle G. Heinonen, Giulia Galli, Francois Gygi, Juan J. De Pablo, Jonathan K. Whitmer
(685f) Multiscale Modeling of Multicompartment Micelle Nanoreactors 978
Connor Callaway, Parveen Sood, Seung Soon Jang
(685g) Reaction Ensemble Monte Carlo: Applications to Ionic Liquids 979
Ryan Gotchy Mullen, Edward J. Maginn
(685h) Temperature-Dependent Physicochemical Properties of Nitrotoluenes from Solvation Free Energies 980
Alauddin Ahmed, Stanley I. Sandler
(688a) Predicting the Solubility and Diffusivity of Gases (CO2, CH4, H2, Noble Gases) in Nano-Confined Interlayer Water and Bulk Water Using Molecular Dynamics Simulations 981
Greeshma Gadikota, Ian Bourg
(688b) Simulation of Tracer Particle Diffusion in Attractive and Repulsive Glassy Matrices 982
Ryan Roberts, Ryan Poling-Skutvik, Jacinta C. Conrad, Jeremy C. Palmer
(688f) [Invited Talk] Predictive Modeling of Adsorption and Reaction Equilibria in Nanoporous Materials 983
J. Ilja Siepmann, Evgenii Fetisov, Mansi S. Shah, Michael Tsapatsis
(688d) Characterizing Protein Hydration to Inform its Interactions 984
Erte Xi, Amish Patel
(688e) Confinement-Induced Supercriticality and Phase Equilibria of Hydrocarbons in Nanopores 985
Sheng Luo, Hadi Nasrabadi, Jodie L. Lutkenhaus
(688c) [Invited Talk] Simulations of Biomolecular Assembly Processes at Interfaces 986
Jeetain Mittal
(688g) Nucleation of Capillary Bridges and Bubbles 987
Jerome Delhommelle, Caroline Desgranges
(688h) Modeling Single Component Transport in Mesoporous Membranes Under Non-Equilibrium Conditions 988
Ashutosh Rathi, Eustathios Kikkinides, David M. Ford, Peter A. Monson
(704a) Optimizing the Formation of Solid Solutions with Colloids of Different Shapes 989
Fernando Escobedo
(704b) Programmed Assembly of Anisotropic Patchy Colloids By Nonlinear Learning and Landscape Engineering 990
Andrew W. Long, Andrew L. Ferguson
(704c) Evaporation-Induced Assembly of Colloidal Crystals 991
Michael P. Howard, Wesley F. Reinhart, Arash Nikoubashman, Athanassios Z. Panagiotopoulos
(704d) Amphiphile-Induced Reorganization of Liquid Crystals at Aqueous Interfaces 992
Hadi Ramezani-Dakhel, Mohammad Rahimi, Joel Pendery, Nicholas L. Abbott, Benoit Roux, Juan J. De Pablo

(704e) Investigating Thermodynamics and Kinetics of a Î²(16-22) Peptide Aggregation Using Coarse-Grained Simulations 993
Yiming Wang, Stefan Auer, Sergei V. Krivov, Carol K. Hall
(704f) Inverse Materials Design from Phase Transitions in Shape Space 994
Rose Cersonsky, Greg Van Anders, Paul Dodd, Sharon C. Glotzer
(704g) Concentration Effects in Simulations of Non-Ionic and Ionic Surfactant Micellization 995
Andrew P. Santos, Athanassios Z. Panagiotopoulos
(704h) Self-Assembly Simulations of Stratum Corneum Lipid Mixtures 996
Timothy C. Moore, Christopher R. Iacovella, Remco Hartkamp, Annette L. Bunge, Clare McCabe
(704i) Self-Assembly of Multi-Flavored DNA-Functionalized Particles into Binary Superlattices 997
Hasan Zerze, Nathan A. Mahynski, Evan Pretti, Vincent K. Shen, Jeetain Mittal
(708a) Accelerating Force Field Parameterization to Improve the Quantitative Predictability of Thermophysical Properties 998
Richard A. Messerly, Andrei Kazakov
(708b) Validation of Trimethylamine-N-Oxide (TMAO) Force Fields Based on Thermophysical Properties of Aqueous Tmao Solutions 999
Daniel Markthaler, Johannes Zeman, Jorg Baz, Jens Smiatek, Niels Hansen
(708d) Optimized Mie Potentials for Phase Equilibria: Application to Noble Gases, Alkanes, Alkynes and Their Mixtures 1008
Mohammad Barhaghi, Jason R. Mick, Jeffrey J. Potoff
(708e) Evaluation of Virial Coefficients and Their Temperature Derivatives for Multibody Potential Models 1009
Navneeth Gokul, Andrew J. Schultz, David A. Kofke, Hainam Do, Richard J. Wheatley
(708f) Prediction of Interaction Parameters for Coarse-Grained Models Using Ab Initio Calculations 1010
Jennifer A. Clark, Erik E. Santiso
(708g) Modifying Nonplanar Vibrational Modes of Aromatic Rings in Biomolecular Modeling 1011
Faramarz Joodaki, Lenore M. Martin, Michael L. Greenfield
(708h) A Polarizable Force Field of Inorganic Phosphates and Hydroxyapatite Based on the Classical Drude Oscillator 1012
Hadi Ramezani-Dakhel, Hui Li, Nader Taheri Qazvini, Juan J. De Pablo, Benoit Roux
(708i) Using Minimal Biasing Methods in Plumed and Lammps 1013
Andrew White
(713a) A General Route for Nanoemulsion Synthesis Using Low Energy Methods at Constant Temperature 1014
Ankur Gupta, Abu Zayed Md Badruddoza, Patrick S. Doyle
(713b) Advanced Emulsion-Templating of Microreactors for the Scalable Production of Semiconductor Nanowires 1015
Maritza Mujica, Michael A. Filler, Victor Breedveld, Sven H. Behrens
(713c) The Discovery of Novel Hydrophobic DES-Water Emulsions 1022
Dannie J. G. P. Van Osch, Nicole M. W. Van Der Heijden, Jaap Van Spronsen, A. Catarina C. Esteves, Remco Tuinier, Mark Vis
(713d) Controlling Foaming during Thermal Cracking - a Non-Silicon Based Antifoaming Agent 1023
Amaka Waturuocha, Michael Volk, Glixon Mavarez Nava, Dwijen Banerjee, Keith Wisecarver
(713e) Effect of Salt on Drainage Via Stratification in Micellar Foam Films 1024
Subinuer Yilixiati, Rabees Rafiq, Yiran Zhang, Vivek Sharma
(713f) Asphaltenes Adsorption at Water/Oil Interface, a Classical Surfactant Approach 1025
Shaghayegh Darjani, Fang Liu, Nelya Akhmetkhanova, Vincent Pauchard
(713g) Rapid Demulsification of Water-in-Oil Emulsions Using Silica Nanoparticles 1026
Ashwin Kumar Yegya Raman, Clint P. Aichele
(713h) Linking Adhesive Emulsions Behaviour in Microfluidic Devices to Direct Force Measurements between Drop Pairs 1027
Christopher Fewkes, Emily Jamieson, Joe Berry, Raymond R. Dagastine
(713i) Interfacial Routes to Gelation in Solid-Stabilized Emulsions 1028
Max Kaganyuk, Ali Mohraz
(713j) Predicting the Microstructure of an Interface Laden with Anisotropic Particles 1029
Sri Harsha Nuthalapati, Christopher L. Wirth
(718a) Controlling the Polymer/Gas Interface of the Ionic Polymer Phase of a PEM Fuel Cell Catalyst Layer during Membrane Electrode Assembly Fabrication 1030
Regis Dowd Jr., Trung Van Nguyen
(718b) Modeling the Effect of Charged Lithium-Metal Anode on Electrolyte Decomposition 1031
Luis E. Camacho-Forero, Perla B. Balbuena

(718c) Dynamic Measurement Method of the Emulsified Water Content Based on Interfacial Polarization 1032
Meiyi Qing, Huaqing Liang, Jinjun Zhang, Min Wei, Chenbo Ma
(718d) Stretchable and Soft Electroadhesion Enhanced By Liquid-Metal Subsurface Microstructures 1041
Sungjune Park, Jun Shintake, Ishan Joshipura, Michael D. Dickey, Kunal Mondal
(718e) Giant Electret Polarization in Electrochemically Deposited Hydroxyapatite Ceramic Coatings 1042
Matthew Yates, Xuefei Zhang, Wanaruk Chaimayo, Benjamin Miller
(718f) Electrochemical Charging of CdSe Quantum Dots: Effects of Adsorption Versus Intercalation 1043
Ajinkya Puntambekar, Qi Wang, Vidhya Chakrapani
(718g) Microstructured Conducting Polymer Swabs for Enhanced Trace Explosive Detection 1044
Jennifer S. Laster, Bryan W. Boudouris, Stephen P. Beaudoin
(718h) Atomistic Modeling of Solvent Decomposition in Magnesium Batteries 1045
Jeffrey S. Lowe, Donald J. Siegel
(719b) Understanding the Reduction Reaction Mechanisms of Sulfur-Based Cathodes: A Theoretical Approach from DFT and Reaxff Molecular Dynamics 1046
Saul Perez Beltran, Perla B. Balbuena
(719c) Reversible Aluminum Intercalation in Transition Metal Sulfides 1047
Linxiao Geng, Juchen Guo
(719d) Highly Structured Titanium Nitride As Novel Cathode Materials of Lithium Sulfur Batteries 1048
Wenduo Zeng, Mark Cheng, Simon Ng
(719e) Elucidating the Solvation Structure and Dynamics of Lithium Polysulfides Using Coupled High Throughput Simulations and Experiments 1058
Nav Nidhi Rajput, Vijayakumar Murugesan, Karl Mueller, Kristin Persson
(719f) Solvation Structure and Behavior of Lithium Polysulfide Species in Electrolytes of Lithium-Sulfur Batteries 1059
Ethan P. Kamphaus, Perla B. Balbuena
(719g) Design Stable Room-Temperature Metal-Sulfur Batteries 1060
Shuya Wei, Lynden A. Archer
(749a) Diverse Colloidal Crystals from DNA-Grafted Spheres Via Self-Assembly 1061
Yifan Wang, Ian Jenkins, James T. McGinley III, Talid Sinno, John C. Crocker
(749b) Material Design By DNA-Mediated Interactions between Colloids 1071
Runfang Mao, Jeetain Mittal
(749c) High Throughput Acoustically Driven Self-Assembly of Microfluidic Colloidal Crystals 1072
Meghana Akella, Jaime Juarez
(749d) The Role of Interaction Heterogeneity in Colloidal Crystallization 1073
Ian Jenkins, John C. Crocker, Talid Sinno
(749e) Self-Assembly of Open Structures Using Depletion 1074
Jens Glaser, Sharon C. Glotzer
(749f) Controlled Self-Assembly of Colloidal Discotic Liquid Crystals 1075
Zhengdong Cheng
(749g) Stratification Dynamics in Drying Colloidal Mixtures 1076
Michael P. Howard, Arash Nikoubashman, Athanassios Z. Panagiotopoulos
(749h) Theoretical and Experimental Investigation of Phase Separation in Noble Metal Nanoparticle Monolayers 1077
Steven Merz, Zachary Farrell, Sergei Egorov, David Green
(749i) Bending-Induced Buckling Instabilities in Self-Assembled Elastomeric Composite Films 1078
Peng Jiang, Sin-Yen Leo, Zhuxiao Gu
(749j) Molecular Modeling of Microstructure, Solubilization and Micro-Emulsion of Block Copolymer Micelles By iSAFT 1079
Shun Xi, Le Wang, Walter G. Chapman
(754a) Spreading of Nanoscale Droplets of Ionic Liquids on the Mica Surface 1080
Lei Li
(754b) Long Range Electrostatic Screening in Ionic Liquids 1081
Matthew A. Gebbie
(754c) Mesopore Confinement Effects on Ionic Liquid Hydration 1082
Yuxin He, Daudi Saang' Onyo, Folami Ladipo, Barbara L. Knutson, Stephen E. Rankin
(754d) Water Sorption in Ionic Liquids Characterized Using a Dynamic Vapor Sorption Analyzer (IGASorp) and High Pressure X-Ray Photoelectron Spectroscopy (APXPS) 1083
M. Alejandra Rocha, Alicia Broderick, John T. Newberg, Mark B. Shiflett
(754e) Property Modeling of Ionic Liquids for Gas Separation Processes 1084
Xinyan Liu, Xiaodong Liang, Xiangping Zhang, Suojiang Zhang, Rafiqul Gani

(754f) An Investigation of Cellulose Solubility in Ionic Liquids with Added Cosolvents 1085
Brooks D. Rabideau
(754g) Effects of Ionic Liquids and Liposomes on Enzymatic Cellulose Hydrolysis Process 1086
Kazuhiko Tanimura, Yoshiko Ooe, Keishi Suga, Yukihiro Okamoto, Makoto Yoshimoto, Hiroshi Umakoshi
(754i) Micellization of Poloxamer Block Copolymers in Ionic Liquids and Their Mixtures with Water 1087
Zhiqi He, Yingzhen Ma, Paschalis Alexandridis
Author Index

Some abstracts in this book refer to images that were not made available by the publisher

488292 Practical Aspects in Using Distillation VLE and Enthalpy Models

Abstract:

Distillation is a very common Unit Operation today and various simulation tools make it very easy to run a simple distillation tower. Behind that simulation is the "heart" of that tower. That "Heart" is the VLE and enthalpy model that the user has chosen. Frequently, users are inexperienced at understanding the background and the limitations of the model they have chosen. It is the intent of this presentation to take the authors practical experiences, with some common distillation applications, and show where today's user can benefit from these experiences. Model choices will be discussed as well as some practical methods to vet a model properly before it is rigorously employed.

486892 Bubble Pressure Measurement and Prediction for n-Alkane + Naphthenic Hydrocarbon Binary Mixtures

Abstract:

Cubic equations of state are widely used for phase equilibrium calculations due to their simplicity and reliability, especially for hydrocarbon mixtures. Unfortunately, common choices such as the Peng-Robinson and Soave-Redlich-Kwong equations of state systematically predict both qualitatively and quantitatively incorrect phase behaviors for long chain n-alkane + aromatic and naphthenic mixtures unless negative binary interaction parameter (k_{ij}) values are used [1, 2]. Binary interaction parameters are typically obtained by fitting vapour-liquid equilibrium (VLE) data for binary mixtures. Experimental VLE data for these mixtures are scarce. In this study, VLE data for representative binary mixtures of naphthenes with long chain n-alkanes are presented. The selected n-alkanes range from n-C_{20} to n-C_{28}. These compounds are paired with naphthenes including cyclohexane, alkyl-cyclohexanes and decahydronaphthalene. These experimental data are compared with computed bubble pressures using the Peng-Robinson, Soave-Redlich-Kwong and PC-SAFT equations of state in order to evaluate the accuracy of predictions and to obtain custom k_{ij} values. In a parallel companion works, bubble pressure data and the corresponding fitted k_{ij} values for 15 representative long chain n-alkanes + aromatic binary mixtures are reported [3] and strategies for estimating k_{ij} in the absence of experimental data are discussed [4]. Expected applications of these contributions include improved phase behavior model accuracy for hydrocarbon production, transport and refining environments, and improved corresponding state based transport property predictions, particularly for viscosity, that require accurate phase behaviors and phase compositions as inputs.

References: [1] Ahitan, S., Satyro, M. A., Shaw, J. M. Systematic misprediction of n-alkane + aromatic and naphthenic hydrocarbon phase behavior using common equations of state. J. Chem. Eng. Data 2015, 60, 3300-3318.

[2] Ahitan, S., Shaw, J. M. Quantitative comparison between predicted and experimental binary n-alkane+ benzene phase behaviors using cubic and PC-SAFT EOS. Fluid Phase Equilib. 2016, 428, 4-17.

[3] Liu, Q. MSc. Thesis, University of Alberta (in progress 2017).

[4] Ahitan, S., Luo, M., McLaughlin, J., Shaw, J. M. On cubic eos interaction parameter estimation for long chain n-alkane + aromatic binary mixtures (in preparation).

503938 When Is It Appropriate to Kiss? a Case Study on the Adverse Consequences of Overfitting GE Models

Abstract:

In his presentation at the 19th International Conference on Chemical Thermodynamics (ICCT-19, Boulder, July/August 2006), Marco Satyro emphasized the importance of developing effective models guided by experience: "In this paper, I describe the different ways in which incomplete data and models are used to do process engineering." He opened his presentation by illustrating the negative consequences of a heat-capacity model where the 24 data points were fit to machine precision using a 24-coefficient model. The "perfect model" was impossible to converge due to absurd temperature derivatives. This presentation shows an analogous case study where the simulation of a simple system - purification of methanol from a mixture containing C2 to C4 alcohols - gave highly inaccurate results compared to established licensor data. It turned out that the two asymmetric NRTL parameters for the various binaries came from a database that fit the near-ideal binary data "exactly," but unintended consequence was that the model predicted strong - and utterly wrong - departures from ideality for the multicomponent system. This is a case where the KISS (Keep it Simple and Straightforward) principle is a superior approach to blind nonlinear fitting of the data. The same principle applies when activity-coefficient parameters are determined from estimation methods such as UNIFAC.

503827 Photochemical Reduction of Enzymes for Light-Driven Chemical Reactions

Abstract:

Photosynthesis powers living cells through the photogeneration of electron-hole pairs that drive energy-demanding chemical reactions. The principles that enable photosynthesis and biological energy conversion provide a foundation for advancing the development of artificial, light-driven systems for catalytic production of chemical fuels. Effective coupling of catalysts with light-harvesting requires understanding how molecular interfaces and electron-proton transfer processes control the overall reaction dynamics. To address these challenges we have been investigating molecular complexes that integrate reduction-oxidation enzymes with light-harvesting materials to understand the mechanisms controlling photocatalysis. A combination of transient spectroscopy, and steady-state infrared and paramagnetic techniques are being used to provide insights on how photoexcited electrons couple to enzymes to drive multi-step reduction reactions. A summary of our research progress on these topics will be presented.

486940 Termite-Inspired Electrochemical Processing of Lignocellulose to Chemicals and Fuels

Abstract:

Lignocellulosic biomass is the most abundantly available raw material on the planet and has great promise as a feedstock for the production of biofuels. Unfortunately, lignocellulose is difficult to break down by fermentation and other conventional approaches and man-made engineered systems have yet to truly harness the energy in wooden waste. Therefore, there is a significant need to find alternative pathways to convert this vital resource into next-generation energy carriers.

The most efficient known system to process lignocellulose to useful chemicals is the digestive track of termites [1-2]. The reason termites are able to efficiently break down these woody feedstocks is because of the diverse community of microbes inhabiting their hindguts, which includes protists, bacteria and archea. These microbes work symbiotically to decompose the lignocellulose, each performing a different role in the digestion process. They self-assemble into a complex community with specific structure and function, resulting in a radial and axial oxygen gradient.

The first objective of this work is to develop a microfluidic habitat that mimics the structure and function of the hindgut of a common lower termite, *Reticulitermies flavipes* [3]. The microfluidic habitat will serve as a well-controlled system to study how chemical and physical composition impact the function of the microbial community. Long-term we intend to use this information to scale up the production of biofuels from wooden materials.

The second objective of this work recognizes that the most common byproduct of the termite digestion is acetate. Though the global market for acetate/acetic acid is limited, in the future it may be widely available since acetate is a very common fermentation product. Therefore, our goal is to find an efficient catalytic pathway for the conversion of acetate to useful commodity chemicals, i.e. methanol.

This talk will highlight our groupâ€™s most recent progress towards both objectives as well as a general discussion of electrochemical pathways for fuels and commodity chemical production.

References

1. Brune, A., D. Emerson, and J.A. Breznak, The Termite Gut Microflora as an Oxygen Sink: Microelectrode Determination of Oxygen and pH Gradients in Guts of Lower and Higher Termites. *Appl Environ Microbiol*, 1995. 61, 2681-2687.

2. Brune, A., Termite Guts: The World's Smallest Bioreactors. *Trends Biotechnol.*, 1998. 16, 16-21.

3. Kadilak, A. L., Liu, Y., Shrestha, S., Bernard, J. R., Mustain, W. E., and Shor, L. M., Selective Deposition of Chemically-bonded Gold Electrodes onto PDMS Microchannel Side Walls. *J. Electroanal. Chem.*, 2014. 727, 141-147.

492284 Nature-Inspired Flow Fields for PEM Fuel Cells

Abstract:

Nature-Inspired Flow Fields for PEM Fuel Cells

Jason In Sung Cho[1,2], Toby Neville[1,2], Panagiotis Trogadas[1], Billy Wu[3], Dan Brett[1,2] and Marc-Olivier Coppens[1]

[1] *Department of Chemical Engineering, University College London, London WC1E 7JE, United Kingdom*

[2] *Electrochemical Innovation Lab, Department of Chemical Engineering, University College London, London WC1E 7JE, United Kingdom*

[3] *Department of Mechanical Engineering, Imperial College London, UK*

Ensuring uniform gas distribution across the catalyst layer remains one of the on-going challenges impeding broader commercialisation of polymer electrolyte membrane fuel cells (PEMFCs). This is an inevitable consequence within the framework of the conventional flow field designs, such as serpentine channel, predominantly due to reactant depletion along the channel and over the active area of the electrode [1].

The novel design of the proposed nature-inspired flow field draws inspiration from the fractal structure and functionality of the upper respiratory tract of the human lung to address this fundamental shortcoming. The fractal geometry of the human lung has been shown to ensure uniform distribution of air from a single outlet (trachea) to multiple outlets (alveoli). The convection driven air flow rate in trachea is reduced until it is equal to the diffusion driven air flow rate in alveoli (Pe ≈ 1)[2], resulting in uniform distribution of entropy production in both regimes[3,4].

By employing a 3D fractal structure as flow field inlet channel, we aim to yield similar benefits from replicating these characteristics of the human lung. The fractal pattern comprises repeating "H" shapes and a fractal dimension of 2, where daughter "H's" are half the parent's size, leading to a plane-filling geometry. Fractal geometry has previously been invoked in flow field designs. However, this design is fundamentally different, as the three-dimensional branching structure allows only the outlets of the fractal inlet channel to be exposed to the membrane-electrode assembly (MEA), eliminating the reactant depletion along the channel, and thus providing uniform local conditions on the surface of the catalyst layer.

Numerical simulations were performed to obtain a detailed insight into the effect of the number of fractal

branching generations on the reactant distribution in the catalyst layer. Based on the simulation results, three fractal flow field prototypes (10 cm^2surface area), each representing a branching generation of N = 3, 4 and 5, were manufactured via selective laser sintering (SLS), and experimentally validated against conventional serpentine flow fields. Results indicate that the N = 4 and N = 5 fractal flow fields outperform the serpentine flow fields at high current densities (at 50% and 75% RH) due to more uniform reactant distribution across the catalyst layer. At higher humidity levels (100% RH), though, the performance of N = 5 fractal flow fields significantly deteriorates; the reduced air flow rate within the fractal flow fields hampers effective gas diffusion within the porous medium resulting in insufficient convective liquid water removal.

Another important characteristic of the fractal flow field design is that uniform gas distribution across the catalyst layer is preserved when these fractal flow fields are scaled-up (27 cm^2 surface area). The fuel cell performance of the large-scale fractal flow field remained almost unchanged on a per area basis, compared to its smaller active area counterpart. On the contrary, larger, conventional serpentine flow fields (27 cm^2surface area) exhibited what appears to be improved performance as a result of an order-of-magnitude higher pressure drop than that of a fractal flow field, because this induces faster reaction kinetics and better liquid water removal. However, such excessive pressure drop renders the use of a large scale serpentine flow field prohibitive, thus favouring the fractal flow field. Implementation of effective water management strategies could circumvent the remaining problems of high-generation fractal flow fields.

References

[1] Tüber, K., Oedegaard, A., Hermann, M., and Hebling, C., *Investigation of fractal flow-fields in portable proton exchange membrane and direct methanol fuel cells.* Journal of Power Sources, **2004**. 131(1–2): p. 175-181.

[2] Coppens, M.-O., A nature-inspired approach to reactor and catalysis engineering. Current Opinion in Chemical Engineering, **2012**. 1(3): p. 281-289.

[3] Kjelstrup, S., Coppens, M.-O., Pharoah, J., and Pfeifer, P., *Nature-inspired energy-and material-efficient design of a polymer electrolyte membrane fuel cell.* Energy & Fuels, **2010**. 24(9): p. 5097-5108.

[4] Trogadas, P., Ramani, V., Strasser, P., Fuller, T.F., and Coppens, M.-O., *Hierarchically Structured Nanomaterials for Electrochemical Energy Conversion.* Angewandte Chemie International Edition, **2016**. 55(1): p. 122-148.

[5] Coppens, M.-O., A nature-inspired approach to reactor and catalysis engineering. Current Opinion in Chemical Engineering, **2012**. 1(3): p. 281-289.

489776 Numerical Analysis of the Flow, Transport, and Interfacial Phenomena Associated with Growth of Crystalline CZT Under Crucible Rotation

Abstract:

Cadmium zinc telluride (CZT) crystals are widely used for advanced gamma ray detectors due to their room temperature operability and high band gap. In order to achieve the high carrier mobility and lifetime properties necessary for its functionality, precise crystal growing techniques are required. CZT is traditionally grown via the gradient freeze method, where molten CZT is directionally solidified in a cylindrical ampoule by drawing it through a temperature gradient, thus encouraging growth of single crystalline material. Directional solidification methods such as this result in complex transport phenomena, as the translating solid-liquid interface induces thermal-buoyancy driven flows, and a non-unity segregation coefficient leads to inhomogeneous concentration gradients. One problem that commonly arises during gradient freeze growth of CZT is the accumulation of thermodynamically supercooled liquid near the solidification interface. This occurs due to local compositional and thermal inhomogeneity created by the growing conditions, which results in constitutional supercooling. Such supercooling promotes an unfavorable morphological instability, changing the solid-liquid interface from planar to cellular. The subsequent cellular morphology can result in the entrainment and entrapment of melt, resulting in micron-sized tellurium-rich inclusions that reduce detector performance and require lengthy annealing process to remove.

The accelerated crucible rotation technique (ACRT), which repeatedly rotates the crucible back and forth, has been shown to result in a lower density of these tellurium-rich particles and dramatically improve detector performance. Under suitable operating conditions, ACRT dramatically changes the patterns of flow in the melt, inciting new rotational phenomena such as Ekman flows and Taylor-Gšrtler instabilities. Enhanced convection also impacts the thermal and solutal fields. Conventional wisdom asserts that ACRT works by homogenizing the composition in the melt. However, little is actually understood from a thermodynamic and mechanistic perspective on how the ACRT works to reduce constitutional supercooling, and, in turn, tellurium-rich inclusions in the crystal. Additionally, little is known on how to select the optimal rotation schedule to achieve high-quality material.

We aim to address these questions using a comprehensive computational model of the flow, heat, and mass transfer within the CZT growth system. We employ an in-house, Galerkin finite-element code with elliptic mesh generation that allows for the computation of flow patterns, thermal field, composition, and interface shape. Using the results from our simulations, we are able to track the quantity and locations of locally supercooled melt. We will show that, contrary to the conventional wisdom, ACRT does not necessarily homogenize the melt, but rather disrupts the local compositional field, which, in turn, affects the thermodynamic stability of the melt. Moreover, we have developed a thermodynamically-consistent metric that allows for the systematic evaluation and optimization of ACRT schedules. Optimal ACRT schedules and generalized ACRT recommendations for this geometry will be presented.

Figures

Fig. 1: a) One example of an ACRT scheme, with four distinct regimes: acceleration, constant, deceleration, and stopped, repeated in the opposite direction, and b) a typical result of the streamlines (left) and thermodynamically undercooled regions (right) in the system during the deceleration regime.

This work is supported by the Department of Energy, National Nuclear Security Administration, under Award DE-NA0002565; no official endorsement should be inferred.

MODELING THE TRANSIENT FLOW AND PREDICTING LAOS FLOW OF A THERMOREVERSIBLE GEL USING SCALAR STRUCTURUAL PARAMETER THIXOTROPIC MODEL

31OCT17

Matthew J. Armstrong[1], Ryan Murphy[2], Antony Beris[2], Norman Wagner[2]
[1]*Department of Chemistry and Life Science, United States Military Academy, West Point, NY 10996*
[2]*Department of Chemical and Biomolecular Engineering, University of Delaware, Newark, DE 19716*

OUTLINE

- Motivation and objectives
- Thermoreversible gel system (old and new)
- Modified Delaware Thixotropic Model for TEVP
- Improvements in fitting thermoreversible gel system?
- Predicting LAOS?
- Future model improvements
- Questions

2

MOTIVATION & OBJECTIVES

Motivation:

-Complex material modeling for TEVP system with different rheological properties at different temperatures

-Develop new/more predictive simple models

-Can we fit LAOS better by first fitting model parameters to series of step up/down experiments

3

THERMOREVERSIBLE GEL SYSTEM

21vol% sticky spheres

- 30nm spheres
- thermo-reversible gel
- visco-elastic
- thixotropic

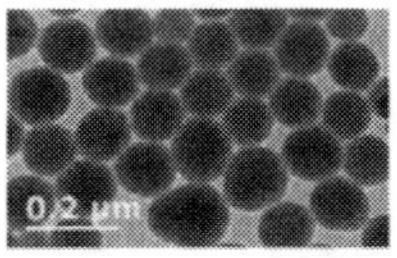

21vol% octadecyl coated silica spheres in n-tetradecane

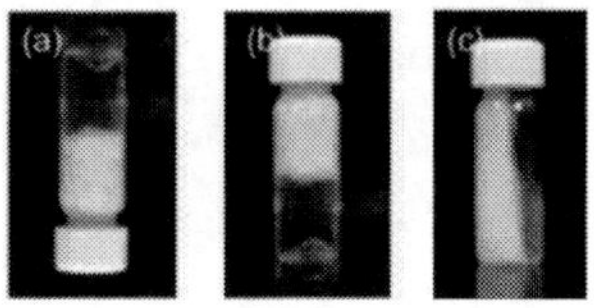

(a) A gelled sample near "gel T"
(b) 15 minutes after tipping
(c) 1 day after tipping the vial

1. Kim JM (2013) PhD Thesis, UDEL; 2. Gurnon AK (2014) PhD Thesis, UDEL; 3. W. M. Keck microscopy Facility.

4

KIM & GURNON ET AL. DATA

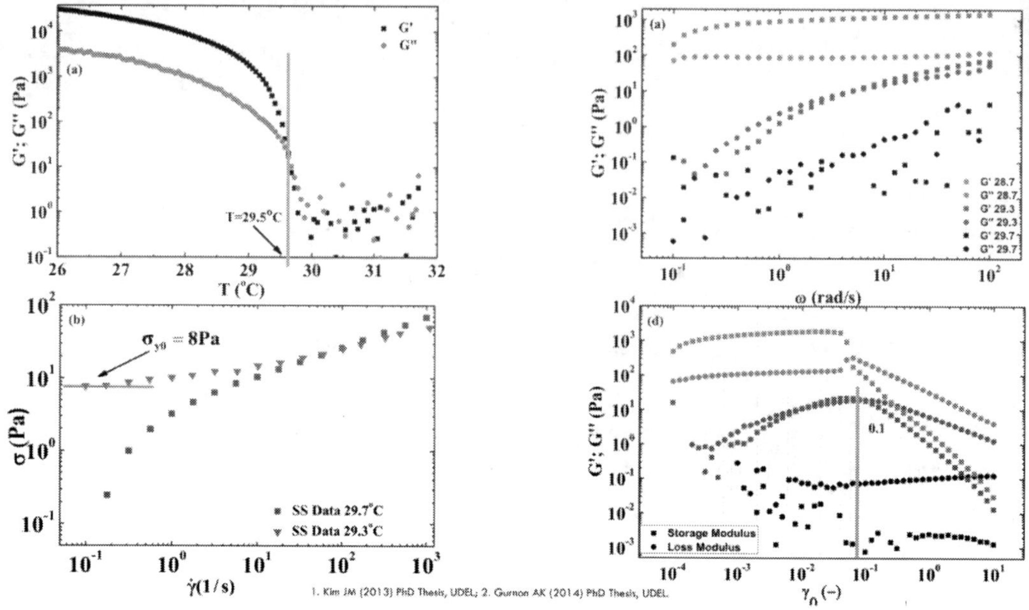

1. Kim JM (2013) PhD Thesis, UDEL; 2. Gurnon AK (2014) PhD Thesis, UDEL.

5

MURPHY & ARMSTRONG ET AL. DATA (1/2)

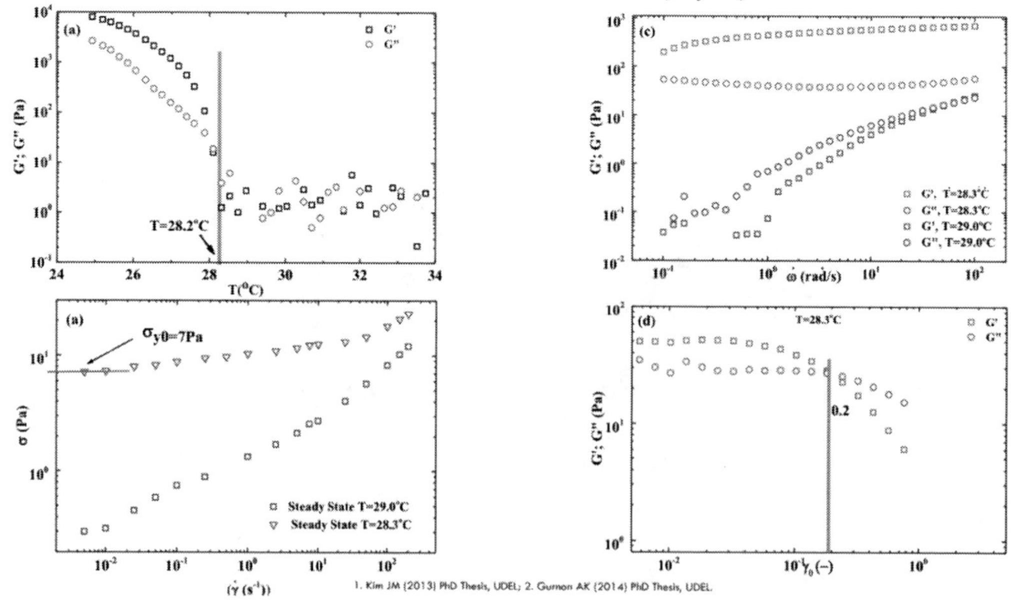

1. Kim JM (2013) PhD Thesis, UDEL; 2. Gurnon AK (2014) PhD Thesis, UDEL.

6

MURPHY & ARMSTRONG ET AL. DATA (2/2)

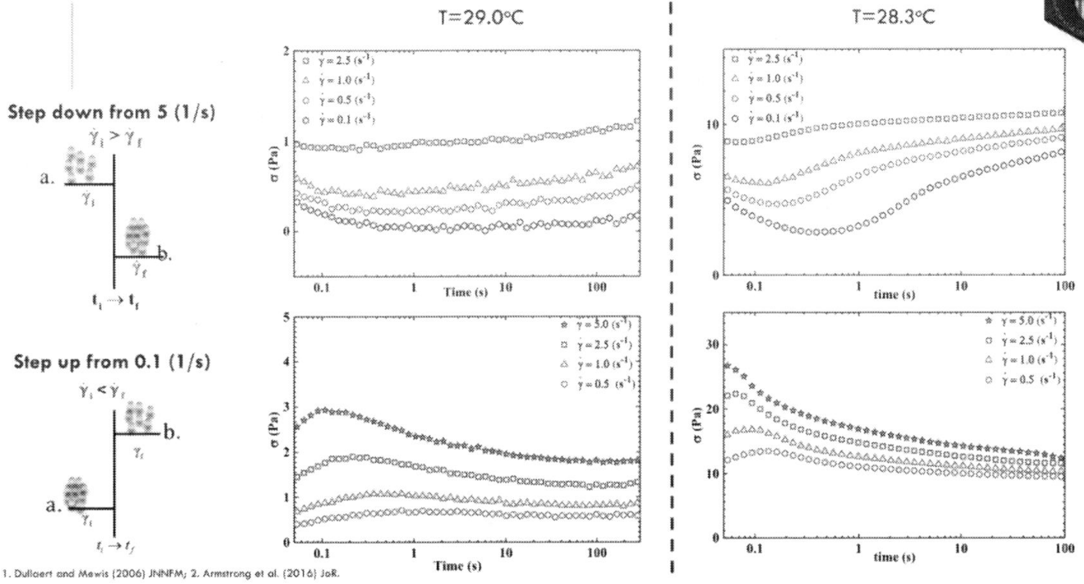

1. Dullaert and Mewis (2006) JNNFM; 2. Armstrong et al. (2016) JoR.

7

MODIFIED DELAWARE THIXOTROPIC MODEL (1/2)

Structural Evolution Equation:

$$\frac{d\lambda}{dt} = -k_{break}\left|\dot{\gamma}\right|^{a}\lambda + k_{aggr}\left|\dot{\gamma}\right|^{0.5}(1-\lambda) + k_{Brown}(1-\lambda)$$

Shear breakage term Shear aggregation term Brownian build up term

$\lambda = 1$

$\lambda ...[0,1]$

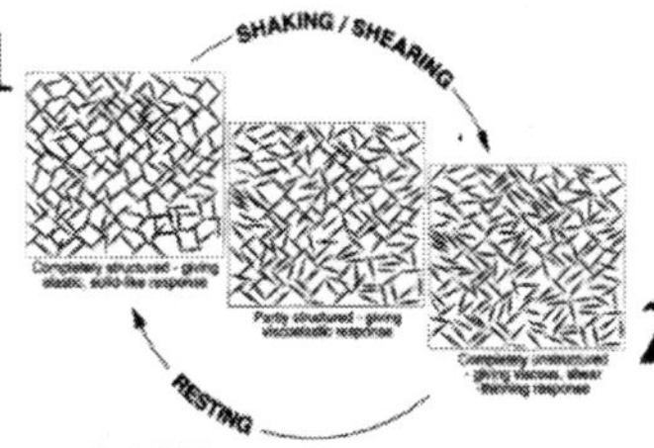

$\lambda = 0$

[1] Mewis and Wagner. Colloid Suspension Rheology. (2012). [2] Dimitriou, Ewoldt, and McKinley J. Rheol (2013).
[3] Mujumbdar, Beris, and Metzner.JNNFM (2002). [4] GoodeveTrans. Faraday Soc. (1938). [5] Dullaert and Mewis, JNNFM (2005).
[6] https://www.google.com/search?q=Maxwell. [7] Wagner and Brady. Physics Today (2009).

8

MODIFIED DELAWARE THIXOTROPIC MODEL (2/2)

Constitutive Equation:

Red=elastic component
Blue=viscous component

$$\sigma = G_f \gamma_e + (\lambda K_{ST}) \dot{\gamma}_p^{n2} + \eta_\infty \dot{\gamma}_p^{n1} \quad (1)$$

$$\gamma = \gamma_e + \gamma_p \Leftrightarrow \dot{\gamma} = \dot{\gamma}_e + \dot{\gamma}_p \quad (2)$$

$$\frac{dG_f}{dt} = k_G \left(G_0 \lambda - G_f \right) \quad (3)$$

$$\frac{d\gamma_e}{dt} = \dot{\gamma}_p - \frac{\gamma_e}{\left(\gamma_{CO} / \lambda^m \right)} \left| \dot{\gamma}_p \right| \quad (4)$$

13 parameters*

$$G_0, \gamma_{CO}, \eta_\infty, \gamma_\infty, m, n_1, n_2, K_{ST}, k_{Brown}, k_{aggr}, k_{break}, a, k_G$$

[1] Mewis and Wagner, Colloid Suspension Rheology. (2012). [2] Dimitriou, Ewoldt, and McKinley J. Rheol (2013). [3] Mujumbdar, Beris, and Metzner.JNNFM (2002).
[4] GoodeveTrans. Faraday Soc. (1938). [5] Dullaert and Mewis, JNNFM (2005). [6] https://www.google.com/search?q=maxwell

9

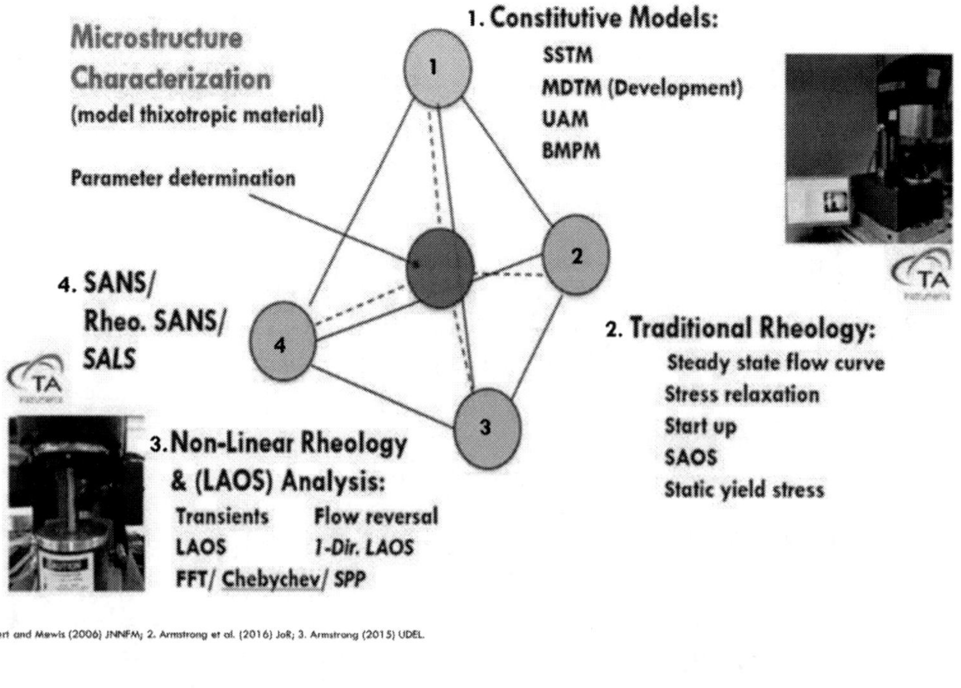

1. Dullaert and Mewis (2006) JNNFM; 2. Armstrong et al. (2016) JoR; 3. Armstrong (2015) UDEL.

10

NEW FITTING RESULTS (28.3^{0}C)

Steady state

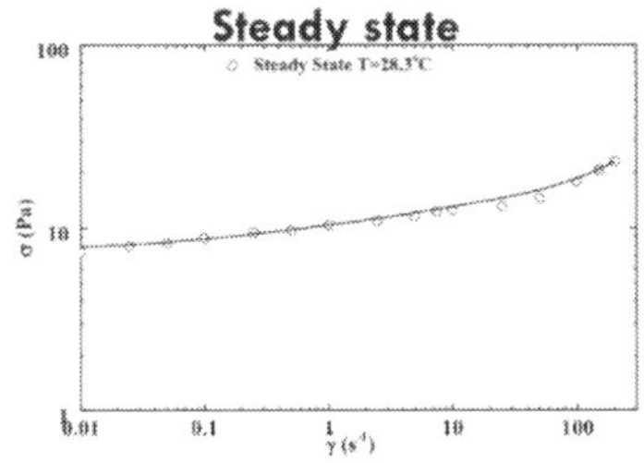

- -MDTM fit
- -Steady state data
- -8 sets of step/down data
- -fit with Par. Temp.

...predict LAOS?

MDTM Model Paramters (28.3°C)

τ_{y0}	7.5
G	1000
η_∞	0.115
m	-1.5
n_1	0.856
n_2	0.6147
K_{ST}	6.98
a	1.05
d	0.4066
t_{r1}	0.3889
t_{r2}	0.7409
k_{rev}	0.4368
k_G	0.0228
γ_∞	1

Step ups from 0.1 (1/s)

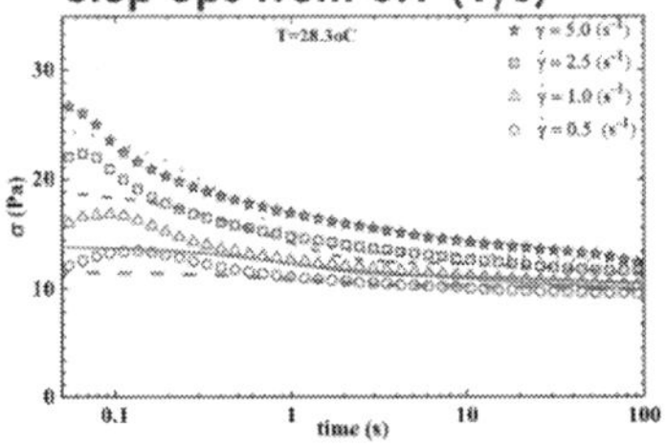

Step downs from 5 (1/s)

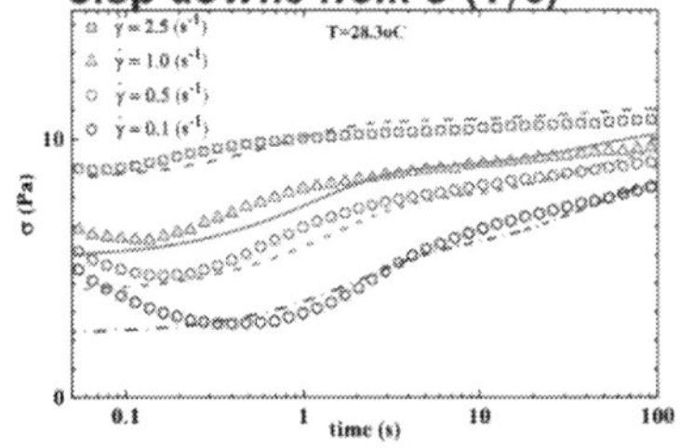

11

NEW FITTING RESULTS (29.0^{0}C)

Steady state

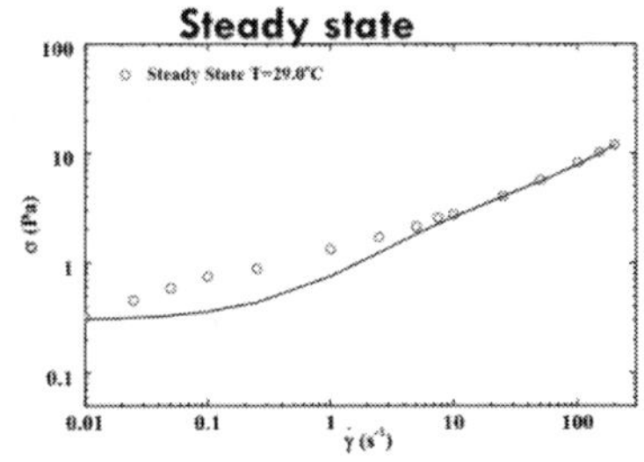

- -MDTM fit
- -Steady state data
- -8 sets of step/down data
- -Fit with Par. Temp.
- -Not as much 'structure'

...predict LAOS?

MDTM Model Paramters (29.0°C)

τ_{y0}	0.3
G	30
η_∞	0.03554
m	-1.5
n_1	0.99
n_2	0.97
K_{ST}	0.6468
a	1.55
d	0.744
t_{r1}	0.817
t_{r2}	2.89
k_{rev}	0.0032
k_G	8.32
γ_∞	1

Step ups from 0.1 (1/s)

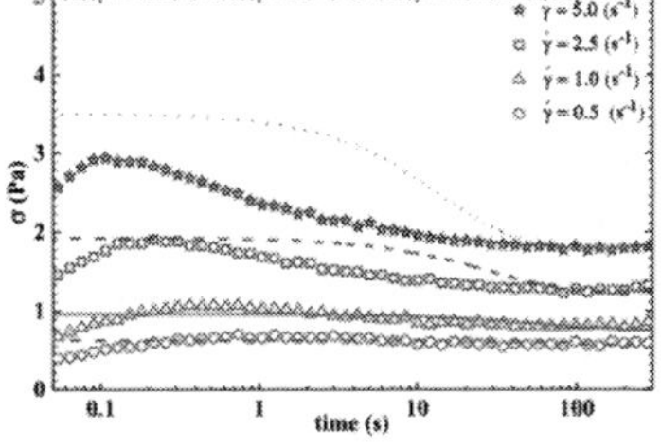

Step downs from 5 (1/s)

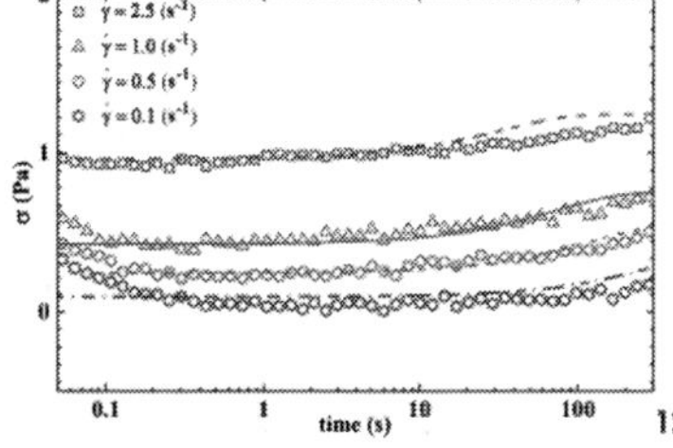

12

LISSAJOUS-BOWDITCH PROJECTIONS (W/ DATA*)

$$\gamma_0 = 1.0; \omega = 1\,(rad/s)$$

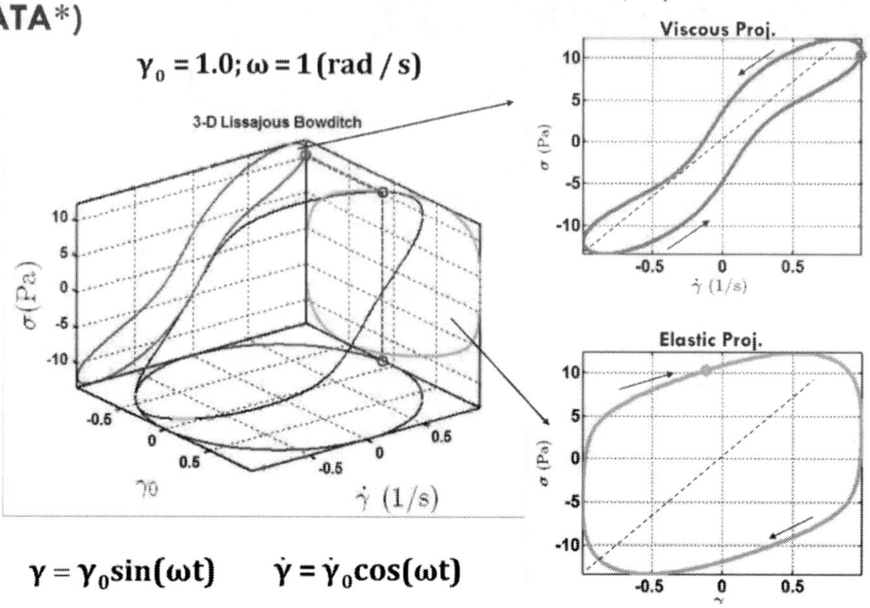

$$\gamma = \gamma_0 \sin(\omega t) \qquad \dot{\gamma} = \dot{\gamma}_0 \cos(\omega t)$$

13

PREDICTING LAOS (28.3°C):

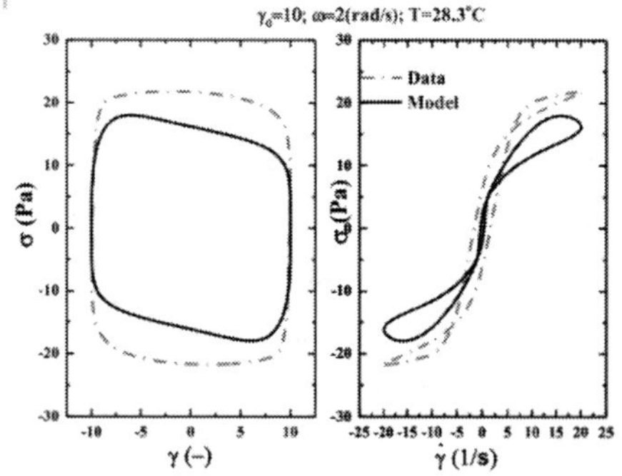

Elastic projection Viscous projection

γ_0=10; ω=2(rad/s); T=28.3°C

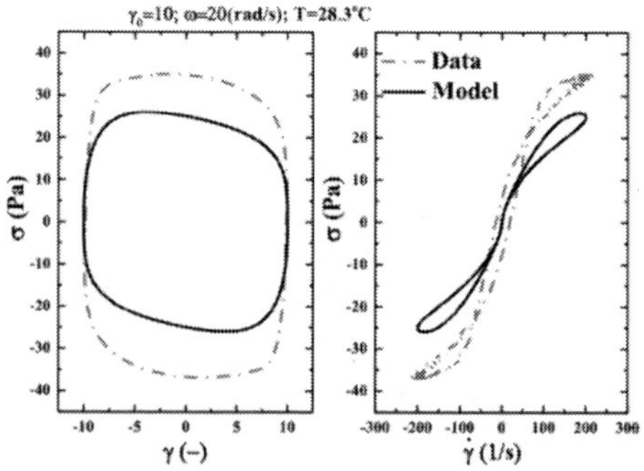

Elastic projection Viscous projection

γ_0=10; ω=20(rad/s); T=28.3°C

14

PREDICTING LAOS (29.0^{0}C):

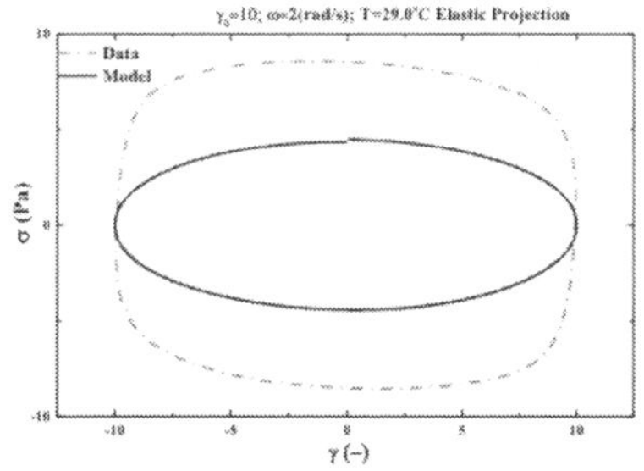

Elastic projection

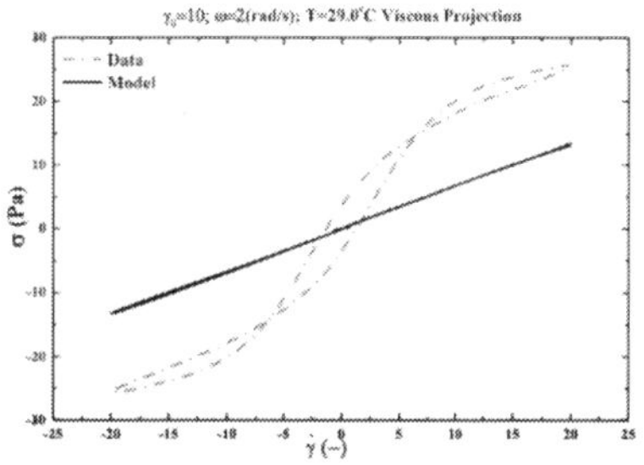

Viscous projection

$\gamma 0=10$; $\omega=2$(rad/s); T=29.0°C

15

CONCLUSIONS

1. **Where structure is relevant (lower T) the LAOS is better predicted by fitting step up/down.**
2. **We improved LAOS predictions by fitting to step up/down first.**
3. **Try to add a more "visco-elastic" term to better fit and predict the viscous time scale at beginning of step up/down.**

1. Armstrong et al (2016) JoR; 2 Conversation with Dr. Antony Beris; Dr. Norm Wagner.

16

FUTURE WORK: MODEL IMPROVEMENTS

Red=elastic component
Blue=viscous component
Purple=viscoelastic

Old MDTM:

$$\sigma = \sigma_e(\lambda, \dot{\gamma}) + \sigma_p(\lambda, \dot{\gamma}) \quad (1)$$

$$\boldsymbol{\sigma} = G_f \gamma_e + (\lambda K_{ST}) \dot{\gamma}_p^n + \eta_\infty \dot{\gamma}_p \quad (2)$$

New MDTM w/ viscoelastic:

$$\sigma = \sigma_e(\lambda, \dot{\gamma}) + \sigma_{ve}(\lambda, \dot{\gamma}) \quad (3)$$

$$\sigma_{ve} + \frac{\eta(\lambda)}{G(\lambda)} \dot{\sigma}_{ve} = \eta(\lambda)\dot{\gamma} \quad (4a)$$

OR

$$\sigma_{ve} + \lambda_{relax} \dot{\sigma}_{ve} = \eta(\lambda)\dot{\gamma} \quad (4b)$$

1. Armstrong et al (2016) JoR; 2 Conversation with Dr. Antony Beris; Dr. Norm Wagner.

17

THANK YOU

Collaborators:

Jeff Horner, Ryan Murphy, Dr. Norm Wagner, Dr. Antony Beris,

Dr. Simon Rogers

18

QUESTIONS

PREVIOUS RESULTS (1/2) (SOR 2013)

Fitting *similar model to Kim and Gurnon SS and LAOS data**

Delaware Model	SS	LAOS
$\sigma_{y0}^{*}(8)$	8	1.000
t^{*}_{1} (0.319, 15.1 s)	0.319	1.000
η^{*}_{ST}	2.990	0.021
η^{*}_{INF}	5.17E-03	1.09E-04
k_1	0.319	15.100
t_2/t_1	0.065	3.217
G^*	0.131	0.66125
n	0.84	0.93
m	0.45	0.46
γ_{CO}	7.63	0.1988

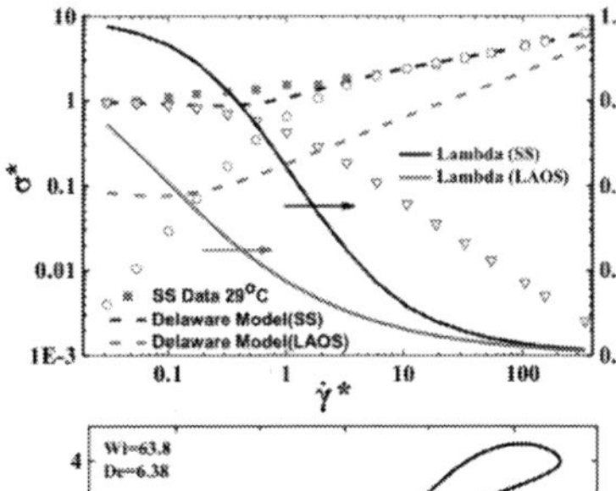

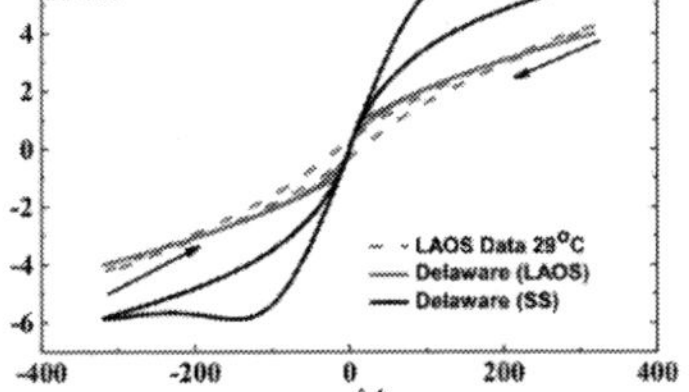

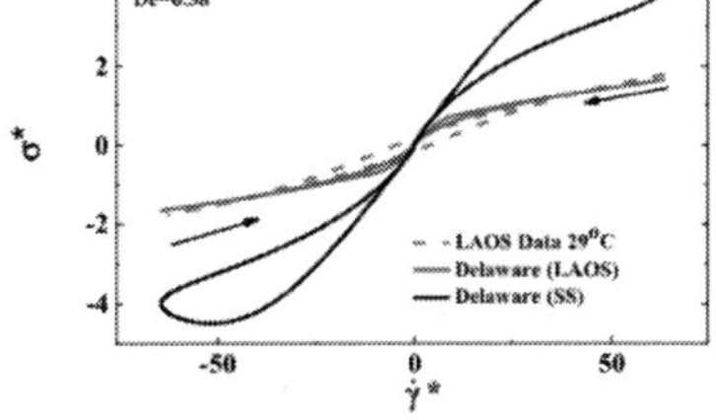

1. Kim JM (2013) PhD Thesis, UDEL; 2. Gurnon AK (2014) PhD Thesis, UDEL; 3. W. M. Keck microscopy Facility.

PREVIOUS RESULTS (2/2) (PhD Thesis)

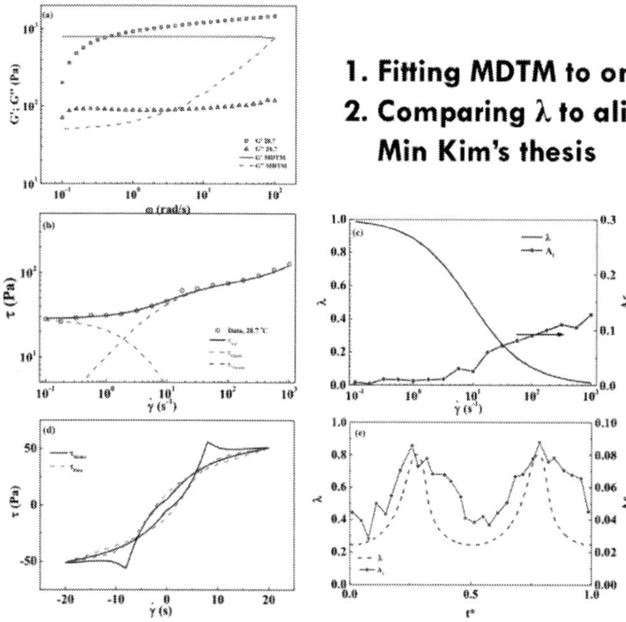

1. Fitting MDTM to one set SS & LAOS (Kim et al. data)
2. Comparing λ to alignment factor from Dr. Min Kim's thesis

Parameter	Units	Optimal	μ(ave)	σ(std)
τ_{y0}	Pa	28.1	(-)	(-)
G	Pa	1600	(-)	(-)
η_∞	Pa s	0.042	(-)	(-)
m	(-)	-1.5	(-)	(-)
n_1	(-)	1	(-)	(-)
K_{ST}	Pa s^{n2}	11.42	11.47	4.19
n_2	(-)	0.87	0.87	0.29
a	(-)	1.09	1.12	0.46
d	(-)	0.27	0.58	0.5
t_{r1}	s^{n1}	0.29	0.28	0.102
t_{r2}	s^{n2}	1.26	1.19	0.638
k_{rev}	s	50	(-)	(-)
k_G	s	10	(-)	(-)

21

1. Kim JM (2013) PhD Thesis, UDEL; 2. Gurnon AK (2014) PhD Thesis, UDEL; 3. W. M. Keck microscopy Facility.

492787 Modeling and Uncertainty Quantification of Vapor Diffusion and Reactions in Polymer

Abstract:

Modeling and uncertainty quantification of vapor diffusion and reactions in polymers

Yunwei Sun, Hom Sharma, and Elizabeth Glascoe

Lawrence Livermore National Laboratory

April 14, 2017

Characterizing the reactive transport of vapor through porous materials is important in predicting the aging and compatibility of polymers that are often used to repel moisture and protect more sensitive materials within. For example, water intrusion into an electronic device (e.g., photovoltaic cells) through a polymer layer may result in corrosion, shorting, and other types of electronic failure. A variety of mathematical models, together with statistical methods and computer algorithms, have been developed to understand and quantify vapor diffusion and sorption in polymers. However, most of those models are limited to steady-state diffusion and equilibrium sorptions and fail to address the dynamic behavior that is directly associated with material aging and compatibility. In this study, we develop a high-fidelity diffusion model coupled with triple-mode sorptions with joint and iterative experimental and modeling capability to characterize and quantify vapor sorption and diffusion through polymeric materials.

When Langmuir adsorption is treated as a kinetic reaction, the corresponding ordinary differential equations (ODEs) are usually solved using ODE solvers. The computational expense for solving those ODEs of the dynamic adsorption has limited the applicability of the high-fidelity model. To accelerate the simulation of coupled diffusion and sorption processes, we developed a statistical approximation method using computationally cheap surrogate models (e.g., algebraic polynomials) that replace the ODE solutions of adsorption and are coupled with the diffusion equations. Since the polynomial presentation of the adsorption term is obtained in a standard format prior to modeling coupled sorption-diffusion, the adsorption operator can be expressed as input data in the transport code. Compared to conventional operator-splitting methods, the polynomial approximation of adsorption offers better computational efficiency. The methodology is demonstrated and validated using a dynamic Langmuir adsorption model that is coupled to diffusion and absorption models and applied to a water vapor sorption-diffusion process in polydimethylsiloxane polymers.

To avoid over parameterizing the physical model, we are conducting global sensitivity analysis to establish a cost-effective model form and identify the most sensitive parameters in the model. Using the developed model, we quantify uncertainty propagation from uncertain parameters to material performance measures.

This work performed under the auspices of the U.S. Department of Energy by Lawrence Livermore National

Laboratory under Contract DE-AC52-07NA27344.

497896 Modeling of Transport and Reaction in a Novel Hydride Vapor Phase Epitaxy System

Abstract:

The research interests of applying hydride vapor phase epitaxy (HVPE) in the field of III-V photovoltaics have recently been revived due to the capability of this process in growing semiconductors of excellent quality at higher growth rate and lower cost compared with its competing epitaxial growth methods [1, 2]. However, the intrinsically high growth rate and near-thermodynamic-equilibrium nature of this process bring the difficulty of producing large-area uniform films in a controlled way, which necessitates the thorough understanding of underlying interactions between transport phenomena and the properties of films grown by this method.

We have proposed a novel HVPE configuration potentially having significantly greater throughput and lower per unit production cost. In this talk, three-dimensional transient transport model of this design coupling a chemical model applicable to HVPE growth of various III-V materials is presented. Finite element methods have been employed to solve this model, the result of which relates process outputs (including materials growth rate, uniformity, and composition, etc.) with process inputs (including reactor geometry, operating conditions, etc.). Besides the ordinary design variables like flow rate and temperature, the influences of dynamic operating factors on process characteristics and corresponding product properties are studied in this work. Through the insights gained in the analysis of transport and reaction within the constraints of different parameter regions, sensitive variables and parameters are identified, based on which optimal reactor geometry and process operating strategies are further determined. This model-based approach could be used to guide experimental design in a much more effective way. The methodology of problem formulation and numerical techniques used in this work may also be extended to design and solve a wider range of chemical vapor deposition systems.

[1] K. L. Schulte, W. L. Rance, R. C. Reedy, A. J. Ptak, D. L. Young, and T. F. Kuech, "Controlled formation of GaAs pn junctions during hydride vapor phase epitaxy of GaAs," Journal of Crystal Growth, vol. 352, no. 1, pp. 253–257, Aug. 2012.
[2] J. Simon, K. L. Schulte, D. L. Young, N. M. Haegel, and A. J. Ptak, "GaAs Solar Cells Grown by Hydride Vapor-Phase Epitaxy and the Development of GaInP Cladding Layers," IEEE Journal of Photovoltaics, vol. 6, no. 1, pp. 191–195, Jan. 2016.

500572 Discretized Modeling of a Simple Motionless 3D Printer Based on Retarded Bending Motion and Electrically Controlled Patterning of Fiber Deposition

Abstract:

The electrospinning process is widely used to produce fibers with diameters in the sub-micron range. In the electrospinning process, a polymer solution is slowly extruded out of a needle and fibers are formed as the solution is accelerated by an electric field. The electrically driven fibers have been of interest in the area of 2D and 3D printing to fabricate ordered fiber mats. These fiber mats have applications in filtration, catalysis, tissue engineering, drug delivery, and energy storage systems. In printing, the nozzle system or the substrate moves to form patterns on the substrate. For this, there needs to be a proper alignment of the nozzle and the substrate to avoid any errors while printing. Therefore, we developed a motionless 3D printer in which the print pattern was formed by electrically controlling the fiber deposition. In our approach, the nozzle system is stationary and the fiber is electrically guided for printing on the substrate. As there are no degrees of freedom in this system, this approach prevents the alignment issues and the print resolution is limited only by the material delivery system.

The 3D printer in this study is developed using the immersed electrospinning process, in which the fiber is electrospun in a liquid medium rather than in air or vacuum. The wet spinning medium improves the collection thickness of the nonwoven fiber and controls its porosity. In general, several instabilities are observed on the polymer fiber in the electrospinning process. In particular, the bending instability results in a non-axisymmetric behavior of the polymer fiber, causing the fiber to undergo a helical motion instead of travelling in a straight line. This bending motion makes it difficult to control the deposition of the fiber. As the immersed spinning setup uses liquid spinning medium, the drag force by this surrounding liquid inhibits the bending motion of the fiber, which makes it easy to electrically control the fiber trajectory. In this paper, we first studied the bending instability of fiber in the system followed by a subsequent study examining the pattern formation for 3D printing with the use of an immersed spinning process.

First, the bending instability of the fiber was studied from simulations and experiments to understand the feasibility of the immersed electrospinning process for 3D printing. In the immersed electrospinning setup, the spinning dope was comprised of polyacrylonitrile (PAN) in the dimethylformamide (DMF) solvent. The PAN/DMF solution was spun in a coagulation bath containing a mixture of chloroform and mineral oil as the spinning medium. With this liquid spinning medium, it was much easier to visualize the bending motion because of the reduced chaotic motion of the fiber. This fiber trajectory obtained from the experiments was used to test the validity of the simulations. For the simulation, we used the discretized model, in which, the fiber is assumed to be a series of beads connected with springs. In this model, the beads are introduced at regular intervals in the nozzle, and the trajectory of the fiber is obtained by applying Newtonâ€™s Second Law of Motion to each bead. The drag force by the liquid spinning medium was incorporated into the discretized model. To obtain the bending motion of the fiber, normal mode perturbations at different frequencies and amplitudes were applied to the beads when they are introduced in the system. The rate of increase in the helical radius of the fiber trajectory is the instability growth rate. This growth rate

was calculated at different frequencies and the frequency with the maximum growth rate is responsible for the onset of the instability. In addition, from the simulations and experiments, the growth rate of the bending instability increases with voltage due to larger Coulombic repulsions between the surface charges of the fiber. Therefore, the bending instability is shown to be in the conducting mode. The radius of the fiber from the simulation is similar to the experimental observation. Overall, the discretized model provided a good prediction of the fiberâ€™s bending motion. Thus, this model was also used to predict the electrically controlled pattern formation of the fiber on the collector. The process conditions for lower bending instability growth rate were used for our next study with simulations and experiments on the electrically controlled pattern formation.

In the second study, the collector in the immersed electrospinning setup was modified to form a square pattern. The collector was designed by attaching four copper plates in a square shape to a non-conducting material. These plates are the lattice sites on the collector, and the voltage is conducted only at these sites. The fiber is guided to that lattice point that has the voltage switched ON. As the fiber reaches one lattice point, its voltage is switched OFF, and based on the desired pattern the voltage at the next lattice point is switched ON. This process was repeated until the end of the printing time. The size and the structural arrangement of the lattice points determine the accuracy of the pattern. The formation of a square pattern was also studied with the discretized model by using a temporal boundary condition for the voltage on the collector. The pattern formed from the discretized model was compared with the experiments. This approach can be developed further by increasing the lattice sites to improve the print resolution. For this, the electrode arrays can be used as a collector, and this will increase the accuracy of the pattern. In addition, the switching of the voltage at the lattice points can be controlled with MOSFETs, which can reduce the delay between the switching of the voltage among the lattice sites. Furthermore, the predictions from the discretized model can be used to decide the process conditions for the 3D printing of various other patterns.

A forward modeling approach for the inverse estimation of transient local heat fluxes

Jiu LUO[1], Ya-Qiao WANG[1], Jia-Li LUO[1], Dong-Chuan MO[1],
Yuan-Xiang FU[1], Shu-Shen LYU[1], Hiroyuki OZOE[2], Yi HENG[1*]

Speaker: Jiu Luo

1. Sun Yat-sen University, China
2. Kyushu University, Japan

Minneapolis Convention Center, Minneapolis, America

October 31, 2017

OUTLINE

- Motivation
- Mathematical formulation
- Results and discussion
- Summary & outlook

1. Motivation

Inverse Heat Conduction Problems (IHCP)

- Extensive applications in industry and academia
- Mathematically severely ill-posed
- Hard to obtain physically stable solutions
- Computationally expensive for 3D transient applications

- A forward modeling approach for the stable and efficient solution.

1. Motivation

Applications for inverse estimation of heat fluxes

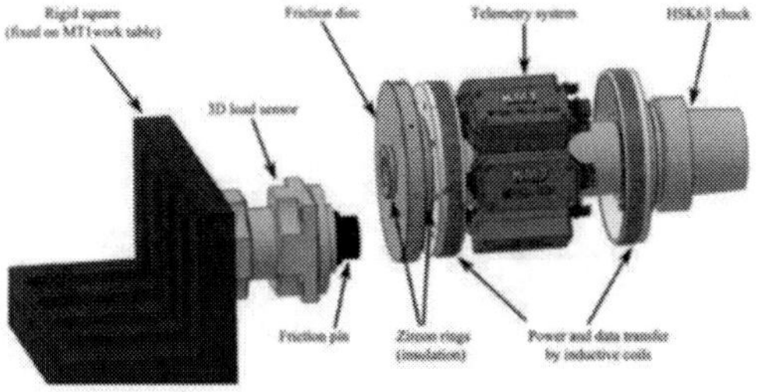

High Speed Tribometer (D. Meresse et al., Applied Thermal Engineering, 2012)

A photograph of the test furnace.
(P. Wikström et al., Applied Thermal Engineering, 2007)

- **In a braking process:** Improving braking efficiency
- **In a steelmaking process:** Making optimal contral of temperatures
- **In an aerospace field:** Choosing heat insulating materials

1. Motivation

Applications for inverse estimation of heat fluxes

Plate structure

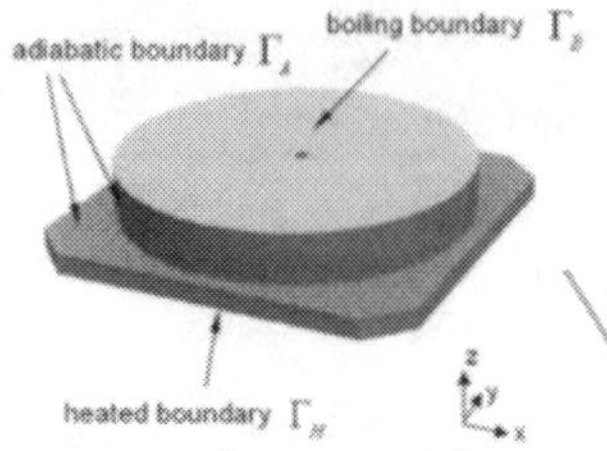

The real heater for pool boiling (Heng et al., Inverse Problems, 2009)

Micro/nano porous structure

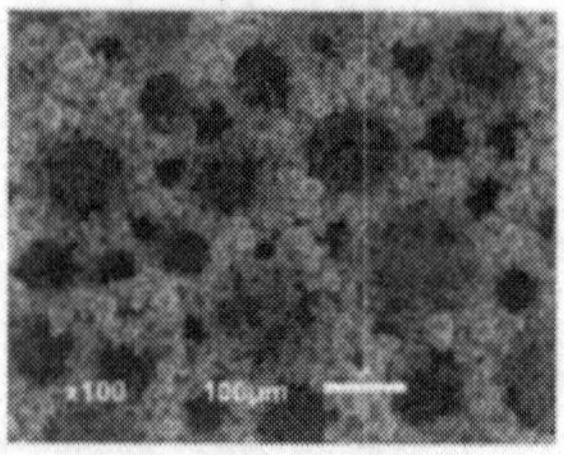

Modified micro/nano porous copper surface (Wang. Y. Q. et al., ASME 2016 5th International Conference)

- **In cooling systems:** Enhancement of boiling heat transfer

Few studies addressed the transient local heat fluxes from available temperature readings for micro and nano porous structure.

1. Motivation

Available solution techniques

Optimized-based inversion method

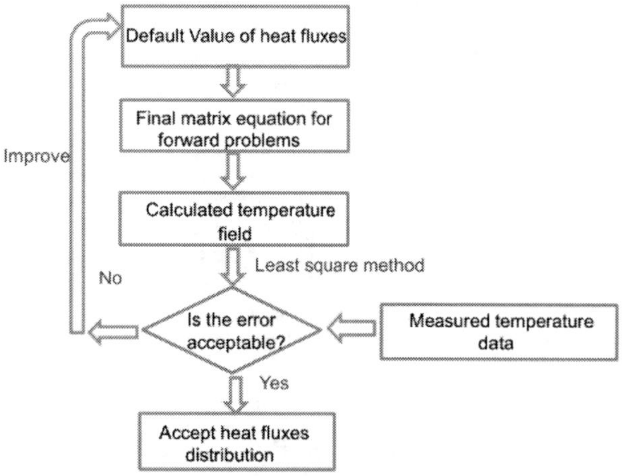

Solution procedure for optimized-based inversion method

Non-iterative method

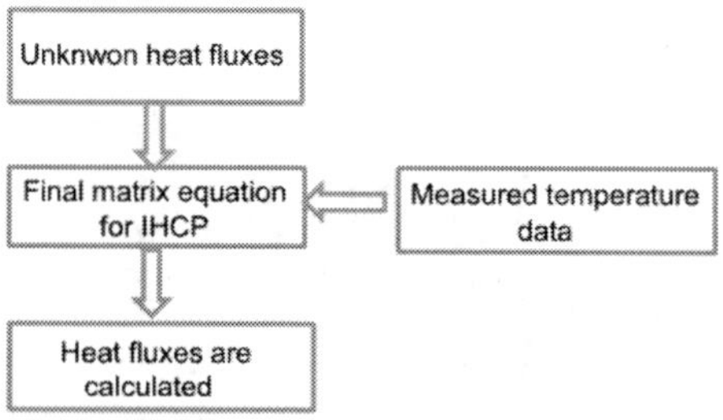

Solution procedure for non-iterative method (Ling et al., INT J NUMER METH ENG, 2003)

More efficient than optimized-based inversion method !!!

2. Mathematical formulation

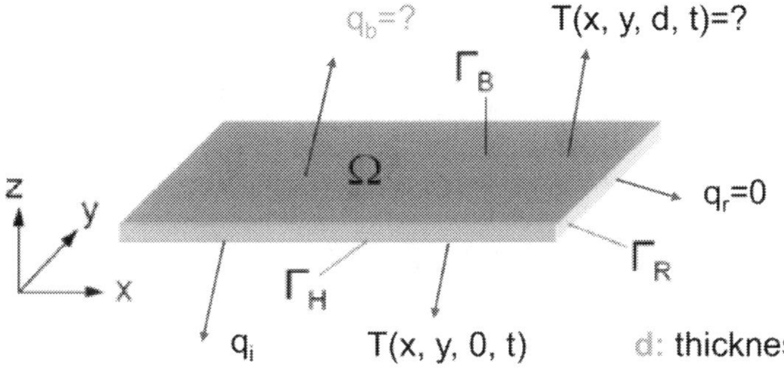

Problem description for IHCP

d: thickness of plate

k: heat conductivity of material

qi: heat fluxes of heated boundary (controlled by current of heater)

Problem formulation

2. Mathematical formulation

$$T(x,y,z,t) = T(x,y,0,t) + \frac{\partial T(x,y,z,t)}{\partial z}\Big|_{z=0} z + \frac{1}{2}\frac{\partial^2 T(x,y,z,t)}{\partial z^2}\Big|_{z=0} z^2 + \cdots + \frac{1}{n!}\frac{\partial^n T(x,y,z,t)}{\partial z^n}\Big|_{z=0} z^n + 0(z^n)$$

$$T_{appro}(x,y,z,t) = T(x,y,0,t) + \boxed{\frac{\partial T(x,y,z,t)}{\partial z}\Big|_{z=0} z} + \frac{1}{2}\frac{\partial^2 T(x,y,z,t)}{\partial z^2}\Big|_{z=0} z^2 + \cdots + \frac{1}{n!}\frac{\partial^n T(x,y,z,t)}{\partial z^n}\Big|_{z=0} z^n$$

$$T_{appro}(x,y,d,t) = T(x,y,0,t) \boxed{-\frac{q_i}{k} z} + \frac{1}{2}\frac{\partial^2 T(x,y,z,t)}{\partial z^2}\Big|_{z=0} z^2 + \cdots + \frac{1}{n!}\frac{\partial^n T(x,y,z,t)}{\partial z^n}\Big|_{z=0} z^n$$

Mathematical derivation

2. Mathematical formulation

$$\begin{cases} \rho c_p \frac{\partial T}{\partial t} = \nabla \cdot (k \nabla T), & \text{in } \Omega \times (0, t_f), \\ T(\cdot, 0) = T_0(\cdot), & \text{on } \Omega, \\ k\frac{\partial T}{\partial n} = q_i, & \text{on } \Gamma_H \times (0, t_f), \\ k\frac{\partial T}{\partial n} = q_r, & \text{on } \Gamma_R \times (0, t_f), \\ k\frac{\partial T}{\partial n} = q_b, & \text{on } \Gamma_B \times (0, t_f), \end{cases}$$

Forward modeling approach →

$$\begin{cases} \rho c_p \frac{\partial T}{\partial t} = \nabla \cdot (k \nabla T), & \text{in } \Omega \times (0, t_f), \\ T(\cdot, 0) = T_0(\cdot), & \text{on } \Omega, \\ k\frac{\partial T}{\partial n} = q_i, & \text{on } \Gamma_H \times (0, t_f), \\ k\frac{\partial T}{\partial n} = q_r, & \text{on } \Gamma_R \times (0, t_f), \\ T(x, y, d, t) = T_{appro}(x, y, d, t), & \text{on } \Gamma_B \times (0, t_f). \end{cases}$$

IHCP when q_b is unknown

Forward Heat Conduction Problem

It is easy to be calculated accurately and efficiently.

Mathematical derivation

2. Mathematical formulation

(Third step) q_b(x, y, d, t) ← finite-difference formulae — Calcuating temperature field for 3D plate (Second step)

(Estimated heat flux) — Compared

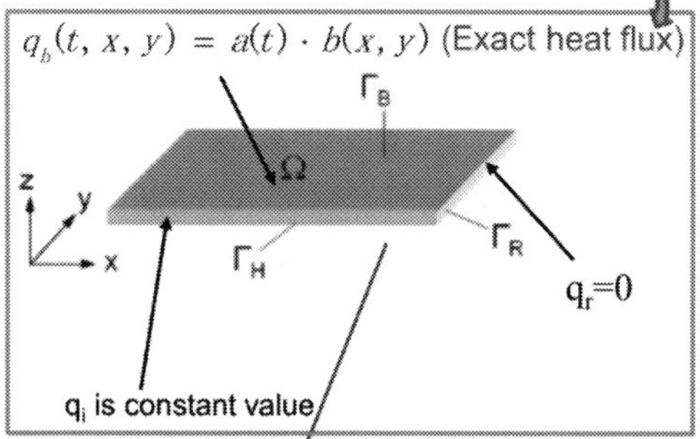

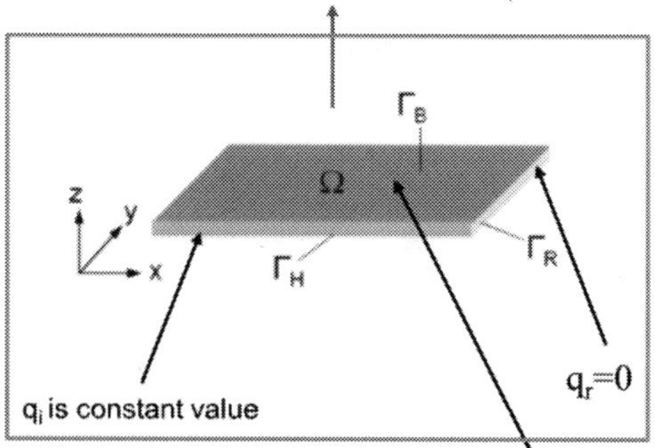

Calcuating temperature field on the heated boundary (Obtain artificial data) (First step) — Forward modeling approach → Derive temperature field on the boiling boundary

The solution precedure

3. Results and discussion

Case study 1

Making online optimal control of temperatures in a reheating furnace

The heat flux is assumed to be a linear and nondifferentiable function (Cui et al., J IRON STEEL RES INT, 2012)

$$q = \begin{cases} 12.5\tau & 0 \le \tau \le 4000 \\ 100000 - 12.5\tau & 4000 < \tau \le 8000 \end{cases}$$

- The heat flux function in this case study is time dependent and space independent.
- The non-differentiable heat flux function can be estimated accurately and efficiently.

First order approximation

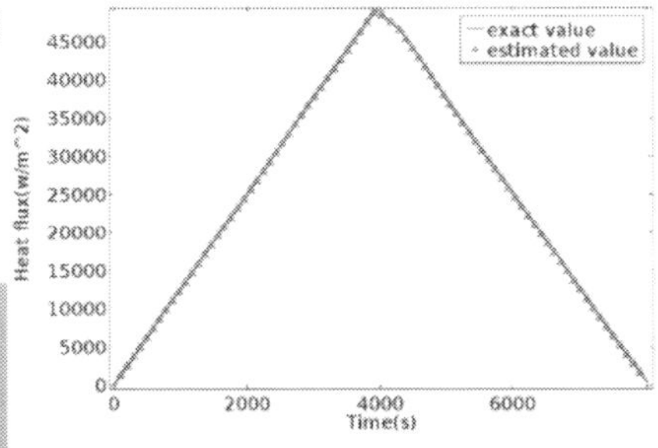

Exact and estimated heat fluxes in the present work

3. Results and discussion

Case study 2

The heat flux $q(S_1,t)$ with leaps on S_1

$$q_1(I,J,t) = 60 \times \sin\left(\frac{t}{t_f}\pi\right), \quad \begin{matrix} 1 \le I \le 12 \\ 1 \le J \le 12 \end{matrix} \quad 24 \ge t \ge 0$$

$$q_2(I,J,t) = 60 \times \sin\left(\frac{t}{t_f}\pi\right), \quad \begin{matrix} 3 \le I \le 10 \\ 3 \le J \le 10 \end{matrix} \quad 24 \ge t \ge 0$$

$$q_3(I,J,t) = 40 \times \sin\left(\frac{t}{t_f}\pi\right), \quad \begin{matrix} 5 \le I \le 8 \\ 5 \le J \le 8 \end{matrix} \quad 24 \ge t \ge 0$$

$$q(I,J,t) = (q_1 + q_2 + q_3), \quad \text{in } \Omega \quad 24 \ge t \ge 0$$

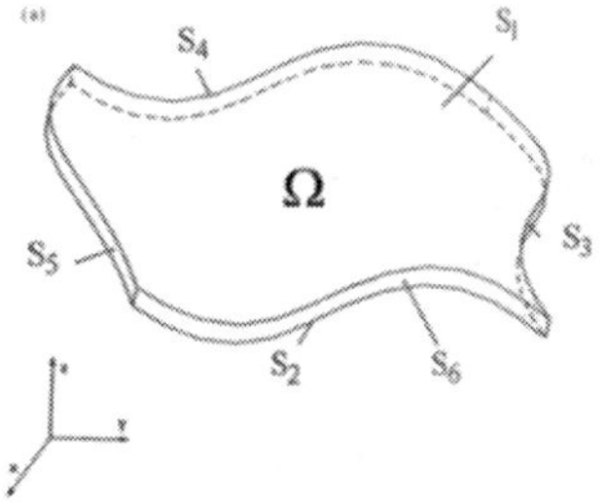

The geometry and coordinates (Hung et al., INT J HEAT MASS TRAN, 1999)

- The estimated heat flux is a smoothed version of the original one by applying the optimized-based inversion method.

3. Results and discussion

Case study 2

Exact and estimated heat fluxes

First order approximation

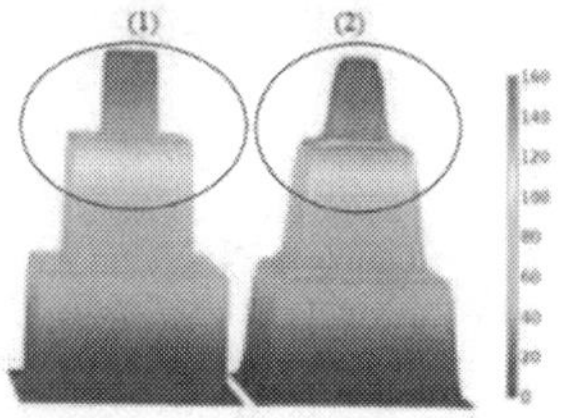

(1) Exact heat flux distribution (t = 12s)
(2) Estimated heat flux in the present work (t = 12s)

Simulated result by optimized-based inversion method (Huang et al.)

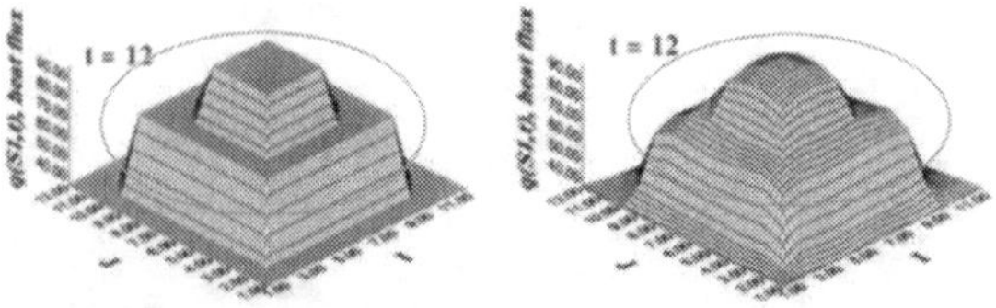

(1) Exact instantaneous heat flux by Huang et al.(Hung et al., 1999)
(2) Estimated instantaneous heat flux by Huang et al.(Hung et al., 1999)

- It yields more precise estimation results compared with those obtained by applying optimize-based inversion method.

3. Results and discussion

Case study 3

Enhancement of boiling heat transfer

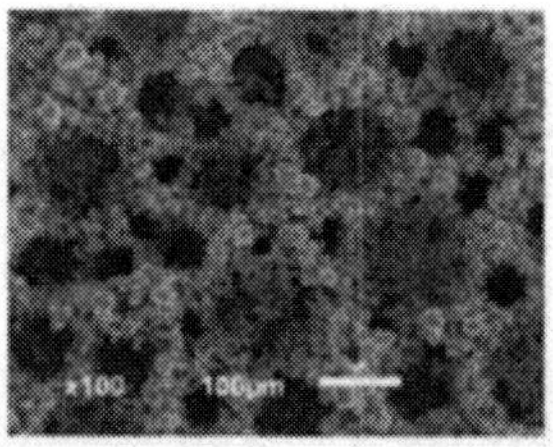

Modified micro/nano porous copper surface (Wang. Y. Q. et al., ASME 2016 5th International Conference)

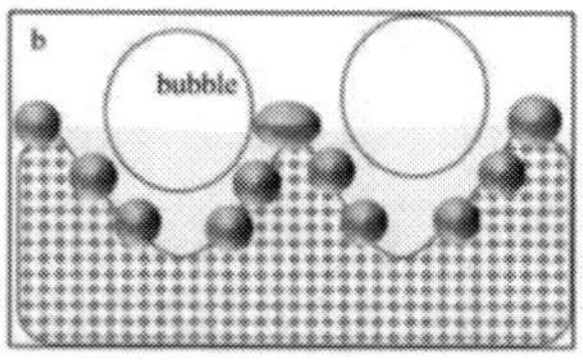

Mechanism for the bubble departure (Wang. Y. Q. et al., ASME 2016 5th International Conference)

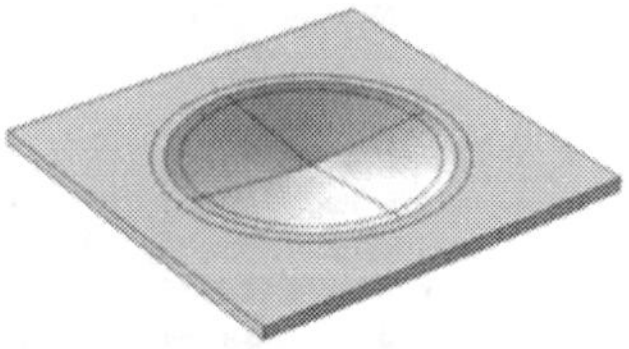

Geometrical model for the micro/nano structure

Hard to build precisely model for micro and nano porous structure !!!

3. Results and discussion

Case study 3

Modeling for micro/nano porous surfaces

$$q_b(t,x,y) = a(t)\cdot b(x,y),$$

$$a(t) = \begin{cases} 0.1\sin(100\pi(t-0.02)) + 0.1 & t \in [0.015, 0.035]\ (t[s]) \\ 0 & t \in [0, 0.015) \cup (0.035, 0.05] \end{cases}$$

$$b(x,y) = \begin{cases} 1, (x,y) \in \boldsymbol{\Omega}_1 \cup \boldsymbol{\Omega}_2 \\ 0, (x,y) \in \boldsymbol{\Omega} \end{cases}$$

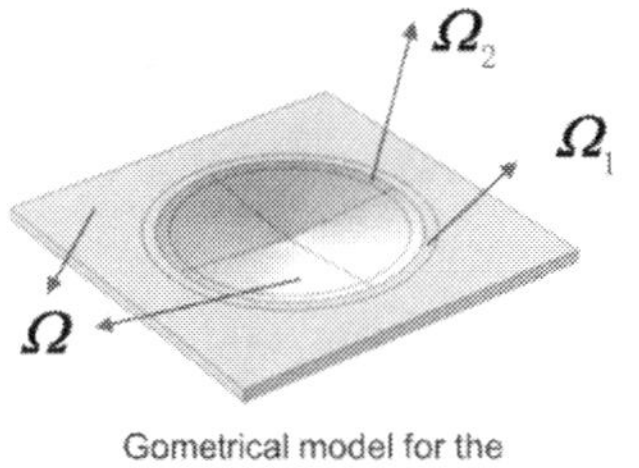

Gometrical model for the micro/nano structure

$\boldsymbol{\Omega}_1$: $0.05 \le \sqrt{(x-0.075)^2+(y-0.075)^2} \le 0.055,\ z = 0.005(x[mm]; y[mm]; z[mm])$

$\boldsymbol{\Omega}_2$: The surface area bounded by two circles of

$\sqrt{(x-0.075)^2+(y-0.075)^2} = 0.05,\ z = 0.005$

$\sqrt{(x-0.075)^2+(y-0.075)^2} = 0.045,\ z = 0.0045625$

$(x[mm], y[mm], z[mm])$

$\boldsymbol{\Omega}$: The rest of the upper suface

Inspired by the microlayer theory, heat during boiling is mainly transferred in the local area of the three-phase contact line by evaporation

Size of plate: $Length \times Width \times Height = 0.15[mm] \times 0.15[mm] \times 0.005[mm]$

Size of inverted cone: $Radius \times Height = 0.05[mm] \times 0.004375[mm]$

3. Results and discussion

Case study 3

Exact and estimated heat fluxes

- The main characteristics can be captured by using first order approximation method.
- The accuracy of estimated result is much better in this case by using second order approximation method.

Exact

Exact surface heat flux at time instant t=0.025s

First order approximation

Estimated surface heat flux using first order approximation at time instant t=0.025s

Second order approximation

Estimated surface heat flux using second order approximation at time instant t=0.025s

3. Results and discussion

Case study 3

Considering measured error for plate structure

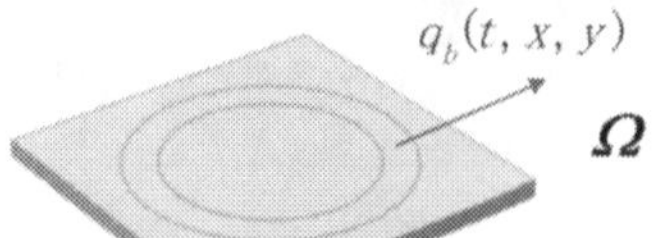

Exact

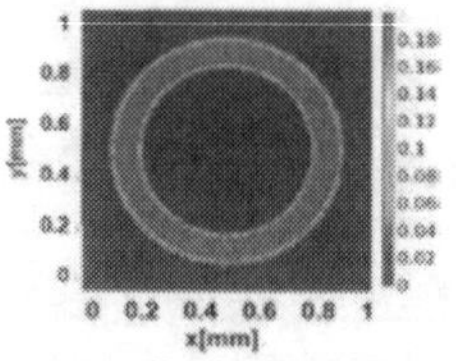

First order approximation(Left: δ=0; Right: δ=0.01K)

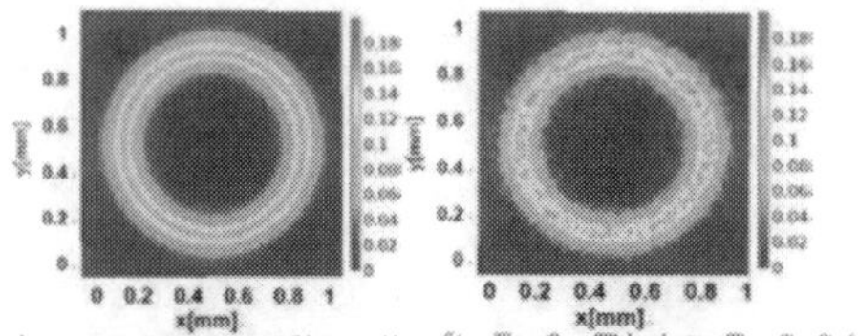

- **Measured error** δ=0.01K

$$T_{appro}(x, y, d, t) = T(x, y, 0, t) + \frac{q_i}{k} d + \boxed{rn(x, y)}$$

$$(+ \frac{1}{2} \frac{\partial^2 T(x, y, z, t)}{\partial z^2} \Big|_{z=0} d^2)$$

Random function (δ=0.01)

Second order approximation (Left: δ=0; Right: δ=0.01K)

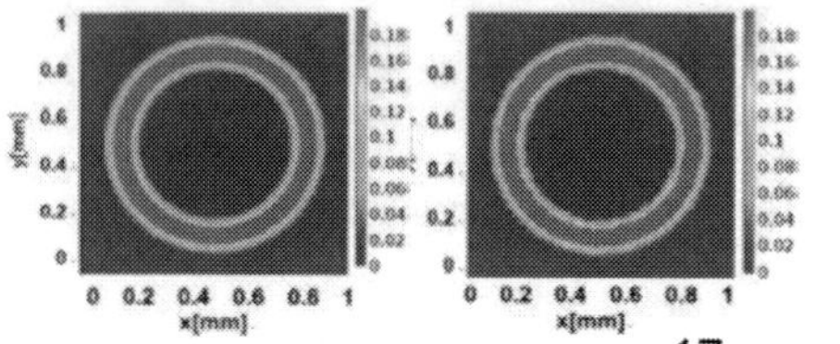

Size of plate:

$$Length \times Width \times Height = 1[mm] \times 1[mm] \times 0.05[mm]$$

$$\Omega : 0.3 \le \sqrt{(x-0.5)^2 + (y-0.5)^2} \le 0.4,$$

$$z = d = 0.05[mm]\ (x[mm];\ y[mm])$$

4. Summary & Outlook

Summary

- Presents a general approach that works for the IHCP of boundary contidions
- Shows its efficiency compared to the optimized-based inversion method.
- Shows its great superiority in estimating non-differentiable and non-continuous heat-flux functions.
- Shows its extensive applications by giving three case studies

4. Summary & Outlook

Outlook

- Develop a more precise model for the micro/nano porous surface
- Validate the method with experimental data
- Integrate the present approach with the optimized-based inversion method

Acknowledgements

Financial support provided by the Chinese Thousand Talents Program for Young Scholars

Thanks for your attention!

AIChE Annual Meeting, 31.Oct.2017

21

Supplementary

Modeling for plate surfaces

$$q_b(t, x, y) = a(t) \cdot b(x, y),$$

$$a(t) = \begin{cases} 0.1\sin(100\pi(t - 0.02)) + 0.1 & t \in [0.015, 0.035] \\ 0 & t \in [0, 0.015) \cup (0.035, 0.05] \end{cases}$$

$$b(x, y) = \begin{cases} 1, (x, y) \in \boldsymbol{\Omega} \\ 0, otherwise \end{cases}$$

$$\boldsymbol{\Omega} : 0.3 \le \sqrt{(x - 0.5)^2 + (y - 0.5)^2} \le 0.4, z = d[mm] \quad (x[mm]; y[mm])$$

Size of plate:

$$Length \times Width \times Height = 1[mm] \times 1[mm] \times d[mm]$$

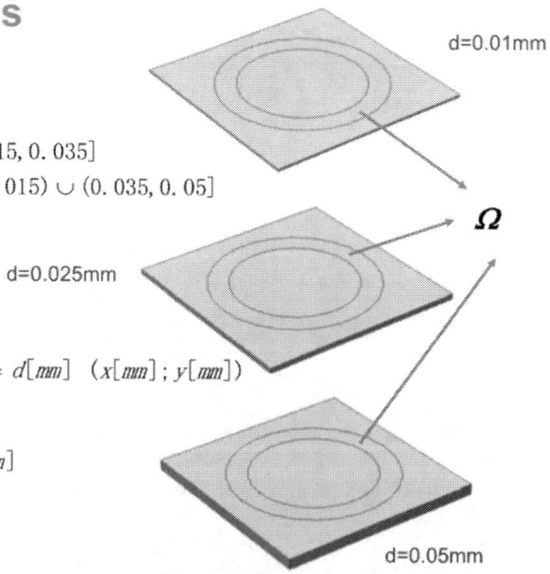

The plate with respect to different thickness

AIChE Annual Meeting, 31.Oct.2017

22

Supplementary

How to produce artificial data

$q_b(t, x, y) = a(t) \cdot b(x, y)$

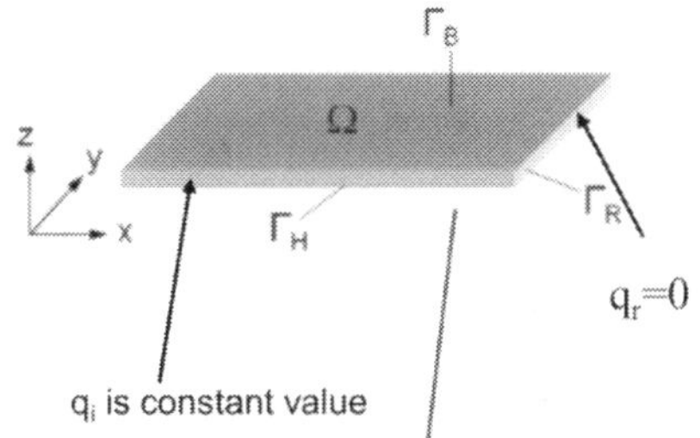

Calcuating temperature field through the heated boundary

(First step)

The first step of solving steps about forward modeling approach for IHCP

- **Measured error** δ=0

$$T_{appro}(x, y, d, t) = T(x, y, 0, t) + \frac{q_i}{k} d \left(+ \frac{1}{2} \frac{\partial^2 T(x, y, z, t)}{\partial z^2}\Big|_{z=0} z^2\right)$$

- **Measured error** δ=0.01K

$$T_{appro}(x, y, d, t) = T(x, y, 0, t) + \frac{q_i}{k} d + \boxed{rn(x, y)} \left(+ \frac{1}{2} \frac{\partial^2 T(x, y, z, t)}{\partial z^2}\Big|_{z=0} z^2\right)$$

Random function (δ=0.01)

First order approximation

Supplementary

Measured error δ=0 δ=0.01K

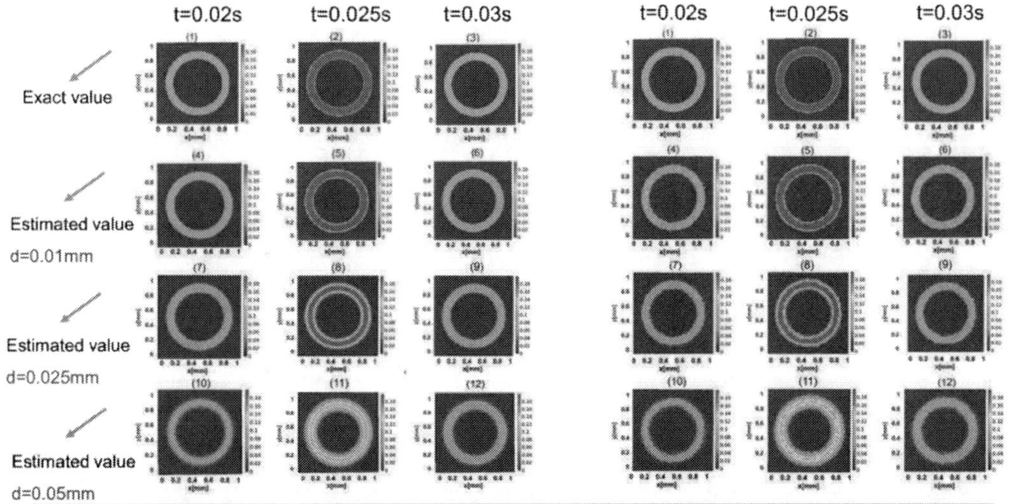

The accuracy for estimated heat flux is decreasing with increasing of thickness

Second order approximation

Supplementary

d=0.05mm

t=0.02s t=0.025s t=0.03s

Exact value

Estimated value
(δ=0)

Estimated value
(δ=0.01 K)

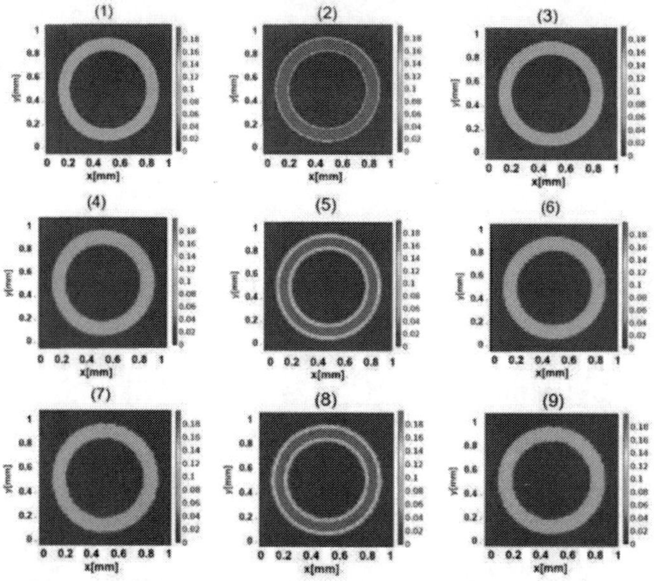

The accuracy is acceptable when the second order approximation is applied

AIChE Annual Meeting, 31.Oct.2017

25

Modeling of the evaporation of the polymer slurry in the porous media

Dr. Shuji Hironaka, Department of Chemical Engineering, Kyushu University, Japan
Dr. Gen Inoue, Department of Chemical Engineering, Kyushu University, Japan
Dr. Jun Fukai, Department of Chemical Engineering, Kyushu University, Japan
Dr. Yoshifumi Tsuge, Department of Chemical Engineering, Kyushu University, Japan

1. Introduction

There are many processes to manufacture the porous media with the wet process such as the catalyst layer in the polymer electrolyte fuel cell, separator in the lithium-ion battery and porous electrode[1–3]. **Fig 1** shows the example of the manufacturing process based on the wet process. The porous structure is formed through the drying of the slurry containing the particle matter of solid material.

The porous property will be important, such as the porosity or tortuosity, in many scenes porous materials are used. Nevertheless, the behavior of particles in drying has not been well understood, therefore, the manufacturing process is empirically designed.

In order to estimate the porous structure after drying out, the numerical simulation can be proper. It is important for the drying simulation to model the free-surface, particle interaction, internal flow and evaporation of the liquid in sub-micro (10^{-6}m ~ 10^{-9}m) scale. **Table 1** shows main factors and currents status of this study. This paper is intended to report the current state of our study to achieve the estimation of the porous structure in the wet process.

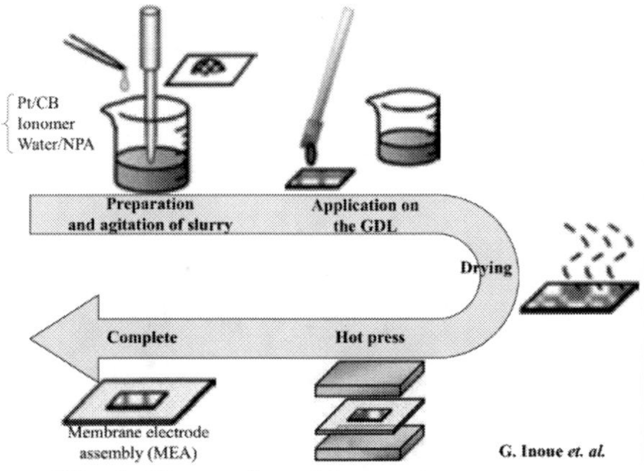

Fig 1. Example of the wet process

Table 1. Modeling factors

Object	Modeling factor	Status
Solid (Particles)	Particle collision and motion	Done
	Interaction between particles	Done
	Surface tension	Not yet
	Shear stress	Not yet
	Buoyancy	Not yet
	Fluid drag	Not yet
Liquid	Surface shape	Done
	Evaporation	Not yet
	Internal flow	Not yet

2. Numerical Modeling

As a first step of modeling, following three factors have been currently included.

· Particle collision and motion
· Particle interaction in 10^{-6}~10^{-9} m scale.
· Surface shape affected by the surface tension

The liquid surface shape is considered as a surface of the minimal area and discretized within the framework of the Finite Element method (FEM). Sub-micro scale particles are treated as the group of rigid spheres with the impulse-based dynamic simulation (IBDS).

The static orthogonal FEM element is used in this study because the particle motion can make the generation of elements fitted with the particle distribution hard. Therefore, particles on the FEM element are treated as the occupancy (**Fig 2**).

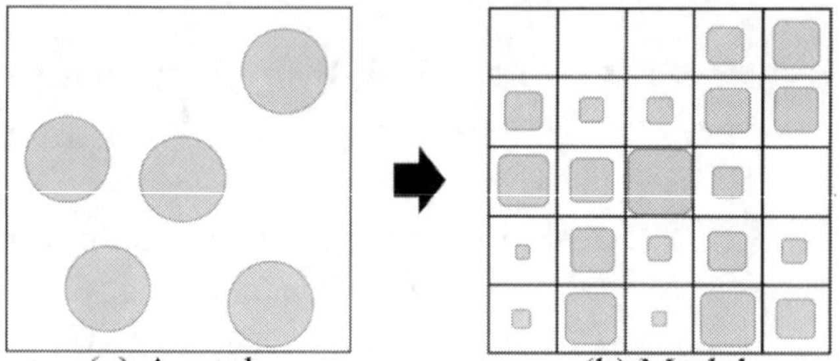

(a) Acutal (b) Model

Fig 2. Modeling of particles on static elements

2.1. Modeling of the surface shape

The surface shape affected by the surface tension is of the minimal area (**Fig 3**). Therefore, when the surface is represented with its height z as a function of x and y, the calculation of the stable surface shape is treated as a problem of the conditional extreme value based on the following equation[4,5].

$$f(z,\lambda)=\int_S \sqrt{1+z_x^2+z_y^2}\,\alpha dxdy-\lambda\left(V_0-\int_S z\alpha dxdy\right) \tag{1}$$

Eq. (1) is the evaluation function of the surface area under the condition of the constant volume. When the surface becomes the shape of the minimal area, eq. (1) must be extreme for all elements. Thus, it is necessary for the surface height z and undetermined multiplier λ to satisfy the following equation.

$$\begin{cases}\dfrac{\partial f}{\partial z}=\displaystyle\int_S \frac{z_x+z_y}{\sqrt{1+z_x^2+z_y^2}}\alpha dxdy+\lambda\int_S \alpha dxdy=0\\[2ex] \dfrac{\partial f}{\partial \lambda}=V_0-\displaystyle\int_S z\alpha dxdy=0\end{cases} \tag{2}$$

Since z is defined on each nodal point and it is shared with neighbor elements (**Fig 4**), the eq. (2) z is simultaneous equations for z at nodal points and λ.

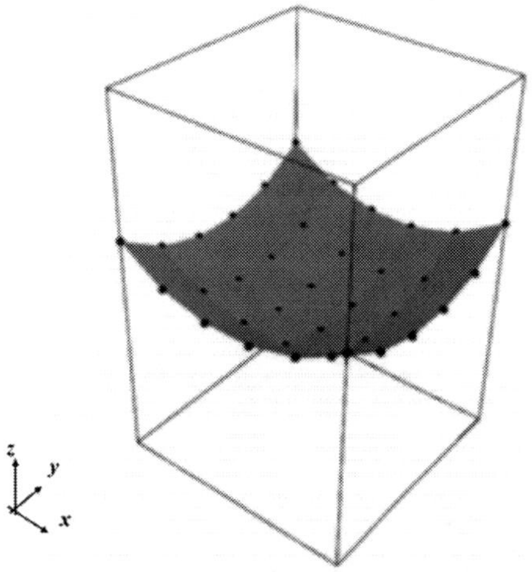

Fig 3. Example of the surface with the minimal area

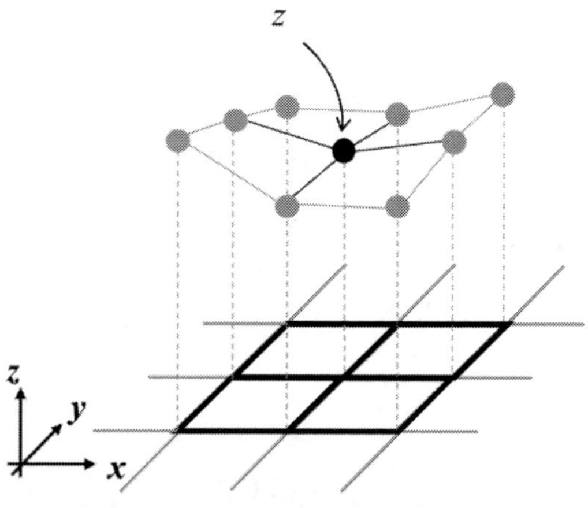

Fig 4. Schematic of the z on nodal points shared with 4 elements

2.2. Modeling of particles

The interaction in two particles' collision is based on the impulse based dynamic simulation (IBDS)[6]. In IBDS, the particle velocity after collision becomes inverted depending on the coefficient of restitution e.

Two particles (p and q) are regarded to be colliding only in the case that they are approaching and contacting. In other words, it is assumed that the distance between p and q particle r_{pq} is less than average diameter and velocity from p to q $\mathbf{v}_{pq}$ is positive in colliding.

At the moment of the collision, the impulse $\mathbf{j}_{pq}$ between p and q particle is obtained by the eq. (3). $\mathbf{j}_{pq}$ and $\mathbf{j}_{qp}$ have the relationship represented by the eq. (4).

$$\mathbf{j}_{pq} = m_{red,pq}(1+e)(\hat{\mathbf{n}}_{pq} \cdot \mathbf{v}_{pq})\hat{\mathbf{n}}_{pq} \quad ; \hat{\mathbf{n}}_{pq} \cdot \mathbf{v}_{pq} > 0 \text{ and } r_{pq} < d_{ave,pq} \tag{3}$$

$$\mathbf{j}_{pq} + \mathbf{j}_{qp} = \mathbf{0} \tag{4}$$

Interaction based on Derjaguin-Landau-Verwey-Overbeek (DLVO) theory is currently considered[7]. According to DLVO theory, particles with the diameter less than 10^{-6}m scale are affected by the electric and inter-molecular force in the slurry. This force between p and q particle $\mathbf{f}_{c,pq}$ are obtained by the gradient of the following potential gradient.

$$V_{T,pq} = V_{R,pq} - V_{A,pq} = \frac{\varepsilon\zeta^2 a^2}{r_{pq}+2a} e^{-\kappa r_{pq}} - \frac{A_H a}{12 r_{pq}} \tag{5}$$

$$\mathbf{f}_{c,pq} = -\nabla V_{T,pq} = \left[\frac{\varepsilon\zeta^2 a^2}{r_{pq}+2a}\left(\frac{1}{r_{pq}+2a}+\kappa\right)e^{-\kappa r_{pq}} - \frac{A_H a}{12 r_{pq}^2} \right] \cdot \frac{\mathbf{r}_{pq}}{r_{pq}} \tag{6}$$

The time evolution of the particle position $\mathbf{r}_p$ and velocity $\mathbf{v}_p$ are calculated by following equations for each particle.

$$\frac{d\mathbf{r}_p}{dt} = \mathbf{v}_p \tag{7}$$

$$m_p \frac{d\mathbf{v}_p}{dt} = \sum_{q \neq p} (\mathbf{f}_{c,pq} + \mathbf{j}_{pq}) \tag{8}$$

2.3. Boundary condition

z on the particle surface z_w will be given by the following equation.

$$z_w = z_{ref} + \frac{\Delta(|\hat{n}_x| + |\hat{n}_y|)}{\tan\theta_p} \tag{10}$$

The gradient of the liquid surface in the normal direction to the particle face is assumed to be same as the reciprocal of the $\tan\theta_p$ (**Fig 5**).

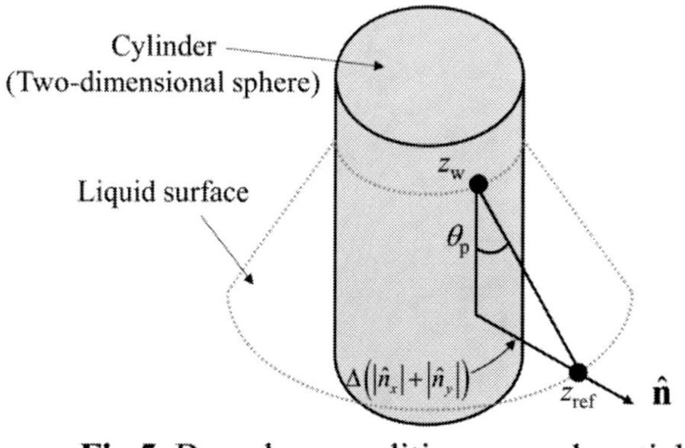

Fig 5. Boundary condition around particle

3. Calculation

Currently, it has been included the effect of the free surface and particle interaction. The following contents are examined in this paper

- Grid-size dependency and comparison with the theoretical solution for the one-dimensional free surface problem
- Stability of the calculation for the time evolution of surface shape induced by moving particles

3.1. One-dimensional problem

As **Fig 6** shows, in order to compare the calculation with the theoretical result for the surface shape, the one-dimensional problem is conducted. The surface shape between two vertical walls is obtained. Each wall has different contact angles and is located at L_x intervals in x direction.

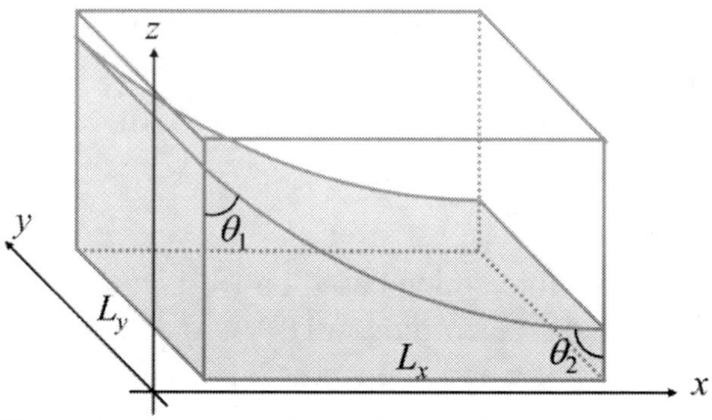

Fig 6. Schematic of one-dimensional problem

Table 2 shows calculation cases. The calculation is conducted in four cases with different element size Δ. **Table 2** also refers to the comparison for the surface position at $x = 0$ z_1 between the calculation and theoretical value. The accuracy will be within 3% when Δ is smaller than 1/40 of L_x.

The theoretical surface shape will be a part of the circle under the condition without gravity. The radius and center position of the circle are given by equations in **Table 3**.

Table 2. Condition and result (One-dimensional problem)

No	Condition				Result	
	Δ [µm]	L_x [µm]	θ_1 [°]	θ_2 [°]	z_1 [µm]	$\left\|\frac{z_1 - z_{1,\mathrm{th}}}{z_{1,\mathrm{th}}}\right\|$ [-]
1	0.40	4.0	45	0	1.86	1.2×10^{-1}
2	0.20				1.76	5.9×10^{-2}
3	0.10				1.71	2.8×10^{-2}
4	0.05				1.68	1.3×10^{-2}

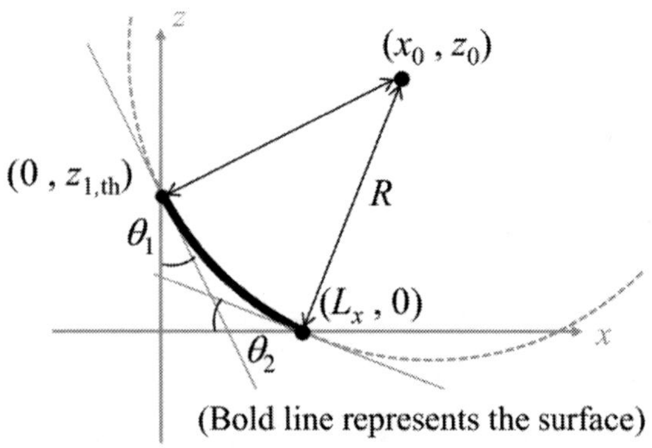

(Bold line represents the surface)

Table 3. Theoretical surface shape

	Equation	Value
R [µm]	$R = \dfrac{L_x}{\cos\theta_1 - \sin\theta_2}$	5.66
x_0 [µm]	$x_0 = L_x + R\sin\theta_1$	4.00
z_0 [µm]	$z_0 = R\cos\theta_2$	5.66
$z_{1,\mathrm{th}}$ [µm]	$z_{1,th} = z_0 - \sqrt{R^2 - x_0^2}$	1.66

Fig 7 shows the calculated surface shape. According to **Fig 7** (b), the surface in all cases is qualitatively similar to the theoretical solution. The theoretical surface position at $x= 0$ $z_{1,\mathrm{th}}$ is 1.66 as **Table 3** shows. Calculation tends to overestimate the surface position at $x= 0$. It is same as the tendency that the surface curvature is overestimated in the coarse element.

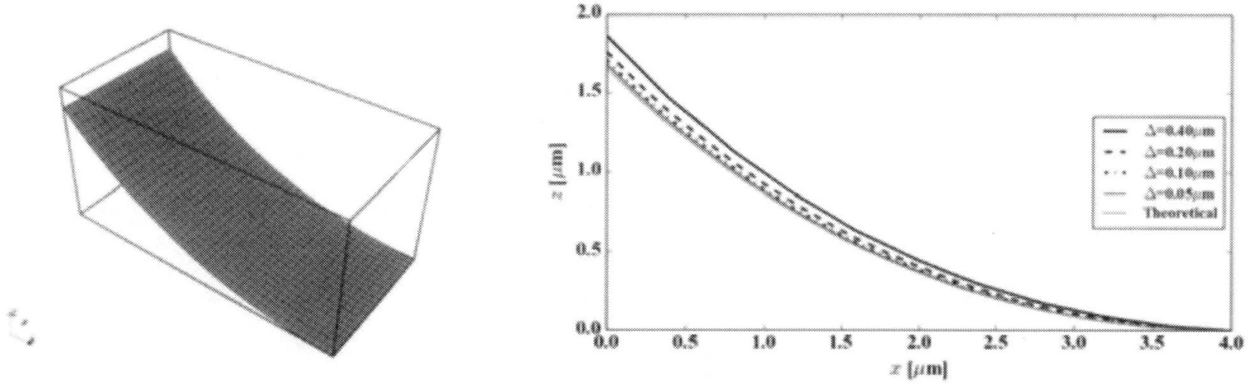

(a) Calculation result ($\Delta = 0.1$ μm) (b) Surface shape in each Δ and theory

Fig 7. Calculated surface shape

3.2. Time evolution of the surface

This calculation is conducted to confirm the calculation stability when moving particles are coupled. **Table 4** shows the calculation condition. Three particles are attracted by the inter-molecular force. They are assumed exactly inelastic. **Fig 8** is the time evolution of the surface shape and particle position. Although particles are assumed as sphere in this calculation, their motion is currently restricted to the two-dimensional. Therefore, they are represented as cylinders in **Fig 8**. The surface elevation around particle is moving together with them and stably calculated.

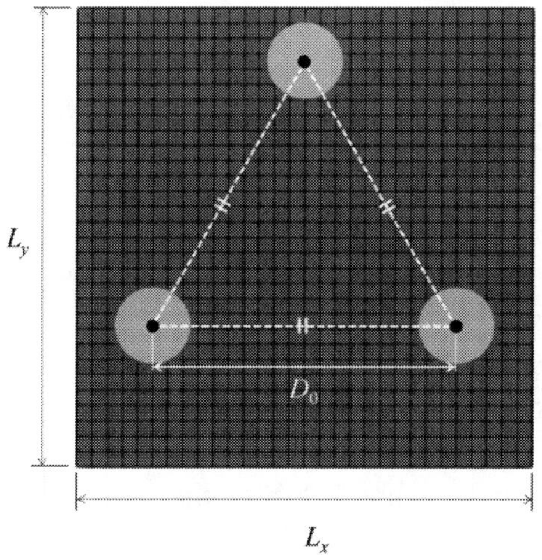

Table 4. Condition (Time evolution of the surface)

Property	Notation	Value	Unit
Domain length in x	L_x	3.0	μm
Domain length in y	L_y	3.0	μm
Element size	Δ	0.1	μm
Time step	Δt	2.0	μs
Contact angle on the particle	θ_p	45	°
Particle diameter	d_p	0.5	μm
Initial particle distance	D_0	2.0	μm
Particle density	ρ_p	2000.0	kg/m^3
Coefficient of the restitution	e	0.0	-
Hamaker constant	A_H	5.8×10^{-19}	J
Primary particle diameter	a	0.1	μm

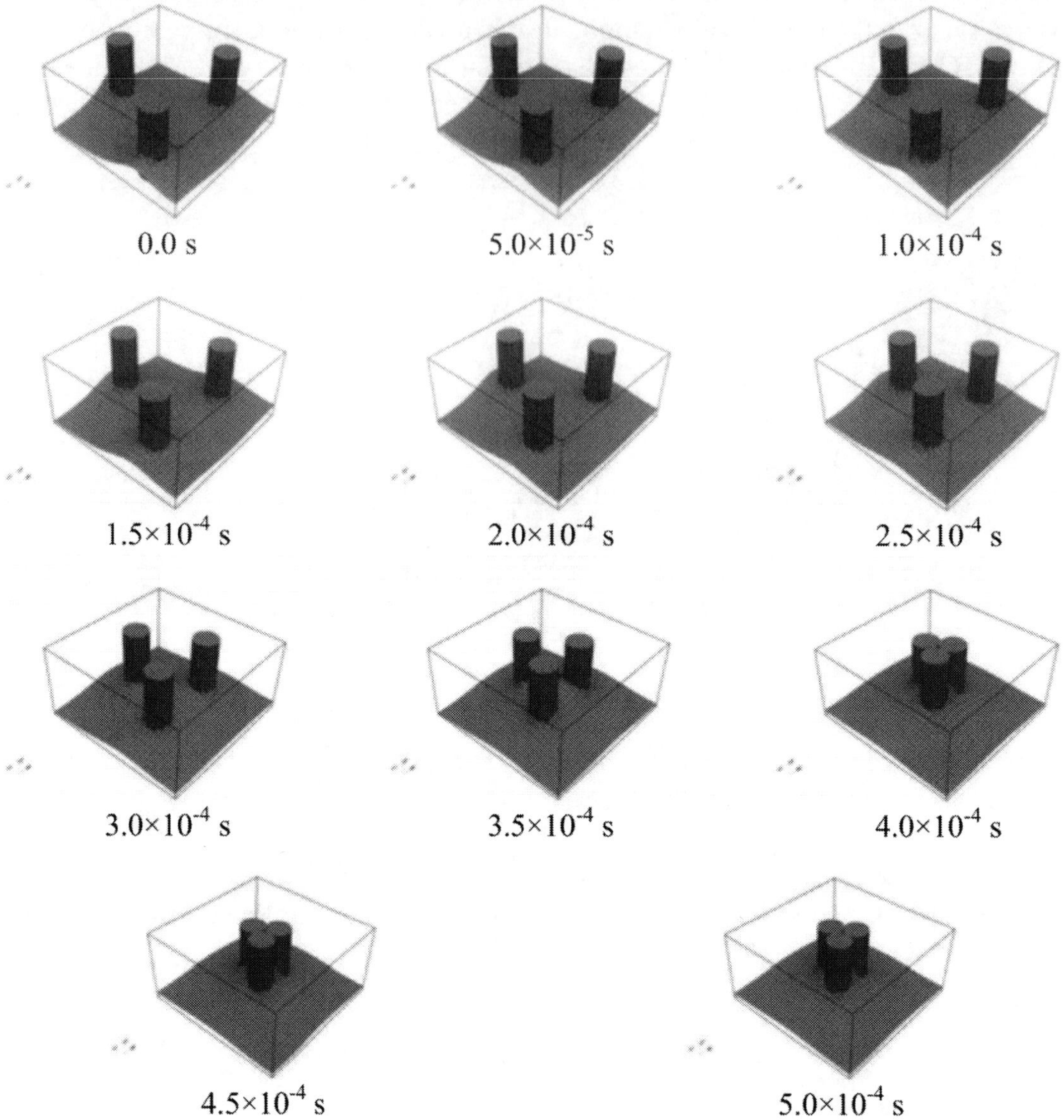

Fig 8. Time evolution of the free surface

4. Conclusion

In order to estimate the porous structure after drying the slurry, the numerical model of the sub-micro scale liquid-solid system including the drying has been considered. As the current work, the particle interaction and surface shape affected by the surface tension are included.

The accuracy of the calculated surface shape is discussed by comparing with the theoretical solution. It was revealed that the error will be less than 3% when the element size is smaller than 1/40 of the distance between two walls. The current method can calculate the surface shape stably when the particle motion is considered together.

Nomenclature

A_H	=	Hamaker constant [J]	a	=	Primary particle diameter [m]
D_0	=	Initial particle distance [m]	d_{ave}	=	Average particle diameter [m]
e	=	Coefficient of the restitution [-]	f	=	Evaluation function of the liquid surface [m^2]
$\mathbf{f}_c$	=	Center force between particles [N]	$\mathbf{j}_{pq}$	=	Impulse between p and q particle [kg·m/s]
L_x	=	Length of the calculation domain in x [m]			
L_y	=	Length of the calculation domain in y [m]			
m_p	=	Particle mass [kg]	$\hat{\mathbf{n}}$	=	Unit normal vector of the particle surface [-]
$\hat{\mathbf{n}}_{pq}$	=	Unit normal vector directed from p to q particle [-]			
$\hat{n}_x$	=	x component of the unit normal vector [-]			
$\hat{n}_y$	=	y component of the unit normal vector [-]			
$\mathbf{r}_p$	=	Position vector of p particle [m/s]	$\mathbf{r}_{pq}$	=	Relative position vector from p to q particle [m]
r_{pq}	=	Distance between p and q particle [m]			
R	=	Radius of the theoretical surface circle [m]			
V_0	=	Liquid volume [m^3]	V_A	=	Attractive potential [J]
V_R	=	Resilience potential [J]	V_T	=	Total potential [J]
$\mathbf{v}_p$	=	Velocity of p particle [m/s]	$\mathbf{v}_{pq}$	=	Relative velocity between p and q particle [m/s]
x_0	=	Center x of theoretical surface circle [m]			
z	=	Vertical surface position [m]	z_0	=	Center z of the theoretical surface circle [m]
z_1	=	Surface position at the left wall [m]			
$z_{1,th}$	=	Theoretical value of the surface position at the left wall [m]			
z_{ref}	=	Reference surface position [m]	z_w	=	Surface position on the particle [m]
z_x	=	Partial derivative of z in x [-]	z_y	=	Partial derivative of z in y [-]
α	=	Particle occupancy in the element [-]	Δ	=	Element size [m]
Δt	=	Size of the time step [s]	ϵ	=	Permittivity [F/m]
ζ	=	Particle zeta potential [J]	θ_1	=	Contact angle at the left wall [°]
θ_2	=	Contact angle at the right wall [°]	θ_p	=	Contact angle on the particle [°]
κ	=	Reciprocal of Debye length [1/m]	λ	=	Undetermined multiplier [1/m]
ρ_p	=	Particle densit [kg/m^3]			

References

1. G. Inoue. and M Kawase. (2016), "Understanding Formation Mechanism of Heterogeneous Porous Structure of Catalyst Layer in Polymer Electrolyte Fuel Cell", *International Journal of Hydrogen Energy*, 41-16, December, pp. 21352-21365.
2. E. Saiz, L. Gremillard, G. Menendez, P. Miranda, K. Gryn and A. P. Tomsia (2006), "Preparation of porous hydroxyapatite scaffolds", *Materials Science and engineering*, C-27, June, pp. 546-550
3. S. Otoishi, Y. Tange and H. Matsuda (2000), "Preparation of Porous Alumina Bodies by W/O-Type Emulsion Method (in Japanese)", *Journal of the Ceramic Society of Japan*, 108-5, pp. 487-491
4. A. Lawal. and R. A. Brown (1982), "The Stability of an Inclined Pendent Drop", *Journal of Colloid and Interface Science*, 89-2, October, pp. 332-345

5. T. Tsukada, M. Hozawa, N. Imaishi, M. Hozawa and K. Fujinawa (1981), "Static Drop Formation on an Inclined Flat Plate", *Journal of chemical engineering of Japan*, 15-6, November, pp. 421-426

6. B. V. Mirtich (1996), "Impulse-based Dynamic Simulation of Rigid Body Systems", *A dissertation submitted in partial satisfaction of the requirements for the degree of Doctor of Philosophy in Computer Science in the Graduate Division of the University of California at Berkeley*, Fall, pp. 58-62

7. D. Myers (1999), "Surfaces, Interfaces, and Colloids Principles and Applications Second Edition", John Wiley & Sons ISBN 0-471-22111-2, pp. 244-246

492481 Flow of a Power-Law Fluid across an Asymmetrically Confined Rotating Cylinder

Abstract:

The present study is motivated by the potential application of a rotating cylinder in the area of drying, filtration of slurries, and calendering process in paper making and polymer industries, cylindrical cooling devices in the glass and plastics industries etc. In addition, such model configurations also advance our understanding of the underlying fluid mechanical issues like wake characteristics, hydrodynamic forces etc. The flow field far away from the cylinder is influenced very little by its rotation and/or by the severity of confinement of the cylinder. Thus, in this work, the combined effects of rotation and asymmetric confinement on the momentum transfer aspects in power-law fluids are investigate numerically.

In particular, this work endeavours to study the momentum transfer characteristics for the flow of shear-thinning fluids, modelled here as a power-law fluid, across a confined, rotating cylinder in between two parallel walls. The cylinder rotational velocity is maintained same as the bulk fluid velocity. The flow is assumed to be laminar, 2-dimensional and incompressible. The flow field is visualized in terms of streamlines, hydrodynamic forces and torque acting on the rotating cylinder. In general, for fixed values of the Reynolds number, power-law index and asymmetry ratio, the drag and lift coefficients and torque are seen to increases steeply with the rising confinement. On the other hand, the shear thinning fluid shows a negative effect on the drag and lift coefficients and torque for a particular value of Reynolds number and cylinder confinement and asymmetry parameter. For the symmetric case, the effective rate of shearing is likely to be high close to the cylinder and close to the walls (due to the no-slip condition) whereas these two regions of low viscosity are separated by a fluid of effectively higher viscosity. For asymmetric positioning, the high viscosity region shrinks and finally disappears for the case when the rotating cylinder is situated close to the channel wall. The velocity field near the cylinder surface is strongly influenced by the Reynolds number, power-law index and degree of confinement and asymmetry due to the low and high viscosity regions depending upon the conditions. Therefore, it is likely that the flow behaviour will exhibit non-monotonic dependence on some of these variables.

For symmetrically placed highly confined cylinder, lubrication analysis holds truc. Thc geometrical and dynamic conditions for the lubrication approximation are generalized for shear-thinning fluids. For Newtonian fluids, an analytical expression for the velocity profile and the torque correction factor in the narrow gap are obtained within the framework of lubrication approximation. For shear-thinning fluids, the resulting velocity profile is more complex due to the existence of non-linear viscous term in the momentum equations. The analytical results in the narrow gap are matched perfectly with the numerical results for all values power-law index

Binary gas mixture in a high speed channel

S. Pradhan

Department of Chemical Engineering, Indian Institute of Science, Bangalore-560 012, India

The viscous, compressible flow in a 2D wall-bounded channel, with bottom wall moving in positive x direction, simulated using the direct simulation Monte Carlo (DSMC) method, has been used as a test bed for examining different aspects of flow phenomenon and separation performance of a binary gas mixture at Mach number $\mathrm{Ma} = (U_w/\sqrt{\gamma k_B T_w/m})$ in the range $0.1 < \mathrm{Ma} < 30$, and Knudsen number $\mathrm{Kn} = (1/(\sqrt{2}\pi d^2 n_d H)$ in the range $0.1 < \mathrm{Kn} < 10$. Here, H is the channel width, U_w is the wall velocity, ρ_m is the volume-averaged gas density, T_w is the wall temperature, m, and d are the molecular mass and molecular diameter, n_d is the number density, μ_w is the molecular viscosity based on the temperature at the isothermal wall, and k_B is the Boltzmann constant. The generalized analytical model is formulated which includes the fifth order differential equation for the boundary layer at the channel wall in terms of master potential (χ), which is derived from the equations of motion in a 2D rectangular ($x - y$) coordinate. The starting point of the analytical model is the Navier-Stokes, mass, momentum and energy conservation equations in the 2D rectangular ($x - y$) coordinate, where x and y are the streamwise and wall-normal coordinates. The linearization approximation is used, where the equations of motion are truncated at linear order in the velocity and pressure disturbances to the base flow, which is a compressible Couette flow. Additional assumptions in the analytical model include high aspect ratio (length of the channel L is large compared to the width H), constant temperature in the base state (isothermal condition), and low Reynolds number. In this limit, the gas flow is restricted to a boundary layer of thickness ($\mathrm{Re}^{-1/2}H$) at the wall of the channel. Here, the Reynolds number $\mathrm{Re} = (\rho_m U_w H/\mu_w)$. The solutions of the generalized analytical model in a high-speed channel are compared with direct simulation Monte Carlo (DSMC) simulations. The comparison reveals that the boundary conditions in the simulations and analysis have to be compared with care. The commonly used 'diffuse reflection' boundary conditions at solid walls in DSMC simulations result in a non-zero slip velocity as well as a 'temperature slip' (gas temperature at the wall is different from wall temperature). These have to be incorporated in the analysis in order to make quantitative predictions. When these precautions are taken, there is excellent agreement between analysis and simulations, to within 10%.

Key words: High speed channel flow, DSMC Simulation, rarefied gas flow.

1. Introduction

A generalized analytical model was developed for calculating the gas flows generated by applied forces in a micro-channel. The base flow is a compressible couette flow. The linearization approximation is used for the perturbed flow, where the equations of motion are truncated at linear order in the velocity and pressure disturbances to the base flow. Additional assumptions in the analytical model include high aspect ratio (length of the channel is large compared to the height), constant temperature in the base state (isothermal condition), and low Reynolds number. In this limit, the gas flow is restricted to a boundary layer of thickness ($\delta = \sqrt{H}$) in the wall-normal coordinate, where H is the height of the channel.

There are two objectives of the present analysis of the rarefied gas flow in a micro-channel. The first is to develope a generalized analytical solutions applicable from very high speed (Mach number = 30) to low speed flows where the Mach number may be $O(1)$. Secondly, we would like to compare the predictions with molecular simulations based on the direct simulation Monte Carlo (DSMC) method for rarefied gas flows, in order to quantify the errors resulting from the approximations at different aspect ratios and Reynolds number. In this study, we have obtained analytical and numerical solutions for the gas flows generated inside the micro-channel due to the moving bottom plate with the external gas in- and out-flow at the boundaries. The results of the analysis are compared with direct simulation Monte Carlo simulations. These studies show that the generalized analytical model (for the boundary layer at the channel wall) is in agreement with numerical solutions to within 10 % for Reynolds number as low as 100 and for channel with length/height ratio as low as 5, provided the wall-slip in both the flow velocity and temperature are incorporated in the analytical solutions.

The main assumptions taken into account in deriving the generalized analytical equation are that the flow field is a small perturbation about an isothermal compressible Couette flow. The inflow/outflow of the gas is small enough compared to the bulk of the fluid present inside the channel. These assumption enables us to linearize the equations of viscous compressible flow, neglect terms that are small in the limit of high Reynolds numbers and aspect ratio, and obtain a fifth-order equation in the cross-streamwise direction for the 'master-potential', which is related to the density, velocity and temperature fields.

As detailed by Chapman & Cowling (1970), the Boltzmann equation can be solved by the ChapmanEnskog approach, which is a series expansion with the Knudsen number as the perturbation parameter. At first-order in Knudsen number, this method retrieves the NavierStokes equations; to second and third order, the Burnett and super-Burnett equations, respectively. An alternative to the ChapmanEnskog method of solving the Boltzmann equation was proposed by

Xu04 presented the gas-kinetic scheme based on the Bhatnagar-Gross-Krook (BGK) model for the microflow simulations in the near continuum flow regime. Their method includes (i) implementing a general non-equilibrium state based on the ChapmanEnskog expansion of the BGK model up to the Knudsen number squared in the gas distribution function, (ii) the compatibility condition to evaluate all high-order time derivative terms in the ChapmanEnskog expansion, and (iii) implementing the kinetic boundary condition for the gas distribution function to obtain 'slip' boundary automatically. As a result, the gas-kinetic BGKBurnett scheme improves the validity of the BGKNS solver for the microchannel flow simulations even in the slip flow regime, where the NavierStokes equations with the slip boundary conditions are considered to be legitimately valid. Owing to the correction to the heat transport in the energy flux, the Prandtl number in the gas-kinetic BGKBurnett method can take any value to capture both viscous and heat conduction effects. Since the current method is based on the direct evaluation of the gas distribution function and captures its time evolution, it is different from those methods that are based on the macroscopic Burnett or extended hydrodynamic equations. The second part of this paper is about the application of the newly developed gas-kinetic BGK Burnett method in the microchannel flows. First, they verify the method in the pressure and external-force-driven Poiseuille flows, where the reliable direct simulation Monte Carlo (DSMC) results are available. In the study of Poiseuille flow with Kn = 0.1, the qualitative differences in the pressure distribution in the cross-stream direction between the NavierStokes and the DSMC results are resolved by the gas-kinetic BGKBurnett scheme. They demonstrates that the BGKBurnett method could give a more realistic description of flow motion than the NavierStokes method even in the slip flow regime. After that, the current method they have used to simulate the microchannel flows, where the experimental data are available. In this study, the similarity in the pressure distribution along the straight microchannel was verified. They have compared the mass flow rates for different gases, such as argon, helium and nitrogen, in the long microchannel of submicron height and further compared with the experimental measurements.

The Burnett equations constitute a set of higher-order continuum equations. These equations are obtained from the ChapmanEnskog series solution of the Boltzmann equation while retaining second-order-accurate terms in the Knudsen number Kn. The set of higher-order continuum models is expected to be applicable to flows in the slip and transition regimes where the NavierStokes equations perform poorly. However, obtaining analytical or numerical solutions of these equations has been noted to be particularly difficult. In the first part of their work (Singh, Gavasane, & Agrawal (2014)), they presented the full set of Burnett equations in cylindrical coordinates in three-dimensional form. The equations were reported in a generalized way for gas molecules that are assumed to be Maxwellian molecules or hard spheres. In the second part, a closed-form solution of these equations for isothermal Poiseuille flow in a microtube was derived. The solution of the equations was shown to satisfy the full Burnett equations up to Kn $\leq$ 1.3. The mass flow rate obtained analytically was shown to compare well with available experimental and numerical results. Comparison of the stress terms in the Burnett and NavierStokes equations was presented. The significance of the Burnett normal stress and its role in diffusion of momentum was brought out by the analysis. An order-of-magnitude analysis of various terms in the equations was presented, based on which a reduced model of the Burnett equations was provided for flow in a microtube.

Based on the fast spectral approximation to the Boltzmann collision operator, Wu, Reese & Zhang (2014) presented an accurate and efficient deterministic numerical method for solving the Boltzmann equation. First, the linearized Boltzmann equation is solved for Poiseuille and thermal creep flows, where the influence of different molecular models on the mass and heat flow rates was assessed, and the OnsagerCasimir relation at the microscopic level for large Knudsen numbers was demonstrated. Experimental measurements of mass flow rates along a rectangular tube with large aspect ratio were compared with numerical results for the linearized Boltzmann equation. Then, a number of two-dimensional microflows in the transition and free-molecular flow regimes were simulated using the nonlinear Boltzmann equation. The influence of the molecular model was discussed, as well as the applicability of the linearized Boltzmann equation. For thermally driven flows in the free-molecular regime, it was found that the magnitudes of the flow velocity are inversely proportional to the Knudsen number. The streamline patterns of thermal creep flow inside a closed rectangular channel were analysed in detail: when the Knudsen number was smaller than a critical value, the flow pattern was predicted based on a linear superposition of the velocity profiles of linearized Poiseuille and thermal creep flows between parallel plates. For large Knudsen numbers, the flow pattern was determined using the linearized Poiseuille and thermal creep velocity profiles at the critical Knudsen number. The critical Knudsen number was found to be related to the aspect ratio of the rectangular channel.

Poiseuille and thermal transpiration flows of a highly rarefied gas were investigated on the basis of the linearized Boltzmann equation by Takata & Funagane (2011), with a special interest in the over-concentration of molecules on velocities parallel to the walls.

An iterative approximation procedure with an explicit error estimate was presented, by which the structure of the over-concentration was clarified. A numerical computation on the basis of the procedure was performed for a hard-sphere molecular gas to construct a database that promptly gives the induced net mass flow for an arbitrary value of large Knudsen numbers. An asymptotic formula of the net mass flow was also presented for molecular models belonging to Grad/s hard potential. Finally, the resemblance of the profiles between the heat flow of the Poiseuille flow and the flow velocity of the thermal transpiration was pointed out.

The flow of binary gaseous mixtures between two parallel plates driven by gradients of pressure, temperature and concentration was studied by Naris, Valougeorgis, Kalempa, & Sharipov (2003), based on the McCormack model of the Boltzmann equation. The coupled kinetic equations are solved numerically by the discrete velocity method. The mass flow, the heat flux and the diffusion flux , which are the mixture quantities of practical importance, are expressed in terms of the so-called thermodynamic fluxes. The latter are written in a form that allows to verify the OnsagerCasimir reciprocity relations. In addition, the analytical expressions for these quantities are derived in the limiting case of the hydrodynamic regime. Their numerical solution together with these expressions provides the solution in the whole range of the gas rarefaction. The influence of the intermolecular interaction potential was also investigated by comparing the results for the rigid sphere model with those for a realistic potential. Finally, the numerical results were presented for two binary mixtures of noble gases ($NeAr$ and $HeXe$) for various values of the molar concentrations.

A binary mixture of rarefied gases between two parallel plates was analyzed by Kosuge & Takata (2008). The Poiseuille flow, thermal transpiration (flow caused by a temperature gradient of the plates) and concentration-driven flow (flow caused by a gradient of concentration of the component species) are analyzed on the basis of the linearized model Boltzmann equation with the diffuse reflection boundary condition. The analyses are first performed for mixtures of virtual gases composed of the hard-sphere or Maxwell molecules and the results are compared with those of the original Boltzmann equation. They performed the analyses for noble gases ($HeNe$, $HeAr$ and $NeAr$) assuming more realistic molecular models (the inverse power-law potential and Lennard-Jones 12, 6 models). By using the results, flux databases covering the entire ranges of the Knudsen number and of the concentration and a wide range of the temperature are constructed. They prepared the databases for the use in the fluid-dynamic model for mixtures in a stationary nonisothermal microchannel derived, and also incorporated in the generalized Reynolds equation in the gas film lubrication theory.

A fully instrumented microscale shock tube was used to study the transmission of a shock wave, produced in a large (37 mm) shock tube, into a 34 m hydraulic diameter and 2 mm long microchannel by Mirshekari, Brouillette, Giordano & Hbert (2013). Pressure microsensors of a novel design, with gigahertz bandwidth, were used to obtain pressure-time histories of the microchannel shock wave at five axial stations. In all cases the transmitted shock wave was found to be weaker than the incident shock wave, and was observed to decay both in pressure and velocity as it propagates down the microchannel. These results were compared with various analytical and numerical models, and the best agreement was obtained with a NavierStokes computational fluid dynamics computation, which assumes a no-slip isothermal wall boundary condition; good agreement was also obtained with a simple shock tube laminar boundary layer model. It was found that the flow developing within the microchannel was highly dependent on conditions at the microchannel entrance, which control the mass flux entering into the device. Regardless of the micrometre dimensions of the present facility, shock wave propagation in a microchannel of that scale exhibits a behaviour similar to that observed in large-scale facilities operated at low pressures, and the shock attenuation was explained in terms of accepted laminar boundary models.

Rarefied gas flow in micro-channels/pipes which height/radius was measured in micro- or nanometers was treated by Djordjevic (2004, 2005, 2008). Knudsen number was varied from very low values present in the continuum flow to very large values pertinent to the free molecular flow. In order to cover the entire Knudsen number range with a single physical model they model the wall slip boundary condition by means of a fractional derivative, whereby its order conveniently depends on the local value of the Knudsen number in the channel flow, or on an average value of this number in the pipe flow. They found excellent agreement between the experimental and numerical results. In this paper they made another attempt to cover the entire Knudsen number range by modeling the slip boundary condition via a fractional derivative. For this purpose they defined a version of Caputo derivative whose order α is either a function of the local value of the Knudsen number in the channel, or a function of the appropriately defined average value of the Knudsen number in the pipe. For $\alpha = 0$ boundary conditions were classical, no-slip boundary conditions in the continuum model, while as α-¿1 the Knudsen number approaches infinity and the flow becomes a free molecular flow. According to their knowledge such an application of a fractional derivative was specific, because they were mostly used for modeling the rheological properties of different viscoelastic materials and in modeling various phenomena in many branches of physics by fractional order differential equations. Thus, the analysis presented in the paper adds to the vast variety of applications of fractional derivatives and opens a new ground for describing complex flow phenomena in rarefied gas dynamics.

In very long micro-channels, or pipes exhausting to a low-pressure environment, with the inlet pressure comparable with atmospheric conditions, gas flow may happen to pass through all regimes mentioned above. In such a case it is highly desirable to have a model for the slip boundary condition, which would cover all regimes of flow, from continuum one, to free molecular flow. The model was introduced on a purely empirical basis in the form of a "rarefaction coefficient", by means of which the expression for the mass flow rate, obtained for relatively small values of the Knudsen number in the slip-flow regime, is "corrected" so as to embody the entire Knudsen number regime. The results obtained for volumetric and mass flow rate as well as for the pressure distribution are fitted well with the results of the direct-simulation Monte Carlo (DSMC) method and the solutions of linearized Boltzmann equation.

Higher order slip boundary conditions have been proposed to stretch the applicability of continuum models over a larger range of Knudsen numbers, but, they have been typically closed with slip coefficients that may themselves vary with Knudsen number. Behaviour of isothermal rarefied gases can be described by the Boltzmann equation Chapman & Cowling (1970). Recently, it has been suggested by several researchers that the constitutive assumption in-built with the standard forms of the Navier-Stokes-Fourier (NSF) equations need to be extended to render them applicable for fluid flows with strong local gradients of density and temperature. A microscopic volume-based kinetic approach for analyzing the molecular spatial distribution was proposed; this led to a set of continuum equations incorporating a volume diffusive flux, which also behaves like a mass diffusion flux. These kind of extended continuum set of equations improve predictions of certain flows in the finite Knudsen number regime. They assessed the implication of a mass diffusion correction for isothermal gas flow along a micro channel. They showed that it can predict the actual pressure distributions in these flows. Axial pressure profiles at various pressure ratios were validated with experimental data, and the effects of Knudsen number on the curvature of the pressure profiles was analyzed. Their results indicate that the mass diffusion correction captures the 'Knudsen paradox' (i.e. the minimum in flow rate) without the need to set higher-order slip boundary conditions. Moreover, the mass diffusion model offers a more reliable explanation of the Knudsen paradox.

An analytical version of the discrete-ordinates method was used by Garcia (2000) to establish a concise and particularly accurate solution to the problem of plane Couette flow for a binary gas mixture described by the McCormack kinetic model. The solution yields, for the general specular-diffused case of Maxwell boundary conditions for each of the two species, the velocity, heat-flow, and shear-stress profiles for both types of particles, as well as the particle-flow and heat-flow rates associated with each of the two species of gas particles. Highly accurate numerical results were reported for the case of a heliumargon mixture confined between molybdenum and tantalum plates.

The flow of binary gaseous mixtures through rectangular microchannels due to small pressure, temperature, and molar concentration gradients over the whole range of the Knudsen number was also studied. The solution was based on a mesoscale approach, formally described by two coupled kinetic equations, subject to diffuse scattering boundary conditions. The model proposed by McCormack substitutes the complicated collision term and the resulting kinetic equations are solved by an accelerated version of the discrete velocity method. Typical results are presented for the flow rates and the heat fluxes of two different binary mixtures (Ne Ar) and (He Xe) with various molar concentrations, in two-dimensional microchannels of different aspect (height to width) ratios. The formulation was very efficient and was used instead of the classical method of solving the NavierStokes equations with slip boundary conditions, which is restricted by the hydrodynamic regime.

Lockerby, Reese & Gallis (2005) made two new propositions regarding the modelling of rarefied (non-equilibrium)isothermal gas flows at the microscale. The first was a new test case for benchmarking high-order, or extended, hydrodynamic models for these flows. This standing time-varying shear-wave problem does not require boundary conditions to be specified at a solid surface, so was useful for assessing whether fluid models can capture rarefaction effects in the bulk flow. They assessed a number of different proposed extended hydrodynamic models, and they found the (R-13) equations perform the best in this case. Their second proposition was a simple technique for introducing non-equilibrium effects caused by the presence of solid surfaces into the computational fluid dynamics framework. By combining a new model for slip boundary conditions with a near-wall scaling of the NavierStokes constitutive relations, they obtained a model that was much more accurate at higher Knudsen numbers than the conventional second-order slip model. They showed that this provides good results for combined Couette/Poiseuille flow, and that the model can predict the stress/strain-rate inversion that was evident from molecular simulations. The model's generality to non-planar geometries was demonstrated by examining low-speed flow around a micro-sphere. It showed a marked improvement over conventional predictions of the drag on the sphere, although there were some questions regarding its stability at the highest Knudsen numbers.

Pressure-driven,implicit boundary conditions are implemented in an open source direct simulation Monte Carlo (DSMC) solver by White, Borg, Scanlon & Reese (2013), and they have benchmarked against simple micro-channel flow cases. DSMC simulations are then carried out of gas flows for varying degrees of rarefaction along micro-channels with both one and two 90-degree bends. The results are compared to those from the equivalent straight micro-channel geometry. They found that away from the immediate bend regions, the pressure and Mach number profiles do not differ greatly from those in straight channels, indicating that there were no significant losses introduced when a bend is added to a micro-channel geometry. It was found that the inclusion of a bend in a micro-channel can increase the amount of mass that a channel can carry, and that adding a second bend produces a greater mass flux enhancement. This increase happens within a small range of Knudsen number ($0.02 < \mathrm{Kn} < 0.08$). Velocity slip and shear stress profiles at the channel walls are presented for the Knudsen showing the largest mass flux enhancement.

Microchannel based systems have emerged as a critical design trend in development of precise control and maneuvering of small devices. In microelectronics, space propulsion and biomedical areas, these systems are especially useful. Nanoscale pores are recently becoming of great interest due to their beneficial drag and heat transfer properties. However it is difficult to predict the flow performance of these microsystems and nanosystems numerically since the standard assumptions of using NavierStokes equations break down at micrometer scales, while the computational times of applicable molecular-dynamics codes become exorbitant. A two-dimensional finite-element based microscale flow model is developed to efficiently predict the overall flow characteristics up to the transition regime for reasonably high Knudsen number flow inside microchannels and nanopores were presented by Roy, Raju, Chuang, Cruden, & Meyyappan (2003). They presented two-dimensional numerical results for Poiseuille flow of a simple fluid through the microchannel and compared with the numerical and experimental data. The nanopore solutions were also validated with experimental data.

Technological advancements in Micro-Electro-Mechanical Systems (MEMS) have rekindled interest in fundamental scientific investigations concerning physical phenomena such as boundary slip of gas flows at small length scales. In microscale rarefied gas flows, deviations from local thermodynamic equilibrium are significant, and the extent of this deviation can be quantified in terms of the Knudsen number (Kn), which is the ratio of the gas molecular mean free path (λ) to the characteristic system length scale (H). Experimental investigations have confirmed that the conventional Navier-Stokes-Fourier (NSF) equations, in their traditional form with no slip boundary condition, underpredict experimental mass flow rates in microscale gas flows, and deviations become more significant with increase in Knudsen number. The phenomenon of enhanced mass flow rates is typically addressed by applying the Maxwell slip boundary condition in conjunction with the continuum equations. However, Maxwell-type first order slip models are accurate only up to small values of the Kn, and also fail to predict the well established Knudsen paradox phenomenon.

In §2, the linearized mass, momentum and energy conservation equations for the flow in a two dimensional channel are presented and different dimensionless numbers are defined. The equations for perturbations to the base state (isothermal compressible couette flow) are then used to get the equation for the master potential In §3. In §4, the boundary layer equations are solved to obtain the generalized analytical equation in the channel. In §5, a brief summary on molecular simulation based on direct simulation Monte Carlo (DSMC) method is presented, and 2D-DSMC simulation results are validated against the 2D finite volume method (FVM) results at subsonic flow conditions (Mach number $\mathrm{Ma} = 0.1, 0.2, 0.3$). In §6, flow-field due to inflow/outflow of gas along the boundaries are studied at high Mach ($\mathrm{Ma} = 1, 3, 5, 10, 30$) and Knudsen numbers ($\mathrm{Kn} = 0.1, 1, 10$) and analytical results are compared with those of DSMC simulations. The important conclusions of the analysis are summarized in §7.

2. Formulation

The configuration consists of a two dimensional channel of length L and height H. A cartesian coordinate system with origin at the left bottom-corner of the channel, with streamwise, and wall-normal coordinates x, and y respectively, is used. Throughout this analysis, we assume that the flow is two dimensional, and there is no dependence on the spanwise coordinate z. The compressible mass, momentum and energy conservation equations in the cartesian coordinate system are provided in the appendix for reference. There are five variables in these equations, the density ρ, the two components of the velocity v_x, v_y, the pressure p and the temperature T.

The pressure is related to the density by the ideal gas equation of state,

$$\rho = \frac{pm}{k_B T}, \tag{2.1}$$

where m is the mass of a molecule and k_B is the Boltzmann constant.

If the temperature is uniform throughout the channel, and there is no inflow or outflow of gas, a steady solution for the equations of motion is one of the isothermal compressible couette flow. The temperature $\bar{T}$ and the components of the velocity $\bar{\mathbf{v}}$ in the base state in this steady state are,

$$\bar{v}_x = \frac{U}{H}(H - y), \quad \bar{v}_y = 0, \quad \bar{v}_z = 0, \quad \bar{T} = T_0, \tag{2.2}$$

where T_0 is the constant temperature. From x and y-momentum equations,

$$\frac{d\bar{p}}{dx} = 0, \tag{2.3}$$

and

$$\frac{d\bar{p}}{dy} = 0. \tag{2.4}$$

Therefore, for isothermal compressible couette flow, there is no pressure gradients in either x or y direction, and pressure is constant throughout the channel. The total mass of gas in the channel is related to $\bar{\rho}$ by,

$$\begin{aligned} \text{Mass of gas} &= HL\left(\frac{1}{H}\int_{-H/2}^{H/2} \rho\, dy\right) \\ &= HL\bar{\rho}, \end{aligned} \tag{2.5}$$

where H and L are the height and length of the channel, and the volume-averaged gas density (total mass divided by volume) is defined by

$$\bar{\rho} = \frac{1}{H}\int_{-H/2}^{H/2} \rho\, dy. \tag{2.6}$$

It is useful to discuss the different dimensionless numbers used in the analysis. The Mach number is defined by

$$\text{Ma} = \frac{U}{\sqrt{\gamma k_B T_0/m}} = \frac{U}{c_s}, \tag{2.7}$$

where U is the wall velocity, and $c_s = \sqrt{\gamma k_B T_0/m}$ is the speed of sound. The Reynolds number, used a little later, is defined by

$$\text{Re} = \frac{\bar{\rho} U H}{\bar{\mu}}, \tag{2.8}$$

where $\bar{\mu}$ is the viscosity of the fluid at the temperature corresponding to the wall temperature T_0, which is the uniform temperature of the fluid in the unperturbed state.

The viscosity of a dilute gas scales as

$$\bar{\mu} \sim \frac{\sqrt{mk_B T_0}}{d^2}, \tag{2.9}$$

where m is the molecular mass and d is the molecular diameter, and T_0 is the temperature in the unperturbed state. Using the above expression for the viscosity, it can easily be verified that the Reynolds number can be rewritten as

$$\mathrm{Re} \sim \frac{\mathrm{Ma}}{\mathrm{Kn}}, \tag{2.10}$$

where the Knudsen number is defined by

$$\mathrm{Kn} = \left(m/\sqrt{2}\pi\rho_w d^2 H\right) = (\lambda/H), \tag{2.11}$$

where $\lambda = \left(m/\sqrt{2}\pi\rho_w d^2\right)$ is the mean free path at the wall. The other dimensionless groups used are the Prandtl number and the ratio of specific heats,

$$\mathrm{Pr} = \frac{C_p\bar{\mu}}{\bar{K}}, \tag{2.12}$$

$$\gamma = \frac{C_p}{C_v}, \tag{2.13}$$

which are both $O(1)$ for an ideal gas.

3. Equations of motion in cartesian coordinates

The disturbance to the base state due to a small mass flux through the channel, is incorporated by perturbing the base velocity and pressure profile as follows:

$$\begin{aligned}
\rho &= \bar{\rho} + \hat{\rho}, \\
p &= \bar{p} + \hat{p}, \\
T &= \bar{T} + \hat{T}, \\
v_x &= (U/H)(H-y) + \hat{v}_x, \\
v_y &= \hat{v}_y.
\end{aligned} \tag{3.1}$$

The above forms for the dynamical variables are inserted into the mass, momentum and energy conservation equations, and linearized in the perturbation (hatted) quantities. Steady state is assumed, so that all the time derivatives are zero. After subtracting the corresponding base state equations, we obtain the following equations for the perturbations:

$$\frac{\partial}{\partial x}(\bar{\rho}\hat{v}_x) + \frac{\partial}{\partial y}(\bar{\rho}\hat{v}_y) + \frac{U}{H}(H-y)\frac{\partial\hat{\rho}}{\partial x} = 0, \tag{3.2}$$

$$\frac{\bar{\rho}U(H-y)}{H}\frac{\partial\hat{v}_x}{\partial x} - \frac{\bar{\rho}U}{H}\hat{v}_y = -\frac{\partial\hat{p}}{\partial x} + \bar{\mu}\left(\frac{\partial^2\hat{v}_x}{\partial x^2} + \frac{\partial^2\hat{v}_x}{\partial y^2}\right), \tag{3.3}$$

$$\frac{\bar{\rho}U(H-y)}{H}\frac{\partial\hat{v}_y}{\partial x} = -\frac{\partial\hat{p}}{\partial y} + \bar{\mu}\left(\frac{\partial^2\hat{v}_y}{\partial x^2} + \frac{\partial^2\hat{v}_y}{\partial y^2}\right), \tag{3.4}$$

$$\bar{\rho}C_v\frac{U}{H}(H-y)\frac{\partial T}{\partial x} = -\bar{p}\left(\frac{\partial\hat{v}_x}{\partial x} + \frac{\partial\hat{v}_y}{\partial y}\right) + \bar{K}\left(\frac{\partial^2\hat{T}}{\partial x^2} + \frac{\partial^2\hat{T}}{\partial y^2}\right). \tag{3.5}$$

Using the mass conservation equation (3.2), the above energy equation (3.5), can be written as

$$\bar{\rho} C_v \frac{U}{H}(H-y)\frac{\partial T}{\partial x} = \frac{T_0}{m}\frac{U(H-y)}{H}\frac{\partial \hat{\rho}}{\partial x} + \bar{K}\left(\frac{\partial^2 \hat{T}}{\partial x^2} + \frac{\partial^2 \hat{T}}{\partial y^2}\right). \tag{3.6}$$

Finally, the ideal gas equation of state is

$$\bar{\rho} = \frac{\bar{p} m}{k_B T_0}. \tag{3.7}$$

Here, $\bar{\mu}$ and $\bar{K}$ are the viscosity and thermal conductivity of the gas in the base state, where the temperature is T_0. Note that in the ideal gas limit, the viscosity and thermal conductivity are functions of the temperature alone, and they do not depend on the density. In equation (3.3), the term on the left is the contribution to the x momentum equation due to the inertial force, similarly in equation (4.2), the left side represents the contribution to the y momentum equation due to the inertial force. In the energy equation (3.6), the term on the left is the pressure work done due to the compression or expansion of the gas, and we have used the equation of state (3.7) and the mass conservation equation (3.2) to simplify this term. The perturbation to the equation of state is

$$\hat{\rho} = \left(\frac{m}{k_B T_0}\right)\hat{p} - \bar{\rho}\left(\frac{\hat{T}}{T_0}\right). \tag{3.8}$$

4. Streamwise and wall-normal flow fields

First, we eliminate the pressure in the y momentum equation as follows. If we take the derivative of the y momentum equation (4.2) with respect to the x coordinate, we obtain

$$\frac{\bar{\rho} U(H-y)}{H}\frac{\partial^2 \hat{v}_y}{\partial x^2} = -\frac{\partial}{\partial y}\left(\frac{\partial \hat{p}}{\partial x}\right) + \bar{\mu}\frac{\partial}{\partial x}\left(\frac{\partial^2 \hat{v}_y}{\partial x^2} + \frac{\partial^2 \hat{v}_y}{\partial y^2}\right). \tag{4.1}$$

Equation (3.3) is used to substitute for $(\partial \hat{p}/\partial x)$ in the above equation, to obtain

$$\begin{aligned}\frac{\bar{\rho} U(H-y)}{H}\frac{\partial^2 \hat{v}_y}{\partial x^2} = \frac{\partial}{\partial y}\Bigg(&\frac{\bar{\rho} U(H-y)}{H}\frac{\partial \hat{v}_x}{\partial x} - \frac{\bar{\rho} U}{H}\hat{v}_y \\ &-\bar{\mu}\left(\frac{\partial^2 \hat{v}_x}{\partial x^2} + \frac{\partial^2 \hat{v}_x}{\partial y^2}\right)\Bigg) \\ &+\bar{\mu}\frac{\partial}{\partial x}\left(\frac{\partial^2 \hat{v}_y}{\partial x^2} + \frac{\partial^2 \hat{v}_y}{\partial y^2}\right).\end{aligned} \tag{4.2}$$

After rearrangement we obtain

$$\begin{aligned}&\frac{\partial^3 \hat{v}_x}{\partial y^3} + \frac{\partial^3 \hat{v}_x}{\partial x^2 \partial y} - \frac{\bar{\rho} U(H-y)}{\bar{\mu} H}\frac{\partial^2 \hat{v}_x}{\partial x \partial y} \\ -\frac{\partial^3 \hat{v}_y}{\partial x^3} - \frac{\partial^3 \hat{v}_y}{\partial x \partial y^2} - &\frac{\bar{\rho} U(H-y)}{\bar{\mu} H}\frac{\partial^2 \hat{v}_y}{\partial x^2} - \frac{\bar{\rho} U}{\bar{\mu} H}\frac{\partial \hat{v}_y}{\partial y} = 0.\end{aligned} \tag{4.3}$$

Here, the viscosity $\bar{\mu}$ in the base state is independent of position, because in the base state the temperature is uniform and the viscosity depends only on temperature.

The velocity components $\hat{v}_x$ and $\hat{v}_y$ can be expressed in terms of a stream function using the mass conservation equation (3.2),

$$\bar{\rho}\hat{v}_x = \frac{\partial\psi}{\partial y} - \bar{v}_x\hat{\rho}, \tag{4.4}$$

$$\bar{\rho}\hat{v}_y = -\frac{\partial\psi}{\partial x}, \tag{4.5}$$

where the difference in the stream function between two wall-normal locations gives the streamwise mass flux in the region between these two locations,

$$\psi(H,x) - \psi(y,x) = \int_y^H (\bar{\rho}\hat{v}_x + \bar{v}_x\hat{\rho})\ dy. \tag{4.6}$$

Expressing in terms of the stream function, the equation (4.3) is

$$\frac{1}{\bar{\rho}}\frac{\partial^4\psi}{\partial y^4} + \frac{1}{\bar{\rho}}\frac{\partial^4\psi}{\partial x^2\partial y^2} - \frac{U(H-y)}{\bar{\mu}H}\frac{\partial^3\psi}{\partial x\partial y^2}$$
$$+\frac{1}{\bar{\rho}}\frac{\partial^4\psi}{\partial x^4} + \frac{1}{\bar{\rho}}\frac{\partial^4\psi}{\partial x^2\partial y^2} + \frac{U(H-y)}{\bar{\mu}H}\frac{\partial^3\psi}{\partial x^3} + \frac{U}{\bar{\mu}H}\frac{\partial^2\psi}{\partial x\partial y} = 0. \tag{4.7}$$

The stream function ψ is expressed in terms of the 'master potential' χ:

$$\psi = \frac{\partial\chi}{\partial y}. \tag{4.8}$$

In terms of 'master potential', the streamwise mass flux is as follows:

$$(\bar{\rho}\hat{v}_x + \bar{v}_x\hat{\rho}) = \frac{\partial^2\chi}{\partial y^2}. \tag{4.9}$$

This is inserted into the momentum conservation equation (4.3) to obtain

$$\frac{1}{\bar{\rho}}\frac{\partial^5\chi}{\partial y^5} + \frac{1}{\bar{\rho}}\frac{\partial^5\chi}{\partial x^2\partial y^3} - \frac{U(H-y)}{\bar{\mu}H}\frac{\partial^4\chi}{\partial x\partial y^3}$$
$$+\frac{1}{\bar{\rho}}\frac{\partial^5\chi}{\partial y\partial x^4} + \frac{1}{\bar{\rho}}\frac{\partial^5\chi}{\partial x^2\partial y^3} + \frac{U(H-y)}{\bar{\mu}H}\frac{\partial^4\chi}{\partial x^3\partial y} + \frac{U}{\bar{\mu}H}\frac{\partial^3\chi}{\partial x\partial y^2} = 0. \tag{4.10}$$

The streamwise and wall-normal coordinates are expressed in terms of the scaled streamwise coordinate x^*, and scaled wall-normal coordinate y^*, and defined by

$$x^* = \frac{x}{L}, \tag{4.11}$$

$$y^* = \frac{y}{H}. \tag{4.12}$$

With these transformations, it can easily be verified that

$$\frac{\partial}{\partial x} = \frac{1}{L}\frac{\partial}{\partial x^*}, \tag{4.13}$$

$$\frac{\partial}{\partial y} = \frac{1}{H}\frac{\partial}{\partial y^*}, \tag{4.14}$$

where $\delta = (\sqrt{H})$ is the boundary layer thickness in the wall-normal coordinate. When the equation for the master potential is expressed in terms of these scaled coordinates, we obtain

$$\frac{1}{\bar{\rho}H^5}\frac{\partial^5\chi}{\partial y^{*5}} + \frac{1}{\bar{\rho}L^2H^3}\frac{\partial^5\chi}{\partial x^{*2}\partial y^{*3}} - \frac{U(H-y)}{\bar{\mu}H^4L}\frac{\partial^4\chi}{\partial x^*\partial y^{*3}}$$

$$+\frac{1}{\bar{\rho}L^4H}\frac{\partial^5\chi}{\partial y^*\partial x^{*4}}+\frac{1}{\bar{\rho}L^2H^3}\frac{\partial^5\chi}{\partial x^{*2}\partial y^{*3}}+\frac{U(H-y)}{\bar{\mu}L^3H^2}\frac{\partial^4\chi}{\partial x^{*3}\partial y^*}+\frac{U}{\bar{\mu}LH^3}\frac{\partial^3\chi}{\partial x^*\partial y^{*2}}=0 \quad (4.15)$$

The above equation can be reduced to dimensionless form by dividing by the prefactor of the first term on the left, to obtain

$$\frac{\partial^5\chi}{\partial y^{*5}}+\frac{H^2}{L^2}\frac{\partial^5\chi}{\partial x^{*2}\partial y^{*3}}-\frac{\bar{\rho}UH(H-y)}{\bar{\mu}L}\frac{\partial^4\chi}{\partial x^*\partial y^{*3}}$$
$$+\frac{H^4}{L^4}\frac{\partial^5\chi}{\partial y^*\partial x^{*4}}+\frac{H^2}{L^2}\frac{\partial^5\chi}{\partial x^{*2}\partial y^{*3}}+\frac{\bar{\rho}UH^3(H-y)}{\bar{\mu}L^3}\frac{\partial^4\chi}{\partial x^{*3}\partial y^*}+\frac{\bar{\rho}UH^2}{\bar{\mu}L}\frac{\partial^3\chi}{\partial x^*\partial y^{*2}}=0. \quad (4.16)$$

After simplification, we obtain

$$\frac{\partial^5\chi}{\partial y^{*5}}+\frac{H^2}{L^2}\frac{\partial^5\chi}{\partial x^{*2}\partial y^{*3}}-\frac{\mathrm{Re}(H-y)}{L}\frac{\partial^4\chi}{\partial x^*\partial y^{*3}}$$
$$+\frac{H^4}{L^4}\frac{\partial^5\chi}{\partial y^*\partial x^{*4}}+\frac{H^2}{L^2}\frac{\partial^5\chi}{\partial x^{*2}\partial y^{*3}}+\frac{\mathrm{Re}H^2(H-y)}{L^3}\frac{\partial^4\chi}{\partial x^{*3}\partial y^*}+\frac{\mathrm{Re}H}{L}\frac{\partial^3\chi}{\partial x^*\partial y^{*2}}=0, \quad (4.17)$$

where the Reynolds number Re, defined as $\mathrm{Re} = (\bar{\rho}UH)/\bar{\mu}$.

The following terms in equation (4.16) are neglected to formulate the generalized analytical equation.

(*a*) The second and fifth terms on the left side of equation (4.16) are neglected in comparison to the first term on the left side, because the parameter $(H^2/L^2) \ll 1$. Clearly, this is valid for high aspect ratio channels, $L \gg H$.

(*b*) The third term on the left side of equation (4.16) is also neglected in comparison to the first term, because the parameter $(\mathrm{Re}(H-y)/L) \ll 1$. Clearly, this is valid for high aspect ratio channels, $L \gg (H-y)$, and also when the Reynolds number (Re) is small.

(*c*) The fourth term on the left side of equation (4.16) is also neglected in comparison to the first term, because the parameter $(H^4/L^4) \ll 1$. Clearly, this is valid for high aspect ratio channels, $L \gg H$.

(*d*) The sixth term on the left side of equation (4.16) is also neglected in comparison to the first term, because the parameter $(\mathrm{Re}H^2(H-y)/L^3) \ll 1$. Clearly, this is valid for high aspect ratio channels, $L^3 \gg (H^2(H-y))$, and also when the Reynolds number (Re) is small.

(*e*) The seventh term on the left side of equation (4.16) is also neglected in comparison to the first term, because the parameter $(\mathrm{Re}H/L) \ll 1$. Clearly, this is valid for high aspect ratio channels, $L \gg H$, and also when the Reynolds number (Re) is small.

Therefore, in the limits $L \gg H$, and $\mathrm{Re} \ll 1$, we formulate the 'generalized analytical equation':

$$L_5(\chi) = 0, \quad (4.18)$$

where the linear differential operator $L_5(\chi)$ is defined by

$$L_5(\chi) = \frac{\partial^5\chi}{\partial y^{*5}}. \quad (4.19)$$

It is easily verified that this operator is self-adjoint, when the inner product is defined as,

$$(A_1, B_1) = \int_0^1 dy^* A_1(y^*)B_1(y^*), \quad (4.20)$$

when the boundary conditions are appropriately specified as discussed next. Therefore, the eigenvalues of this operator are real, and the eigenfunctions are orthogonal.

Since the equation (4.18) is a fifth order differential equation, it is necessary to specify three boundary conditions at the top wall $y = H$, and two boundary conditions at the bottom wall $y = 0$. Here we will consider the homogeneous form of the generalized analytical equation and we will investigate the streamwise flow due to inflow/outflow of gas along the boundaries. The following expressions for the velocity components and the streamwise flow rate are useful for specifying the boundary conditions:

$$\bar{\rho}\hat{v}_x = \frac{1}{H^2}\frac{\partial^2\chi}{\partial y^{*2}}, \tag{4.21}$$

$$\bar{\rho}\hat{v}_y = -\frac{1}{LH}\frac{\partial^2\chi}{\partial x^*\partial y^*}, \tag{4.22}$$

$$\begin{aligned} F_x &= \int_0^H dy\, \bar{\rho}\hat{v}_x \\ &= \frac{1}{H^2}\int_0^1 dy^*\, \frac{\partial^2\chi}{\partial y^{*2}} \\ &= \frac{1}{H^2}\frac{\partial\chi}{\partial y^*}\bigg|_0^{y^*}, \end{aligned} \tag{4.23}$$

where F_x is the total flow rate in the streamwise direction.

The streamwise velocity condition at the bottom wall $(y = 0)$ can be written as

$$\frac{\partial^2\chi}{\partial y^{*2}} = V^*_{x,\, bw}, \tag{4.24}$$

where $V^*_{x,\, bw}$ is the scaled streamwise velocity at the bottom wall.

The zero wall-normal velocity boundary condition at the bottom wall $(y = 0)$ requires that

$$\frac{\partial\chi}{\partial y^*} = 0. \tag{4.25}$$

The streamwise velocity boundary condition at the top wall $(y = H)$ gives,

$$\frac{\partial^2\chi}{\partial y^{*2}} = V^*_{x,\, tw}, \tag{4.26}$$

where $V^*_{x,\, tw}$ is the scaled streamwise velocity at the top wall.

The zero wall-normal velocity boundary condition at the top wall $(y = H)$ requires that

$$\frac{\partial\chi}{\partial y^*} = 0. \tag{4.27}$$

At the top wall $(y = H)$, the condition for the total streamwise mass flow rate (equation 4.23) can be used to obtain a boundary condition for the master potential χ. Since $(\partial\chi/y^*) = 0$ at the top wall $(y = H)$, the total flow rate condition reduces to,

$$\frac{\partial\chi^*}{\partial y^*} = F^*, \tag{4.28}$$

where $F^* = (FH^2)$ is the scaled streamwise flow rate.

Analytical solutions can be obtained for the generalized analytical equation (4.18), in the limit $\mathrm{Re} \ll 1$, when there is a moving bottom wall and a net streamwise flow of material. The most general solution for the equation, (4.18), is,

$$\chi = \left(\frac{C_0}{24} y^{*4} + \frac{C_1}{6} y^{*3} + \frac{C_2}{2} y^{*2} + C_3 y^* + C_4 \right). \tag{4.29}$$

5. DSMC simulation

The direct simulation Monte Carlo method (DSMC) is a numerical method for modeling rarefied gas flows. Here, the Boltzmann equation for the probability distribution function for gas molecules is solved using a probabilistic (Monte Carlo) simulation technique. The Boltzmann equation is more fundamental than the Navier-Stokes equations, since it describes the evolution of the velocity distribution function of the gas molecules. The Navier-Stokes equations are obtained by taking the first, second and third moments of the Boltzmann equation with respect to the molecular velocities. While taking moments, the assumption of 'local equilibrium' (velocity distribution function close to the Maxwell-Boltzmann distribution) is made, thereby restricting the Navier-Stokes equations to near-equilibrium flows. Since there is no assumption of local equilibrium in the Boltzmann equation, it is valid even for far-from equilibrium flows, provided the molecular chaos approximation is satisfied.

The DSMC method uses 'simulated molecules' which represent a large number of real gas molecules. A basic assumption is that the inter-molecular collisions and molecular motion between collisions can be decoupled over time scales comparable to the time between collisions. The translation of molecules between collisions is deterministically modeled, while the inter-molecular collisions and molecular-surfaces are incorporated using probabilistic models. The domain is divided into a large number of cells, and properties are evaluated in these cells. The cells are further divided into 'sub-cells' for collision localization.

In the DSMC simulations, inter-molecular collisions are executed at random from all molecules within 'sub-cells' within the simulation cell, whose linear dimension is much smaller than the mean free path. The probability of collisions between a pair of molecules within a sub-cell is selected equal to that for a dilute gas with the same temperature as that in the cell. In this manner, the temperature (mean square velocity), mean free path and collision frequency for the simulated molecules are equal to those for the real molecules, provided the sub-cell dimension is much smaller than the mean free path.

The most common model is the hard-sphere model, but other models such as the variable hard sphere and variable soft sphere models are also used in order to accurately reproduce the density and temperature dependence of the viscosity. The hydrodynamic fields (density, velocity, temperature) are obtained by averaging over the molecular distribution functions in the DSMC simulations.

At the boundaries, it is necessary to prescribe interaction laws between the molecules and the solid surfaces. Accurate specification of the interaction laws requires a detailed knowledge of the molecule-surface interaction. However, there are standard interaction laws based on the surface velocity and temperature which are sufficient for predicting the hydrodynamic fields accurately. The most commonly used is the 'diffuse reflection' conditions, where a molecule incident on the surface gets reflected with a Gaussian velocity distribution with the same mean velocity and temperature as that of the surface. Another commonly used condition is the specular reflection condition, that the velocity normal to the surface is reversed, while that tangential to the surface is unchanged. It is important to note that these boundary conditions are different from the no-slip conditions in the continuum models. In high Knudsen number flows, the diffuse and specular reflection conditions do predict a slip velocity at the surface. Therefore, these simulations could be used even in situations where there is a slip at the surface.

From kinetic theory of dilute gases Chapman & Cowling (1970), the viscosity of a hard-sphere gas in the dilute limit is given by,

$$\mu = \frac{5\sqrt{mk_BT}}{16\sqrt{\pi}d^2}, \tag{5.1}$$

where k_B is the Boltzmann constant, while the speed of sound c_s is given by the relation

$$mnc_s^2 = \gamma p = \gamma n k_B T, \tag{5.2}$$

where m is the molecular mass, n is the number density, and p is the gas pressure. From this the Mach number (Ma), Reynolds number(Re), channel height (H), molecular diameter(d) and volume fraction of the molecules(ϕ) are related by the following equation,

$$\mathrm{Re} = \frac{96}{5} \times \frac{1}{\sqrt{\pi}} \mathrm{Ma}\, \phi \, \frac{H}{d}, \tag{5.3}$$

where the volume fraction of the molecules is given by

$$\phi = \frac{\pi}{6} d^3 n, \tag{5.4}$$

where n is the number density of the molecules. The important point to note, from 5.3, is that the Reynolds number is proportional to the Mach number times $\phi(H/d)$, where H is the macroscopic scale. Therefore if we are keeping the volume fraction (density) a constant, increasing (decreasing) the Reynolds number at constant Mach number corresponds to increasing (decreasing) the system size H in relation to the molecular diameter.

The Reynolds number and Mach number are also related to the Knudsen number, which is the ratio of the macroscopic scale H and the mean free path $(1/\sqrt{2}\,\pi\, n_d\, d^2)$, as follows,

$$\mathrm{Kn} \propto \frac{\mathrm{Ma}}{\mathrm{Re}}. \tag{5.5}$$

Therefore, the flow is in the continuum limit (Knudsen number is small) if the Reynolds number is large compared to the Mach number. In the present analysis, we will be working in the small to large Knudsen number limit (Kn in the range (0.1 − 10). Therefore the non-continuum effects are negligible in the present analysis.

The details of the DSMC simulation procedure are discussed in (Bird (1963) and Bird (1994)), and so we do not repeat the discussion here, but restrict attention to the parameters in the simulations. The number of cells along the streamwise direction is 200, while there are 100 cells along the wall-normal direction. The cell width is no greater than one-third of the local mean free path. The number of subcells per cell is taken as 4 for both streamwise and cross-streamwise directions. The number of simulated particles per cell is taken as 100. In our simulations, the number of real molecules per simulated particles varied between 10^{16} and 10^{18} so that the statistical fluctuations not becomes large.

The direct simulation Monte Carlo method (DSMC) is used here to simulate the gas flow in the two-dimensional ($x-y$) plane inside a channel. The initial binary gas mixture contains 50% mole-fraction of each component with $\epsilon_a = (2\Delta m/(m_1+m_2)) = 0.01$, and 0.5. The fluid is a hard sphere gas mixture in which both species have equal molecular diameter $d = 5.5\times10^{-10}$ m, with average molecular mass $m_{av} = (m_1+m_2)/2 = 58.4426\times 10^{-26}$ kg/molecule, specific heat ratio $\gamma = 1.066$, and Prandtl number Pr $= 0.95$. It is

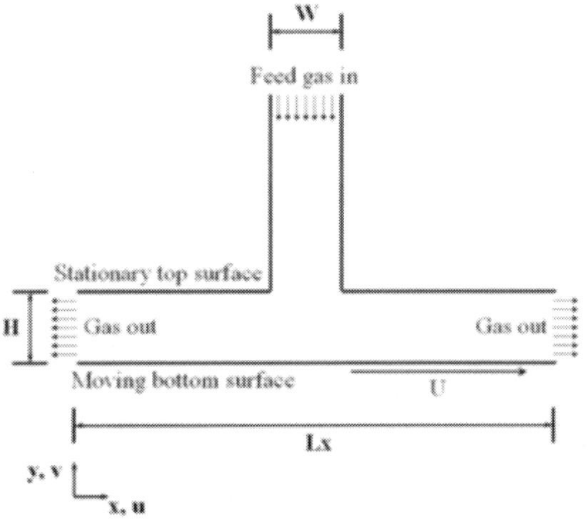

FIGURE 1. Computational domain and coordinate system for 2D channel flow simulations. The stationary walls are at $y = \pm H$. Periodic boundary condition is used in the streamwise direction.

assumed that the gas molecules undergo diffuse reflection from all the solid surfaces with the thermal accommodation coefficient equal to 1. The wall temperature is maintained at 300 K.

The time step in the simulation is chosen to be small compared to the time between collisions. The mean collision time is calculated as, $t = (\lambda / c_{av})$, where λ is the mean free path and c_{av} is the average molecular velocity. The minimum value of mean free path at the rotating sidewall, calculated from the gas density at the wall, is $\lambda = 2.316742 \times 10^{-7}$ m. The average molecular velocity is calculated as $\sqrt{k_B T/m} = 84.17717$ m/sec. This gives the mean collision time (t) $= 2.752221 \times 10^{-9}$ sec. The time step used in the simulations is 5.0×10^{-10} sec, which is less than mean collision time. The flow geometry and coordinate system are shown in figure 1.

First, we validate the DSMC simulation results at subsonic flow conditions (Ma = 0.1, 0.2, 0.3), and Kn = 0.1, with the finite volume method (FVM) results as shown in figure 2. This shows that the present DSMC simulations are in agreement with the finite volume method (FVM) results for the low-Mach number and low-Knudsen-number limit.

In order to check the convergence of the simulations, we have varied both the time step and the number of simulated molecules per cell, keeping the number of real molecules a constant. In all cases, the time step is decreased by a factor of 5 and the number of simulated molecules increased by a factor of 2, and verify that there is less than 1 % change in the resulting flow profiles. The comparisons for a streamwise velocity profile with Ma = 30, and Kn = 0.1 is shown in figure 3. Different time steps (1×10^{-10}, 5×10^{-10}, 1×10^{-9} sec) and different numbers of simulated molecules per cell (100, 80, 60) are used to test the convergence. It is found that the results are insensitive to the time step and the number of simulated molecules per cell. This clearly shows the convergence of the simulation results. Similar results are obtained for low Mach numbers also (Ma = 1, 3, 5, 10), but we do not provide the details.

Next, we compare the DSMC predictions for the mean free path and the collision frequency, which are molecular properties, with those predicted by theory. It is important to verify that the predictions of the DSMC simulations are in agreement with theoretical predictions. The mean free path, which is the average distance traveled by a molecule

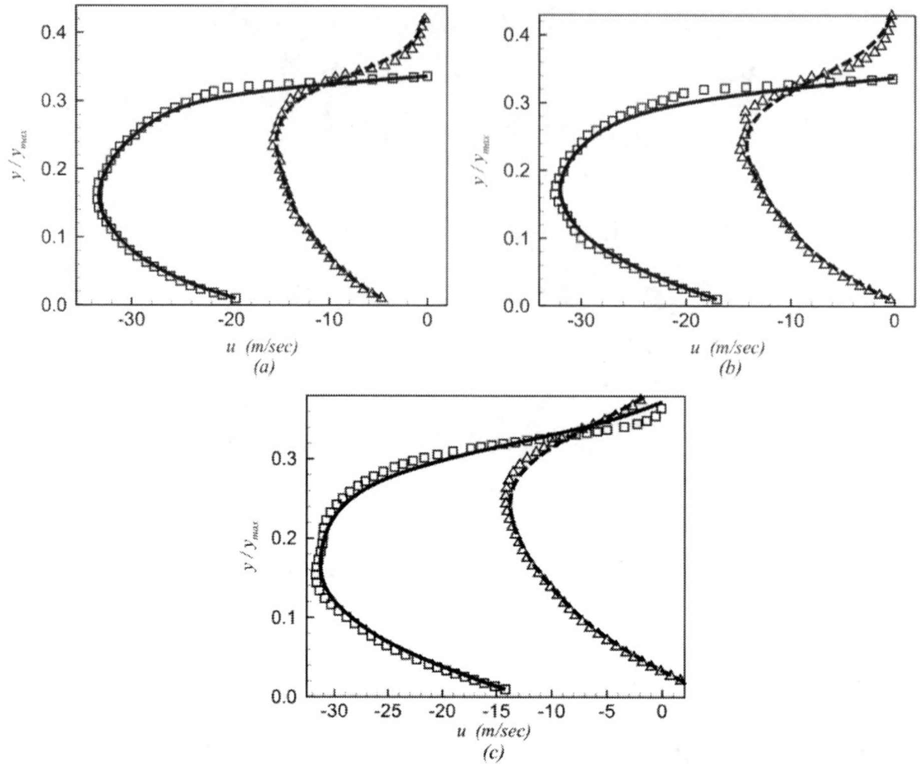

FIGURE 2. The streamwise velocity (u), scaled by wall velocity (U), as a function of y/y_{max}, for Mach numbers Ma = 0.1 (a); Ma = 0.2 (b); and Ma = 0.3(c), at $Kn = 0.1$, for $\epsilon_a = 0.5$. The initial binary gas mixture contains 50% mole-fraction of each component. The symbols show the DSMC results for : ($x/x_{max} = 0.2$ (□); $x/x_{max} = 0.4$ (△)). The lines show the 2D-FVM results for : $x/x_{max} = 0.2$ (—); $x/x_{max} = 0.4$ (– –)).

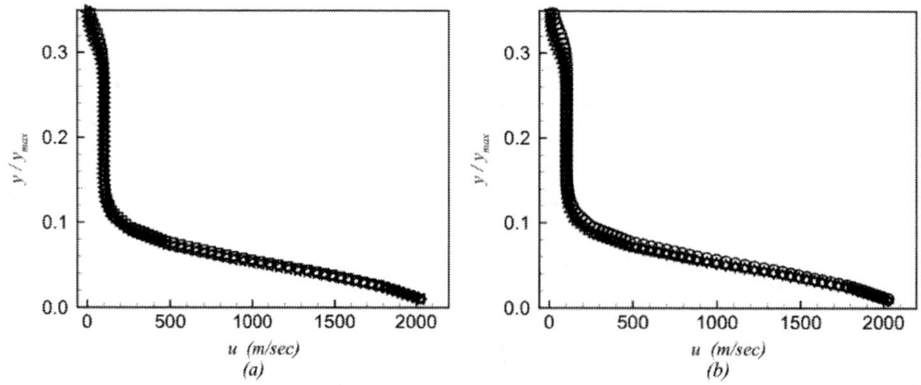

FIGURE 3. The streamwise velocity (u), scaled by wall velocity (U), as a function of y/y_{max}, for Mach number Ma = 30, $Kn = 0.1$, $\epsilon_a = 0.5$. The initial binary gas mixture contains 50% mole-fraction of each component. Sub-figure (a) for different time steps (t) in sec. The symbols show the DSMC results for : ($t = 5 \times 10^{-10}$(□), $t = 1 \times 10^{-9}$(◁), $t = 1.5 \times 10^{-9}$(▷)). Sub-figure (b) for different number of simulated molecules per cell (n_s). The symbols show the DSMC results for : ($n_s = 100$ (○), $t = 80$ (▽), $t = 60$ (△)).

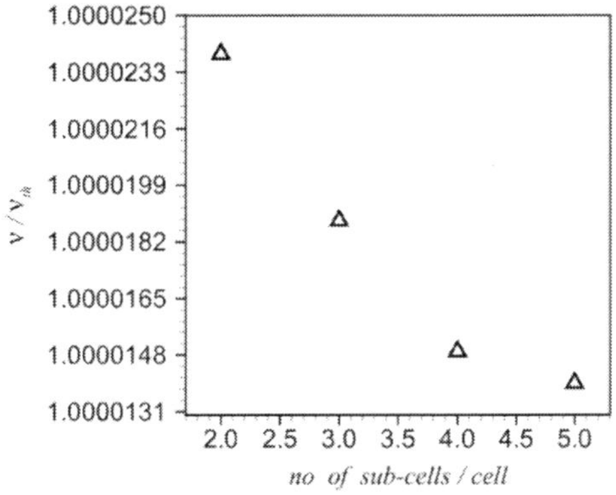

FIGURE 4. The ratio of collision frequency obtained from DSMC simulations (ν) and the theoretical value (ν_{th}), as a function of the number of sub-cells per cell at the location $((x/x_{max}) = 0.8)$, $((y/y_{max}) = 0.1)$, for Kn = 0.1 and Ma = 30 ($\triangle$). The grid resolution used is $(N_x : N_y) = (200 : 100)$.

between two successive collisions, is defined in a frame of reference moving with the mean velocity of the gas. For a hard sphere gas, the mean free path is given by $\lambda = (\sqrt{2}\pi n d^2)^{-1}$, where n is the number density and d is the molecular diameter. In the simulations, it is found that the collision frequency and theoretical value of the mean free path based on the Maxwell-Boltzmann molecular velocity distribution are in close agreement with the actual values based on the distance traveled by the molecule between two successive collisions. However, it is important to note that the collision probabilities in the DSMC simulation have to be calculated based on the local mean temperature, and not the global average temperature or the wall temperature. For the sub-sonic laminar flows, where the temperature is nearly a constant, the average temperature is used for calculating the collision probabilities. In our high Mach number (Ma = 30) DSMC simulations, it is found that the mean free path is not in agreement with the theoretical value if the global average temperature is used for calculating the collision frequency. However, good agreement is obtained if the local mean temperature is used.

The ratio of the actual collision frequency ν and the theoretical collision frequency $\nu_{th} = 4 n_d \, d^2 \sqrt{\pi k_B T/m}$ is shown in figure 4. In this figure, the ratio is shown as a function of the number of equally sized sub-cells per cell within which the collision partners are chosen. Increasing the number of sub-cells per cell localises the collision to a smaller region within the cell. Clearly, from figure 4, increasing the number of sub-cells does not alter the collision frequency significantly, and so it is sufficient to use 2^2 sub-cells per cell in our simulations to capture the collision frequency accurately for laminar flows.

In figure 5, the ratio of theoretical mean free path ($\lambda_{th} = 1/\sqrt{2}\pi d^2 n_d$) and mean free path based on actual distance travel by molecules between collisions obtained in the simulations (λ) is plotted as a function of Mach number. There is some ambiguity in assigning the local mean free path in the simulations, since the mean free path for a molecule between successive collisions could be assigned to either the beginning, the end or the mid-point of the trajectory of the molecule. All three choices are shown in figure 5, and it is observed that there is little difference between the three choices even for the most extreme case simulated here, which is a laminar flow at Kn = 0.1 and Ma = 30. Further the actual mean free path in the DSMC simulations is in excellent agreement with the theoretical value based on the local number density. Therefore, we calculate the

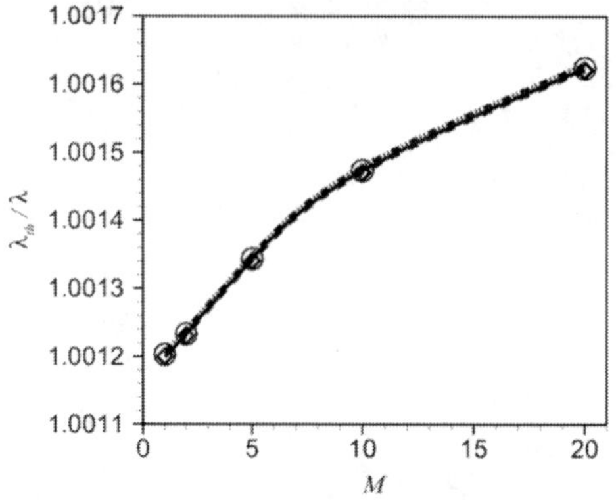

FIGURE 5. The ratio of the theoretical mean free path (λ_{th}) and average distance traveled by the molecules between collision in the simulations (λ) as a function of Mach number for a simulation with 16 sub cells per cell at the location $((x/x_{max}) = 0.8)$, $((y/y_{max}) = 0.1)$, for Mach number in the range $1 - 30$. The grid resolution used is $200 : 100$. For Kn $= 0.1$, Ma $= 30$. The lines show the DSMC results for assigning the mean free path to the cell at the center of the particle trajectory (—), at the start of the trajectory (– –), and at the end of the trajectory (· · – · ·).

mean free path based on the theoretical formula for a Maxwell-Boltzmann distribution in the remainder of the analysis.

6. Results

We consider the binary gas mixture for $\epsilon_a = (2\Delta m/(m_1 + m_2)) = 0.01$, and 0.5. Here, Δm is the difference in the molecular masses of the two species (m_1-m_2). The fluid is a hard sphere gas mixture in which both species have equal molecular diameter $d = 5.5 \times 10^{-10}$ m, with average molecular mass $m_{av} = (m_1 + m_2)/2 = 58.4426 \times 10^{-26}$ kg/molecule, specific heat ratio $\gamma = 1.066$, and Prandtl number Pr$= 0.95$. It is assumed that the gas molecules undergo diffuse reflection from all the solid surfaces with the thermal accommodation coefficient equal to 1. Simulations were carried out for Mach number in the range $(0.1 - 30)$, and Knudsen number in the range $(0.1 - 10)$. The flow geometry and coordinate system are shown in figure 1.

6.1. *Two-dimensional flow at low Mach number*

Before proceeding to simulate the high Mach number flows, first, the DSMC simulations are carried out at subsonic flow conditions (Ma $= 0.1, 0.2, 0.3$), at Kn $= 0.1$, for $\epsilon_a = (2\Delta m/(m_1 + m_2)) = 0.01$, and 0.5. The gas is introduced to the channel from the top at $y = H$. The gas flows in the streamwise direction due to the motion of the bottom channel wall, and splits into two streams, one going along positive x direction and the other going along the negative x direction. At the two end ($x = 0$, and L), The gas is taken out at two locations along the streamwise direction, one at the ($x = 0$) and other at ($x = L$). Here, L is the total length of the channel.

In the case of inflow/outflow of gas along the channel boundaries, the dimensionless scaled feed rate is defined as,

$$F^* = \frac{F}{(H_g/t_c)}, \tag{6.1}$$

where F is the feed rate, H_g is the total gas holdup and t_c is the characteristic time $t_c = L/U$, where, L is the total length of the channel and U is the bottom wall velocity.

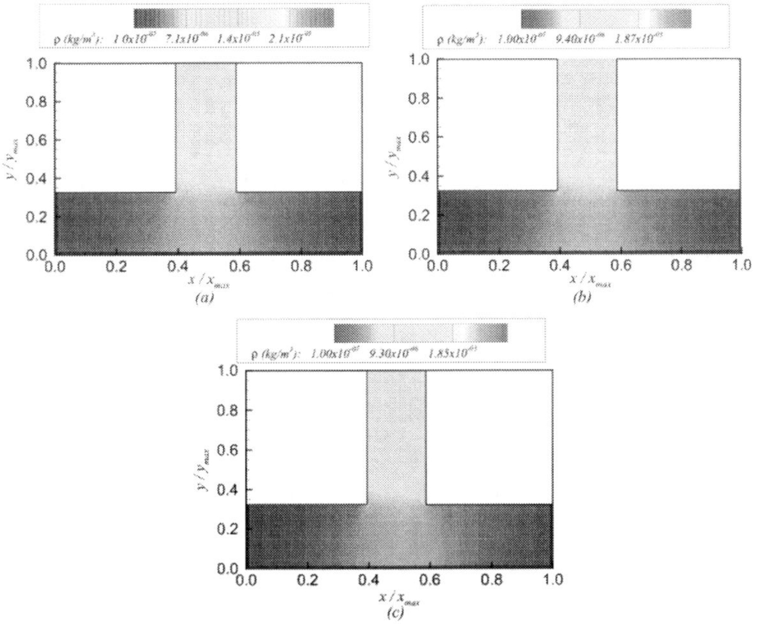

FIGURE 6. The density contours for Mach numbers Ma = 0.1 (a), Ma = 0.2 (b), Ma = 0.3 (c); at Kn = 0.1, for $\epsilon_a = 0.01$.

The scaled feed rate is a measure of the perturbation to the base flow due to external feed. Here, we have carried out simulations for $F^* = 0.2$. The actual feed rate (F) is varied in such a way that the desired scaled feed rate is obtained at different Mach and Knudsen numbers.

The contours of the steady state density, streamwise ($x - component$), and wall-normal ($y - component$) velocity components, streamwise mass flux (ρu), streamlines, and velocity vectors in the flow domain at low Mach numbers Ma = 0.1, 0.2, 0.3, at Kn = 0.1, for $\epsilon_a = 0.01$, are shown in figures 6, 7, 8, 9, and 10 respectively. We find that the gas density decreases as the Mach number is increased. Increasing the Mach number also leads to a increase in the gas temperature across the viscous sub-sayer due to viscous heating.

Next, we study the steady state density profiles along the streamwise direction for sub-sonic flow conditions (Ma = 0.1, 0.2, 0.3), at Kn = 0.1, $\epsilon_a = 0.01$, and 0.5 as shown in figure 11 (a, b, c). It is clear that the density increases significantly, by a factor of 6, when we move along streamwise direction from the ends to the feed location. Figure 11 shows sharp peak in the density at the feed location, due to the high density of the feed gas entering into the channel. We have also studied the grid-size dependence of the results; there is a negligible difference in the results when the grid resolution is increased from (140 : 70) to (200 : 100).

Next, we study the steady state streamwise and wall-normal velocity profiles along the streamwise direction for sub-sonic flow conditions (Ma = 0.1, 0.2, 0.3), at Kn = 0.1, $\epsilon_a = 0.01$, and 0.5 as shown in figure 12, and figure 13. Figure 12 shows a gradual increase in the streamwise velocity along the streamwise direction, and the velocity profile is expected to be the same as that for an isothermal incompressible flow. We find that the streamwise velocity increases as the mass difference between two species in a binary gas

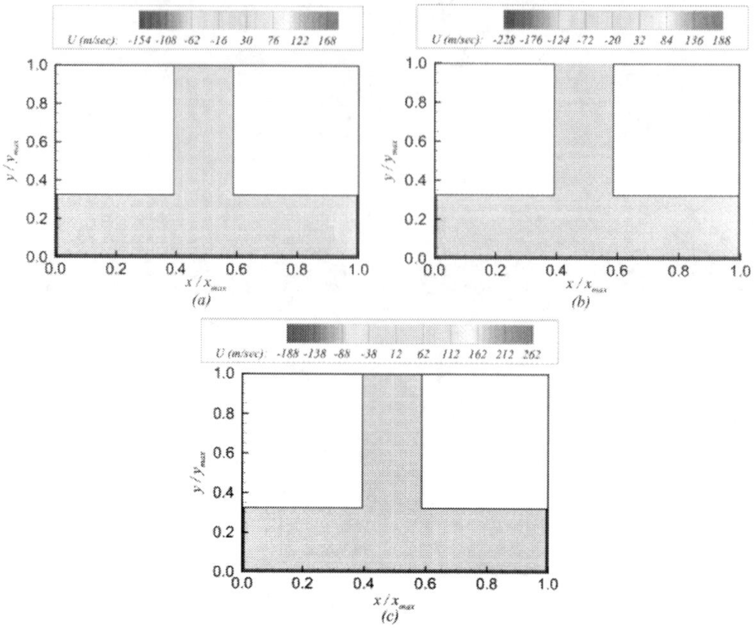

FIGURE 7. The streamwise velocity contours for Mach numbers Ma = 0.1 (a), Ma = 0.2 (b), Ma = 0.3 (c); at Kn = 0.1, for $\epsilon_a = 0.01$.

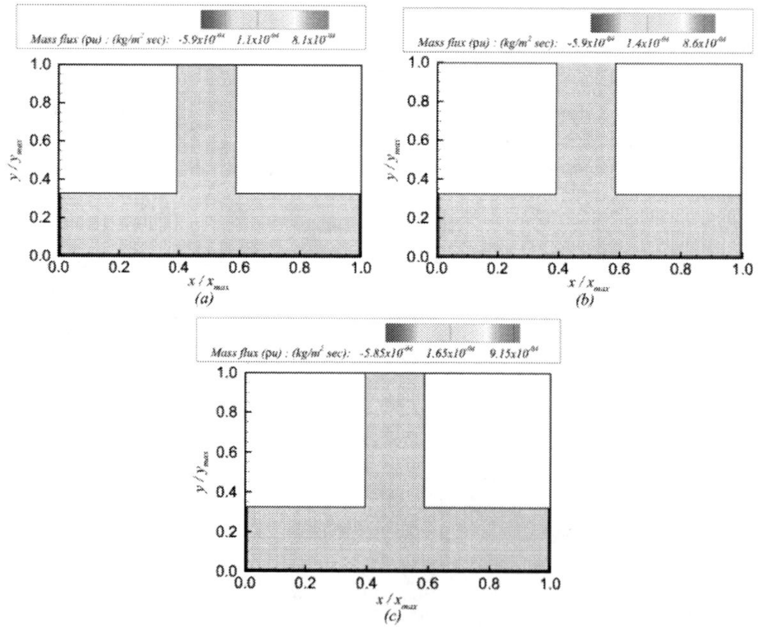

FIGURE 8. The streamwise mass flux contours for Mach numbers Ma = 0.1 (a), Ma = 0.2 (b), Ma = 0.3 (c); at Kn = 0.1, for $\epsilon_a = 0.01$.

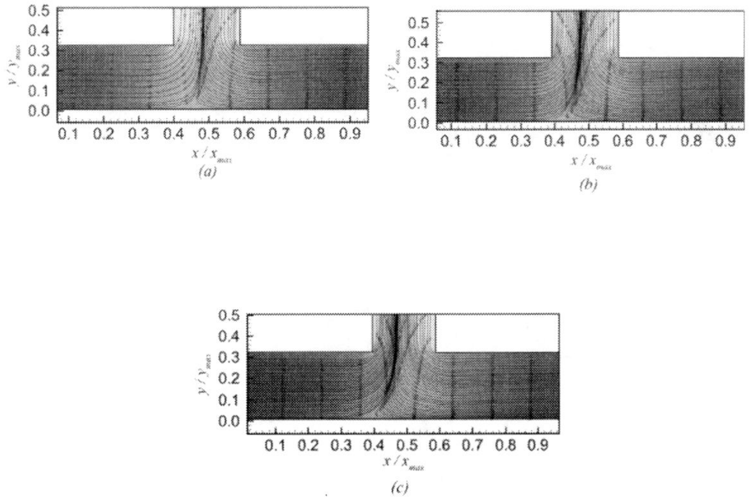

FIGURE 9. The streamlines in the flow domain for Mach numbers Ma = 0.1 (a), Ma = 0.2 (b), Ma = 0.3 (c); at Kn = 0.1, for $\epsilon_a = 0.01$.

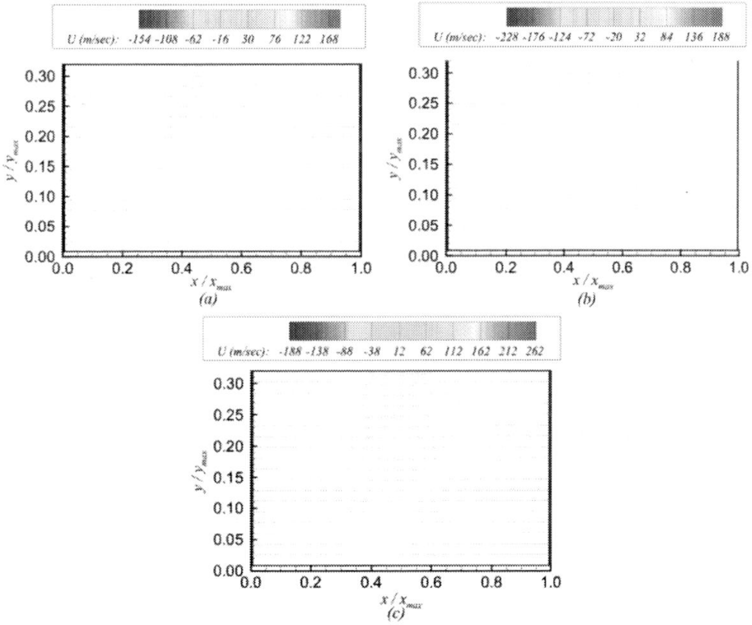

FIGURE 10. The velocity vectors in the flow domain for Mach numbers Ma = 0.1 (a), Ma = 0.2 (b), Ma = 0.3 (c); at Kn = 0.1, for $\epsilon_a = 0.01$.

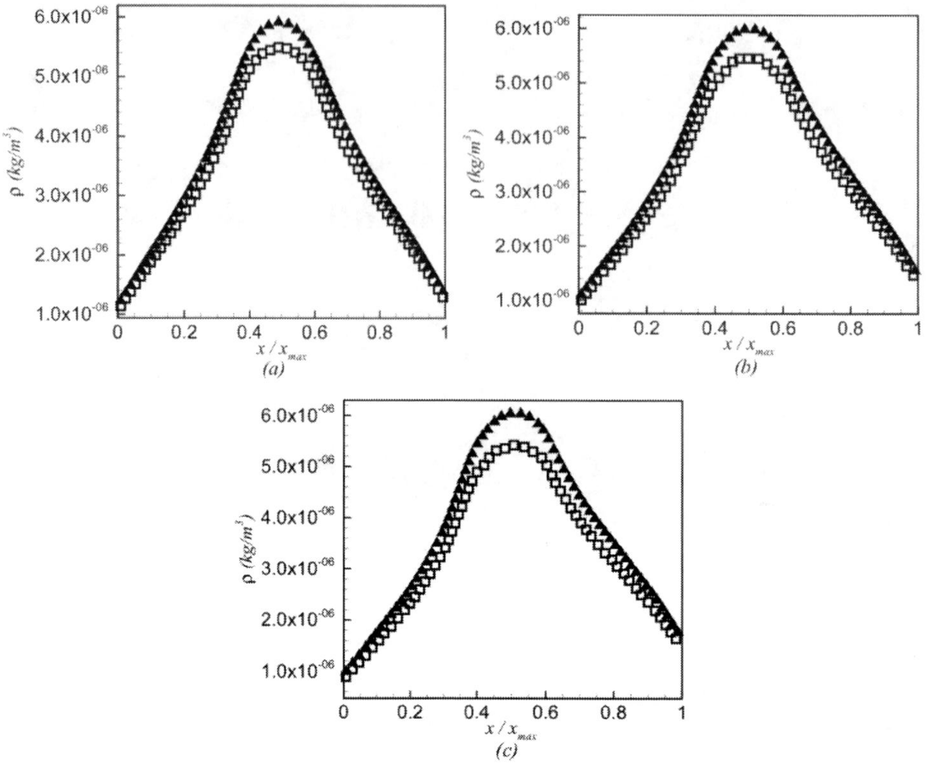

FIGURE 11. The variation of density (ρ) as a function of x/x_{max} at $y/y_{max} = 0.15$, for Kn $= 0.1$, and for Mach numbers Ma $= 0.1$ (a), Ma $= 0.2$ (b), Ma $= 0.3$ (c). The symbols show the DSMC results for : ($\epsilon_a = 0.01$: (□) $\epsilon_a = 0.5$: (▲)).

mixture increases. Figure 13 shows a dip in the wall-normal velocity at the feed location, owing to the inflow of the feed gas into the channel.

The streamwise mass flux profiles (ρu) along the streamwise direction is studied for the sub-sonic flow conditions (Ma $= 0.1, 0.2, 0.3$), at Kn $= 0.1$, $\epsilon_a = 0.01$, and 0.5 as shown in figure 14. Figure 14 shows a sharp increase of the streamwise mass flux (ρu) along the streamwise direction for $x/x_{max} = 0.35 - 0.65$, and is nearly constant for $x/x_{max} = 0 - 0.35$, and $x/x_{max} = 0.65 - 1$. We find that the streamwise mass flux increases as the mass difference between two species in a binary gas mixture increases.

Next, we study the steady state number densities of component-1 (n_1) and component-2 (n_2) along the streamwise direction for sub-sonic flow conditions (Ma $= 0.1, 0.2, 0.3$), at Kn $= 0.1$, $\epsilon_a = 0.01$ as shown in figure 15 (a, b, c). It is clear that the number density increases significantly, by a factor of 12, as we move along streamwise direction from $x/x_{max} = 0$, to $x/x_{max} = 0.5$. Figure 11 shows a peak in the number density at the feed location, due to the high number density of the feed gas entering into the channel.

6.2. *Two-dimensional flow at high Mach number*

Next, DSMC simulations are carried out at high Mach numbers (Ma = 1, 3, 5, 10, 30) for Knudsen number Kn $= 0.1$, $\epsilon_a = (2\Delta m/(m_1 + m_2)) = 0.01$, and 0.5. At the wall, the molecules undergo diffuse reflection with thermal accommodation coefficient equal to 1. This implies that the temperature of the reflected molecules is equal to the

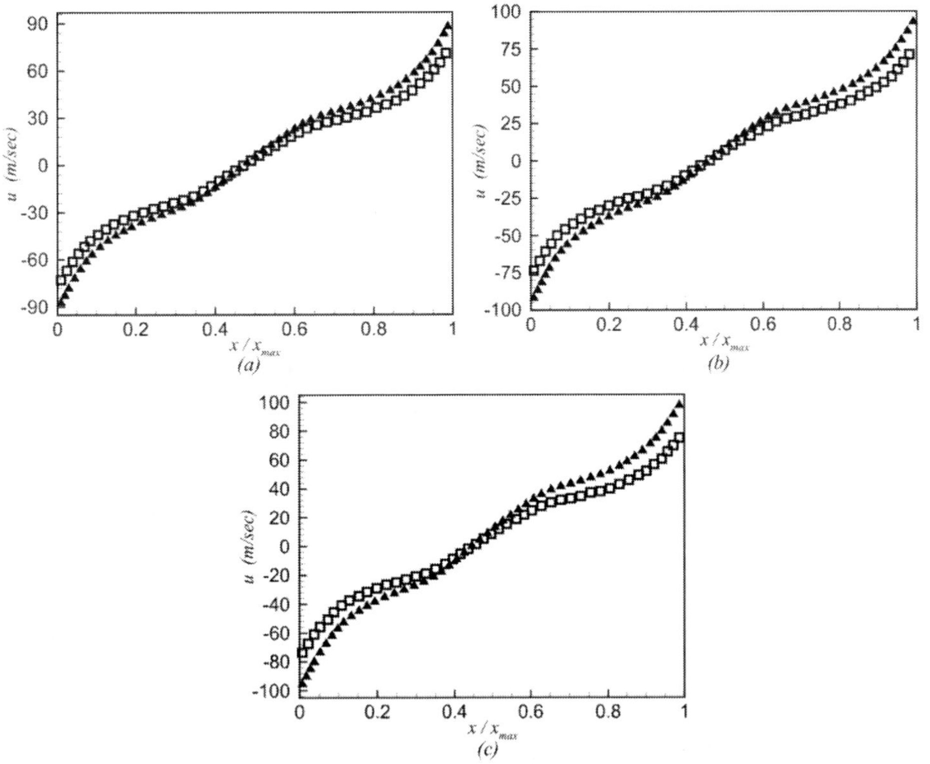

FIGURE 12. The variation of streamwise velocity u) as a function of x/x_{max} at $y/y_{max} = 0.15$, for Kn = 0.1, and for Mach numbers Ma = 0.1 (a), Ma = 0.2 (b), Ma = 0.3 (c). The symbols show the DSMC results for : ($\epsilon_a = 0.01$: (□) $\epsilon_a = 0.5$: (▲)).

wall temperature. However, the gas temperature at the wall (the average of the incident and reflected molecules) is not equal to the wall temperature in the high Mach number limit, since the temperature of the incident molecules could be different. This results in a temperature 'slip' at the wall, similar to the velocity slip, because the velocity of the incident molecules is different from the wall velocity, though that of the reflected molecules is equal to the wall velocity (Pradhan & Kumaran (2011)). We find that the velocity slip increases as the Mach number is increased, as shown in figure 17, and is at a maximum for Mach number Ma = 30.

The existence of slip at the wall is shown in figure 16, where the variation along the walls of the y-component of velocity (v) along x, and x-component of velocity (u) along y, are shown. The velocity slip at the walls (($x/L_x = 0.5$, $y/L_y = 1$) and ($x/L_x = 1$, $y/L_y = 0.5$)), shown in figure 16 (c) increase approximately linearly with Mach number at high Mach number.

Next, we study the variation of shear strain rate ($\partial u/\partial y$) with Mach number Ma at a fixed location ($x/x_{max} = 0.5$, $y/y_{max} = 0$), on the moving wall for Kn = 0.1. There is a monotonic increase in the shear strain rate ($\partial u/\partial y$) as the Mach number is increased, as seen in figure 18.

The contours of the steady state density, streamwise, and wall-normal velocity components, streamlines, and velocity vectors in the flow domain at high Mach numbers (Ma = 1, 3, 5, 10, 30), at Kn = 0.1, for $\epsilon_a = 0.01$, and 0.5 are shown in figures 19, 23, 22,

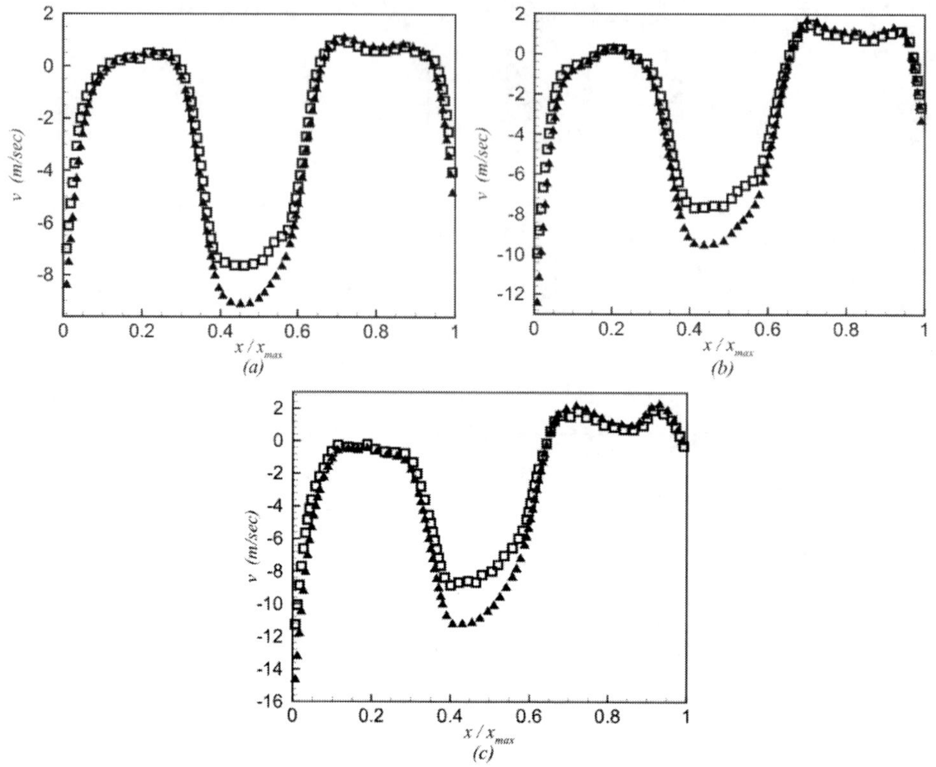

FIGURE 13. The variation of wall-normal velocity (v) as a function of x/x_{max} at $y/y_{max} = 0.15$, for Kn = 0.1, and for Mach numbers Ma = 0.1 (a), Ma = 0.2 (b), Ma = 0.3 (c). The symbols show the DSMC results for : ($\epsilon_a = 0.01$: (□) $\epsilon_a = 0.5$: (▲)).

21, and 22 respectively. We find that the gas density decreases as the Mach number is increased. Increasing the Mach number also leads to a increase in the gas temperature across the viscous sub-sayer due to viscous heating.

Next, we study the steady state density profiles along the streamwise direction at high Mach numbers (Ma = 1, 3, 5, 10, 30), for Kn = 0.1, and $\epsilon_a = 0.01, 0.5$ as shown in figure 24 (a, b, c). It is clear that the density increases significantly, by a factor of 6, when we move along streamwise direction from the ends to the feed location. Figure 24 shows sharp peak in the density at the feed location, due to the high density of the feed gas entering into the channel. We have also studied the grid-size dependence of the results; there is a negligible difference in the results when the grid resolution is increased from (140 : 70) to (200 : 100).

Next, we study the steady state streamwise and wall-normal velocity profiles along the streamwise direction at high Mach numbers (Ma = 1, 3, 5, 10, 30), for Kn = 0.1, and $\epsilon_a = 0.01, 0.5$ as shown in figure 25, and figure 26. Figure 25 shows a gradual increase in streamwise velocity along the streamwise direction, and the velocity profile is expected to be the same as that for an isothermal compressible flow. We find that the streamwise velocity increases as the mass difference between two species in a binary gas mixture increases. Figure 26 shows a dip in the wall-normal velocity at the feed location, owing to the inflow of the feed gas into the channel.

The streamwise mass flux profiles (ρu) along the streamwise direction is studied at high

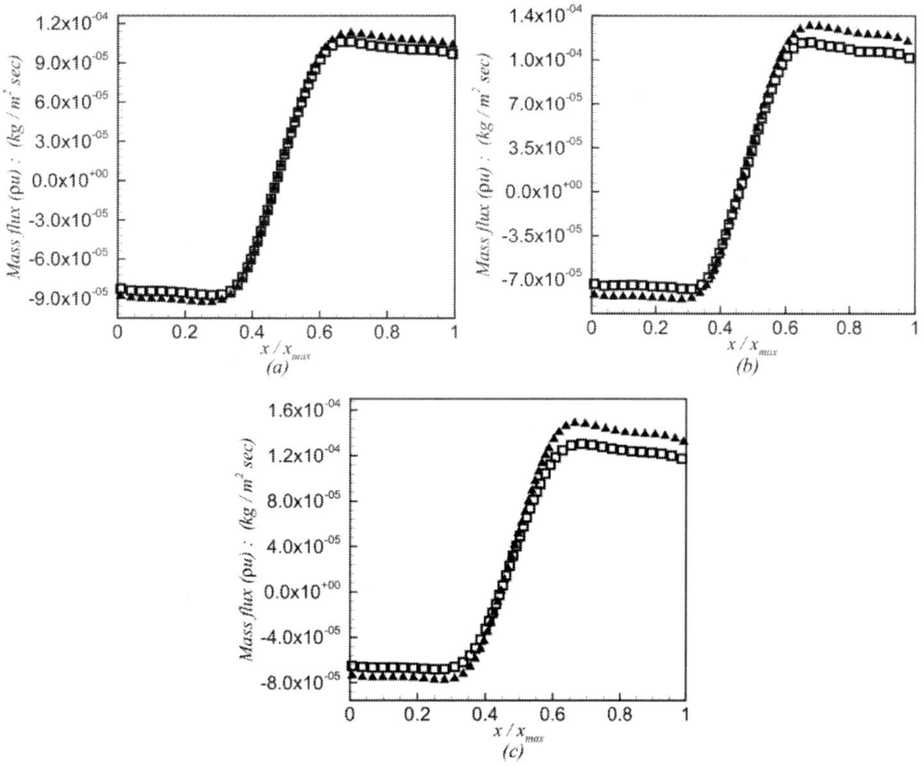

FIGURE 14. The variation of streamwise mass flux (ρu) as a function of x/x_{max} at $y/y_{max} = 0.15$, for Kn = 0.1, and for Mach numbers Ma = 0.1 (a), Ma = 0.2 (b), Ma = 0.3 (c). The symbols show the DSMC results for : ($\epsilon_a = 0.01$: (□) $\epsilon_a = 0.5$: (▲)).

Mach numbers (Ma = 1, 3, 5, 10, 30), for Kn = 0.1, and $\epsilon_a = 0.01$, as shown in figure 27. Figure 27 shows a sharp increase of the streamwise mass flux (ρu) along the streamwise direction for $x/x_{max} = 0.35 - 0.65$, and is nearly constant for $x/x_{max} = 0 - 0.35$, and $x/x_{max} = 0.65 - 1$. We find that the streamwise mass flux increases as the mass difference between two species in a binary gas mixture increases.

The steady state streamwise velocity (u), and streamwise mass flux (ρu), along the wall-normal direction y/y_{max} is plotted at high Mach numbers Ma $= 1, 3, 5, 10, 30$, for $Kn = 0.1$, and $\epsilon_a = 0.01$, 0.5, in figures 28, and 30 respectively. We find that the streamwise velocity (u), and streamwise mass flux (ρu), increases as the mass difference between two species in a binary gas mixture increases.

The streamwise mass flux, (ρu), is plotted as a function of wall-normal coordinate y/y_{max} for different Mach numbers (Ma = 1, 3, 5, 10, 30) and at different streamwise locations (x/x_{max} = 0.2, 0.4, 0.6, 0.8) as shown in figure 30(a, b, c, d, e), for Kn = 0.1, $\epsilon_a = 0.01$. There is quantitative agreement (with a difference of less than 5%) between the predictions of the analytical model and the simulations for Mach numbers Ma = 1, 3, 5, 10, 30, and for Knudsen number Kn = 0.1.

Next, we study some qualitative features of the flow in the central region of the channel. At the outset, it is important to note that there is a significant variation of the gas viscosity with Mach number, due to the variation in the temperature. The variations in

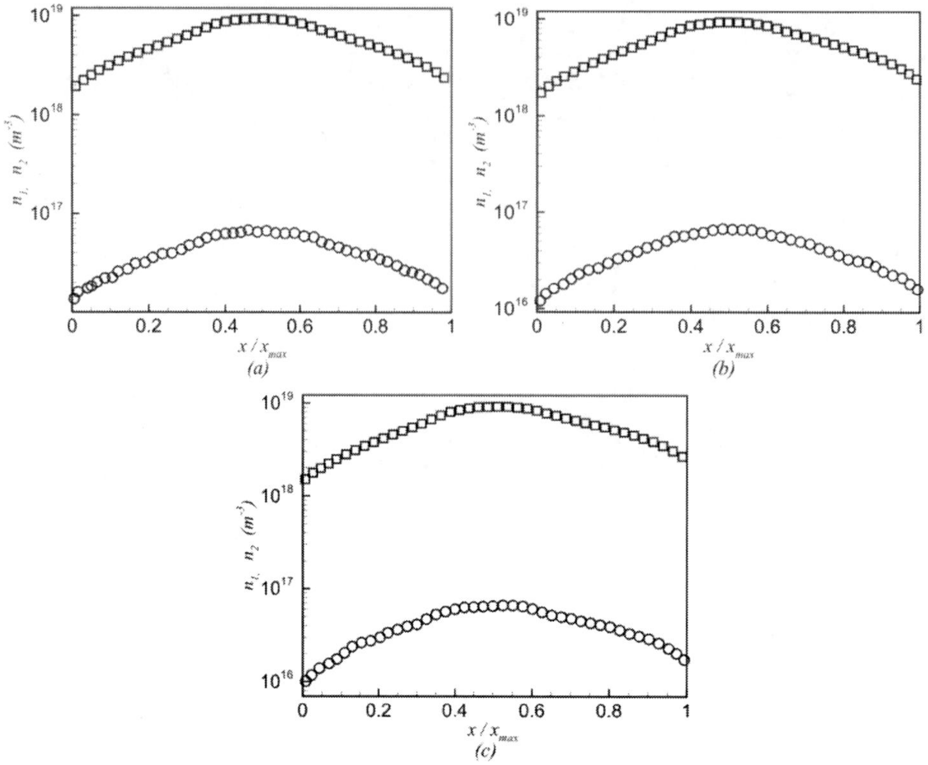

FIGURE 15. The variation of the number density profiles as a function of x/x_{max} at y/y_{max} = 0.15, for Mach numbers Ma = 0.1 (a), Ma = 0.2 (b), Ma = 0.3 (c); at Kn = 0.1, for $\epsilon_a = 0.01$. The symbols show the DSMC results for : ($Component - 1\ N_1$: (○); $Component - 2\ N_2$: (□)).

the time-averaged viscosity, scaled by the corresponding volume averages, are shown in figure 31.

Next, we study the steady state number densities of component-1 (n_1) and component-2 (n_2) along the streamwise direction at high Mach numbers (Ma = 1, 3, 5, 10, 30), for Kn = 0.1, ϵ_a = 0.01, as shown in figure 32 (a, b, c, d, e). It is clear that the number density increases significantly, by a factor of 12, as we move along streamwise direction from $x/x_{max} = 0$, to $x/x_{max} = 0.5$. Figure 32 shows a peak in the number density at the feed location, due to the high number density of the feed gas entering into the channel.

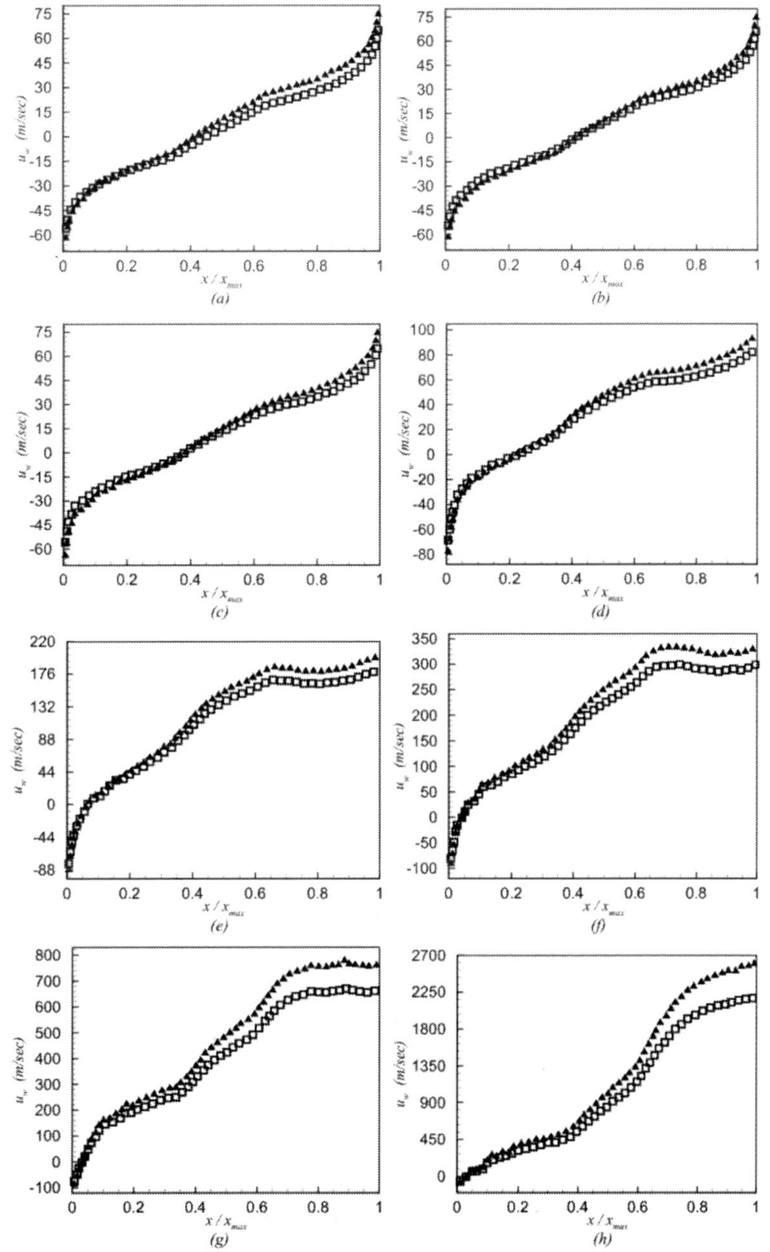

FIGURE 16. Streamwise velocity slip (u_w) at the bottom wall ((y/y_{max}) = 0), as a function of x/x_{max}, for Kn = 0.1, and for : Ma = 0.1 (a); Ma = 0.2 (b); Ma = 0.3 (c); Ma = 1 (d); Ma = 3 (e); Ma = 5 (f); Ma = 10 (g); Ma = 30 (h). The symbols show the DSMC results for : (ϵ_a = 0.01 : (□) ϵ_a = 0.5 : (▲)).

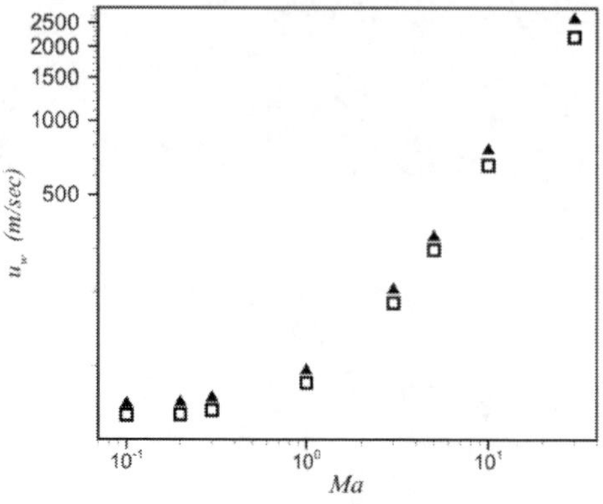

FIGURE 17. Streamwise velocity slip (u_w) at the bottom wall ($(y/y_{max}) = 0$), as a function of Mach number Ma, at $(x/x_{max}) = 1$, for Kn = 0.1. The symbols show the DSMC results for : ($\epsilon_a = 0.01$: (□) $\epsilon_a = 0.5$: (▲)).

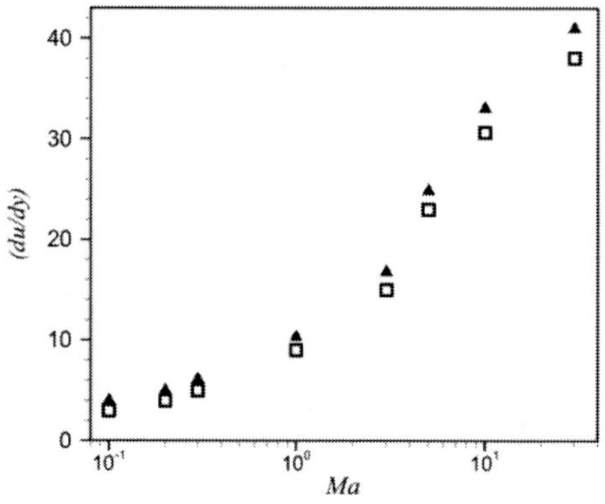

FIGURE 18. Variation of shear strain rate ($\partial u/\partial y$) with Mach number Ma on the moving wall at ($x/x_{max} = 0.5$, $y/y_{max} = 0$), for Kn = 0.1. The symbols show the DSMC results for : ($\epsilon_a = 0.01$: (□) $\epsilon_a = 0.5$: (▲)).

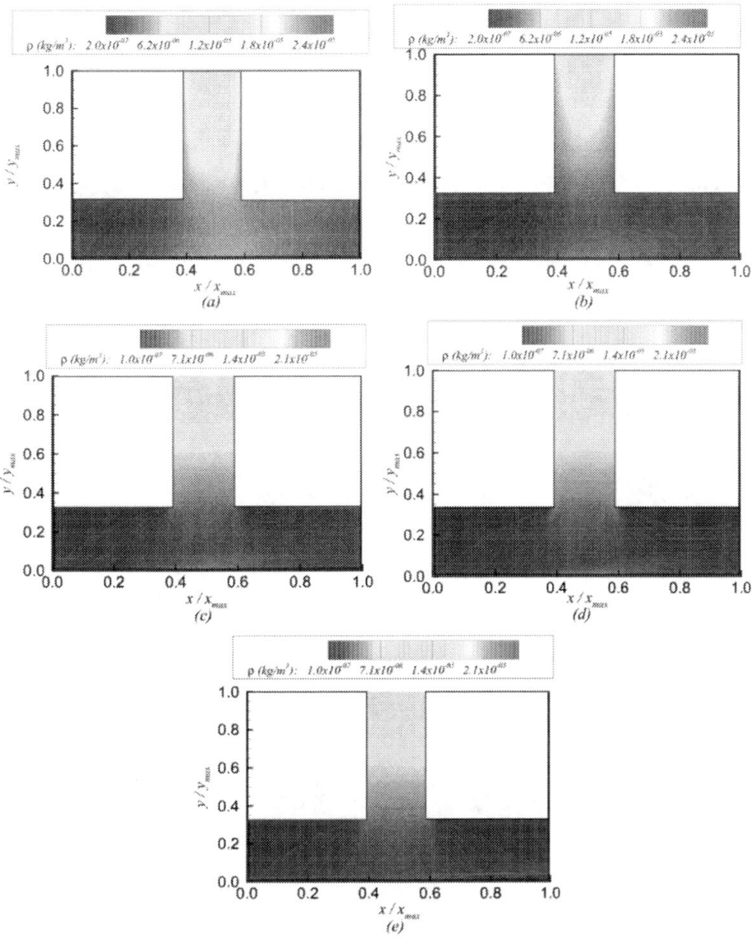

FIGURE 19. The density contours for Mach numbers Ma = 1 (a), Ma = 3 (b), Ma = 5 (c), Ma = 10 (d), Ma = 30 (e); at Kn = 0.1, for $\epsilon_a = 0.01$.

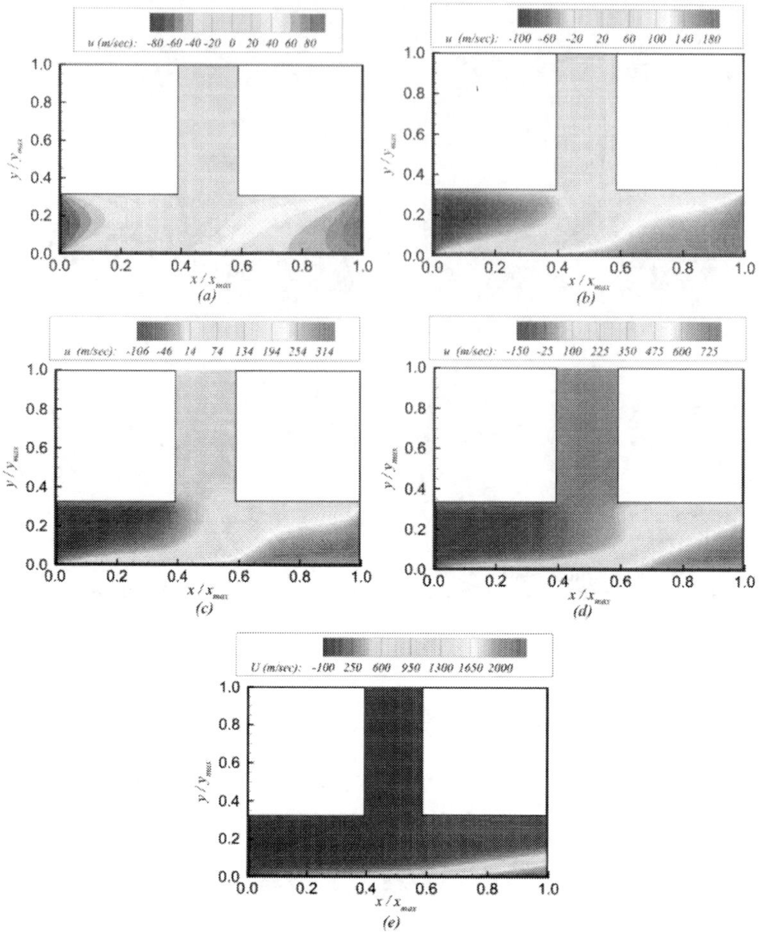

FIGURE 20. The streamwise velocity contours for Mach numbers Ma = 1 (a), Ma = 3 (b), Ma = 5 (c), Ma = 10 (d), Ma = 30 (e); at Kn = 0.1, for $\epsilon_a = 0.01$.

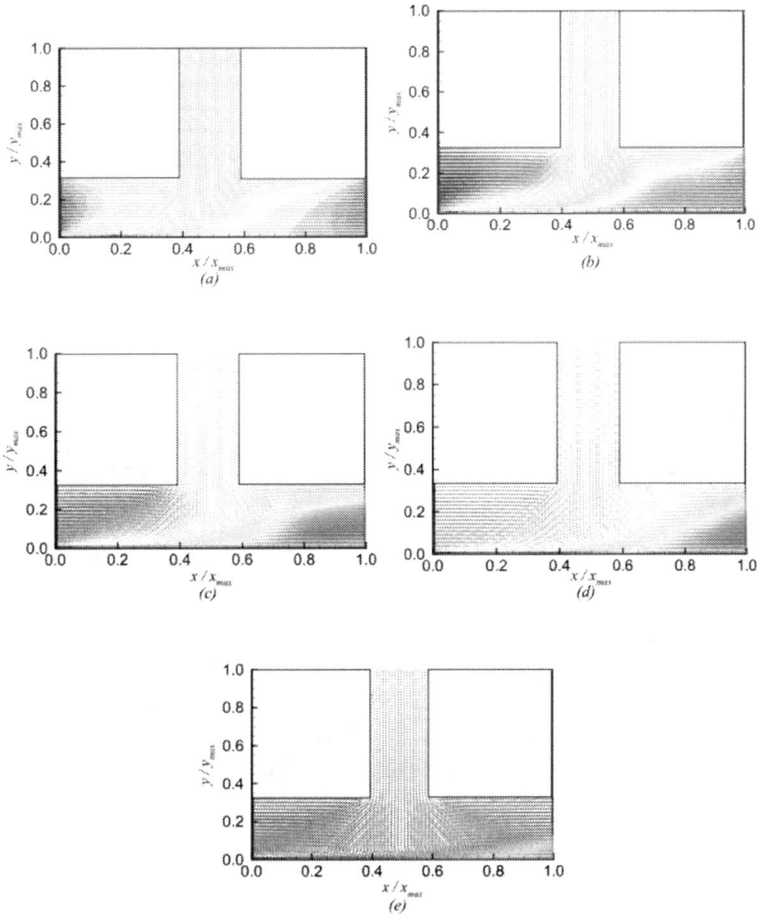

FIGURE 21. The velocity vectors in the flow domain for Mach numbers Ma = 1 (a), Ma = 3 (b), Ma = 5 (c), Ma = 10 (d), Ma = 30 (e); at Kn = 0.1, for $\epsilon_a = 0.01$.

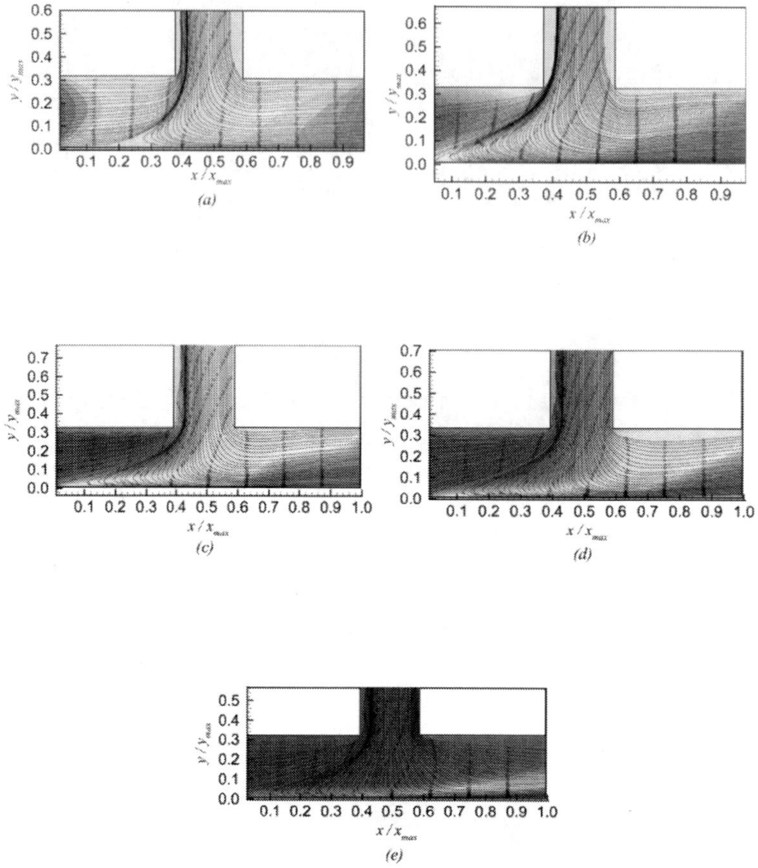

FIGURE 22. The streamlines in the flow domain for Mach numbers Ma = 1 (a), Ma = 3 (b), Ma = 5 (c), Ma = 10 (d), Ma = 30 (e); at Kn = 0.1, for $\epsilon_a = 0.01$.

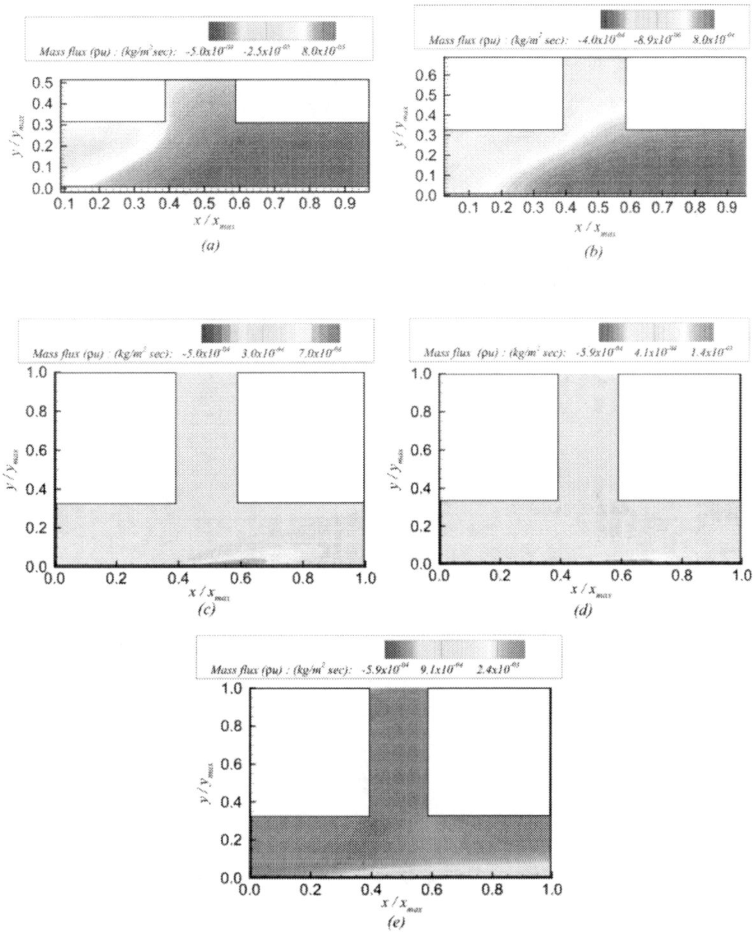

FIGURE 23. The streamwise mass flux (ρu) contours for Mach numbers Ma = 1 (a), Ma = 3 (b), Ma = 5 (c), Ma = 10 (d), Ma = 30 (e); at Kn = 0.1, for $\epsilon_a = 0.01$.

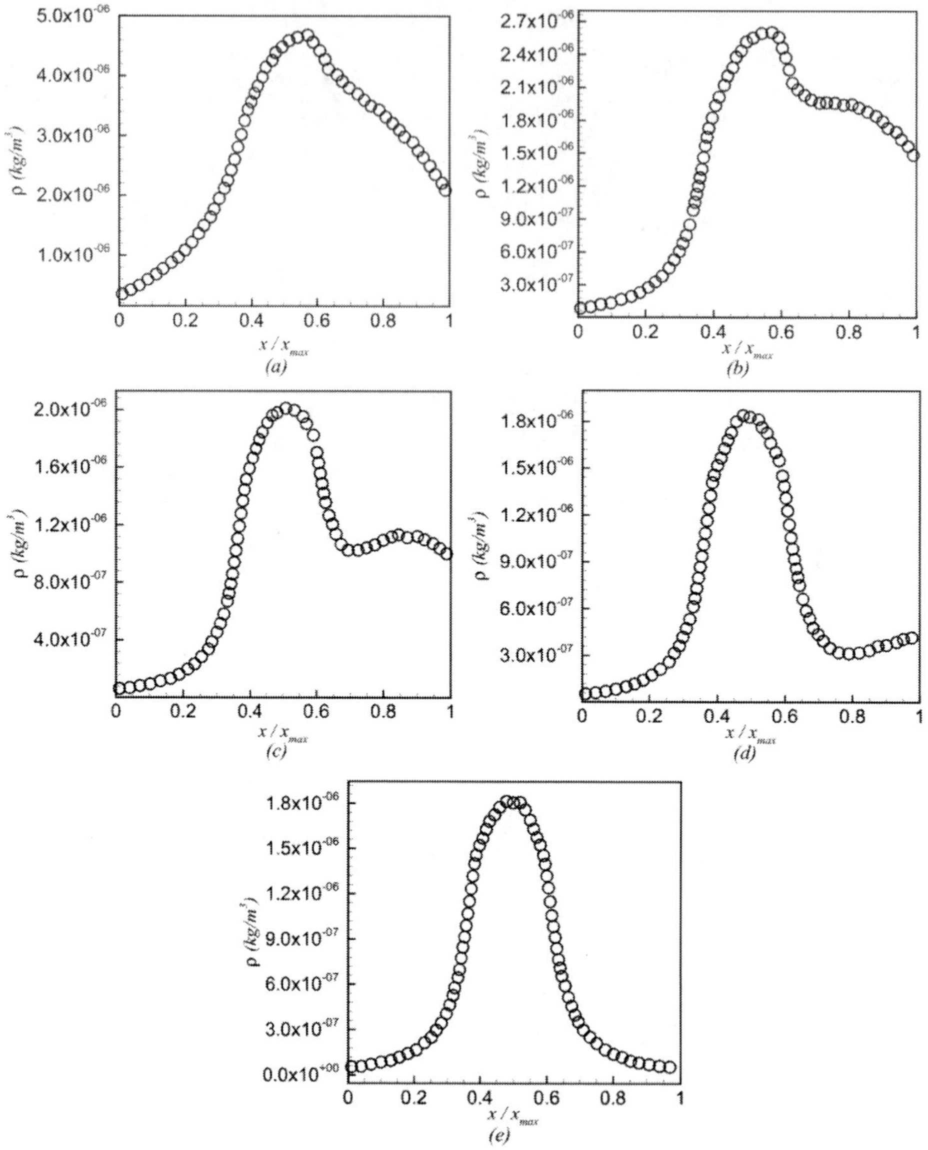

FIGURE 24. The variation of the density (ρ) as a function of x/x_{max} at $y/y_{max} = 0.15$, for Kn = 0.1, and for Mach numbers Ma = 1 (a), Ma = 3 (b), Ma = 5 (c), Ma = 10 (d), Ma = 30 (e). The symbols show the DSMC results for : ($\epsilon_a = 0.01$: ($\circ$)).

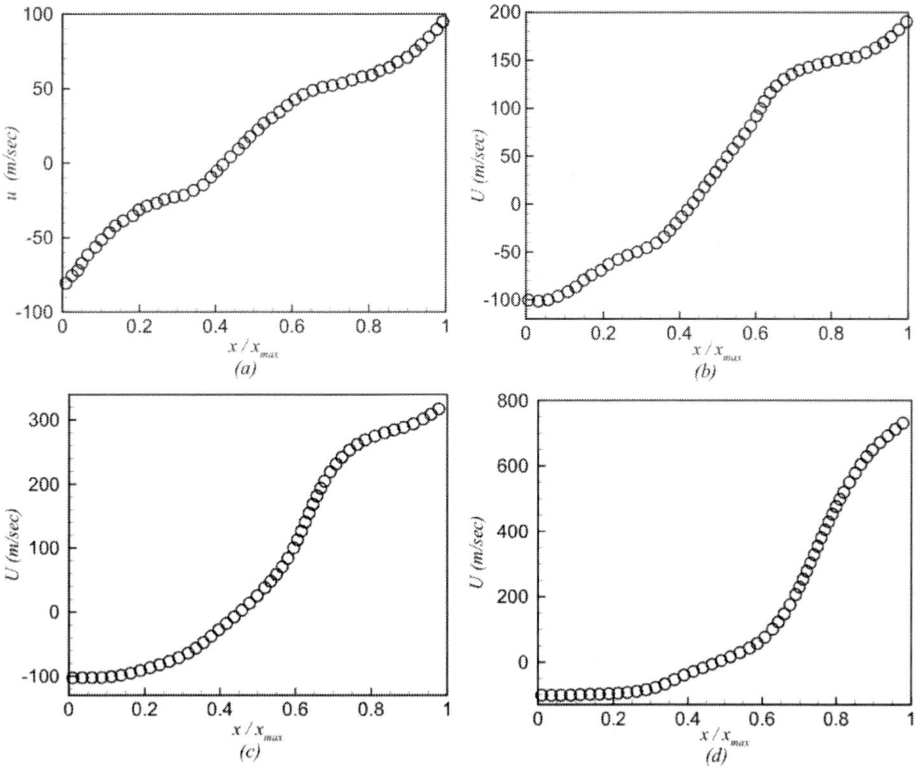

FIGURE 25. The variation of the streamwise velocity (u) as a function of x/x_{max} at $y/y_{max} = 0.15$, for Kn = 0.1, and for Mach numbers Ma = 1 (a), Ma = 3 (b), Ma = 5 (c), Ma = 10 (d), Ma = 30 (e). The symbols show the DSMC results for : ($\epsilon_a = 0.01$: ($\circ$)).

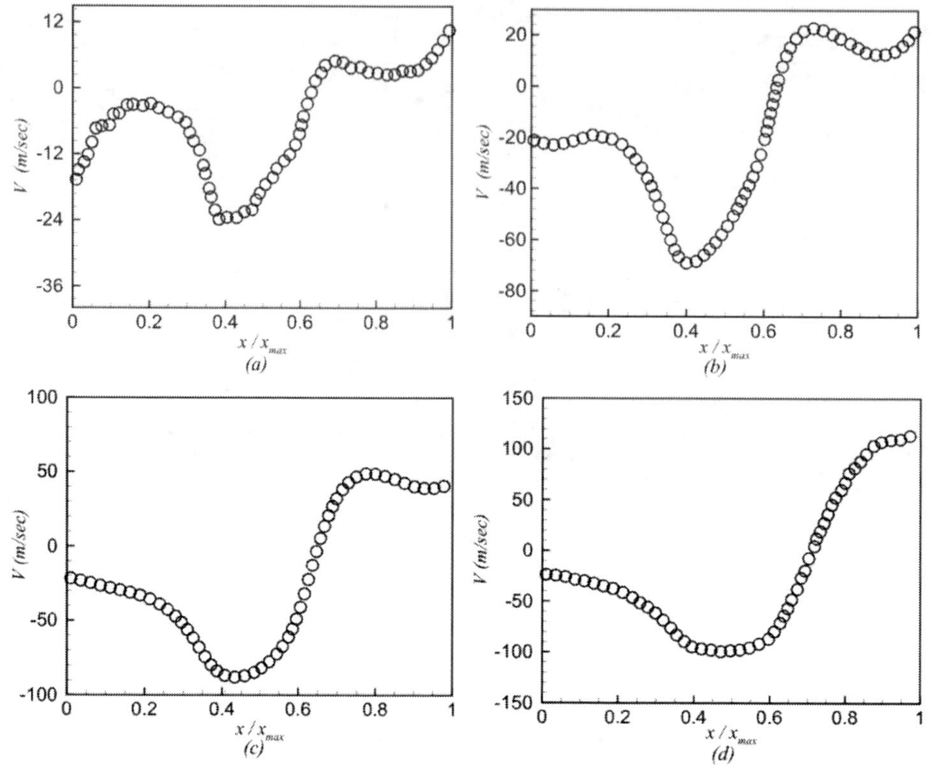

FIGURE 26. The variation of the wall-normal velocity (v) as a function of x/x_{max} at y/y_{max} = 0.15, for Kn = 0.1, and for Mach numbers Ma = 1 (a), Ma = 3 (b), Ma = 5 (c), Ma = 10 (d), Ma = 30 (e). The symbols show the DSMC results for : ($\epsilon_a = 0.01$: ($\circ$)).

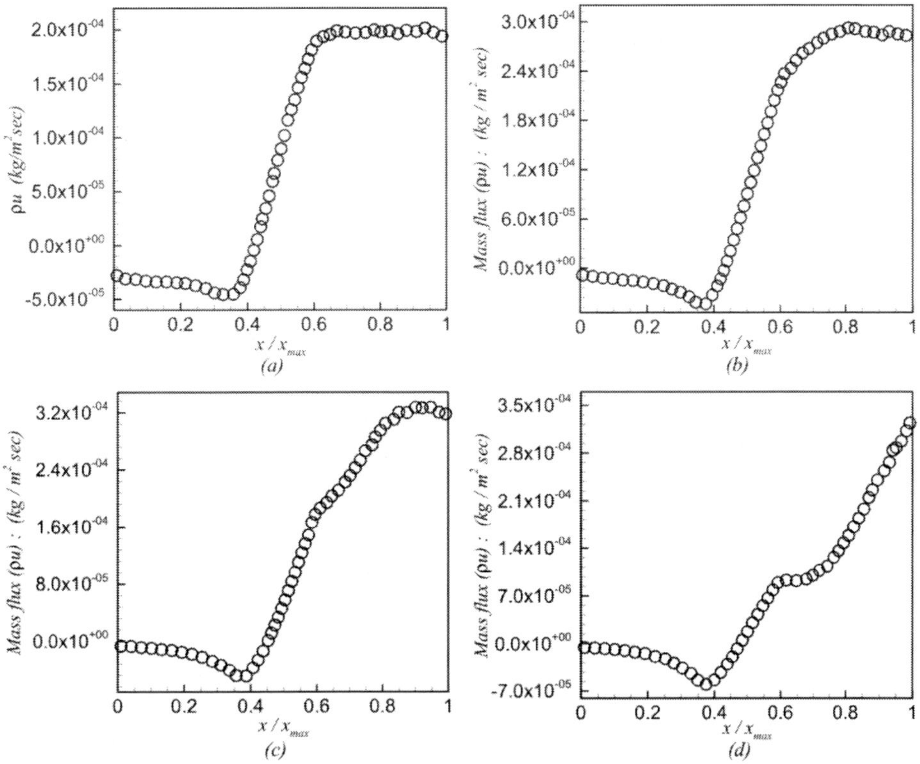

FIGURE 27. The variation of the streamwise mass flux (ρu) as a function of x/x_{max} at y/y_{max} = 0.15, for Kn = 0.1, and for Mach numbers Ma = 1 (a), Ma = 3 (b), Ma = 5 (c), Ma = 10 (d), Ma = 30 (e). The symbols show the DSMC results for : ($\epsilon_a = 0.01$: ($\circ$)).

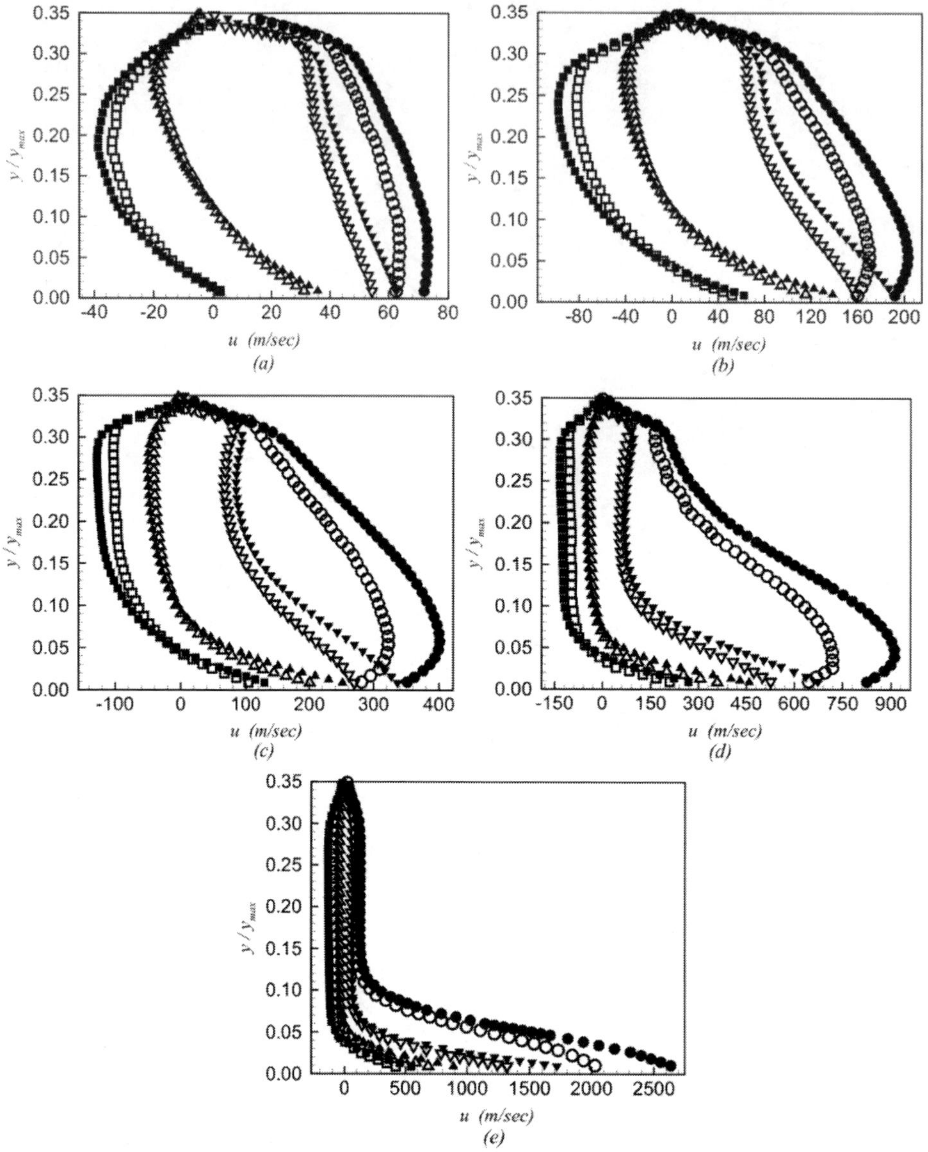

FIGURE 28. The variation of the streamwise velocity (u) as a function of y/y_{max} for Kn = 0.1, and for Mach numbers Ma = 1 (a), Ma = 3 (b), Ma = 5 (c), Ma = 10 (d), Ma = 30 (e). The symbols show the DSMC results for $\epsilon_a = 0.01$: ($x/x_{max} = 0.2$ (□), $x/x_{max} = 0.4$ (△), $x/x_{max} = 0.6$ (▽), $x/x_{max} = 0.8$ (○)). The symbols show the DSMC results for $\epsilon_a = 0.5$: ($x/x_{max} = 0.2$ (■), $x/x_{max} = 0.4$ (▲), $x/x_{max} = 0.6$ (▼), $x/x_{max} = 0.8$ (●)).

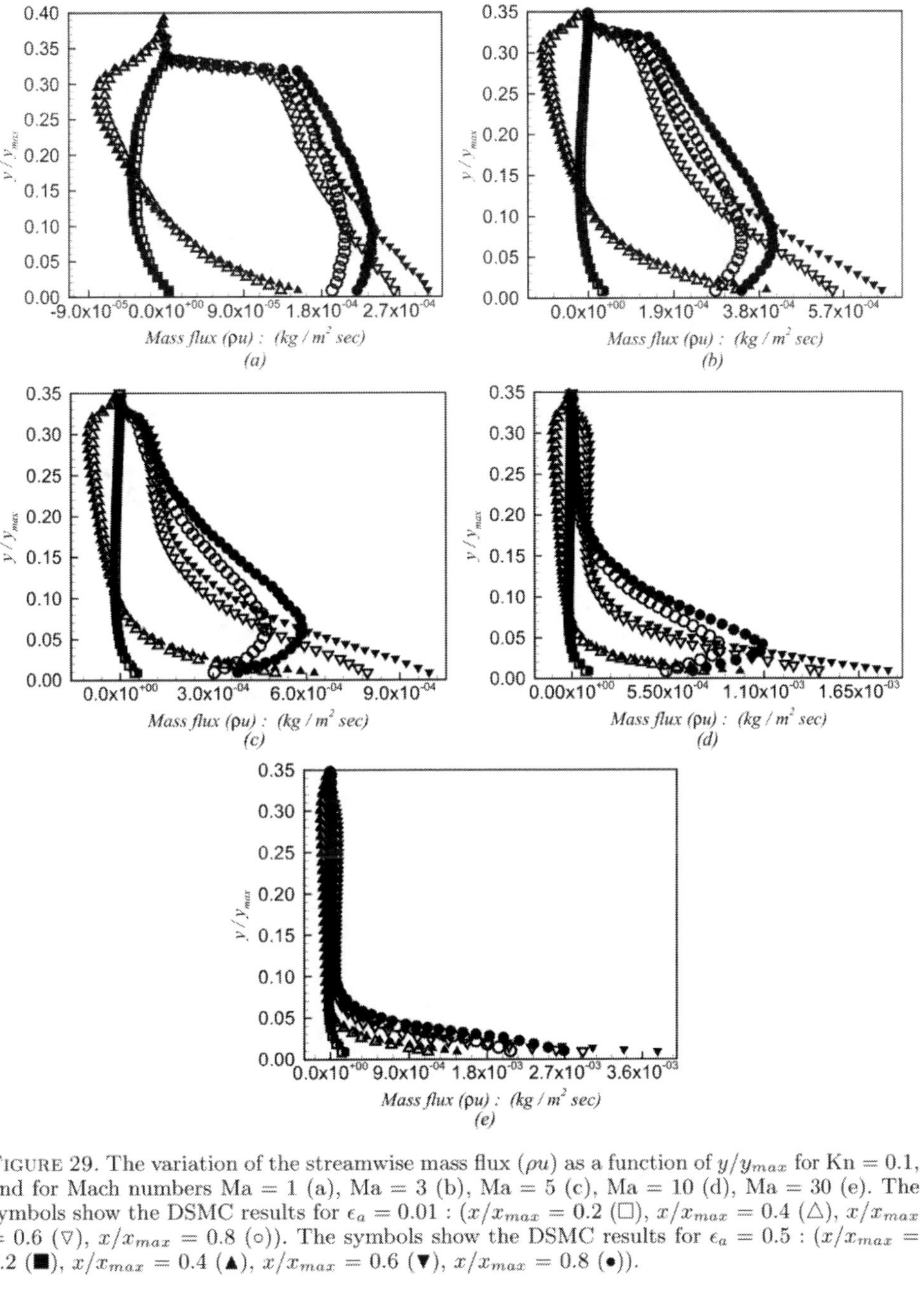

FIGURE 29. The variation of the streamwise mass flux (ρu) as a function of y/y_{max} for Kn = 0.1, and for Mach numbers Ma = 1 (a), Ma = 3 (b), Ma = 5 (c), Ma = 10 (d), Ma = 30 (e). The symbols show the DSMC results for $\epsilon_a = 0.01$: (x/x_{max} = 0.2 (□), x/x_{max} = 0.4 (△), x/x_{max} = 0.6 (▽), x/x_{max} = 0.8 (○)). The symbols show the DSMC results for $\epsilon_a = 0.5$: (x/x_{max} = 0.2 (■), x/x_{max} = 0.4 (▲), x/x_{max} = 0.6 (▼), x/x_{max} = 0.8 (●)).

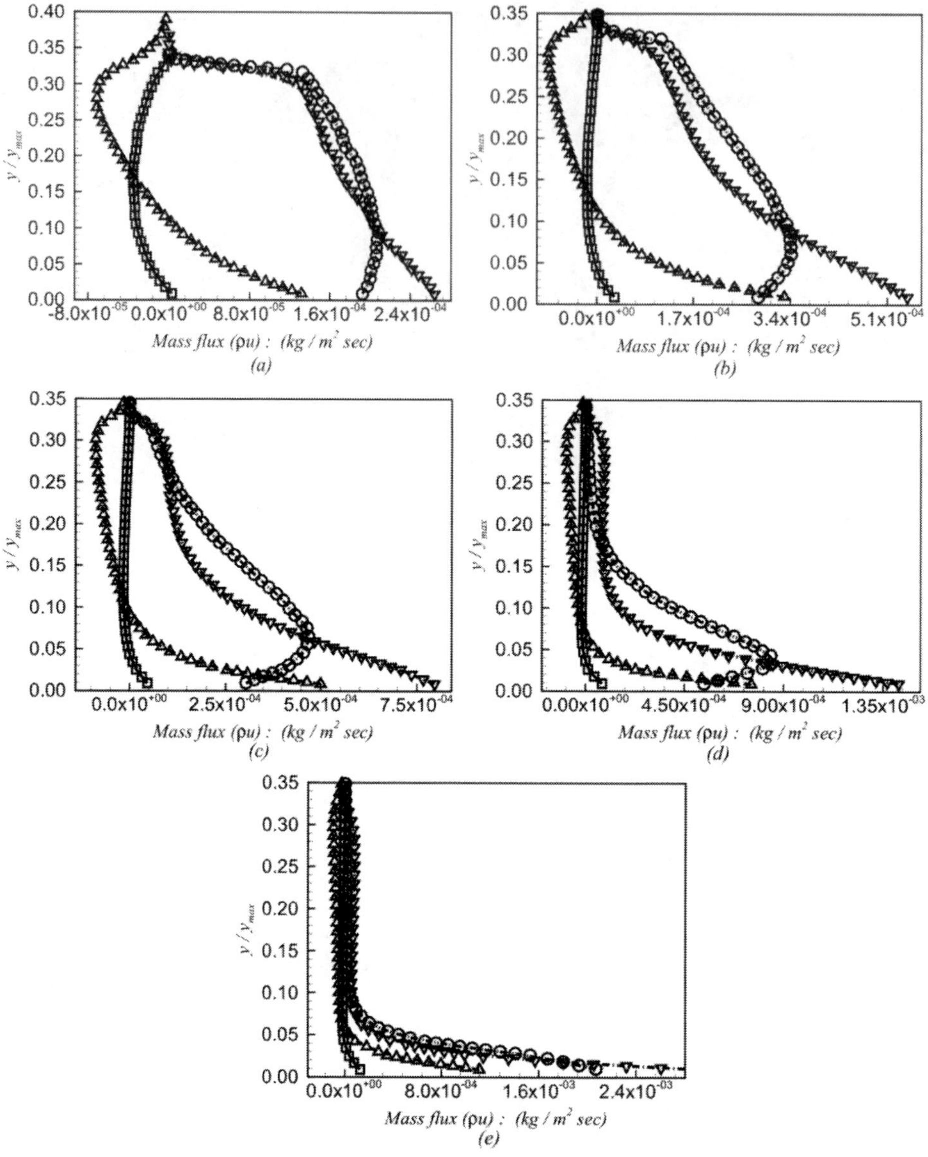

FIGURE 30. The variation of the streamwise mass flux (ρu) as a function of y/y_{max} for Kn $= 0.1$, and for Mach numbers Ma $= 1$ (a), Ma $= 3$ (b), Ma $= 5$ (c), Ma $= 10$ (d), Ma $= 30$ (e); for $\epsilon_a = 0.01$. The symbols show the DSMC results for : ($x/x_{max} = 0.2$ ($\square$); $x/x_{max} = 0.4$ ($\triangle$); $x/x_{max} = 0.6$ ($\triangledown$); $x/x_{max} = 0.8$ ($\circ$)). The lines show the generalized analytical model with slip for : ($x/x_{max} = 0.2$ (– –), $x/x_{max} = 0.4$ (—), $x/x_{max} = 0.6$ ($\cdot - \cdot$), $x/x_{max} = 0.8$ ($\cdot\cdot - \cdot\cdot$)).

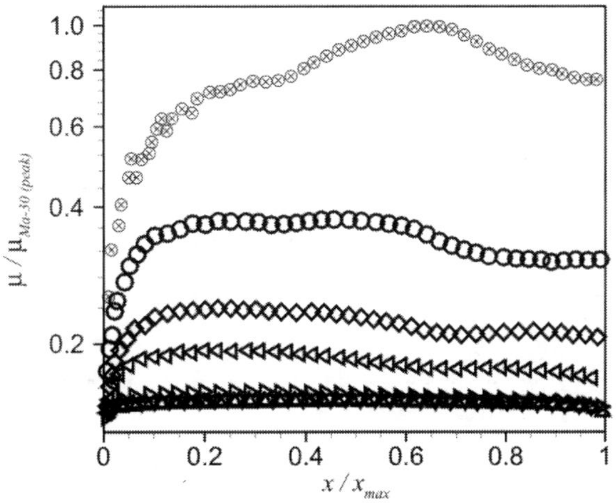

FIGURE 31. Viscosity μ, scaled by the peak value (at Mach number Ma = 30), as a function of x/x_{max} at $y/y_{max} = 0.15$, at Kn = 0.1, $\epsilon_a = 0.01$. The symbols show the DSMC results for : (Ma = 0.1 (□); Ma = 0.2 (△); Ma = 0.3 (▽); Ma = 1 (▷); Ma = 3 (◁); Ma = 5 (◊); Ma = 10 (○); Ma = 30 (⊗)).

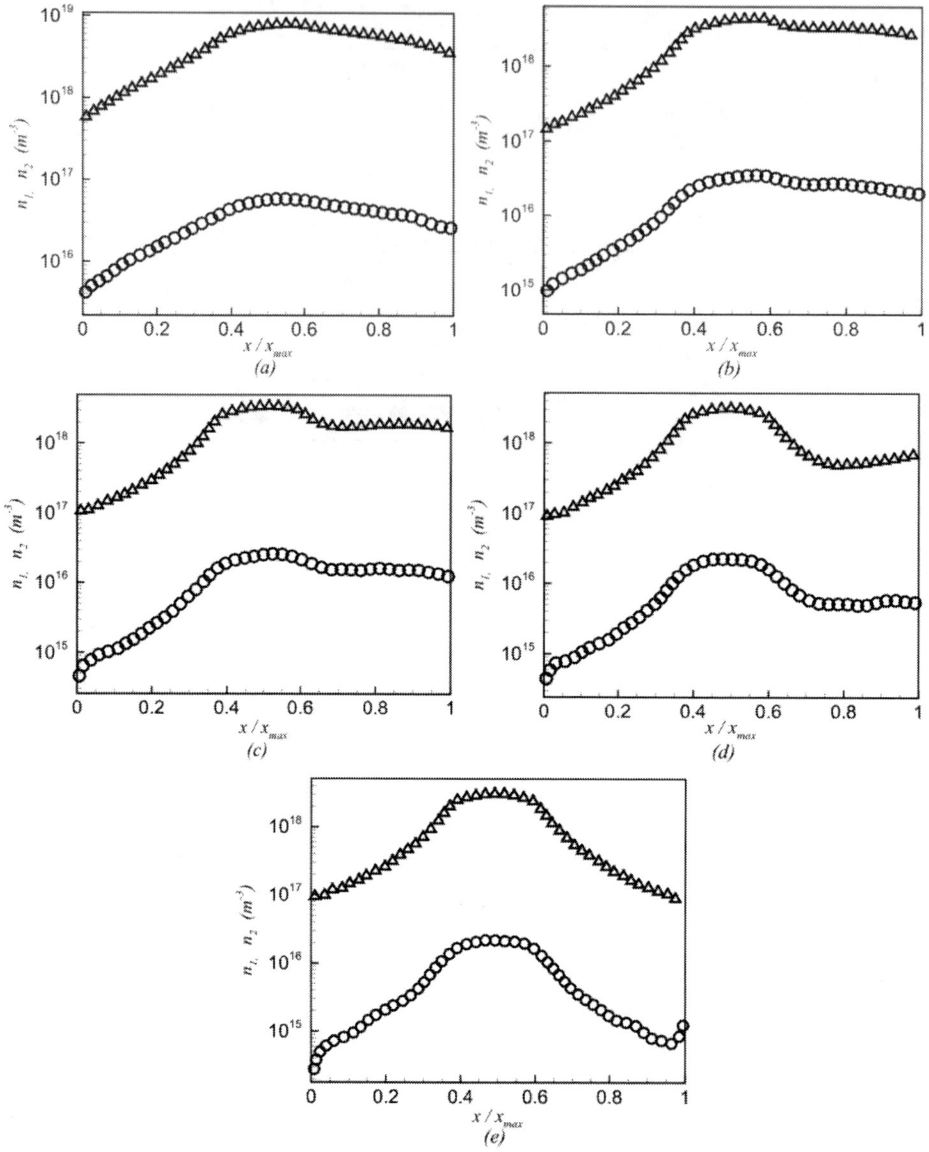

FIGURE 32. The variation of the number densities for Mach numbers Ma = 1 (a), Ma = 3 (b), Ma = 5 (c), Ma = 10 (d), Ma = 30 (e); at Kn = 0.1, for $\epsilon_a = 0.01$. The symbols show the DSMC results for : (component-1: n_1 (○); component-2: n_2 (△)).

7. Conclusions

The viscous, compressible flow in a 2D wall-bounded channel, with bottom wall moving in the positive x direction, studied using generalized analytical model and numerical simulations based on direct simulation Monte Carlo (DSMC) method, has been used as a test bed for examining different aspects of flow phenomenon and separation performance at Mach number $\mathrm{Ma} = (U_w/\sqrt{\gamma k_B T_w/m})$ in the range $0.1 < \mathrm{Ma} < 30$, and Knudsen number $\mathrm{Kn} = (1/(\sqrt{2}\pi d^2 n_d H)$ in the range $0.1 < \mathrm{Kn} < 10$ for a binary gas mixture. Here, H is the channel height, U_w is the wall velocity, ρ_m is the volume-averaged gas density, T_w is the wall temperature, m, and d are the molecular mass and molecular diameter, n_d is the number density, μ_w is the gas viscosity based on the temperature at the isothermal wall, and k_B is the Boltzmann constant. The generalized analytical model includes the fifth order differential equation for the boundary layer at the channel wall in terms of master potential (χ), which is derived from the equations of motion in a 2D rectangular $(x - y)$ coordinate. The starting point of the analytical model is the compressible Navier-Stokes, mass, momentum and energy conservation equations in the 2D rectangular $(x - y)$ coordinate, where x and y are the streamwise and wall-normal coordinates. The linearization approximation is used, where the equations of motion are truncated at linear order in the velocity and pressure disturbances to the base flow, which is a isothermal compressible Couette flow. Additional assumptions in the analytical model include high aspect ratio (length of the channel L is large compared to the height H), constant temperature in the base state (isothermal condition), and low Reynolds number (laminar flow). In this limit, the gas flow is restricted to a boundary layer of thickness ($\mathrm{Re}^{-1/2} H$) at the channel wall. Here, the Reynolds number $\mathrm{Re} = (\rho_m U_w H/\mu_w)$.

The main assumptions taken into account in deriving the generalized analytical equation are that the flow field is a small perturbation about an isothermal compressible Couette flow. The inflow/outflow of the gas is small enough compared to the bulk of the fluid present inside the channel. These assumption enables us to linearize the equations of viscous compressible flow, neglect terms that are small in the limit of high Reynolds numbers and aspect ratio, and obtain a fifth-order equation in the cross-streamwise direction for the 'master-potential', which is related to the density, velocity and temperature fields.

The major advantage of the present formulation is that it is not restricted to the asymptotic limit of high Mach number. Due to this, the model is not restricted to just boundary layer flows at the surface of the channel, but is even applicable for situations where the flow domain extends all the way to the center of the channel. Therefore, the model can be used for much slower flows, that is for incompressible flows where the incompressible Navier-Stokes equations can be used.

The solutions of the generalized analytical model for the channel flow are compared with DSMC simulations. The comparison reveals that the boundary conditions in the simulations and analysis have to be matched with care. The commonly used 'diffuse reflection' boundary conditions at solid walls in DSMC simulations result in a non-zero slip velocity as well as a 'temperature slip' (gas temperature at the wall is different from wall temperature). These have to be incorporated in the analysis in order to make quantitative predictions. It is also necessary to ensure that the homogeneous temperature boundary conditions are accurately satisfied in the simulations. When these precautions are taken, there is excellent agreement between analysis and simulations, to within 10%.

Appendix A. Conservation equations for compressible flow

The mass, momentum and energy conservation equations are,

$$\frac{\partial \rho}{\partial t} + \frac{\partial (\rho u_j)}{\partial x_j} = 0, \tag{A 1}$$

$$\frac{\partial (\rho u_i)}{\partial t} + \frac{\partial (\rho u_i u_j)}{\partial x_j} = -\frac{\partial p}{\partial x_i} + \frac{\partial \tau_{ij}}{\partial x_j} + \rho f_i, \tag{A 2}$$

$$\frac{\partial \rho T}{\partial t} + \frac{\partial (\rho T u_j)}{\partial x_j} = \frac{1}{c_v}\left[-p\frac{\partial u_j}{\partial x_j} + \tau_{ij}\frac{\partial u_i}{\partial x_j} - \frac{\partial q_j}{\partial x_j}\right], \tag{A 3}$$

where the viscous stress tensor,

$$\tau_{ij} = \mu\left(\frac{\partial u_i}{\partial x_j} + \frac{\partial u_j}{\partial x_i} - \frac{2}{3}\frac{\partial u_k}{\partial x_k}\delta_{ij}\right), \tag{A 4}$$

and the heat flux

$$q_j = -K\frac{\partial T}{\partial x_j}. \tag{A 5}$$

The bulk density ρ_m and the bulk velocity U_m are defined as,

$$\bar{\rho} = \frac{1}{H}\int_{-H/2}^{H/2} \rho \, dy, \tag{A 6}$$

$$U = \frac{1}{H\bar{\rho}}\int_{-H/2}^{H/2} \rho u \, dy. \tag{A 7}$$

The non-dimensional equation of state is

$$p = \frac{1}{\gamma \mathrm{Ma}^2}\rho T. \tag{A 8}$$

Equation A 1 is the mass conservation equation, equations A 2, are the momentum conservation equations and equation A 3 is the energy conservation equation.

Appendix B. Conservation equations for incompressible flow

The mass, momentum and energy conservation equations are,

$$\frac{\partial u_j}{\partial x_j} = 0, \tag{B 1}$$

$$\frac{\partial u_i}{\partial t} + \frac{\partial (u_i u_j)}{\partial x_j} = -\frac{1}{\rho}\frac{\partial p}{\partial x_i} + \frac{\mu}{\rho}\frac{\partial^2 u_i}{\partial x_j \partial x_j}, \tag{B 2}$$

$$\frac{\partial T}{\partial t} + \frac{\partial (T u_j)}{\partial x_j} = \frac{K}{\rho c_p}\frac{\partial^2 T}{\partial x_j \partial x_j} + Q, \tag{B 3}$$

where Q is the heat source. Equation B 1 is the mass conservation equation, equations B 2, are the momentum conservation equations and equation B 3 is the energy conservation equation.

REFERENCES

BIRD, G. A. 1963 Approach to translational equilibrium in a rigid sphere gas. *Phys. Fluids* **6**, 1518.

BIRD, G. A. 1994 Molecular gas dynamics and the direct simulation of gas flows. *Clarendon Press, Oxford.*

BIRD, G. A. 1998 Recent advances and current challenges for DSMC. *Comput. Maths Applics.* **35 (1-2)**, 1.

BRADSHAW, P. & FERRISS, D. H. 1971 Calculation of boundary layer development using the energy equation: compressible flow on adiabatic walls. *J. Fluid Mech.* **46**, 83.

BURNETT, D. 1971 The distribution of molecular velocities and the mean motion in a non-uniform gas. *Proc. Lond. Math. Soc.* **40**, 382.

CERCIGNANI, C. & DANERI, A. 1963 Flow of a rarefied gas between two parallel plates. *J. Appl Phys.* **34**, 3509.

CERCIGNANI, C. 1990 Mathematical Methods in Kinetic Theory. *Plenum.*

CERCIGNANI, C. 2000 Rarefied gas dynamics. From basic concepts to actual calculations. *Cambridge University Press.*

CHAPMAN, S. & COWLING, T. G. 1970 The Mathematical Theory of Non-Uniform Gases,2nd edition. *Cambridge University Press.*

CHANG, W. & UHLENBECK, G. E. 1948 On the transport phenomena in rarified gases. *Tech Rep. DTIC Document.*

CHUN, J. & KOCH, D. L. 2005 A direct simulation Monte Carlo method for rarefied gas flows in the limit of small Mach number. *Phys. Fluids.* **17 (10)**, 107107.

CLAUSING, P. 1971 The flow of highly rarefied gases through tubes of arbitrary length. *J. Vac. Sci. Technol.* **8 (5)**, 636.

DJORDJEVIC, V. 2004 Modeling of the slip boundary condition in rarefied gas micro-channel flow via fractional derivative *Proc. 1st IFAC Workshop on Fractional Differentiation and its Applications, ENSEIRB, Bordeaux*, 363.

DJORDJEVIC, V. 2005 On the rarefied gas flow in pipes. *Proc. ASME Int. Eng. Conference DETC2005* **6B**, 1591.

DJORDJEVIC, V. 2008 Modeling of the slip boundary condition in Micro-Channel/Pipe flow via fractional derivative. *Advances in Nonlinear Sciences II - Sciences and Applications, Belgrade* **2**, 136.

EWART, T. & PERRIER, P. & GRAUR, I. A. & MEOLANS, J. G. 2007 Mass flow rate measurements in a microchannel, from hydrodynamic to near free molecular regimes. *J. Fluid Mech.* **584**, 337.

FAN, J. & SHEN, C. 2001 Statistical simulation of low-speed rarefied gas flows. *J. Comput Phys.* **167**, 393.

GARCIA. 2000 Time step truncation error in direct simulation Monte Carlo. *Phys. Fluids* **12**, 2621.

GRAD, H. 1949 On the kinetic theory of rarefied gases. *Commun. Pure Appl. Maths* **2**, 331.

KOSUGE, S. & TAKATA, S.2008 Database for flows of binary gas mixtures through a plane microchannel. *European Journal of Mechanics B/Fluids* **27**, 444.

KUMARAN, V. & PRADHAN, S. 2014 The generalized Onsager model for a binary gas mixture. *J. Fluid Mech.* **753**, 307.

KUNDU, P. & COHEN, I. 2007 Fluid Mechanics. *Academic Press.*

LOCKERBY, D. A. & REESE, J. M. & GALLIS, M. A. 2005 Capturing the Knudsen layer in continuumfluid models of nonequilibrium gas flows. *AIAA J.* **43**, 1391.

MANELA, A. & HADJICONSTANTINOU, N. G. 2007 On the motion induced in a gas confined in a small-scale gap due to instantaneous boundary heating. *J. Fluid Mech.* **593**, 453.

MIRSHEKARI, G. & BROUILLETTE, M. & GIORDANO, J. & HBERT, C. 2013 Shock waves in microchannels. *J. Fluid Mech.* **724**, 259.

NANBU. 1980 Direct simulation scheme derived from the Boltzmann equation I. Multicomponent gases. *J.Phys.Soc. of Japan* **45**, 2042.

NARIS, S. & VALOUGEORGIS, D. & KALEMPA, D. & SHARIPOV, F. 2003 Gaseous mixture ow between two parallel plates in the whole range of the gas rarefaction. *Physica A.*

NASSIOS, J. & SADER, J. E. 2012 Asymptotic analysis of the BoltzmannBGK equation for oscillatory flows. *J. Fluid Mech.* **708**, 197.

OHWADA, T. & SONE, Y. & AOKI, K. 1989 Numerical analysis of the Poiseuille and thermal transpiration flows between two parallel plates on the basis of the Boltzmann equation for hard-sphere molecules. *Phys. Fluids.* **A 1**, 2042.

PARK, J. H. & BAHUKUDUMBI, P. & BESKOK, A. 2004 Rarefaction effects on shear driven oscillatory gas flows: a direct simulation Monte Carlo study in the entire Knudsen regime. *Phys. Fluids.* **16 (2)**, 317.

PRADHAN, S. & KUMARAN, V. 2011 The generalized Onsager model for the secondary flow in a high-speed rotating cylinder. *J. Fluid Mech.* **686**, 109.

RADTKE, G. A. & HADJICONSTANTINOU, N. G. & WAGNERHA, W. 2011 Low-noise Monte Carlo simulation of the variable hard sphere gas. *Phys. Fluids.* **23 (3)**, 030606.

ROY, S. & RAJU, R. & CHUANG, HELEN F. & CRUDEN, BRETT A. & MEYYAPPAN M. 2003 Modeling gas flow through microchannels and nanopores. *J. Applied Physics.* **93**, 4870.

SCHLICHTING, H. T. & GERSTEN, K. 2000 Boundary-Layer Theory. *McGrawHill.*

SINGH, N. & AGRAWAL, A. 2014 The Burnett equations in cylindrical coordinates and their solution for flow in a microtube. *J. Fluid Mech.* **751**, 121.

SINGH, N. & GAVASANE, A. & AGRAWAL, A. 2014 Analytical solution of plane Couette flow in the transition regime and comparison with direct simulation Monte Carlo data. *Comput. Fluids.* **97**, 177.

SONE, Y. & OHWADA, T. & AOKI, K. 1989 Temperature jump and Knudsen layer in a rarefied gas over a plane wall: numerical analysis of the linearized Boltzmann equation for hard-sphere molecules. *Phys. Fluids.* **A 1**, 363.

SREEKANTH, A. K. 1968 Slip flow through long circular tubes. In Rarefied Gas Dynamics. *Ed. L. Trilling H. Y. Wachman, Academic Press.*

STRUCHTRUP, H. & TORRILHON, M. 2003 Regularization of Grad's 13-moment equations: derivation and linear analysis. *Phys. Fluids* **15**, 2668.

TAKATA, S. & FUNAGANE, H. 2011 Poiseuille and thermal transpiration flows of a highly rarefied gas: over-concentration in the velocity distribution function. *J. Fluid Mech.* **669**, 242.

VARADE, V. & AGRAWAL, A. & PRADEEP, A. M. 2014 Behaviour of rarefied gas flow near the junction of a suddenly expanding tube. *J. Fluid Mech.* **739**, 363.

VINCENTI. 1967 Introduction to Physical Gas Dynumics. *Wiley, New York.*

VON KARMAN, T. 1963 From Low-speed Aerodynamics to Astronautics. *Pergamon.*

WAGNER, W. 2008 Deviational particle Monte Carlo for the Boltzmann equation. *Monte Carlo Meth. Applic.* **14 (3)**, 191.

WHITE, C. & BORG, K. MATTHEW & SCANLON, J. & REESE, M. 2013 A DSMC investigation of gas fows in micro-channels with bends. *Computers & Fluids* **71**, 261.

WU, L. & REESE, J. M. & ZHANG, Y. 2014 Solving the Boltzmann equation deterministically by the fast spectral method: application to gas microflows. *J. Fluid Mech.* **746**, 53.

XU, K. & LI, Z. 2004 Microchannel flow in the slip regime: gas-kinetic BGKBurnett solutions. *J. Fluid Mech.* **513**, 87.

ZOHAR, Y. & LEE, S. Y. K. & LEE, W. Y. & TONG, P. 2002 Subsonic gas flow in a straight and uniform microchannel. *J. Fluid Mech.* **472**, 125.

498259 The Shape Evolution of Pendant Droplets in Miscible Environments

Abstract:

The spreading of liquids is a classical problem in interfacial fluid mechanics and, historically, the examination has been limited to immiscible systems. We have reported previously on our experimental studies and observations of the spreading of sessile drops in miscible environments, which have distinctly different shape evolution and power law dynamics from sessile drops that spread in immiscible environments. We have extended this experimental work to include the shape evolution of pendant drops existing in a miscible environment. By examining pendant drops, the need to account for surface energies arising from a solid-fluid interface, as in the case of a sessile drop, is eliminated. We have complemented these experimental studies with a theoretical scaling analysis as well as numerical simulation.

As time evolves, diffusion across the miscible liquid-liquid boundary proceeds due to the chemical potential difference between the two initially distinct, homogeneous phases. Diffusion, in turn, imparts a time-dependence to the properties of the liquids in the diffusive region – notably the density, viscosity, and interfacial tension – that influence the shape evolution. It was found for these droplets in miscible environments that gravitational forces dominate the shape evolution process. The presence of diffusion sets up a fluid flow of free convection in the case of a pendant drop in a miscible environment.

A series of liquid pairs (corn syrup-water, silicone oil-silicone oil) and volumes of droplets have been studied in the experimental study for pendant droplets. Solving the convection-diffusion and Stokes equations numerically and in tandem has been used to simulate these systems, which quantitatively match observations of the experiments. This talk will present our pendant drop analyses, spanning the experimental, theoretical, and numerical work.

Figure 1: Image sequence taken in time of a corn syrup pendant drop immersed in water. A strand emanates from the apex of the drop and continues to flow as the entire drop descends and elongates.

Figure 2: Numerical simulation of a pendant drop in a miscible environment evolving in time.

Figure 3: Volume of pendant drops of corn syrup in water. Volume scaled by maximum volume; time scaled by radius of pendant drop at maximum volume squared divided by diffusion coefficient. Symbols are experimental results. Solid lines are numerical results.

499528 Shape Evolution of Miscible Drops with Arbitrary Viscosity Ratio: Lagrangian-Eulerian Swarms of Stokeslets and Subgrid Resolution

Abstract:

Gravity/buoyancy driven motion, deformation, coalescence and breakup of miscible drops is central to geophysical flows of mantle plumes, and also enables an emerging, patented technology of fluid-phase self-assembly of drug delivery particles (toroidal-spiral particles or TSP) by competitive-kinetics. Essentially unchecked by (negligible) interfacial forces, viscous effects lead to evolution of extremely intricate drop shapes and their interfaces, which can frustrate Eulerian-volumetric and boundary-element methods.

In this talk we present a Lagrangian/Eulerian hybrid method based on a swarm of Stokeslets that accurately tracks complex shape evolutions. The case of unit viscosity ratio is interesting in its determination of intricate local structures by global hydrodynamic interactions, These are efficiently treated with a particle-mesh (PM) scheme in which cell-cell interactions are computed in O(N log N) operations.

The new element presented here is a subgrid-accurate scheme for treating different viscosities of the drop versus bath. A particle-particle / particle-mesh (P3M) scheme is combined with iterative treatment of piecewise constant viscosity (and discontinuities at the interfaces) as an effective body force. Computer simulations will be compared with drop visualization experiments using a high-speed camera.

The particulate character of the method opens an avenue for tracking strain histories and thereby extending the treatment to viscoelastic drops as well.

498166 Coalescence of Drops Due to a Constant Force Interaction in a Viscous Quiescent Fluid

Abstract:

A Cantilevered-Capillary Force Apparatus is used to study the time scale for the coalescence of two droplets compressed together with a constant force. Power-law trends for the coalescence time as a function of droplet radius and compression force are measured. The measurements are compared against several different scaling theories from the literature. One of the existing theories is found to correctly predict the dependence on the droplet radius, but all of the theories over-predict the dependence on the force. A transition is also measured in the coalescence process from a predominately deterministic to a predominately stochastic process. A qualitative explanation for this transition is provided via scaling arguments.

499106 ÂNanoemulsion Formation: Controlling and Predicting Droplet Size

Abstract:

Nanoemulsions contain droplets of oil in water or water in oil with sizes on the order of 100 nm, and due to attractive properties such as high surface area, robust stability and tunable rheology, they are used in drug delivery, cosmetics, pharmaceutical formulations and development of smart food drinks. However, the literature uses relations for droplet size correlations developed for large emulsion droplets (or macroemulsions) in an ad-hoc manner without considering the effects of small droplet sizes. Here, we highlight that the droplet size correlations for macroemulsions (droplet size ~ 100 um) cannot be extrapolated to nanoemulsions, and hence derive a new criterion for predicting droplet size of nanoemulsions. The new scaling relation takes into account the effect of viscosity of droplet phase as well as continuous phase. We validate our scaling theory with a wide range of experimental droplet size data. Further, building on the proposed scaling theory, we propose a new breakage model for predicting the frequency of successful droplet breakup event, the rate-controlling step in nanoemulsion formation. A population balance analysis with the proposed breakage model provides insights about the kinetics of droplet size change and successfully captures the effect of droplet viscosity, unlike the earlier breakage models. The predictions of the proposed breakage model are supported by the experimental droplet size data. The proposed scaling theory and models are useful for rational design of nanoemulsions.

499638 Using a Novel Approach to Model Drop Size Distribution: The Adaptive Multi-Size Group Method

Abstract:

Liquid-Liquid dispersions and emulsions occur in many applications in the chemical, food and cosmetics industry in equipment such as stirred vessels, rotor-stator devices and colloid mills. The behavior of such two phase systems is strongly governed by how one phase disperses in the continuous phase. Such behavior can understood only when a good description of the distribution of droplet size is available. The droplet size distribution depends on the coalescence of small droplets in to larger droplets and break-up of large droplets in daughter droplets. Coalescence and break-up is a complex phenomenon governed by the rheology of the fluids involved as well as the fluid dynamics effect such as turbulence and shear.Many researchers have investigate these behavior both experimentally and numerically [1].The mathematical models used to simulate such systems include simple correlation based models [1], moments based methods such as S-gamma model [2] and quadrature based models such as QMOM/DQMOM

The simpler models can increase the accuracy in dispersed flow but are not sufficient to capture all the details, especially if the size distributions do not follow log normal type of distributions. The quadrature based models generally give higher accuracy but at a disproportionately higher compute cost. This is due to them computing size bins which may be empty and/or cannot adapt to the changing distribution shape. In this paper we present the results from a recent model termed as AMuSiG or adaptive multi-size group model. This model is expected to give higher accuracy comparable to quadrature based models but at a much lower compute cost. A basic description of the model with validation examples will be provided.

[1] H. Steiner, R. Teppner, G. Brenn, N. Vankova, S. Tcholakova, N. Denkov, Numerical simulation and experimental study of emulsification in a narrow-gap homogenizer, Chemical Engineering Science 61, 2006, 5841 – 5855

[2] S. Lo, A .Splawski & B.J. Yun, "The importance of correct modeling of bubble size and condensation in predicting sub-cooled boiling flows", JThe Journal of Computational Multiphase Flows pp299-308, 2012.

[3] A. Vichansky, A. Splawsky, Adaptive multiply size group method for CFD-population balance modelling of polydisperse flows, The Candadian Journal of Chemical Engineering, Volume 93, Issue 8 August 2015, Pages 1327–1334

494888 Describing Emulsions in Terms of Pseudo-Potential Lattice Boltzmann Modeling

Abstract:

Liquid-liquid dispersions or emulsions are usually described in terms of densities and interfacial tension, while their mean droplet size and their droplet size distribution are assumed to depend on either shear rate (Capillary number) or (average) energy dissipation rate in the system. This phenomenological (or engineering) approach, however, does not explicitly and accurately represent the intermolecular interactions that lead to phase separation. On the contrary, the pseudopotential lattice Boltzmann model is excellently geared to describe such interactions between the constitutive components of a system, which lead to phase separation, and in this method the interfaces between phases are captured automatically. We use a pseudopotential lattice Boltzmann method for small density ratios between the two liquid phases to study systems with three non-ideal components (described by the Carnahan-Starling EoS), one of which models a surfactant. We report a systematic study of the effects of the interaction strengths between the components and their reduced temperatures on the number and composition of the phases present at equilibrium. These results allow us to perform simulations of emulsions at prescribed density ratios between the two liquid phases. We also study how the surfactant component, which is present in small amounts relative to the other two components and repelled by both of them, is distributed between the interfacial region and the bulk phases. We determine how this third component alters the surface tension of the interfaces between the fluid phases.

494018 Comparison of the Rise Dynamics of Standard Bubbles and Oil-Coated Bubbles: Experiments and Simulations

Abstract:

Ordinary air bubbles are routinely used in froth flotation, a highly efficient separation method utilized in many industrial processes; and the rise dynamics of standard air bubbles has also been extensively studied for many years. There is also an emerging interest in replacing the standard bubbles with oil-coated bubbles to separate hydrophilic particles with low affinity for the air-water interface, but the rise behavior of oil-coated bubbles has not been well explored yet. In our work, we systematically investigated the shape, velocity and trajectory of bubbles engulfed in a layer of oil, and experimental results were compared to computer simulations to confirm a pronounced effect of the oil coat on the bubble dynamics. We consistently found that oil-coated bubbles display a more spherical shape, lower rise velocity and straighter trajectory than the standard air bubbles, which can benefit the flotation process.

496607 A Quadrature-Based Model for Polydisperse Laminar Bubbly Flows

Abstract:

A quadrature-based moment method for the simulation of laminar polydisperse bubbly flow is developed and implemented into OpenFOAM® as part of the OpenQBMM project (OpenQBMM, 2017; Passalacqua et al., 2017). The proposed approach aims at addressing one key limitation of previously adopted quadrature-based methods coupled to gas-liquid systems (Sanyal et al., 2005, 1999): the use of the local mean velocity of the gas phase to advect bubbles of different sizes. The proposed approach represents an evolution of the mass-velocity moment method proposed in Yuan et al., (2014), which presented numerical difficulties when the continuous phase disappeared and the disperse phase became continuous.

In the proposed approach, the disperse phase is assumed to be mono-kinetic, due to the moderate Stokes number of bubbles. Because of this assumption, only the first-order velocity moments are considered, while p moments in size are transported. The major novelty of this contribution consists in the solution algorithm, in which the moment method is formulated as a function of the mean gas-phase velocity, and of the deviation of the bubble size velocity from the mean. This allows a scheme that ensures the boundedness of the numerical solution, and, in particular, of the volume fraction, to be formulated. Key element of the algorithm is the treatment of the moment advection term, which is split into two stages. First moments of the disperse phase are advected with respect to the deviation velocity with respect to the mean using kinetic fluxes. Then, a continuity and momentum equation for the disperse phase, similar to those used in traditional two-fluid solvers are solved and coupled to the equations for the liquid phase. These equations are solved with an iterative procedure (Spalding, 1980), supplemented by a bounded scheme based on the Multi-dimensional Universal Limiter for Explicit Solution (MULES) (Weller, 2006) for the volume fraction. Notable difference, with respect to the traditional two-fluid model is the definition of the momentum exchange term, which is function of the bubble size distribution in the proposed approach, while it relies on a mean bubble size diameter in the traditional two-fluid model. Once the updated mean volume fraction and mean velocity are found, size and joint size-velocity moments are updated as a consequence of the mean transport. The effect of the force term on velocity moments is included by directly integrating the ordinary differential equation for the velocity abscissae, as discussed in Fox (2008).

The proposed numerical approach was tested considering two bubble columns studied in the literature. A mono-disperse case (Pfleger et al., 1999) was used to verify the approach in the limit case of monodisperse bubbles, to ensure the solution reproduces the results of the equivalent two-fluid model. Results are in agreement with the predictions of the two-fluid model implementation available in OpenFOAM 4 (OpenFOAM, 2016). A polydisperse case (Díaz et al., 2008) was also considered, obtaining results in agreement with their experimental data for what concerns the gas mean axial velocity and hold-up.

Acknowledgments

The research discussed in this work was supported by the National Science Foundation of the United States,

under the SI2 – SSE award NSF – ACI 1440443.

References

Díaz, M.E., Montes, F.J., Galán, M.A., 2008. Experimental study of the transition between unsteady flow regimes in a partially aerated two-dimensional bubble column. Chemical Engineering and Processing: Process Intensification 47, 1867–1876. doi:10.1016/j.cep.2007.10.012

Fox, R.O., 2008. A quadrature-based third-order moment method for dilute gas-particle flows. Journal of Computational Physics 227, 6313–6350. doi:10.1016/j.jcp.2008.03.014

OpenFOAM, 2016. The OpenFOAM Foundation [WWW Document]. OpenFOAM. URL http://openfoam.org/

OpenQBMM, 2017. An open-source implementation of Quadrature-Based Moment Methods [WWW Document]. URL https://www.openqbmm.org/

Passalacqua, A., Laurent, F., Madadi-Kandjani, E., Heylmun, J.C., Fox, R.O., 2017. An open-source quadrature-based population balance solver for OpenFOAM.

Pfleger, D., Gomes, S., Gilbert, N., Wagner, H.G., 1999. Hydrodynamic simulations of laboratory scale bubble columns fundamental studies of the Eulerian-Eulerian modelling approach. Chemical Engineering Science 54, 5091–5099. doi:10.1016/S0009-2509(99)00261-4

Sanyal, J., Marchisio, D.L., Fox, R.O., Dhanasekharan, K., 2005. On the Comparison between Population Balance Models for CFD Simulation of Bubble Columns. Industrial & Engineering Chemistry Research 44, 5063–5072. doi:10.1021/ie049555j

Sanyal, J., Vásquez, S., Roy, S., Dudukovic, M.P., 1999. Numerical simulation of gas–liquid dynamics in cylindrical bubble column reactors. Chemical Engineering Science 54, 5071–5083. doi:10.1016/S0009-2509(99)00235-3

Spalding, D.B., 1980. Numerical Computation of Multi-phase fluid flow and heat transfer, in: Taylor, C. (Ed.), Recent Advances in Numerical Methods in Fluids. Pineridge Press.

Weller, H.G., 2006. Bounded Explicit and Implicit Second-Order Schemes for Scalar Transport.

Yuan, C., Kong, B., Passalacqua, A., Fox, R.O., 2014. An extended quadrature-based mass-velocity moment model for polydisperse bubbly flows. The Canadian Journal of Chemical Engineering 92, 2053–2066. doi:10.1002/cjce.22006

506177 Stability of Stagnation Point Flows of Newtonian and Complex Fluids

Abstract:

We present a summary of our recent results examining flow instabilities arising in stagnation point flows generated in cross-slot-type geometries. Our studies span purely inertial flows of simple Newtonian fluids, through inertio-elastic flows of highly dilute polymer solutions, to purely-elastic instabilities in more concentrated polymeric systems. In Newtonian fluids, axially-aligned spiral vortices are formed along the outlet channels of the cross-slot for flow above a modest Reynolds number, $Re_c \sim 40$. The addition of small quantities of polymer has a striking effect on the inertial instability, causing reductions in both the value of Re_c and the strength of the spiral vortex that forms. For higher polymer concentrations, the vorticity is completely suppressed and a purely-elastic instability mode becomes dominant above a critical Weissenberg number, Wi_c. Potential mechanisms for the various flow instabilities are also discussed.

497974 Prediction of Mean Ionic Activity Coefficient of NaCl-H_2OÂ System at High Concentrations Using Molecular Dynamics Simulations

Abstract:

Aqueous electrolyte thermodynamic models are extensively use to correlate and predict the liquid phase non-ideality for many industrial applications such as hydraulic fracturing, desalination, electrodialysis, e-waste and nuclear waste treatment, etc. Among the electrolyte thermodynamic models, Pitzer model [1] and electrolyte non-random two liquid (eNRTL) model [2] are the two most widely used models for aqueous electrolytes as the thermodynamic models for NaCl-H_2O to predict mean ionic activity coefficient. The Pitzer model is valid up to 6 molal whereas the eNRTL model claims to extrapolate up to pure salt. Specifically, for NaCl in water, the Pitzer model predicts a very sharp increase in mean ionic activity coefficient of NaCl after 6 molal. In contrast, the eNRTL model suggests the NaCl mean ionic activity coefficient should increase to a maximum and then decline with further increase in NaCl concentration. In this work, we use Molecular Dynamics simulations to calculate the mean ionic activity coefficient of aqueous NaCl system at 298.15 K and compare the predictions with those from the two thermodynamic models. The mean ionic activity coefficient is determined using the Kirkwood-Buff method [3]. The Kirkwood-Buff Force Field (KBFF) [4] parameters for NaCl are optimized to reproduce the NaCl lattice dimensions, ion to water oxygen contact distances, and Kirkwood-Buff Integral (KBI) which is related to mean ionic activity coefficient of aqueous NaCl solution. The predictions of the NaCl mean ionic activity coefficient beyond 6 molal from the Kirkwood-Buff method are found to be in line with the eNRTL model extrapolation and differ from the Pitzer model extrapolation.

References

[1] K.S. Pitzer, J.C. Peiper, and R.H. Busey, "Thermodynamic Properties of Aqueous Sodium Chloride Solutions," *Journal of Physical and Chemical Reference Data*, vol. 13, pp. 1-102, 1984.

[2] Y. Yan and C.-C. Chen, "Thermodynamic representation of the NaCl + Na_2SO_4 + H_2O system with electrolyte NRTL model," *Fluid Phase Equilibria*, vol. 306, pp. 149-161, 2011.

[3] J.G. Kirkwood and F.P. Buff, "The statistical mechanical theory of solutions. I," *Journal of Chemical Physics*, vol. 19, pp. 774-777, 1951.

[4] S. Weerasinghe and P.E. Smith, "A Kirkwoodâ€"Buff derived force field for sodium chloride in water," *Journal of Chemical Physics*, vol. 119, pp. 11342-11349, 2003.

500806 Thermodynamic Modeling of Lithium Salt Systems with the Electrolyte NRTL Model

Abstract:

Thermodynamic models are essential to facilitate the advancement of process design, optimization, and operation range of absorption heat pumps and refrigeration units that utilize lithium salt systems. The objective of the present work is to continue working toward a comprehensive thermodynamic representation of lithium salt systems by implementing the symmetric electrolyte NRTL (eNRTL) activity coefficient model. This solubility model correlates the nonideality of the solution with the composition dependency by utilizing two binary interaction parameters for each of the electrolyte-electrolyte and molecule-electrolyte pairs in the system. Each parameter is determined by applying up to three temperature dependent coefficients and then regressed from experimental data. This allows the parameters to establish the accuracy of the model for concentrations up to saturation and temperatures between 273.15-473.15 K. The lithium chloride and water binary system has already been completed, therefore this work will focus on the binary systems containing lithium bromide, lithium iodide, and lithium nitrate with water.

500165 Why Tetrabutylphosphonium Hydroxideâ€“Water Mixtures Can Dissolve Cellulose at Extremely High Water Mass Fractions

Abstract:

In past research work, mixtures of the ionic liquid tetrabutylphosphonium hydroxide (TBPH) and water have shown a unique ability to dissolve a high percentage of cellulose at near-room temperatures.[1] This mixture only showed the high dissolution rate of cellulose for high mole fractions of water, well above the allowed range for other ionic liquids such as the imidazolium series.[1] This makes TBPH a potential solvent of interest to study to for the biofuel extraction process.

In this work we determine the thermodynamic properties of the ionic liquid TBPH in water. The density of the mixtures, over whole range of water percentages, were calculated from the simulation data. The reason for the high cellulose dissolution rate is examined, from the simulations visualized data and the radial distribution functions (RDFs). It appears that the physical location and structure of the water may highly influence the ability of this mixture to dissolve cellulose.

By calculating other thermodynamic properties (heat capacities, excess molar volumes, etc.) from the water/TBPH mixture, as well as including cellulose, we can better compare the performance of TBPH to other cellulose-dissolving ionic liquids.

[1] Mitsuru Abe, Yukinobu Fukaya and Hiroyuki Ohno, 2011, Fast and facile dissolution of cellulose with tetrabutylphosphonium hydroxide containing 40 wt% water

484101 Phase Behavior of Compressed Gases in Ionic Liquids at the Liquid-Solid Transition Point

Abstract:

Ionic liquids (ILs) are a unique class of molecular salts that exhibit melting points below 100°C. The functionality of ILs provide this class of molecules numerous advantages for applications in reactions, separations, and materials processing due to their molecular tunability through cation and anion selection. Ionic liquids also exhibit negligible vapor pressures which may lead to more sustainable processes and new engineering applications. For these reasons, ILs are being considered for a wide range of industrial processes.

Select ionic liquids demonstrate significant absorption capacity for compressed gases including carbon dioxide (CO_2). Our laboratory is setup to measure and model the solubility of gases and vapors in ionic liquids using state of the art microbalance instruments. This presentation will discuss the absorption of compressed gases such as CO_2 in ionic liquids at temperatures approaching the freezing point of the pure IL. The behavior of an absorbed gas in an ionic liquid as the IL transitions between liquid and solid states will be presented for the first time.

500593 Is Difference in Hydrogen Bond Ability of Anions an Exclusive Descriptor of Nonideality in Binary Ionic Liquid Mixtures?

Abstract:

Ionic liquids (ILs) are molten salts composed entirely of ions that exist as liquids under ambient conditions. The traditional approach for tuning the property behavior of ionic liquid has taken advantage of a proper ion selection. Recently, the use of ionic liquid mixtures has been proposed as a strategy to obtain task-specific properties of ionic liquid solvent systems. As a result of which, a thorough understanding of the structure-property relationships is only now beginning to be probed for binary ionic liquid mixtures. One of outstanding questions in the binary ionic liquid mixture research is whether there are molecular descriptors that can be used to predict nonideal behavior of such mixture systems. In an effort to answer this question, atomistic molecular dynamics (MD) simulations were carried out to evaluate if the difference in hydrogen bonding ability of the anions can serve as a primary descriptor for nonideal in binary ionic liquid systems. For this study, three sets of binary IL systems containing the common cation 1-*n*-butyl-3-methylimidazolium ($[C_4mim]^+$) in combination with chloride (Cl^-) and acetate ($[CH_3COO]^-$), Cl^- and trifluoroacetate ($[CF_3COO]^-$), and $[CH_3COO]^-$ and $[CF_3COO]^-$

assessing the nonideal behavior of these systems. The Cl-$[CH_3COO]$ system exhibited nearly ideal behavior as the hydrogen bonding ability of the anions were similar. On the contrary, positive deviation was observed for the Cl-$[CF_3COO]$ system which is not consistent with difference in the hydrogen bonding ability of anions. Further, the $[CH_3COO]$-$[CF_3COO]$ system displayed negative excess molar volume despite very small difference in the hydrogen bonding ability of the two anions and the fact that considerable structural changes were observed in first solvation shell of the cation. Distribution of specific interaction energies, combination of van der Waals and electrostatic interaction experienced by cation and anions individually were evaluated and will be presented to rationalize the excess volume behavior for these binary IL systems. The results obtained for the three systems will be compared to those reported in our previous study[1] where nonideal behavior was accounted for by differences in the hydrogen bonding strength that leads to changes in structural organization of ions.

Overall, our results from MD simulations of binary IL mixtures indicate that changes in local microstructure of ILs can provide only a partial understanding of the nonideality in the binary ionic liquid mixtures, and a delicate balance between hydrogen bonding, electrostatic and dispersion interactions is required to explain the origin of nonideality.

1. Kapoor, U. and Shah, J. K., â€œPreferential ionic interactions and microscopic structural changes drive nonideality in binary ionic liquid mixtures as revealed from molecular simulationsâ€ *Ind. Eng. Chem. Res.*, 55, 13132-13146 (2016).

496974 Polymer–Solvent Phase Behavior of Lignin with Hot Aqueous Solvent Systems

Abstract:

Lignin is one of the world's most abundant biopolymers, comprising almost 30% of all woody biomass; only cellulose is more common. Furthermore, it is unique among renewable biopolymers in having significant aromatic content, giving it the potential for polymeric systems applications ranging from electrodeposited coatings to polyurethane foams to an inexpensive precursor for carbon fibers. We have discovered that lignins exhibit some of the classic features of polymer–solvent phase behavior when combined with selected aqueous organic solvent systems at elevated temperatures; additionally, this liquid-liquid phase behavior has been found to be useful for solvating, fractionating, and even cleaning lignins. For example, when raw alkali lignins are combined with hot acetic acid–water mixtures, two equilibrium liquid phases are formed, one solvent-rich and one polymer (lignin)-rich. The metal salts in the raw lignin (typically 10,000 ppm) are extracted into the solvent phase, resulting in a purified "ultrapure" lignin as the polymer phase. Furthermore, the lignin also partitions between the two phases according to mol wt, with the distribution of the lignin being controlled by the organic/water ratio in the solvent mixture. Finally, the lignin polymer phase is highly solvated, rendering it easily processable.

Determining the temperature at which the lignin–solvent mixture undergoes the phase transition from solid–liquid equilibrium to liquid–liquid equilibrium can be difficult, as visual observation is impeded by the opalescence of the system. Thus, a detection method based on electrochemical impedance spectroscopy (EIS) has been developed and was used to locate SLE to LLE phase boundaries for two lignin polymer–solvent systems over a range of organic/water solvent ratios. We expect that this EIS method can be applied to other polymer–solvent systems when standard techniques such as differential scanning calorimetry (DSC) fail.

Liquid–liquid equilibrium (LLE) phase behavior has been measured for the acetic acid–water–lignin system at 70 and 95 °C, with the results being presented as ternary phase diagrams depicting all regions of both fluid and solid phase behaviour. Measurements for at least one other lignin–solvent–water system will also be presented. A number of important thermodynamic properties relevant to the processing of lignin can be obtained from these diagrams, including the degree of solvation in the lignin (polymer)-rich phase as a function of solvent composition, the size of the LLE region available for polymer processing, and the distribution of lignin between the polymer-rich and solvent-rich phases. Our ability to both control and model the size of the LLE region will also be discussed, along with the implications of the observed polymer–solvent phase behavior for the development of a viable purification and fractionation process for raw, unpurified lignins.

483612 Application of Thermodynamic Models in the Context of Liquid Chromatography

Abstract:

Applying the assumption of dilute conditions, the use of thermodynamic models in the context of liquid chromatography is most often avoided. Chromatographic models are commonly based on component mass balances which consider a single convective phase and a thermodynamic equilibrium between adsorbed phases and liquid phases which is described through an isotherm, being a function of liquid phase concentrations. In some cases, however, the common assumption of dilute conditions is not valid, and one has to account for highly concentrated, non-ideal liquid phases, which might even result in a phase split and subsequent multiphase flow [1,2]. Examples for such non-ideal behavior are certain reactions in chromatographic reactors, or systems where interactions of adsorbing components lead to a considerable enrichment of these components. Under these non-dilute conditions, the thermodynamic equilibrium between adsorbed phase and convective phase(s) is consistently described as a function of liquid phase activities (rather than liquid phase concentrations), and, in the case that solubility limits are exceeded, the phase split and compositions of the convective phases have to be determined. For these problems, a thermodynamic model for the liquid phase(s) is required.

In this contribution, we consider the quaternary system phenetole (PNT), 4-tert-butylphenole (TBP), methanol and water at 296 K and ambient pressure, and its interaction with the adsorbent Zorbax 300SB-C18. In order to describe the dynamic, chromatographic behavior of this system under non-dilute conditions, accounting for adsorption and possible phase separations/ two-phase flow, we have to characterize the thermodynamic behavior of this system.

In a first step, we characterize the thermodynamic behavior of the convective phase(s), i.e. of the quaternary system in the absence of the adsorbent, through liquid-liquid, solid-liquid, and solid-liquid-liquid equilibrium experiments. A UNIQUAC model is used to quantitatively describe the thermodynamic properties of the system (see Figure 1), with model parameters fitted to the experimental data.

Secondly, the thermodynamic equilibrium between the convective phase(s) and the adsorbed phase has to be analyzed. For this purpose, adsorbed phase concentrations of the adsorbing components PNT and TBP are determined experimentally through dynamic column experiments as a function of the liquid phase composition (compare Figure 2a). Liquid phase activities can be calculated from the liquid phase composition applying the established UNIQUAC model. Determined adsorbed phase concentrations can be related to liquid phase activities through simple isotherms (see Figure 2b). The established isotherms describe the adsorption behavior for a very broad range of conditions (diluted to highly concentrated and multi-phase conditions), while common isotherms, being a function of liquid phase concentrations, are only valid in a low concentration range of the adsorbing component and for a constant solvent (methanol : water) ratio.

Application of the established UNIQUAC model and adsorption isotherms in the chromatographic model provides an accurate description of the chromatographic behavior for a broad range of conditions.

Figure 1: Thermodynamic model of the quaternary system PNT-TBP-methanol-water (prediction by UNIQUAC model)

(a)

Figure 2: (a) experimental adsorption data for PNT at different solvent ratios r (methanol content in solvent in mass fractions), over the liquid phase concentration of PNT, (b) adsorption data for PNT, with liquid phase compositions converted to liquid phase activities by applying the UNIQUAC model.

References

[1] Jermann S., et al., (2015), *J. Chrom. A, 1425, 116-128*

[2] Ortner F., Mazzotti M., (2017), *in preparation*

Solubility and Freezing point depression (FPD) measurements of Na-1-ethanethiolate, Na-1-propanethiolate, Na-2-propanethiolate, Na-1-butanethiolate and Na-2-methyl-2-propanethiolate salts in pure water.

Qazi Mohammed Omar[1], Salman Haroon[1], Jean-Noël Jaubert[2], **Javeed A. AWAN[1]****

[1]*Institute of Chemical Engineering & Technology, University of the Punjab, Lahore, Pakistan.*
[2]*Université de Lorraine, Ecole Nationale Supérieure des Industries Chimiques, Laboratoire Réactions et Génie des Procédés (UMR CNRS 7274), 1 rue Grandville, 54000 Nancy, France.*

ABSTRACT

In this work we have measured the solubility and the freezing point depression (FPD) of Na-1-ethanethiolate, Na-1-propanethiolate, Na-2-propanethiolate, Na-1-butanethiolate and Na-2-methyl-2-propanethiolate salts in pure water within concentration range of 0-49 wt/wt % at 1 atm. The freezing point curves and solubility curves of five mercaptide salts have been measured at various concentrations of mercaptide salts and further used to find eutectic points.

Key Words: *Freezing Point Depression, Thiol salts, Sodium mercaptide salts, Sodium mercaptide salts + water system.*

*Corresponding authors: i) Dr. Javeed A. Awan (Associate Professor), Email: javeedawan@yahoo.com , Tel: +92 42 992 304 62; Fax: +92 42 992 311 59 ii) Prof. Dr. Jean-Noël Jaubert (Head and Professor), Email: jean-noel.jaubert@univ-lorraine.fr,

Introduction

The colligative property of solutions such as freezing point depression is a concept that demonstrates how various solutes affect colligative properties with the change in temperature and composition. Phase equilibria, such as solid–liquid equilibria in electrolyte solutions are of crucial importance in many industrial processes like wastewater treatment, desalination, and crystallization (Bodzewski *et al.*,2016; Marliacy *et al.*, 1998; Mohs *et al.*, 2011). The prediction of the freezing point is especially important for freeze concentration (FC) and eutectic freeze concentration (EFC). The eutectic freeze concentration (EFC) is a novel and promising separation technique by concentration to recover simultaneously moisture in the form of ice and salt crystals with high purities (Van der Ham et al., 1999). There are numerous calculations in chemical and petroleum engineering where reliable predictions of phase behavior of electrolyte solutions are required, from desalination units to gas hydrate inhibition in offshore/arctic drilling operations or in oil/ gas pipelines. Accurate models describing kinetic behavior of thiols are therefore necessary for analyzing, designing, and optimizing processes in the chemical and petroleum industry. The effect of electrolyte solutions on phase behavior of electrolyte solutions is important for a better understanding of the effect of salts on gas hydrate stability conditions. (Haghighi et al., 2008) In oil and gas processing industry, it is important to know the properties of SLE (Solid liquid equilibrium) of mercaptide salts because an increase in concentration or a decrease in temperature could result in an unexpected precipitation of solid. Also the addition of mercaptide salts proved helpful for the leakage detection of natural gas. Accumulation, precipitation of solids, freezing and possible clogging of pipes and pumps can be avoided by knowing the physical or chemical properties of these components. Solid equilibrium data is more difficult to obtain as compared to vapor-liquid and liquid-liquid data. The most frequently used methods consist of following i.e. the indirect determination of phase diagram by means of cooling/or heating curves and by the direct measurement of activity coefficients in concentration cells (Null, 1970) . Mercaptan belongs to thiol-group of compound which contain an –SH group bound to a radical R. Mercaptan properties are governed to a large extent by the length of this radical. (Awan *et al.*, 2011; *Awan et al.*, 2009) The sulfur-containing thiols act as gases in their acid form. This study will focus on the sodium salts of thiols. Unfortunately limited or no information about the kinetics and phase behavior on mercaptide salt-water solution is available in open literature.

The experimental solid liquid equilibrium data are required for optimization of the theoretical models, with adjustable parameters. In this work, the freezing point depressions of five electrolyte aqueous solutions (H_2O+Na 1-ethanethiolate, H_2O+Na-1-propanethiolate, H_2O+Na-2-propanethiolate, H_2O+Na-1-butanethiolate and H_2O+Na-2-methyl-2-propanethiolate) has been measured by using the apparatus provided by PHYWE Germany, according to the standard equipment conditions provided by PHYWE. The eutectic points and the solubility temperatures has also been studied in this work. The measured results of freezing point depressions were then compared with calculated depression of mercaptide salts based on molecular weight.

Methods and Materials

Materials

The chemicals used in this work are Na-1-ethanethiolate (purity ≥90%), Na-1-propanethiolate (purity ≥95%), Na-2-propanthiolate (purity ≥90%), Na-1-butanthiolate (purity ≥95%), Na-2-methyl-2-propanethiolate (purity ≥90%) and Ethanol (purity ≥99.9% by volume) were provided by Sigma-Aldrich Germany. No further purification of these chemicals was made prior to their use in experimental work. The chemicals used along with their CAS no. are shown in Table 1. The NaCl (purity ≥99.5%) has been purchased from GATT-KOLLER Germany and was used for calibration of the FPD apparatus. The deionized water has been prepared at ICET through alfa-pore machine (WAP-4).

Experimental Setup

Experimental boundary conditions

The Na mercaptide salts were dissolved in deionized water with a conductivity of 0.6 μS/cm. The mercaptide salts has been weighed in grams (g) by using analytical weighing balance provided by Anton Par having an accuracy of ± 0.0001g. The concentrations of mercaptide salts + deionized water solution were calculated by weight percent (wt %) by the following formula.

$$wt\% = \frac{\text{m of mercaptide salt}}{\text{m of mercaptide salt+m of water}} \quad (1)$$

To measure the freezing point depression of the mercaptide salts, it is appropriate to find the saturation point first so that we can identify the eutectic point. As it is impossible to predict the eutectic point of a solution having an unknown substance, it is assumed that it is in a level that is approximately half of the saturation point at 298.15K. The freezing points of five mercaptide salts were measured at 1wt%, 2 wt%, 5 wt%, 8 wt%, 10 wt%, 15 wt% and 23 wt%. The solubility measurements of mercaptide salts were performed at 23 wt%, 28 wt%, 31 wt%, 34 wt%, 37 wt%, 40 wt%, 42 wt%, 47 wt% and 49 wt%.

Freezing-point depression method

Eutectic points of the mercaptide salts were determined by the two methods *i.e.* freezing point of two component system, by means of lowering the freezing point and the precipitation of the solid using a method similar to the FPD called SLE by FPD.

Apparatus

Freezing-point depression apparatus has been supplied by PHYWE (Germany). Actually freezing point depression has been found by using a Beckmann apparatus with few modifications as shown in Fig.1. The method to find freezing point depression is already described in literature (Marzzacco *et al, 1980)* In all the experiments, the temperature was controlled by ethanol-salt cooling bath system (1) through circulation. The temperature probe (2) was connected to a temperature meter device (3), making it possible to observe data for the duration of the experiment. A sample tube (4) is provided with a lid (5) which is mounted on a rubber stopper (6). Rubber stopper acts as a lid for the temperature-controlled cooling bath (7). A manual stir spring (8) and the temperature probe (2) is directed through the rubber stopper (6) and the lid (5) of the sample tube (4), which makes it possible to maintain a homogeneous temperature in the sample tube during the experiment. The cooling jacket (9) and sample tube (4) are both equipped with a magnetic stirrer tablet (10) and cooling jacket (9) is placed on a magnetic stirrer (12). A beaker (13) containing ethanol is placed on a second magnetic stirrer (14) which is used for heating the sample after achieving freezing point of sample solution. The temperature meter device has been synchronized with central data acquisition unit (15) through cobra port 4.

For each experiment, the whole apparatus like probe sleeves, magnetic stirrers, sample tube of FPD apparatus and beakers were washed properly with chromic acid solution, tap water, distilled water and finally with de-ionized water to get accurate results. Chromic acid solution was prepared with chromic acid powder and concentrated sulfuric acid (98% pure, commercial grade), as it removes all dirt and sticky material from inner tube.

Ethanol has been used in outer tube of freezing point depression apparatus which acts as a heat transfer media. It transfers cooling to target solution present in the inner tube. The small quantity (two to three drops) of ethanol has been added in temperature probe sleeves for uniform heat measurements. The level of coolant (Ethanol) in the temperature-controlled cooling bath was maintained higher than the liquid level of the sample in the sample tube for sufficient cooling of the sample, but level of the sample should be under the lid of the sample tube so that cooling liquid would not enter into the sample tube and contaminate the sample.

In order to avoid super cooling, it is important that the temperature of the cooling bath should be 7K below the supposed or calculated freezing point of the sample. During the experiment, it is important to carry out the stirring continuously with magnetic stirrer to ensure a homogeneous temperature distribution in the test tube. Freezing points measured for each concentration was based on ten measured points. The average of the ten measurements for each point were considered and shown in the tables. After the measurements of each freezing point, the sample was heated again in the beaker containing ethanol or in this case water, until the sample melted and only a small amount of crystals remained in the test tube. Then sample was cooled down till frozen. The remaining amount of crystals in the sample tube was used as seeds for the formation of ice at the next measurement. This is also the reason why cooling in the subsequent tests is considered to be negligible. The freezing points of NaCl + water were measured at different concentrations as shown in Table 2 and were compared with Kaj Thomsen database (Thomsen et al., 1996) as shown in Fig. 2 for purpose of calibration.

Solubility tests

In order to determine the solubility of the mercaptide salts, solutions were cooled until precipitation occurred. Subsequently the test tube was heated in a beaker with water or air, or by holding the tube by hand until the sample was again visually assessed to be in solution. Temperature at which the last crystal dissolved was recorded by a temperature probe and that is the equilibrium temperature of that solid – liquid system. These experiments were repeated 5 to 6 times with each sample till we get the consistent results. The solubility and saturation point of the mercaptide salts has been measured by dilution method and evaporation method. In dilution method the solubility was found by naked eye whether the mercaptide salt is completely soluble in liquid or not. It is therefore necessary to have a certain minimum amount of the solvent in aqueous solution so that it will be possible to observe the phases in the mixture.

Procedure

The solubility of mercaptide salts has been measured in 1g of deionized water. The amount of salt added were limited, therefore the saturation point of solubility has been observed as the range because it is difficult to hit the accurate values. In this work the mean value of saturation point of solubility range has been reported. The samples prepared for solubility tests were centrifuged to separate the precipitated solids from the mercaptide salt-water solution. The solution sample was drawn up and placed in a bowl in a vacuum chamber at 308 K until all the water is evaporated leaving behind a dry matter. The amount of dry matter is weighed and thus it is possible to calculate the concentration of the solution.

Results and discussion:

In this work two methods were used to find the saturation point and solubility of five mercaptide salts. The solubility and saturation point obtained from dilution or evaporation methods have close correlation based on concentration. The solubility of mercaptide salts in water by dilution is shown in **Table 3** and the solubility of mercaptide salts in water by evaporation is shown in **Table 4**. The freezing point depression measurements of mercaptide salts and solubility points of Na-1-ethanethiolate, Na-1-propanethiolate, Na-2-propanethiolate, Na-1-butanethiolate and Na-2-methyl-2-propanethiolate have been measured at room temperature and presented in **Table 5 and 6** respectively.

Freezing point depression and solubility measurements

Measurement of Na-1-ethanethiolate in the concentration range i.e. 1% to 49wt%

The freezing points and solubility for *Na-1-ethanethiolate* in aqueous solution has been measured in this work. The eutectic point is found to be approximately 23.1 wt% solution at 252.1K as shown in Fig. 3. These results are in accordance with the assumption that the eutectic point would be half of the solubility point. The standard deviation of Freezing point depression curve is lower than the solubility curve located to the right of the eutectic point. The last point *i.e* 49wt% solution must be considered as an outlier.

Measurement of Na-1-propanethiolate in the concentration range 1% to 37 wt%

The eutectic point of Na-1-propanethiolate is found slightly higher than the previous one at a level of about 24.2 wt% solution as shown in Fig. 4. Furthermore, the temperature of the eutectic point was also slightly lower *i.e* 253K. The standard deviation of Freezing point depression curve is lower than the solubility curve located to the right of the eutectic point. Here we have again an outlier at the highest concentration i.e. 37 wt% of mercaptide salt-solution.

Measurement of Na-2-Propanthiolate in the concentration range 1 wt% to 47 wt%

For Na-2-Propanthiolate, both temperature and solubility rate further decreased in relation to *Na-1-ethanethiolate* and *Na-1-propanethiolate* as shown in Fig. 5. The eutectic point is found to be approximately 26.6 wt% solution at 250.3K. The standard deviation of Freezing point depression curve as well as solubility curve is low in this case.

Measurement of Na-1-Butanethiolate in the concentration range 1 wt% to 42 wt%

Na-1-butanethiolate achieved a eutectic concentration of approximately 24 wt% solution at 261.5K as shown in Fig. 6. The standard deviation of Freezing point depression curve is lower than the solubility curve located to the right of the eutectic point. However, here we have again an outlier at the highest concentration of mercaptide salt-solution *i.e* 42 wt%.

Measurement of Na 2-methyl 2-propanethiolate in the concentration range 1 wt% to 34 wt%

A declining freezing point depression curve for Na 2-methyl 2-propanethiolate with relatively low standard deviation is shown in Fig. 7. Again, we have eutectic point at which the concentration of mercaptide salt has increased to about 27.5 wt% at 253K. Standard deviation of the both freezing point depression and solubility curve is very low.

Comparison of the five measured Mercaptide salts based on wt/wt%

The FPD of five mercaptide salts decreases with concentration as shown in Fig. 8. It is interesting to note that the Na 1-butanethiolate is the only mercaptide salt for which the freezing point depression curve is linear with concentration. It is also observed that FPD of other mercaptide salts are much lower at higher concentrations irrespective of the fact that all five mercaptide salts have the same functional group. The solubility of all five mercaptide salts are measured and shown in Fig. 9. The reason why the five points are located at the same temperature is because it is obtained from the maximum solubility measured (saturation points) at room temperature *i.e.* 298.15K.

Comparison of the five measured Mercaptide salts corrected for purity wt/wt%

It is not correct just to compare on the basis of weighed amounts because the five mercaptide salts have different purity so it is necessary to add correction for purity prior to compare the mercaptide salts with each other. The purity was calculated by using formula, where the mass (m) of the mercaptide salts is multiplied by the percentage purity (R%) of the product:

$$wt\% = \frac{R\,\% \times m \text{ of Mercaptide salt}}{(R\,\% \times m \text{ of Mercaptide salt}) + m \text{ of water}} \qquad (2)$$

The measured concentrations corrected for purity for comparison of five mercaptide salts in terms of freezing point depression and solubility are shown in Table 7 and 8 and plotted in Fig. 10 and 11 respectively. The correction in purity had not greatly affected on the trends of five mercaptide salts as shown in Fig. 10. Na-1-butanethiolate is still at the top of the five mercaptide salts and still Na-1-propanethiolate showed the lowest and steepest temperature curve.

Fig. 11 shows that the individual behavior of the mercaptide salts in the plot is a little different, but it is very little and it is clear that the trend for each mercaptide salt is not greatly changed.

Comparison of the five measured Mercaptide salts in terms of molality and corrected for purity

For comparison of mercaptide salts it is not enough just to apply correction for purity. As all the five mercaptide salts represent different molecules, it also means that they have different molar mass (MW) and molecular structure. So, it would be appropriate to further compare the individual mercaptide salts based on their molality.

Molality (b) is calculated as follows:

$$b = \frac{\text{m Mercaptide salt (adjusted for purity)}}{\text{MW of Mercaptide salt x 1 kg of m of water}} \quad (3)$$

The freezing point depression for all five mercaptide salts adjusted for weight % purity and molality are presented in Table 9 and plotted in Fig. 12. Whereas the measured solubility for all five mercaptide salts adjusted for both weight% and molality are presented in **Table 10** and plotted in Fig. 13. As clearly shown in Fig. 12, measurements in terms of molality have changed the results; it means that the mercaptide salts mole to mole properties are slightly closer to each other than when only measured on weight measurements. This trend also seems logical since been adjusted for the relatively large fluctuations in molecular mass, from 84.12g / mole to 112.17g / mole. However, five mercaptide salts have not changed their position but moved a little bit closer to each other and it can be seen that the trend lines for each of the five Mercaptide salts approach one another becomes more linear.

Theoretical calculations of the freezing point depression

The activity coefficient based models like Pitzer equation and three characteristic parameter correlation (TCPC) models have not been used in this investigation, due to the fact that mercaptide salts has been considered as weak electrolyte salts. In this work the freezing point depression has been calculated and considered to be dependent on the solute concentrations only. It has been estimated by simple linear relationship with the cryoscopic constant (kf) also known as Balagden's law which has been reported in (Atkins, 2009).

$$\psi_F = k_f \times b \times i \quad (4)$$

Where ψ_F is the freezing point depression, k_f is cryoscopic constant which is dependent on the properties of the solvent b is the molality of salt and $\boldsymbol{i}$ is the vanthoff factor. The cryoscopic constant k_f=1.86 K kg mol^{-1} **(Atkins, 2009)** and van't hoff factor which shows the number of ions included in mercaptide salt has been taken from literature.

$$\psi_{FS} = \psi_{DW} - \psi_F \tag{5}$$

Where ψ_{FS} is the freezing point of solution and ψ_{DW} is the freezing point of deionized water. The calculated freezing points for mercaptide salts are shown in table 11.

Comparison of measured and calculated freezing point for Mercaptide salt

Comparison of measured and calculated freezing points of mercaptide salts (Na-1-ethanethiolate, Na-1-propanethiolate, Na-2-propanethiolate, Na-1-butanethiolate and Na-2-methyl-2 propanethiolate) corrected for purity and molality are plotted in Figs from 14 to 18 respectively. Fig. 14 highlights that the prediction of freezing points of Na-1-ethanethiolate at small concentrations is promising although measured values start to deviate from predicted values for molal concentration higher than 0.75. However, it is to be expected that measured values of Na salts have a tendency to lie at a level below the ideal values of the calculated values. Fig. 15 highlights that the prediction of freezing points of Na-1-propanethiolate at small concentrations is promising although measured values start to deviate from predicted values for molal concentration higher than 0.71. Again it could be seen that the measured values lie below the ideal curve. Again, Fig. 16 highlights that the prediction of freezing points of Na 2-propanethiolate at small concentrations is promising although measured values start to deviate from predicted values for molal concentration higher than 0.62. However, the deviation was not so strong as in the case with Na 1-propanethiolate, where there was comparatively a great difference between calculated and measured values at 2.5 molal concentration. Na-2-propanethiolate has comparatively small temperature difference at 2.5 molal concentrations. Again it could be seen that the measured values lie below the ideal curve. There is approximately a strong correlation between the measured and calculated values for all concentrations in case of Na-1-butanethiolate as shown in Fig. 17. However, the other mercaptide salts deviate from ideal curve at approximately 0.7 molal concentration.

As shown in Fig. 18, a prediction of freezing points of Na-2-methyl-2-propanethiolate is promising at small concentrations; however measured values started to deviate from predicted values for molal concentration higher than 0.7 . The temperature difference grows to about 9K at approximately 2.3 molal concentrations.

The prediction of freezing points of mercaptide salts at lower concentrations is promising however measured values deviate at high molal concentration for all the five mercaptide salts. This might be due to the colligative properties of the solution as the solution is both under the influence of the quantity of a dissolved mercaptide salt and the type of salt that is dissolved. However, Na-1-butanethiolate behaves almost ideally across the measured process where the other mercaptide salts confirmed that the amount also influences whether a solution behaves ideally. As mercaptide salts used in this work are not 100% pure. So impurities can also affect freezing points and solubility measurements. That's why error bars of 5% are applied in all measurements.

The measurements of solubility and freezing point depression of five mercaptide salts shows that they have different physical and chemical properties attributed by the molality, and purity of the substance but the complexity of the individual mercaptide salt molecules has no influence on Na-2-methyl-2-propanethiolate as it carries more branches and has comparatively large molecular structure.

Conclusion

The freezing point depressions and solubility points for the five mercaptide salts Na-1-ethanethiolate, Na-1-propanethiolate, Na-2-propanthiolate, Na-1-butanthiolate and Na-2-methyl-2-propanethiolate in water has been measured and compared with the calculated freezing points in terms of molality and purity. The measured values are in excellent agreement with the calculated values only in case of Na-1-butanthiolate. The prediction of freezing points of five mercaptide salts at small concentrations are promising, however measured values start deviating from a molal concentration larger than approximately 0.50. This simple model for the measured data in this work has its place at low concentrations of a solution. At high concentrations, intermolecular forces and water activity becomes more pronounced which has not been included in this investigation. It is further concluded that there is no significant relationship between freezing point depressions, solubility points with the molar masses of Na-mercaptide salts. The measured data is very important for downstream gas processing as thermodynamic behavior and kinetic datum are highly important for separation process design.

References

Atkins, P. (2009). "De Paula Julio. Atkins' Physical Chemistry." Table 6.3, 135.

Awan, J. A. (2009). *Vapor-Liquid Equilibria of Organic Sulfur Species-Alkanolamine Aqueous Solutions in the Presence of Acid Gases* (Doctoral dissertation, PhD Thesis, Mines Paris Tech, France).

Awan, J. A., Usman, M. R., Khan, R. U., & Richon, D. (2011). Electrical Conductivity of n-Propyl mercaptan (n-PM) in N-Methyldiethanolamine (MDEA) Aqueous Solutions at 303K, J. PICHE., *39*(1), 29-33.

Bodzewski, K. Y., Caylor, R. L., Comstock, A. M., Hadley, A. T., Imholt, F. M., Kirwan, K. D., Oyama K.S., Wise, M. E. (2016). Investigating Freezing Point Depression and Cirrus Cloud Nucleation Mechanisms Using a Differential Scanning Calorimeter, J. Chem. edu., *93*(4), 729-732.

Haghighi, H., Chapoy, A., & Tohidi, B. (2008). Freezing point depression of electrolyte solutions: experimental measurements and modeling using the cubic-plus-association equation of state, Ind Eng Chem Res., *47*(11), 3983-3989.

https://www.phywe.com/en/freezing-point-depression.html#tabs4.

Marliacy, P., Hubert, N., Schuffenecker, L., & Solimando, R. (1998). Use of Pitzer's model to calculate thermodynamic properties of aqueous electrolyte solutions of Na 2 SO 4+ NaCl between 273.15 and 373.15 K, Fluid Phase Equilib., *148*(1), 95-106.

Marzzacco, C., & Collins, M. (1980). Convenient freezing point depression apparatus, J. Chem. edu., 57(9), 650.

Mohs, A., Decker, S., & Gmehling, J. (2011). The solid–liquid equilibrium of the binary system H 2 O–DMSO and the influence of a salt (NaCl, KCl) on the thermodynamic behavior: Correlations using a revised LIQUAC model, *Fluid Phase Equilib.*, *304*(1), 12-20.

Null, H. R. (1970). *Phase equilibrium in process design*. John Wiley & Sons Inc.

Thomsen, K., Rasmussen, P., Gani, R. (1996). Correlation and prediction of thermal properties and phase behaviour for a class of aqueous electrolyte systems, Chemical Engineering Science, 51(14), 3675-3683.

Van der Ham, F., Witkamp, G. J., De Graauw, J., & Van Rosmalen, G. M. (1999). Eutectic freeze crystallization simultaneous formation and separation of two solid phases, J. Cryst. Growth, *198*, 744-748.

FIGURES

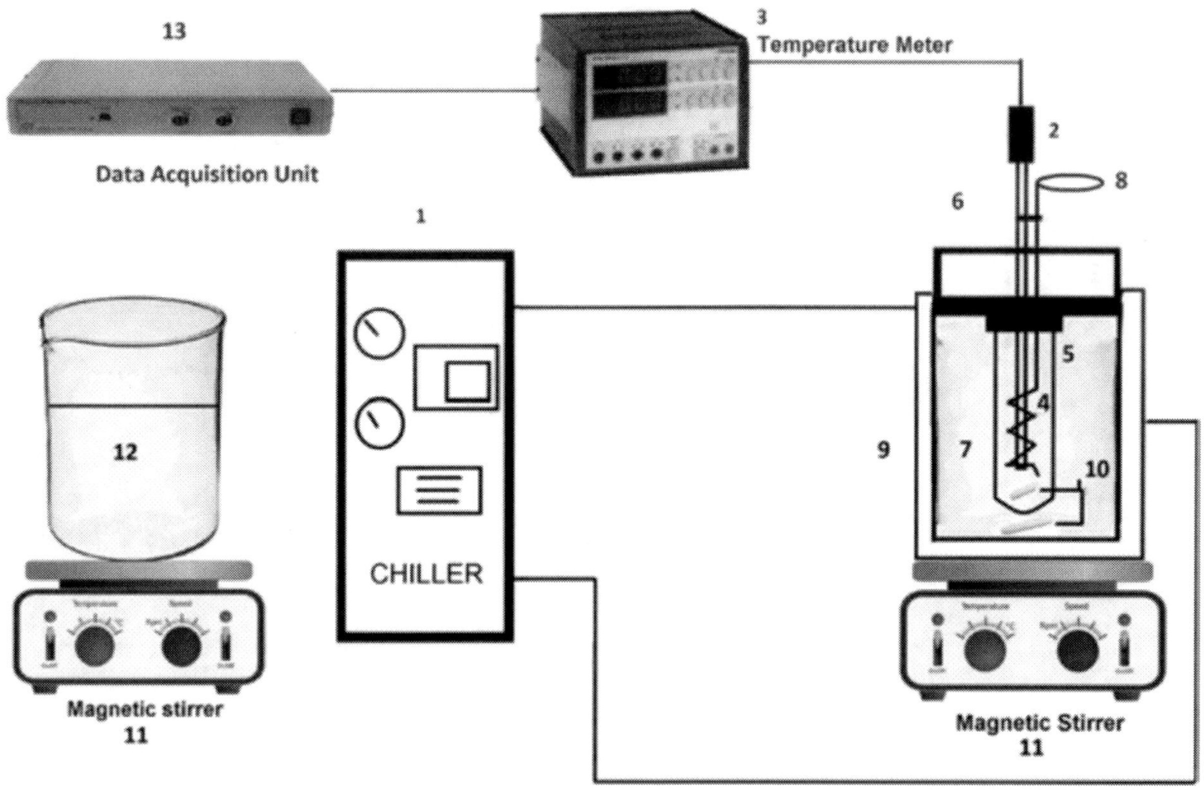

Fig. 1. Schematic diagram of experimental setup: cooling bath (1), temperature probe (2), temperature meter device (3), sample tube (4), lid (5), rubber stopper (6), temperature-controlled cooling bath (7), manual stir spring (8), cooling jacket (9), Magnetic stirrer tablet (10), magnetic stirrer (11), beaker (12), Data Acquisition unit (13)

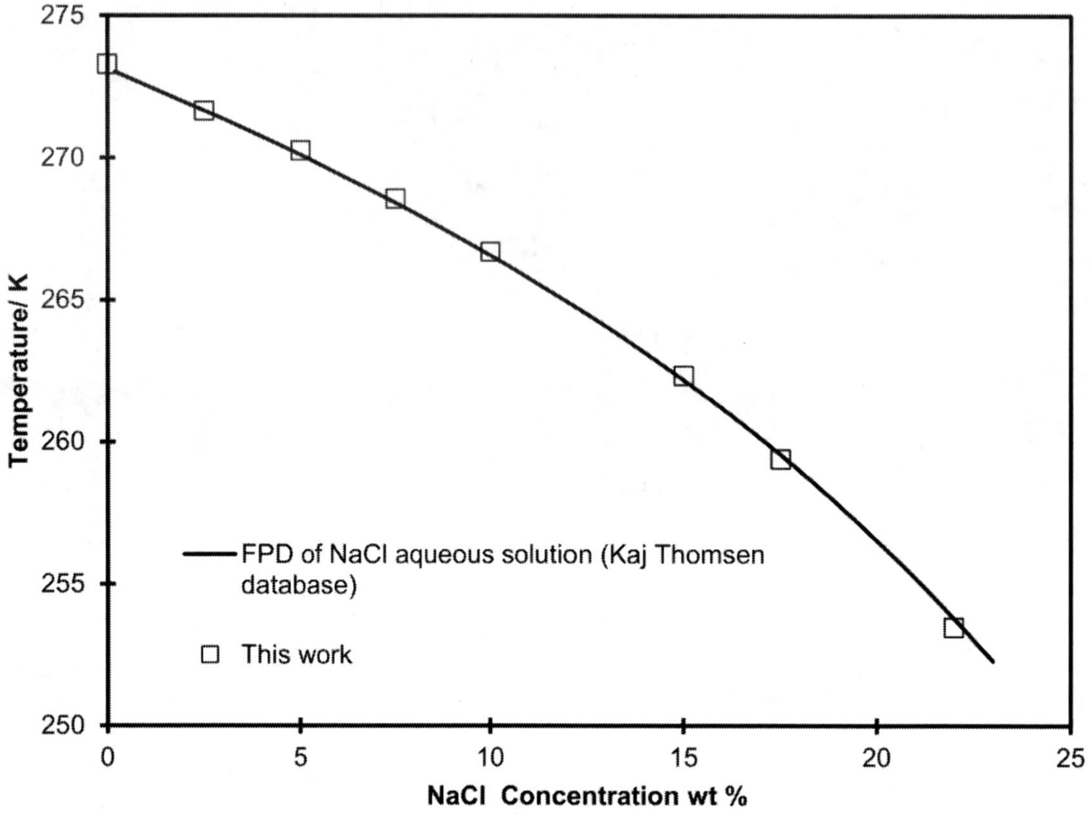

Fig. 2. Freezing point depression of NaCl + H_2O solution for calibration.

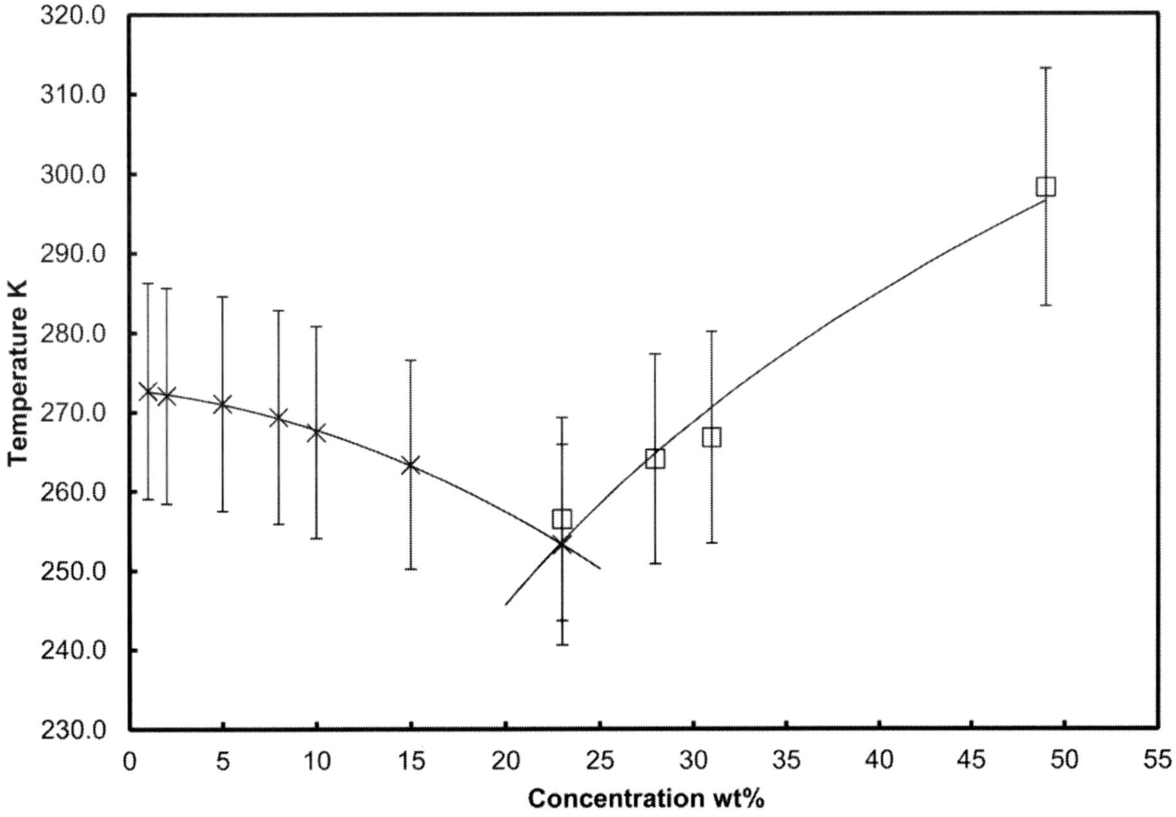

Fig. 3. Freezing point depression and solubility measurement of Na-1-ethanethiolate in water. Error band shows experimental uncertanity ±5%.

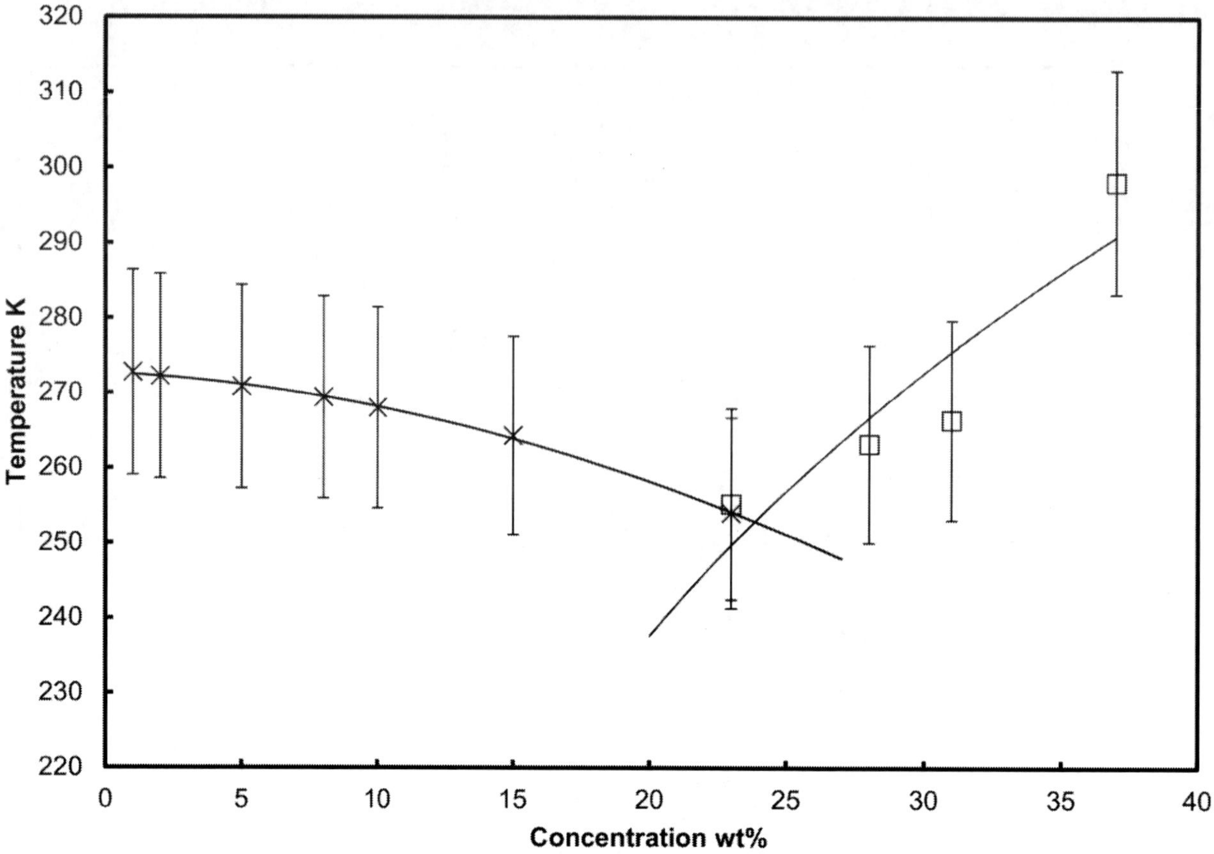

Fig. 4. Freezing point depression and solubility measurement of Na-1-propanethiolate in water. Error band shows experimental uncertanity ±5%.

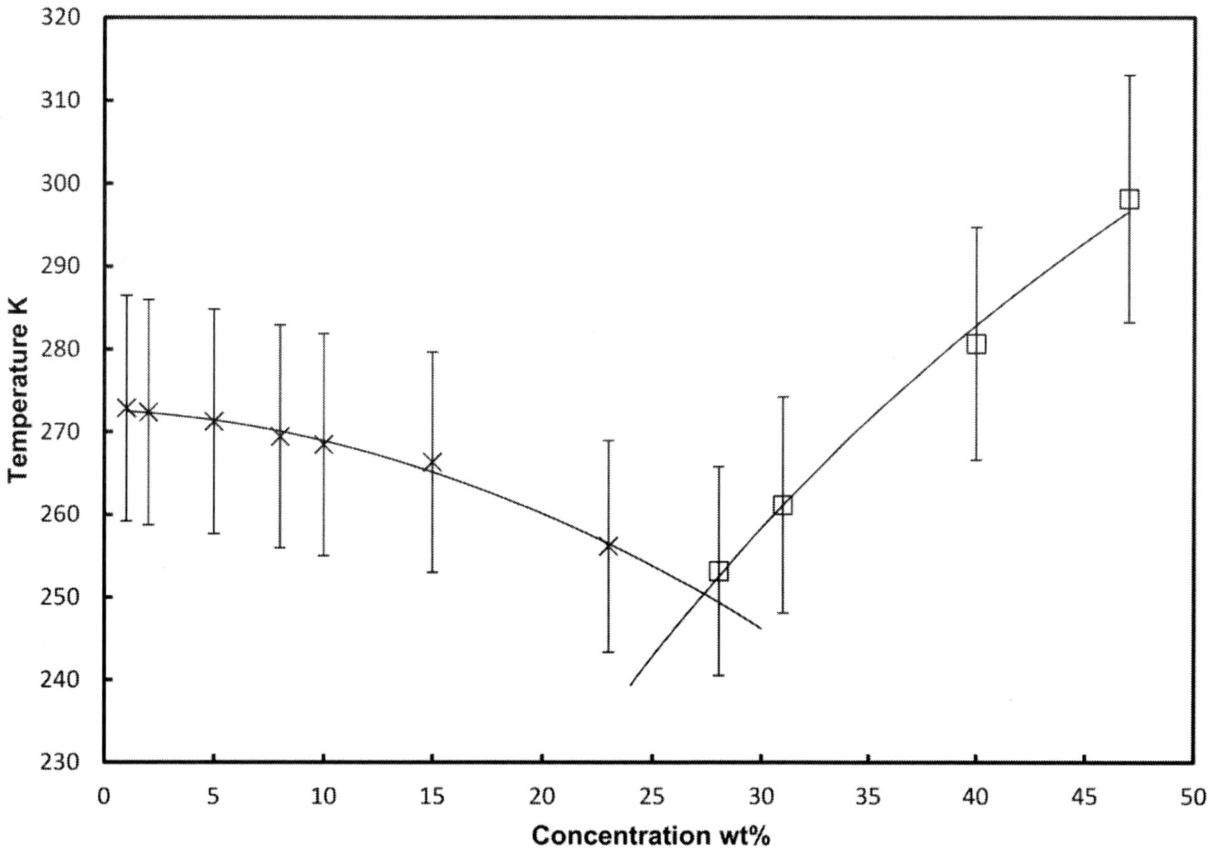

Fig. 5. The freezing point depression and solubility measurement of Na-2- propanethiolate in water. Error band shows experimental uncertanity ±5%.

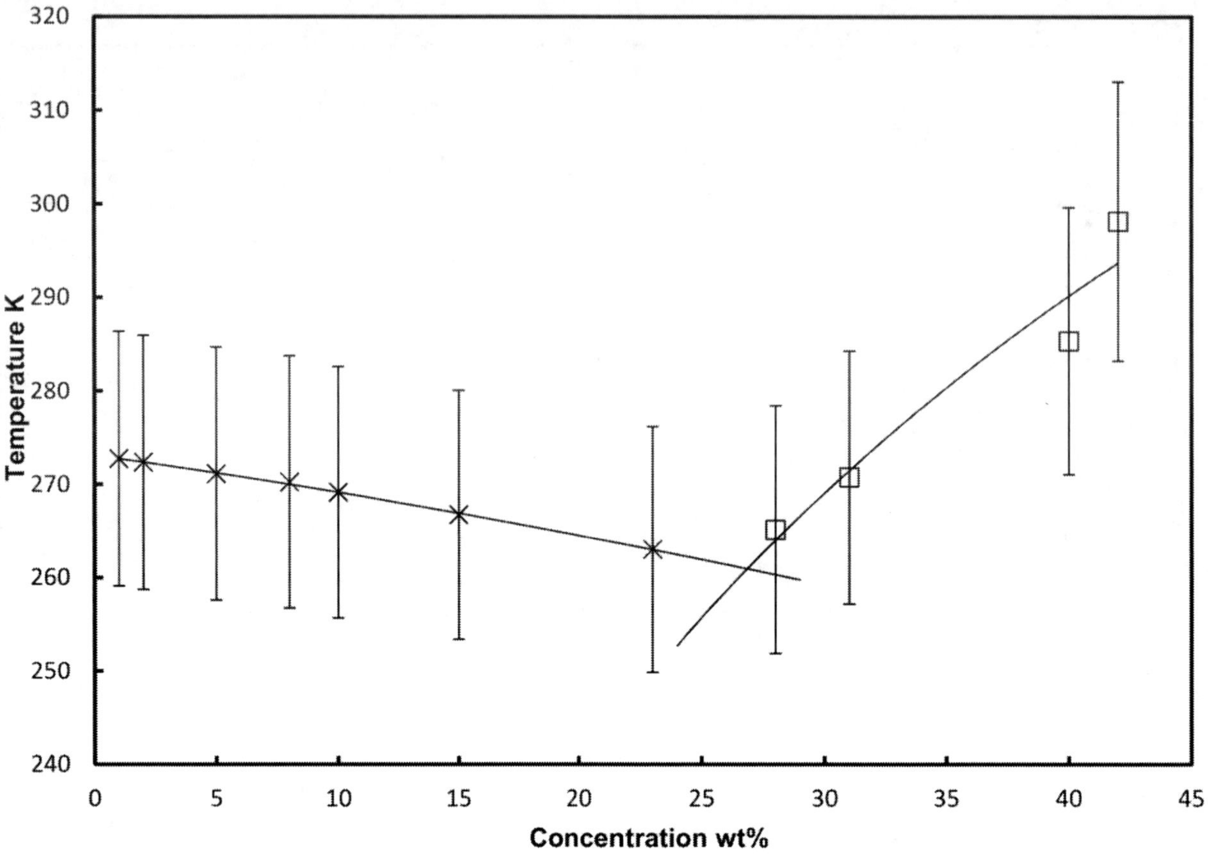

Fig. 6. The freezing point depression and solubility measurement of Na-1-butanethiolate in water. Error band shows experimental uncertanity ±5%.

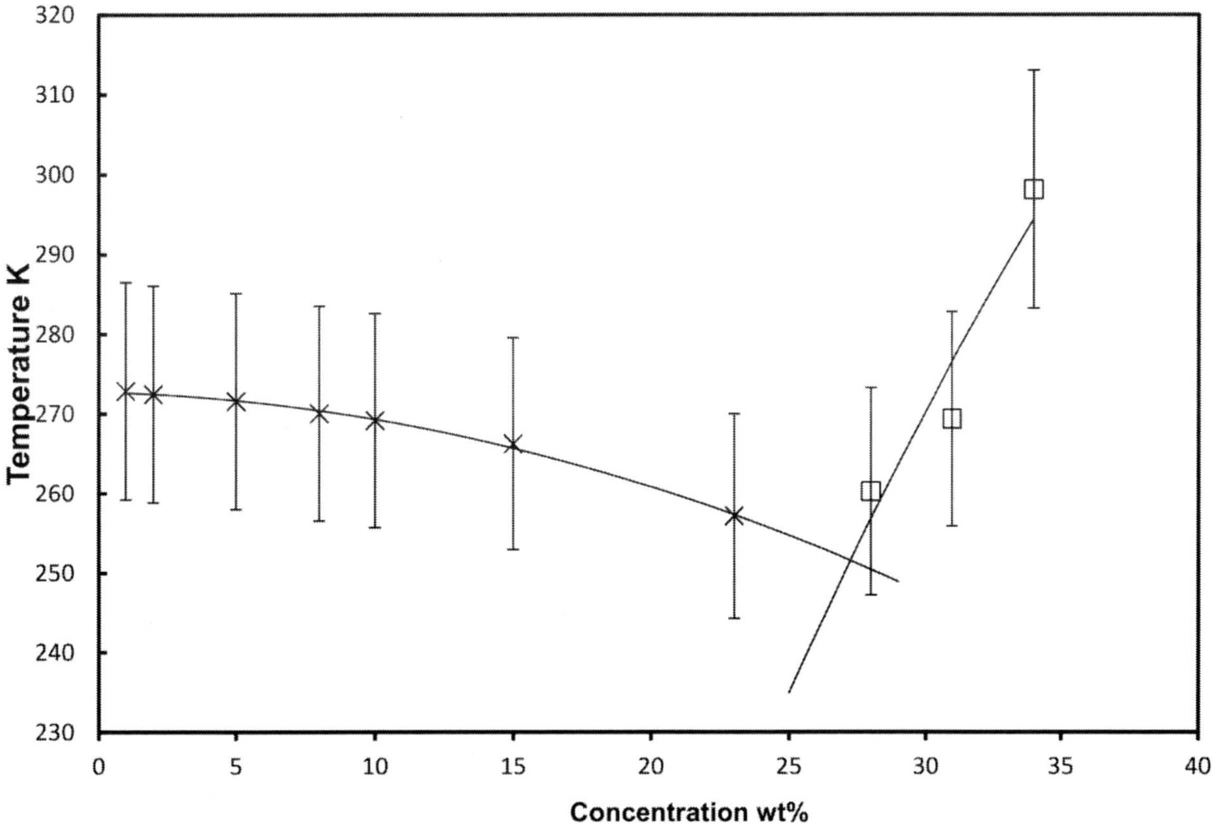

Fig. 7. The freezing point depression and solubility measurement of Na-2-methyl-2-propanethiolate in water. Error band shows experimental uncertanity ±5%.

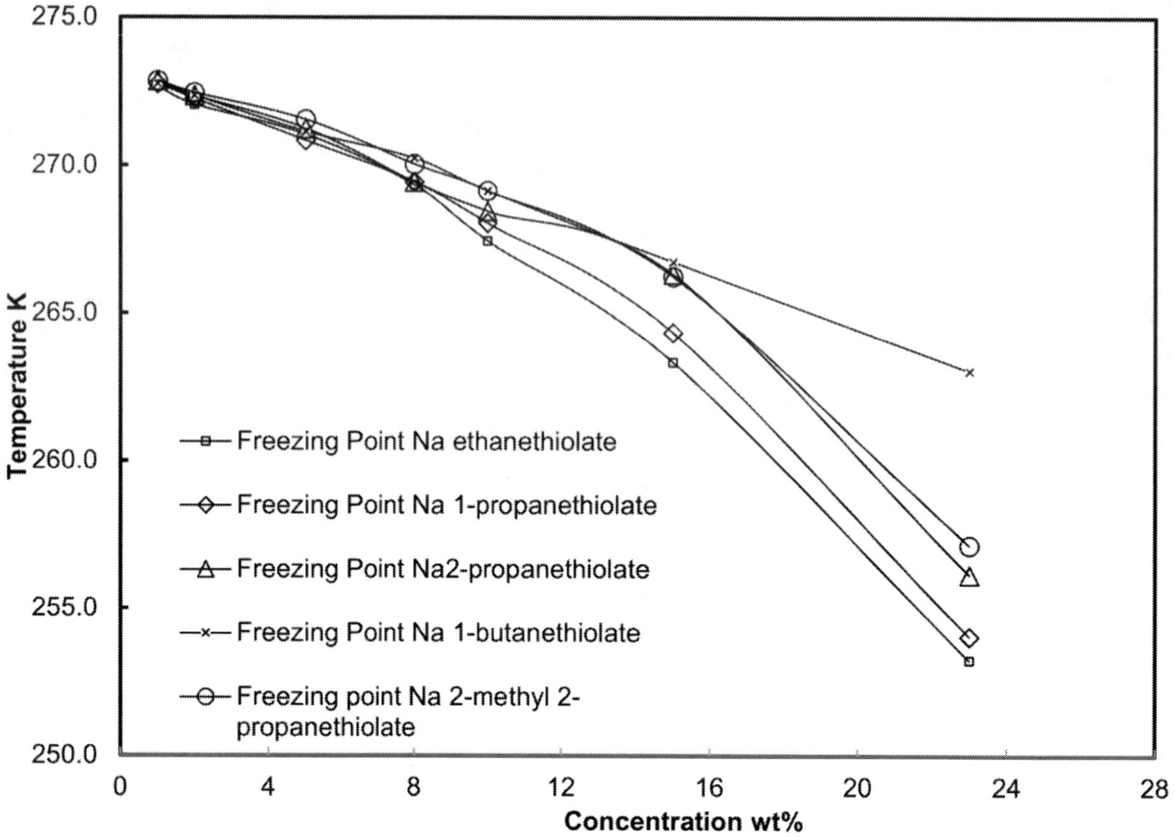

Fig. 8. The comparison of measured freezing point depression of the Mercaptide salts (Na-1-ethanethiolate, Na-1-propanethiolate, Na-2-propanethiolate, Na-1-butanethiolate and Na-2-methyl-2-propanethiolate) in water.

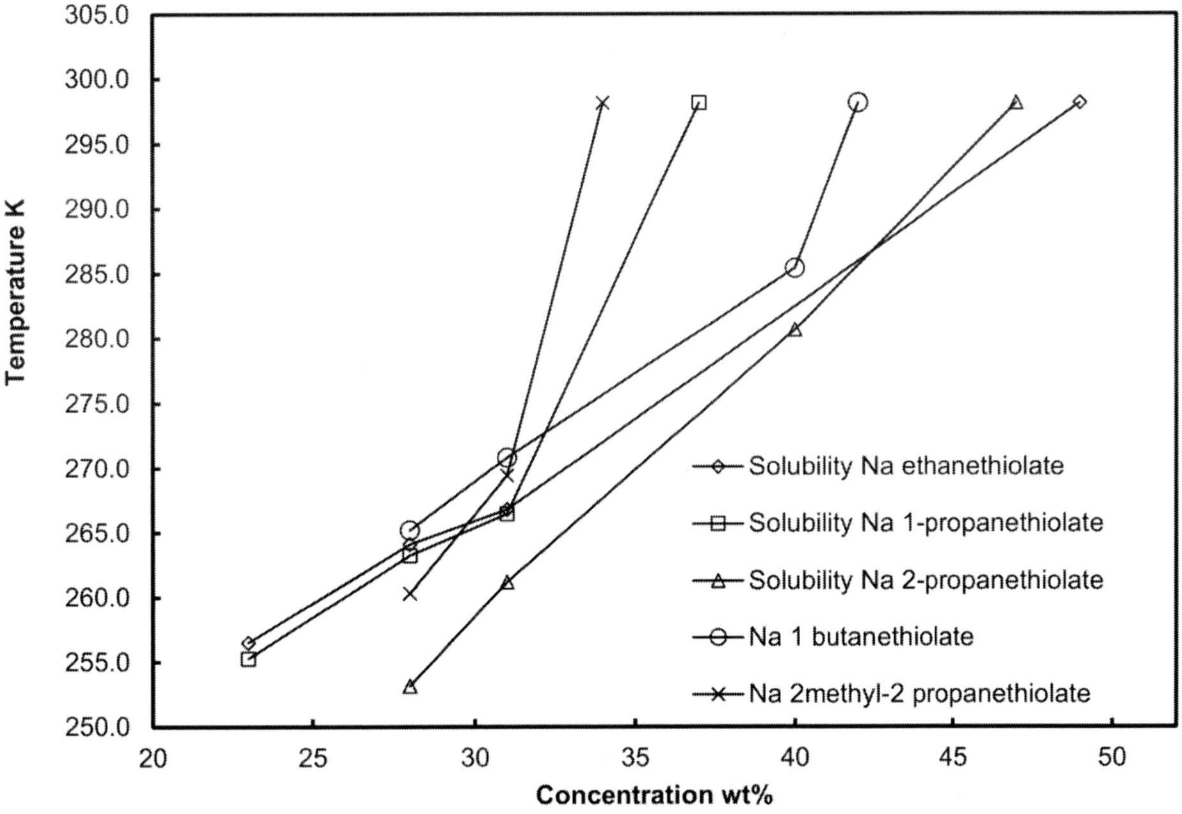

Fig. 9. The measured Solubility of the Mercaptide salts (Na-1-ethanethiolate, Na-1-propanethiolate, Na-2-propanethiolate, Na-1-butanethiolate and Na-2-methyl-2-propanethiolate) in water.

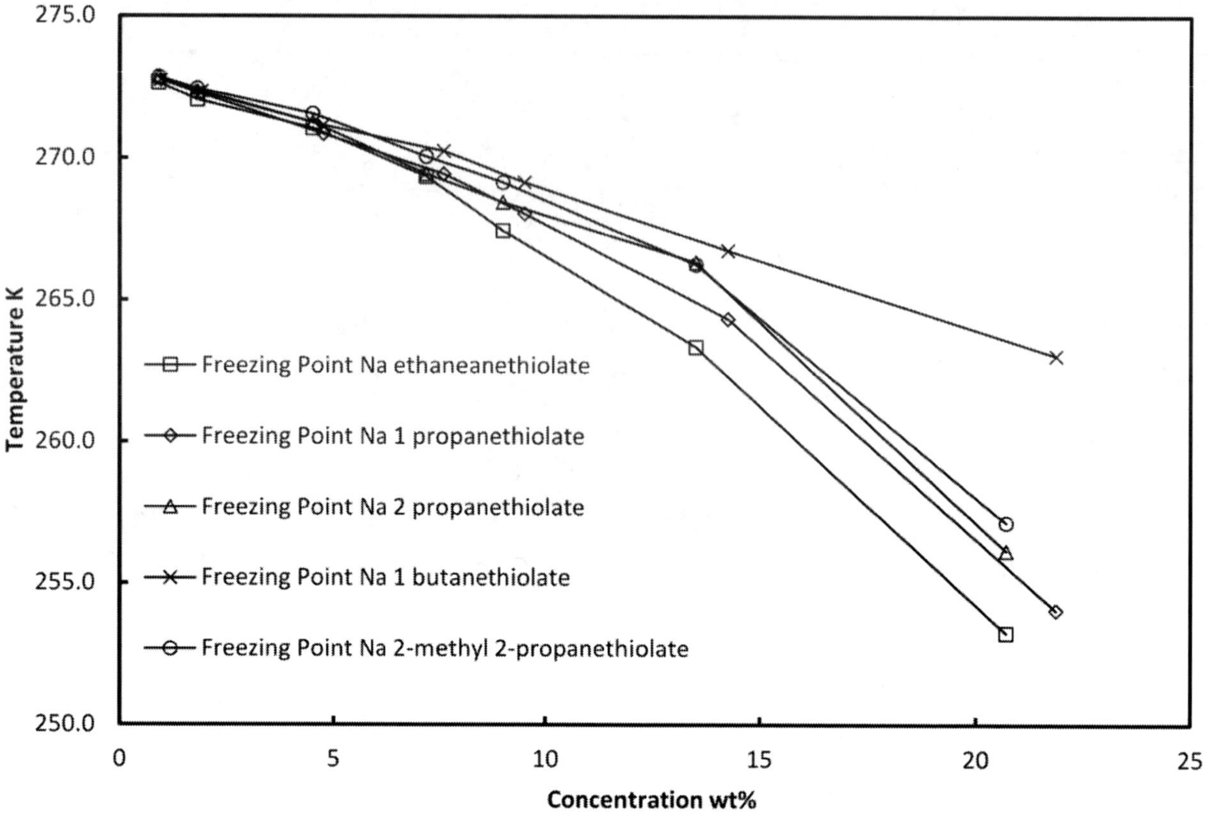

Fig. 10. The measured freezing point depression of the five mercaptide salts adjusted for purity of the mercaptide salts (Na-1-ethanethiolate, Na-1-propanethiolate, Na-2-propanethiolate, Na-1-butanethiolate and Na-2-methyl-2-propanethiolate) in water.

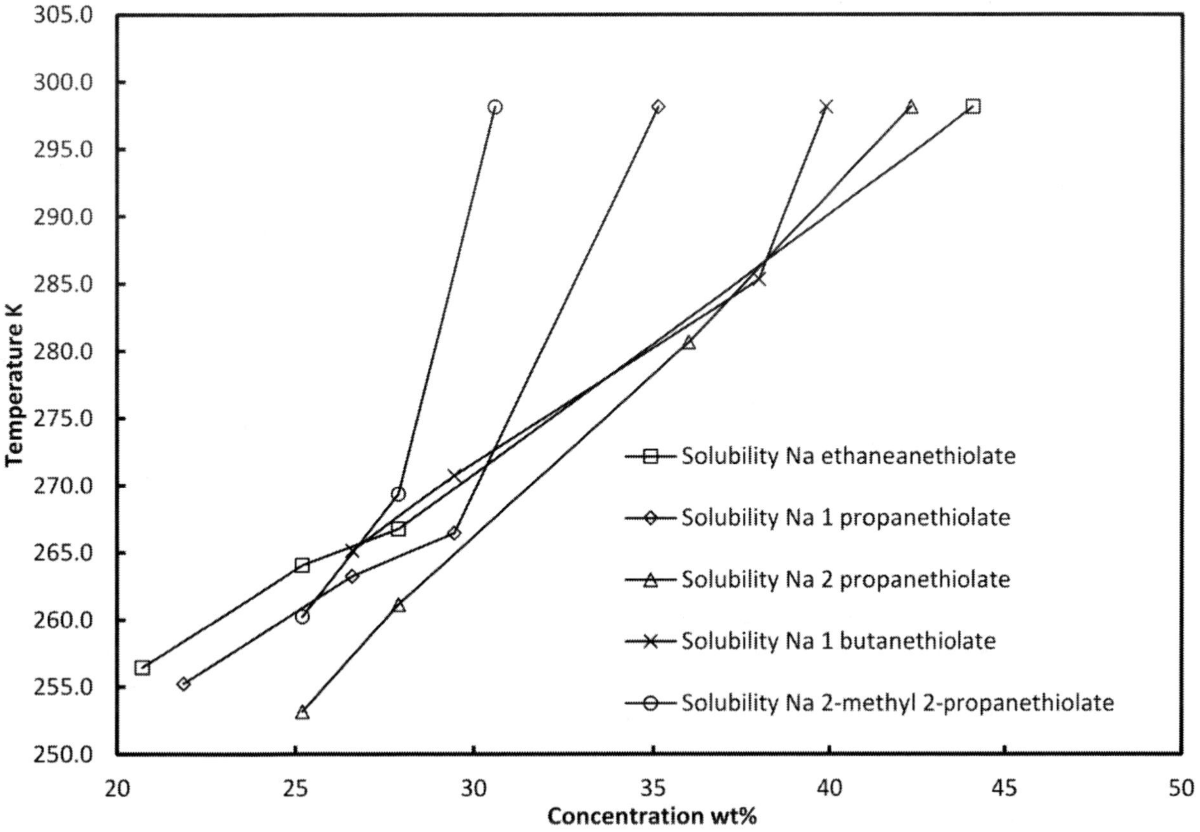

Fig. 11. The measured solubility of the five Mercaptide salts adjusted for the purity of mercaptide salts (Na-1-ethanethiolate, Na-1-propanethiolate, Na-2-propanethiolate, Na-1-butanethiolate and Na-2-methyl-2-propanethiolate) in water.

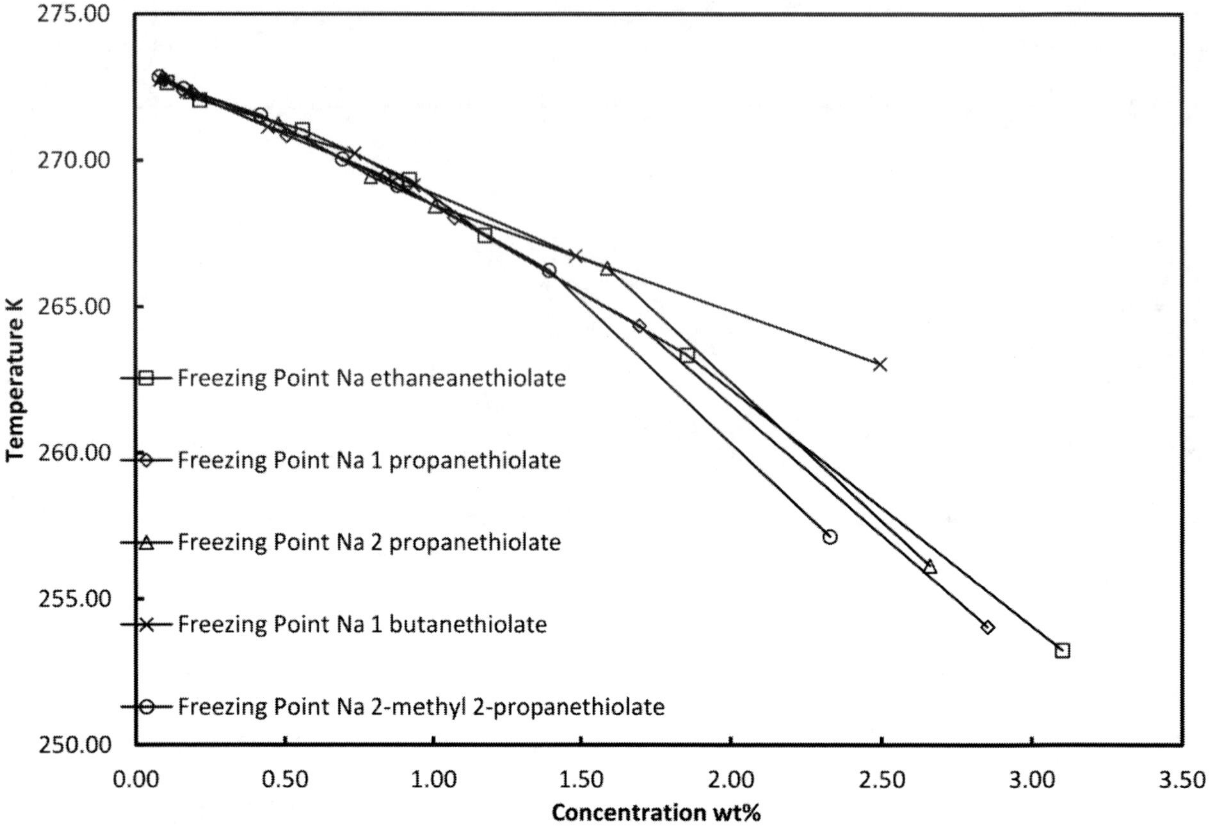

Fig. 12. The measured freezing points of mercaptide salts (Na-1-ethanethiolate, Na-1-propanethiolate, Na-2-propanethiolate, Na-1-butanethiolate and Na-2-methyl-2-propanethiolate) in water adjusted for purity and molality

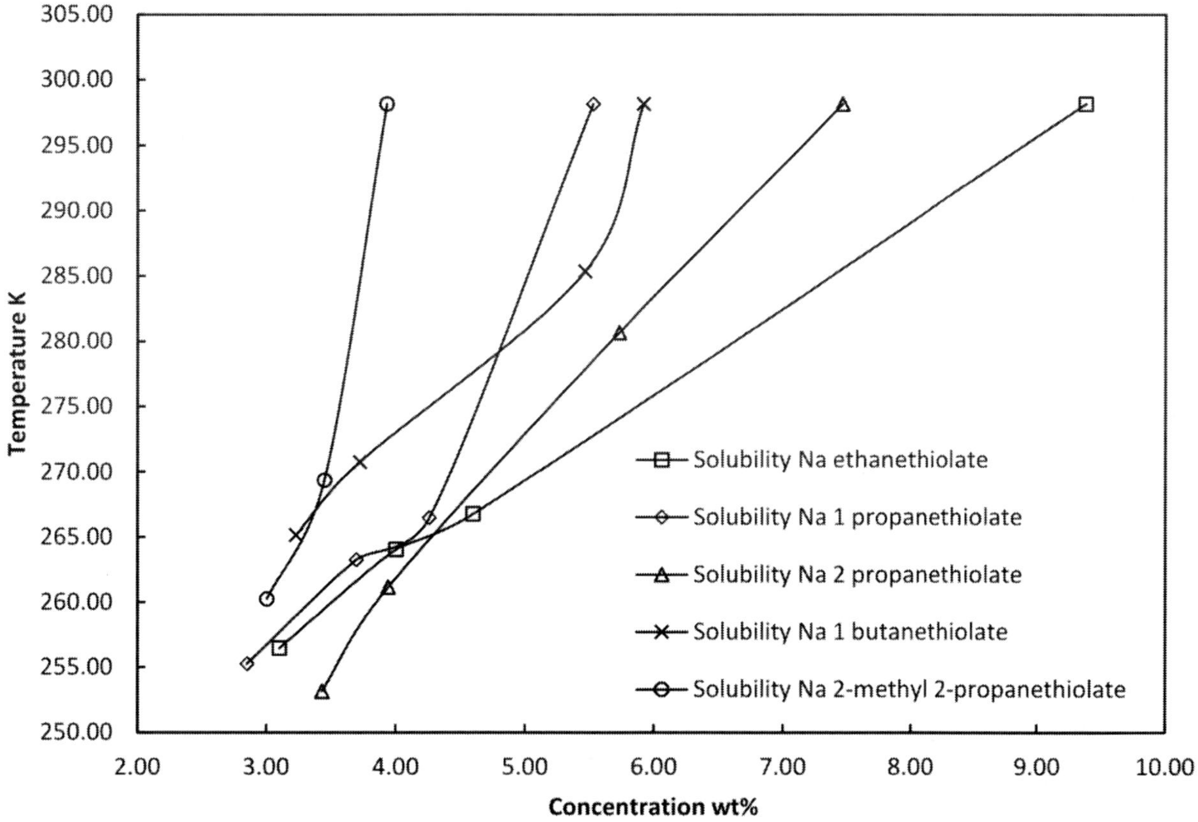

Fig. 13. The measured sloubility of mercaptide salts (Na-1-ethanethiolate, Na-1-propanethiolate, Na-2-propanethiolate, Na-1-butanethiolate and Na-2-methyl-2-propanethiolate) in water adjusted for purity and molality.

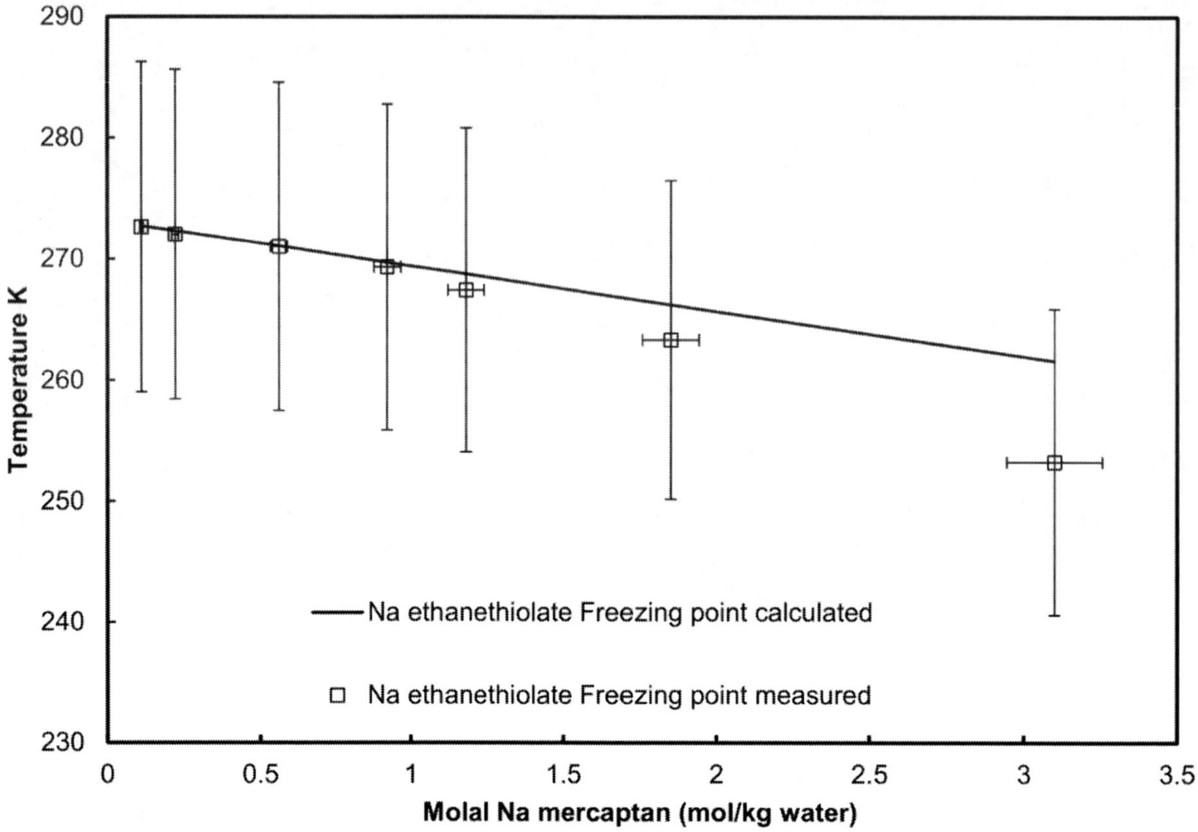

Fig. 14. Measured and calculated values of Na-1-ethanethiolate in water. Error band shows experimental uncertanity ±5%.

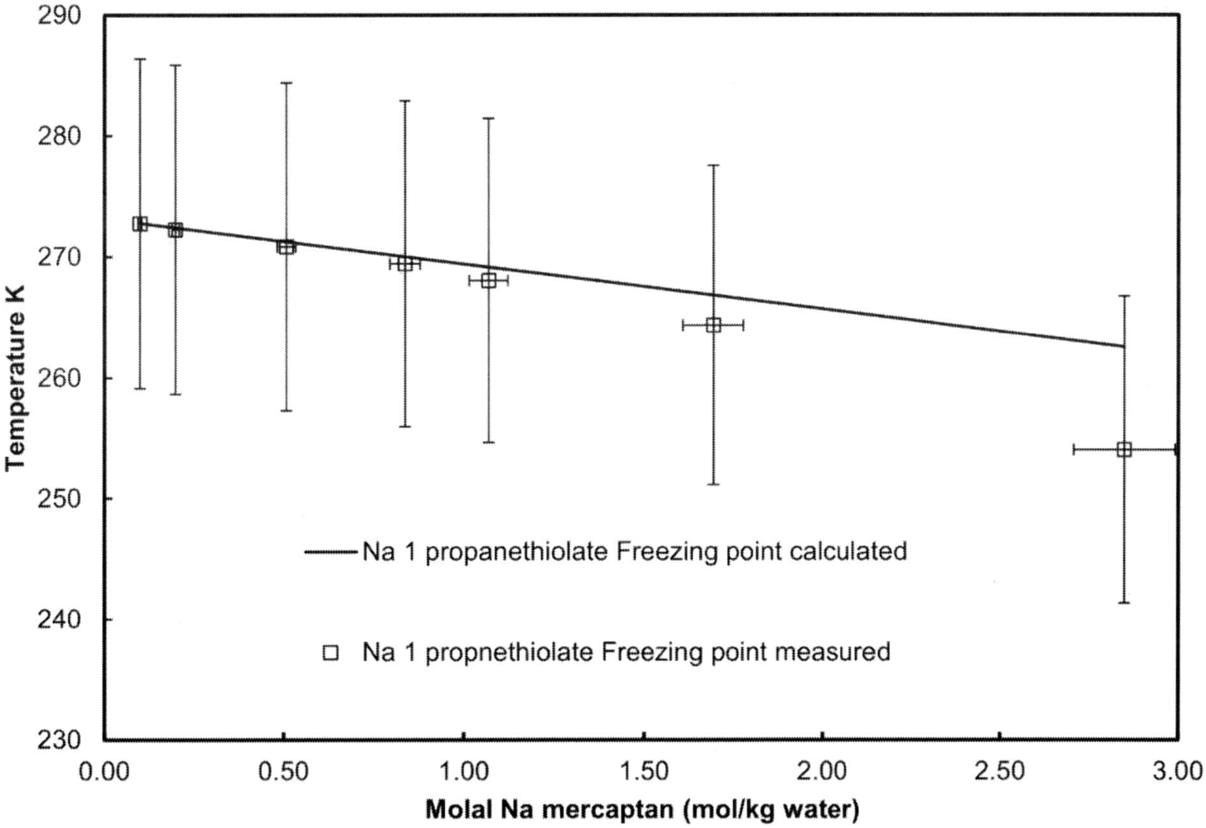

Fig. 15. Measured and calculated values of Na-1-propanethiolate in water. Error band shows experimental uncertanity ±5%.

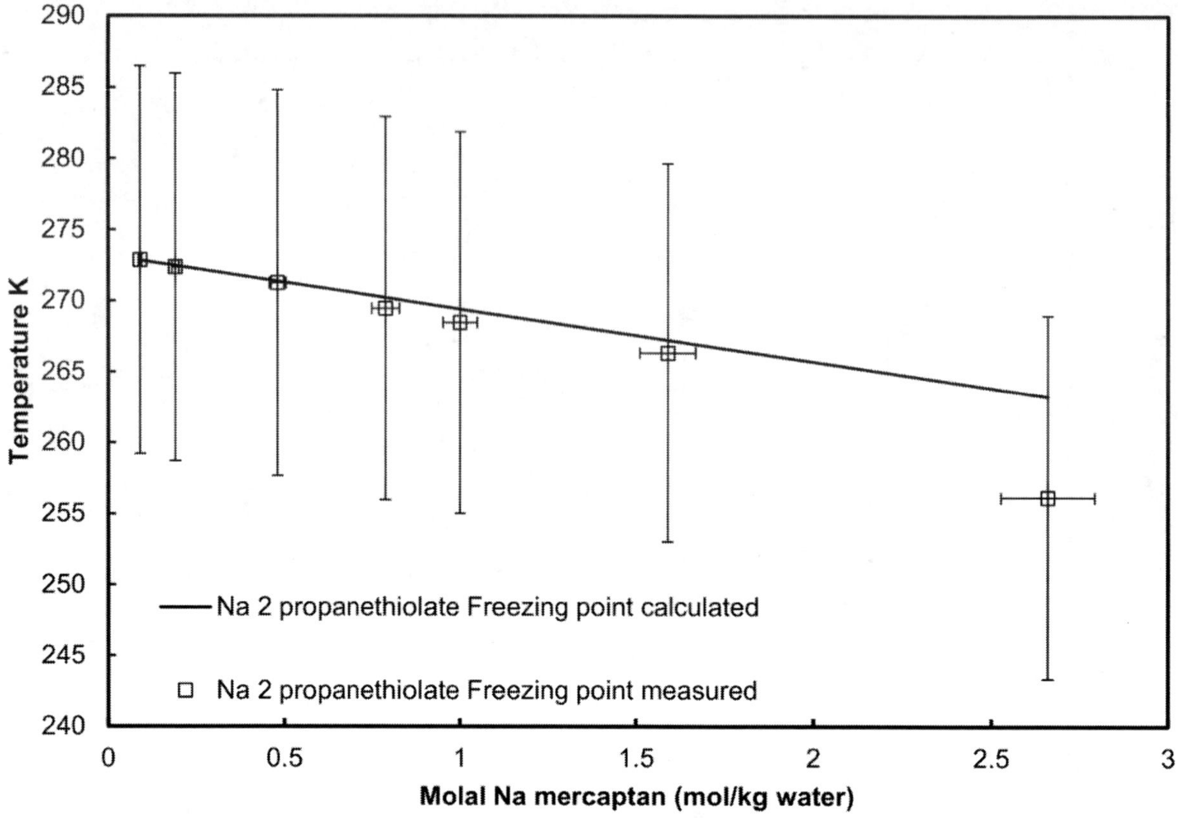

Fig. 16. Measured and calculated values of Na-2-propanethiolate in water. Error band shows experimental uncertanity ±5%.

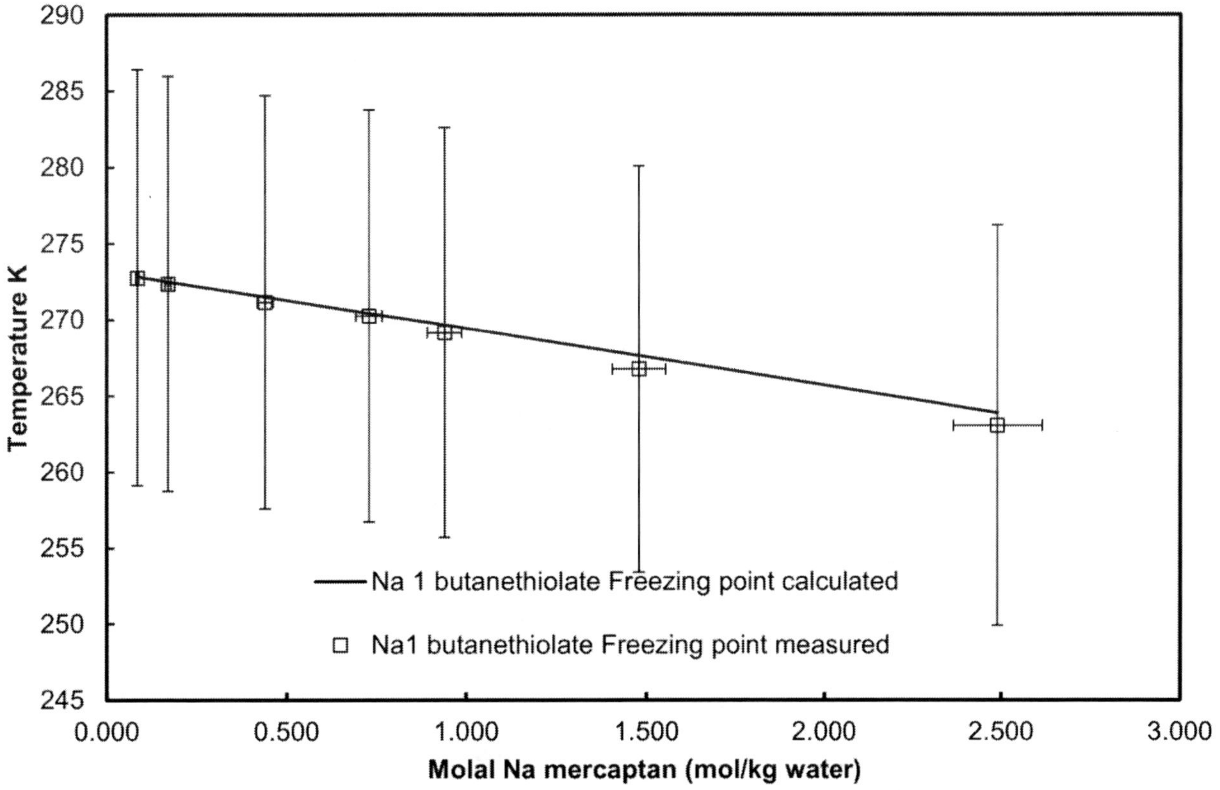

Fig. 17. Measured and calculated values of Na-1-butanethiolate in water. Error band shows experimental uncertanity ±5%.

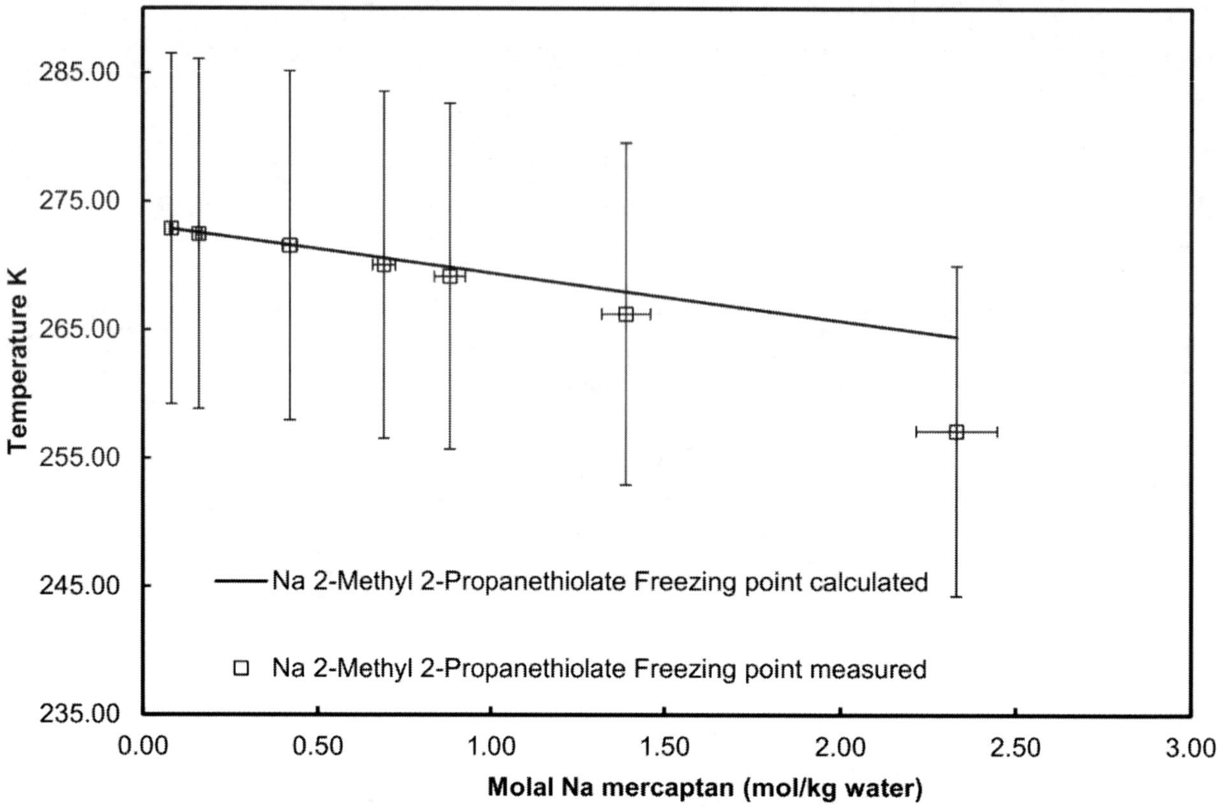

Fig. 18. Measured and calculated values of Na 2-methyl 2-propanethiolate in water. Error band shows experimental uncertanity ±5%.

Tables:

Table 1

CAS numbers, purities, and suppliers of materials

Chemical Name	CAS No.	Purity %	Supplier
Sodium ethanethiolate	811-51-8	≥ 90%	SIGMA-ALDRICH
Sodium 1-propanethiolate	6898-84-6	≥ 95%	SIGMA-ALDRICH
Sodium 2-propanethiolate	20607-43-6	≥ 90%	SIGMA-ALDRICH
Sodium 1-butanethiolate	107-03-9	≥ 95%	SIGMA-ALDRICH
Sodium 2-mathyl 2-propanethiolate	29364-29-2	≥ 90%	SIGMA-ALDRICH
Ethanol	64-17-5	≥99.9%	SIGMA-ALDRICH
Sodium Chloride	7647-14-5	≥99.5%)	GATT KOLLER

Table 2

The freezing point depression of NaCl + water solution were found at different concentrations

Concentration (wt%) of NaCl	Freezing Point (K)
0	273.15
2.5	275.65
5	278.15
7.5	280.65
10	283.15
15	288.15
17.5	290.65
22	295.15

Table 3

The solubility of mercaptide salts in water by dilution method at 298.15K.

Salt	water g	Mercaptide salt g	Concentration wt %
Sodium ethanethiolate	1.00	0.96	0.49
Sodium propanethiolate	1.03	0.60	0.37
Sodium 2-propanethiolate	1.04	0.92	0.47
Sodium 1-butanethiolate	1.01	0.73	0.42
Sodium 2-mathyl 2-propanethiolate	1.01	0.52	0.34

Table 4

The solubility of mercaptide salts in water at 298.15K after evaporation in oven at 308.15 K.

Salt	water g	Mercaptide salt G	Concentration wt %
Sodium ethanethiolate	0.71	0.66	0.48
Sodium 1-propanethiolate	0.68	0.40	0.37
Sodium 2-propanethiolate	0.71	0.60	0.46
Sodium 1-butanethiolate	0.69	0.48	0.41
Sodium 2-mathyl 2-propanethiolate	0.69	0.34	0.33

Table 5

The measured freezing point depression (K) of mercaptide salts **(This work)**.

	Concentration wt %						
Salt	1%	2%	5%	8%	10%	15%	23%
Sodium ethanethiolate	272.65	272.05	271.05	269.35	267.45	263.35	253.25
Sodium 1-propanethiole	272.75	272.25	270.85	269.45	268.05	264.35	254.05
of Sodium 2-propanethiolate	272.85	272.35	271.25	269.45	268.45	266.33	256.15
of Sodium 1-butanethiole	272.75	272.35	271.15	270.25	269.15	266.75	263.05
Sodium 2-methyl 2-propanethiolate	272.85	272.45	271.55	270.05	269.15	266.25	257.15

Table 6

The measured solubility temperature (K) of mercaptide salts **(This work)**.

	Concentration wt %								
Salt	23%	28%	31%	34%	37%	40%	42%	47%	49%
Sodium ethanethiolate	256.45	264.05	266.75	-	-	-	-	-	298.15
Sodium 1-propanethiolate	255.25	263.25	266.45	-	298.15	-	-	-	-
Sodium 2-propanethiolate	-	253.15	261.15	-	-	280.65	-	298.15	-
Sodium 1-butanethiolate	-	265.15	270.75	-	-	285.35	298.15	-	-
Sodium 2-methyl-2-propanethiolate	-	260.25	269.35	298.15	-	-	-	-	-

Table 7

The measured freezing point depression (K) of mercaptide salts corrected for purity of salt **(This work)**.

	Corrected Concentration and Freezing point depression							
Sodium ethanethiolate	Concentration (wt%)	0.90%	1.80%	4.52%	7.20%	9.0%	13.5%	20.7%
	Freezing Point (K)	272.65	272.05	271.05	269.35	267.45	263.35	253.25
Sodium 1-propanethiole	Concentration (wt%)	0.95%	1.90%	4.75%	7.6%	9.50%	14.25%	21.85%
	Freezing Point (K)	272.75	272.25	270.85	269.45	268.05	264.35	254.05
Sodium 2-propanethiol	Concentration (wt%)	0.90%	1.80%	4.50%	7.20%	9.0%	13.5%	20.70%
	Freezing Point (K)	272.85	272.35	271.25	269.45	268.45	266.33	256.15
Sodium 1-butanethiole	Concentration (wt%)	0.95%	1.90%	4.75%	7.60%	9.50%	14.25%	21.85%
	Freezing Point (K)	272.75	272.35	271.15	270.25	269.15	266.75	263.05
Sodium 2-methyl 2-propanethiolate	Concentration (wt%)	0.90%	1.80%	4.50%	7.20%	9.00%	13.50%	20.70%
	Freezing Point (K)	272.85	272.45	271.55	270.05	269.15	266.25	257.15

Table 8

The measured solubility temperature (K) of mercaptide salts corrected for purity of salt. **(This work)**

	Corrected Concentration and solubility									
Sodium ethanethiolate	Concentration (wt%)	20.70%	25.20%	27.9%	-	-	-	-	-	44.10%
	Solubility Temperature (K)	256.45	264.05	266.75	-	-	-	-	-	298.15
Sodium 1-propanethiolate	Concentration (wt%)	21.85%	26.60%	29.45%	-	35.15%	-	-	-	-
	Solubility Temperature (K)	255.25	263.25	266.45	-	298.15	-	-	-	-
Sodium 2-propanethiolate	Concentration (wt%)	-	25.2%	27.90%	-	36.00%	-	-	42.30%	-
	Solubility Temperature (K)	-	253.15	261.15	-	280.65	-	-	298.15	-
Sodium 1-butanethiolate	Concentration (wt%)	-	26.60%	29.45%	-	-	38.00%	39.9%	-	-
	Solubility Temperature (K)	-	265.15	270.75	-	-	285.35	298.15	-	-
Sodium 2-methyl-2-propanethiolate	Concentration (wt%)	-	25.20%	27.90%	30.60%	-	-	-	- -	-
	Solubility Temperature (K)	-	260.25	269.35	298.15	-	-	-	- -	-

Table 9

The measured freezing point depression (K) of mercaptide salts adjusted for molality and purity of salt **(This work).**

		Molality and Freezing point depression						
Sodium ethanethiolate	Molality (mol/kg of water)	0.11	0.22	0.56	0.92	1.18	1.85	3.10
	Freezing Point (K)	272.65	272.05	271.05	269.35	267.45	263.35	253.25
Sodium 1-propanethiolate	Molality (mol/kg of water)	0.10	0.20	0.51	0.84	1.07	1.69	2.85
	Freezing Point (K)	272.75	272.25	270.85	269.45	268.05	264.35	254.05
Sodium 2-propanethiolate	Molality (mol/kg of water)	0.09	0.19	0.48	0.79	1.00	1.59	2.66
	Freezing Point (K)	272.85	272.35	271.25	269.45	268.45	266.33	256.15
Sodium 1-butanethiolate	Molality (mol/kg of water)	0.09	0.17	0.44	0.73	0.94	1.48	2.49
	Freezing Point (k)	272.75	272.35	271.15	270.25	269.15	266.75	263.05
Sodium 2-methyl 2-propanethiolate	Molality (mol/kg)	0.08	0.16	0.42	0.69	0.88	1.39	2.33
	Freezing Point (K)	272.85	272.45	271.55	270.05	269.15	266.25	257.15

Table 10

The measurd solubility temperature (K) of mercaptide salts adjusted for molality and purity of salt **(This work)**.

		Molality and measured Solubility points			
Sodium ethanethiolate	Molality (mol/kg of water)	3.10	4.00	4.60	9.38
	Solubility Temperature (K)	256.45	264.05	266.75	298.15
Sodium 1-propanethiole	Molality (mol/kg of water)	2.85	3.69	4.25	5.52
	Solubility Temperature (K)	255.25	263.25	266.45	298.15
Sodium 2-propanethiol	Molality (mol/kg of water)	3.43	3.94	5.73	7.47
	Solubility Temperature (K)	253.15	261.15	280.65	298.15
Sodium 1-butanethiole	Molality (mol/kg of water)	3.23	3.72	5.46	5.92
	Solubility Temperature (K)	265.15	270.75	285.35	298.15
Sodium 2-methyl 2-propanethiolate	Molality (mol/kg of water)	3.00	3.45	3.93	-
	Solubility Temperature (K)	260.25	269.35	298.15	-

Table 11

The calculated freezing point depression (K) of mercaptide salts adjusted for molality and purity of salt by using eq. (4).

Molality and Calculated Freezing point depression								
Sodium ethanethiolate	Molality (mol/kg of water)	0.11	0.22	0.56	0.92	1.18	1.85	3.10
	Freezing point (K)	272.75	272.34	271.10	269.72	268.78	265.25	261.61
Sodium 1-propanethiole	Molality (mol/kg of water)	0.10	0.20	0.51	0.84	1.07	1.69	2.85
	Freezing point (K)	272.79	272.42	271.26	270.03	269.17	266.85	262.55
Sodium 2-propanethiol	Molality (mol/kg of water)	0.09	0.19	0.48	0.79	1.00	1.59	2.66
	Freezing point (K)	272.81	272.46	271.36	270.21	269.40	267.23	263.26
Sodium 1-butanethiole	Molality (mol/kg of water)	0.09	0.17	0.44	0.73	0.94	1.48	2.49
	Freezing point (K)	272.83	272.50	271.50	270.42	269.67	267.64	263.88
Sodium 2-methyl 2-propanethiolate	Molality (mol/kg of water)	0.08	0.16	0.42	0.69	0.88	1.39	2.33
	. Solubility Temperature (K)	272.85	272.45	271.55	270.05	269.15	266.25	257.15

502874 Computation of Amino Acid-Mineral Surface Thermodynamics and Interaction Modes via Importance Sampling

Abstract:

Understanding interactions between biomolecules and minerals is essential to the development of biomimetics and the treatment of biomineralization diseases. Yet few force fields are validated for the interactions of biomolecules at mineral interfaces, and there are few rigorous benchmarks of the effectiveness of the conformational sampling in this context. Toward those ends, we used new detailed-balance Monte Carlo methods in the Rosetta molecular modeling suite to generate equilibrium canonical ensembles of the twenty naturally occurring amino acids interacting with stable, flat surfaces of calcite, hydroxyapatite, calcium oxalate monohydrate, and brushite. We tested the convergence of the algorithms and found that all converge energetically. In several cases of strong adsorption, there are multiple structural modes of interaction with similar binding energies. Adsorption free energies range from –40 to 0 kcal/mol depending on the surface and amino acid. Negatively charged amino acids consistently adsorb most strongly, with the electrostatic term dominating the interactions. Hydroxyapatite binds amino acids tightest, and calcite the weakest. We compare results to published molecular dynamics studies and experimental adsorption and single molecule force spectroscopy studies. Our findings have implications for studies of protein-mineral interactions and design of biomolecules to control crystal growth.

491177 Research on the Mechanism of Spontaneous DNA Extension on Grooved Surface Covered with Cationic Lipid Bilayers

Abstract:

We have recently developed a novel DNA optical gene mapping platform. Its working principle relies on the phenomenon that DNA can spontaneously extend along grooves covered with cationic lipid bilayers. Although the physical gene map can be readily obtained from the extended DNA, the mechanism causing DNA extension is not clear. Since DNA extends only along the grooves, it was suggested that there exists an electrostatic energy well for DNA. The origin of the energy well was postulated to be the uneven lipid distribution due to the steric effect that the shape of lipids will affect their distribution on a curved surface. Moreover, we suspect a simple geometry effect may also contribute to the energy well. The geometry effect comes from that fact that DNA on a surface with positive curvature can interact more strongly with lipid molecules. To examine both postulations, we investigated DNA behavior on three sets of bilayers with different composition: (A) DOPC/DOTAP(+), (B) DOPC/EPC(+) and (C) DOPE/EPC(+). The lipids used here have the same tails but different head groups. The experimental results show both the steric and the geometry effect are important. However, the geometry effect is always in favor of DNA extension while the steric effect can either enhance or undermine the phenomenon. We have also investigated the effect of the concentration of the positively charged lipids and found higher concentration always help DNA to extend. In order to quantify the geometry effect, we have derived the electrostatic potential of DNA on curved surface and found the theoretical prediction is in good agreement with the experimental results.

494798 Effect of Interchain Associations on Hybridization Activity of DNA Brushes

Abstract:

Hybridization in DNA brushes is commonly encountered in diagnostic and materials science applications. Typically, the brush chains have lengths of ten or more nucleotides to provide sufficient sequence specificity to uniquely identify a partner to pair with. At these lengths, coincidental occurrence of complementary regions between two brush chains is expected to lead to associations due to base pairing between the brush chains. This talk examines how such associations impact the hybridization activity of DNA brushes with analyte molecules. DNA brushes were prepared from 20mer chains with 4-nucleotide long "adhesive regions" through which neighboring chains could interact. The affinity of association, as well as the position of the adhesive region along the chain backbone, were varied. The impact of associations between the brush chains on hybridization thermodynamics between the brush and analyte strands was assessed from melting transitions. Higher affinity adhesive regions were found to more extensively suppress hybridization with analyte, and associations near the middle of the chains were found to be more penalizing than those at the immobilized or the free end of the chains. Comparison of the brush results with data extracted from commercial single-nucleotide-polymorphism (SNP) microarrays revealed agreement in the impact of probe associations on hybridization.

501315 Understanding Marine Mussel Adhesion: Interfacial Energy from Microscopic to Macroscopic Length Scales

Abstract:

The marine mussel adheres to organic and inorganic surfaces under water through a combination of physical and chemical bonds, and translates these interactions into energy dissipation and high detachment forces through macroscopic geometry. This work explores protein-surface interactions and the behavior of a mussel-inspired single-component coacervate adhesive in a confined geometry. The mussel delivers its adhesive proteins to surfaces in a protein-rich liquid phase called a coacervate that readily spreads on and adheres to surfaces. The interaction energy often depends on the molecular-scale geometry between the adhesive proteins and the surface (determined by the primary and secondary structure of the protein). For example, the proximity of cationic residues to the aromatic residue 3,4-dihydroxyphenylalanine (Dopa)â€"nearly ubiquitous in marine adhesionâ€"impacts the ability of Dopa to form long-lived bidentate hydrogen bonds with the surface. Geometry is also crucial on the macroscopic scale: the force of detachment and energy dissipation of the mussel holdfast (the byssus) results from its unique structure. Inspired by the success of mussel adhesion, a random copolymer incorporating Dopa was designed that forms a polymer-dense liquid phase in aqueous environments over a wide pH range without precipitating (a single-component coacervate). The cohesion of this material as a function of pH and Dopa content was investigated between molecularly smooth mica surfaces. An understanding of the adhesion mechanisms of marine mussels has motivated the development of this and many other adhesives for use in aqueous environments, and will continue to guide the development of adhesives for medical, dental, and marine applications.

500876 Tuning Underwater Adhesion with Cation-Pi Interactions

Abstract:

Cation–π interactions feature prominently in the self-assembly and adhesion of many biological molecules, including the adhesion proteins of several marine organisms. Although the origin of cation–π bonds in isolated pairs has been extensively studied, the mechanism of cation–π-interactions in molecular films remains uncharted. Here we discuss nanoscale force measurements demonstrating that the cohesive properties of simple aromatic and lysine-rich peptides rival those of the strong adhesion exhibited by adhesion proteins of the marine mussel. More broadly, we demonstrate that interfacial confinement fundamentally alters the energetics of cation–π-mediated assembly, and we discuss how this insight can be used to tune self-assembly and adhesion in peptide-based biomaterials.

497977 Influence of the Surface Charge on the Brownian Motion of Colloidal Particles Near Surfaces with Biomimetic Characteristics

Abstract:

Many important biological phenomena, such as bacterial invasion, absorption of food particles or cleansing of impurities are set up at the interface between biological fluids and tissues. Such processes commence with non-specific interactions between surfaces (mucosa, epithelia, endothelia) and colloids (mucus, bacteria, dirt particles), when the colloids are in the vicinity of the surfaces. Knowledge of these phenomena is relevant for the engineering of drug delivery systems or the modification of the adhesion properties of tissues and biomaterials.

In this work, we used fluorescent colloidal particles suspended in proximity of surfaces functionalized with biomolecules. The colloidal particles were tracked three-dimensionally by digital video microscopy. From their trajectories we computed their mean square displacement to calculate the diffusion coefficient in directions parallel and perpendicular to the surface. We used both negatively-charged and positively-charged colloidal micro-particles on surfaces functionalized with poly-L-lysine, hyaluronic acid and phospholipid monolayers and found differences in the parallel and the perpendicular diffusions highly dependent on the interfacial characteristics of the colloidal particles and the surfaces.

496105 Charged Nanoparticles Interacting with Giant Vesicles Fabricated from Inverted-Headgroup Lipids

Abstract:

The surface chemistry of the cell membrane plays an important role in how cells interact with particulate species. These interactions are dictated in large part by lipid headgroup charge. To investigate the nature of electrostatic interactions between lipid bilayers and nanoparticles in solution, we investigated nanoparticles interacting with DOPC (1,2-dioleoyl-glycero-3-phosphocholine) and DOCP lipids, in which the zwitterionic head group was inverted from its natural configuration in DOPC.

These interactions were investigated by fabricating giant unilamellar vesicles (GUVs) with DOPC lipids and DOCP lipids respectively, and introducing nanoparticles to suspensions of both. In the presence of charged polystyrene nanoparticles, GUVs displayed various degree of deformation depending on the charge sign and size of the nanoparticles. These differences illuminate how the relative positions of phosphate group and choline group in the headgroup determine the interaction with charged particles. This work provides fundamental insights regarding the nature of this interaction, which has implications for the development of new drug delivery vehicles as well as understanding the environmental toxicity of nanomaterials.

499622 Modulation of Phase Morphology of Surfactant Monolayer with Interfacial Curvature

Abstract:

For the past 100 years, surfactant monolayers have been studied on planar interface of the Langmuir trough while most physiological, such as, alveolar interface in mammals, and technological interfaces are curved at the micron to millimeter scale. We present the design and engineering of a unique instrument assembly and methodology for creating curved monolayer on air-aqueous interfaces of different background curvatures (25-500 μm radius). Our unique approach combines 3-D imaging capabilities of confocal fluorescence microscopy with this setup to investigate curvature-dependent changes in the phase morphology of monolayer on interface. Using a biologically relevant multi-component surfactant system that forms coexisting ordered, 'solid-like' domains in the disordered, 'liquid-like' phase on a planar interface, we show that as the background curvature of interface approaches a critical range the monolayer yields a contrasting equilibrium phase-coexistence pattern where size, shape and connectivity of the domains dramatically change. Time-lapsed investigation of early events at interfaces reveals curvature sensitive conversion of adsorbed surfactant MLVs into monolayer and subsequent evolution of phase pattern. Surprisingly, the background curvature-mediated dramatic shift in phase-coexistence pattern impacts the dynamic properties of monolayer as well, in particular the complex dilatational modulus. Modulation of local phase structure and dynamics with the background curvature is found reversible up to certain extent. We present a theoretical model to explain how electrostatic (dipole:dipole) interaction energy between phases and within domains compete with the line tension energy on a curved interface and yield distinct phase morphology in the monolayers.

497134 Recognition of Target Cells By *Vibrio cholerae* outer Membrane Vesicles

Abstract:

*Vibrio cholerae*Â is the etiologic agent of cholera, a disease characterized by severe, often fatal diarrhea. Cholera toxin (CT), the major virulence determinant of disease, binds a specific ganglioside, GM1, on intestinal epithelial cells, allowing internalization of the toxin. CT is secreted by the bacteria both in a water-soluble â€œfreeâ€ form, as well as associated with outer membrane vesicles (OMVs). OMVs are spherical, lipid-encapsulated vesicles, on the order of 100 nm in diameter, that are secreted by Gram negative bacteria. These vesicles contain bacterial virulence factors, implicated in disease, as well as DNA and other molecules. OMV-associated CT has been demonstrated to be physiologically active; however, the mechanism by which delivery of OMV-associated CT occurs has not yet been demonstrated. In the present study, we sought to characterize the delivery of CT-containing OMVs with the goal of identifying novel therapeutic targets to prevent disease. We hypothesized that CT on the OMV surface might act as a targeting agent, allowing the OMV to bind to the CT receptor, GM1. We have demonstrated that *V. cholerae* OMVs contain CT both inside and on the surface. At neutral pH, the toxin has no affinity for the anionic lipid components of the bacterial membrane, with the exception of lipid A. This affinity suggests that the CT is located on the OMV membrane surface, where it is obscured by the polysaccharide chains of lipopolysaccharide (LPS) component of the OMV membrane so that the OMV-associated CT is unable to bind GM1 on the cell surface. As a result, *V. cholerae* OMVs do not bind GM1, and soluble GM1 does not inhibit binding or toxicity of *V. cholerae* OMVs. Additionally, OMVs purified from a strain of *Vibrio cholerae* that does not produce CT are trafficked in an identical method as CT-containing OMVs. These results indicate that *V. cholerae* OMVs recognize target cells in a CT-independent mechanism.

492471 Applying Computational Tools of Polymer Field Theory to out-of-Equilibrium Polymer Solutions in Flow

Abstract:

In polymer solutions, the relationship between hydrodynamic interactions (HI) and out-of-equilibrium polymer conformation has been an area of intense interest. This is an area of rheology that is fundamentally important for a wide array of polymer processing methods, especially those involving polymer solutions in flow; for example, fiber spinning, coating, and printing processes can lead to materials with properties that are sensitive to how molecular structure is controlled by fluid flow. This talk will focus on our efforts to expand the role of simulation in understanding the nature of the coupling between polymer conformation and hydrodynamics. First, we show that there are non-trivial connections between ring polymer architectures and HI, where the intramolecular HI effects can dominate conformation. This motivates the need for resolved simulations of polymer solutions. Inspired by simulation methods that have become standard in polymer field theory, we revisit early ideas in polymer dynamics and develop an iterative procedure for capturing HI. This method circumvents some standard computational bottlenecks in Brownian dynamics simulations, enabling large simulations of polymer solution dynamics. We demonstrate the advantages and limitations of this method, and show how we are beginning to apply these ideas to challenges at the forefront of molecular polymer rheology.

500781 Capillary Thinning and Pinch-Off Dynamics and Extensional Rheology of Polyelectrolyte Solutions

Abstract:

Biological macromolecules like proteins, DNA and polysaccharides, and many industrial polymers, are classified together as polyelectrolytes for in solution, the repeat units in their backbone are decorated with disassociated, charge-bearing ionic groups, surrounded by counter-ions. In diverse applications like inkjet printing, sprayable cosmetics and insecticides, paints and coatings that involve formation of fluid columns or sheets that undergo progressive thinning and pinch-off into drops, the dominant flow within the necking filament is extensional in nature. The extensional rheology response of the charged macromolecular solutions is not as well understood as that of their uncharged counterparts. Here focus on the characterization of capillary thinning and pinch-off dynamics, extensional rheology and printability of two model systems: sodium (polystyrene sulfonate) and poly(acrylic acid) by using dripping-onto-substrate (DoS) rheometry technique. Both the measured extensional relaxation times and the extensional viscosity values show salt- and polymer concentration-dependent behavior that is not expected or anticipated from the typical shear rheology response.

491652 Relaxation of Knotted Polymers

Abstract:

In recent years, there has been interest in understanding the dynamics of knotted DNA due to the fact that such molecules have been observed in nanotechnology applications such as single-molecule gene-mapping and sequencing. In this talk, we will discuss our latest experiments and simulations to characterize the relaxation behavior of knotted chains from a highly extended initial state. The single-molecule experiments we perform repeatedly stretches and relaxes a knotted DNA in a microfluidic T-junction via an extensional electric field. We observe several findings: (1) the knots swell at a time scale that is comparable to the end-to-end relaxation of the polymer chain, and (2) the swelling behavior is fairly insensitive to the chain topology, at least for the large knots we sample. These results are corroborated by Brownian dynamic simulations, which are able to sample a wide range of knot topologies and several chain sizes. We note that in the experiments performed, the DNA contour length is much larger than the contour stored in a tight knot (75 microns to 2 microns). In the last part of the talk, we will discuss how knots alter the stress relaxation of shorter polymer chains. We perform Brownian dynamic simulations and develop scaling theories to understand how a finite-size knot at the center of the chain alters the polymerâ€™s entropic elasticity and effective drag.

ON SYNTHETIC DRILLING FLUID RHEOLOGICAL MODELLING

SILVIO BALDINO[1], REZA E. OSGOUEI[2], EVREN OZBAYOGLU[1], STEFAN Z. MISKA[1] AND ROLAND MAY[3]

(1) TUDRP, Universay Of Tulsa, Tulsa, OK; (2) Baker Hughes, a GE company, Houston, TX; (3) Baker Hughes, a GE company, Celle, DE

INTRODUCTION – DRILLING FLUIDS IN THE OIL & GAS

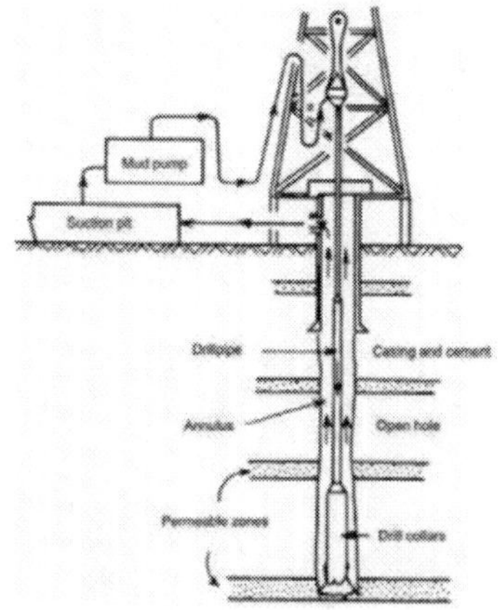

The well-fluid system [from Bourgoyne et al. (1991)].

API defines a drilling fluid as: *A circulating fluid used in rotary drilling to perform any or all of the various functions required in a drilling operation.* Some of them are:

- Cool and lubricate the bit
- **Transport drilled cuttings to the surface**
- Control drilling rates
- Exert sufficient hydrostatic pressure to stabilize the wellbore

Water Based | Oil Based | Synthetic Oil Based

2

INTRODUCTION – SYNTHETIC-BASED DRILLING FLUIDS (SBFs)

In general, drilling fluid should **NOT**:

- **Be harmful to personnel nor the environment**
- Chemically react with the formation
- Damage the production rate from the formation
- Cause corrosion of the drilling equipment and subsurface tubulars

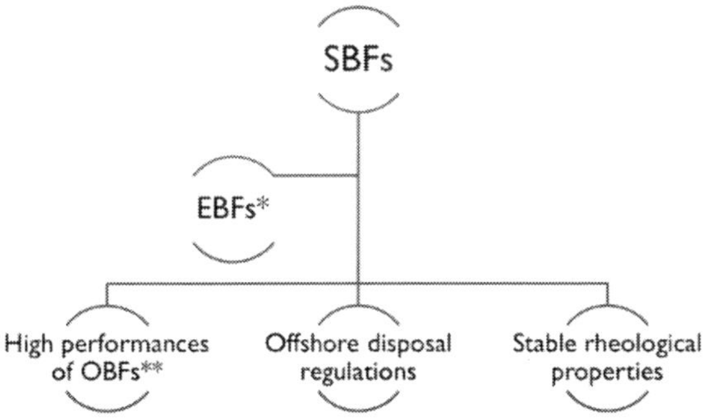

*Emulsion-Based Drilling Fluids (EBFs)
**Oil-Based Drilling Fluids (OBFs)

3

PROBLEM STATEMENT – YIELD STRESS DETERMINATION

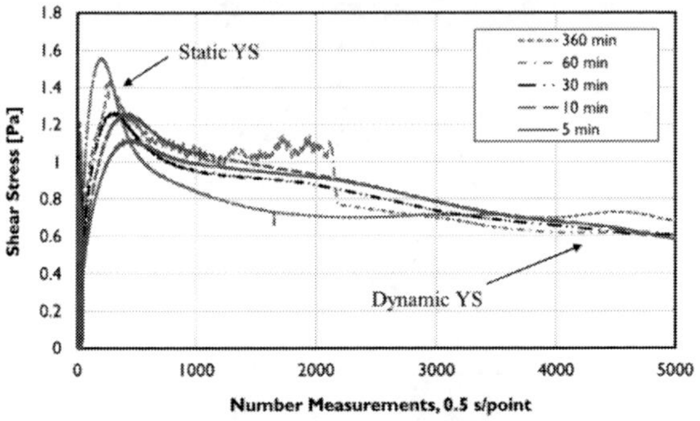

Static and Dynamic yield stresses evaluated with a Stress over Shoot Test, under constant 0.1[s⁻¹] shear rate and 21.11 °C, as a function of aging time. ***The fluid is clearly thixotropic.***

The most usual rheological model for these materials is the Herschel–Bulkley model:

$$\tau = \tau_y + K\gamma^n$$

CHALLENGES

- Lack of a unique established method
- Dependence on the experimental protocol
- Dependence on the available instrumentation
- Two yield stresses – one static and one dynamic

- K = Consistency Index [Pa.s]
- n = Generalized flow behavior index

4

QUEMADA RHEOLOGICAL MODEL FOR THIXOTROPIC FLUIDS

Structural model based on the generalization of the hard sphere (HS) model and on a revised Effective Volume Fraction (EVF or ϕ_{eff}) concept.

$$\eta = \eta_\infty \left[\frac{1 + \Gamma^p}{\chi + \Gamma^p}\right]^2$$

$$\eta_\infty = \eta_F \left(1 \quad \frac{\phi}{\phi_\infty}\right)^{-2}$$

$$\eta_0 = \eta_F \left(1 \quad \frac{\phi}{\phi_0}\right)^{-2}$$

$$\chi(\phi) = \frac{1 \quad \phi/\phi_0}{1 \quad \phi/\phi_\infty} = \pm\left(\frac{\eta_\infty}{\eta_0}\right)^{05}$$

- Γ = Dimensionless shear rate = $\Gamma = \gamma/\gamma_c$
- γ_c = characteristic shear rate [s^{-1}]
- η_F = suspending fluid viscosity [Pa.s]
- ϕ = Particle volume fraction (packing)
- ϕ_0 = Packing at zero shear rate
- ϕ_∞ = Packing at infinite shear rate

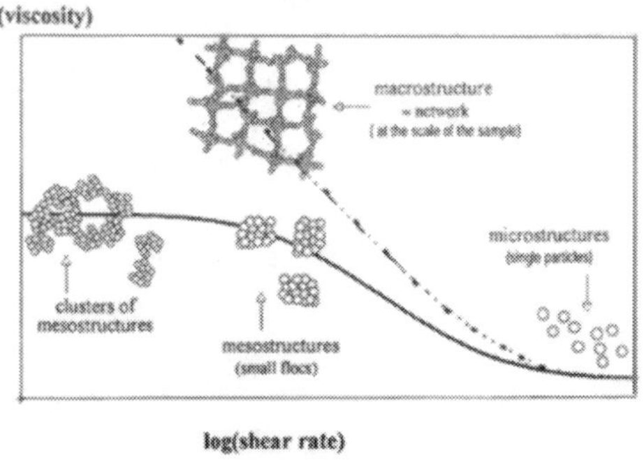

Structural interpretation of shear-thinning and plastic fluids, after Quemada (1998).

5

EXPERIMENTAL PROTOCOL – RHEOMETER AND INDUSTRIAL SBF

Baker Hughes Inc. synthetic drilling fluid before and after mixing.

Approximate formulation of the SBF

Property	Value
Density	10.5 [lb/gal]
Oil/Water Ratio	85/15
Oil Viscosity	4 cP
Components	
Base Oil	Isomerized Olefin
Internal Phase	21% Calcium Chloride
Polyamide Emulsifier	8 [lb/bbl]
Organophilic Clay	4 [lb/bbl]
Lime	0.5 [lb/bbl]
Fatty Acid Rheological Modifier	0.2 [lb/bbl]
Barite	138 [lb/bbl]
Low Gravity Solids	48 [lb/bbl]

Anton Paar rheometer, TUDRP Lab

Model parameters are evaluated by means of the **Stress over Shoot** testing technique (SOT). The latter requires the application of a constant shear rate over time.

6

QUEMADA MODEL – MODIFICATION AND RESULTS

7

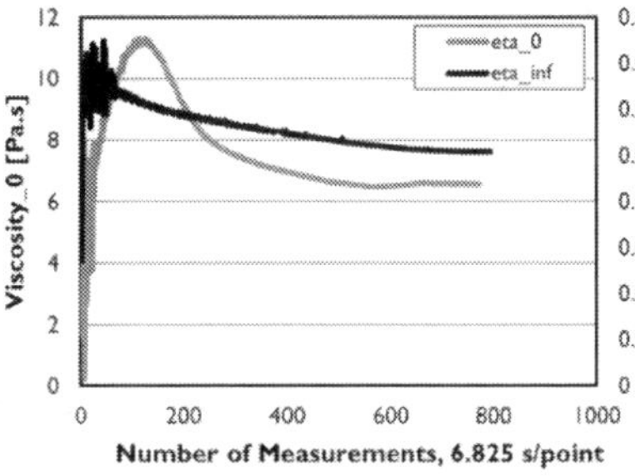

SBF Asymptotic Viscosities measured at 0.1 s-1 and 1000 s-1constant shear rate. T=21.11 °C.

$$t_c \approx \frac{\eta_{\text{eff}} a^3}{W_1} \approx \frac{\eta_{\text{eff}} a^3}{kT}$$

$$2a = 2215\ \mu\text{m}$$

- W_1 = Particle interaction energy [J]
- k = Boltzam's constant
- η_{eff} = suspension viscosity [Pa.s]

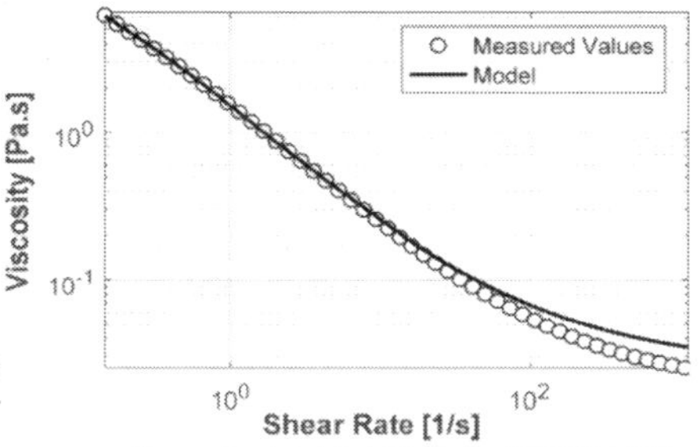

Measured and modelled data for the SBF. T=21.11 °C.

$$\eta = \eta_\infty \left[\frac{1 + \Gamma^p}{\left(\frac{\eta_\infty}{\alpha \eta_0}\right)^{0.5} + \Gamma^p} \right]^2$$

Optimal parameters for the SBF at T=21.11 °C

Parameter	Value
p	0.5548
t_c	0.02144 [s]
α	5.751

TEMPERATURE EFFECT ON SBF ASYMPTOTIC VISCOSITIES

8

During drilling operations, the drilling fluid is exposed to **temperature variation** because of the geothermal gradient (temperature rises by 25°C per km of depth).

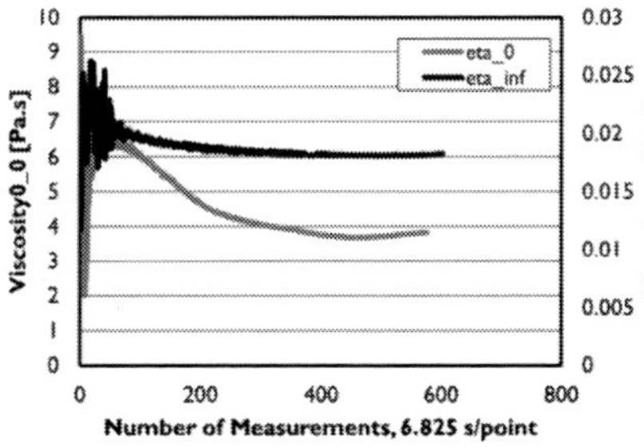

SBF Asymptotic Viscosities measured at 0.1 s-1 and 1000 s-1 constant shear rate. T=40.55 °C.

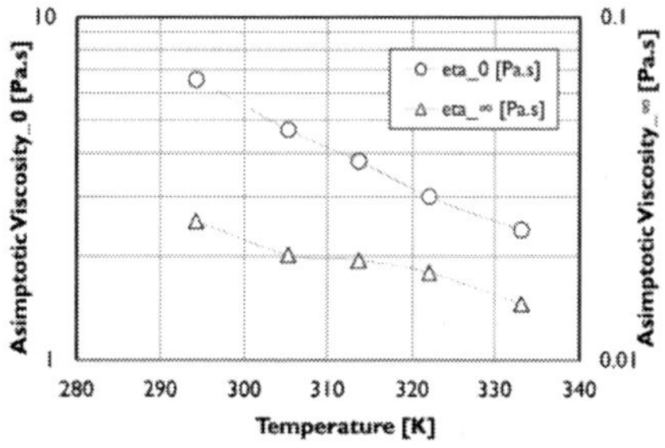

SBF Asymptotic Viscosities measured at 0.1 s-1 and 1000 s-1 constant shear rate, as a function of Temperature [K].

Measured and modelled data for the SBF.T=32.22 °C.

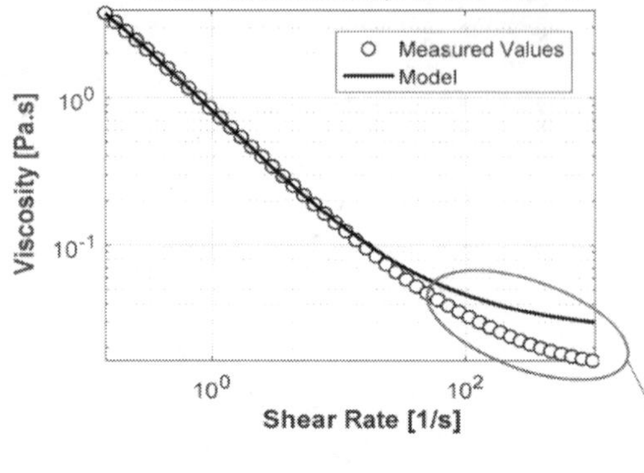

Measured and modelled data for the SBF.T=60 °C

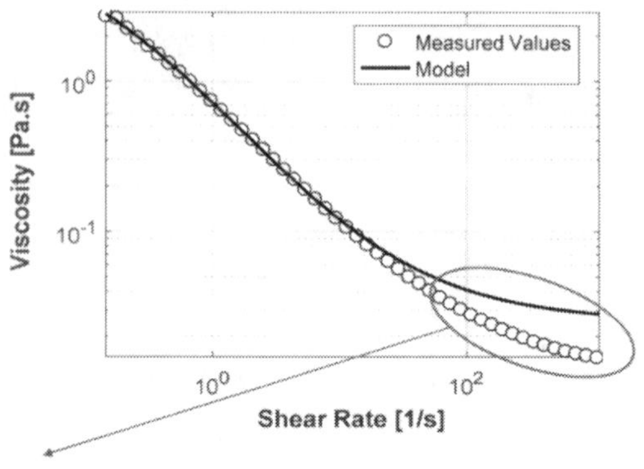

Discrepancies at high shear rate increase with increasing temperature.This suggests that an s-type curve would me more appropriate for high values of temperature. S-type curve could be found to predict the non- existence of a yield stress (Hodne et al., 2007), which is expected at high temperatures.

TEMPERATURE EFFECT - RESULTS

9

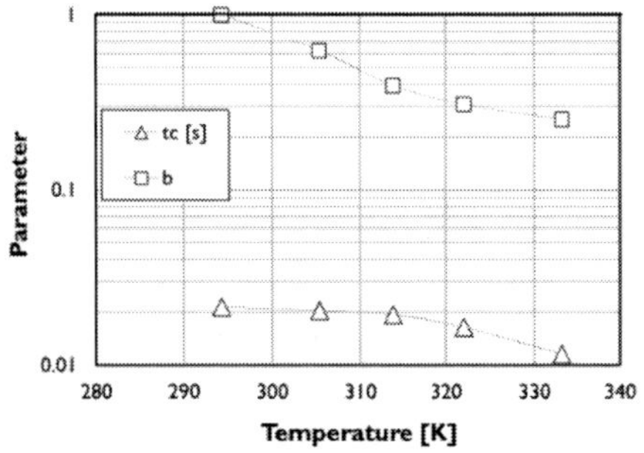

$$\eta = \beta(T)\eta_\infty \left[\frac{1 + (t_c(T)\gamma)^p}{\left(\frac{\eta_\infty}{\alpha\eta_0}\right)^{05} + (t_c(T)\gamma)^p} \right]^2$$

REMARKS

- As expected, the value of β decreases almost linearly with temperature
- For an increase in temperature, t_c decreases
  - η_{eff} of the suspension decreases
  - particles interaction energy increases, as the reference thermal energy, kT, increases

Optimal parameters for the SBF as a function ofT [K].

Parameter	Value			
$T\ [K]$	305.37	313.7	322.04	333.15
p	0.5626	0.5975	0.6102	0.5819
t_c	0.02045	0.01944	0.01643	0.01155
β	0.6261	0.3967	0.3087	0.254

TEMPERATURE EFFECT ON MODEL'S PARAMETERS

10

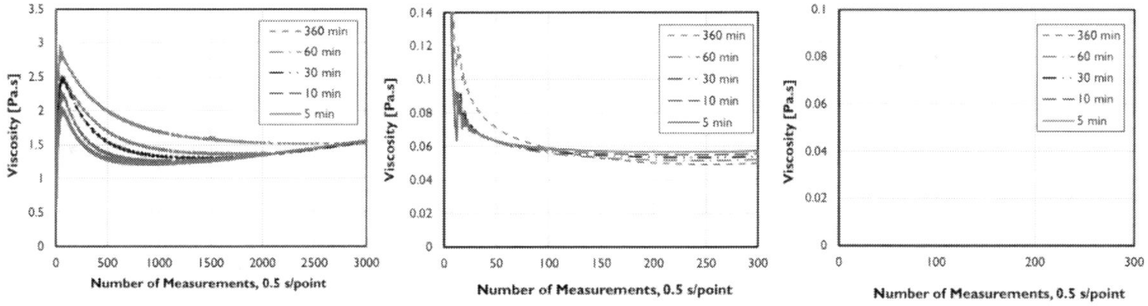

SOT results for the SBF. The viscosity is plotted versus time for constant applied shear (1-100-511s-1) as a function of resting time.

SBF AGEING TIME EFFECT

11

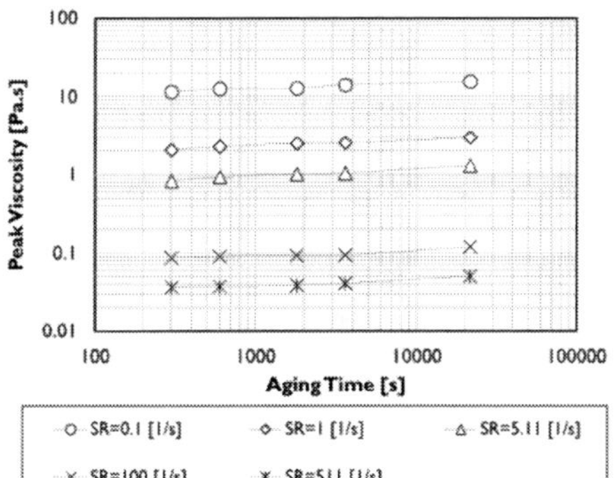

Values of the Viscosity Peak for the SBF at different shear rates and resting time. T=21.11 °C.

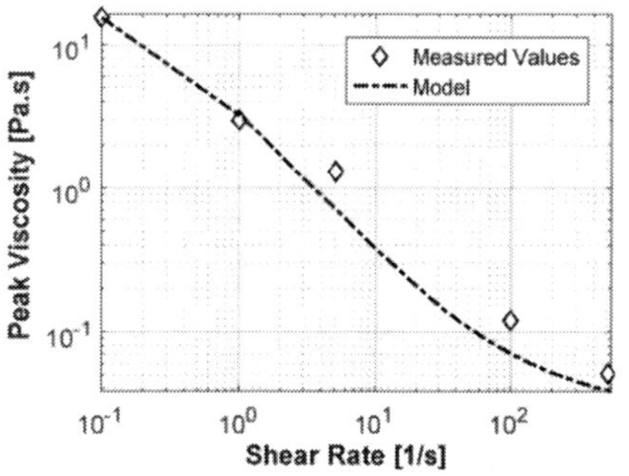

Measured and modelled data for the SBF peak viscosity at t=360 min, as a function of shear rate. T=21.1 °C.

Optimal parameters for the SBF as a function of t [min].

Parameter	Value				
t [min]	360	60	30	10	5
p	0.689	0.663	0.641	0.638	0.62

AGEING TIME EFFECT - RESULTS

12

CUTTINGS SETTLING VELOCITY IN THE SBF

13

Retrieved drilled cuttings - before and after cleaning

The modified Quemada model is used in the Concha and Barrientos (1982) correlation for irregularly shaped particle settling.

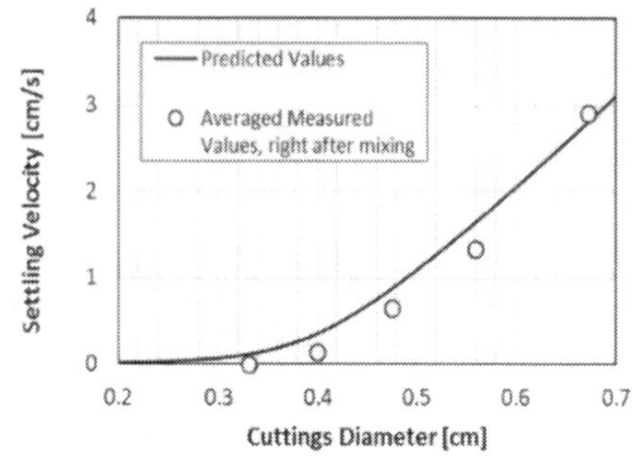

Measured and modelled data of settling velocity in SBF, T=21.11°C and right after mixing.

CUTTINGS SETTLING VELOCITY IN THE SBF

14

Retrieved drilled cuttings - before and after cleaning

The modified Quemada model is used in the Concha and Barrientos (1982) correlation for irregularly shaped particle settling.

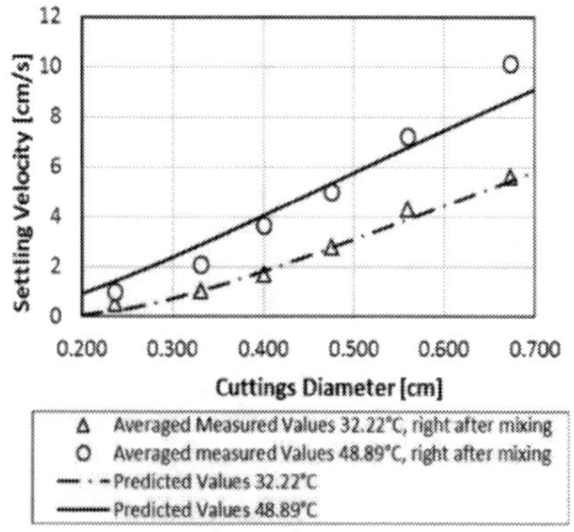

Measured and modelled data of settling velocity in SBF, T=32.22°C and T=48.89°C, right after mixing.

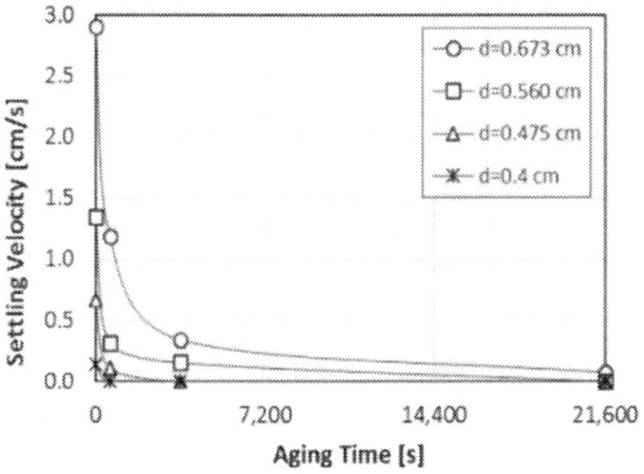

Measured settling velocity in SBF versus resting time, as a function of cuttings size. T=21.11°C.

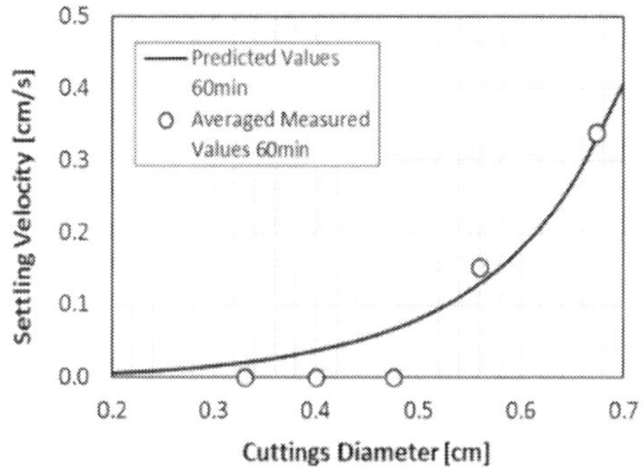

Measured and modelled data of settling velocity in SBF, T=21.11°C and 60 min resting time.

CUTTINGS SETTLING VELOCITY IN THE SBF – AGEING EFFECT

15

CONCLUDING REMARKS

- In this work we have proposed a new view on Synthetic Drilling Fluids exhibiting a yield stress.
- Most of the difficulties encountered in accounting for yield stress behavior disappear when thixotropy is considered by means of *structural* rheological models, such as the one of Quemada (1998).
- By applying an adequate experimental protocols and some minor modifications, we have been able to predict ageing and shear rejuvenation of our SBF for different values of temperature and resting time.
- The proposed methodology reduces the number of parameters to be fitted with measured data, as compared to other experimental procedures (Hodne et al., 2007).
- Both the experimental observations and the model indicate that changes in viscosity and yield point occur simultaneously. They derive from the same physical mechanism, the existence of Structural Units in the fluid.
- The magnitude of the change is affected by temperature, as well as shear rate and resting time, and it is driven by the response of the structure's parameters.
- A further validation of the model was obtained by implementing it into a settling velocity correlation and successfully estimate the sedimentation velocity of drilled cuttings, retrieved from a real well.

16

ACKNOWLEDGEMENT

The University of Tulsa Drilling Research Projects

BAKER HUGHES a GE company

The authors thank Roland May, Dennis Clapper and **Baker Hughes** for the donation of the synthetic drilling fluid and the retrieved cuttings used to perform the experiments. The authors are also thankful to the **University of Tulsa Drilling Research Projects** member companies for their technical and financial supports.

17

17 AIChE Annual Meeting, Minneapolis, MN

Thank You! Any Questions?

Silvio Baldino, PhD Candidate, Petroleum Engineering, University of Tulsa, sib202@utulsa.edu

488098 Experimental Evidence for Shear Banding in Large Amplitude Oscillatory Shear of Entangled DNA Solution.

Abstract:

Combining a planar Couette shear cell with high-resolution confocal microscopy, we investigate the shear profiles of highly entangled DNA solutions undergoing large amplitude oscillatory shear (LAOS). The linear shear profiles of low-concentration solutions transition to wall-slip dominant profiles at high DNA concentrations and ultimately evolve into shear-banded profiles with increasing shear rates or Weissenberg numbers (Wi). We analyze the dynamics of embedded tracer particles in the shear frame, which provide important insights into the microscopic structural changes of shear-banded polymeric solutions. Interestingly, the tracer particles display unusual transient super-diffusivity and enhanced dynamic heterogeneity in the high-shear-rate band. The probability distribution function of particle displacements indicates a Levy-flight-type of motion with power-law decays at large displacements. Such dynamics have also been observed in other shear-banded complex fluids, suggesting a universal feature of shear-banded complex fluids under time-dependent shear.

502425 PadÃ© Approximants for Shear Stress in Large-Amplitude Oscillatory Shear Flow

Abstract:

Analytical solutions for shear stress in large-amplitude oscillatory shear flow, both for continuum or molecular models, often take the form of the first few terms of a power series in the shear rate amplitude. We get this series using the Goddard-Miller integral expansion. Our previous work shows that the Padé approximants for this truncated series, and specifically for the corotational Maxwell model, can agree closely with the corresponding exact solution. We observe this close agreement, even for the Padé approximant of the truncation that contains only the fifth harmonic of the shear stress response. Here we begin with our recent extension of the power series to the next order in the shear rate amplitude [*Phys. Fluids* **29**, 043101 (2017)]. We then use this extension to explore its Padé approximants. We uncover its best approximant, compare it with the Goddard-Miller integral expansion, and also with the exact solution [*Macromol. Theory Simul.* **24**, 352 (2015)]. We use Ewoldt fingerprints to show the stunning accuracy of our new Padé approximant. We quantify this accuracy with the objective functions introduced in this paper. Our worked examples illustrate how researchers can use our new approximant reliably.

492753 The Effect of Branching on Shear Band Formation and Evolution in Wormlike Micelles (WLMs)

Abstract:

Self-assembled wormlike micelles (WLMs) are of particular scientific and technological interest due to their ability to branch, break, and reform under shear, which can lead to nonlinear flow phenomena such as shear banding. Nonlinear rheology methods like shear startup have shown to be useful in determining the molecular architecture and chain branching [1], but quantitative relationships are lacking. Thus, understanding the coupling between flow behavior and molecular topology requires experimental methods that combine time- and spatially-resolved small angle neutron scattering (flow-SANS) measurements of structure with rheometry for nonlinear deformations. Here, we explore the relationship between branching, shear-induced microstructure and nonlinear responses using a series of well-characterized, branched WLM solutions [2,3]. The degree of branching in the mixed cationic/anionic surfactant (CTAT/SDBS) solutions is controlled via the addition of the hydrotropic salt sodium tosylate [2,3]. The shear-induced micellar alignment is spatially and temporally characterized under steady shear and shear startup by flow-SANS. Segmental orientation and alignment (A_f) is a complex function of the branching level and deformation type. Using spatially-resolved 1-2 plane measurements during steady shear, shear banding is verified in low and mildly branched solutions, but is not observed in highly branched solutions. Shear startup SANS measurements are then used to explore the mechanism of shear band formation. A long transient is observed in the A_f response in the low and mildly branched solutions during banding; no transient is observed in the highly branched solutions, confirming shear thinning. While the mildly branched WLMs appear to follow the `disentangle, re-entangleâ€™ mechanism [4], a new mechanism of shear band formation is identified in the low branched WLMs. Here, after the micelles re-entangle, the low shear rate band `re-alignsâ€™ and becomes more oriented before steady state is achieved. This multi-technique approach, combining nonlinear rheology and spatially-and temporally-resolved SANS, has enabled us to link the micellar microstructure and topology to the macroscopic flow properties of WLM solutions, and is part of a broader effort to characterize branching in chemical polymers and self-assembled systems.

[1] F. Snijkers, et al., *J. Rheol.,* **57**, 4 (2013).

[2] M. A. Calabrese, et al., *J. Rheol.,* **59**, 5 (2015).

[3] B. A. Schubert, N.J. Wagner, and E.W. Kaler, *Langmuir* **19**, 10 (2003).

[4] C.R. Lopez-Barron, et al., Physical Review E **89**(4), 2014.

491997 Predictions for Non-Linear Flows of Polydisperse Blends Based on a Differential-Constitutive Analogue of the Double Reptation Model

Abstract:

The flows of industrially relevant polymeric materials are notoriously difficult to model due in part to the broad distribution of molecular weights. While double reptation theory provides a simple and useful description of polydisperse blends in the linear flow regime, at present there is no comparable model in the non-linear flow regime. A true tube-based model is possible in the bi-disperse limit [1], but unfortunately this model is computationally intractable for spatially resolved flows. For studies of spatially resolved flows, multimode constitutive models have greater practical value. Multimode models describe a blendâ€™s overall rheology as the superposition of responses from a set of simple constitutive models, fit to experimental data. However, the multimode approach can be unreliable: for example, a model fit to simple shear flow data can yield poor predictions for extensional flows.

As an improvement to existing multimode models of polymer blends, we have developed a differential constitutive analogue to double reptation theory. Double reptation theory can be viewed as a temporary network model; entanglements are approximated as mutual, binary, topological constraints between chains, and the stress at an entanglement relaxes when one of the two chains reptates through the constraint [2]. In a differential constitutive framework, we apply the micromechanical model of double reptation to all stress relaxation mechanisms so that entanglements can also relax their stress by chain retraction and convective constraint release. The resulting multimode constitutive model is consistent with classical double reptation in the linear flow regime, but also incorporates the non-linear stress relaxation couplings for strong flow conditions. For well-entangled bi-disperse blends, our multimode model yields predictions that are qualitatively consistent with the more detailed (but computationally challenging) tube-based theory. We believe that the differential constitutive analogue to double reptation theory offers a compelling balance between reliability and practical utility for predicting flow behavior in industrially relevant polymer blends.

[1] D. J. Read, K. Jagannathan, S. K. Sukumaran, and D. Auhl, Journal of Rheology 56, 823 (2012)

[2] J. des Cloizeaux, Europhysics Letters, 5 (5), pp 437 â€“ 442 (1998)

494129 Marangoni Instability during the Non-Solvent Induced Phase Separation of a Ternary Polymer Solution

Abstract:

Non-solvent induced phase separation (NIPS) processes are widely used by industry and by Nature as a means of controlling the microstructure of polymer materials. While the general principles of the NIPS process have been known for a long time, a detailed quantitative description has proven difficult, in part due to a lack of robust theories and numerical models. We recently addressed this need by developing a multi-fluid model of the NIPS process that combines continuum fluid mechanics with a field-theoretic description of polymer thermodynamics. Using our model, we show that in certain cases non-solvent transport between a polymer-lean and polymer-rich domain can lead to a hydrodynamic instability driven by local surface tension gradients. We will discuss the prerequisites for this instability and its potential impacts on the microstructure of materials made by the NIPS process.

497240 Highly Stable Mn-Based Nanocarbon: A New Class of Bifunctional Electrocatalyst for Oxygen Reduction and Evolution Reactions

Abstract:

Carbon has many advantages over transitional metal oxide for being an oxygen reduction reaction (ORR) electrocatalyst due to high activity, low cost, excellent electrical conductivity, and high surface area. Nevertheless, its intrinsic instability in oxygen evolution reaction (OER) potential regime (>1.23 V *vs.* RHE) has largely inhibited the potential for bifunctional ORR and OER applications in reversible electrochemical energy devices. Here, we reported a new class of highly stable carbon nanocomposite bifunctional catalysts consisting of substantial carbon tubes growing from a very thick and dense graphitic substrate, which is derived from an inexpensive dicyandiamide as carbon/nitrogen precursor simultaneously catalyzed by Fe, Co, Ni, and Mn during carbonization. The dense substrate leverages the characteristics of high degree of graphitization carbon derived from Mn; active nitrogen-doped carbon tubes are stemmed from FeCoNi during carbonization process. The bifunctional stability and activity of the carbon nanocomposite catalysts were found greatly dependent on content of Mn. The existences of various metal/metal oxides and effective nitrogen doping was responsible for its high bifunctional activities for the ORR and the OER. The unprecedented stability of the bifunctional catalyst was verified by using various accelerated stress tests up to 60,000 potential cycles (0-1.9 V *vs.* RHE) at 25°C and 100-hour life test at constant current density of 10 mA cm^{-2} at 60°C. Instead of degradation, ORR and OER activity was considerably increased during the long-term potential cycling tests, indicating more active sites gradually exposed. The OER current was verified to be a sole outcome of oxygen evolution, rather than carbon or metal oxidations. The remarkable electrochemical stability of this catalyst was primarily attributed to its unique hybrid carbon structures featured with a very high graphitization degree, thick graphitic layers, and abundances of carbon tubes grown from a dense substrate. More importantly, the irreversible formation of Î²-MnO_2 phase and FeCoNi based metals oxides provide crucial protection roles in enhancing carbon corrosion resistance.

494824 Density Functional Theory Study of the Oxygen Reduction Reaction in a Graphene Surface Modified with S or N

Abstract:

Due to the high cost and scarcity of platinum resources, the design and synthesis of new materials for reducing or replacing platinum materials used in the cathode side of low temperature fuel cells has become imperative. In the last decades, novel materials to reduce or replace Pt or Pt alloys have been used as candidates in fuel cells electrodes. Carbon nanomaterials show good performance when used as supports of the active phase. On the other hand, carbon nanomaterials modified with heteroatoms such as sulfur or nitrogen has been also evaluated for the oxygen reduction reaction (ORR) in alkaline media as a promising alternative. To characterize this alternative, first-principles computational methods are useful tools to explain the influence of heteroatom doped-carbon materials on the adsorption/reaction phenomena associated with the ORR. In this context, we present the electrochemical behavior of the surface reactions on graphene doped with S or N, by computational modeling using density functional theory (DFT). We discuss the adsorption/reaction phenomena occurring during catalysis processes on graphene structures modified with dopant S or N atoms. The analysis is related to the electrochemical performance of these materials under adsorption of different functional groups that can describe the behavior of ORR in alkaline media.

497349 Highly Stable Nanocarbon Supports for Pt Cathode Catalysts in Polymer Electrolyte Fuel Cells

Abstract:

Polymer electrolyte membrane fuel cells (PEMFCs) have been widely considered one of environmentally friendly and highly efficient electric power sources for a wide variety of power applications ranging from portable and stationary power supplies to transportation. However, some of the most critical barriers to their commercialization are the intrinsic high cost and insufficient long-term performance durability. Among others, the spherical carbon black particle support used for Pt catalyst in MEAs have a number of intrinsic structural disadvantage. It is partially responsible for current insufficient performance and durability of MEAs. They include (1) inefficient O_2transport due to significant tortuosity in carbon black-based cathode, (2) non-uniform distribution of ionomer due to significant microporosity of carbon blacks, (3) catastrophic corrosion during start-up/shut-down events due to low corrosion resistance, (4) low electrical conductivity, and (5) weak binding sites with Pt nanoparticles.

University at Buffalo, SUNY has developed a few types of highly stable large size nitrogen-diped graphene tubes (NGT) in a highly controlled manner and at high yields. The diameters of these graphene tubes can be tuned from 50 to 500 nm, which is much larger than conventiona carbon nanotubes (20 nm). It should be noted that the graphene tube is low cost using inexpensive dicyandiamide as precursors ($1-2/kg) and the synthesis method is easily scaled up. Apart from the obvious advantage of high electronic conductivity and enhanced corrosion resistance, the large tube size provided better Pt nanoparticle dispersion and more facil mass transfer including O_2 and H_2O. Importantly, the NGT exhibited oxygen reduction reaction (ORR) activity in acidic media comparable with state of the art PGM-free catalsyts. Thus, the novel ORR active nitrogen-doped graphene tubes are able to not only offer remarkable support effects by geometrically and electronically modifying the deposited Pt nanoparticles, but also provide a large amount of complementary nonprecious ORR active sites, thereby significantly improving the overall activity and durability of the Pt catalysts. Thus, relative to conventional carbon blacks, the highly graphitized and active N-GT is an ideal support for low-PGM catalysts for high-performance MEAs with enhanced mass transport and durability along with ideal interfaces with ionomer.

491429 Approaching 2 W·cm^{-2} Aemfcs through Electrode Engineering and Controlling the Cell Water Content and Balance

Abstract:

In recent years, advances in alkaline exchange membrane fuel cells (AEMFCs) with anion exchange membrane (AEM) solid polymer electrolytes have gained traction due to their distinct – and potentially game-changing – advantages over proton exchange membrane fuel cells. However, AEMs and AEMFCs are at a significantly less mature stage in their developmental than proton exchange membrane fuel cells (PEMFCs), and have experienced limitations specifically in the area of stability, carbonation, and achievable current and power densities, exhibiting a sizable performance gap vs PEMFCs.[1] This talk will focus on several fundamental and engineering advances that have enabled the creation of AEMFCs that are able to achieve up to 1.9 W·cm^{-2} peak power (> 2 W·cm^{-2} iR-corrected) and 100's of hours of stable operation, bringing AEMFCs much closer to the incumbent PEMFC technology, and opening the way to overcome the cost barrier that has slowed the growth and large scale market implementation of fuel cells in the transportation sector.[2]

One of the largest contributors to the challenges faced in AEMFCs is the cell water content and balance, where the role of water has an impact on the anode (generation through the hydrogen oxidation reaction), cathode (consumption through the oxygen reduction reaction), and movement from the cathode to anode through electro-osmotic drag. This results in a significantly larger current-driven movement of water than PEMFCs, where less water required for the transport of the protons and there is no electrochemical consumption, and a need to provide adequate water to maintaining proper ionomer and membrane hydration while avoiding catalyst layer flooding or dry-out. Back diffusion of water from anode to cathode in the AEMFC, which is coupled to the membrane conductivity, is a promising way to decrease the extreme water gradient, and simplify the water balance problem in AEMFCs.[3]

In this study, we will discuss the influence of the membrane, ionomer and gas diffusion layer as well as the flow rate and dew points of the anode and cathode gases on AEMFC performance. Using a hydrophilic gas diffusion layer without a microporous layer increases membrane hydration, but also increases the possibility for flooding. Catalyst layer engineering – through adjusting the carbon/ionomer/catalyst loadings, adjusting hydrophilicity/hydrophobicity, manipulating electrode thickness and porosity, etc. – is investigated. We will also discuss controlling the total amount of water fed to the AEMFC through the gas feeds, as well as the water balance between the anode and cathode. Manipulating the dew points led to the counter-intuitive discovery that the cell performs better with the humidity higher at the anode than the cathode, despite water generation and electro-osmotic drag towards that electrode. In fact, removing too much water from the anode caused instability in the cell, while increasing the water at the anode decreased the membrane resistivity. Water back diffusion from the anode to the cathode likely plays an important role in membrane hydration and hydroxide transport through the membrane. Finally, we will also discuss the behavior of the

AEM and powder anionomer in the presence of air both with and without CO_2.

References

1. J. R. Varcoe, P. Atanassov, D. R. Dekel, A. M. Herring, M. A. Hickner, P. A. Kohl, A. R. Kucernak, W. E. Mustain, K. Nijmeijer, K. Scott, T. Xu and L. Zhuang, *Energy Environ. Sci.*, **7**, 10 (2014).

2. L. Q. Wang, E. Magliocca, E. L. Cunningham, W. E. Mustain, S. D. Poynton, R. Escudero-Cid, M. M. Nasef, J. Ponce-González, R. Bance-Souahli, R. C. T. Slade, D. K. Whelligan and J. R. Varcoe, *Green Chem.*, **19** (2017).

3. T. D. Myles, A. M. Kiss, K. N. Grew, A. A. Peracchio, G. J. Nelson and W. K. Chiu, *J. Electrochem. Soc.*, **158**, 7 (2011).

498743 Design and Engineering of Membraneless Electrolysis Devices

Abstract:

Although the sun is a promising energy source to renewably meet societyâ€™s demands, the storage of solar energy during intermittent hours remains a significant challenge. One possible solution is to electrolyze water into storable fuels, but the present capital costs are still too high for this to be viable. The current state of the art technology relies on the use of ion selective membranes, which are expensive and prone to degradation and failure. The extreme sensitivity of membranes to impurities can incur significant costs associated with installation, maintenance, and electrolyte purification. Our group has focused on creating low cost devices by designing electrolyzers which can operate safely without a membrane. In particular, we have investigated devices based on mesh flow-through electrodes. In these studies, we were able to demonstrate safe and reliable separation of H_2 and O_2 product gasses using forced convection.[1] By selectively depositing the electrocatalyst on only one side of the mesh substrate, we were even able to separate the product gasses with high purity in an un-pumped, free-standing electrolyte.[2] In this talk, we discuss the present challenges for membraneless electrolysis as well as the opportunities going forward.

[1] Oâ€™Neil G D, *et al.* 2016 Hydrogen Production with a Simple and Scalable Membraneless Electrolyzer *J. Electrochem. Soc.* **163** F3012â€"9

[2] Davis, *et al.* 2017 Floating Membraneless PV-Electrolyzer Based on Buoyancy-Driven Product Separation *Manuscript in Submission.*

500751 Design of a Novel Electrochemical Membrane Reactor for Hydrogen Production *Via* the S-NH_3 Cycle

Abstract:

The solar-driven hybrid sulfur-ammonia (S-NH_3) water splitting cycle is a novel approach to produce hydrogen through a large-scale design. Originally proposed by T-Raissi *et al.* (2006), this cycle involves the photochemical oxidation of ammonium sulfite as a solar-photocatalytic hydrogen production step. Later, the resulting ammonium sulfate is decomposed into ammonia, sulfur dioxide and oxygen through a high-temperature solar thermochemical oxygen evolution step. Finally, ammonium sulfite is regenerated by means of chemical absorption of ammonia and sulfur dioxide in aqueous phase. Thereby, only water is needed to release hydrogen and oxygen in this cycle. This process has many advantages like high-purity hydrogen production at ambient temperature; generation of value-added by-products such as ammonium sulfate; exploitation of the entire solar spectrum and consumption of greenhouse gases (*i.e.* sulfur dioxide). However, even though the hydrogen production step has been studied, there is a lack of information on aspects which have strong impact on hydrogen production rates like pH, kinetic parameters, and/or efficient reactor design. Moreover, new approaches have been proposed to enhance the large-scale viability of the S-NH_3 cycle, including an electrochemical hydrogen production step established by Taylor *et al.* (2014). This electrochemical approach still makes use of solar energy resources, but via thermodynamic cycles which take advantage of the high temperature gases during the oxygen evolution step to produce electricity. Thereby, the electrooxidation of ammonium sulfite is carried out with the same advantages as described above. Nevertheless, a more detailed analysis is needed in order to consolidate a high-performance hydrogen production step.

In this work, we propose a novel electrochemical reactor for the hydrogen production step for the S-NH_3 cycle, which uses an ionic exchange membrane to achieve pH control. The reactor is designed in such a way that allows a suitable and dynamic analysis of hydrogen production at laboratory-scale.

In order to support this proposal, a preliminary pH-control analysis was carried out. Figure 1 illustrates the chronoamperograms resulting from the electrooxidation of ammonium sulfite at room temperature applying 700 mV (*vs* Ag/AgCl). The best two conditions of reaction rates were using a buffer and an Anion Exchange Membrane (AEM), as shown. Thus, the latter was selected as the most practical pH-control method for this reaction. Note that such experimental conditions enforced hydrogen production *via* electrooxidation, while prevented water electrolysis in contrast to those experimental conditions earlier reported, 125°C and 1.00 V by Taylor *et al.* (2014). Whereas hydrogen production was verified by Gas Chromatography with Thermal Conductivity Detector (GC-TCD), production of ammonium sulfate was proven by Raman spectroscopy, as shown in the inset in Figure 1. Based on these results, a two-chamber electrochemical membrane reactor separated with an AEM was developed to allow the hydrogen production analysis, as shown in Figure 2.

The reactor performance was later analyzed using a Design of Experiments (DOE) approach through a Central Composite Design (CCD), to establish optimal operating conditions. Ammonium sulfite concentration (mM) and

applied voltage (mV) were selected as the studied factors. The response variable was the normalized half-life time, a parameter linked directly to the reaction rate and calculated through chronoamperometric data of each ammonium sulfite electrooxidation. Extensive experimental tests were carried out for each condition and a second order (quadratic) model for the response variable was obtained. Through this model, suggested optimal operating conditions are 350 mV (*vs* Ag/AgCl) and a concentration of ammonium sulfite at 55 mM. Finally, hydrogen production rate at optimal operating conditions was verified. Hydrogen gas produced during reaction was stored inside a collection chamber over the cathode. Continuous samples were taken and measured by GC-TCD. A calibration curve was used to obtain the amount of hydrogen produced. To compare hydrogen production levels, tests at 800 mV / 50 mM were conducted as well. Currently, kinetic and mass transport studies are being carried out to discuss a better reaction design for this electrochemical approach.

Figure 1. Electrooxidation kinetics of $(NH_4)_2SO_3$.

Figure 2. Conceptual principle of operation of the proposed electrochemical membrane reactor.

499822 Portable Smartphone-Enabled Capillary-Based Viscometer and Tensiometer

Abstract:

Here we present a simple capillary-based approach for the measurement of viscosity and surface tension of liquids. A smartphone is used to record videos (at up to 240 frames per second) of liquid imbibition inside horizontally placed capillaries of both uniform (100-400 micrometers in diameter) and tapered circular cross-sections. Automated image analysis of the videos is then performed to detect fluid displacement and retrieve the meniscus position, L, as a function of time, t. Because liquid imbibition is measured over relatively short lengths of the capillary (typically less than 6 cm) and could possibly involve high imbibition speeds (there is no gravity slowing down the fluid as in capillary rise), the effect of a dynamic contact angle is expected in our measurements. Therefore, to enable broader application of this approach, we have developed an analytical model for the capillary-driven imbibition of a Power-Law fluid in capillaries of uniform circular cross-section, while accounting for dynamic contact angle variations in the viscous regime. For a Newtonian fluid and neglecting contact angle dynamics, the mathematical expression obtained simplifies to the well-known Lucas-Washburn equation. A systematic fitting using our analytical model is then employed to retrieve fluid properties from the L vs. t data collected with the smartphone. Results from our viscometer are in quantitative agreement with known fluids used as references. In addition, an equilibrium condition achieved by fluids inside capillaries of tapered circular cross-sections, also imaged with a smartphone, has been explored as a convenient method for measuring the surface tension of small amounts of liquids (less than 10 microliters). The materials needed (capillaries and a smartphone camera) and the analysis performed (implemented as a user-friendly software) enable viscosity and surface tension measurements to be performed rapidly and without requiring special knowledge or training. All of these advantages make this approach low cost and portable, which facilitates the study of the rheology and surface tension of various fluids in point-of-use and resource-limited settings.

493058 An Experimental Study of a Liquid Drop Impinging on a Liquid Pool

Abstract:

An experimental study of a liquid drop impinging on a liquid pool

Manas Ranjan Behera [a], Anirvan Dasgupta [b], Sudipto Chakraborty [a]

[a] *Department of Chemical Engineering, Indian Institute of Technology Kharagpur, India.*

[b] *Department of Mechanical Engineering, Indian Institute of Technology Kharagpur, India.*

E-mail: manaswebmail@gmail.com, anir@mech.iitkgp.ernet.in, sc@che.iitkgp.ernet.in

The coalescence of a drop into a liquid pool has been studied in the context of a vortex ring[1], foam[2], emulsion[2], and microfluidic device[3]. The research on the drop coalescence is motivated due to their relevance to various industrial applications; viz. transport of oxygen across air-water interface[4], and spray cooling[5], as well as its fundamental importance in fluid mechanics. Generally, when a drop impinges on the surface of the liquid pool having similar properties at a low velocity, a vortex ring is formed. A vortex ring is a locale where the fluid mostly turns around an imaginary axis line that forms a closed loop. The formation of the ring is a commonly occurring phenomenon which is seen in both natural occurring processes (e.g. volcano[6]) as well as artificial processes (e.g. blast of atomic bomb[7], generated at wings of the aeroplane).

The current work attempts to study the formation of vortex ring for the miscible liquid system by impinging a liquid drop on a liquid pool. It is the simplest and relatively inexpensive method to generate vortex ring. A schematic diagram of the experimental setup for the generation of vortex ring has been shown in Figure 1. A rectangle tank [Plexiglas, Volume: 25x25x45 cm^3] was used for all these experiments. Distilled water was filled up to 35 cm of total tank volume to create the liquid pool. The microsyringe assembly was attached to a stepper motor system for dispensing drops in a controlled manner. A fine vertical adjustment system was used for impingement of liquid at different heights. The impingement height was varied from 10-40 mm. The size of the drop was maintained 2.96 mm. A high-speed digital camera (manufacturer: Phantom, Vision Research Inc. (USA), model: V 7.3) was used to capture the post impact drop dynamics. The video was recorded in 800x600 resolutions at 1500 frames per second with an exposure time of 150 μs.

The formation and growth of the ring are described and presented in a sequence of images at the different time interval in Figure 2. When a particle laden drop impinges on a liquid pool, it creates a crater which is marked as the onset point for vortex ring generation. The dynamics of crater influences the characteristics of the ring. The impact of crater dynamics on vortex ring formation is investigated by measuring the variation of depth and velocity of the crater with time. The effect of drop impact height on the formation and characterization of vortex ring is also studied. How the impingement height affects the eccentricity of the drop, crater dynamics, translational velocity, and diameter of the vortex ring is investigated. From the observation, it is concluded that good vortex rings are formed over 10-30 mm height. The effect of drop eccentricity on the characterization of the ring is also investigated. It is concluded that the maximum penetrating vortex rings are achieved when the drop shape is prolate at the time of impact whereas minimum penetrations are obtained for the oblate shaped drop. From our observation, it can be concluded that the eccentricity of the drop also affects the crater dynamics. The shape and motion of the ring are measured and the variations are depicted with respect to time. The effects of various forces on penetration and formation of the ring were characterized by different dimensionless numbers such as Reynolds number, Weber number, Ohnesorge number, and Froude number.

Keywords: Interfacial Transport Phenomena, Droplet Dynamics, Crater Dynamics, Vortex Ring, High-speed imaging.

Figure 1. Experimental setup for generation of vortex ring.

Figure 2. Formation and growth of vortex ring.

References:

1. Lee, J. S., Park, S. J., Lee, J. H., Weon, B. M., Fezzaa, K. and Je, J. H. (2015), "Origin and dynamics of vortex rings in drop splashing", *Nature Communications*, Vol. 6, pp. 8187-8195.

2. Behera, M. R., Varade, S. R., Ghosh, P., Paul, P. & Negi, A. S. (2014), "Foaming in Micellar Solutions : Effects of Surfactant, Salt, and Oil Concentrations", *Industrial & Engineering Chemistry Research*, Vol. 53, pp. 18497-18507.

3. Velev, O. D., Prevo, B. G. and Bhatt, K. H. (2003), "On-chip manipulation of free droplets", *Nature*, Vol. 426, pp. 515–516.

4. Ho, D. T., Asher, W. E., Liven, L. F. B., Schlosser, P. and Gordan, E. L. (2000), "On mechanisms of rain-induced air-water gas exchange", *JOURNAL OF GEOPHYSICAL RESEARCH*, Vol. 15, pp. 24,045–24,057.

5. Manzello, S. L. and Yang, J. C. (2002), "An experimental study of a water droplet impinging on a liquid surface", *Experiments in Fluids*, Vol. 32, pp. 580–589.

6. Fuentes, O. V. (2014), "Early observations and experiments on ring vortices", *European Journal of Mechanics, B/Fluids*, Vol. 43, pp. 166–171.

7. Sigurdson, L. (1991), "Atom Bomb/Water Drop", *Physics of Fluids A: Fluid Dynamics*, Vol. 3, pp. 2034–2034.

500232 Arresting Dissolution By Interfacial Rheology Design

Abstract:

A strategy to halt dissolution of particle coated air bubbles in water based on interfacial rheology design will be presented. Whereas previously a dense monolayer was believed to be required for such an "armored bubble" to resist dissolution, in fact engineering a two-dimensional yield stress interface suffices to achieve such performance at sub-monolayer particle coverages. We use a suite of interfacial rheology techniques to characterize the behavior of spherical PS-PVP particles at an air-water interface as a function of surface coverage. Spherical bubbles with varying particle coverages are made and their resistance to dissolution evaluated using a microfluidic technique. Whereas a bare bubble only has a single pressure at which a given radius is stable, we find a range of pressures over which bubble dissolution is arrested for armored bubbles. This pressure window is dictated by the particle surface coverage and the yield stress of the interface. The link between interfacial rheology and macroscopic dissolution of ~100μm bubbles coated with ~1μm PS-PVP particles is presented and discussed. The results imply a generic design rationale that can be applied to successfully inhibit Ostwald ripening in a multitude of foam and emulsion applications.

487819 The Importance of the Amorphous Material Interface in the Measurement of Thermodynamic Parameters, Using Inverse Gas Chromatography

Abstract:

Hansen Solubility Parameters (HSP) are thermodynamic quantities of crucial importance for the understanding of the behavior of polymeric materials with respect to different solvent systems.[1] Traditionally, the measurement of HSP was based on lengthy solubility/swelling experiments, followed by extensive statistical analysis of the data. Inverse Gas Chromatography (IGC), a technique employed in polymer characterization since mid-70's,[2,3] was introduced as a tool for the determination of HSP about 25 years ago.[4,5] Since then, it has gained a lot of momentum in the field, as it provides faster measurements, it is less labor intensive and requires much less amount of materials, such as polymer and solvents.

IGC enables the determination of the infamous Flory-Huggins χ solubility parameter[5,6,] for the interaction of the material with different solvents.[7,8] The values of χ obtained can be then used to determine quantities such as Hansen Solubility Parameters (HSP) and diffusion coeffcients.[9] Considering the chromatographic nature of the technique and the amorphous character of much of the materials of interest, it is evident that the flow rate of the carrier gas in the chromatographic column is of high importance, even for amorphous materials below their glass transition temperature (T_g). The interface between the solvent and the solute gives rise to capillary effects.[10-12] Thus, for accurate measurements, enough time should be given for the solvent to overcome these limitations. Up to now, the influence of this interesting interfacial phenomenon has been omitted for such types of measurements. In the majority of the published literature the authors do not provide any rational for the selection of the carrier gas flow rate in their measurements and they seem to ignore its importance.

This work addresses the importance of interfaces in the measurement of HSP using a series of simple experiments on some model compounds. The HSP of crystalline materials (P-Monoclinic Carbamazepine and Monoclinic Acetaminophen) were measured at two different flow rates. The results are identical for all the flow rates suggesting that in crystalline materials the influence of diffusion is limited. Following that, Copovidone, a common polymeric pharmaceutical excipient, was employed for more advanced studies. Its T_g was determined both with DSC and IGC and it was found to be around 100 °C. Then its solubility parameters were measured at 30, 50, 70, 90 and 100 °C for different carrier gas flow rates. For measurements conducted well below the T_g, the value of HSP increases with decreasing flow rate whereas for the measurements on the T_g the HSP value appears to be independent of the flow rate. These findings highlight that at high temperatures the limitations induced by capillarity diminish, as expected from the analysis of these systems in terms of Fickian diffusion. Analysis of the χ parameters obtained, for different solvents and at different temperatures, reveal that the influence of entropic phenomena scales with

decreasing flow rate and decreasing temperature. The findings of these work couple well with the theory proposed by de Gennes, which, in fact, encompasses the influence of local capillary phenomena; the importance of which have been predicted by J.W. Gibbs in his groundbreaking work "On the equilibrium of heterogeneous substances", published in 1874.

Overall, this work addresses the influence of interfacial phenomena in the measurement of the bulk properties of amorphous materials; revealing the overwhelming importance of entropic interactions. It highlights the importance of experimental conditions, proposing a road-map of good experimental practice. This would provide the investigators with a magnifying glass enabling them to look the very fine details of the literature data.

References

1. Braun J, Guillet JE. Determination of crystallinity of olefin copolymers and polyolefin powders by inverse gas chromatography. *Polymer Chemistry*. 1975, 13, 1119-1131.

2. Hudec P. Determination of crystallinity of polyethylene powder by inverse chromatography. *Macromolecular Chemistry and Physics*. 1977, 78, 1187-1193.

3. Voelkel A, Janas J. Solubility parameters of broad and narrow distributed oxyethylates of fatty alcohols. *Journal of Chromatography*. 1993, 645, 141-151.

4. Adamska K, Bellinghausen R, Voelkel A. New procedure for the determination of Hansen solubility parameters by means of inverse gas chromatography. *Journal of Chromatography A*. 2008, 1195, 146-149.

5. Flory PJ. Thermodynamics of high polymer solutions. *Journal of Chemical Physics*. 1941, 9, 660-661.

6. Huggins LM. Solutions of long chain compounds. *Journal of Chemical Physics*. 1941, 9, 440.

7. DiPaola-Baranyi G, Guillet JE. Estimation of polymer solubility parameters by gas chromatography. *Macromolecules*. 1978, 11, 228-235.

8. Ito K, Guillet JE. Estimation of solubility parameters for some olefin polymers and copolymers by inverse gas chromatography. *Macromolecules*. 1979, 12, 1163-1167.

9. Hansen CM. *The three dimensional solubility parameter and solvent diffusion coefficient and their importance in surface coating formulation.* [PhD]. The Technical University of Denmark. 1967.

10. de Gennes PG. Dynamics of fluctuations and spinodal decomposition in polymer blends. *Journal of Chemical Physics*. 1980, 72, 4756-4763.

11. Forrest BM, Toral R. The phase diagram of the flory-huggins-de gennes model of a binary polymer BlendF. *Journal of Statistical Physics*. 1994, 77, 473-489.

12. Geoghegan M, Krausch G. Wetting at polymer surfaces and interfaces. *Progress in Polymer Science*. 2003, 28, 261-302.

485085 Mechanisms of Dynamic Wetting Failure in the Presence of Soluble Surfactants

Abstract:

A hydrodynamic model and flow visualization experiments are used to understand the mechanisms through which soluble surfactants can influence the onset of dynamic wetting failure. In the model, a Newtonian liquid displaces air in a rectangular channel in the absence of inertia. A Navier-slip boundary condition and constant contact angle are used to describe the dynamic contact line (DCL), and surfactants are allowed to adsorb to the interface and moving channel wall (substrate). The Galerkin finite element method is used to calculate steady states and identify the critical capillary number at which wetting failure occurs. It is found that surfactant solubility weakens the influence of Marangoni stresses, which tend to promote the onset of wetting failure. Adsorption of surfactants to the substrate can delay the onset of wetting failure due to the emergence of Marangoni stresses that thicken the air film near the DCL. The experiments indicate that the critical capillary number increases with surfactant concentration. For the more viscous solutions used, this behavior can largely be explained by accounting for changes to the mean surface tension and static contact angle produced by surfactants. For the lowest viscosity solution used, comparison between the model predictions and experimental observations suggests that other surfactant-induced phenomena such as Marangoni stresses may play a more important role.

485086 Electrostatic Assist of Liquid Transfer in Printing Processes

Abstract:

Transfer of liquid from one surface to another plays an important role in many printing processes. Incomplete liquid transfer can produce defects that are detrimental to the operation of printed electronic devices, and one strategy for minimizing these defects is to apply an electric field, a technique known as electrostatic assist (ESA). However, the underlying physical mechanisms of ESA remain a mystery. To better understand these mechanisms, slender-jet models for both perfect dielectric and leaky dielectric Newtonian liquid bridges with moving contact lines are developed. Nonlinear partial differential equations describing the time- and axial-evolution of the bridge radius and interfacial charge are derived, and then solved using finite-element methods. For perfect dielectrics, it is found that application of an electric field enhances transfer of liquid to the more wettable surface. For leaky dielectrics, application of an electric field can augment or oppose the influence of wettability differences, depending on the direction of the electric field and the sign of the interfacial charge. The physical mechanisms underlying these observations will be discussed, and the results of complementary flow-visualization experiments will be presented.

489035 Analysis of Temperature Gradient Zone Melting and Annealing for Mitigation of Second-Phase Particles in Single Crystals

Abstract:

Interfacial transport plays a predominant role in the temperature gradient zone melting (TGZM) technique, whereby small, second-phase particles embedded within a crystalline matrix can be induced to migrate by heating to slightly above the eutectic temperature and applying a temperature gradient across the sample. Under such conditions, the now-liquid particle dissolves on the hot side and re-solidifies on the cool side, with a net effect of the particle migrating toward the hotter region. Historically, the discovery of these effects was motivated by the experience of arctic explorers, who found that newly formed sea ice was not potable due to brine inclusions. However, after 2-3 years of exposure to sunlight, this sea ice could be used for drinking water. In the aging process, brine inclusions migrate to the surface by diurnal heating, thus purifying the ice.

In a more modern application, TGZM has been applied to crystals of cadmium zinc telluride (CZT) grown from liquid phases. This material typically exhibits significant populations of large (10 micron and above) tellurium-rich particles that are deleterious to the performance of semiconductor radiation detectors. While the specific formation mechanisms of these particles are not well understood, their presence is unavoidable due to the supersaturation of tellurium controlled by near-equilibrium thermodynamics during growth.

In this presentation, we present the formulation of a mathematical model for particle migration via TGZM that represents solute transport across interfaces defined by conditions of local thermodynamic eqilibrium. We demonstrate that an approximate analytical solution to this model in one spatial dimension well describes the general behavior of particle migration. The analytical solution shows that steady-state migration is not possible, and that, under a constant thermal gradient, the particle velocity and size increase continually with time. We demonstrate that this analytical solution provides good estimates for the overall outcomes of TGZM experiments.

We also describe the implementation of a moving-boundary, finite-element method that solves for particle position and shape, with no simplifying assumptions about mechanistic interactions. We account for the Gibbs-Thompson effect, whereby equilibrium melting temperature is affected by local curvature of the solid-liquid interface. By varying the Gibbs-Thompson parameter, we are able to gauge its role in maintaining the particleâ€™s interface stability during migration. We observe evolution of particle velocity, size, and shape over time and compare model results with experimental observations of tellurium-rich particles in CZT. We also consider how annealing under a cadmium-rich atmosphere can reduce the size of second-phase particles. Finally, we consider the combined strategy of TGZM and cadmium-overpressure annealing.

This work has been supported in part by U.S. Department of Homeland Security, 2012-DN-077-ARI066-06, and no official endorsement should be inferred.

497928 *In situ* Partitioning Measurements of Multiple Fermentation End-Products into Lipid Membranes

Abstract:

Product inhibition during fermentation is driven in part by solvent partitioning into bacterial lipid membranes. Solvent partitioning leads to membrane fluidization, or a reduction in lipid ordering, that alters transmembrane permeability and the function of membrane proteins. Membrane engineering continues to be used to enhance bacterial resistance to membrane fluidization and create robust bacteria for biofuel production. A key aspect to membrane engineering is understanding how solvent partitioning varies with lipid composition in realistic, complex environments. We present a new approach for measuring the partitioning behavior of metabolic end-products produced during butanol fermentation using quantitative nuclear magnetic resonance (qNMR) spectroscopy. Butanol is a viable alternative fuel to gasoline; however, it is lipophilic and inhibits fermentation at low concentrations. The qNMR technique facilitates rapid *in situ* measurements of multiple species partitioning simultaneously and can be performed as a function of temperature and solution chemistry. This presents clear advantages over calorimetric techniques. Butanol, ethanol, 1,3-propanediol, acetic acid, and butyric acid have been examined individually and in mixtures as a function of lipid tail saturation using mixtures of dipalmitoylphosphatidylcholine (DPPC, saturated) and dioleoylphosphatidylcholine (DOPC, unsaturated), as well as reconstituted membrane extracts from *Clostridium pastuerianum*. qNMR and dynamic light scattering results show that butanol partitioning is minimally affected by the presence of other metabolic end-products and membranes swell to accommodate butanol based on their ability to interdigitate at high butanol concentrations. The qNMR technique should be broadly applicable to fermentation processes and can enable rapid screening of membrane partitioning.

493239 Synchronization and Traveling Waves within Arrays of Electrostatic Oscillators

Abstract:

In Nature, large numbers of molecular-scale actuators work together to achieve macroscopic functions such as locomotion. The pursuit of artificial systems capable of similar dynamic functions requires strategies by which to coordinate the motion of many active components. One such approach relies on the spontaneous synchronization of mechanical oscillators coupled together within spatially distributed arrays. This talk describes how linear arrays of electrostatic oscillators can synchronize to create traveling waves that persists over distances and times much larger than those of the individual oscillators. The motion of each oscillator relies on contact charge electrophoresis (CCEP) of a conductive particle between two parallel electrodes. Electrostatic interactions between neighboring oscillators guide the selection of a preferred wavelength, which depends on the geometry of the array. Detailed simulations that account for the many body electrostatic interactions among the charged, polarizable particles reproduce the key experimental observations and offer insights into the mechanisms underlying traveling wave synchronization. We demonstrate how these traveling waves can be harnessed to do mechanical work, with potential opportunities in the areas of colloidal machines and soft robotics.

497576 Probing Viscosity of Complex Emulsions with Micro-Rheometry

Abstract:

Rapid in-situ characterization of emulsions is of direct relevance to the chemical processing industry. Conventional surface science and rheological tools require large sample volumes and possess well-known limitations. For instance, a pendant drop tensiometer is inaccurate for fluids with a low density and optical contrast and a rheometer is chiefly developed for probing viscosities at a macro scale. Microfluidics provide the opportunity to probe rheological and interfacial properties of emulsions at the single drop level.

Current microfluidic methods focus on using pressure driven flow through various geometries to investigate shear response of single phase fluids (Newtonian and non-Newtonian). We extend this approach to design lab on a chip devices for simultaneous measurement of viscosity and interfacial tension in two phase droplet flows. A PDMS based device is micro-fabricated and coupled to a pressure driven Labview controlled system to generate emulsions with high mono-dispersity. The dispersed phase droplets are systematically perturbed by tuning the flow field in the dispersed phase. High speed camera imaging is used to record the drop deformations. Drop shapes are compared to analytical models for drop shape relaxation (after an initial well-defined deformation) and viscosity and interfacial tension are obtained by direct comparison of experiments with theory.

494223 The Study on Mechanism of Flow Distribution and Its Optimization in a Spray Column

Abstract:

The study on mechanism of flow distribution and its optimization in a spray column

Spray column is widely used as an absorption apparatus in wet flue gas desulfurization (WFGD) technology due to its efficiency and reliability[1]. Because of the stirred slurry tank in the bottom of the column, flue gas enters the column from the side typically in a radial direction. As a result, the gas velocity is unevenly distributed across the cross section since the turning and expanding gas flow. With increasing attention of the environment problem, WFGD spray column has the trend of very large-scale, extremely high efficiency and low resistance. Especially a spray column with large diameter will deteriorate the maldistribution of the velocity[2], which will seriously affect the removal rate. For engineering practice, it is necessary to research flow distribution in a spray column. In spray column without internals, the only factor to adjust gas distribution is falling droplets, while the mechanism of the adjustment of droplets to gas distribution is not studied in the literatures. In spray column with rod bank internal, gas flows around the rod bank under the countercurrent condition between droplets and gas, and less attention was paid to this flow field. In this paper, the theoretical analysis and numerical simulation for both mechanism of flow distribution is combined. Firstly, the self-adjustment effect with single spraying layer is proposed. Then explorations are conducted for a spray column with a rod bank internal and a spray column with 4 spraying layers, and an optimization scheme for flow distribution and mass transfer is also obtained.

Different from drag distributed mechanism[3] for traditional internal, the adjustment of droplets to gas is derived from the momentum transfer between two phases. During the momentum transfer, the motion of droplets tend to optimize the flow distribution, so we call it droplet self-adjustment effect. This effect equalizes the gas velocity across whole column, however, it only works under the countercurrent condition.

In the spray column with single layer, an internal is necessary to assist the flow equalization if droplets are too small to make the droplet self-adjustment effect work. The adjustment effect of a rod bank internal to gas velocity is based on drag distributed mechanism which has no precondition, but it works only within a limited range near the rod bank. When these two effects work together, a rod bank internal can firstly mitigate the gas maldistribution and reduce the potential of the concurrent condition in the lower part of the column, then the gas velocity profile will be affected by the droplet self-adjustment effect across the whole column. A synergistic effect exists between these two effects which is shown in Fig.1 and Tab.1.

Fig.1 Distributions of gas velocity in a spray column (plane y=0)

Table 1 nonuniform coefficient in the upper part of a spray column

The spray column usually has 4 spraying layers. The flow distribution in the column with 4 layers becomes better from bottom to top, hence the distribution problem across whole column turns into the problem in the lowest layer. Reactive absorption is the key step for spray column, so a dilemma appears: droplets in small diameter are beneficial for larger interfacial area, while droplets in large diameter are beneficial for droplet self-adjustment effect. Since spray column with multilayer can regulate operating conditions for each layer, a scheme to optimize the flow distribution and mass transfer is proposed. A small amount of spraying liquid with droplets in large diameter in lowest layer can be used to achieve the flow distribution requirement, while the rest of spraying liquid with small droplets in the upper layers can optimize the mass transfer. Through numerical simulation about the influence of droplet diameter and spraying rate for flow distribution, an optimal diameter and spraying rate is obtained at d=3mm in the lowest layer, and d=2mm in the rest layers. Compared with the spray column with uniform diameter d=2.5mm (under the given L/G, it is the smallest droplet diameter for uniform distribution), the optimization scheme increases 15% in liquid hold-up and 52% in interfacial area.

1. Neveux, T. and Y. Le Moullec, *Wet industrial flue gas desulfurization unit: model development and validation on industrial data.* Industrial & Engineering Chemistry Research, 2011. **50**(12): p. 7579-7592.

2. Nygaard, H.G., et al., *Full-scale measurements of SO 2 gas phase concentrations and slurry compositions in a wet flue gas desulphurisation spray absorber.* Fuel, 2004. **83**(9): p. 1151-1164.

3. Idelchik, I.E., N.A. Decker, and M. Steinberg, *Fluid Dynamics of Industrial Equipment: Flow Distribution Design Methods.* Hemisphere Publishing Corporation, USA, 1991.

496914 Transition between Turbulence Regimes in Particle-Laden Channel Flows

Abstract:

In this talk we will present recent findings from high-resolution numerical simulations on two distinct turbulence regimes observed in particle-laden channel flows. At low mass loading, it is well established that the majority of the underlying carrier-phase turbulence manifests from classical mean-shear production. In this limit, particles tend to preferentially concentrate in high-strain regions of the flow near the channel wall. When the mass loading is order one or larger, the relative motion between the phases leads to additional sources of instabilities as a result of interphase coupling. Particles are known to spontaneously self organize into densely packed clusters. At mass loading of 10, our previous work demonstrated that fluid-phase turbulent kinetic energy (TKE) arises almost entirely from drag production past clusters, giving rise to a separate class of multiphase turbulence referred to as cluster-induced turbulence (CIT). Surprisingly, it remains to be seen how wall-bounded particle-laden flows transition from the dilute limit in which fluid-phase TKE arises from classical mean shear production, to the high-mass-loading regime dominated by drag production. In this study, we examine this transition as a function of mass loading and Stokes number. One-point and two-point statistics will be presented, and the and the separate mechanisms contributing to fluid-phase TKE energy balance will be reported.

498563 Analysis of Interfacial Forces of a Slurry Bubble Column

Abstract:

Slurry bubble columns are multiphase contactors widely used in industry applications, due to high transfer rates with low operating and maintenance costs. However, their fluid dynamics is very complex, and several studies are still in progress concerning interfacial forces present in such equipment, manly when a third solid phase is added in the flow. Solids particles holdup, velocity and physical properties affect directly flow parameters as bubble distribution, liquid velocity and gas holdup, which are directly related to the fluid/fluid and solid/fluid interaction. As in the most of application, solids act as catalyst, which influences on mixing, mass transfer and chemical reaction, emphasizing the importance of their correct dynamics prediction. Despite the extensive research involving bubble columns, the way that the three phases interact with each other still continues as an open subject. In this context, this study proposes a numerical simulation of a three phase bubble column for evaluating the influence of the interaction forces between gas/liquid, gas/solid and liquid/solid phases. Numerical simulations were performed with a CFD commercial code and results were compared with the experimental data available in literature.

499005 Semi-Batch Liquid-Liquid Taylor Vortex Flow: Comparison of 3-D and 2-D Eulerian-Eulerian CFD Simulations and Experimental Measurements for Fluid Flow Patterns and Droplet Holdup

Abstract:

In this paper, semi-batch liquid-liquid Taylor-Couette flow regimes have been studied in a vertically-oriented reactor using both experiments and Eulerian-Eulerian multiphase computational fluid dynamic (CFD) models. This flow features competition between wall-driven azimuthal flow and buoyancy-driven axial flow of the droplet phase. The present study utilizes a water as the continuous phase and n-hexane as the droplet phase, while two key operating parameters (inner cylinder rotation speed and feed rate of hexane) are varied in order to modulate the azimuthal and axial Reynolds numbers. Fluid flow regimes are delineated as a function of these parameters. Other quantities considered here include dispersed phase droplet size and size distribution, droplet holdup, droplet spatial distribution. Both three-dimensional and axisymmetric two-dimensional CFD simulations are compared with experimental measurements, and the limitations of two-dimensional simulations are discussed.

492636 Experiments on Bubble Behaviors in Three Phase Internal-Loop Airlift Reactor Using Optical Fiber Probe Measurements

Abstract:

Experiments on bubble behaviors in three phase internal-loop airlift reactor using optical fiber probe measurements

Yongxiang Gao, Haoran Lu, Youwei Cheng, Lijun Wang and Xi Li.*

Department of Chemical and Biological Engineering, Zhejiang University, Hangzhou 310027, PR China

The internal-loop airlift reactor has been gathered widely interests because of its simple structure, high efficiency of mass and heat transfer, good solid suspension and good mixing. A gas-liquid-solid three phase internal-loop airlift reactor combined with a conical bottom and a nozzle inlet was investigated deeply in term of its fluid dynamics in bubble-scale, besides in column-scale. The bubble behaviour was implemented by optical probe method. The image analysis was also carried out to validate the results. The data collected by the optical probe was processed in time and frequency domain analysis via MATLAB algorithms. The time domain analysis obtained dispersion parameters, such as the local gas holdup, the bubble rise velocity, and the bubble chord length distribution, and the frequency domain analysis revealed chaotic characteristics. The maximum entropy method was utilized to transfer the bubble chord length distribution (CLD) to bubble size distribution (BSD). The effect of the superficial gas velocity and solid concentration on the bubble behaviours in the riser and downcomer regions were discussed carefully. The reasonable agreement between the two measured methods certified and supplied a feasible method to predict the bubble behaviours in gas-liquid-solid three phase fluidized beds. The bubble behaviours in this work can be a benchmark for CFD simulation verifications for airlift reactor, and these in return can provide important guidance for reactor design and scale-up.

* Corresponding author. E-mail address: ywcheng@zju.edu.cn (Y. Cheng)

491174 Flow Properties of Gas Hydrate Slurry in Oil-Dominated Transportation Systems

Abstract:

As oil & gas industry develops into deep water, gas hydrate has been one of the major hazards to the deep sea flow assurance. Hydrate risk management strategy is a novel method for hydrate control, in which hydrate is allowed to form and the petroleum product is transported in the form of hydrate slurry. In this work, hydrate slurry flow properties in w/o emulsion systems were studied using a high pressure flow loop. The w/o emulsion was formed by deionized water and diesel oil in three different water cut (10%, 20% and 30%). Civil natural gas was used as the hydrate former and 1% AA was also used to avoid the blockage after hydrate formation. For each water cut condition, hydrate slurry flow experiments were carried out with four different velocities (0.39m/s, 0.58m/s, 0.97m/s and 1.29m/s). It was found that the flow friction factor increased obviously after hydrate formation, and would keep increasing with the increasing volume fraction of hydrates formed. The increase degree of the friction factor decreased with the increasing flow velocity and increased with the increasing water cut. But when the velocity was larger than 0.97m/s, the increase degree of the friction factor would no longer change with the velocity. Then, a model was proposed to estimate the hydrate-caused friction factor increase, which could well predict the flow fraction factor of hydrate slurry flow. In addition, based on the force analysis of the suspended hydrate particles, the stable suspension region of hydrate particles in different flow velocities were calculated. It was found that there was a critical suspension height in the hydrate slurry flow pipeline. If the hydrate particles suspended above the critical suspension height, the hydrate-caused friction factor would be small and would not change with the flow velocity; while if the hydrate particles suspended under the critical suspension height, the hydrate-caused friction factor would be large and would decrease obviously with the increasing flow velocity.

Paper Number: 494188

RELATIONSHIP BETWEEN FLOW PATTERN IN CYCLONE CLASSIFIER AND CLASSIFICATION PROCESS

Zhanpeng Sun[1], Guogang Sun[1 2], Yan Du[1], Jiyang Chao[1]

([1]College of Chemical Engineering, China University of Petroleum, Beijing 102249, China; [2]Beijing Key Laboratory of Process Fluid Filtration and Separation, Beijing 102249, China)

Abstract: The cyclone classifier was divided into three feeding regions (A, B, C. from center to wall) according to the characteristics of the gas velocity distribution. Effects of the three representative feeding positions on particle trajectory and classifying accuracy were investigated using Reynolds Stress Model (RSM) with the discrete phase model (DPM). The results showed that feeding at the near-wall region C aggravates the fine particle entrainment into the coarse powder, and then impairs the classifying accuracy. The bypass of coarse particle reporting to the fines is high when the material was released from the central region A. The residence time of particles via the position 2 at the region B with high force varies from 0.304s to 0.444s, which is averagely shortened 0.103s and 0.031s than that via the position 1 and 3. This helps to achieving a rapid classification. With increasing the inlet air velocity from 10m/s to 22m/s the classifying accuracy of the cyclone classifiers was averagely increased by 25.3% and 45.7%. Those provide another strategy to improve the performance of the various centrifugal air classifiers.

Key words: centrifugal air classifier; multiphase flow; feeding position; numerical simulation; particle trajectory

Size classification is an important unit in various industries such as mineral processes, chemical, cement, power generation, agriculture and resources recycling[1-5]. To date, both the dynamic turbo- and static centrifugal air classifiers are widely used for dry classification[5-9]. Its principle of size classification lies in the use of centrifugal force generated by way of a flow with vortex.

With the rapid development of the computational fluid dynamics (CFD), optimizing the flow field of the centrifugal air classifier attracts more attention. To eliminate the harmful secondary vortex in a static centrifugal air classifier, Sun et al[7] optimized the inserted depth and diameter of fine particle outlet. Some researchers have been devoted to optimize the velocity distribution between the rotor cage blades of the dynamic turbo air classifiers[10-13]. In recent studies, more attentions have been paid on design improvement of the guide vane to achieve a uniform flow in the annual region [14-15].

There is no doubt that the flow field distribution has a significant influence on the performance of the air classifier. Otherwise, the feeding type cannot be ignored because it largely determines the material dispersion. The mechanical spreading plate is widely used for the dynamic turbo air classifiers to improve the material dispersion[16-18]. Wu et al[19]recently designed a new double layer spreading plate and the classification sharpness can be increased up to 52.47%.

Besides the flow field distribution and the material dispersion, the feeding position may be another factor affecting the particle classification efficiency that not mentioned yet. The main gas swirling flow pattern can be described as Fig 1 after analyzing the flow field distribution in various types of centrifugal air classifiers. From the wall to the centre, the tangential velocity magnitude increases from 0 to the maximum and then decreases to 0 again. When the materials are fed from positions A~C in Fig 1 respectively, the particle flow pattern may differs greatly and thus will the particle classification efficiency.

The objective of this research is to make clear the relationship between the effects of the feeding position (or particle initial position) and the flow field distribution. To better understand the particle classification behavior in the classifiers, the turbulence Reynolds Stress Model (RSM) with an unsteady discrete phase model (DPM) was used.

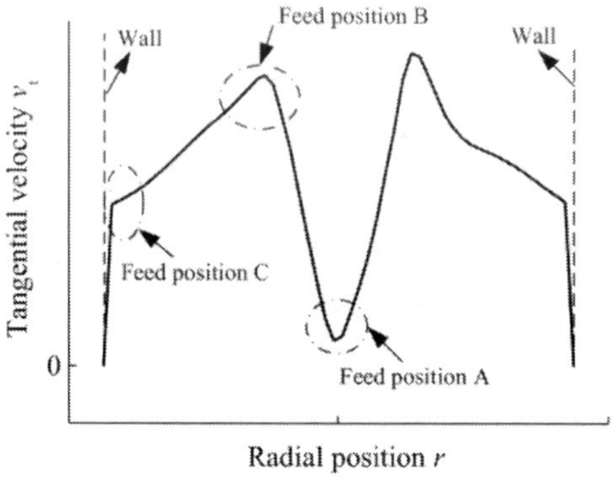

Fig.1 Schematic of the feed position and tangential velocity distribution

1 Model descriptions

1.1 The cyclone classifier model

A schematic of the cyclone classifier with its geometric parameters is shown in Fig. 2a. The origin of the coordinates is set at the top of the classifier, and the positive direction of Z-axis is upward. Two tangential-inlets were designed to keep the gas flow symmetry in the classifier. The separating air stream is generated by an external fan and sucked into the classifier. The powder material fed is classified into coarse and fine fractions under influences of the air currents. Centrifugal force due to the air rotation tends to push the coarse particles towards the wall, and the coarse particles are collected by the discharge hopper finally. Fine particles are carried out from the upper outlet by the air and then transported to the bag filter.

Three feeding positions 1~3 are choose from the regions C, B, A (discussed in section 2.2) respectively as shown in Fig. 2b~2d. The material was fed into the classifier from a thin pipe during the classification experiments. In numerical simulations, the thin pipe was neglected.

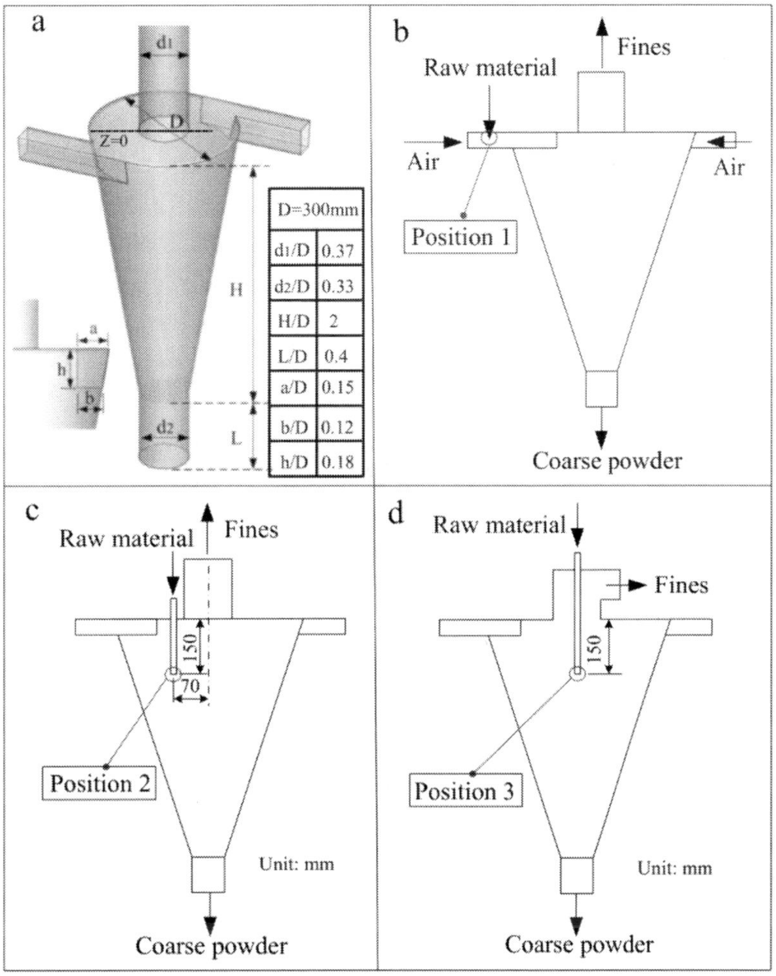

Fig.2 Sketch of the cyclone classifier and different feeding positions

1.2 **CFD Model descriptions and simulation conditions**

Three-dimensional simulation of the cyclone classifier was performed using FLUENT-ANSYS 14.0. A reasonable selection for the fluid dynamic models is crucial. To predict the turbulent air flow inside the classifier, the Reynolds Stress Model has been proved to be suitable in previous study[20]. Thus, the RSM model was adopted to make the equations closed in all calculations. The equations used to solve the continuity, momentum, and Reynolds stress were reported in the literature[21]. The pressure equation was discretized by PRESTO (Pressure Staggering Option) scheme, which is suitable for high-speed rotating flows. The SIMPLEC algorithm was adopted for the pressure-velocity coupling and QUICK difference scheme was used for the convection and diffusion.

After solving the steady fluid flow field, the motion of particles is calculated by the unsteady RSM with a transient time step of 0.0001s. The particle feed time was set 0.2s with the feed rate m'_p of 0.00001kg/s. The simulation of the particle flow in the classifier has been performed by discrete phase model (DPM) based on Newton' second law considering the coupling between the gas phase and particle phase. The particle track method is used to calculate the particle classification efficiency.

A no-slip boundary condition was assumed on the wall boundary, which reflects the physical operation. The near wall treatment was a standard wall function. A "velocity inlet" boundary condition was used at the two inlets. The inlet gas velocity was set as 14 m/s, and its direction is normal to the inlet boundary. The hydraulic diameter and turbulent intensity are 50mm and 4.1% respectively. The boundary condition at the gas outlet was prescribed as a fully developed pipe flow and treated as "outflow" with the length of the gas outlet pipe ten times its diameter.

1.3 Mesh-independent solutions

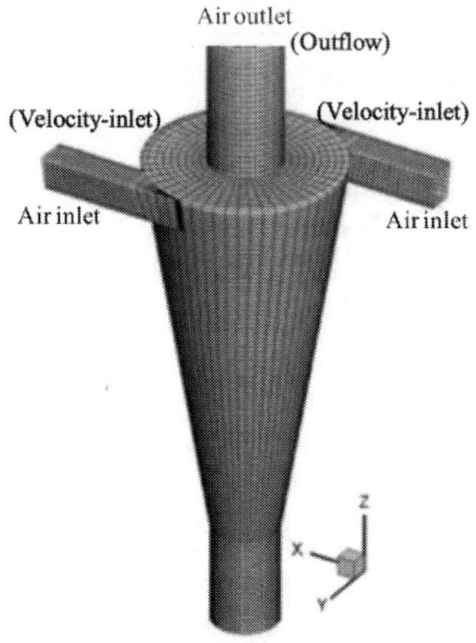

Fig.3 CFD grid and boundary conditions of the cyclone classifier

Grid partitioning method was adopted to mesh the cyclone classifier using the Gambit2.3.16. Fig. 3 shows a representative grid arrangement in the computational domain. The whole classifier is divided by the high-quality structured grids. To ensure that the solution is not dependent on the grid, the grid refinement tests were carried out by changing the grid numbers from 64,318 to 376,834. The predicted pressure drops were very close to each other when the grid number beyond 229140, and in good agreement with the experimental date.

2 Simulation results and analysis

2.1 Flow field pattern

The tangential velocity is responsible for the swirling motion of the fluid and it leads to centrifugal action on particles. As seen from Fig. 4a, the tangential velocity profile resembles a so-called Rankine vortex with its peak value a 1.2 ~ 1.3 times the air velocity at the inlet. The peak value of tangential velocity almost keeps unchanged along the axial direction. However, the position of the peak value approaches gradually to the geometric center of the classifier from top to bottom.

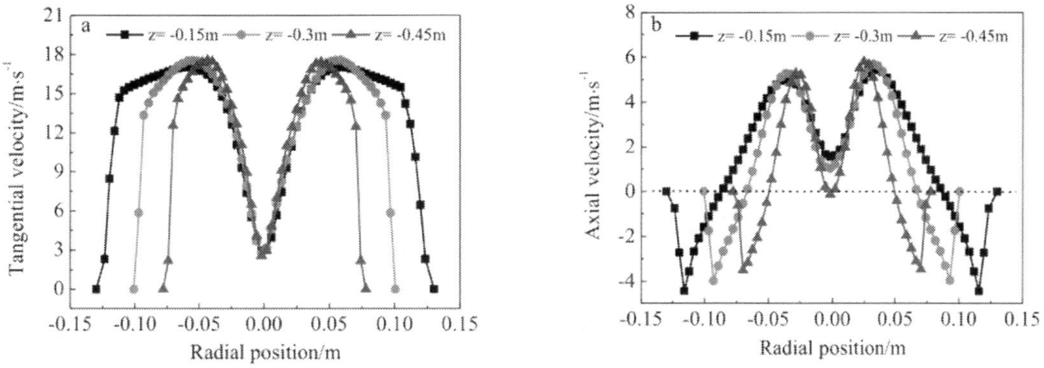

Fig.4 Distributions of the tangential and axial velocity in different sections

The axial velocity profiles in the axial heights shows quantitative and qualitative (M shape) similarities in the cyclone classifier, as shown in Fig. 4b. This figure also confirms the presence of a descending flow in the periphery and an ascending flow in the central part of the cyclone classifier. A part of airflow in the central region moves upward but with a relatively small value. The peak value of the ascending flow of the axial velocity is 0.4 ~ 0.46 times the inlet air velocity.

2.2 Space division for feeding

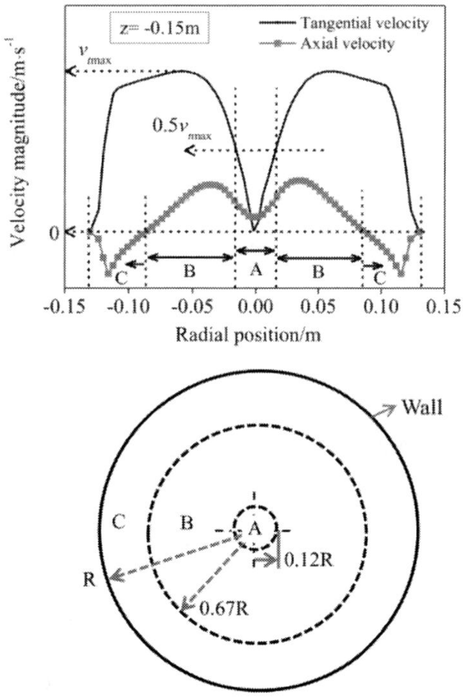

Fig.5 Feeding space division of the cyclone classifier

According to the distribution of the tangential and axial velocities, the space for feeding in the classifier can be divided into three zones from inner to outer, as shown in Fig. 5. The swirling downwind gases dominates the region C (0.67R<r<R). The particles were pushed outward and downward in this region, which may lead to fine particles mixed into the coarse fraction. At the region B (0.12R<r<0.67R), there exists a strong vortex flow. The gas tangential and axial velocity reaches maximum value in this region. The particles may be rapidly classified due to the strong centrifugal and drag forces, which also decrease the probability of particle collision and aggregation. In the central part of the classifier named region A (r<0.12R), both the tangential and axial velocity are relatively small. Particle velocity becomes small at the region A, thus the particle residence time is lengthened and probability of aggregation for small particles is increased.

2.3 Particle flow

Fig. 6 shows a series of particle trajectories from different feeding positions. It shows that the particle flow patterns are sensitive to the particle size. The particles of 60, 45, 35, 25 and 15μm have the obvious moving orbit, while the positions of 8, 3 and 0.5μm particles fluctuate stochastically in the radial direction. Numerous fine particles are carried downward and captured by the bottom surface when material was fed via the position 1. Therefore, there is a higher tendency for small particles to mix into the coarse product. This phenomenon also accounts for the low classification accuracy of the conventional cyclone classifier. In addition, there exists a relatively high concentration rotating particle ring in the top space of the classifier during the particle classification process.

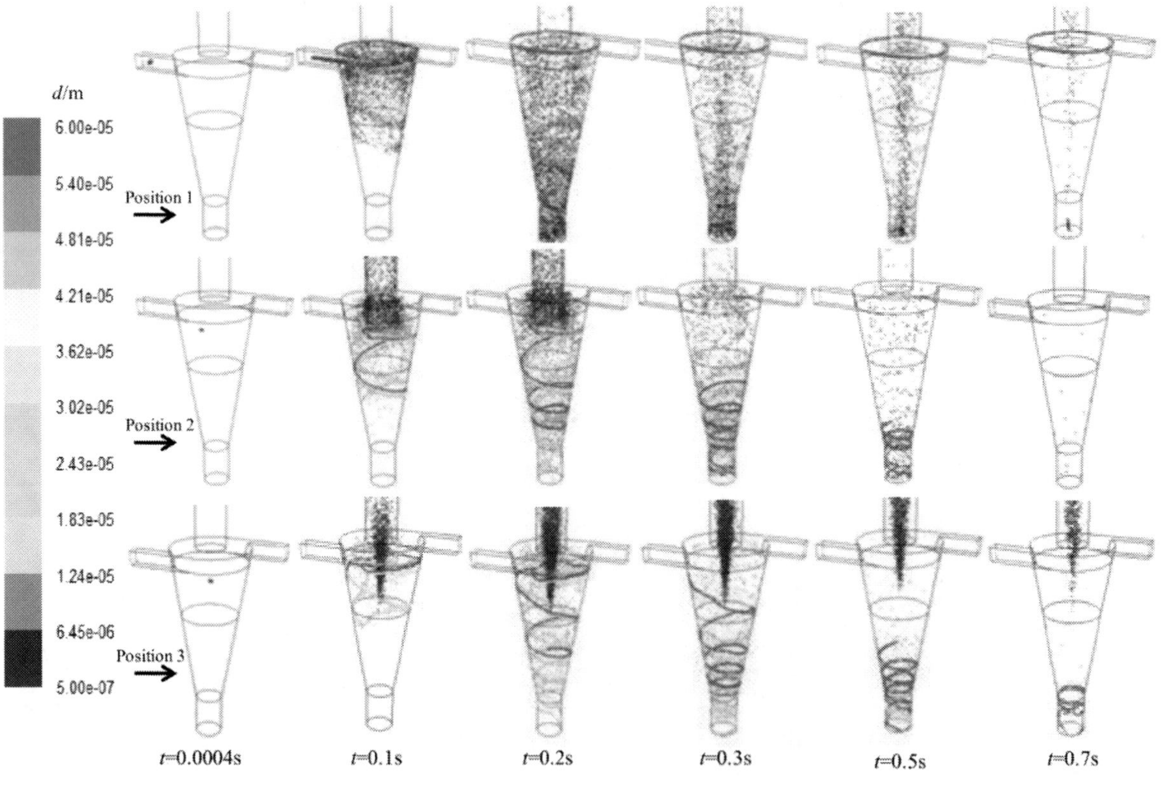

Fig.6 Comparison of the particle flow pattern in the three classifiers

Comparing the particle swarm position at t=0.7s in Fig. 6, the residence time of the particles via the position 2 is shortened greatly. That means that the interaction time between particles and the fluid phase eddy is shortened and decrease the particle collision and aggregation. Thus, the raw material can obtain the rapid classification.

Small particles of 0.5μm and 3μm gather slowly and form a vertical narrow particle band when powder enters the classifier via the position 3. This increases the particle concentration in the classifying zone and causes inadequate powder dispersion, which will reduce the classification efficiency. Large particles spend more time to reach the coarse fraction outlet. In addition, some large particles directly escaped from the air outlet due to the short circuit flow as seen from Fig. 6 at t=0.1s.

2.4 Particle residence time

The shorter residence times can result in a significant carry-over of particles and a concomitant loss in gas cyclones' separation efficiency[22]. However, shortening the residence time decreases the particle aggregation and back-mixing, which has positive effect on improving the particle classification accuracy.

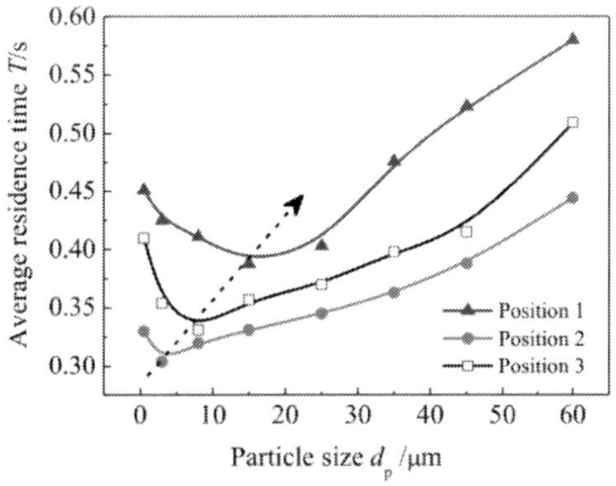

Fig.7 Comparison of particle residence time of the 8 groups of particles

Fig. 7 shows the average particle residence time in the classifier quantitatively. It can be seen that the particle residence time of the 8 groups of particles firstly decreases and then sees an increase trend with increasing the particle size from 0.5μm to 60μm on the whole. This is because the large particles rotate more laps and the movement distance is lengthened due to the particle-wall collision. The movement of the fine particles has a strong stochastic characteristic (shown in Fig. 6) due to the gas velocity fluctuation and turbulent effect, which leads to a smaller settling velocity and an increased residence time. The particles of 15, 3 and 8μm take least time to be classified via the positions 1~3. The residence time of particles via the position 2 varies from 0.304s to 0.444s, which is averagely shortened 0.103s and 0.031s than that via the position 1 and 3. The above results suggest that particles via the position 2 are classified quickly due to the strong classification forces at the feeding point.

3 Powder classification experiments

The dust-laden air is suck into the cyclone classifier by a fan. The coarse particles are collected by the hopper of the classifier, and then the gas flows through a filter, which removes the remaining fine fraction from the gas. The inlet air velocity was measured using a standard L-type pitot tube. The gas flow rate is adjusted by a biplane butterfly valve equipped on the outlet channel. The pressure drop was measured as the static pressure between the air inlet and outlet using a U-type tube. Experiment was performed at the inlet air velocities of 10, 14, 18 and 22 m/s. The feed material has the actual density of 1500 kg/m^3, with a volume median diameter of 40.13μm.

The raw material was fed at a concentration of 0.1 kg/m^3 by controlling the feeding time.

The characteristic particle sizes D10 and D90 were taken from the size distribution curve of the coarse and fine fraction as shown in Fig. 8. The classification size ratio (CSR) was calculated according to the formula CSR = D10/D90[23]. Obviously, bigger CSR value indicates that less fine particle mixes into the coarse product and the classifier works better.

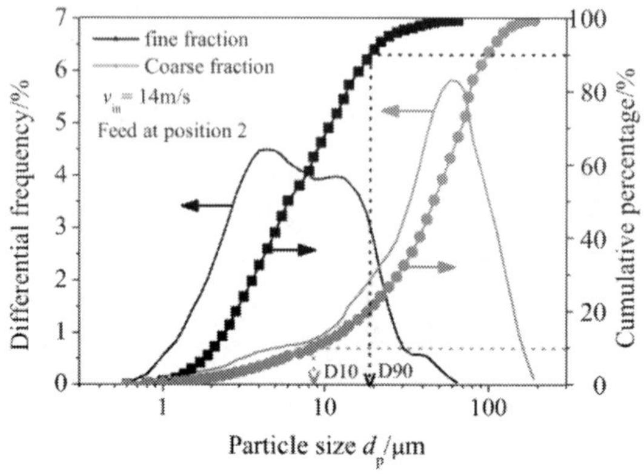

Fig.8 Particle size distribution of the coarse and fine powder

Fig. 9 shows that the experimental CSR value first increases to a maximum and then decreases, as the inlet air velocity increases from 10m/s to 22m/s. The cyclone classifier behaves worst when the particle enters via the position 1. Note that the CSR value via the position 2 increases averagely by 25.3% compared to that via the position 1, suggesting that the performance of the cyclone classifier has been greatly improved. Moreover, the classifier in this study works in the optimal condition when the inlet air velocity reaches 18 m/s.

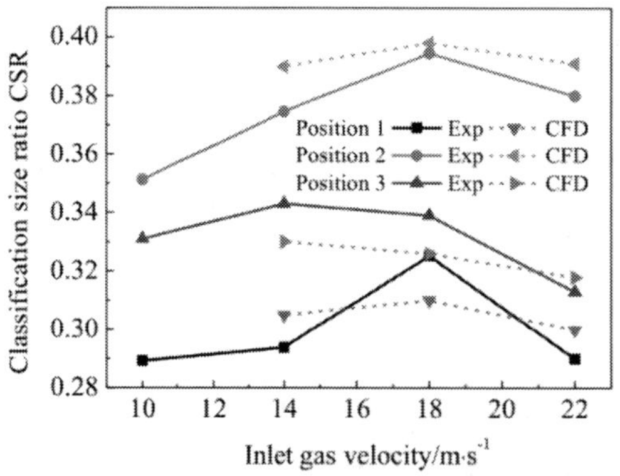

Fig.9 Comparison of the classification size ratio for different feed positions

On the other hand, the predicted CSR value is not agreed well with the experimental one. This can be explained by a number of reasons as the particle interaction that the DPM model does not take into account. However, it can be found that the overall trend of CSR value is represented well by the simulations. Thus, the CFD with the discrete phase model is still useful to study the effect of the feed position on the particle motion in the cyclone classifier, although it cannot precisely predict the CSR value.

4 Conclusions

Effects of the feeding position on particle trajectory and classification efficiency were investigated in the cyclone classifiers numerically and experimentally. The underlying mechanisms have been analyzed in terms of particle flow pattern and residence time. The flowing conclusions could be obtained:

(1) According to the characteristics of the gas velocity distribution, the cyclone classifier can be divided into three regions for feeding. Feeding at the near-wall region results in a large amount of fine particle reporting to the coarse product and impairs the classifying accuracy. But this is very appropriate when the key problem is gas-solid separation.

(2) The centrifugal force and the axial drag force the dominant driving force to make the coarse and fine particles separated in the cyclone classifier. In the radial direction, however, the drag force has less influence on the particle motion. Feeding at the region with high force shortens the particle residence time in the cyclone classifier, which helps to improve the classification efficiency. According to the classification experiments, when the inlet air velocity varies from 10m/s to 22m/s, the classifying accuracy was averagely increased by 25.3%.

Acknowledgments

This work was financially supported by the National Natural Science Foundation of China (No. 21276274) and National Key Project of Basic Research of the Ministry for Science and Technology of China (No. 2014CB744304).

References

[1] Shapiro M, Galperin V. Air classification of solid particles: a review. Chemical Engineering and Processing: Process Intensification, 2005, 44(2): 279-285.

[2] Karunakumari L, Eswaraiah C, Jayanti S, Narayanan SS. Experimental and numerical study of a rotating wheel air classifier. AIChE J. 2005, 51(3):776-790.

[3] Lanzerstorfer C. Application of air classification for improved recycling of sinter plant dust. Resources Conservation and Recycling. 2015, 94: 66-71.

[4] Routray S, Rao R B. Classification studies in an advanced air classifier. Journal of The Institution of Engineers (India): Series D, 2016: 1-6.

[5] Johansson R. Evertsson M. CFD simulation of a centrifugal air classifier used in the aggregate industry. Mineral Engineering. 2014, 63: 109-116.

[6] Li H, He Y, Shi F, et al. Performance of the static air classifier in a Vertical Spindle Mill. Fuel. 2016, 177: 8-14.

[7] Sun Z, Sun G, Yang X, et al. Effects of fine particle outlet on performance and flow field of a centrifugal air classifier. Chemical Engineering Research and Design, 2017, 117: 139-148.

[8] Guizani R, Mokni I, Mhiri H, et al. CFD modeling and analysis of the fish-hook effect on the rotor separator's efficiency. Powder Technology. 2014, 264: 149-157.

[9] Altun O, Benzer H. Selection and mathematical modelling of high efficiency air classifiers. Powder Technology, 2014, 264: 1-8.

[10] Silva S R, Walsh D C, Johansen S T, et al. Air classification: computer models and experimental results. KONA POWDER PART J. 1991, 9(0): 131-138.

[11] Galk J, Peukert W, Krahnen J. Industrial classification in a new impeller wheel classifier. Powder Technology. 1999, 105(1): 186-189.
[12] Wang Q, Melaaen M C, De Silva SR. Investigation and simulation of a cross-flow air classifier. Powder Technology. 2001, 120: 273-280.
[13] Ren W, Liu J, Yu Y. Design of a rotor cage with non-radial arc blades for turbo air classifiers. Powder Technology. 2016, 292: 46-53.
[14] Huang Q, Liu J, Yu Y. Turbo air classifier guide vane improvement and inner flow field numerical simulation. Powder Technology. 2012, 226: 10-15.
[15] Liu R, Liu J, Yu Y. Effects of axial inclined guide vanes on a turbo air classifier. Powder Technology. 2015, 280: 1-9.
[16] Li G, Yu Y, Liu J X. Simulation experimental study of material distribution of turbo air classifier. Chemical Engineering (China), 2011, 39(5):55-58.
[17] Eswaraiah C, Angadi S I, Mishra B K. Mechanism of particle separation and analysis of fish-hook phenomenon in a circulating air classifier. Powder technology, 2012, 218: 57-63.
[18] Diao X, Li S, Huang P. Flow field numerical analysis and structure optimization for feeding pipe of SCX superfine classifier. Chemical Engineering (China), 2012, 40: 31-35
[19] Wu S, Liu J, Yu Y. Design of a new double layer spreading plate for a turbo air classifier. Powder Technology, 2017, 312: 277-286.
[20] Gao L, Yu Y, Liu J. Study on the cut size of a turbo air classifier. Powder Technology. 2013, 237: 520-528.
[21] Hu L, L Zhou X., Zhang J, Shi X.. Studies on strongly swirling flows in the full space of a volute cyclone separator, AIChE J, 2005, 1: 740-749.
[22] Hoffmann A C, Stein L E. Gas cyclones and swirl tubes: principles, design and operation. Second ed., Springer-Verlag, Berlin, Heidelberg, New York, 2007.
[23] Lai W H, Lu W F, Chen C C. The new expression of the effectiveness of powder classification. Advanced Powder Technology, 2005, 16(6): 611-620.

494826 Computational Study of the Stability of the Miniprotein Trp-Cage in Extreme Environments

Abstract:

The native state of a protein is an ensemble of compactly folded conformations in which the protein is stable and functional. While perturbations, such as changes in temperature, pressure or cosolvent concentration, often destabilize the protein leading to the formation of highly populated denatured states, it is known that these perturbations can have stabilizing effects as well. Fundamental understanding of the relationship between protein stability and deviations from physiological conditions is important due to its evolutionary implications and relevance to pharmaceutical applications, but the molecular mechanism of stabilization/destabilization by these external conditions is yet to be fully unveiled. Heat denaturation of globular proteins is widely known to be manifested by loss of secondary and tertiary structures due to the gain of configurational entropy. Cold-denaturation at ambient pressure has recently been found to be accompanied by compact, partially-folded denatured states in which the secondary structure is mainly preserved. The application of high pressure is known to shift the equilibrium of conformational states promoting denaturation near room temperature. We use extensive replica-exchange molecular dynamics simulations and thermodynamic analysis to investigate the effects of high pressure in the low temperature range and ionic liquid on the folding/unfolding thermodynamics of the Trp-cage miniprotein. We find that Trp-cage is stabilized when the pressure is increased at subzero temperatures and conformational states in the low temperature range show strong resemblance to room temperature states. We also present that mechanisms of cold- and ionic-liquid-induced denaturations exhibit similarities where ionic liquids show signs of stabilizing secondary structures.

499568 Engineering Robust Activity in Extremophilic Enzymes

Abstract:

Enzymes are highly specific and environmentally friendly biological catalysts with a potential multibillion dollar market. Examples of their applications include a wide array of industries, such as food and beverages, textiles, detergents, pharmaceuticals, and production of biofuels. The broad range of operating temperatures of these industries require the catalytic activity of enzymes to be less sensitive to temperature. Accordingly, the physical and chemical basis for the temperature adaptability of enzymes from psychrophiles and thermophiles are extensively studied. These studies indicate that the cold adaptability of psychrophilic enzymes is due to their high flexibility, specifically near their active site. In contrast, the loss of flexibility helps thermophilic enzymes to retain their stability at high temperatures.

We hypothesize that the incorporation of flexibility near the active site of thermophilic enzyme can increase its activity at low temperatures without compromising its overall stability. We test this hypothesis by mutating glutamic acid to glycine near the active site of Geo thermocatenulatus (GTL) â€“ a thermophilic enzyme. We performed two such mutations, Glu316Gly and Glu361Gly and observed an increase in the specific activity at both lower and higher temperatures compared to the wild type (WT) GTL. We use all-atom molecular dynamics (MD) simulations to explore the local and global flexibilities and understand the mechanisms through which the mutations have increased the catalytic activity of GTL. In our presentation, we will discuss these results and comment on their implications for designing enzymes with a broader range of stability and activity.

500155 Effect of Manufacturing Conditions on the Fabrication of Protein/Polysaccharide Biocompatible Nanotubes (BNTs)

Abstract:

Effect of Manufacturing Conditions on the Fabrication of Protein/Polysaccharide Biocompatible Nanotubes (BNTs)

Abstract

The Layer-by-Layer (LbL) technique with the assistance of a template is used to allow the formation of single or multilayer nanotubes (Decher, 2002). The polycarbonate (PC) track-etched nanoporous membranes are one of the most convenient template materials because they are easily dissolved allowing the extraction and dispersion (in aqueous solution) of virtually intact nanotubes (Azzaroni & Lau, 2011; Sadeghi et al., 2013; Saghazadeh et al., 2015). The PC membranes are available in different pore sizes allowing the fabrication of nanotubes with different diameters. The pore size and thickness of the template determines the final dimensions of the nanotubes, including wall thickness (Alem, Blondeau, Glinel, Demoustier-Champagne, & Jonas, 2007; Cho, Lee, & Hong, 2014; Roy, Dupont-Gillain, Demoustier-Champagne, Jonas, & Landoulsi, 2010). The biopolymers used in the fabrication of BNTs should be easily dispersible in the solvent of choice, water being the solvent most conveniently used, and should possess large opposite charges and be able to form secondary bonds (electrostatic, ionic or hydrogen bonds) with the template membrane and subsequent layers.

LbL deposition technique has been successfully used to fabricate edible and biodegradable nanotubes with well-defined and controlled structure and properties. To date, there is insufficient understanding of the biopolymer deposition process, especially for edible polyelectrolyte molecules. To address this gap, this research has studied the interactions and operating conditions that lead to the formation of biocompatible nanotubes (BNTs) of bovine serum albumin (BSA) and sodium alginate (SA) through the template-assisted LbL technique, and their application for encapsulation of curcumin as a model if hydrophobic bioactive compound. Dynamic light scattering (DLS) and isothermal titration calorimetry (ITC) were used to study the ability of a protein and polysaccharide to interact and form nanotubes. These methods also offered insight into the types of interactions occurring between these two biopolymers. ITC measurements indicated that interaction between BSA and SA was predominant at pH 3-4, while no interaction occurred at pH 7. This was also correlated by zeta potential measurements that showed opposite charges for these two biopolymers at acidic pH and similar charges at neutral pH. Other assembly parameters, including ratio and concentration of biopolymers, rates of addition, and stability at different pH values were also studied. The optimum concentrations for BSA/SA nanotubes were BSA at 0.8 mg-mL^{-1} and SA at 0.6 mg-mL^{-1} with a rate of addition of 1 mL.min^{-1}. The assembly of BNTs was studied with polycarbonate templates of 200, 400, 600 and 800 nm pore sizes. The wall thickness that led to the most stable BNTs was 4 bilayers $(BSA/ALG)_4$. The morphology, outer diameter, thickness of the wall and length of the freeze dried BNTs were characterized with SEM and TEM. The SEM images showed that $(BSA/ALG)_1$ and $(BSA/ALG)_2$ yield a weaker BNT. The application of these nanotubes for encapsulation and release of curcumin was also studied. Curcumin concentrations used were 10, 15 and 20μg.mL^{-1}. The types of interactions occurring between the BNTs and curcumin were also characterized with ITC. The development of optimal biocompatible nanotubes, which enhances the encapsulation and bioavailability of poorly water-soluble curcumin can be used as a delivery system for other bioactive compounds.

Keywords: Biocompatible nanotubes; layer-by-layer technique, polyelectrolyte.

References

1. Alem, H., Blondeau, F., Glinel, K., Demoustier-Champagne, S., & Jonas, A. M. (2007). Layer-by-Layer Assembly of Polyelectrolytes in Nanopores. *Macromolecules*, *40*(9), 3366–3372. https://doi.org/10.1021/ma0703251

2. Azzaroni, O., & Lau, K. H. A. (2011). Layer-by-layer assemblies in nanoporous templates: nano-organized design and applications of soft nanotechnology. *Soft Matter*, *7*(19), 8709–8724. https://doi.org/10.1039/C1SM05561E

3. Cho, Y., Lee, C., & Hong, J. (2014). Pore size effect on the formation of polymer nanotubular structures within nanoporous templates. *Colloids and Surfaces A: Physicochemical and Engineering Aspects*, *443*, 195–200. https://doi.org/10.1016/j.colsurfa.2013.11.013

4. Decher, G. (2002). Polyelectrolyte Multilayers, an Overview. In G. Decher & J. B. Schlenoff (Eds.), *Multilayer Thin Films* (pp. 1–46). Wiley-VCH Verlag GmbH & Co. KGaA. Retrieved from http://onlinelibrary.wiley.com/doi/10.1002/3527600574.ch1/summary

5. Roy, C. J., Dupont-Gillain, C., Demoustier-Champagne, S., Jonas, A. M., & Landoulsi, J. (2010). Growth Mechanism of Confined Polyelectrolyte Multilayers in Nanoporous Templates. *Langmuir*, *26*(5), 3350–3355. https://doi.org/10.1021/la903121e

6. Sadeghi, R., Kalbasi, A., Emam-jomeh, Z., Razavi, S. H., Kokini, J., & Moosavi-Movahedi, A. A. (2013). Biocompatible nanotubes as potential carrier for curcumin as a model bioactive compound. *Journal of Nanoparticle Research*, *15*(11), 1–11. https://doi.org/10.1007/s11051-013-1931-8

7. Saghazadeh, S., Zhang, S., Lefèvre, D., Le Beulze, A., Jonas, A. M., & Demoustier-Champagne, S. (2015). Universal Method to Transfer Membrane-Templated Nano-Objects to Aqueous Solutions. *Langmuir*, *31*(26), 7264–7273. https://doi.org/10.1021/acs.langmuir.5b01648

496300 Towards a Thermodynamic Model for Predicting Coiled-Coil Protein Structures

Abstract:

A coiled-coil protein structure consists of two or more interacting α-helical strands that together form a supercoil structure. Coiled-coil structures entail unique mechanical properties critical to the function and integrity of various motor proteins, cytoskeletal filaments and extra-cellular matrix proteins.[1,2]

The main objective of our work is to develop a statistical mechanical model to predict the propensity of a protein sequence to form an isolated coiled-coil structure and to study its structural response to the mechanical load. We develop a partition function for coiled-coil dimerization to estimate the probability of a given amino acid sequence to form various dimeric coiled-coil structures. We use our model's predictions to reexamine the assumptions of the thermodynamic models employed to obtain various free energy terms from the experimental measurements. The model parameters will be optimized based on comparisons with our experimental data of coiled-coil forming sequences designed to quantify various free energy effects. We will combine the above coiled-coil dimerization model with our previously developed model on the tensile mechanics of α-helical polypeptide[3]. This results in a model that predicts the structural response of a given coiled-coil motif to a mechanical tension.

(1) Rose, A.; Meier, I. *Cell. Mol. Life Sci.* **2004**, *61*.

(2) Burkhard, P.; Stetefeld, J.; Strelkov, S. V. *Trends cell Biol.* **2001**, *11*, 82–88.

(3) Torabi, K.; Schatz, G. C. *Macromolecules* **2013**, *46*, 7947–7956.

489019 Effect of Residual Secondary Structure Propensity on Liquid-Liquid Phase Separation of Tdp-43

Abstract:

TDP-43 is a partially disordered protein involved in RNA metabolism with implications in the disease pathology of amyotrophic lateral sclerosis, and frontotemporal dementia (ALS/FTD). It is a 414-residue protein containing multiple folded and disordered domains, capable of assembling into liquid droplets composed primarily of protein. Most notable is its disordered 141-residue C-terminal prion-like domain, where most ALS/FTD-associated mutations occur.This region also undergoes transient helix formation in solution which may contribute in some way to the self-assembly behavior. Our goal is to elucidate the role of the disordered domain in the formation and stability of these assemblies, and how transient helicity may promote, or inhibit such behavior. Using a newly developed coarse-grained model for multiprotein assembly, we are able to compute thermodynamic phase behavior of TDP-43 and how it is perturbed by changes in residual helicity due to disease or designed mutations. Our results highlight important functional role for helical structures populated in the TDP-43 equilibrium ensemble in mediating interpeptide contacts and stabilizing liquid droplets.

494022 Solvation Contribution to the Conformational Preferences of the Hydrated Peptides

Abstract:

Previous studies have shown that disordered polar peptides such as polyglycine prefer collapsed structures in an aqueous solution, suggesting that water is a poor solvent even for a protein backbone lacking hydrophobic sidechain residues. This raises several intriguing possibilities with regard to the role of water-mediated â€œhydrophobicâ€ interactions in the behavior of solvated proteins. We use all-atom simulations, coupled with advanced sampling methods, of solvated model peptides to delineate various free energy contributions that govern their behavior. Our results highlight important role of favorable direct protein-water interactions in counterbalancing the unfavorable indirect cavity contribution in case of model peptides. Furthermore, the presence of secondary structures can favor the stabilization of peptide conformations that are not necessarily the most compact ones. Most importantly, our results are found to be largely insensitive to the variations in the protein and solvent force fields.

Influence of Backbone Hydrogen Bonding on the Stability of the WW Domain of the Protein Pin1

Daniel Markthaler, Niels Hansen

Institute of Thermodynamics and Thermal Process Engineering
University of Stuttgart

November 1, 2017

The WW Domain

- 11 backbone hydrogen bonds distributed over 3 β-strands

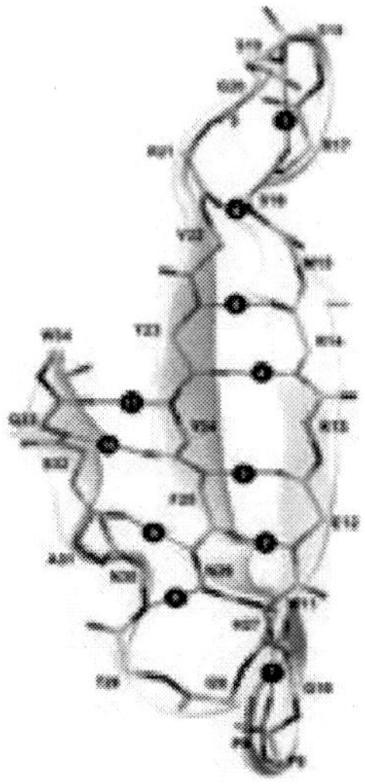

Thermodynamic contribution of the individual backbone H-bonds to the protein stability?

Amide-to-Ester (A-to-E) Mutation

- Enable perturbation of specific backbone H-bonds [Kelly, 2004]

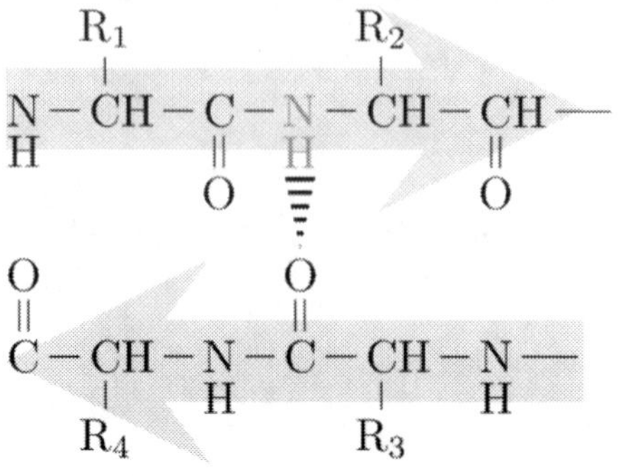

Wild Type protein

3

Amide-to-Ester (A-to-E) Mutation

- Enable perturbation of specific backbone H-bonds [Kelly, 2004]

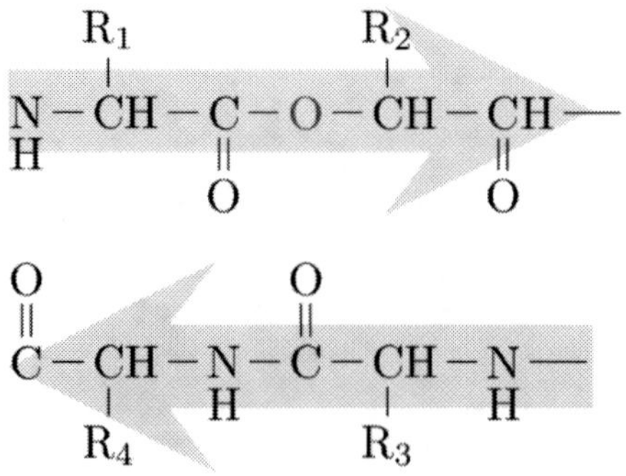

A-to-E mutant

→ 20 A-to-E mutants for the 11 backbone H-bonds

4

Computational Approach

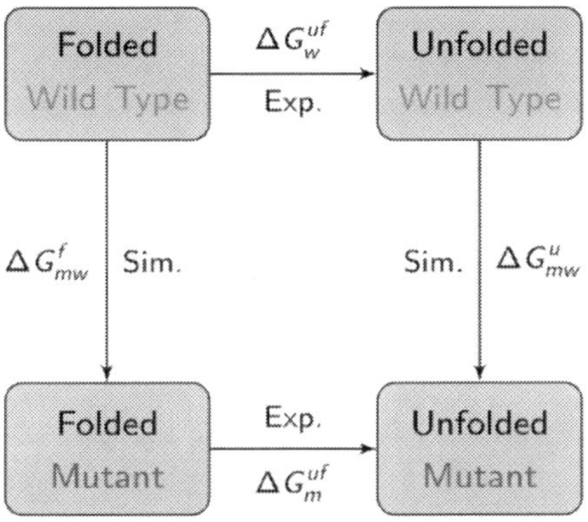

$$\underbrace{\Delta G_w^{uf} - \Delta G_m^{uf}}_{\text{Experiment}} = \Delta\Delta G = \underbrace{\Delta G_{mw}^{f} - \Delta G_{mw}^{u}}_{\text{Simulation}}$$

$$\Delta\Delta G \begin{cases} > 0 : \text{mutant less stable} \\ < 0 : \text{mutant more stable} \end{cases}$$

Free-Energy-MD-Simulations:

- $\Delta G_{mw} = \int_0^1 \langle \frac{\partial H(\lambda)}{\partial \lambda} \rangle_\lambda d\lambda$
- Constant p,T in SPC water
- GROMOS 53A6 / 53A6$_{\mathrm{OXY}}$ FF for wild type protein / ester linkages

Representation of unfolded state:

- Capped tripeptides ACE-X$_{i-1}$-X$_i$-X$_{i+1}$-NME [Hansen, 2015]

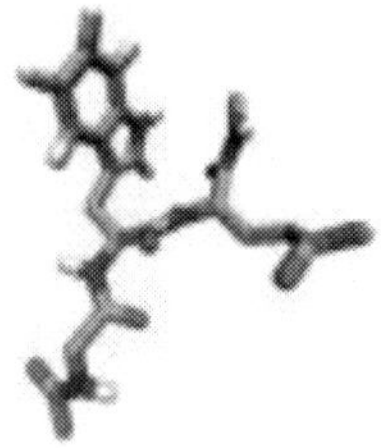

5

Simulation vs. Experiment

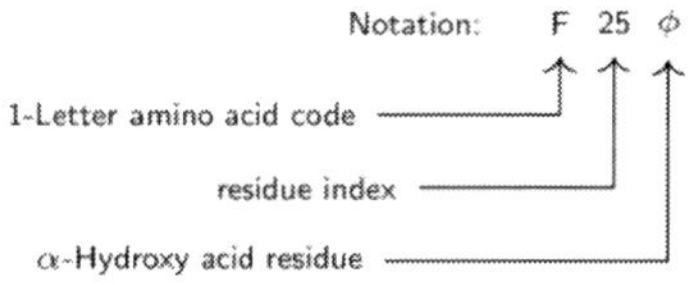

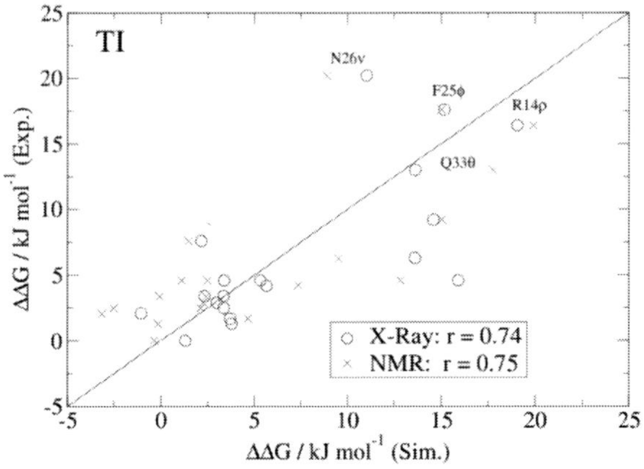

R14ρ, *F25ϕ*, *N26ν*, *Q33θ* only stable upon addition of TMAO [Kelly, 2004]

Deviations between ...

1. ... simulations using different starting structures in folded state
2. ... simulation and experiment
   - Inadequate representation of the unfolded state?
   - Insufficient sampling?
   - Difference in thermodynamic state points?

→ Limiting correlation should be purely dictated by the force field

6

Representation of the Unfolded State

Notation: F 25 ϕ

1-Letter amino acid code → F

residue index → 25

α-Hydroxy acid residue → ϕ

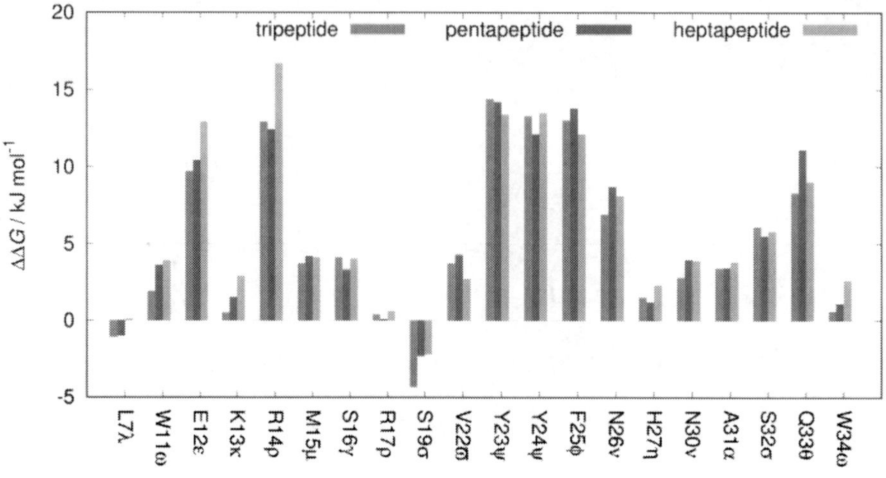

→ Effect of increased chain length not systematic and in most cases below 2 kJ mol^{-1}

Influence of the Starting Structure

Notation: F 25 ϕ

1-Letter amino acid code → F

residue index → 25

α-Hydroxy acid residue → ϕ

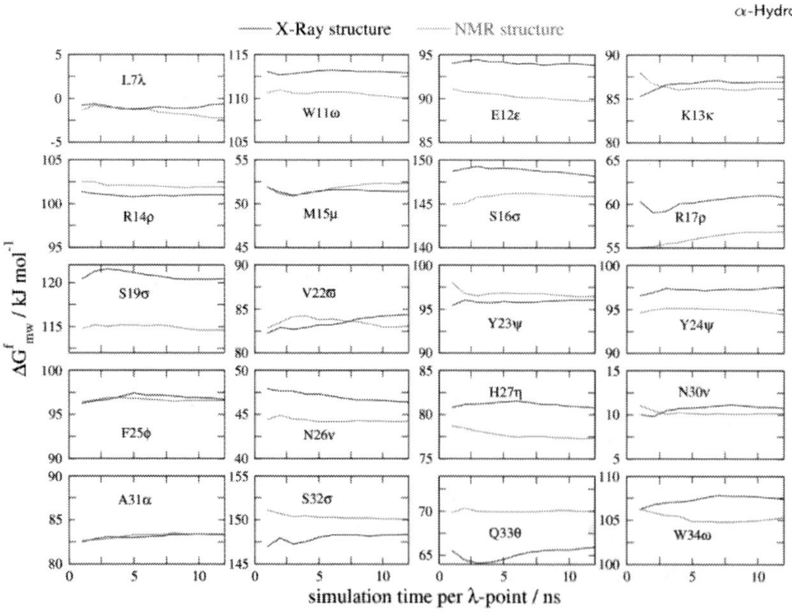

→ Some cases for which the 2 structures do not approach

Influence of the Starting Structure

Notation: F 25 ϕ

1-Letter amino acid code → F

residue index → 25

α-Hydroxy acid residue → ϕ

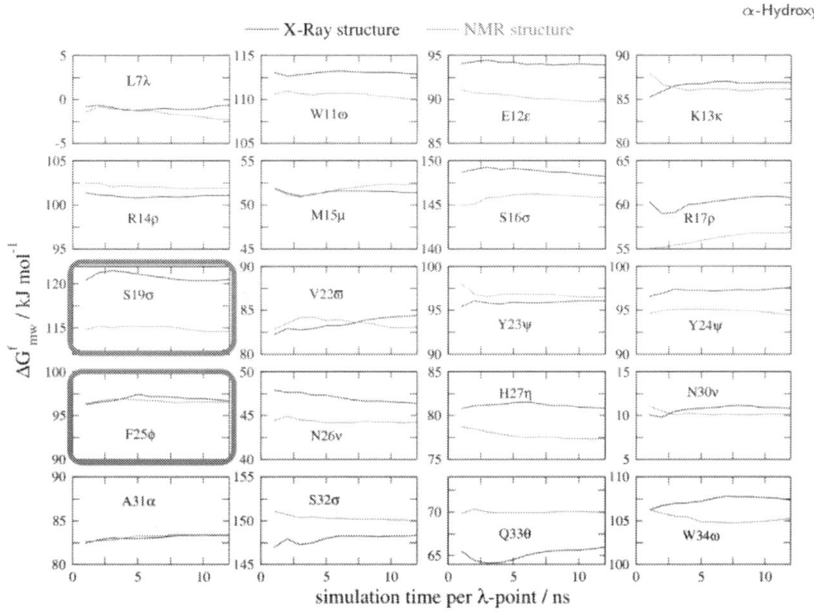

→ Some cases for which the 2 structures do not approach

Sampling of Backbone Dihedrals

Example for similar $\Delta\Delta G$: F25ϕ

φ : C-N-Cα-C

ψ : N-Cα-C-N

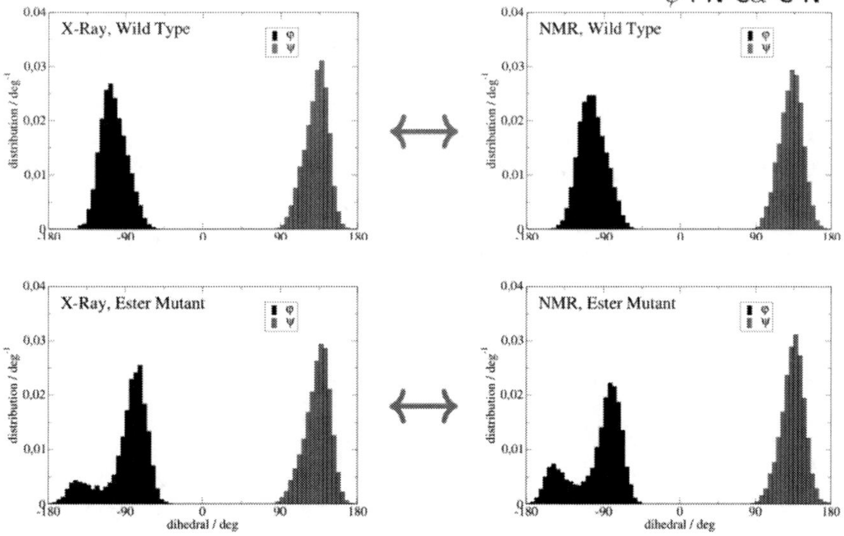

→ Very similar distributions for both structures

Sampling of Backbone Dihedrals

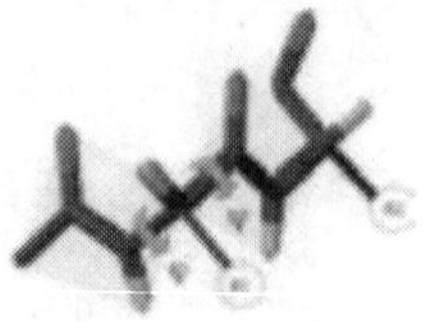

Example for deviating $\Delta\Delta G$: S19σ

φ : C-N-Cα-C
ψ : N-Cα-C-N

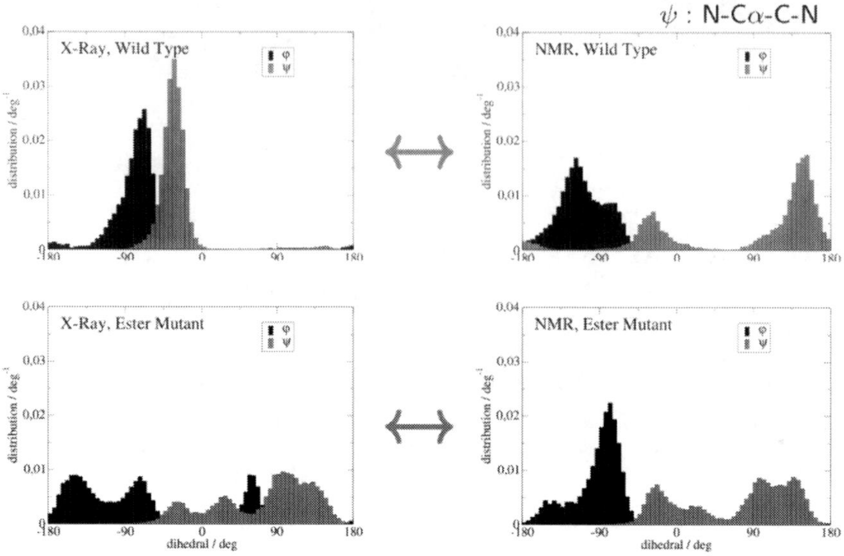

→ Sampled distributions differ considerably between the two structures

Summary and Conclusion

- Most cases differ less than 4 kJ mol^{-1} towards experimental values.
- ... BUT correlation coefficient still clear less than 1.
  → Even better correlation possible with the applied force field? [1]
- Representation of the unfolded state:
  → Increasing peptide length has only minor (unsystematic) effect
- Influence of the starting structure (folded state):
  → Caused by insufficient sampling of backbone dihedrals
  → Replica Exchange between λ-points for improved sampling

[1] correlation coefficient between different measurements based on a set of 2156 single mutations: 0.86 [Potapov, 2009]

Outlook

- Interaction of the WW domain with rest of protein Pin1
- Comparison with Differential Scanning Calorimetry (DSC)

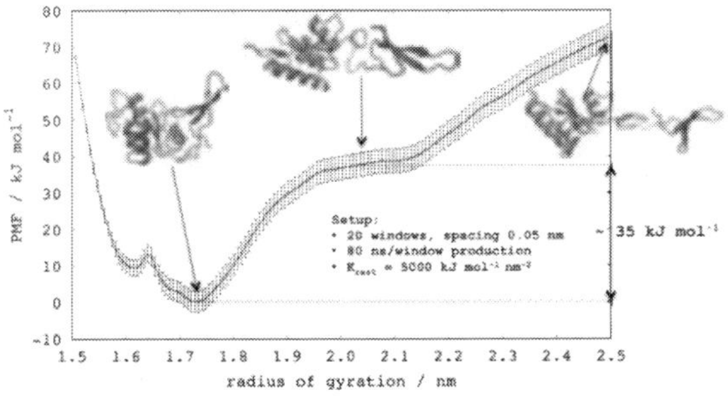

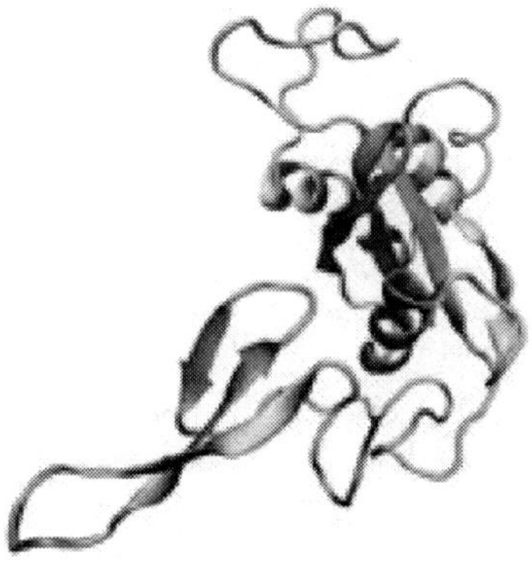

→ Could serve as model system for studying cooperativity effects

Acknowledgements

DFG Deutsche Forschungsgemeinschaft

DAAD
Deutscher Akademischer Austausch Dienst
German Academic Exchange Service

SimTech
Cluster of Excellence

References

S. Deechongkit, P. E. Dawson and J. W. Kelly
Toward Assessing the Position-Dependent Contributions of Backbone Hydrogen Bonding to β-Sheet Folding Thermodynamics Employing Amide-to-Ester Perturbations
JACS 126 (2012), 16762.

A. P. Eichenberger, W. F. van Gunsteren, S. Riniker, L. von Ziegler and N. Hansen
The key to predicting the stability of protein mutants lies in an accurate description and proper configurational sampling of the folded and denatured states
Biochim. Biophys. Acta 1850 (2015), 983 – 995.

V. Potapov, M. Cohen and G. Schreiber
Assessing computational methods for predicting protein stability upon mutation: good on average but not in the details
Protein Eng. Des. Sel. 22 (2009), 553–560.

Folded state simulations (EDS) Sampli

$$V_R = -\frac{1}{\beta s}\ln\left[e^{-s\beta\left(V_A - E_A^R\right)} + e^{-s\beta\left(V_B - E_B^R\right)}\right]$$

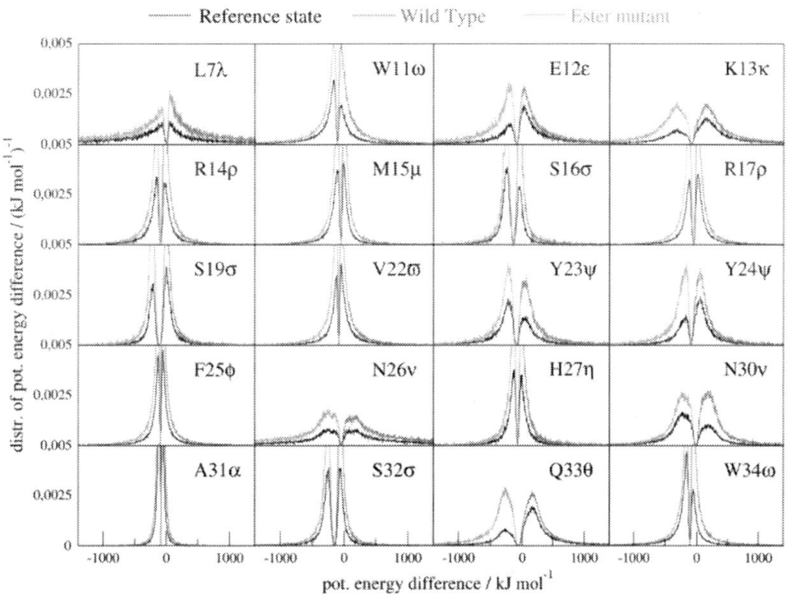

→ Some cases for which sampling does not seem to be sufficient

Comparison of TI with EDS

- Unfolded state and folded state simulations:

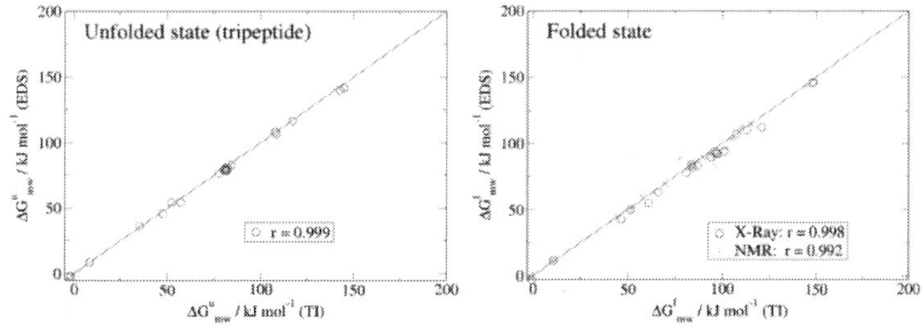

- Correlation with experimental data:

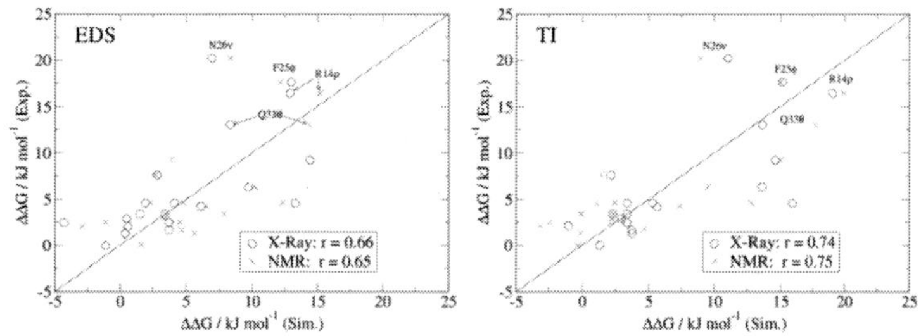

TI results closer to experiments

491824 Molecular Simulations of the Force Spectroscopy of a Biomimetic Polymer

Abstract:

It has become a common practice of probing various aspects of biological polymers via force spectroscopy. Interestingly, a crossover in the unfolding kinetics of the ubiquitin protein has been observed in molecular simulations: The inverse of the unfolding time (i.e., the rate coefficient of transitioning from a collapsed state to an expanded state) is practically constant at low forces and essentially exponential at high forces [1]. This behavior has been noted in other proteins, suggesting in turn that the crossover may be a universal signature of force spectroscopy. In this work, we consequently examine such a possibility via molecular simulations of a biomimetic polymer: Although this homopolymer is solely based on a bead-spring model with a square-well potential, it is capable of universally capturing the protein-like unfolding of any heteropolymer [2]. Foremost, via the Wang-Landau procedure [3], we calculate at zero force the free energy as a function of an order parameter of interest [4]. We continue via perturbation theory, determining the free energy at nonzero force. We then proceed by examining the unfolding kinetics of the biomimetic polymer as a function of force. For this purpose, we invoke "Transition Path Sampling" with replica exchanges [5], and we especially focus on the evluation of the reaction coordinate [6]. Importantly, except that we vary the magnitude of the force, we also apply the force on different sets of monomeric sites. Overall, it appears that the reaction coordinate might have significant ramifications on the crossover observed for the rate coefficient.

[1] R. B. Best, E. Paci, G. Hummer, and O. K. Dudko. Pulling direction as a reaction coordinate for the mechanical unfolding of single molecules. Journal of Physical Chemistry B 112:5968-5976, 2008.
[2] M. P. Taylor, W. Paul, and K. Binder. All-or-none proteinlike folding transition of a flexible homopolymer chain. Physical Review E 79:050801, 2009.
[3] F. Wang and D. P. Landau. Efficient, multiple-range random walk algorithm to calculate the density of states. Physical Review Letters 86:2050-2053, 2001.
[4] C. Leitold and C. Dellago. Folding mechanism of a polymer chain with short-range attractions. The Journal of Chemical Physics 141:134901, 2014.
[5] T. S. van Erp. Reaction rate calculation by parallel path swapping. Physical Review Letters 98:268301, 2007.
[6] B. Peters and B. L. Trout. Obtaining reaction coordinates by likelihood maximization. The Journal of Chemical Physics 125:054108, 2006.

500283 Insights into the Binding of β-Wrapins to Amyloidogenic Proteins Using Simulations and Experiments

Abstract:

Amyloidogenic proteins aggregate into β-sheet-rich amyloid states associated with several diseases. Pancreatic islet amyloids formed by islet amyloid polypeptide (IAPP), senile plaques formed by amyloid-β (Aβ), and Lewy bodies formed by α-synuclein (α-syn) are pathological features of types II diabetes, Alzheimer's disease, and Parkinson's disease, respectively. Increasing evidence suggests possible molecular links between the three amyloid peptides in the aforementioned diseases [1-3], and thus, the discovery of molecules inhibiting amyloid formation of one or combinations of the amyloid proteins may be a promising therapeutic approach. β-wrapins (β-wrap proteins) are a class of proteins that bind, sequester, and thus inhibit amyloid formation by the three amyloidogenic proteins. Subtle changes of one to four mutations at the sequence level of β-wrapins can considerably change their potencies and inhibitory activities for Aβ, α-syn, or IAPP, and specific engineered β-wrapin variants bind one or combinations of the aforementioned protein monomers with submicromolar affinities [4,5,6].

In our studies, we combine computational and experimental methods to gain biophysical insights into the binding and specificity of a series of β-wrapins in complex with the three disease-associated amyloidogenic proteins Aβ, α-syn, and IAPP. In our investigation of β-wrapins in complex with both Aβ and α-syn, we identify the presence of common interactions between the current most promising dual, Aβ and α-syn, binding β-wrapin AS10 and corresponding Aβ / α-syn residues leading to dual-targeted properties, we depict the interactions which potentially act as switches diminishing the potency of β-wrapins for Aβ or α-syn, and we provide insights into the binding of the current most active β-wrapin (ZSYM73 [7]) in complex with Aβ [8], as well as the key role of polar electrostatic interactions contributing to an enhanced binding of β-wrapins to amyloidogenic proteins. We are currently investigating the interaction of β-wrapins with IAPP, and our preliminary results confirm our previous findings, and overall suggest that the use of computational association free energy calculations can be introduced to differentiate between active versus inactive engineered β-wrapins, as well as to identify the most active β-wrapins. We suggest that our studies can constitute the basis for the computational design of novel highly potent single or multi-targeted β-wrapin variants as potential therapeutics.

1. K. Jackson, G. A. Barisone, E. Diaz, L. W. Jin, C. DeCarli, F. Despa. Amylin deposition in the brain: A second amyloid in Alzheimer disease? Ann Neurol. 2013; 74(4) : 517-526.

2. I. Horvath, P. Wittung-Stafshede. Cross-talk between amyloidogenic proteins in type-2 diabetes and Parkinson's disease. Proc. Natl. Acad. Sci. U. S. A. 2016; 113(44) : 12473-12477.

3. L. K. Clinton, M. Blurton-Jones, K. Myczek, J. Q. Trojanowski, F. M. LaFerla. Synergistic Interactions between Abeta, tau, and alpha-synuclein: acceleration of neuropathology and cognitive decline. J Neurosci. 2010; 30(21) : 7281-9.

4. W. Hoyer, C. Grönwall, A. Jonsson, S. Ståhl, T. Härd. Stabilization of a β-hairpin in monomeric Alzheimer's amyloid-β peptide inhibits amyloid formation Proc. Natl. Acad. Sci. U. S. A. 2008,;105(13) : 5099-5104.

5. E. A. Mirecka, H. Shaykhalishahi, A. Gauhar, Ş. Akgül, J. Lecher, D. Willbold, M. Stoldt, W. Hoyer. Sequestration of a β-hairpin for control of α-synuclein aggregation. Angew. Chem. Int. Ed. Engl. 2014; 53(16) : 4227-4230.

6. E. A. Mirecka, L. Gremer, S. Schiefer, F. Oesterhelt, M. Stoldt, D. Willbold, W. Hoyer. Structural Characterization of Fibrils from Recombinant Human Islet Amyloid Polypeptide by Solid-State NMR: The Central FGAILS Segment Is Part of the β-Sheet Core. J. Biotechnol. 2014; 191 : 221–227.

7. H. Lindberg, T. Härd, J. Löfblom, S. Ståhl. A truncated and dimeric format of an Affibody library on bacteria enables FACS-mediated isolation of amyloid-beta aggregation inhibitors with subnanomolar affinity. Biotechnol J. 2015; 10(11) : 1707-18.

8. A. A. Orr, M. M. Wördehoff, W. Hoyer, P. Tamamis. Uncovering the Binding and Specificity of β-Wrapins for Amyloid-β and α-Synuclein. J Phys Chem B. 2016; 120(50) : 12781-12794.

493394 Monte Carlo Simulations Probing Interfacial Adsorption and Uptake in Surfactant Bilayers

Abstract:

Configurational-bias Monte Carlo simulations in variants of the Gibbs ensemble are used to investigate (a) the adsorption of gas molecules at the gas-liquid surface of aqueous solutions, (b) the spatial and orientational distributions near the liquid-liquid interface of water/oil mixtures, and (c) the uptake of additives in surfactant bilayers. The talk will emphasize the computational methodologies employed for the study of these complex interfacial systems.

489066 Phase and Interfacial Behaviour of Aqueous-Two Phase Systems Based on Hyperbranched Polymers

Abstract:

Extraction which is based on the miscibility gap of aqueous two phase systems (ATPS) is a principal possibility to concentrate proteins. These systems are composed by adequate polymers, salts. The fundamental idea of this work was the extension of the material base by the application of hyperbranched polyesteramid in combination with dextran to form an ATPS. The characteristics of this ATPS were compared with a classical ATPS composed of PEG8000 and dextran. In contrast to linear polymers hyperbranched polymers with different functional groups and different architectures which influence the thermodynamic behaviour and the miscibility gap can be synthesized. A further advantage of hyperbranched polymers in fluid separation is the low viscosity compared to linear polymers. In both ATPS the partitioning of L-serin on the two coexisting phases was experimentally investigated. In addition to the experimental analysis, both ATPS were modelled using the Lattice Cluster Theory (LCT) combined with Wertheim Association theory. The LCT was used to take especially the branched architecture of the polymers into account. By the incorporation of the Wertheim theory, the influence of polar interactions is considered. With help of the binary phase behaviour of PEG 8000 + water and of HB + water the binary interaction parameters of the used model were determined. With the introduced model both ATPS were calculated in good agreement with experimental data. The partitioning behavior of L-serine was also experimentally and theoretically investigated in both ATPSâ€™s in order to compare the extraction performance. To obtain interaction parameters for L-serine the solubility of L-serine was investigated in water and aqueous polymer solutions; whereas the experimental data for the solubility in water. Because of a miscibility gap, the solubility of L-serine could not be estimated in hyperbranched polyesteramide containing solutions. For that reason, binodal curve was estimated and the model was fitted to this liquid-liquid equilibrium. With the thermodynamic model, the partitioning coefficient for L-serine was calculated in both ATPSâ €™s in good agreement with experimental data.

Next to phase equilibria calculations also the interfacial properties of polymeric ATPS were calculated. To calculate the interfacial properties of polymeric ATPS the density gradient theory (DGT) was used in combination with the developed thermodynamic model. The DGT allows the calculation of interfacial tension as well as of mass transfer across the interface. At first, the DGT was used to calculate the interfacial tension of investigated ATPS. By the use of one parameter fitted on one experimental data point, the interfacial tension was calculated in good agreement with experimental data. Moreover, the concentration profile across the interface was calculated. It can be stated, that the polymers are not accumulated at the interface. Based on this calculation the DGT was used to calculate the mass transfer across the interface. At first the system ATPS PEG8000+ dextran T40 + water was experimentally and theoretically investigated. The experiments were conducted in a Nitsch-cell for two tie lines. By the DGT the mass transfer across the interface could be modelled by adjusting the mobility coefficient of each component pair in one experiment. This mobility coefficient could be used to predict mass transfer in this ATPS for the other tie line in good agreement of experimental data. Based on the mobility coefficients estimated for Dextran and PEG, the mass transfer in the quaternary system PEG8000 + dextran T40 + L-Serine + water was examined. It can be stated, that the mass transfer in this system can be modelled with the DGT combined with the developed thermodynamic model. But it has to be considered, that L-Serine is accumulates at the interface, when the experiment starts. Next to this system also the mass transfer in the system HB + dextran T40 + L-serine + water was examined. It can be stated, the mass transfer in this system can be modelled with DGT combined with LCT+Wertheim.

492572 From Metallic Hydrogen to Polymeric Sulfur: Peculiar Thermodynamics of Polyamorphic Fluids

Abstract:

"Fluid polyamorphism" is the existence of two alternative amorphous phases in a single-component fluid. Fluid polyamorphism is found in a broad group of physically different materials, such as, silicon, silicon dioxide, sulfur, phosphorous, cerium, and hydrogen, usually at extreme conditions. In particular, this phenomenon is hypothesized in metastable, deeply supercooled water, inaccessible for direct bulk-water experiments because it is presumably located a few degrees below the empirical limit of homogeneous ice formation. We present a generic phenomenological approach, based on the Landau theory of phase transitions and the concept of the order parameter, to describe fluid polyamorphism, applicable regardless of the difference in the microscopic origin of the phenomena. To specify this approach, we consider two alternative scenarios: (a) "discrete" - a fluid with "chemical reaction" equilibrium between two competing interconvertible molecular structures and (b) "continuous" – one possible example is liquid-liquid separation caused by a "two-scale" (soft repulsion ramp) intermolecular potential. We discuss the applicability of these scenarios to dissociation/polymerization reactions (hydrogen, sulfur, phosphorous) and to the degree of tetrahedrality in fluid structures (silicon, silicon dioxide, water).

494274 Computational Studies of Mixtures of Refrigerants, Ionic Liquids and Deep Eutectic Solvents for Absorption Refrigeration Systems

Abstract:

Heating and cooling buildings in the U.S. consumes an enormous amount of energy (>10 quadrillion BTU), and is responsible for adding ~1 billion metric tons of CO_2 in the earthâ€™s atmosphere every year. Much of this energy is used as electricity in vapor-compression systems; however, this technology is mature and only evolutionary improvements are expected in the near future. Several recent studies[1],[2],[3]have shown that several common ionic liquids (ILs) can be combined with standard fluorocarbon refrigerants for use in absorption refrigeration systems that use waste heat at relatively low temperatures (~100 Â°C). Nevertheless, there is limited understanding (and data) on the VLE behavior of these systems, and only one working example of an absorption system using this type of mixture. Furthermore, deep eutectic solvents (DESs), a relatively new class of solvents, share many of the properties of ILs while being considerable cheaper and mostly nontoxic. A fundamental understanding of how the chemical structure of the different species affects the solubility of fluorocarbons in a DES is crucial to design mixtures suitable for use in absorption refrigeration systems that use solar energy or waste heat. In this work, we performed molecular dynamics simulations of mixtures of a conventional fluorocarbon refrigerant, 1,1,1,2-tetrafluoroethane (R134a) with three common deep eutectic solvents (1:2 choline chloride/urea, 1:2 choline chloride/glycerol and 1:2 choline chloride/ethylene glycol). We report and discuss a number of properties for these systems, including Henryâ€™s law constants of R134a in the three DESs as a function of temperature, radial distribution functions, and diffusion coefficients. We will also describe our recent efforts aimed at computationally screening a large number of mixtures of standard fluorocarbon refrigerants with ILs and DESs. Here we used the quantum chemistry-based conductor-like screening model for real solvents (COSMO-RS) and other theoretical tools, to quickly predict values of several key properties (e.g., solubility, viscosity, phase equilibrium and thermochemical data), and perform thermodynamic evaluations of absorption refrigeration cycles using these working fluid mixtures.

---

[1] Shiflett, M. B.; Yokozeki, A. *AIChE J.* **2006**, *52*, 1205-1219; Shiflett, M. B.; Yokozeki, A. *J. Chem. Eng. Data* **2007**, *52*, 2007-2015.

[2] Kim, S.; Kim, Y. J.; Joshi, Y. K.; Kohl, P. A.; Fedorov, A. G. *J. Electron. Packaging* **2012**, *134*, 031009; Kim, S.; Patel, N.; Kohl, P. A. *Ind. Eng. Chem. Res.* **2013**, *52*, 6329-6335.

[3] Zheng, D.; Dong, L.; Huang, W.; Wu, X.; Nie, N. *Renew. Sust. Energ. Rev.* **2014**, *37*, 47-68

494373 Oil Desulfuriztion Using Deep Eutectic Solvents Via Liquid-Liquid Extraction

Abstract:

Catalytic hydrodesulfurization is the common process that is used in the industry to achieve low sulfur concentrations in oil fuels [1]. However, it is usually conducted at relatively high pressure and temperature that results in an expensive and energy intensive process. In this paper, two deep eutectic solvents (DESs) were evaluated for their extraction properties of sulfur derivatives from two aliphatic hydrocarbons via liquid-liquid extraction. The mixtures thiophene/*n*-hexane and thiophene/*n*-octane were selected as oil models. The selected DESs were *(i)* tetrahexylammonium bromide:ethylene glycol with molar ratio = 1:2 (DES 1) and *(ii)* tetrahexylammonium bromide:glycerol with molar ratio = 1:2 (DES 2). To confirm the suitability of the DESs for the system {aliphatic+thiophene+DES}, the binary solubility of thiophene/DES and aliphatic/DES was tested at T/K = 298 and atmospheric pressure. Then, the liquid−liquid equilibrium (LLE) data of the ternary systems {*n*-hexane + thiophene + DES 1}, {*n*-hexane + thiophene + DES 2}, {*n*-octane + thiophene + DES 1} and }*n*-octane + thiophene + DES 2} were determined at T/K = 298 and atmospheric pressure. Also, the solute distribution ratios and the selectivities were calculated from the experimental LLE data and compared to the literature. Moreover, the DES were successfully regenerated by means of vaccum evaporation. Finally, the experimental LLE data were correlated using the nonrandom two-liquid (NRTL) thermodynamic model. It was found that the DESs are potential solvents for desulfurization procesess.

References

[1] S. Brunet, D. Mey, G. Perot, C. Bouchy, and F. Diehl, “On the hydrodesulfurization of FCC gasoline: A review,” *Appl. Catal. A Gen.*, vol. 278, no. 2, pp. 143–172, 2005.

Relevance of Thermophysical Property Data in Rate-Based Gas Treating Simulations

Jacob M. Crosthwaite[1], Diego E. Cristancho[2], Allen Day[2]
Clare Worley[2] and John Dowdle[2]

[1] Analytical Sciences, The Dow Chemical Company, Midland, Michigan 48667, United States
[2] Dow Oil, Gas, & Mining (DOGM), The Dow Chemical Company, Freeport, Texas 77541, United States

2017 AIChE Annual Meeting

Thermophysical Properties and Phase Behavior III: Complex Molecules and Mixtures

11/1/2017

The Dow Chemical Company

1

Acid Gas Removal with Liquid Solvents

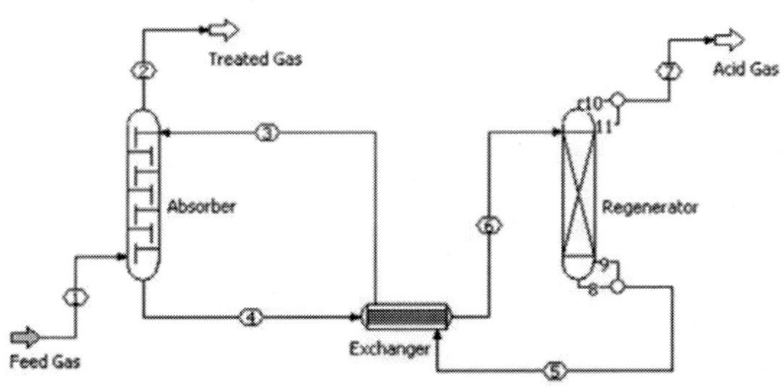

- **Physical Solvents**
  - Glycol ethers
  - Good acid-gas removal (RSH)
  - Higher hydrocarbon removal
- **Chemical Solvents**
  - Aqueous amine solutions (MDEA)
  - Deep acid-gas removal (low loadings)
  - Poor RSH
  - Lower hydrocarbon removal

- Remove acid gases (CO_2, H_2S, RSH, etc.) from feed gas
- Separation occurs in the absorber during gas-liquid contact
- Solvent is regenerated and cycled back through the process
- Key challenges in modeling:
  - Electrolyte thermodynamics
  - Mass transfer rate modeling in columns

2

Why Simulation is Important?

- Process simulations enable design, operation and troubleshooting of gas treating plants.
- Complex thermophysical processes interplay during plant operation in gas treating facilities:
  - Electrolyte chemistry (Thermodynamics and reaction kinetics)
  - Mass transfer in gas-liquid interphases
  - Internal configuration of process equipment (Bed and trays columns)
- A good simulation tool should be able to capture the correct behavior of process variables.
- Allow customized modeling for specific solvents and plant configurations:
  - Unique modeling capabilities
  - Expert knowledge of performance with different solvents
  - Tailor formulation for specific applications

Dow

3

Development of Simulator Capabilities

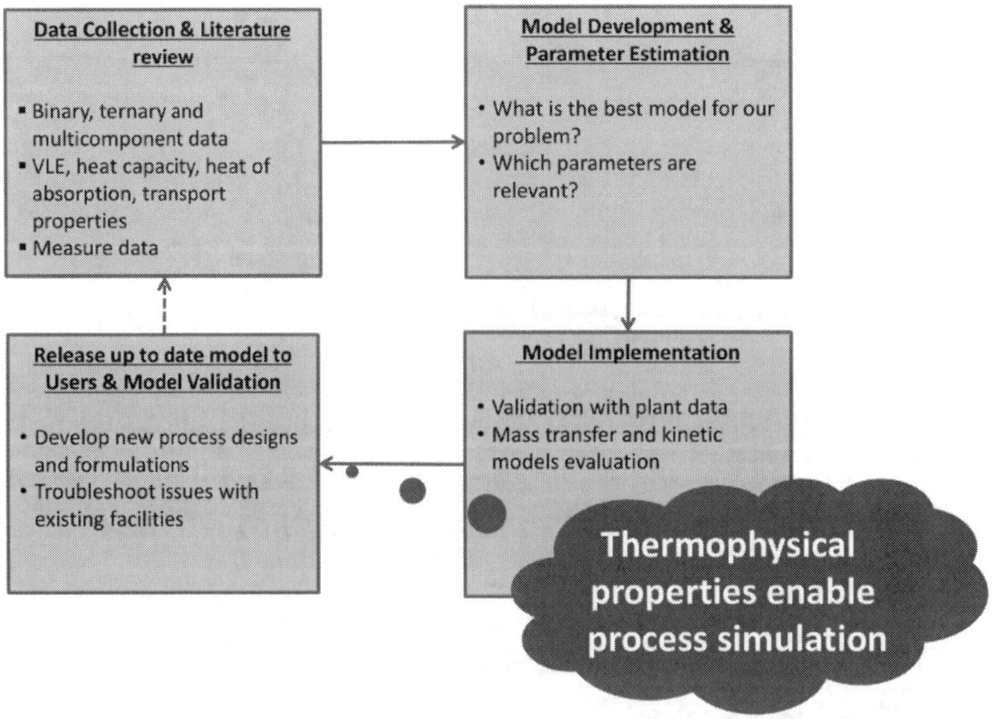

Dow

4

Rate-based Mass Transfer vs. Equilibrium Stage Model

Classical Equilibrium Stage models for column design→

- Rely on concept of equilibrium stage
- Empirical tray efficiencies or HETP required for estimating actual stages

Fails for gas absorption where systems are often very far from equilibrium.

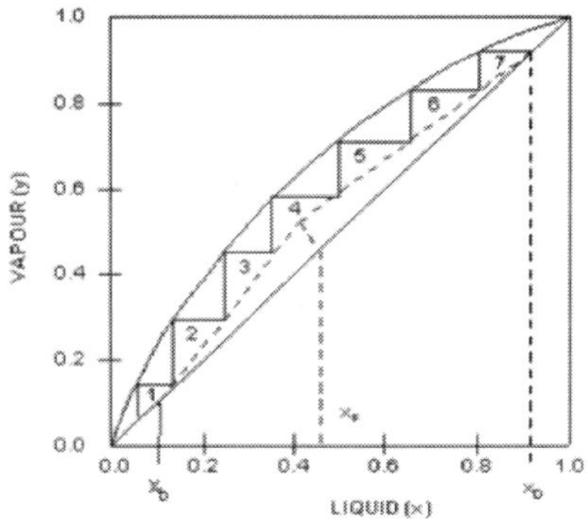

5

Rate-based Mass Transfer vs. Equilibrium Stage Model

Classical Equilibrium Stage models for column design→

- Rely on concept of equilibrium stage
- Empirical tray efficiencies or HETP required for estimating actual stages

Fails for gas absorption where systems are often very far from equilibrium.

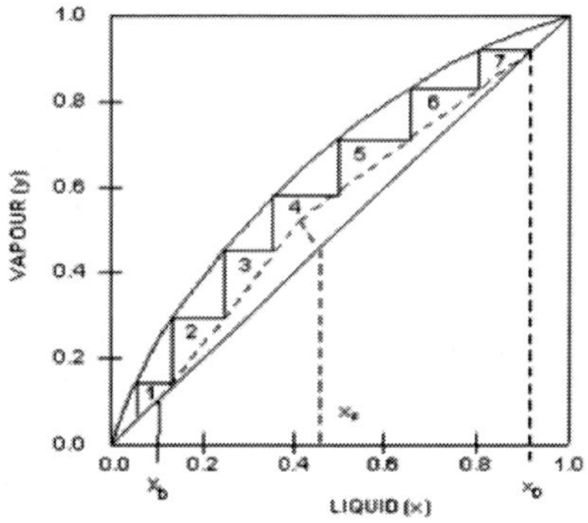

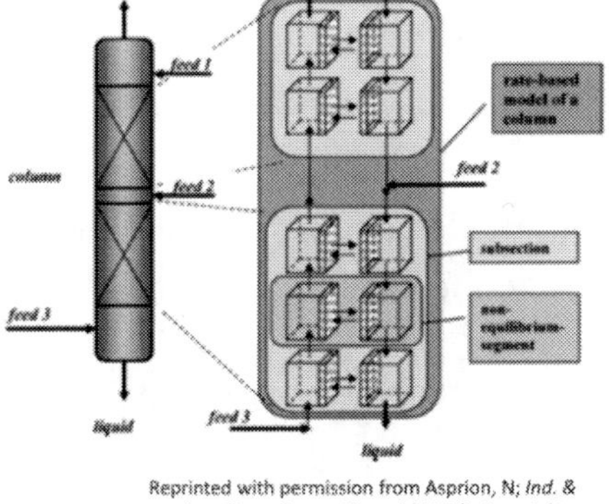

Reprinted with permission from Asprion, N; *Ind. & Eng. Chem. Res.* v. 45 no. 6 (2006). Copyright 2006 American Chemical Society

←Rate-based Mass Transfer Models

- Columns are discretized into a number of segments
- Rigorous mass transfer equations are solved on each segment
- Goal is to eliminate empirical parameters like HETP through physics-based modeling

Now the status quo for chemical absorption simulations!

6

Segment Model: Gas-Liquid Mass Transfer With Chemical Reaction

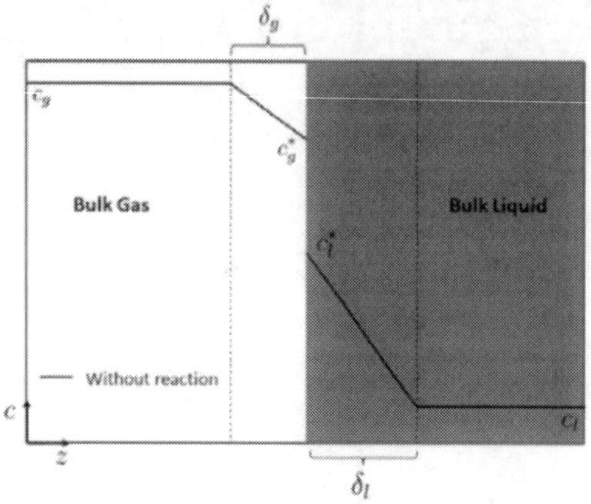

$$N = K_{OV} a_e \left(c_g - c_l / m\right)$$

Overall mass transfer coefficient

Effective gas-liquid contact area

Thermodynamic driving force

Segment Model: Gas-Liquid Mass Transfer With Chemical Reaction

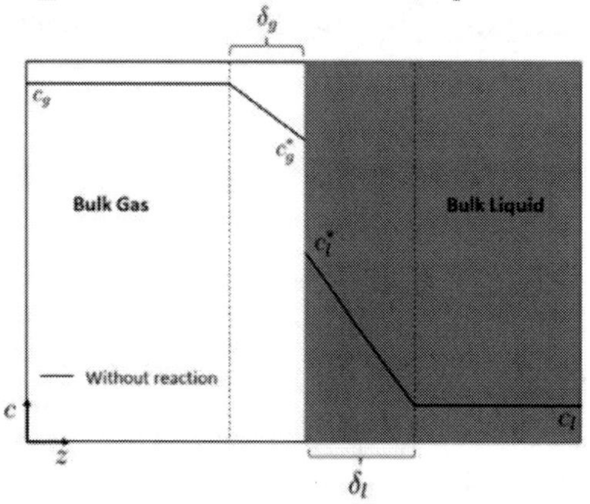

$$N = K_{OV} a_e \left(c_g - c_l / m\right)$$

Overall mass transfer coefficient

Effective gas-liquid contact area

Thermodynamic driving force

$$K_{OV} = \frac{1}{\frac{1}{k_g} + \frac{1}{m k_l E}} \qquad E \equiv \frac{N_{\text{with reaction}}}{N_{\text{without reaction}}}$$

Terms:

- Vapor film coefficient (k_g)
- Liquid film coefficient (k_l)
- Chemical enhancement factor (E)
- Equilibrium partition coefficient ($m = c_l^*/c_g^*$)

For CO_2 removal, mk_lE generally dominates

- *E ~O(1)* for slow reacting solvents
- ***m* estimated by analogy for aq. solvents**
- **k_l is a function of the diffusivity**

$$k_l \sim \sqrt{D}$$

Implications for Design

- Consider a solvent that "slips" CO_2, *e.g.*, tertiary amine-based
    - Then $k_g >> mk_l E$ and E is $O(1)$, so
      $$J_{CO_2} \approx mk_l(c^* - c_b)$$
    - Also note that typically $k_l \sim \sqrt{D_{CO_2}}$
- If initial driving forces are similar the difference in absorption rate will be proportional to $m\sqrt{D_{CO_2}}$

- For a real application, enhanced absorption is predicted by the model of the absorption tower.
- However, at high physical solvent mass fractions, the driving force diminishes and CO_2 capture decreases

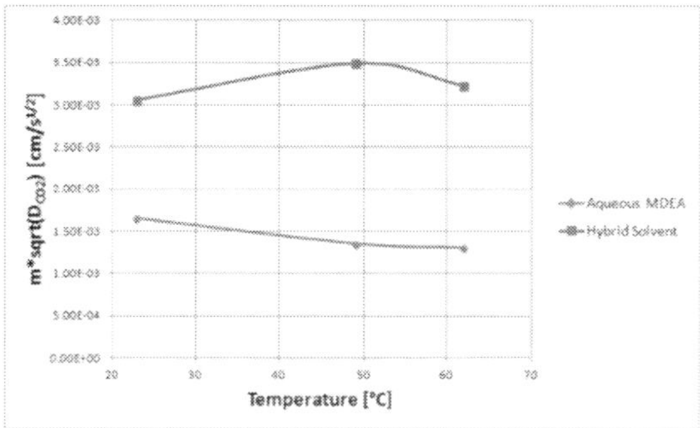

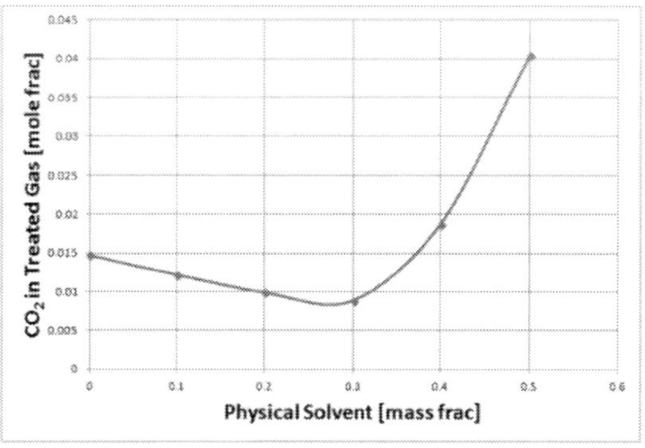

Dow

9

Types of Data to Incorporate in Simulator

- Density (Neat, CO_2-Loaded Solutions)
- Heat Capacity (Neat, CO_2-Loaded Solutions)
- Heat/Kinetics of Absorption/Desorption (C80/BT Calorimeters)
- Vapor-Liquid Equilibrium
    - BTEX
    - Gas Solubility (CO_2, H_2S, Propane, N_2O)
        - » Including low loading regime, speciation[4,5]

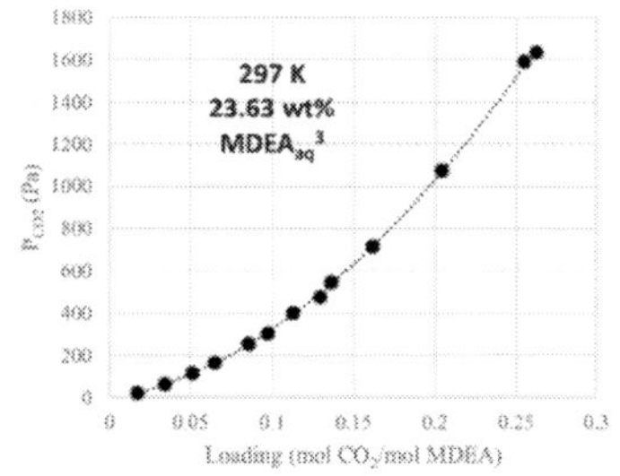

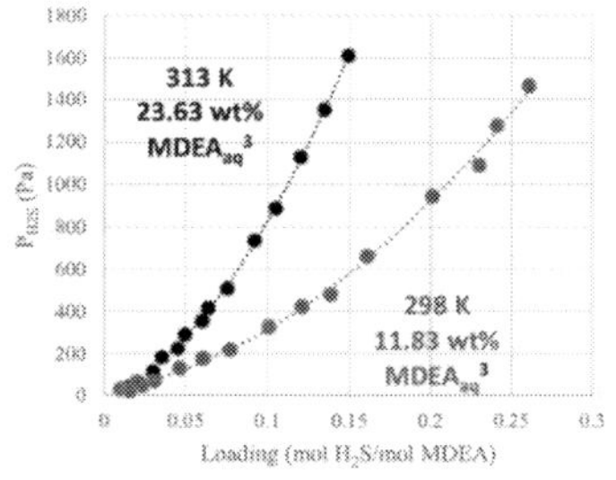

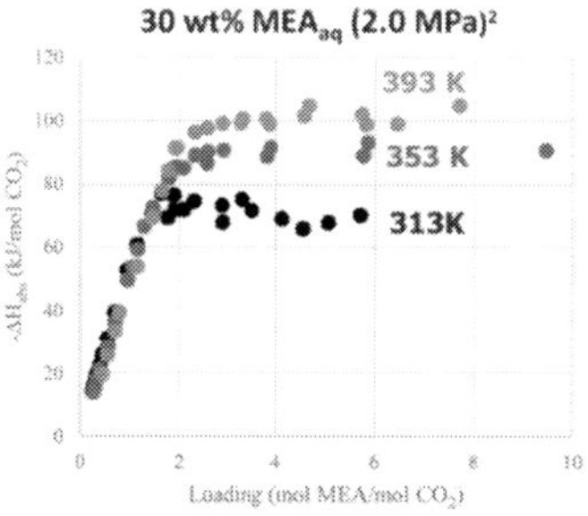

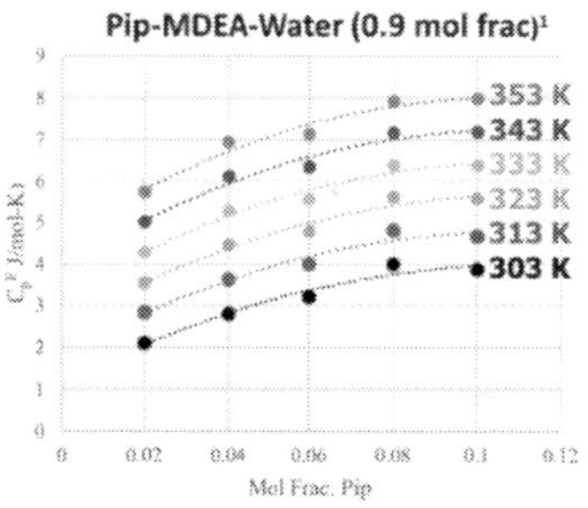

[1] Chen et al., *J. Chem. Thermo.*, **42**, 54-59 (2010)
[2] Mathonat, *Ind. Eng. Chem. Res.*, **37**, 4136-4141 (1998)
[3] Lemoine et al., *FPE*, **172**, 261-277 (2000)
[4] Hamorg et al., *J. Chem. Eng. Data*, **52**, 2491-2502 (2007)
[5] Bottinger et al., *Ind. Eng. Chem. Res.*, **47**, 7917-7926 (2008)

Dow

10

CASE I: SELECTIVE H_2S REMOVAL

11

Existing Data (Low Lean Regime)

50% MDEA in Water

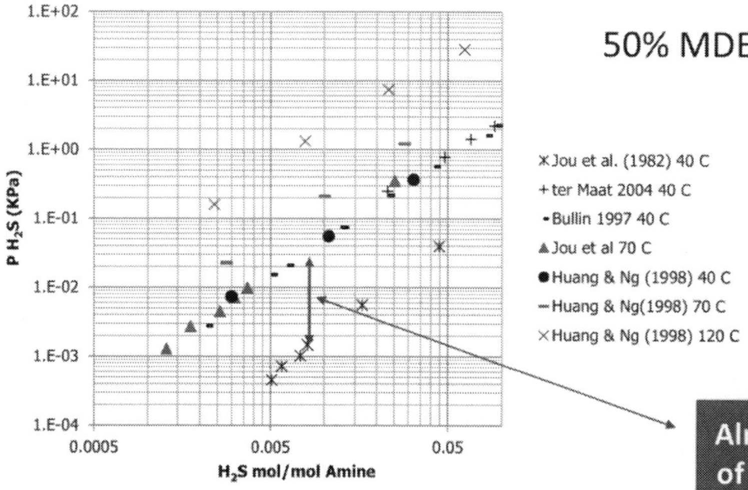

Almost an order of magnitude in discrepancy!

- Selective and deep H_2S removal require high accuracy VLE data in the low lean loading range

- *Jou, F-Y, Carroll, J.J., Mather, A.E., Otto, F.D., Canadian Journal of Chemical Engineering, Vol. 71, April 1993, pp 264-268.*
- *Bullin, J.A., Davison, R.R., Rogers, W.J., Gas Processors Association Research Report RR-165, March 1997.*
- *Jou, F-Y, Mather, A.E., Otto, F.D., Ind. Eng. Chem. Process Des. Dev., 21, pp 539-544, 1982.*
- *Huang, S.H., Ng, H.-J, Research Report RR-155 to the Gas Processors Association and GRI, Sept. 1998.*
- *Ter Maat H, Praveen S, Ijben P, Arslan D, Wouters HF, Huttenhuis JG, Hogendoorn JA, Versteeg GF (2004) GPA-RR-186, RR 186.*

12

VLE Rig for Low Lean Loading Experiments

Unique experimental capability that allows measuring of vapor-liquid equilibrium (VLE) at very low acid gas concentrations in formulated solvents.

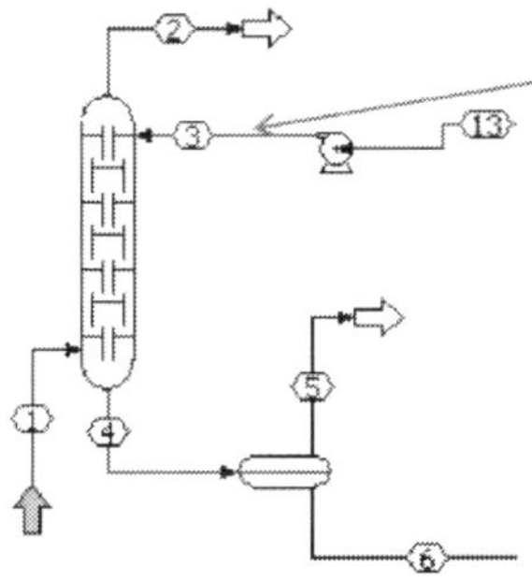

- Applications where acid gas concentration in the treated gas is controlled by absorber lean VLE.
- Performance removal for H_2S applications (Selective H_2S removal) is dictated by the lean amine concentration.
- Stringent regulations for H_2S emission: Less than 10 ppm for many applications.
- Acid gas concentration strongly influenced heat stable salts, particularly in refineries.

- Produce valuable information to define optimal CAPEX and OPEX in process design.
- Data variability impacts process design and risk management in engineering companies and chemical companies.
- Generates data required for the construction of robust thermodynamics models.
- ***Helps engineers manage risk in process design and troubleshooting (minimize over/under design)***

Dow

13

Selective H_2S Removal

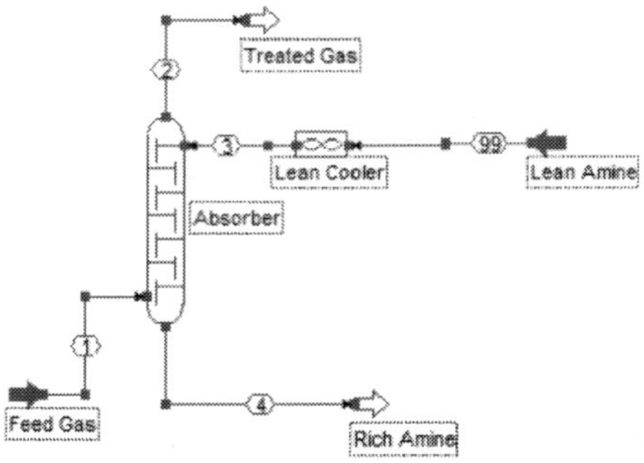

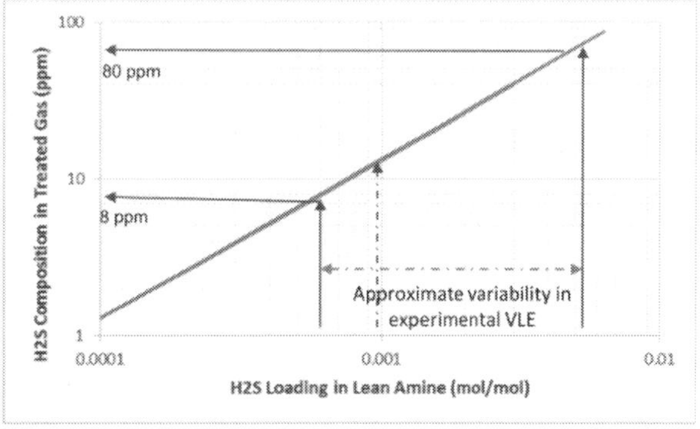

Analysis based upon common Selective H_2S Applications

- Variability in the low lean VLE for H_2S tracks proportionally composition predictions of H_2S in the treated gas

Dow

14

CASE II: HYBRID SOLVENTS

Dow

15

Hybrid Solvents

- Mixtures of Chemical and Physical Solvent
- Advantages
  - Fine tuning of physical/chemical properties of solvent
  - Energy savings (vs. chemical solvents)
  - Lower hydrocarbon losses (vs. physical solvents)
  - Achieve high removal of trace components (mercaptans, COS, BTEX)
- Rigorous understanding of physical/chemical behavior lacking for these solvents

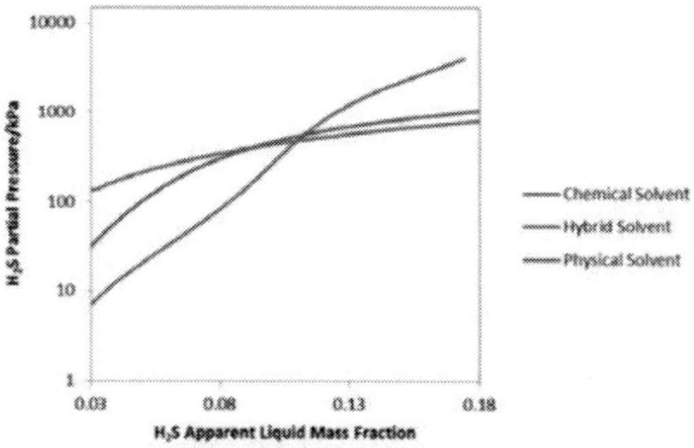

Comparison of H_2S partial pressure vs. apparent liquid mass fraction of H_2S in three representative solvents type

Dow

16

Thermodynamic Framework

Vapor phase described by a Fugacity model

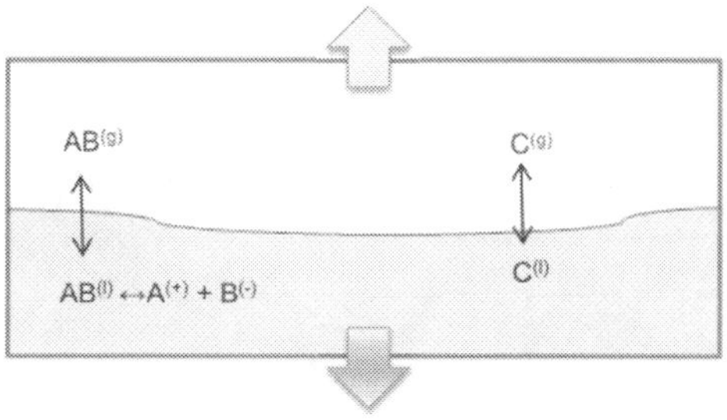

- Chemical equilibrium present in liquid phase → Physical and Chemical solubility for Acid Gases.
- Challenge: Electrolyte Systems → Speciation, Electroneutrality, dielectric constant of solvents.
- Activity coefficient model has to be able to describe short range interactions, long range interactions as well as the effect of mixed solvents.

$$G^{*,\mathrm{ex}} = G^{*,\mathrm{SR}} + G^{*,\mathrm{LR}} + \Delta G^{*,\mathrm{Born}}.$$

Excess Gibbs Energy Model

DOW

17

Mercaptans Approach

- Mercaptans exhibit properties of both acid gases and hydrocarbons.
- Mercaptans are less acidic than CO_2 and H_2S, this reduces their solubility in liquid phase with loading.

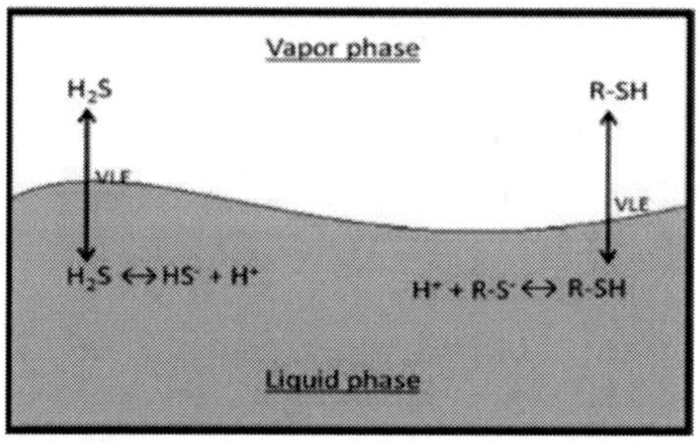

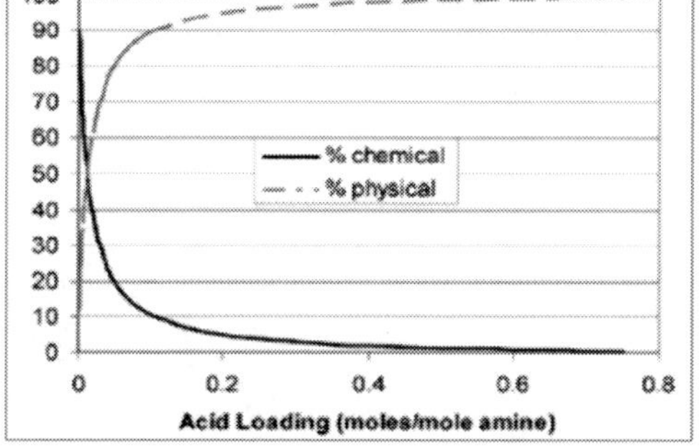

Degree of physical and chemical solubility of methyl mercaptan in 50% MDEA at 40 ° C with various acid loadings.
Bedell, S.; Miller; Eng. Chem. Res, 2007, 46, 3729-3733

Physical Solubility vs. Chemical Solubility

- At low loadings, the chemical solubility dominates the total solubility. As the acid loadings in the solution increase, chemical solubility is substantially reduced and physical solubility dominates.
- Hybrid solvents allow for the same lean amine reactions plus sustained greater physical solubility at the higher loadings.

DOW

18

Model Implementation: Plant Data Comparison

UCARSOL HYBRID™ 701

ABSORBER

Feed Gas

Component	Actual Plant (molar)
CH_4	82.20%
C_2H_6	5.20%
C_3H_8	2.05%
CO_2	5.60%
H_2S	3.90%
CH_4S	121 ppm

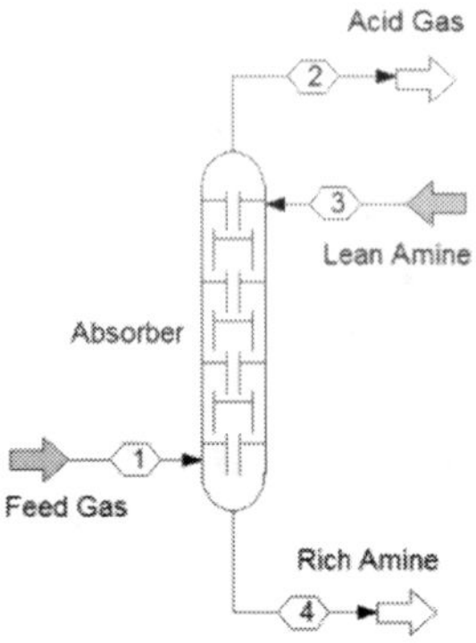

Treated Gas

Component	Actual Plant (molar)	Simulated (molar)
CH_4	87.40%	88.10%
C_2H_6	5.55%	5.55 %
C_3H_8	2.10%	2.19 %
CO_2	2.61%	2.60%
H_2S	8.50 ppm	8.37 ppm
CH_4S	13 ppm	19 ppm
Flow rate (kmol/h)	4256	4262

Simulator reproduces Plant Data

19

Summary & Conclusions

- Dow has developed a simulator for acid gas removal processes, accounting for different types of phenomena with mixed solvents
  - Thermodynamics
  - Mass Transfer
  - Kinetics of Adsorption/Desorption

- Hybrid solvents have been developed as an alternative to physical and chemical solvents for acid gas removal
  - Deep Mercaptans, H_2S removal

- Simulator for hybrid solvent applications matches plant data
  - Case studies for different process conditions, feed gas compositions

20

Acknowledgements

Analytical Sciences

- Florin Dan
- Lynn Stiehl
- Michael Bishop
- Suzanne Lehr
- Binghe Gu
- Clifford Todd

Oil, Gas, & Mining

- Diego Ortiz-Vega
- Prameela Susarla
- Ajay Badhwar
- Norm Byrne

Dow

21

Questions??

Dow

22

499907 Identification of the Equilibrium Parameters for the Modeling of the Hydrogen Solubility in Heavy Crude Cuts Using the Matlab-Aspen Connection

Abstract:

Identification of the equilibrium parameters for the modeling of the hydrogen solubility in heavy crude cuts using the Matlab-Aspen connection.

Camilo MONROY-PEÑA[1], Johana ORJUELA[1]

1 Universidad Nacional de Colombia Universidad de Cundinamarca, Colombia

At present there is an increase in the use of heavy crude oil and bitumen, which must be converted into clean fuels. The conversion is carried out in processes such as hydrotreatment and hydrorupture. The design and operation of these equipment requires knowledge of the solubility of hydrogen in heavy crude blends. Various investigations have shown that thermodynamic methods such as Grayson Streed (GS)have an acceptable approximation to the experimental behavior for pure organic compounds, but does not adequately represent the solubility in heavy cuts. The present work implements the identification of the parameters of various thermodynamic models using Aspen Properties ® software as a tool for thermodynamic modeling and Matlab as a tool for optimization and programming of the NSPSO bioinspired algorithm in order to reduce the error between the experimental results and the model. Figure 1 compares the experimental data obtained by Torres et al. 2013 (black), with the GS (blue) and AGS (red) models developed by Torres et al. 2013 and the Peng Robinson (PR) model with the optimized parameters (yellow) in this work. We found that the thermodynamic models of Peng Robinson (PR), PengRobinson modified by Bostion Matias, Benedith Webb Rubin Straling reduced by more than 20% the approximation of the solubility in pure hydrocarbons as well as in the heavy cuts compared to the work developed by Torres et al. 2013.

Torres, Roger, JC.de Hemptinne, and I. Machin. "Improving the modeling of Hydrogen solubility in heavy oil cuts using an augmented Grayson Streed (AGS) Approach." Oil & Gas Science and Technology–Revue d'IFP Energies nouvelles 68.2 (2013): 217233.

500064 Molecular Perspective on Protein-Protein Interactions at the Tight Junctions Interface

Abstract:

The epithelial and endothelial tissues serve as the physiological barriers against chemical and biological insurgents by constituting the tight junctions. Tight junctions are specialized adhesion constructs formed by the assembly of claudin family proteins. Membrane-embedded claudin proteins are the functional determinants of the tight junctions. In the paracellular space between two adjacent cells, claudins form size and charge selective pores that impede free diffusion of these solutes. The vital role of tight junctions in various tissue barriers such as the blood-brain barrier, renal barrier, and gut barrier, etc. are currently active fields of research, yet a complete molecular understanding of the tight junctions is elusive. Here we have employed multiscale molecular dynamics simulations to gain insight into the underpinnings of claudin assembly at the tight junctions. From these simulations, a clear picture of how the membrane milieu influences the tight junctionâ€™s assembly is evident. Both cholesterol and ceramide lipids limit claudin-claudin interaction in the lipid bilayer, due to the reduced fluidity. Further, we were able to demonstrate that claudins preferentially assemble into distinct dimers that are the primary building blocks of the paracellular pores. The molecular-level insights gained from this work are critical in unraveling the physiological roles of tight junctions.

496422 Pore Formation By Aggregates of Antimicrobial Peptides in Dmpc Bilayers

Abstract:

Antimicrobial peptides (AMP) inactivate microorganisms by forming pores in cell membrane. Since their mechanism of action differs from that of antibiotics, they could be very useful for combating drug resistant microbes, for treatment of microbial infections and also for prevention. Elucidation of mechanism of pore formation will help in the design of synthetic AMP for specific microorganism. Pore formation in DMPC liposomes by AMP melittin, its mutants G1I (higher hydrophobicity) and I17K (higher net charge) and Cecropin P1 as monitored by leakage of fluorescent dye calcein, encapsulated within DMPC liposomes indicated that the leakage rates and extent of leakage were in the order Cecropin P1 > melittin > G1I > I17K. A mathematical model for the evaluation of energy barrier for formation of pore consisting of peptide aggregates of different size interspersed with phospholipid head was proposed which considers detailed pore structure as well as intermolecular interactions. Estimated free energy barriers for insertion of melittin into an ideal paraboloid pores of different sizes were compared with those evaluated by potential mean force calculations. A mathematical model was proposed for the rates of formation of pores by AMP by adsorption and surface diffusion. The rate of dissociation of peptides from pores was evaluated by first passage time analysis. The model was employed for the prediction of rates of nucleation of pores, which were compared with experimental data.

498558 Single-Molecule Characterization of Protein Adsorption to Multivalent Glycan Polymers

Abstract:

Protein adsorption to multivalent carbohydrate (or glycan) derived ligands plays a critical, but poorly understood, role in both fundamental cellular (e.g., cell glycocalyx and viral protein binding) and industrial processes (e.g., cellulosic biofuel production). However, we currently have a poor understanding of the molecular mechanisms driving protein adsorption to complex glycan polymers. We also lack suitable experimental tools and predictive adsorption models that can explain the heterogeneous and high affinity multivalent protein-carbohydrate binding interactions. Here, we study the adsorption of native carbohydrate-binding proteins to model polysaccharides using complementary bulk and single-protein based adsorption techniques. Our single-protein adsorption method is an optical tweezers based bond lifetime assay following a modified methodology as first reported by Brady et al. This technique has allowed us to probe protein-glycan binding interactions at a range of rupture forces. Detailed analysis of the single-protein assay dataset revealed underlying trends that help explain the bulk ensemble observations. The range of bond lifetimes and forces reflect the underlying heterogeneity of available binding sites for multivalent carbohydrate-based ligands. These results provide an underlying molecular basis for using multi-site adsorption models to characterize protein adsorption to multivalent polysaccharides.

Brady, S. K., Sreelatha, S., Feng, Y., Chundawat, S. P. S. & Lang, M. J. Cellobiohydrolase 1 from Trichoderma reesei degrades cellulose in single cellobiose steps. Nature Communications. 6, 10149 (2015).

500042 Controlling Local Hydrophobicity in Poly(ethylene glycol) Brushes with Poly(sulfobetaine) to Mediate the Conformation of Fibronectin on Biomaterial Surfaces

Abstract:

Polymer brushes consisting of poly(ethylene glycol) (PEG) have been extensively studied as coatings to minimize protein adsorption and ultimately the foreign body response to biomaterials *in vivo*. While investigating the mechanistic connection between the properties of PEG brushes and non-specific protein adsorption and denaturation, we recently showed that increasing grafting density led to the stabilization of unfolded fibronectin (FN) on the brush surface. Based on this observation, we hypothesized that this effect was due to an increase in local regions with low hydration within the brush as grafting density increased. In this work, we investigated the extent to which the addition of poly(sulfobetaine) (PSB), a zwitterionic polymer that is highly hydrophilic, disrupted these local hydrophobic "hotspots" as means to overcome this effect. This extent was investigated using novel single-molecule methods, which permit unprecedented insight into dynamic, spatial, and population heterogeneity associated with protein-brush interactions. Such methods included combining high-throughput single-molecule fluorescence tracking with intramolecular Förster resonance energy transfer. Our results indicated that, as the fraction of PSB in the brush increased, the rate of FN adsorption decreased. However, interestingly, an optimum fraction of PSB for inhibiting the stabilization of unfolded FN as well as minimizing the denaturation of FN was observed. Additionally, in single-molecule experiments using an environmentally-sensitive fluorophore, we showed that the relative fraction of hydrophobic "hotspots" decreased with increased PSB. These results suggest that the optimum ratio of PSB-to-PEG arises from the balance between the density of hydrophobic "hotspots" and electrostatic interactions at high PSB fractions. The enhancement of electrostatic interactions between FN and the brush at high PSB fractions may also be denaturing. We are currently exploring the molecular basis for the impact of mixing PSB with PEG to prevent denaturation and the stabilization of unfolded FN via molecular dynamics simulations. Ultimately, these findings implicate the potential utility of mixed PEG/PSB brushes as a novel coating for biomaterials with improved biocompatibility over conventional polymer brushes.

491808 Adsorption of Recombinant Human Interleukin-1 Receptor Antagonist to Silicone Oil-Water Interfaces Leads to Gel Formation and Subsequent Surface-Induced Aggregation

Abstract:

Therapeutic proteins formulated in prefilled syringes (PFS) lubricated with silicone oil contact silicone oil-water interfaces for their entire shelf lives. Protein adsorption to silicone oil-water interfaces may induce aggregation and particle formation which potentially may in turn result in reduced product efficacy, or worse, unwanted immune responses in patients. We hypothesize that the formation of aggregated recombinant human interleukin-1 receptor antagonist (rhIL-1ra) in PFS stems primarily from the formation of viscoelastic gel-like protein layers at the silicone oil-water interface and that conditions fostering stronger interfacial gels lead to increased protein aggregation.

1 mg mL^{-1} rhIL-1ra was incubated in siliconized and unsiliconized PFS at room temperature both quiescently and with end-over-end rotation (3.5RPM). Formulation buffer was varied from 60, 150mM, 500mM, to 800mM NaCl in 10mM MES, at pH 6.5. The extent of protein monomer loss at those ionic strengths was monitored using size exclusion chromatography. We employed a static light scattering instrument to measure bulk protein-protein interactions through measurements of the osmotic second virial coefficient B_{22}. A magnetic rod interfacial shear rheometer was used to determine interfacial gelation times and interfacial gel strengths by measuring the elastic and viscous moduli. To probe for interfacial aggregation, front-face fluorescence of Thioflavin T was used as a marker of intermolecular beta-sheet formation in gelled layers of rhIL-1ra adsorbed at silicone oil-water interfaces.

Up to 15% monomer loss was observed in silicone oil-coated PFS that underwent rotation, whereas less than 4% monomer loss was detected in quiescently incubated siliconized PFS and in unsiliconized PFS that were either quiescently incubated or that were rotated. Protein monomer loss was most pronounced in low ionic strength buffer and diminished with increasing ionic strength. These results correlated well with static light scattering B_{22} measurements where net attractive forces were measured at 60mM and at 150mM and net repulsive forces were measured at 500mM and 800mM NaCl. To determine the cause of aggregation inside PFS, rotation was stopped in siliconized PFS after 14 days of incubation and no further monomer loss was observed over a 7 day period indicating that protein aggregation occurred at the silicone oil-water interface. Shear rheometer measurements showed that rhIL-1ra formed strong gels at the silicone oil-water interface under low ionic strength conditions while increasing NaCl concentration led to five times weaker gels. From front face fluorescence measurements, it was possible to determine that those strong gel layers were formed by interfacial intermolecular beta sheets as fluorescence intensity of Thioflavin T increased by up to 600% in rhIL-1ra samples incubated with silicone oil in 60mM buffer.

We showed that agitation-induced exposure to silicone oil-water interfaces caused rhIL-1ra aggregation in siliconized PFS. This aggregation was greatly increased at low ionic strengths where strong gels formed and interfacial intermolecular beta sheet structures predominated for protein adsorbed at the silicone oil-water interface.

501194 Fibrinogen Adsorption Onto Phospholipid Monolayers: Â Aging and Stiffening

Abstract:

All mammals, including humans, utilize lung surfactant (LS) to reduce surface tension at the alveolar interface and facilitate respiration. Serum proteins, including fibrinogen, have been implicated in the inactivation of LS, and in the progression of acute respiratory distress syndrome (ARDS). Leakage of fibrinogen into the lung can occur following traumatic injury and quickly develop into ARDS, the results of which are frequently fatal. Motivated by the mechanical role fibrinogen may play in the progression of ARDS, we study mixed systems of fibrinogen and dipalmitoylphosphatidylcholine (DPPC), the main consituent of LS. As fibrinogen adsorbs to air/water interfaces, the interfacial rheology increases dramatically, well before the surface pressure changes in any measurable way. We find that DPPC is ineffective at displacing preadsorbed fibrinogen monolayers, and the resulting mixed monolayer has a strongly elastic shear repsonse. In contrast, a pre-existing DPPC monolayer is effective at preventing fibrinogen adsorption in its liquid condensed (LC) state, at relatively high surface pressures. For DPPC in the coexistence region between LC and liquid expanded (LE) states, fibrinogen does adsorb to the interface, increasing the interfacial viscoelastic moduli. Furthermore, the monolayer domain structures are qualitatively different when fibrinogen is added to DPPC and provide insight into ARDS processes.

501236 Morphological and Mechanical Studies of Multicomponent Phospholipid Monolayers Containing 27-Hydroxycholesterol

Abstract:

27-Hydroxycholesterol (27OH) is structurally similar to cholesterol with the addition of a hydroxyl group located at the 27th carbon of cholesterol. This oxysterol is part of the biosynthetic pathway for the synthesis of bile salts. 27OH is also implicated in a host of pathologies including Chronic Obstructive Pulmonary Disease (COPD) where airways are found to have elevated levels of 27OH. The biological relevance of this sterol is paired with unique phase behavior in model lipid membranes making 27OH an attractive molecule to investigate. Like cholesterol, 27OH induces liquid-liquid immiscibility in phospholipid monolayers. However, the morphology of these systems is strikingly different from cholesterol/phospholipid systems. These monolayers can exhibit regions of miscibility at low surface pressures as well as high surface pressures. Additionally, the size distribution of liquid domains is significantly sharper than in other sterol/phospholipid systems. Here we will utilize traditional Langmuir film balance techniques to extract thermodynamic parameters and molecular packing densities. We report the elastic and viscous modulus for these monolayers measured using the oscillatory mode of our pendant drop tensiometer. Finally, we report morphological changes as the monolayer undergoes a symmetric compression to collapse.

495071 Understanding the Effect of Engineered Carbon Nanodiamonds on the Reversible Collapse of Lung Surfactant Monolayers

Abstract:

Lung surfactants [LSs] are a complex mixture of lipids and proteins lining the air-water interface of the alveolar wall of our lungs. LSs help in reducing the energy required for breathing by lowering the surface tension of water. This makes it easier for the alveoli to expand with inspiration. LSs also act as a barrier against any foreign particle capable of entering the alveoli of the lungs. At present, Engineered Carbon Nanodiamonds [ECNs] are being considered for drug delivery and the respiratory tract is a lucrative route for this purpose. This in turn raises the question of their compatibility with LS mixtures, which would certainly come in contact with the nanoparticles. Therefore, any possible changes in the properties of the LSs because of its interaction with ECNs must be carefully scrutinized. In this report, we try to understand the effect of ECNs on the reversible collapse of the LSs at high compressions. Reversible collapse is a particularly important property of LSs that allows material to be retained near the surface at high compressions. This can be observed in the form of bright folded structures with fluorescent microscopy. The retained material is then capable of spreading back on the surface and minimizing the loss of surfactants. Here, multiple compositions were tested with fluorescent microscopy to evaluate the effect of electrostatics and lipid tail saturation on the collapse mechanism of LS. Under the microscope, model mixtures of dipalmitoyl phosphatidyl choline (DPPC)-dipalmitoyl phosphatidyl glycerol (DPPG) and DPPC-palmitoyl oleoyl glycerol (POPG), tagged with Texas red dye, revealed the formation of folded collapse features. Upon immediate exposure to ECN, this feature was still retained in both the samples. However, the DPPC-DPPG mixture collapsed irreversibly after four days of exposure to ECNs. This was observed as bright specs on the surface that are believed to be vesicles and are incapable of spreading back to the surface. Interestingly, this was not observed in the case of the DPPC-POPG mixture, which continued to collapse with the formation folded features. Therefore, unsaturation in the lipid mixture may be a crucial component, capable of countering any potential negative impact by the ECNs on the LS mixture at higher compressions.

500290 High Resolution Optical and Electrical Recording of Free Standing Lipid Bilayers

Abstract:

Cellular membranes possess a transmembrane electric potential of â^¼100 mV but this fact, which is generally conserved across all cell and organelle types, is often neglected in model phospholipid bilayer studies. The cause of this potential is due to a variety of factors operating in tandem, namely the action of ion pumps to maintain an osmotic pressure difference, the presence of charged membrane proteins, membrane curvature, and asymmetry of charged lipid composition.Â This talk will highlight recent work evaluating the electromechanical properties of model lipid bilayers which dictate how cell membranes respond to electrical stress. Simultaneous optical and capacitive monitoring of a large-area free standing model membrane provides unique insight into the dynamic response as the bilayer compresses and stretches due to the applied field. Measurements of the thickness-voltage relationships for monocomponent bilayers and lipid mixtures containing different degrees of saturated lipids and anionic lipids will be presented to elucidate the factors that govern membrane elastic properties. These properties play an important role many biological processes, namely the function of membrane proteins, and as an outlook initial experiments incorporating model membrane proteins will be discussed.

498756 Dynamics of Lipid Lamellae and Lamellar Vesicles in Shear and Extensional Flows

Abstract:

Mechanics and dynamics of lipid bilayers and vesicles play a crucial role in cellular processes and cell functions that range from cell division to targeted drug delivery. Liposomes, spherical vesicles made up of lipids, have been widely used in drug delivery and medical diagnostics and, as a model structure for studying cell mechanics. We use coarse grained molecular dynamics (MD) simulations to study the influence of shear and extensional flow deformations on vesicle shape and dynamics as well as experimentally observed shear-induced structure transition from bilayers to multi-lamellar vesicles.

Specifically, we simulate self-assembly of dodecyldimethylammoniumbromide (DDAB) lipids in water using a coarse-grained Martini force field. Large-scale MD simulations are performed using GROMACS and LAMMPS packages on the XSEDE supercomputing platform. MD simulations have revealed various equilibrium morphologies such as micelles, uni-lamellar and multi-lamellar vesicles as well as layered structures with increasing lipid concentration. In this presentation, we will focus on the following points.

Dynamics of vesicles in shear flow: We have developed a numerical method to precisely calculate the reduced volume of a vesicle by expanding the vesicle-water interface topology obtained from MD simulations in terms of spherical harmonics. We then tracked the orientation of non-spherical (ellipsoidal) vesicles by calculating their principal axes from the inertia tensor. Our simulations suggest that the reduced volume of vesicles is a key determinant of their shear flow dynamics. A deflated vesicle shows tumbling dynamics while a nearly spherical vesicle is likely to exhibit tank-treading. We will discuss how the tumbling frequency changes with the applied shear rate.

Lamellar Structures in Shear Flow: We will discuss the mechanical response of lamellar structures in shear flow. As the deformation rate increases, we see a buckling instability leading to a morphology transition from lamellae to vesicles which is accompanied by an increase in the shear viscosity. Our results suggest that the linear dimension of the vesicles thus formed is approximately equal to the wavelength of the primary buckling mode.

Uniaxial Extension of Lamellae: We will discuss the response of lamellar structures under uniaxial extension. We have developed a method to calculate the Poisson's ratio of lamellae. We will discuss the effect of electrostatic screening induced by co-amphiphilic molecules stacked parallel to the lipids on their mechanical properties. Specifically, our simulations show that stiffness and Poisson ratio of the lamellae increases and decreases respectively with increasing concentration of co-surfactants. We will discuss the mechanisms underlying this observation.

Probing the Structure of High Viscosity Complex Fluids at Extreme Shear Rates

Javen Weston[1,2], Steve Hudson[2], Katie Weigandt[3]

[1]Georgetown University, Department of Physics
[2]NIST Materials Science and Engineering Division
[3]NIST Center for Neutron Research

17AIChE
Annual Meeting, Minneapolis, MN

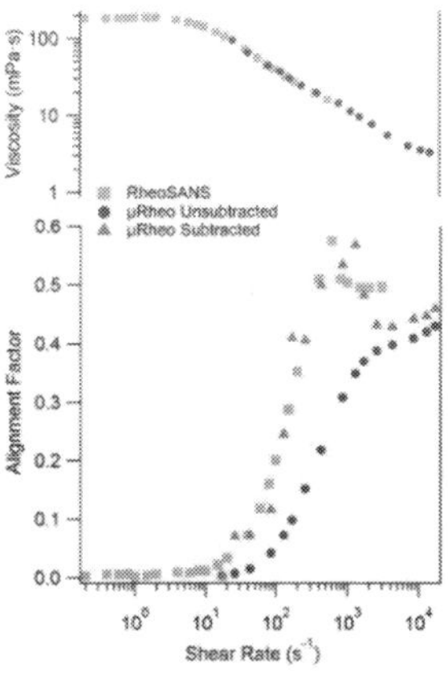

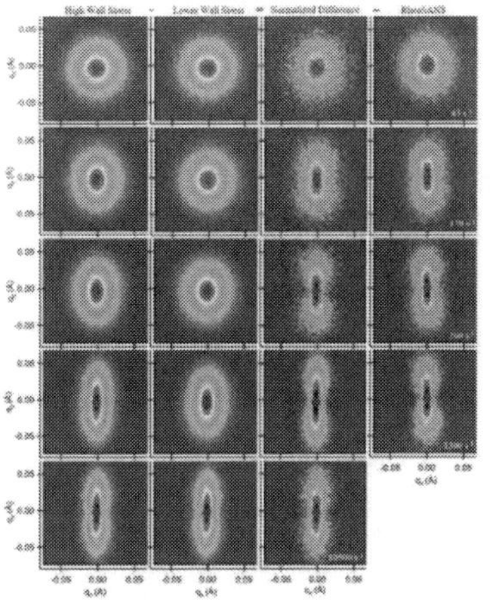

RheoSANS – Introduction

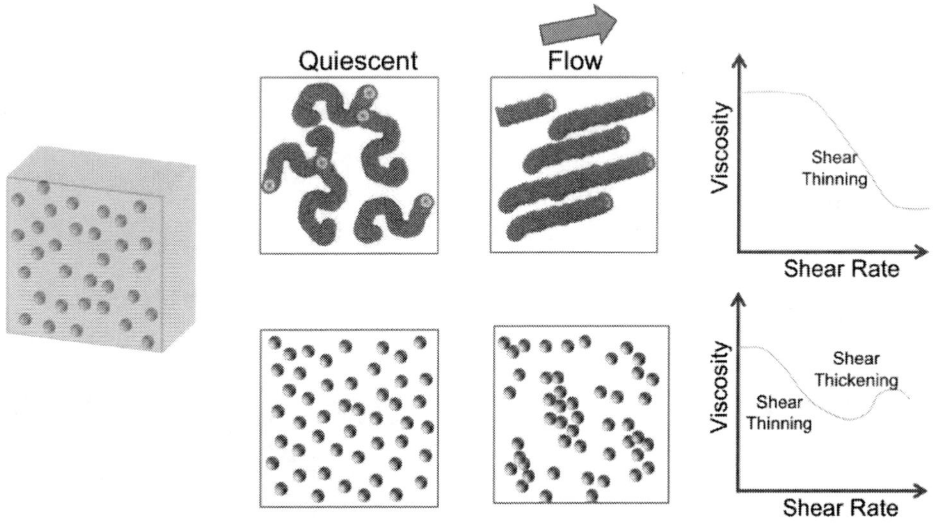

Current Opinion in Colloid & Interface Science 17 (2012) 33–43

Flow-SANS and Rheo-SANS applied to soft matter

Aaron P.R. Eberle [a], Lionel Porcar [b,*]

[a] NIST Center for Neutron Research, National Institute of Standards and Technology, Gaithersburg, MD, 20899, USA
[b] Large Scale Structure Group, Institut Laue Langevin, Grenoble, France

RheoSANS – Status Quo

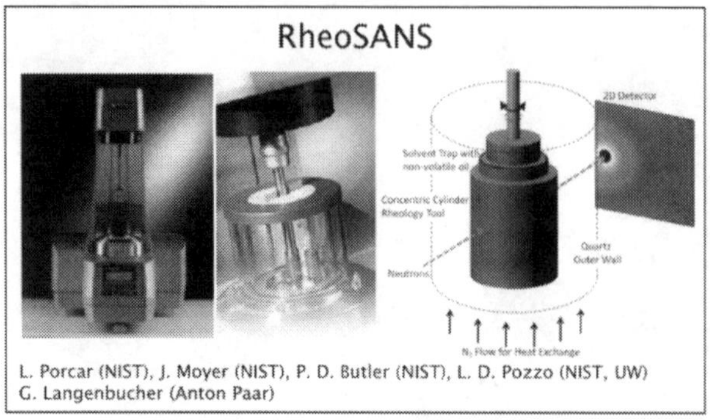

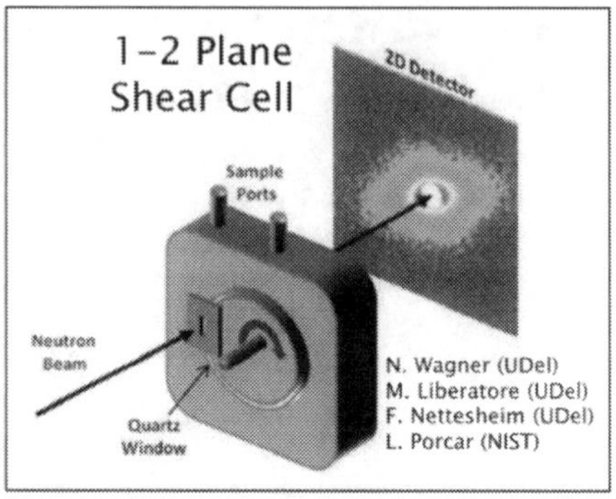

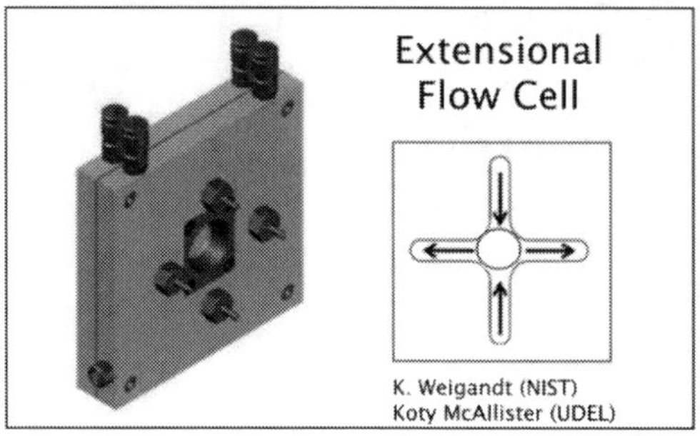

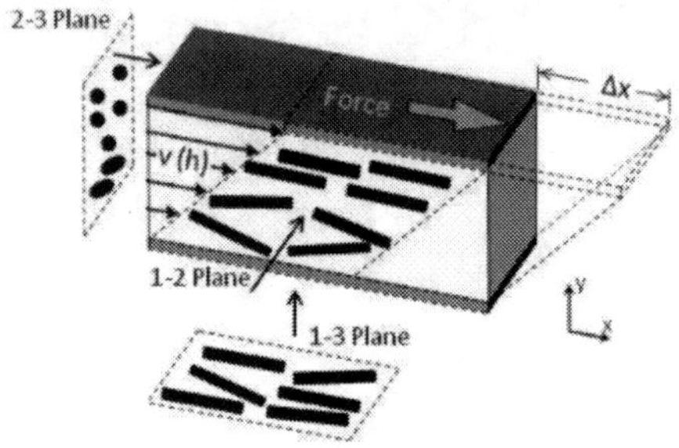

It is often very difficult to measure the nanostructure of complex fluids in dynamic and industrially relevant environments!

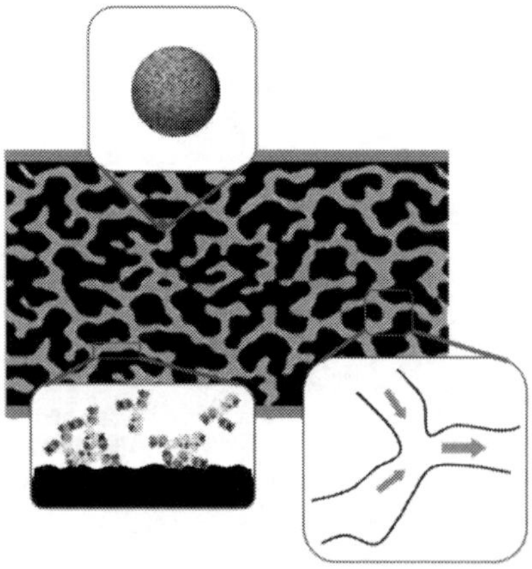

High Strain/Flow Rates
(~1,000,000 s^{-1})

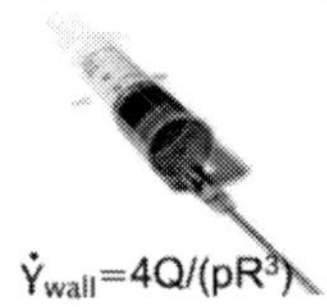

Non-Ambient Conditions

High Temperatures (>200°C) | High Pressures (up to 70 MPa)

and

μRheoSANS of Anionic Wormlike Micelles

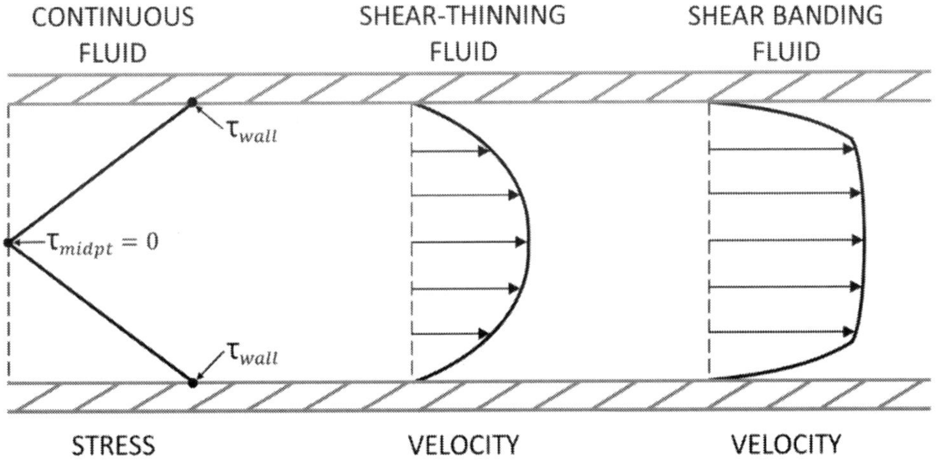

μRheoSANS of Anionic Wormlike Micelles

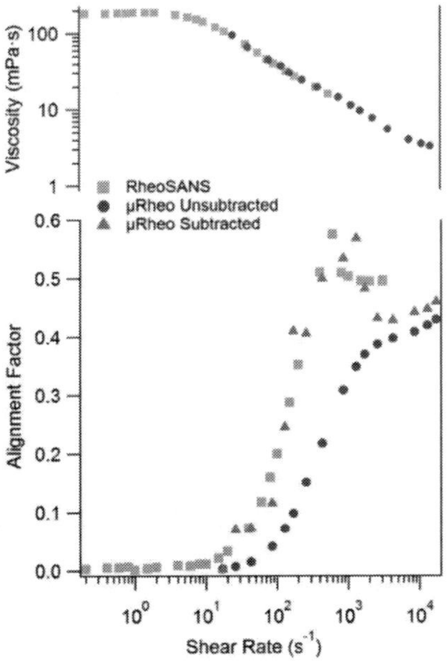

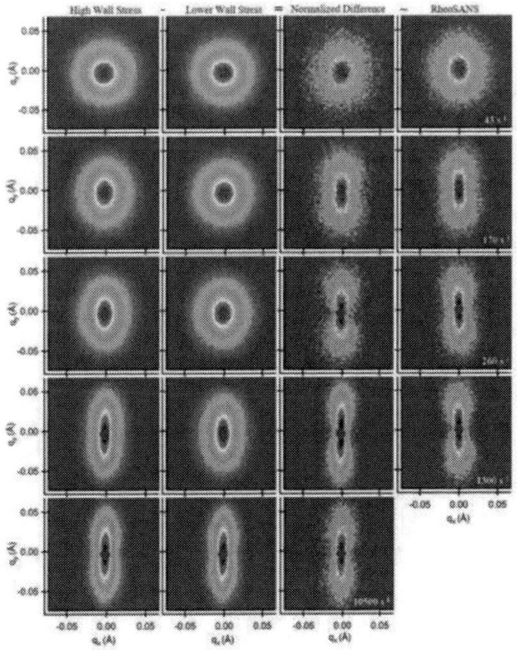

High Pressure Upgrade

Full Apparatus Schematic

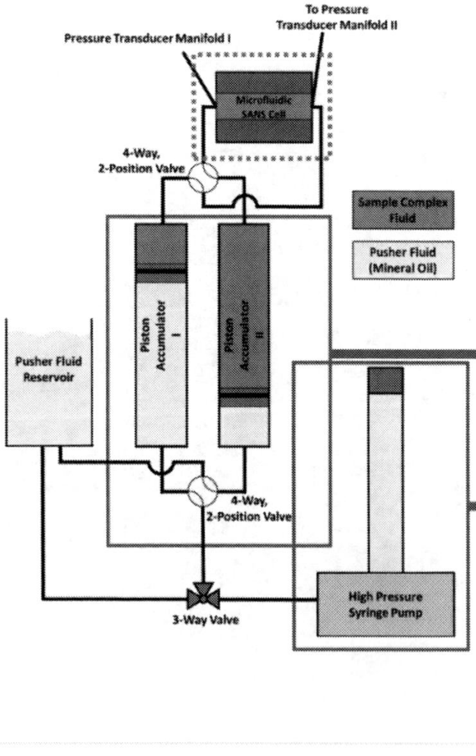

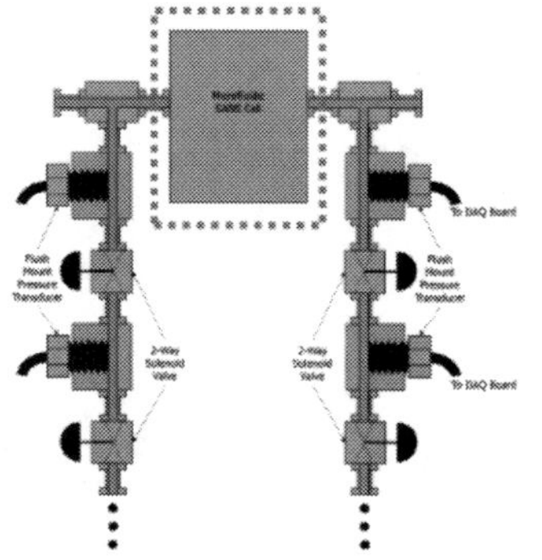

HiP Tubular Reactor w/Piston
- 125 mL each
- 10 000 psi – 700 bar
- 304 Stainless Steel
- 0 – 125°C

DCI VPA Syringe Pump
- 400 mL/min
- 5000 psi – 345 bar
- Stainless Steel Cylinder
- 12 nL volume resolution

Modular High Pressure SANS Flow Cell

Modular High Pressure SANS Flow Cell

Modular High Pressure SANS Flow Cell

Modular High Pressure SANS Flow Cell

Cross-Section View

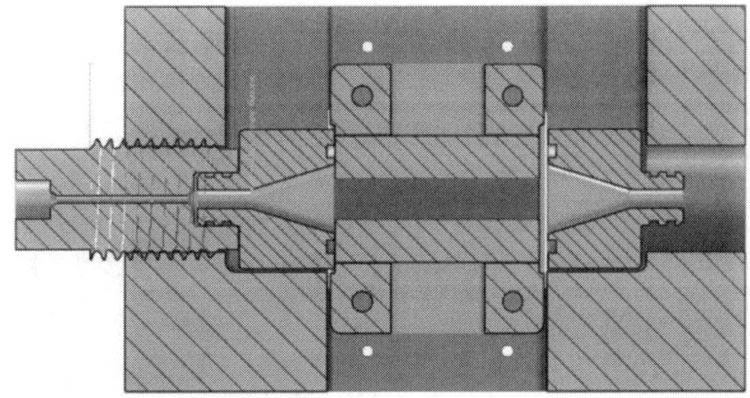

Machining is *Almost* Completed

First Flow Cells Are Ready for Testing

Full Apparatus and Expected Performance

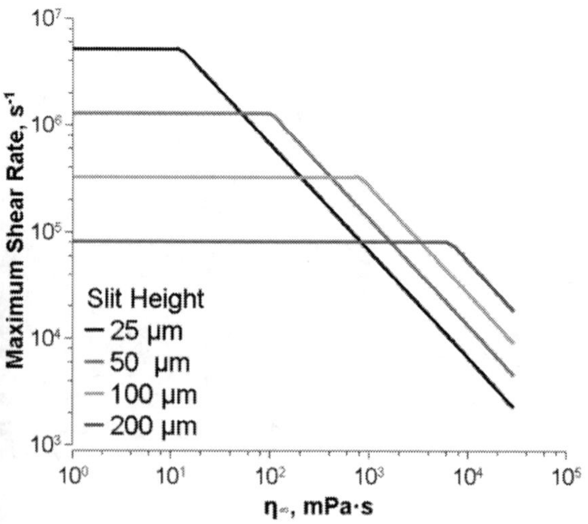

Modularity

Model Porous Media

Novel Synthesis of Model Porous Media via Self-Assembly

Ludox-TM, 22nm, 30 w/w%, 4 ml + 4 ml lutidine

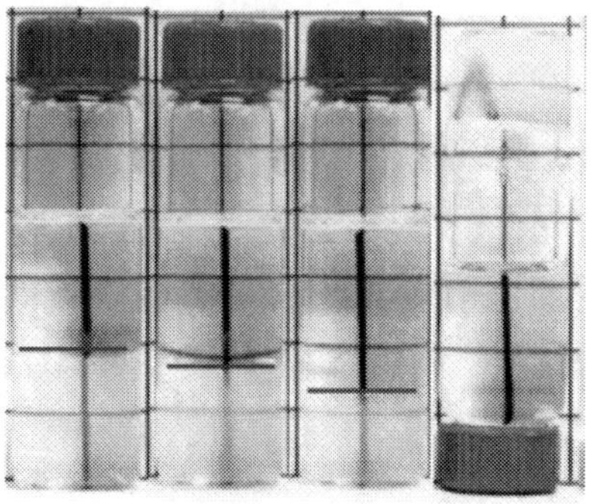

1 hr 1 day 6 days

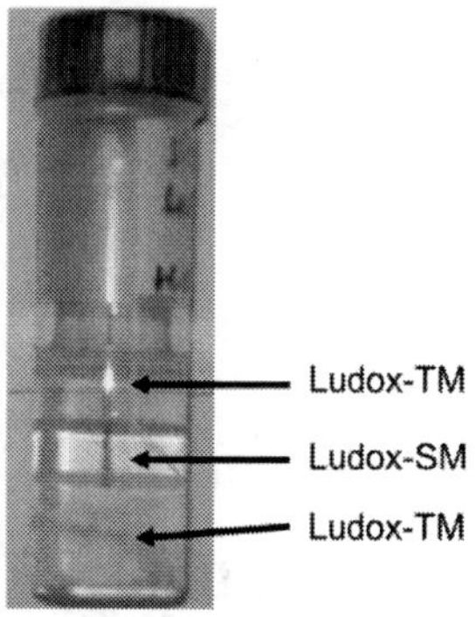

Hongyu Guo, Raul Lobo, Yun Liu, "Novel Method to Synthesize Porous Materials with Controlled Pore Sizes Based on the Self-Assembly of Colloidal Silica Particles", in preparation, 2017

Novel Synthesis of Model Porous Media via Self-Assembly

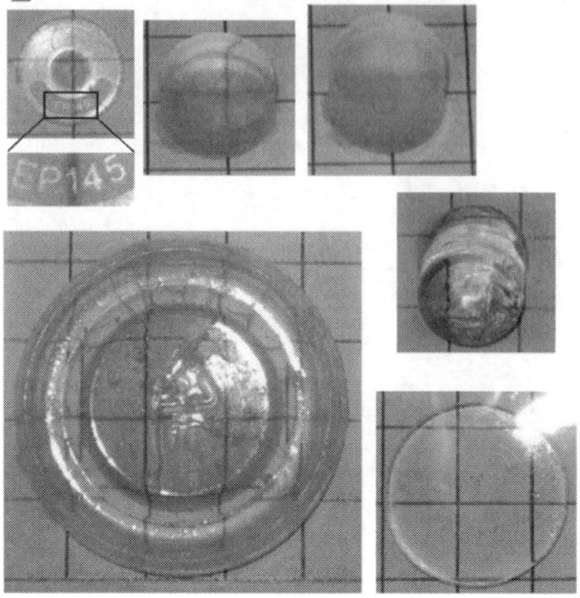

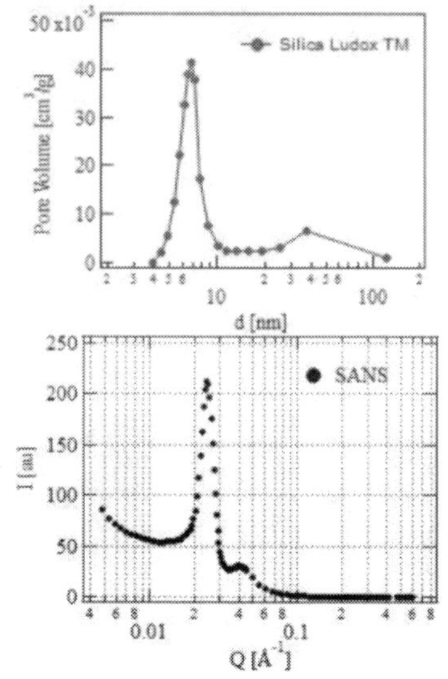

Hongyu Guo, Raul Lobo, Yun Liu, "Novel Method to Synthesize Porous Materials with Controlled Pore Sizes Based on the Self-Assembly of Colloidal Silica Particles", in preparation, 2017

In Situ Porous Media Synthesis?

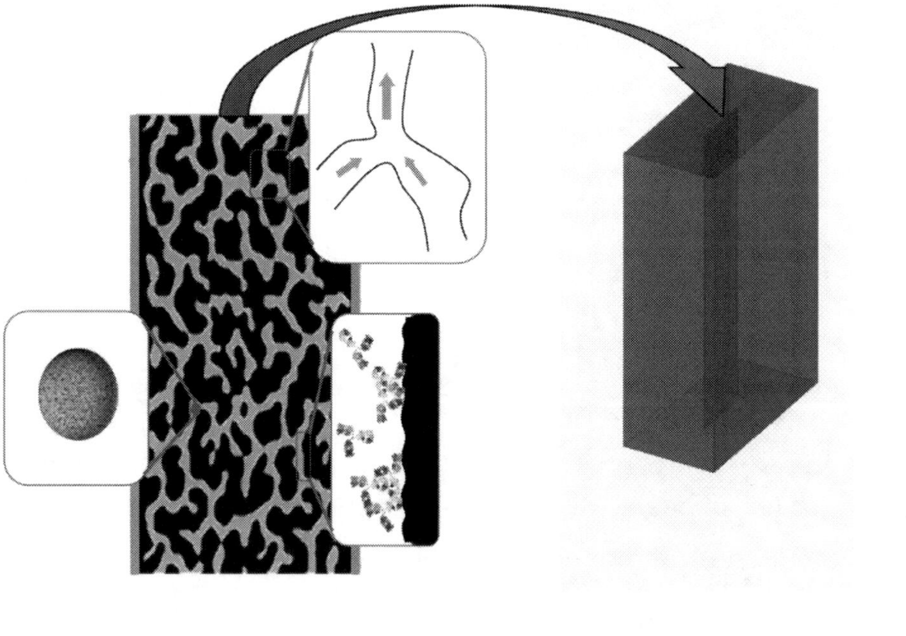

Thanks to Collaborators – Questions?

-NIST Staff Collaborations:
Cedric Gagnon (Technical Drawings)
Chris Amigo (Machining)
Yun Liu (Model Porous Media)

-NIST Postdoc Collaborations:
Wei-Shan Chiang (High Pressure)
Hongyu Guo (Model Porous Media)
Paul Salipante (Fluid Dynamics)

-Academic Collaborations:
Daniel Blair, Georgetown

Postdoctoral Positions Available
javen.weston@nist.gov

Extra Slides

High shear scattering in rectangular channels

- Depth sectioning method demonstrated by Fernandez-Ballester et al., JoR '09 (WAXS).
    - Isolate high stress region

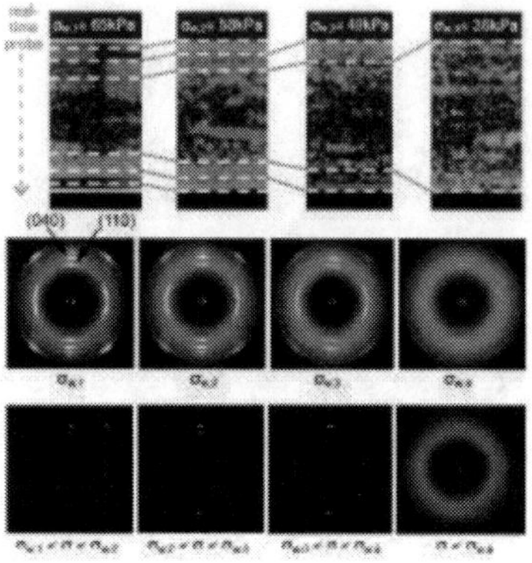

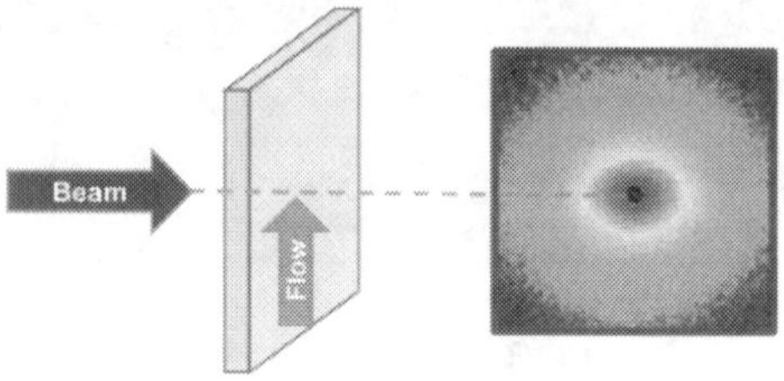

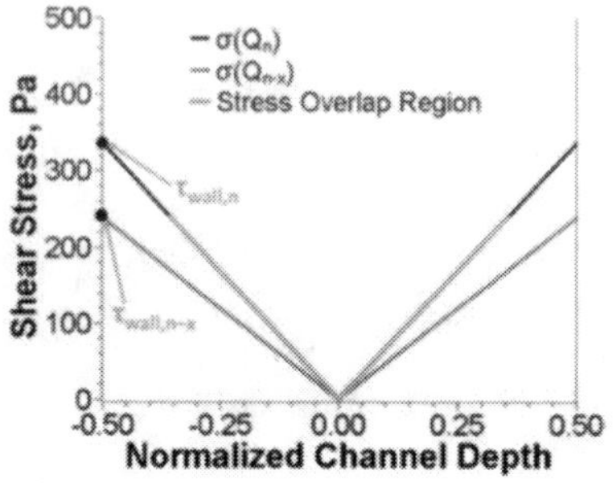

$$I_{q_x,q_y}(wall) = \frac{1}{1-\frac{\tau_a}{\tau_b}}\left[I_{q_x,q_y}(\tau_b) - \frac{\tau_a}{\tau_b} I_{q_x,q_y}(\tau_a)\right]$$

μRheoSANS Limitations

- Depth dependent scattering analysis:
  - Approach will fail if the scattering depends on position independent of stress
    - No concentration gradient
    - Continuous sample (no sample fracture)
  - Slice depth relative to scattering object dimensions
  - Time…
- Pressure drop or flow rate limited to reach high shear rates…

Evaluation of Stress-Ratio Importance

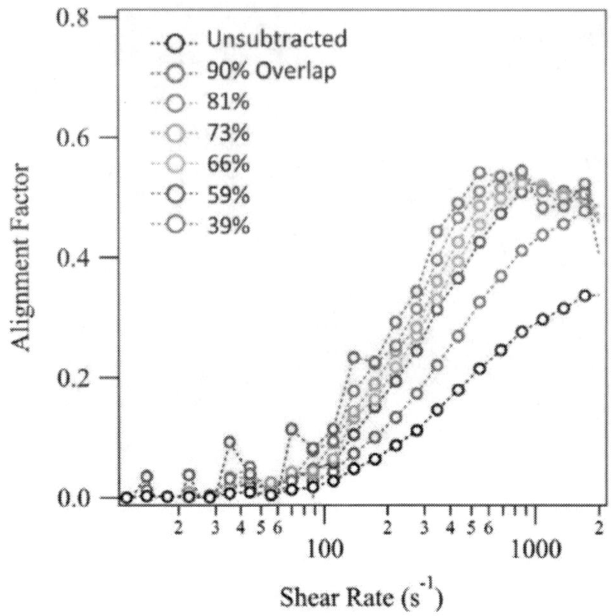

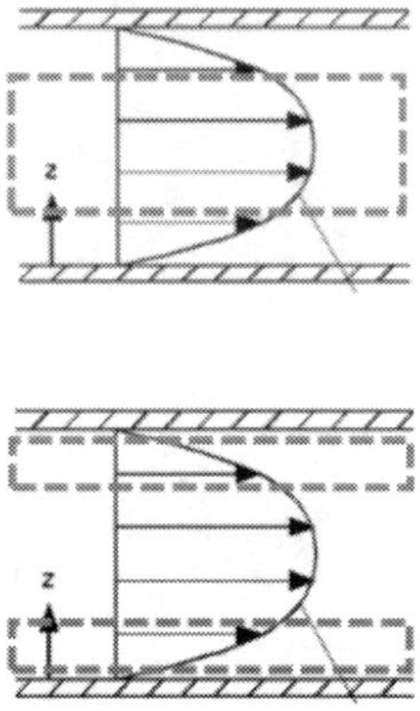

Direct Measurement of Scattering Across Slit

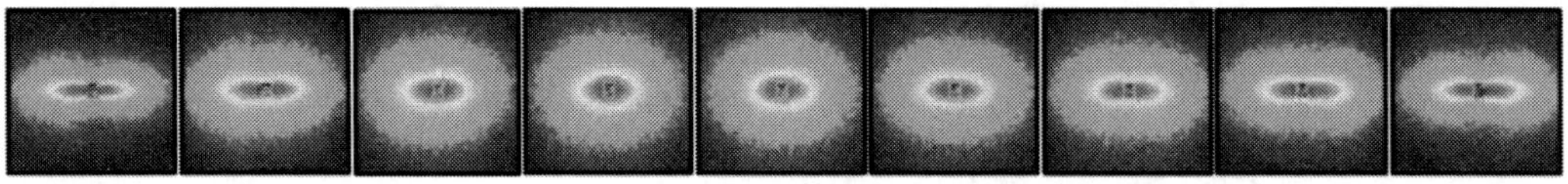

- Is asymmetry shear-induced or inertial?
  Re ≈ 7
- Use symmetric inlets in future.

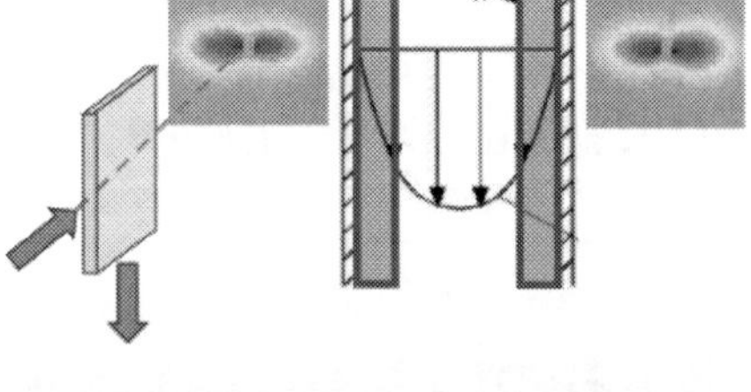

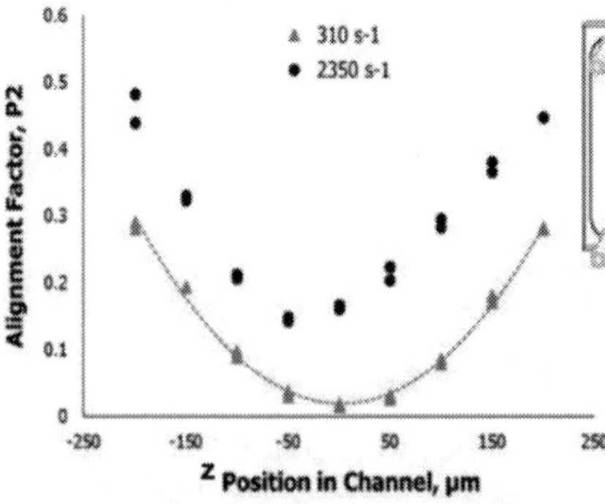

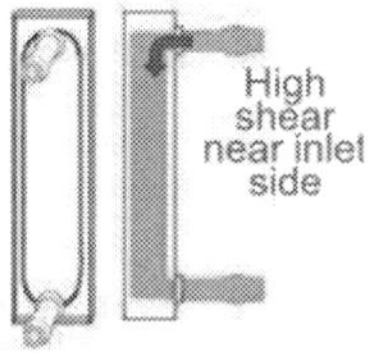

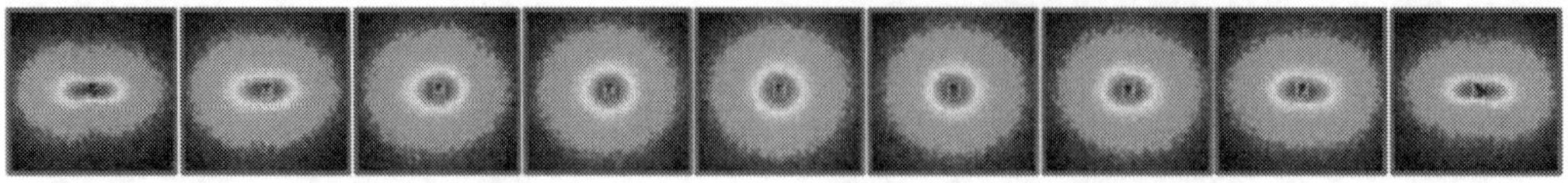

491806 Mechanistic Constitutive Model for Wormlike Micelle Solutions with Flow-Induced Structure Formation

Abstract:

We present a tensor constitutive model for predicting stress and flow-induced structure formation in dilute wormlike micellar solutions. The micellar solution is treated as a dilute suspension of rigid Brownian rods whose length varies dynamically. Consistent with the mechanism presented by Turner and Cates [J. Phys.:Condens. Matter 4, 3719 (1992)], flow-induced alignment of the rods is assumed to promote increase of rod length that corresponds to the formation of flow-induced structures observed in experiments. At very high deformation rate, hydrodynamic stresses causes the rod length to decrease. These mechanisms are implemented in a phenomenological equation governing the evolution of rod length, with the number density of rods appropriately modified to ensure conservation of surfactant mass. The model leads first to an increase in both shear and extensional viscosity as deformation rate increases and then to a decrease at higher rates. If the rate constant for flow-induced rod growth is sufficiently large, the model predicts a multivalued relation between stress and deformation rate in both shear and uniaxial extension. Predictions for shear and extensional flow at steady state are in reasonable agreement with experimental results. The model is simple enough to serve as a tractable constitutive relation for computational fluid dynamics studies.

Rheology and microstructure of thermoreversible micellar crystals in ionic liquid

Ru Chen, Carlos R. López-Barrón, Norman J. Wagner
Department of Chemical and Biomolecular Engineering, University of Delaware, Newark, Delaware, USA

1. Introduction

Block copolymer (BCP) solutions are fascinating complex fluids because they can self-assemble into a wide range of nanoscale structures in selective solvents, which can be functionalized for a variety of potential applications in the fields as diverse as consumer products, food additives and involved in polymer processing. The macroscopic viscoelastic properties of those solutions are determined by the spatial organization of the constituent micelles, usually referred to as the microstructure. It has been demonstrated via rheological measurements in BCP aqueous solution that controlling the relative corona/core size of the constituent micelles one can obtain a wide range of viscoelastic response, from polymeric to colloidal. The colloidal character is typically attributed to the hard core, and the polymeric character to the softer corona.[1] However, a remaining challenge is understanding BCP self-assembly in ionic liquids (ILs), where the ultimate goal is to link the microstructure to the macroscopic flow properties at various micelle phases. In this work, the structural and rheological properties of a model series of binary mixtures of Pluronic BCPs dissolved in a protic IL are studied as a strategy to modulate and control the soft solid behavior of amphiphilic BCPs in ionic liquids. The inter-particle softness of the soft solids is controlled via the ratio of Pluronic BCPs mixtures self-assembled in deuterated ethylammonium nitrate (dEAN). Equilibrium microstructures are examined by linear viscoelasticity and small angle neutron scattering (SANS) measurements, while the shear induced microstructures are probed by state of the art *in situ* rheo-SANS in the radial direction (1-3 plane of flow) under steady shear flow.[2]

2. Experimental section

Sample preparation. Deuterated EAN (dEAN) was synthesized by solvent exchange with D_2O following by drying. Solutions with total Pluronic® composition 27.5 wt. % at mass ratio of Pluronic P123/Pluronic F127: 100/0, 10/90, 25/75, 50/50, 75/25, 90/10, and 0/100 in dEAN were prepared by strong mechanical mixing and centrifugation until clear solutions were obtained.

Steady rheology-small angle neutron scattering (rheo-SANS). SANS measurements were performed at 15 and 40 °C during steady shear flow (shear rates = 0.1 to 1000 s^{-1}) using a stress controlled MCR501 rheometer (Anton-Paar) mounted in the SANS instrument NGB30m at the NIST NCNR in Gaithersburg, MD. SANS measurements were performed in the 1-3 flow-vorticity plane of shear with the neutron beam parallel to the radial direction in a titanium Couette geometry.

3. Results and discussion

At higher polymer concentration and system temperature, Pluronic® P123 and F127 spherical micelles spontaneously packed into cubic crystal lattices, body-centered cubic phase (or BCC) for P123 and face-centered cubic (or FCC) phase for F127. Furthermore, both systems show a thermoreversible transition from a liquid to a gel.[3] The effect of micelle softness on the microstructure and rheological properties of 27.5 wt. % PFdE solutions were measured using rheo-SANS on flow-vorticity plane of shear under steady shear flow. Fig. 1 showed the 2D scattering pattern and corresponding rheology response for a characteristic mixture, 27.5 wt. % P50/F50dE solution, abbreviation for sample comprising 50 wt. % P123 and 90 wt. % F127 in total polymer weight composition. Compared our experimental data to the Ackerson Loose model prediction,[4,5] at quiescent state, the scattering is mostly isotropic, indicating the micelles are oriented in all possible orientations. When the sample is sheared at 0.1 s^{-1} to 0.316 s^{-1}, the development of six-fold symmetry was overserved, which indicates the micelles start to orient themselves and stack into hexagonal-closed packed (HCP) layers. As the shear rate increases to 100 s^{-1}, the six-fold symmetry becomes more defined and the intensity of spots on vorticity direction decreases. This reflects micelles layers sliding in a zig-zag fashion and losing the long-range order in the velocity direction. The formation of the two-dimensional HCP layers and the zig-zag motion comply easily with the flow, which is signaled by the observed shear-thinning behavior. However, a small viscosity overshoot is overserved above 100s^{-1}, followed by shear thinning. This behavior has been previously reported for transition with a weak order-disorder transition. As the shear rate increases further, the powder scattering profile re-appears and the system shear thins further, indicating system melting.

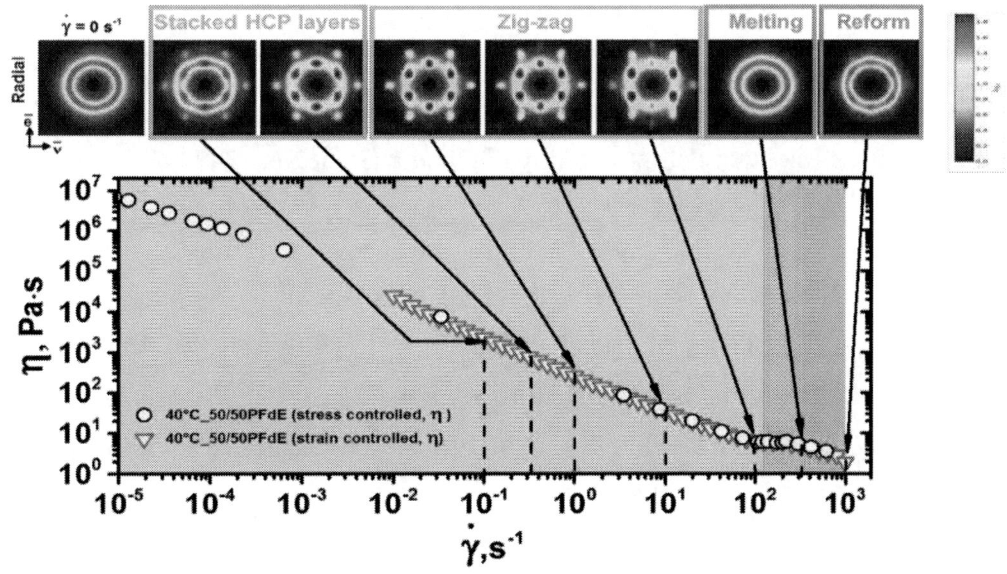

Fig. 1 Steady rheo-SANS data from 27.5 wt. % P50/F50dE solution measured at 40 °C. Lower panel: Steady shear viscosity versus stress and versus shear rate. Both strain-controlled measurements and stress-controlled measurements are shown. Upper panel: 2D SANS patterns measured at the shear rates and indicated by the arrows in the radial direction.

4. Conclusion

The block copolymers mix at the molecular level and their mixtures exhibit two different types of crystalline ordering depending on composition: Pluronic P123 rich mixtures show a body centered cubic (BCC) ordering – characteristic of pure Pluronic P123 micelles, while Pluronic F127 rich mixtures exhibit face centered cubic (FCC) ordering – characteristic of pure Pluronic F123 micelles. Both structures crystallize from liquid-like micellar solutions into cubic phases by quiescent heating, *i.e.*, inverse melting. Under steady shear flow, increasing shear rate induces three structural transitions: (1) twinning of the FCC lattice at low shear rates, (2) formation of hexagonal close packed (HCP) layers at intermediate shear rates, and (3) shear-melting at high shear rates. This research identifies a strategy for modifying the structural and rheological properties of Pluronic BCPs in solutions via mixing. It also provides insights useful for comparing the properties of self-assembled amphiphilic BCPs in molecular versus ionic liquid solvents.

5. Acknowledgement

We acknowledge the support of the National Institute of Standards and Technology (NIST), U.S. Department of Commerce, in providing the neutron research facilities used in this work. The statements, findings, conclusions and recommendations are those of the author(s) and do not necessarily reflect the view of NIST or the U.S. Department of Commerce. N.J.W and R.C. acknowledge support of cooperative agreements 70NANB12H239 and 70NANB15H260 from NIST, U.S. Department of Commerce. R.C. also acknowledges support from the National Science Foundation Graduate Research Fellowship Program under Grant No. 1247394.

References:

(1) Vlassopoulos, D.; Cloitre, M. Tunable Rheology of Dense Soft Deformable Colloids. *Curr. Opin. Colloid Interface Sci.* **2014**, *19* (6), 561–574.

(2) Eberle, A. P. R.; Porcar, L. Flow-SANS and Rheo-SANS Applied to Soft Matter. *Curr. Opin. Colloid Interface Sci.* **2012**, *17* (1), 33–43.

(3) López-Barrón, C. R.; Li, D.; Wagner, N. J.; Caplan, J. L. Triblock Copolymer Self-Assembly

in Ionic Liquids: Effect of PEO Block Length on the Self-Assembly of PEO-PPO-PEO in Ethylammonium Nitrate. *Macromolecules* **2014**, *47* (21), 7484–7495.

(4) Loose, W.; Ackerson, B. J. Model Calculations Structures for the Analysis of Scattering Data from Layered Structures. *J. Chem. Phys.* **1994**, *101* (9), 7211–7220.

(5) López-Barrón, C. R.; Wagner, N. J.; Porcar, L. Layering, Melting, and Recrystallization of a Close-Packed Micellar Crystal under Steady and Large-Amplitude Oscillatory Shear Flows. *J. Rheol. (N. Y. N. Y).* **2015**, *59* (3), 793–820.

503468 Use of Microfluidics and Advanced Microscopy to Study Shear-Induced Structuring in Wormlike Micellar Solutions

Abstract:

In the last decades, a growing attention has been devoted to wormlike micellar solution in industrial applications. Indeed, thanks to their viscoelastic properties, these systems are widely used to tune the rheological response of home-care and personal-care products. On the other side, academic research is showing a great interest in studying wormlike micelles, thanks to their unique morphology and flow behavior, characterized by peculiar phenomena, as the occurrence of flow instabilities [1] at high shear rates, with the appearance of flow-induced structures [2]. Although this great effort, there are still many open questions, mainly concerning the dynamics and the physical origin of these instabilities.
In this scenario, microfluidics is a powerful tool to get a deeper insight into the flow behavior of a wormlike micellar solution. Indeed, the high confinement that characterizes a microfluidic device facilitates the onset of the flow instabilities; on the other side, its small dimensions allows a direct visualization of flow structuring phenomena, by coupling the microfluidic device with optical microscopy.
Here, we couple a glass capillary with mPIV techniques in order to study the flow of a widely used wormlike micellar solution, and we show the relation between rheological behavior, velocity profiles and flow induced structures.

[1] E. K. Wheeler *et al.*, J. Non-Newtonian Fluid Mech. **75** (1998)

[2] J. J. Cardiel et al., Proc. Natl. Acad. Sci. **110** (2014)

498675 A Multiscale Tensor-Based Model for the Rheology of Aggregating Thixotropic Colloidal Suspensions

Abstract:

Concentrated colloidal suspensions, in which the primary particles associate at rest into aggregates, or flocs, which, however, are broken apart upon an applied flow deformation, often exhibit a complex thixotropic rheological behavior characterized by yield stress and a deformation history dependent rheology[1-4]. Recently, well defined rheological measurements on a model thixotropic suspension [2,6-7] have been extended to include large amplitude oscillatory shear (LAOS) flow, shear flow reversal, and a novel unidirectional LAOS flow to provide an extended rheological data set for testing constitutive models [5]. In [5], this extended data set has been used to test a new structure-based model developed by extending the Delaware thixotropic model [8]. A similar, scalar structural parameter-based, model has also been recently developed to describe the thixotropic behavior of another aggregating colloidal system, human blood [9]. However, the LAOS-based work [8] has shown the inherent limitation of the new as well as other, phenomenological, single structural-parameter based models to unidirectional shear flows, in that they cannot account for the reversal of flow directionality inherent in LAOS flow. The present work attempts to circumvent that limitation by systematically developing a new, multiscale thixotropy model where a structural scalar parameter model, developed from the model coarsening of a microscopic population balance analysis [10], is combined with two internal tensor structure parameters that are introduced through a thermodynamically consistent formulation [11] to describe the orientation and deformation under flow of free and interacting aggregates, respectively. The resultant model is demonstrated to quantitatively capture the key rheological signatures observed in the experimental data on thixotropic suspensions with distinctly superior predictions when compared to current thixotropy models. More specifically, the multiscale tensor model arising from this work is shown to offer substantially better predictions to the model thixotropic suspension of Dullaert and Mewis [6] than the single scalar parameter models that have been tested before[7,9]. An additional advantage of the new formulation is the capability to model the systemâ€™s rheology under general flow deformation.

In summary, we have developed a multiscale tensor model that incorporates microscopic information at the particle level into a two structural tensors-based, thermodynamically consistent, macroscopic description of soft aggregating colloidal particles in a thixotropic suspension that displays a yield stress. The good agreement with rheological experiments on a model thixotropic suspension provides additional confidence in the results arising from the multiscale framework adopted in this modeling effort.

References:

[1] J. Mewis, J. *J. Non-Newtonian Fluid Mech.*, 6, 1 (1979).

[2] K. Dullaert, K., J. Mewis. *J. of Colloid and Interface Science*, 287, 542 (2005).

[3] J. Mewis, N. J. Wagner, *Adv. Colloid and Interface Science*, 147-148, 214 (2009).

[4] J. Mewis, N. J. Wagner, *Colloidal Suspension Rheology*, Cambridge U. Press, New York, 2012.

[5] M. J. Armstrong, A. N. Beris, S. A. Rogers and N. J. Wagner, *J. Rheol.*, in press (2016).

[6] K. Dullaert, J. Mewis, *Rheol. Acta*, 45, 23 (2005).

[7] K. Dullaert, J. Mewis, *J. Non-Newtonian Fluid Mech.*, 139, 21 (2006).

[8] A. Mujumbdar, A. .N. Beris , A. B. Metzner, *J. Non-Newtonian Fluid Mech.*, 102, 157 (2002).

[9] A. J. Apostolidis, M. J. Armstrong, A. N. Beris, *J. Rheol.*, 59, 275 (2015).

[10] P.M. Mwasame, A.N. Beris, R.B. Diemer and N. J. Wagner, *AIChE Journal*, *63*(2), 517-531, (2017).

[11] A. N. Beris, B. J. Edwards, *Thermodynamics of Flowing Systems*, Oxford UP, New York, 1992.

496639 Effect of Shear Stress on Rheological Properties of Fibrous Colloidal Gels during Gel-Sol Transitions Induced By Osmotic Pressure Gradients

Abstract:

The manufacturing of creams and colloidal suspensions requires a variety of processing steps including dilutions and mixing that can cause permanent changes to the microstructure of the product. This work investigates the effects of shear, such as from mixing, on the material properties of a colloidal gel after repeated phase transitions. Hydrogenated castor oil (HCO) is a fibrous colloid that is a stable suspension in the presence of a surfactant, and has large aspect ratios ranging from 50-2500. A gel-sol phase transition is induced by an osmotic pressure gradient. Material properties are measured using both bulk rheology and Î¼^{2}rheology, the combination of microfluidics and multiple particle tracking microrheology (MPT). Experiments are performed with two starting materials, a non-sheared 4 wt% HCO gel and a sheared 0.125 wt% HCO sol. Bulk rheology measures the degradation of the 4 w% gel and determines that osmotic forces are not large enough to completely degrade the gel into a sol. This is evidenced in the data because there is no crossover point between the storage and loss moduli. Gelation of the 0.125 wt% sol does show a crossover point during a change of phase, but the final material properties are a weaker gel than the equilibrated 4wt% sample. In MPT, fluorescent 0.5 Î¼m probe particles are embedded into a sample and video microscopy records their Brownian motion. Particle trajectories are then related to material properties using the Generalized Stokes-Einstein Relation. The microfluidic device used in Î¼^{2}rheology enables repeated phase transitions on a single sample while minimizing the amount of shear during fluid exchange. Î¼^{2}rheology results conclude that the material can undergo repeated phase changes without changes to microstructure, but again determine that the material properties are dependent on the shear history of the sample. This work shows the importance of sample handing during processing, and the ways that different processing steps can permanently alter material properties.

496265 Microstructural Basis for Colloidal Gel and Glass Aging

Abstract:

The study of out-of-equilibrium systems is of fundamental scientific interest and industrial applicability, and, as such, characterizing and controlling these systems is considered by the Department of Energy as one of the five grand challenges facing scientists today [1]. Colloidal dispersions are usually out-of-equilibrium systems that are of particular relevance as they are found in a wide variety of natural and synthetic consumer products, such as foods, pharmaceuticals, paints and cosmetics [2]. Moreover, colloidal dispersions are often employed as model systems to probe and understand non-equilibrium behavior in other systems of interest, such as atomic or molecular systems, because colloidal dispersions have more experimentally accessible length scales and time scales [3]. Importantly, the properties of colloidal systems are known to change with gel age, but the processes and mechanisms by which aging occurs is not well understood and limits our ability to predict macroscopic behavior in these systems. Due to its fundamental significance and industrial applicability, we investigate the microstructural basis of aging in a well-studied, model adhesive hard sphere (AHS) system, consisting of silica nanoparticles grafted with an octadecyl brush dispersed in tetradecane [4, 5]. In particular, we quantitatively relate rheological aging to structural aging using concurrent rheometry and time-resolved small angle neutron scattering (Rheo-SANS) measurements in both glasses and gels [6]. Moreover, by probing the system with different thermal profiles and shear rates, we investigate the effect of thermal and shear history on the mechanical properties and microstructure during aging to assess the efficacy of shear as a rejuvenation method. Our results showed that after mechanical rejuvenation, the evolution of the microstructure and bulk properties progressed more rapidly compared to the thermal rejuvenation, indicating that either the microstructure may not be completely erased by shear or that the microstructure continues to evolve under shear. While the formation and aging trajectory in the both the gel and glass samples are significantly affected immediately following a temperature quench or the application of stress, the samples appear to eventually evolve along a similar aging trajectory, independent of the thermal and shear history at long times.

References:

[1] "Directing Matter and Energy: Five Challenges for Science and the Imagination," Basic Energy Sciences Advisory Committee Report, U.S. Department of Energy (2007).

[2] Joshi, Y. M., "Dynamics of Colloidal Glasses and Gels," Annual Review of Chemical and Biomolecular Engineering, Vol 5 **5**, 181-202 (2014).

[3] Lu, P. J., and Weitz, D. A., "Colloidal Particles: Crystals, Glasses, and Gels," Annual Review of Condensed Matter Physics, Vol 4 **4**, 217-233 (2013).

[4] Eberle, A. P. R., Castaneda-Priego, R., Kim, J. M., and Wagner, N. J., "Dynamical Arrest, Percolation, Gelation, and Glass Formation in Model Nanoparticle Dispersions with Thermoreversible Adhesive Interactions," Langmuir **28**, 1866-1878 (2012).

[5] Eberle, A. P. R., Wagner, N. J., Akgun, B., and Satija, S. K., "Temperature-Dependent Nanostructure of an End-Tethered Octadecane Brush in Tetradecane and Nanoparticle Phase Behavior," Langmuir **26**, 3003-3007 (2010).

[6] Gordon, M.B., Kloxin, C.J., and Wagner, N.J., "Rheology and Microstructure of an Aging Thermoreversible Colloidal Gel," Journal of Rheology (2017).

495237 The Mechanism for Shear Thickening in Viscoelastic Suspensions

Abstract:

Suspensions of solid particulates in viscoelastic fluids are ubiquitous in engineering applications, yet their material functions under shear flow are difficult to predict due to the nonlinearity of the stress in the suspending fluid. Previous experiments in the literature on suspensions in Boger fluids have suggested that elasticity in the suspending fluid causes shear thickening in the suspension averaged viscosity and the first normal stress coefficient. Upon examining these results, authors were unable to determine whether this effect arose from particle-particle interactions or particle-fluid interactions. Previously we performed simulations of shear flow past a sphere and identified that the extra fluid stress induced by polymer stretching in the disturbance flow of each particle in a dilute suspension leads to shear-thickening which is at least in qualitative agreement with experimental results. However, quantitative differences with the experimental data suggest a closer examination of the mechanism of shear-thickening is necessary. Thus, we first use numerical simulations to analyze the local disturbance flow field surrounding a single particle as well as its effect on the polymer conformation to determine the regions that most contribute to the average suspension viscosity thickening. We identify the importance of Lagrangian, time-dependent extensional flow in creating large polymer stretching in closed streamlines around the particles. The insights from these analyses imply particle-particle interactions are relatively unimportant for predicting the shear-thickening of the viscosity but are important for accurately predicting the zero-shear viscosity as a function of volume fraction. Moreover, our results suggest that the effective extra viscosity (beyond the zero shear viscosity) of the suspension per particle is correlated with the suspension shear stress. We verify these findings by analyzing literature experiments and new, multi-particle simulations using Immersed Boundary Methods.

503557 Homogenization of Self-Organizing Colloidal Particles in a Microscopy Chamber

Abstract:

Colloidal particles are typically observed to self assemble in a thin chamber under a compound light microscope. It is desirable to initiate observations with a homogeneous suspension of particles. There are two ways of homogenizing or "mixing; one is "chaotic", and the other is "separation and recombination" or SAR, which forms the basis of in-line static mixers functioning at low Reynolds' number. Another term applied to this method is "interfacial surface generation" (ISG), which is a good description of what actually happens. Most electroosmosis based mixers in microfluidics tend to be of the chaotic type, introducing localized regions of backflow at the low Reynolds' numbers that characterize microfluidics. In our mixing module the objective is actually to separate particles that had once reacted with one another via long-range forces so that a self-assembly experiment can be repeated (at a modified temperature, for example). It is assumed that any non-uniform fluid motion will break up the investigator's colloidal assembly. In order to achieve the generation of new interfacial surfaces small finite elements of a static fluid need to be moved with respect to each other. Electroosmosis accomplishes this by the application of an electric field in the y direction causing a reactive force in the diffuse double layer of the chamber wall if the wall is charged and there are ions in solution. Usually the wall carries fixed negative charges and the fluid adjacent to the wall drifts toward the cathode. In a closed chamber this flow is balanced by flow in the opposite direction along the midplane of the chamber. With the Smoluchowski slip velocity at the wall given by $v_y = \mu_{eo}E$ and assuming we want an element of fluid at the wall to move ½ the length of a 2 cm chamber in 10s to get $v_y = 0.1$ cm/s we will need a potential of the order of 100 V based on the generous assumption of electroosmotic mobility $\mu_{eo} = 0.001$ cm^2/V-s. For the desired mixing effect we should repeat this process so that each element of fluid is brought to a different location with every repetition. Thus a pulse train starting with 100V in one direction for 10 s followed by -100V for a shorter time followed by multiple subsequent alternating square waves of progressively shorter duration should produce satisfactory micromixing in 2-3 minutes. This process was implemented in a 0.7 x 7.0 x 40 mm glass chamber with microscope observation. Video images demonstrated that all particles in the viewing field were relocated. This research was supported by NASA contract NNX12CE76P.

498946 Elevated Pressure Interfacial Dynamics of LS Surfactants at the Water-CO_2Â Interface

Abstract:

There are several examples of surfactants employed under extreme temperatures and pressures, e.g. the Deep Horizon oil spill, cleaning and recycling techniques, and fuel recovery and transport. Understanding the effect of these conditions, particularly pressure, on the fundamental behavior of surfactants has yet to be fully explored. In this work, a novel high pressure microtensiometer setup is detailed which allows for dynamic interfacial tension measurements at elevated temperatures and pressures at microscale interfaces. We will present the effect of pressure and temperature on the governing surfactant transport mechanisms and overall equation of state for multiple LS surfactants, which are commonly used in high pressure applications.

494000 Atomistic Simulations of Micellization and Adsorption of Imidazoline-Based Surfactants Near Metal-Water Interfaces

Abstract:

Imidazoline-based surfactants are popular corrosion inhibitors in the oil and gas industry. These molecules are known to strongly adsorb onto the metal-water interfaces, as well as form micelles in the aqueous phase. High degree of corrosion inhibition is observed at concentrations close to critical micelle concentration (CMC). However, molecular underpinnings of the relationship between micellization and surface adsorption are not completely understood. By employing atomistic representation of imidazoline-based surfactants in explicit water, we have studied micellization and transport properties of these molecules as a function of concentration, size of the alkyl tail and ionic strength. Furthermore, we have investigated the adsorption characteristics of these molecules at the metal-water interface at different concentrations. We find that the micelle forming tendency of these molecules depends strongly on the length of the alkyl tail and the ionic strength of the solution. The dynamic equilibrium between adsorption at the metal-water interface and micellization will be discussed.

499522 Surfactant Chemistry and Counter-Ion Clouds in Bile Salt Stabilized Swcnts

Abstract:

Enrichment and separation of SWCNTs based on chirality has been previously achieved by subjecting bile salt stabilized aqueous suspension of SWCNTs to analytical centrifugation. The ability of bile salts to amplify relatively subtle differences in SWCNTs based on chirality to significant differences in hydrodynamic behavior can be attributed to the structure and dynamics of the counter-ion cloud surround the stabilized SWCNT in water. The counter-ion cloud is known to create an "ion wall" between surfactant coated SWCNTs, which is key for stabilizing the SWCNTs in water. Experiments on bile salts have shown that minor chemical differences (such as point mutations) in surfactant backbone or "tail" chemistry lead to drastic changes in both, the solution properties as well as their ability to stabilize SWCNTs. Here, we use atomistically detailed molecular dynamics simulations to study the free energy of separation of SWCNTs stabilized by different bile salts. Understanding the molecular mechanism underlying the interplay between the chemistry of the bile salt backbone, the structure and dynamics of the counter-ion cloud, and the hydrodynamic behavior of the surfactant stabilized-SWCNT is important for separation of SWCNTs based on chiral fractions.

499524 Tailoring Inhibitors of Pathological Crystallization: New Platforms for Drug Design

Abstract:

Pathological mineralization arises from the precipitation of inorganic and organic constituents in the human body. Their occurrence is driven by supersaturated environments where dissolved solute undergoes a phase transformation from solution via processes consisting of nucleation, growth, aggregation and retention on tissues. One such physiological disorder is kidney stone formation, primarily composed of calcium salts and other minor constituents. About 80% of kidney stones are composed of calcium oxalate monohydrate (COM).[1] The current therapy is citrate supplementation to reduce stone growth; however, stone recurrence remains a concern for patients and clinicians, thereby stressing the need for alternative therapies. In order to address this issue, we have investigated a series of crystal growth inhibitors, which are molecules that adsorb on crystals and block solute incorporation.[2-3] Inhibitors have different modes of action to suppress crystal growth, knowledge of which is integral to the design of effective drugs.

Here, we present the results of bulk, kinetic, and interfacial studies of COM growth inhibition. We identify modifiers with binding specificity to COM crystal surfaces,[4-5] leading to either the inhibition or promotion of growth. Modifier interactions with crystal surfaces are instigated using *in situ* atomic force microscopy (AFM) to capture dynamic events with sufficient spatiotemporal resolution. AFM reveals unique modes of actions for different modifiers; and from these studies we assess modifier efficacy as a platform to design therapeutics for pathological crystallization.

1. Wesson, J. A.; Ward, M. D., Pathological Biomineralization of Kidney Stones. *Elements* **2007,** *3* (6), 415-421.

2. Alamani, B. G., Rimer, J. D., Molecular modifiers of kidney stones. *Curr Opin Nephrol Hy* **2017**.

3. Olafson, K. N.; Li, R.; Alamani, B. G.; Rimer, J. D., Engineering Crystal Modifiers: Bridging Classical and Nonclassical Crystallization. *Chemistry of Materials* **2016,** *28* (23), 8453-8465.

4. Chung, J.; Granja, I.; Taylor, M. G.; Mpourmpakis, G.; Asplin, J. R.; Rimer, J. D., Molecular modifiers reveal a mechanism of pathological crystal growth inhibition. *Nature* **2016,** *536* (7617), 446-450.

5. Farmanesh, S.; Alamani, B. G.; Rimer, J. D., Identifying alkali metal inhibitors of crystal growth: a selection criterion based on ion pair hydration energy. *Chemical Communications* **2015,** *51* (73), 13964-13967.

490492 Directed Crystal Nucleation and Growth of Blue Phases By Chemically Patterned Surfaces

Abstract:

Blue phase liquid crystals (BPs) are states of matter that exhibit fluid properties and a configurational structure that produces ordered networks of topological defects. The kinetics of the crystal nucleation and growth processes of such networks are central to the understanding and manipulation of the properties of BPs. In this work, the nucleation and epitaxial growth of BPs are examined with exquisite detail by relying on chemically nano-patterned surfaces to direct their assembly. The use of nano-patterns enables formation of single-crystal BP domains, which grow from BP nuclei that arise when the material is heated up from a cholesteric phase, or when it is cooled down from the isotropic phase. Through a combination of field-theoretic simulations and experiments, it is shown that once a single-crystal BP is formed, the transition into another BP with a different defect network proceeds through a local re-organization of the crystalline structure, which does not require a nucleation stage and exhibits little hysteresis. The transition between single-crystal BPs is shown to be weak first order, with the characteristics of a martensitic-like transformation which is remarkably fast and is the first example of a diffusion-less transformation of this type in soft materials, this opens intriguing possibilities for development of optical devices. From a fundamental perspective, perfectly ordered BPs are shown to offer fertile grounds for the study of directed crystal-nucleation and growth of soft matter.

500532 Transient Reorientation of Liquid Crystals Induced By Water Transport

Abstract:

Liquid crystals (LCs) present an ordered environment at interfaces that enables new functionalities in light focusing devices and biosensors. Numerous applications of LCs rely on instantaneous reorientation of the molecules at an interface which can propagate well into the bulk. In this contribution, we show that the spontaneous flux of water across the interface between a thermotropic liquid crystal and an aqueous phase gives rise to a transient reorientation of the liquid crystal; a new principle that differs from that of explicit liquid crystal-solute interactions. A critical water flux is created by changing the water activity of an aqueous electrolyte solution using specific salts; water is then transported through the liquid crystal phase and across the interface. The orientational response of the liquid crystal phase is then imaged by polarized optical microscopy. The magnitude of the spontaneous water flux is controlled by the concentration and type of solutes, as well as the rate of salt addition. These results provide unprecedented fundamental principles that can be exploited to design biological sensors, drug delivery systems, and molecular machines.

500852 DNA-Micropost Adsorption and Enhanced Transport in More Crowded Micropost Arrays

Abstract:

Crowded environments typically are expected to slow molecular transport. However, recent experiments found enhanced transport of DNA molecules in more crowded nanoslit micropost arrays. The unexpected behavior has been attributed to DNA attraction to the microposts, leading to weak adsorption and cross-post translation. We investigated the physical competition between polymer-obstacle attraction, polymer conformational entropy, and simultaneous polymer interactions with multiple obstacles.

We found that enhanced transport occurs for weak polymer-obstacle attractions under which the polymer conformation can extend to interact with other obstacles. In this regime, polymer diffusivity increases as the inter-post gap size decreases to smaller than the free polymer radius of gyration, then the polymer diffusivity decreases as the inter-post gap further decreases to nearly the polymer width. For very strong polymer adsorption and thin adsorption layer thickness, polymer transport is very weakly affected by changes in the gap size. Sufficiently long polymers are able to span across multiple obstacles for small gap sizes, resulting in reptation-like behavior.

500725 A Miniaturized, Radial Langmuir Trough for Simultaneous Visualization and Dilatational Deformation of a Complex Fluid-Fluid Interface

Abstract:

The viscoelastic properties of monolayers of amphiphilic molecules at a fluid-fluid interface determine, in part, the stability of the foams and emulsions found in cosmetics, pharmaceutical formulations, foodstuffs, and petroleum refining. In addition to foam and emulsion stability, the mechanical properties of fluid-fluid interfaces are relevant to the study of biological membranes, especially lipid bilayers, where viscoelastic properties affect the transport of molecules tangentially through the membrane, and mechanical properties provide biological cells with requisite structural integrity. Unlike most bulk fluids, fluid-fluid interfaces are highly compressible and display a rich variety of viscoelastic behavior that depend on the area-density of the interfacial molecules. As a consequence of compressibility, a complete rheological description requires measurements of both shear and dilatational properties; thus, interfacial rheological devices are typically designed to generate either pure shear deformations, or pure dilatational deformations. However, isolating purely dilatational deformations in a geometry amenable to simultaneous direct visualization of the interfacial structure, e.g. with fluorescence microscopy, has always been an experimental challenge. We present a novel design, and data from, a miniaturized radial Langmuir trough that generates purely dilatational deformations with a 9:1 compression ratio, and operates on the platform of a conventional fluorescence microscope. This equipment, which was produced almost entirely from basic additive manufacturing techniques, is intended to broaden access to simultaneous dilatation-visualization measurements that otherwise require specialized assemblies.

485620 Electrocatalytic CO_2 Reduction for Enhanced C-C Coupling

Abstract:

An appealing option for the conversion of solar energy to fuels is the (photo)-electrochemical reduction of CO2 (CO2RR) using water as the source of hydrogen. It has been known that polycrystalline Cu is prone to produce C1 products, mainly methane in aqueous solution. The product selectivity of such conversion can be affected by the electrolyte composition including anions and cations and catalyst structure. Nanostructuring of polycrystalline copper surfaces via electrochemical cycling in the presence of different halide anions in the electrolyte allows for a suppression of methane and an enhancement of ethylene formation. On the other hands, cations have direct influence on CO2RR in which increasing the size of mono-valent cations can increase the selectivity and activity to ethylene. In addition, controlling an atomic-level catalyst structure accelerates electrocatalytic activity and selectivity toward the formation of C-C bonds.

496446 Reaction Path Discovery Under Potential Bias

Abstract:

Electrocatalysis may form the basis for a fully renewable means to convert the greenhouse gas carbon dioxide into fuels or value-added chemicals. Since the conversion efficiency depends intimately on the mechanism of reactionâ €"which is not fully understoodâ€"ab initio quantum chemical calculations may provide molecular details that allow for the rational design of electrocatalysts. This work extends the method of Goodpaster, et al.[1], which treats surfaces under applied potential and implicit solvent, to the growing string method[2] for reaction path discovery. The growing string method is capable of finding the activation energy and reaction energy from a starting reactant and a set of driving coordinates for bond breaking or forming. Several vital surface reactions are conducted to demonstrate the efficacy of this method for CO2 reduction under potential bias on a Cu(100) surface. Furthermore, the same reaction is investigated on a Ag(100) surface to explain the experimentally observed behavior of silver to switch between carbon monoxide or hydrogen production as a function of potential bias. The reactions path discovery examples on Cu and Ag will demonstrate the power of the growing string method for investigating surface reactions under electrochemical bias, and provide foundational insights into the rational design of electrocatalysts.

[1] Goodpaster, J. D.; Bell, A. T.; Head-Gordon, M. Identification of Possible Pathways for C-C Bond Formation during Electrochemical Reduction of CO2: New Theoretical Insights from an Improved Electrochemical Model. *Phys. Chem. Lett.* **7** (2016) 1471-1477.

[2] Jafari, M.; Zimmerman, P. M. Reliable and Efficient Reaction Path and Transition State Finding for Surface Reactions with the Growing String Method. *J. Comput. Chem.* 38 (2017) 645-658.

500514 Electrocatalytic Carbon Fixation on Molecular-Material Hybrids

Abstract:

Our research focuses on the development of catalysts for CO_2 reduction as a route towards the storage of intermittent renewable energy sources. New catalysts for carbon-neutral energy conversion processes are essential to addressing climate change and rising global energy demands. We are taking a unified approach to this small-molecule activation problem by developing molecularly functionalized materials catalysts for CO_2 reduction that can be used in parallel under environmentally green, aqueous conditions. Molecular materials derived from molecular-nanoparticle composites and molecular-bulk materials hybrids, complement heterogeneous catalyst designs. In particular, we are inspired by recent advances in organometallic chemistry, where careful design and modification of ligand structures has led to novel reactivity of small molecule organometallic complexes for homogeneous catalysis of organic reactions. As such, these molecular-material hybrids can be thought of as heterogeneous analogs to homogeneous organometallic complexes. Our latest work on the development of systems capable of CO_2 and CO reduction to C-C coupled products will be discussed in this presentation.

500879 Investigation of the Morphological Factors That Drive Selectivity on Polycrystalline Cu during CO_2 Electroreduction

Abstract:

Investigation of the morphological factors that drive selectivity on polycrystalline Cu during CO_2 electroreduction

Alexandros N. Karaiskakis, Elizabeth J. Biddinger*

Department of Chemical Engineering, City College of New York, CUNY, New York, NY 10031 (USA)

*ebiddinger@ccny.cuny.edu

Intensive human activity through industrial processes, combustion and agriculture produce more carbon dioxide (CO_2) than the ability of nature to absorb, leading to disruption of the natural carbon cycle and accumulation in the atmosphere. Carbon dioxide is a known greenhouse gas responsible for the contribution to climate change due to global warming. Electroreduction of CO_2 (CO2ELR) is a promising process that combines the utilization of CO_2 residual gas with renewable electricity and water to form a variety of useful carbon neutral chemicals and fuels such as methane (CH_4), ethylene (C_2H_4), ethanol, and formic acid. A key component of the CO2ELR process is the performance of the catalyst in terms of current efficiency, selectivity, and stability[1, 2] which continue to limit the broader use of the process. Cu was found to be the only metal that catalyzes the reduction towards valuable products (hydrocarbons) in comparison with other metals[3].

Recent studies have presented the importance of Cu surface morphology on product selectivity and activity[4-6]. The oxidation state of Cu along with several morphological parameters, such as roughness factor, particle sizeand crystal orientation, have been evaluated and related as dominant factors that influence the selectivity and the activity of CO2ELR on Cu-based catalysts. Oxidized Cu (CuO and Cu_2O) and higher roughness factor have been related with C_2H_4 enhancement in comparison with smooth Cu, which produces mainly methane[5, 7] Particle size has been related with product selectivity, and in particular 2nm diameter nanoparticles were observed to enhance H_2 formation, rather than CO2ELR seen on smooth Cu under similar CO2ELR conditions[8]. The crystal orientation of Cu, on single crystal work, has been related with the selectivity between CH_4 and C_2H_4[3]. We recently reported that polycrystalline Cu under CO_2 reduction conditions undergoes reconstruction that influences the morphology and the crystal orientation[9]. This work examines the oxidation state and morphological aspects on rough polycrystalline Cu (roughness factor, particle size, and crystal orientation) provides insights regarding which factors seem to control CO2ELR selectivity. A current distribution analysis on the surface due to the morphology of each catalyst was also performed. Electrodeposition of copper was used as the synthesis technique for the catalysts studied for its ability to control catalyst morphology. The evaluation of the synthesized catalysts involved the examination of their morphology, current efficiency, and product selectivity. The morphological aspects on the surface of each catalyst were evaluated with scanning electron microscopy (SEM), X-ray diffraction (XRD), X-ray photoelectron spectroscopy (XPS), atomic force microscopy (AFM) and capacitance measurements with cyclic voltammetry. Micro GC was used for gaseous product analysis. The results obtained present a direct relation between roughness, particle size and current distrubution on the surface with the product selectivity. In particular, lower roughness catalysts, with a 1.6 roughness factor and 3μm particle size form methane as the main hydrocarbon product, whereas higher surface roughness catalysts, with a 7.8 roughness factor and 300nm particle size were associated with the higher formation of ethylene. The current distribution analysis illustrates lower current intensity on lower roughness surfaces in comparison with higher roughness surfaces under the same potential.

References

1. Qiao, J.; Liu, Y.; Hong, F.; Zhang, J., *Chemical Society Reviews* **2014,** *43* (2), 631-675.

2. Hori, Y., Electrochemical CO_2 reduction on metal electrodes. In *Modern Aspects of Electrochemistry*, Vayenas, C. G.; White, R. E.; Gamboa-Aldeco, M. E., Eds. Springer: New York, 2008; Vol. 42, pp 89-189.

3. Hori, Y.; Murata, A.; Takahashi, R., *Journal of the Chemical Society, Faraday Transactions 1: Physical Chemistry in Condensed Phases* **1989,** *85* (8), 2309-2326.

4. Tang, W.; Peterson, A. A.; Varela, A. S.; Jovanov, Z. P.; Bech, L.; Durand, W. J.; Dahl, S.; Norskov, J. K.; Chorkendorff, I., *Physical Chemistry Chemical Physics* **2012,** *14* (1), 76-81.

5. Kas, R.; Kortlever, R.; Milbrat, A.; Koper, M. T. M.; Mul, G.; Baltrusaitis, J., *Physical Chemistry Chemical Physics* **2014,** *16* (24), 12194-12201.

6. Chi, D.; Yang, H.; Du, Y.; Lv, T.; Sui, G.; Wang, H.; Lu, J., *RSC Advances* **2014,** *4* (70), 37329-37332.

7. Mistry, H.; Varela, A. S.; Bonifacio, C. S.; Zegkinoglou, I.; Sinev, I.; Choi, Y.-W.; Kisslinger, K.; Stach, E. A.; Yang, J. C.; Strasser, P.; Cuenya, B. R., *Nature Communications* **2016,** *7*, 12123.

8. Reske, R.; Mistry, H.; Behafarid, F.; Roldan Cuenya, B.; Strasser, P., *Journal of the American Chemical Society* **2014,** *136* (19), 6978-6986.

9. Karaiskakis, A. N.; Biddinger, E. J., *Energy Technology* **2016,** *DOI: 10.1002/ente.201600583.*

496963 Modulation of Charge Transfer Kinetics By Two Orders of Magnitude at Back-Gated Monolayer MoS_2 Electrodes

Abstract:

Charge transfer at the electrode/electrolyte interface is an elementary step in electrochemical reactions and is essential to many electrochemical applications. For a specific reaction, the charge transfer kinetics are largely dependent on the electronic structure at the electrode surface. We demonstrate a continuous and reversible physical method to modulate the charge transfer kinetics at nanometer thick two-dimensional semiconductor electrodes in which a transverse electric field (produced by a gate bias) applied from the back side of the electrode shifts the electrodeâ€™s band alignment on the front side. This gate tunable electrode structure is analogous to a metalâ€“oxideâ€“semiconductor field-effect transistor (MOSFET). We find that at back-gated monolayer MoS_2 electrodes, the charge transfer kinetics can be modulated from irreversible to nearly reversible by the back-gate bias. In particular, we were able to tune the charge transfer rate constant for MoS_2 with the ferrocene/ferrocenium redox couple by two orders of magnitude, from 4×10^{-6} to 9.7×10^{-4} cm/s, with varying back-gate biases. As a general approach, the field effect, with its potential to alter the electrochemical properties of two-dimensional semiconductors over a large range, opens up new possibilities for fundamental studies of gate-tunable electron occupation and Fermi level alignment on electrochemical reactions.

487069 Predicting Vapor-Liquid Equilibria with Augmented *Ab Initio* Potentials

Abstract:

Understanding the vapor-liquid equilibria of fluids is important for many key chemical engineering processes [1]. The topic is also challenging from a scientific perspective because it involves simultaneously predicting two diverse phases of matter. Historically, the prediction of vapor-liquid equilibrium has been addressed by proposing a suitable equation of state [2]. However, as the governing mechanism for phase separation is directly related to intermolecular interactions, the use of molecular simulations [3] that involve a suitable intermolecular potential can be particularly insightful. Simple intermolecular potentials such as the Lennard-Jones model have been used widely, but considerable progress has been [4,5] achieved in developing accurate potentials from first principles. Although these *ab initio* potentials are currently restricted to simple atomic systems [4] or small polyatomic molecules such as water [6], greater use of such models is likely in the future. In this work, we examine the accuracy of some recent models to predict vapor-liquid equilibria. An advantage of this approach is that it can be systematically refined to improve the quality of predictions. We illustrate this approach by combining two-body ab intio potentials with a computationally simple model of three-body interactions. Comparison with experimental data indicates that the quality of predictions is improved considerably.

1. R.J. Sadus, *High Pressure Phase Behaviour of Multicomponent Fluid Mixtures*, Elsevier, Amsterdam, 1992.
2. Y.S. Wei and R. J. Sadus, AIChE J. **46**, 169 (2000).
3. R.J. Sadus, *Molecular Simulation of Fluids: Theory, Algorithms and Object-Orientation, Elsevier*, Amsterdam, 1999.
4. B. Jaeger, R. Hellmann and E. Vogel, J. Chem. Phys. **144**, 114304 (2016).
5. M. Vlasiuk, F. Frascoli and R.J. Sadus, J. Chem. Phys. **145**, 104501 (2016).
6. R. Bukowski, K. Szalewicz, G. C. Groenenbnoom and A. van der Avoird, Science **315**, 1249 (2007).

484589 New Vapor Pressure Prediction with Improved Thermodynamic Consistency Using the Riedel Equation

Abstract:

New Vapor Pressure Prediction with Improved Thermodyamic Consistency using the Riedel Equation

Joseph W. Hogge[1], Neil F. Giles, Richard L. Rowley, Thomas A. Knotts, W. Vincent Wilding

[1] *Department of Chemical Engineering, Brigham Young University, USA*

**Joseph Hogge: joehogge@byu.edu*

Scientific topic: Thermodynamics and Transport Properties

Thermophysical Properties and Phase Behavior

Keywords: Multi-Property Optimization, Vapor Pressure, Heat Capacity, Heat of Vaporization

Abstract

Vapor pressure, enthalpy of vaporization, liquid heat capacity, and ideal gas heat capacity for pure compounds between the triple point and critical point are important properties for process design and optimization. Unfortunately, the literature is filled with poor and inconsistent data or is void of experimental data for leading-edge compounds common in todayâ€™s chemical industries. As such, there is high demand to develop methods to assess the accuracy and consistency of experimental data and correctly predict properties when such data are not available.

This work focuses on improving the thermodynamic consistency and accuracy of vapor pressure, enthalpy of vaporization, liquid heat capacity, and ideal gas heat capacity since these can drastically affect the safety and profitability of chemical processes. This is done by improving the temperature correlation of the vapor pressure because it is related to the other properties through rigorous thermodynamic relationships involving temperature derivatives. For many years, the Riedel equation has been considered an excellent and simple choice among vapor-pressure correlating equations [1], but this work shows that it requires modification of the final coefficient to provide thermodynamic consistency with thermal data [2]. Specifically, Riedelâ€™s equation set the final coefficient to a constant value of 6, but this severely restricts the ability of the model to produce accurate liquid heat capacity data below the normal boiling point.

This presentation will show 1) newly created predictive correlations where this final coefficient is changed from 1 to 6 in integer steps, and 2) how these result in improved prediction of vapor pressures for compounds with limited or no experimental data. The methodology followed is that originally proposed by Riedel [3] but the last coefficient was chosen based on its ability to simultaneously represent vapor pressure and liquid heat capacity. This procedure improves the fit to liquid heat capacity data by 5-10% (average absolute deviation), while maintaining the fit of vapor pressure data similar to other prediction methods. This procedure also greatly improves the ability to predict low-temperature vapor pressure by using liquid heat capacity data. This feature of the work suggests that previous understanding of the temperature dependence of low-temperature vapor pressure was incorrect and that new paradigms need to be developed.

References

1. Velasco, S., F.L. Roman, J.A. White, and A. Mulero, *A predictive vapor-pressure equation.* Journal of Chemical Thermodynamics, 2008. **40**(5): p. 789-797.

2. Hogge, J.W., N.F. Giles, T.A. Knotts, R.L. Rowley, and W.V. Wilding, *The Riedel vapor pressure correlation and multi-property optimization.* Fluid Phase Equilibria, 2016. **429**: p. 149-165.

3. Riedel, L., *Eine neue universelle Dampfdruckformel Untersuchungen über eine Erweiterung des Theorems der übereinstimmenden Zustände. Teil I.* Chemie Ingenieur Technik, 1954. **26**(2): p. 83-89.

490262 Solvent Screening and Pure Component Thermophysical Property Prediction for Pharmaceutical Process Design

Abstract:

The selection of optimal solvent or a mixture of solvents is essentially useful to achieve pharmaceutical production with desirable product and production qualities as well as a minimum cost of production and a maximum space-time yield. Despite the availability of high-throughput technologies, experimental solvent screening is most often expensive and laborious and sometimes simply not possible because of lack of sufficient amount of API or relevant impurities. This is especially the case when many combinations of design needs to be tested for identifying an optimal mixture of binary or ternary solvents made up among several common solvents in pharmaceutical industry. To this end, application of predictive thermodynamic models for estimation of phase behaviors in pure and mixed solvent systems have been shown to be a supporting and complementary tool to design pharmaceutical processes. In this work, different predictive thermodynamic models such as UNIFAC, COSMO-SAC, COSMO-RS and NRTL-SAC were applied for solubility prediction of a steroid-like structure antibiotic. The predicted solubility data were compared with experimental solubility data in n-butanol, isobutanol, methyl tert-butyl ether, methyl isobutyl ketone, ethyl acetate, hexane, methanol, acetone and toluene by calculating the root mean square error (RMSE) in order to evaluate the models capability. Therefore, a model which gives more accurate solubility prediction can be selected. For this specific case study, NRTL-SAC is capable of predicting solubility in a good agreement with experimental solubility with RMSE of 4.5%. Moreover, the application of such predictive thermodynamic models for the design of separation/purification unit operations is sometimes limited by the lack of pure component thermophysical properties like melting point and enthalpy of fusion. This mostly happens in the design of separation/purification unit operations to separate analytically identified but not commercially available impurities. Therefore, in addition to the application of those predictive thermodynamic models for solubility prediction, a group-contribution$^+$ (GC^+) based model is applied for pure component thermosphysical property prediction. However, the GC^+ based model has a limited capability for a large and complex organic molecules that pure component thermosphysical property prediction is prone to uncertainty. In such circumstance, a reference solvent approach can be applied to predict solubility[1]. The reference solvent approach requires no pure component thermosphysical properties but an experimental solubility data in a single solvent. The experimental solubility data of a noncommercially available impurity can be measured by concentrating a sample containing the impurity. This work concludes by critical examination of the performance and the applicability of the above mentioned four property models under different pure component data constraints especially for solubility calculations in life science product (API) separation applications.

Acknowledgment: We would like to thank the Danish Council for Independent Research (DFF) for financing the project with grant ID: DFF-6111600077B.

---

[1] Jens Abildskov, John P. O'Connell, *Ind. Eng. Chem. Res.*(2003) 5622 -5634

A Two-Structure Equation of State for Super Cooled and Stretched Light and Heavy Water

Michal Duška[1,2]
Jan Hrubý[2], Frédéric Caupin[3] and Mikhail A. Anisimov[1]

[1]Department of Chemical & Biomolecular Engineering and Institute for Physical Science & Technology, University of Maryland, College Park, U.S.A.
[2]Institute of Thermomechanics of the Czech Academy of Sciences, 182 00Prague 8, Czech Republic
[3]Université de Lyon, Université Claude Bernard Lyon 1, CNRS, Institut Lumière Matière, F-69622, Lyon, France

AIChE Meeting 2017

1

Highlights

- Thermodynamic hypotheses about the origin of water anomalies
- New Two-Structure EOS to explaint connection between the hypotheses
- Difference between heavy and ordinary water

2

Scenarios behind water anomalies

Speedy's stability limit conjecture

Poole et al.'s second critical point

Critical point scenarios

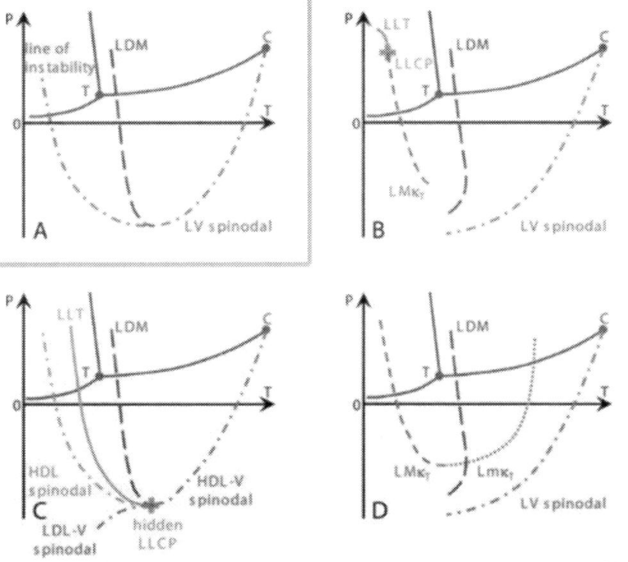

Poole et al.'s "weak bond"-modified van der Waals model, now the critical-point-free scenario

Sastry et al.'s singularity-free scenario

Published in: Paola Gallo; Katrin Amann-Winkel; Charles Austen Angell; Mikhail Alexeevich Anisimov; Frédéric Caupin; Charusita Chakravarty; Erik Lascaris; Thomas Loerting; Athanassios Zois Panagiotopoulos; John Russo; Jonas Alexander Sellberg; Harry Eugene Stanley; Hajime Tanaka; Carlos Vega; Limei Xu; Lars Gunnar Moody Pettersson; *Chem. Rev.* **2016,** 116, 7463-7500.
DOI: 10.1021/acs.chemrev.5b00750

Two state approach to criticality - chemical reaction

$$\frac{G(\hat{T},\hat{p})}{k_B\hat{T}} = \frac{G^A}{k_B\hat{T}} + x\frac{(G^B - G^A)}{k_B\hat{T}} + x\ln(x) + (1-x)\ln(1-x) + \frac{\omega}{k_B\hat{T}}x(1-x)$$

$$\omega = \omega_0\left(1 + \omega_1\hat{p} + \omega_2\hat{T}\right)$$

$$\frac{G^B - G^A}{k_B\hat{T}} = -\ln K = \lambda\left(\frac{1}{\hat{T}} + a_1\frac{\hat{p}}{\hat{T}} + a_2 + a_3\hat{T} + a_4\frac{\hat{p}^2}{\hat{T}} + a_5\frac{\hat{p}^3}{\hat{T}} + a_6\hat{p}\hat{T}\right)$$

Water with broken hydrogen bonds

Two state approach to criticality - chemical reaction

Vincent Holten and John Biddle proved it as an effective theory to explain anomalies in experimental and modeled water

Can criticality explain the re-entering feature of the vapor-liquid spinodal predicted by Speedy?

What if water has a normal vapor-liquid spinodal and the re-entering feature was predicted because of the curved density surface as a consequence of the existence of the critical point. Can we unite all the scenarios?

5

State A as Speedy EOS "spinodal equation"

Speedy's EOS was developed in 1982 to find a vapor liquid spinodal of water

$$1-\frac{p}{p_S}=B\left(\frac{V_S}{V}-1\right)^2 \longrightarrow \left(\frac{\partial G_A}{\partial \hat{p}}\right)_{\hat{T}}=\frac{\hat{V}_S(\hat{T})\sqrt{B(\hat{T})}}{\sqrt{1-\frac{\hat{P}}{\hat{p}_S(\hat{T})}}+\sqrt{B(\hat{T})}}$$

- Taylor expansion of the pressure as a function of density about the limit of stability along each isotherm, and neglecting all but the first term
- fits the very precise density data well, with physical parameters p_S as the pressure at spinodal, V_S the volume at spinodal and B the second derivative of pressure with respect to density at spinodal
- It was integrated to get G_A Will be implemented later

6

State A of water with a “normal” spinodal

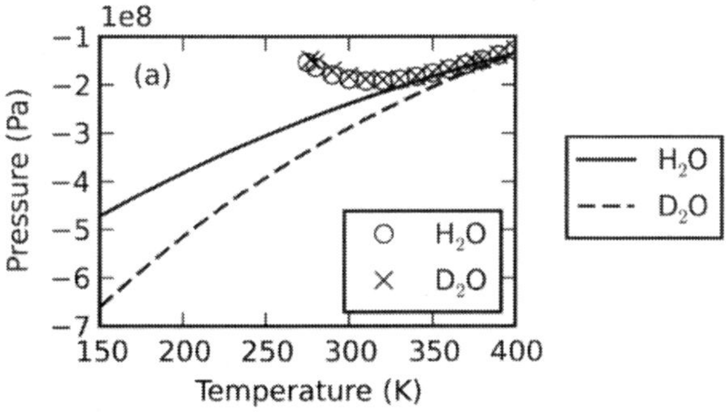

Spinodal pressure calculated directly from Speede’s equation (points), and spinodal pressure of state A (curves) represented in model by polynomial function of temperature

7

Density

at positive pressure was used for model development only

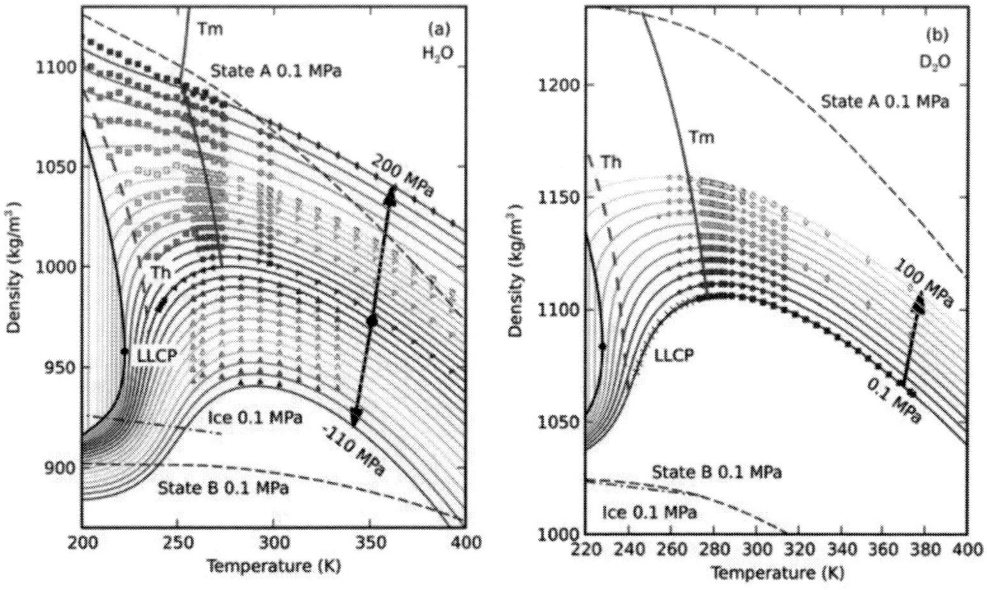

8

Isothermal compressibility

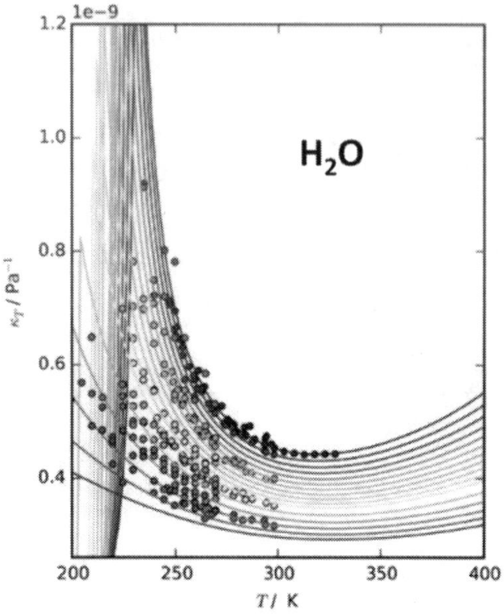

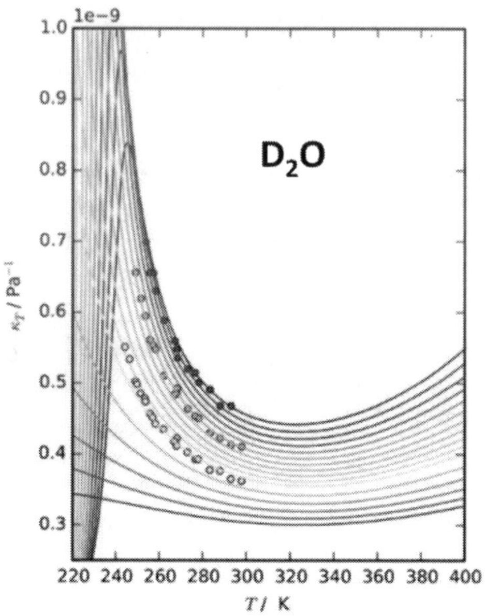

9

Heat capacity

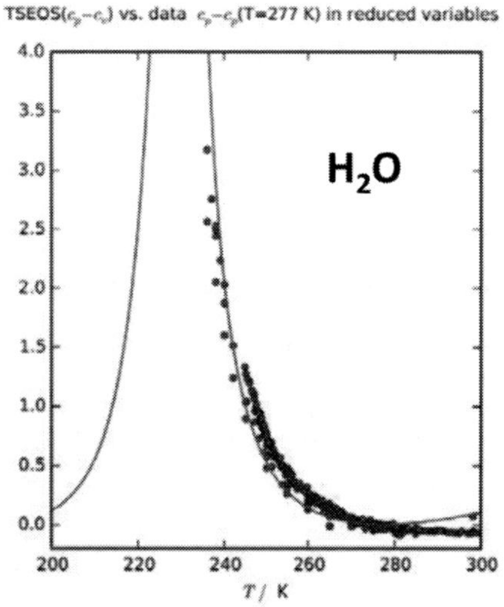

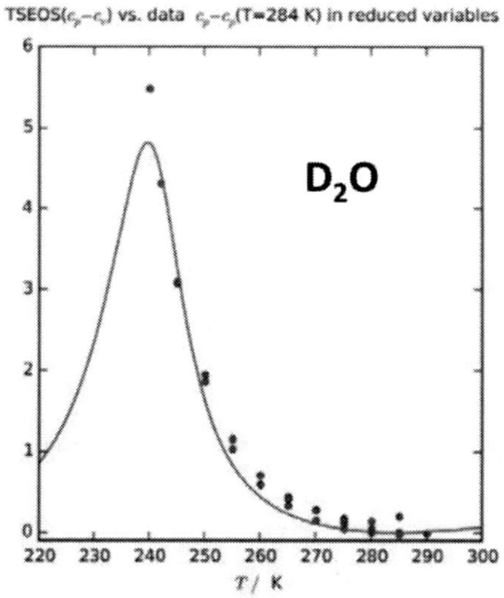

Divergence based on density data is not far from divergence of heat capacity. The position of the second critical point after taking into account both measurements is unlikely to be very different from what we have.

10

Phase diagram

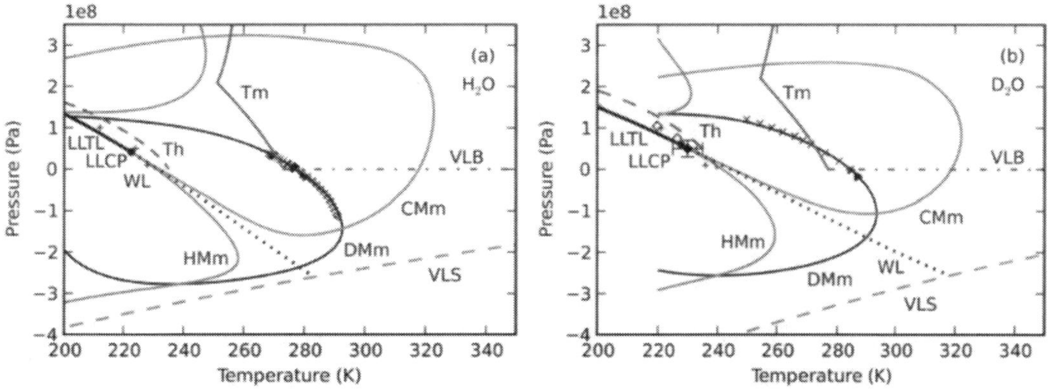

Widom line of heavy water is more pronounced at higher temperatures with respect to the homo-geneous nucleation limit, crossing it at positive pressure (22 MPa), and also position of maximum of density with respect to the melting line. We see this as a consequence of stronger tetrahedral structures in heavy water.

11

Conclusions

Critical phenomena within our reach:

alleged presence of the Widom line of heavy water in an accessible region should be experimentally investigated

Re-entering spinodal as another anomaly explained by existence of a critical point proved by new Two Structure EOS with state A as Speedy's EOS – there are no two hypotheses but one critical point scenario

12

Thank you for your attention

The study has been supported by the Young Scientist IAPWS Project, the Czech Science Foundation grant GJ15-07129Y, and the institutional support RVO:61388998.

13

Expansivity

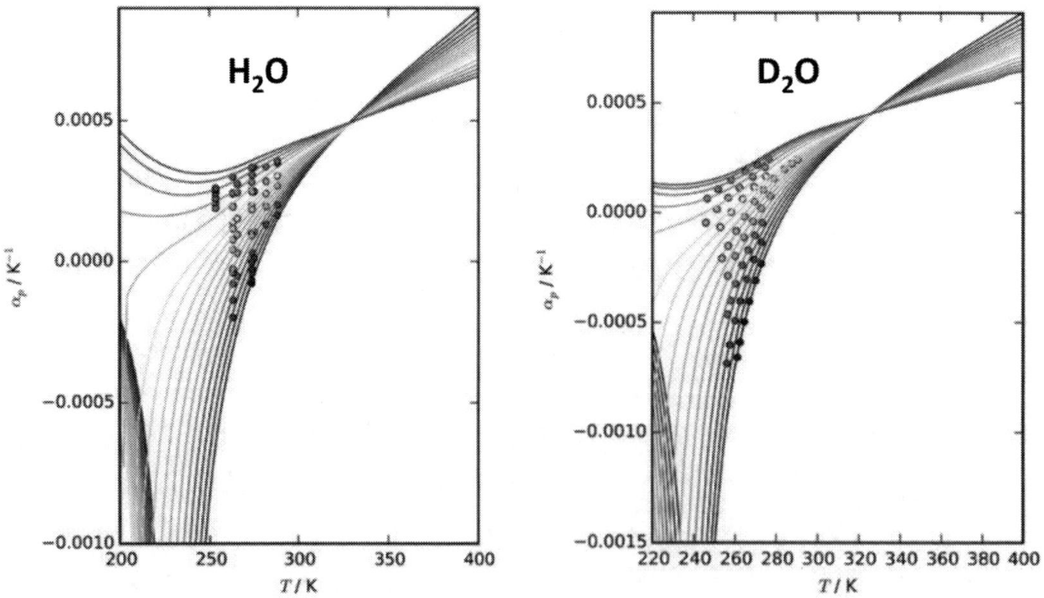

14

State A of water with a "normal" spinodal

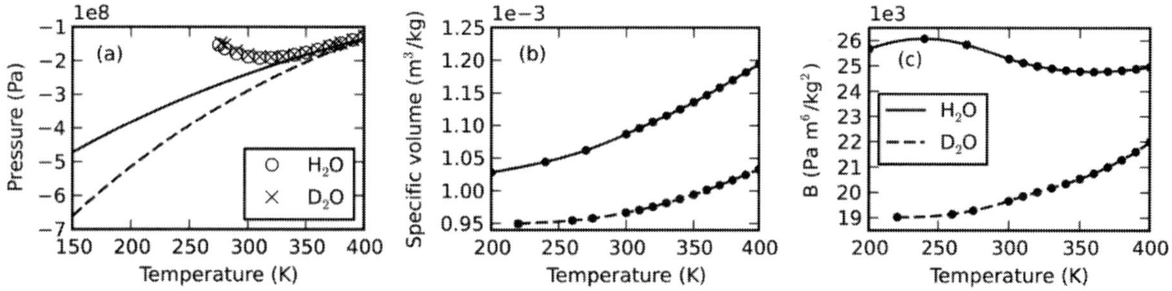

(a) **Spinodal pressure** calculated directly from Speede's equation (points), and spinodal pressure of state A (curves) represented in model by polynomial function of temperature;

(b) **volume at spinodal**;

(c) and **second derivative of pressure** with respect to density at constant temperature at spinodal both represented by cubic spline of the points .

15

497978 Can Molecular Simulations Predict the Binary Interaction Parameters of the Activity Coefficient Models?

Abstract:

Ability for predicting the thermodynamic and phase equilibrium properties of liquid mixtures is vital for the design of industrial chemical processes. In practice, molecular thermodynamic models such as non-random two liquid (NRTL) and UNIFAC models are widely applied for predicting these properties. The binary interaction parameters that are the input to these thermodynamic models, quantify the nature of interactions between different components of the mixture. These binary interaction parameters are usually calculated by regressing experimental data. In this work, we determine these interaction parameters by combining molecular simulations with theory, as such an approach is predictive and does not depend on the availability of experimental data. The thermodynamic model of interest in this work is the non-random two liquid (NRTL) activity coefficient model. Binary interaction parameters are calculated for the water-methanol, methanol-methyl acrylate, and water-methyl acrylate mixtures. For each of the binary mixtures, the effective interaction strength between the components is obtained from molecular dynamics simulations, which are in turn used in the calculation of the binary interaction parameters. We show that the binary interaction parameters calculated from molecular simulations agree reasonably well with those estimated from regressing the experimental data.

490610 A Wertheim Activity Coefficient Model for Associating Mixtures

Abstract:

Accurate, robust thermodynamic models are essential for the design of separation equipment. Models must represent highly non-ideal solutions often encountered in industrial applications. With global interest in biofuels on the rise, these types of solutions are now more ubiquitous than ever and hence there is a demand for more accurate models to handle solutions containing polar components. Traditional models, such as NRTL[1], do not capture hydrogen bonding (or association) explicitly but instead use large dispersion parameters to empirically mimic its effects. Methods that do explicitly account for the association often do so through chemical theory or Wertheim's theory[2–5]. In our previous work[6], we relate the constants in these two methods and develop a new generalized expression for self-associating systems.

Wertheim's theory has been incorporated into a number of models as a means of calculating non-ideality due to association. Equations of state (EOS) have traditionally been the approach of choice and models such as SAFT[3], CPA[8] and ESD[9] have been used to represent a range of industrially important systems. However, the accuracy of an EOS hinges on its ability to simultaneously fit both vapor pressure and excess Gibbs energy data. In contrast, activity coefficient model parameters are fitted only to excess Gibbs and are therefore more favored by industry.

In the present work, an activity coefficient model combining NRTL and Wertheim's theory is used to model two classes of systems. Class 1 systems, such as alcohols in alkanes, are those in which one component self-associates and the others are "inert" or inactive in hydrogen bonding. In contrast, class 2 systems involve components that only cross-associate and do not have any hydrogen bonding in their pure states. The proposed model is used to fit binary and ternary liquid-liquid equilibria data and the parameters are evaluated using infrared spectroscopy data.

[1] H. Renon, J.M. Prausnitz, On the thermodynamics of alcohol-hydrocarbon solutions, Chem. Eng. Sci. 22 (1967) 299–307. doi:10.1016/0009-2509(67)80116-7.

[2] M.S. Wertheim, Fluids with highly directional attractive forces.1. Statistical thermodynamics, J. Stat. Phys. 35 (1984) 19–34. doi:10.1007/bf01017362.

[3] M.S. Wertheim, Fluids with highly directional attractive forces. 2. Thermodynamic perturbation-theory and integral-equations, J. Stat. Phys. 35 (1984) 35–47. doi:10.1007/bf01017363.

[4] M.S. Wertheim, Fluids with highly directional attractive forces. 3. Multiple attraction sites, J. Stat. Phys. 42 (1986) 459–476. doi:10.1007/bf01127721.

[5] M.S. Wertheim, Fluids with highly directional attractive forces. 4. Equilibrium polymerization, J. Stat. Phys. 42 (1986) 477–492. doi:10.1007/bf01127722.

[6] A.M. Bala, C.T. Lira, Relation of Wertheim Association Constants to Concentration-based Equilibrium Constants for Mixtures with Chain-forming Components, Fluid Ph Equilib. 430 (2016) 47–56. doi:10.1016/j.fluid.2016.09.015.

[7] W.G. Chapman, K.E. Gubbins, G. Jackson, M. Radosz, New reference equation of state for associating liquids, Ind. Eng. Chem. Res. 29 (1990) 1709–1721. doi:10.1021/ie00104a021.

[8] G.M. Kontogeorgis, E.C. Voutsas, I.V. Yakoumis, D.P. Tassios, An Equation of State for Associating Fluids, Ind. Eng. Chem. Res. 35 (1996) 4310–4318. doi:10.1021/ie9600203.

[9] J.R. Elliott, S.J. Suresh, M.D. Donohue, A simple equation of state for nonspherical and associating molecules, Ind. Eng. Chem. Res. 29 (1990) 1476–1485. doi:10.1021/ie00103a057.

494545 Understanding the Role of Intermolecular Forces in Molecular-Based Equations of State: 20 Years of the Soft-SAFT Equation

Abstract:

More than twenty five years have passed since the first paper published on the development of what it is known today as the SAFT equation of state or the SAFT approach[1-2] (from the Statistical Associating Fluid Theory). SAFT was developed, from the very beginning, to describe the thermodynamic properties and phase behavior of associating systems, naturally extended to oligomers, polymers and other systems with highly directional and short range attractive forces, for which the cubic equations of state fail to provide an adequate description. The original equation was developed using a hard fluid as a reference term, while dispersion was added as a perturbation. However, it was very soon discovered that more refined reference terms were needed for accurate predictions of complex systems, leading to the development of soft-SAFT-LJ[3], soft-SAFT[4-6] and SAFT-VR[7] in the mid-late 90's, in which the reference term take into account repulsion and dispersion forces. In 2001 PC-SAFT[8] provided a different approach in which the dispersion term was added as a perturbation to a chain reference term. SAFT-gamma-Mie[9] was recently developed aiming at having a soft, flexible, reference term, generalizing the SAF-LJ and soft-SAFT EoS. During these years different version of SAFT have been implemented by different authors and the approach has been widely used to model highly non ideal systems. There are several excellent reviews on this topic.[10-12]

This talk will first provide an overview on how having the right physics with the adequate level of approximations has made soft-SAFT one of the most accurate, yet simple, versions of SAFT, twenty years after the first paper on soft-SAFT was published. In addition, the built in theory in the equation allows its extension to calculate different properties of the systems, delivering a global equation for the characterization of thermophysical properties. Today, soft-SAFT is a robust tool for calculating critical, phase, interfacial, transport and thermodynamic derivatives properties of associating fluids and polymers. Examples on the relationship between the intermolecular forces description in the equation and the performance of the equation will be provided, including systems such as ionic liquids, aqueous solutions and deep eutectic solvents, highlighting the robutness of the equation. We will also tackle some of remaining challenges and the opportunities on moving SAFT from academia to industry and backwards.

However, there are still remaining challenges and opportunities for developments before soft-SAFT (as any SAFT equation) can be truly considered a standard tool in industrial process development and design and these will also be addressed in this presentation.

1. W.G. Chapman, K.E. Gubbins, G. Jackson, M. Radosz, Fluid Phase Equilib. 52 (1989) 31–38.
2. W.G. Chapman, K.E. Gubbins, G. Jackson, M. Radosz, Ind. Eng. Chem. Res. 29 (1990) 1709–1721.
3. J.K. Johnson, E.A. Müller, KE Gubbins, Journal of Physical Chemistry 98 (1994) 6413-6419.
4. F.J. Blas, L.F. Vega, Mol. Phys. 92 (1997) 135–150.
5. F.J. Blas, L.F. Vega, Ind. Eng. Chem. Res. 37 (1998) 660–674.
6. J.C. Pàmies, L.F. Vega, Ind. Eng. Chem. Res., 40 (2001) 2532-2543.
7. A. Gil-Villegas, A. Galindo, P.J. Whitehead, S.J. Mills, G. Jackson, A.N. Burgess, J. Chem. Phys. 106 (1997) 4168-4186.
8. J. Gross, G. Sadowski, Ind. Eng. Chem. Res., 40 (2001) 1244–1260.
9. T. Lafitte, A. Apostolakou, C. Avendaño, A. Galindo, C. S. Adjiman, E. A. Müller, G. Jackson, J. Chem. Phys. 139 (2013) 154504.
10. S.P. Tan, H. Adidharma, M. Radosz, Ind. Eng. Chem. Res. 47 (2008) 8063–8082.
11. C. McCabe, A. Galindo, SAFT associating fluids and fluid mixtures, in: A.R.H. Goodwin, J.V. Sengers, C.J. Peters (Eds.), Applied Thermodynamics of Fluids, Royal Society of Chemistry, UK, 2010, pp. 215–279.
12. L.F. Vega, F. Llovell, Fluid Phase Equilib. 416 (2016) 150-173.

487812 Coarse-Grained Modeling of Polymer Materials in Molecular Simulations with SAFT-γ Mie Corresponding States Correlation Specified Potentials

Abstract:

Polyester polyol polymer materials are utilized in various forms such as in coatings, adhesives, sealants, polyurethane flexible and rigid foams, and in elastomers. These polymers can be synthesized with many different monomer molecules that are either solely hydroxy or carboxylic acid difunctional or multifunctional to give multiple ester bonds and hydroxy ends in the polyester polyol. Incorporation of this polymer for applications commonly involves crosslinking the polyols to make a network structure. To simulate this material, we have made top-down coarse-grained models of the monomers involved using the corresponding states correlation to the SAFT-γ Mie EoS.[1–3] Using this theory, we obtained Mie potential parameters from monomer critical temperatures, acentric factors, and densities at Tr=0.7[4] to use in molecular simulations. Joining the monomer models together appropriately constructs the polymer model. The simulations are designed to model specific experimental polyester polyol resins, matching experimental composition and characterization of molecular weight distribution. Key properties of the polyester polyol resin agree well between simulation and experiment including density, glass transition temperature (Tg) and change in heat capacity across the glass transition. Additional simulations designed to explore the Mie model parameter space led to a fit equation for Tg that predicted the Tg of experimental resin simulations. Our results with polyester polyols suggest that the properties of polymer resins can be estimated with reasonable accuracy from a relatively small amount of experimental data. Crosslinker and solvent molecules are also being fit with Mie models. Further simulations involve heuristic crosslinking of the polyester polyols and characterization of structural and mechanical properties.

References

1. Mejia, A., Herdes, C. & Mueller, E. A. Force Fields for Coarse-Grained Molecular Simulations from a Corresponding States Correlation. *Ind. Eng. Chem. Res.* **53,** 4131–4141 (2014).

2. Lafitte, T. *et al.* Accurate statistical associating fluid theory for chain molecules formed from Mie segments. *J. Chem. Phys.* **139,** (2013).

3. Lymperiadis, A., Adjiman, C. S., Galindo, A. & Jackson, G. A group contribution method for associating chain molecules based on the statistical associating fluid theory (SAFT-gamma). *J. Chem. Phys.* **127,** 234903 (2007).

4. Design Institute for Physical Properties, S. by Aic. *DIPPR Project 801 - Full Version. Design Institute for Physical Property Research/AIChE* (Design Institute for Physical Property Research/AIChE, 2005).

495611 New SAFT Eos for Electrolyte Solution

Abstract:

In this study, the electrolyte Statistical Associating Fluid Theory (SAFT) equation of state (EOS) has been considered to represent solid liquid equilibrium of aqueous electrolyte solutions. In this regard, the SAFT-VR Morse and ePC-SAFT models were developed and modified respectively for this purpose. In the modified electrolyte PC-SAFT model, the modified version of static permittivity was utilized to consider the electrostatic effect of charged particles in solvent dielectric constant. In the case of electrolyte SAFT-VR Morse, the robust variable range potential was used to consider the short-range interaction between ion-ion and ion-solvent. For aforementioned models, the ion-specific parameters were obtained from the fitting of mean ionic activity coefficient and solution density (or apparent molar volume) data of several 1:1, 1:2 and 2:1 types binary electrolyte, simultaneously. Using single electrolyte parameters solid liquid equilibrium were predicted in some simple mixed electrolyte solution, however to have an accurate estimation of thermodynamic properties of mixed electrolyte solutions an additional parameter was needed to improve the cross dispersion energy between unlike particles. Therefore, the binary interaction parameters (BIPs) have been introduced to improve the model prediction results. The results showed that, for the case of SAFT-VR Morse model smaller absolute value of BICs compared to ePC-SAFT is required to predict the solid-liquid equilibrium (SLE) of mixed electrolyte solutions. It is indicated that, the suitable short-range interaction between charged and uncharged particles can improves the model capability especially in mixed electrolyte solutions. In addition, the Gibbs free energy of hydration and hydration enthalpy were predicted at 298.15 K and 1bar. The results were in good agreement with experimental data. The results showed that in the case of Gibbs free energy and enthalpy of hydration, two mentioned models give comparable values but in SLE of mixed electrolyte solutions, the SAFT-VRE Morse EOS gives accurate results using smaller correction parameters (BIPs).

501606 Controlling Vesicle Shape By Adsorption of a Semiflexible Polymer

Abstract:

Interactions between membranes and polymers are ubiquitous in biology, with examples including membrane-cytoskeleton interactions, the membrane-facilitated assembly of some virions, and drug delivery by polymeric carriers. Despite the importance to biology, many aspects of polymer-membrane interactions remain to be addressed, in part because the flexibility of membranes can lead to nontrivial coupling between configurations of the polymer and of the membrane. In this work, we study the adsorption of a semiflexible polymer onto a fluid membrane vesicle using Monte Carlo computer simulations. We find that polymer adsorption can induce vesicle shape changes, with configurations that depend on the polymer-membrane adsorption strength, the polymer persistence length, and the membrane bending rigidity. Stiff membranes maintain approximately spherical shapes, with the adsorbed polymer adopting configurations on the surface that are controlled by its persistence length. At smaller values of the bending rigidity, sufficiently strong adsorption of the polymer can cause invagination of the vesicle, resulting in inward budding of the membrane around a collapsed polymer or a cylindrical deformation that envelops a rod-like polymer. For small vesicles, the polymer can induce a dumbbell-like vesicle shape with a narrow membrane neck that is suggestive of protein-mediated scission mechanisms. By characterizing the shapes and statistical properties of the membrane-polymer system, we construct a pseudophase diagram that provides insight into the underlying physical interactions governing polymer adsorption onto the vesicle and the influence on vesicle shape. Analytical calculations of the energetics of idealized configurations provide additional insight into the interplay of adsorption strength, polymer persistence length, and membrane stiffness.

489321 Polymer-Induced Restructuring in the Gut

Abstract:

The gut governs digestion and nutrient absorption, is a promising target for drug delivery, and teems with micro-organisms that can have remarkably strong effects on host health. Despite its importance, however, little is known about how the structure and function of the gut are influenced by many of the materials that transit through it regularly. By combining experiments on mice, biophysical processing and characterization, and microscopy, we show how polymers abundant in the gut â€" such as secreted mucins, dietary fibers, or administered therapeutics â€" can alter the structure of the gut and its contents. We compare our results to the predictions of a simple thermodynamic model, showing how aspects of this complex behavior can be understood in the framework of polymer solution theory. Our results provide a foundation for disentangling the interactions between polymers and the gut; more broadly, our work highlights how physical principles can be used to understand and possibly control how polymers regulate structures in the body.

Multicomplex Formation in the Presence of Polymer Crowders Mimicking in-Cell Environment

Vahid Rahmanian, Young C. Kim, Jeetain Mittal

Department of Chemical and Biomolecular Engineering
Lehigh University, Bethlehem, PA

Crowded environment in the cell

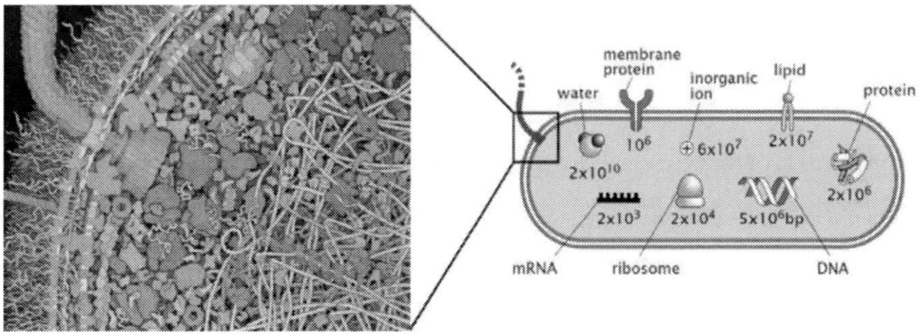

Macromolecular Crowding Effects:

Change in the properties of biomolecules due to this dense environment.

From Goodsell, D.S., The Machinery of Life (2nd ed.), Springer (2009).

Protein binding interactions

- Protein-protein interactions are important in many biological functions
  - DNA transcriptional regulation
  - Signal transduction
  - Nuclear pore complex
  - Membrane-protein trafficking
  - Viral entry and release

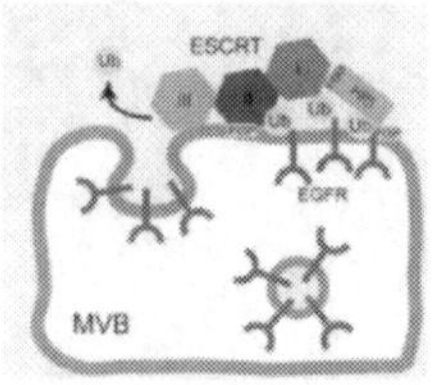

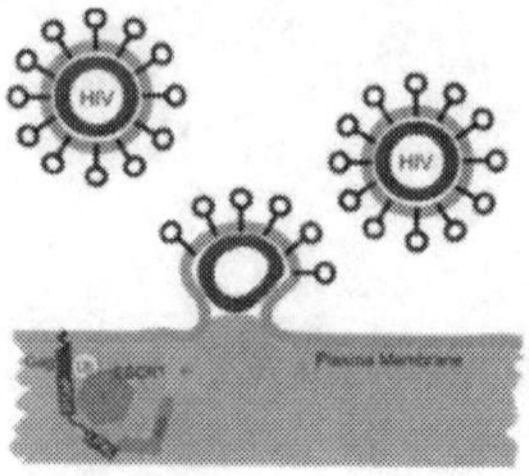

Protein model: Ubiquitin/UIM1

Representation of in-cell crowders

Most previous works had focused on mimicking the crowded environment with rigid spherical crowders.

⇩

Rigid spherical crowders are devoid of intramolecular conformational fluctuations that are commonly present in cellular biomolecules acting as crowding agents.

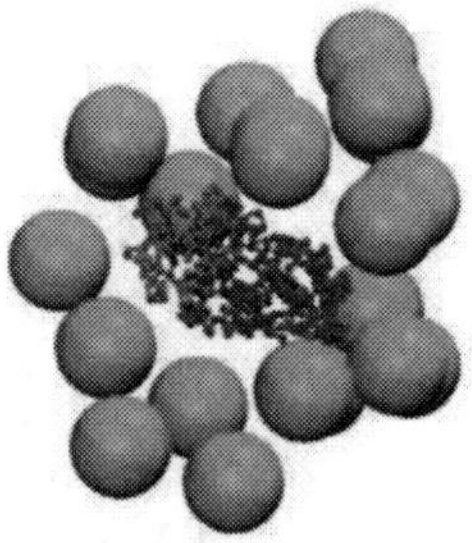

Representation of in-cell crowders

Most previous works had focused on mimicking the crowded environment with rigid spherical crowders.

Rigid spherical crowders are devoid of intramolecular conformational fluctuations that are commonly present in cellular biomolecules acting as crowding agents.

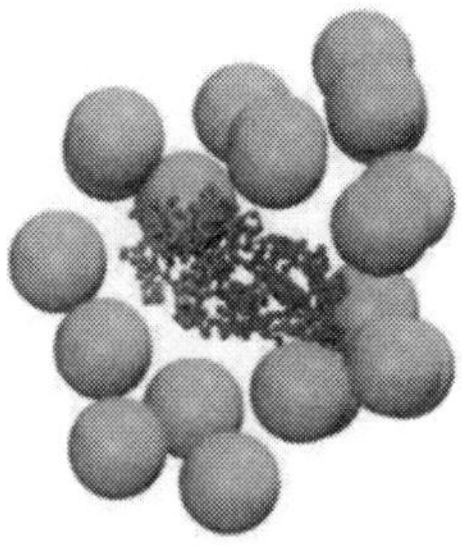

A more faithful representation of crowders in the crowded environment is polymeric chains

Simulation method

All Atom Model

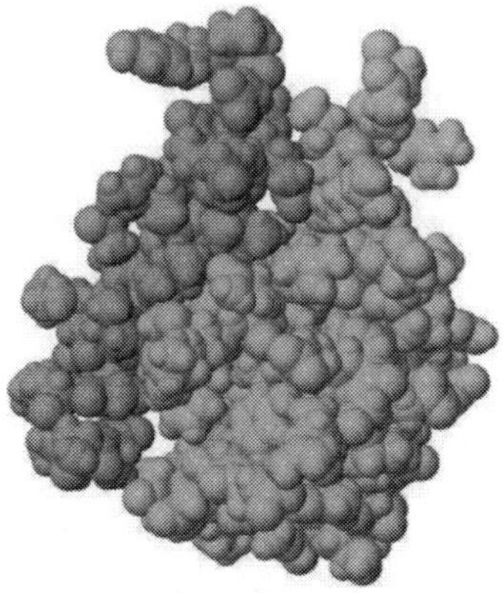

Ubq/UIM1 complex

Coarse-Grained Model

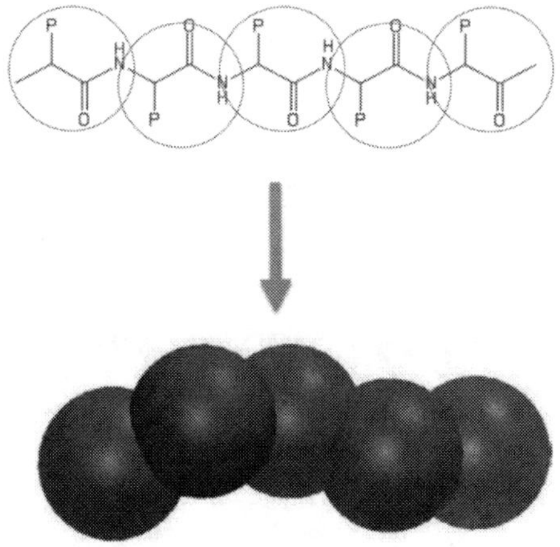

Simulation method

All Atom Model

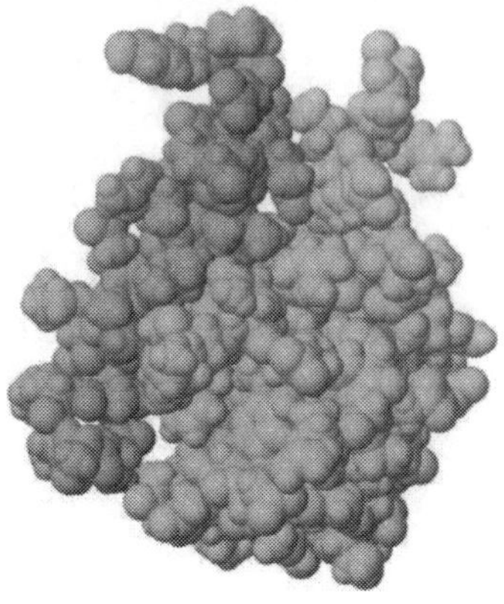

Ubq/UIM1 complex

Coarse-Grained Model

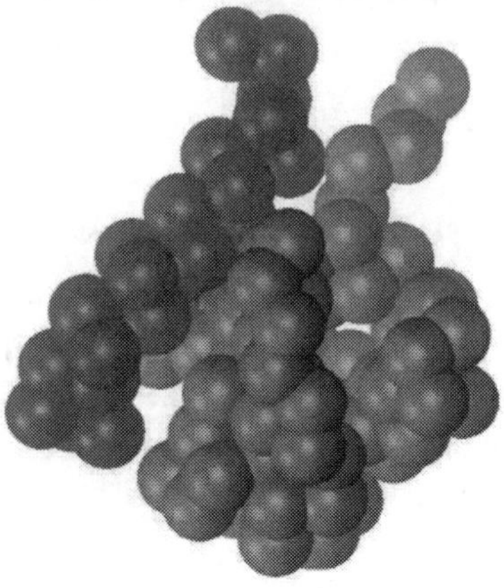

Simulation method

Coarse-Grained Model for Protein-Protein Interactions

$$\varphi_{ij}(r) = \varphi_{ij}^{LJ}(r) + \varphi_{ij}^{el}(r)$$

$$\varphi_{ij}^{LJ}(r) = 4|\varepsilon_{ij}|[(\frac{\sigma_{ij}}{r})^{12} - (\frac{\sigma_{ij}}{r})^{6}]$$

$$\varphi_{ij}^{LJ}(r) = \begin{cases} 4\varepsilon_{ij}[(\frac{\sigma_{ij}}{r})^{12} - (\frac{\sigma_{ij}}{r})^{6}] + 2\varepsilon_{ij}, & \text{if } r < r_{ij}^{0} \\ -4\varepsilon_{ij}[(\frac{\sigma_{ij}}{r})^{12} - (\frac{\sigma_{ij}}{r})^{6}], & \text{if } r \geq r_{ij}^{0} \end{cases}$$

$$\varphi_{ij}^{el}(r) = q_i q_j \frac{\exp(-r/\xi)}{4\pi D r}$$

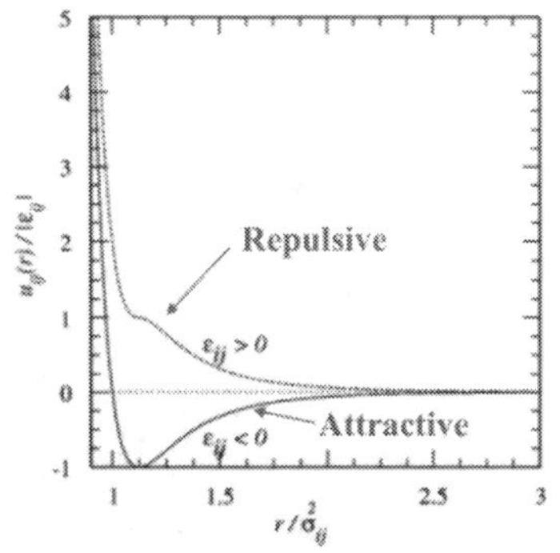

Coarse-Grained Model for Protein-Crowder and Crowder-Crowder Interactions

$$u_{ij}(r) = \varepsilon[\frac{\sigma_{ref}}{r - \sigma_{ij} + \sigma_{ref}}]^{12}$$

Kim, Y.C.; Hummer, G. *Journal of molecular biology* *2008*, *375*(5), 1416-1433.

Model validation: binding affinity

Ubiquitin-ubiquitin binding domains(UBDs)

- Ub/CUE
- Ub/UIM1
- Ub/UBA
- Ub/GAT
- Ub/DUIM
- Ub/UIM2
- UbL/UBA
- UbL/UIM
- UbL/UIM2

Fully transferable!

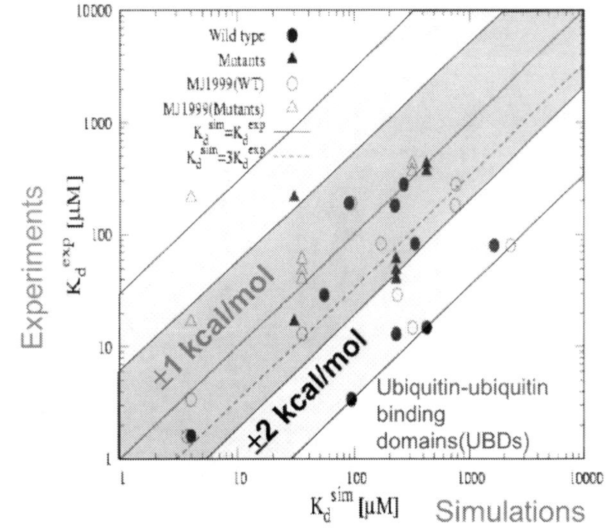

Kim, Y.C.; Hummer, G. *Journal of molecular biology 2008, 375*(5), 1416-1433.

Model validation: native structure

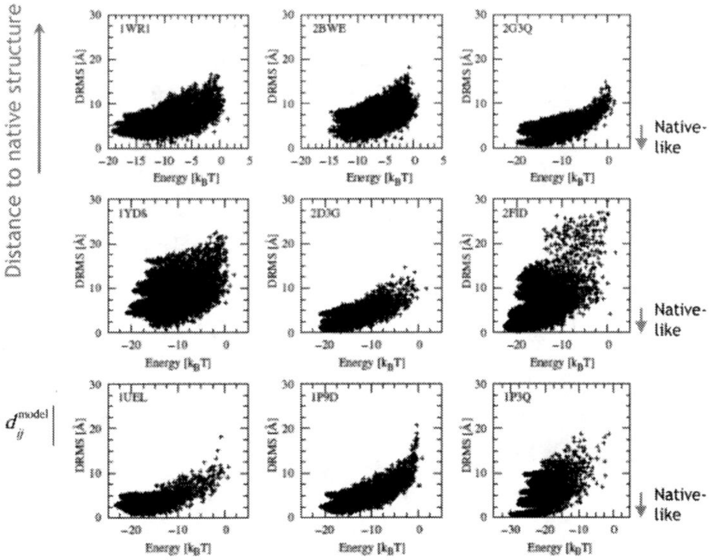

$$\mathrm{DRMS} = \frac{1}{N}\sum_{i,j}\left|d_{ij}^{\exp} - d_{ij}^{\mathrm{model}}\right|$$

Kim, Y.C.; Hummer, G. *Journal of Molecular Biology 2008, 375*(5), 1416-1433.

Key factors in protein binding interactions

Translational diffusion: Mutual diffusion search in solution.

Rotational diffusion: Finding the best binding conformation.

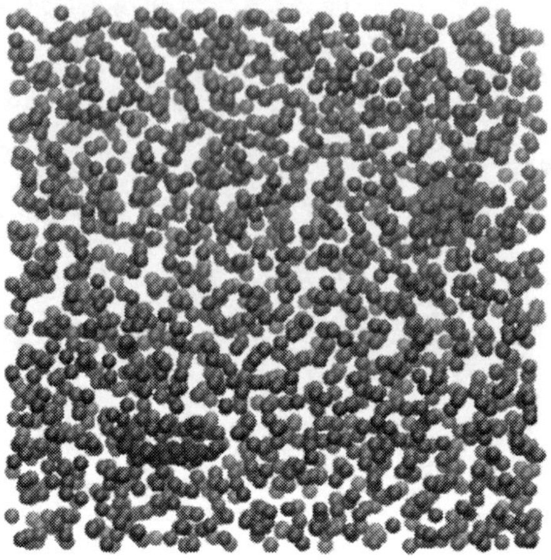

Translational diffusion coefficient analysis

Distance distributions as a function of time

$$P(r,t) = \frac{4\pi r^2}{(4\pi Dt)^{3/2}} \exp\left(\frac{-r^2}{4Dt}\right)$$

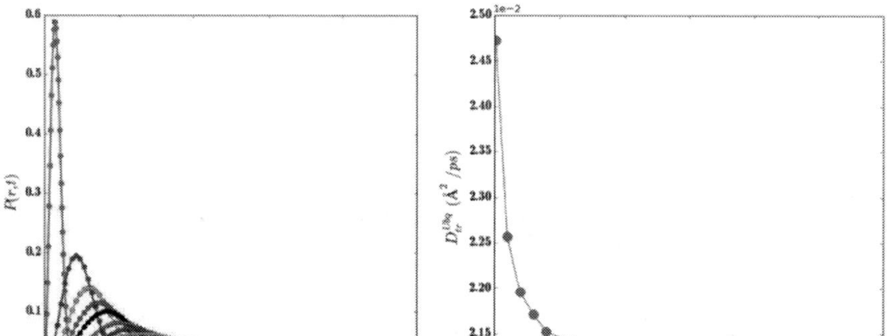

Chechkin, A.V., Seno, F., Metzler, R.,Sokolov, I.M.. *Physical Review X 2017*, 7(2), p.021002.

Translational diffusion coefficient analysis

Distance distributions

$$P(r,t) = \frac{4\pi r^2}{(4\pi Dt)^{3/2}} \exp\left(\frac{-r^2}{4Dt}\right)$$

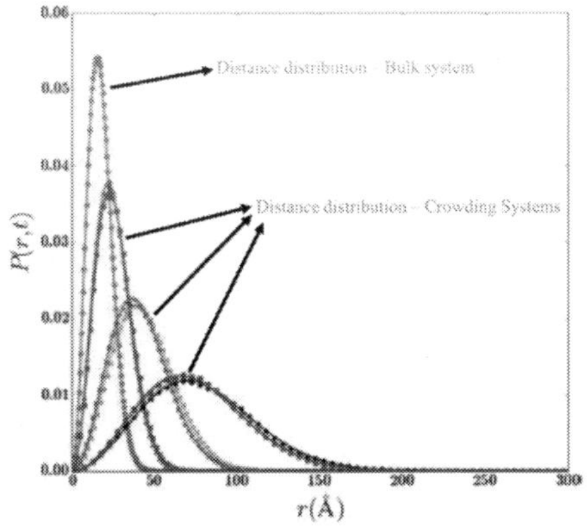

Mean square displacement

$$\langle r^2(t) \rangle = 2dDt$$

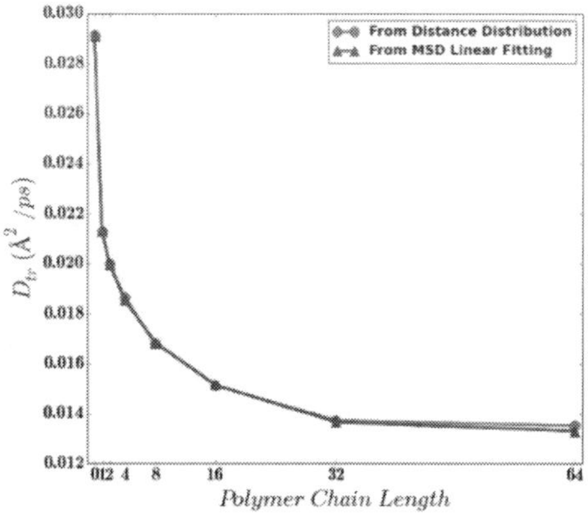

Chechkin, A.V., Seno, F., Metzler, R., Sokolov, I.M., *Physical Review X 2017*, 7(2), p.021002.

Results: translational diffusion (Ubq)

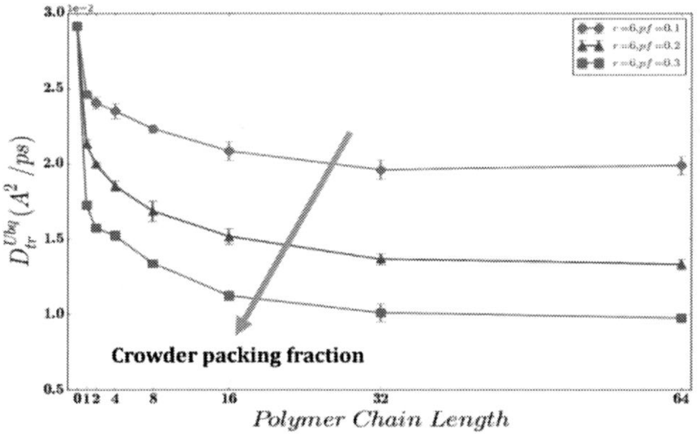

Translational diffusion coefficient of Ubq in the presence of polymeric crowders at fixed radius of monomers but different packing fractions.

Results: translational diffusion (Ubq)

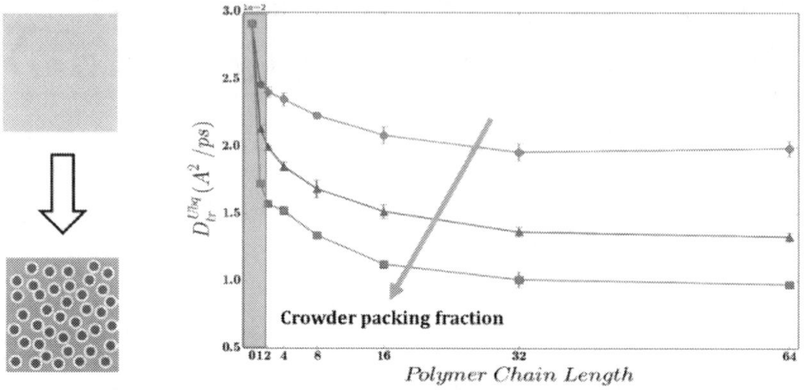

Translational diffusion coefficient of Ubq in the presence of polymeric crowders at fixed radius of monomers but different packing fractions.

Results: translational diffusion (Ubq)

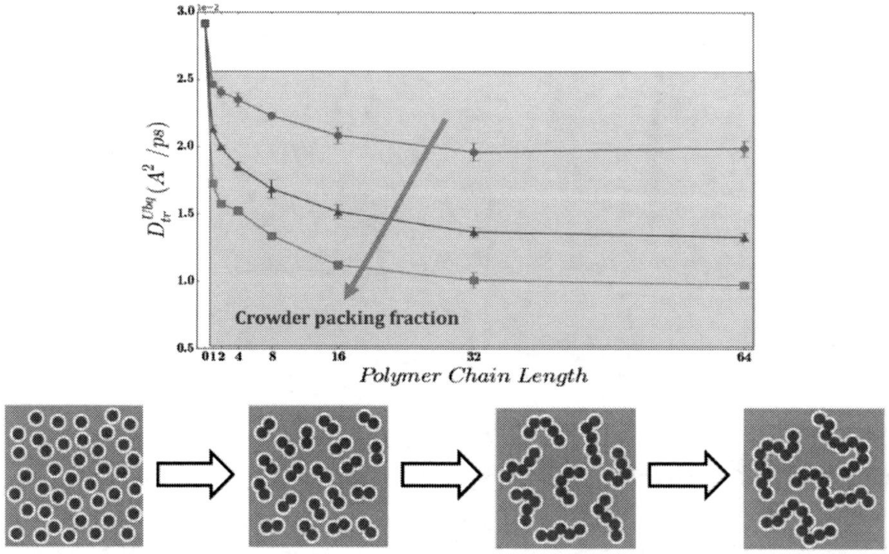

Results: crowding medium viscosity

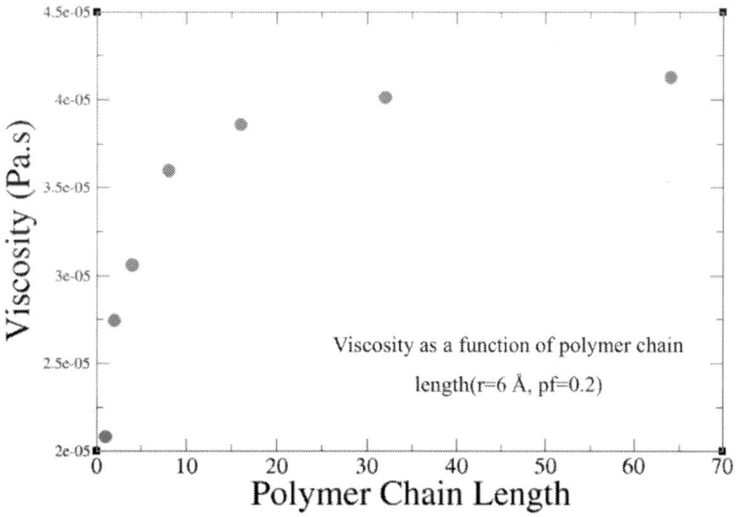

Results: Translational Diffusion (Ubq)

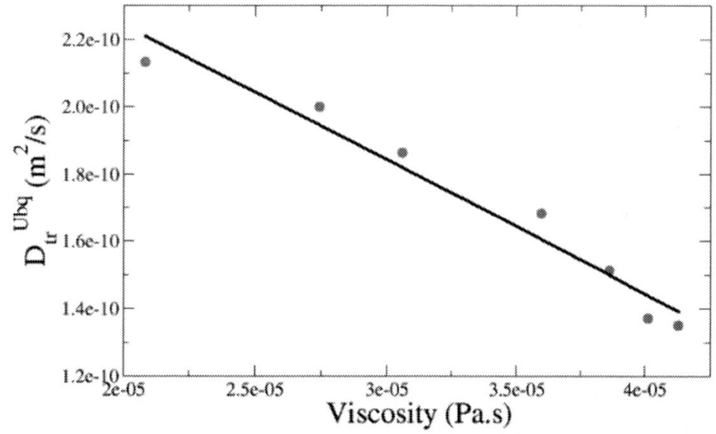

So, increase in viscosity due to the reduction of diffusion of polymeric crowders is the reason for decreasing the translational diffusion of proteins.

Rotational Diffusion Analysis

$$\frac{\partial c(\mathbf{n},t)}{\partial t} = D\nabla^2 c(\mathbf{n},t) = -D_{\text{rot}}\hat{I}^2 c(\mathbf{n},t)$$

$$\begin{aligned} &C_1(\tau) \\ &= sin^2(\theta)cos^2(\varphi)\exp[-(D_y + D_z)\tau] \\ &+ cos^2(\theta)\exp[-(D_x + D_y)\tau] \\ &+ sin^2(\theta)cos^2(\varphi)\exp[-(D_x + D_z)\tau] \end{aligned}$$

Fitting the rotational correlation functions of as many random unit vectors chosen to the following equation

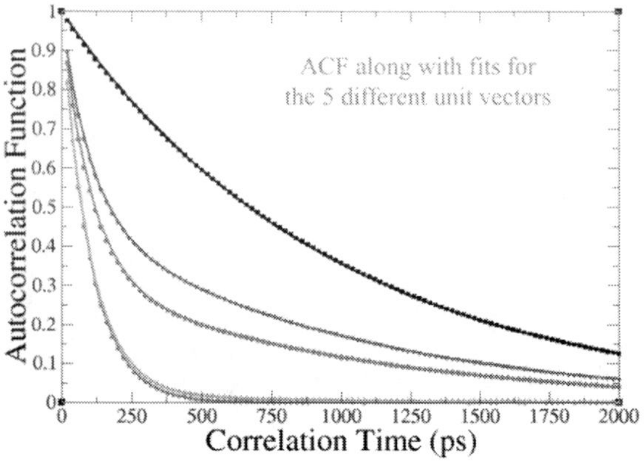

Results: Rotational Diffusion (Ubq)

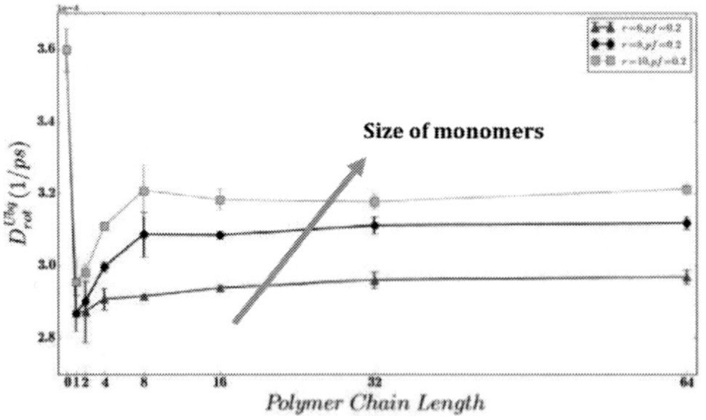

Rotational diffusion coefficient of Ubq in the presence of polymeric crowders at fixed packing fractions and different radius of monomers.

Results: Rotational Diffusion (Ubq)

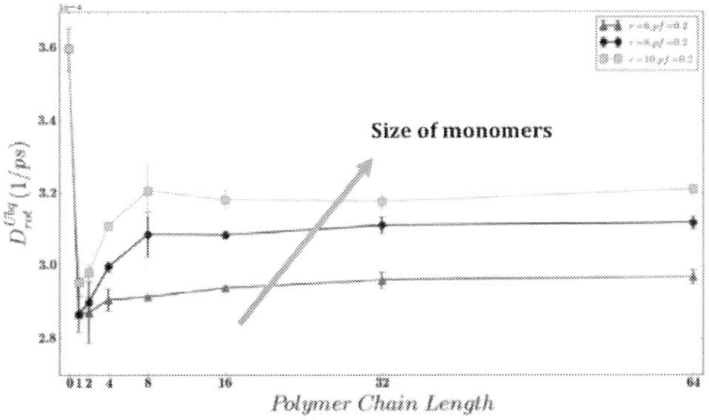

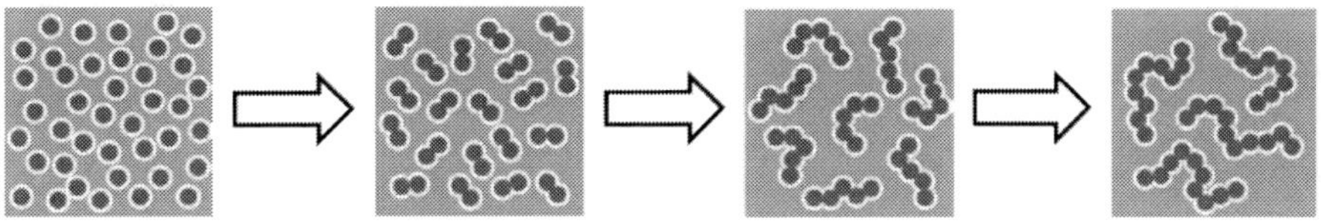

Results: Rotational Diffusion (Ubq)

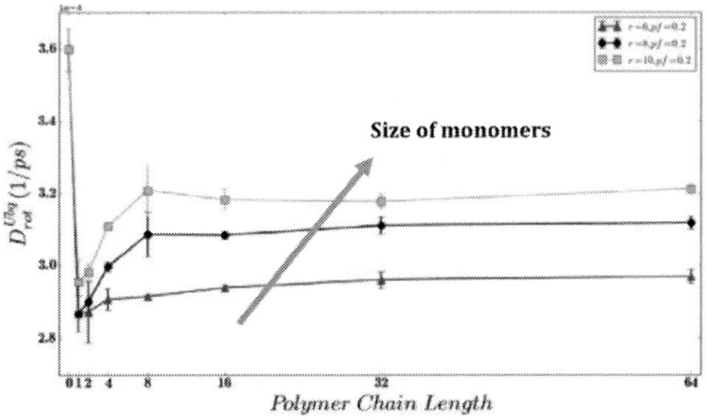

We guess that the excluded volume effects is responsible for increasing in the rotational diffusion coefficient of proteins.

Rate Analysis

Binding Interaction is considered as a 2nd order reaction.

$$A + B \underset{k_{off}}{\overset{k_{on}}{\rightleftharpoons}} AB \quad , \qquad K_d = \frac{k_{off}}{k_{on}}$$

We construct a coarse master equations for Ubq/UIM1 binding interaction dynamics.

$$\dot{P}_i(t) = \sum_{j=1}^{N} [k_{ij}P_j(t) - k_{ji}P_i(t)] \qquad\qquad \mathbf{KP}_{eq} \equiv 0$$

The rate constants computed using the mean-first-passage time as follows:

$$k_{on} = \frac{1}{(MFPT_{off \to on})C} \qquad\qquad k_{off} = \frac{1}{MFPT_{on \to off}}$$

Results: Rate Constants

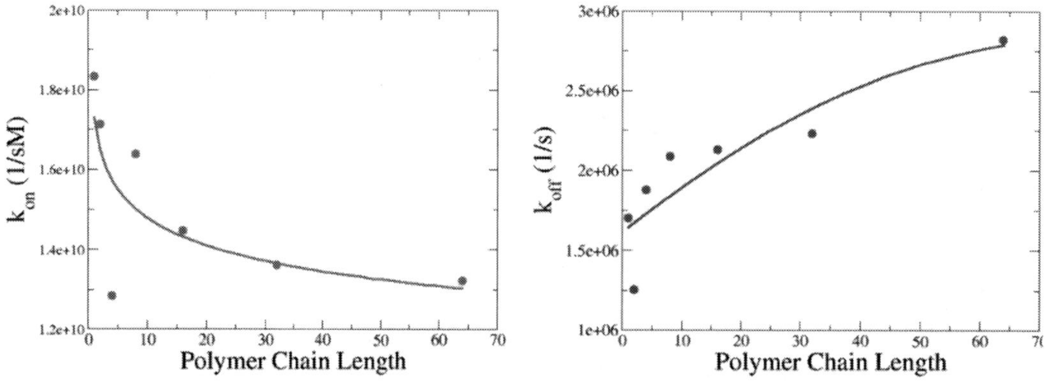

Conclusions

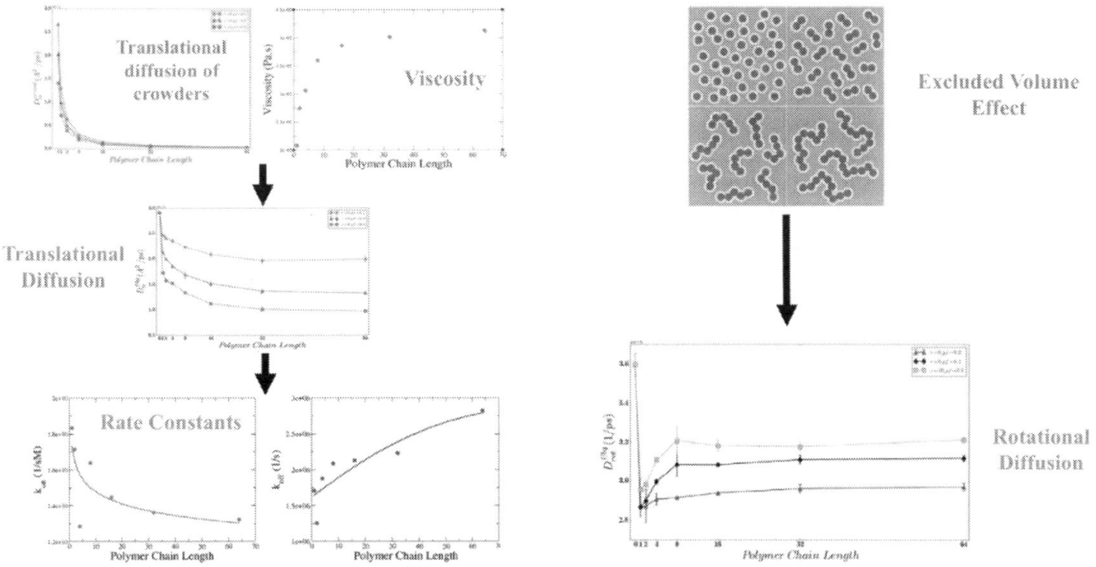

Acknowledgments

Biophysics and soft matter group @ Lehigh

Prof. Jeetain Mittal

Members:

Gregory Dignon
Tiarra Ann Maula
Runfang Mao
Brandon Horan
Samira Anbari
Elif Irem Senyurt
Evan Pretti
Huilai Gu
Eddie McGowan

Funding Resources

NIH National Institutes of Health

U.S. DEPARTMENT OF ENERGY
Office of Science

Computing Resources

XSEDE
Extreme Science and Engineering Discovery Environment

NERSC

498279 Effect of Palmitoylation in Membrane Proteins at the Blood-Brain Barrier Interface

Abstract:

Claudin family of tetraspanin transmembrane proteins form the building blocks of tight-junction barriers that regulate the diffusion of molecules through the paracellular spaces of epithelial and endothelial cells. In the lipid environment, the claudin proteins can undergo a reversible post-translational modification where long chain fatty acids covalently bond with the specific cysteine sites on the protein. Palmitoylation is implicated in the change in membrane-associated functions such as protein trafficking, cell signaling, and raft formation, among others. Studies on claudin-18 indicate that palmitoylation of claudin is essential for tight junction localization, wherein the claudins whose palmitoylation sites were mutated showed reduced localization due to poor membrane trafficking efficiency.[1] In this study, we investigate the role of palmitoylation in claudin-5 in cholesterol-rich complex membranes using coarse-grained molecular dynamics simulations. Through signature patterns in the protein sequence, four potential palmitoylation sites have been identified for claudin-5, which include the residues Cys-104, 107, 182, and 183.[2]Comparison of protein-lipid interactions and claudin-claudin self-assembly interaction patterns both with and without palmitoylation will be presented. The results of these molecular simulations highlight the impact of palmitoylation on protein trafficking, protein assembly, subcellular localization, and intracellular signaling.

References:

1. **Van Itallie, C. M., Gambing, T. M., Carson, J. L., and Anderson, J. M.**, "Palmitoylation of claudins is required for efficient tight-junction localization," Journal of Cell Science **118**, pp. 1427-1436 (2005).
2. **Awan, F. M., Anjum, S., Obaid, A., Ali, A., Paracha, R. Z., and Janjua, H. A.**, "In-silico analysis of claudin-5 reveals novel putative sites for post-translational modifications: Insights into potential molecular determinants of blood–brain barrier breach during HIV-1 infiltration," Infection Genetics and Evolution, **27**, pp. 355–365, (2014).

498601 Determining the Limitations of Processive Polysaccharide Deconstruction By Glycoside Hydrolases

Abstract:

Processive glycoside hydrolases (GHs) deconstruct polysaccharides (e.g., cellulose and chitin) by associating with the substrate and repeatedly hydrolyzing glycosidic bonds without dissociating. The repetitive hydrolytic process – the processive cycle – is thought to occur in a stepwise fashion; the GH cleaves the bond, expels the dimeric product, and moves forward along the polymer chain to the next available glycosidic bond. The cycle continues until the enzyme reaches the end of a chain or a barrier and dissociates. Each step likely has an associated free energy barrier, but the entire process must be energetically favorable overall. For two processive cellulases, *Hypocrea jecorina* Cel7A and Cel6A, free energy calculations have previously demonstrated that the hydrolytic event is rate-limiting within the processive cycle. However, similar studies on the energetic barriers for other processive glycoside hydrolases do not exist, and it is not clear that the rate-limiting step must be the same for different GH families. As such, we do not have sufficient knowledge of processive GH reaction coordinates to provide a generalized description of processive functionality. Accordingly, we elucidated the free energy barriers along the processive cycle of a GH family 18 chitinase, *Serratia marcescens* ChiA, extending our knowledge to a third GH family. Molecular dynamics simulations and umbrella sampling free energy calculations were used to determine the energetic barriers of forming the 'presliding,' 'sliding,' and 'Michaelis complex' intermediate states in the processive cycle. Comparing these barriers to experimental kinetic rate data, we show that ChiA hydrolytic step is also rate-limiting within the *S. marcescens* ChiA processive cycle, temptingly suggesting this may be a general limitation of the processive GH cycle. Our findings reveal another piece of the puzzle that will enable us to purposefully manipulate enzyme function for improved enzymatic hydrolysis of biomass and oligosaccharide synthesis.

499813 Computational Design of Novel Self-Assembling Peptide Biomaterials Based on an Amyloid Forming Motif from the Adenovirus Fiber Shaft

Abstract:

The properties of self-assembling peptides can be tuned through changes at their sequence level, and thus amyloid-forming peptides are increasingly gaining interest as potential biomaterials for a series of applications (1-5).Â Naturally occurring sequences from amyloid proteins or Î²-sheet rich proteins, for example sequence NSGAITIG from the adenovirus fiber shaft, self-assemble into amyloid fibril peptide nanostructures. Previous computational studies have suggested that the GAITIG motif is the key amyloidogenic region of the peptide with sequence NSGAITIGC (6), and experiments exploited these for the fabrication of biomaterials with several biomedical and technological applications. Here, we introduced a combination of computational and experimental studies, and investigated the addition of the cell-adhesive motif as well as single residues at the two termini of GAITIG motif, aiming at discovering novel bifunctional self-assembling amyloid peptide biomaterials with advanced biomedical applications.

In the first part of the study, we used replica exchange MD simulations and free energy calculations, as in ref. (7,8) in CHARMM (9) to design a self-assembling amyloid peptide biomaterial with sequence RGDSGAITIGC (1). According to analysis of the simulations, the GAITIG motif is at the core of the fibril and is the motif leading to self-assembly (1,6,7,10), the cell-adhesive motif RGD at the N-terminal end is flexible and adequately solvent exposed to promote cell-adhesion, and the cysteine residue at the C-terminal domain is not part of the amyloid peptideâ€™s amyloid zipper region, thus providing the biomaterial with metal binding properties. Experiments confirmed the bifunctional properties of the designed self-assembling amyloid peptide biomaterial which can be exploited for tissue engineering applications.

In the second part of the study, we introduced a novel in-house computational design protocol, which innovatively combines biophysical and optimization principles, and aimed at further stabilizing and potentially functionalizing the aforementioned self-assembling biomaterial by computationally and experimentally investigating additional mutations at the C-terminal end of the peptide. The mutations suggested by the computational protocol were investigated using replica exchange MD simulations and free energy calculations, and preliminary results from simulations and experiments confirm that the introduction of a specific amino acid substitution leads to a cell-adhesive self-assembling amyloid peptide biomaterial with promising advanced mechanical properties which can be further exploited in future tissue engineering applications.

1. Deidda G, Jonnalagadda SVR, Spies JW, Ranella A, Mossou E, Forsyth VT, Mitchell EP, Mitraki A, Tamamis P (2016) Self-Assembled Amyloid Peptides with Arg-Gly-Asp (RGD) Motifs As Scaffolds for Tissue Engineering. ACS. Biomater. Sci. Eng. Article ASAP. DOI: 10.1021/acsbiomaterials.6b00570

2. Loo Y, Goktas M, Tekinay AB, Guler MO, Hauser CA, Mitraki A (2015) Self-Assembled Proteins and Peptides as Scaffolds for Tissue Regeneration. Adv. Healthc. Mater. 4(16):2557-86.

3. Terzaki K, Kalloudi E, Mossou E, Mitchell EP, Forsyth VT, Rosseeva E, Simon P, Vamvakaki M, Chatzinikolaidou M, Mitraki A, Farsari M (2013) Mineralized self-assembled peptides on 3D laser-made scaffolds: a new route toward 'scaffold on scaffold' hard tissue engineering. Biofabrication 5(4):045002.

4. Kasotakis E, Mossou E, Adler-Abramovich L, Mitchell EP, Forsyth VT, Gazit E, Mitraki A (2009) Design of metal-binding sites onto self-assembled peptide fibrils. Biopolymers 92(3):164-72.

5. Kasotakis E, Mitraki A (2012) Silica biotemplating by self-assembling peptides via serine residues activated by the peptide amino terminal group. Biopolymers 98(6):501â€"509.

6. Tamamis P, Kasotakis E, Mitraki A, Archontis G (2009) Amyloid-like self-assembly of peptide sequences from the adenovirus fiber shaft: insights from molecular dynamics simulations. J. Phys. Chem. B. 113: 15639-15647.

7. Tamamis P, Kasotakis E, Archontis G, Mitraki A (2014) Combination of Theoretical and Experimental Approaches for the Design and Study of Fibril-forming Peptides. In Protein Design: Methods and Applications. Methods Mol. Biol. 1216: 53-70.

8. Tamamis P, Terzaki K, Kassinopoulos M, Mastrogiannis L, Mossou E, Forsyth VT, Mitchell EP, Mitraki A, Archontis G (2014) Self-Assembly of an Aspartate-Rich Sequence from the Adenovirus Fiber Shaft: Insights from Molecular Dynamics Simulations and Experiments. Journal of Physical Chemistry B, 118 (7): 1765-1774.

9. Brooks BR, Brooks CL 3rd, Mackerell AD Jr, Nilsson L, Petrella RJ, Roux B, Won Y, Archontis G, Bartels C, Boresch S, Caflisch A, Caves L, Cui Q, Dinner AR, Feig M, Fischer S, Gao J, Hodoscek M, Im W, Kuczera K, Lazaridis T, Ma J, Ovchinnikov V, Paci E, Pastor RW, Post CB, Pu JZ, Schaefer M, Tidor B, Venable RM, Woodcock HL, Wu X, Yang W, York DM, Karplus M. (2009) CHARMM: the biomolecular simulation program. J Comput Chem 30(10):1545-614.

10. Tamamis P, Archontis G (2011) Amyloid-like Self-Assembly of a Dodecapeptide Sequence from the Adenovirus Fiber Shaft: Perspectives from Molecular Dynamics Simulations. Journal of Non-Crystalline Solids, 357(2): 717-722.

492802 Conformational Changes of 2'-Hydroxybiphenyl-2-Sulfinate Desulfinase

Abstract:

Biodesulfurization using the 4S enzymatic pathway can effectively remove sulfur from refractory thiophenic molecules at ambient conditions. 2'-hydroxybiphenyl-2-sulfinate (HBPS) desulfinase (DszB), one of four enzymes in the 4S pathway, catalyzes the cleavage of the carbon-sulfur bond to produce 2-hydroxybiphenyl (HBP) and sulfite; this final step in the 4S pathway exhibits the slowest reaction rate, partially as a result of product inhibition. Here, we examine the effect of solvent on a large-scale conformational change, which may have implications for understanding the mechanism of product inhibition. Molecular Dynamics (MD) simulation was used to analyze how the HBPS substrate and inhibitory product, HBP, bind in DszB active site and influence protein conformational changes in two solutions: aqueous buffer and a biphasic hexane-water (0.5 vol/vol) solution. Three loops, including residues 50-60 (loop 1), 135-150 (loop 2), and 180-200 (loop 3), were identified as participating in the dominant conformational change; they are located close to active site and appear to form a "gate" that may be responsible for ligand ingress/egress. In aqueous solution, both HBPS and HBP remained bound in a stable position in the active site. HBPS-bound DszB displayed large fluctuations in a region distal to the active site (residues 220-240), and the entire protein backbone was less stable in comparison to HBP-bound DszB. Structural studies suggest the catalytic residues, Cys27, His60, and Arg70, remain close to the HBPS substrate and maintain substrate interactions; though in the HBP-bound DszB, the catalytic cysteine moved away from the active site along the simulation. In biphasic solution, HBPS-bound DszB showed improved rigidity distal from catalytic center. Due to the high affinity of HBP for the organic phase, HBP escaped the active site to the hexane-aggregated phase at the nonpolar DszB surface region, leaving the gate open facilitating the next reaction. Umbrella sampling provided a quantitative assessment of the energetic penalty for the gate conformational change. The energy cost associated with the gate opening from the close position was 3 kcal/mol. This value was independent of bound ligand and solvent conditions.

492818 Chromatin Folding Via Coarse-Grained Multi-Scale Simulation

Abstract:

The compaction of eukaryotic DNA in to chromatin represents an essential, yet poorly understood mechanism for controlling gene expression. Errors during compaction are associated with numerous diseases, and therefore elucidating the molecular factors that control compaction is a central goal of genetics. A significant challenge in studying chromatin compaction is the many length scales involved, typically ranging from angstroms to microns. In this work, we present a multi-scale approach that couples a detailed molecular model of the nucleosome to a coarse-grained mesoscale model of chromatin. We show that this approach can reproduce the dynamics and thermodynamics of available experimental measurements, and is computationally efficient enough to examine the self-assembly of large regions of chromatin. Notably, we show that subtle features of the chromatin fiber, such as the underlying DNA sequence, can have a significant impact on the stability and accessibility of different chromatin structures. The work presented here represents important steps toward understanding the molecular processes that dictate the dynamics and structure of chromatin.

498746 Quantitative Prediction of Industrial Turbulent Processes in Stirred Tank Reactors- Invited Presentation in Turbulent and Reactive Flows

Abstract:

CFD has been used to model flows in stirred tanks for many years. Different levels of success have been achieved for this purpose. For example, it has been widely accepted that CFD can predict mean velocity field of a turbulent flow in a stirred tank with reasonable accuracy using RANS model. More recent publications have shown that LES turbulence models provide more accurate prediction of time dependent turbulent flows. Although, good prediction of flow field is important, we should keep in mind that the ultimate goal for an industrial application is to predict process results, such as mixing, reaction, mass transfer, etc. For process industry, two basic measures for the performance of a stirred tank reactor are mixing time and mixing quality. Both require quantitative accuracy from CFD solutions. Quantitative predictions for reactive systems and multiphase systems are more challenging. Are today's CFD codes and computers ready to satisfy such quantitative requirements for industrial applications? What are preventing us from obtaining accurate mixing results? In this talk, we will first reaffirm what process industry is looking for from CFD solutions. Then we will discuss some key challenges/obstacles in our way to achieve such solutions.

498087 Passive Scalar Mixing in Anisotropic Turbulence from Line Sources

Abstract:

Mixing in turbulent flows is important both in nature and in engineering. Convection effects along with the motion of three-dimensional coherent structures in turbulent flow disperse a substance more efficiently than molecular diffusion does on its own. While the first effect increases the length scale of mixing and reduces its time scale, the latter works locally and enhances mixing at molecular level, which is of significant importance in chemical reactions and turbulent combustion applications. Several studies have been devoted to molecular mixing in homogeneous, isotropic turbulence. In this study, however, we explore the effects of anisotropic turbulence, in addition to molecular diffusion, on passive scalar mixing. Simulation of a turbulent channel flow is conducted by direct numerical simulation (DNS), followed by the Lagrangian Scalar Tracking (LST) method, where the motion of particles that correspond to markers of passive scalars with different Schmidt numbers (Sc) is calculated in the flow field [1-3]. Mass markers are released from several instantaneous line sources, starting at the channel wall and elevated to different distances from the channel wall to the center of the channel. The combination of diffusion, differences in the mean fluid velocity, and the effects of the coherent flow structures in the viscous wall region lead to the observation of different concentrations of particles downstream from the source. We then explore in details the effects of molecular diffusion and turbulence advection on mixing, which is observed to reach an upper limit at long time. We quantify mixing by introducing a "mixing factor" that is an indication of the quality of mixing and it represents the number of particles from one source that are present with particles from the other source at a specific location in the channel. Results from numerical simulation at friction Reynolds number of 300 will be discussed for Schmidt number scalars ranging from 0.7 to 2,400.

ACKNOWLEDGEMENTS

The computational support of XSEDE (CTS090017) is gratefully acknowledged.

REFERENCES

1. Nguyen, Q. and D.V. Papavassiliou, A statistical model to predict streamwise turbulent dispersion from the wall at small times, *Physics of Fluids,* 2016, **28**(12), Art. 125103
2. Nguyen, Q., Srinivasan, C., and D.V. Papavassiliou, Flow induced separation in wall turbulence, *Phys Rev E,* 2015, **91**, Art. 033019
3. Srinivasan, C., and D.V. Papavassiliou, Heat transfer scaling for wall bounded turbulent flows, *Applied Mechanics Review,* 2013, **65**(3), Art. 031002

491876 Dynamics and Structures of Viscoelastic Turbulence in Transitional Channel Flow

Abstract:

Introducing a trace amount of polymer into liquid turbulent flows can result in substantial reduction of friction drag. This phenomenon has been widely used in fluid transport; however, the mechanism is not fully understood. In this work we perform direct numerical simulations of viscoelastic turbulence in channel flow, using a high-resolution scheme that guarantees the positive-definiteness of polymerconformation tensor without artificial diffusion. It is found that at low to medium Weissenbergnumbers (Wi) the statistics from the high-resolution scheme show no difference from the statistics obtained from spectral code with global artificial diffusion. However the high-resolution scheme is able to resolve a finer spatial structure of the polymer conformation, which exhibits thin sheet structures. Intermittency of low- and high-drag events as well as the anti-correlation between the velocity and polymer stretch are observed in both. Frequency analysis shows that the natural frequency and the burst frequency are similar between Newtonian and low-Wi cases.

499359 Polymer Effects on the Development and Bursting of Turbulent Vortices: Implication on High-Extent Drag Reduction

Abstract:

Two major problems in viscoelastic turbulence, the effects of polymers on the laminar-turbulent transition dynamics and the origin of the maximum drag reduction asymptote, are both better understood in the regime near the margin of turbulence. Direct numerical simulation trajectories initiated from the edge state are used to follow its unstable manifold into the turbulent basin. In Newtonian flow, the growth of turbulence starts with the intensification of velocity streaks and a sharp rise in the Reynolds shear stress. It is followed by a quick breakdown into high-intensity small-scale fluctuations before entering the core of turbulence. This breakdown process not only occurs during the transition but also shows up in steady-state turbulence as intermittent bursting events. Adding drag-reducing polymers does not affect the initial growth of turbulence but stabilizes the primary streak-vortex structure. As a result, the breakdown stage is circumvented. Throughout the process, polymers act in reaction to the growing turbulence and do not drive the instability. Before the MDR regime, there is also a transition between low-extent and high-extend drag reduction (LDR vs. HDR) where qualitative differences are observed in flow statistics. Its existence indicates that there are at least two primary mechanisms for drag reduction. Our research shows that the second one, which does not occur until the LDR-HDR transition, results from the suppression of bursting and the corresponding shift of vortex-sustaining mechanism. Understanding this transition is important for achieving high-extend drag reduction in flow control practices.

486654 Macroscale Insensitivity of Type B Drag Reduction By Two Biopolymers

Abstract:

Turbulent drag reduction by distilled water solutions of two polysaccharide biopolymers, Diutan and Xanthan, was studied in three electropolished pipes of 4.57, 7.82 and 15.8 mm ID. The experiments encompassed solvent-based Reynolds numbers from 2000 to 200000, pipe wall shear stresses from 1 to 900 Pa, and biopolymer concentrations from 2 to 2000 weight parts per million. Solutions of both biopolymers exhibited Type B drag reduction, characteristic of extended macromolecules, in all three pipes. Low concentrations, $c < 200$ wppm, yielded turbulent flow segments roughly parallel to, but displaced upwards from, the Prandtl-Karman law, resembling ladder rungs that ascended with increasing concentration. High concentrations, $c > 500$ wppm, exhibited a continuous transition from shear-thinning laminar flow at low Re to turbulent flow along the maximum drag reduction asymptote at high Re. At fixed high $Re\surd f = 5000$, low concentration flow enhancements relative to solvent $S' = [(1/\surd f)_p - (1/\surd f)_n]_{Re\surd f}$ increased almost linearly with increasing concentration, while the highest concentration solutions all exhibited flow enhancements $S' \sim 20$, approaching the asymptotic maximum drag reduction. At the same fixed $Re\surd f = 5000$, intrinsic flow enhancements $[S'] = \mathrm{Lim}_{c\rightarrow 0}[S'/c]$ ($1/\surd f$ units/wppm) in each of the 4.57, 7.82, 15.85 mm ID pipes were, respectively, for Diutan [0.12±0.01, 0.12±0.03, 0.12] and for Xanthan [0.12±0.02, 0.13±0.02, 0.11±0.03]. The invariance of intrinsic slip [S'] with respect to pipe diameter for each biopolymer implies, more generally, that Type B drag reduction is insensitive to turbulence macroscale. This newly-found macroscale insensitivity combined with long-known microscale insensitivity suggest that Type B drag reduction is insensitive to turbulence parameters, being mainly controlled by parameters of the extended macromolecular additive. It is also curious that both biopolymers induced virtually identical intrinsic slips $[S'] = 0.12 \pm 0.03$, despite Diutan having the higher molecular weight M = 5.0E6 vs 2.5E6 and longer formal contour length $L_c \sim 11.7$ mm vs 3.2 mm relative to Xanthan, suggesting an influence of other skeletal attributes, perhaps stiffness and aspect ratio.

498362 Euler-Euler and Euler-Lagrange Simulations of Heavy Particles Clustering in Homogeneously Sheared Gas

Abstract:

Particle-laden flows of sedimenting heavy solid particles or droplets in a carrier gas have strong inter-phase coupling even at low volume fractions. The slip velocity between phases lead to sustained clustering that can strongly modulate the overall flow. Simulations of homogeneously sheared flow are conducted in the Euler-Lagrange and Euler-Euler formalisms. We show that the formation of particle clusters involve three fundamental mechanisms: (1) the preferential concentration of inertial particles in the stretching regions of the flow (2) particle-trajectory crossing (PTC) and (3) a Rayleigh-Taylor instability due to the vertical stacking of particle-rich and particle-depleted regions. The Euler-Lagrange simulation method, based on particle tracking, capture all three effects but suffer from lack of scalability to industrial scales. Euler-Euler methods offer better scalability but require extreme care in the presence of PTC.

499376 Breakage of Single Drops and Bubbles in a Turbulent 2-D Orifice Flow

Abstract:

Much of the existing literature on droplet break-up deals with either non-inertial (Stokes flows) or turbulent flows with droplets in the inertial or viscous subranges. The current work examines the break-up of single macroscale droplets and bubbles in turbulent flows due to short-term high-intensity deformation events. Information from both experimental observations and computational simulations are combined to develop break-up criteria and determine the probability of droplet break-up.

The experimental apparatus consists of a parallel plate channel with a 25%-open slit orifice, creating a 2-D planar orifice flow in the center plane. Water was used as the continuous phase, with Reynolds numbers ranging from approximately 8,000 to 19,000. Oil droplets and air bubbles were injected upstream of the orifice using a variety of capillary tubes, generally resulting in droplet diameters between 200 and 1200 microns, much larger than the Kolmogorov microscale. The path and behavior of the droplets and bubbles were recorded using high speed digital imaging.

To understand the local conditions of the droplet prior to break-up, computational simulations of the orifice were performed for each Reynolds number case. The channel flows were simulated using 3-D RANS equations with the realizable k-Îµ turbulence model. The droplet path from the experimental images was then compared to the simulated deformation field to construct the deformation history of the droplet as it passed through the orifice. The synthesis of this experimental and computational information led to the development of breakage probability and regime plots that can be correlated using an appropriately defined turbulent Weber number.

(577i) ON THE INFLUENCE OF SPANWISE BOUNDARY CONDITIONS ON THE LARGE EDDY SIMULATION OF THE FLOW PAST A SINGLE CYLINDER

Bárbara L. da Silva, Matheus R. Barbieri, Jonathan Utzig, Waldir P. Martignoni, Henry F. Meier,
University of Blumenau, Blumenau, SC, Brazil

Abstract

The flow past a single cylinder has been the focus of scientific investigations due to two major reasons: first, it is a simplified model to study unsteady forces and consequent flow-induced vibrations in many practical applications, for example marine petroleum risers, offshore structures, chimneys and as a component of tubular heat exchangers. Second, as in other bluff body flows, complex phenomena such as shear layer separation and transition to turbulence take place, which make it a classical benchmarking case for the validation of mathematical models. The tridimensional solution of turbulent flows past a single cylinder usually requires approaches of higher computational costs, such as the Large Eddy Simulation (LES). Therefore, simplifications are commonly adopted in the computational domain to allow for reasonable simulation times. One of these simplifications is the use of periodic conditions in the spanwise boundaries of the domain, limiting the spanwise length into an arbitrary value, sometimes guided by the estimated wavelength of the flow structures. This study aims to investigate the validity of the periodic simplification, comparing it to other spanwise boundary conditions for a low aspect ratio cylinder and confronting the resulting velocity profiles with experimental data obtained by a two-component Laser Doppler Anemometry system. Simulations were carried out in a commercial Computational Fluid Dynamics (CFD) code with the LES approach at a subcritical Reynolds number. Force and pressure coefficients and the Strouhal number were also compared between the investigated cases, revealing that periodic boundary conditions have considerable influence in the evaluated parameters. Results show that a symmetry boundary condition is more appropriate than the periodic one in the investigated case, and it should be prioritized for moderate to low aspect ratio cylinders if computational and time resources allow for it.

Introduction

Vibrations are present in any flow due to the transient feature of the fluid dynamic forces. They occur from the two-way interaction between the reaction from the equipments' surfaces and the fluid forces (SILVA *et al.*, 2016; BLEVINS, 1990). These reactions depend on many factors such as natural vibration frequency and Reynolds number (BLEVINS, 1990; BEARMAN, 2011). This phenomenon is of special interest in the cross-flow past structures such as tube bundles, once it can cause noise and damage to the structure of industrial equipments, which threatens operational safety and generates financial loss.

The study of the fundamentals and mechanisms of flow-induced vibrations involves numerical and physical experimentation, which are conducted in cross-flow past bluff bodies, typically using a single circular cylinder as an initial simplified model. In the numerical field, the choice of an appropriate turbulence model over the computational domain, such as Reynolds-Averaged Navier-Stokes (RANS) and Large Eddy Simulation (LES) models (MOUKALED, MANGANI, DARWISH, 2015) is of great importance, as complicated phenomena such as flow separation and transition to turbulence take place.

Complex flow patterns are caused by different flow conditions, showing a strong dependency on geometrical characteristics. One of these characteristics is the aspect ratio (AR), which one is the ratio between the cylinder height and its diameter.

The effect of end-wall conditions along the spanwise direction transforms the flow pattern, resulting in considerable influence on drag coefficient, base-pressure coefficient, and on fluctuating and mean pressure coefficient distributions (WEST, APELT, 1982; LAM, ZOU, 2010). The effects caused by different aspect ratios become smaller for AR > 20, but small influences can still be detected up to aspect ratios of 40 (WEST, APELT, 1982).

Studies have shown that only models that resolve at least part of the turbulent eddies can simulate well this type of flow (SILVA, UTZIG, MEIER, 2016). The proximity between the time scale of the greatest turbulent vortices and the period of the shedding vortices compromise the RANS models. Therefore, despite its higher computational cost due to the fine mesh required to solve at least the large turbulent eddies, the LES model is a common choice to resolve the tridimensional turbulent bluff body flow (PRSIC *et al.*, 2014; LAM *et al.*, 2010).

Accordingly, to obtain acceptable simulation times, simplifications are adopted usually aiming to reduce the computational domain extension. One of these simplifications is related to the spanwise boundary condition, such as the periodic and symmetry conditions, that are widely applied in numerical studies. When the physical geometry of interest has a mirror symmetry, a symmetry condition can be used to reduce the computational domain to a symmetric subsection of the overall system, while when the physical geometry and the solution pattern have a periodically repeating nature, the periodic condition can be used. Therefore, in this work, the effects of simplifications in the spanwise direction were evaluated in comparison to the simulation of the complete domain of our test section. The cylinder used had an aspect ratio equal to 10.7, which, according to West, Apelt (1982), can be considered low. In addition to it, time-averaged velocity data were experimentally acquired in our test section to serve as a reference. The results of this work were used to support further numerical and experimental investigations on the cross-flow around tube bundles.

Materials and Methods

Both experimental and numerical approaches were adopted in this study to confront the simulation results using different spanwise boundary conditions with experimental data obtained in our test section. The methodology of each approach is described separately in the following subsections.

Experimental details

Experiments were conducted in a 1 m long closed-circuit water channel built in acrylic with an internal width of 8 cm. The channel height was kept at 10.7 cm by using a segmented lid, to avoid the formation of a free surface and provide an accurate boundary condition for the simulations. The middle section of the lid is perforated to allow the insertion of a single polyacetal cylinder with a diameter of 1 cm, as in Figure 1a. Therefore, the cylinder model had an aspect ratio of 10.7 and a blockage ratio (cylinder diameter divided by the width of the channel) of 12.5%. A fixed flow condition was adopted, corresponding to an average inlet flow velocity V_{in} = 0.12 m/s and a Reynolds number based on the cylinder diameter $Re_D = \rho V_{in} D/\mu$ = 1200. The average turbulence intensity was measured 15 cm upstream from the cylinder position, yielding a value of 10.1%. The flow rate was controlled using a PID algorithm written in LabVIEW, from National Instruments AS.

Velocity measurements were carried out in the test section using a two-component Laser Doppler Anemometry system (LDA) from Dantec Dynamics AS, model FlowExplorer DPSS 150 2D. Aluminosilicate particles with a volume-weighted mean diameter of 25.4 μm and average density of 1500 kg/m³ were used to seed the flow. Tracing behavior was assured by providing a Stokes number far smaller than 1. The LDA measurement volume position was adjusted with a three-dimensional traversing unit along the cylinder wake centerline at a fixed normal distance of 4.8 cm from the bottom of the channel. Figure 1b shows the measurement points positions.

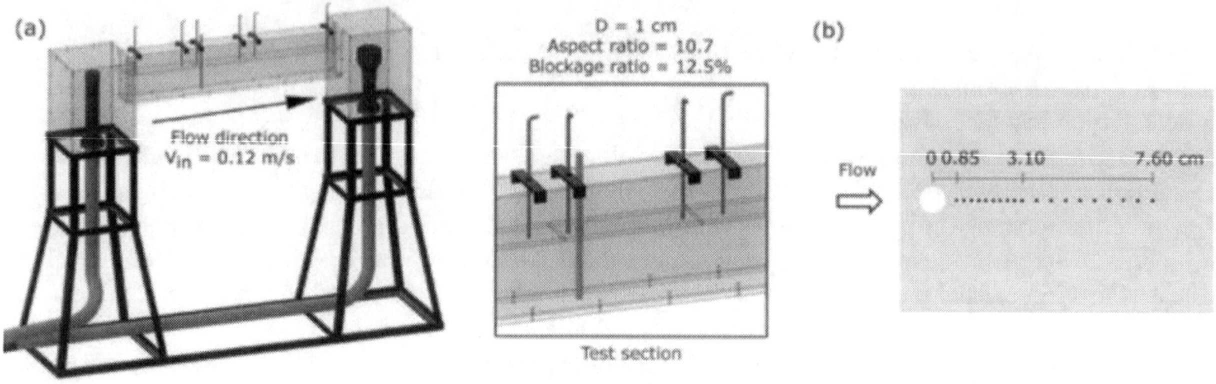

Figure 1. (a) Scheme of the experimental setup and (b) LDA measurement points.

Experimental reproducibility was assured by conducting two runs for each measurement point. Up to 5000 validated samples were acquired for each run and data point with average frequencies in the order of 10^1 to 10^2 Hz. A single mean velocity and its fluctuation at each data point was obtained by adding the samples acquired in the two validated runs.

Simulations setup

The finite volume method implemented in the commercial code ANSYS Fluent 14.0 with the physical properties of liquid water (ρ = 998.2 kg/m³; ν = 1e-6 m²/s) was used to solve the filtered three-dimensional incompressible Navier-Stokes equations for a Newtonian flow:

$$\frac{\partial \bar{\boldsymbol{u}}}{\partial \boldsymbol{x}} = 0,$$

$$\frac{\partial \bar{\boldsymbol{u}}}{\partial t} + \frac{\partial \bar{\boldsymbol{u}}\bar{\boldsymbol{u}}}{\partial \boldsymbol{x}} = \frac{\partial}{\partial \boldsymbol{x}}\left[\nu\left(\frac{\partial \bar{\boldsymbol{u}}}{\partial \boldsymbol{x}} + \frac{\partial \bar{\boldsymbol{u}}}{\partial \boldsymbol{x}}\right)\right] - \frac{1}{\rho}\frac{\partial \bar{P}}{\partial \boldsymbol{x}} - \frac{\partial \boldsymbol{\tau}}{\partial \boldsymbol{x}},$$

where $\bar{\boldsymbol{u}}$ are the filtered velocity components $(\bar{u}, \bar{v}, \bar{w})$, $\boldsymbol{x}$ are the cartesian coordinates (x, y, z), ρ is the density of the fluid, ν is the kinematic viscosity of the fluid and $\bar{P}$ is the filtered pressure. The subgrid-scale Reynolds tensor $\boldsymbol{\tau}$ includes the effects of small scales in the flow, and it was modeled with the Dynamic LES turbulence approach (GERMANO et al., 1991) for all cases. Four geometries were considered, with different extensions in the spanwise direction: the complete immersed cylinder span of 10.7D, half of the immersed span (5.35D), πD and 2πD. The two last extensions were chosen because they are commonly adopted by other authors (LUO *et al.*, 2014; LAM *et al.*, 2010). In all four geometries the inlet boundary condition was set 15 cm upstream of the cylinder (15D), and the outlet boundary condition 30 cm downstream of the cylinder (30D). The full channel width of 8 cm (8D) was kept.

At the inlet boundary, a uniform velocity of 0.12 m/s was prescribed at all cases, with no turbulence perturbations despite of the high turbulence intensity verified experimentally. At the outlet, the static pressure P was set as 0 Pa. For the "Complete" case with the full cylinder span, remaining boundaries were set as no-slip wall conditions. This case was simulated as a reference for the other cases, which have geometry simplifications in the spanwise direction and also different spanwise boundary conditions. In the case with half of the immersed span, a no-slip wall boundary condition was set at the bottom face of the channel, while a symmetry boundary condition was adopted at the top, designating this case as "Symmetry" in Figure 2. The spanwise boundaries at the cases with πD and 2πD extensions were set as periodic, so these cases are referred to respectively as "Periodic - πD" and "Periodic - 2πD" as illustrated in Figure 2. These boundary conditions are summarized in Table 1.

Hexahedral grids were generated for all cases with 192 cells in the cylinder perimeter. The node closest to the cylinder and channel walls is located at a distance of 0.03 mm, which provided y^+ values inferior to 1. The maximum cell size is of 1.25 mm within 3 cm (3D) from the cylinder, expanding up to 2.49 mm at the inlet and outlet faces. The mesh sizes ranged from 1,088,019 nodes to 4,755,046 nodes.

Table 1. Boundary conditions of the simulations setup

Simulation setup	Inlet	Outlet	Side walls	Top wall	Bottom wall
Complete	$u = 0.12$ m/s $v = 0$ $w = 0$	$P = 0$ Pa	$\boldsymbol{u} = 0$	$\boldsymbol{u} = 0$	$\boldsymbol{u} = 0$
Symmetry	$u = 0.12$ m/s $v = 0$ $w = 0$	$P = 0$ Pa	$\boldsymbol{u} = 0$	$v\|_{y=top} = 0$ $\left.\frac{\partial u}{\partial y}\right\|_{y=top} = 0$ $\left.\frac{\partial w}{\partial y}\right\|_{y=top} = 0$	$\boldsymbol{u} = 0$
Periodic - πD Periodic - 2πD	$u = 0.12$ m/s $v = 0$ $w = 0$	$P = 0$ Pa	$\boldsymbol{u} = 0$	$\boldsymbol{u}\|_{y=top} = \boldsymbol{u}\|_{y=bottom}$	

The SIMPLEC pressure-velocity coupling algorithm was used, with the bounded central differencing momentum discretization scheme and second order implicit transient formulation. The fixed time step of 2e-4 s provided maximum Courant numbers inferior to 1 in all cases. The drag and lift coefficients were monitored during the simulations until their behavior was quasi-stationary, then unsteady statistics were acquired during 13 s of flow time.

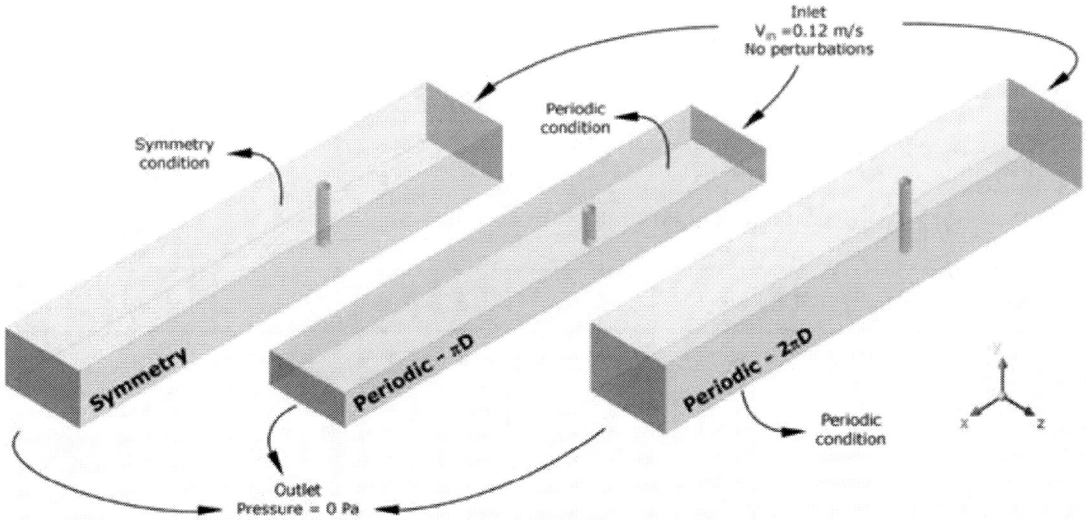

Figure 2. Boundary conditions used in the cases with geometry simplifications.

Results and Discussion

Figure 3 shows the mean streamwise velocity component U, mean streamwise velocity fluctuations u' and mean crosswise velocity fluctuations w' normalized by the inlet velocity V_{in} for the LDA data and the four simulation cases described. Table 2 presents characteristic lengths for all cases: the formation length L_f/D, measured from the center of the cylinder to the position of the minimum mean streamwise velocity U_{min}, the recirculation length L_r/D, measured from the center of the cylinder to the position of the zero mean streamwise velocity, $L_{u'}/D$, measured from the center of the cylinder to the position of the maximum streamwise fluctuations u'_{max}, and $L_{w'}/D$, measured from the center of the cylinder to the position of the maximum crosswise fluctuations w'_{max}.

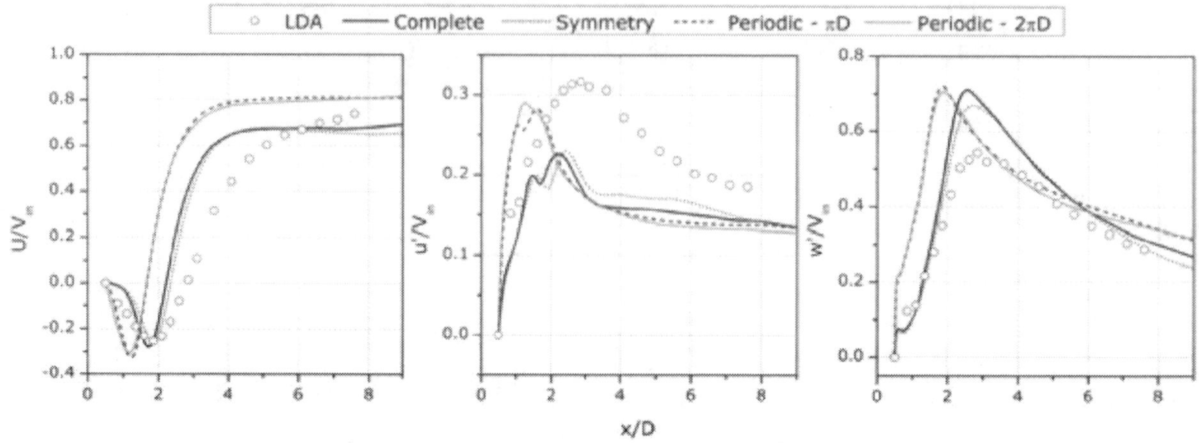

Figure 3. Velocity distributions in the wake centerline of the cylinder.

The Complete case followed the experimental data trends, but with notable differences such as a shorter wake, lower fluctuations in the streamwise direction and higher fluctuations in the crosswise direction, numerically pointed out in Table 2. These differences may be due to the absence of turbulent perturbations at the inlet boundary condition of the simulations. This hypothesis is reinforced by the two peaks shown in distribution of u', observed by Norberg (1998) only at cases with $Re_D < 8000$ for a turbulence intensity lower than 0.1%. The Symmetry case results were very close to the Complete case, aside from minor variations related to a slightly longer wake.

Both periodic cases featured recirculation lengths 25% shorter with reference to the Complete case, which contradicts the wake shortening effect due to lowering the aspect ratio of the cylinder reported in the literature (SUMNER, HESELTINE, DANSEREAU, 2004). In spite of it, this behavior was also observed by Lam and Zou (2010) in their numerical simulation of four cylinders in an in-line square configuration with the periodic condition. The maximum value observed for u' was closer to the one obtained from the LDA measurements, with only the Periodic - πD case presenting two peaks in the distribution.

Table 2. Relevant lengths and velocities in the wake centerline

Case	L_f/D	U_{min}	L_r/D	$L_{u'}/D$	u'_{max}	$L_{w'}/D$	w'_{max}
LDA	1.85	-0.25	2.85	2.85	0.32	2.85	0.54
Complete	1.74	-0.28	2.22	2.18	0.23	2.52	0.71
Symmetry	1.86	-0.25	2.34	2.44	0.23	2.70	0.67
Periodic - π	1.22	-0.33	1.68	1.58	0.28	1.86	0.72
Periodic - 2π	1.16	-0.31	1.67	1.24	0.29	1.88	0.70

The time history of the drag and lift coefficients is shown in Figure 4 for all simulated cases. and the mean value of each coefficient $\bar{C}_D$ and $\bar{C}_L$ as well as its fluctuations C_D' and C_L' are organized in Table 3. The drag and lift coefficients of the cylinder are calculated as $C_D = 2F_D/(\rho V_{in}^2 A)$ and $C_D = 2F_D/(\rho V_{in}^2 A)$, where F_D and F_L represent the drag and lift forces, respectively, and A is the projected area of the cylinder.

Again, the Complete and Symmetry cases presented a different behavior from the periodic cases. The mean drag coefficient was about 10% higher with the periodic condition, with twice the magnitude of the drag fluctuations and more than four times the magnitude of the lift fluctuations of the Complete case. This is in accordance with the results obtained by Lam and Zou (2010) with the periodic condition. The increase in the drag coefficient is related to the decrease of the pressure coefficient at the rear of the cylinder, also presented in Table 3.

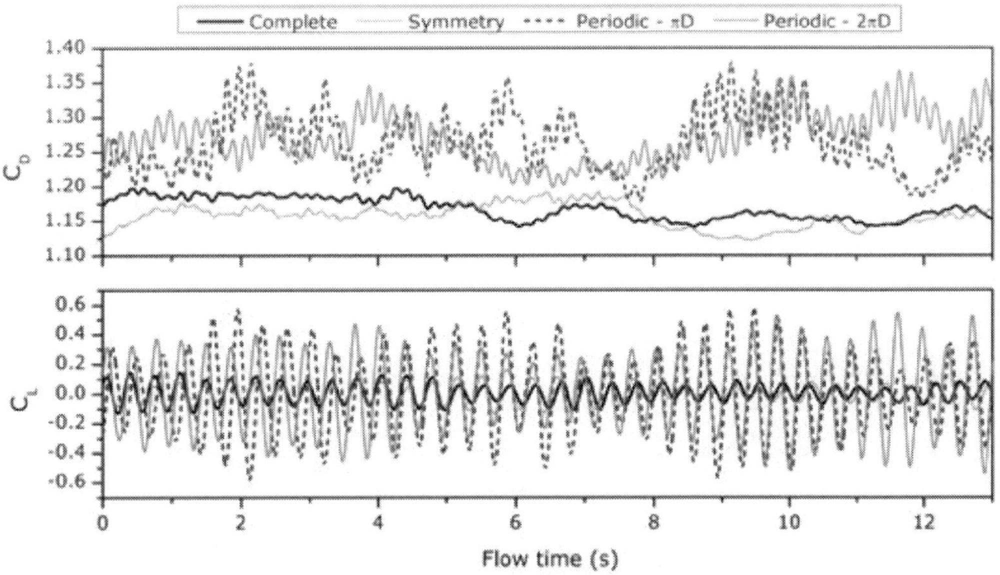

Figure 4. Time history of the drag and lift coefficients obtained numerically.

The mean pressure coefficient, usually taken at the back of the cylinder, is also presented in Table 3 and is related to the base suction of the wake recirculation region. It is calculated as $Cp = 2(P - P_\infty)/(\rho V_{in}^2)$, where P_∞ is the reference static pressure adopted as zero. The Strouhal number was also obtained for the simulated cases. It is related to the vortex shedding frequency and defined as $St = fD/V_{in}$, where f is the frequency, obtained from the Fast Fourier Transform of the lift coefficient time history. As shown in Table 3, all cases have a Strouhal number of 0.23, close to the value of 0.21 reported by Blevins (1990) for this subcritical Reynolds number.

Table 3. Force and pressure coefficients and Strouhal number

Case	$\overline{C}_D$	C'_D	$\overline{C}_L$	C'_L	$\overline{Cp}_{back}$	St
Complete	1.168	0.017	0.004	0.061	-0.696	0.23
Symmetry	1.157	0.018	0.000	0.058	-0.675	0.23
Periodic - πD	1.263	0.043	-0.003	0.275	-0.878	0.23
Periodic - 2πD	1.270	0.035	0.002	0.258	-0.897	0.23

Conclusions

The influence of spanwise geometry simplifications and associated boundary conditions was investigated numerically in the Large Eddy Simulation of the flow past a single cylinder with a limited aspect ratio of 10.7. The full domain simulation results were compared with time-averaged velocity data acquired by the Laser Doppler Anemometry technique. It served as a reference for the evaluation of simplifications with the symmetry and periodic boundary conditions, although the inclusion of an appropriate turbulent inlet boundary condition for the Large Eddy Simulations could have improved the numerical results against the experimental data.

The time-averaged velocity distributions and the wake formation and recirculation lengths revealed that the periodic simplification resulted in a much shorter wake, with a higher maximum value of the streamwise velocity fluctuations. This was reflected in the time-averaged force coefficients, with higher values for the mean drag coefficient and for the drag and lift coefficients fluctuations. The symmetry simplification case, on the other hand, provided an outcome close to the full domain simulation, concerning all evaluated parameters for this investigated case. This shows that, for moderate to low aspect ratio cylinder cases, it should be prioritized over the periodic condition simplification if computational and time resources allow for it. This study supports a broader research on the flow-induced vibration of tube bundles with variable tube spacing ratios, by means of physical experimentation and numerical simulation.

Acknowledgements

The authors would like to acknowledge the support of FAPESC (grant term number 2016TR2219) and CNPq (process number 308714/2016-4).

References

1. Bearman, P.W. (2011), "Circular cylinder wakes and vortex-induced vibrations", *Journal of Fluids and Structures*, v. 27, p. 648-658.
2. Blevins, R.D. (1990), "Flow-Induced Vibration", 2nd edition, *Van Nostrand Reinhold*, New York.
3. Germano, M., Piomelli, U., Moin, P., Cabot, W.H. (1991), "A dynamic subgrid- scale eddy viscosity model", *Physics of Fluids A*, v. 3, p. 1760-1765.
4. Lam, K., Lin, Y. F., Zou, L., Liu, Y. (2010), "Experimental study and large eddy simulation of turbulent flow around tube bundles composed of wavy and circular cylinders", *International Journal of Heat and FLuid Flow*, v. 31, p. 32-44.
5. Lam, K., Zou, L. (2010), "Three-dimensional numerical simulations of cross-flow around four cylinders in an in-line square configuration", *Journal of Fluids and Structures*, v. 26, p. 482–502.
6. Luo, D., Yan, C., Hongkang, L., Zhao, R. (2014), "Comparative assessment of PANS and DES for simulation of flow past a circular cylinder", *Journal of Wind Engineering and Industrial Aerodynamics*, v. 134, p. 65-77.
7. Moukalled, F., Mangani, L., Darwish, M. (2015), "The Finite Volume Method in Computational Fluid Dynamics An Advanced Introduction with OpenFOAM(R) and Matlab (Fluid Mechanics and Its Applications) ", 1st edition, v. 113 of Fluid Mechanics and its applications, *Springer - Verlag*, Switzerland.
8. Norberg, C. (1998), "LDV-measurements in the near wake of a circular cylinder", in *Proceedings of the 1998 ASME Fluids Engineering Division Summer Meeting*, C. H. K. Williamson and P. W. Bearman, ed. Washington DC, p. 1–12.

9. Prsic, M, A., Ong, M. C., Pettersen, B., Myrhaug, D. (2014), "Large Eddy Simulations of flow around a smooth circular cylinder in a uniform current in the subcritical flow regime". *Ocean Engineering*, v. 77, p. 61-73.

10. Silva, B.L., Santos, C.M., Bianchi Neto, P., Meier, H.F., Oliveira, E., Martignoni, W.P. (2016), "Flow simulations solve WHRSG vibration issues", *Oil and Gas Journal*, v. 114, p. 72–83.

11. Silva, B.L., Utzig, J. and Meier, H.F. (2016), "CFD-based design of experiments for the study of flow-induced vibrations in a tube bundle", *Proceedings of the 10th ABCM Spring School on Transition and Turbulence*, São José dos Campos, Brazil.

12. Sumner, D., Heseltine, J.L., Dansereau, O.J.P. (2004), "Wake structure of a finite circular cylinder of small aspect ratio", *Experiments in Fluids*, v. 37, p. 720-730.

13. West, G.S., Apelt, C.J. (1982), "The effects of tunnel blockage and aspect ratio on the mean flow past a circular cylinder with Reynolds numbers between 10^4 and 10^5", *Journal of Fluid Mechanics*, v. 114, p. 361-377.

499727 Out-of-Equilibrium Phase Behavior of Dielectric/Paramagnetic Nanoparticle Suspensions in Toggled Electric/Magnetic Fields

Abstract:

Structured materials self-assembled from dispersions of nanoparticles and colloids have numerous demonstrated applications including wave guides for optical computing, microlens arrays for energy harvesting, and porous electrodes for energy storage. One simple, tunable way to direct the self-assembly of these materials is by using electric and magnetic fields to induce dipole and higher order moments in dielectric or paramagnetic nanoparticles. As the particles self-assemble under a steady field, a thermodynamic constraint on the phase separation kinetics forces a tradeoff between quality of the self-assembled microstructure and its rate of formation. This is a key difficulty preventing adoption of self-assembled nanoparticle materials at scale. Dynamically self-assembling dispersions with interactions that vary in time do not have this constraint and offer a promising method to suppress kinetic arrest while accelerating growth of ordered nanostructures. In the case of dielectric/paramagnetic dispersions, varying the particle interactions in time is easily achieved by toggling the external electric/magnetic field on and off cyclically in time. Because of the cyclic driving force, materials formed by toggled self-assembly are dissipative and out-of-equilibrium, and therefore are classified as active matter, along with particles that swim, grow, or self-rotate. As toggled self-assembly cannot be understood in terms of equilibrium thermodynamics, new theories must be developed before it can be used to reliably fabricate nanomaterials, but there are several theoretical and computational challenges.

A fundamental model of these dispersions requires solving a many-bodied problem for the electric or magnetic field. Exact solutions are computationally expensive, so numerical simulations and thermodynamic calculations typically make the simplifying approximation that each particle behaves as a point dipole of fixed strength and direction. These models are inconsistent with experimental conditions, where particle moments are strong functions of configuration, and lead to phase predictions that fail to satisfy thermodynamic coexistence criteria. We have developed a new method to compute the multipole moments and forces to desired accuracy in suspensions of dielectric/paramagnetic nanoparticles. The method uses a moment expansion of the periodic, Fourier space representation of the Greenâ€™s function for Poissonâ€™s equation. Our method is spectrally accurate, scales nearly linearly with particle number, and is parallelized on graphics processing units, allowing for rapid simulation of large dielectric/paramagnetic dispersions. We use our method to perform thermodynamic calculations predicting the equilibrium phase diagram of hard, dielectric/paramagnetic nanoparticles in steady electric/magnetic fields, which agree with results from dynamic simulations. Next, we use our simulation method to show that cyclically toggling the electric/magnetic fields on and off can avoid kinetic barriers and yield well-ordered crystalline domains in these dispersions. The rate of phase separation, local and global quality of the self-assembled structures, and range of tunable parameters leading to acceptable self-assembly are all enhanced with toggled fields compared to steady fields. The growth mechanism and terminal structure of the dispersion are easily controlled by parameters of the toggling protocol, allowing for selection of processes that yield rapidly self-assembled, low defect crystals as well as out-of-equilibrium structures that cannot be stabilized with steady fields.

500111 Investigation of the Motion of Patchy Particles at Liquid/Fluid Interfaces

Abstract:

The majority of swimming microorganisms involved in human functions and diseases are found in confined fluidic environments. Inspired by the motion of such swimmers, many studies have been carried out to understand the principles underlying their motion and propulsion mechanism. Among these studies is the emerging field of active colloids including micro- and nanometer scale self-propelling particles. Active colloids have the ability to move, carry and transport components in a fluidic environment. They have a wide variety of applications, including medical and biomedical applications, lab-on-chip transport, targeted drug delivery and cancer fight-agent delivery. In many cases, such active particles are suspended in a solution bound by a liquidâ€"fluid interface, which raises interesting questions regarding the interaction of swimmers with a liquid/fluid boundary. To better understand the interactions in such soft mater systems, we study the dynamic motion of patchy and Janus particles with varying patch sizes trapped at liquid/fluid interfaces.

In this context, we have studied the Brownian and active motion of Janus and patchy particles trapped at the oil/air and water/air interfaces. Patchy particles are prepared using the GLAD deposition technique. Upon exposure to hydrogen peroxide, patchy particles display a self-induced phoretic motion by decomposing hydrogen peroxide into water and oxygen. The asymmetric distribution of the product molecules around the particles leads to the particlesâ€™ autonomous motion. The phoretic velocity and diffusivity of patchy particles of varying patch size trapped at a liquid/fluid interface will be analyzed and compared with the case of particles immersed in the bulk solution.

500432 Preparation and Characterization of Janus Dumbbells Nanomotors

Abstract:

The research on Janus nanoparticles has been gaining substantial momentum in the last years. Their non-homogeneously functionalized surface renders them much more interesting than their isotropic counterparts. One unique possibility offered by these particles is their transformation into nanomotors. In this work, we will discuss the synthesis, functionalization, characterization and self-assembly of Janus Dumbbells asymmetrically functionalized to become nanomotors.

We start by synthesizing polystyrene dumbbell-shaped nanoparticles in a three-step process, as has already been reported [1]. We then performed silane chemistry to selectively bring amino groups on the surface of one hemisphere of the dumbbells only. This is possible thanks to presence of silanol groups on the surface of the first hemisphere. One hemisphere of the dumbbells has then been selectively covered with gold nanocrystals. We prepared negatively-charged gold nanocrystals (coated with citric acid) by a Turkevich-like method. Due to the electrostatic interactions between the amino groups on the polystyrene dumbbells and the citrate groups on the gold nanocrystals, mixing the Janus dumbbells with the gold nanocrystals leadsÂ the latter selectively attached on one hemisphere of the former particles, while the other hemisphere remains naked. On top of interesting plasmonic properties, gold nanoparticles are also known to act as â€œseedsâ€ for the growth of platinum nanoparticles. An asymmetric distribution of such catalytic nanoparticles leads to self-propulsion of the dumbbells in the presence of hydrogen peroxide as fuel.

In this work, Janus nanomotors with different sizes, from 100nm to several hundred of nanometers have been prepared, and their self-propelling ability in the presence of hydrogen peroxide have been tested using single-point fluorescent correlation spectroscopy.Â Additionally, their self-assembly in the presence of electrolytes with/without fuel has been investigated.

[1] F. Guignard, M. Lattuada, Langmuir 2015, 31(16):4635-43.

500946 Magnetic Microlassos for Reversible Cargo Capture, Transport, and Release

Abstract:

Most microbot propulsion schemes powered by external fields or chemical reaction lack the capability to load and unload contents as may be necessary for applications such as drug delivery. To address this shortcoming, we introduce a reversible method that employs flexible and linked superparamagnetic chains that form closed rings or â€œlassosâ€ in the presence of a planar rotating magnetic field. We show that these magnetic lassos can roll at substantial velocities with precise spatial control by manipulating both field strength and phase lag. These rings can be used to grab onto cargo then translate based on a wheel-type mechanism, all without the need for chemical attachment or disengagement. In addition, we demonstrate that once the target destination is reached, cargo can be easily released upon field removal leaving the lasso ready for reuse.

492534 Fly Larvae Mix to Increase Eating Rates

Abstract:

Black soldier fly larvae are edible maggots that can transform tons of food waste into sustainable protein per day. Although they are known to have a collective motion around and inside food sources, a physical understanding of their high eating rates is missing. We take an active matter approach to studying larvae to show how their feeding behavior allows large aggregations to eat. Larvae form a dry, polar active system that exhibits either turbulent or jammed behavior. A combination of chemotaxis and quorum sensing allows larvae to find the easiest path to a food source. Active self-mixing allows hungry larvae to push full larvae from food, increasing their eating rates tenfold compared to hypothetical stationary larvae. We show that there is a boundary layer of mixing larvae near food, while the rest of the aggregation exhibits collective burrowing behavior using video tracking and compression force experiments.

493076 Programming the Dynamics of Active Colloids in 3D

Abstract:

Directing particle-motion at low Reynolds number is the key to engineer future biomimetic active materials. Presently used methods of introducing active motion in colloids lack in directional control over their non-linear migration, and preprograming the particle-trajectory is a scientific challenge. Here we study the spatial propagation of metallodielectric patchy microspheres by Induced Charge Electrophoresis (ICEP) in the presence of AC-electric field. Application of the high frequency electric field to a patchy colloidal dispersions leads to a local force imbalance around the particle, resulting into its direction motion. The surface-force imbalance and torque acting on the particle determines the motion of particle in 3D space. Here we demonstrate that the particleâ€™s velocity, and its 3D trajectory can be programed by engineering the patch size and shape. We show that by introducing chirality into the surface patch of the particle, we can program the motion of individual particle into helices of desired handedness. Our approach introduces a new method of engineering the motion of microparticles in three dimensional space, which could lead to the development of advanced micromotors and miniature robots.

494420 Tuning the Collective Behavior of Active Electrohydrodynamic Motors

Abstract:

As demonstrated in biological systems, breaking the symmetry of surrounding hydrodynamic flow is the key to achieve autonomous locomotion of microscopic motors. In recent years, a variety of synthetic motors have been developed based on different propulsion mechanisms. Although collective motion of those motors has been observed, fine control of the hydrodynamics of individual motors and the resulting assembly behavior is chanlleging. In this talk, we will report our recent observation of collective assembly of asymmetric colloidal motors under an applied aleternating current electric field. Although the motors experience strong dipolar repulsion from each other and are highly active, surprisingly, they assemble into a variety of close-packed and stable structures. We show that, through combined experimental and numerical investigations, it is the electrohydrodynamic interaction between specific parts of the asymmetric motors that drives the assembly. By changing both the interfacial and geometric properties of the colloidal motors via a chemical approach, we can fine-tune their assembly patterns resulting from collective motion of the motors.

496301 Heterogeneous Active Matter Systems

Abstract:

Active systems are composed of self-propelled (i.e. active) particles that locally convert energy into motion and exhibit emergent collective behaviors, such as fish schooling and bird flocking. Most work so far have focused on monodisperse, one-component active systems. However, real systems are heterogeneous, and consist of several active components. We perform molecular dynamics simulations of multi-component active matter systems and report on emergent behavior. We discuss the phase diagram of dynamic states demonstrating accessible behaviors for mixed systems following basic rules of motion. We present potential utility of heterogeneous active matter systems in the context of medicinal, natural, and synthetic applications.

491392 Directed Motion of Metallodielectric Particles By Contact Charge Electrophoresis

Abstract:

Active colloidal particles harness energy from their environment to power directed motions relative to their fluid surroundings. This talk will describe a mechanism for colloidal propulsion based on contact charge electrophoresis (CCEP) of metallodielectric Janus particles between two parallel electrodes. CCEP uses a constant voltage to repeatedly charge and actuate conductive particles within a dielectric fluid, resulting in rapid oscillatory motion. In addition to particle oscillations, micrometer-scale Janus particles move perpendicular to the field at high speeds (up to 600 Î¼m/s) and over large distances. We characterize these directed motions and propose a mechanism based on the rotation-induced translation of the particle following charge transfer at the electrode surface. The propulsion mechanism is supported both by experiments with fluorescent particles that reveal their rotational motions and by simulations of CCEP dynamics that capture the relevant electrostatics and hydrodynamics. We also show that interactions among multiple particles can lead to repulsion, attraction, and/or cooperative motions depending on the position and phase of the respective particle oscillators. Our results demonstrate how particle asymmetries can be used to direct the motions of active colloids powered by CCEP. By expanding the repertoire of active colloids, we aim to discover new dynamical behaviors and enable applications such as cargo delivery.

495388 Molybdenum Dioxide Functionalized CoP Porous Nanocatalyst for Highly Efficient Hydrogen Evolution

Abstract:

The development of highly active and stable catalysts based on low-cost materials for hydrogen evolution reaction (HER) is crucial to catalytic water splitting. In this work, we report on using molybdenum oxide as an intermediary (pore-forming agent and molybdenum dioxide precursor) to synthesis molybdenum dioxide functionalized CoP nanosheets for HER. By virtue of modulating hydrogen adsorption/desorption energies on the central CoP, together with the in situ formation of the porous structure, the as-prepared porous CoP/MoO_2 hybrid nanosheets exhibit an excellent HER catalytic performance, only requiring small overpotential of 41 mV to support a current density of 20 mA cm^{-2}, which is comparable catalytic performance to Pt/C benchmark and exceeds all reported transition-metal phosphides HER catalysts.

495753 Bijel-Derived Materials for Electrochemical Energy Storage and Conversion

Abstract:

Pressing energy demands for national security, distributed power, and portable technologies require the development of next-generation batteries and supercapacitors that can simultaneously deliver high power and energy densities. In order to achieve this goal, the kinetics of ion and electron transport within the electrodes must be enhanced while maintaining a large capacity for energy storage. Therefore, the idealized electrode microstructure in batteries has been envisioned as a three-dimensional, co-continuous arrangement of percolating passages for efficient ion and electron transport that would also enable a large volumetric density of active materials. We report a novel method to produce this distinctive microstructure based on colloidal self-assembly into bicontinuous interfacially jammed emulsion gels (bijels), which are a new class of soft materials with a co-continuous microstructure where two interpenetrating fluid domains are kinetically arrested through jamming of particles at the fluid-fluid interface during spinodal decomposition. Bijels can be converted to electrolytically active composites with co-continuous internal microstructure through simple chemical post processing steps, maintaining the unique morphological characteristics, optimum transport properties, and tunable internal morphology of the original emulsion. In this talk, we will present microstructured composite electrodes that have been synthesized by this technique and demonstrate how their electrochemical characteristics can be tuned over a wide range, allowing for concurrent delivery of high energy and power. The broader applications of this new class of soft materials in electrochemical energy conversion and storage, sensing, and catalysis will also be discussed.

496015 Electrochmical Lithiation-Delithation of Sulfur in Sub-Nano Confinement

Abstract:

Despite the complexity of lithium–sulfur (Li–S) electrochemical reactions, most of the challenges of Li-S batteries are originated from the polysulfide solution phase reactions. In this study, we investigate a solid-state Li-S reaction in liquid electrolytes enabled by sub-nano confinement of sulfur. We investigate two factors that play decisive roles in Li–S electrochemical processes: the size of the sulfur confinement and the type of electrolyte solvents. To precisely capture the subtle changes in Li–S electrochemical behavior due to the different sulfur confinement size, a series of porous carbon hosts with different pore sizes (1, 2, 2.5, 3 nm) are selected. In addition, two electrolytes, ether-based electrolyte (TEGDME) and carbonate-based electrolyte (EC/DEC) with very different polysulfides solubility are utilized for the study. The results demonstrate a clear correlation between the size of sulfur confinement and the resulting Li–S electrochemical mechanisms. In particular, when sulfur is confined within sub-nano pores, we observe identical lithium–sulfur electrochemical behavior, which is distinctly different from conventional Li–S reactions, in both ether and carbonate electrolytes. We propose that Li ions can only enter the sub-nano pores through desolvation, therefore, the Li-S electrochemical reaction in the sub-nano pores is in solid state. To further study the effect of Li-ion solvation on the Li-S reaction in the sub-nano confinement, electrolytes based on a series of ethereal solvents (dimethoxyethane, diglyme, triglyme, tetraglyme, and 15-crown-5) are studied. With combination of electrochemical analysis, spectroscopic characterization and DFT-based computation, our results revealed the clear effect of solvation energy of different solvent molecules to Li-ion on the Li-S electrochemical reactions in the sub-nano confinement.

496071 Durable High Capacity Li-O_2 Cathode Composed of Iron-Nitrogen-Doped Mesoporous Core-Shell Carbon Loaded with RuO_2 Nanoparticles

Abstract:

The reversible lithium oxygen battery has generated a lot of interests for its high energy density. Limitations of oxygen reduction and evolution (ORR/OER) kinetics in aprotic solvents, lack of reversible lithium peroxide deposition/dissolution, and degradation problems have hintered development of Li-O_2 technology. There are few reports of durable high capacity cathode performance. Nitrogen-doped mesoporous carbon has been succesfully deployed for ORR in aqueous electrolytes[1]. Here, we report Fe-N-doped hierarchical mesoporous carbon loaded with RuO_2 nanoparticles (Fe-N-HCMS-RuO_2) for application as a Li-O_2 cathode. Promising results are observed with capacity reaching 1/3 the theoretical energy density of Li-O_2 (3,500 Wh/kg). The Fe-N-HCMS-RuO_2 electrode demonstrates excellent performance which can be attributed to three features of the composite. A well-defined core-shell mesoporous structure provides excellent mass transport and limits the Li_2O_2 deposition to be dispersed in nanoscale. The Fe-N doping of carbon provides the corrosion resistance, high conductivity, and active sites for favorable interaction with RuO_2. Small and amorphous RuO_2 provides highly catalytic activity for decomposition of Li_2O_2. Discharge capacity of 10,000 mAh/g is demonstrated with discharge potential of 2.4 V and recharge potential of 3.6 V maintained at 2 A/g based on total mass of catalyst and carbon support and maintained at 2.5 V and 3.5 V at 1 A/g over 50 cycles.

501279 Magnetic Field-Induced Fabrication of Fe_3O_4@Graphene Nanocomposites for Enhanced Electrode Performance in Lithium-Ion Batteries

Abstract:

We report a novel magnetic field-assisted approach for the fabrication of nanoporous, non-agglomerated, wrinkled Fe_3O_4@reduced graphene oxide (RGO) anode material for lithium ion batteries (LIBs). The RGO sheets provide protection to Fe_3O_4 nanoparticles by preventing Fe_3O_4 agglomeration and facilitating the accommodation of the drastic volume change of Fe_3O_4 during lithiation/delithiation, thus enhancing the structural integrity and interfacial stabilization of the electrode. The Fe_3O_4@RGO anode with porous/wrinkled architecture does not only facilitate the Li ion and electron transport kinetics, but also simplifies the electrode fabrication process, as it does not require a polymer binder and a carbon conductive additive. As a result, at relatively low specific currents (157 mA/g), such Fe_3O_4@RGO anode shows high reversible specific capacities (up to 903 mAh/g at 157 mA/g). At high discharge rate (1.57 A/g), the specific capacity of such Fe_3O_4@RGO anode stays at 345 mAh/g (with capacity retention of 90%) after 100 discharge/charge cycles compared to the rapid capacity fading associated with nonporous /flat Fe_3O_4@RGO anode with a specific capacity of 178 mAh/g after same number of cycles. These results demonstrate that magnetic field treatment is a promising approach for fabricating desired nanocarbon structures with magnetic nanoparticles.

490885 Elucidating Mechanisms of Substrate Transport in Membrane TransportersÂ

Abstract:

Understanding the exchange of biological molecules across the phospholipid bilayer is of fundamental interest to the scientific community. In particular, transporters from the major facilitator superfamily (MFS) are involved in the transport of a wide spectrum of substrates across biomembranes and play a pivotal role in multiple physiological processes. Despite intense investigation, only few MFS proteins from six subfamilies have been structurally elucidated. Computational investigations have provided tremendous insights in the structure and function of these transporters. However, large scale unbiased simulations that could provide a molecular view of the complete transport cycle have not been reported. Here, we report development and application of novel computational methods for (a) enhanced unbiased sampling of the conformational landscape of membrane transporters and (b) optimal design of experimental probes for membrane protein dynamics. We demonstrate the application of these methods to investigate the transport cycle of the following membrane transporters: peptide transporter $\mathrm{PepT_{SO}}$, nitrate transporter *At*NRT1.1 and Sugar Transporter *Os*SWEET2b. These proteins have only one available crystal structure. Our results provide for the first time, the complete transport cycle of these key transporters including key conformational states along the transport pathway using large scale (100's of microseconds) molecular dynamics simulations.

494867 Grafting Charged Species to Membrane-Embedded Scaffolds Dramatically Increases the Rate of Bilayer Translocation

Abstract:

Functionalized nanoparticles (NPs) are versatile materials with heterogeneous surface properties that can be engineered to mimic typical biological macromolecules. Recently, a particular class of charged, amphiphilic NPs were shown to embed within lipid bilayers to obtain configurations similar to transmembrane proteins. This behavior is surprising because the NPs must also transport charged groups across the hydrophobic bilayer core within short timescales to stably insert into the bilayer, a process that is conventionally thought to occur on timescales far too long (hours to days) to be experimentally relevant. However, the rapid translocation of charged species across lipid bilayers is not unique to these NPs, but is also observed in disparate membrane protein systems, suggesting that some shared mechanism may lead to short (sub-second) translocation timescales. Here, we use atomistic molecular dynamic simulations with enhanced sampling techniques to gain a mechanistic understanding of this rapid charge transport. We show that charged species grafted to a bilayer-embedded scaffold â€“ such as the NPs or multispanning transmembrane proteins â€“ translocate across the bilayer significantly more rapidly than isolated ions. We find that rapid translocation is enabled by the partial dehydration of charged species in their initial configurations, and is further assisted by cooperative interactions between multiple hydrophilic end groups. This work suggests design guidelines for synthetic materials capable of transporting charged, soluble small molecules across bilayers, which may be useful for ferrying therapeutic molecules to the interior of cells for drug delivery applications.

495147 Molecular Simulations of Nanoparticles Permeating Lipid Membranes for Drug Delivery Applications

Abstract:

The study of effectiveness of surface-decorated nanoparticles is important for developing targeted drug delivery systems where functionalized nanoparticles act as carriers for therapeutic agents delivered to specific sites in the body. Computational modeling of the transport of functionalized nanoparticles across membranes and nanoparticle-membrane interactions is valuable to experimentalists exploring a variety of nanoparticles as practical candidates for biological applications such as cancer therapeutics or imaging. In this study, we use molecular dynamics simulations to investigate permeation of lipid bilayer membranes by surface-functionalized nanoparticles and to provide molecular-level detailed mechanisms that are otherwise not readily available via most experimental means. Permeation of small molecules through lipid bilayer membranes is a fundamental biological process that is important to understand in developing therapeutic applications. Nanoparticles have recently become important carriers for drug delivery in such therapeutic applications. We describe coarse-grained simulation methods applicable to such large systems and use examples where the permeants are bare gold nanocrystals, gold-core nanoparticles with hydrophobic ligands (alkanethiol ligands of various lengths), and gold-core nanoparticles with hydrophilic ligands (methyl-terminated polyethylene glycol ligands of various lengths). In addition to spherical gold nanoparticles, we also examine the characteristic permeation mechanisms of the differently shaped gold nanorods with polyethylene glycol ligands, where the aspect ratio different from 1 makes the permeation event dependent on the angle of the nanorod axis relative to the membrane surface. We have examined the phenomena associated with the interaction of these various permeants with the phospholipid bilayer that serves as our model membrane. We consider adsorption at the interface, the permeant within the top lipid leaflet, in the middle of the membrane within the lipid tail region, and finally exiting the membrane on the way to recovery, for the various permeants. As a consequence of nanocarrier permeation, we observe the formation of a water pore, occasional transport of ions across the lipid bilayer, lipid translocation, and lipid displacement from the membrane. We investigate the rotational behavior of PEGylated nanorods during the permeation process and the available pathways for these nanorods in completing the permeation. These events differ depending on the chemical nature (hydrophobic or hydrophilic) of the ligands, their length, the surface coverage density on the gold surface, and the aspect ratio of the gold nanocarrier core. Our work, which highlights comparisons of permeation of nanoparticles with size, shape, and functional diversity, should aid experimentalists who are designing desirable candidates for drug delivery applications.

501666 Transport Characteristics and Stability of Artificial Water Channels in Lipid and Biomimetic Membranes

Abstract:

Cell membranes owe their high solute or water permeability to various membrane proteins (MPs) that are an essential component of the cellular machinery for exchanging materials, energy, and information. Many MPs such as channels and pumps possess transport characteristics with high specificity that is unmatched in synthetic systems. For example, MPs can sense and transport small solute or solvent molecules both along and against concentration gradients at a rate of billions of molecules per second. Despite these outstanding properties, it is not possible to employ cellular membranes containing MPs in industrial scale separation processes due to their poor chemical and mechanical stability. Although biomimetic membranes made up of self-assembling block copolymers are emerging as a highly promising alternative, the interactions and transport characteristics of biological or bioinspired artificial channels embedded in block copolymer membranes are not well understood at the molecular scale. In this talk, we will present results from long time-scale all-atom molecular dynamics (MD) simulations of a synthetic peptide-appended pillar[5]arene (PAP) channel incorporated in lipid and block-copolymer membranes. Our results suggest that the membrane environment can significantly influence the conformational dynamics of PAP channels and potentially their diffusive as well as transport characteristics.

499644 Solute Transport through Outer Membrane Porins of Gram-Negative Bacteria

Abstract:

Porins facilitate the uptake of antibiotic and nutrients across the outer membrane of Gram-negative bacteria like Escherichia coli. In this work, we focus on solute transport through the porin channels by systematically generating the complex outer membrane description of E. coli in molecular-level detail. In general, cell envelope of gram-negative bacteria contains lipopolysaccharide (LPS) rich outer membrane that acts as the first line of defense for bacterial cells in adverse physical and chemical environments. The LPS macromolecule has negatively charged oligosaccharide domain that limits the permeability of charged chemical agents through the membrane. Elucidating the interplay of these contributing macromolecular components and their role in the transport of solutes through porins remains a challenge. To bridge the gap in our current understanding, we have developed a coarse-grained force field to quantify the transport properties through OmpF channels in E. coli outer membrane. The model is computationally affordable for simulating dynamical processes over physiologically relevant timescales. The force field was benchmarked against available experimental and atomistic simulations data for properties such as membrane thickness, density profiles of the residues, the area per lipid, gel to liquid-crystalline phase transition temperatures, and order parameters. A comparison of simulated structural and dynamical properties with corresponding experimental data shows that the developed force field reproduces the overall physiology of solute transport through LPS rich E. coli outer membranes. The simulations provide the molecular mechanisms for permeation of antibiotics through the outer membrane and the strategies that can be employed to develop a new class of antimicrobial agents.

500020 Could Class IIb Bacteriocins Induce Pore Formation? Investigation through Microsecond Long Atomistic Molecular Dynamic Simulation

Abstract:

Antibiotic resistance poses an alarming threat to public health and modern society at large. In our times, it is an unfortunate reality that resistance has been found to all commercially available antibiotics, even to the so-called "last resort" ones [1,2]. Different government-issued reports warn that we are approaching a "post-antibiotic era", where even minor bacterial infections could be deadly [1,2]. As the number of cases of antibiotic resistance is increasing, the effective utilization of conventional antibiotics is rapidly decreasing. To overcome this challenge, alternative drugs should be examined. One promising alternative source of new antibiotic drugs that has been investigated extensively of late is the use of antimicrobial peptides (AMPs) [3]. Bacteriocins are antimicrobial peptides (AMPs) produced by bacteria that show great potential as novel antibiotic agents [4]. They are active at lower concentrations and they exhibit higher specificity against their targets, compared to other AMPs [5]. In order to most powerfully utilize the full potential of bacteriocins and use them as a platform to develop new antibacterial agents, we must develop an understanding of their mechanism of action.

We study Plantaricin EF (PlnEF), a two-peptide class IIb bacteriocin, that is comprised of Plantaricin E (PlnE) and Plantaricin F (PlnF). It has been well-documented that class IIb bacteriocins induce cell death through membrane leakage [6]. Although, the complete mechanism of action of class IIb bacteriocins is still unclear, it has been suggested that these peptides act through a receptor-mediated mechanism of action. To elucidate the initial interaction of the peptides with the membrane, we have previously performed molecular dynamics simulations of the two-peptide bacteriocin PlnEF on the surface of a model lipid bilayer [7]. Next, we employed an extensive mutational analysis and created fusion peptides that blocked the activity of each of the terminai of each peptide in order to identify the orientation of the peptides in a transmembrane conformation [8]. These experiments suggested that PlnE and PlnF create an antiparallel dimer when located inside a membrane. We were then able to design a transmembrane model of the dimer embedded in the bilayer.

The dimer showed a remarkable stability over a 1 μs long atomistic molecular dynamics simulation. We analyzed the dimer structure and identified important amino acids that are responsible for the interaction between the peptides and the anchoring of the dimer in the membrane. We further investigated the impact of the of the dimer's presence to the rest of the system (the membrane, water, and ions). Most importantly, we showed that the transmembrane PlnEF dimer can form a small torroidal pore that allows water permeation and suggests possible ion conduction. This is the first time (to our knowledge) that it has been demonstrated in atomistic detail that a LAB bacteriocin with narrow antimicrobial activity range, can form pores on its own. We believe this finding could be of immense importance to the designing of new antibiotic agents, as it would steer the search for better bacteriocins toward peptides that form stable pores, interact more strongly with the membrane in specific regions, and increase water or ion permeability.

References:

[1] S. Reardon, WHO warns against "post-antibiotic" era, Nature. (2014).

[2] Antimicrobial Resistance: Tackling a Crisis for the Future Health and Wealth of Nations, in: Rev. Antimicrob. Resist. Chaired by Jim O'Neill, 2014.

[3] D.H. Lloyd, Alternatives to conventional antimicrobial drugs: a review of future prospects., Vet. Dermatol. 23 (2012) 299–304, e59-60.

[4] P.D. Cotter, R.P. Ross, C. Hill, Bacteriocins - a viable alternative to antibiotics?, Nat. Rev. Microbiol. 11 (2013) 95–105.

[5] J. Nissen-Meyer, I.F. Nes, Ribosomally synthesized antimicrobial peptides: their function, structure, biogenesis, and mechanism of action, Arch. Microbiol. 167 (1997).

[6] J. Nissen-Meyer, C. Oppegård, P. Rogne, H.S. Haugen, P.E. Kristiansen, The Two-Peptide (Class-IIb) Bacteriocins: Genetics, Biosynthesis, Structure, and Mode of Action, in: Prokaryotic Antimicrob. Pept., Springer New York, New York, NY, 2011: pp. 197–212.

[7] P.K. Kyriakou, B. Ekblad, P.E. Kristiansen, Y.N. Kaznessis, Interactions of a class IIb bacteriocin with a model lipid bilayer, investigated through molecular dynamics simulations, Biochim. Biophys. Acta - Biomembr. 1858 (2016) 824–835.

[8] B. Ekblad, P.K. Kyriakou, C. Oppegård, J. Nissen-Meyer, Y.N. Kaznessis, P.E. Kristiansen, et al., Structure–Function Analysis of the Two-Peptide Bacteriocin Plantaricin EF, Biochemistry. 55 (2016) 5106–5116.

501013 Host-Pathogen Interactions at the Tight Junctions: Molecular Insights from Membrane Guided Self-Assembly

Abstract:

Transmembrane proteins of the claudin family assemble in the paracellular space to from tight junctions. Clostridium perfringens is a spore forming opportunistic pathogen that releases enterotoxins (CPE) that disrupt intestinal tight junctions. *C. perfringens* related enteric infections are the second leading cause of foodborne illness in developed countries. Several members in the claudin family act as receptors for the CPE, and they specifically bind to the C-terminal domain of the toxin (cCPE). CPE initially forms a small complex with receptor claudins that consequently initiates the assembly of a ~450 kDa hexameric pore complex CH-1. The CH-1 complex causes Ca2+ influx and consequently apoptosis. The interaction between the cCPE and claudins are active topics of research for potential translational applications that range from cancer treatment and opening the blood-brain barrier. Crystal structures of the full-length toxin show them to form solvent stabilized trimers in C3 symmetry and more recently, the cCPE complex with claudin-4 was resolved, yet still, the molecular and mechanistic details of the pathogenesis remain to clarified. In this study using multiscale self-assembly simulations, we show that many claudins (including CPE receptor claudin-3/-4), form C3 mediated symmetric interaction in the membrane, which match the receptor orientation for the trimeric CPE toxin. The trimer arrangement presents an electrostatic interaction site that complements the binding pockets in the CPE. We further performed molecular docking and atomistic molecular dynamics to unravel the mechanism of Claudin-CPE interactions. The results from this work will be instrumental in designing translational therapeutics utilizing CPE.

501149 Structure of Multi-Component, Gel-Phase Lipid Bilayers

Abstract:

The composition of lipid membranes has a strong impact on the membraneâ€™s structure and transport properties. While fluid-phase phospholipid bilayers have been studied extensively, far less is known about how lipid composition and packing influence the behavior of dense, gel-phase bilayers. Molecular dynamics (MD) simulations of multi-component, gel-phase bilayers of 1,2-distearoyl-sn-glycero-3-phosphatidylcholine (DSPC), free fatty acids (FFA), and alcohol (OH) molecules of varying chain lengths and composition are used to elucidate the role of lipid composition on gel-phase membranes. Specifically, binary and ternary systems are studied, and the effect of molecular substitution, chain length and DSPC fraction on the bilayer properties are examined. In agreement with previous work[1,2], large headgroups are found to increase the area-per-lipid, area-per-tail, and tilt angle, while longer chain lengths increase the bilayer height. Interdigitation is found to depend on the relative fraction of the protruding tails and tail length asymmetry between molecular groups. Both binary and ternary systems demonstrate placement of molecular groups at different depths within the bilayer as a result of the competing molecular interactions. The relative depths of molecules can subsequently affect further displacement within the bilayer due to steric repulsions, tail interactions, and electrostatic forces from multiple components. Alcohol molecules located at a depth between two molecular groups push the deeper group deeper into the bilayer due to steric repulsions overcoming any electrostatic attraction. Free fatty acid molecules located at a depth between two molecular groups attract the deeper group closer to the bilayer surface due to electrostatic attractions overcoming steric repulsions. Finally, depending on the length of tails, interdigitation can drive molecular groups deeper within the bilayer or closer to the surface. The balance between steric repulsions from head groups, van der Waals interactions from tails, and electrostatic interactions affects the structure of multi-component, gel-phase bilayers in complex, yet rational ways.

Works Cited:

[1] Hartkamp, R.; Moore, T. C.; Iacovella, C. R.; Thompson, M. A.; Bulsara, P. A.; Moore, D. J.; McCabe, C. Journal of Physical Chemistry B, 2016, 120, 12863-12871

[2] Hartkamp, R; Moore, T. C.; Iacovella, C. R.; Thompson, M.; Bulsara, P; Moore, D. J.; McCabe, C. Biophysical Journal, 2016, 111, 813-823

502399 Multiplicity of States and Internal Structure of Supported Lipid Bilayers

Abstract:

Lipid bilayers attached to solid surfaces play important role in biomimetic materials and devices and serve as model systems for studies of interactions of cell membranes with particles and biomolecules. Despite active experimental and theoretical studies, the interactions of lipid membranes with solid substrates are still poorly understood. In this work, we explore, using atomistic molecular dynamics simulations, equilibrium and stability of phospholipid DMPC membrane supported on hydroxylated amorphous silica. We reveal two distinct types of thermodynamically stable states, characterized by the different width of the water layer between the membrane and the substrate. In Î±-states, the membrane is closely attached with the lipid head groups interacting directly with surface hydroxyls, however due to the molecular level roughness of amorphous silica surface, there exists an inhomogeneous water layer trapped between the substrate and the membrane. In Î²-states, the membrane is separated from the silica surface by a water film of ~2.5 nm in thickness. The thermodynamic equilibrium is quantified in terms of the disjoining pressure isotherm as a function of membrane-substrate separation, which has a double sigmoidal shape with two minima and one maximum, which correspond to the limits of stability ofÂ Î± and Î² states. The thermodynamic properties and bilayer structure are compared with experimental findings and simulation results for relevant systems.

485531 Line Tension and Lipid Sorting Modulate Dynamics of Hemifusion Diaphragm Dissipation

Abstract:

Lipid bilayer fusion is commonly believed to proceed through the stalk-pore mechanism. This hypothesis includes an intermediate in which two distal monolayers of merging bilayers come together to form a new bilayer, called the hemifusion diaphragm (HD). Several factors aid in the formation and stabilization of an HD, including the presence of non-bilayer lipids. Lipids with negative intrinsic curvature (NIC) have been shown to enhance the probability of HD formation and effect the stability and formation of the resulting HD structure. Despite a theoretical basis for the necessity of NIC lipids to the HD, lipid contributions to the dynamics of HD formation are still not fully understood. We studied the effects of intrinsic curvature and lipid concentration on protein-free HD dynamics through the use of molecular dynamics (MD) and a coarse-grained lipid model. Large HD's (r_{HD} 10 nm, 20 nm) were artificially produced. Binary lipid mixtures were added by randomly seeding the HD's with lipids of varying strengths of NIC (low, moderate, and high) at 0.05, 0.10, 0.17, 0.25, 0.50, 0.75, 1.00). Systems were run for a total of 2 μs in an $NP_{xx}P_{yy}L_zT$ ensemble.

HD's are found to decay to three different metastable end states dependent on intrinsic curvature and composition of the constituent lipids. These end states include a non-fusogenic double bilayer and fusogenic fusion pore and stable HD. Systems decay to a double bilayer at low compositions of NIC lipids or when NIC lipids are weak. Moderate NIC lipids or moderate compositions of strong NIC lipids decay to a metastable fusion pore while high compositions of strong NIC lipids stabilize the HD. Systems which decay to a double bilayer are found to have similar dynamics regardless of lipid composition while in fusogenic systems, lipid compositions are correlated to the rate of HD dissipation. In addition, the method of dissipation appears different between fusogenic and non-fusogenic systems. Non-fusogenic systems formed a network of interconnected bicelles in the stalk region with large pores in the region. In some of these systems, the network of bicelles aligned in such a way that a single flat bilayer formed and the previously joined bilayer was left with a large pore ($r_{pore} \approx 8$ nm) in the center. The fusogenic systems maintained a connected stalk region throughout the dissipation process with only small pores present around the three-junction. The difference in decay rates and method of dissipation alludes to a fundamentally different driving force in fusogenic compositions and non-fusogenic compositions.

We conclude that the metastable end state is dependent on the mode of HD dissipation. When systems do not have sufficient NIC lipids to mediate the line tension of the three-junction, the stalk region forms a network of free edges. In this case, the stalk region harbors a line tension and a compressive force on the interior HD bilayer exists. This compressive force can be measured through a progressively decreasing apparent area per lipid in the HD bilayer as the HD dissipates. In the fusogenic system, no such compression exists. The fusogenic systems are found to sort NIC lipids towards the three-junction to effectively minimize the line tension in the stalk region. In these systems, though no significant compressive force on the HD bilayer exists, dissipation is driven by lipid movement into the three-junction from the HD bilayer. As the HD dissipates, the secondary radius dictated by the HD size decreases which effectively increases the necessary composition of NIC lipid required to minimize the line tension. NIC lipids respond by flipping out of the HD and outer bilayer and into the three-junction. The movement of NIC lipids out of the HD bilayer simultaneously decreases the NIC lipid concentration in the HD bilayer and the radius of curvature of the three-junction. These two factors drive the HD to shed more NIC lipids to mediate the increased line tension in the three-junction and increased effective area per lipid in the HD bilayer. When the NIC lipids have stabilized the three-junction and sorting towards the junction halts, the driving force vanishes and the system assumes a stable HD state.

This work aims to elucidate the effect of NIC lipids in fusion events and further understand the dynamics of the fusion process. In the two proposed dissipation methods, the HD is dependent on lipid flip-flop to shrink the HD. Flippases have previously been indicated as a functional component of the fusion process. Our work implicates flip-flop as a mechanism by which the bilayer may control the fusion process and sheds light on the role of flip-flop in HD formation and dissipation. Though this work uses a highly coarse-grained lipid model, it is believed that these findings may have implications in biological fusion events.

494505 Understanding the Removal of Pharmaceuticals from Water Effluents By Adsorption in Activated Carbons – a Molecular Simulation Approach

Abstract:

There is an increased concern in recent years on the removal of pharmaceuticals from water effluents. Many of these compounds are at concentrations of mg/L or below, are hardly biodegradable even at the outlet of wastewater treatment plants and much on their potential impact on environment and human health are still not completely known. Among the main removal technologies reported in the literature, removal by adsorption is one of the most promising techniques because of its convenience when applied in current water treatment processes. Adsorption using activated carbon is ideal for removing organic compounds due to the availability of high surface area and combination of well-developed pore structure and surface functional group properties [1]. Hence, the identification of the mechanisms governing the adsorption phenomena of pharmaceuticals in aqueous solutions is of fundamental importance.

The main objectives of this study was to evaluate, by molecular simulations, the competitive adsorption of pharmaceutical compounds as pollutants contented in water on activated carbons of varied characteristics, and to compare the removal capacities from equilibrium simulations on single component and aqueous mixtures. Molecular simulations using the Grand Canonical Monte Carlo (GCMC) method have been performed in order to understand the interaction of pharmaceutical molecules (ibuprofen, diclofenac, naproxen, paracetamol and amoxicillin, all with similar dimensions, except amoxicillin) on activated carbons in both, pure and aqueous mixtures. "Realistic" nanoporous carbon models based on units of polyaromatic molecules with different number of rings, defects and polar-oxygenated sites have been employed for this purpose [2].

The modeling approach we have used for the activated carbons, inspired by the early work of Gubbins and co-authors (see, for instance, reference [3]) represents a middle ground between the slit pore model and the fully atomistic models of carbon; it is an extension of the model packing idealized structures first developed by Segarra and Glandt (SG) [4]. We have modeled graphene planes as platelets, but including polar groups and defects, not explicitly considered in the SG model, and in a similar fashion of the extended Liu-Monson model [5]. Furthermore, in order to create imperfections on the surface, increased surface roughness were created by deleting up to 10–20% of the atoms in the inner carbon layers. Details on the calculations can be found in [2].

Our methodology allowed us to obtain a successful prediction of the drug adsorption properties. Comparison of the simulated capacities with experimental reported values is highly consistent. The model is sufficiently flexible to tune its parameters in a systematic way and introduce additional features to reflect specific properties of the material of interest, *i.e.*, effects of platelet sizes and geometries, surface area, concentration and nature of surface groups, among others. Results confirm the role of surface sites in the enhancement of water adsorption and reduction of drug adsorption as the oxygenated sites exceeds 10%wt. Reduction of the surface functionality gives a higher adsorption capacity of pharmaceuticals in aqueous mixture, since there is not competition with water molecules.

This work is a further step in order predict macroscopic performance of activated carbons in drug removal applications, providing useful molecular-level insights into the drug/water adsorption mechanism and allowing finding the most favorable material.

References:

1. L. Nielsen, T. J. Bandosz, Analysis of the competitive adsorption of pharmaceuticals on waste derived materials, Chemical Engineering Journal 287 (2016) 139–147.

2. D. Bahamon, L. Carro, S. Guri, L.F. Vega, Computational study of ibuprofen removal from water by adsorption in realistic activated carbons, Journal of Colloid and Interface Science, 498 (2017), 323–334.

3. J. Pikunic, C. Clinard, N. Cohaut, K.E. Gubbins, J.M. Guet, R.J.M. Pellenq, I. Rannou, J.-N. Rouzaud, Structural modeling of porous carbons: constrained reverse Monte Carlo method, Langmuir 19 (2003) 8565–8582.

4. E.I. Segarra, E.D. Glandt, Model microporous carbons: microstructure, surface polarity and gas adsorption, Chem. Eng. Sci. 49 (1994) 2953–2965.

5. J.C. Liu, P.A. Monson, Monte Carlo simulation study of water simulation study of water adsorption in activated carbon, Ind. Eng. Chem. Res. 45 (2006) 5649–5656.

504009 Insights into the Adsorption and Phase Behaviour of Fluids in Nanoporous Materials with Hierarchical Pore Structure:Â Towards an Advanced Textural Characterization

Abstract:

In recent years, major synthetic efforts have been focused on the introduction of hierarchical pore structures into many different materials such as zeolites, carbons, silicas, MOFs, hybrid layered and pillared structures. Nanoporous materials which consist of hierarchical pore networks with pore widths spanning the entire micro-mesopore range (i.e. up to 50 nm) allow one to overcome accessibility and diffusion limitations associated with classical microporous materials for catalysis and separations.

In this talk recent developments in the structural characterization of these materials will be described. An accurate textural characterization is crucial not only within the discovery process of novel hierarchically ordered nanoporous materials but also contributes to advance their application in a variety of areas including heterogenous catalysis, separation, battery devices, drug delivery. Among a variety of techniques, gas adsorption is widely applied because it assesses a wide range of pore sizes, spanning the entire micro- and mesopore range. Within the last two decades major progress has been achieved in physical adsorption characterization also because of the development of advanced approaches based on statistical mechanics such as molecular simulation and density functional theory (DFT). This progress, coupled with the availability of high resolution experimental methodologies for the adsorption of various subcritical fluids (also allowing one to couple adsorption techniques with complimentary techniques such as small angle scattering), has led to major advances in the structural characterization of nanoporous materials [1]. However, major challenges still exist concerning an in-*depth* characterization of complex pore networks. Application of characterization techniques to hierarchical materials has revealed new phenomena, such as cavitation processes during desorption, and led to significant refinement of these techniques [2].

[1] M. Thommes, K. Kaneko, A.V. Neimark, J.P. Olivier, F. Rodriguez Reinoso, J. Rouquerol and K.S.W Sing, (IUPAC Technical Report), *Pure Appl. Chem*. **87**, 1051 (2015)

[2] K. A. Cychosz, R. Guillet-Nicolas, J. Garcia-Martinez, J., M. Thommes, M. *Chem. Soc. Rev*. **46**, 389, (2017)

504010 Molecular Studies of Supercapacitors: Ionic Liquids Adsorbed into Porous Carbon Electrodes

Abstract:

Energy storage continues to attract the attention of the scientific and engineering research communities. Renewable energy sources, such as solar and wind, require so-called grid-based energy storage so that energy generated during periods of high energy production can be made available during periods of low or zero energy production. Mobile energy storage (for everything from mobile phones to automobiles and beyond) is clearly increasing exponentially in importance. Among the many choices for energy storage devices, electrical double layer capacitors (EDLCs), also called supercapacitors, are attracting considerable attention. Supercapacitors store electrical energy via ion electrosorption directly in the EDLs at the electrolyte-electrode interface, suggesting that such liquid-solid interfaces play a dominant role in the underlying energy storage mechanism and the resulting device performance. Because electrical energy in supercapacitors is stored based on physical phenomena rather than chemical reaction (as in batteries), supercapacitors have fast rates of charge/discharge and a virtually limitless number of charge cycles (unlike batteries, which are often limited to 10^4or less cycles). Much of the goal of supercapacitor research is aimed at increasing the amount of energy stored (energy density is the strong point in favor of batteries), which in turn focuses attention on the electrolyte, the nature of the electrode, and the electrode-electrolyte interactions.

To date, ionic liquids (ILs) have become emerging candidates for electrolytes used in supercapacitors, due to their exceptionally wide electrochemical window, excellent thermal stability, nonvolatility, and relatively inert nature; meanwhile carbons are the most widely used electrode materials in supercapacitors, due to their high specific surface area, good electrical conductivity, chemical stability in a variety of electrolytes, and relatively low cost. To improve the energy density and the transport properties of the charge carriers in supercapacitors, carbons have been developed in diverse forms such as activated carbons, carbon nanotubes (CNTs), onion-like carbons (OLCs), carbode-derived carbons and graphene. Using molecular modeling combined with molecular experimental probes, such as SAXS, SANS, NMR, and AFM, we report on our investigations into the interfacial phenomena occurring between the IL electrolytes and electrodes of varying geometries to understand the energy storage mechanism of supercapacitors that rely on EDLs established at IL-electrode interfaces.

504011 Structure of Ice in Confinement; Water in Mesopores

Abstract:

We report X â€"rays diffraction studies of water adsorbed in nanoporous activated carbon fibres (ACFs) and CMK-3 and CMK-8 carbon mesopores of different pore sizes. The fibres are built of turbostratic nanoparticles separated by quasi two-dimensional voids, forming narrow slit-shaped pores; CMK-3 and CMK-8 are the reverse carbon replica of silica SBA-15 and KIT-6 porous matrices. In order to determine the structure of water within the pores and its influence on the fibresâ€™ structure, mean interatomic and intermolecular distances have been estimated from the positions of the maxima of the normalized angular distribution functions obtained by X-ray diffraction [1]. We observe significant changes in the interlayer distance of the carbon nanoparticles; the results suggest that very high pressures arise within the pores,of the order a few hundreds MPa, as has been observed in molecular simulations [1,2]. Such results are confirmed by observation of high pressure forms of ice in cylindrical nanocarbons , using neutron and X â €"ray diffraction [3]. For water in multi-walled carbon nanotubes (MWCN), below the pore melting point cubic ice was observed. This structure of ice can be obtained during re-crystallisation from high pressure phases at low temperature for bulk water. For CMK-3 and CMK-8 mesopores we observed the existence of a stacking-disordered ice, I_{sd} [4,5]. This metastable ice is neither cubic nor hexagonal, and is not a simple mixture of the two, but a combination of cubic sequences intertwined with hexagonal sequences, which was identified as having the space group p3m1 [4]. Moreover, the stacking disorder can vary in complexity depending on the way the ice is formed and on the prevailing thermal conditions during this process. An analysis of the kind of I_{sd} formed in CMK-3 and CMK-8 of different pore sizes will be presented. These crystal forms, which occur in bulk water only at temperatures below 180 K in the case of cubic ice, and at pressures of hundreds or thousands of MPa are stabilized by the confinement.

References

[1] M.Sliwinska-Bartkowiak, M.Drozdowski, M.Jazdzewska, Y.Long, J.Palmer, K.E.Gubbins, Phys. Chem. Chem. Phys., **14**, 71454, 2012.

[2] Y. Long, J.Palmer, B.Coasne, M.Sliwinska-Bartkowiak, K.E.Gubbins, Phys. Chem. Chem. Phys , **14** 17163, 2011 .

[3] M.Jazdzewska , M. Sliwinska-Bartkowiak, A.I. Beskrovnyy, S.G. Vasilovskiy, S.W. Ting, K.Y. Chan, L.L. Huang, K.E. Gubbins, Phys. Chem. Chem. Phys., **13** ,9008 ,2011

[4] Malkin, T. L., Murray, B. J., Salzmann, C., Molinero, V., Pickering, S. J., & Whale, T. F. Phys. Chem. Chem. Phys., **17(1)**, 60-76, 2014; A.Haji-Akbari, P.G.Debenedetti, PNAS, **112**, 10582, 2015

[5]K.Domin,K.Y.Chan,H.Yung,A.Sterczynska,M.Jurek,K.E.Gubbins, M.Sliwinska-Bartkowiak, J.Chem..Eng. Data, 61, 4252-4260, 2016

504012 Towards Understanding the Role of Microstructure in Energetic Material Response: Coarse-Grain Modeling and Simulation

Abstract:

Mechanical and thermal loading of energetic material (EM) composites can incite responses over a wide range of spatial and temporal scales due to inherent microscale features. Many energy transfer processes within these materials are atomistically governed, yet the overall material response is manifested at the micro- and mesoscale â€“ scales beyond those amenable to atomistic simulation techniques. Moreover, continuum level approaches rely on field-based formulations that are empirically based, which lack sufficient fidelity to capture microstructure effects.

Particle-based microscale simulation methods that utilize coarse-grain models offer a promising route for extending the attributes of atomistic modeling toward the mesoscale. We have developed such microscale methods based upon the constant-energy dissipative particle dynamics method that includes chemical reactions. Coarse-grain models of EMs have been built using the Multiscale Coarse-Graining approach. Time-resolved depictions of an EM responding to insult has been simulated for a variety of microstructures, including samples with intra- and inter-granular voids of varying shape, size and relative orientation, varying grain shapes and sizes, and polymer binder. A sampling of these results will be presented.

504014 Adsorption and Transport in Multiscale Porous Media

Abstract:

Hierarchical porous materials such as hierarchical zeolites, which combine several porosity scales, are widely used in industry (adsorption, separation, catalysis) to overcome slow diffusion in microporous solids (< 2 nm) and enhance access to their large surface area. Available modeling approaches for adsorption and transport in such multiscale porous media are limited to empirical parameters which cannot be derived from molecular coefficients. In particular, existing approaches do not offer the ground for a bottom up model of adsorption/transport in multiscale materials as (1) they describe empirically the adsorption/transport interplay and (2) they do not account for the breakdown of hydrodynamics at the nm scale.

In this talk, I will present a multiscale model of adsorption and transport in hierarchical materials obtained by adding mesopores (~ few nm) and macropores (> 10 nm) to existing microporous crystals [1]. I will first show how adsorption, permeance, and transport in such media can be described without having to rely on macroscopic concepts such as hydrodynamics [2,3]. Using fundamental parameters and coefficients available to simple experiments, we will see how transport coefficients can be rigorously obtained from simple models in the framework of Statistical Mechanics. Then, I will present a multiscale model of adsorption and transport in hierarchical materials [4]. This approach consists of upscaling accurate molecular simulations in a lattice model. Thanks to the use of atom-scale simulations, which capture the different adsorption and transport regimes upon varying the temperature, pore size, pressure, etc. this bottom-up model does not rely on hydrodynamics and, hence, does not require assuming a given adsorption or flow type. I will also discuss NMR experimental results on transport in hierarchical zeolites [5].

[1] B. Coasne et al., Langmuir 29, 7864 (2013).

[2] K. Falk et al. Nature Comm. 6, 6949 (2015).

[3] T. Lee et al. Nature Comm. 7, 11890 (2016).

[4] A. Botan et al. Phys. Rev. E 91, 032133 (2015).

[5] A. Galarneau et al., J. Phys. Chem. C 120, 1562 (2016).

494064 Redox-Triggered Mixing and Demixing of Surfactants within Assemblies Formed in Solution and at Surfaces

Abstract:

This presentation will report experiments that test the hypothesis that redox-triggered changes in the architectures of surfactants permit control of mixing of surfactants within assemblies. Specifically, I will describe surface tension, light scattering, atomic force microscopy, and quartz crystal microbalance measurements that characterize the redox-dependent behaviors of cationic surfactants with a ferrocene group located either at the surfactant terminus (11-ferrocenylundecyl-trimethylammonium bromide; FTMA) or head (*N,N*-dimethylferrocenylmethyldecyl ammonium bromide; DMFA). In bulk solution, we find that reduced and oxidized FTMA do not mix within micellar assemblies but that reduced and oxidized DMFA do form mixed micelles. Because oxidized FTMA has the architecture of a bolaform surfactant whereas oxidized DMFA has a conventional surfactant architecture with a divalent head group, these results suggest that redox-triggered changes in molecular architecture permit control of the extent of mixing of surfactants in micellar assemblies in bulk solution. This conclusion receives further support from measurements performed with mixtures of dodecyltrimethylammonium bromide and FTMA, with FTMA in either reduced or oxidized states, and was found to extend to hemimicellar assemblies formed at hydrophobic solid surfaces but not to mixed monolayers formed at the surface of water. The latter is attributed to differences in the conformations of surfactants within monolayers and micellar assemblies. Overall, these results provide insight into the design of surfactant assemblies within which mixing can be controlled reversibly using redox processes.

492777 Static and Dynamic Signatures of Branching in Wormlike Micelles (WLMs) Via Advanced Techniques in Rheology and Neutron Scattering

Abstract:

Self-assembled wormlike micelles (WLMs) are of scientific and technological interest due to their ability to branch, break, and reform under shear. Controlling the level of branching in WLMs allows the rheological properties to be tuned for specific applications. Dynamic measurements like neutron spin echo have proven useful in differentiating between topologies in polymers and WLMs [1,2] due to differences in chain dynamics. Branching in WLMs provides additional stress relief mechanisms due to the presence of sliding branch points, and in contrast to linear micelles, an array of relaxation modes may be expected in highly branched or network-like micelles. Here, we use multiple rheological and neutron techniques to explore the relationship between branching, dynamics, microstructure and nonlinear responses using a model WLM series [3,4], where the degree of branching is controlled via the addition of sodium tosylate. The linear viscoelastic rheology (LVE) of the WLMs shows deviations from Maxwellian behavior with branching, similar to branched polymers. These spectra are then connected to two distinct relaxation modes identified by DLS and neutron spin echo (NSE). In NSE, solutions with low branching exhibit wormlike chain behavior, whereas highly branched solutions approach the behavior of flexible membranes, indicative of the network morphology. Steady shear and startup results indicate that shear banding disappears at high branching levels, and that the stress overshoot behavior upon shear startup is reflective of the topology. Orthogonal superposition rheology (OSP) is then used to quantify the changes in the LVE spectra under shear, which are distinct based on branching level. The normalized orthogonal plateau modulus and crossover frequency as a function of shear rate (Wi) decrease more rapidly with branching, indicating a breakdown of network-like structures. This breakdown of branched structures identified in OSP is confirmed using 1-3 plane flow-SANS, by analyzing the WLM orientation distribution function (ODF). The ODF analysis indicates that WLM branches breakage past a critical shear rate, at which point the branched WLMs recover the linear WLM behavior. The combination of rheology and advanced neutron scattering techniques enables distinct differences in the rheological response, flow-induced microstructure, and solution dynamics of WLMs with branching to be identified. This research is part of a broader effort to characterize branching in chemical polymers and self-assembled systems, and provides a complete data set for the development and improvement of constitutive models that incorporate branching.

[1] F. Snijkers, et al. *Macromolecules* **46** (2013).

[2] S. A. Rogers, M. A. Calabrese and N. J. Wagner, *Curr Opin Coll Int Sci* (2014).

[3] M. A. Calabrese, et al., *J. Rheol.,* **59**, 5 (2015).

[4] B. A. Schubert, N.J. Wagner, and E.W. Kaler, *Langmuir* **19**, 10 (2003).

496018 Self-Assembly of Achiral Surfactants Can Conduct Enantioselective C-C Bond Formation of Amino Acid Derivative

Abstract:

The use of achiral surfactant assemblies as a reaction platform for an alkylation reaction resulted in a high enantiomeric excess. Dilauryldimethylammonium bromide (DDAB) vesicles were modified with cholesterol to promote alkylation of N-(diphenylmethylene)glycine tert-butyl ester (DMGBE) with benzyl bromide, resulting in high conversion (~90%) and high enantioselectivity (up to 80%) [1]. The R-enantiomer was formed on using the DDAB vesicles, whereas the use of phospholipid liposomes prepared from 1,2-dioleoyl-sn-glycero-3-phosphocholine (DOPC) produced an excess of the S-enantiomer [2]. Considering the chemical structures of the reaction substrates and amphiphiles as well as the membrane structures and properties of DDAB vesicles and DOPC liposomes, it is suggested that the enantiomeric excesses result from the location of the quaternary amine of the amphiphiles and the DMGBE at the outer surface of the membrane. We show that the enantioselective reaction at the surface of the self-assembly could be regulated by adjusting the chemical structures and resulting membrane properties of the self-assembly.

[1] F. Iwasaki *et al.*, Enantioselective C—C Bond Formation Enhanced by Self-Assembly of Achiral Surfactants. *ACS Omega*, **2017**, *2* (4), 1447–1453. [2] F. Iwasaki *et al.*, Liposomes Can Achieve Enantioselective C–C Bond Formation of an α-Amino Acid Derivative in Aqueous Media. *ACS Omega*, **2017**, *2* (1), 91–97.

501800 A Statistical Associating Fluid Theory (SAFT) Framework for Aqueous Nonionic Surfactant Systems

Abstract:

An established molecular thermodynamic model of micellization and phase behavior of aqueous nonionic systems is incorporated into a statistical associating fluid theory (SAFT) framework to obtain a combined model capable of describing both microscopic aspects of solution behavior (i.e., micelle shape and size distribution) and bulk phase behavior including phase separation and osmotic pressure. The original form of the model advanced by Puvvada and Blankschtein [1] considers three distinct contributions to the Gibbs energy of the surfactant solution: 1) formation of the micellar distribution, 2) ideal mixing of solution species including monomers, micelles and water molecules, and 3) interaction between micellar aggregates. The strength of this model is in the detailed accounting for factors that account for micellization including transfer of hydrocarbon tails from water to the micellar core, formation of a curved hydrocarbon-water interface, hydrocarbon chain packing within the core, and steric head group interactions. However, the Blankschtein phenomenological model of micellization uses a mean field expression with an empirical parameter for the Gibbs free energy of interaction.

The SAFT approach [2] has been successfully applied to correlate and/or predict thermodynamic properties and phase behavior of a wide range of pure fluids and mixtures. This work draws on the SAFT methodology and develops a theoretically-based expression for the free energy of species interaction that replaces the mean field expression in Puvvada and Blankschteinâ€™s original formulation [2]. A main feature of the new model is a recently proposed equation of state [3] for the repulsive free energy contribution. We show in this talk several examples of the combined modelâ€™s predictive ability for micellar and bulk aqueous solution properties of surfactants from the polyoxyethylene glycol monoether and glucoside families.

References

[1]. Puvvada, S. and D. Blankschtein, Molecular-Thermodynamic Approach to Predict Micellization, Phase Behavior and Phase Separation of Micellar Solutions. I. Application to Nonionic Surfactants, *J. Chem. Phys.* **92**, 3710-3724 (1990).

[2]. Muller, E. A. and K. E. Gubbins, Molecular-Based Equations of State for Associating Fluids: A Review of SAFT and Related Approaches, *Ind. Eng. Chem. Res.* **40**, 2193-2211 (2001).

[3]. Gow, A. S., S. Alkhaldi and S. Demir, Cubic and Quartic Hard-Sphere and Lennard-Jones Chain Equations of State as Foundations for Complex Fluid Modeling, *Fluid Phase Equilibria* 399, 1-15 (2015).

492708 First Proof of Self-Assembly of Block Copolymers in a Deep Eutectic Solvent

Abstract:

It was in 2003 when a new class of innovative solvents was presented in literature.[1] These so called deep eutectic solvents (DESs) are a combination of two or more solids, which undergo interactions via hydrogen bonding and presumably Van der Waals forces upon the formation of the liquid DES.[2] DESs can be seen as the successors of ILs with many advantages over them. The main advantage is their easy preparation, which is simply mixing the two components at moderate temperatures to form the liquid DES. Furthermore, DESs are considered as sustainable solvents, an aspect that can be tuned by the selection of the components.

From the start, the search for news DESs mainly focused on using solid components with a high amount of hydrogen bonding groups, leading to hydrophilic DESs. In 2015 hydrophobic DESs were discovered, which were used for the removal of volatile fatty acids from water.[3] These hydrophobic DESs were composed of decanoic acid and quaternary ammonium salts. Recently, more applications regarding hydrophobic DESs were presented in the literature. These include the removal of biomolecules from an aquatic environment, the removal of metal ions from pure water and the capture of CO_2.[4,5]

An area that is not explored yet in the literature is the self-assembly of block copolymers in DESs, while this can be of interest for many applications such as drug delivery, nanomaterial synthesis and separation processes, where the sustainable character and cheap production costs of DESs can offer a major advantage over ILs and conventional solvents. The first step towards studying self-assembly of polymers in DESs was measuring the solubilities of hydrophobic and hydrophilic polymers in DESs. Results that gave an insight towards which block copolymers would self-assemble in the DESs. Subsequently, a block copolymer was designed, which was self assembled in the DESs for the first time.

References

[1] A. P. Abbott, G. Capper, D. L. Davies, R. K. Rasheed, and V. Tambyrajah, *Chemical Communications*, 2003, 70–71.

[2] M. Francisco, A. van den Bruinhorst, and M. C. Kroon, *Angew. Chem. Int. Ed.*, 2013, **52**, 3074–85.

[3] D. J. G. P. van Osch, L. F. Zubeir, A. van den Bruinhorst, M. A. A. Rocha, and M. C. Kroon, *Green Chemistry*, 2015, **17**, 4518–4521.

[4] B. D. Ribeiro, C. Florindo, L. C. Iff, M. A. Coelho, and I. M. Marrucho, *ACS Sustainable Chemistry & Engineering*, 2015, **3**, 2469–2477.

[5] D. J. G. P. van Osch, D. Parmentier, C. H. J. T. Dietz, A. van den Bruinhorst, R. Tuinier, and M. C. Kroon, *Chemical Communications*, 2016, **52**, 11987–11990.

495468 Fast and Slow Dynamical Processes in Simulations of Block Copolymer Micelles

Abstract:

Experimental studies of dynamics in micellar solutions of small molecule surfactants show evidence of "fast" and "slow" processes with widely disparate time . The "fast" process is associated with insertion and expulsion of individual surfactant molecules, while the "slow" process is associated with creation and destruction of entire micelles. We discuss a simulation study of rates and mechanisms of both fast and slow processes in a simple coarse-grained model of "hairy" diblock copolymer micelles of highly asymmetric copolymers. Quantitative study of rates of insertion and expulsion show that insertion rates in this model are essentially diffusion controlled, and that the micelle corona does surprisingly little to hinder exchange of individual molecules. Study of the rates of the slow process focus on a comparison of predicted rates for micelle creation and destruction by postulated competing pairs of processes of (1) step-wise association and dissociation and (2) fission and fusion of entire micelles, in order to clarify which of these processes controls relaxation of the micelle size distribution.

501318 Synthesis of Hydrophobically Modified Polybetaines (HMPB) and Study of Their Self-Assembly By Molecular Dynamics Simulations

Abstract:

The presentation discusses reversible addition fragmentation chain-transfer (RAFT) polymerization of hydrophobically modified polybetaines (HMPB) and their self-assembly in aqueous solutions using a combination of experiments and molecular dynamics (MD) simulations. RAFT was used to synthesize HMPB for the first time and MD simulations were used to illustrate the self-assembled morphologies.

As a first step in the RAFT polymerization of HMPB, the desired enamine form of alkylaminocrotonate (monomer) was synthesized by thermal bulk reaction method. The synthesis was followed by purification of enamine alkylaminocrotonate, from the impurities of reactants and the imine form of alkylaminocrotonates, using silica gel chromatography. The monomer was characterized by FTIR and H-NMR. FTRP and C13-NMR, combining with GPC, were used to confirm polymer compositions and molecular weight distribution. Molecular weight of around 80,000 g/mol and a narrow molecular weight distribution of 1.214 was achieved.

The self-assembly behavior of RAFT-HMPB was investigated by both experimental characterization techniques and MD simulations. Both TEM characterization and MD simulations suggest that, at low concentration HMPB forms spherical aggregates. MD simulations show that at high concentrations, worm-like micelles are formed. The stability of these micelles was also investigated using MD simulations. Furthermore, DLS and zeta potential measurements indicated pH responsive behavior of HMPB aggregates. Concentration and pH responsive behavior of HMPB combined with its stability suggest that HMPB has potential for applications in drug delivery and enhanced oil recovery.

500564 A Hexagonal Columnar Liquid Crystal Phase Formation in Dilute Solutions of Carbon Nanotubes

Abstract:

Solutions of carbon nanotubes (CNTs) in chlorosulfonic acid (CSA) form liquid crystals at high concentrations. From a fundamental perspective, CNTs are an ideal candidate system for testing theories of liquid crystals for rod-like systems with high degree of polydispersity. From an application point of view, these liquid crystal solutions are of special interest for the fluid-phase self-assembly approach for processing CNT-based macroscopic materials. While CNT-CSA solutions exhibit a phase behavior consistent with the Onsager prediction for non-interacting rigid rods, the ordering nature of the solution and how it depends on the CNT properties and the concentration are not completely understood. Here, we use small angle x-ray scattering to characterize the morphology of the liquid crystal phase and get more insights on the ordering of CNTs in solutions. Our results show that the at high concentration (~5% in volume), our system self assembles into a hexagonally packed columnar phase. Theoretically, for a length polydisperse system with p > 0.18 (polydispersity is defined as p = â^š(<L^2>/<L>^2 â^' 1) where L is the length of molecules), the system will exhibit a direct nematic-columnar phase transition at high concentrations (about 50% by volume). Surprisingly, CNTs show the nematic-columnar transition at much lower concentration (~ 5 % by volume). We attribute this difference in behavior to steric forces that could be the result of electrostatic repulsions between CNTs. Such steric forces confine the CNTs to a larger area that increases the effective volume fraction and results in a nematic-columnar transition at lower concentrations. Beside its fundamental importance, this finding should have important implications for processing of multifunctional CNT fibers at high concentrations where the alignment of the CNTs affects the mechanical properties of the final product.

499122 Computational Phase Space Screening of Isotropic Multi-Well Pair Potentials

Abstract:

While the assembly of intricate crystals is often attributed to an inherent complexity in building block geometry or interaction, soft-matter scientists have found surprisingly complex structures in many very simple systems in both experiment and simulation in recent years. Inspired by the emergence of complexity from several seemingly simple systems, we screened the phase spaces of multiple isotropic pair potentials and found a wealth of structures with varying degrees of complexity and order. The results of our systematic analysis of the assembled crystals can be used to engineer the assembly of desirable and novel structures in future studies.

493187 Methanol Self-Association and Preferential Solvation of Chelating Agents for the Extraction of Nuclear Fission Products in Supercritical CO_2

Abstract:

While supercritical carbon dioxide (sCO_2) is a promising solvent in heavy metal extractions due to its low toxicity, surface tension and viscosity, many chelating molecules exhibit limited solubility in sCO_2. The addition of volatile polar molecules to sCO_2, such as methanol (MeOH), improves the solubility of chelating molecules and metal-ligand complexes. MeOH self-assembly and preferential solvation of 2-ethylhexylphosphonic acid mono-2-ethylhexyl ester (HEHEHP), a chelating ligand, in sCO_2 at 19.7 MPa and 323 K was investigated with diffusion-ordered nuclear magnetic resonance spectroscopy (DOSY-NMR) and molecular dynamics simulations. DOSY-NMR pulse sequences contain a pair of pulsed field gradients that attenuate the NMR resonance and enable calculation of diffusivity for each NMR active component in the mixture. Diffusometry results were analyzed using a model that integrates the Stokes-Einstein relationship with self-association models to approximate the size-dependent, energy of cluster formation landscape. DOSY-NMR results are consistent with molecular dynamics simulations and suggest the formation of small clusters of MeOH limited to a composition of under 10 molecules in a cluster at high concentrations of MeOH. Next, self-assembly of HEHEHP into dimers was quantified with DOSY-NMR in sCO_2 at 19.7 MPa and 323 K. The propensity of HEHEHP to self-associate into dimers in sCO_2 was evaluated with the integrated Stokes-Einstein, self-association model. The results suggest that HEHEHP is primarily composed of dimers at concentrations of over 40 mM in sCO_2. Finally, we compare the extent of dimerization of HEHEHP in sCO_2 containing 4 mole % of MeOH to the dimerization of HEHEHP in neat sCO_2 with a series of DOSY-NMR experiments. The experiments indicate that MeOH preferentially solvates HEHEHP, increasing both the solubility of HEHEHP in the supercritical fluid and the weight fraction of monomeric HEHEHP. Our results suggests that the increased amount of monomeric chelators may be an additional, synergistic factor in the enhanced heavy metal extraction in MeOH modified sCO_2.

491505 Effect of Salt Concentration on the Ability to Form Stable Close-Packed Vesicular Dispersions for Stabilizing Suspensions of Dense Particles Against Sedimentation

Abstract:

For many applications of colloidal dispersions or suspensions, such as inks and paints, the particles must be suspended for long times. While this is often accomplished by preventing agglomeration, dense particles may still settle rapidly even without agglomeration. Stabilizing dense particles with low bulk viscosity of the dispersion is challenging and has received little attention. We have presented a general method for preventing dense particles from settling by using close-packed vesicular dispersions of the cationic double-chain surfactant DDAB (didodecyldimethyl-ammonium bromide) (Yang et al., 2015, Langmuir, 31, 8802). To understand the feasibility of the vesicular stabilization mechanism under various salinities, we examined the effect of sodium bromide (NaBr) concentration on the stability of dense particle suspensions against sedimentation.

DDAB forms lamellar liquid microcrystallites, or â€œliposomes,â€ with stirring in water at 25 C, and it also easily forms dispersions of unilamellar vesicles at low concentrations. Such vesicular dispersions were found to be quite effective for stabilizing settling suspensions of 1-10 wt% titania particles, with diameters of 300 ± 200 nm and a density of 4.2 g/cm^3. From cryo-TEM and Dynamic Light Scattering measurements it was found that the vesicles are deformable with diameters of 400-600 nm. At 0.5 wt% DDAB and an estimated volume fraction of less than 0.2, the vesicles are mobile. At 2.0 wt% DDAB, the vesicles have an estimated volume fraction of over 0.5 and are quite immobile, forming a close-packed vesicular dispersion. Both dispersions are highly shear-thinning, and free-flowing at high shear rates. Only the close-packed vesicles provide strong resistance against the settling of the dense particles. At 0.5 wt% DDAB, the titania particles settled completely by 1 cm in two months, while at 2.0 wt%, the particles remained suspended for over 18 months. We extended the study to silica particles, which can also be easily stabilized against sedimentation for long times.

To better understand the effect of salt on the use of DDAB vesicular dispersions to minimize sedimentation, the stabilization of titania and silica particles against sedimentation was studied for NaBr concentrations ranging from 1 to 1000 mM. At 1.0 wt% DDAB with NaBr concentrations below 10 mM, the suspensions were again found to be quite stable. At higher NaBr concentrations, the suspensions were unstable, evidently because no liposomes or vesicles were formed. Various thermophysical properties and the phase behavior of DDAB solutions as a function of salinity were also studied to gain insight into the impact of NaBr concentration on vesicle formation.

498953 Toward a Better Understanding of Nanoassembly: Importance of Molecular Granularity on Colloidal Forces

Abstract:

To obtain fundamental understanding of nanoassembly processes such oriented attachment (OA), an important mechanism of non-classical crystal growth, one would need an in-depth analysis of the coupling between colloidal interactions and molecular details owing to intrinsic length scales involved in the processes; a continuum-based Derjaguin-Landau-Verwey-Overbeek (DLVO) theory needs to be revised. Utilizing the spatial density response of a solvent to a surface combined with simple correlating schemes, we related the discrete nature of solvent to electrostatic and dispersion interactions. It was found that colloidal interactions are sensitive to the spatial variation of solvent density, demonstrating significant deviations in the interactions from the conventional formulation (e.g., with uniform solvent density). A comparison to force measurements from surface force apparatus (SFA) and atomic force microscopy (AFM)-based dynamic force spectroscopy (DFS) indicates that such molecular granularity of solvent response would be critical to capture a physical picture of the assembly.

498963 Structural Insights into DNA-Stabilized Silver Clusters through Analytical Ultracentrifugation

Abstract:

Noble-metal nanoparticles with sizes approaching their Fermi wavelengths (~0.5 nm for Ag) display remarkable optical properties owing to the discretization of their electronic states. Among these, silver clusters stabilized by DNA (Ag_NDNA) stand out for their exceptional stability and tunability in optical properties. Selection of the stabilizing DNA or RNA sequence guides the formation of Ag_NDNA clusters, controlling the synthesized particle size to including between ~10-24 Ag atoms. These particles are then excellent candidates for a number of applications including sensing schemes and for use as novel nanophotonic elements in devices. However, applications involving Ag_NDNAs remain limited by the lack of mechanistic understanding regarding their photo-physical and folding pathways in response to stimuli, and this lack of understanding is largely due to challenges in obtaining structural information about the particles in solution.

In this work we address the structural characterization of the particles through in situ characterization with analytical ultracentrifugation (AUC). In AUC, the sedimentation rate of the AgNPs is monitored during centrifugation and structural information can be extracted from modelling of the concentration boundary of the particles as a function of time. This information is amplified through the use of density contrast variation experiments, in which a deuterated solvent is utilized to enable precise determination of the particle density in the given solvent system. For AG15DNA NPs this is of particular interest for aqueous - salt buffer and water-ethanol mixtures in which significantly different NP optical properites have been hypothesized as being due to structural rearrangement of the particle. Data and analysis of the sedimentation behavior measured of purified Ag15DNA NPs will be presented for both as produced and aged particles in both solvent systems, shedding new light on the differential behavior and structures of the particles in these systems.

499107 Stabilization of Colloidal Suspensions with a Bimodal Distribution of Microparticles and Nanoparticles

Abstract:

Charged nanoparticles have been found to enhance the stability of colloidal suspensions by forming a non-adsorbing nanoparticle layer surrounding larger neutral colloidal particles that induces an effective electrostatic repulsion. This stabilization mechanism is known as nanoparticle haloing. It has been previously shown that silica particles could be stabilized by this mechanism using highly charged zirconia nanoparticles at a low range of volume fractions. In this study we examined whether binary systems of silsesquioxane-zirconia and silsesquioxane-silica can develop nanohaloing. One of the advantages of using silsesquioxane compared to silicon is that they can be easily functionalized by fluorescent markers which is extremely valuable for imaging in confocal microscopy. The other advantage is that isoelectric point of silsequioxane is near the neutral pH while isoelectric point of silica is around pH 1.5. These advantages make silsesquioxane more favorable for conditions where highly acidic solutions cannot be used due to materials or safety concerns.

The charging behavior of the microspheres and the effect of the nanoparticles on the overall charge of the system have been studied. Ground results showed that silsesquioxane-zirconia system was stable at pH=4.5 where particles were highly charged and it was unstable near the isoelectric point of microspheres. Our results were in agreement with the recent results obtained from similar experiments performed onboard the ISS. Considering that both silsesquioxane microspheres and zirconia nanoparticles had similar electrostatic behavior, nanohaloing was not observed in this system. Based on these results, this study was aimed to identify an optimized binary system to be used in a future experiments on board the International Space Station.

500915 Rheological and Electrical Percolation Behavior of Carbon Black Suspensions in Polar Aprotic Solvents

Abstract:

In this work, the microstructural origin of the rheo-electric behavior of carbon black gels and suspensions is studied. These materials find widespread use as conductive fillers in composites and slurry-based electrochemical energy storage technologies. In these applications, both the viscosity and electrical conductivity are key design parameters and are determined by the microstructure.Â Using small angle neutron scattering (SANS) and electron microscopy, we perform detailed structural analysis on carbon black suspensions as a function of volume fraction for two commonly used conductive additives â€“ KetjenBlack and Vulcan. From these measurements, the static structure factor is determined and characterized at low-Q by a power-law scaling $I(Q) \sim Q^{-3}$. We characterize the onset of gelation using oscillatory rheological measurements and identify the percolation threshold associated with the jamming transition. While the onset of mechanical percolation is commonly associated with the electrical percolation threshold for these materials, results from impedance spectroscopy measurements show that the electrical percolation threshold occurs at far lower volume fractions. Further, analysis of the critical scaling parameters associated with the rheology are inconsistent with those extracted from the conductivity. These results suggest a rich relationship between electrical conductivity and the fluid properties of carbon black dispersions that could guide the design of future low-viscosity, high conductivity conductive additives for electrochemical flow applications.

501651 Functional Hybrids Based on Assembly of 2D Materials and Polymeric Nanoparticles

Abstract:

The increasing demand for systems with multiple functionalities requires developing new hybrid materials containing multiple nanocomponents. Despite this demand, the question of how the interactions in these systems can be engineered to control their bottom-up self-assembly into scalable and stable materials remains essentially unexplored. Here we show that self-assembly via different pathways can integrate graphene oxide, nanosilicates and gelatin nanoparticles into macroscopic hybrid structures with tunable properties and tissue compatibility. Depending on the overall concentration, charge, shape, and aspect ratio of the 2D components, the investigated hybrid systems exhibit distinct rheological behaviors and signatures from a viscoelastic liquid to viscoelastic soft solid and from a gel-like to glass-like dynamics. The structural evolution and the arrested dynamics are explained via the interaction potential among the components. The hybrid soft solids reveal a noticeable shear-induced failure but immediately recover their elasticity upon removal of the shear strain. Complementary single-step stress relaxation experiments provide a comprehensive description of the nonlinear rheological response of these materials. The tunability of the rheological properties, demonstrated by our hybrid system, is valuable for emerging technologies such as 3D printing. Further, we show that upon drying under controlled conditions, these hybrid soft materials can be converted to preformed dense structures, self-standing films, coatings on metallic or glass substrates, and 3D porous scaffolds. The prepared scaffolds are biocompatible and synergistically promote osteogenic differentiation of mesenchymal stem cells.

501875 Convective Fluid Motions in Droplets Driven By Global and Local Chemical Gradients

Abstract:

Colloidal migration in chemical gradients are important to both scientific studies and industrial applications. Chemical gradients can generate interfacial flows between no-slip particle surfaces and bulk fluids, driving diffusiophoretic motion. Meanwhile, interfacial tension gradients may develop due to non-uniform distribution of chemicals, giving rise to Marangoni flows that yield solutocapillary motion. Here, we present studies on these two types of motions associated with droplets by creating global and local chemical environments in dispersed droplet systems.

First, we show diffusiophoretic transport of droplets in external electrolyte gradients. In addition, gradients of surface-active chemicals may direct droplet migration oppositely compared with solid particles, due to different boundary conditions at particle surfaces. Second, we demonstrate spontaneous flows inside sessile drops (immersed in oil) driven by local chemical gradients. In a ternary water-oil-butanol system, coupled interfacial transport of butanol between two adjacent droplets yield fluid flows. Patterns of these flows depend on the direction of interfacial mass transport, which can be understood in terms of induced Marangoni stresses at droplet interfaces, owing to the inhomogeneous transport. Such effects appear important in a variety of flows of both scientific and practical interest. Moreover, we find that convective flows even exist in single droplets when butanol is present. Internal flows of droplets are directed radially inward at the plane near the supporting substrate, regardless of interfacial mass transport direction.

485429 Characterizing Gelatin Hydrogel Viscoelasticity with Diffusing Colloidal Probe MicroscopyÂ

Abstract:

In this study, we investigate viscoelasticity in gelatin hydrogels using diffusing colloidal probe microscopy (DCPM) to directly measure the elastic potential energy interaction between colloidal probes and the underlying viscoelastic media. Gelatin samples are prepared in four different concentrations between 0.3 wt% to 0.6 wt% to examine changes in viscoelasticity with concentration. A force balance describing the interaction between the colloidal probes and the hydrogel as a spring-damper system lead to a simple model for mean square displacement. A histogram of locations sampled by the colloidal probes is directly related to the elastic potential energy and the effective spring constant of the gelatin hydrogels. The effective spring constant is a fixed parameter used in the mean square displacement model to find effective viscosity. These parameters are comparable to viscoelastic parameters obtained by a microrheology analysis of two-dimensional mean square displacements. These results can serve as a guide for assessing hydrogel systems where viscoelastic properties are an important factor in biomaterial design.

490324 Self-Immobilized Cross-Linked Enzyme AggregatesÂ for Biocatalysis

Abstract:

Enzymes have many advantages over inorganic catalysts but they are very sensitive to temperature and pH. Recently, we showed that enzymes can be immobilized as cross-linked enzyme aggregate (CLEAs) without using any solid carriers to improve their stability. Firstly, millifluidic reactors with two laminar flows of an enzyme solution and an organic solvent were employed to prepare uniform CLEAs of cellulase and laccase. CLEAs of cellulase had a size between 200 and 400 nm and were highly uniform. They can be collected on silica gels or entrapped in a membrane as practical biocatalysts to hydrolyze cellulose with an activity up to 86% of the free cellulase. Because of their large size, CLEAs can also be encapsulated in calcium alginate beads without any leaching problems. This features allowed CLEAs of cellulase to be reused and recycled up to 10 times for hydrolysis of cellulose. Similarly, CLEAs of laccase were prepared and entrapped in a membrane for oxidization of phenolic compounds. Trypan blue, a dye molecule, can be degraded by using membrane-bound CLEAs continuously for 96 h. Finally, glucose oxidase (GOx) and horseradish peroxidase (HRP) were co-immobilized as combi-CLEAs by using a coaxial-flow millifluidic reactor. The combi-CLEAs were used to catalyze cascade chemical reactions and minimize the decomposition of hydrogen peroxide in the presence of catalase. As a result, the overall reaction rate increased by 14 times compared to free enzyme mixtures in the presence of 6.3 mM of catalase. The combi-CLEAs were further exploited to develop colorimetric glucose assays with a limit of detection of 0.5 mM and a linear range up to 27.8 mM. In the future, millifluidic reactors can be used to prepare CLEAs with tunable sizes and properties for various industrial applications.

502889 Impact of Molecular Structure on Surface and Thermal Properties of Amido-Amine Cationic Gemini Surfactants

Abstract:

One of the great challenges of surfactant flooding is the stability of surfactants in harsh reservoir conditions such as high salinity and high temperature. The less stable surfactants easily degrade under these condition and loss their property of lowering the interfacial tension between oil and water. In order to overcome the drawbacks of surfactant flooding, a variety of gemini cationic surfactants were synthesized. The structures of the synthesized surfactants were confirmed using NMR (^{1}H and ^{13}C), FTIR and elemental analysis. Thermogravimetric analysis was used to identify the short-term thermal stability and structure analysis technique was used to investigate the long-term thermal stability of the synthesized surfactants. Pendant drop technique was used to measure the surface properties of the surfactant. The synthesized surfactants exhibited excellent short-term and long-term thermal stabilities and no changes in structure were detected. The critical micelle concentration and the corresponding surface tension of the surfactants showed excellent surface properties and these values are comparable with the commercially available surfactants. The critical micelle concentration of gemini surfactants was decreased with increasing length of the spacer. Similarly, the critical micelle concentration and corresponding surface tension decreased by increasing tail length. The rigidity and spacer conformation of the spacer double bond also affect the surface properties. The surface area per molecules decreased with increasing tail length, spacer flexibility, and spacer length.

485574 Surfactant Effect on Hydrate Crystallization at Oil-Water Interface

Abstract:

Clathrate hydrates are solid crystalline compounds with cage-like structures of water molecules. Hydrates have found important applications in many areas, including flow assurance of oil and gas lines, potential sources of natural gas from permafrost and deep-sea hydrate deposits, and use as a medium for energy storage and transportation. Plugging of transport pipelines due to hydrate formation is a major concern in the oil and gas industry which spends over US $200 M annually to prevent hydrate formation and aggregation to maintain flow assurance. Blockage of the oil and gas lines not only affects production but is also a major safety concern due to possible pressure buildup upon hydrate agglomeration. Anti-agglomeration is perhaps the most effective approach for flow assurance. Anti-agglomerates are surfactants which prevent the formation of the hydrate crystalline structure. Over the past several decades, even though significant knowledge has been gained regarding properties of hydrates, many aspects still remain puzzling or simply unknown. Among these aspects is the mechanism for destabilization of hydrates by surfactants. In this study a novel approach using interfacial rheology was applied to understand cyclopentane clathrate hydrate formation in the presence of nonionic surfactants. The hydrate inhibiting performance of low(<CMC), medium (~CMC), and high (>CMC) concentrations of Span 20, Span 80, Pluronic L31, and Tween 65 at 2°C on a 2Î¼L water droplet showed a morphological shift in crystallization mechanism from planar shell growth to conical growth. Monitoring the internal pressure of a droplet undergoing planar hydrate crystallization provided a strong correlation of decreasing interfacial tension to the shrinking area of the water cyclopentane interface. In the case of conical growth, interfacial tension measurements showed oscillations. These oscillations result from the cycling growth and release of the crystals from the interface to the bulk. This study provides a first hard evidence of the formation of different crystallization morphologies depending on the concentration and structure of the surfactant molecules. The results of this study can lead to the design of more effective, eco-friendly surfactants which will have broad applications in offshore natural gas production and seabed oil capture.

491601 Surface Tension Behavior of Aqueous Solutions of a Propoxylated Surfactant and Interfacial Tension Behavior Against an Un-Preequilibrated Crude Oil

Abstract:

Equilibrium interfacial tensions (EIFTs) are important in determining the oil recovery efficiency of surfactant-based enhanced oil recovery (EOR) processes. The dynamic IFTs (DIFTs), and the adsorption mechanisms are also important for the initial stages of EOR and for screening various surfactants for potential field uses. We report on the DIFT and EIFT behavior of a commercial anionic surfactant, S13D, which is a single-chain propoxylated sodium sulfate salt. The solubility in water is over 25 wt%, and the solubility in the brine used is over 20 wt%. The synthetic brine used is similar to the one present in an actual oil reservoir. It contains 0.95 wt% salts, primarily NaCl, and $NaHCO_3$ with smaller concentrations of $CaCl_2$, $MgCl_2$, $BaCl_2$, KCl, Na_2SO_4, and $MnCl_2$. The low-viscosity crude oil samples used were prepared from oil produced from a reservoir in the Illinois Basin.

For comparison purposes, the dynamic and equilibrium surface tensions, DSTs and ESTs, and DIFTs and EIFTS, were measured, with either the emerging bubble or drop method, and often with the spinning bubble or drop method. Only the latter method was used for the ultralow IFT measurements. DST has a simpler adsorption mechanism than DIFT. DSTs were measured in water and in brine at surfactant concentrations ranging from 0.00001 wt% to 1.0 wt% (0.1 ppm to 10,000 ppm). The DST data revealed long equilibration times, ranging from 24,000 s to 1 s. The times were shorter in brine than in water. Several surface tension relaxation tests after surface area perturbation were done, for testing the stability of the STs, and establish the validity of the ESTs. The EST data reveal a cmc of about 12 ppm in water and 1 ppm in brine. The molecular areas of the surfactant in the adsorbed monolayer were estimated as 1.2 nm^2 in water and 1.0 nm^2 in brine.

The un-pre-equilibrated DIFTs and EIFTs in water and brine against crude oil drops were also measured for the same concentration range with similar area perturbations to test for the stability and accuracy of the EIFTs. In water, the lowest observed EIFTs were 1 mN/m. In brine, the EIFTs were much lower above the cmc, reaching ultralow values, less than 0.01 mN/m. The timescales of IFT equilibration were slightly longer than that for ST equilibration. The results suggest that the low and ultralow IFTs are associated with adsorbed monolayers at the aqueous/crude oil interface.

497533 Eco-Friendly Sacrificial Amphiphiles As Chemical Herders for Oil Spill Remediation

Abstract:

Currently, the two principal methods for responding to a maritime oil spill are the use of floating barrier booms to encircle and contract the spill, so that it may be removed by mechanical skimming or in-situ burning, and the aerial spraying of surfactants onto the spill to cause the oil slick to emulsify into small droplets that are carried away from the site or are broken down by microbes. Taken together they are an inadequate toobox: Booms are difficult to deploy, and dispersants do not really remove the oil from the marine biota, and the effect of currently utilized dispersants on marine life is under dispute.

The object of this proposal is to develop eco-friendly chemical herders for removing maritime oil spills. In this method, surfactant is aerially sprayed around the spill, to lower the large surface tension of the air/sea surface uncontaminated by the oil. The reduction in the surface tension causes the floating oil layer to retract and thicken to the point where it can be ignited, and the spill burned away. The most effective surfactant herders that have emerged from laboratory and field testing are silioxane polymers, which do not easily or rapidly degrade in the marine environment and are ineffective in rough seas where wave action can disintegrate the herding monolayer.

The overall goal of this proposal is to develop a new class of effective surfactant herders which are molecularly designed to be eco-friendly and to be effective even on rough sea surfaces. Our central idea is to design the herder as a mixture consisting of two surfactant components: The first is a glycolipid/oligosaccharide surfactant, which, because of its sugar moiety can complex with polysaccharides in the sea surface to form a monolayer with high surface dilatational viscosity. High surface viscosities dampen surface waves. The second is a surfactant with isoprenoid (i.e. branched) aliphatic chains (e.g. phytanic acid) which intercalate into surfactant monolayers through their branched structure to achieve the significant tension reduction necessary for herding.

We report measurements of the surface pressure isotherms and dilatational viscosities of monolayers of a glycolipid (MGDG), phytanic acid and their mixtures on the surface of pure water and artificial seawater with natural polysaccharides. In addition AFM images of Langmuir-Blodgett films of the monolayers are obtained to understand the molecular structure. From these results we identify mixtures suitable for effectively herding maritime oil spills in rough seas.

495098 Interfacial Interactions in Oil/Brine Emulsions Stabilized By Combinations of Cellulose Nanocrystals and Emulsion Stabilizers

Abstract:

Oil/Brine emulsions were prepared and stabilized using a combination of negatively charged cellulose nanocrystals (CNC) and emulsion stabilizers of various charges. Dodecane was used as the oil phase, while two high salinity brines were used as aqueous phases, namely, synthetic sea water (SSW) and American Petroleum Institute (API: 2wt% $CaCl_2$ + 8wt% NaCl) brine. Emulsions prepared using either nanoparticles or surfactants have been extensively characterized and the interfacial phenomenon involved in stabilization of these emulsions are well documented. Emulsions and emulsion based products that use a combination of both nanoparticles and surfactants have various applications in oil and gas production, as well as environmental cleanup; understanding interactions between nanoparticles and emulsion stabilizers in such emulsion systems is thus very important. This study was designed to characterize mechanisms by which CNCs and emulsion stabilizers interact to stabilize oil-water emulsions. Both CNCs (1 wt.%) and surfactants (at concentrations just below their critical micelle concentration) were dispersed in the aqueous phase prior to emulsion formation. The critical micelle concentrations of the stabilizers were measured in each brine, both with and without nanocrystals, to study adsorption behavior of surfactant molecules onto the CNC surface using a force tensiometer. Representative samples with volume ratios of 25:75 and 90:10 aqueous to oil phase were prepared using high shear mixing. The oil rich emulsions have potential applications for oil/gas production, while the water rich emulsions are relevant for oil spill cleanup. These emulsions were then characterized as either O/W (oil in water) or W/O (water in oil) based on electrical conductivity measurements. It was found from the measurements that at low oil to water volume ratio all emulsions were of O/W type but as the relative volume of the oil phase increased they switched to W/O. The stabilities of the prepared emulsions were studied in terms of macroscopic creaming rates over the duration of 24 hours and by quantifying the size distribution of the dispersed droplets with an optical microscope. Two formulations, guar gum with CNC and CNC alone, stabilized the emulsions for longer times compared to charged surfactants mixed with CNC. Droplets for the most stable emulsions tested were smaller in size and uniform in distribution, indicating that they are able to withstand coalescence. Interfacial tension measurements of the brine/dodecane system were used to estimate the adsorption of CNC and surfactants onto the interface using the pendant drop method. The results indicate that using non-ionic stabilizer and CNC with minimal interaction between them can be used to significantly improve stability of dodecane/brine emulsion.

RICE

Surface Complexation Modeling of Calcite Zeta Potential in Mixed Brines for Carbonate Wettability Characterization

Jin Song, Yongchao Zeng, Le Wang, Xindi Duan, Maura Puerto, Lisa Biswal, George Hirasaki

Rice University

1

Carbonate Wettability in Oil Recovery

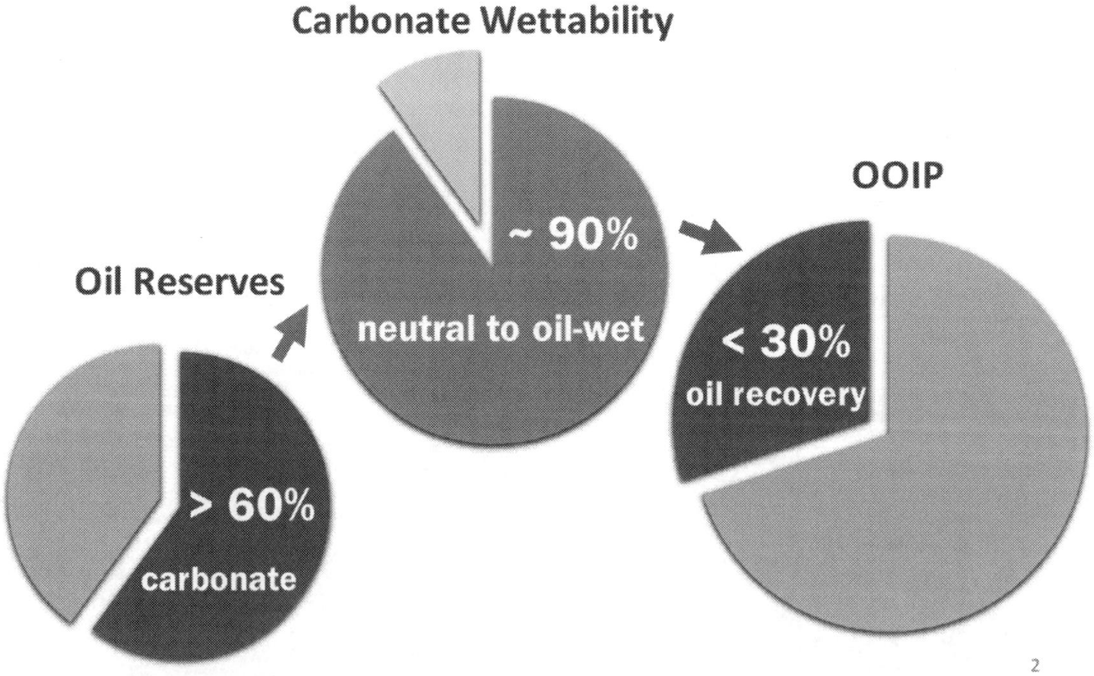

Strand, et al. 2005. Colloids and Surfaces A

2

Role of Electrostatic Forces

Smart water: brines with designed ionic composition that enhance oil recovery.

→ Possible reason: divalent ions affect carbonate surface charge

$$\cos\theta(h) = 1 + \frac{\int_h^\infty \Pi(\xi)d\xi + h\Pi(h)}{\sigma}$$

$$\Pi(h) = \Pi_{vdw}(h) + \boxed{\Pi_{electrostatic}(h)} + \Pi_{structural}(h)$$

Oil/brine surface charge characterization

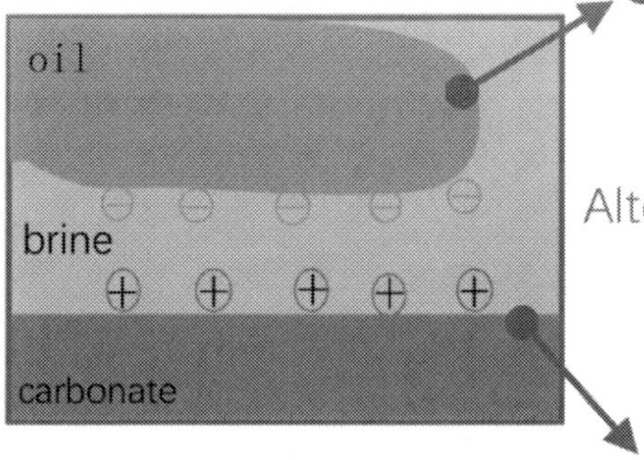

Alter wettability by manipulating surface charges
Divalent ions were found to be important

Carbonate/brine surface charge characterization

Hirasaki, 1991. SPE Formation Evaluation

3

Goal of Current Work

- Understand the role of brine composition in determining carbonate surface charge
  - Measure synthetic calcite (model for carbonate) zeta potential in various brines with divalent potential determining ions (PDIs)
  - Develop a model to fit for the measured zeta potential
  - Quantify the individual role of ions using the model

4

Zeta Potential Measurement of Synthetic Calcite

- **Rock:** synthetic calcite (>99.5%)
- **Brines**: I=0.1M $CaCl_2$, $MgCl_2$, Na_2SO_4, Na_2CO_3 & their mixtures with NaCl (controlled ionic strength, single potential determining ion (PDI))
- **Gas**: $P_{CO2} = 10^{-3.4}$ atm or 1 atm
- **Equilibrium**: 3 phases pre-equilibrated for 1~4 days until pH matches with PHREEQC calculation.

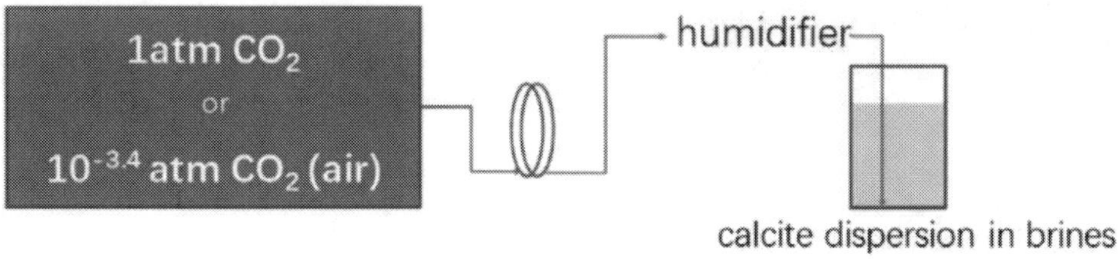

5

Zeta Potential Measurement of Synthetic Calcite

- Mixed brines of NaCl w/ $CaCl_2$, $MgCl_2$, Na_2CO_3, Na_2SO_4 to different ratios (I=0.1M)

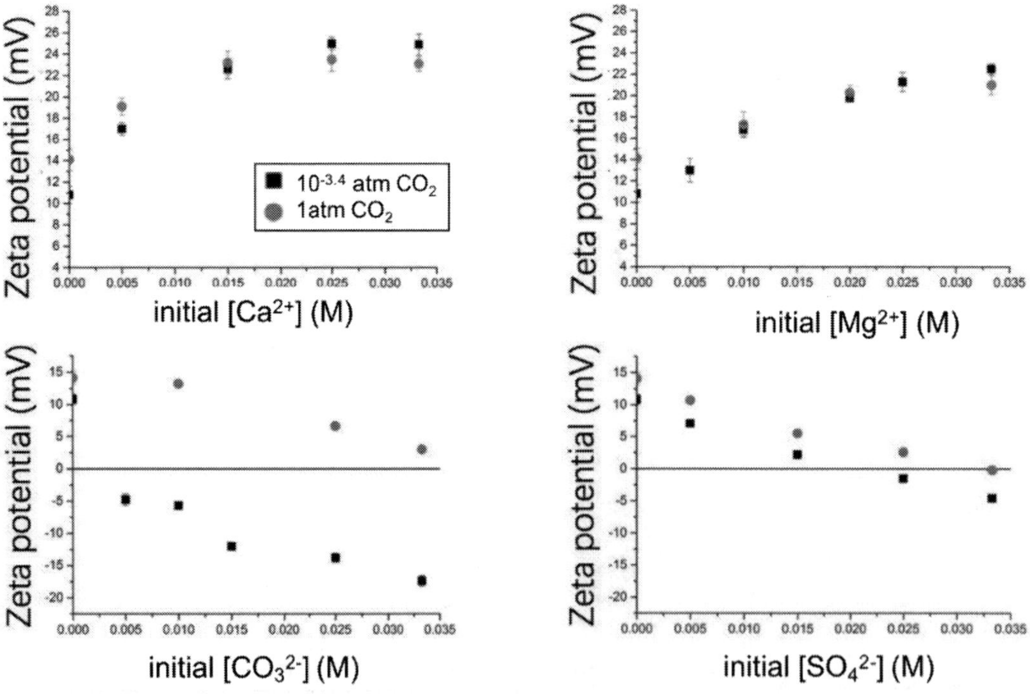

6

Zeta Potential Measurement of Synthetic Calcite

- Ionic strength = 0.1M for all brines.
- Changing pH by P_{CO2} does not have dramatic impacts on zeta potential.

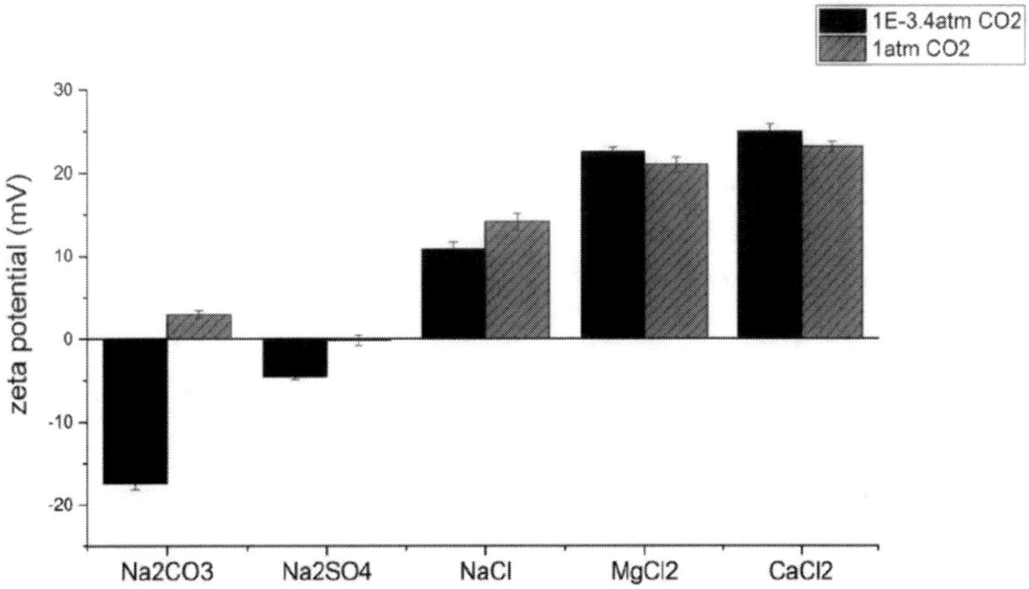

equilibrium pH	9.6	6.6	8.4	6.1	8.3	6.0	8.3	6.0	7.5	5.8

7

Origin of Calcite/Brines Surface Charge

- **Surface reactions** between surface sites & ions in brine

Major surface sites(**$>CaOH^{-0.75}$**, **$>CO_3H^{+0.75}$**) attract ions with **opposite** charge

Example: $>CaOH^{-0.75} + Ca^{2+} \rightleftharpoons >CaOH\cdots Ca^{+1.25}$

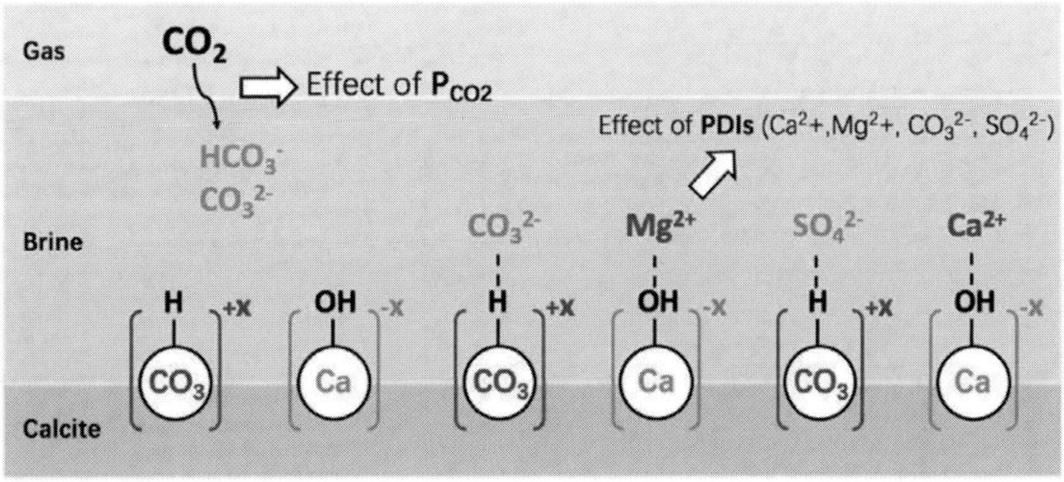

Song, J. et al. 2017 J Colloid Interface Sci

8

Calcite Surface Complexation Modeling (SCM)

- **Goal**: Predict calcite surface charge in given brine composition & P_{CO2}

surface reaction	intrinsic equilibrium constants
$>CaOH^{-0.75} + H^{+} \rightleftharpoons >CaOH_2^{+0.25}$	K_1
$>CO_3H^{+0.75} + OH^{-} \rightleftharpoons >CO_3^{-0.25} + H_2O$	K_2
$>CaOH^{-0.75} + Ca^{2+} \rightleftharpoons >CaOH \cdots Ca^{+1.25}$	IB_1
$>CaOH^{-0.75} + Mg^{2+} \rightleftharpoons >CaOH \cdots Mg^{+1.25}$	IB_2
$>CO_3H^{+0.75} + CO_3^{2-} \rightleftharpoons >CO_3H \cdots CO_3^{-1.25}$	IB_3
$>CO_3H^{+0.75} + SO_4^{2-} \rightleftharpoons >CO_3H \cdots SO_4^{-1.25}$	IB_4
$>CO_3H^{+0.75} + HCO_3^{-} \rightleftharpoons >CO_3H \cdots HCO_3^{-0.25}$	IB_5
$>CaOH^{-0.75} + Na^{+} \rightleftharpoons >CaOH \cdots Na^{+0.25}$	IB_6
$>CO_3H^{+0.75} + Cl^{-} \rightleftharpoons >CO_3H \cdots Cl^{-0.25}$	IB_7

$[>CO_3H^{+0.75}]_0 = [>CaOH^{-0.75}]_0 = 8.2\ \mu mol/m^2$

9 adjustable parameters

Phase equilibrium calculation

↓

Surface charge balance

$\sigma = f(charged\ surface\ species)$

↓

Grahame equation

$\psi_0 = f(\sigma)$

↓

Debye-Huckel equation

$\zeta = \psi_0\ e^{-\kappa x_d}$

↓

Zeta potential (ζ) measurement

optimize

9

SCM Fitting with Experimental Measurements

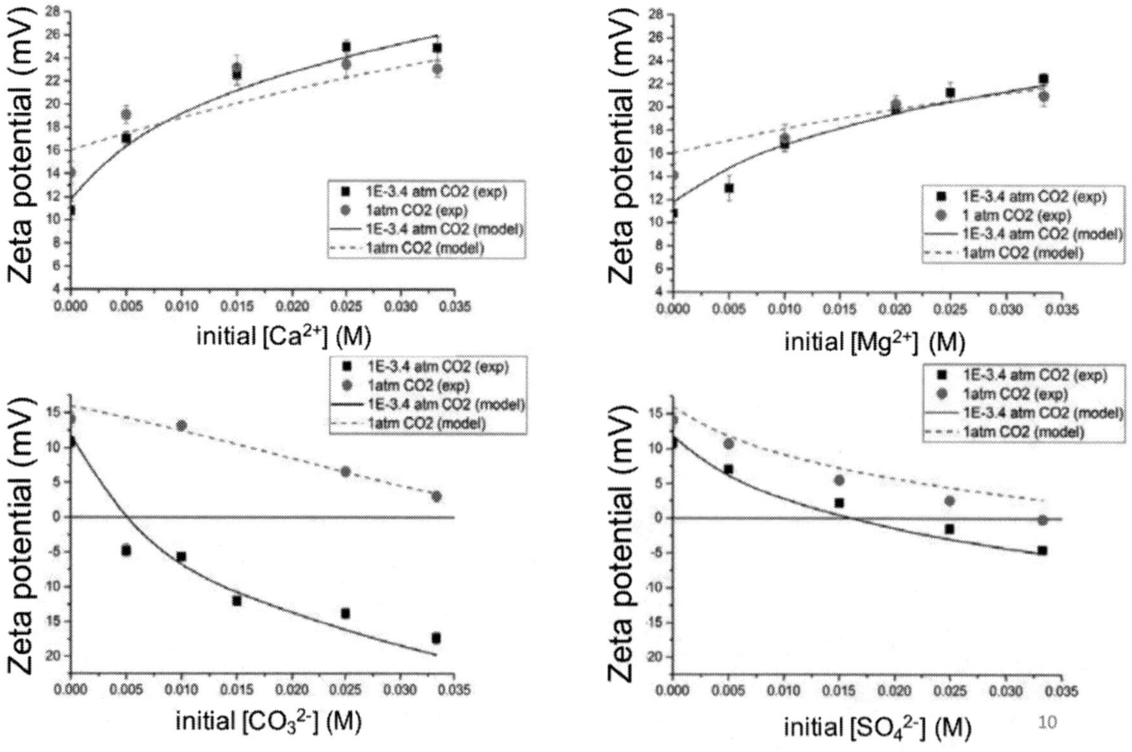

10

Calcite Surface Complexation Modeling (SCM)

- IB_i are negatively correlated to the hydrated ionic radius.
- IB (divalent) are at least 1 order of magnitude greater than IB(monovalent).

parameter ($\log_{10}K$)	value	surface reaction
K_1	0.30	$>CaOH^{-0.75} + H^+ \rightleftharpoons >CaOH_2^{+0.25}$
K_2	0.50	$>CO_3H^{+0.75} + OH^- \rightleftharpoons >CO_3^{-0.25} + H_2O$
IB_1	1.74	$>CaOH^{-0.75} + Ca^{2+} \rightleftharpoons >CaOH \cdots Ca^{+1.25}$
IB_2	1.62	$>CaOH^{-0.75} + Mg^{2+} \rightleftharpoons >CaOH \cdots Mg^{+1.25}$
IB_3	2.23	$>CO_3H^{+0.75} + CO_3^{2-} \rightleftharpoons >CO_3H \cdots CO_3^{-1.25}$
IB_4	1.00	$>CO_3H^{+0.75} + SO_4^{2-} \rightleftharpoons >CO_3H \cdots SO_4^{-1.25}$
IB_5	0.09	$>CO_3H^{+0.75} + HCO_3^- \rightleftharpoons >CO_3H \cdots HCO_3^{-0.25}$
IB_6	0.14	$>CaOH^{-0.75} + Na^+ \rightleftharpoons >CaOH \cdots Na^{+0.25}$
IB_7	-0.64	$>CO_3H^{+0.75} + Cl^- \rightleftharpoons >CO_3H \cdots Cl^{-0.25}$

11

Successful Prediction in $MgCl_2/Na_2SO_4$ Mixed Brines

- Mixed brine of $MgCl_2/Na_2SO_4$ to different ratios (I=0.1M) under $P_{CO2} = 10^{-3.4}$ atm
- These zeta potential measurements are NOT included in the parameter fitting.

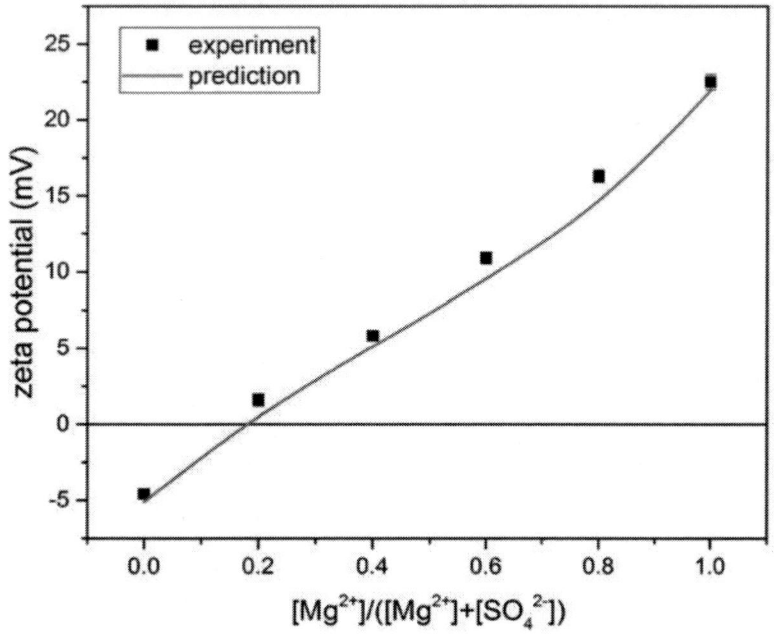

12

Successful Prediction in Modified Seawater

- Iceland spar is pure natural calcite.
- Zeta potential was measured in closed system with no gas environment.
- These zeta potential measurements are NOT included in the parameter fitting.

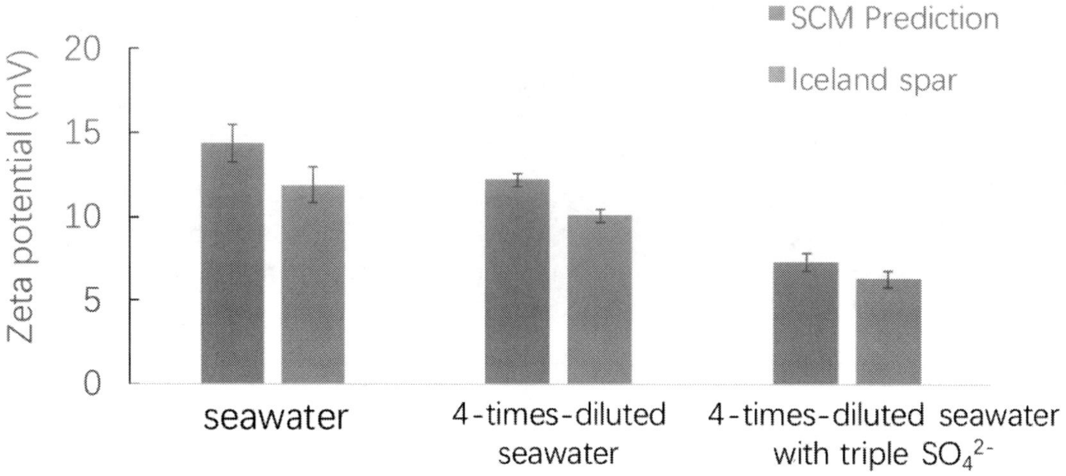

Example of Species Contribution Analysis

- distinguish the contribution of each ion in specific condition

$$contribution(i) = frac(i) \cdot z_i$$

$$frac(i) = \frac{[surface\ specie\ i]}{[>CO_3H^{+0.75}]_0} \qquad (anionic\ surface\ species)$$

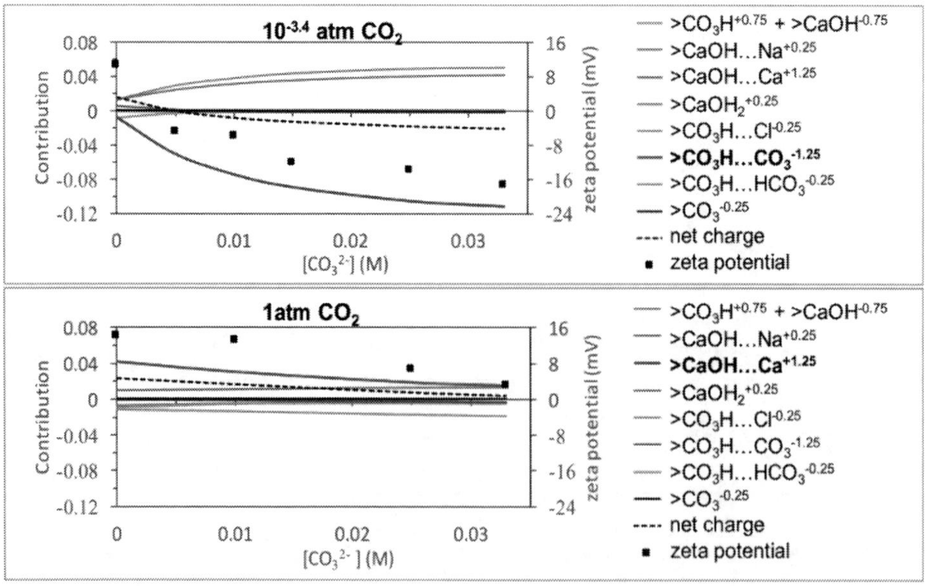

Conclusions

- Divalent ions (Ca^{2+}, Mg^{2+}, CO_3^{2-}, SO_4^{2-}) have surface binding constants (IB) at least one order of magnitude higher than monovalent ions.

- The calcite zeta potential is sensitive to CO_2 partial pressure in CO_3^{2-} brines

- Our SCM has the ability to predict calcite zeta potential in mixed brines with four PDIs (Ca^{2+}, Mg^{2+}, SO_4^{2-}, CO_3^{2-}).

15

RICE

أدنوك
ADNOC

THANK YOU!
Q&A

16

497065 Dissolution and Restructuring of Calcite Surfaces in Contact with Electrolyte Solutions - Implications for Enhanced Oil Recovery

Abstract:

Injecting diluted brine into carbonate (e.g., calcite) oil reservoirs has been shown to enhance oil recovery; yet, the mechanism(/s) behind this 'dilution effect' remains elusive. Using a Surface Forces Apparatus (SFA), we studied the dissolution and restructuring of calcite surfaces at 75°C that are sequentially exposed to electrolyte solutions of decreasing concentration, starting from highly concentrated brine (e.g., formation water with composition similar to that is found in oil reservoir) via seawater and different dilutions of seawater, to diluted monovalent solutions. Calcite surfaces are observed to restructure and roughen (100s of nanometers) within seconds, and persisted over several hours, after contact with electrolyte solutions and after refluxing with diluted brine solutions. This restructuring of the calcite surfaces due to dilution of the injection brine can have equally large – or even larger – effect as the changes in the physical/colloidal (electric double layer, van der Waals, and hydration) forces on the water contact angle, wettability, and – by implication – the oil recovery.

494435 Marine Bacteria Adhesion to Oil/Water Interfaces

Abstract:

Alcanivorax Borkumensis **is a hydrocarbon-degrading bacterium that dominates hydrocarbon-degrading bacterial communities around many oil spills.** ***A. Borkumensis*** **cells were grown either with dissolved organic carbon (no-spill condition) or with hexadecane layer (oil spill condition) as the sole energy source. Bacterial ability to adhere to the oil/water interface post-growth was monitored using interfacial tension, bacteria cell hydrophobicity measurements and fluorescence microscopy. Our experiments show that** ***A. Borkumensis*** **grown with dissolved organic carbon as the sole energy source remain hydrophilic and do not show significant adhesion to the oil/water interface. The bacteria grown under the hexadecane layer become partially hydrophobic and adhere to oil/water interfaces, where they reduce interfacial tension and form the early stages of a biofilm. We show that the biosurfactant produced by** ***A. Borkumensis*** **after exposure to hydrocarbons stay bound to bacterial cell membranes. Thus biosurfactant synthesis is vital for the bacteria to able to adhere to oil/water interfaces, which subsequently permits the bacteria to utilize hydrocarbons as an energy source. This study provides key insights into how** ***A. Borkumensis*** **accesses and degrades oil.**

497837 Oil Spill Dispersion with Food-Grade Amphiphiles: Formation of Oil-Bacteria Agglomerates during Oil Biodegradation

Abstract:

Food grade amphiphilic molecules, such as soybean lecithin and polyethoxylated surfactants, have been shown to be effective emulsifiers and capable as substitutes for conventional surfactants (e.g. dioctyl sulfosuccinate or DOSS) in the design of new dispersant systems. A key requirement of a dispersant system is that it facilitate bacterial adhesion to dispersed oil droplets to promote oil biodegradation. We have employed a microplate-based approach coupled with respirometry and microscopy studies to examine the interfacial and colloidal behavior associated with the adhesion of *Alcanivorax borkumensis*, an oil-eating bacterium commonly found near oil spills, to oil dispersed by lecithin/Tween-80 mixtures. Using hexadecane as a model oil, we show that increasing oil concentration (50 to 1000 mg/L) and bacteria concentration drive the formation of oil-bacteria agglomerates that sediment over several days. Microscopy analysis shows that the bacteria remain viable (alive) once adhered to the oil-water interface and form a dense biofilm with a fibrous network of exopolymeric substances. This work complements our research on the dispersion effectiveness of lecithin and lecithin/Tween-80 mixtures, and confirms that such mixtures can promote oil biodegradation.

499539 The Role of Bacterial Biofilm in Particle-Stabilized Oil-in-Water Emulsions: Implications to the Biodegradation of Oil Spills

Abstract:

The formation of oil-in-water emulsions from spilled oil and produced water has been attributed to the presence of natural suspended particles that stabilize the oil-water interface. While this phenomenon has been well-studied, the role of biofilm produced by oil degrading bacteria in these systems is yet to be fully understood. We use a model system of flat sheet clays (kaolinite) to form particle stabilized emulsions and contact the oil droplets with a hydrocarbon degrading bacterial organism (*Alcanivorax borkumensis*).Â Optical microscopy and cryo-SEM imaging of the microbial biofilm in the presence of particles shows a dense aggregate of cells embedded in the biofilm and evidence of biofilm bridging between particles. Surface engineering of the clay sheets through carbonization to optimize the contact angle, improves emulsion stability. It is found that the biofilm bridging that lead to particle and droplet aggregation leads to the natural phenomenon of â€œmarine snowâ€ and a partial sinking of oil that has been reported in the environment of oil spills. We thus conclude that natural and engineered particulates adhere at oil-water interface resulting in significant and rapid sedimentation only upon microbial activity, specifically due to biofilm formation on oil surface decorated by particles. The results of our work provide insight into interactions between oil degrading biofilm producing marine organisms and clay particles at solid stabilized oil interfaces.

501025 Novel Cathode Materials for Rechargeable Aluminum Batteries

Abstract:

Aluminum metal is one of the most coveted electrode materials for rechargeable batteries because it is inexpensive, globally abundant, non-flammable, non-toxic, and has a volumetric energy capacity almost four times greater than lithium metal. However, its widespread use in rechargeable batteries has thus far remained elusive, in part due to the limited number of cathode and electrolyte materials that can reversibly electrochemically cycle with aluminum metal anodes. Here, novel cathode materials for rechargeable aluminum batteries will be presented that are based on graphene and sulfur-containing materials (e.g., transition metal sulfides). Characterization of the cathode materials will be discussed as a function of state-of-charge and cycle number, emphasizing molecular-scale understanding of the reaction mechanisms via multi-dimensional NMR spectroscopy, X-ray diffraction, electron microscopy, and electrochemical methods. The results yield insights into design strategies for synthesizing next-generation cathode materials with improved energy densities, rate capabilities, and cycle lifetimes.

501329 Reversibility of Zinc Metal Anode: Fundamentals and Engineering

Abstract:

Zinc metal anode provides 3 times the volumetric capacity of Li metal anode, and is compatible with non-flammable aqueous electrolyte. Indeed, primary zinc-air batteries have about the same volumetric energy density of Li-S battery (3 times the one of $LiCoO_2$/graphite combination), and have long been used for electronic devices requiring extremely high energy density (e.g. hearing aid). However, rechargeable zinc-air batteries have not been successful, limiting their applications in larger markets such as cell phones, laptops, and electric vehicles. In this presentation, I will talk about our effort towards making reversible zinc-air battery, as a new research group at Georgia Tech. We find the main hurdle to reversibility lies in the zinc anode side. We developed in situ microscopy to understand the fundamental reason for irreversibility, and designed nanostructured zinc anode to overcome these challenges.

501554 An Air Breathing Lithium-Oxygen Battery

Abstract:

The Li-air batteries possess the highest theoretical energy density compared to any other battery chemistries[1,2]. However, these batteries suffer from a poor cycle life in the presence of real air components such as nitrogen, oxygen, carbon dioxide and moisture because of anode degradation, electrolyte instability, and slow kinetics of oxygen reduction and evolution in the air cathode[3–5].

Here, we report a new Li-Air battery system working with the actual air components (N_2, O_2, CO_2 and H_2O) with very high cycle life and energy efficiency that can work up to 550 continues cycles with the capacity of 500 mAh/g. This Li-Air battery benefits from a protected Li-anode, an air cathode based on MoS_2 catalyst and a mixture electrolyte. Different characterizations including differential electrochemical mass spectroscopy (DEMS), X-ray photoelectron spectroscopy (XPS), and Raman spectroscopy on the cathode reveal that the only discharge product in our system is lithium peroxide (Li_2O_2) without any evidence of side products e.g., lithium hydroxide (LiOH) and lithium carbonate (Li_2CO_3). Nuclear magnetic resonance (NMR) results also show the stability of the electrolyte after 550 cycles. DFT calculations reveal that the combination of MoS_2 cathode and mixture electrolyte make the system kinetically favorable for Li_2O_2 formation hindering the formation of other possible products. The Li-air system studied here is a key step toward the development of next generation of lithium batteries with higher energy density.

References:

1. Aurbach, D., McCloskey, B. D., Nazar, L. F. & Bruce, P. G. Advances in understanding mechanisms underpinning lithium–air batteries. *Nat. Energy* **1,**16128 (2016).

2. Bruce, P. G., Freunberger, S. A., Hardwick, L. J. & Tarascon, J.-M. Li–O2 and Li–S batteries with high energy storage. *Nat. Mater.* **11,**19–29 (2011).

3. Ye, L. *et al.* Lithium-Air Batteries: Performance Interplays with Instability Factors. *ChemElectroChem* **2,**312–323 (2015).

4. Asadi, M. *et al.* Cathode Based on Molybdenum Disulfide Nanoflakes for Lithium-Oxygen Batteries. *ACS Nano* **10,**2167–75 (2016).

5. Geng, D. *et al.* From Lithium-Oxygen to Lithium-Air Batteries: Challenges and Opportunities. *Adv. Energy Mater.* **201502164,** 1–14 (2016).

499594 Appropriate Characterization Techniques for Lithium-Oxygen Batteries, and Implications for Understanding 2e⁻ Vs. 4e⁻ Oxygen Reduction Processes

Abstract:

The lithium-oxygen battery is an intensely studied “beyond lithium-ion” battery technology due to its high theoretical specific energy. Commonly consisting of a lithium metal negative electrode, a gaseous oxygen-saturated nonaqueous electrolyte, and a porous carbon positive electrode, the battery operates via the formation (discharge) and decomposition (charge) of lithium peroxide at the porous carbon cathode. Currently, fundamental electrochemical challenges, including electrolyte instability, porous carbon cathode instability, and cathode passivation due to lithium peroxide's insulative nature, prevent the realization of the battery's high theoretical specific energy.[1] Therefore, researchers have focused on developing electrolyte and material engineering approaches to address these fundamental electrochemical challenges.

To assess the impact of electrolyte and material engineering approaches in lithium-ion batteries, researchers have commonly used galvanostatic cycling and cyclic voltammetry. However, differences between the lithium-oxygen battery and the lithium-ion battery, namely the use of gas and a surface reaction forming a solid discharge product, make these typical battery techniques insufficient for studying the lithium-oxygen battery. This talk will focus on critical advances in experimental characterization techniques for metal-air batteries, including differential electrochemical mass spectrometry, titrations of extracted cathodes, impedance spectroscopy, and simply the use of a small ratio of electrolyte to cathode surface area. We will show that these techniques are critical to understanding what is going on inside the battery, with less rigorous characterization resulting in incorrect conclusions.

Highlighted will be our recent work on characterizing the ability of lithium iodide and water to form lithium hydroxide rather than lithium peroxide as the primary discharge product, and how our characterization techniques elucidate the interesting but irreversible nature of this reaction.[2]

(1) McCloskey, B. D.; Burke, C. M.; Nichols, J. E.; Renfrew, S. E. *Chem. Commun.* **2015**, *51*, 12701.

(2) Burke, C. M.; Black, R.; Kochetkov, I. R.; Giordani, V.; Addison, D.; Nazar, L. F.; McCloskey, B. D. *ACS Energy Letters* **2016**, *1*, 747.

493014 Mechanistic Insights into the Sodium-Oxygen Battery Cathode Electrochemistry

Abstract:

The nonaqueous lithium-oxygen ($Li-O_2$) battery has received increasing attention over the past decade, owing largely to its high theoretical specific energy density. Prior study has demonstrated that the primary $Li-O_2$ discharge product is produced via a two-electron oxygen reduction reaction at the cathode, though the efficiency of O_2 conversion on discharge and the evolution of O_2 on charge is less than ideal.[1] A similar system, the nonaqueous sodium-oxygen ($Na-O_2$) battery, has been reported as a possible alternative, where the primary discharge product is produced via a one-electron oxygen reduction reaction.[2] Despite its lower theoretical specific energy, $Na-O_2$ batteries have been reported to provide improved cyclability and energy efficiency compared to their Li counterparts.[2-4] Sodium's natural abundance and lower cost also make large-scale implementation an attractive possibility.[4] However, prior studies have shown that $Na-O_2$ battery performance can be highly sensitive to operating conditions and contaminants, leading to changes in capacity, discharge product morphology, and overall cell performance, the fundamental causes of which are not all well understood.[4-6]

In this work, we present research concerning the operation of nonaqueous $Na-O_2$ cells while varying key parameters, with a focus on the oxygen pressure in the headspace on discharge and the current density employed on discharge and charge. We characterize the dependence of cell performance on these conditions, including the cell capacity and the discharge and charge overpotentials. In particular, we examine the causes of the "sudden death" phenomena, a decrease in potential on discharge and an increase in potential on charge that limit cell capacity and efficiency.[7] The one-electron oxygen reduction and evolution reactions at the cathode are verified quantitatively by pressure decay/rise measurements and differential electrochemical mass spectrometry (DEMS). Further qualitative study is performed using scanning electron microscopy to examine the morphology and deposition of the discharge product. Through this study, we elucidate the limitations of $Na-O_2$ cell capacity and the mechanistic origins of the "sudden death" phenomena.

References
1. McCloskey, B. D.; Burke, C. M.; Nichols, J. E.; Renfrew, S. E. *Chem Commun* **2015,** *51* (64), 12701-12715.
2. Hartmann, P.; Bender, C. L.; Vračar, M.; Dürr, A. K.; Garsuch, A.; Janek, J.; Adelhelm, P. *Nat Mater* **2013,** *12* (3), 228-232.
3. McCloskey, B. D.; Garcia, J. M.; Luntz, A. C. *J Phys Chem Lett* **2014,** *5* (7), 1230-1235.
4. Adelhelm, P.; Hartmann, P.; Bender, C. L.; Busche, M.; Eufinger, C.; Janek, J. *Beilstein J Nanotechnol* **2015,** *6*, 1016-1055.
5. Knudsen, K. B.; Nichols, J. E.; Vegge, T.; Luntz, A. C.; McCloskey, B. D.; Hjelm, J. *J Phys Chem C* **2016,**

120 (20), 10799-10805.
6. Xia, C.; Black, R.; Fernandes, R.; Adams, B.; Nazar, L. F. *Nat Chem* **2015,** *7* (6), 496-501.
7. Nichols, J. E.; McCloskey, B. D. *J Phys Chem C* **2017,** *121* (1), 85-96.

496474 Mechanistic Evaluation of Thermal Runaway in Potassium-Ion Batteries

Abstract:

Despite widespread commercial use, safety concerns of rechargeable lithium-ion batteries continue due to their susceptibility to thermal runaway. Exothermic reactions between the organic electrolyte, electrode materials, and the anode solid electrolyte interphase (SEI) lead to thermal runaway at higher temperatures (> 80°C), resulting in eventual cell rupturing and combustion. Recent interest in potassium-ion batteries (KIBs) as an inexpensive energy storage technology has arisen due to its natural abundance and compatibility with graphite anodes. The thermal stability of this K-ion graphite system, however, is completely unknown in terms of runaway onset temperature, enthalpy of reactions, and degradation mechanisms. We have explored thermal runaway of graphite and carbon anodes in KIBs through differential scanning analysis (DSC) of charged electrodes to evaluate the practicality of this new system. By selectively studying cell components (e.g. electrolyte, SEI, binder), we identified the sources of exothermic peaks and found that the profile is strongly dependent on the cell preparation and material properties. Interestingly, compared to the Li-ion battery case, less total heat generation occurs from 50-450°C (415 J g^{-1} vs. 1170 J g^{-1}) for a standard graphite anode, and the system overall shows great potential for improved safety by materials engineering.

498844 Molten Salt Batteries: Mechanics and Electrolyte Transport

Abstract:

Molten salt batteries are unique power sources with large energy and power densities per unit mass. The electrolyte used to transport charged ions between electrodes is a solid at room temperature which eliminates self-discharge over long storage times. To activate a molten salt battery, the electrolyte is melted by heat generated from an exothermic, pyrotechnic reaction. Because the battery will only remain active while the electrolyte is liquid, insulation is used on all sides of the cell stack.

To model the performance of molten salt batteries, mechanical behavior of the cell (cathode, separator, and anode) and the insulation must be understood. For example, the insulation acts like a spring to maintain a compressive force on the cell stack during activation, yet this material is subject to stress relaxation during battery aging at room temperature. At the electrolyte melting transition, the initially porous separator compresses significantly, redistributing the electrolyte and also determining the ionic resistance across the cell. The initial separator composition affects this behavior greatly, including electrolyte concentration and separator porosity. The insulation at this transition expands in thickness nonlinearly, determining the stress state of the activated battery, which is also investigated. The effects of these mechanical changes on electrolyte transport and battery performance will be discussed. To understand implications of design variables such as materials choice or initial stress state, these mechanical behaviors are included in a comprehensive thermal, electrochemical, and mechanical model for molten salt battery activation.

Sandia is a multi-mission laboratory operated by Sandia Corporation, a Lockheed Martin Company, for the United States Department of Energy under contract DE-AC04-94AL85000.

493867 Entropy-Driven, Two-Step Crystallization of Colloidal Clathrate Crystal

Abstract:

Non-classical crystallization, such as two-step crystallization, is the primary mechanism of solidification for many substances. Despite its importance, aspects of the system responsible for determining the crystallization mechanism are generally poorly understood. Identifying idealized model systems that exhibit non-classical crystallization is therefore desirable. Here, we report two-step crystallization in a three-dimensional system of hard shapes. Unlike previous reports of two-step nucleation and growth, in our system it is driven by entropy alone. Monte Carlo simulations reveal that crystallization is preceded by the appearance of a dense amorphous phase.The thermodynamics of crystallization is investigated by determination of the equation of state and via free energy calculations.

499352 Unraveling the Coupling between Demixing and Crystallization in Mixtures

Abstract:

Crystallization has remained an incredibly complex and mysterious process. This is for instance the case on the nanoscale where our ability to control nanoparticle properties is limited by our lack of understanding of how nanoparticles nucleate and grow. This is however key for many applications as metallic nanoparticles are increasingly used as protein markers, in high-density magnetic recording and in catalysis. Using molecular simulation, we shed light on the mechanisms underlying crystal nucleation and growth in metallic and bimetallic systems. In particular, we unravel the competition between the two nonequilibrium processes of phase separation and crystallization in bimetallic systems and discuss its impact on structure and composition on the nanoscale.

500651 La Mer Burst Nucleation and Growth: Assumptions, Models, and Data

Abstract:

Recent efforts to synthesize monodispersed nanoparticles have employed in situ reactions to steadily and homogeneously generate a mounting solute supersaturation. As the supersaturation climbs past a critical level in solution, the nucleation rate spikes giving birth to a large population of nuclei over a very short interval of time. Subsequent growth of these nuclei lowers the supersaturation and thereby suppresses further nucleation. The result is a large population of nearly monodisperse particles. Classic experiments by T. Sugimoto and others have been modeled with deterministic models having separate stages of supersaturation build-up, nucleation, growth, and sometimes ripening. We examine assumptions in these models within the context of chalcogenide nanoparticle synthesis (Prof. J. S. Owen) and gold nanoparticle synthesis (Prof. S. L. Scott). We present a fully non-equilibrium stochastic model for LaMer burst nucleation with comparisons to results from simpler models. The results should clarify whether fit parameters from simple models accurately resolve the microscopic nucleation and growth kinetics.

489027 A Stochastic Model of Primary Nucleation of Polymorphs

Abstract:

Polymorphism is a crucial feature of crystals, with particular importance in the fine chemical and pharmaceutical industry, since different polymorphs have different physical and chemical properties, such as solubility and crystal structure. Crystal nucleation is known to be a stochastic phenomenon; however, the implications of its statistical nature for the order of formation of the polymorphs (Ostwaldâ€™s rule of stages) have not yet been explored. Similarly, the role of stochastic nucleation in polymorphs transformation has not been thoroughly investigated either.

Based on large experimental evidence, and building upon previous theoretical work, we have developed a stochastic model of primary nucleation of multiple polymorphs from a clear solution, accounting also for growth, dissolution, and secondary nucleation. With this model, one can simulate and qualitatively study the interplay of all phenomena acting in the system - often coupled and not directly measurable - and their role in determining detection times and polymorphs transformation. Assessing such relationships and the mechanisms underlying them would allow to not only gain a deeper understanding of polymorphism, but also to better design and optimise the production processes of polymorphs.

492589 Step Velocity of a Crystal Edge with Alternating Rows of Growth Units

Abstract:

The morphology of a crystal impacts bioavailability, catalytic activity, processability, etc. Mechanistic growth models have accurately predicted crystal shapes for a number of organic molecules. [1-3] A critical parameter in these models is the step velocity, i.e., the rate at which a spiral or 2D nucleus edge spreads across the surface. The step velocity depends on the density of kink sites along an edge and the net attachment rate to a kink site; kinks are favorable surface sites for attachment of growth units (atoms, molecules, etc.). [4] Determining these quantities is complicated for non-centrosymmetric crystals, as their multiple growth units can lead to complex step edge configurations.

In this work, we examine an edge with alternating rows of A and B growth units (A-B edge), which can exhibit stable and unstable behavior depending on the supersaturation and anisotropic interactions between A and B. [5] For example, the B row can actually dissolve under supersaturated conditions. We capture these phenomena in a stochastic 1D nucleation model which employs a Fokker-Planck equation to treat formation of an AB nucleus (new A and B rows). This model is used to determine the supersaturation-dependent kink density by balancing kink creation (1D nucleation) with kink annihilation (kink collisions). The resulting step velocity predictions yield great agreement with kinetic Monte Carlo simulations for various supersaturations and anisotropic interactions. We find that the step velocity undergoes a cross-over in supersaturation dependence as the edge changes from unstable to stable (via increasing supersaturation). This suggests the possibility of a supersaturation-dependent aspect ratio of spirals and 2D nuclei with an A-B edge. The step velocity model captures important changes in growth at higher supersaturations and will improve crystal shape predictions.

[1] J. Li, C.J. Tilbury, S.H. Kim, M.F. Doherty, *Prog. Mater. Sci.* **2016**, *82*, 1-38.

[2] M.A. Lovette, A.R. Browning, D.W. Griffin, J.P. Sizemore, R.C. Snyder, M.F. Doherty, *Ind. Eng. Chem. Res.* **2008**, *47*, 9812-9833.

[3] P. Dandekar, Z.B. Kuvadia, M.F. Doherty, *Annu. Rev. Mater. Res.* **2013**, *43*, 359-386.

[4] P.G. Vekilov, *Cryst. Growth Des.* **2007**, *7*, 2796-2810.

[5] C.J. Tilbury, M.N. Joswiak, B. Peters, M.F. Doherty, *Cryst. Growth Des.* **2017**, *17*, 2066-2088.

497948 Multiscale, Multiphysics, Mechanistic Model for Computation of Face-Specific Growth Rates

Abstract:

The morphology or shape of the crystalline material is a critical determinant of their physical properties such as stability, flowability, wettability, bulk density, and dissolution rates. For example, the plate- or needle-like pharmaceutical crystals are difficult to granulate and compress into a tablet form. Identification of theoretical models to predict the effect of crystallization conditions on crystal morphologies is imperative to develop control strategies for preventing such undesired morphologies to grow. The dynamic and steady-state prediction of crystal morphology requires the knowledge of face-specific growth rates of crystals. The current models to predict crystal growth rates assumes the regimes of crystal growth mechanism (spiral, 2D nucleation and rough) based on the empirical calculations of interfacial free energies of solute and solvent molecules. The growth rate is then estimated using the height, spacing, and velocity of layers spreading on the crystal surface. However, a single face of the crystal can have multiple growth mechanisms on the surface and there need not be a strict barrier between different crystal growth mechanisms. This could be due to different processes â€“ diffusion, desolvation and surface integration, and topography evolution - occurring simultaneously on the nucleated crystal surface. There is a need for a multiscale model which accounts for all these processes. In this talk, I will describe the multi-scale model for determining crystal growth rate using Kinetic Monte-Carlo method. This multi-scale model captures the most fundamental mechanism of crystallization â€“ the association and dissociation of molecules â€“ and find the most probabilistic realization of the surface. Our method considers probabilities of events such as nucleation of 2D islands and screw dislocations and spreading of the layer based on the energy barrier for integration of molecules onto the crystal surface. The prediction of growth rates of Glutamic acid crystals will be shown as an example.

486860 Anisotropic Growth Kinetics of Triblock Janus Colloids

Abstract:

Anisotropic colloidal particles have been shown to exhibit equilibrium crystal structures of vastly different density and coordination number depending upon surface geometry and thermodynamic conditions. However, due to very small free energy differences between accessible structures, it is not known whether self-assembly into large single crystals is feasible. We use molecular dynamics simulations to calculate interface propagation velocities for different faces of these anisotropic crystals. Our results show that the growth kinetics are strongly influenced by geometric factors which are difficult to predict from equilibrium properties. Developing a better understanding of these effects could unlock a new set of tools for directed self-assembly by leveraging growth kinetics to select between structures with similar free energies.

501528 Enhanced Sampling Methods for Modulating Density Fields

Abstract:

Driven by hydrophobic effects, aqueous surfactant solutions self-assemble into mesostructures with distinct symmetries, such as micelles, vesicles and bilayers. Molecular simulations play an important role in connecting the molecular characteristics of the surfactants to the properties of the mesostructures, the thermodynamic stability of the mesophases, and transitions between them. However, studying these mesostructures using straightforward simulations can be challenging, due to the large separation in timescales between the relaxation of individual molecules and that of the mesostructures. Here, we present an enhanced sampling simulation technique to address this challenge. Our approach involves organizing the molecules of interest into structures with well-defined symmetries, through the application of an external potential, which biases the molecular density field. The response of the system to the biasing potentials allows us to estimate mesostructure properties, as well as the thermodynamic stability of a particular mesophase with respect to another.

496866 Mapped Averaging Methods for Accurate and Precise Evaluation of Free Energies and Other Properties By Molecular Simulation

Abstract:

"Mapped averaging" is a recently published scheme that provides alternative, rigorous expressions for the ensemble averages that underlie molecular simulation.[1-3] The approach introduces knowledge from approximate theories while relying on the molecular simulation to measure the deviation from the theory. The approach also exploits the knowledge of the forces and the Hessian of the potential. In this manner, calculation of the mapped averages by molecular simulation can proceed without contamination by noise produced by behavior that has already been captured exactly by the theory, thereby yielding results of unprecedented precision with minimal computational effort. The approach can be applied in principle to any property that can be expressed as a derivative of the free energy. It does not affect sampling, so it can be used for many properties in the same simulation. We describe our recent advances in formulating and applying these methods, focusing in particular on two broad cases: (1) the free energy, which can be obtained with great efficiency via thermodynamic integration of first-derivative properties; and (2) second-derivative properties such as the dielectric constant, which are of great practical interest, but which suffer particularly from stochastic uncertainty because their evaluation is based on averaging ensemble fluctuations when performed using conventional methods.

(1) Schultz, A. J.; Moustafa, S. G.; Lin, W.; Weinstein, S. J.; and Kofke, D. A. Reformulation of Ensemble Averages via Coordinate Mapping, *Journal of Chemical Theory and Computation* 2016.

(2) Moustafa, S. G.; Schultz, A. J.; and Kofke, D. A. Very fast averaging of thermal properties of crystals by molecular simulation, *Phys. Rev. E* 2015, *92*, 043303.

(3) Moustafa, S. G.; Schultz, A. J.; and Kofke, D. A. Harmonically Assisted Methods for Computing the Free Energy of Classical Crystals by Molecular Simulation: A Comparative Study, *Journal of Chemical Theory and Computation* 2017, *13*, 825-834.

489235 Predicting the Free Energy Landscape of Multicomponent Fluids

Abstract:

We derive a statistical mechanical framework for predicting the grand canonical free energy landscape of multicomponent fluids obtained from flat-histogram Monte Carlo simulations at chemical potentials and temperatures other than those used during a simulation itself. This is accomplished by expanding the landscape in a multidimensional Taylor series at each macrostate on the surface defined by the sampling order parameter; we illustrate how all equations necessary to precisely obtain the coefficients in these series may be obtained using classical statistical mechanics, and may be computed using data trivially available during the simulation. While each individual series may be relatively simplistic, when used in concert they are capable of producing feature-rich landscapes out of relatively simple ones. This approach allows us to obtain the thermodynamic behavior of multicomponent fluid systems, for instance, at low temperatures using only a single simulation at a much higher one. For multicomponent systems, we illustrate how a small number of different simulations may be rationally combined in a self-consistent fashion that yields a continuous thermodynamic phase phase for the fluid. This enables the equation of state for multicomponent fluid systems to be easily computed, and phase diagrams to be obtained over a broad range of conditions. After surveying a variety of different fluid systems, we identify a range of temperature and chemical potential differences over which the extrapolation of simulations tends to quantitatively predict various thermodynamic properties accurately. Beyond this range, extrapolation still provides a reasonable estimate of the free energy landscape; this requires less computational effort to refine via additional simulation(s) at the desired conditions than construction of the surface without any initially informed estimate. In either case, this method significantly increases the computational efficiency of flat-histogram Monte Carlo simulations when investigating thermodynamic properties of multicomponent fluids over a wide range of conditions.

498180 A Novel Molecular Simulation Method for Liquid-Liquid Equilibria Predictions and *in silico* Screening of Desalination Solvents

Abstract:

Conventional desalination technologies include energy-intensive distillation or pressure driven membrane processes. Recently, directional solvent extraction (DSE) has emerged as a viable low-temperature and low-energy alternative which utilizes a solvent capable of selectively dissolving water by rejecting the salt ions and at the same time having negligible solubility in water phase. The dissolved water is then extracted either by heating or cooling the water-solvent phase. Thus far, the proof-of-concept has been demonstrated with decanoic acid for which the extraction efficiency, defined as amount of pure water produced per 100 kg of solvent, is only 0.5% suggesting that improvements are needed. This can be achieved by *in-silico* screening of a large number of molecular structures that can extract substantially higher amounts of water than decanoic acid. Success of such a screening critically depends on the ability to calculate solubility of water in solvents. As is widely known, liquid-liquid equilibria (LLE) calculations from molecular simulations continue to remain challenging due to the challenges associated with the transfer of molecules between two dense phases. In this presentation, we will describe the development of a novel molecular simulation methodology based on computing the fugacity of water as a function of concentration to caclulate the extent of water dissolution in a range of solvents. First, the applicability of the LLE method is shown considering that the solubility of the solvent in water is negligible. The calculated solubilities are shown to give good agreement with experimental data for carboxylic acids with six to ten alkyl chain lengths over a range of temperatures. This method is also extended and demonstrated for systems exhibiting substantial mutual solubility unlike directional solvents and water. In addition, the structural properties elucidating the role of the hydrogen bonding interactions between water and directional solvents governing the solubility will be presented.

492659 Ssages: A Comprehensive Platform for Enhanced Sampling Simulations

Abstract:

Molecular simulations are an essential tool in modern research, guiding experimental efforts and providing fundamental insights into underlying phenomena. Well-designed simulations offer a molecule-scale microscope with exquisite control over the physical phenomena of interest. While immensely powerful, the simulation toolbox has some important limitationsâ€"the accuracy of computations requires proper sampling of the system. While this may be achieved by simulating long trajectories or multiple independent copies of a system, the size of the systems and the time-scales that are accessible are still restricted by hardware. Importantly, systems that have rough energetic landscapes can become â€œstuckâ€, impeding proper exploration of configurations.

A palette of techniques has evolved to overcome these difficulties through static and dynamic modifications of the statistical ensemble underlying the sampling; these may be restriction based (as in umbrella sampling and replica exchange techniques) or on-the-fly adaptations (as in the powerful adaptive biasing force and metadynamics methods). Regrettably, the implementation of these techniques is not always straightforward and can be unattainable to users. This has often constrained newly developed, ostensibly general methods to problem or software specific applications, preventing the use of these techniques across platforms and research.

With these concerns in mind, we have developed SSAGES: Software Suite for Advanced General Ensemble Simulations. SSAGES is a multi-platform software that allows the use of enhanced sampling techniques on molecular simulations, which allows users to seamlessly apply enhanced sampling techniques to a variety of systems using widely used simulation engines such as LAMMPS, GROMACS, QBox, and OpenMD. We leverage SSAGES to study a variety of systems using multiple methods and discuss relative merits of flat histogram and string methods. In this talk, we will demonstrate the powerful methods, such as metadynamics, string methods, forward flux sampling and adaptive biasing force to several example systems, including dissociation of NaCl, isomerization of alanine dipeptide, mechanical properties of liquid crystals, self-assembly of porous materials, and discuss applications to related classes of problems.

501120 Multiscale Modeling of Multicompartment Micelle Nanoreactors

Abstract:

In recent years, research in industrial applications of polymeric materials has begun to explore the field of immobilized catalysis. In particular, the idea of catalysts bound to a micelle backbone, creating a nanoscale molecular reactor (commonly referred to as nanoreactor), has become an area of great interest. From a computational perspective, investigating the potential of micelles as nanoreactors requires analyzing the miscibility of block copolymers, both on a fully atomistic and on a mesoscale basis.

Extensive work has already been completed in this group to establish a robust method of estimating the Flory-Huggins Ï‡ interaction parameter for a given pair of molecules in order to probe polymer-polymer miscibility; this information is necessary for preparing coarse-grained molecular simulations (e.g., micellization simulations). In our present work, we apply miscibility analyses to a relatively nascent technology in immobilized catalysis science, viz. the multicompartment micelle nanoreactor. This technology offers a way to harness both the enhanced reactivity of homogeneous catalysis *and*the ease of separation traditionally enjoyed by heterogeneous catalysis.

Through the use of mesoscale calculations, we will study the feasibility of a three-compartment micelle nanoreactor composed of hydrophilic, lipophilic, and fluorophilic blocks. Both aromatics and alkyls will be explored for the fluorophilic block, in order to find the polymer architecture which allows for maximum phase segregation. Additionally, we will investigate reactant and product diffusion through the micelle compartments in order to explore potential applications of this technology to polymer production. We hope to demonstrate the viability of a micelle nanoreactor capable of reaction compartmentalization and tandem catalysis, two hugely promising capabilities for highly selective multistep-catalyzed reactions in polymer science and manufacturing.

499342 Reaction Ensemble Monte Carlo: Applications to Ionic Liquids

Abstract:

Reaction ensemble Monte Carlo (RxMC) has been successfully used to study equilibrium compositions in many non-ideal environments, e.g. in nanopores or at high temperatures and pressures. In classical simulations, a reaction move is accomplished by deleting reactant molecules and inserting product molecules. Random insertions and deletions in condensed phases are rarely accepted. In simple liquids, this problem can be overcome using continuous fractional component (CFC) Monte Carlo with Wang-Landau (WL) biasing. But in ionic liquids, the extremely low vapor pressure and heterogeneous nanostructure cause even CFC-WL to fail. We present a simple adaptation of the original reaction move to align both the position and orientation of inserted product molecules to the deleted reactant molecules. We discuss applications to absorption of CO_2 in the ionic liquid triethyloctylphosphonium 2-cyanopyrrolide, [P2228][cnp] and the dissociation of the acid-base pair acetic acid and N-methylpyrrolidine.

502781 Temperature-Dependent Physicochemical Properties of Nitrotoluenes from Solvation Free Energies

Abstract:

The temperature-dependent vapor pressure, solubility, Henryâ€™s law constant, and partition coefficients are important in many industrial applications and in environmental modelling. However, the reliable data for these properties are not always available, especially, for hazardous and exothermic compounds, such as the nitrotoluenes. Here we demonstrate how force field-based solvation and self-solvation free energies as functions of temperature can be used to compute a wide range of physicochemical solvation properties. The temperature-dependent solvation free energies in water, 1-octanol, and in the pure liquid or subcooled phases for a set of nitrotoluenes are computed using expanded ensemble molecular simulation method over the temperature range of 273 K to 330 K. TraPPE force field parameters with CHELPG charges were used to model the nitrotoluenes. TIP3P and TraPPE-UA force field parameters were used to model water and 1-octanol, respectively. Subcooled and liquid-phase vapor pressures, solubilities, Henryâ€™s law constant, and partition coefficients (air-water, air-octanol, and water-octanol) are predicted and compared with available experimental data. The changes in entropy, enthalpy, and heat capacities are estimated both for the solvation of the solute in water and in 1-octanol, and for the transfer of the solute from water to 1-octanol. The effects of functional groups and molecular structure on physicochemical and thermodynamic properties are also discussed. The protocol followed here can be used for compounds for which temperature-dependent data are scarce or not accessible due to hazardous or exothermic nature.

499272 Predicting the Solubility and Diffusivity of Gases (CO_2, CH_4, H_2, Noble Gases) in Nano-Confined Interlayer Water and Bulk Water Using Molecular Dynamics Simulations

Abstract:

The chemistry of water films with dissolved gases confined between clay mineral surfaces has applications for a number of areas in energy and environment. For example, gas adsorption in solid-water interfaces impacts the formation of CH_4 hydrates in fine-grained sediments, the behavior of CH_4 in shale, CO_2 leakage across caprocks of geologic CO_2 sequestration sites, H_2 leakage across engineered clay barriers of high-level radioactive waste repositories, and the use of noble gases as tracers for subsurface hydrocarbon migration. Here, we present molecular dynamics (MD) simulations of the solubility and diffusivity of a range of gases (CH_4, CO_2, H_2, noble gases) in bulk water and in nano-confined interlayer water in clays. Our results indicate that the affinity of dissolved gases for clay mineral surfaces has a non-monotone dependence on the hydrated radius of the gas molecules. This non-monotone dependence arises from a combination of two effects: the enthalpic affinity of the gas solutes for the surrounding species and the entropic penalty of hydrophobic solvation.

500822 Simulation of Tracer Particle Diffusion in Attractive and Repulsive Glassy Matrices

Abstract:

Anomalous transport in confined systems is most commonly associated with particle sub-diffusion. Characterization of confined particle dynamics has been extended to matrices of different topologies, including ordered and disordered particles; the effects of slow relaxations of the confining matrix, however, have received comparatively little attention. Physical systems in which this scenario arises include colloidal gels[1] and polymer nanocomposites,[2]which are used in a variety of technological applications. A matrix of particles exhibiting glassy dynamics provides a convenient model in which to explore the effect of slow structural relaxations on the dynamics of strongly confined tracer particles. In the limit of infinite tracer dilution, the motions of confined tracers minimally affect the matrix relaxation; hence characterization of tracer dynamics may provide insights into the structural relaxation processes of glassy matrices, whose underlying physical mechanisms remain the subject of vigorous debate.

We use event-driven molecular dynamics to scrutinize the diffusion of tracer particles confined within "attractive" and "repulsive" glassy matrices formed from a well-characterized, bi-disperse system of colloidal spheres with hard cores.[3] Fixing the volume fraction of both the tracer particles and glassy matrix, the temperature is modulated to confine tracer particles within 'repulsive' high-temperature and 'attractive' low-temperature glassy matrices. These two broad types of glasses are dominated by different relaxation dynamics and mechanisms.[4]Here, we show that these distinct relaxation processes strongly influence the dynamics and trajectories of the tracer particles. By varying the size of the tracer relative to that of a matrix particle, we investigate the role of the size asymmetry on the particle trajectories and dynamics. Because dispersing particles within slowly-relaxing matrices with varying interactions appear in settings ranging from the crowded cytoplasm inside cells to natural soils in the environment to artificial fiber nanocomposites, these results provide insight into the coupling between particle transport and matrix dynamics across a wide range of scientifically and technologically relevant processes.

1. Babu, S., Gimel, J. C. & Nicolai, T. Tracer diffusion in colloidal gels. J. Phys. Chem. B 112, 743–748 (2008).
2. Kalathi, J. T., Yamamoto, U., Schweizer, K. S., Grest, G. S. & Kumar, S. K. Nanoparticle diffusion in polymer nanocomposites. Phys. Rev. Lett. 112, 108301 (2014)
3. Zaccarelli E. and Poon W., (2009) Colloidal glasses and gels: The interplay of bonding and caging, PNAS. 106: 36
4. K. N. Pham, et. al., Multiple glassy states in a simple model system. Science, 296:104–106, 2002.

493411 [Invited Talk] Predictive Modeling of Adsorption and Reaction Equilibria in Nanoporous Materials

Abstract:

Force-field-based and first principles Monte Carlo simulations are used to explore (a) the adsorption of complex mixtures in microporous and hierarchical zeolites (including multicomponent sour gas mixtures and aqueous solutions of alcohols and biorefinery compounds), (b) the adsorption of gas molecules in metal-organic frameworks with undercoordinated metal sites, and (c) the influence of nanoscale confinement and preferential adsorption sites on reaction equilibria. The talk will highlight simulation methodologies and microscopic-level origins of the observed thermodynamic behavior.

501336 Characterizing Protein Hydration to Inform its Interactions

Abstract:

The extent to which the inherent structure of water is perturbed by complex molecules, such as proteins, peptides, and surfactants, influences the thermodynamics and the kinetics of their assembly. However, accurately characterizing this perturbation is challenging, because the manner in which proteins disrupt the inherent structure of water depends not only on the chemistry of the underlying protein surface, but also on the precise topography and chemical pattern of amino acids. Nevertheless, understanding the role of water in protein interactions is essential to understanding, predicting, and eventually controlling such interactions, which play a crucial role in the development of therapeutic strategies and in protein separations. In this presentation, I will discuss our recent efforts on quantitatively characterizing this disruption of water structure, with the goal of using this information for efficiently predicting protein interaction interfaces as well the design of ligands with high affinity and specificity.

492704 Confinement-Induced Supercriticality and Phase Equilibria of Hydrocarbons in Nanopores

Abstract:

For over a century, the phase behavior of bulk fluids has been described as *PVT* (pressure–volume–temperature) three-dimensional properties, but it has become increasingly clear that the liquid–vapor phase behavior in confined geometries is significantly altered from the bulk. Efforts have been devoted to accessing confined phase transitions using sorption, molecular simulations, and theoretical methods. However, a comprehensive picture of *PVT* relationships for confined hydrocarbons remains uncertain. Herein, we introduce *d* (confining pore diameter) as a fourth dimension, and we present *PVT-d* behavior of confined fluids in nanopores. A *T–d* phase diagram is presented for *n*-hexane, *n*-octane, and *n*-decane under multiple confinement scales (37.9, 14.8, 9.8, 6.0, 4.1, 3.3, and 2.2 nm cylindrical pore diameter) using experimental differential scanning calorimetry and *PVT-d* equation of state theory at atmospheric pressure. As pore diameter decreases from 37.9 to 4.1 nm, the bubble point increases by as much as 15 K above bulk, until we observe behavior consistent with a supercritical state, pointing to confinement-induced supercriticality. Remarkably, experimental and theoretical findings overlap very well, showing that this approach effectively captures the phase boundaries between the liquid, vapor, and supercritical fluid regions. The model and completed EOS are additionally extended to calculation of isothermal capillary adsorption, and its validity is discussed.

501017 [Invited Talk] Simulations of Biomolecular Assembly Processes at Interfaces

Abstract:

Although biomolecular folding, binding and assembly are usually conceived of as occurring in a bulk aqueous solution, there are numerous instances of such processes occurring near interfaces within cells. Examples of these interfaces include very large macromolecules, membranes bounding intracellular compartments and the cell membrane itself. Interfacial behavior of biomolecules is also important in several technological applications such as DNA-based nanomaterials, biosensors, and microarrays. In this talk, I will discuss several ongoing research problems in my group that illustrate rich behavior exhibited by biomolecules (protein, DNA) at interfaces.

499504 Nucleation of Capillary Bridges and Bubbles

Abstract:

Using molecular simulation, we analyze the capillary condensation and evaporation processes. For this purpose, we define the entropy of the adsorbed fluid as a reaction coordinate, and determine the free energy associated with both processes. We find capillary condensation to proceed through the nucleation of a liquid bridge across the nanopore, followed by its expansion throughout the pore to give rise to the stable phase of high density. For capillary evaporation, we also find a multi-staged process, starting with the initial destabilization of the fluid structure followed by the formation of a bubble in the nanopore.

500052 Modeling Single Component Transport in Mesoporous Membranes Under Non-Equilibrium Conditions

Abstract:

Mesoporous membranes have potential for energy efficient removal of pollutants like CO_2 and volatile organic compounds from their mixture with air in the form of industrial effluents. Logical design of mesoporous materials requires in-depth knowledge of transport of both condensable vapors and gases in pure and mixture form through the pores. Characterization of these materials through measurement of pore size distribution as well as application to gas separation, require partial pressure gradients and are thus inherently nonequilibrium. Modeling of these systems with classical modeling techniques such as computational fluid dynamics require predetermined models for various phenomenon such as Knudsen diffusion, capillary condensation and surface adsorption while molecular simulations require computational effort too large to be used on regular basis. Dynamic mean field theory (DMFT), a coarse-grained lattice based theory, computationally efficient technique has been used to model mesoporous membranes. We used this approach to study permporometry, a characterization technique where permeation of a light gas in presence of condensable vapor is used to determine pore size distribution and deviation from equilibrium is small. We have further investigated separation process associated with use of significant pressure gradient to separate condensable vapor from its mixture with light gas. This represented a large deviation from equilibrium. An interesting phenomenon was revealed as a result where capillary condensation is confined to high pressure side of the system and pore filling is incomplete at steady state. This phenomenon has been previously hypothesized in the literature but concrete understanding is still lacking. We used a simple slit pore with pure fluid to understand the transport in partial capillary condensation state using both DMFT and dual control volume grand canonical molecular dynamics.

498151 Optimizing the Formation of Solid Solutions with Colloids of Different Shapes

Abstract:

A key challenge to engineer ordered solids from the co-assembly of two differently-shaped building blocks is to predict the key particle characteristics that lead to maximal mutual ordered-phase compatibility (MaxOC). While both entropy disparity, as captured by the relative size of the components, and energetic inter-species selectivity affect MaxOC, it is the former whose effect is less intuitive and the main focus of this work. Such MaxOC predictive rules are formulated and validated by using Monte Carlo simulation results for hard-core mixtures of octahedra and spheres and of other previously studied mixtures. Specifically, it is proposed that component size ratios should maximize their â€œsubstitutional symmetryâ€ and hence minimize the combined free-energy cost associated with mutating a host-particle into a guest-particle in each of the solid phases. For the hard-core mixtures examined, packing entropy stabilizes substitutionally disordered solid solutions but not stoichiometric compounds. Additional molecular simulations were hence used to demonstrate, consistent with recent experimental findings, that such compounds can be formed by strengthening the inter-species compatibility via orientation-dependent attractions.

492003 Programmed Assembly of Anisotropic Patchy Colloids By Nonlinear Learning and Landscape Engineering

Abstract:

The inherently many-body nature of assembly makes it a challenge to identify the underlying collective mechanisms and assembly pathways, and has hindered the rational design of building blocks to produce aggregates with desired structure and function. We have combined tools from statistical mechanics, nonlinear machine learning, and molecular simulation to develop an approach to determine low-dimensional assembly landscapes preserving the important thermodynamics (*what* stable aggregates can assemble?) and mechanisms (*how* do they assemble?) of assembly. We have applied this approach to understand and predict the assembly of anisotropic patchy colloids, Janus particles, "digital colloids", and asphaltene molecules.

In this work we build on these foundations to perform inverse design of colloidal building blocks to spontaneously self-assemble into aggregates with desired structure. We engage this challenge through the principle of "landscape engineering" â€“ rational sculpting of the low-dimensional assembly landscape to maximally favor the target morphology. Implementing this concept in practice, we couple our nonlinear machine learning approach with state-of-the-art optimization engines to iteratively optimize building block properties to maximize the thermodynamic stability and kinetic accessibility of the target structure. We demonstrate our approach in the inverse design of anisotropic patchy colloids engineered to assemble hollow Platonic polyhedra with applications as biomimetic models of viral capsids and in the hierarchical assembly of 3D photonic crystals. In an application to icosahedral target aggregates, our approach discovers a colloid architecture that produces ~45% improved yields than that designed by expert knowledge and geometric concerns alone. Starting from this optimized icosahedral design and selecting octahedra as our target, our approach is capable of efficiently mutating the building block design from this poor starting guess to engineer a new colloid programmed to assemble into octahedra with high yield and selectivity.

487728 Evaporation-Induced Assembly of Colloidal Crystals

Abstract:

Colloidal crystals have promising applications for technologies such as photonics, catalysis, and sensing. One convenient method to create such crystals is to disperse the colloids in solution and evaporate out the solvent, leaving behind a close-packed structure. In most situations, however, the crystal structures that form after evaporation are polycrystalline and highly sensitive to the processing conditions. For example, Cheng and Grest [*J. Chem. Phys.* **138**, 064701 (2013)] used computer simulations to show that in-plane crystalline order at the air-solvent interface undergoes a maximum with respect to evaporation rate. The influence of evaporation on crystal structure at larger scales, e.g, across multiple crystalline layers, is not fully understood.

We performed massive-scale, explicit-solvent molecular dynamics simulations to study the evaporation-induced nucleation and growth of a colloidal crystal. We classified the structure of crystallizing colloids using a machine-learning approach, and showed how the crystal nucleates and grows from the drying air-solvent interface. Complementary implicit solvent simulations were performed to demonstrate the nontrivial role that solvent plays in controlling the crystallization process. Our work has important implications for the processing of colloidal crystals from solution and nonequilibrium molecular modeling.

500639 Amphiphile-Induced Reorganization of Liquid Crystals at Aqueous Interfaces

Abstract:

It is well-known that monolayer assemblies of amphiphiles at planar interfaces between thermotropic liquid crystals (LCs) and aqueous phases give rise to configurational transitions of the liquid crystals. The general assumption is that hydrophobic tails of surfactants interdigitate into the liquid crystal molecules at the interface to trigger reconfiguration of LC films. A different mechanism is introduced here, whereby a monolayer decoration of surfactants disturbs antiparallel orientation of 4-cyano-4â€²-pentylbiphenyl (5CB) molecules to cause a nematic-to-isotropic phase transition of LC within a ~2 nm layer at the interface. Upon formation of this isotropic layer, the anchoring strength of 5CB at the aqueous interface reaches zero to spawn a transition of a hybrid configuration to a uniform configuration. This configuration transition releases the stored elastic energy of the system. These results present a previously unknown mechanism that can be exploited for rational design of biological sensors, and drug delivery systems.

489931 Investigating Thermodynamics and Kinetics of a β(16-22) Peptide Aggregation Using Coarse-Grained Simulations

Abstract:

Aβ(16-22) is an intrinsically disordered and amyloidogenic fragment of full length Aβ protein which is associated with Alzheimer's disease. In the present study, we perform computer simulations of Aβ(16-22) peptide oligomerization and fibrillation by applying discontinuous molecular dynamics (DMD) simulation combined with an intermediate resolution protein model, PRIME20. We first construct a thermodynamics phase diagram of Aβ(16-22) peptide in the plane of temperature and concentration. The coexistence line between five types of peptide aggregates (native oligomer, disordered oligomer, 2-, 3- and 4-sheet fibrils) and the solution phase are determined by measuring the equilibrium soluble peptide concentration Ce(T) at which the peptide aggregate neither grows nor dissolves. The latent heat of soluble peptide aggregation into fibril is calculated by fitting lnCe-1/T data to the Van't Hoff equation. The values are found to be close to the binding energy per peptide within the fibril. Next we investigate the effect of various degrees of supersaturation on cluster size and helix and β sheet content of the Aβ(16-22) aggregates formed at various temperatures and concentrations. The kinetically observed aggregates are structurally different from those predicted from the equilibrium phase diagram. In addition, we reveal the major energy barriers and intermediate structures along the Aβ(16-22) oligomerization (dimer to octamer) pathway by calculating the free energy profiles as a function of an optimized reaction coordinate.

498459 Inverse Materials Design from Phase Transitions in Shape Space

Abstract:

Since the development of the theory of atoms, packing arguments have been used in an attempt to explain and predict the structure of condensed matter. For atomic and even surfactant systems packing provides a useful heuristic rationalizing structure. However, studies prompted by the recent explosion of new nanoparticles show that packing arguments often fail in two ways. First packing arguments fail to predict the emergence of ordered mesophases. Second, packing arguments fail to predict particles likely to assemble dense packing structures at finite pressure. Though the formation of local dense packing motifs in order to maximize shape entropy has been proposed to explain why global packing arguments fail, there is no quantitative measure of how and when packing arguments fail. Here we use a recently proposed digital alchemy framework to distinguish between finite pressure assembly behavior and the onset of infinite pressure packing behavior through Alchemical Monte Carlo (Alch-MC) simulation. We study families of FCC-, BCC-, and SC- forming symmetric convex polyhedra and find for FCC a first-order shape-shape phase transition in shape space from an "assembly" regime to a "packing" regime. In the assembly regime we determine optimal shapes, as well as the most and least sensitive aspects of shapes for crystal formation. In the packing regime, we find that the optimal packing shape fails to predict the thermodynamically preferred shape up to packing fractions of 99% and that vertex truncation of these convex polyhedra generally optimizes the particle shape. We report parallel results for a family of simple cubic formers and body-centered cubic formers.

491859 Concentration Effects in Simulations of Non-Ionic and Ionic Surfactant Micellization

Abstract:

Surfactants are well known for self-assembling into micellar aggregates from free surfactants at the critical micelle concentration (cmc). Although some micellar properties above the cmc are assumed to be constant, such as the free surfactant concentration and the degree of co-ion association, experiments and simulations have shown that the free surfactant concentration decreases above the cmc. We have performed histogram reweighting grand-canonical ensemble Monte Carlo simulations supplemented with cluster moves to study the micellization of implicit-solvent, non-ionic lattice[1] and ionic off-lattice[2] surfactants. Molecular dynamics simulations of the ionic surfactants were also performed for large systems. We calculate the cmc, osmotic pressure, concentration of free surfactants and aggregation number distributions[3]. For ionic surfactants, we also obtain the degree of co-ion association. Above the cmc, the free surfactant concentration decreases for both non-ionic and ionic surfactants. From simulation configurations, we offer an explanation for the decrease of free surfactants using a methodology developed for non-ionic surfactants to calculate the excluded volume[3] and theory developed for ionic surfactants[4]. Our finding suggest that simulation studies that use the free surfactant concentration at high surfactant loadings as a proxy for the cmc provide underestimates, potentially by large factors, of the true cmc. Models are proposed for reliable extrapolations of results from simulations at high loadings back to the cmc.

[1] Larson, R. G. "Monte-Carlo Lattice Simulation of Amphiphilic Systems in 2 and 3 Dimensions", *J. Chem. Phys.* 1988, 89, 1642-1650

[2] Jusufi, A., Hynninen, A.P., Panagiotopoulos, A.Z., "Implicit Solvent Models for Micellization of Ionic Surfactants", *J. Phys. Chem. B*, 2008, 112, 13783-13792

[3] Santos, A.P., Panagiotopoulos, A.Z., "Determination of the critical micelle concentration in simulations of surfactant systems", *J. Chem. Phys.*, 2016, 144, 044709

[4] Jusufi, A., LeBard, D.N., Levine, B.G., Klein, M.L., "Surfactant Concentration Effects on Micellar Properties", *J. Phys. Chem. B*, 2012, 116, 987-991

500035 Self-Assembly Simulations of Stratum Corneum Lipid Mixtures

Abstract:

The barrier function of human skin is localized to its outermost layer, the stratum corneum (SC); specifically, the lipid matrix which surrounds dead, flat corneocytes is thought to be the main barrier to chemical penetration through the skin. The chemical makeup of the SC lipids is known, but the molecular level details of its structure are not, in part due to the fact that 14 unique lipids have been identified in the SC. As such, experiments often consider simplified mixtures of synthetic lipids whose composition can be precisely controlled, however, even simple systems lead to different hypotheses (e.g., whether two-tailed ceramide lipids adopt a hairpin or extended conformation in lamellar structures). Molecular simulation is an important complement to these experimental studies, as simulation provides a direct 3-D molecular-level resolution and precise control over composition. Thus far, most simulation studies of SC lipids have focused on pre-assembled structures. While such studies have offered much insight into lipid-lipid and lipid-water interactions, it is possible that the preassembled structures are not representative of the molecular structures in the skin, given the low diffusion coefficient of the lipids and the fact that single bilayers are considered, rather than experimentally/biologically relevant multilayer structures. As such, self-assembled structures would provide a more realistic structure of SC lipid systems, but the timescales required to access self-assembly, and the system sizes needed to model multilayer structures, are likely prohibitive for atomistically detailed models, necessitating the use of coarse-grained (CG) models.

In this work, we present the results of the derivation and validation of CG models of the lipids present in the SC: ceramides, cholesterol and free fatty acids. The potentials were derived via multi-state and multi-ensemble iterative Boltzmann inversion [1], based on simulations of fully atomistic models [2]. Validation is first performed on preassembled bilayers and multilayers, where results are compared to corresponding atomistic simulations. Self-assembly into lamellar structures was observed, and the details of the process are elucidated. We find that the intramolecular conformations of the ceramides are dependent on the lipid composition of the system, with a greater number of different components giving rise to a greater fraction of extended or linear ceramide conformations, compared to less diverse systems which show a greater fraction of hairpin ceramide conformations. Thus, these self-assembled structures provide clear insight into the experimentally proposed

References

[1]. Moore, T. C., Iacovella, C. R. & McCabe, C. Derivation of coarse-grained potentials via multistate iterative Boltzmann inversion. J. Chem. Phys. 140, 224104 (2014).

[2]. Moore, T. C., Iacovella, C. R., Hartkamp, R., Bunge, A. L. & McCabe, C. A Coarse-Grained Model of Stratum Corneum Lipids: Free Fatty Acids and Ceramide NS. J. Phys. Chem. B 120, 9944â€“9958 (2016).

500158 Self-Assembly of Multi-Flavored DNA-Functionalized Particles into Binary Superlattices

Abstract:

DNA base-pairing can be exploited to induce programmable interactions between colloidal particles via anchoring of complementary DNA-strands on their surface. This allows one to design colloidal systems which can crystallize into a variety of superlattice structures. A common strategy has been to functionalize two particle types by complementary DNA strands so that the same particle types are repulsive whereas the opposite particles are attractive. A more recent scheme, so-called multiflavoring, is to blend complementary DNA strands on the same particle to achieve systems in which like-particles also attract each other. In this work, we show that control over interparticle interactions in such systems can allow control over the resulting binary crystal structures in a way that is not feasible by other methods.

491659 Accelerating Force Field Parameterization to Improve the Quantitative Predictability of Thermophysical Properties

Abstract:

A plethora of force fields for organic compounds exist in the literature. These models differ in the underlying molecular assumptions and the data used in the parameterization process, which are typically liquid density, vapor pressure, and/or heat of vaporization. The traditional parameterization approach rarely considers the entire parameter space due to the relatively high computational cost for molecular simulations, although significant progress has been made in recent years. Instead, the force field parameters are usually optimized in a trial and error manner or with gradient based algorithms. With the traditional optimization approach it is difficult to assure that a global minimum is obtained, to assess the degree of nonlinear coupling between parameters, and to quantify the uncertainty in the optimal set of parameters. Furthermore, the optimization process typically includes only a small set of experimental data (to limit the number of molecular simulations required) for a few compounds (to reduce the number of nonbonded parameters that are optimized simultaneously). This is significant because the reliability and transferability of the resulting force field is intimately related to the quality and quantity of data to which it is parameterized. For this reason, we present a parameterization approach designed to overcome these deficiencies and, thereby, develop a more quantitatively predictive force field for thermophysical properties.

The novel methodology employed in this work reduces the computational cost of force field parameterization by up to three orders of magnitude. The optimization is accelerated by using simulated configurations for an initial parameter set to predict physical property values for unsampled parameter sets. This acceleration enables the simultaneous optimization of several parameters to the large amount of experimental data available at the National Institute of Standards and Technology (NIST) Thermodynamics Research Center (TRC). In addition, this approach facilitates the use of Bayesian inference to assign uncertainties to the force field parameters. This is important as meaningful uncertainties are essential for molecular simulation to be a quantitative tool in thermophysical property prediction. Furthermore, advanced Bayesian methodologies are employed to ensure that the parameters are not overfit to the training data set by simultaneously investigating several nonbonded potential functional forms (e.g. Lennard-Jones 12-6, Mie n-m, Buckingham exponential-6, etc.). This improves the transferability of the force field when compared to data at different state points.

Universität Stuttgart

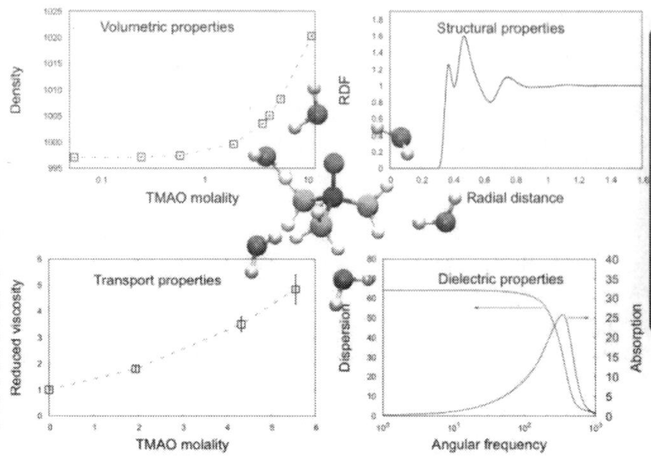

Validation of Trimethylamine-N-Oxide (TMAO) Force Fields Based on Thermophysical Properties of Aqueous TMAO Solutions

AIChE Annual Meeting 2017

Daniel Markthaler,
Jörg Baz,
Johannes Zemann,
Jens Smiatek,
Niels Hansen

itt

INSTITUTE FOR COMPUTATIONAL PHYSICS

Introduction

Why is TMAO interesting to study?

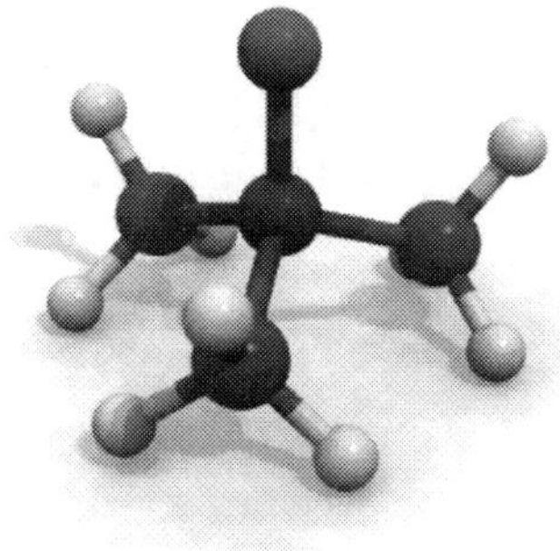

- TMAO is commonly used as protein stabilizer
- TMAO counteracts destabilizing effects of:
  - Elevated p and T (pH dependent)
  - Denaturants like urea
- Different stabilization mechanisms proposed:
  - Preferential exclusion from protein surface
  - Direct interaction TMAO-protein

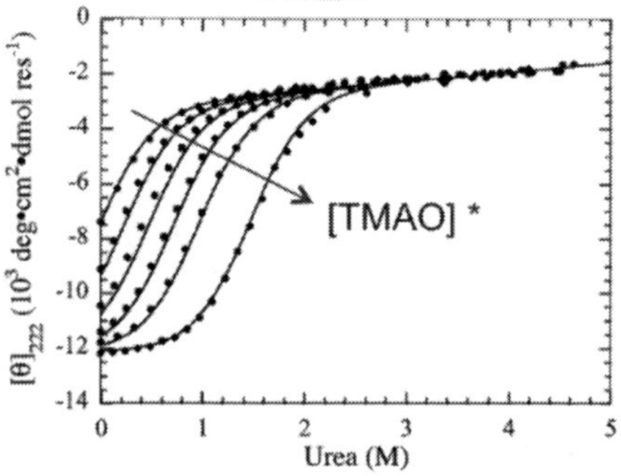

Proper description of binary aqueous solution properties essential for interpretation of TMAO behaviour

* Taken from C. C. Mello et al., *Protein Sci.* (2003) 12

1

Overview: Simulated Systems

Tested TMAO models

- 4/5 tested models derived from Kast 2003 force field (FF) [1]: same intramol. parameters but optimized nonbonded parameters
- Models differ in water model, comb. rules, 1-4 interactions, sim. param.

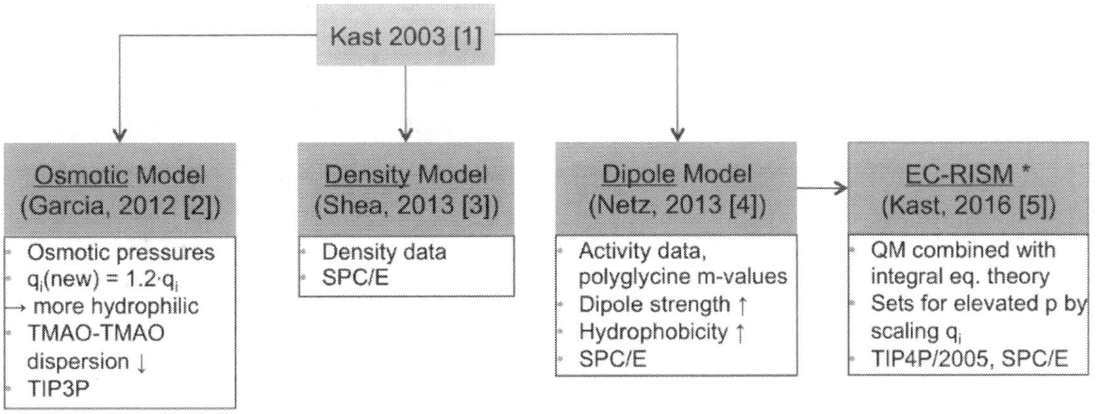

* **E**mbedded **C**luster-**R**eference **I**nteraction **S**ite **M**odel

2

Overview: Simulated Systems

Tested TMAO models

- 4/5 tested models derived from Kast 2003 force field (FF) [1]: same intramol. parameters but optimized nonbonded parameters
- Models differ in water model, comb. rules, 1-4 interactions, sim. param.

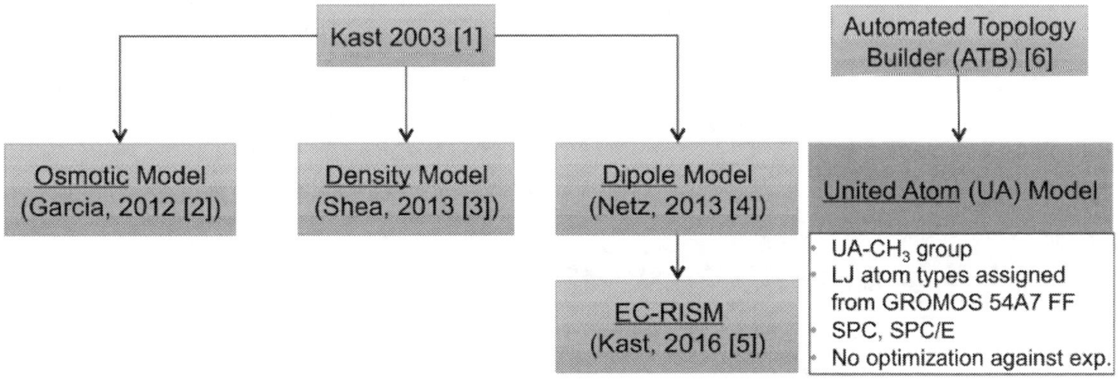

3

Universität Stuttgart

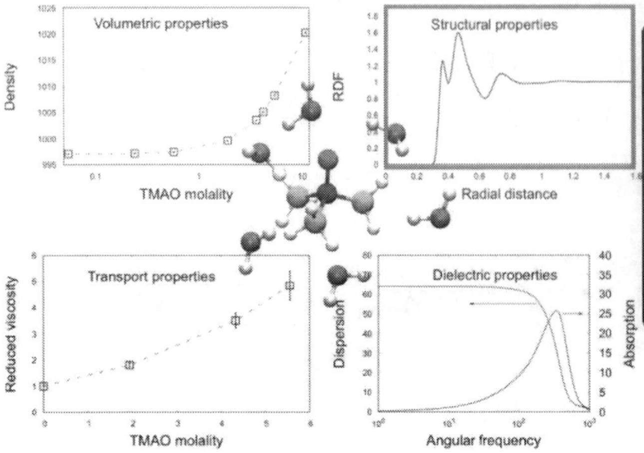

Structural Properties

Structural Properties

Results: Hydrogen bonding

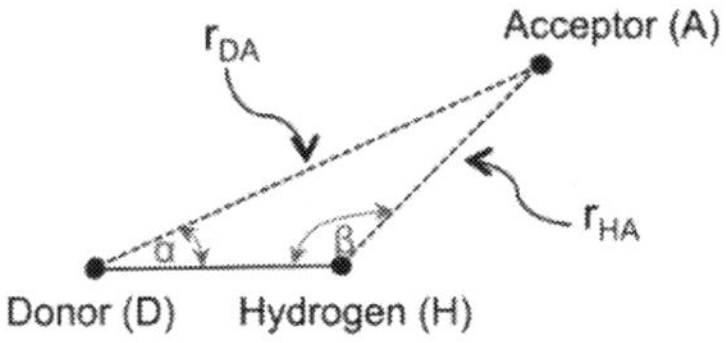

- Tested H-bond existence criteria:
  1. $r_{HA} < 0.25$ nm & $\beta > 135°$
  2. $r_{DA} < 0.35$ nm & $\alpha < 30°$
  3. $r_{HA} < -0.171\ \text{nm} \cos(\beta) + 0.137$ nm *

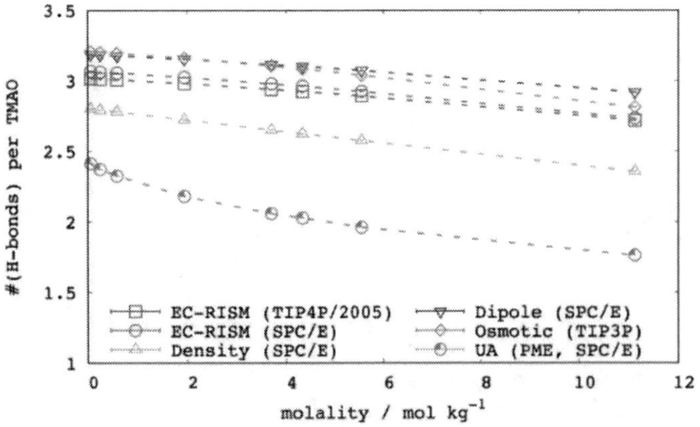

- Experiments propose clusters with TMAO bound to 2-3 H_2O
- UA most hydrophobic
- Osmotic and Dipole most hydrophilic

* S. Imoto et al., *Phys. Chem. Chem. Phys.* (2015) 17

4

Universität Stuttgart

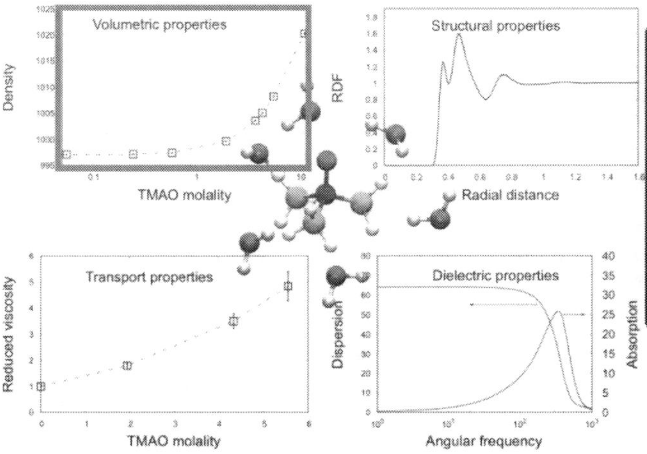

Volumetric Properties

Volumetric Properties

Results: Density

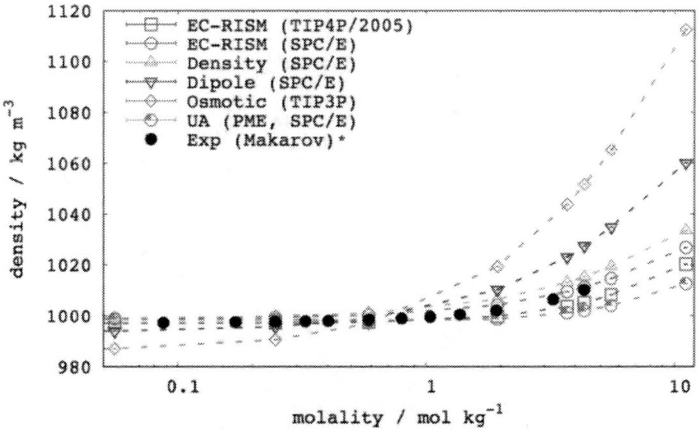

- Best agreement for EC-RISM and Density model
- Strong overestimation for Osmotic and Dipole model

* D. M. Makarov et al., *J. Chem. Eng. Data* (2015) 60

Volumetric Properties

Results: Partial molar volume at infinite dilution

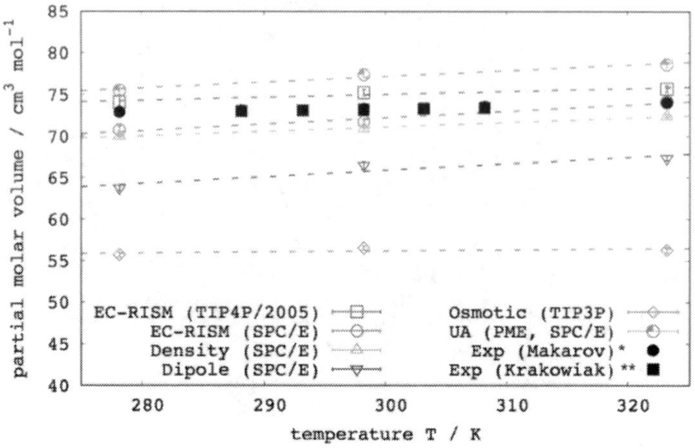

- Best agreement for EC-RISM and Density model
- UA represents too bulky TMAO
- Osmotic and Dipole model represent too hydrophilic TMAO

** J. Krakowiak et al., *J. Chem. Thermodyn.* (2013) 58

* D. M. Makarov et al., *J. Chem. Eng. Data* (2015) 60

Universität Stuttgart

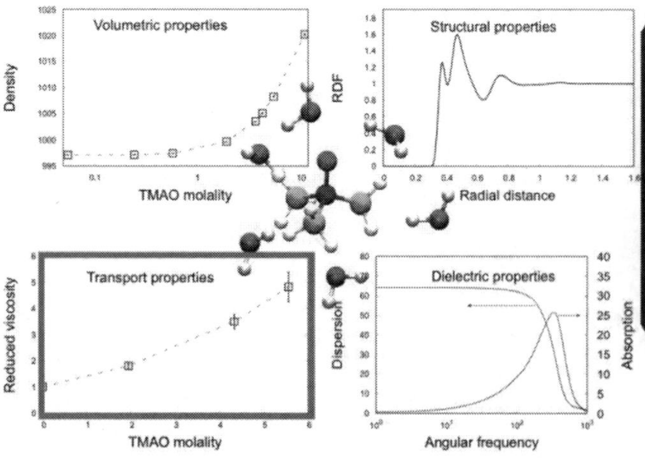

Transport Properties

Transport Properties

$$\lim_{t\to\infty} \langle (\mathbf{r}_i(\tau+t)-\mathbf{r}_i(\tau))^2\rangle_{i,\tau} = 6D_{\mathrm{self,i}}\, t + \mathrm{const.}$$

Results: Self-diffusion coefficient of water

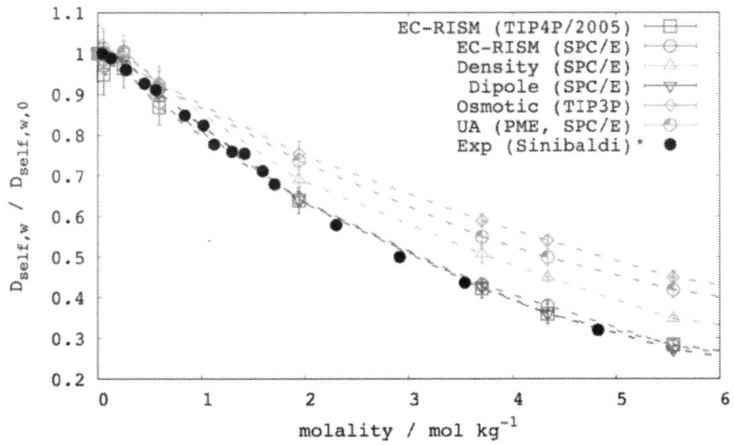

- Best agreement for EC-RISM and Dipole model
- High deviations in case of UA and Osmotic model

* R. Sinibaldi et al., *J. Phys. Chem. B* (2006) 110

7

Universität Stuttgart

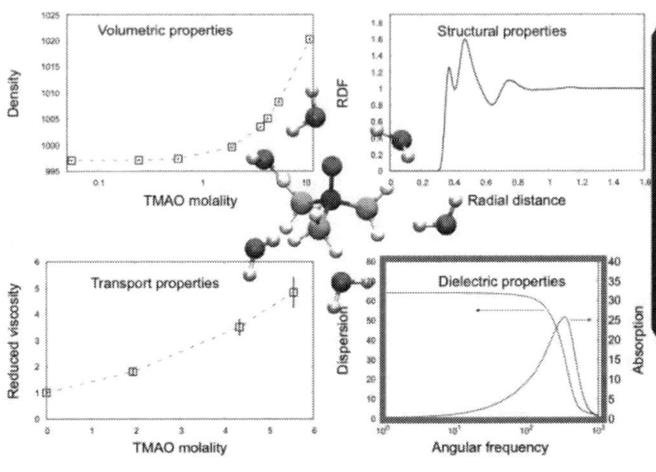

Dielectric Properties

Dielectric Properties

Theory

- The complex frequency-dependent permittivity $\varepsilon_\mathrm{r}(\omega) := \boxed{\varepsilon_\mathrm{r}'(\omega)} - i\boxed{\varepsilon_\mathrm{r}''(\omega)}$ ("dispersion", "loss") ("dielectric spectrum") is calculated from the complex total conductivity $\sigma(\omega)$ following a "Green-Kubo" approach *:

$$\boldsymbol{J}_\mathrm{tot}(t) = \sum_n q_n \boldsymbol{v}_n(t)$$

$$\sigma(\omega) = \frac{1}{3Vk_\mathrm{B}T}\int_0^\infty \langle \boldsymbol{J}_\mathrm{tot}(0)\boldsymbol{J}_\mathrm{tot}(t)\rangle e^{i\omega t}dt$$

$$\varepsilon_\mathrm{r}(\omega) = 1 + \frac{i}{\varepsilon_0\omega}(\sigma(\omega) - \sigma(0))$$

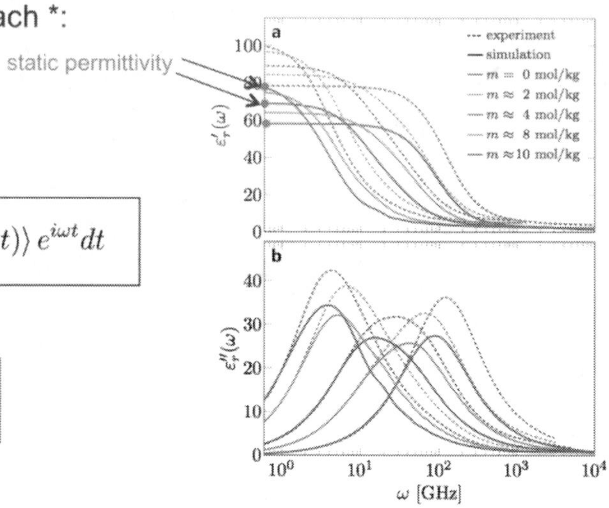

* A. N. Krishnamoorthy et al., *Phys. Chem. Chem. Phys.* (2016) 18

8

Dielectric Properties

Results: Static permittivity

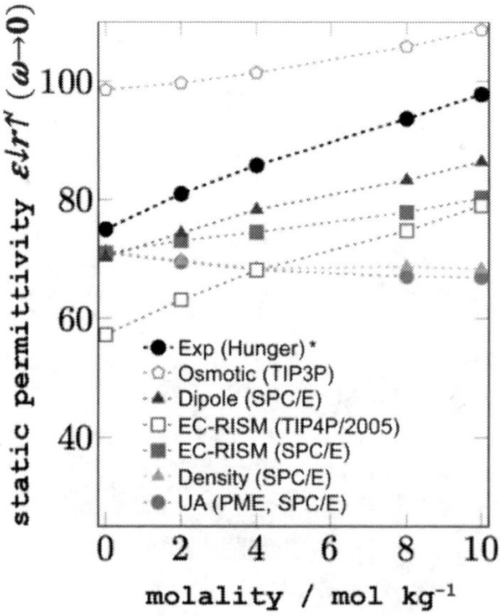

- Best overall agreement for Dipole model
- EC-RISM and Osmotic model good qualitative agreement.
- UA and Density model even fail qualitatively

* J. Hunger et al., *J. Phys. Chem. B* (2012) 116

9

Thermodynamic Properties

Results: Rational activity coefficient of TMAO

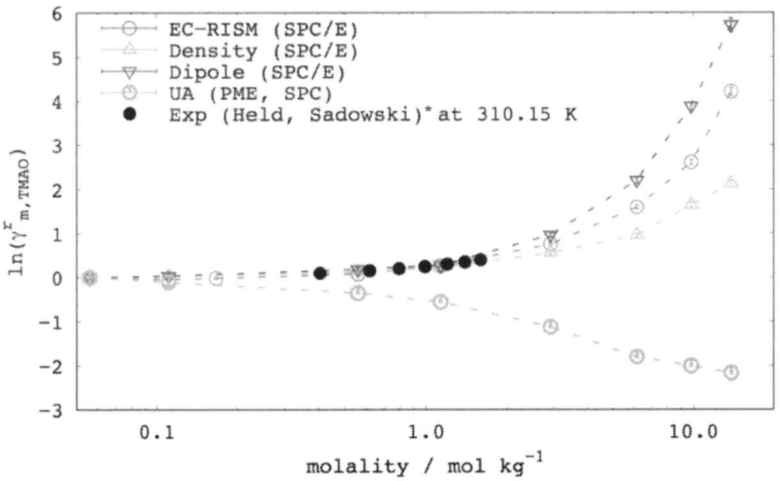

- EC-RISM, Density and Dipole model yield similar agreement with experiments
- UA model even fails qualitatively

Computational Details:

- Stepwise decoupling of 1 TMAO (18 states, LJ/Electrostatics separately)
- Solvation free energy from Multistate Bennett's Acceptance Ratio (MBAR) estimator

* C. Held et al., *FPE* (2016) 407

10

Summary

- Aqueous mixtures of 5 TMAO models were compared in terms of:
  - Thermodynamic and volumetric properties
  - Structural properties
  - Transport properties
  - Dielectric properties

- EC-RISM performed best in nearly all regarded properties
  -> good balance between solute-solute and solute-solvent interactions
  -> probably best model to gain molecular interpretation of stabilizing effect on proteins

11

Acknowledgements

DFG Deutsche Forschungsgemeinschaft

DAAD
Deutscher Akademischer Austausch Dienst
German Academic Exchange Service

SimTech
Cluster of Excellence

12

References

Literature TMAO models

- **[1]** K. M. Kast; J. Brickmann; S. M. Kast; R. S. Berry.
  Binary Phases of Aliphatic N-Oxides and Water: Force Field Development and Molecular Dynamics Simulation.
  J. Phys. Chem. A **2003**, 107, 5342–5351.
- **[2]** D. R. Canchi; P. Jayasimha; D. C. Rau; G. I. Makhatadze; A. E. Garcia.
  Molecular Mechanism for the Preferential Exclusion of TMAO from Protein Surfaces.
  J. Phys. Chem. B **2012**, 116, 12095–12104.
- **[3]** L. Larini; J.-E. Shea.
  Double Resolution Model for Studying TMAO/Water Effective Interactions.
  J. Phys. Chem. B **2013**, 117, 13268–13277.
- **[4]** E. Schneck; D. Horinek; R. R. Netz.
  Insight into the Molecular Mechanisms of Protein Stabilizing Osmolytes from Global Force-Field Variations.
  J. Phys. Chem. B **2013**, 117, 8310–8321.
- **[5]** C. Hölzl; P. Kibies; S. Imoto; R. Frach; S. Suladze; R. Winter; D. Marx; D. Horinek; S. M. Kast.
  Design principles for high-pressure force fields: Aqueous TMAO solutions from ambient to kilobar pressures.
  J. Chem. Phys. **2016**, 144, 144104.
- **[6]** A. K. Malde; L. Zuo; M. Breeze; M. Stroet; D. Poger; P. C. Nair; C. Oostenbrink; A. E. Mark.
  An automated force field topology builder (ATB) and repository: Version 1.0.
  J. Chem. Theory Comput. **2011**, 7, 4026–4037.

13

494312 Optimized Mie Potentials for Phase Equilibria: Application to Noble Gases, Alkanes, Alkynes and Their Mixtures

Abstract:

Transferable united-atom force fields, based on n-6 Mie potentials are presented for noble gases, branched alkanes, and alkynes. By tuning the repulsive exponent, it is possible to simultaneously reproduce experimental saturated liquid densities and vapor pressures with high accuracy, from the normal boiling point to the critical point. Vapor-liquid coexistence curves for pure fluids are calculated using histogram reweighting Monte Carlo simulations in the grand canonical ensemble, while Gibbs ensemble Monte Carlo is used to calculate binary and ternary mixture phase behavior. All calculations are performed with GOMC[1]. A semi-automated process for optimization of non-bonded force field parameters is presented. For each compound, vapor-liquid-coexistence curves, vapor pressures, heats of vaporization, critical properties, and normal boiling points are predicted and compared to experiment.

For all noble gases, saturated liquid densities and vapor pressures are reproduced to within 1% and 4% of experiment, respectively. For alkynes, experimental saturated liquid densities are reproduced to within 2% average absolute deviation (AAD), except for 1-hexyne, which are reproduced with 3% AAD. Experimental saturated vapor pressures are reproduced to within 3% AAD, except for 1-pentyne, 2-pentyne, and 2-hexyne, where deviations from experimental up to 20% AAD were observed. For branched alkanes, simulations are performed on branched isomers of butane, pentane, hexane, heptane, and octane to assess the transferability of potential parameters. Experimental saturated liquid densities and critical temperatures are reproduced with a median absolute average error of 0.6%, while vapor pressures are reproduced with a median absolute average error of 2.2%. Binary phase diagrams predicted by the Mie potentials for a variety of systems are in close agreement with experiment. Comparisons are made to the predictions of existing force fields, including SPEAD[2] and 2CLJQ[3], NERD[4, 5] and TraPPE[6].

References:

1. *GOMC: http://gomc.eng.wayne.edu.*

2. Unlu, O., N.H. Gray, Z.N. Gerek, and J.R. Elliott, *Transferable Step Potentials for the Straight-Chain Alkanes, Alkenes, Alkynes, Ethers, and Alcohols.* Industrial & Engineering Chemistry Research, 2004. **43**(7): p. 1788-1793.

3. Vrabec, J., J. Stoll, and H. Hasse, *A set of molecular models for symmetric quadrupolar fluids.* Journal of Physical Chemistry B, 2001. **105**(48): p. 12126-12133.

4. Nath, S.K. and J.J. De Pablo, *Simulation of vapour-liquid equilibria for branched alkanes.* Molecular Physics, 2000. **98**(4): p. 231-238.

5. Nath, S.K. and R. Khare, *New forcefield parameters for branched hydrocarbons.* The Journal of Chemical Physics, 2001. **115**(23): p. 10837-10844.

6. Martin, M.G. and J.I. Siepmann, *Novel configurational-bias Monte Carlo method for branched molecules. Transferable potentials for phase equilibria. 2. United-atom description of branched alkanes.* Journal of Physical Chemistry B, 1999. **103**(21): p. 4508-4517.

496956 Evaluation of Virial Coefficients and Their Temperature Derivatives for Multibody Potential Models

Abstract:

The most widely used molecular models are pairwise-additive and have been formulated to describe properties of condensed phases. As a result they do not describe the true interactions between molecule pairs, but rather are effective potentials that attempt to include multibody interactions and other temperature- and density-dependent effects in the pairwise form. Models such as TraPPE family are formulated to be more broadly applicable, aiming to describe vapor-liquid phase equilibria in particular, but they are still pairwise in nature, and thus are effective potentials.

With improvements in simulation algorithms and computer speed, it is becoming increasingly viable to perform simulations using potentials that more accurately describe the true interactions between/among two, three, or more molecules. This includes polarizable models such as GCPM, and explicit pair and three-body potentials fit to ab initio potential surfaces. A crucial test of any model that purports to describe true intermolecular interactions is its ability to reproduce experimental virial-coefficient data, which rigorously characterize interactions of two, three, etc. molecules.

Wheatley recently introduced a very efficient recursive algorithm for evaluation of the integrand needed for virial coefficient calculations. Wheatleyâ€™s derivation assumes a pairwise additive potential, but we show that it may be extended for application to multibody potentials, and moreover it provides a route for calculation of temperature derivatives. This allows calculation of high-order virial coefficients for such models. Coefficients determined this way can be compared to experimental data where available (which is typically no higher than the 3rd-order coefficient), thereby providing an assessment of the accuracy of the model. It can also be used to formulate equations of state that are fully predictive and accurate over much of the supercritical region, for both pure substances and mixtures; these models may also be used to compute critical properties. In this manner the usefulness of the accurate molecular model is greatly extended. We discuss these issues and demonstrate applications.

499370 Prediction of Interaction Parameters for Coarse-Grained Models Using *Ab Initio* Calculations

Abstract:

Coarse Grained (CG) models enable the simulation of larger systems, and phenomena occurring on longer time scales, than traditional atomistic simulation methods. However, determining CG model parameters is usually a time-consuming task, which involves carrying out atomistic-level simulations and fitting CG parameters to the results. An alternative approach is top-down coarse graining, which obtains CG parameters by fitting directly to experimental data. The SAFT-Î³-Mie force field, where the parameters of a Mie potential representing interactions between coarse-grained beads are fit to phase-equilibrium data, is an example of this approach[1,2]. A challenge in developing such CG models is the determination of unlike-pair interaction parameters. If the number of beads in the force field is large, it becomes impractical to fit all the unlike interactions due to the combinatorial explosion in the number of parameters. The commonly used Lorentz-Berthelot mixing rules work well for systems where the main interactions are dispersion forces, but they do not account for electrostatics or molecular polarizability. In this work, we derive these mixing rules for the SAFT-Î³-Mie force field to include these interactions. In the resulting model, the unlike-pair interactions account for the effect of charges, dipole moments, quadrupole moments, and polarization. We then show how, by estimating these properties for the CG molecular fragments using *ab initio*Â calculations, it is possible to obtain realistic cross-interaction energies without the need to use extensive experimental data.

1. Dufal, S. *et al.* Prediction of Thermodynamic Properties and Phase Behavior of Fluids and Mixtures with the SAFT-Î³-Mie Group-Contribution Equation of State. *Journal of Chemical & Engineering Data* **59,** 3272â€“3288 (2014).
2. Papaioannou, V. *et al.* Group contribution methodology based on the statistical associating fluid theory for heteronuclear molecules formed from Mie segments. *The Journal of Chemical Physics* **140,** 054107 (2014).

499394 Modifying Nonplanar Vibrational Modes of Aromatic Rings in Biomolecular Modeling

Abstract:

Several biomolecular studies have stated the important role of aromatic amino acids such as Tryptophan (TRP) and Tyrosine (TYR) in the folding process of proteins at water-membrane interfaces. The accuracy of biomolecular modeling results is based in part on force field (FF) parameters. In this study, molecular dynamics (MD) simulations were conducted on TRP and TYR amino acids in solution by applying the most recent version of Charmm 36. The results from these simulations show that the structures of aromatic rings have deviations from planarity. Considering improper torsion in MD simulation brings the aromatic rings in TYR and TRP closer to planar structures. On the other hand, inherent vibrations of atoms in molecules might be the reason for this nonplanarity. Hence, a concern arises about the effects of imposing improper torsion parameters on the nonplanar vibrational modes of atoms in the aromatic ring. Addressing this concern requires calculating these vibrational modes on the basis of FFs and comparing these obtained vibrational modes to experimental data. All-atom Normal Mode Analysis (NMA) provides the vibrational frequencies of those rings on the basis of diverse FFs. Analytical and numerical NMA were applied on the configuration results of MD simulations to calculate the average nonplanar vibrational frequencies and normal modes of atoms in those rings. The results of NMA were compared to Raman spectra and to results of quantum calculations for TRP and TYR. The nonplanar mode frequencies that are obtained for both TRP and TYR on the basis of the standard Charmm36 are higher compared to Raman spectra and to the results of quantum calculations. These frequencies increase even more when improper torsions are added to Charmm36. Increasing the improper torsion force constants in the potential equation shifts these frequencies to even higher amounts. Beside the improper torsions, torsion angles that include atoms in the ring have a direct effect on nonplanar motion. In order to obtain proper nonplanar frequencies, NMA has been applied on different sets of amended torsion angle parameters with and without considering improper torsion angles. The results of these calculations demonstrate that by applying a specific set of amended torsion angle parameters and considering the improper torsion with specific force constants, the pattern and frequencies of nonplanar normal modes can match properly with Raman spectra and the results of quantum calculations.

498820 A Polarizable Force Field of Inorganic Phosphates and Hydroxyapatite Based on the Classical Drude Oscillator

Abstract:

Calcium phosphate (CaP) is a ubiquitous mineral with a major contribution to the formation of human bone and tooth enamel. The biomedical applications of calcium phosphate include reconstructive surgeries and porous scaffolds for tissue engineering. Despite its fundamental significance, the early-stage nucleation mechanism of CaP particles from a saturated solution remains speculative because the critical nanoscale details pertaining to the nanometer-sized â€œpre-nucleationâ€ clusters remain inaccessible to characterization techniques. In this contribution, we develop a polarizable force field of phosphate ions, in various protonation states, and hydroxyapatite (HAP) using the classical Drude oscillator model. The force filed reproduces the key properties of ions in solution as well as the properties of the HAP crystal. The quantum mechanical electrostatic potential maps, atomic partial charges, molecular polarizabilities, and molecular dipoles are well reproduced. The experimental properties of the solvated ions and HAP crystals including solvation free energy of ions, lattice constants and density of crystal, cleavage energies, and mechanical properties of HAP are reproduced in agreement with experiments. The force filed is further applied to study nucleation process of calcium phosphate particles. The results show instantaneous growth of branched polymeric assemblies prior to the formation of spherical particles.

498132 Using Minimal Biasing Methods in Plumed and Lammps

Abstract:

Minimal biasing methods are a class of methods which restrain one or more collective variables to match a particular set-point. These techniques can be used to create molecular dynamics trajectories consistent with experimental data. This presentation will focus on the experiment directed simulation (EDS) and metadynamics (EDM) techniques. EDS allows biasing one or more collective variable to match in a set-point expectation. EDM allows biasing a PMF of a collective variable to match a target distribution (EDM). Specific examples will be presented for biasing coordination number to improve the structure of water in ab-initio molecular dynamics and matching a 2D density profile in a Lennard-Jones fluid. The EDS/EDM methods result in a minimally biased simulation consistent with the target values. That simulation can be used to study dynamic or static properties with the now experimentally consistent force-field. This presentation will utilize the Plumed free energy plugin, LAMMPS, and open-source plugins developed by us.

499147 A General Route for Nanoemulsion Synthesis Using Low Energy Methods at Constant Temperature

Abstract:

The central dogma of nanoemulsion formation using low energy methods at constant temperature â€“ popularly known as the emulsion inversion point (EIP) method â€“ is that the continuous phase should be added to a mixture of dispersed phase and surfactant. Here, we demonstrate that the above order of mixing is not universal and a reverse order of mixing could be superior depending on the choice of surfactant and liquid phases. We propose a more general methodology to make nanoemulsions by studying the variation of droplet size with surfactant HLB for several model systems. Based on our understanding and experimental results, we utilize the reverse order of mixing for two applications: crystallization and formulation of pharmaceutical drugs with faster dissolution rates, and synthesis of alginate-based nanogels. The general route provides insights into nanoemulsion formation through low energy methods and also opens up possibilities that were previously overlooked in the field.

Advanced Emulsion-Templating of Microreactors for the Scalable Production of Semiconductor Nanowires

Maritza Mujica,

Michael A. Filler, Victor Breedveld and Sven H. Behrens

School of Chemical & Biomolecular Engineering

Georgia Institute of Technology®

Nanowire Applications

Flexible Electronics

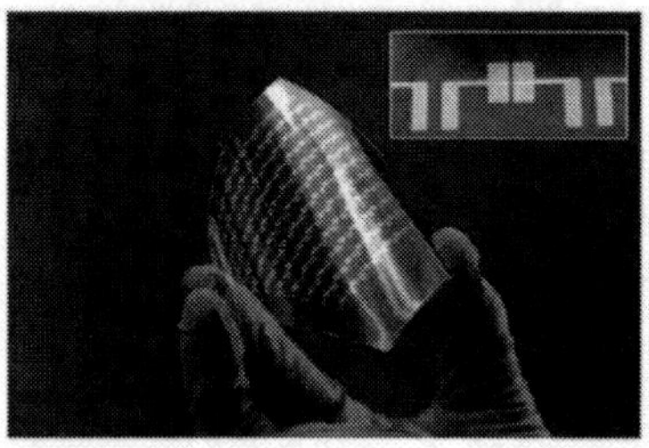

Min et al. *Nat. Commun.* 2013, *4*, 1773.

Solar Cells

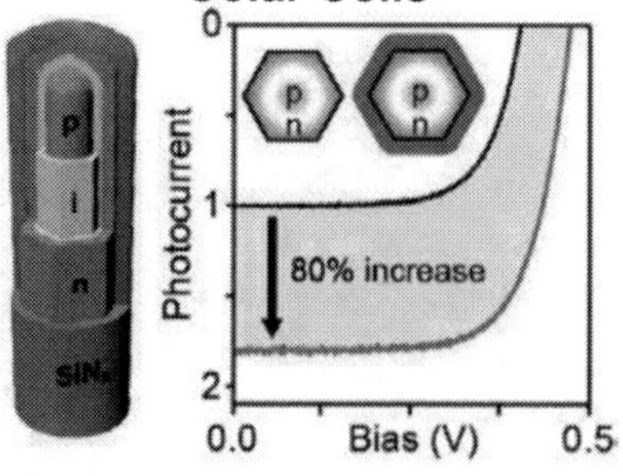

Kim et al. *Nano Lett.* 2015, *15*, 753.

Bioelectronics

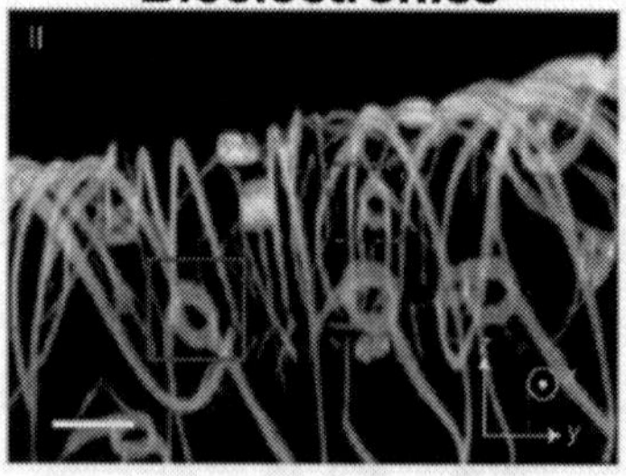

Tian et al, *Nat. Mater.* 2013, *11*, 986.

Georgia Tech

1

Nanowire Growth Process

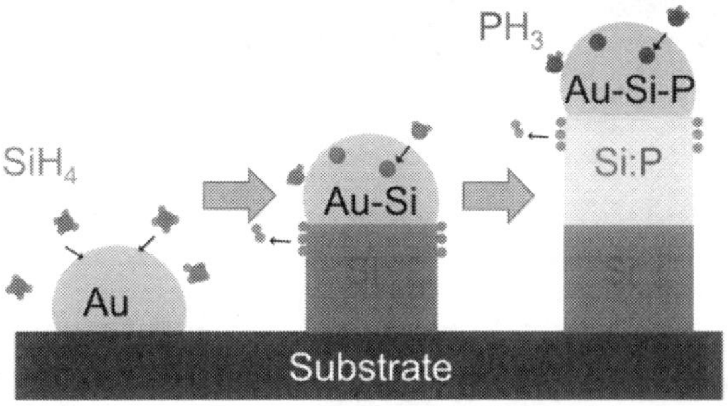

Heterostructures

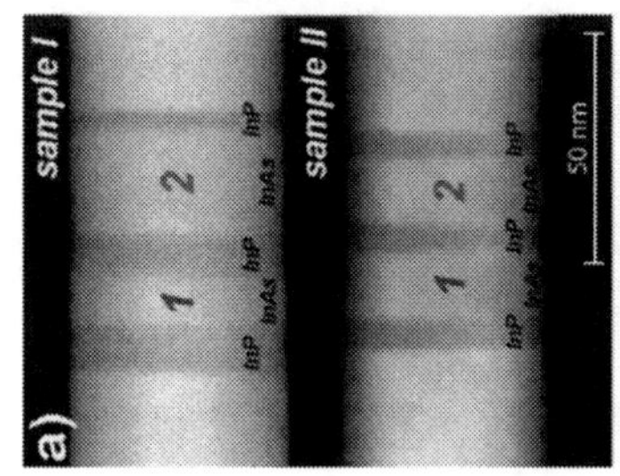

Fuhrer et al., *Nano Lett.* 2007, 7, 243

Doped and Selectively Etched

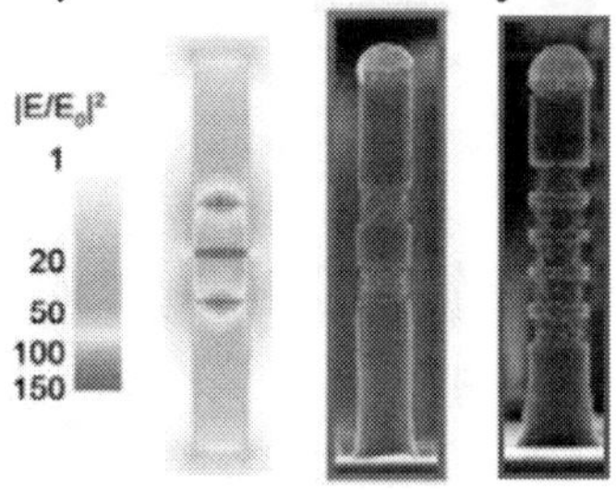

Boyuk *et al.*, *ACS Photon.* 2016, *3*, 184

Kinked Structures

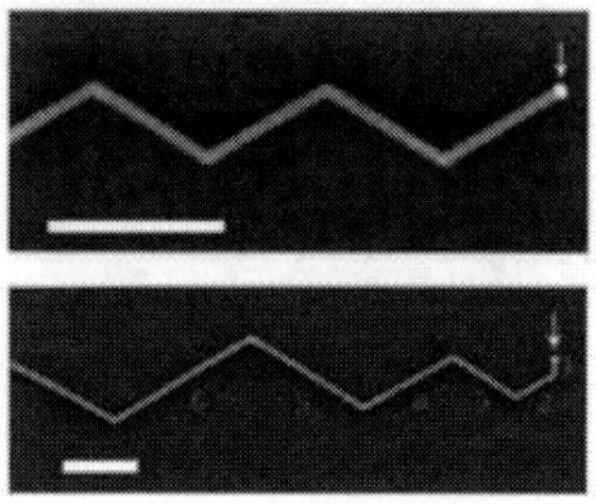

Tian et al. *Nature Nanotechnology*, 2009, 4, 824.

Georgia Tech

2

Nanowire Scale-up

Solution Phase

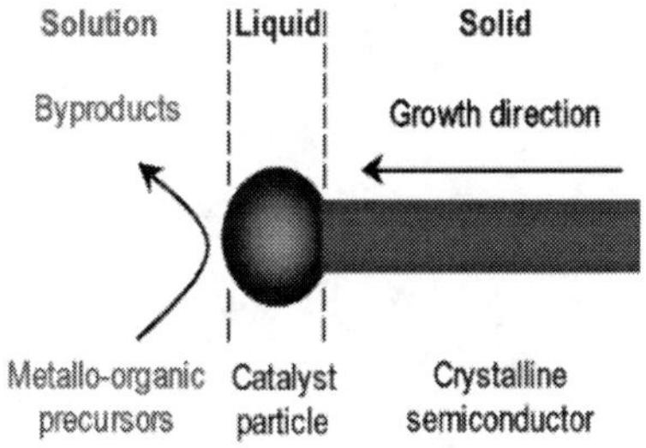

Pros:
- ✓ Low temperature

Cons:
- × Slow transition between precursors
- × Solvent and surfactant difficult to remove

Gas Phase

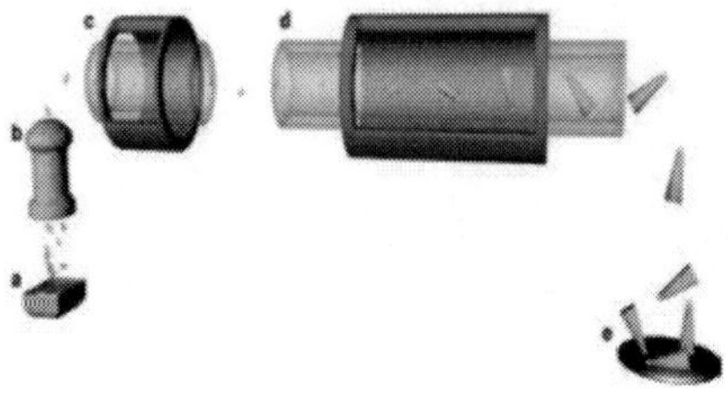

Pros:
- ✓ Rapid switching between different precursors

Cons:
- × Nanowire shape non-uniformity
- × Low productivity due to gold loss and agglomeration

Li et al. *J. Mater. Chem.*, 2012, *22*, 22821-22831.
Heurlin, et al. *Nature* 2012, *492*, 7427, 90-94.

Georgia Tech

3

The "Geode" Process

Emulsion-templated Microcapsule

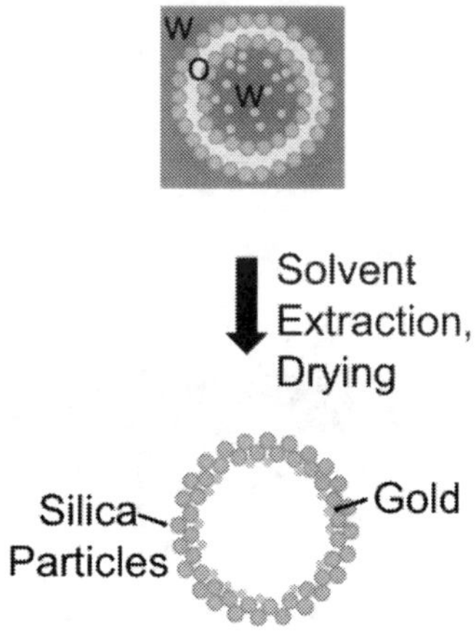

Nanowire Growth

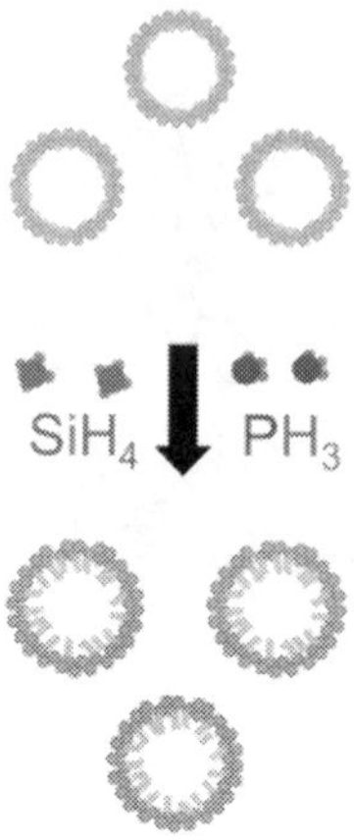

Functional Nanowires

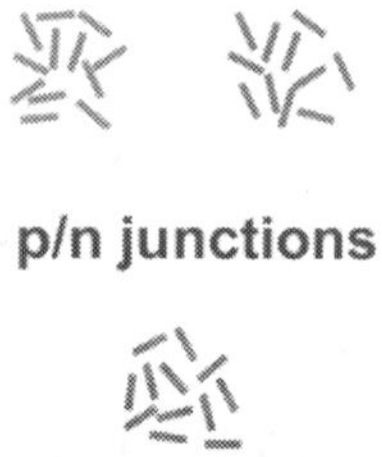

Nanowires are harvested from capsules in liquid phase after dissolution of sacrificial capsule substrate.

Georgia Tech

4

Previous Microcapsules

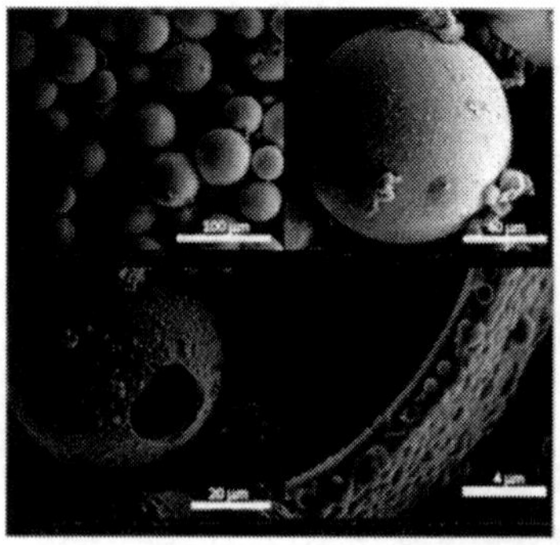

van Wijk et al. *J. Mater. Chem. B*, 2015, *3*, 7745.

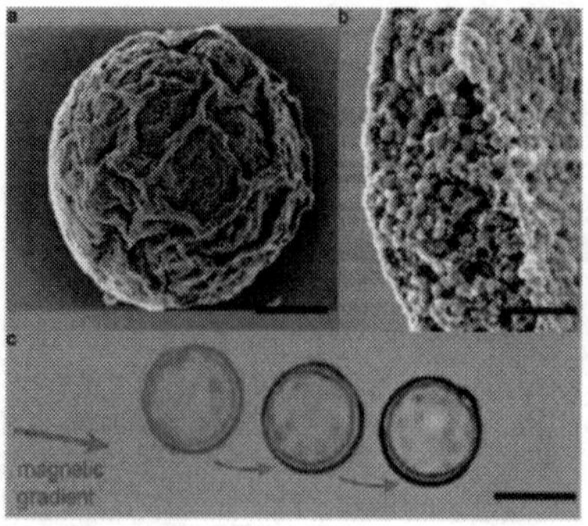

Sander and Studart. *Soft Matter, 2014, 10, 60.*

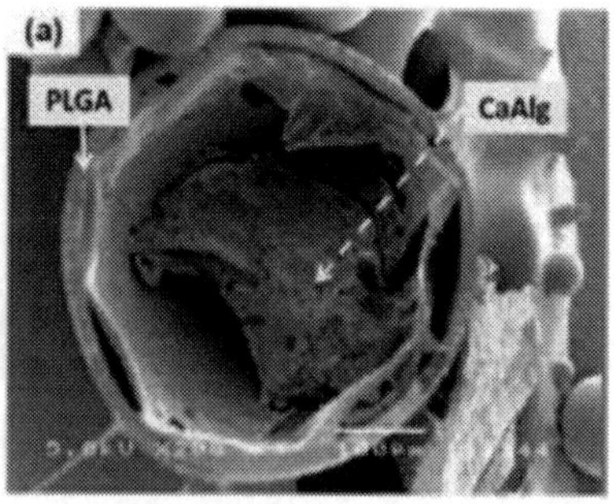

Lim et al. *Biomater. Sci.*, 2013, *1, 486*

Georgia Tech

5

Our Process

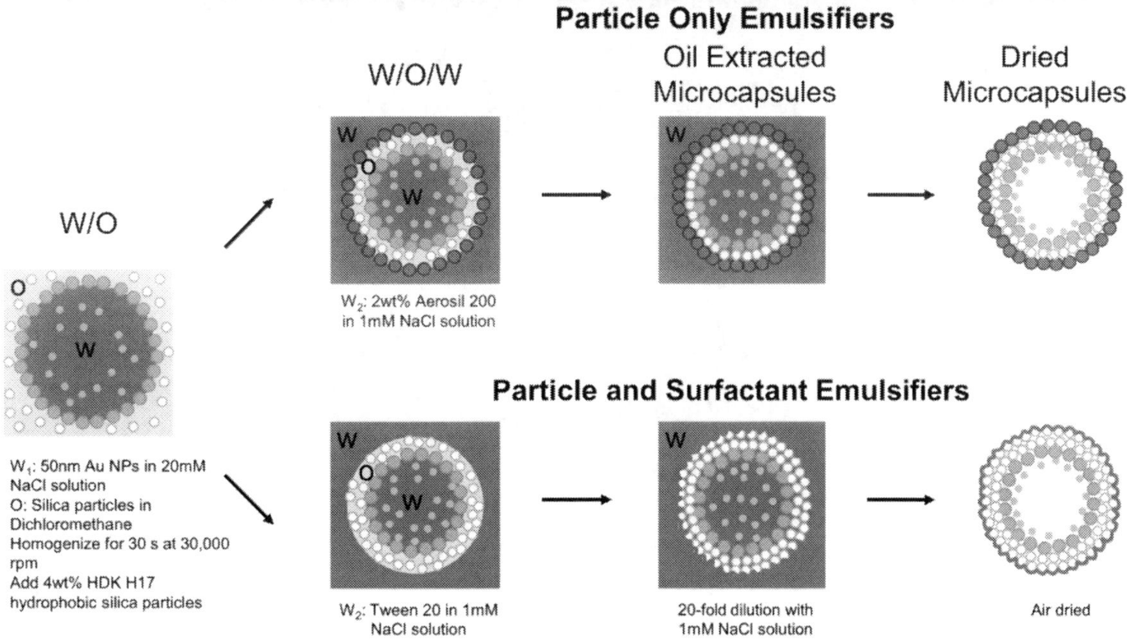

6

Georgia Tech

Water-in-Oil Emulsion

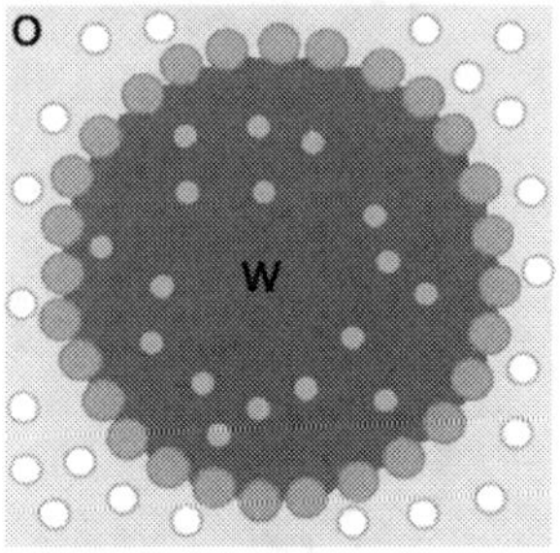

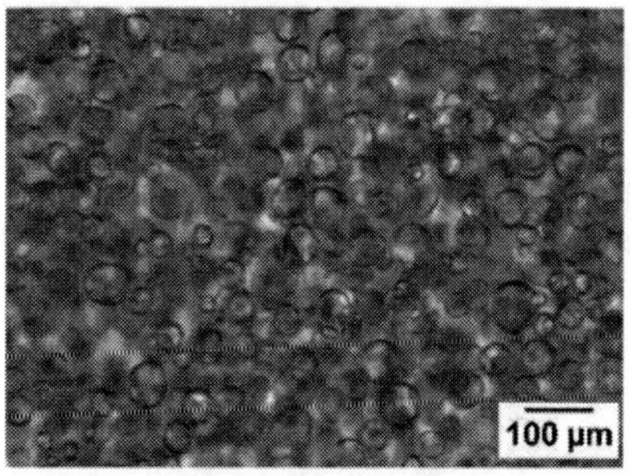

W1: 50nm Au nanoparticles in 20mM NaCl solution
O: 0.5wt% partially hydrophilic silica particles in dichloromethane

Oil Selection:

Solvent	Solubility in H_2O @ 25˚C (mg/g)	P_{vap} @ 25^{0}C (mmHg)
Ethyl Acetate	73.72	92.52
Dichloromethane	19.38	435.5
Toluene	0.54	28.31

7

Georgia Tech

Microcapsule Design: Second Emulsifier

Hydrophilic Silica Particle (Aerosil 200)

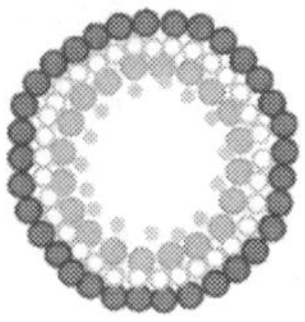

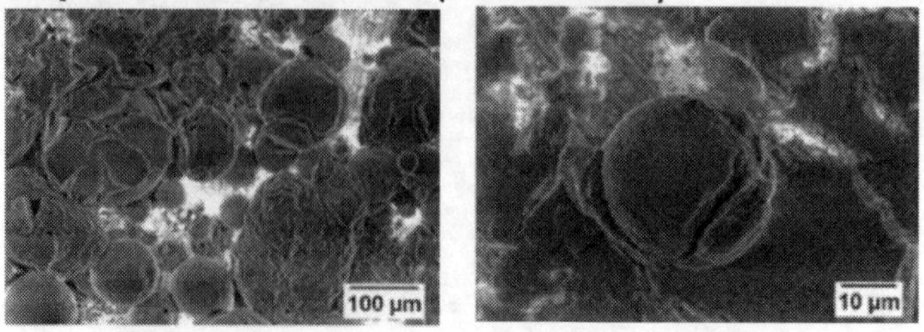

Surfactant (Tween 20)

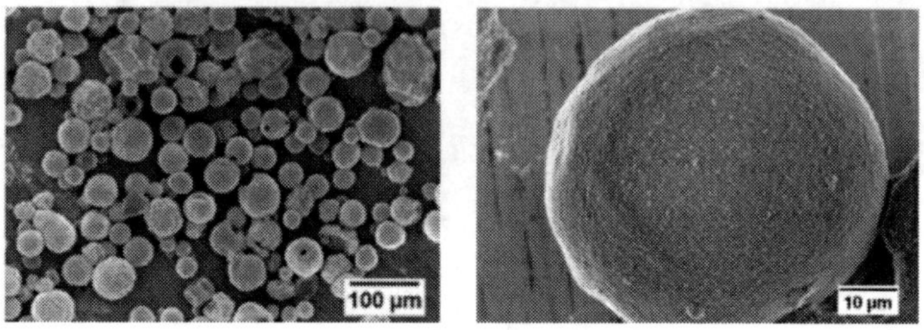

Osmotic Swelling

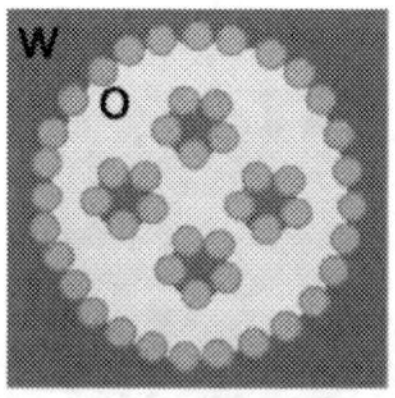

[Salt]$_{internal}$ > [Salt]$_{external}$

Osmotic swelling and coalescence →

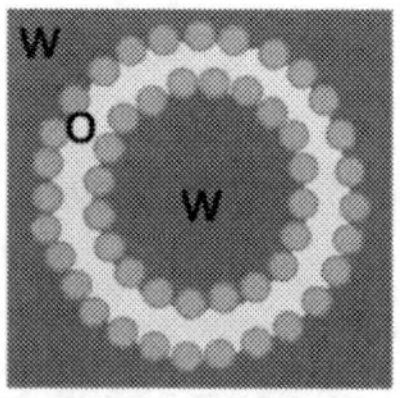

[Salt]$_{internal}$ ~ [Salt]$_{external}$

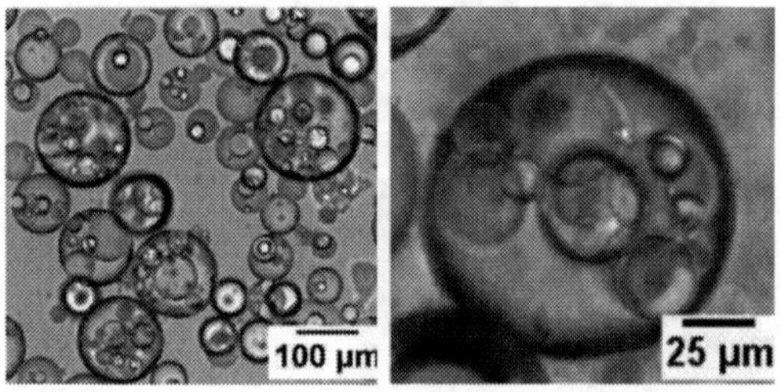

→

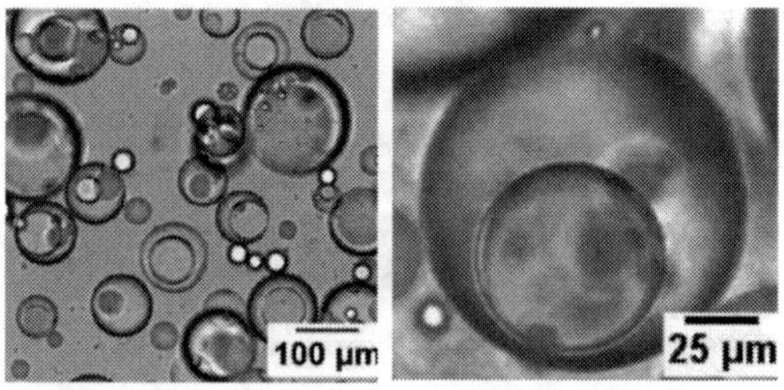

Osmotic swelling tunes internal morphology

Complete Procedure

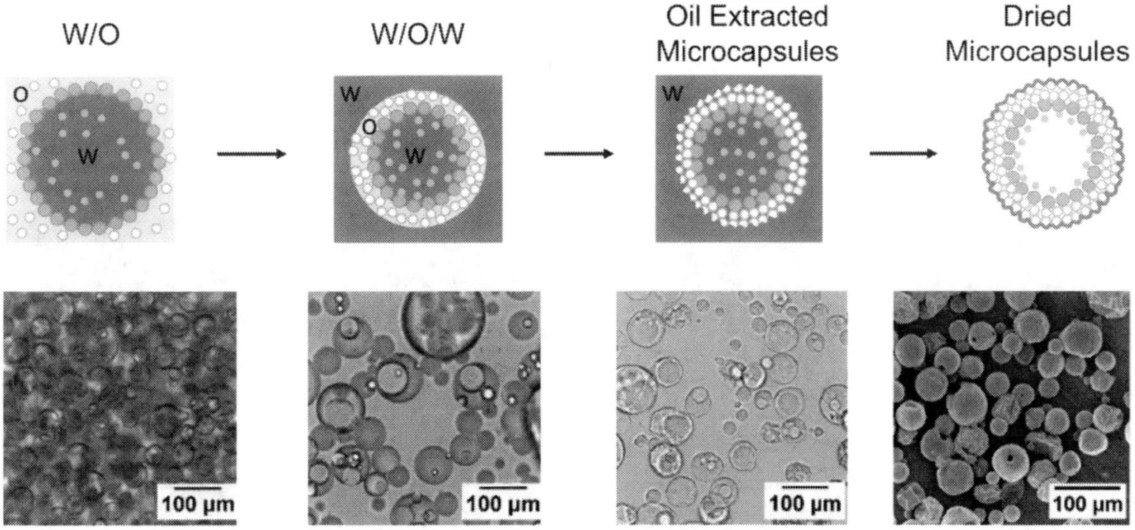

W_1: 50nm Au nanoparticles in 20mM NaCl solution
O: Silica particles in Dichloromethane
Homogenize for 30 s at 30,000 rpm

W_2: Tween 20 in 1mM NaCl
Vortexed

20-fold dilution with 1mM NaCl solution

Air dried

Microcapsules achieved through double emulsion scheme

Georgia Tech

10

Nanowire Growth

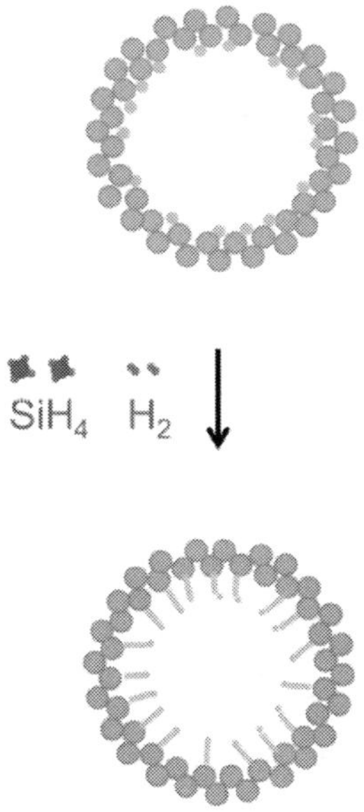

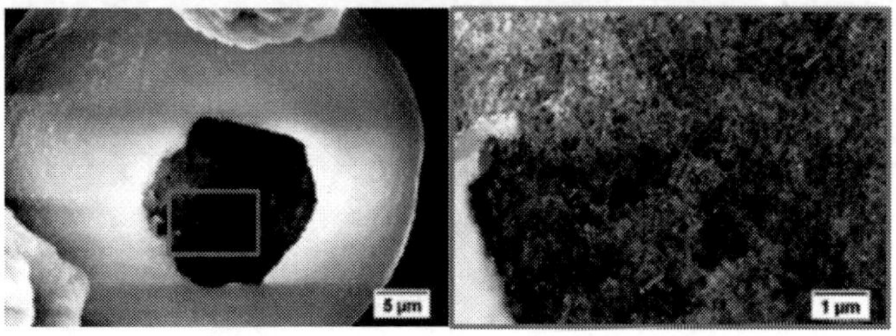

Si Nanowire Growth Conditions:
T=550˚C; P=20 Torr
SiH_4 flow=50 SCCM; H_2 flow = 0.60 SLPM

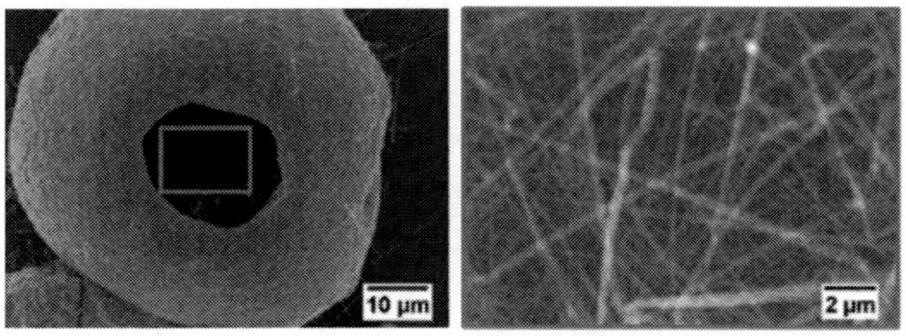

Nanowire growth achieved in porous microcapsule

Georgia Tech

11

Summary

- Developed double emulsion scheme for catalyst-lined microcapsules
- Demonstrated nanowire growth on the interior

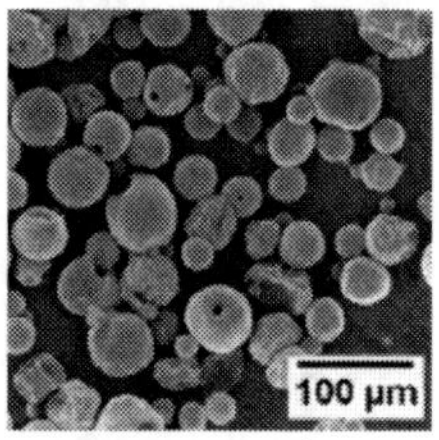

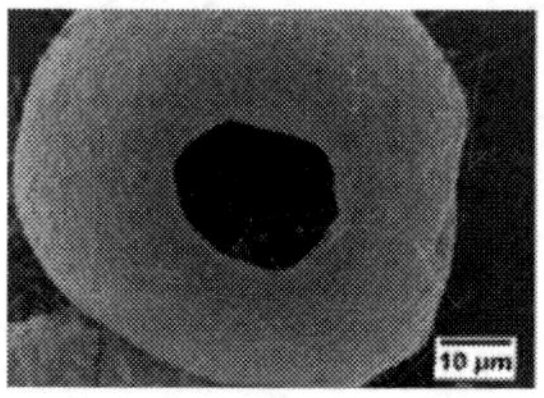

Future Outlook

- Improve microcapsule production yield and explore spray drying
- Understand mass transport of precursor gases across microcapsule wall
- Demonstrate functional nanowire encoding (e.g, pn junction nanowires)

Acknowledgements

- Filler, Behrens, and Breedveld Groups

12

Georgia Tech

492742 The Discovery of Novel Hydrophobic DES-Water Emulsions

Abstract:

We present the first experimental realization of an emulsion of a deep eutectic solvent (DES) in water. DESs are a new, innovative class of solvents presented in the literature for the first time in 2003.[1] A DES is the combination of two or more solid components, which self-associate via hydrogen bonding and presumably Van der Waals forces forming a liquid phase upon the formation of the DES.[2] Compared to regular solvents, DESs have the advantage that the physical properties of the solvent can be tuned through the selection of the components. Compared to ionic liquids, DESs can be prepared easily by mixing the two solid components at moderate temperatures and can be constructed from readily-available biorenewable compounds.

Until 2015 all DESs found in literature were hydrophilic, due to the high amount of hydrogen bonding groups present in the components investigated to form DESs. It was this year when a number of us reported hydrophobic DESs for the first time. These hydrophobic DESs consisted of decanoic acid and a range of quaternary ammonium salts and were used for the removal of fatty acids from water.[3] A few months later in 2015 another publication reported the use of hydrophobic DESs for the extraction of biomolecules from an aquatic environment.[4]

Recently, more publications regarding hydrophobic DESs were presented. One of these investigations focused on the removal of metal ions from water with a DES composed of decanoic acid and lidocaine in a 2:1, 3:1 and a 4:1 ratio,[5] while other research focused on the capture of CO_2 with the hydrophobic DESs composed of decanoic acid and quaternary ammonium salts.

A disadvantage of many extraction processes is, however, the slow extraction due to limited interfacial area between DES and water. We show that this can be tremendously improved by employing emulsions of DES in water, stabilized by suitable surface active agents. Additionally, we investigate the fundamental properties of these novel emulsions, detailing droplet size, interfacial tension, and adsorption density.

References

[1] A. P. Abbott, G. Capper, D. L. Davies, R. K. Rasheed, and V. Tambyrajah, *Chemical Communications*, 2003, 70–71.

[2] M. Francisco, A. van den Bruinhorst, and M. C. Kroon, *Angew. Chem. Int. Ed.*, 2013, **52**, 3074–85.

[3] D. J. G. P. van Osch, L. F. Zubeir, A. van den Bruinhorst, M. A. A. Rocha, and M. C. Kroon, *Green Chemistry*, 2015, **17**, 4518–4521.

[4] B. D. Ribeiro, C. Florindo, L. C. Iff, M. A. Coelho, and I. M. Marrucho, *ACS Sustainable Chemistry & Engineering*, 2015, **3**, 2469–2477.

[5] D. J. G. P. van Osch, D. Parmentier, C. H. J. T. Dietz, A. van den Bruinhorst, R. Tuinier, and M. C. Kroon, *Chemical Communications*, 2016, **52**, 11987–11990.

493229 Controlling Foaming during Thermal Cracking - a Non-Silicon Based Antifoaming Agent

Abstract:

Foaming in hydrocarbon systems is a dynamic process that occurs when gas bubbles travel to the surface of a liquid without breaking or coalescing. The life of foam is related to its stability. The stability of foam in turn is related to forces such as surface tension, gravity, viscosity and capillary forces acting on the liquid film around the bubble. During thermal cracking of heavy oils, the presence of cracked vapors traveling through a liquid phase can cause foaming. The foam produced during this process can be stabilized by natural surfactants present in heavy oils such as asphaltenes and resins. Foaming can pose serious issues in the refinery if it overflows from the reacting vessel such as in a delayed coke drum. Silicon-based antifoams (AF) are commonly used to mitigate foaming. These antifoams contribute significantly to operating costs and can result in contamination of catalysts downstream of the delayed coking unit and in some cases; it leads to undesirable coke morphology that can pose risk to operators when decoking the reactor. An antifoaming agent that effectively controls foam without contaminating the liquid stream with silicon is desirable.

This paper shows how a non-silicon based antifoam is used to control foaming during thermal cracking in a pilot unit delayed coker drum. A vacuum resid with known properties was fed to the 3-inch diameter coker equipped with a gamma densitometer that allows for a transient analysis on foam growth and collapse during the experiment. This non-silicon antifoam; Tire oil, is an aromatic compound that does not contain any contaminants which can affect the downstream units or catalyst. Two modes of AF injection are tested; intermittent and continuous injections (at 0.36 cc/min and 0.18 cc/min after initial injection). Both modes were found to be effective at controlling the foam in the coke drum. The continuous foaming case resulted in a morphology change and liquid yield increased. Two AF concentrations are also tested (a neat injection and a mix with a hydrocarbon carrier) using the intermittent injection mode. Both concentrations were effective at controlling the foam.

The processing capacity of an industrial coke drum utilizing this non-silicon AF may be increased by decreasing the zone used to impede foamovers as a result of the eliminated risk of catalyst poisoning typically caused by foamover of silicon containing liquid. This increases the profitability of the delayed coker process.

500534 Effect of Salt on Drainage Via Stratification in Micellar Foam Films

Abstract:

Understanding and controlling the drainage kinetics of thin films is an important problem that underlies the stability, lifetime and rheology of foams and emulsions. Foam films containing micelles, colloidal particles or polyelectrolyte-surfactant mixtures exhibit step-wise thinning or stratification, due to the influence of non-DLVO forces, including supramolecular oscillatory structural forces. In this study, we use Interferometry, Digital, Imaging, Optical Microscopy protocols to investigate the drainage and stratification in micellar foam films (\textless 100 nm) with high spatial (thickness < 10 nm) and temporal resolution (< 1 ms). We determine how the concentration of surfactants and added salt influences the nanoscopic topography, stratification kinetics and step size of foam films formed using micellar sodium dodecyl sulfate (SDS) solutions.

498239 Asphaltenes Adsorption at Water/Oil Interface, a Classical Surfactant Approach

Abstract:

Time consuming gelation of asphaltenes covered interfaces has persistently been called to explain both slow evolution of interfacial tension and crude oil emulsion stability, even if the latter can be obtained after short mixing times. To resolve this inconsistency (and some others), asphaltenes adsorption has been re-investigated from scratch. Irrespective to adsorption time, asphaltenes appear to behave as surfactants insofar they follow an equation of state which can be approximated by a Langmuir model. The associated molecular area corresponds to the average asphaltenes polyaromatic core parallel to the water surface, a conformation in line with previously published Sum Frequency Generation spectroscopy experiments and with recent quantum based Self Consistent Charge Density Functional Tight Binding simulations. The coupling of the Langmuir model with a distribution in adsorption coefficient among the asphaltenes solubility class further enables matching quantitatively both dynamic interfacial tension and dilatational rheology measurements with diffusional models. The slow evolution of interfacial properties between water and asphaltenes solution then appears to be governed by the diffusion of a tiny sub-fraction of extreme surface activity. The Langmuir adsorption model is however a gross approximation insofar it assumes surfactants to be smaller than adsorption sites (here water molecules). This can be improved with a Lattice Gas model of hexagons covering 3 triangular adsorption sites (a choice dictated the size and shape of asphaltenes aromatic cores). This model not only matches the experimental equation of state extremely well but it also predicts a fluid to solid transition at the very surface pressure value for which a contracted pendant droplet loses its Laplacian shape. Around phase transition the lattice gas model also exhibits some dynamic frustration that would explain the observed Soft Glass (shear) rheology features of asphaltenes laden interfaces. In turn, the reported onset of emulsion stability at a critical mass coverage can be interpreted in terms of jamming of the interface upon partial coalescence. This would resolve the discrepancy between fast emulsion stabilization and slow evolution of interfacial properties in quiescent experimental setups. Finally the observed transition to multilayer adsorption at high bulk asphaltenes concentration seems to be fairly well captured by a BET model using the same molecular area as the previously mentioned Langmuir and Lattice gas models. The corresponding ratio in adsorption energies between first and subsequent layers is fairly large. This would be consistent with a first monolayer adsorption through strong pi interactions with water and additional layers adsorption through weak van der Waals interactions with alkyl dangling chains.

495000 Rapid Demulsification of Water-in-Oil Emulsions Using Silica Nanoparticles

Abstract:

Our research focuses on investigating the efficiency of fumed silica nanoparticles as potential demulsifiers for water-in-oil emulsions stabilized using a surfactant. From our experimental results, the hydrophobicity of silica nanoparticles determined the efficiency of demulsification. Partially hydrophobic silica nanoparticles had the highest demulsification efficiency followed by hydrophilic silica nanoparticles. Hydrophobic silica nanoparticles did not aid in complete phase separation of water-in-oil emulsions. The observed phenomenon was attributed to the difference in dynamic interaction mechanisms between silica nanoparticles and surfactants. Hydrophilic and partially hydrophobic silica nanoparticles adsorb the surfactants via hydrogen bonding that in turn led to depletion of surfactants at the oil-water interface. Upon addition of hydrophilic silica nanoparticles to the water-in-oil emulsion, the preferential distribution of silica nanoparticles in the water phase led to lower adsorption of surfactants as compared to partially hydrophobic silica nanoparticles, which have higher affinity towards the oil phase. Weak hydrophobic interactions between highly hydrophobic silica nanoparticles and surfactants led to poorer demulsification efficiencies for water-in-oil emulsions stabilized using a non-ionic surfactant. Furthermore, we present observations regarding the effect of salinity on demulsification efficiency of water-in-oil emulsions stabilized using a non-ionic surfactant upon addition of fumed silica nanoparticles of different wettabilities.

496083 Linking Adhesive Emulsions Behaviour in Microfluidic Devices to Direct Force Measurements between Drop Pairs

Abstract:

Adhesive emulsions are systems with attractive interactions between drops in which aggregation, rather than coalescence, is prevalent. Adhesive interactions between drops control processes such as gel formation, fluid microstructure and the deposition of coatings as well as play a crucial role in governing responsive complex fluids (*e.g.* formulated products in food, personal care products, pharmaceutical formulations). This talk will focus on polymer surfactant complexes, due to their ability to control both solution and surface properties as well as form aggregated emulsions where drops stick or adhere rather than coalesce within a system. To understand these interactions one can employ direct force methods such as Atomic Force Microscopy (AFM) that offer exceptional accuracy, but are often time-consuming or limiting when attempting to examine a large number of parameters or solution conditions. Whereas microfluidics is a commonly used technique for the generation and investigation of drops behaviour at many solution conditions, but drop collision behaviour is often dictated by hydrodynamic interactions, not interfacial forces. In this work, we compare the adhesive interactions between polymer-surfactant coated drops measured via AFM to the behaviour of a novel microfluidic device capable of forming chains of drops based on the surface forces acting between drops under dynamic flow conditions. The implications of correlating adhesive interactions with solution conditions and molecular composition at a higher throughput scale using this combination of methods will be discussed.

495730 Interfacial Routes to Gelation in Solid-Stabilized Emulsions

Abstract:

Colloidal gelation is loosely defined as the emergence of arrested dynamics in a particulate suspension, and is typically due to attractive interparticle interactions that lead to clustering and the formation of a percolating particle network. Such transitions are heavily exploited in a myriad of technological applications to enhance the microstructure, texture, rheology, or shelf life of suspensions and emulsions. In this talk, new routes to gelation in three-phase mixtures will be discussed, which are based on colloidal assembly at liquid-liquid interfaces. Two different systems will be presented, where gelation results from the formation of a percolating colloid network. However, particle trapping at fluid interfaces fundamentally changes the route to network formation and results in starkly dissimilar microstructures in the two systems. In the first example, gelation is due to droplet bridging, which leads to the formation of a tenuous particle/droplet network in a continuous third phase. In the second system, gelation is the result of particle jamming at fluid interfaces during spinodal decomposition, resulting in a co-continuous arrangement of uniform fluid domains separated by a jammed monolayer of colloidal particles. Motivating the need to understand the rheology of these multiphase materials for their modern technological applications, we will discuss how the different kinetic pathways to gelation mediate the resultant microstructures and, in turn, the rheology of the mixtures.

498757 Predicting the Microstructure of an Interface Laden with Anisotropic Particles

Abstract:

Micrometer scale particles regularly interact with and adsorb to planar and curved interfaces found in a variety of consumer products, foodstuffs, industrial and coating processes. Often times, interfaces stabilized with solid particulates are more stable than interfaces stabilized by molecular surfactants. The ultimate properties of the interface depend on the interfacial microstructure, which in turn depend on interparticle interactions. We conducted Monte Carlo simulations for canonical ensemble (NVT ensemble) to predict the microstructure of spheroids, ranging from spheres to ellipsoids, at a liquid/liquid interface. In these simulations, particles experienced a superposition of interactions arising from electrostatics and capillarity. Results summarized in this presentation will compare microstructure through parametric change guided by experimental data. The experimental and simulation data showed that capillary energy expressions are accurate, while our data suggests the need for a more sophisticated expression for the electrostatic interaction between ellipsoids pinned to an interface.

487698 Controlling the Polymer/Gas Interface of the Ionic Polymer Phase of a PEM Fuel Cell Catalyst Layer during Membrane Electrode Assembly Fabrication

Abstract:

During high current density operation, water produced in the cathode catalyst layer of a PEM fuel cell negatively impacts performance by lowering mass transport of oxygen into the cathode. In this presentation, a novel heat treatment process for controlling the ionic polymer phase surface properties in the fuel cell catalyst layer is incorporated into the fabrication process of the membrane electrode assembly (MEA). XPS is used to characterize the catalyst layer ionomer-gas interface in order to verify specific heat treatment conditions lead to a hydrophobic or hydrophilic ionomer surface. The MEA fabrication procedure is modified to ensure the catalyst layer is subjected to suitable heat treatment exposure conditions in order to create a hydrophobic ionomer-gas interface inside the cathode catalyst layer. Fuel cell performance results will be presented using the new MEAs under different operating temperatures, flow field types, and air humidification conditions.

Acknowledgement

The work presented was partially funded by the National Science Foundation under Grant No. EFRI-1038234 and CBET-1518755.

493272 Modeling the Effect of Charged Lithium-Metal Anode on Electrolyte Decomposition

Abstract:

The increasing demand of high-performance energy storage devices for commercial applications has accelerated the search for the next generation of battery systems. One of the most promising candidates to replace the conventional lithium-ion battery, which is approaching its limits, is the Lithium-Sulfur (Li-S) technology, where the lithium-metal anode is an essential component of the cell. Although Li-metal is considered an ideal candidate for anode materials due to its low density, extremely high theoretical specific capacity (3860 mA h $g^{\text{âˆ'1}}$), and very low negative electrochemical potential (âˆ'3.04 V *vs.* SHE), several challenges have stopped Li-metal anodes from becoming feasible to be used in commercial batteries. These issues are mainly related to the continuous decomposition of the electrolyte as a result of contact with the Li surface. These reaction products play an important role in the formation of the solid-electrolyte interphase (SEI).Â Thus, a rigorous understanding of the electrolyte stability as well as the reaction mechanisms that can occur on it may help us elucidate improvements in performance of the Li-S system and provide some guidelines for enhanced materials to extend battery life. Recent studies have shown that nature and composition of the electrolyte may improve the battery performance. In addition, computational studies have revealed new insight on electrolyte decomposition at Li-metal anode. However, the effect of charged electrode has not been incorporated yet. In this work, we use density functional theory (DFT) and ab initio molecular dynamics (AIMD) methods in order to model and investigate the effect of charged electrode-electrolyte interface on electrolyte mixtures composed by 1, 2-dimethoxyethane (DME) and salt. Here, two different salts: LiFSI (Lithium bis(fluorosulfonyl)) and LiTFSI (Lithium bis(trifluoromethanesulfonyl)imide) are studied at 1 M concentration. The dynamics, reaction pathways, structure of the products, and charge distribution are examined and characterized in detail. Finally, some insights on thermodynamics and kinetics of the reactions are presented.

DYNAMIC MEASUREMENT METHOD OF THE EMULSIFIED WATER CONTENT BASED ON INTERFACIAL POLARIZATION

Abstract ID: 494216

1. *Qing Meiyi, State Key Lab of Petroleum Resources and Prospecting, China University of Petroleum-Beijing, Beijing, China*
2. *Liang Huaqing, State Key Lab of Petroleum Resources and Prospecting, China University of Petroleum-Beijing, Beijing, China*
3. *Zhang Jinjun, National Engineering Laboratory for Pipeline Safety/MOE Key Laboratory of Petroleum Engineering/Beijing Key Laboratory of Urban Oil & Gas Distribution Technology, China University of Petroleum-Beijing, Beijing, China*
4. *Ma Chenbo, National Engineering Laboratory for Pipeline Safety/MOE Key Laboratory of Petroleum Engineering/Beijing Key Laboratory of Urban Oil & Gas Distribution Technology, China University of Petroleum-Beijing, Beijing, China*

Abstract

Water content in static crude emulsions has been extensively studied. However, the emulsion in the flowing conditions has higher emulsified water content, which affecting the flow characteristics of oil-water two-phase in gathering lines. This study focuses on exploring the electrical characteristics associated with water content of the flowing emulsions.

There are interfaces films between oil and water phase in crude emulsions because of the existence of some natural surfactants including asphaltenes, resins, and naphthenic acids et al. In the alternating electric field, the charged particles will accumulate on interface film, which caused the interfacial polarization. According to the difference between the conductivity of oil phase and water phase, and the polarization of the phase boundary film, the emulsion system in the alternating electric field can be equivalent to a circuit model composed of resistors and capacitors in series or in parallel. Based on the circuit model, the complex impedance formula capable of describing of emulsion system is thus deduced. There exists a maximum value of the response curve of imaginary part of complex impedance due to the occurrence of interfacial polarization of emulsion. The corresponding frequency of imaginary maximum is defined as the interfacial polarization frequency $f_0(w)$, which is strongly correlated to the water content of the emulsion. A dynamic testing apparatus of complex impedance of emulsion is developed in our work based on the principles stated above. To realize the dynamic testing and improve the sensitivity of the measurements, the sample cell is made of non-metal material with two half-cylindrical surface electrodes. The measuring circuit is based on a Frequency Response Analyzer PSM1700, which is used to measure the impedance of emulsions in the frequency range of 50Hz-1MHz. Using this apparatus, different stirring rates and test temperatures were controlled for the oil-water mixtures with different water contents in experiments to prepare emulsions with varied amount of emulsified water content.

The experiment results show that in the observed range, the interfacial polarization frequency of impedance has the linear relation with the water content of the crude oil, which is well consistent with the theoretical model established. The linear dependence can be applied to the measurements of the water content of flowing crude oil.

Keywords: Water content, Interfacial polarization, Dynamic mearsurement, Testing apparatus, RC equivalent circuit model

Introduction

Oil-water two-phase system in the flow shear conditions tends to form a certain stable emulsion, affecting the flow characteristics of oil-water two-phase in gathering lines. Compared to the quiescent state, the emulsion in the flowing state has higher emulsified water content. Conventional emulsified water content studies are usually under static conditions [1-4]. The objective of this study is to propose an effective method for measuring the emulsified water content of oil-water two-phase system under flowing conditions.

Because of the existence of asphaltenes, wax, solid particles and other natural surface active substance, there is a certain thickness of the interfacial film between oil and water phases in the W/O emulsions. In the AC field the charges will gather at oil-water boundary film, which leads to the interfacial polarization. Therefore, there are two electrical processes of emulsion in the AC electric field: conduction and polarization. Due to the certain time required for interfacial polarization, the emulsion will exhibit different electrical characteristics at different frequencies. In the recent study of water content in crude oil emulsion, some researches have already noticed and explored the influence of frequency dispersion on the dielectric constants of emulsions[5~7]. However, these theories are often based on the geometric structure of emulsions droplets to establish theoretical models, which is limited in explaining the electrical characteristics in unstable, heterogeneous emulsions. And there is fewer investigations focus on the impacts of frequency dispersion on the frequency characteristics of the emulsions system under flowing conditions.

Impedance spectroscopy is a method of electrical measurement with small amplitude sinusoid potential as perturbation signal, which provides the amplitude and phase of the impedance through the ratio between the response signal and the perturbation signal at different frequencies in a certain frequency range. The analysis generally makes considerable use of equivalent circuits to represent experimental frequency response [8]. Impedance spectroscopy has been widely used in the chemical materials fields to measure and analyze the materials in which ionic conduction strongly predominates (EIS) [9~11]. The method also be applies to dielectric materials: solid or liquid nonconductors whose electrical characteristics involve dipolar rotation, and to materials with predominantly electronic conduction. Thus the impedance spectroscopy has been employed to obtain information of materials by many investigators and efforts were made to apply impedance spectroscopy approach to detecting water content and distribution in various materials this work, such as cement paste, soil, fuel cell, biological tissue [9~13]. In the measurement of the water content of crude oil emulsions, some studies have adopted high-frequency (GHz-THz) [14~15] impedance spectroscopy or dielectric spectroscopy to analyze the relationship between the water content and electrical characteristics. But the laboratory implementation is complex and costly, and can't realize the measurement of the water content of emulsions under flowing conditions.

In this paper, the technique of impedance spectroscopy was applied to characterize the water contents of crude oil emulsions. The equivalent electrical model was established to describe the impedance response of W/O emulsion system. A flowing test device was developed to measure the AC impedance of W/O emulsions with water ratios of 0 to 0.9. The experiment results show that the interface polarization frequency of AC impedance varies linearly with the water content of the emulsion, which is consistent with the established model.

Theoretical

In low-frequency AC field, the current in the emulsion mainly go through the continuous phase and bypass the dispersed phase droplets. With the frequency increasing, the charged particle moved and gathered on the interfacial film, resulting in the charge redistribution and interfacial polarization [16~18]. When the frequency increased to a certain value, the interfacial capacitance decreases and a part of the current goes through the films and the droplets. Therefore, the emulsions exhibit different impedance characteristics at different frequencies, i.e., frequency dispersion of emulsions.

According to the impedance characteristics of the emulsion, the established RC equivalent circuit is shown in Figure 1(a). During the observed frequency range, the capacitances of the dispersed phase and continuous phase are small, while the resistance of the interfacial film is large, thus the circuit model can be simplified as shown in Figure 1 (b), where R_d and C_d are the resistance and capacitance of the dispersed phase; R_i and C_i are the resistance and capacitance of the interfacial film; R_c and C_c are the resistance and capacitance of the continuous phase.

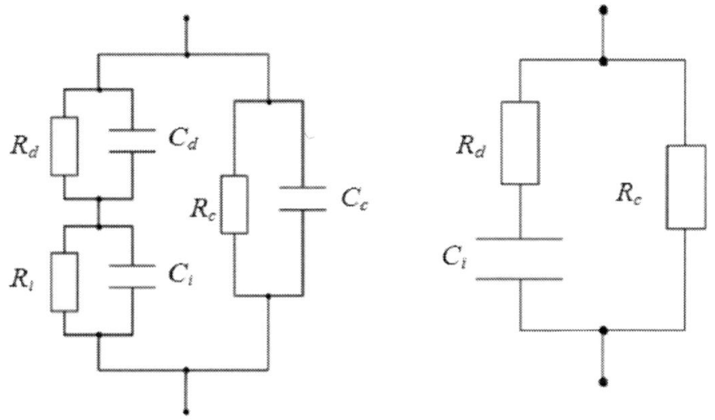

(a) (b)

Fig. 1 Equivalent circuit model of emulsion
(a) Theoretical RC equivalent circuit model
(b) Simplified RC equivalent circuit model

Based on the equivalent circuit model shown in Figure 1(b), the impedance of an ideal emulsion system is derived:

$$Z = \frac{R_c(1+jwR_dC_i)}{1+jwC_i(R_c+R_d)} = \frac{R_c + w^2C_i^2R_cR_d(R_c+R_d)}{1+w^2C_i^2(R_c+R_d)^2} - j\frac{wC_iR_c^2}{1+w^2C_i^2(R_c+R_d)^2} \tag{1}$$

The real and imaginary part of impedance Z_a、Z_b of emulsion can be obtained from Formula (1):

$$Z_a = \frac{R_c + w^2 C_i^2 R_c R_d (R_c + R_d)}{1 + w^2 C_i^2 (R_c + R_d)^2}$$

$$Z_b = \frac{w C_i R_c^2}{1 + w^2 C_i^2 (R_c + R_d)^2} \tag{2}$$

It can be seen that the Z_b gets the maximum value Z_{bmax} at f_0:

$$Z_{b\max} = \frac{w C_i R_c^2}{2}$$

$$f_0 = \frac{1}{2\pi\tau} = \frac{1}{2\pi (R_c + R_d) C_i} \tag{3}$$

The f_0 is defined as the interfacial polarization frequency. Basing on the Formula (1) and (2), the frequency characteristic curve of the RC equivalent circuit model is drawn as Figure 2.

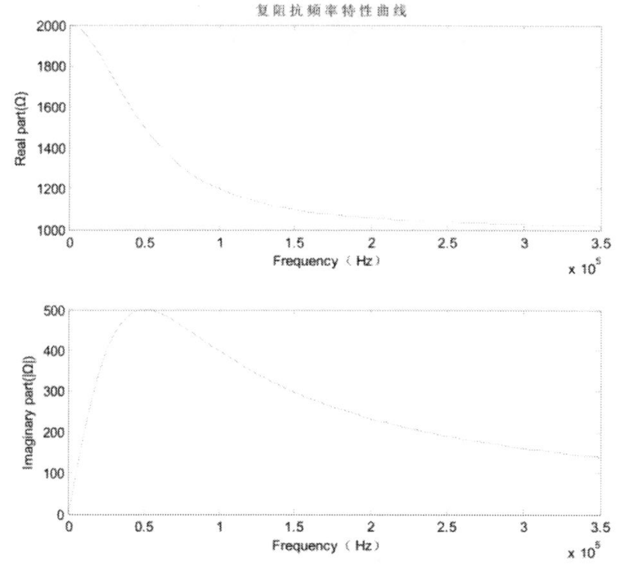

Fig. 2 Frequency characteristic curve of equivalent circuit model

Experimental

Laboratory preparation

Dehydrated oil of Xinjiang Tarim Oilfield in China was used to prepare oil-water mixtures by adding distilled water. Oil-water mixtures with water ratios of 0~0.9 were obtained. The laboratory flowing test device was shown as Figure 3.1. The sample cell was a non-metal cylindrical container, with a pair of double cylindrical metal electrodes attaching to its inside surface. The AC impedance spectroscopy of emulsions from 50Hz to 1MHz was measured by a PSM1700 Frequency Response Analyzer made in Bri N4L Co., Ltd. A stirring system was applied to construct a shear filed in the sample cell to simulate the actual flow conditions, which can help the oil-water mixture to emulsify and hold a steady state during the measurement.

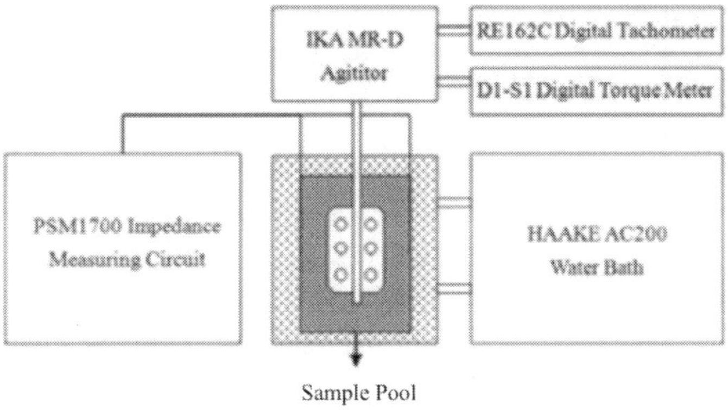

Fig.3 The laboratory flowing test device

Experimental procedure

In experiments, different water phase ratio (W), stirring rate (S) and temperature (T) of oil-water mixtures were controlled. To determine impacts brought about by a single parameter, only one parameter was adjusted during one test procedure. Test parameters of samples were listed in Table 1.

In experiments, W was adjusted from 0 to 0.9 while S increased from 0 to 500 rpm and then decreased to 100 rpm. T was adjusted from 20℃ to 40℃ respectively.

After keeping stirring for 5min at each rate, the AC impedance of flowing emulsions were measured by PSM1700 in the frequency range from 50Hz to 1MHz.

Table 1 Test parameters of oil-water mixture sample

Parameter	**W**	**T(℃)**	**S(rpm)**
	0 0.1 0.2 0.3 0.4 0.5 0.6 0.7 0.8 0.9	20 30 40	0 100 200 300 400 500

Results and Discussions

When the stirring rate reached 500rpm under 30℃, the oil-water mixtures with water ratios of 0-0.7 had formed stable and homogeneous emulsions, which showed no stratification and maintained the consistency of viscosity of each position in the sample cell. Thus we considered that the oil-water mixtures had been fully emulsified and maintained a stable amount of emulsified water content W. On the contrary, with obvious stratification at each stirring rate, the oil-water mixture with water ratio of 0.9 could not form stable emulsions.

The impedance spectroscopes of flowing emulsions with different water contents (0-0.9), stirring rates and temperatures were successively measured. Figure 4.1(a), (b) and (c) shows the curves of the impedance modulus(|Z|), the real part of the impedance (Z_a) and the imaginary part of the impedance (Z_b) vs. frequency at S=500rpm, T=30℃, respectively.

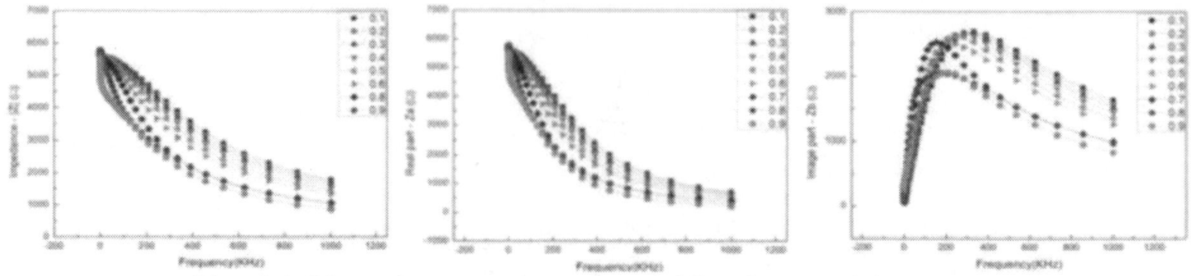

Fig. 4 AC impedance spectroscopes of flowing emulsions(0-0.9)

It can be seen that the values of impedance modulus (|Z|), the real part of the impedance (Z_a) and the imaginary part of the impedance (Z_b) of various emulsions decreased as the water content increased throughout the whole frequency range, which were well consistent with the established theoretical model. Therefore, the AC impedance spectroscopes can be used to characterize the water content of emulsions.

The relationship between |Z| and water content

Figure 4 shows the curve of |Z| at 1MHz vs. water content (S=500rpm, T=30℃). The data exhibit that the |Z| value linearly decreases when increasing the water content of emulsions. From the mechanism analysis, the increase of water content leads to the increase of the conductivity of the emulsion system, which causes the decrease of the impedance modulus.

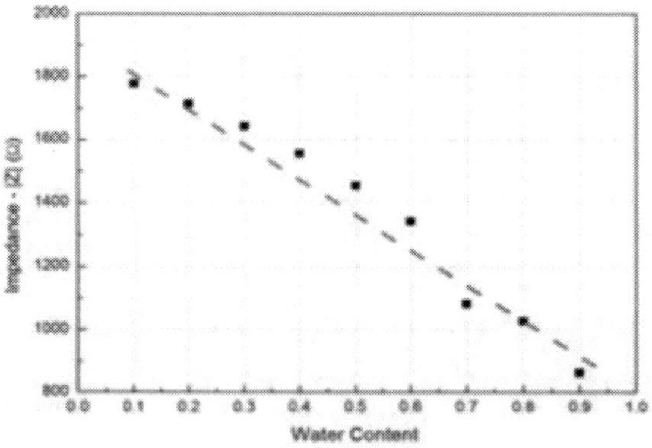

Fig. 5 The impedance modulus |Z| at 1MHz vs. water content W

The relationship between f_0 and water content

Figure 6 shows the interfacial polarization frequency f_0 changes with the water content of flowing emulsion at S=500rpm. At the water contents range of 0-0.7, the f_0 linearly increases as the water content increases. However, when the water content exceeds 0.7, the f_0 no longer increases with the water content increasing. Therefore, f_0 can be used to determine the water content of emulsions with medium or low water content. It also indicates the increased emulsified water content leads to the linearly increase of polarization time because of the increased dispersed phase resistance.

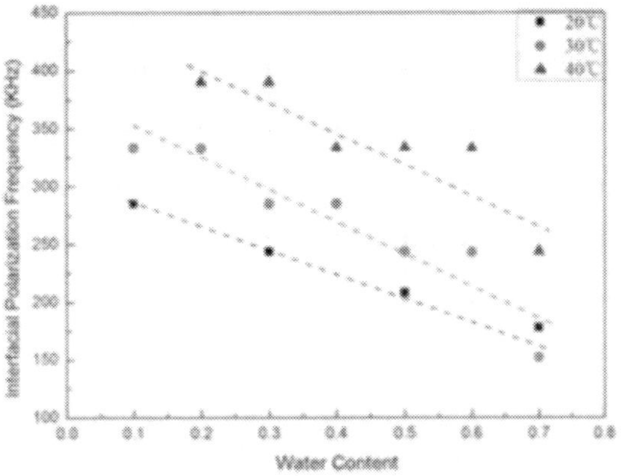

Fig. 6 The interfacial polarization frequency f_0 vs. water content W

It can be observed that the oil-water mixture with a water phase ratio of 0.9 can't form stable emulsion at any stirring rates, thus the degree of emulsification can be described by the stirring rate S in the unstable emulsions. As shown in Figure 7, there is a linear relationship between S and f0 at different temperatures (W=0.9). It can be explained that the increase of stirring rate results in the increase in the number and the decrease in the volume of dispersed phase droplet, which causes the decrease of dispersed phase capacitance.

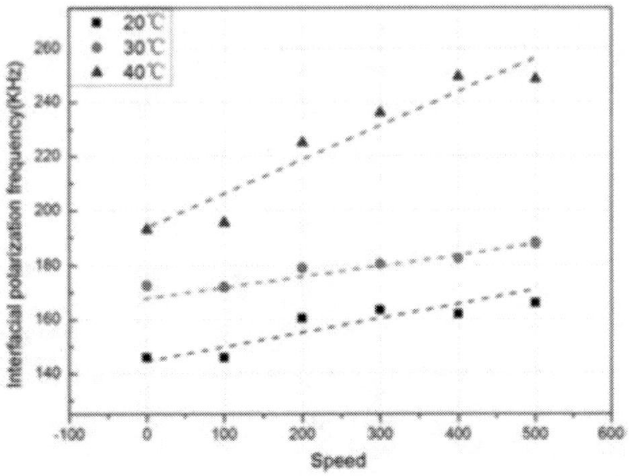

Fig. 7 The interfacial polarization frequency f_0 vs. stirring rate S

The results suggest that the impedance spectroscope can be used not only to determine the water content in crude oil emulsions, but also to evaluate the degree of emulsification of oil-water mixture by measuring whether the interfacial polarization frequency varies with the rate.

Conclusions

We investigated the potential of impedance spectroscope to accurately determine the water content in crude oil emulsion. The equivalent circuit model based on the frequency dispersion provides a theoretical basis for the impedance measurements. By employing a stirring system, the flowing emulsions with water content from 0-0.9 were measured.

The experimental results suggest that both the impedance modulus and interfacial polarization frequency can be used to calculate the water content in the emulsions. The impedance modulus and water content in flowing emulsions are strongly and linearly correlated in the examine range. In addition, the interfacial polarization frequency exhibits a marked linear decrease with the increase of water content in stable emulsion and a linear increase with the increase of degree of emulsification in unstable emulsion. This linear dependence greatly reduces the complexity of accurate measurement of water content of crude oil emulsion. Further investigations should be undertaken to establish the method for characterizing water content in emulsions.

Reference

1. Nädler M, Mewes D (1997), "Flow induced emulsification in the flow of two immiscible liquids in horizontal pipes." *International Journal of Multiphase Flow*, 23, pp. 55-68.
2. Grassi B, Strazza D, Poesio P (2008), "Experimental validation of theoretical models in two-phase high-viscosity ratio liquid-liquid flows in horizontal and slightly inclined pipes," *International Journal of Multiphase Flow*, 34(10), pp. 950 - 965.
3. Wang W, Gong J, Angeli P (2011), "Investigation on heavy crude-water two phase flow and related flow characteristics," *International Journal of multiphase Flow*, 37(9), pp. 1156-1164.
4. Klink IM, Phillips RJ, Dungan SR (2011), "Effect of emulsion drop-size distribution upon coalescence in simple shear flow: A population balance study," *Journal of Colloid & Interface Science*, 353(3), pp. 467-475.
5. T Skodvin, J Sjöblom(1996), "Models for the Dielectric Properties of Flocculated W/O-Emulsions," *Journal of Colloid & Interface Science*, 182 (1), pp. 190-198.
6. S Pirouzpanah, B Erdoğan, GL Morrison(2014), "Study of hysteresis effects and emulsion properties in water cut measurement using high speed multi-frequency impedance sensor," *Flow Measurement & Instrumentation*, 40, pp. 1-8.
7. S Gorenflo, U Tauer, I Hinkov, A Lambrecht, R Buchner(2006), "Dielectric properties of oil–water complexes using terahertz transmission spectroscopy", *Chem. Phys. Lett.*, 421, pp. 494 - 498.
8. JR Macdonald(2003), "Impedance spectroscopy, Annals of Biomedical Engineering," 20(3), pp. 703-715.
9. N. Perini, A.R. Prado, C.M.S. Sad, E.V.R. Castro, M.B.J.G. Freitas(2012), "Electrochemical impedance spectroscopy for in situ petroleum analysis and water-in-oil emulsion characterization," *Fuels*, 91, pp. 224–228.
10. Smiechowski MF, Lvovich VF(2005). "Characterization of non-aqueous dispersions of carbon black nanoparticles by electrochemical impedance spectroscopy," *J. Electroanal Chem*, 577, pp. 67–78.
11. F Lisdat, D Schäfer(2008), "The use of electrochemical impedance spectroscopy for biosensing," *Analytical & Bioanalytical Chemistry*, 391(5), pp. 1555.
12. K Takehara, T Takehara(2007-6-27), "Body water amount condition judging apparatus by multi-frequency bioelectric impedance measurement," EP 1177760B1[P].
13. S Perron, J.J Beaudoin(2002), "Freezing of water in portland cement paste – An AC Impedance Spectroscopy Study," *Cement & Concrete Composites*, 24 (5), pp. 467 - 475.

14. WJ Jin, K Zhao, C Yang, et al (2013). "Experimental measurements of water content in crude oil emulsions by terahertz time-domain spectroscopy," *Applied Geophysics* 10(4), pp. 506-509.
15. YZ Li, SX Wu, XL Yu, et al (2017). "Optimization of pyrolysis efficiency based on optical property of semicoke in terahertz region," *Energy*, pp. 126.
16. J Jiao, DJ Burgess(2002), "Multiple Emulsion Stability: Pressure Balance and Interfacial Film Strength," *Journal of Colloid & Interface Science*, 250(2), pp. 444-450.
17. A Bhardwaj, S Hartland(1994). "Dynamics of Emulsification and Demulsification of Water in Crude Oil Emulsions," *Industrial & Engineering Chemistry Research* , 33(5), pp. 1271-1279.
18. T Hanai, T Imakita, N Koizumi(1982), "Analysis of dielectric relaxations of w/o emulsions in the light of theories of interfacial polarization," *Colloid and Polymer Science*, 260 (11), pp. 1029-1034.

496934 Stretchable and Soft Electroadhesion Enhanced By Liquid-Metal Subsurface Microstructures

Abstract:

Electroadhesion is an attractive mechanism for robotic components to electrically modulate adhesion to surfaces for gripping, climbing, and locomotion. Previously, electroadhesion has been implemented on flexible substrates, but not on soft or stretchable substrates that are compatible with, for example, soft robotics or stretchable electronics. Here, electroadhesion was generated and enhanced in a soft and stretchable elastomeric substrate by embedded subsurface liquid metal-filled electrodes. Applying potential between these interdigitated electrodes modulates the adhesion of these devices to surfaces in a reversible manner. In addition to creating sot and stretchable constructs, the use of liquids intrinsically enhances the adhesion by the surface stresses caused by the capillary pressure inside the fluid-filled microchannels. This approach to electroadhesion can be implemented in ultra-stretchable and soft substrates, including those used in robotics, due to the inherently soft and deformable electrical conductivity of the liquid metal electrodes.

501273 Giant Electret Polarization in Electrochemically Deposited Hydroxyapatite Ceramic Coatings

Abstract:

Calcium phosphate coatings synthesized electrochemically were discovered to retain extremely large quasi-permanent electrical polarization. The coatings are synthesized onto a cathode surface from an aqueous electrolyte solution under a low temperature galvanostatic process. Calcium cations are concentrated near the cathode surface due to Coulombic attraction, while hydroxide ions are generated at cathode surface due to water splitting reactions. Changes in pH and ionic strength near the cathode surface result in the nucleation and growth of a thin calcium phosphate coating. X-ray diffraction spectra show that the only crystal phase observed in the coating is hydroxyapatite, although other amorphous phases are likely present. The coating grows within the electrical double layer at the cathode/electrolyte interface, where ions experience extreme electric fields and ion concentration gradients. As the coating grows on the cathode surface, ions become trapped in the dielectric calcium phosphate layer. These trapped charges charges result in an internal electric field in the coating oriented in the opposite direction as the applied field during synthesis. Since the trapped charges are immobilized at room temperature in the solid coating, the calcium phosphate is an electret with stored charge among the highest measured for any material. The unique polarization mechanism will be discussed, and biomedical applications of the hydroxyapatite coatings in implants will be demonstrated. In addition, synthesis of metal and semiconductor nanoparticles on the coatings will be described. When acting as a support for metal and semiconductor particles, the coatings may impact catalytic processes due to the strong internal electric field.

502187 Electrochemical Charging of CdSe Quantum Dots: Effects of Adsorption *Versus* Intercalation

Abstract:

Effects of electrochemical charging of quantum dots have been reported previously, wherein optical and electrical properties could be modulated through cation adsorption and electron injection into the quantum-confined $1S_e$ states. In this presentation, we report two different modes of electrochemical double layer charging in CdSe quantum dots and their effects on the electronic and optical properties. We show that the charging mechanism at the interface involves cation intercalation for smaller ions such as Li^+, Na^+ or K^+, and cation adsorption for larger bulky ions such as tetrabutylammonium ions, where steric hindrance precludes intercalation. As a result, while cation adsorption leads to increase in the absorbance in the mid-infrared spectral range, cation intercalation into CdSe core results in an absorbance increase from visible-to-infrared spectral range, an enhancement in radiative lifetime of $e^{âˆ'}$, an increase of 158% in the intensity of band-edge photoluminescence, and strong emission in the near-infrared spectral range as a result of the formation of Se vacancies. The nature of charging mechanisms is discussed using the results of combined photoluminescence, radiative lifetime and X-ray photoemission studies. The cation-coupled electronic and optical modulation reported here in CdSe QDs have important implications for electrochromic smart windows, photovoltaic and other devices.

498716 Microstructured Conducting Polymer Swabs for Enhanced Trace Explosive Detection

Abstract:

The detection of trace amounts of explosive materials is critical to national security, and ion mobility spectrometer (IMS)-based analysis of trace explosives collected using contact sampling continues to be the most common explosives and chemical agent detection method used in airports and high security checkpoint locations. In this process a swab (or particle trap) is used to probe a surface of interest (e.g., the hands and luggage of passengers) to collect and transfer particulate residue to an IMS for analysis. However, the collection of residue from a surface remains the limiting step in this process. As such there is a significant need to develop new materials with the improved ability to interrogate surfaces and to increase the adhesion to trace explosive residues of interest.

In this work, we present a novel swabbing material for the enhanced removal of explosive particles from surfaces. Polypyrrole (PPy) microstructured films were synthesized with a variety of form factors through a templated electropolymerization process. The removal of representative fluorescent particles from aluminum surfaces was examined as a function of polymer microstructure. PPy-based microstructured films displayed an increase in particle removal properties when compared to PPy-based non-microstructured films and commercial of-the-shelf (COTS) swabs. This increase in removal is due to an increase in particle-swab contact from the increased surface area of the microstructured films. Furthermore, these polymer films were easily functionalized with surface groups with increased adhesion to explosive compounds such as trinitrotoluene (TNT) to develop next generation materials for trace explosive detection.

505790 Atomistic Modeling of Solvent Decomposition in Magnesium Batteries

Abstract:

The lithium-ion (Li-ion) battery has revolutionized the consumer electronics industry in the past two decades, achieving moderate battery lifetimes and higher energy storage densities. However, we have begun to reach the limits to improvements in Li-ion batteries. Further, in order to reach the energy demands posed by the transportation industry, batteries that possess energy densities much larger than current state-of-the-art Li-ion technologies are highly desirable. This research investigates one â€˜beyond Li-ionâ€™ battery technology that boasts a theoretical energy density more than five times larger than Li-ion batteries: metallic magnesium (Mg) batteries. We used density functional theory to probe atomistic interactions on the metallic Mg anode surface. Specifically, we identified likely reactions occurring between the anode surface and the battery solvent. The composition of the model Mg anode surface was modified to mimic realistic electrode surfaces. We found that solvent decomposition products and reaction energies were highly dependent on the model anode surface composition, with the pristine Mg anode displaying a greater tendency for solvent decomposition. This study provides a more complete picture of the atomic-scale factors governing the successful implementation of metallic Mg anodes for beyond Li-ion batteries.

493844 Understanding the Reduction Reaction Mechanisms of Sulfur-Based Cathodes: A Theoretical Approach from DFT and Reaxff Molecular Dynamics

Abstract:

The next generation of batteries must help to do the jump from portable consumer electronics to high demanding energy applications like electric vehicles (EVs) and smart-grids. In this regard, as the energy density of the Li-ion batteries is approaching to its theoretical limit, Lithium/Sulfur (Li/S) batteries come up as one of the most promising technologies able to close the gap between the capacity of the current electrode materials and the future energy requirements. Besides the natural abundance, non-toxicity and low-cost of sulfur, the sulfur-based cathode materials are theoretically able to deliver a specific capacity of 1675 Ah/kg and an energy density of 2600 Wh/kg, which happens to be up to 10 times higher than the commonly-used $LiCoO_2$ cathodes.

The successful commercialization of Li/S batteries requires, however, critical improvements regarding cycle-life, stability, and electrochemical activity. Since its initial introduction around 30 years ago, the sulfur-based cathode still suffers from the insulating nature of sulfur and the formation of soluble lithium sulfide intermediates. The most followed strategy to overcome these two issues is the sulfur trapping within porous conductive carbon-based materials. This way, the carbonaceous structure serves as the electrical conductor increasing the cathode's conductivity and also as the container diminishing the loss of active material out of the cathode. So far, there is a vast and always growing list of structured sulfur-carbon composites proposed by different authors: hollow carbon spheres, carbon nanotubes, carbon nanofibers, and graphene, are just a few examples of them.

Complete trapping/isolation of sulfur inside the carbonaceous structures is not the key for sulfur-based cathodes though. It is also known the dissolution of the polysulfides (PS) created during sulfur reduction is essential for the performance of a Li/S battery. The non-conductive nature of sulfur and its reduction products implies the sulfur reduction can only take place on the surfaces of conductive carbon. The dissolution of PS guarantees then the remaining sulfur is exposed so the reduction process can progressively go on. The incomplete understanding of these processes is still a bottleneck for the commercial development of Li/S batteries. The exclusive use of experimental techniques does not allow to get this deeper understanding, but now the emergence of the computational methods opens a new window at the very atomic level for the description of the reduction processes taking place during the discharge/charge of Li/S batteries.

In this work, we describe at the atomic level the reduction reactions taking place during discharging. Insights about the different structural arrangements during reduction along with the sulfur/polysulfide-carbon interaction were studied. For doing so, we used in conjunction state-of-the-art density functional theory (DFT), ab initio molecular dynamics (AIMD), and reactive molecular dynamics (ReaxFF). The sulfur/carbon composite cathode is represented with different arrangements of slabs of orthorhombic sulfur along the 110 plane coupled with graphene layers (~60 %-wt S). These structures were set in contact with liquid DOL and liquid mixtures of DOL/DME with Lithium (Li^+) ions embedded in it for studying the early stages of sulfur reduction with the ReaxFF method. The diffusion of Li^+ ions from the bulk liquid towards the sulfur slabs was observed along with the formation of Li/S bonds breaking the outer sulfur rings. We also observed the further formation of Li-S bonds, which gave us insights on the solid-liquid two-phase reaction mechanisms happening at the onset of cathode reduction. Formation of lithium/sulfur bulk structures around the graphene layers also gave us insights on the sulfur/carbon interactions and the role of carbon hindering the massive detachment/diffusion of the formed PS into the bulk liquid. DFT and AIMD calculations were also performed to corroborate, at a smaller scale, the mechanisms of polysulfide formation upon sulfur contact with the liquid electrolyte. Hints on the Li solvation mechanisms and its effects on the polysulfide diffusion into the bulk liquid electrolyte were drawn.

The studying of the liquid-liquid single-phase reduction from the dissolved Li_2S_8 to low-order PS is also addressed.
The studying of the liquid-liquid single-phase reduction from the dissolved Li_2S_8 to low-order PS is also addressed.
The molecular dynamics calculations have shown that the Li_2S_8 reduction is influenced by the graphene layers present. When the graphene layers are present, the insoluble reduction products of Li_2S_8 tend to agglomerate around them. This way, we observed that carbon not only hinders the diffusion of polysulfides into the bulk liquid but also influences in the precipitation of Li_2S_8 into insoluble Li_2S_2/Li_2S like structures, helping in the complete reduction of sulfur. These theoretical calculations greatly contribute to the understanding of the reduction reactions of sulfur-based, helping in the engineering of Li/S batteries with longer cycle-life, better stability, and higher electrochemical activity.

496041 Reversible Aluminum Intercalation in Transition Metal Sulfides

Abstract:

Rechargeable aluminum batteries are promising technologies for large scale energy storage. One of the main obstacles hindering the development of rechargeable aluminum batteries is the lacking of understanding of cathode materials. We hereby demonstrate Chevrel phase molybdenum sulfide (Mo_6S_8) as the intercalation-type of cathode material with aluminum chloride/1-butyl-3-methyimidazolium chloride ($AlCl_3$-[BmIm]Cl) ionic liquid as the electrolyte. Electrochemical characterizations including cyclic voltammetry and galvanostatic charge-discharge demonstrate unambiguous Al intercalation-extraction in the Mo_6S_8. The X-ray diffraction analysis further indicates the crystal structure change of Mo_6S_8 after Al intercalation. In addition, we report the synthesis and electrochemical properties of layered TiS_2 and cubic Ti_2S_4 as intercalation-type cathode materials for rechargeable aluminum battery at both room temperature and 50 °C. We further confirmed the aluminum intercalation in the TiS_2 and Ti_2S_4 crystal structure using combined electrochemical analysis and ex-situ XRD. The proposed titanium sulfide cathode materials showed decent reversible capacity and a higher working potential than Mo_6S_8. Furthermore, our study suggests the low Al ion diffusivity in the titanium sulfide framework induced by strong coulombic intercalation is the key obstacle for fast discharge-charge. This finding could provide guideline for design and synthesis of future titanium sulfides electrode materials.

College of Engineering

Novel Nano-structure Cathode Materials of Titanium Nitride for Lithium Sulfur Battery

Wenduo Zeng, Zhao Wang, Mark Cheng, and K. Y. Simon Ng
Department of Chemical Engineering
Wayne State University
Detroit, Michigan

Lithium Sulfur

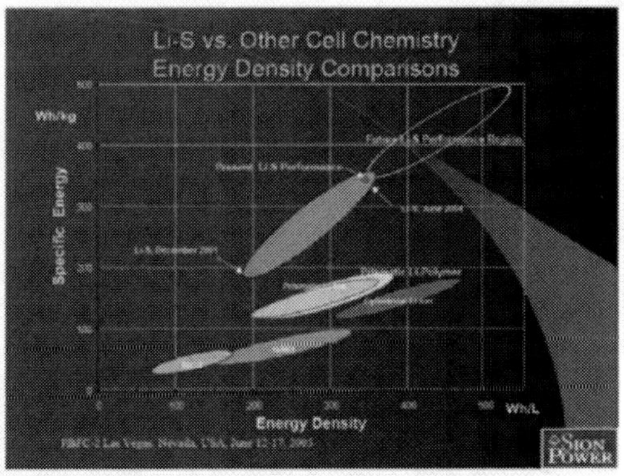

L-S vs other cells energy density comparison

- Advantages
  - High theoretical capacity
  - Low cost
  - Safe
  - Easily rechargeable

- Technical barriers
  - Cycleability
  - Volumetric Density

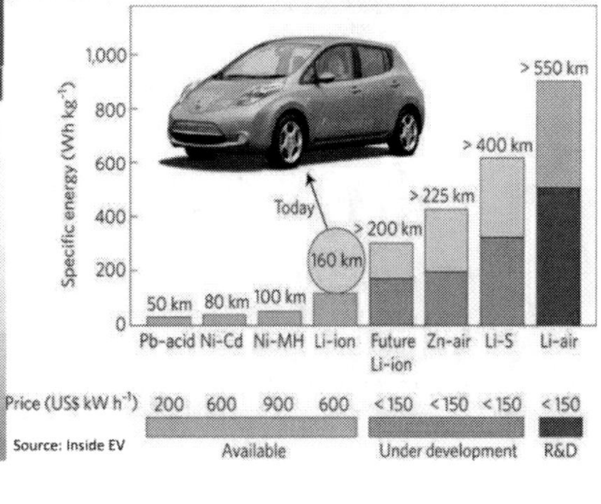

Voltage Profile of Li-S battery

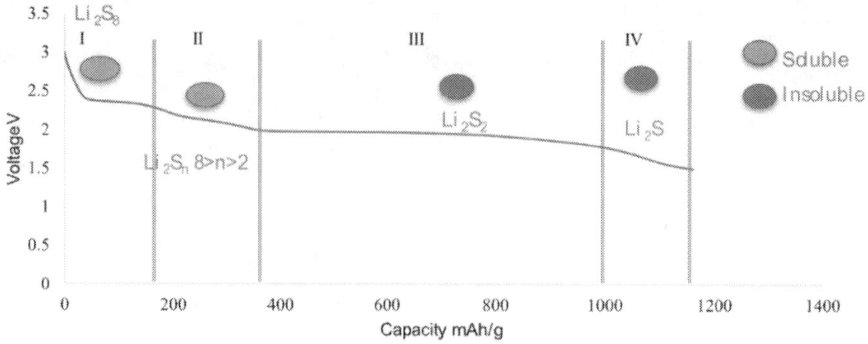

Discharge profile of Li-S battery

Region I

- First plateau (2.3-2.5) V
- $S_8+2Li \leftrightarrow Li_2S_8$

Region II

- Reduction (Li_2S_8 to low order polysulfide)
- $Li_2S_8+2Li \leftrightarrow Li_2S_{8-n}+ Li_2S_n$

Region III

- Second lower plateau (1.9-2.1) V
- $2Li_2S_n+2n-4Li \leftrightarrow nLi_2S_2$
- $2Li_2S_n+2n-2Li \leftrightarrow nLi_2S$

Region IV

- Reduction of Li_2S_2 to Li_2S
- $Li_2S_2+2Li \leftrightarrow 2Li_2S$

Zhang, Sheng S. Liquid electrolyte lithium/sulfur battery: Fundamental chemistry, problems, and solutions

Technical Challenges of Li-S battery

Insulating active materials [Dai et al., 2011]

- Higher impedance for charge transfer

Volume expansion [Barai et al., 2016]

- Lower performance stability
- Rapid capacity deactivation

Shuttle Reactions [Bauer et al., 2014]

- High self-discharge

Non soluble lithium sulfide [Nazar et al., 2015]

- Increases the charge transfer
- Increases the sulfur active material loss

Proposed Solutions to Current Issues

Novel cathode materials with:

- Lower electron transfer impedance (Cheng et al., 2016..)
- High surface area to provide higher reaction kinetics and accommodate structural change
- Ability to adsorb/absorb and accommodate polysulfides by physisorption or chemisorption

Reported Li-S Battery Results

Materials	C-rate	Cycles	Capacity (mAh/g)	Reference
Porous hollow carbon	0.5 C	100	974	Jayaprakash et al., 2011
Carbon nanotube	1 C	100	915	Manthiram et al., 2012
Graphene	0.2 C	80	670	Wang et al., 2012
$Mg_{0.6}Ni_{0.4}O$ additives	1 C	50	~1000	Lee et al., 2004
Mesoporous TiN	0.5 C	500	644	Goodenough et al., 2016

The best has been ever get is around 850 mAh/g after 100 cycles under 0.5 C-rate.

Why Nanostructured TiN ?

Materials Property

- Exceptional electrical conductivity (in the order of 10^6 S m^{-1}), compared with graphite (in the order of 10^5 S m^{-1})
- Good thermal conductivity as 28.84 W/(m*K)
- High thermodynamic stability and corrosion resistance

Nanotube structure

- Higher surface area
- More flexibility to accommodate with volume change
- Spatial confinement (vs. nanoparticles)

Objectives

- Elucidate the effects of preparative parameters on the size and morphology of TiN nanotube structures.
- Understand the effects of morphological and electronic properties on the electrochemical performance of TiN nanotube.

Anodization Fabrication and Chemical Synthesis

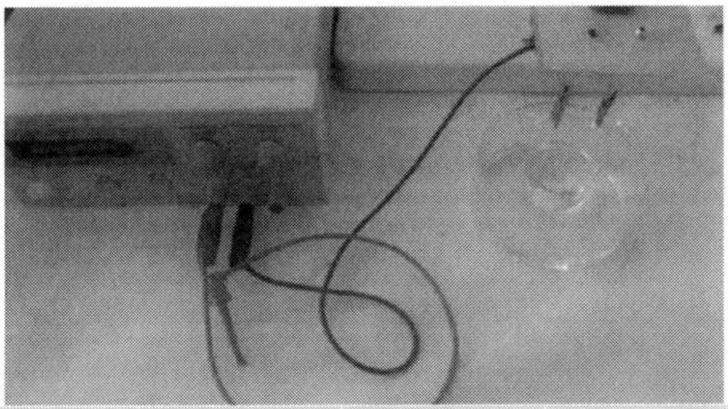

Anodization Voltage	***45 V***	***60 V***	***75 V***
Diameter of TiO_2 nanotube	***35 nm***	***65 nm***	***100 nm***
Anodization Time	***10 min***	***5 h***	***10 h***
Length of TiO_2 nanotube	***4.5 um***	***15 um***	***30 um***

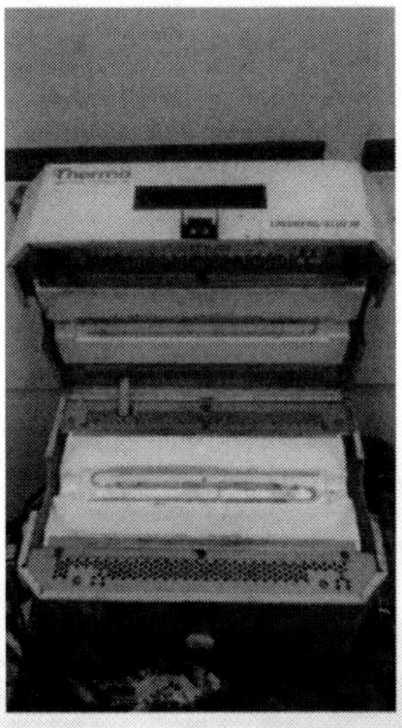

Nitridation procedure under 800 ℃

Structure and Morphology of TiN Nanotube

XRD spectra of TiN nanoparticle and nanotube

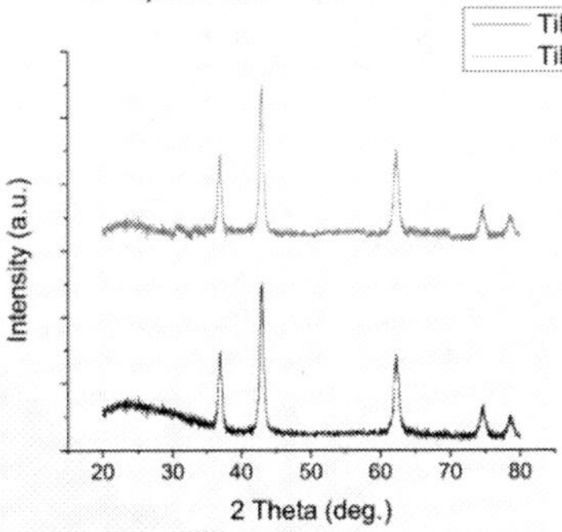

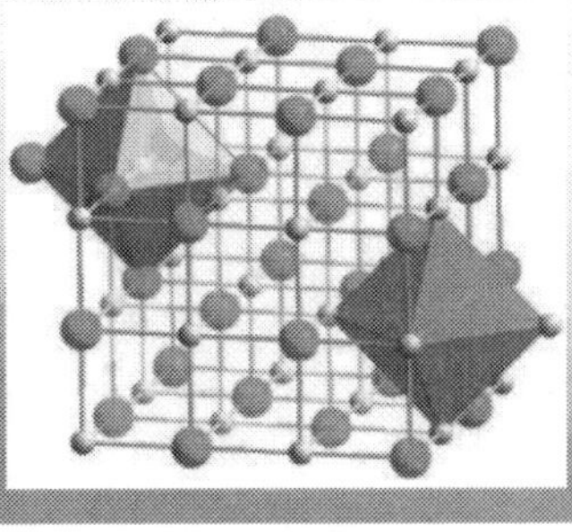

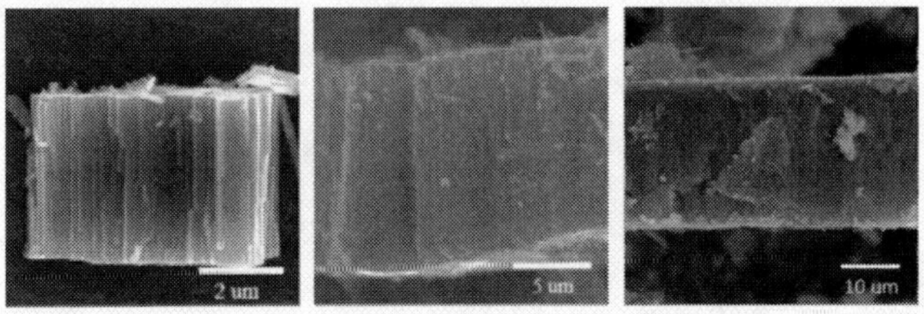

Length: 4.5 um 15 um 30 um

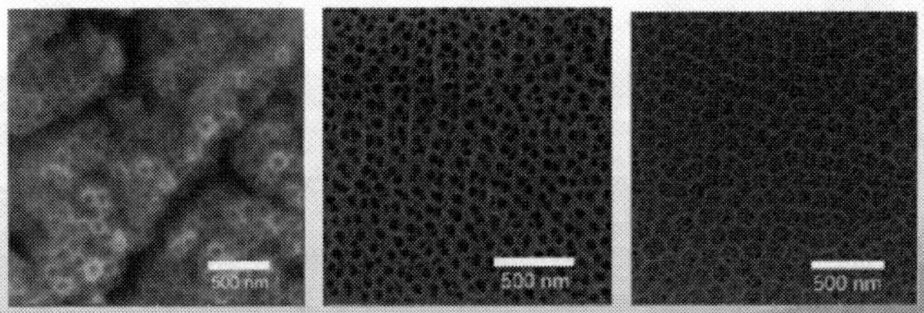

Diameter: 100 nm 65 nm 35 nm

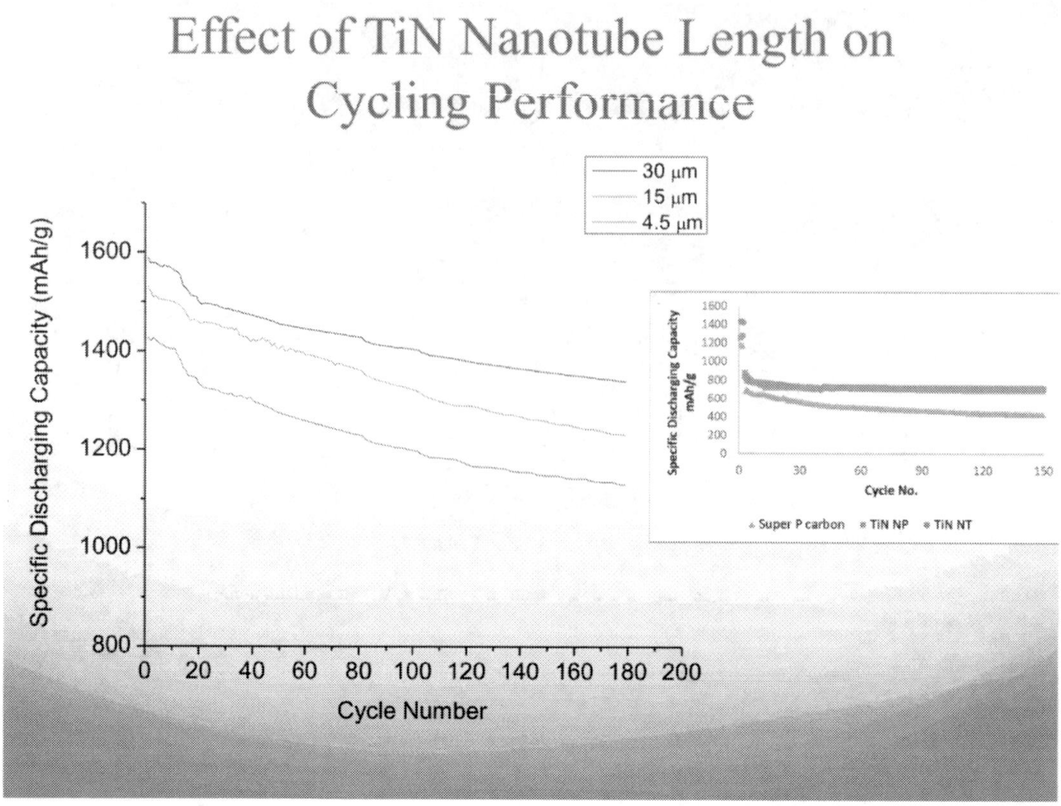
Effect of TiN Nanotube Length on Cycling Performance
30 μm
15 μm
4.5 μm
Specific Discharging Capacity (mAh/g)
1600
1400
1200
1000
800
0
20
40
60
80
100
120
140
160
180
200
Cycle Number
Specific Discharging Capacity mAh/g
Cycle No.
Super P carbon
TiN NP
TiN NT

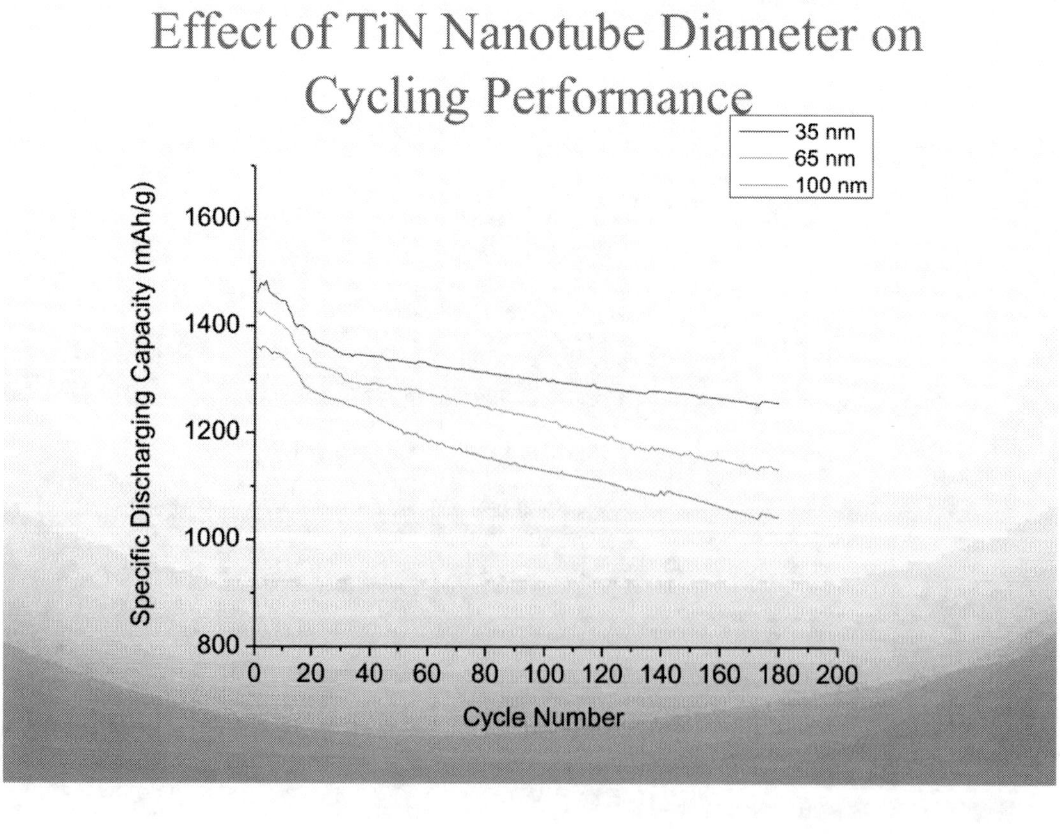
Effect of TiN Nanotube Diameter on Cycling Performance
35 nm
65 nm
100 nm
Specific Discharging Capacity (mAh/g)
1600
1400
1200
1000
800
0
20
40
60
80
100
120
140
160
180
200
Cycle Number

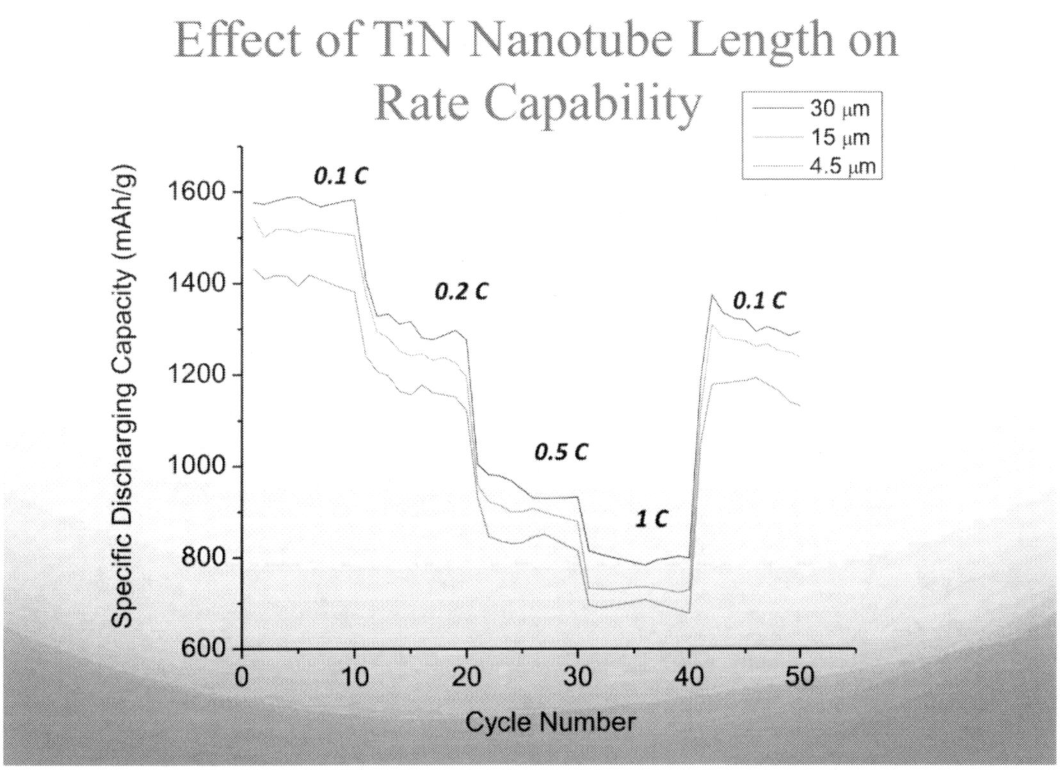
Effect of TiN Nanotube Length on Rate Capability
30 μm
15 μm
4.5 μm
0.1 C
0.2 C
0.5 C
1 C
0.1 C
Specific Discharging Capacity (mAh/g)
Cycle Number

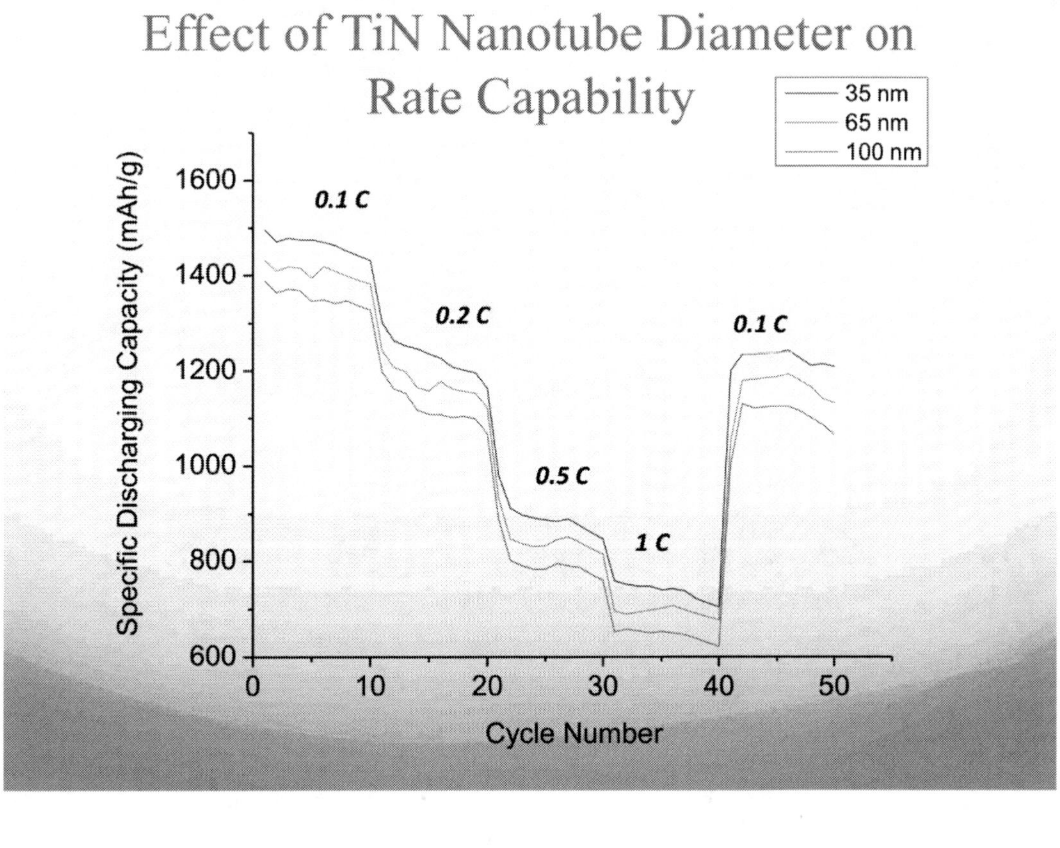
Effect of TiN Nanotube Diameter on Rate Capability
35 nm
65 nm
100 nm
0.1 C
0.2 C
0.5 C
1 C
0.1 C
Specific Discharging Capacity (mAh/g)
Cycle Number

Effect of TiN Nanotube Length on Charge Transfer Impedance

Fresh cell

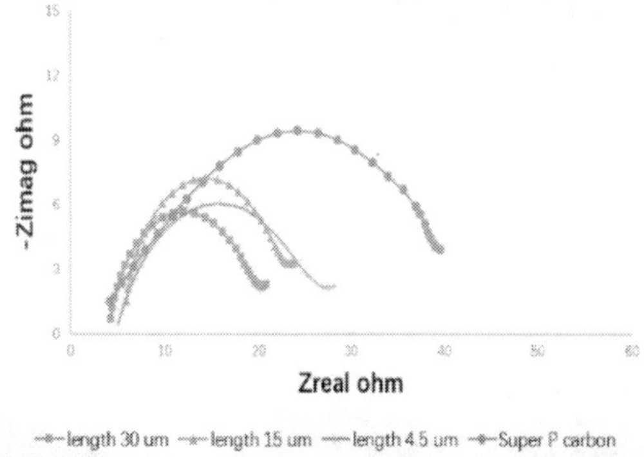

After 180 cycles

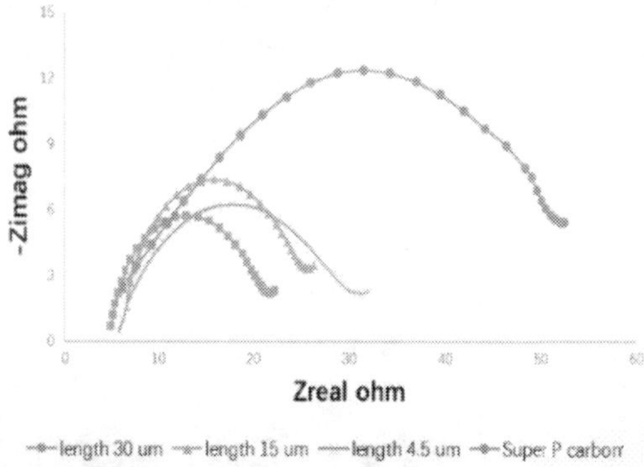

Length	Before cycles / ohm	After cycles / ohm	Percentage of increase
30 um	20	21	5%
15 um	23	26	13%
4.5 um	27	31	15%

Effect of TiN Nanotube Diameter on Charge Transfer Impedance

Fresh cell

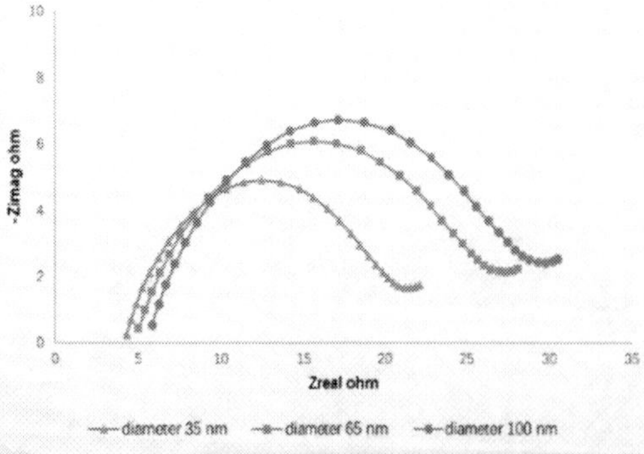

After 180 cycles

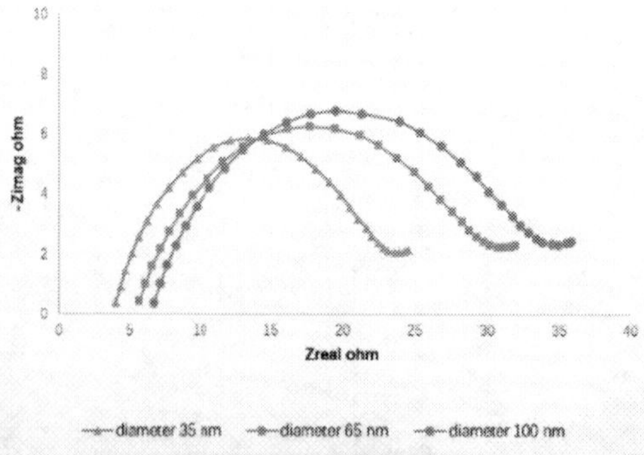

Diameter	Before cycles / ohm	After 180 cycles / ohm	Percentage of increase
35 nm	17	20	15%
65 nm	22	26	18%
100 nm	24	29	21%

BET Specific Surface Area of TiN nanotubes

Effect of Diameter

Nanotube with length of 4.5 um	35 nm	65 nm	100 nm
Specific surface area ($m^2 g^{-1}$)	45.15	33.14	21.76

Effect of Length

Nanotube with diameter of 65 nm	4.5 um	15 um	30 um
Specific surface area ($m^2 g^{-1}$)	34.14	33.97	35.32

Discussion / Mechanism

It has been observed that Li-S battery performance improves by introducing TiN with varying size.

Smaller diameter:

- One of the factors determining electrocatalytic activity is the accessible surface area
- Fine spatial separation to induce less internal stress

Larger length:

- More inhibition for soluble polysulfide to escape from the host material
- More effective conductive matrix constructed by TiN nanotube as electron transfer channel

Summaries

- TiN nanotube is a promising cathode material for advanced Li-S batteries showing a capacity of 1370 mAh/g after 180 cycles at 0.1 C rate.

- Demonstrated high stability of 0.064 % decay per cycle on average.

- Increase in nanotube length from 4.5 to 30 um and decrease in nanotube diameter from 100 to 35 nm enhance capacity by 19.1% and 24.7% after discharge/charge of 180 cycles, respectively .

Acknowledgements

- **Financial support of this research from DOE is gratefully acknowledged**

498000 Elucidating the Solvation Structure and Dynamics of Lithium Polysulfides Using Coupled High Throughput Simulations and Experiments

Abstract:

Fundamental molecular level understanding of functional properties of liquid solutions provides an important basis for designing optimized electrolytes for numerous applications. In particular, exhaustive knowledge of solvation structure, stability and transport properties is critical for developing stable electrolytes for fast charging and high energy density next-generation energy storage systems. Here we report the correlation between solubility, solvation structure, and translational dynamics of a lithium salt (Li-TFSI) and polysulfides species using well-benchmarked classical molecular dynamics simulations combined with nuclear magnetic resonance (NMR). It is observed that the polysulfide chain length has a significant effect on the ion-ion and ion-solvent interaction as well as on the diffusion coefficient of the ionic species in solution. In particular, extensive cluster formation is observed in lower order polysulfides (S_x2-; xâ ‰¤4), whereas the longer polysulfides (S_x^{2-}; x>4) show high solubility and slow dynamics in the solution.[1] It is observed that optimal solvent/salt ratio is essential to control the solubility and conductivity as the addition of Li salt increases the solubility but decreases the mobility of the ionic species. We also observed that counter anion (such as TFSI, TfO, FSI & TDI) and solvent interaction strength with Li^+ is critical in controlling polysulfide solubility.[2] Also, high mobility of counter anions can cause faster capacity loss due to enhanced polysulfide solubility and SEI layer formation. This work provides a coupled theoretical and experimental study of bulk solvation structure and transport properties of multi-component electrolyte systems, yielding design metrics for developing optimal electrolytes with improved stability and solubility.

References

1. Rajput, N. N.; Murugesan, V.; Shin, Y.; Han, K. S.; Lau, K. C.; Chen, J.; Liu, J.; Curtiss, L. A.; Mueller, K. T.; Persson, K. A., Elucidating the Solvation Structure and Dynamics of Lithium Polysulfides resulting from Competitive Salt and Solvent Interactions. *Chem. Mater.* **2017**.

2. Vijayakumar, M.; Govind, N.; Walter, E.; Burton, S. D.; Shukla, A.; Devaraj, A.; Xiao, J.; Liu, J.; Wang, C.; Karim, A., Molecular structure and stability of dissolved lithium polysulfide species. *Phys. Chem. Chem. Phys.* **2014,** *16* (22), 10923-10932.

492405 Solvation Structure and Behavior of Lithium Polysulfide Species in Electrolytes of Lithium-Sulfur Batteries

Abstract:

Batteries with large amounts of energy storage are needed to meet the requirements of modern technology and electronics like electric vehicles or cell phones. Current lithium ion batteries are unable to meet these modern energy demands; new battery technology is required. The lithium-sulfur (Li-S) battery is promising technology that has the theoretical energy capacity to meet societal needs. However, several large issues plague Li-S battery performance such as the polysulfide shuttle reaction which results from migration to the anode side of soluble Li polysulfide species formed at the cathode surface. The dissolution of polysulfide species in the electrolyte is the fundamental problem with this parasitic shuttle reaction. However, the solvation structure and fundamentals are not well known due to the complex interactions and chemistry in Li-S batteries. The molecular processes are difficult to study experimentally and therefore are not well understood.

Quantum density functional theory (DFT) and ab-initio molecular dynamics (AIMD) were used to investigate and screen different electrolyte compositions in order to identify the structure and behavior of lithium and polysulfide solvation. Promising and interesting electrolyte interactions can be identified by computational simulations to help with experimental design of batteries. Electrolytes consisting of dimethoxyethane, dioxolane, and fluorinated ethers were investigated. We determined thermodynamically favorable solvation structures, degree of dissociation of lithium polysulfides, and established the basis for a rational design that may improve overall Li-S battery performance by reducing the shuttle redox reaction.

502771 Design Stable Room-Temperature Metal-Sulfur Batteries

Abstract:

High-energy and inexpensive rechargeable battery systems based on earth-abundant materials are important for both mobile and stationary energy storage technologies. Rechargeable sodium-sulfur (Na-S) batteries that are able to operate stably at room temperature are among the most sought-after of these platforms because these cells take advantage of a two-electron-redox process to yield high storage capacity from inexpensive electrode materials. Realization of practical Na-S batteries has been fraught with multiple stubborn problems ranging from unstable electrodeposition of sodium during battery recharge to rapid loss of the active cathode material by dissolution into the electrolyte. In this study, we develop a room temperature Na-S battery that uses a sodium metal anode, a microporous carbon-sulfur composite cathode, and a liquid electrolyte containing a functional imidazolium-based ionic liquid as a deposition stabilizer. We show that the Na-S cells with this configuration can cycle stably for over 100 cycles at 0.5C (1C = 1675 mAh/g) with 600 mAh/g reversible capacity and nearly 100 percent Coulombic efficiency. By means of spectroscopic and electrochemical analysis, we find that the high stability and reversibility of the cells stem from at least two sources related both to the cathode and anode. First, the functional ionic liquid spontaneously forms a Na-ion conductive film on the anode, which appears to stabilize deposition of sodium by reducing the electric field near the electrode and prevent electrolyte decomposition and depletion. Second, on the cathode side, microporous carbon materials play a key role that can constrain the electrochemical reaction between sodium ion and sulfur to the solid state without the formation of the intermediate soluble sodium polysulfide species. This combination of electrolyte and carbon substrate are shown to provide sufficiently strong association of sulfur in the cathode and at the same time stabilize the surface of the highly reactive sodium metal anode.

17AIChE
Annual Meeting, Minneapolis, MN

Penn
UNIVERSITY of PENNSYLVANIA

Diverse Colloidal Crystals from DNA-grafted Spheres Via Self-assembly

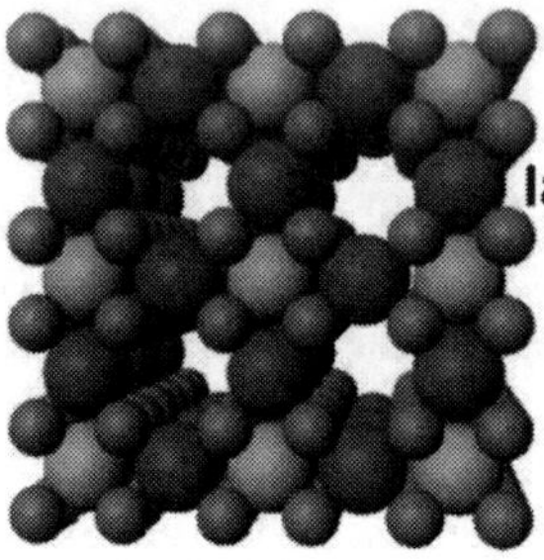

Yifan Wang
Ian Jenkins, Talid Sinno, John Crocker
Dept. of CBE
University of Pennsylvania
Advisor: Prof. John Crocker

AIChE Annual Meeting
November 2nd , 2017

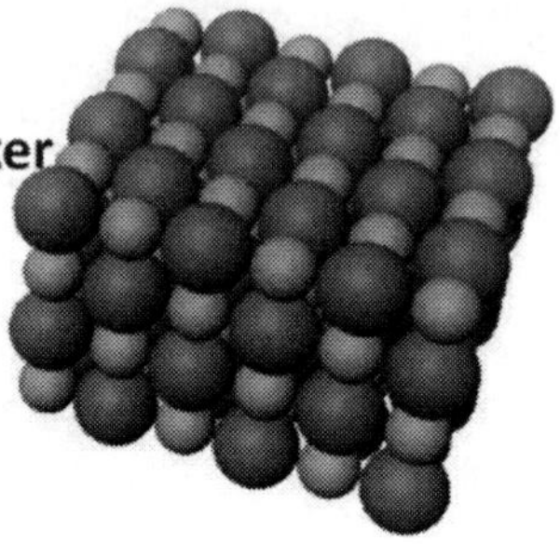

Motivation

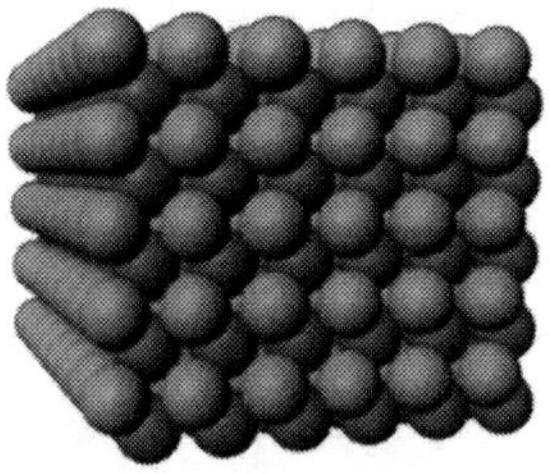
CsCl

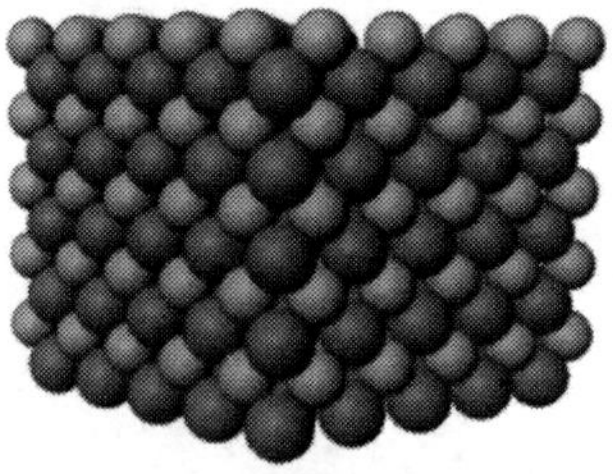
FCC

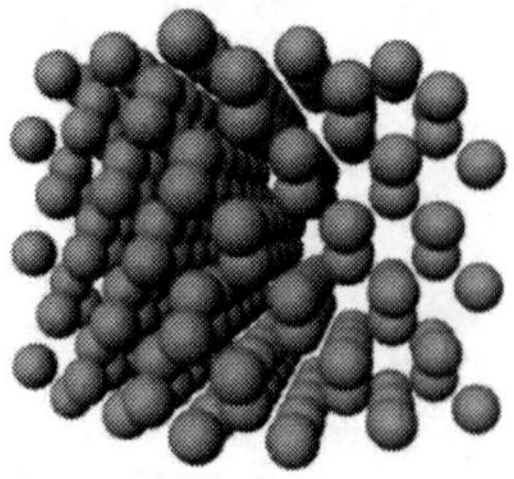
Diamond

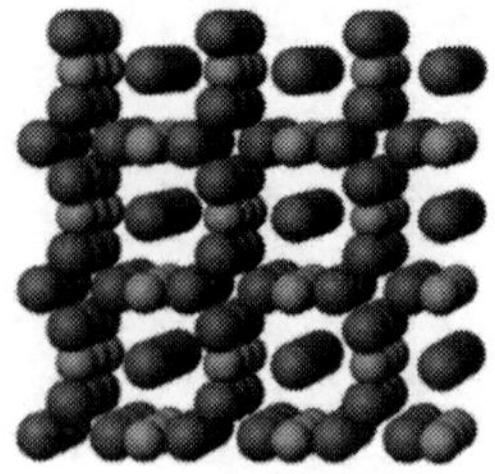
Cr3Si

Can novel crystal structures be made of colloids?	Yes!
How to connect each species particles?	DNA!
Are they stable? What do they transform to?	2 step self-assembly

DNA Binding

Rapid Prototyping!!!

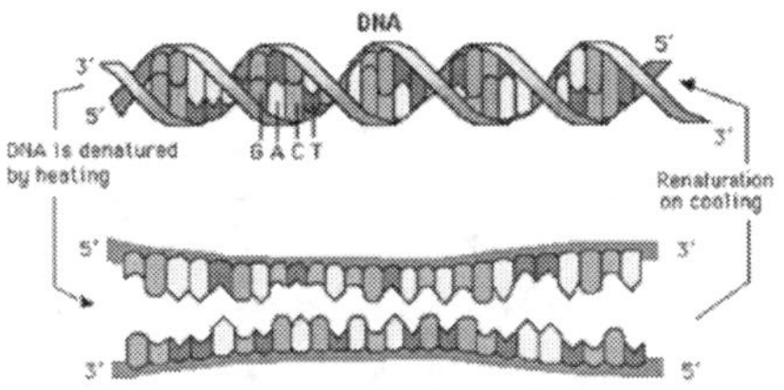

Through DNA linker

TGACGTTCAT
||||||||||
ACTGCAAGTA

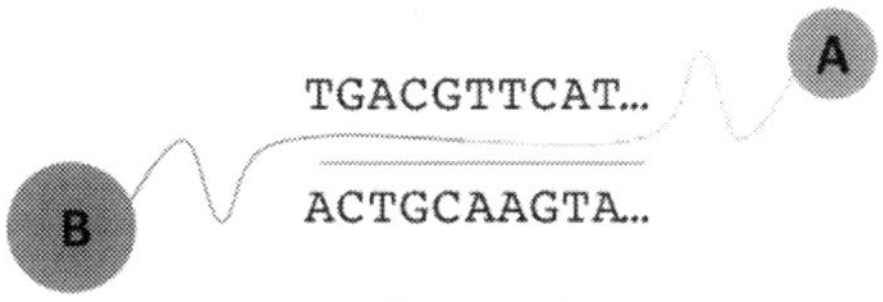

Polystyrene Colloids

Base Pairs Binding

A–T
T–A
C–G
G–C

5′-ATGGCTACGT-3′
||||||||||
3′-TACCGATGCA-5′

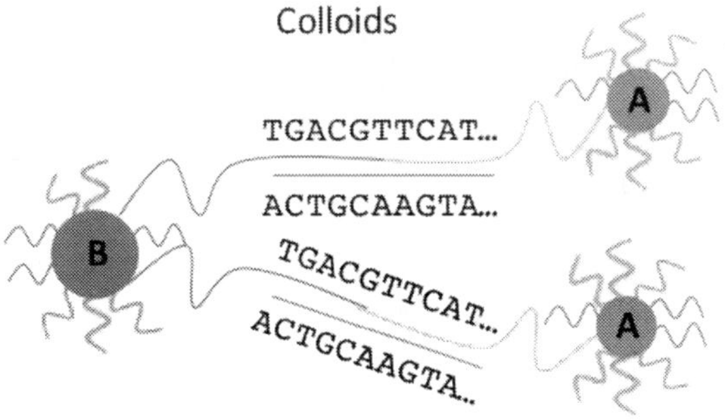

Multiple Particles

3

Chemistry Version I

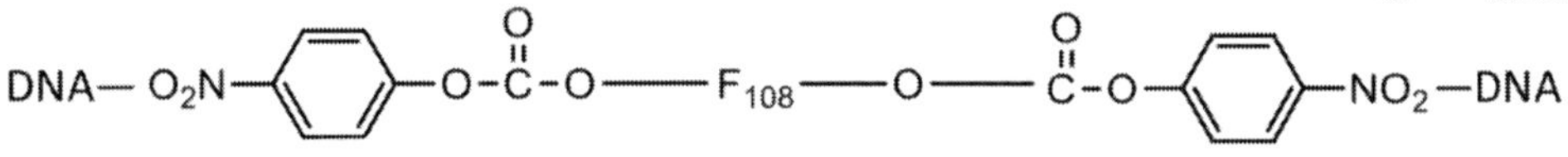

4-NPCF

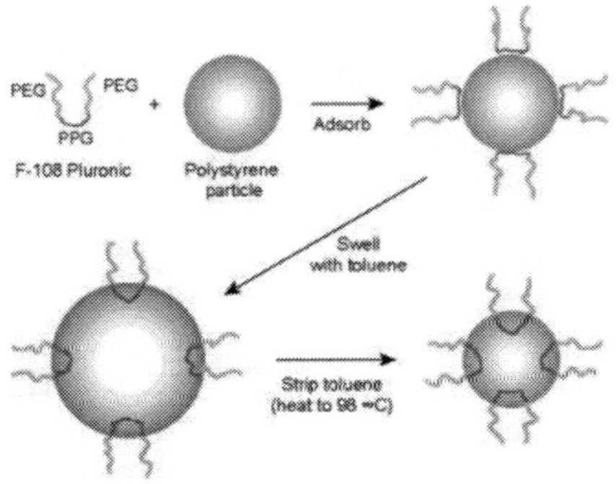

Particle formation is important, particles could **roll and slide** using our method after swell/deswell

Swelling-Based Method for Preparing Stable, Functionalized Polymer Colloids, John Crocker, JACS, 2005

High DNA concentration obtained!!

An Improved chemistry right now!

On the scope, crystals are formed in less than 1 hour

4

2 Step Self-Assembly

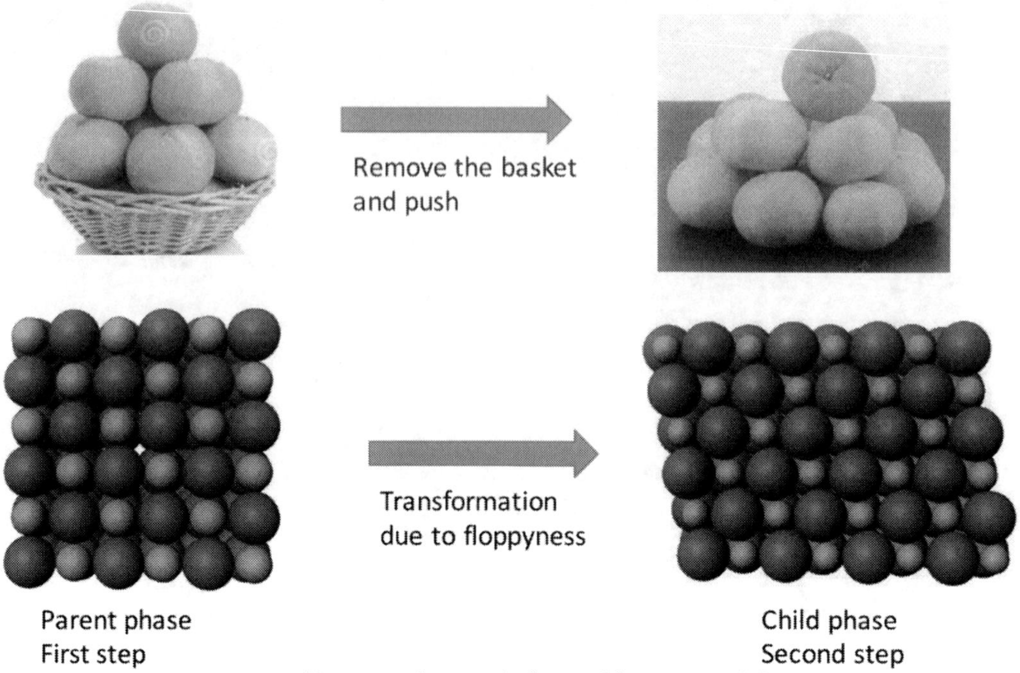

2 step directed particle self-assembly

5

Basic Idea

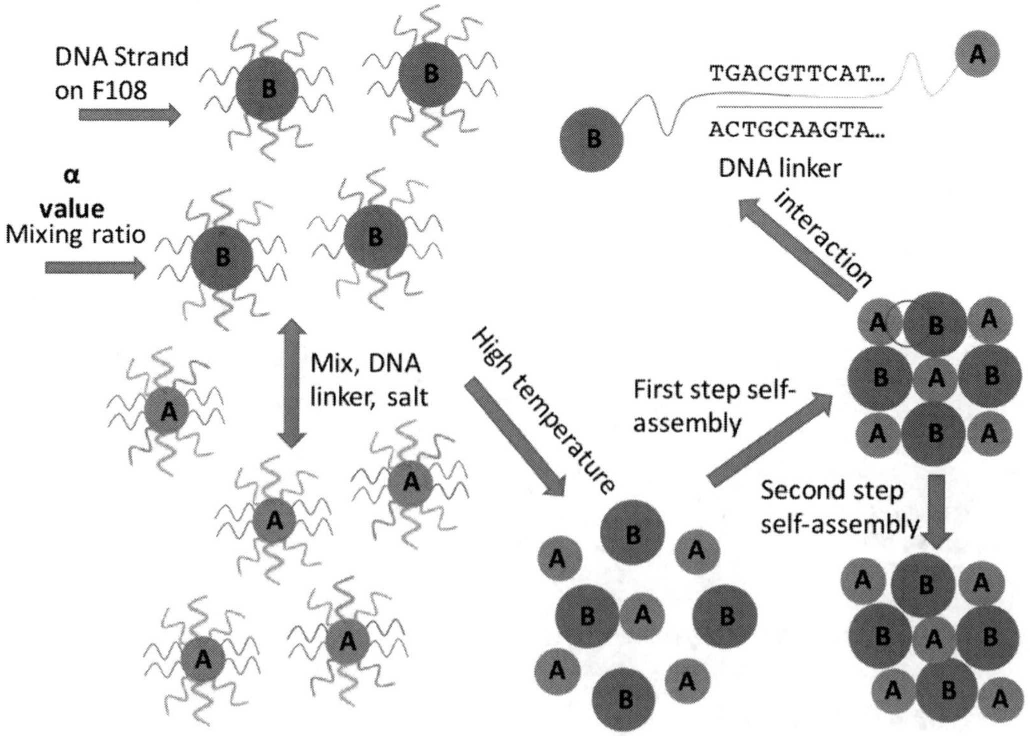

6

Parameters

A	245	378	392	378	378	(nm)
B	445	445	445	392	378	(nm)
Size Ratio:	**0.551**	**0.849**	**0.881**	**0.964**	**1**	

Stoichiometry:

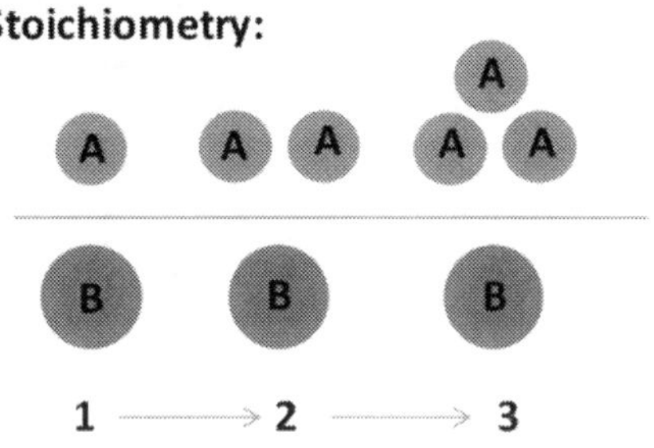

Interaction/Attraction

Only A-B	A	B
A		+
B	+	

A-A A-B	A	B
A	+	+
B	+	

Matrix and mixing ratio

B-B A-B	A	B
A		+
B	+	+

A-A B-B A-B	A	B
A	+	+
B	+	+

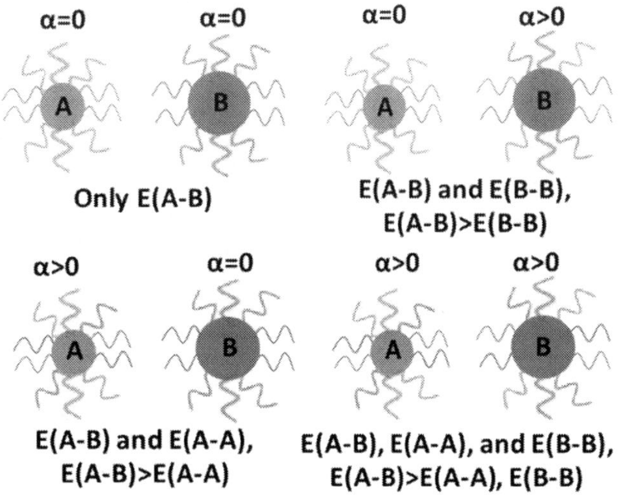

7

Crystallography

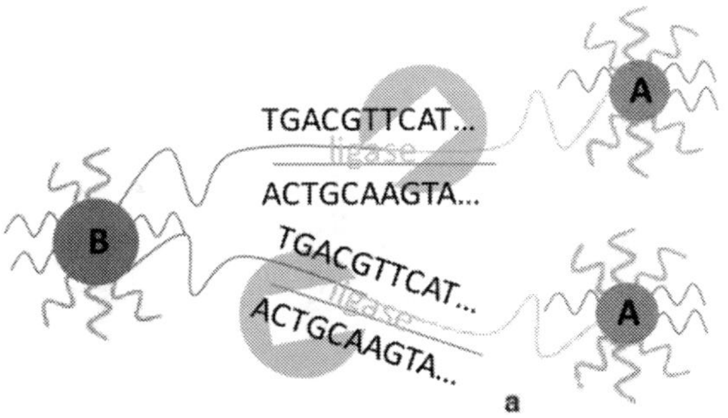

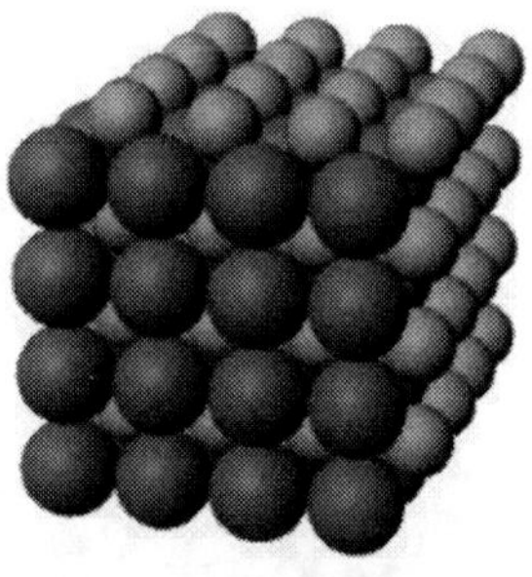

Obtain stable crystal structure

Z-Stack measurement using index match oil, see through the structure, compare different 2-D planes.

Resolvable

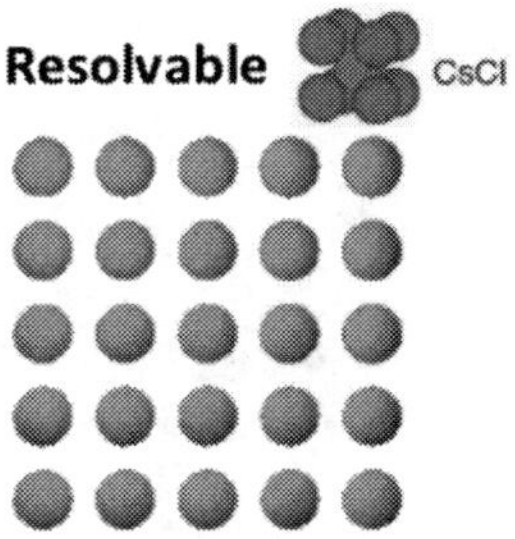

Compare lattice space and angle

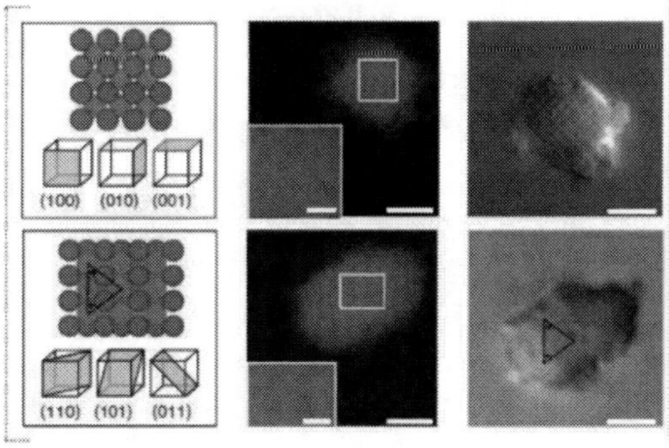

Driving diffusionless transformations in colloidal crystals using DNA handshaking, M. Casey, Sinno, Crocker, et al., Nature Communications, 2012

Experimental Summary

Parent Phase:
a, b, h

Child Phase:
c, d, e, f, g, i, j

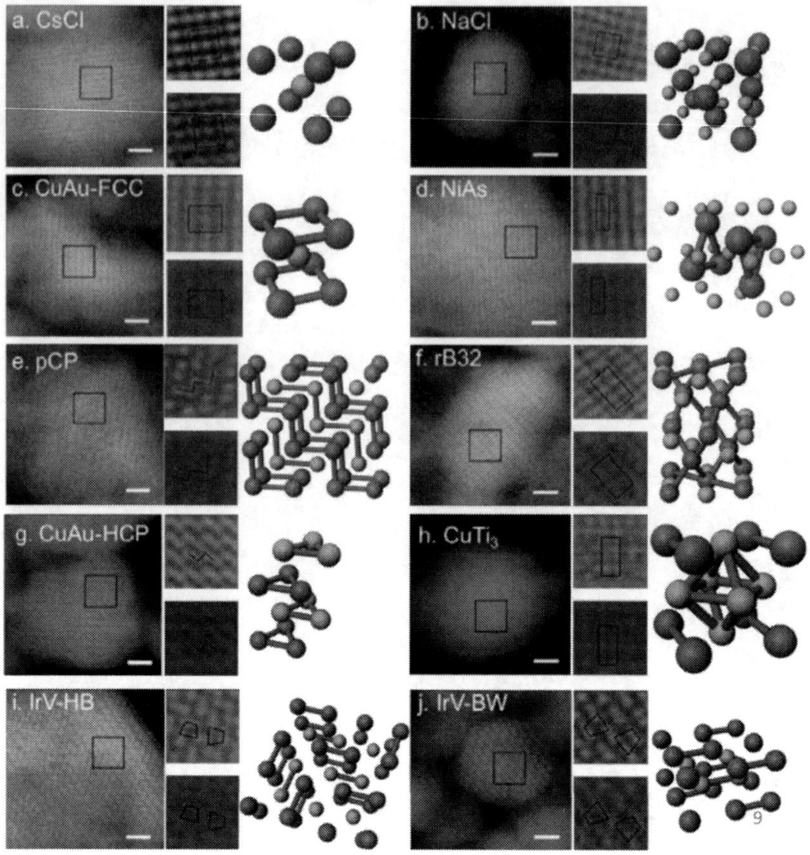

Transformation Paths

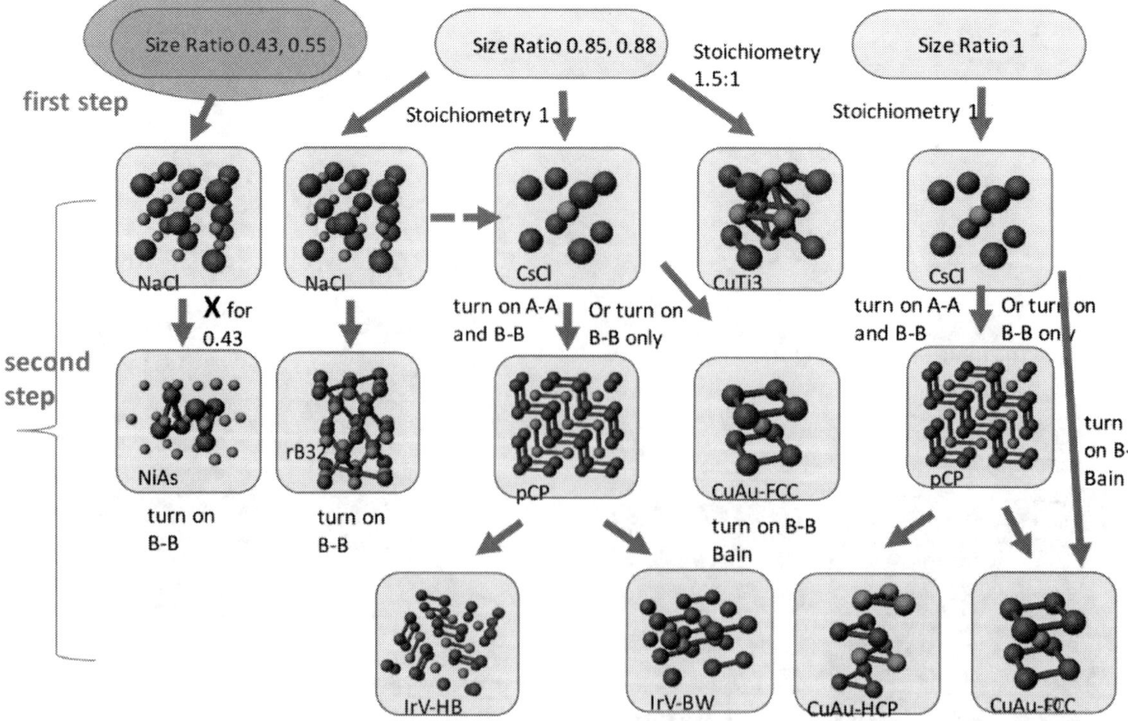

NaCl through first step self-assembly

A 245nm

B 445nm

Size Ratio: 0.551

Stoichiometry: 1

Only A-B	A	B
A		+
B	+	

Perspective view of NaCl

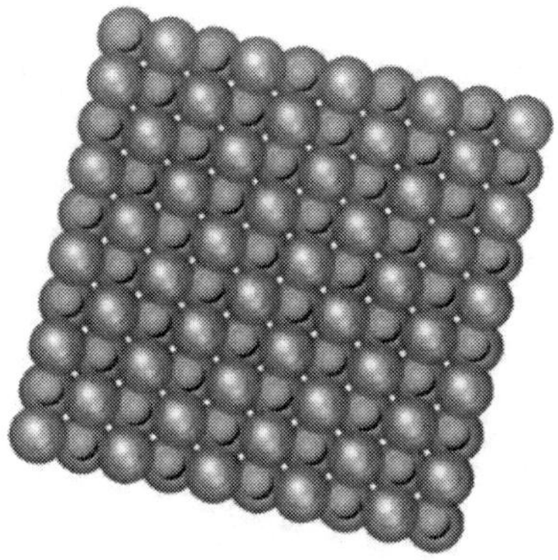

MAYA image of NaCl 100 Plane

Lattice spacing ~380nm

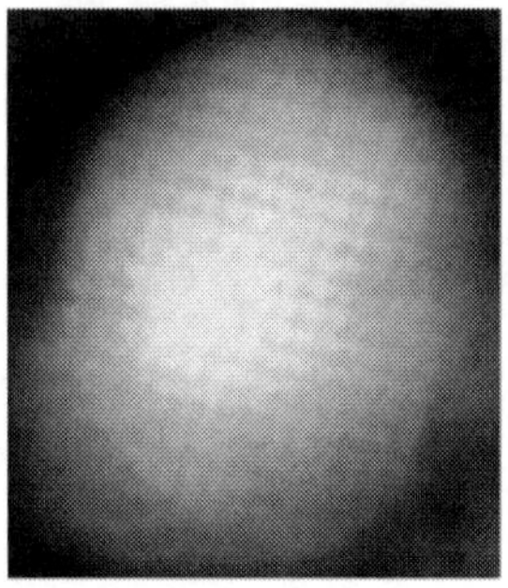

Confocal Image of NaCl 100 Plane

Lattice spacing ~400nm

11

Trigonal NaCl (NiAs) through second step self-assembly

A 245nm

B 445nm

Size Ratio: 0.551

Stoichiometry: 1

Only A-B	A	B
A		+
B	+	+

Perspective view of sheared NaCl

100 80 ~375nm

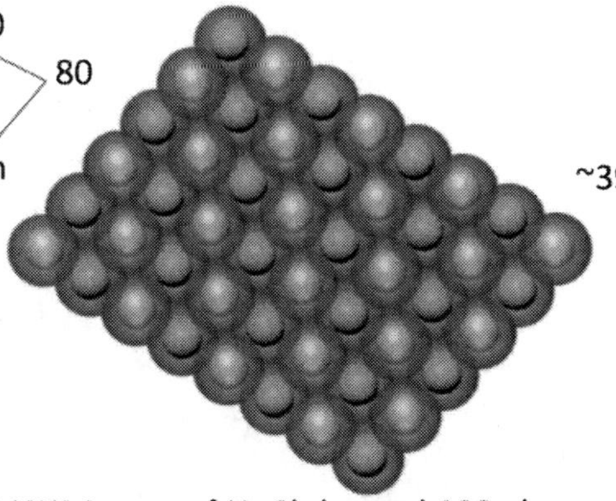

MAYA image of NaCl sheared 100 plane

98 82 ~390nm

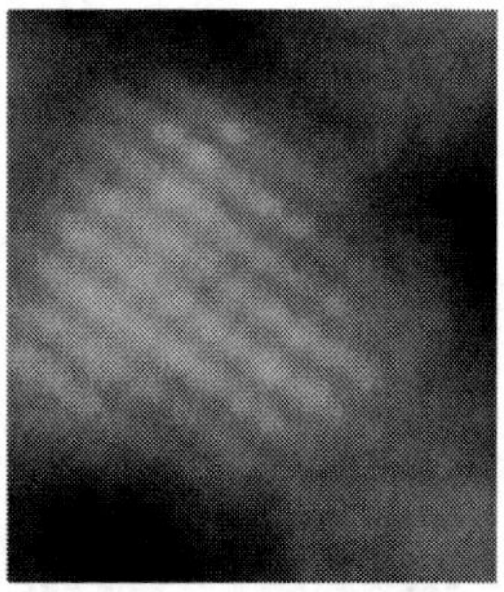

Confocal Image of NaCl sheared 100 Plane

12

Transformation Paths

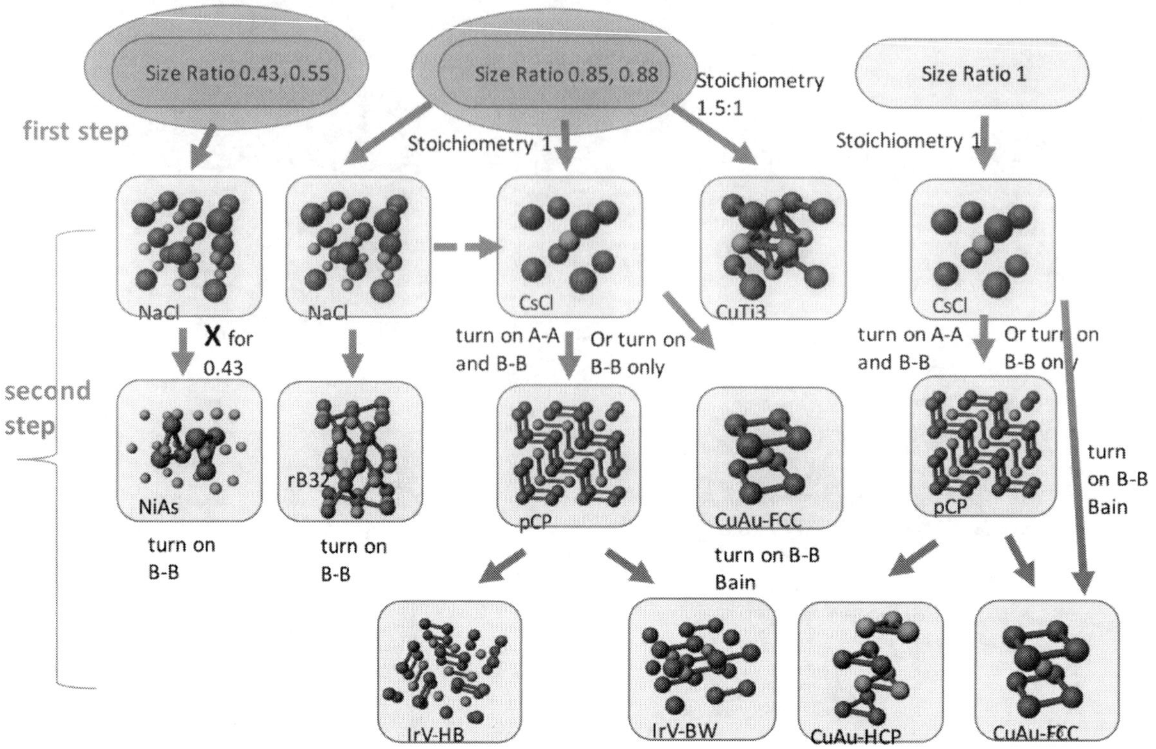

IrV Reconstruction

A 378nm

B 445nm

Size Ratio: 0.849

Stoichiometry: 1

A-A B-B A-B	A	B
A	+	+
B	+	+

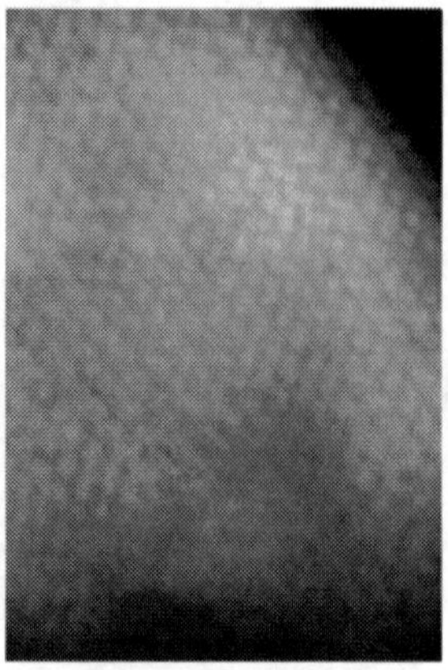

IrV_herringbone

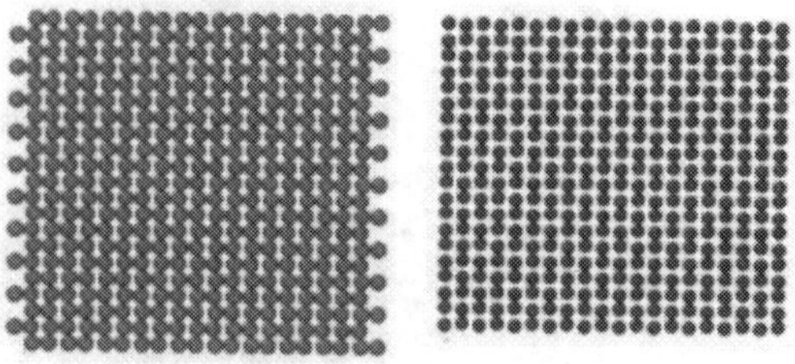

IrV_Basket weave

Stacking Difference

14

Pseudo Close Pack (PCP) Structure as an intermediate

Credit to Ian and Talid

A 392nm

B 445nm

Size Ratio: 0.88

Stoichiometry: 1

A-A B-B A-B	A	B
A		+
B	+	+

Confocal of PCP structure at 0.88 size ratio, green channel

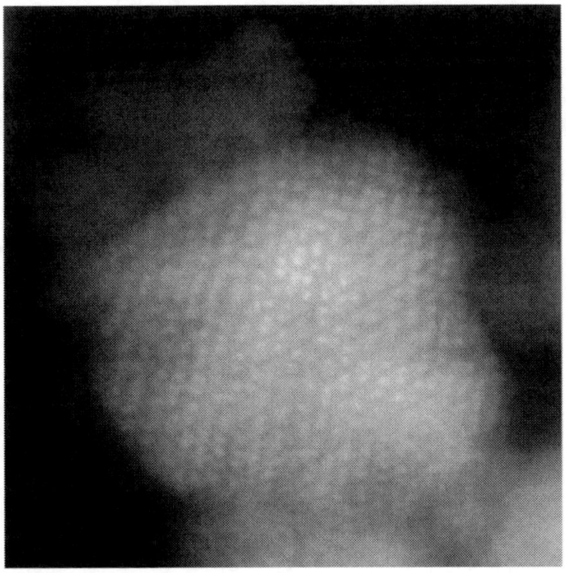

15

Double Diamond (rB32)

A 392nm

B 445nm

Size Ratio: 0.88

Stoichiometry: 1

B-B A-B	A	B
A		+
B	+	+

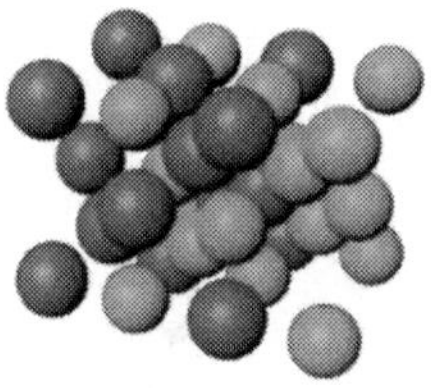

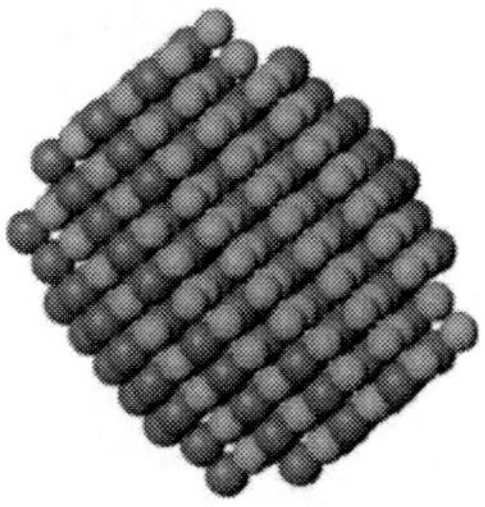

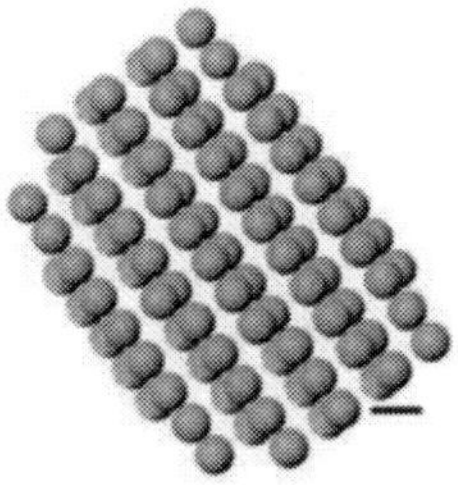

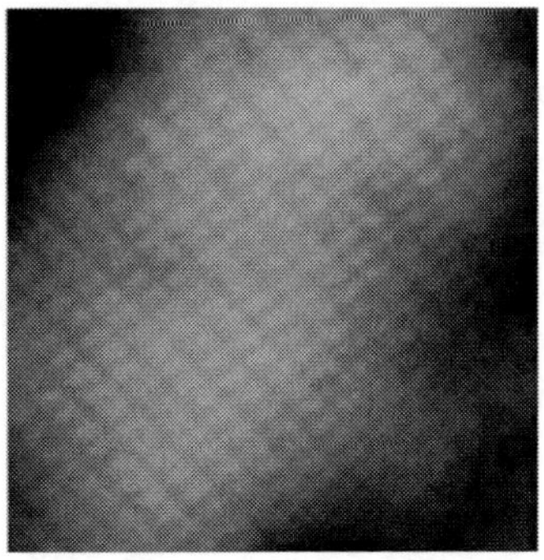

Double Diamond (211) plane

16

Double Diamond Crystal Proof

While the mechanism of the Double Diamond formation is still yet to study(rB32)
We think it came from NaCl

Colloidal crystals with diamond symmetry
at optical lengthscales, Yifan Wang et al., Nature Communication, 2017

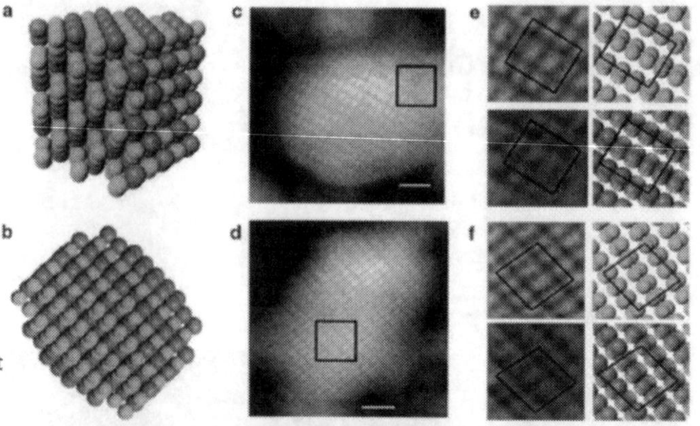

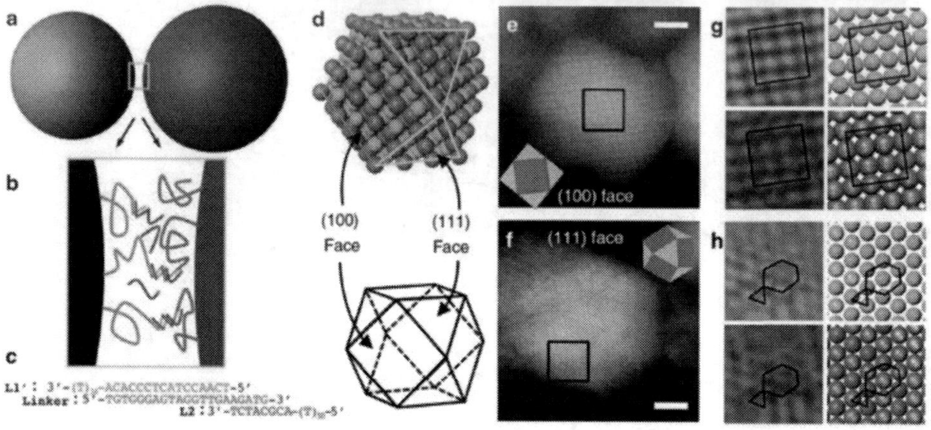

17

Transformation Paths

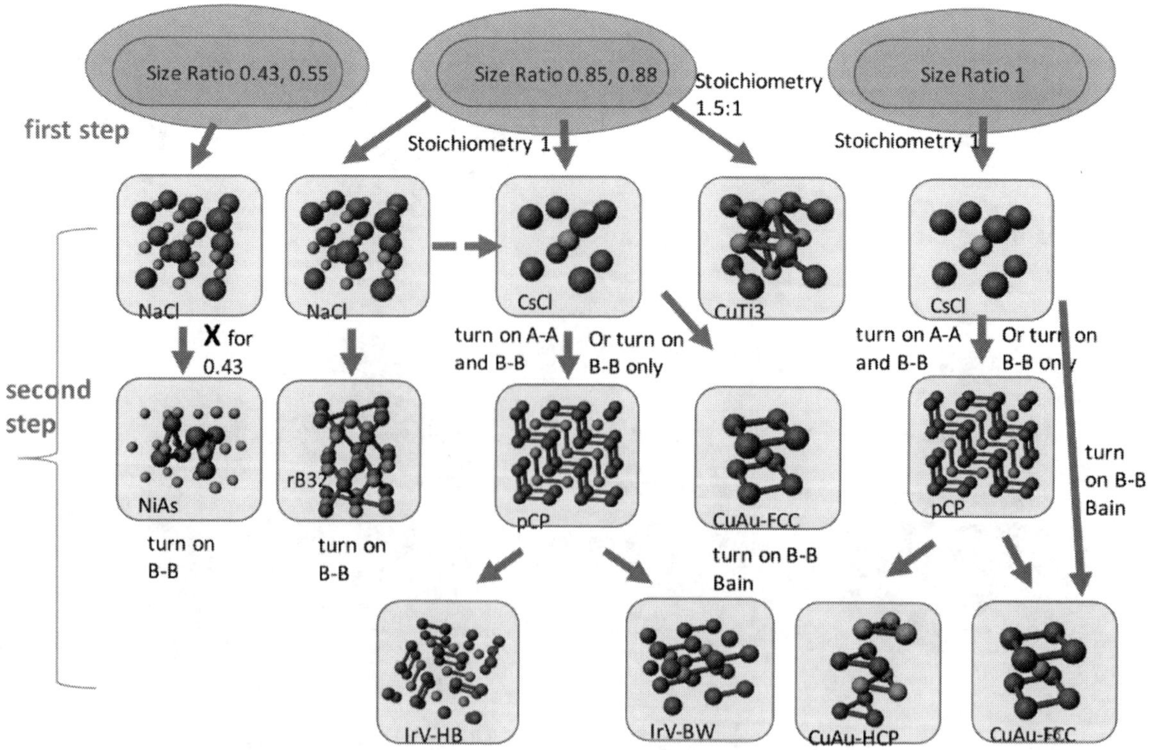

Conclusion and Future Plans

Conclusion:

1) We obtained DNA colloidal crystal from DNA-grafted spheres with high DNA concentration of different size ratios, imaged them and measured crystallography to determine the structure.
2) We applied different interaction matrix and α value studies to different size ratio systems and found the properties of each.
3) We proposed and further demonstrated different parent structures and their transformation mechanisms and paths.

Future Plans:

1) Keep on trying different parameters to build a library of DNA colloidal crystals made from various size ratios and shapes.
2) Apply on different interaction matrix and study on possible transformations.
3) Find ways to make different building blocks through new chemistry and DNA design and make 2nd order crystals through DNA cluster assembly.

19

Acknowledgement

Professor John Crocker

Other Crocker lab Members:
Jennifer Hwang
Chris Porter
Mehdi Molaei
Ehsan Atefi

Doctor James McGinley

Doctor Taesoup Shim

Thank you all for your attention and patience.

Any Questions?

yifanw@seas.upenn.edu

20

500871 Material Design By DNA-Mediated Interactions between Colloids

Abstract:

The ability to rationally design and synthesize complex nano-, micro-, and hierarchically structured materials with tunable functionality represents a holy grail that holds exciting promise for revolutionizing materials applications spanning catalysis, molecular sensing, drug delivery, molecular and charge manipulation, and beyond. Such applications hinge upon precise control over morphology, topology, material connectivity, and function, and demand accommodation of a wide range of materials compositions spanning organic, inorganic, and hybrid structures. One of the most promising approach to achieve such unprecedented control over colloidal self-assembly is to use DNA-mediated interactions between particles functionalized with partially complementary DNA sequences. In general, DNA functionalized particles have several adjustable parameters: the hybridization interaction strength, temperature, solution stoichiometry and DNA grafting density. Using molecular dynamics simulations of a coarse-grained model developed specifically to study such systems, we show that variations in these key factors provide selectivity between different lattice structures such as body-centered cubic (BCC) and close-packed (FCC and HCP) structures. In addition, we are able to program structures that interconvert via diffusionless transformations via external stimuli such as temperature changes.

485046 High Throughput Acoustically Driven Self-Assembly of Microfluidic Colloidal Crystals

Abstract:

Manufacturing defect-free ordered materials is immensely important for producing phononic and photonic crystals for energy applications or to serve as waveguides for optical computing. External fields (e.g., electric, magnetic, flow) that drive self-assembly are a form of bottom-up manufacturing, where micro- or nanoscale building blocks assemble spontaneously to form complex shapes. Drawbacks to current approaches include low throughput (~10 particles per second or less), the formation of permanent defects (e.g., grain boundaries, point defects, line defects) during assembly and the high cost associated with fabricating devices to produce external fields. In this presentation, we demonstrate a high throughput (~300 particles per second) self-assembly process based on acoustically-driven assembly of microparticles using of-the-shelf components without the need to access a clean room.

Fifteen micron polystyrene particles dispersed in water are continuously introduced to glass-capillary microfluidic flow cell with a syringe pump. A piezoelectric element attached to the size of the flow cell is driven to generate an acoustic standing wave that drives particles to assemble at acoustic nodes. An experimental state diagram shows the effect that fluid flow rate and applied acoustic pressure have the assembled microstructure. Highly ordered, continuous crystals are observed to assemble in less than a minute with a throughput rate of several hundred particles per second were observed under various pressure and flow rates. Particle tracking software is used to identify the location of microparticles assembled in ordered structures. This data is used to as part of a distribution analysis to understand the compressive influence that the acoustic pressure has on the observed structures. The flow cell throughput is quantified using a micro-particle image velocimetry analysis of optical video microscopy data. The degree of order within the assembled structures is quantified by comparing the number of nearest neighbors to theoretical expectations for an ideal ordered structure.

Future work is focused on mitigating defect formation in our ordered crystal structures. Acoustic assembly is dynamically controllable, allowing us to understand the kinetic effects that fluid flow and acoustic pressure have on dislocation dynamics. We aim to develop real-time annealing techniques to remove defects during crystal formation.

499771 The Role of Interaction Heterogeneity in Colloidal Crystallization

Abstract:

Heterogeneity in a population of particles is generally considered to be a detrimental factor for the self-assembly of ordered structures. While this is well-established for certain types of heterogeneity such as size polydispersity, here we show, using a combination of equilibrium and non-equilibrium simulations, that heterogeneity in the pairwise interaction strength among a collection of particles may in fact be useful for nucleation of crystalline phases. In particular, we consider interaction heterogeneities that may arise, for example, from particle-to-particle variations in the density of DNA oligomer brushes grafted on the surface of micron-scale spherical particles. In this setting, it is now well-known that the multivalent nature of DNA-mediated particle binding introduces constraints for assembly by increasing the temperature dependence of the resultant interparticle potential and consequently narrowing the operating window for high quality crystallization. For a given areal density of DNA strands, this effect becomes particularly pronounced for larger particles and represents a key challenge for DNA-driven assembly of micron-scale particles.

The beneficial impact of population-wide interaction heterogeneity is shown to arise from a synergistic combination of two effects. First, we employ equilibrium umbrella sampling simulations to show that heterogeneity lowers the free energy barrier associated with the nucleation of crystals by the formation of strongly-bound small clusters, even if the overall average binding energy between pairs of particles is held constant. This finding is confirmed by direct, non-equilibrium crystal growth simulations, which demonstrate that crystallization proceeds over a wider range of average interaction energy. On the other hand, these simulations also show that variations in the interaction strength between particles inhibit gelation and polycrystallinity by keeping the number of stable nuclei low, allowing individual nuclei to grow unhindered. In combination, these effects lead to a â€˜spreading-outâ€™ of the nucleation process and a concomitant widening of the crystallization window. These results are used to investigate potential avenues for intentionally controlling the nucleation rate in colloidal crystallization experiments.

500901 Self-Assembly of Open Structures Using Depletion

Abstract:

Depletion is a universal mechanism to induce order in colloidal self-assembly. In the absence of explicitly attractive forces or patchy interactions, anisotropic colloids usually crystallize into close-packed structures. Open structures are of high interest for applications for example as photonic crystals. Here we investigate the depletion-driven self-assembly of faceted colloids using Monte Carlo simulations. We find that such colloids can form open structures purely from hard shape and depletant entropy, as well as complex, close-packed unit cells. Our highly efficient simulation algorithm can readily be used to study the self-assembly of (almost) arbitrary convex and concave shapes with depletion.

499296 Controlled Self-Assembly of Colloidal Discotic Liquid Crystals

Abstract:

Colloidal anisotropic platelets are model systems for discotic liquid crystal investigation. We have systematically studied the ZrP monolayers due to the capability to control the size and surface chemistry. I will discuss our recent research on liquid crystal phase formation in colloidal discotic suspensions, particularly the controlled self-assembly under external fields, such as gravity, electric and magnetic field, and temperature gradient.

487724 Stratification Dynamics in Drying Colloidal Mixtures

Abstract:

Drying colloidal mixtures can undergo a spontaneous, nonequilibrium vertical separation (stratification) into layers enriched in particles of a specific size, an effect which can be exploited to create functional coatings. It was recently observed in experiments and simulations that for large particle size ratios and high drying rates, smaller colloids migrate to the top of the film, while big colloids are pushed to the bottom, creating a peculiar and counterintuitive â €œinvertedâ€ stratification. To investigate this behavior, we performed implicit-solvent molecular dynamics simulations of binary colloidal mixtures for a wide range of particle size ratios and drying rates, and quantified the stratification dynamics in detail. We found that the stratified layer of small colloids grew faster and to larger thicknesses for larger size ratios. Interestingly, inverted stratification was observed even at moderate drying rates, but the thickness of the stratified layer decreased. We propose a model based on dynamical density functional theory which captures the salient features of our simulations and can help to explain the observed phenomena.

493106 Theoretical and Experimental Investigation of Phase Separation in Noble Metal Nanoparticle Monolayers

Abstract:

Controlling the interfacial area of nanoparticles is a crucial component to controlling their properties due to their high surface area to volume ratio. One of the most common ways to control this interfacial area is with the use of a monolayer. Therefore, careful design of nanoparticle monolayers allows for a wide range of control of nanoparticle properties. Our work looks at the design rules that influence the morphology of a monolayer. To do this we use a combination of experimental and computational techniques to explore the effect of physical and chemical changes in the nanoparticle monolayer systems. Experimentally we use Transmission Electron Microscopy (TEM) to assess nanoparticle shape and size and Matrix-Assisted Laser Desorption and Ionization (MALDI) to assess the degree of order present in nanoparticle monolayers. In addition, we model the systems using both Self-Consistent Field Theory (SCFT) and atomistic simulations to model these nanoparticle monolayers. We are then able to calculate predicted MALDI spectrum from these simulations which allows us to directly compare theory and experiment. This direct comparison allows us to have greater confidence in the insights provided by our computational simulation.

499798 Bending-Induced Buckling Instabilities in Self-Assembled Elastomeric Composite Films

Abstract:

Buckling instabilities induced by various mechanical stresses are universal phenomena observed at all length scales in a wide range of materials (e.g., polymers, metals, and ceramics) in both natural and manmade systems. Bilayer buckling systems consisting of rigid skin films attached to thick compliant polymer foundations have been extensively investigated. Unfortunately, it is challenging to create a universal bilayer buckling platform that can cover a broad range of length scales, while simultaneously can achieve high structural stability and system durability. Additionally, bending-induced buckling instabilities of bilayer composite systems possessing discrete rigid phases, which are of great technological and scientific importance in developing next-generation wearable and flexible devices and new non-linear instability models, are far from being well studied. Here we report a new and universal bilayer buckling system by applying scientific principles drawn from two disparate fields that do not typically intersect â€“ the mature polymer buckling and colloidal self-assembly techniques. Self-assembled colloidal crystals embedded in elastomers provide a model buckling system exhibiting many unique properties, such as high effective elastic moduli of the composite skin layers, well-controlled skin-layer thicknesses spanning at least two orders of magnitude (from 100 nm to 10 Âµm), unconventional buckling microstructures comprising discrete rigid phases dispersed in a continuous elastic matrix, and high mechanical durability. The fundamental mechanisms of bending-induced buckling instabilities exhibited by the self-assembled elastomeric composite films have been investigated by mechanical finite element analysis. The numerical simulations match well with the experimental results.

501561 Molecular Modeling of Microstructure, Solubilization and Micro-Emulsion of Block Copolymer Micelles By iSAFT

Abstract:

Block copolymer micelles have been the focus of much interest for both academia and industry over decades for extensive applications in enhanced oil recovery, cosmetic products, and drug delivery. The immiscibility of constituents causes micro-phase separation for high density block copolymer melts, and micellar aggregates when it is dissolved in a selective solvent. A few chemical potential models have been developed to predict sizes, aggregation number, solubilization ability, critical micelle concentration(CMC) etc. Most of the theories are phenomenological equation of state(EoS) types of theories and thus cannot fully capture the physics at a molecular level. We study the microstructures using inhomogeneous statistical associating fluid theory(iSAFT) a classical density functional theory iSAFT takes into account a range molecular interactions, including hard-sphere repulsion, chain connectivity, van der Waals attraction, and association.

iSAFT is demonstrated to predict different influences of hydrophobic tail and hydrophilic head of block copolymer surfactants on the size and CMCs of micelles. iSAFT showed that radii and CMCs of micelles are strongly dependent on the length of hydrophobic tail but not for the length of hydrophilic head. This agrees with experimental finding of Pluronic micelles. iSAFT also predicts the radii of triblock surfactant micelles is smaller than of diblock copolymer surfactants with the same head and tail groups. Similar agreement between theory and experiment results is also found for the CMCs of micelles of triblock and diblock copolymer surfactants.

Structure of solubilized micelles is similar to a surfactant micro-emulsion. By predicting the distribution of solute, solvent and surfactants within micelles by iSAFT, we are able to show the enhanced solubility of hydrocarbon in water with surfactants. iSAFT predicts the different solubility of normal alkanes and branched alkanes with different carbon numbers when using different surfactants.

491598 Spreading of Nanoscale Droplets of Ionic Liquids on the Mica Surface

Abstract:

Extensive studies have been conducted on the molecular arrangement of ionic liquids (ILs) next to a solid surface. In these previous studies, mica has been utilized as a model solid surface due to its atomic-level smoothness and availability. The solid-like layering has been reported for many ILs on the mica surface. What remains unknown is the mobility of ILs confined to the solid mica surface: Are these solid-like layers mobile? There have been much less study on this topic and the previous results are inconsistent. In this paper, we report our recent AFM study on the spreading of the nanodroplets of an IL, 1-butyl-3-methylimidazolium tris(pentafluoroethyl) trifluorophosphate (BMIM-FAP), on the mica surface. The results indicate that the solid-like layers of IL are not real solid though their mobility is significantly lower than in bulk IL. As a result, the morphology of the nano-confined ILs on the solid surface could change significantly with time. The governing key parameters and the underlying mechanisms are discussed.

500947 Long Range Electrostatic Screening in Ionic Liquids

Abstract:

As liquids composed solely of ions, the scientific consensus has been that ionic liquids have exceedingly high ionic strengths and very short Debye screening lengths. However, recent experiments from several independent laboratories reveal remarkable long range forces that appear to be electrostatic in origin. We discuss temperature-dependent surface force measurements that yield insight into the mechanism of these long range forces and propose a general model of long-range electrostatic screening in ionic liquids, where thermal fluctuations, either free dissociated ions or correlated domains, take on the role of ions in traditional dilute electrolyte solutions. More broadly, our work provides an alternative way of envisioning highly concentrated electrolytes, with implications for diverse areas of inquiry, ranging from designing electrochemical devices to rationalizing electrostatic interactions in biological systems.

498793 Mesopore Confinement Effects on Ionic Liquid Hydration

Abstract:

Room temperature ionic liquids (ILs) have been widely investigated as solvents due to their low volatility, good thermal stability, and tunable solvent properties, they have been employed as media for catalytic dehydration of glucose into 5-hydroxymethylfurfural (HMF). [1][2][3] However, their high price, unknown toxicity and difficulty of solute recovery inhibit their large-scale application. Supported ionic liquid systems (SILPs) are believed to be the key to overcome the disadvantages of ILs and allow their properties to be utilized in heterogeneous catalysis. [4][5] The long-term goal of the present study is to understand the interfacial science behind the functionalization of mesoporous silica films with IL and metal-based catalyst systems for heterogeneous catalysis, with dehydration of glucose to HMF as a representative reaction.

Here, *in situ* transmission FT-IR study is developed as a technique to characterize the interactions of water with IL (1-butyl-3-methylimidazolium chloride (BMIM-Cl)) supported by mesoporous silica films. The approach is to use thin silicon wafers as supports to allow direct transmission monitoring of ionic liquid response to solvent vapors. The development starts with investigation of structural changes of IL thin films on unmodified silicon wafers with respect to relative humidity (RH). This establishes the amount of water absorbed and the mode of interaction with the ionic liquid, as well as the reversibility of hydration upon changes in relative humidity. [6] Next, the effects of confining IL in silica films with mesopores of varying diameters is performed to isolate the effects of confinement on hydration of BMIM-Cl. The base case for this study uses silica films with pore diameter of 8-9 nm prepared using pores templated with Pluronic surfactant P123 that are vertically oriented by chemically modified using a crosslinked layer of P123. [7] Hydration of BMIM-Cl will be compared in films with smaller pores templated with CTAB, and larger pores formed by swelling the P123 templates with triisopropylbenzene. [7]

References

[1] Tariq, Mohammad, et al. "Ionic Liquids in Bulk and at an Interface." *Ionic Liquid-Based Surfactant Science: Formulation, Characterization, and Applications* (2015)

[2] Zhao, H., J. E. Holladay, H. Brown, and Z. C. Zhang. "Metal Chlorides in Ionic Liquid Solvents Convert Sugars to 5-Hydroxymethylfurfural."*Science* 316.5831 (2007)

[3] Stefanopoulos, Konstantinos L., George E. Romanos, Olga C. Vangeli, Konstantina Mergia, Nick K. Kanellopoulos, Alexandros Koutsioubas, and Didier Lairez. "Investigation of Confined Ionic Liquid in Nanostructured Materials by a Combination of SANS, Contrast-Matching SANS, and Nitrogen Adsorption." *Langmuir* 27.13 (2011)

[4] Stefanopoulos, Konstantinos L., et al. "Investigation of confined ionic liquid in nanostructured materials by a combination of SANS, contrast-matching SANS, and nitrogen adsorption." *Langmuir* 27.13 (2011)

[5] Gupta, Krishna M., and Jianwen Jiang. "Cellulose dissolution and regeneration in ionic liquids: a computational perspective."Â *Chemical Engineering Science*Â 121 (2015)

[6] Maiti, Amitesh, Arvind Kumar, and Robin D. Rogers. "Water-clustering in hygroscopic ionic liquidsâ€"an implicit solvent analysis."Â *Physical Chemistry Chemical Physics*Â 14.15 (2012)

[7] Das, Saikat. "Fundamental Studies of Surfactant Templated Metal Oxide Materials Synthesis and Transformation for Adsorption and Energy Applications." Ph.D. Thesis, University of Kentucky, 2015.

498724 Water Sorption in Ionic Liquids Characterized Using a Dynamic Vapor Sorption Analyzer (IGASorp) and High Pressure X-Ray Photoelectron Spectroscopy (APXPS)

Abstract:

Ionic liquids (ILs) are commonly defined as salts composed of distinct cations and anions with melting points below 100 °C, and many are liquid at ambient temperature. IL research has grown rapidly over the past two decades due to the realization that these materials have many unique properties, such as negligible vapor pressure and outstanding solvation potential. Ionic liquids have been further emphasized by the fact that their physical and chemical properties can be finely tuned by varying both the cation and anion.

The presence of water in ionic liquids can significantly affect physical properties, as well as chemical properties such as polarity and reactivity. Water is considered the most common impurity in ILs as most ionic liquids are hygroscopic to a certain extent, and preventing their exposure to moisture is quite difficult. Due to the array of conditions used in the numerous IL applications, it is of interest to observe the water uptake of ionic liquids at different temperatures and relative humidities. This study compares bulk and interfacial water uptake in imidazolium-based ILs over a range of temperatures and humidity levels at ambient pressure. Important experimental details on using a gravimetric microbalance (Hiden IGAsorp) for bulk measurements are discussed, and ambient pressure X-ray photoelectron spectroscopy (APXPS) results for water-IL interface are reported. Additionally, this study determines and compares the absorption enthalpies of water in ionic liquids using bulk and interfacial water sorption measurements.

488895 Property Modeling of Ionic Liquids for Gas Separation Processes

Abstract:

With increasing calls for "green" and sustainable technologies, ionic liquids (ILs) have been receiving much attention as a potential solvent for gas separation tasks due to their unique properties. Therefore, for a sustainable design point of view, IL-gas absorption has been identified as a replacement for the energy intensive distillation based gas separation.

One of the promising advantages of ILs is their negligible vapor pressure, which eliminates the risk of contamination of the gas stream and the potential loss of ILs. Many researchers have evaluated the possibility of ILs as a gas absorption solvent by measuring the solubility of various gases in different ILs. The most common studies concentrated on the CO_2 capture capacities of different ILs, which points to the potential of an IL-based gas decarbonization technology (Liu et al, 2016[1]). Encouraged by good absorption of CO_2 with IL, new studies have focused on the solubility of other gases in ILs. As the potential number of ILs that can be generated by combining cations and anions is enormous, experimental studies of gas solubility in ILs only provide a fraction of the data-knowledge needed to search over a wide spectrum of ILs to find the most appropriate for gas separation tasks. Not only the solubility of the selected gas in the solvent (IL or other chemicals) is important but for a gas separation task, the selectivity of the IL solubility for the gases to be separated is also very important, together with other pure component properties of the IL, such as viscosity, stability, density, etc. Therefore, before a selection-screening method for ILs for use in absorption based gas separation tasks can be developed, the used properties must be available. The objective is to identify those ILs that would match, a priori, the desired properties of the gas absorption solvent and thereby lead to an efficient gas separation process.

In this work, a property modeling tool has been developed for use in screening method for selection of the most appropriate IL for specific absorption based gas separation tasks. The property models are available in two scales. At the lower scale, the solubility data is generated in the form of Henry's constants of gases in various ILs through the COSMO-RS model (Klamt et al. 2000[2]) and/or molecular simulations (Huang et al.2013[3]). After evaluating their accuracy, they are included in a database of solubility of gases in ILs together with IL properties. These data are then used in a mezo-scale predictive engineering models (like the group contribution methods) for the same properties. Note that the role of the smaller-scale property modelling is to enlarge the potential search space for the ILs. Solubility, Henry's constant, selectivity, viscosity, density, surface tension, IL-stability, vapor pressure, for a given set of desired (target) properties, the screening method is then able to find the ILs that match the target properties using a search algorithm similar to that of Harper et al. 2000[4].

The presentation will highlight the contents of the database, the validation of the generated data and the developed property models together with applications of the screening-selection method for IL as solvents for various absorption based gas separation problems. More specifically, separations of gases like H_2, CO_2, CH_4, C_2H_4, C_2H_6, with IL based absorption will be illustrated.

References

1. Liu, X., Huang, Y., Zhao, Y., Gani, R., Zhang, X., & Zhang, S. (2016). Ionic Liquid Design and Process Simulation for Decarbonization of Shale Gas. Industrial & Engineering Chemistry Research, 55(20), 5931-5944.
2. Klamt, A., & Eckert, F. (2000). COSMO-RS: a novel and efficient method for the a priori prediction of thermophysical data of liquids. Fluid Phase Equilibria, 172(1), 43-72.
3. Huang, Y., Dong, H., Zhang, X., Li, C., & Zhang, S. (2013). A new fragment contribution-corresponding states method for physicochemical properties prediction of ionic liquids. *AIChE Journal*, *59*(4), 1348-1359.
4. Harper, P. M., & Gani, R. (2000). A multi-step and multi-level approach for computer aided molecular design. Computers & Chemical Engineering, 24(2), 677-683.

503947 An Investigation of Cellulose Solubility in Ionic Liquids with Added Cosolvents

Abstract:

The development of processes to convert cellulose, the most abundant organic substance on earth, into useful products that are currently made from nonrenewable resources is of the utmost importance for a green and sustainable future. Cellulose is an energy-rich resource but itâ€™s relatively difficult to process since few organic solvents are capable of breaking apart its dense hydrogen bond network. In the past few years, ionic liquids (ILs) have been shown to be efficient at solubilizing cellulose and shown a lot of potential due to the ability to modify the properties of the solvent by varying the constituent cations and anions. Even so, the adoption of ILs in processes has been slowed by their prohibitive costs and mass transport issues. Recently, however, it was shown that the addition of polar aprotic cosolvents can substantially alleviate these mass transport issues while reducing the amount of the costly IL required for dissolution. Furthermore, it was shown that in some cases these cosolvents can significantly improve cellulose solubility and even the water tolerance of the solvents. The precise mechanisms by which this solubility enhancement occurs and however remains unclear. In this study, molecular dynamics simulations are used to study the role that cosolvents play in the enhancement of cellulose solubility. Separate simulations employing either crystalline or dissolved strands of cellulose with ionic liquid, cosolvent and, in some cases, water were performed to understand the underlying molecular driving forces of dissolution and how varying the proportions of these substituents can influence the thermodynamics of the process. Using these two state simulations we quantify the changes in hydrogen bonding, interfacial energy and the free energy of dissolution using the two phase thermodynamic method. We find that the calculated changes agree quite well with the experimentally determined phase diagram and provide a detailed physical picture for how the addition of moderate amounts of cosolvent enhance cellulose solubility in neat and aqueous ionic liquids.

496126 Effects of Ionic Liquids and Liposomes on Enzymatic Cellulose Hydrolysis Process

Abstract:

The heterogeneous system of cellulose reaction suffers from irreversible adsorption of the enzyme molecules to solid substrate, which is recognized as one of the mechanisms responsible for the enzyme deactivation and decreasing accessible surface area of the substrate. Ionic liquids (ILs) decrease the crystallinity of cellulose, whereas IL treatment usually decrease the enzymatic activity of cellulose. Here, the effect of liposome membranes on cellulase from Trichoderma viride was studied in the presence or absence of ILs. The liposomal cellulase and free enzyme were examined in their hydrolytic activities to insoluble cellulose powder CC31 in the acetate buffer solution (pH 4.8) of 15 w/w% [Bmim][Cl] (1-butyl-3-methylimidazolium chloride). The mean diameter and size distribution of cellulase-containing liposome were practically unchanged under the above condition. In the saccharification reaction of insoluble cellulose, the localization of cellulase and liposome was found to be important, because the cellulase dispersed in liposome suspensions resulted higher activity than that conjugated into liposomes. The activities of liposomal and free cellulase to cellobiose as soluble substrate were less susceptible to [Bmim][Cl] than their cellulolytic activities to CC31, meaning that Î²-glucosidase is relatively stable among the three enzyme components of cellulase. The rate of glucose production could be appreciably improved by the pretreatment of CC31 with [Bmim][Cl] alone at 120Â°C for 30 min followed by the liposomal cellulase-catalyzed hydrolysis of the substrate at 45Â°C at the [Bmim][Cl] concentration of 15 w/w% [1]. In addition, the coexistence of liposomes prevented deactivations of enzymes in the presence of [Bmim][Cl].

[1] K. Tanimura et al., Hydrolysis of insoluble cellulose to glucose catalyzed by cellulase-containing liposomes in an aqueous solution of 1-butyl-3-methylimidazolium chloride. *Biotechnol. Prog.* **2013**, 29, 1190â€"1196.

500821 Micellization of Poloxamer Block Copolymers in Ionic Liquids and Their Mixtures with Water

Abstract:

Ionic liquids (ILs), organic salts that are fluid at ambient conditions, are a novel class of compounds with a combinatorially great chemical diversity and unique properties. The very low volatility and high thermal and chemical stability that many ILs exhibit render them promising as solvents. We are interested in the self-assembly properties in ionic liquids of block polymers, in particular those of the poly(ethylene oxide)-poly(propylene oxide)-poly(ethylene oxide) (PEO-PPO-PEO) family, commercially available as Pluronics or Poloxamers. We consider here solutions of a relatively hydrophobic PEO-PPO-PEO block copolymer (Pluronic P123) in the ionic liquid ethylammonium nitrate (EAN) and its mixtures with water. We report on the onset of micelle formation (cmc) and on micellization thermodynamics, and discuss the underlying interactions between P123 and EAN, P123 and water, and EAN and water. We compare the effects on aqueous Pluronic P123 micellization of ethylammonium nitrate to those of ammonium nitrate. We further compare the effects of the protic IL EAN to those of imidazolium ILs.

AUTHOR INDEX

Abbasi, Pedram 428, 961
Abbott, Nicholas L. 45, 465, 468, 834, 923, 992
Abedin, Rubaiyet 780
Abedini, Asghar 53
Abel, Steven M. 858
Abildskov, Jens 844
Abraham, Nathan 497
Abrams, Cameron F. 916
Abruna, Hector D. 119
Acharya, Prashant 364
Acosta, Edgar 34
Adams, Ryan A. 964
Agah, Shaghayegh 102
Aglave, Ravindra 644
Ahitan, Sourabh 537, 551
Ahmed, Alauddin 980
Ahmmed, Shamim 39
Ahn, Gwangnoh 334
Ahn, Yun-Ho 381
Aichele, Clint P. 466, 1026
Akano, Koichi 644
Akella, Meghana 1072
Akhmetkhanova, Nelya 47, 1025
Al Ghafri, Saif Zs. 316
Alamani, Bryan G. 832
Al-Dahhan, Muthanna 323, 324
Alexandridis, Paschalis 359, 1087
Alicke, Alexandra 736
Alkire, Richard 419
Allen, Andrew J 91
Allen, Kyle 132
Alpak, Faruk O. 221
Alvarez, Nicolas J. 219, 829
Amama, Placidus B. 906
Amrhein, Lauren E. 779
Amundsen, Ted J. 58
Andersen, Simon 536
Anderson, Ryan 457
Anisimov, Mikhail A. 779, 845
Annaland, Martin Van Sint 781
Antila, Hanne 181
Antoniuk-Pablant, Antaeres 9
Antony, Lucas 295
Aponte-Rivera, Christian 512
Arauz-Lara, B. Jose Luis 706
Archer, Lynden A. 433, 438, 1060
Arlt, Wolfgang 230
Armstrong, Matthew 276, 332, 333, 557
Asadi, Mohammad 428, 961
Ashbaugh, Hank 416
Asthagiri, Dilip 92
Athas, Jasmin C. 89
Au, Sam 35
Auer, Stefan 993
Awan, Javeed 656
Ayappa, K. G. 234
Back, Tyson C. 906
Badruddoza, Abu Zayed Md 1014
Baek, Seungjun 381
Bagusetty, Abhishek 52
Bahadur, Vaibhav 411
Bahamon, Daniel 917
Bai, Peng 294
Bai, Xue 881
Bajdich, Michal 16
Bala, Aseel M. 854
Balarama Sridhar, Dwadasi 275
Balbuena, Perla B. 728, 1031, 1046, 1059
Baldea, Michael 345
Baldino, Silvio 713
Baled, Hseen O. 208
Ball, Madelyn 121
Banerjee, Atiya 233
Banerjee, Dwijen 1023
Banerjee, Sanjoy 453
Bao, Nanqi 465
Bara, Jason E. 53
Barakat, Joseph M. 33
Barbieri, Matheus R. 885
Barhaghi, Mohammad 1008
Barringer, David A. 534
Bastos, Jaci Carlo Schramm Camara 342, 746
Bates, Frank S. 358
Baz, Jorg 184, 999
Bazant, Martin Z. 152, 175, 406
Bazyleva, Ala 477
Beaudoin, Stephen P. 463, 1044
Becerra, Mildred 472
Beckham, Gregg T. 874
Behdani, Behrouz 322
Behera, Manas Ranjan 734
Behrens, Sven H. 3, 462, 527, 646, 1015
Bell, Alexis T. 11
Beltramo, Peter J. 736, 802
Beltran, Saul Perez 1046
Benck, Jesse D. 436
Benz, Gregory 462
Berg, John C. 153
Berger, Manuel 331
Berghout, Pieter 645
Beris, Antony N. 135, 515, 557, 824
Berry, Joe 222, 1027
Besser, Ronald S. 122, 525
Bevan, Michael A. 1, 503
Bezik, Cody 977
Bezinge, Leonard 969
Bhagavatula, Dinesh N V S S R 412
Bharti, Bhuvnesh 898
Biddinger, Elizabeth J. 840
Biggs, Simon 4

AUTHOR INDEX

Bilir, Taner 436
Bischofberger, Irmgard 405
Bishop, Kyle J. M. 742, 901
Biswal, Sibani Lisa 173, 326, 343, 517, 947
Biswas, Shaurjo 430
Bizarri, Gregory A. 73
Blake, Diane A. 958
Blas, Felipe J 855
Blenner, Mark 761
Bliatsiou, Chrysoula 320
Bobbitt, N. Scott 296
Bognet, Brice 478
Bohm, Lutz 320, 456
Boles, J. 955
Bolton, Christopher 5
Bonnecaze, Roger T. 411, 469, 513
Bose, Arijit 956
Bothun, Geoffrey D. 89, 741, 957, 958
Boudouris, Bryan W. 463, 944, 1044
Bourg, Ian 981
Bourque, Alexander 765
Bourret, Edith D. 73
Brady, Sonia K 796
Braun, Trevor M 437
Breedveld, Victor 70, 1015
Brennan, John K. 921
Brett, Dan 555
Brewer, Catherine E. 59
Broderick, Alicia 1083
Brown, Angela C. 709
Brun, Pierre 407
Bucior, Benjamin 296
Bui, Justin 731
Bull, Geoffrey 276
Bulsara, Pallav 914
Bunge, Annette L. 996
Burke, Colin M. 962
Burns, Mark A. 174
Burshtein, Noa 648
Butler, Paul 937
Butz, Julian 90
Cadirov, Nicholas 955
Caiazza, Carla 823
Calabrese, Michelle A. 724, 924
Calabrese, Richard V 884
Callaway, Connor 978
Camacho-Forero, Luis E. 1031
Camaioni, Donald M. 522
Campbell, Charlie 522
Campbell, Charlton 747
Campos, Christian A. T. 238
Cao, Jinrong 106
Cao, Lixia 530
Cao, Ping 368
Cao, Zhi 15, 839
Capecelatro, Jesse 745
Carbone, Paola 49
Carneiro, Juliana S. A. 523
Carter, Emily A. 426
Cartier, Charles A. 742, 901
Carvalho, Marcio S. 485, 487, 738
Casali, Dominic M. 28
Castaneda-Priego, Ramon 417
Castro-Arellano, Jose J. 241, 382
Caupin, Frederic 779, 845
Celik, Fuat E. 120
Cersonsky, Rose 994
Chaimayo, Wanaruk 1042
Chaimovich, Aviel 24, 775
Chakraborty, Aishik 801
Chakraborty, Sudipto 734
Chakrapani, Vidhya 1043
Chambers, Scott A. 429
Champagne, Jean-Yves 458
Chan, Karen 16
Chan, Kwong-Yu 50, 905
Chandran, Prashanth 976
Chandran, Vishnu Deep 145
Chang, Christopher 15, 839
Chang, Hsueh-Chia 170
Chapman, Walter G. 44, 92, 182, 221, 476, 947, 1079
Chauvel, Paul 186
Chavez-Flores, David 732
Cheluget, Eric L. 854
Chen, Chau-Chyun 649, 650, 853
Chen, Jiayi 296
Chen, Jingyi 777
Chen, Leanne 16
Chen, Mengjie 729
Chen, Qiao 235
Chen, Ru 819
Chen, Szu-Ying 955
Chen, Wan-Ting 56
Chen, Xue 321
Chen, Yeng-Long 35, 835
Chen, Yusi 436
Cheng, Feng 59
Cheng, Mark 1048
Cheng, Xiang 336, 400, 403, 486, 722
Cheng, Zhengdong 1075
Chhabra, R. P. 592
Chiang, Hao-Chun 703
Chiarot, Paul R. 535
Chicaroux, Andres Kulaguin 778
Chien, Fan-Tso 835
Chien, Wei 835
Cho, Jason 555
Choi, Siyoung 470, 480
Choi, Won Tae 70
Choudhary, Madhuresh K. 72

AUTHOR INDEX

Choudhury, Pabitra 52
Choudhury, Snehashis 433
Chun, Jaehun 491, 934
Chundawat, Shishir 796
Chung, Jaeyub 944
Churchill, Stuart W. 165
Clark, Aurora E. 932
Clark, Ezra L. 11
Clark, Jennifer A. 1010
Clark, Thomas 321
Cloitre, Michel 513
Co, Anne 365
Coasne, Benoit 922
Cogswell, Kyle 27, 29
Colina, Coray M. 127
Collins, Charles 10
Collins-Martinez, Virginia 732
Colon, Yamil J. 295, 977
Comer, Benjamin 424
Conrad, Jacinta C. 490, 982
Contreras-Naranjo, Jose C. 733
Coppens, Marc-Olivier 555
Coquelet, Christophe 312
Cordoba, Andres 877
Cordova-Figueroa, Ubaldo M. 506
Corson, Elizabeth 427
Corti, David . S. 933
Coutinho, Joao A. P. 855
Cowman, Jonathan 144
Cox, Kenneth R. 476
Crabtree, Ellis 53
Crawford, Brad 651
Creel, Erin 427
Criscenti, Louise J. 232
Cristancho, Diego E. 782
Crocker, John C. 518, 1061, 1073
Croell, Arne 38
Crosthwaite, Jacob M. 782
Csernica, Peter 119
Cummings, Peter T. 42, 104, 180, 919
Curtis, Jennifer Sinclair 147
Curtiss, Larry A. 961
Cussler, Edward L. 439
Dagastine, Raymond R. 5, 222, 464, 1027
Dai, Gance 408, 744
Dalal, Indranil Saha 327
Dallas, Andrew J. 176
Damasceno, Pablo F. 931
Danahy, Brooks B. 652
Dani, Archit 65, 743
Daniel, Susan 357
Dann, Kevin 943
Darjani, Shaghayegh 1025
Das, Soumen 528
Das, Soumik 411, 469
Dasetty, Siva 761
Dasgupta, Anirvan 734
Dasgupta, Sunando 528
Datta, Sujit S. 396, 859
Daux, Virgile 478
David, Allan E. 509
Davis, Jonathan 731
Davis, Robert H. 172
Day, Allen 782
De Pablo, Juan J. 103, 133, 295, 398, 479, 833, 834, 877, 938, 977, 992, 1012
De Pablo, Luis X. 103
De Yoreo, James J. 934
Debenedetti, Pablo G. 123, 760
Deegan, Michael 333
Defever, Ryan 367
Degen, George 704
Dehghani, Mohammad Reza 857
Deidda, Graziano 875
Delacruz-Araujo, Ronal A. 506
Delhommelle, Jerome 967, 987
Dellago, Christoph 775
Derby, Jeffrey J. 38, 73, 484, 556, 740
Derksen, Bridget 52
Desgranges, Caroline 967, 987
Deshpande, Parag Arvind 283
Desikan, Rajat 234
Desjardins, Olivier 745, 883
Deskins, N. Aaron 287
Devarajan, Dinesh Sundaravadivelu 68
Dhakal, Subas 803
Dhar, Prajnaparamita 801
Diamond, Scott L. 518
Dickey, Michael D. 1041
Dighe, Anish V. 971
Dignon, Gregory L. 764
Diky, Vladimir 475, 477
Ding, Junhuan 654
Ding, Liang-Xin 902
Ding, Lin 749
Dinic, Jelena 329, 711
Dinivahi, Mangalaa 832
Disalvo, Francis J. 119
Divecha, Mia S. 556
Divvela, Mounica Jyothi 570
Do, Hainam 1009
Dobbs, Howard 955
Dodd, Paul 502, 505, 994
Doherty, Michael F. 970
Donahue, Melissa 345
Dong, Xiaorui 733
Dorfman, Kevin D. 722
Dou, Yong 742, 901
Dowd Jr., Regis 1030
Dowdle, John 782

AUTHOR INDEX

Doyle, Patrick S. 177, 643, 712, 1014
Dshemuchadse, Julia 502, 505, 931
Du, Chrisy Xiyu 500, 505
Du, Yan 750
Du, Yingge 429
Duan, Xindi 947
Dumesic, James A. 121
Dunn, Alexander 129
Duska, Michal 779, 845
Dutcher, Cari S. 37, 176, 516, 743
Dutta, Sarit 818
Dyatkin, Boris 42
Dybeck, Eric 497
Edgar, James H. 231
Egger, Torben 355
Egorov, Sergei 1077
El Hajem, Mahmoud 458
Eldridge, R. Bruce 345
Elizalde-Solis, Octavio 227
Emanuelsson, Emma 346
Enders, Sabine 90, 501
Engel, Michael 931, 966
Englert, Derek 454
Enick, Robert M. 208
Eppinger, Thomas 644
Errington, Jeffrey 498
Errington, Jeffrey R. 41, 71, 975
Escobedo, Fernando 989
Espin, Leonardo 409
Espinoza, Said E. Aranda 706
Esposito, Daniel V. 521, 731
Esquivel-Mora, Pedro 382
Esteves, A. Catarina C. 927, 1022
Fagan, Jeffrey A. 935
Fan, Tzu-Hsien 702
Fang, Bing 368
Farrell, Zachary 1077
Favero, Claudio Vilas Boas 168
Fawzi, Nicolas 764
Feaster, Jeremy T. 9
Fei, Wenjie 901
Fell, James 971
Ferguson, Andrew L. 990
Fernandes, Robert L. 239
Ferrier, Nicola J. 977
Fetisov, Evgenii 983
Fewkes, Christopher 1027
Fialoke, Suruchi 973
Fieg, Georg 355
Filler, Michael A. 1015
Fiore, Andrew 61
Flake, John C. 780
Follette, Marissa 906
Ford, David M. 499, 988
Ford, Katie 325, 401
Fouad, Wael A. 476
Fox, Rodney O. 647, 745
Francis, Lorraine F. 487
Frank, Florian 221
Frank, Timothy C. 854
Franses, Elias I 933, 944
Frechette, Joelle 69
Fredrickson, Glenn H. 725, 726
Freysinger, Wolfgang 331
Friedrich, Jochen 38
Frisbie, C. Daniel 841
Frityanti, Maya 795
Frostad, John M. 642
Fu, Chengyin 904
Fu, Yuan-Xiang 571
Fujigaya, Tsuyohiko 12
Fukai, Jun 584
Fuller, Gerald G. 335, 640
Fulton, John L. 522
Funazukuri, Toshitaka 236, 240
Furst, Eric M. 150
Gadikota, Greeshma 91, 981
Gajula, Kishore 275
Galicia-Luna, Luis A. 226, 241, 382
Galli, Giulia 977
Gallup, Jennifer 118
Gamble, Jordan A. 654
Gamwo, Isaac 208
Gani, Rafiqul 1084
Gao, Jingsi 151
Gao, Xi 747
Gao, Yongxiang 748
Garbin, Valeria 39
Garcia, Robert M. 534
Gardel, Margaret L. 398
Gardeniers, J. G. E. 171
Gardner, Jasmine 916
Garg, Aaron 121
Gebbie, Matthew A. 705, 1081
Geißelbrecht, Michael 230
Geng, Linxiao 1047
Geng, Yina 502
Ghosh, Parthasarathi 356, 461
Giacomelli, Jason J. 318
Giacomin, A. Jeffrey 723
Giberti, Federico 977
Gilbert, William J. R. 188
Gilchrist, James F. 493
Giles, Neil 843
Giles, Stephen A. 520
Glascoe, Elizabeth 568
Glaser, Jens 19, 1074
Glatz, Brittany 43
Glotzer, Sharon C. 19, 164, 500, 502, 505, 931, 966, 994, 1074

AUTHOR INDEX

Gochez, Roque ... 31
Godfrin, Michael ... 956
Godfrin, P. Douglas ... 417
Goel, Sachin ... 34
Gogotsi, Yury ... 42
Gokul, Navneeth ... 1009
Goltl, Florian ... 292
Gong, Jing ... 749
Gong, Ming ... 15
Goodpaster, Jason ... 286
Gopalan, Arun ... 296
Gordillo, Leonardo ... 336
Gordon, Melissa B. ... 826
Gosavi, Shachi ... 234
Goshayeshi, Bijan ... 237
Gow, Arthur S. ... 926
Goyal, Akshara ... 974
Graham, Michael D. ... 402, 818, 880
Graham, Trent R. ... 932
Gray, Jeffrey J. ... 701
Greeley, Jeffrey ... 282
Green, David ... 1077
Greenfield, Michael L. ... 183, 1011
Greenlee, Lauren F. ... 364
Grenville, Richard K. ... 318
Grillet, Anne M. ... 534
Grime, John M. A. ... 17
Gu, Xiang-Kui ... 523
Gu, Zhuxiao ... 1078
Gu, Zongyu ... 175
Guala, Diego ... 832
Guay, Martin ... 723
Gubbins, Keith E. ... 93
Guida, Vincenzo ... 823
Guignard, Florian ... 895
Guio, Carlos Morales ... 9
Guo, Ashley ... 295, 977
Guo, Hou-Jun ... 702
Guo, Juchen ... 904, 1047
Guo, Shuo ... 403
Guo, Wenjing ... 71
Gupta, Ankur ... 177, 643, 1014
Gupta, Manish ... 736
Gupta, Rakesh ... 275
Gupta, Shiva ... 727
Gutierrez, Oliver ... 522
Gygi, Francois ... 977
Ha, Harry Z. ... 538
Hadjittofis, Eftychios ... 532, 737
Haghighat, Arya ... 747
Hahn, Christopher ... 9, 11
Hall, Carol K. ... 18, 124, 993
Hamasaki, Yuki ... 12
Han, Endao ... 479
Handleton, Rachel M. ... 28
Hanley, Alan ... 509
Hann, Sarah ... 415
Hansen, Jesper S. ... 183
Hansen, Niels ... 184, 766, 999
Harris, James ... 436
Hartkamp, Remco ... 996
Hartman, Ryan L. ... 169
Hasebe, Shinji ... 334
Hassan, Yassin A. ... 36
Hatton, T. Alan ... 177, 643
Hatzell, Marta ... 424
Hauser, Brad G. ... 176
Haward, Simon ... 335, 640, 648
Hayashi, Keita ... 217
He, Peng ... 343
He, Yuxin ... 1082
He, Zhiqi ... 1087
Headen, Thomas ... 339
Headley, Alexander ... 965
Heidarian, Sharareh ... 780
Heinonen, Olle G. ... 133, 977
Helfferich, Julian ... 977
Helton, Tyler ... 332
Hendrix, Marco M. R. M. ... 927
Heng, Jerry ... 532, 737
Heng, Yi ... 571
Herceg, Eldad ... 186
Hernandez-Ortiz, Juan ... 133, 833
Herrera-Peraza, Eduardo ... 732
Hess, Dennis W. ... 70
Heusinger, Ferdinand ... 230
Heylmun, Jeffrey C ... 647
Higgins, Drew ... 9
Hill, James C. ... 165
Hinkle, Kevin R. ... 21
Hinton, Zachary R ... 219, 829
Hirasaki, George J. ... 947
Hironaka, Shuji ... 584
Hirose, Masanori ... 218
Ho, Tuan ... 232
Hoepfner, Michael P. ... 339
Hoffmann, Kyle ... 295
Hofmann, Jennifer L. ... 489
Hoga, Heloisa E. ... 224, 238, 239
Hogge, Joseph W. ... 843
Horner, Jeffrey S. ... 135
Horsch, Martin T. ... 25
Hoss, Darby J. ... 463
Hossain, Md. Tashfin Zayed ... 526
Hossain, Nazir ... 649
Howard, Michael P. ... 991, 1076
Hoyer, Wolfgang ... 776
Hrenya, Christine ... 494
Hruby, Jan ... 845
Hsiao, Lilian ... 66

AUTHOR INDEX

Hsieh, An-Hsuan 933
Hsieh, Chieh-Ming 223
Hsieh, Chih-Chen 115, 702
Hu, David L. 897
Hu, Jianguang 408
Hu, Sheng 360
Hu, Yue 379
Huang, Chung-Hsuan 739
Huang, Liangliang 93
Huang, Min 280
Huang, Tan 117
Huang, Wenxin 12
Huang, Yun-Ru 467
Hudgins, Jesse 276
Hudson, Steven D. 804
Hung, Francisco R. 105, 780
Hunt, Sean T. 121
Huo, Yijie 436
Hussain, S. M. Shakil 942
Huynh, Hang 430
Hwang, Hyun-Tae 454
Hwang, Luke 480
Iacovella, Christopher R. 180, 914, 996
Ilias, Shamsuddin 526
Inoue, Gen 584
Intikhab, Saad 118
Irudayanathan, Flaviyan Jerome 794, 873, 913
Ismagilov, Rustem 859
Ismail, Ahmed E. 651
Israelachvili, Jacob 704, 705, 955
Iwasaki, Fumihiko 925
Iyemperumal, Satish 287
Jackson, James E. 854
Jadhav, Ankur 959
Jaeger, Heinrich 479
Jain, Deeksha 365
Jalilvand, Zohreh 894
Jamali, Vida 930
Jameson, Cynthia J. 909
Jamieson, Emily 1027
Jana, Suvamay 874
Jang, Seung Soon 978
Jaramillo, Thomas F. 9, 11, 436
Jasim, Ahmed 524
Jauss, Thomas 38
Jena, Umakanta 59
Jenkins, Ian 1061, 1073
Jhong, Molly 12
Jia, Jieyang 436
Jiang, Peng 1078
Jiang, Shaoyi 51
Jiang, Xikai 133, 977
Jiang, Zhitong 973
Jiao, Feng 10
Jimenez, Leidy N. 329, 711
John, George 945
John, Vijay T. 89, 957, 958
Johnson, J. Karl 52
Jokar, Mojtaba 763
Jones, James 492
Jones, Matthew 346
Jonnalagadda, Sai Vamshi R. 776, 875
Joo, Yong Lak 570
Joodaki, Faramarz 1011
Joshi, Kedar 493
Joshipura, Ishan 1041
Joswiak, Mark 970
Juarez, Jaime 940, 1072
Kaar, Joel L. 797
Kadilak, Andrea 554
Kadri, Olufemi 145
Kaganyuk, Max 1028
Kale, Amullya 327
Kamal, Muhammad Shahzad 942
Kamphaus, Ethan P. 1059
Kamyabi, Nabiollah 130
Kang, Yijin 116
Kapoor, Utkarsh 653
Karaiskakis, Alexandros N. 840
Karamad, Mohammadreza 16
Karim, Alireza Mohammad 487
Kasbaoui, Mohamed H. 883
Kaufman, Yair 955
Kawaji, Masahiro 453
Kazakov, Andrei 475, 998
Kaznessis, Yiannis N. 912
Kelsey, Jarred 466
Kenis, Paul J. A. 12
Kenny, Dermot 144
Khabaz, Fardin 513
Khair, Aditya S. 6
Khalaf, Nidal 723
Khan, Aliza 911
Khan, Saad A. 856
Khare, Ketan S. 831
Khare, Rajesh 68, 649, 853, 929
Khomami, Bamin 360
Khraisheh, Majeda 384
Kian, Kourosh 189
Kikkinides, Eustathios 988
Killian, William G. 854
Kim, Chaerin 12
Kim, Doyoung 832
Kim, Hye Hyun 797
Kim, Kibum 428
Kim, Kyuhan 470, 480
Kim, Sang Beom 760
Kim, Young C. 764, 860
Kim, Youngsang 427
King, Jerry W. 187

AUTHOR INDEX

King, Paul 553
Kinoshita, Motohiro 30
Kiran, Bandaru 344
Kirk, Charlotte 16
Kirkes, Toni 650
Kitchens, Christopher L. 31
Kjelstrup, Signe 368
Klapp, Sabine H. L 124
Klein, Christoph 180
Klett, Adam S. 654
Klotz, Alexander 712
Kloxin, Christopher J. 826
Knapp, Ellen M. 508
Knehr, Kevin 430
Knott, Brandon C. 874
Knotts, Thomas A. 843
Knutson, Barbara L. 1082
Knuutila, Hanna K. 312
Ko, Derrick I. 884
Koch, Donald L. 883
Kofke, David A. 498, 974, 1009
Koga, Hiroaki 106
Kokini, Jozef 762
Kolb, Thomas 900
Kolomeisky, Anatoly 102
Komives, Claire F. 277
Kong, Bo 647, 747
Koplik, Joel 397, 510
Kostecki, Robert 427
Kowalik, Mikolaj 742
Kral, Florian 331
Kramer, Peter R. 399
Kraume, Matthias 320, 456
Kraxner, Michael 331
Kremer, Kurt 24
Kretzschmar, Ilona 506, 508, 510, 894, 901
Krishnamoorthy, Senthil 550
Kristiansen, Kai 955
Krivov, Sergei V. 993
Kroenlein, Kenneth 475, 477
Krohl, Patrick 134
Kroon, Maaike C. 781
Kuczera, Zachary 523
Kudaibergenov, Sarkyt 929
Kuech, Thomas F. 569
Kuei, Steve 326, 343, 517
Kuhn, Simon 496
Kulkarni, Ambarish R. 284
Kumar, Manjesh 72
Kumar, Nitin 398
Kumar, Satish 409, 485, 486, 738, 739
Kumar, Vimal 454
Kundan, Akshay 443
Kun-Lin, Yang 941
Kurapati, Yathish 830
Kurdziel, Sophia 521
Kurihara, Kiyofumi 229, 236
Kurk, Michael (Andy) 828
Kurtenbach, Khia 103, 834
Kwon, Youngkook 837
Kyriakou, Panagiota 912
Labrador, Natalie 521
Lacassagne, Tom 458
Ladipo, Folami 1082
Lakkaraju, Rajaram 461
Lakshmanan, Hari Hara Sudhan 340
Lamie, Willam 26
Landers, Alan 9
Lang, Matthew J. 796
Lapeyri, Corey 277
Larson, Ronald G. 40, 168
Laster, Jennifer S. 1044
Lattanzi, Aaron 494
Lattuada, Marco 895
Laursen, Siris 422
Lavey, John 747
Lawal, Akanni S. 190, 209, 228, 298
Le Gac, S. 171
Leal, L. Gary 642, 725
Le-Doux, Travis 59
Lee, Daeyeon 415
Lee, Huen 381
Lee, Hyundo 177
Lee, Jae W. 381
Lee, Jingyun 898
Lee, Jong-Min 117
Lee, Ka Yee 480
Lee, Sangmin 966
Lee, Young Ki 518
Legg, Meesha 747
Lehmann, Marcus 134
Leitold, Christian 775
Lele, Pushkar 325, 401
Lemarchand, Claire A. 183
Leo, Sin-Yen 1078
Lequieu, Joshua 877, 977
Lercher, Johannes A. 522
Levicky, Rastislav 703
Li, Bing 858
Li, Chien-Yi 223
Li, Chi-Ying Vanessa 905
Li, Dongsheng 934
Li, Hui 1012
Li, Jiyuan 133, 977
Li, Ke 504
Li, Lei 530, 1080
Li, Ting 915
Li, Weihua 485
Li, Xi 514
Li, Xiao 833

AUTHOR INDEX

Li, Xue....493
Li, Yanxiang....530
Li, Yawei....361
Li, Yunzi....174
Li, Zheng....288
Liang, Huaqing....1032
Liang, Xiaodong....1084
Liang, Xiaoyu....41
Liang, Yunfeng....106, 383
Liedtke, Aleesha M.....328
Lin, John....9
Lin, Nancy J.....935
Lin, Weisong....974
Lin, Yu-Jiun....173, 343
Lindberg, Seth....319, 341, 511, 825
Linic, Suljo....425
Linico, Audrey....362
Lira, Carl T.....854
Liu, Bin....231
Liu, Chen-Yu....738
Liu, Fang....47, 1025
Liu, Honglai....46, 93, 128, 533
Liu, Jinlu....92
Liu, Jiuxu....55
Liu, Kunlei....274
Liu, Minye....878
Liu, Nian....960
Liu, Song....231
Liu, Tianfei....513
Liu, Wei....15
Liu, Xinyan....1084
Liu, Yang....749
Liu, Ying....554, 641
Liu, Yun....417
Livingston, Dana A.....186
Llovell, Felix....855
Lo, Simon....644
Loewenberg, Michael....444
Lone, Sohail Rasool....454
Long, Andrew W.....990
Long, Thomas....510
Lopez-Barron, Carlos R.....819
Lowe, Jeffrey S.....435, 1045
Lowe, Jennifer....321
Lu, Hang....32
Lu, Li....535
Lu, Shawn....12
Lu, Xiaohua....126
Lubers, Alia M.....362
Luc, Wesley....10
Lunt, Richard....148
Luo, Jia-Li....571
Luo, Jiu....571
Luo, Miao....537
Luo, Sheng....985
Luo, Yimin....2
Lutkenhaus, Jodie L.....985
Lyu, Shu-Shen....571
Lyu, Yuan....795
Ma, Anson....478
Ma, Chenbo....1032
Ma, Huilian....504
Ma, Huilin....911
Ma, Yingzhen....1087
Maag, Alex....58
Mackintosh, Fred....930
Maddala, Jeevan....340
Madsen, Jesper J.....17
Magee, Joseph W.....477
Maggioni, Giovanni Maria....969
Maginn, Edward J.....185, 293, 979
Maguire, Megan....362
Mahadevan, L.....483
Mahmoodi, Seyed Reza....122, 525
Mahynski, Nathan A.....975, 997
Maktabi, Sepehr....535
Maldarelli, Charles....65, 397, 945
Maldonado, Luis F.....762
Maldonado-Camargo, Lorena....62, 132
Malmstadt, Noah....707
Maloney, Ryan C.....124
Mamontov, Eugene....104
Mamtani, Kuldeep....365
Maneval, James E.....457
Mao, Runfang....1071
Mao, Yating....67
Markthaler, Daniel....766, 999
Marquez, Diego....277
Marquez-Montes, Raul....732
Marr, David W. M.....896
Marruecos, David Faulon....797
Marshall, Kristin A.....328
Marson, Ryan L.....40
Marthelot, Joel....407
Martignoni, Waldir Pedro....885
Martin, Lenore M.....1011
Martinez, Adam D.....534
Martinez-Gonzalez, Jose....103, 833
Masuda, Yoshihiro....106, 383
Mathias, Paul M.....538, 552, 854
Matiazzo, Tatiana....342, 746
Matin, Naser S.....274
Matsuda, Hiroyuki....229, 236
Matsuoka, Toshifumi....106
Matthews, Michael A....28
Maula, Tiara Ann....179
Mavrikakis, Manos....45, 292, 465
Mayer, Matthew....122, 525
Mazzotti, Marco....655, 969
McCabe, Clare....180, 914, 996

AUTHOR INDEX

McCloskey, Bryan D. 420, 427, 962, 963
McCormick, Alon 89, 928, 957
McDevitt, Kyle 903
McGinley III, James T. 1061
McHugh, Mark A. 208
McKenna, Gregory B. 514
McKone, James R. 119
McLaughlin, John B. 537
McNeary, William 362
Medford, Andrew 424
Mees, Bruna L 342, 746
Mehrle, Andreas 331
Meier, Henry F. 342, 746, 885
Melendez, Brandon 832
Mena, Sarah E. 174
Meredith, J. Carson 462, 527, 646
Merker, David 456
Merz, Steven 1077
Messerly, Richard A. 998
Messinger, Robert J. 959
Metaxas, Athena E. 516
Mettu, Srinivas 222, 464
Meza, Juan Manuel Hernandez 706
Mick, Jason R. 1008
Middleton, Chadwick E. 946
Milina, Maria 121
Miller, Benjamin 1042
Miller, Juanita 59
Min, Juwon 381
Ming-Yi, Chang 115
Minkara, Mona 777
Minnick, David L. 188, 652
Miranda, Michael 466
Mirecka, Ewa A. 776
Mironenko, Alexander V. 291
Mirri, Francesca 930
Mirzadeh, Mohammad 406
Mishra, Arpit 356, 461
Miska, Stefan Z. 713
Mitraki, Anna 875
Mittal, Jeetain 179, 764, 765, 860, 986, 997, 1071
Mo, Dong-Chuan 571
Mohanty, Ritesh P. 519
Mohraz, Ali 903, 1028
Molla, Getachew S. 844
Moller, Joshua 877, 977
Mondal, Kunal 1041
Monjezi, Saman 492
Monroy-Pena, Camilo 793
Monson, Peter A. 499, 988
Montiel-Macias, Elizabeth 728
Montoya, Joseph H. 289, 423
Moore, David J. 914
Moore, Timothy C. 914, 996
Moradi, Marzieh 936
Moravec, Davis B. 176
Morgan, Nathan T. 740
Morin, Samuel 103
Morris, Jeffrey F. 491
Morrish, Rachel 279
Morse, David 928
Mortazavi-Manesh, Sepideh 472
Moustafa, Sabry G. 974
Moyers, Scott 828
Mozaffari, Ali 397
Mu, Xiaoqun 44, 221
Mueller, Karl 432, 1058
Mujica, Maritza 1015
Mukherjee, Dibyendu 360
Mukherjee, Sanjoy 463
Mukherjee, Siddhartha 645
Mullen, Ryan Gotchy 293, 979
Muller, Erich A. 125
Muller, Karsten 230
Mumm, Daniel R. 903
Mundy, Christopher J. 934
Municchi, Federico 495
Murad, Sohail 909
Murphy, Ryan P. 557
Murugesan, Vijayakumar 432, 1058
Mustain, William E. 554, 730
Mwasame, Paul M. 515, 824
Mysona, Joshua 928
Nacy, Ayad 363
Naegele, Gerhard 417
Naganawa, Shigemi 383
Nakamura, Hidemi 217
Nakashima, Naotoshi 12
Nan, Yue 55
Nandy, Lucy 37
Nangia, Shikha 794, 873, 911, 913
Narayan, Shweta 176, 743
Narsimhan, Ganesan 795
Narsimhan, Vivek 712
Nasrabadi, Hadi 985
Nava, Glixon Mavarez 1023
Navarro, Rocio 172
Nealey, Paul F. 833
Neeves, Keith B. 134, 896
Neilsen, John 89
Neimark, Alexander V. 915
Neitz, Tracy 536
Nelson, Zachary 364
Nemer, Martin B. 965
Nemmaru, Bhargava 796
Neville, Tobias P. 555
Newberg, John T. 1083
Newcomb, Ken 185
Newman, Richmond S. 500
Ng, Jia Wei Desmond 436

AUTHOR INDEX

Ng, Samson 34
Ng, Simon....1048
Nguyen, Frank....402
Nguyen, Le Truc 941
Nguyen, Quoc P. 469
Nguyen, Quoc T. 166, 879
Nguyen, Thao T. T. 443
Nguyen, Trung 20
Nichols, Jessica E. 963
Nikolla, Eranda 363, 523
Nikoubashman, Arash 991, 1076
Ning, Fulong 368
Nitopi, Stephanie....9
Nitsche, Ludwig C 641
Norskov, Jens 16, 284, 285
Nourian, Pouria....68
Nunes, John 741
Nuraje, Nurxat 929
Nuthalapati, Sri Harsha 1029
Obermeier, Jonas 230
O'Brien, David 536
Oettle, Shawn 492
Ofoli, Robert Y. 148
Oglesby, Irene K. 144
Ohashi, Yuki....229
Okada, Kazuya....404
Okamoto, Yukihiro....217, 218, 925, 1086
Olbricht, William L. 131
Olivieri, Gustavo V. 225
Olsen, Michael....747
Omarova, Marzhana....958
Omasta, Travis J. 730
Onyo, Daudi Saang' 1082
Ooe, Yoshiko 1086
Orjuela, Johana....793
Ornithopoulou, Eirini 875
Oropeza, Maria Veronica Carranza....330
Oroskar, Priyanka 909
Orozco-Mena, Raul....732
Orr, Asuka....776, 875
Ortiz, Camilla U. 531
Ortner, Franziska 655
Osgouei, Reza E....713
Osuji, Chinedum O. 488
O'Sullivan, Denis....823
Otsi, Naresh C. 104
Ouaknin, Gaddiel 22
Ousley, Evan....333
Overney, Rene M. 153
Ozbayoglu, Evren....713
Ozkan, Umit S. 365
Ozoe, Hiroyuki 571
Pacella, Michael S. 701
Palmer, Jeremy C....982
Pan, Zehao 170
Panagiotopoulos, Athanassios Z. 972, 991, 995, 1076
Pandey, Shashank 742, 901
Papavassiliou, Dimitrios V. 63, 166, 879
Parajuli, Sanjiv 946
Parihar, Vartika 528
Park, Joontaek....322, 492
Park, Nayoung....490
Park, Sungjune 1041
Parker, Morghan....422
Parsi, Nikhila 711
Pasquali, Matteo....60, 102, 930
Passalacqua, Alberto....647
Patanwala, Huseini S. 478
Patel, Amish 973, 984
Patel, Anjli M. 284
Patel, Navin....854
Patton, Steven 220
Pauchard, Vincent....47, 1025
Paul, Alexandra 642
Paulechka, Eugene 477
Paulsen, Alex D. 58
Payne, Christina M. 874, 876
Pednekar, Sidhant 491
Peles, Yoav 167
Pellenq, Roland J.-M. 48
Pendery, Joel....992
Peng, Xiaoguang....514
Peng, Xiong....554, 730
Peng, Yi 400, 403
Peng, Yunhu 66
Pennington, Ashley M. 120
Pera-Titus, Marc 533
Perera, Dinal....457
Perrodin, Didier....73
Persson, Kristin 423, 432, 1058
Pervaje, Amulya 856
Pesek, Joseph 277
Peter, Christine....24
Peters, Baron 968, 970
Peters, Cor J. 777, 781
Peterson, Jeffrey H. 73, 484
Peterson, Joseph 725
Pham, Ngoc Hong....63
Pham, Truong....486
Phelan Jr., Frederick R. 21, 831
Phillips, Carolyn L. 931
Pillei, Martin....331
Pimentel-Rodas, Alfredo....226, 241
Pinho, Bruno....169
Pinto, Diego D. D. 312
Plata, Paola Leon....641
Plawsky, Joel L. 167, 443
Pol, Vilas G. 964
Poling-Skutvik, Ryan 982
Pope, Daniel J. 932

AUTHOR INDEX

Porcar, Lionel....724, 924
Porter, Christopher....518
Potoff, Jeffrey J....1008
Poudyal, Samiksha....422
Pourjafar, Sara....57
Pourtousi, Mohammad....645
Pradhan, Sahadev....242, 272, 593
Prajapati, Aditya....14
Prasad, Ajay K....520
Prashant, Ishan....327
Pretti, Evan....997
Preuster, Patrick....230
Preziosi, Valentina....823
Prieve, Dennis....146
Priye, Aashish....36
Pulikkathara, Merlyn....536
Puntambekar, Ajinkya....1043
Purohit, Apoorva....498
Pushpavanam, S....412
Qazvini, Nader Taheri....938, 1012
Qi, Qin M....144
Qian, Yuzhou....399
Qing, Meiyi....1032
Qiu, Fen....427
Qu, Xiaohui....432
Quevedo, Miguel Gonzalo Arenas....227
Quevillon, Michael....977
Quinn, Joseph....425
Quirke, Nicholas....50
Rabideau, Brooks D....1085
Raciti, David....13
Radke, Clayton J....149
Radl, Stefan....495
Radler, Mike....321
Rafiq, Rabees....1024
Raghavan, Srinivasa R....89, 957
Rahimi, Mohammad....834, 977, 992
Rahmanian, Vahid....860
Rai, Beena....275
Rai, Prabin....45
Rajagopal, Nandhini....873
Rajput, Nav Nidhi....432, 1058
Ramachandran, Arun....34
Raman, Ashwin Kumar Yegya....466, 1026
Ramezani-Dakhel, Hadi....834, 977, 992, 1012
Ramirez-Saito, Angeles....706
Ramos-Pallares, Francisco....474
Ramos-Sanchez, Victor....732
Randolph, Theodore W....798
Rankin, Stephen E....1082
Rao, Hemanth Giri V....234
Rashidi, Aidin....507
Rasti, Elnaz S....709
Rathee, Vikramjit....977
Rathi, Ashutosh....988
Rathore, Prerana....178
Rattan, V. K....236
Ravichandran, Ashwin....649, 853, 929
Ravisankar, Vijay....36
Rawlings, James B....569
Razavi, Sepideh....901
Reboucas, Rodrigo....444
Recheis, Wolfgang....331
Redlinger-Pohn, Jakob D....495
Reid, Daniel....977
Reimann, Christian....38
Reinhart, Wesley F....972, 991
Reis, Pedro....407
Remias, Joseph E....274
Ricco, Antonio J....144
Richards, Jeffrey J....937
Riehm, David....89
Riera-Ortiz, Jaime A....226
Riest, Jonas....417
Rimer, Jeffrey D....72, 832
Rinaldi, Carlos....62, 132
Ripken, R. M....171
Rivera, Rafael G. Henriquez....641
Roberts, Christine Cardinal....965
Roberts, Mark E....654
Roberts, Ryan....982
Roberts, Scott A....965
Robinson, Jacob....521
Rocchio, Joseph....957
Rocha, Alejandra M....188
Rocha, M. Alejandra....380, 1083
Rodriguez, Nerea R....781
Roericht, Margarete....501
Rogers Jr., Reginald E....281
Rogers, Simon....724, 924
Roman-Leshkov, Yuriy....121
Root, Addison....318
Rosch, Justin....131
Rosenbaum, Amanda....779
Rosenfeld, Liat....943
Rosete-Barreto, Jose M....226
Ross, Jennifer....398
Roussell, William L....180
Roux, Benoit....834, 992, 1012
Rowley, Richard L....843
Roy, Arnab....461
Rudra, Apoorva....453
Ruppli, Chantal....655
Ruszczynski, Lukasz....844
Rutkowski, David M....124
Sabri, Laith....323, 324
Sachan, Amit Kumar....220, 480, 481, 708, 800
Sadati, Monirosadat....103, 833, 834
Sadus, Richard J....842
Saeedfar, Amin....536

AUTHOR INDEX

Saengow, Chaimongkol 723
Sahu, Kirti Chandra 646
Sakabe, Junichi 240
Salehi-Khojin, Amin 428, 961
Salunkhe, Aditya A. 153
Samaniuk, Joseph R. 836
Samanta, Devranjan 403
Samira, Samji 363
Sammalkorpi, Maria 181
Sanchez, Francisco Javier Veronico 227
Sanchez, Javier 741
Sandler, Stanley I. 980
Santiso, Erik E. 856, 1010
Santos, Andrew P. 995
Santos-Serena, Raul A. 382
Sanyal, Udishnu 522
Sarupria, Sapna 43, 367, 761
Satoh, Akira 404
Saunders, Steven R. 932
Sayahpour, Baharak 961
Schaffer, C. B. 131
Schenter, Gregory K. 934
Schertzer, Jeffrey W. 535
Schideman, Lance 56
Schieber, Natalie 497
Scholz, Carmen 741
Schrader, Alex 955
Schultz, Andrew J. 498, 974, 1009
Schultz, Danielle 935
Schultz, Kelly M. 319, 341, 511, 825
Schwartz, Brian 938
Schwartz, Daniel K. 797
Schwartz, Daniel T. 437
Schwartzentruber, T. E. 439
Scurto, Aaron M. 189
Seames, Wayne S. 57
Seay, Jeffrey 454
Seebeck, Jan 73
Seitz, Linsey C. 436
Selvam, Balaji 907
Sen, Ramkrishna 356
Senftle, Thomas P. 426
Seo, Dongjin 955
Serra, Francesca 2
Sevgen, Emre 977
Seymour, Joseph D. 457
Shabaniverki, Soheila 940
Shah, Jindal K. 233, 653, 976
Shah, Mansi S. 983
Shahriari, Arjang 411
Shahryari, Reza 857
Shahzad, Khurram 496
Shao, Qing 18
Shapiro, Harrison 103
Shaqfeh, Eric S. G. 33, 144, 827
Shardt, Orest 645
Sharma, Hom 568
Sharma, Sumit 830
Sharma, Vivek 178, 329, 337, 338, 471, 482, 531, 711, 1024
Shaw, John M. 472, 537, 551
Shaykhalishahi, Hamed 776
Shekar, Ashwin 880
Shen, Amy 335, 640, 648
Shen, Vincent K. 975, 997
Shen, Yan 105
Sherman, Zachary 893
Shi, Bohui 749
Shi, Nan 939
Shiflett, Mark B. 188, 380, 652, 1083
Shimoyama, Yusuke 30, 315
Shin, Seunghwan 722
Shin, Young Hwan 56
Shintake, Jun 1041
Shirts, Michael 497, 797
Shishkov, Olga 897
Shivaprasad, Parimala 346
Shor, Leslie M. 554
Shrestha, Sujan 554
Shukla, Diwakar 907
Siahrostami, Samira 16, 285
Sides, Paul J. 146
Sidky, Hythem 977
Siegel, Donald J. 435, 1045
Siepmann, J. Ilja 777, 983
Sihlabela, Maswazi 956
Sikora, Benjamin J. 977
Silva, Barbara L. 885
Silva, Marcela Kotsuka 342, 746
Simoã«ns, Serge 458
Sin, Gurkan 844
Sing, Charles E. 710
Singh, Meenesh R. 14, 971
Singh, Narendra 439
Singh, Nirala 522
Singh, Preet M. 70
Sinno, Talid 518, 1061, 1073
Siperstein, Flor R. 49
Skuntz, Matthew 457
Skylogianni, Eirini 312
Sliwinska-Bartkowiak, Malgorzata 920
Smiatek, Jens 999
Smith, Kevin 317
Smyrl, William H. 418
Snurr, Randall Q. 296, 414
Snyder, Joshua 118, 361
Solich, Ryan 965
Song, Jin 947
Song, Kyoo 476
Song, Shangfei 749

AUTHOR INDEX

Songcuan, Eva 521
Sood, Parveen 978
Sorgenfrei, Tina 38
Sorret, Lea L. 798
Soto, Bernardo Yanez 706
Squires, Todd M. 7, 220, 470, 480, 799, 939
Srinivasan, Chiranth 166
Srinivasan, Siddarth 483
Stavig, Mark 965
Stebe, Kathleen J. 2, 415
Steen, Paul 410
Stefanov, Zdravko 186
Steinberg, Asher Preska 859
Steingart, Daniel 430
Stoerzinger, Kelsey A. 429
Storer, Jackson A. 854
Stottrup, Benjamin 800
Strong, Elizabeth 407
Su, Xiao 434
Suga, Keishi 217, 218, 925, 1086
Sultan, Abbas 323, 324
Sultan, Abdullah S. 942
Sultana, Khondker 526
Sum, Amadeu K. 379
Summers, Daniel R. 550
Sun, Guogang 750
Sun, Ting-Pi 336
Sun, Yunwei 568
Sun, Zhanpeng 750
Sunol, Aydin K. 27, 29
Sureshkumar, Radhakrishna 803
Suszynski, Wieslaw 487
Suteria, Naureen 39
Swan, James 61, 489, 893
Swann, Britany M. 67
Swientoniewski, Lauren T 958
Szilvasi, Tibor 45, 465
Tabrizi, Kayvon 701
Taguchi, Minori 240
Taguchi, Shogo 217
Tai, Yishu 400
Tam, Wai-Ming 536
Tamamis, Phanourios 776, 875
Tan, Mingyang 67
Tanaka, Hiroyuki 106
Tang, Maureen H. 118
Taniguchi, Satoshi 334
Tanimura, Kazuhiko 1086
Tao, Yutao 38
Tasci, Tonguc Onur 896
Tasker, Alison 4
Taufertshofer, Kirstin 501
Tavlarides, Lawrence L. 55
Taylor, Ross 473
Taylor, Shawn D. 474, 536
Tcheimou, Stephane 462
Tenma, Norio 383
Thakur, Pooja 592
Thapar, Vikram 977
Thies, Mark C. 26, 654
Thomas, John A. 317
Thommes, Matthias 918
Thompson, Michael 914
Tilton, Robert D. 467
Timko, Michael T. 58
Tindall, Graham W. 654
Tirrell, Matthew V. 938
Tiwari, Naveen 592
Tochigi, Katsumi 229, 236
Todd, Paul W. 828
Todt, Anika H. 328
Tomaiuolo, Giovanna 823
Tomoya, Aoyama 334
Toner, Mehmet 35
Tonomura, Osamu 334
Torabi, Korosh 763
Torne, Jordi 855
Torres, Ricardo B. 224, 225, 238, 239
Torres-Diaz, Isaac 503
Tosuji, Yuta 383
Tree, Douglas 726
Treftz, Brian 59
Tremsin, Anton S. 73
Tripathi, Anubhav 956
Tripathi, Anurag 327
Tripathi, Manoj Kumar 646
Trogadas, Panagiotis 555
Truskett, Thomas M. 413
Trusler, J. P. Martin 316
Tsao, Joanna W. 3
Tsapatsis, Michael 983
Tsuge, Yoshifumi 584
Tu, Raymond 508
Tuinier, Remco 927, 1022
Turner, C. Heath 53
Twieg, Robert 45
Ugaz, Victor M. 36, 733
Umakoshi, Hiroshi 217, 218, 925, 1086
Underhill, Patrick T. 399
Unni, Mythreyi 132
Uralcan, Betul 760
Urban, Jeffrey 427
Urena-Benavides, Esteban E. 946
Usui, Keishi 383
Utzig, Jonathan 885
Val, Natalia D. F. 224
Valadez, Nestor 417
Van Aeken, Wouter 496
Van Anders, Greg 500, 502, 505, 994
Van Den Akker, Harry E. A. 645

AUTHOR INDEX

Van Der Heijden, Nicole M. W. 927, 1022
Van Der Schoot, Paul 930
Van Gerven, Tom 496
Van Lehn, Reid 908
Van Nguyen, Trung 1030
Van Osch, Dannie J. G. P. 927, 1022
Van Spronsen, Jaap 927, 1022
Van Tassel, Paul R. 181
Vanapalli, Siva A. 39, 130
Vangala, Sai Phani Kumar 283
Varga, Zsigmond 489
Varma, Arvind 964
Vashisth, Harish 910
Vega, Lourdes F. 855, 917
Vega-Bellido, Gabriel 506
Velasquez, Sandy Candiotti 330
Velegol, Darrell 8
Velev, Orlin D. 124
Venegas Sr., Luis Carrasco 330
Venegas, Alexis 277
Verde-Gomez, Ysmael 728
Verma, Sumit 12
Vermant, Jan 736, 802
Vichansky, Alexander 644
Vigeant, Margot 278
Vigil, R. Dennis 747
Vilaseca, Oriol 855
Virk, Preetinder S. 455, 882
Vis, Mark 1022
Vishnyakov, Aleksey 915
Vlachos, Dionisios G. 291, 520
Vlasiuk, Maryna 842
Vlugt, Thijs J. H. 368
Vogel, Sven 73
Volk, Michael 1023
Vora, Sahil R. 478
Voronov, Roman 145
Voss, Johannes 16
Voth, Gregory A. 17
Vreeland, Thomas 31
Vu, Dung T. 854
Wagner, Norman J. 64, 135, 151, 417, 515, 557, 724, 819, 824, 826, 924, 937
Waite, J. Herbert 705
Walker, Eric 290, 838
Walker, Travis W. 67, 328
Walls, Dan 335, 640
Wang, Chao 13
Wang, Cheng 322
Wang, Ching-Kuan 115
Wang, Haihui 902
Wang, Huan 906
Wang, Jialun 514
Wang, Jian-Guo 54
Wang, Jun 902
Wang, Kerry 740
Wang, Le 947, 1079
Wang, Lei 9
Wang, Liang 520
Wang, Lu 707
Wang, Nan 913
Wang, Penghui 744
Wang, Qi 1043
Wang, Rui 297
Wang, Songcheng 462, 646
Wang, Sung-Ning 880
Wang, Weigao 761
Wang, Xiaoyi 913
Wang, Yan 841
Wang, Ya-Qiao 571
Wang, Yifan 1061
Wang, Yifeng 232
Wang, Yiming 993
Ward, Kevin R. 174
Warrag, Samah E. E. 781
Warshavsky, Vadim B. 499
Wasserscheid, Peter 230
Waturuocha, Amaka 1023
Wayner Jr., Peter C. 443
Webb, Michael 977
Wehrman, Matthew 319, 341, 511, 825
Wei, Min 1032
Wei, Shuya 438, 1060
Wei, Zhiyan 483
Weigandt, Kathleen 804
Weimer, Alan W. 362
Werner, Yannick 230
Weston, Javen 804
Wheatley, Richard J. 1009
Whitcomb, Kevin 64
White, Andrew 1013
White, Jeff 466
White, Joshua 276
Whitmer, Jonathan K. 977
Wieseneck, Stacey 898
Wilding, W. Vincent 843
Wilkinson, Nikolas A. 516
Williams, Christopher 49
Williams, Ian 220, 799
Williams, Stuart J. 936
Willing, Gerold A. 936
Winiarski, Aubrey 926
Wirth, Christopher L. 507, 1029
Wisecarver, Keith 1023
Wojcik, Ewelina 338
Woodcock, Cory 167
Woolf, Scott 59
Wordehoff, Michael M. 776
Worku, Dereje 526
Worley, Clare 782

AUTHOR INDEX

Woulfe, Donna S. 135
Wozniak, Justin 977
Wu, Billy 555
Wu, Chu 222, 464
Wu, Gang 366, 727, 729
Wu, Jianyang 368
Wu, Jinglan 235
Wu, Ning 896, 899
Wu, Yanyang 46
Wyczalkowski, Wojciech 318
Xi, Erte 984
Xi, Li 881
Xi, Shun 1079
Xia, Yi 410
Xiao, Junyin 46, 533
Xie, Hanguang 408
Xie, Jingyi 906
Xie, Liangxu 50
Xin, Hongliang 288
Xing, Yangchuan 524
Xu, Bicheng 521
Xu, Jeffrey 959
Xu, Xinliang 403
Yadav, Gautam G. 431
Yadav, Geetanjali 356
Yamamoto, Minoru 240
Yan, Yushan 421, 520
Yang, Alexander 914
Yang, Chuanfang 530
Yang, Mengfei 827
Yang, Pengpeng 235
Yang, Tao 896
Yang, Xingfu 899
Yang, Yuan 339
Yang, Yung-Jih 933
Yao, Min 569
Yarranton, Harvey W. 474
Yasue, Masahiro 383
Yates, Matthew 1042
Yelvington, Paul E. 58
Yezer, Benjamin A. 146
Yilixiati, Subinuer 337, 338, 471, 1024
Yin, Xiaolong 494
Ying, Hanjie 235
Yokozaki, Yuta 315
Yoshimoto, Makoto 1086
Young, Charles 478, 710
Yu, Huaizhe 45
Yu, Jiaheng 443
Yu, Xiangfei 167
Yu, Xiaochen 93
Yu, Yue 876
Yu, Yuncheng 36
Yuan, Hao 148
Zambrano, Andres E. Rodriguez 946
Zane, Kylie 938
Zasadzinski, Joesph A. 220, 470, 480, 481, 708, 800
Zeiner, Tim 778
Zeman, Johannes 999
Zeng, Wenduo 1048
Zeng, Yongchao 947
Zeng, Zhenhua 282
Zeng, Zhiqiao 493
Zerze, Gul H. 764, 765
Zerze, Hasan 997
Zhang, Chang 73
Zhang, Geoff G. Z. 532, 737
Zhang, Hanguang 366
Zhang, Jinjun 1032
Zhang, Junshe 381
Zhang, Kechun 400
Zhang, Peng 56
Zhang, Qian 530
Zhang, Qing 405
Zhang, Rui 23, 398, 479, 833, 834
Zhang, Suojiang 1084
Zhang, Xiangping 1084
Zhang, Xuefei 1042
Zhang, Yi 462, 527
Zhang, Yiran 337, 338, 471, 482, 1024
Zhang, Yu 42, 104
Zhang, Yuanhui 56
Zhang, Yuchong 182
Zhang, Zhuqing 343
Zhao, Guolin 46, 533
Zhao, Ruiyang 462
Zhao, Shuangliang 46, 93, 128, 533
Zhao, Xiao 929
Zhao, Xujun 977
Zheng, Weiqing 520
Zhou, Hao 945
Zhou, Jiarun 43, 462
Zhou, Ming 905
Zhou, Mingxia 231
Zhou, Ran 322
Zhou, Ye 103, 833
Zhu, Lu 881
Zhu, Xiao 795
Zhuang, Wei 235
Zia, Roseanna N. 512, 514, 519
Zimmerman, Paul M. 838
Zinchenko, Alexander 172
Zou, Fengxia 235